Proceedings of the 10th International Conference on

Electrorheological Fluids and Magnetorheological Suspensions

Proceedings of the 10th International Conference on

Electrorheological Fluids and Magnetorheological Suspensions

Lake Tahoe, USA June 18 – 22, 2006

Edited by

Faramarz Gordaninejad
Olivia A. Graeve
Alan Fuchs
David York

University of Nevada, Reno

World Scientific

NEW JERSEY • LONDON • SINGAPORE • BEIJING • SHANGHAI • HONG KONG • TAIPEI • CHENNAI

Published by

World Scientific Publishing Co. Pte. Ltd.
5 Toh Tuck Link, Singapore 596224
USA office: 27 Warren Street, Suite 401-402, Hackensack, NJ 07601
UK office: 57 Shelton Street, Covent Garden, London WC2H 9HE

British Library Cataloguing-in-Publication Data
A catalogue record for this book is available from the British Library.

ELECTRORHEOLOGICAL FLUIDS AND MAGNETORHEOLOGICAL SUSPENSIONS
Proceedings of the 10th International Conference on ERMR 2006

ISBN-13 978-981-277-119-3
ISBN-10 981-277-119-0

Printed in Singapore by World Scientific Printers (S) Pte Ltd

ERMR 2006
June 18-22, 2006
Embassy Suites Hotel
Lake Tahoe, USA

The 10th International Conference on Electrorheological Fluids and Magnetorheological Suspensions

Foreword

The 10th International Conference on Electrorheological Fluids and Magnetorheological Suspensions ("ERMR 2006") took place in Lake Tahoe, USA on June 18-22, 2006. The purpose of this Conference was to provide a venue for scientists and engineers to present their findings and to interact in a variety of subjects in the ER and MR materials area. ERMR 2006 comprised a wide range of research topics, which includes chemical compositions, physical mechanisms, material processing and characterization, device development, theoretical modeling, and system integration. The contributed papers offer advances in the field of ER and MR materials structure and systems.

ERMR 2006 included invited speakers, technical presentations, poster presentations, the student paper competition, and the banquet address by Dr. David Carlson of Lord Corporation. One of the unique features of ERMR Conferences is that they offer a wide range of issues from chemistry to engineering applications of ER and MR materials held in a single general session to enhance the interaction between the scientists and engineers. The sessions in ERMR 2006 were organized based on: a) *Materials* subjects which includes: mechanisms, preparation, and characterization of ER and MR materials, and b) *Applications* subjects which includes: ER and MR devices, control systems, system Integration, and applications. This structure was implemented in order to enhance the interaction between the ER and MR materials research issues and investigators.

ERMR 2006 was co-sponsored by the University of Nevada, Reno, Lord Corporation, BASF Company, Anton Paar Corporation, and Advanced Materials and Devices, Inc. Their support in partially sponsoring this event are gratefully acknowledged and appreciated. Finally, we wish to acknowledge the great effort put forth by the local organizing committee, and particularly, Ms. Azita Madani of Advanced Materials and Devices, Inc.

Faramarz Gordaninejad
Olivia Graeve
Alan Fuchs
David York

University of Nevada, Reno
Reno, Nevada, USA

ERMR 2006
June 18-22, 2006
Embassy Suites Hotel
Lake Tahoe, USA

The 10th International Conference on Electrorheological Fluids and Magnetorheological Suspensions

Organization

Conference Chair

Faramarz Gordaninejad — Director, Composite and Intelligent Materials Laboratory
University of Nevada, Reno, USA, faramarz@unr.edu

Conference Secretariat

Olivia A. Graeve — Director, Nanomaterials Processing Laboratory
University of Nevada, Reno, USA, oagraeve@unr.edu

Local Organizing Committee

Faramarz Gordaninejad	Foundation Professor University of Nevada, Reno, USA
Olivia A. Graeve	Assistant Professor University of Nevada, Reno, USA
Cahit Evrensel	Associate Professor University of Nevada, Reno, USA
Alan Fuchs	Associate Professor University of Nevada, Reno, USA
Yanming Liu	Senior Engineer Advanced Materials and Devices, Inc., USA
Azita Madani	Chief Operating Officer Advanced Materials and Devices, Inc., USA
Xiaojie Wang	Research Assistant Professor University of Nevada, Reno, USA

International Scientific Committee

Medhi Ahmadian	Virginia Polytechnic Institute and State University, USA
Georges Bossis	University of Nice, France
William A. Bullough	University of Sheffield, UK
J. David Carlson	Lord Corporation, USA
Andrejs Cebers	Institute of Physics, Latvia
Seung-Bok Choi	Inha University, Korea
Hans Conrad	North Carolina State University, USA
Frank E. Filisko	University of Michigan, USA
John Ginder	Ford Motor Company, USA
Xinglong Gong	University of Science and Technology of China, China
Faramarz Gordaninejad	University of Nevada, Reno, USA
Daniel J. Klingenberg	University of Wisconsin, Madison, USA
William Kordonski	QED Technologies, Inc., USA
Evguenia Korobko	A.V. Lykov Heat and Mass Transfer Institute, Belarus
Kunquan Lu	Chinese Academy of Sciences, China
James E. Martin	Sandia National Laboratory, USA
Masami Nakano	Yamagata University, Japan
Pradeep P. Phulé	University of Pittsburgh, USA
Ping Sheng	Hong Kong University of Science and Technology, China
Rongjia Tao	Temple University, USA
Norman M. Wereley	University of Maryland, USA

ERMR 2006
June 18-22, 2006
Embassy Suites Hotel
Lake Tahoe, USA

The 10th International Conference on Electrorheological Fluids and Magnetorheological Suspensions

Sponsoring Organizations

University of Nevada, Reno
http://www.unr.edu/

Advanced Materials and Devices, Inc.
http://www.amadinc.com

Lord Corporation
http://www.mrfluid.com/

BASF, The Chemical Company
http://www.corporate.basf.com/

Anton Paar GmbH
http://www.anton-paar.com/

CONTENTS

TRANSIENT BEHAVIOR OF ELECTRORHEOLOGICAL FLUIDS IN SHEAR FLOW

DAVID KITTIPOOMWONG AND DANIEL J. KLINGENBERG

Department of Chemical and Biological Engineering and Rheology Research Center, University of Wisconsin, 1415 Engineering Dr., Madison, WI 53706, USA
E-mail: klingen@engr.wisc.edu

JEFFREY F. MORRIS

Levich Institute and Department of Chemical Engineering, City College of New York, New York, NY
Email: morris@ccny.cuny.edu

The transient response of electrorheological suspensions in shear flow subjected to a suddenly imposed electric field is investigated experimentally. Barium titanate/silicone oil and alumina/mineral oil suspensions are employed. The evolution of both the rheological properties and suspension structure are investigated. Transient responses appear above a critical field field strength, and the critical Mason number for the onset of a transient rheological response is equivalent to the critical Mason number for the onset of lamella formation, within experimental uncertainty. These results are consistent with previous predictions. However, the shear rate and volume fraction dependence of the critical Mason number deviate from that predicted.

1 Introduction

Electrorheological (ER) and magnetorheological (MR) suspensions exhibit dramatic changes in rheological properties upon the application of electric or magnetic fields, respectively. The suspensions change from viscous liquids to viscoplastic solids in the order of milliseconds.[1,2,3] The dramatic rheological changes are closely associated with a change in microstructure. At zero field strength, the microstructure is isotropic. Upon the application of the external field, the particles aggregate into columnar structures. The rapid initial increase in apparent viscosity caused by the applied field can be followed by a much slower transient increase in the apparent viscosity.[4,5] Such behavior is believed to be intimately connected to the field-induced formation of lamellar structures in shear flow.[4,5,6,7,8]

von Pfeil et al.[9] examined the shear induced structure in ER and MR suspensions by employing a two-fluid continuum model. The evolution of the particle concentration is governed by the competition of field-induced and hydrodynamic particle stresses. The resulting conservation equation for the particle volume fraction yields column-like structures in quiescent suspensions and lamellar structures in shear flow, consistent with experimental observations. Employing a linear stability analysis, the authors predicted three features of the structure evolution under shear that are relevant to this article: (1) lamellar structures should appear only for field strengths greater than a critical field strength, E_c; (2) the critical field strength corresponds to a critical Mason number, $\mathrm{Mn}_c = \eta_c\dot{\gamma}/(2\epsilon_0\epsilon_c\beta^2 E_c^2)$, which should be independent of shear rate, but depends on the particle concentration. Here, η_c is the suspending fluid viscosity, $\dot{\gamma}$ is the shear rate, $\epsilon_0 = 8.8542 \times 10^{-12}$ F/m, ϵ_c is the relative dielectric constant of the the suspending fluid, and β is the relative particle polarizability;[10] and (3) the predicted dependence of Mn_c on the volume fraction is of a specific form, $\mathrm{Mn}_c = 2\phi_{\max}^2 \left(1 - \phi/\phi_{\max}\right)^3 / \left(1 - \beta\phi\right)^3$, where ϕ is the particle volume fraction and $\phi_{\max}$ is the maximum packing fraction. Although these predictions pertain to the evolution of the microstructure, it follows that a transient structure evolution should give rise to a transient rheological response.[11]

Ulicny et al.[5] investigated the transient rheological response of MR suspensions composed of carbonyl iron particles in an aliphatic oil. Fumed silica was added to slow sedimentation; the fumed silica also imparted non-Newtonian character to the suspension in the absence of a magnetic field. For small field strengths, the apparent suspension viscosity rapidly increased to a steady value when the field was applied. However, for sufficiently large field strengths, a transient shear stress was observed. The appearance of a transient response for field strengths above a critical value is consistent with the first prediction described in the previous paragraph. However, Ulicny et al. found that the critical Mason number increased substantially with shear rate, and the values were much smaller than those predicted by von Pfeil et al. The authors attributed the discrepancies between their experimental observations and the predictions in part to effects of the colloidal forces imparted by the fumed silica, which are not considered in the continuum model.

In this report, we describe an experimental investigation of the transient rheological response and microstructure evolution in ER suspensions in shear flow. The rheological response to a suddenly imposed electric field only exhibits a slow transient above a critical field strength, in agreement with the predictions of von Pfeil et al.[9] Experiments in which the rheological and structural responses are monitored simultaneously reveal that the critical Mason numbers marking the onset of a transient rheological response and the appearance of lamellar structures are equivalent within experimental uncertainty. However, the shear rate and volume fraction dependence of the critical Mason number differ from that predicted.

2 Experimental

Alumina suspensions were composed of acidic alumina particles (Al_2O_3) in heavy mineral oil (viscosity $\eta_c = 0.198$ Pa·s) with Brij®30 as a surfactant. The amount of surfactant was 2.5 wt% (based on the particle mass). Barium titanate suspensions were composed of barium titanate particles ($BaTiO_3$) in silicone oil (viscosity $\eta_c = 0.504$ Pa·s) with no surfactant added. Both the alumina and the barium titanate particles were ground with a mortar and pestle and sieved to obtain particle diameters in the range 53–90 μm. The particles were dried for 12 hours at 110 °C. The oils were stored over molecular sieves overnight to remove water.

Rheological measurements were made using a Bohlin VOR rheometer configured with either parallel disk or concentric cylinder geometries. The stainless steel parallel disks are attached to Delrin fixtures to electrically insulate the disks from the rheometer. A potential difference is applied across the disks using a Trek 10/10 amplifier driven with a function generator, connected to the top plate with a wire, and to the bottom plate with a sliding contact. The concentric cylinder cell is composed of a stainless steel bob and transparent quartz cup, each connected to Delrin fixtures. The inner surface of the cup was made electrically conducting with a coating of indium tin oxide.[12] For all experiments reported here, ac electric fields are employed, with frequencies of 300 Hz for alumina suspensions, and 500 Hz for barium titanate suspensions.

Rheological experiments were performed as follows. Suspensions were first sheared at a large shear rate ($\dot{\gamma} \geq 50$ s^{-1}) with no applied field to ensure that the particles were

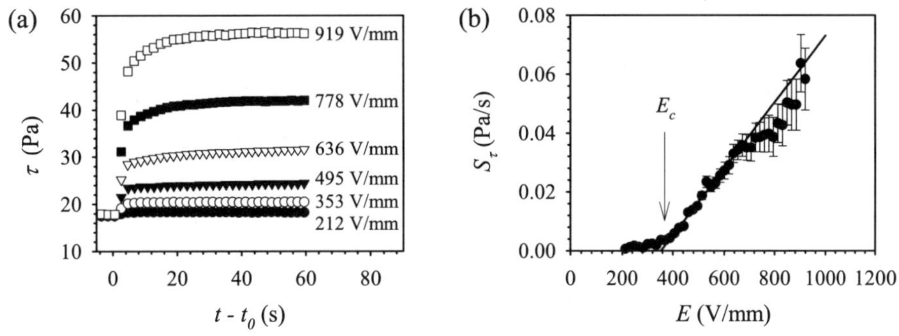

Figure 1. (a) Shear stress of 12.8 vol% barium titanate suspensions as a function of time for different applied field strengths ($\dot{\gamma} = 19.9\ \mathrm{s}^{-1}$). (b) Slope of shear stress vs time as a function of field strength for 12.8 vol% barium titanate suspensions ($\dot{\gamma} = 19.9\ \mathrm{s}^{-1}$).

distributed homogeneously. The shear rate was then set to the desired value and maintained for 30 s with no applied field. An electric field was then applied for 1-3 minutes. The field strength was then set to zero and the shear stress was recorded for another 30 seconds. These measurements were repeated for various field strengths, shear rates and concentrations. All experiments were carried out at room temperature (20–25°C).

The evolution of the suspension structure in the concentric cylinder geometry was examined by recording video images with a CCD video camera. The camera was directed normal to the outer transparent cylinder. Video images were analyzed using NIH image processing software. Raw images were sharpened using an FFT filter.

3 Results

The shear stress τ of a 12.8 vol% barium titanate suspension is plotted in Fig. 1(a) as a function of time after the electric field has been applied, $t - t_0$, for various field strengths (parallel disks; $\dot{\gamma} = 19.9\mathrm{s}^{-1}$). When the field is applied, the shear stress initially increases rapidly, as commonly observed in ER fluids. For sufficiently large field strengths, the stress then continues to increase slowly.

Following Ulicny et al.,[5] we characterize the slow transient rheological response by the slope S_τ of the shear stress vs time data, while the field is applied. This slope is calculated by linear regression of the data collected over the period $2 < t - t_0 < 60$ s; data collected during the first two seconds after the field is applied are ignored in order to avoid potential problems with a lag in the data acquisition. The slope is plotted as a function of electric field strength in Fig. 1(b) for 12.8 vol% barium titanate suspensions at $\dot{\gamma} = 19.9\ \mathrm{s}^{-1}$. There is a clear transition in the behavior of the shear stress. For small E, S_τ is indistinguishable from zero, which indicates steady shear flow. For sufficiently large E, $S_\tau > 0$, which indicates a transient increase in the shear stress.

We define a critical field strength, E_c, as the field strength at which slope begins to deviate from zero. The value of E_c is obtained by linear regression of the data for which $S_\tau > 0$, equating E_c with the intercept on the abscissa as illustrated in Fig. 1(b). Data for all runs were analyzed in this manner. The critical field strength corresponds to a critical

Mason number $\text{Mn}_c = \eta_c\dot{\gamma}/(2\epsilon_0\epsilon_c\beta^2 E_c^2)$, which is plotted as a function of $\dot{\gamma}$ for the barium titanate and alumina suspensions at various concentrations in Fig. 2 (using $\epsilon_c = 2$ and $\beta = 1$). For some of the systems represented in Fig. 2, Mn_c does not vary with $\dot{\gamma}$ within experimental uncertainty; for other systems, Mn_c increases with $\dot{\gamma}$ by an amount comparable with the experimental uncertainty. We thus conclude that Mn_c depends, at most, weakly on shear rate. This differs from that predicted by von Pfeil et al.[9] where Mn_c is independent of shear rate. We note that Ulicny et al.[5] reported that Mn_c increased significantly with shear rate for MR suspensions composed of iron and fumed silica particles in an aliphatic oil.

Next, we consider the behavior of alumina suspensions in shear flow between concentric cylinders. The outer cylinder is transparent so that the evolution of the suspension structure could be observed simultaneously with the rheological measurements. Video images for shear flow of an alumina suspension ($\phi = 0.15$, $\dot{\gamma} = 29.2\ \text{s}^{-1}$) at various field strengths and times are presented in Fig. 3, where the motion of the outer cylinder is from left to right. For sufficiently small electric field strengths, suspensions under shear appear to be homogeneous, with no variation of the particle concentration along the axial direction. For large field strengths, lamellar patterns oriented in the flow direction form after the field is applied.

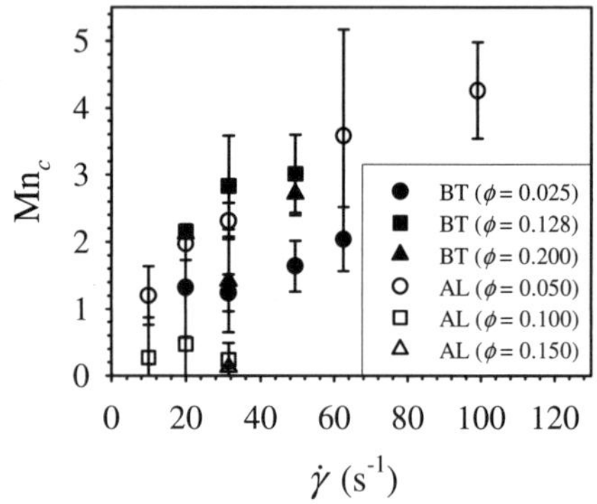

Figure 2. Critical Mason number as a function of shear rate for barium titanate (BT) and alumina (AL) suspensions at different concentrations.

The suspension structure evolves slowly with time, as the lamella coalesce into larger lamella. This is illustrated in Fig. 4(a) where the average stripe thickness h is plotted as a function of time after the application of the electric field for a 15 vol% alumina suspension ($\dot{\gamma} = 29.2\ \text{s}^{-1}$). At a field strength of 612 V/mm, lamella do not form; this is represented in Fig. 4(a) by the symbols at $h = 0$. The large error bars in Fig. 4(a) characterize the variation in lamella thickness throughout the gap, which is also apparent in Fig. 3.

The transient evolution of the lamellar patterns is characterized similarly to the rheological response, via the slope $S_h \equiv dh/dt$, obtained by linear regression of the data $h(t)$ for $2 < t < 180$ s. The slope S_h is plotted as a function of electric field strength in Fig. 4(b) for a 15 vol% suspension of alumina particles ($\dot{\gamma} = 29.2\ \text{s}^{-1}$). Also plotted is the slope S_τ of the transient shear stress. These two measures of transient responses exhibit similar behavior, and similar to that depicted in Fig. 1(b). For small field strengths, the rates of change are indistinguishable from zero (and lamella are not present), while for large field strengths, transient behavior arises.

It is apparent in Fig. 4(b) that the onset of lamella formation occurs at roughly the same field strength as the onset of a transient rheological response, supporting the notion that the transient rheological response is caused by lamella formation. The critical field strengths for the onset of transient behavior are compared quantitatively for alumina sus-

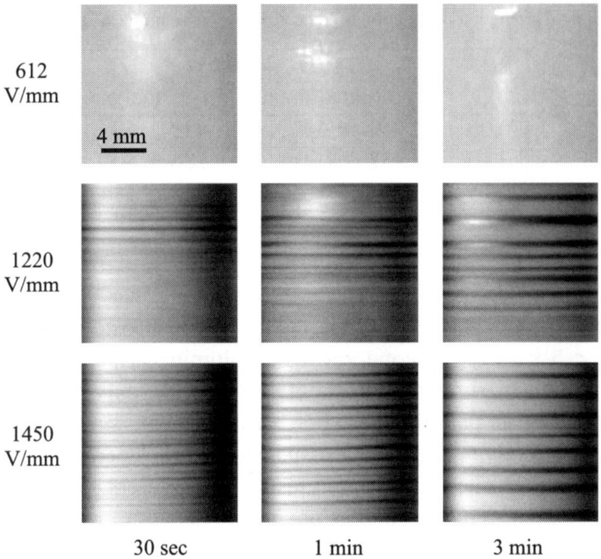

Figure 3. Suspension structure as a function of time for 15 vol% alumina suspensions at various field strengths ($\dot{\gamma} = 29.2\ \mathrm{s}^{-1}$).

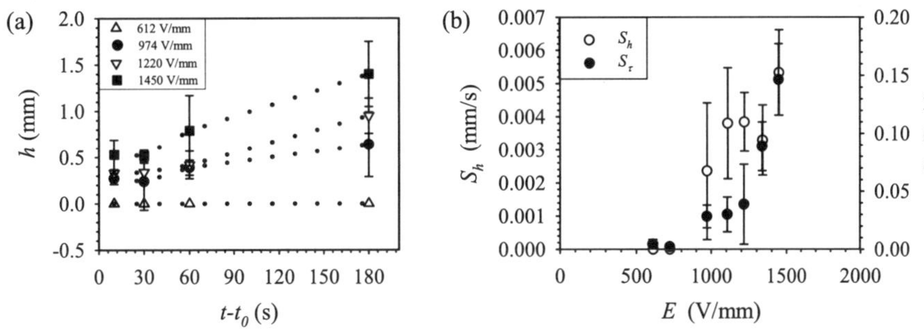

Figure 4. (a) Average stripe thickness as a function of time for 15 vol% alumina suspensions at various field strengths ($\dot{\gamma} = 29.2\ \mathrm{s}^{-1}$). (b) Rates of change of average lamellae thickness (S_h) and shear stress (S_τ) as functions of applied field strength for 15 vol% alumina suspensions ($\dot{\gamma} = 29.2\ \mathrm{s}^{-1}$).

pensions at various concentrations ($\dot{\gamma} = 29.2\ \mathrm{s}^{-1}$) in Table 1. Again, the critical field strengths obtained from rheological and microstructural data agree reasonably well for all concentrations.

In Fig. 5, Mn_c is plotted as a function of concentration for alumina and barium titanate suspensions. The values of Mn_c in this figure were obtained by averaging the results obtained from rheological experiments at different shear rates. For alumina suspensions, results from experiments with both parallel disk and concentric cylinder geometries are included. Also plotted in Fig. 5 is the prediction reported by von Pfeil et al.[9] There are several features to point out in this figure. First, the experimental results are closer to that

Table 1. Critical field strength as a function of concentration for alumina suspensions ($\dot{\gamma} = 29.2\ \mathrm{s}^{-1}$).

ϕ	E_c (V/mm)	
	rheological data	image analysis
0.05	448 ± 44	442 ± 69
0.10	837 ± 106	786 ± 80
0.15	1172 ± 165	1058 ± 114

predicted than the experimental results reported for MR suspensions by Ulicny et al., where $2 \times 10^{-5} < \mathrm{Mn} < 5 \times 10^{-3}$.[5] Second, results for alumina suspensions using different geometries agree within experimental uncertainty. Finally, the experimental results exhibit a dependence of $\mathrm{Mn}_c(\phi)$ on particle type. For barium titanate suspensions, Mn_c increases with concentration initially. For alumina suspensions, Mn_c decreases monotonically with increasing concentration. The experimental results for different particle types only appear to be similar, and similar to that predicted, at small concentrations.

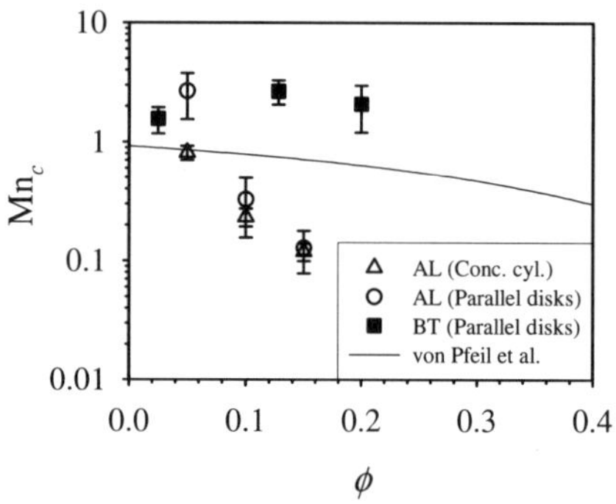

Figure 5. Critical Mason number as a function of particle volume fraction for barium titanate (BT) and alumina (AL) suspensions.

There are several possible explanations for the different values of Mn_c obtained for different materials, which does not seem to be captured by the continuum model. First, Ulicny et al.[5] suggested that colloidal forces may significantly influence the value of Mn_c. We have made no attempt to characterize the colloidal forces for our systems. Second, in the continuum analysis described by von Pfeil et al.,[9] a point-dipole model was employed for the electrostatic stresses. This model approximately captures the many-body electrostatic interactions between particles, producing a dependence of Mn_c on particle type through its dependence on the relative polarizability β (see Sec. 1). However, since we expect β to be near one for both particle types [e.g., for barium titanate, $\epsilon \approx 2000$ at large frequencies, which gives $\beta \approx 0.996$ for silicone oil with dielectric constant $\epsilon_c = 2.75$.[13,14] For alumina particles in silicone oil at $\phi = 0.057$ with 3 wt% Brij®30, Kim reported a dielectric constant at 300 Hz of 3.77[15]; the suspension dielectric constant obtained from Maxwell's formula $\epsilon = \epsilon_c(1 + 2\beta\phi)/(1 - \beta\phi)$ is 3.40 for $\beta = 1$, $\epsilon_c = 2.88$, suggesting that the actual value of β is near 1.0], this dependence is too weak to explain the experimental observations. It is well known that higher-order multipole interactions which were neglected in the continuum model can significantly affect field-induced forces, especially as $\beta \rightarrow 1$.[16] Inclusion of higher-order multipoles should therefore affect the suspension stress, and thus cause the critical Mason number to depend more sensitively on the materials' dielectric properties than predicted by the continuum model of von Pfeil et al.

4 Conclusion

Experimental data for the transient evolution of the rheological properties and the structure of ER suspensions support the hypothesis that a transient rheological response is caused by the formation of lamella. Critical field strengths for onset of transient behavior are similar for both types of experiments. That transient behavior only appears above a critical field strength is consistent with the continuum model of von Pfeil et al.[9] However, the critical Mason number appears to depend weakly on shear rate, whereas von Pfeil et al. predict that the critical Mason number should be independent of shear rate. We also find that the volume fraction dependence of Mn_c differs from that predicted by von Pfeil et al., and different dependences are obtained for the alumina and barium titanate suspensions.

Acknowledgments

The authors thank D. Nest, L. Hsiao and J. Schmidt for helping with the experiments. This work was supported in part by the National Science Foundation under grant CTS-0424087, and by General Motors.

References

1. Tanaka, K., A. Sahashi, R. Akiyama and K.Koyama, Phys. Rev. E., **52**, 3325 (1995).
2. Nava, R., M. A. Ponce, L. Rejón, S. Viquez and V. M. Castano, Smart Mater. Struct., **6**, 67 (1997).
3. Tian, Y., C. Li, M. Zhang, Y. Meng and S. Wen, J. Coll. Int. Sci., **288**, 290 (2005).
4. Vieira, S. L., L. B. P. Neto and A. C. F. Arruda, J. Rheol., **44**, 1139 (2000).
5. Ulicny, J. C., M. A. Golden, C. S. Nanuduri and D. J. Klingenberg, J. Rheol., **49**, 87 (2005).
6. Henley, S. and F. E. Filisko, J. Rheol., **43**, 1323 (1999).
7. Tang, X., W. H. Li, X. J. Wang and P. Q. Zhang, Int. J. Mod. Phys. B, **13**, 1806 (1999).
8. Volkova, O., S. Cutillas and G. Bossis, Phys. Rev. Lett., **82**, 233 (1999).
9. von Pfeil, K., M. D. Graham, D. J. Klingenberg and J. F. Morris, J. Appl. Phys., **93**, 5769 (2003).
10. Parthasarathy, M. and D. J. Klingenberg, Mat. Sci. Eng. R, **R17**, 57 (1996).
11. von Pfeil, K., M.S. Thesis, University of Wisconsin, Madison, WI (2002).
12. Liu, J., E. Radlein and G. H. Frischat, Phys. Chem. Glasses, **40**, 277 (1999).
13. Rankin, P. J. and D. J. Klingenberg, J. Rheol., **42**, 639 (1998).
14. Lan, Y. C., S. Q. Men, X. Y. Xu and K. Q. Lu, J. Phys. D: Appl. Phys., **33**, 1239 (2000).
15. Kim, Y., Ph.D. thesis, University of Wisconsin, Madison, WI (1996).
16. Clercx, H. J. H. and G. Bossis, Phys. Rev. E., **48**, 2721 (1993).

THE INFLUENCES OF THE MATERIAL AND GEOMETRICAL PROPERTIES OF SLIDING PLATE ON ER EFFECT OF GEL-STRUCTURED ERF

SATOKO SAITO[1], YASUHIRO KAKINUMA[1], TOJIRO AOYAMA[1], KATSUTOSHI TANAKA[2] AND HIDENOBU ANZAI[3]

[1] *School of Integrated Design Engineering, Keio University, 3-14-1 Hiyoshi, Kohoku-ku, Yokohama 223-8522, Japan*[2] *Toshiba Machine CO., LTD.*[3] *FUJIKURA KASEI CO.,LTD.*
E-mail: kakinuma@ sd.keio.ac.jp

Electro-rheological Fluid (ERF) is the functional fluid that can change its viscoelastic property by applying electric field. However, ERF shows the sedimentation of the ER particles in long time use, and it caused the unstable ER effect. The Gel-structured ERF (ERG) has been developed to solve this problem. When a metallic flat plate is placed on the ERG sheet, ERG shows the high shear force due to the occurrence of the contact between the gel and the plate in response to the applying electric field (ERG effect). ERG effect can be obtained not only for metallic materials but also non-metallic materials by applying the one-sided electrode, and the ER effect is supposed to be considerably influenced by the material and geometrical properties of flat plate on the ERG sheet. In this study, the influences of the material and geometrical properties of sliding plate are experimentally analyzed from the viewpoint of the relative permittivity, surface roughness and flatness, and the design instruction for the selection of materials and their geometrical issues are proposed.

1 Introduction

Electro-rheological Fluid (ERF) [1, 2] is the functional fluid that can change its viscoelastic property by applying electric field. This property is called the ER effect[3]. However, ERF shows the sedimentation of the ER particles in long time use and its application needs a seal structure for use in machine elements [4]. Electro-rheological Gel (ERG) [5] is developed to overcome these two shortcomings of ERF. The mechanism of ER effect of ERG is different from the case of ERF [2] as illustrated in Fig. 1. Many ER particles with low frictional coefficient are protruded at the ERG surface under no electric field. When the electric field is applied to ERG, the surface frictional property of ERG changes from a slippery to sticky state because the protruding particles retract into the sticky gel. As a result, the surface frictional resistance increases. This phenomenon is called the ERG effect in this study.

To obtain the ERG effect, it is necessary to place the ERG between a pair of plane-parallel electrodes and to apply an electric field to ERG. However, applying ERG to the device with rotating parts such as clutches, the double-sided electrode is need to wire the both rotating electrodes with use of slip ring and the structure of wiring becomes complex. Thus the application of one-sided electrode to ERG is proposed to simplify the wiring structure.

One-sided electrodes have arranged the anode and cathode on the base plate alternately as illustrated in Fig 2 [6]. ERG is placed between the one-sided electrodes and the sliding plate. By using the one-sided electrodes, the ERG effect can be obtained for not

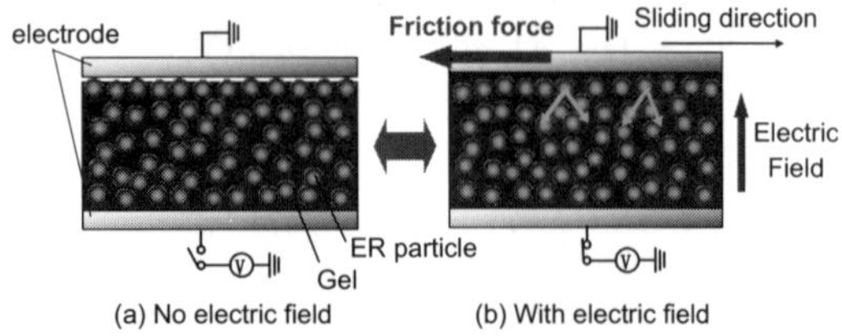

Figure 1. Mechanism of ER effect in ER Gel

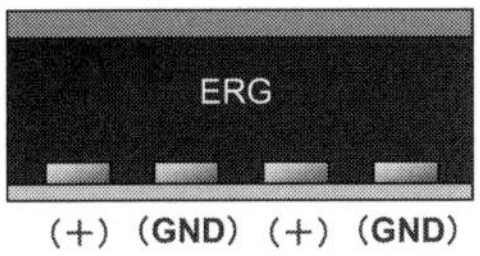

Figure 2. Model of one-sided electrode

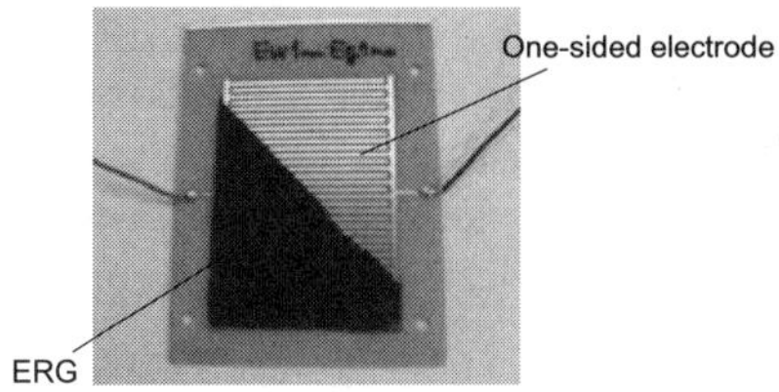

Figure 3. One-sided electrode with ERG

only the conducting materials but also the insulating materials. When insulating materials are used for the sliding plate, some insulating materials shows the ERG effect but the others not.

In order to find the cause of this result, the influence of the properties of the insulating material, such as the relative permittivity, the surface roughness and the flatness on the ER effect is investigated in this study.

2 One-sided Electrode with sliding insulating material

One-sided electrode is made from Printing Wiring Board (PWB) by etching process. PWB is used for manufacturing the electronic circuit in the electronics and computer industry. Fig. 3 shows the photograph of the one-sided electrodes and Fig. 4 shows the structure of the one-sided electrode with ERG sheet. The basic design parameters of one-sided electrode considered in this study is the electrode width Ew and gap Eg. The both dimensions of electrode width (Ew) and gap (Eg) are 1.0mm respectively, and the size of ERG sheet is 50×70×0.3mm (L×W×T).

3 Basic properties of ERG on one-sided electrodes with the sliding insulating materials

3.1 Numerical analysis of electric field

The ERG effect of several different insulating materials is evaluated. The result of the simple test clearly shows that the ERG effect strongly depends on the materials. In order to find the cause of the difference of ERG effect, the distribution of electric field is numerically analyzed for various insulating materials and the influence of the surface

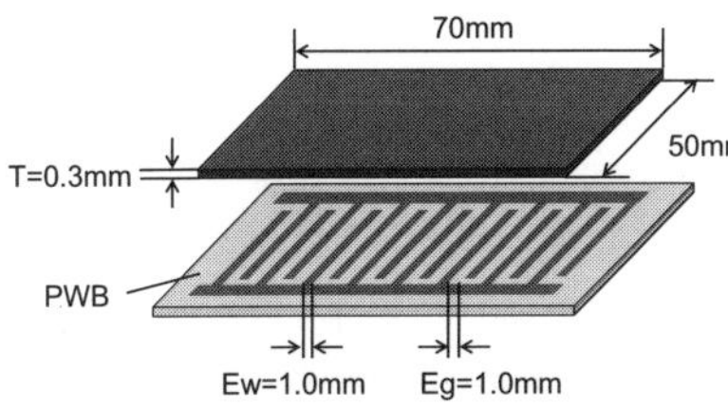

Figure 4. Structure of one-sided electrode

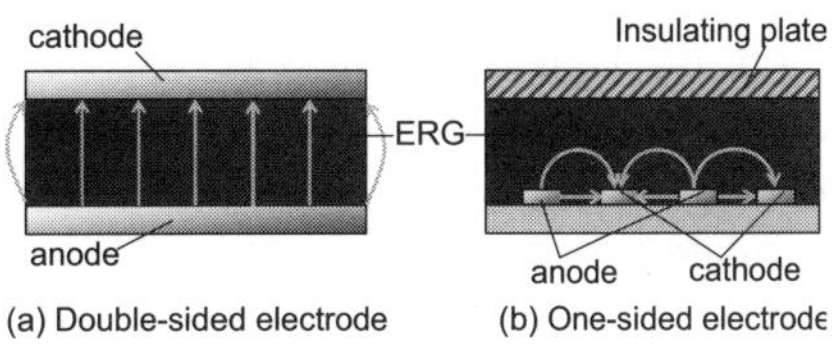

Figure 5. Configuration of electric field

condition of the material on the ERG effect is also investigated by shear test.

ER effect in ERG depends on the electric field intensity at the interface between ERG surface and the sliding plate. As shown in Fig. 5, the configuration of electric field in one-sided electrode with insulating sliding plate is different from the case of double-sided electrode. It is also predicted that the distribution of electric field in ERG is influenced by the permittivity of the insulating materials. In order to investigate the influence of the permittivity of the material on the electric field intensity at the interface, the numerical analysis of electric field is carried out by using the ANSYS 9.0.

The model for numerical analysis is illustrated in Fig. 6. The analysis is carried out for the parameters summarized as follows:
Electrode width (Ew) and gap (Eg) 0.5mm,
ERG thickness (T) 0.3mm,
Applying voltage to electrodes 750V- 0V,
Relative permittivity of PWB (bakelite) 3.0,
Relative permittivity of ERG 30,
Relative permittivity of insulating material ε_s 100 - 0.1

The distribution of the electric field is shown in Fig.7. A vector shows the strength and direction of the electric field at each point. The behavior of electric field strength at the interface between the ERG and the insulating material depends on the relative permittivity of both materials.

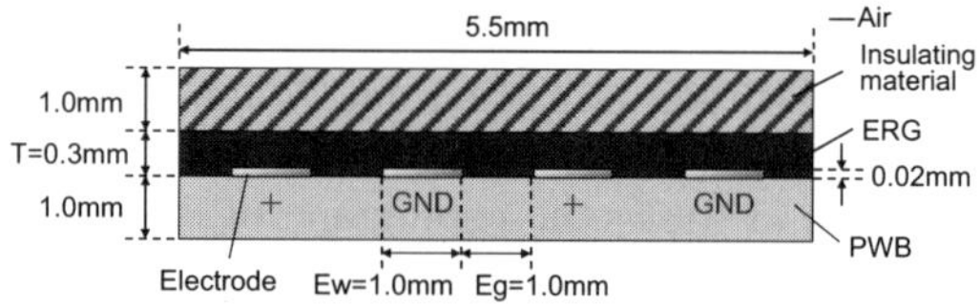

Figure 6. Analysis model of one-sided electrode

Considering that the mechanism of the ERG effect originates in the sinking of the polarized ER particles from the surface into gel, the perpendicular component (y component) of electric field to the surface of ERG greatly influence on the ERG effect.

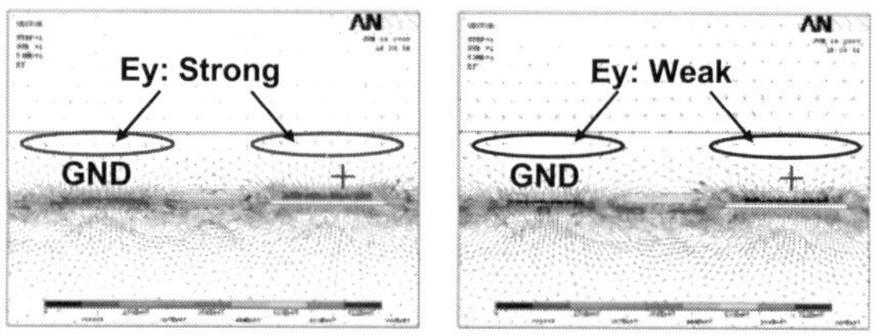

(a) ε_s=100 (b) ε_s= 0.1

Figure 7. Result of the electric field analysis

When the relative permittivity of insulating material is high (Fig.7(a), ε_s=100), the y component of electric field in the ERG sheet close to its surface is large, and therefore the high ERG effect is expected. However the small y component is obtained by using the insulating material with low relative permittivity (Fig.7(b), ε_s=0.1).

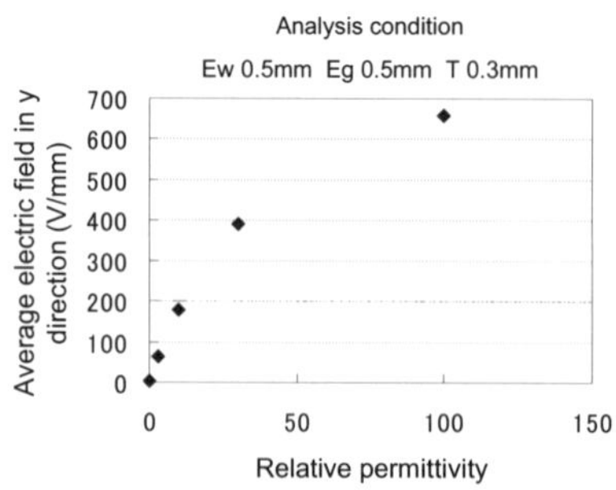

Figure 8. Relationship between relative permittivity of insulating materials and average electric field in y direction

The average value of y components of electric field at the ERG surface is calculated and the result is shown in Fig. 8 as a function of relative permittivity of insulating materials. The average electric field intensity in y direction increases with increase of the relative permittivity of insulating materials.

This result suggests that higher ERG effect can be obtained by using the insulating material with higher permittivity.

3.2 Experimental analysis of the basic properties

The ERG effect of the one-sided electrode with the sliding insulating material is also experimentally investigated. 7 kinds of insulating materials are prepared; the plate of BK7, pyrex glass (PX), blue glass, synthetic silica glass, the polyethylene resin (PE), fluorine resin and ceramics.

3.2.1 Properties of the insulating materials

The relative permittivity of the insulating materials is measured by the capacitance method using a pair of plane-parallel electrodes. Applying the current to the insulating material with varying the frequency from10 to 1~2MHz, the capacitance is obtained and the relative permittivity is calculated. The measured permittivities of the 7 kinds of materials are listed in Table. 1. The sliding plates of different insulating materials have different surface geometrical properties due to their different manufacturing process. The surface roughness average (Ra) and the flatness of the sliding plates of insulating materials are measured. The results are summarized in Table 2. The glass materials especially give the fine surface roughness and better flatness.

Table 1. Relative permittivity of insulating materials

Materials	L×W×D (mm)	Relative permittivity (measured figure)	Relative permittivity (theoretical figure)
BK7	50×70×5	4.440	Glass 3.7~10.0
PX	50×70×5	5.364	
Blue Glass	50×70×5	7.852	
Synthetic Silica Glass	50×70×5	4.396	
Ceramics	50×70×2	3.217	9.8
PE	50×90×5	2.423	2.3~2.4
Fluorine Resin	50×70×5	2.171	18.6

Table 2. Surface roughness and flatness

Materials	Surface roughness (Ra) (μm)	Flatness (mm)
BK7	0.06	0.0048
PX	0.03	0.0032
Blue Glass	0.05	0.0042
Synthetic Silica Glass	0.03	0.0477
Celamics	0.37	0.0478
PE	0.50	0.0757
Fluorine Resin	0.44	0.1067

3.2.2 Experimental setup for shear test

To examine the ERG effect of the one-sided electrode with the sliding insulating material, the behavior of shear stress with applying electric field is measured. The experimental setup is schematized in Fig. 9. The sliding plate made of insulating material is placed on the ERG sheet formed on the one-sided electrode. The behavior of shear force between the sliding plate and the ERG sheet is measured with applying electric field. The shear motion is applied to the sliding plate by the micrometer screw with motor-drive equipment. The shear force is measured by the strain gauge load cell attached to the sliding plate. The displacement of the sliding plate is

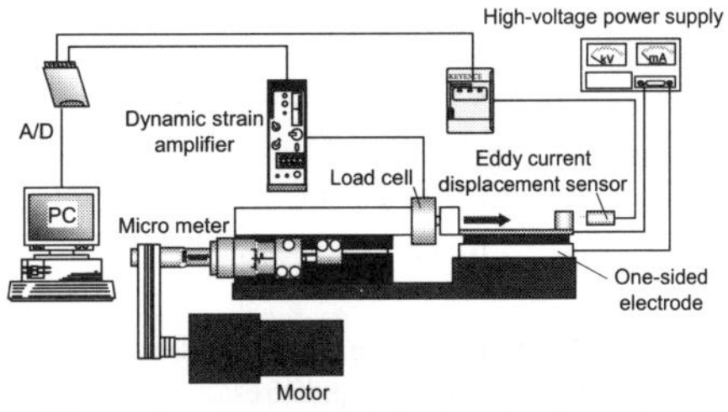

Figure 9. Experimental setup for shear stress measurement

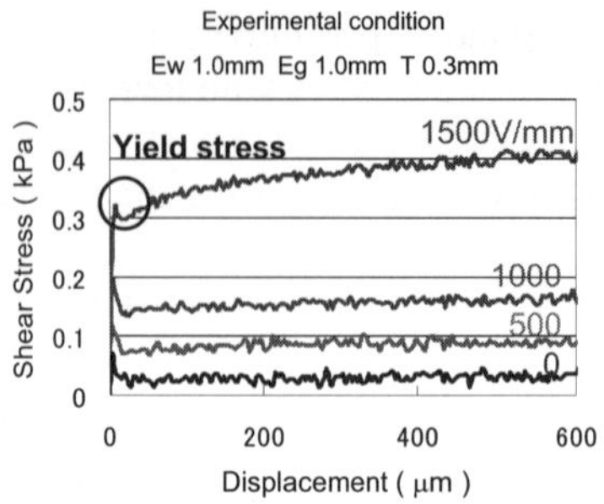

Figure 10. Behavior of the shear stress under electric field (Blue Glass)

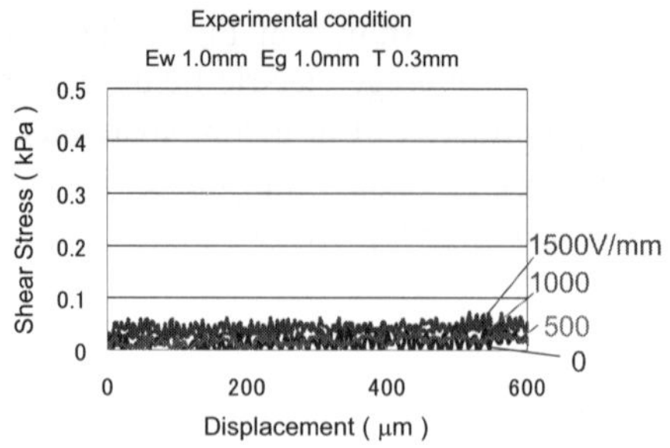

Figure 11. Behavior of the shear stress under electric field (PE)

measured by the eddy current displacement sensor. The behavior of shear force is evaluated by varying the electric field intensity of electrode gap from 0 to1500 V/mm. The shear length was set to 600μm, and the shear speed was adjusted to 30μm/s.

3.2.3 The ERG effect for various insulating materials

When the blue glass is used for shear test, the behaviors of shear stress under a variable electric field, for instance, are shown in Fig. 10. When the displacement of the sliding plate increases, the shear stress shows yielding. Beyond the yield point, the sliding plate begins to slip on the ERG surface. The shear stress at the yield point is defined as yield stress. Higher yield stress was obtained under higher electric field. The yield stress under 1500V/mm is 0.32kPa while the yield stress under 0V/mm is 0.07kPa. It is clear that the blue glass is one of the insulating materials which provide the remarkable ERG effect in the one-sided electrodes.

On the other hand, as shown in Fig. 11, the shear stress in the PE hardly changes even if the electric field is applied. The shear stress vary within only 0.05kPa., and the yield stress can not confirmed clearly.

In order to find the relationship between the sliding insulating plate materials and the ERG effect, the variation of the yield stress is summarized as the function of the relative permittivity, the surface roughness (Ra) and the flatness of the sliding plates of insulating materials. The results are shown in Fig. 12, 13 and 14 respectively.

When the relative permittivity of the materials increases, the generated yield stress

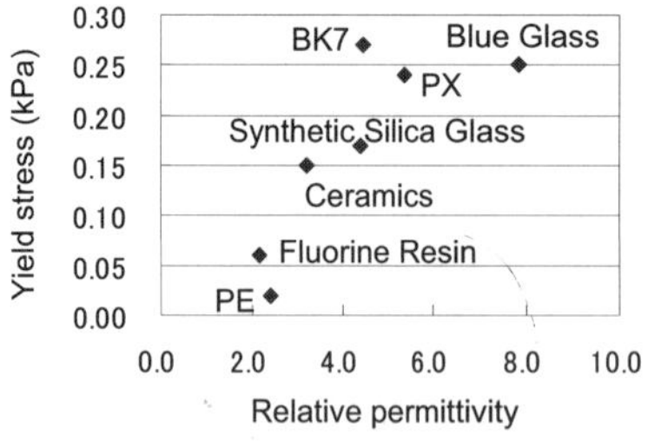

Figure 12. Relationship between relative permittivity and yield stress

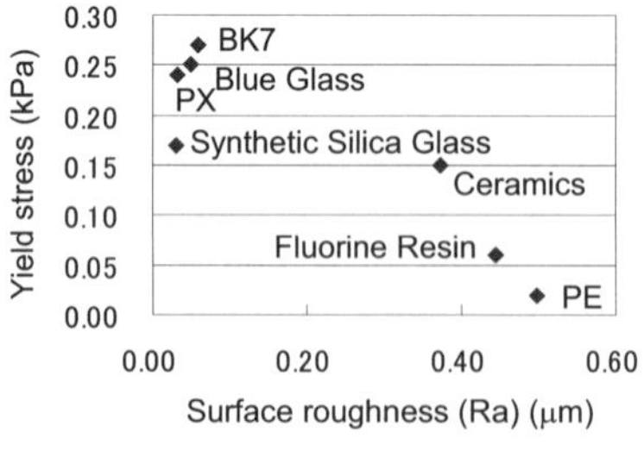

Figure 13. Relation between surface roughness and yield stress

tends to increase. This tendency agrees with the results of numerical analysis shown in Fig. 8.

Also, as for influence of geometrical properties of the sliding plates on the ERG effect, it is suggested that the materials with fine and flat surface provide ERG effect efficiently.

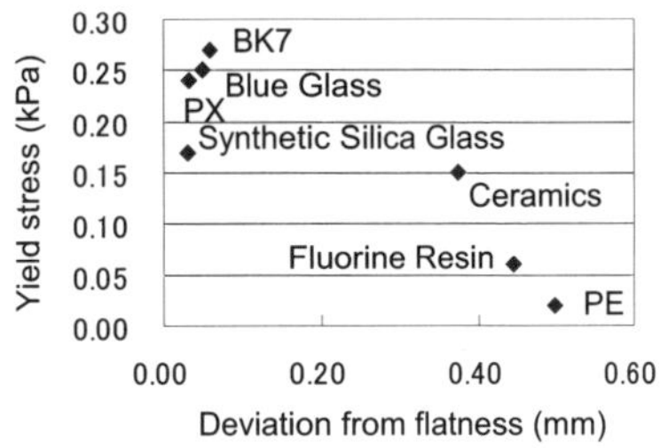

Figure 14. Relation between deviation from flatness and yield stress

4 Conclusion

In this study, the influences of the properties of the sliding insulating material on ERG effect is investigated, especially from the viewpoint of the relative permittivity, the surface roughness and the flatness. From the results of the numerical and experimental analysis, the instruction for the selection of materials and their geometrical issues are pointed out.

It is demonstrated that the high ERG effect can be obtained when the insulating material with high relative permittivity and, fine and flat surface is used as the sliding plate in the one-sided electrode.

References

1. Nakano, M., and Nagata, T., ER Properties and Flow-induced Microstructures of an ER Fluid between Two Parallel Disk Electrodes in Squeeze Flow Mode, International Journal of Modern Physics B, Vol. 16, No. 17 & 18, (2002), pp. 2555-2561.
2. Jolly, M.R., and Nakano, M., Properties and Applications of Commercial Controllable Fluids, 6th International Conference on New Actuators, Bremen, Germany, June 17 - 19, (1998), pp. 411-416.
3. Tomas C. and Halsey, Electrorheological Fluids, Science, Vol. 258, No. 5083, (1992), pp. 761-766.
4. Aoyama, T., Inasaki, I., Application of Electrorheological Fluid Dampers to Machine Tool Elements, Annals of the CIRP, 46/1, (1997), pp. 309-312.
5. Y. Kakinuma, T. Yakoh, T. Aoyama, H. Anzai, K. Isobe, Development of Gel-structured Electro-rheological Fluids and their Application to Mechanical Damper Elements, the 8th IEEE International Workshop on Advanced Motion Control (AMC'04), Kawasaki, Japan, March 25 - 28, (2004), pp.541-545.
6. N. Takesue, J. Furusho, A. Inoue, Electro-rheological Effects of Liquid Crystalline Polymer on One-sided Pattern Electrodes, Journal of Applied Physics, Vol. 91, No. 3, February 1,(2002), pp.1618-1623.

VISCOELASTIC BEHAVIOR OF CONDUCTIVE POLYMER BASED ER DISPERSIONS UNDER SMALL DEFORMATIONS

ANNA KRZTOŃ-MAZIOPA, MARCIN OSUCHOWSKI AND JANUSZ PŁOCHARSKI

Department of Chemistry, Warsaw University of Technology

ul. Noakowskiego 3, 00-664 Warszawa, Poland, E-mail: janusz@ch,pw.edu.pl

ER fluids under shear are often approximated by the Bingham model with the electric field induced yield stress. This approach, however, neglects the properties of the solid phase created. The oscillation measurements taken under field below the yield point and leading to calculation of storage and loss moduli of the electrified solid do not allow for determination of the full stress–strain relationship of the solidified ER suspension. In order to overcome this problem and collect information about the mechanical properties of this solid a dedicated rheometer was constructed capable of measuring small strains under moderate stresses and under electric field. Full shapes of the stress–strain curves were recorded leading to determination of storage modulus, linearity range, maximum deformation and yield point. Looking for relations between the viscoelastic behavior of ER suspensions and properties of their material components a series of fluids comprising silicone oil dispersions of conductive polymers (polyaniline, poly(*p*-phenylene), pyrolysed polyacrylonitrile) and polymer electrolytes (polyacrylonitrile complexes with inorganic salts) was studied. It was found that the viscoelastic responses of ER fluids were affected by material characteristics of the dispersed phase.

1 Introduction

ER effect in a heterogeneous suspension relies on fast and reversible transformation of the liquid into a poorly characterized solid upon application of electric field. The behavior of the electrified ER fluids is often approximated by the Bingham model that is very useful when applied to such fluids under shear flow but the behavior of ER systems is definitely more complex. This model neglects the properties of the electrified solid under stresses below the yield point assuming that the material does not deform. This assumption leads to overlooking the interesting area of properties of ER systems in the pre-yield state which remain almost completely uninvestigated in contrast to the ample literature presenting flow curves of ER fluids in electric field. In fact we do not know whether an ER fluid under electric field really resembles an elastic solid (at least within the human scale of time). The literature related to the pre-yield state of ER fluids is devoted mainly to oscillatory shear experiments capable to determine storage and loss moduli at given electric field strength and oscillation amplitude [1-3]. The authors report on quadratic dependence of storage modulus on electric field strength and its large decrease with increasing a
mplitude of the oscillations. The so called "creep and recovery" studies are also worth to mention [4,5]. The important conclusion is that the created electrified solid seems to be elastic at least under very small deformations but the full stress-strain relations remain undisclosed.

Therefore, believing that the direct measurements would be useful, we performed a series of experiments allowing for direct determination of the stress-strain curves of selected ER fluids in the pre-yield region. We chose two groups of materials as the dispersed phase in the ER fluids – conjugated polymers and polymer electrolytes. According to our earlier investigations [6,7] the ER effect in these materials originates from mobile ionic species in the bulk of dispersed grains (polyphenylene, complexes of polyacrylonitrile with inorganic salts) and/or presence of polar surface groups (pyrolyzed polyacrylonitrile, polyaniline). The differences in electric polarization processes in the studied materials should be reflected in the shape of the stress-strain curves.

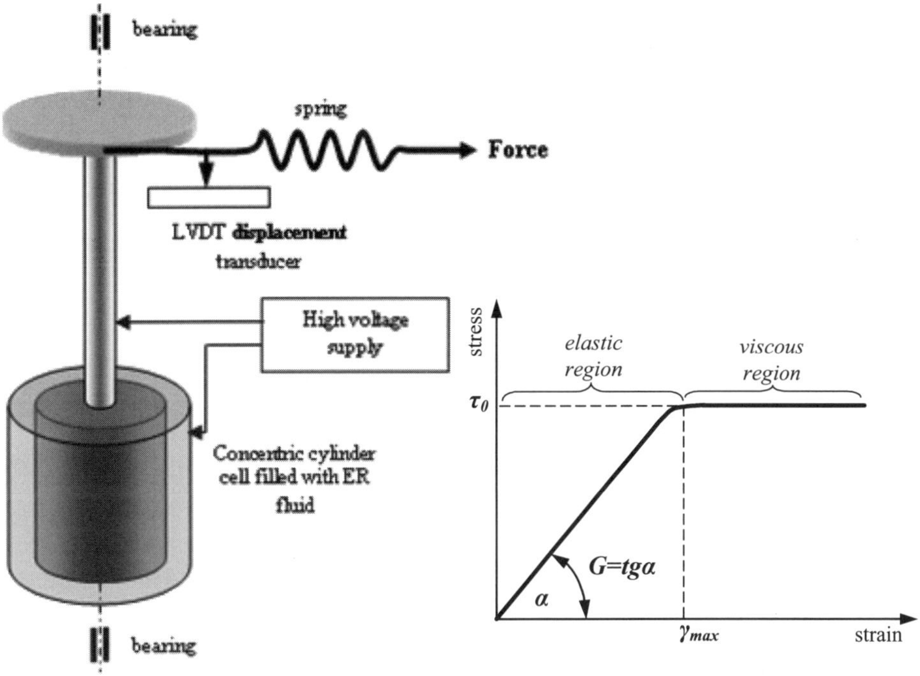

Figure 1. Scheme of deformation measuring system and interpretation of data.

2 Experimental

2.1 Preparation of initial materials and ER fluids

A series of conjugated polymers have been prepared and then used for preparation of ER suspensions. Poly(*p*-phenylene) coded **PPP** was synthesized from benzene according to the modified Hirai method [8,9] and then doped with $FeCl_3$ to obtain iron concentration of 0.47% by weight and conductivity of about $5 \cdot 10^{-6}$ S/cm. Polyaniline coded **PANI** was synthesized by chemical oxidation of aniline, total deprotonation and then protonation with H_2SO_4·aq up to pH=5.7 [10]. Pyrolyzed polyacrylonitrile was obtained from commercial acrylic fibers (Dralon®) by initial chemical oxidative treatment and then pyrolysis in N_2 at 470ºC (coded **PANN**) or in O_2 at 470ºC (coded **PANOX**) or in O_2 at 430ºC (coded **PANOX-430**) [11]. Ionic complexes of polyacrylonitrile were prepared by dissolution of the polymer in DMF together with the inorganic salt. KSCN, NaSCN and lithium triflate were used in the ratio of 10% to the polymer weight. After evaporation of the solvent the foils were ground and screened [7]. They were coded as **PAN/KSCN**, **PAN/NaSCN** and **PAN/LiTF**, respectively.

The prepared powders were then carefully dispersed in silicone oil of 0.1 Pas viscosity. The concentrations studied were 17% by volume for PPP and PANI and 35% by weight for all other suspensions.

2.2 *Measurement of deformation under stress*

Fig.1 presents the scheme of the deformation measuring system. It consisted with a Couette concentric cylinder cell whose inner bob was connected to a disc to which a force could be applied. The Couette cell (20 mm diameter, 1 mm gap) was filled with an ER fluid and high voltage was switched on between the inner bob and the outer cylinder. Application of the force lower than the yield point resulted in a deformation of the solidified ER fluid and when the force became higher the ER fluid yielded. The applied force was increased in small increments so that recording the whole curve up to 30% strain took about 15 minutes. The value of the linear displacement and the applied force were used to calculate the strain and stress values, respectively.

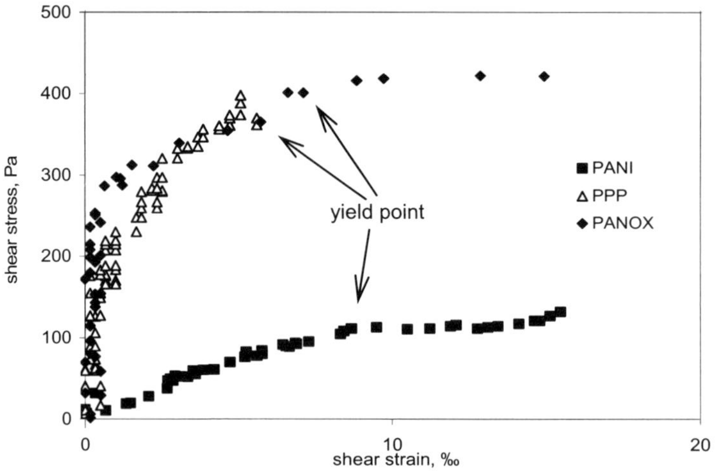

Figure 2. Stress-strain curves in pre-yield region for three ER fluids; E = 2 kV/mm.

3 Results and Discussion

Three examples of the recorded stress-strain curves are shown in Fig.2. The measurement were taken just after pouring a fluid into the measuring cell and switching the field on. The most intriguing feature of this picture is that the shapes of the curves were very much different. For example the deformation of the PANOX suspension was below resolution of our equipment until about 250 Pa. Then deformation started to increase until a brake identifying the yield point what happened at the strain of 0.007 and upon the stress of 400 Pa. The PPP suspensions deformed almost linearly up to about 230 Pa and slightly faster up to the brake that happened at the strain of about 0.005 upon the stress of 400 Pa. The PANI suspension deformed easily from the very beginning until the brake at 0.009 and 110 Pa. One could even say that the PANOX suspension was brittle, the PPP was hard and elastic while the PANI was soft and elastic.

In order to analyze the recorded curves on a quantitative basis three parameters were identified: the maximum deformation or strain at yield point, the yield stress and the elastic modulus. Due to the apparent nonlinearity of the relations the estimation of the modulus was the most difficult. It could not be just the yield stress divided by the maximum deformation. We have calculated it taking into account the slope of approximately linear region close to the 0,0 point but anyhow the obtained values were not very precise.

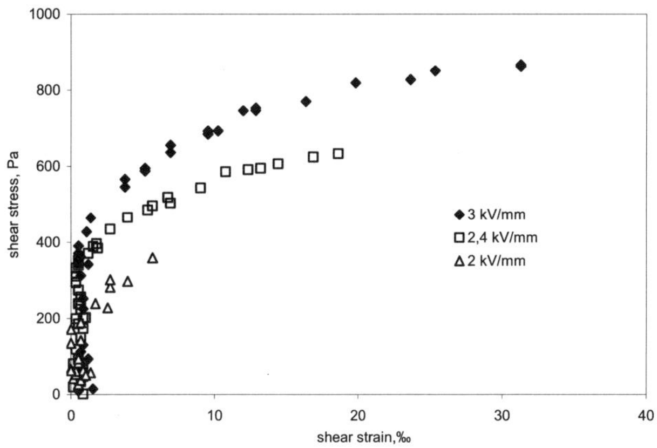

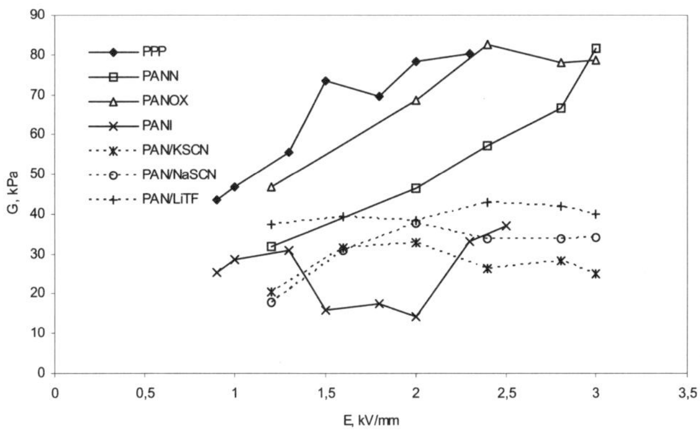

Figure 3. Field strength dependence of PANOX-430 suspension.

Figure 4. Electric field strength dependence of storage modulus of the studied ER suspensions.

The parameters of the curves depended on the applied voltage what was presented in Fig.3 for the PANOX-430 suspension. What is interesting, both the yield stress value and the maximum strain increased with the electric field strength. What is surprising, at least for this particular suspension, the values of the elastic moduli estimated from the initial slope of the three curves are high and seem to be equal. In order to see how this relation looks for other fluids we performed calculations and presented results in Fig.4. It can be seen that the differences between the modulus values of the studied fluids are considerably bigger than just the differences of the yield stress values measured dynamically in a classic rheometer. In addition the presented relations show that the moduli increase linearly with field or even remain constant. Apparently they are not quadratic what stays in contrast with the literature data obtained from the oscillation measurements.

As already seen in Fig.3 the value of deformation at the yield point depends on the field strength. The quantitative correlations are shown in Fig.5. It can be noticed that the solidified state of the ER fluids was not able to withstand deformations exceeding 1% and their maximum values increased linearly with field strength.

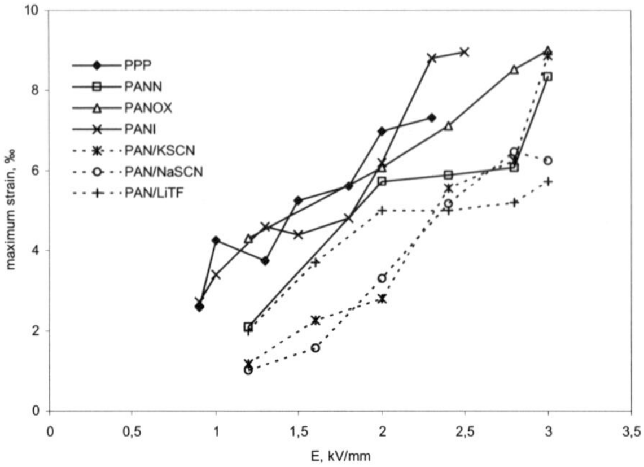

Figure 5. Electric field strength dependence of strain at yield point of the studied ER suspensions.

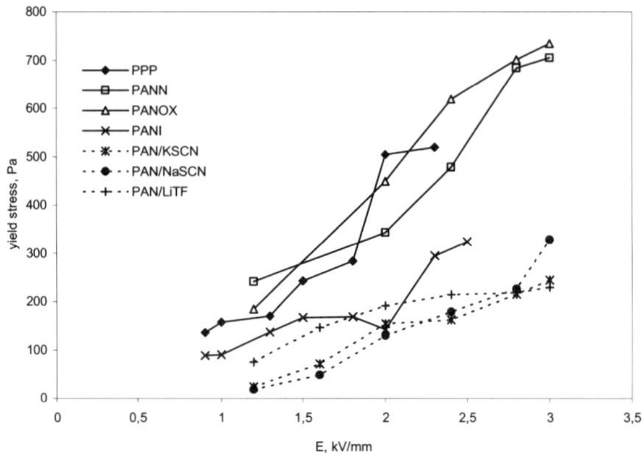

Figure 6. Electric field strength dependence of yield stress estimated from the stress-strain curves.

The next Fig.6 shows the electric field dependence of the yield stress determined by the static method. The results look reasonably and the observed correlations are between linear and quadratic. The absolute values, however, differ from the analogous results obtained from a dynamic measurements taken with a classic rotating rheometer. Fig.7 shows the comparison of such data recorded for two pyrolyzed polyacrylonitrile samples. In this case the static results are considerably higher than the values obtained by extrapolation the flow curves to the zero shear rate. It is, however, not a general rule. We have also studied samples for which this relation was opposite.

The presented data described the basic behavior of the ER suspension studied in the pre-yield phase. Using our equipment it was also possible to extend the range of strain and take measurements in the post-yield region. Two examples of such measurements are shown in Fig.8. In the case of the PAN/KSCN sample the result was just as it could be expected. After the initial increase of the stress causing the small deformation of the electrified solid (not clearly seen in the scale of this plot) the value of deformation increased continuously with the stress value kept at a constant level. However, the other sample comprising pyrolyzed polyacrylonitrile behaved in a significantly different way. After the typical initial increase of stress and small deformation the electrified solid broke and the deformation increased very fast and considerably. During this jump the stress decreased (there was a spring in the force applying system – see Fig.1) to a certain value and the structure of the solid was restored. Then, the stress increased again until the next jump have occurred. During each jump the ER suspension underwent some structural changes resulting in elevated value of the maximum stress at the next jump. After several jumps this process saturated and the stress value remained at the constant level. This observation proves that a kind of superstructure is formed within an ER fluid as a result of combined action of electric and shear fields.

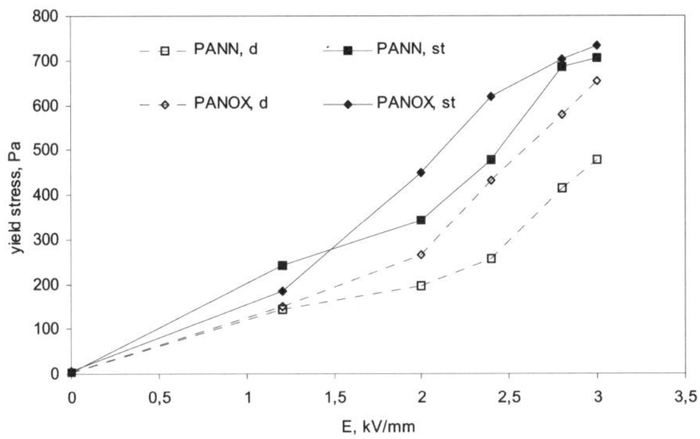

Figure 7. Comparison of yield stress values estimated by static (st) and dynamic (d) methods as function of electric field strength.

These superstructures have been already reported in the literature [12]. They have very strong influence on the reproducibility of the results obtained in our arrangement. As already mentioned the data presented here were recorded on fresh fluids that were not sheared before the measurement. We have observed that quantitatively different results were obtained when prior to the measurement the bob was rotated even by a few degrees with voltage on. At the moment we are not able to correlate the tendency of the continuous or step-like deformations with material properties of the investigated fluids but the steps seem to occur more often in the fluids of higher ER activity.

4 Conclusions

We have constructed a dedicated static rheometer that was used for direct measurements of stress-strain relations in pre-yield stages of electrified ER suspensions. We have found that maximum deformation and shear modulus increased with electric field strength. The modulus value seemed to depend linearly on electric field that, we believe, need not to be surprising if we assume that this value could be approximately estimated as the ratio of the yield stress and the strain at yield point. Since the increase of yield stress is quadratic and the deformation increases linearly the increase of the modulus must be linear too.

We have found also that some superstructure was created within the ER fluid upon combined action of the shear and electric fields. This phenomenon considerably influences the behavior of ER fluids.

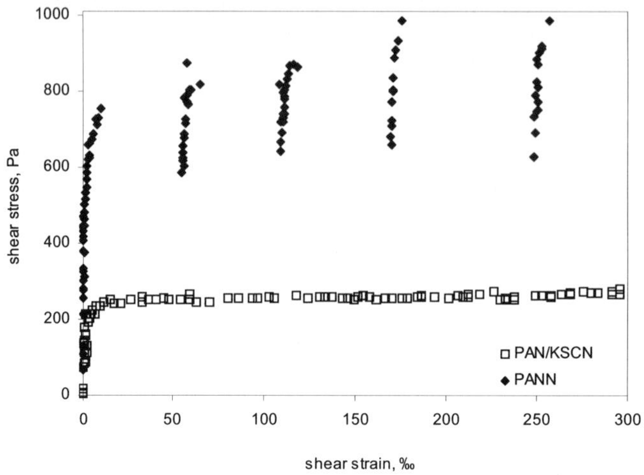

Figure 8. Comparison of stress-strain curves for ER fluids consisting of polymer electrolyte (PAN/KSCN) and conjugated polymer (PANN), E = 3kV/mm.

5 Acknowledgements

The authors are greatly indebted to Mr J. Harasimowicz and Dr Z. Jabłoński who constructed the static rheometer. The work was financially supported by the Faculty of Chemistry, Warsaw University of Technology.

References

1. Parthasarathy M., Klingenberg D.J., *J. Non-Newton. Fluids*, **81** (1999) 83.
2. Lee C., Cheng C., *Int. J. Mech. Sci.*, **42** (2000) 561.
3. Cho M.S., Choi Y.S., Choi H.S., Kim S. G., Jhan M. S., *J. Mol. Liq.*, **75** (1998) 13.
4. Gow J. C., Zukoski, C. F., *J. Coll. Int. Scie.*, **136** (1990) 175.
5. Lau K. C., Shi L., Tam W. Y., Sheng P., *Phys. Rev. E*, **67** (2003) 052502.
6. Krztoń-Maziopa A., Wyciślik H. and Płocharski J., *J. Rheol.*, **49** (2005) 1177
7. Krztoń-Maziopa A., Ciszewska M. and Płocharski J., *Electrochimica Acta*, **50** (2005) 3838.
8. Hirai H., Toshima N., Tanaka K., Koshirai A., *Chem. Lett.* (1987) 1461.
9. Płocharski, J., Kosiek A., Wyciślik H., Łużny W., Wielgus I., *J. New Mater. Electrochem. Syst.* **4** (2001) 161
10. Gozdalik A., Wyciślik H., Płocharski J., *Synthetic Metals*, 109 (2000) 337.
11. Krztoń-Maziopa A., Turowski M. and Płocharski J., *Int. J. Mod. Phys. B*, **19** (2005) 1083.
12. Henley S., Filisko, F. E., *J.Rheol.*, **43** (1999) 1323.

The Physical Mechanism to Reduce Viscosity of Liquid Suspensions

R. TAO

Department of Physics, Temple University, Philadelphia, PA 19122, USA

Reducing the viscosity of liquid suspensions is vital in science and engineering. This paper explores the physical mechanism for the viscosity reduction method in liquid suspension by pulsed electric or magnetic field. The key is that the maximum volume fraction to be available for the suspended particles in the suspension increases with the particle size and the polydispersity in the particle size distribution. Positive experimental results with various liquid suspensions indicate that this method, developed from the basic mechanism of viscosity, is universal and powerful for all liquid suspensions with broad applications.

1. Introduction

Viscosity of liquid suspensions and emulsions is of great importance in all areas of science and engineering. In many cases, we need to lower the viscosity. For example, reducing blood's viscosity improves circulation and prevents cardiovascular events. Lowering the viscosity of crude oil is the key to transporting offshore oil via undersea pipelines *(1)*. Reducing the viscosity of gasoline or diesel improves the fuel atomization, which leads to more efficient combustion and less pollution. Unfortunately, to date there are no effective methods to reduce the viscosity of liquid suspensions except by changing the temperature. In case that changing temperature is not an option, such as in the above blood and off-shore crude oil cases, reducing the viscosity becomes formidable.

In our recent work, we showed that a pulsed electric or magnetic field could significantly reduce the viscosity of crude oil and the viscosity of a magnetorheological (MR) fluid made of nanoscale magnetic particles *(2)*. In this paper, we will discuss the physical mechanism of this viscosity reduction method The key is that the maximum volume fraction to be available for the suspended particles in the suspension increases with the particle size and the polydispersity in the particle size distribution. Positive experimental results with various liquid suspensions indicate that this method, developed from the basic mechanism of viscosity, is universal and powerful for all liquid suspensions with broad applications.

2. Viscosity and the Particle Size Distribution

Einstein first studied a dilute liquid suspension of noninteracting uniform spheres in a base liquid of viscosity η_0 *(3)*. The apparent viscosity η was found,

$$\eta = \eta_0(1+2.5\phi), \qquad (1)$$

for the volume fraction of the spheres $\phi < 0.01$. Batchelor and others derived the second correction for the dilute case *(4,5)*.

For high ϕ, we must consider the maximum volume fraction, ϕ_m, to be available for adding particles. When we add $d\phi$ volume fraction of spheres to a liquid suspension of volume fraction ϕ, the net available volume fraction to add spheres is only $1-\phi/\phi_m$, the viscosity increase *(6)* would be

$$d\eta/\eta = 2.5d\phi/(1-\phi/\phi_m)\,. \tag{2}$$

Hence at high ϕ the viscosity is given by

$$\eta/\eta_0 = (1-\phi/\phi_m)^{-2.5\phi_m}\,. \tag{3}$$

Krieger-Dougherty introduced the intrinsic viscosity, $[\eta]$,

$$\eta/\eta_0 = (1-\phi/\phi_m)^{-[\eta]\phi_m}, \tag{4}$$

which enables us to estimate the viscosity for particles of any shape by choosing a suitable $[\eta]$ *(7)*. For example, $[\eta]=5/2$ for spherical particles and $[\eta]=5.8$ for glass plates *(8)*. While $[\eta]$ changes significantly for particles of different shapes, the product $[\eta]\phi_m$ almost remains as a constant, close to 1.9.

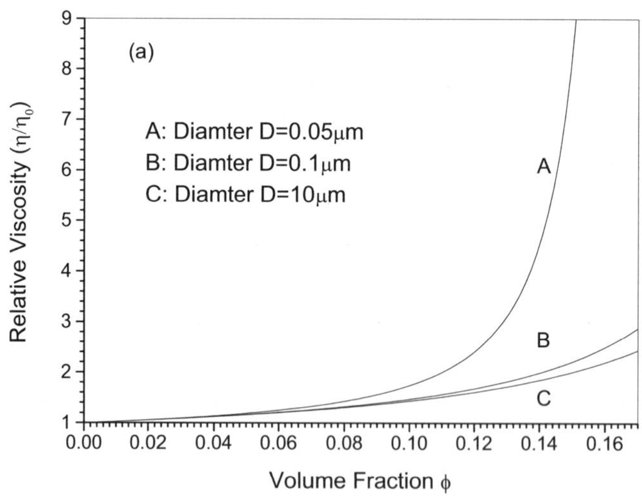

Fig.1 The viscosity versus volume fraction and particle size for mono-dispersed suspensions.

At a high ϕ, the particle size distribution has strong effect on the viscosity. There are substantial experiments on monodispersed suspensions of particles on the order of microns and sub-microns, showing that at constant ϕ, the viscosity goes down as the particle size increases *(9,10)*. As shown in Fig 1, for example, at $\phi = 15\%$, a suspension of 10.0 μm particles has a viscosity only 23.9% of the viscosity of a suspension of $0.05 \mu m$ particles *(9)*.

Generally, the effective viscosity depends on how much freedom the suspended particles have in the suspension. The less freedom for the particles, the faster the energy dissipates and the higher the effective viscosity is. The mean free path of the spheres inside the suspension is given by $a/(3\phi)$, where a is the particle radius. As a gets bigger, the mean free path becomes longer, indicating that the suspended particles have more freedom to move in the suspension. Thus η goes down.

Let us consider a liquid suspension with noninteracting spheres of diameter D suspended in a base liquid of viscosity η_0. Since D is on the order of nanometer or micrometer, the particles are classical ones. Because of the thermal vibration, these particles in the base liquid seem to be larger than their actual physical size D. This can be estimated as follows. From the thermal motion, $\frac{1}{2}mv^2 \approx \frac{3}{2}k_B T$, we have the average thermal velocity $\bar{v} \approx \sqrt{3k_B T / m}$. If the relaxation time is t_0, the length $\lambda = \bar{v} t_0$ is the additional dimension due to the thermal vibration. Therefore, in the base liquid, the particles seem to have a diameter,

$$D_x = D + \lambda . \tag{5}$$

Then, the maximum volume fraction to be available for the particles is given by

$$\phi_m = \phi_{m0}(D / D_x)^3 = \phi_{m0} /(1 + \lambda / D)^3 , \tag{6}$$

where ϕ_{m0} is about 0.64 for random close-packing, smaller than 0.74, the maximum packing fraction of a face-centered cubic (fcc) lattice. It is clear from Eq.(6) that as D increases, the maximum volume fraction ϕ_m is increased and the viscosity η is reduced from Eq.(3) or Eq.(4)

The values of ϕ_m in Eq.(4) also increases with increasing polydispersity *(8)*. For example, when the ratio of the large particles to small particles increases in a suspension of binary particle-size distribution, the viscosity reduces significantly (Fig.1b): At $\phi \geq 50\%$, when this ratio reaches 1:1, the viscosity is reduced more than 50% from the mono-disperse case. A qualitative explanation is as follows: For a binary particle-size distribution, we can consider that the small particles thicken the continuous phase and the next-size-up particles then thicken this phase, and hence

$\eta = \eta_0(1-\phi_1/\phi_{m1})^{-[\eta_1]\phi_{m1}}(1-\phi_2/\phi_{m2})^{-[\eta_2]\phi_{m2}}$, which is lower than that of a suspension of uniform small particles with volume fraction $\phi_1+\phi_2$.

In addition, experiments also show that in flow through capillary tubes the viscosity is further reduced for large suspended particles because of a tendency for large particles to migrate toward the centre of the tube *(11)*.

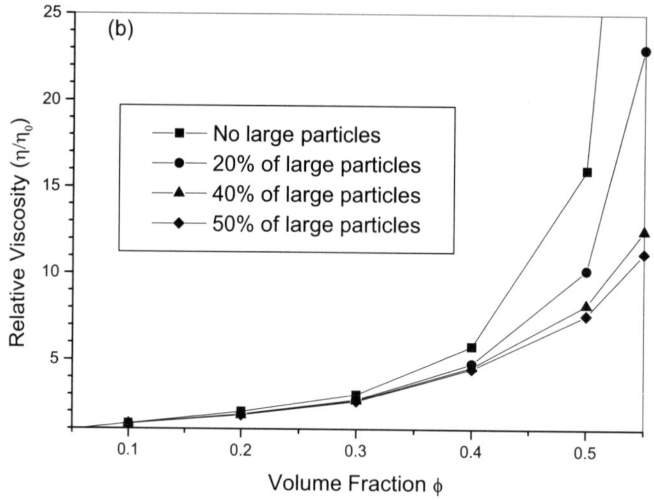

Fig.2 The viscosity versus particle volume fraction and particle size distribution for suspensions of a binary size distribution. The particle-size ratio is 5:1.

3. Reducing the Viscosity

It is clear from the above background that aggregating small particles into large ones in a liquid suspension will reduce the effective viscosity while ϕ remains the same. For most suspensions, this aggregation can be realized with either electric field or magnetic field.

We discuss the magnetic field here, but the same physics applies to the case of electric field. We assume that the magnetic permeability μ_p of the particles is different from μ_f of the base liquid. In a magnetic field, the particles are polarized along the field direction. If the particles are uniform spheres of radius *a*, the dipole moment is $\vec{m} = \vec{H}a^3(\mu_p-\mu_f)/(\mu_p+2\mu_f)$, where $\vec{H}$ is the magnetic field acting on the sphere. The interaction between two induced magnetic dipoles takes

$$U = \mu_f m^2(1-3\cos^2\theta)/r^3, \qquad (7)$$

where r is their distance and θ is the angle between the field and the line joining the two dipoles. If this interaction is strong enough to overcome the Brownian motion, the particles aggregate and align in the field direction. If this interaction is very strong, the particles quickly aggregate into macroscopic chains or columns to jam the liquid flow and increase the viscosity with the magnetic field on, a well-known phenomenon in magnetorheological (MR) fluids and electrorheological (ER) fluids *(12-17)*.

On the other hand, if the applied magnetic field is in such a short pulse that the dipolar interaction does not have enough time to affect particles separated by macroscopic distances, but has enough time to assemble nearby ones together, the assembled clusters are of limited size, say, in micrometer range. Although some particles may not join the aggregation, the aggregated particles have their size increased. During the application of field, the viscosity changes rapidly. However, after the magnetic field is turned off, the suspension has a reduced viscosity. This reflects the fact that while the volume fraction remains the same, the particle size distribution is changed: There are more large particles and the polydispersity is also increased.

It is important to note that this viscosity reduction method does not come from a change of the suspension's temperature. The reduction becomes more pronounced as the volume fraction ϕ increases. The electric or magnetic field pulse is thus more effective in dense suspensions than in dilute ones.

Let us estimate the minimum magnetic field H_c required to form clusters. For particle number density n, the typical separation between two neighbouring particles is about $n^{-1/3}$ and their dipolar interaction is about $m^2 n\mu_f$. This interaction must overcome the thermal Brownian motion in order to pull them together. Then it is required to have $\mu_f m^2 n/(k_B T) \geq 1$, where k_B is the Boltzmann constant and T is the absolute temperature. Hence we obtain the critical field

$$H_c = [k_B T/(n\mu_f)]^{1/2}(\mu_p + 2\mu_f)/[a^3(\mu_p - \mu_f)]. \qquad (8)$$

In order to change the viscosity of the liquid suspension, the applied magnetic field cannot be lower than H_c. It is quite different from ER and MR fluids, where this ratio $\mu_f m^2 n/(k_B T)$ exceeds hundreds. Here we only require the dipolar interaction is not weaker than the thermal motion.

Now let us estimate the required pulse duration. The force between two neighbouring particles is about $6\mu_f m^2 n^{4/3}$. From this force and the Stoke's drag force $6a\pi\eta_0 v$, we estimate the particle's average velocity $v = \mu_f m^2 n^{4/3} /(\pi\eta_0 a)$. The time required for two neighbouring particles to come together is about

$$\tau = n^{-1/3} / v = \pi\eta_0(\mu_p + 2\mu_f)^2 /[\mu_f n^{5/3} a^5 (\mu_p - \mu_f)^2 H^2]. \quad (9)$$

If the magnetic field pulse is much shorter than τ, there is insufficient time for the aggregation. If the pulse lasts much longer than τ, macroscopic chains may be formed to jam the flow, unfavourable for viscosity reduction. Thus, to reduce the viscosity, the pulse duration should be of order τ.

To apply the above equations for the electric field case, we only need to replace the magnetic permeability by the respective dielectric constant. The aggregated particles are usually elongated along the field direction *(18)*. The viscosity can thus be further reduced if the flow and the field direction are parallel.

After the field is turned off, the dipolar interaction disappears and the aggregated particles gradually dissemble under the Brownian motion. Therefore, the viscosity is expected to increase gradually and will return to the original value after all aggregated particles disintegrate. Let us estimate the time interval for the viscosity reduction. In absence of other disturbances, such as in a static or a constant flow state, the particles in the suspension separate diffusively only due to the Brownian motion with a diffusion constant $k_B T /(6\pi a\eta_0)$. Two spheres of radius a which are initially in contact diffuse apart by a distance of a in a time interval $3\pi a^3\eta_0 /(k_B T)$. With $a = 3\mu m$ and $\eta_0 = 1$ poise, this estimated time is about 2 hours at room temperature. Therefore, this dissembling process is slow and the viscosity reduction lasts for several hours, long enough for many important applications.

After all aggregated particles are disintegrated, the suspension returns to the rheological state prior to the magnetic treatment. Thus the viscosity returns to the original value. Reapplication of the magnetic field pulse will again reduce the viscosity. The process is repeatable.

4. Experimental Verification and Applications

The above theory has been verified by our experiments *(2)*. We first conducted experiments with a dilute MR fluid of iron nanoparticles of diameter 35-40 nm in pump oil. These iron particles were well dispersed by addition of small amount of surfactant to coat the particle surface. The volume fraction was $\phi = 9\%$. The MR fluid was in a thermal bath with a constant temperature of 23.5°C and had viscosity 880 cp on our rotational viscometer at 10 rotations per minute (rpm). It is well known that for such a

MR fluid, the viscosity increases in a constant magnetic field and the stronger the magnetic field keeps on, the higher the effective viscosity *(19)*. However, after applying a magnetic field pulse of 0.15 T for 5 minutes, the viscosity of the MR fluid decreased to 496 cp (Fig.3). Then the viscosity gradually went up. After 240 minutes, the viscosity was up to 780cp, but still lower than 880cp. After 12 hours, the viscosity was back to the original value. The viscosity right after application of a magnetic field pulse depended on the pulse duration. From 10rpm to 90 rpm, the 5-min pulse seemed to produce the maximum viscosity reduction. We measured the average size of aggregated magnetic particle cluster, which were similar to ellipsoids *(18)*.After a 1-second pulse of 0.38T, the 30-40 nm particles aggregated in to clusters with average size 9.9 μm; after a 10-second pulse, it is about 15.4 μm, and so on. The increasing average cluster size leads to the viscosity reduction.

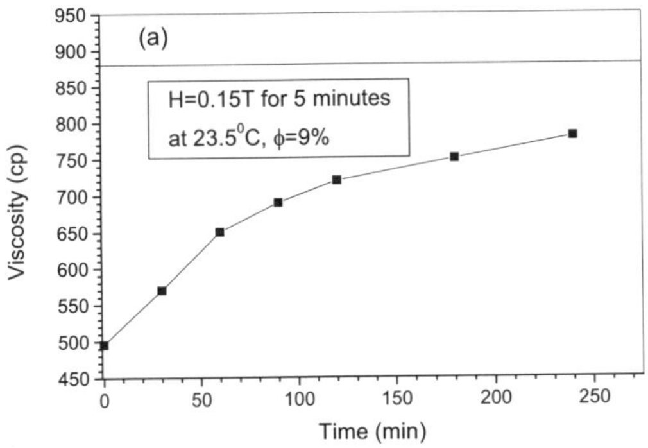

Fig.3. The viscosity of an MR fluid decreased to 496 cp from 880cp after applying a magnetic field of 0.15T for 5 minutes. It then gradually went up to return to the original rheological state.

We also applied this viscosity reduction method to crude oil *(2)*. To be specific, the paraffin-base crude oil can have its viscosity reduced by a pulsed magnetic field and the asphalt-base crude oil can have its viscosity reduced a pulsed electric field. These positive experimental results indicate that this method, developed from the basic mechanism of viscosity, is universal and powerful for all liquid suspensions with broad applications.

References

1. R. Von Flatern, *Offshore* **57**, N3, 28-31 (1997).
2. R. Tao and X. Xu, *Energy & Fuels*, (accepted for publication).
A. Einstein, Ann. Phsik [4] **17**, 549 (1905); **19**, 289, 371 (1906).
3. W. B. Russel, D. A. Saville, and W. R. Schowalter, *Colloidal Dispersion,* (Cambridge Univ. Press, Cambridge, 1991), 456-503.

4. G. K. Batchelor, J. Fluid Mech. 83, 97-117 (1977).
5. M. Mooney, J. of Colloid Sci., 6, 162 (1951).
I. M. Krieger & T. J. Dougherty, *Tans. Soc. Rheol.* **3**, 137-152 (1959).
6. H. A. Barnes, J. F. Hutton & K. Walters, K. An introduction to rheology. Elsevier, Amsterdam, 1989, pp 119-127.
7. S. Matsumoto & P. Sherman, *J. of Colloid and Interface Sci.* ***30*****,** 525-536 (1969).
8. D. G. Thomas, *J. of Coll. Scie.* **20**, 267-277 (1965).
9. G. Serge & A. Silibergerg, *J. Fluid Mech.* **14**, 86 (1951).
10. R. Tao & J. M. Sun, *Phys. Rev. Lett.* **67**, 398-341 (1991).
11. R. Tao, Journal of Physics: Condensed Matter Physics, V13, R979 (2001).
12. M. Whittle & W. A. Bullough, *Nature* **358**, 373 (1992).
13. T. C. Halsey. *Science* **258**, 761-766 (1992).
J. D. Carlson & K. D. Weiss, Magnetorheological materials based on alloy particles. US Patent 5,382,373 (1995).
K. M. Ginder & L. C. Davis, *Appl. Phys. Lett.* **65**, 3410-3412 (1994).
14. J. H. E. Promislow & A. P. Gast, *Langmuir* **12** (17), 4095-4102 (1996).
15. C. Kormann, H. M. Laun, & H. J. Richter. *Int. J. of Modern Phys. B* **10** (23-24), 3167-3172 (1996).

DYNAMIC LIGHT SCATTERING STUDY OF REVERSE MICELLAR SOLUTIONS FOR THE SYNTHESIS OF MAGNETIC NANOPARTICLES

OLIVIA A. GRAEVE AND KAUSTAV SINHA

University of Nevada, Reno
Department of Chemical and Metallurgical Engineering
1664 North Virginia Street – MS 388, Reno, NV 89557, USA
E-mail: oagraeve@unr.edu

Iron nanoparticles dispersed in hydraulic oil were prepared by mixing two microemulsions containing iron (II) sulfate and sodium borohydride at a temperature of 60°C. Six values of ω_o = [water]/[surfactant] were used, namely 5, 10, 15, 20, 25, and 30. Dynamic light scattering measurements of the hydrodynamic radius of the reverse micelles showed that the average size, surfactant shell thickness and water core radius, increased with ω_o. The micelle size distribution for ω_o values of 5, 10, and 15, was in the nanometer regime, while for ω_o values of 20, 25, and 30 it was in the micrometer regime. Scanning electron microscopy showed that the nanoparticle diameters were around 30 nm. In addition, a comparison between the particle size distribution of the iron nanoparticles dispersed in isopropyl alcohol and the iron nanoparticles in the reverse micellar solution was made. It was shown that once the particles are cleaned and separated from the micellar solution, they agglomerate into particles that are about 1 μm in size.

1 Introduction

Magnetic fluids [1] have long been considered a potential smart material due to its myriad applications ranging from magnetic seals in motors, dampers and shock absorbers [2, 3], and biological applications [4]. The strength of these smart fluids depends to a large degree on the properties of the magnetic particles dispersed in the carrier fluid, such as their magnetization strength and particle-particle and particle-fluid interactions. Even though iron-based magnetic fluids are the materials of choice for most MR applications, there are several issues associated with these fluids such as sedimentation and agglomeration of the particles.

Due to their small size, nanosized powders can remain in suspension for long periods. However, they suffer from agglomeration problems as well as high costs. To address the agglomeration problem, several synthesis techniques have been developed in the last twenty years that allow for the preparation of well-dispersed powders. In particular, the microemulsion technique has come a long way as a technique for the production of high quality nanomaterials [5, 6]. Moreover, this technique has proven successful in the preparation of uniform metallic [7, 8] and oxide [9, 10] magnetic nanoparticles. Indeed, Zhou *et al.* [11] compared this technique with conventional precipitation for the formation of Fe_3O_4 nanoparticles. In contrast to the nanoparticles prepared using the microemulsion technique, which were very uniform, the nanoparticles prepared using conventional precipitation appeared irregular in shape and size due to the uncontrolled nucleation and growth that took place in the bulk aqueous phase.

The microemulsion process involves the emulsification of an aqueous phase in an organic solvent. In this type of system, known as a reverse micellar system, the high interfacial tension between the water phase and the oil phase is reduced by the addition of a surfactant and possibly some co-surfactants. The surfactant molecules orient themselves according to the polarities of the constituent water and oil phases. Thus, due to the polarity of water, the polar heads of the surfactant molecules orient themselves towards the water droplets, as shown in Figure 1. The hydrodynamic radius refers to the combined size of the water droplets and the surfactant. If particles form inside the reverse micelles due to a precipitation reaction from solution, then the hydrodynamic radius refers to the combined size of the particles and the surfactant. The billions of reverse micelles, which form upon addition of the surfactant, act as nanoreactors for the synthesis of the desired powders. Control of the reverse micelle size is achieved by changing the water content and the amount of surfactant in the micellar solution, which is known as the water-to-surfactant ratio, ω_o [12].

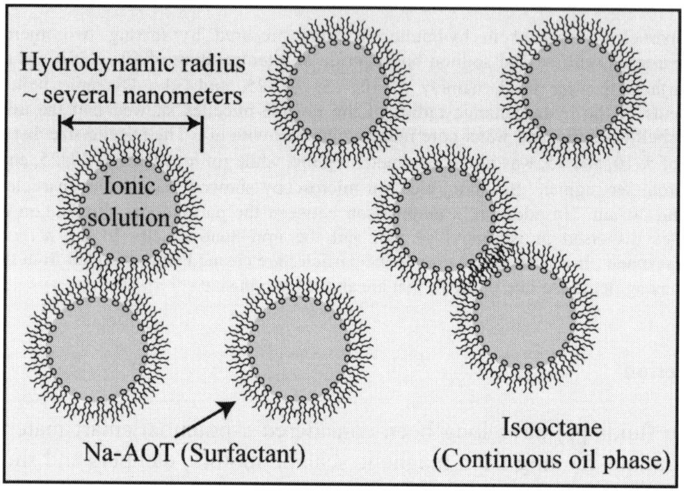

Figure 1. Schematic of a reverse micellar solution.

In this paper, we present and analyze the effect of the water-to-surfactant ratio on the hydrodynamic radius distribution of iron nanopowders using dynamic light scattering. We also present a preliminary analysis of the relationship between the final particle size of the powders and the hydrodynamic radius of the iron nanopowders.

2 Experimental Techniques

A pictorial flow chart of the synthesis process is shown in Figure 2. With some modifications, the synthesis route used in this project is based on the reaction described by Zhang [13]. As an initial step, two mixtures were prepared: (1) Mixture A: an aqueous mixture of $FeSO_4 \cdot 7H_2O$ (99.9%, Sigma Aldrich, St. Louis, MO 63178) in water, with isooctane (99+%, Fisher Scientific, Hampton, NH 03842) and Na-AOT (Sigma Aldrich, St. Louis, MO 63178) to form a microemulsion, and (2) Mixture B: an aqueous mixture of $NaBH_4$ (99.9%, Alfa Aesar, Ward Hill, MA 01835) in water, with isooctane and Na-AOT to form another microemulsion. These solutions were then mixed in a nitrogen (Air Liquide America, Houston, TX 77056) atmosphere at room temperature to form Mixture C. The combined mixture was then stirred for two hours in a nitrogen environment at 60°C. For mixtures A and B, the Na-AOT concentration was fixed at 0.5

M and the number of moles of water required was calculated using the following expression:

$$\omega_o = \frac{[H_2O]}{[AOT]} \quad (1)$$

During these experiments the ω_o ratios were varied between 5 and 30 with precursor concentrations fixed at 0.9 M for $FeSO_4$, 2.7 M for $NaBH_4$, and 0.5 M for Na-AOT. During mixing, collisions between reverse micelles in the solution are frequent with about one collision in every thousand to ten thousand leading to the formation of a short-lived dimer, during which two (or more) reverse micelles exchange the contents of their aqueous core [14], resulting in a chemical reaction. The presumed chemical reaction that occurs during mixing is the following:

$$2Fe^{2+}(aq) + (BH_4)^-(aq) + 3H_2O(\ell) \longrightarrow 2Fe^0\downarrow + (H_2BO_3)^-(aq) + 4H^+(aq) + 2H_2(g) \quad (2)$$

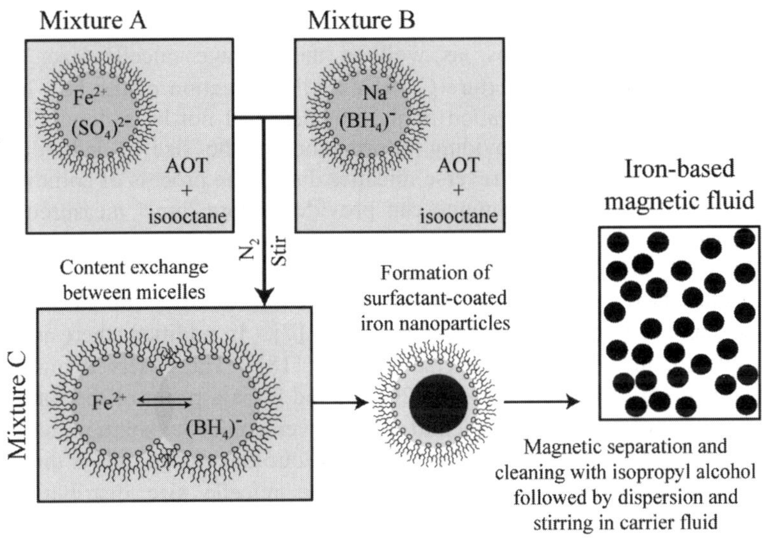

Figure 2. Pictorial flow chart of the reverse micelle synthesis process.

As can be seen from the reaction, the borohydride reduces the Fe^{2+} in the water domains, which then forms iron particles via a nucleation and growth process. Since the precipitation occurs inside the water domains, the growth of the particles is constrained. In addition, the presence of the surfactant isolates the particles so that little agglomeration occurs at this stage. After precipitation, the particles were magnetically separated, washed with isopropyl alcohol (99+%, Fisher Scientific, Hampton, NH 03842) and de-ionized water to remove both organic and inorganic impurities, and dispersed in hydraulic oil (Mobil DTE 25). During this step, agglomeration occurred due to tangling of the surfactant molecules, as has been seen during the centrifugation of ZrO_2 powders [15].

Once preparation of the fluids was completed, the powder particle sizes and micelle sizes were experimentally measured using the dynamic light scattering (DLS) technique on a Microtrac Nanotrac Ultra instrument (Microtrac, Inc., Montgomeryville, PA 18936). This instrument measures particle size distribution in solution, with measurement capability from 0.8 nm to 6.5 μm. It is capable of measuring up to 40% by solids in some cases. Our measurements were run for 45 seconds three times, after which an average was calculated. The measurement of micelle size distribution did not require any special preparation. The measurement of iron particle size distribution required the dispersion of the clean powders in isopropyl alcohol followed by light overhead stirring. Scanning electron microscopy (SEM) samples were also prepared by drop coating the fluid on a Si(111) wafer and allowed to dry. The SEM analysis was done on a Hitachi S-4700 instrument (Hitachi, Ltd., Tokyo 100-8280, Japan).

3 Results and Discussion

In a reverse micellar system, the water-to-surfactant ratio is one of the most important parameters to determine and control the size of the aqueous domains where the particle formation occurs. Controlling the size of the micelles in turn controls the particle size of the final powders. Figure 3 shows the DLS measurements of the micelle size distribution at various ω_o ratios, as well as the average micelle size, d. These measurements were taken on Mixture C. Hence, the reduction of the Fe^{2+} had already occurred, but cleaning and separation of the powders had not been done. In addition, these measurements are not providing information on the size of individual reverse micelles. Instead, the size of the reverse micelles during the process of colliding is being presented. Hence, the DLS technique can provide a more direct measurement of the dynamic behavior of the micelles and a better correlation of the real size of the micelles to the final crystallite size of the powders.

Our results show that there is an increase in the micelle size with an increase in the ω_o ratio, as has been shown in other studies [16, 17]. In addition, there is an evident polydispersity in the size of the reverse micelles [18]. The observed high values of hydrodynamic diameter for ω_o values of 20, 25, and 30 is a result of the formation of a microemulsion, as opposed to a nanoemulsion or reverse micellar solution, which is what forms at ω_o values of 5, 10, and 15. To the best of our knowledge, this is the first report on the in-situ and experimental analysis of the micelle size distribution of iron nanoparticles inside reverse micelles using the DLS technique at varying water-to-surfactant ratios. Lin *et al.* [19] measured the hydrodynamic radius distribution of Fe and Au-coated Fe particles at a value of water-to-surfactant ratio of eight and with cetyltrimethylammonium bromide as surfactant, 1-butanol as a co-surfactant, and octane as the continuous oil phase. Our own study is on a different reverse micellar system. Hence, a direct comparison between the two studies cannot be made.

After synthesis, cleaning, and separation of the powders, the final iron particle size distribution was measured. The particle size distribution was much higher than the initial micelle size due to agglomeration of the particles as a result of the magnetic separation. Figure 4 shows the difference between the in-situ micelle size in a reverse micellar system and the iron particle size after magnetic separation. Both of these size distributions correspond to an ω_o of 10. In regards to the morphology of the particles, it can be seen that very small nanoparticles (Figure 5(a)), which are less than 30 nm, cluster together into larger agglomerates (Figure 5(b)).

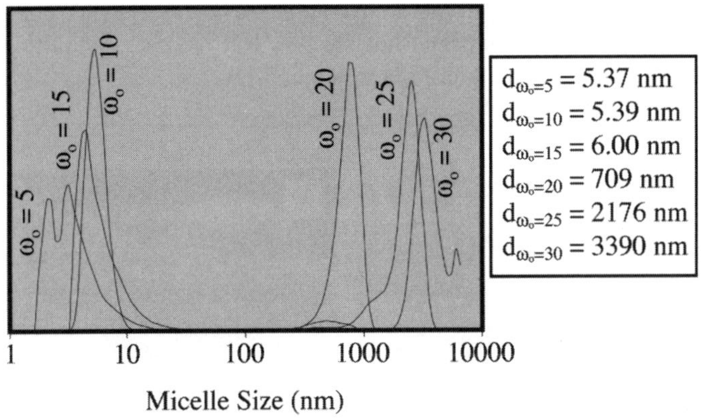

Figure 3. Effect of ω_o on the micelle size.

Currently, x-ray diffraction measurements to determine the exact phase as well as the x-ray crystallite size of the nanoparticles are underway. In addition, measurements of the particle size distribution of the nanopowders at w_o values other than 10 are being completed using DLS. Even though the particles are presumed to be pure iron, we believe there is oxidation both in our powders and the powders prepared by Lin *et al.* [19], as has been shown by Kinoshita *et al.* [20].

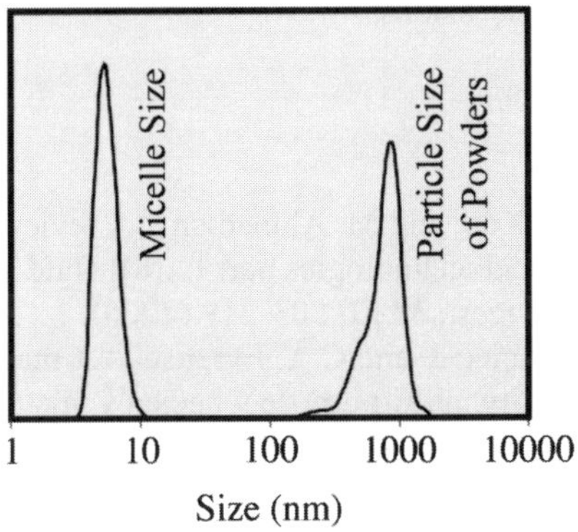

Figure 4. Dynamic light scattering measurements of reverse micelles and iron nanoparticles.

4 Conclusions

A microemulsion synthesis technique was used for the preparation of fluids that consisted of iron nanoparticles in hydraulic oil. The synthesis involved the reduction of Fe^{2+} using $(BH_4)^-$ in the protected environment of reverse micelles. During synthesis, variation of the micelle size with water-to-surfactant, ω_o, ratio was determined using dynamic light scattering measurements. The results showed that increasing the ω_o ratio increased the micelle size. The particle size of the nanopowders in isopropyl alcohol was also measured using dynamic light scattering. The particle size of these powders was in the micrometer range due to agglomeration of the particles. Indeed, observations from scanning electron microscopy indicate that the powders have a crystallite size of around 30 nm with a particle size in the micrometer range.

Figure 5. Electron micrographs of iron nanoparticles.

5 Acknowledgements

The authors gratefully acknowledge the help of Mojtaba Ahmadian-Tehrani for assistance with scanning electron microscopy, as well as Faramarz Gordaninejad and Joseph Whiteley for useful discussions.

References

[1]. F.D. Goncalves, J.-H. Koo, and M. Ahmadian, “A review of the state of the art in magnetorheological fluid technologies part I: MR fluid and MR fluid models,” *Shock and Vibration Digest*, **38** [3] 203-219 (2006).

[2]. U. Dogruer, F. Gordaninejad, and C.A. Evrensel, “A magneto-rheological fluid damper for high-mobility multi-purpose wheeled vehicle (HMMWV),” *Proceedings of the SPIE – The International Society for Optical Engineering*, **5386** [1] 195-203 (2004).

[3]. X. Wang and F. Gordaninejad, “Lyapunov-based control of a bridge using magneto-rheological fluid dampers,” *Journal of Intelligent Material Systems and Structures*, **13** [7-8] 415-419 (2002).

[4]. S.W. Charles, “Alignment of biological assemblies using magnetic fluids – a review,” *Journal of Magnetism and Magnetic Materials*, **85** [1-3] 277-284 (1990).

[5]. M.P. Pileni, “The role of soft colloidal templates in controlling the size and shape of inorganic nanocrystals,” *Nature Materials*, **2** (2003) 145-150.

[6]. B.L. Cushing, V.L. Kolesnichenko, and C.J. O’Connor, “Recent advances in the liquid-phase syntheses of inorganic nanoparticles,” *Chemical Reviews*, **104** [9] 3893-3946 (2004).

[7]. M.P. Pileni, “Reverse micelles as microreactors,” *Journal of Physical Chemistry*, **97** (1993) 6961-6973.

[8]. F. Li, C. Vipulanandan, K.K. Mohanty, "Microemulsion and solution approaches to nanoparticle iron production for degradation of trichloroethylene," *Colloids and Surfaces A*, **223** (2003) 103-112.

[9]. G.X. Cheng, F. Shen, L.F. Yang, L.R. Ma, Y. Tang, K.D. Yao, and P.C. Sun, "On properties and structure of the AOT-water-isooctane reverse micellar microreactor for nanoparticles," *Materials Chemistry and Physics*, **56** (1998) 97-101.

[10]. A.K. Gupta and M. Gupta, "Synthesis and surface engineering of iron oxide nanoparticles for biomedical applications," *Biomaterials*, **26** (2005) 3995-4021.

[11]. Z.H. Zhou, J. Wang, X. Liu, and H.S.O. Chan, "Synthesis of Fe_3O_4 nanoparticles from emulsions," *Journal of Materials Chemistry*, **11** (2003) 1704-1709.

[12]. J.H.R. Clarke, J.D. Nicholson, and K.N. Regan, "Self-diffusion and polydispersity in water+AOT+p-xylene microemuslsion: a dynamic light scattering study," *Journal of the Chemical Society. Faraday Transactions I*, **81** (1985) 1173-1182.

[13]. W.-X. Zhang, "Nanoscale iron particles for environmental remediation: an overview," *Journal of Nanoparticle Research*, **5** (2003) 323-332.

[14]. P.D.I. Fletcher, A.M. Howe, and B.H. Robinson, "The kinetics of solubilisate exchange between water droplets of a water-in-oil microemulsion," *Journal of the Chemical Society. Faraday Transactions I*, **83** (1987) 985-1006.

[15]. H. Singh and O.A. Graeve, "Reverse micelle synthesis of zirconia powders: the use of hydrogen peroxide as washing Solvent," *Mater. Res. Soc. Symp. Proc.*, **879E** (2005) Z10.28.1-Z10.28.6.

[16]. M. Zulauf and H.-F. Eicke, "Inverted micelles and microemulsions in the ternary system H_2O/aerosol-OT/isooctane as studied by photon correlation spectroscopy," *Journal of Physical Chemistry*, **83** (1979) 480-486.

[17]. A.M. Cazabat and D. Langevin, "Diffusion of interacting particles: light scattering study of microemulsions," *Journal of Chemical Physics*, **74** [6] 3148-3158 (1981).

[18]. B.H. Robinson, C. Toprakcioglu, and J.C. Dore, "Small-angle neutron-scattering study of microemulsions stabilized by aerosol-OT," *Journal of Chemical Society, Faraday Transactions 1*, **80** (1984) 13-27.

[19]. J. Lin, W. Zhou, A. Kumbhar, J. Wiemann, J. Fang, E.E. Carpenter, and C.J. O'Connor, "Gold-coated iron (Fe@Au) nanoparticles: synthesis, characterization, and magnetic field-induced self-assembly," *Journal of Solid State Chemistry*, **159** (2001) 26-31.

[20]. T. Kinoshita, S. Seino, K. Okitsu, T. Nakayama, T. Nakagawa, and T.A. Yamamoto, "Magnetic evaluation of nanostructure of gold-iron composite particles synthesized by a reverse micelle method," *Journal of Alloys and Compounds*, **359** (2003) 46-50.

RELAXATION DYNAMICS OF MAGNETIC PARTICLE CHAINS

M. BELOVS
University of Latvia, Zellu-8, Riga, Latvia
E-mail: belovs@lanet.lv

A. CEBERS
Institute of Physics, University of Latvia, Salaspils-1, LV-2169, Latvia
E-mail: aceb@sal.lv

1. Introduction.

Chains of superparamagnetic particles determine the behavior of magnetorheological suspensions[1]. Permanent flexible magnetic chains are created by linking the particles by some linker molecules[2,3]. Description of the properties of the magnetic particle chains is possible by the model of flexible magnetic filaments proposed recently[4,5]. There are several experimental investigations of the magnetic susceptibility of the magnetic chains under the action of an AC magnetic field[6,7], which possess several interesting features. As it was for the first time illustrated in[8], the magnetic filaments possess a spectrum of the magnetic relaxation times, which is determined by the deformation modes of filaments. This shows the possibility to explain the experimentally observed distribution of the relaxation times by the intrinsic properties of the magnetic filaments without any ad hoc hypothesis about the distribution of the particle sizes and other as it usually takes place when considering the dynamic magnetic properties of the magnetic colloid[9]. Efficient way of the exploring of the distribution of the relaxation is the use of the so-called Cole-Cole plots[10]. We are illustrating here that the Cole-Cole plots of the complex magnetic susceptibility of the magnetic filaments have rather interesting features, which experimental resolution can give the valuable information on the properties of the magnetic particle chains.

As it was illustrated in[8] for the ferromagnetic filaments there are two main contributions to the magnetic susceptibility – straightening of the thermal fluctuations of the filaments, which can be considered in the limit of infinite length of the filament, and bending of the free ends of the filaments in the direction of the applied field. Here we consider the magnetic susceptibility of the magnetic filaments with a different nature – the chains of the superparamagnetic particles, when an AC magnetic field is perpendicular or parallel to the constant polarizing field. In the first case the dispersion of the magnetic susceptibility arises due to the bending of the free ends of the filament, in the second – due to the straightening of the thermal fluctuations of the filaments.

2. Magnetic susceptibility in the perpendicular field.

Due to the interaction between the particles, the magnetic susceptibility of the filaments is anisotropic. Its dependence on the orientation of the filament is given by $\kappa_{ik} = \kappa_{\perp}\delta_{ik} + \left(\kappa_{\parallel} - \kappa_{\perp}\right) t_i t_k$, where $\vec{t} = (\cos\vartheta, \sin\vartheta)$ is the tangent to the filament, characterized by its angle with the direction of a constant field H_0 Magnetic torque on the filament per unit length $\left[\vec{M} \times \vec{H}\right]\pi a^2$ (a is the radius of the cross-section of the filament assumed to be circle, $\vec{M}$ is its magnetization per unit length) up to the first order terms in ϑ is $\left(\kappa_{\parallel}-\kappa_{\perp}\right)\pi a^2\left[H_0H_1-\vartheta H_0^2+H_1^2\vartheta\right]$. According to the model of the semiflexible magnetic filament [4, 5] the normal force in the cross-section of the filament due to the distributed torques is $F_n' = -\left[\vec{M}\times\vec{H}\right]_b \pi a^2$ ($\vec{n}$ is a normal to a filament, $\vec{b}=\vec{t}\times\vec{n}$ is a binormal). As a result the equation for the tangent angle in dimensionless form up to the first order term in H_1 is

$$\vartheta_t = -\vartheta_{,llll} + 2\mathrm{Cm}\,\vartheta_{,ll}\,. \tag{1}$$

The boundary conditions correspond to the free and unclamped ends of the filament

$$\vartheta_{,l}\,|_{l=\pm 1} = 0\,,$$
$$-\vartheta_{,ll} + 2\mathrm{Cm}\,\vartheta\,|_{l=\pm 1} = 2\mathrm{Cm}\frac{H_1}{H_0}\,. \tag{2}$$

Here the characteristic length scale is L ($2L$ is the length of the filament), time scale is $\tau = \dfrac{\zeta L^4}{C}$, where C is the curvature elasticity constant of the filament, ζ is the hydrodynamic drag coefficient per unit length, but $\mathrm{Cm} = \dfrac{1}{2}\dfrac{\left(\kappa_{\parallel}-\kappa_{\perp}\right)\pi a^2 L^2 H_0^2}{C}$ is the magnetoelastic number. The magnetic susceptibility in the case of superparamagnetic particles in the perpendicular field has two parts – instantaneous one, which is determined by the intrinsic processes inside the particles $\kappa_{\perp}$, and relaxing part determined by the relaxational processes of the filament. The last one contributes to the magnetization of the suspension

$$m_y = L\int_{-1}^{1}\left(\kappa_{\parallel}-\kappa_{\perp}\right)H_0\vartheta\,\mathrm{d}l\,\pi a^2\,. \tag{3}$$

Solution of the problem (1), (2) allows us to calculate according to the relation (3) the complex magnetic susceptibility in dependence on the frequency of an AC magnetic field and the magnetoelastic number. This gives $m_y = \mathrm{Re}\left[\kappa H_1 e^{i\omega t}\right]$, where

$$\kappa = 4L\pi a^2\left(\kappa_{\parallel}-\kappa_{\perp}\right)\mathrm{Cm}\,\frac{\left(\alpha_2^2-\alpha_1^2\right)\tanh\alpha_1\tanh\alpha_2}{\alpha_1\alpha_2\left[\alpha_2^3\tanh\alpha_2-\alpha_1^3\tanh\alpha_1\right]}\,.$$

Here

$$\alpha_1^2 = \mathrm{Cm} + \sqrt{\mathrm{Cm}^2 - i\omega}\,,$$
$$\alpha_2^2 = \mathrm{Cm} - \sqrt{\mathrm{Cm}^2 - i\omega}\,,$$

but ω is the frequency of AC field in dimensionless units.

We see that real κ' and imaginary κ'' parts of complex susceptibility $\kappa = \kappa' - i\kappa''$ are determined by

$$\chi' = \mathrm{Re}\left[\mathrm{Cm}\,\frac{\left(\alpha_2^2-\alpha_1^2\right)\tanh\alpha_1\tanh\alpha_2}{\alpha_1\alpha_2\left[\alpha_2^3\tanh\alpha_2-\alpha_1^3\tanh\alpha_1\right]}\right],$$

$$\chi'' = -\mathrm{Im}\left[\mathrm{Cm}\,\frac{\left(\alpha_2^2-\alpha_1^2\right)\tanh\alpha_1\tanh\alpha_2}{\alpha_1\alpha_2\left[\alpha_2^3\tanh\alpha_2-\alpha_1^3\tanh\alpha_1\right]}\right].$$

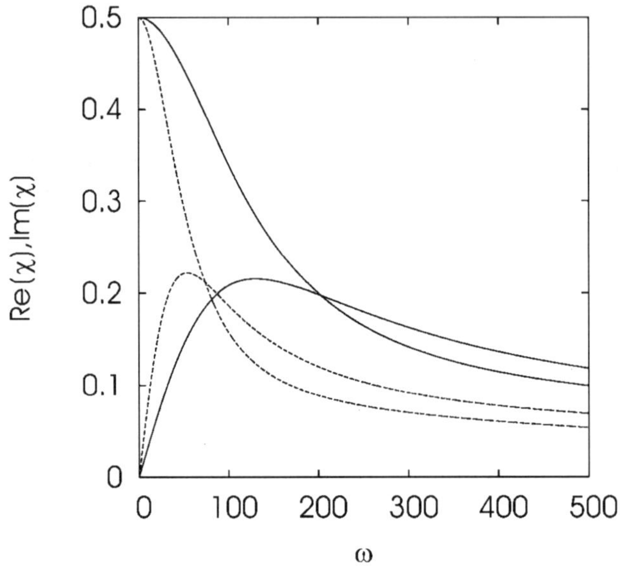

Figure 1. Real and imaginary parts of complex magnetic susceptibility in dependence on the frequency for Cm=10 (dashed line) and Cm=25 (solid line).

Interesting is the behaviour of the complex magnetic susceptibility at high frequencies. Then $\tanh\alpha_{1,2}\approx 1$ and we have $\chi\approx\dfrac{\mathrm{Cm}\left(\alpha_1+\alpha_2\right)}{\alpha_1\alpha_2}$, what gives $\chi\approx\dfrac{\mathrm{Cm}\,e^{-i\pi}\left(1-i\right)}{\omega^{3/4}}$. Thus, we see that the power law $\omega^{3/3}$ is valid for the magnetic susceptibility in the high frequency region. Dependencies of the real and imaginary parts on the frequency for two values of Cm (10; 25) are shown in Fig. 1. We should remark that these values of Cm are characteristic for the superparamagnetic filaments linked by some polymer [5].

Interesting feature of the dependencies is the shift of the frequency at which κ'' has maximum to higher value with the increase of Cm number. This corresponds to the increase of effective stiffness of the chain with increase of constant polarizing field. The Cole-Cole plots has fine structure in the region corresponding to the high-frequency ($\mathrm{Cm}\approx\sqrt{\omega}$) as it is illustrated for two values of Cm in Fig. 2 (Cm=1, 25). By dashed line in Fig. 2 the dependence corresponding to power law is shown.

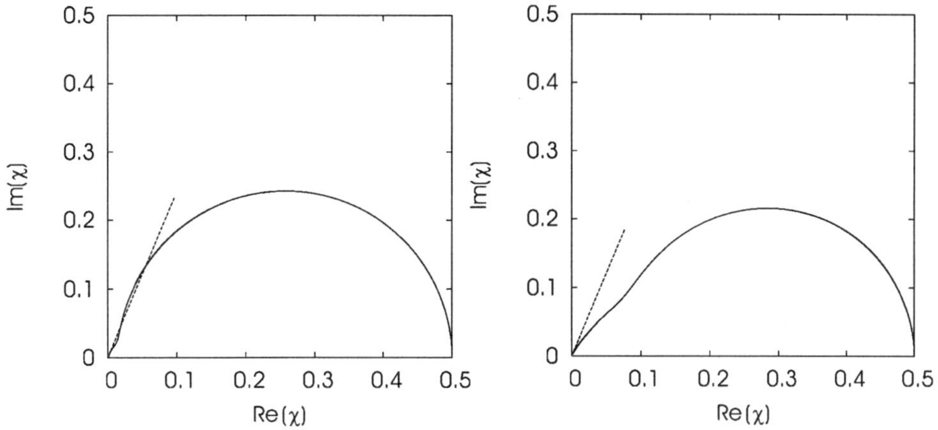

Figure2. Cole-Cole plot Cm = 1 (left) and Cm = 25 (right).

The features described for the magnetic susceptibility of the filaments in the perpendicular AC field are qualitatively similar to these of the magnetic susceptibility of the chains of superparamagnetic particles in the AC field parallel to constant field[7], where the shift of the frequency at which the imaginary part of the susceptibility has a maximum to the higher value at the magnetic field increase is observed. Nevertheless the physics responsible for the magnetic susceptibility in the case when an AC magnetic field is parallel to the constant polarizing field is quite different.

Since when the AC field is parallel to the polarizing field there are not reasons for inducing the bending of the filaments as in the previous case unless they are not exactly aligned along the field lines. The reason why the filaments are not aligned along the field lines may be due to the thermal fluctuations. Supporting evidence for the importance of the thermal fluctuations in this case, is the increase of the magnetic susceptibility with the decrease of the external magnetic field. We believe that the chains formed due to aggregating superparamagnetic particles in the strong field remain unbroken by thermal fluctuations in weaker field. The straightening of the thermal fluctuations by the applied AC field is the reason for the dispersion of the magnetic susceptibility in this case. It in the case when the polarizing field is absent is considered in[8]. Here we generalize this calculation for the case, when external polarizing field is present.

3. Straightening of thermal fluctuations in an AC magnetic field.

The equation for the tangent angle of filament reads

$$\zeta \frac{\partial \vartheta}{\partial t} = - C \frac{\partial^4 \vartheta}{\partial l^4} + \left(\kappa_{\parallel} - \kappa_{\perp}\right) \pi a^2 \left(H_0 + H_1\right)^2 \frac{\partial^2 \vartheta}{\partial l^2} + f^{\vartheta} , \qquad (4)$$

where f_{ϑ} describes the effect of thermal noise. The magnetic moment of the filament along the field per its unit length in the case when thermal fluctuations are small is

$\pi a^2\left(\kappa_{\|}(H_0+H_1)-(\kappa_{\|}-\kappa_{\perp})\vartheta^2(H_0+H_1)\right)$. Therefore the linear in the applied AC field response of the filaments averaged by the thermal noise is

$$\pi a^2\left[\left(\kappa_{\|}-(\kappa_{\|}-\kappa_{\perp})\left\langle\vartheta^2\right\rangle_0\right)(H_0+H_1)-(\kappa_{\|}-\kappa_{\perp})H_0\left\langle\vartheta^2\right\rangle_1\right], \tag{5}$$

where $\left\langle\vartheta^2\right\rangle_1$ is the term of the tangent angle fluctuation proportional to the AC field. It can be calculated according to the equation (4), where we keep only the first order terms in H_1. Representing tangent angle fluctuation by the Fourrier series

$$\vartheta=\frac{1}{L}\sum_k \vartheta_k e^{ikl}$$

and applying fluctuation–dissipation theorem

$$\left\langle f_k^{\vartheta}(t')f_{k'}^{\vartheta}(t'')\right\rangle=2\zeta k_B TLk^2\delta_{k,-k'}\delta(t'-t'')$$

in the steady case we have

$$\left\langle\vartheta^2\right\rangle_1=-\frac{8k_BT(\kappa_{\|}-\kappa_{\perp})\pi a^2H_0}{\zeta^2\omega L}\sum_k k^4\left(\sin\omega t\int_0^\infty(1-\cos\omega\tau)e^{-\tau/\tau_0(k)}\,\mathrm{d}\tau+\right.$$
$$\left.+\cos\omega t\int e^{-\tau/\tau_0(k)}\sin\omega\tau\,\mathrm{d}\tau\right). \tag{6}$$

Here the characteristic relaxation time $\tau_0(k)$ of the mode with the wave number k is introduced

$$\tau_0=\frac{\zeta}{2\left(Ck^4+(\kappa_{\|}-\kappa_{\perp})\pi a^2H_0^2k^2\right)}.$$

Substituting the summation along the wavenumbers by integration according to $\sum_k\rightarrow\frac{L}{2\pi}\int\mathrm{d}k$ and calculating the integrals as a result for the real and imaginary parts of the complex magnetic susceptibility we obtain

$$\kappa'=K\frac{1}{2\pi}\int_0^\infty\frac{k^4\mathrm{d}k}{\left(Ck^4+Hk^2\right)^2+(\omega\zeta/2)^2} \tag{7}$$

$$\kappa''=K\frac{1}{2\pi}\int_0^\infty\frac{k^4\mathrm{d}k(\omega\zeta/2)}{\left(Ck^4+Hk^2\right)}\frac{1}{\left(Ck^4+Hk^2\right)^2+(\omega\zeta/2)^2}. \tag{8}$$

Here $K=4L(\kappa_{\|}-\kappa_{\perp})^2(\pi a^2)^2H_0^2k_BT$, $H=(\kappa_{\|}-\kappa_{\perp})H_0^2\pi a^2$.

Expressions (7, 8) give the general relations for the dispersion of the magnetic susceptibility of semiflexible magnetic filament in an AC magnetic field parallel to the constant magnetic field. In order to obtain useful relations, which could be compared with the available experimental findings[7] we apply the expression for the curvature elasticity constant of the magnetic particle chain $C = M^2/2^5$ determined only by magnetodipolar interactions. Substituting $M = \pi a^2 \kappa_{\|} H_0$, scaling k in relations (7, 8) with a^{-1} and multiplying expressions (7, 8) by a number of particles per unit volume n we obtain the relations

$$\kappa' = K_T \frac{1}{2\pi} \int_0^\infty \frac{k^4 \mathrm{d}k}{\left(k^4 + l^2 k^2\right)^2 + \left(\omega \tau_m\right)^2} \tag{9}$$

and

$$\kappa'' = K_T \frac{1}{2\pi} \int_0^\infty \frac{k^4 \mathrm{d}k \, \omega \tau_m}{\left(k^4 + l^2 k^2\right)\left[\left(k^4 + l^2 k^2\right)^2 + \left(\omega \tau_m\right)^2\right]}. \tag{10}$$

Here $\tau_m = \dfrac{\zeta}{\left(\pi \kappa_{\|} H_0\right)^2}$, $K_T = \varphi \dfrac{16L}{a} \left(\dfrac{\kappa_{\|} - \kappa_{\perp}}{\kappa_{\|}}\right)^4 \dfrac{k_B T}{V\left(\left(\kappa_{\|} - \kappa_{\perp}\right) H_0\right)^2}$.

The prefactor K_T shows that magnetic susceptibility due to the straightening of thermal fluctuations is proportional to volume fraction of particles $\varphi = nV$ multiplied by the number of the particles in the chain L/a and divided by ratio of the characteristic torque due to the polarizing field per particle $V\left(\left(\kappa_{\|} - \kappa_{\perp}\right) H_0\right)^2$ to thermal energy. Due to this last factor the magnetic susceptibility is increasing at the diminishing of the polarizing field.

Parameter l^2 in terms of the physical parameters of the model can be expressed as follows $l^2 = \dfrac{2\left(\kappa_{\|} - \kappa_{\perp}\right)}{\pi \kappa_{\|}^2} = \dfrac{8}{\mu + 1}$.

This shows that the realistic value of the parameter l is about 1. The Cole-Cole plot in terms of the frequency dependent integral on the right part of relations (9) and (10) is shown in Fig. 3. We see that in this case the Cole-Cole plot is the arc of circle which height is about 4 times less than its chord. As we judge from misleading scales of

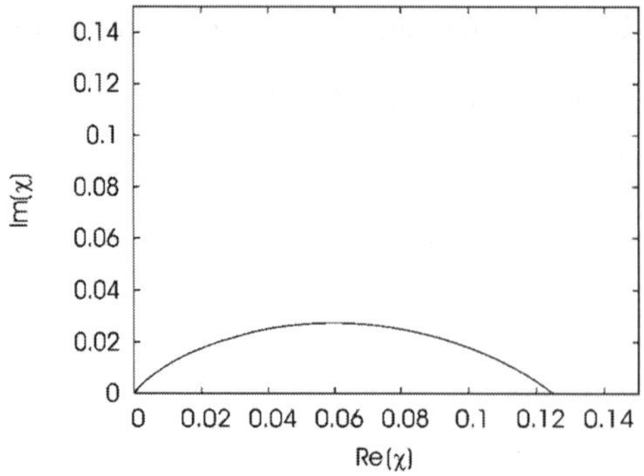

Figure 3. Cole-Cole plot of the complex magnetic susceptibility due to straightening of thermal fluctuations. $l = 1$.

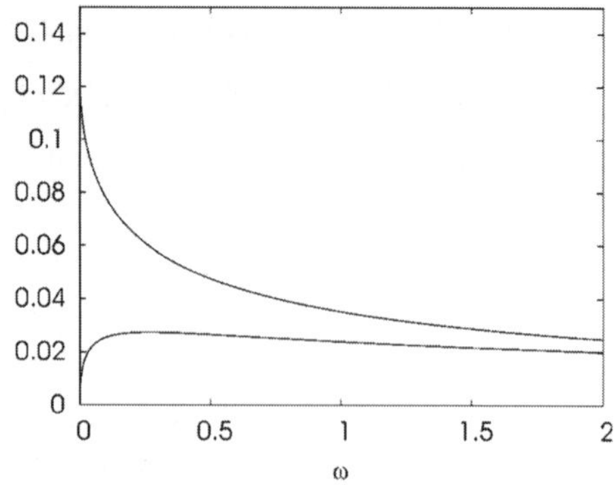

Figure 4. Real and imaginary parts of complex magnetic susceptibility at straightening of thermal fluctuations. $l = 1$.

Fig.4 in[7], it is close to what is observed in experiment. Close to experimental data is also the value of the frequency at which the maximum of the imaginary part of complex susceptibility takes place. As we can see from Fig. 4, where real and imaginary parts of the complex susceptibility are shown at $l = 1$ the maximum take place at $\omega\tau_m \approx 0.27$. This corresponds to the frequency of an AC field $f_m = \frac{\omega}{2\pi} = \frac{0.27}{2\pi\tau_m} = \frac{0.27}{2\pi}\frac{(\pi\kappa_{\|}H_0)^2}{4\pi\zeta}$. Estimating ζ according to $\zeta = 4\pi\eta/[\ln 10 + 1]$, $\eta = 10^{-2}$ P, for f_m in the field $H = 13.6$ kA/m we have 2.04 kHz, which is very close to the value 1.77 kHz given in[7].

It should be noticed that the Cole-Cole plot of the frequency dependence of κ' and κ'' in this case is qualitatively different from the characteristic to Debye model. So at frequency at which imaginary part has maximum it is much smaller than real part. This and rather slow decay of κ'' with frequency also is in qualitative agreement with experimental findings in[7].

4. Conclusion.

On the basis of the model of semiflexible magnetic filaments the characteristic features of the magnetic dispersion of the suspension of the filaments are analyzed. Two characteristic machanisms of the magnetic relaxation are considered – bending of the free ends and staightening of the thermal fluctuations. The results obtained are in qualitative and in some parts also in quantitative agreement with the available experimental data.

References

1. G. Bossis (Ed.), *Electrological Fluids and Magnetorheological Suspensions* (World Scientific Publishing Company, 2002).
2. C. Goubault *et al*, *Phys. Rev. Lett.* **91**, 260802, (2003).
3. S.L. Biswal, A.P. Gast, *Phys. Rev. E* **68**, 021402, (2003).
4. A.Cebers, *J. Phys: Condens. Matter* **15**, S1335, (2003).
5. Cebers, *Current Opinion in Colloid & Interface Science* **10**, 167, (2005).
6. K. Butter *et al, J.Phys.:Condens.Matter* **15**, S1451, (2003).
7. P.C. Fannin *et al*, *JMMM* **300**, e210, (2006).
8. M. Belovs, A. Cebers, *Phys. Rev. E* **73**, 021507, (2006).
9. F.Gazeau *et al, Phys.Rev. E* **56**, 614, (1997).
10. E. Blums *et al*, *Magnetohydrodynamics* (N.Y.) **23**, 45, (1997).

ELECTRORHEOLOGICAL AND DIELECTRIC PROPERTIES OF UREA/SIO_2 NANOCOMPOSITE SUSPENSIONS MODIFIED BY N, N-DIMETHYLFORMAMIDE

T. BELZA, V. PAVLINEK, I. KURITKA AND P. SAHA

Tomas Bata University in Zlin, Polymer Centre, TGM 275, 762 72 Zlin, Czech Republic
E-mail: pavlinek@ft.utb.cz

OTAKAR QUADRAT

Institute of Macromolecular Chemistry, Academy of Sciences of the Czech Republic, Heyrovsky Sq.2, 162 06 Prague 6, Czech Republic

In this study, electrorheological (ER) behaviour of silica nanocomposite suspensions treated with urea and N, N – dimethylformamide (DMF) in DC electric field has been investigated. While the ER effect of the neat silica itself was very low, the modification of silica nanoparticles improved compatibility of the solid and liquid phase and increased considerably ER activity of the system. In contrast to maximum possible concentration about 5 wt.% of neat silica due to particle aggregation 20 wt.% suspension of treated particles with low field-off viscosity could be prepared. The dielectric measurements showed that with increasing amount of urea deposited on the silica particles both the difference between the limit values of the relative permittivities and the relaxation frequency increased. This indicates a great influence of both particle polarizability and the rate of rearrangement of the ER structure in the electric field on the ER intensity. After DMF addition the changes in dielectric properties reflected the higher ER activity. At higher particle loading (25 wt.%) mutual particle interaction increased and field-off viscosity steeply rose. The comparison of the behaviour of 20 and 25 wt.% suspensions of modified particles showed that even if high yield stress at higher particle content under electric field application sets in, its relative increase indicating the ER efficiency due to high field-off value may be much lower than at lower suspension loading.

1 Introduction

At the present time electrorheological (ER) fluids belong to the highly perspective smart materials for application in various technical fields. They are typically prepared by dispersing the solid electrically polarizable particles in a non-conducting liquid of a low relative permittivity. In the absence of electric field, they have the properties of suspensions of neutral solid particles. Upon the application of the electric field their structure undergoes a rapid reversible transition resulting in development of the high yield stress and increased viscosity. The polarized particles form fibrous chains oriented along the streamlines of the electric field and a yield stress is required to break this structure. When the electric field is removed the particles return to random distribution and the system behaves like the original suspension.

Since the discovery of the ER effect [1] many studies have been published. Survey of the main results obtained in this field has been summarized in several reviews [2-7]. Various mechanisms have been proposed to explain the ER phenomenon. It has been concluded that particle polarizability plays the main role in the building of the organized structure in the electric field.

In the preceding study [8] the crucial dependence of the electrorheological efficiency of the suspensions of silica nanocomposites treated with urea and N, N-diemethylformamide (DMF) on the particle concentration has been demonstrated. This paper analyses the relationships between the ER and dielectric properties of these materials.

2 Experimental

2.1 *Modification of silica with urea and sample preparation*

10 g of silica nanoparticles (average particle size ≃ 10 nm, Aerosil A 200, Degussa, Germany) was immersed in 500 ml of distilled water at 40 – 50 °C. After 1 hour of stirring 10 ml of 5, 10 and 20 wt. % of aqueous solutions of urea was added and suspension stirred for another 16 hours. Then the suspension particles were collected by suction filtration, washed with successive small portions of ethanol and dried at 40 °C in vacuum for 24 hours. The amount of urea deposited on the silica particles estimated by thermogravimetric analyzer (Setaram SetSys 1200 Evolution, France) coupled with FTIR spectrometer (Scimitar, Digilab, USA) was about 0.7, 1.2 and 2.8 wt.%.

Suspensions with 5, 10, 20 and 25 wt.% of dry silica powders treated with urea in silicone oil (Lukosiol M15, Chemical Works Kolín, Czech Republic; viscosity η_c = 14.475 mPas, density d_c = 0.965 g.cm^{-1}) were prepared. In the given cases 0.5 wt.% of DMF was added. Suspensions were mechanically stirred before each measurement.

2.2 *Viscometry, conductivity and dielectric measurements*

Measurement of apparent viscosity was carried out in the range of shear rates 0.7 – 50 s^{-1} using a rotational viscometer (Bohlin Gemini, Malvern, UK) with parallel plates 40 mm in diameter and 0.2 mm gap. The plates were connected to a DC power supply, U = 0 – 5 kV (high-voltage source TREK 668B, TREK, USA). The sample temperature during experiments was maintained at 25 °C.

The DC conductivity of suspensions was calculated from the current passing through the sample between the plates in the gap at rest. The maximum current density obtained with 20 wt.% suspension of urea – coated nanoparticles with DMF at the field strength 5 kV mm^{-1} was only 31 μA cm^{-2} (Fig. 1). For suspensions of urea – coated silica particles the current density was lower. Thus, short-circuit between the electrodes did not occur.

The frequency dependence of relative permittivity, ε', and relative dielectric loss, ε'', in the frequency range of 10 – 10^5 Hz were measured with Hioki 3522 RLC HiTester (Hioki, Japan).

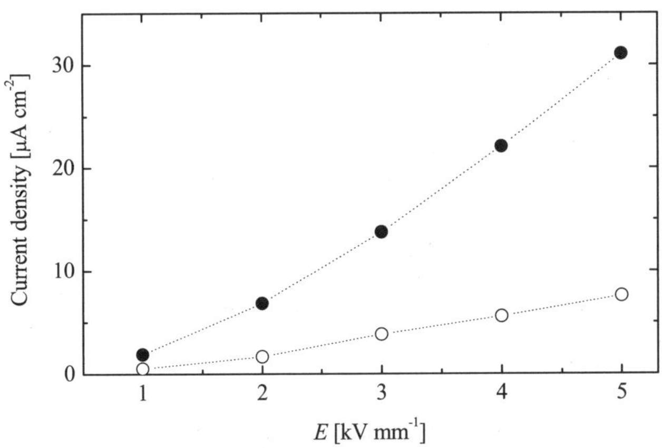

Figure 1. The dependence of the current density on the field strength E for 20 wt.% suspensions in silicone oil. Symbols: ○ silica + urea, ● silica + urea + DMF. Urea concentration (wt.%): 1.2.

3 Results and discussion

3.1 Flow behaviour

It has been shown that bad compatibility of uncoated silica nanoparticles with the oil medium does not allow to prepare homogeneous suspensions with concentrations higher than about 5 wt.%. With higher particle content big aggregates arose (Fig. 2) and, consequently, viscosity steeply increased (Fig. 3a). In contrast, however, after particle coating with urea the maximum particle concentration with low suspension viscosity was 20 wt.%. Then a steep rise started till the maximum possible loading 25 wt.%. In both cases the DMF addition had no influence.

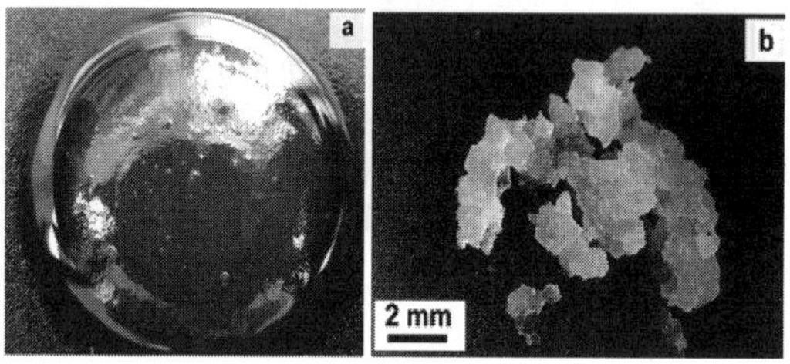

Figure 2. Micrographs of neat silica suspensions a) 5 wt. % and b) 7 wt. % in silicone oil.

The different flow curves of 5 wt. % suspensions (Fig. 3b) suggest considerable lowering of particle interaction and, consequently, weakening their aggregation tendency due to particle modification.

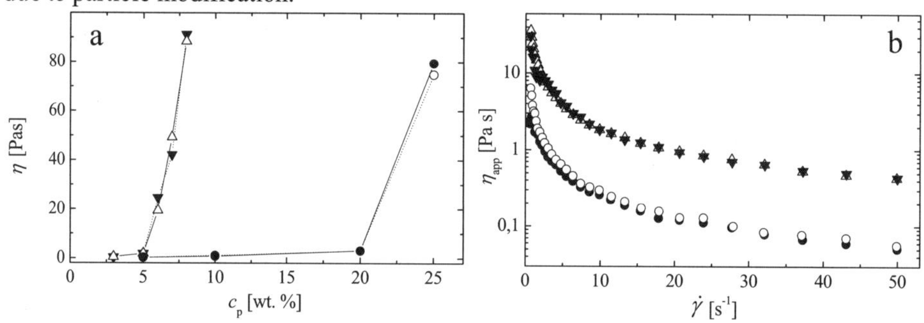

Figure 3. The dependence of suspension apparent viscosity at the shear rate 10 s^{-1} on the particle concentration c_p (a) and the dependence of suspension apparent viscosity η_{app} (b) on the shear rate. Symbols: Δ neat silica, ▼ silica + DMF, ○ silica + urea, ● silica + urea + DMF. Urea concentration (wt.%): 1.2.

After electric field application the viscosity of both uncoated and coated particles increased due to formation of ER chain particle structure in suspensions. A relative increase in apparent viscosities of 5 wt.% suspension of urea – modified particles was much higher than that of neat silica particles (Fig. 4). The flow curves of both systems nearly merged at the maximum field strength (5 $kVmm^{-1}$) in spite of different field off viscosities.

At higher suspension loading with modified particles intensity of the ER effect increased (Fig. 5a). For better evaluation of rigidity of ER structure the dynamic yield stress has been obtained by linear extrapolation of the shear stress – shear rate plots (Fig.

5b). For 20 wt.% suspensions of urea modified particles or after DMF addition it has been shown that while the field – off values of dynamic yield stress were virtually the same (26 Pa), after application the field strength 5 kV mm^{-1} in the former case the dynamic yield stress increased to about 1 kPa and in the latter even more.

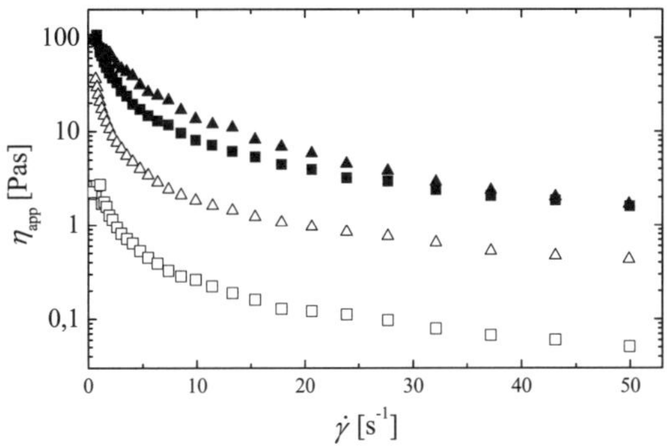

Figure 4. The dependence of apparent viscosity η_{app} of the 5 wt.% suspensions on the shear rate. Symbols/electric field strength [kV mm^{-1}]: △/0 neat silica, □/0 silica + urea, ▲/5 neat silica, ■/5 silica + urea. Urea concentration (wt.%): 1.2.

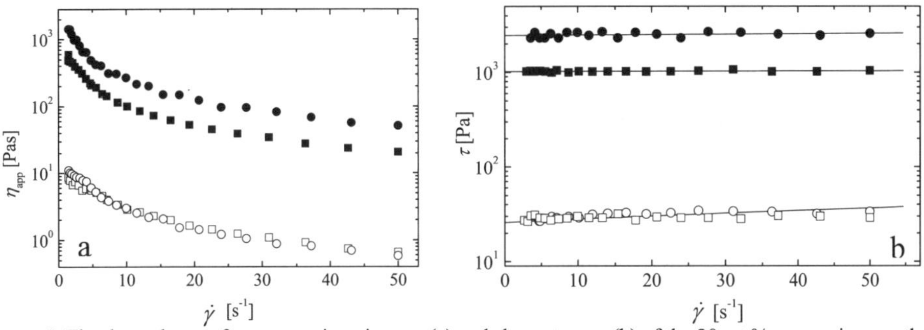

Figure 5. The dependence of apparent viscosity η_{app} (a) and shear stress τ (b) of the 20 wt.% suspensions on the shear rate. Symbols/field strength [kV mm^{-1}]: □/0 silica + urea, ○/0 silica + urea + DMF, ■/5 silica +urea, ●/5 silica + urea + DMF. Urea concentration (wt.%): 1.2.

3.2 Dielectric properties and ER activity

Concerning the mechanism of electrorheology it is beyond doubt that in the electric field dielectric polarization of suspension particles occurs. The extent of charge separation responsible for this effect and the rate at which the polarization proceeds are related to permittivity in its complex form

$$\varepsilon^* = \varepsilon' + \varepsilon'' \tag{1}$$

where ε' and ε'' are respectively the relative permittivity and the dielectric loss factor. It has been assumed that the magnitude of polarization responsible for electrorheological effect depends on the difference between the limit values of the relative permittivities at the frequencies below and above the relaxation frequency f_{crit} for which the dielectric loss factor has a maximum [2, 9, 10]. It has been concluded that if the interfacial polarization of suspension particles as the main cause of ER phenomenon occurs, the relaxation frequency of electrorheological fluid must be in the range between ca 10^2 – 10^5 Hz.

Figure 6 illustrates the frequency spectra of the relative permittivity and the dielectric loss factor of suspensions of particles modified with various amount of urea without and with the DMF addition. The solid line represents the theoretical results calculated according to Havriliak-Negami empirical equation [11]

$$\varepsilon^* = \frac{\varepsilon_\infty + (\varepsilon_0 - \varepsilon_\infty)}{\left((1 + i2\pi f t_c)^{1-\alpha}\right)^\beta} \qquad (2)$$

where t_c is the relaxation time, α the scattering degree of relaxation time and β is related to the asymmetry of the relaxation times spectrum. The large α value means a great scattering of relaxation time. When α and β differ more and more from zero and unity, respectively, the relaxation time spectrum becomes more and more asymmetrical.

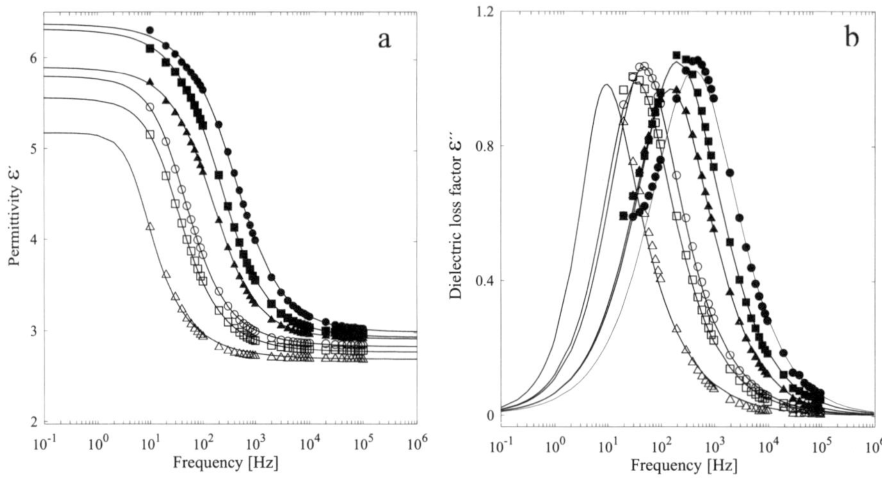

Figure 6. The dependence of the relative permittivity ε'(a) and the dielectric loss factor ε''(b) on frequency f of suspensions with 20 wt.% of modified silica particles. Urea concentration (wt.%): (△/▲) 0.7, (□/■) 1.2, (○/●) 2.8. Solid symbols represent suspensions with DMF addition.

If $2\pi f t_c = 1$, the relaxation frequency $f_c = 1/2\pi t_c$ corresponds to the maximum of the dielectric loss factor. The permittivities ε_0 and ε_∞, the parameters α and β and relaxation time t_c calculated from Eq.2 are presented in Table 1.

Table 1. Parameters of Eq. 2 for 20 wt.% suspensions of silica particles modified with various amount of urea *u* and DMF addition.

u (wt.%)	ε_0	ε_∞	$\Delta\varepsilon$	t_c (s)	α	β
0.7	2.68	5.17	2.49	0.024	0	0.612
1.2	2.76	5.56	2.80	0.0057	0.147	0.768
2.8	2.82	5.79	2.97	0.0042	0.169	0.798
0.7+DMF	2.90	5.89	2.99	0.00102	0.268	1.00
1.2+DMF	2.92	6.31	3.39	0.000711	0.289	0.979
2.8+DMF	2.98	6.37	3.39	0.000332	0.324	1.08

Figure 7 demonstrates that the intensity of the ER effect expressed as the dynamic yield stress of suspensions under the highest field strength used (5 kV mm^{-1}) increases with higher amount of urea deposited on the silica particles. The difference between the relative permittivities $\Delta\varepsilon = \varepsilon_0 - \varepsilon_\infty$ increases too. This means that particle polarizability rises. At the same time a steep decrease in the relaxation time respectively increase in relaxation frequency can be observed. This trend was more pronounced after DMF addition. The effect may be explained as a result of increasing tendency to rearrangement of ER structure in flow due to higher rate of particle polarization [12]. The parameters α and β correspond to moderate broadening and higher asymmetry of relaxation time spectra.

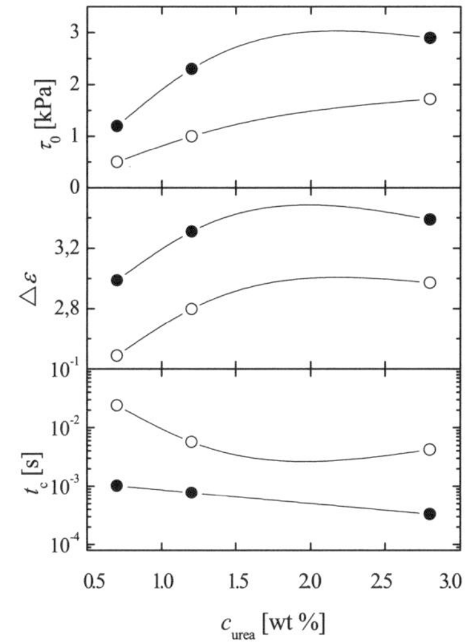

Figure 7. The dependence of the dynamic yield stress, τ_0, the difference between relative permittivities, $\Delta\varepsilon$, and relaxation times t_c on the urea concentration c_{urea}. Suspensions with 20 wt.% of modified silica particles. Urea concentration (wt.%): 1.2. Solid symbols represent suspensions with DMF addition.

3.3 *The efficiency of ER fluids*

Figure 3a demonstrates that field – off viscosities of suspensions between 20 and 25 wt.% of modified particles went through a steep rise. When the electric field is applied, viscosities as well as the dynamic yield stresses of both suspensions increased (Fig. 8a). Even though the dynamic yield stress of 25 wt.% suspension was, at the maximum field strength, three times higher than that at 20 wt.% loading. Its relative increase in electric field proved to be much lower, owing to the high field-off yield stress (Fig. 8b).

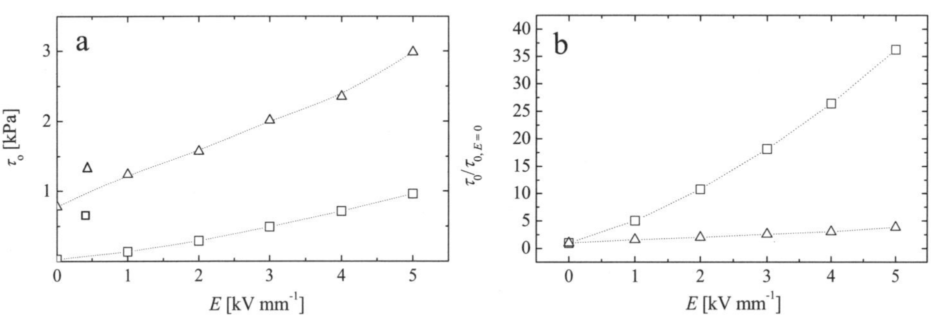

Figure 8. The dependence of dynamic yield stress τ_0 (a) and the relative increase in the dynamic y ield stress $\tau_0/\tau_{0,E=0}$ (b) on the electric field strength E. Suspensions of 20 wt.% (□) and 25 wt.% (Δ). Urea concentration (wt.%): 1.2.

4 Conclusion

It is well known that the nanosize dimension of suspension particles is a significant factor for obtaining high-performance ER system. It appeared that in case of silica nanoparticles their treatment by high permittivity agent such as urea provided material with promising ER efficiency. In this system obviously a firm anchoring of urea by possible bonding with silanyl groups present on the silica particle surface enhances rigidity of particle chain structure in the electric field.

The measurement of frequency dependence of complex permittivity demonstrated an increasing difference between the limit values of the relative permittivities and decreasing relaxation time of suspension particles with the amount of urea or DMF deposited on silica particles. This findings confirmed dielectric properties as crucial factor controlling both particle polarizability and rate of rearrangement the ER structure in the electric field.

The comparison of flow properties of ER structure with field–off viscosity showed that the practical utilization of the ER phenomenon apart from high ER activity calls for the lowest field-off fluidity of ER suspension so that a highest possible relative increase in the low-shear viscosity or yield stress after field strength application may be obtained.

5 Acknowledgement

Authors wish to thank to Ministry of Education, Youth and Sports of the Czech Republic (MSM 7088352101) and Grant Agency of the Czech Republic project No. 202/06/0419 for financial support.

References

1. W. M. Winslow, Induced Fibration of Suspensions, *J. Appl. Phys.* **20** (1949) pp. 1137-1140.
2. H. J. Block, P. Kelly, Electro-rheology, *J. Phys. D: Appl. Phys.* **21** (1988) pp. 1661-1677.
3. T. C. Jordan, M.T. Shaw, Electrorheology, *IEEE Trans. Electric Insul.* **24** (1989) pp. 849-861.
4. K. M. Blackwood, H. Block, Semi-conducting Polymers in Electrorheology: A Modern Approach to Smart Fluids, *TRIP* **14** (1993) pp. 98-104.
5. M. Parthasarathy, D.J. Klingenberg, Electrorheology: Mechanisms and Models, *Mat. Sci. Eng.,* R **17** (1996) pp. 57-103.
6. T. Hao, Electrorheological Fluids, *Adv. Mater.* **13** (2001) pp.1847-1857.
7. T. Hao, Electrorheological Suspensions, *Adv. Colloid Interface Sci.* **97** (2002) pp. 1-35.
8. T. Belza, V. Pavlínek, P. Sáha, O. Quadrat, Electrorheological properties of urea/SiO_2 nanocompositesuspensions modified by N, N – dimethylformamide *Colloids and Surfaces A*, accepted.
9. F. Ikazaki, A. Kakai *et al*, Mechanism of electrorheology: the effect of the dielectric property, *J. Phys. D, Appl. Phys.* **31** (1998) pp. 336-347.
10. H. J. Choi, M. S. Cho, K. To, Electrorheological and dielectric characteristics of semiconductive polyaniline-silicone oil suspensions, *Physica A*, **254** (1998) pp. 272-279.
11. S.Havriliak, S.J. Havrilijak, Dielectric and Mechanical Relaxation in Materials, Hanser/Gradner Pub., Cincinnati (1997) pp. 29.
12. H. Block, J. P. Kelly, A. Quin, T. Watson, Materials and Mechanisms in Electrorheology, *Langmuir*, **6** (1990) pp. 6-14.

VISCOELASTIC PROPERTIES OF SILICONE-BASED MAGNETORHEOLOGICAL ELASTOMERS

H. BÖSE

Fraunhofer-Institut für Silicatforschung ISC, Neunerplatz 2, D-97082 Würzburg, Germany
E-mail: boese@isc.fraunhofer.de

Magnetorheological (MR) elastomers are composite materials consisting of magnetic particles in elastomer matrices, whose mechanical properties can be influenced by applying a magnetic field. Main parameters which determine the behavior of these smart materials are the concentration of the magnetic particles and the mechanical stiffness of the elastomer matrix. The viscoelastic properties of silicone-based MR elastomers are outlined in terms of their storage and loss moduli. The mechanical behavior of the material is also influenced by a magnetic field during the curing of the elastomer matrix, which leads to materials with anisotropic microstructures. The storage and loss moduli of soft elastomer matrix composites can be significantly increased in the presence of a magnetic field by considerably more than one order of magnitude. This increase exceeds that of all previously reported data. A shape memory effect, i. e. the deformation of an MR elastomer in a magnetic field and its return to original shape on cessasion of the magnetic field, is described.

1 Introduction

Magnetorheological (MR) elastomers are a relatively new class of adaptable materials which have attracted some attention in recent years [1-5]. These composite materials generally consist of an elastomeric matrix containing a magnetically polarizable particle dispersion. In the presence of a magnetic field the particles become polarized and exert forces on oneanother due to dipolar interactions. As a consequence, a reversible and continuously controllable increase in mechanical properties such as Young's modulus and shear modulus occurs.

MR elastomers may be regarded as a solid counterpart of MR fluids which are suspensions of magnetically polarizable particles in a carrier liquid. The main difference between MR elastomers and fluids is that the former have a defined shape into which they return on release of the deforming load. Consequently, MR elastomers can be periodically deformed without damage only for limited amplitudes. However, in contrast to MR fluids, they do not require to be contained. Another advantage of MR elastomers is that sedimentation of the magnetic particles does not occur. The exciting properties of these materials are caused by the superposition of internal magnetic and elastic forces which can be tailored by the composition of the composite.

Possible commercial applications of MR elastomers include magnetically controllable vibration dampers, capable of being operated under shear or compression load conditions. Other products may be identified in adaptable mounts, sealings or valves, where the flow of a liquid or gas medium may be controlled by a magnetic field. In addition, haptic surfaces with changeable consistency make interesting potential applications.

In comparison with MR fluids, only relatively few publications have hitherto dealt with the preparation, properties and applications of MR elastomers. Jolly et al. established a model of the mechanical properties of MR elastomers based on a silicone matrix and compared the results with experimental data [1]. The curing of their MR elastomer samples was performed in a magnetic field. A maximum increase in the measured storage modulus of 39 % whilst in the magnetic field was observed. Lokander and Stenberg described the performance of MR rubber materials with different elastomer compositions and achieved a maximum increase in storage modulus of about 25 % [2, 3]. Ginder et al. used natural rubber for their MR elastomers and reported comparable properties [4].

In the following, the viscoelastic properties of MR elastomers are described that are based on relatively soft silicone rubber filled with iron particles. The materials show factors of increase in storage modulus in the magnetic field, which are considerably larger than corresponding data in former publications.

2 Experimental

Samples of MR elastomers used in the investigations were composed of polydimethylsiloxane, which is the most common silicone rubber, and carbonyl iron particles with a mean particle size of about 5 μm. The rubber was synthesized by a reaction of two different types of silicone molecules containing vinyl groups or Si-H groups, respectively. The mechanical stiffness of the silicone rubber can be influenced by changing the density of crosslinks between polymer molecules. The iron particles were dispersed in the liquid precursors in different concentrations (10 and 20 % by volume) prior to the curing reaction. Parts of the samples were cured in a field with a magnetic flux density of 1.3 T generated by an electromagnet.

Measurements of the viscoelastic properties of the MR elastomer samples were conducted using a Paar-Physica MCR 300 rheometer equipped with a magnetorheological cell with parallel plates. The magnetic flux lines were generated across the gap perpendicularly to the plates. For the samples cured in a magnetic field, the orientation of this field was the same as the orientation of the magnetic field applied later in the rheometer measurements.

Oscillation experiments were carried out by varying several parameters. These parameters included the magnetic flux density, the oscillation frequency and the deformation amplitude. Viscoelastic properties such as storage modulus G' (elastic energy storage) and loss modulus G'' (viscous energy dissipation) were derived from the measurement of shear stress and phase shift between the shear stress and deformation. The loss factor is defined as the ratio of loss and storage moduli.

3 Results and Discussion

3.1 Variation of the iron particle content

MR elastomer samples with iron particle contents of 10 and 20 % by volume were compared with a reference sample without particles. Fig. 1 shows the dependence of the storage and loss moduli on the magnetic flux density for the three materials which were cured without the influence of a magnetic field. The storage modulus of the sample with 20 % iron particles is increased by more than 100 % in a magnetic field of 670 mT. This hardening of the composite material is caused by magnetic interactions between dipoles that are polarized in the magnetic field.

A significant increase of the loss modulus of the MR elastomers in the magnetic field becomes visible in Fig. 1, too. This means that the damping behavior is also strongly influenced by the magnetic field. Additionally, if the magnetic field is absent the loss modulus rises with the iron particle content, which is attributed to friction phenomena of the particles during shear motion.

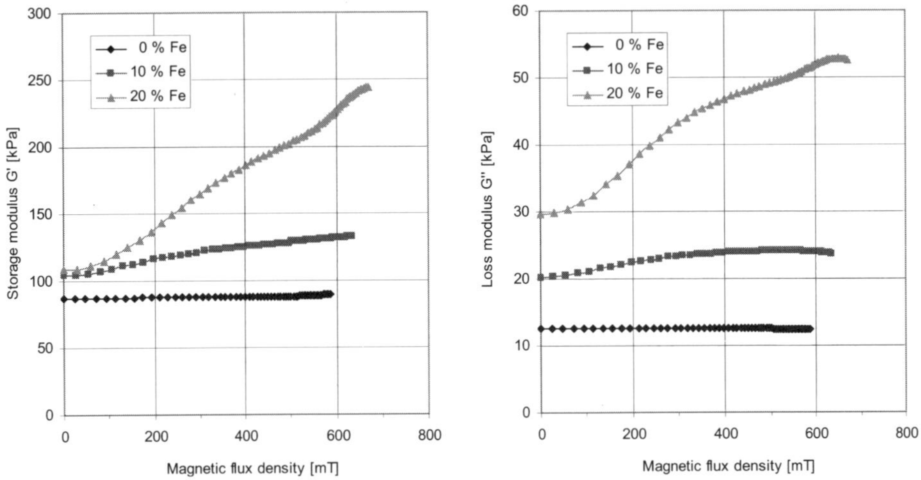

Figure 1. Dependence of storage modulus (left) and loss modulus (right) on magnetic flux density for samples with different iron particles contents in % by volume (frequency 10 Hz, amplitude 1 %)

3.2 *Curing in a magnetic field*

For comparison, corresponding MR silicone elastomer samples were prepared by applying a strong magnetic field during the curing process. Fig. 2 reveals the microstructure and the arrangement of the magnetic particles in the two samples with 20 % iron particles, cured without and with magnetic field, respectively. It can be clearly observed that the particles in the first sample are randomly distributed, whereas in the second sample the particles form chain segments along the magnetic field lines.

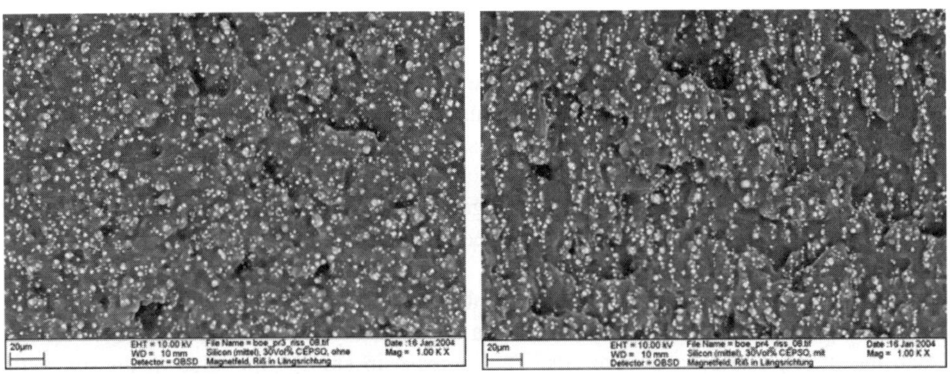

Figure 2. Scanning electron microscope (SEM) graphs of MR elastomer samples cured without (left) and with (right) magnetic field (field lines in vertical direction)

The viscoelastic data of the two groups of MR elastomer materials is shown in Fig. 3. The application of the magnetic field during the curing process enhances the basic storage modulus of the composite with 20 % particles (without a magnetic field being applied during the measurement). In addition, the relative increase in storage modulus in the magnetic field is more pronounced for the sample cured within the magnetic field than for the corresponding sample cured without field (maximal factor of 4.5 vs. 2.5). Furthermore, the loss factor of the MR elastomer materials cured in the magnetic field is increased in comparison to that of samples cured in the absence of a magnetic field. In conclusion, it can be said that the elastic and viscous properties of the MR elastomers cured with or without magnetic field, respectively, can be strongly influenced by a magnetic field after the curing process. This effect is more pronounced for samples cured within a magnetic field.

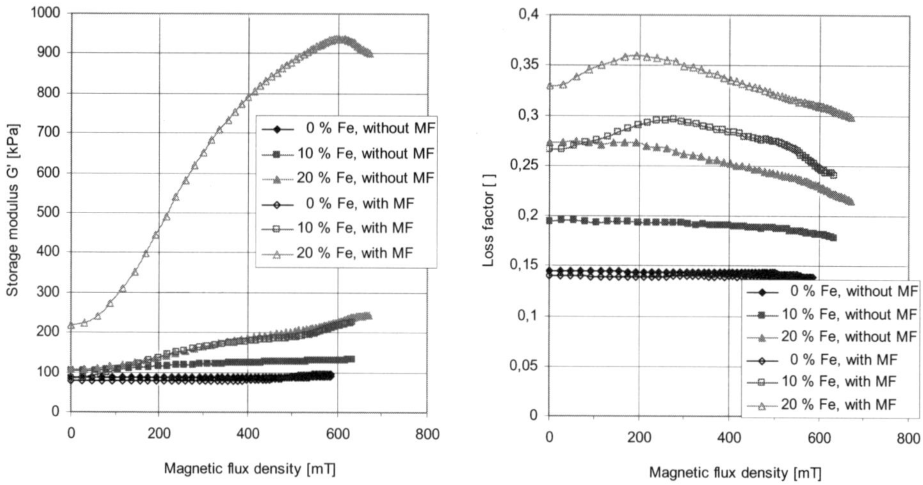

Figure 3. Dependence of storage modulus (left) and loss factor (right) on magnetic flux density for samples with different iron particles contents in % by volume, cured without or with magnetic field, respectively (frequency 10 Hz, amplitude 1 %)

3.3 *Variation of crosslinking*

In addition to the iron particle content and arrangement (random or chained distribution), the mechanical stiffness of the MR elastomer also depends on the density of elastomer matrix crosslinks. In order to investigate this influence, three composite samples with the same iron particle content (20 % by volume) but different crosslink densities were synthesized without the application of a magnetic field. In the following, they are denoted by “soft”, “medium” and “hard”, where “medium” corresponds to the matrices of the MR elastomer samples previously described.

The results of oscillation measurements on these materials are shown in Fig. 4. Apparently the soft MR elastomer exhibits a considerably higher increase of both moduli than the medium and hard materials. The increase in storage modulus in the magnetic field amounts to a factor of about 40, which strongly exceeds that in all previously published data. The corresponding increase in loss modulus is less than a factor of 15. These results demonstrate that, with an increasing magnetic field strength, the elastic behavior of the composite becomes more dominant in comparison with the viscous behavior due to magnetic interaction between particles.

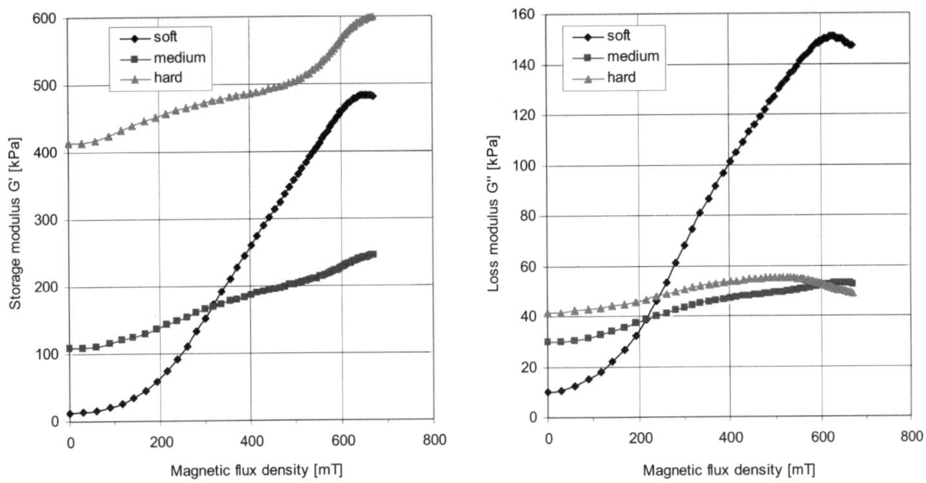

Figure 4. Dependence of storage modulus (left) and loss modulus (right) on magnetic flux density for samples with 20 % by volume iron particles and different crosslinking densities (frequency 10 Hz, amplitude 1 %)

3.4 Variation of frequency and amplitude

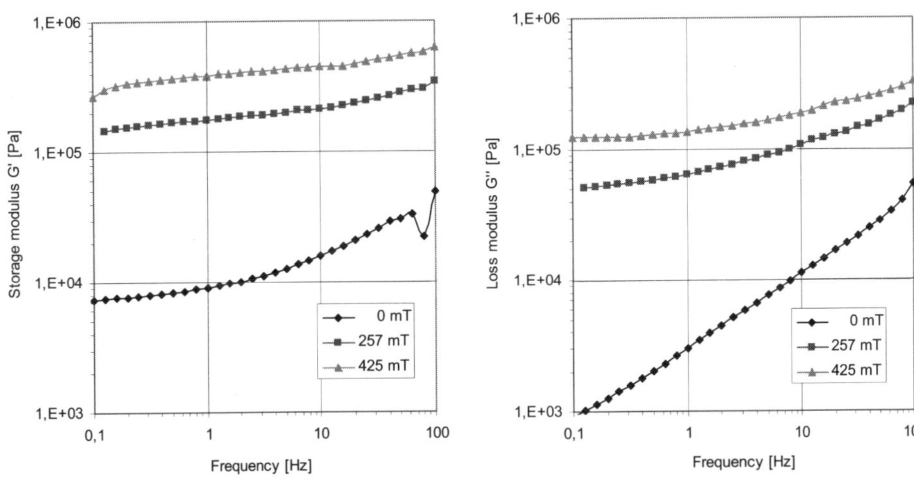

Figure 5. Dependence of storage modulus (left) and loss modulus (right) on oscillation frequency for samples with 20 % by volume iron particles and soft silicone at different magnetic flux densities (amplitude 1 %)

The frequency dependence of the storage and loss moduli of the soft MR elastomer with a content of iron particles of 20 % within magnetic fields of different strengths at a fixed amplitude is shown in Fig. 5. For the sample in the magnetic field the curves have lower slopes than that for the same sample without field, which indicates the behavior of a polymer with a higher degree of crosslinking within the field. Apparently, under these conditions, the material behaves as if it were more strongly crosslinked due to magnetic interactions.

For the same sample, fig. 6 shows the dependence of the viscoelastic properties on deformation amplitude at a fixed frequency and at various field strengths. With rising amplitude the increase in storage and loss moduli by the magnetic field becomes smaller. However, even at unity amplitude (corresponding to 100 %) a remarkable increase of both moduli can be observed.

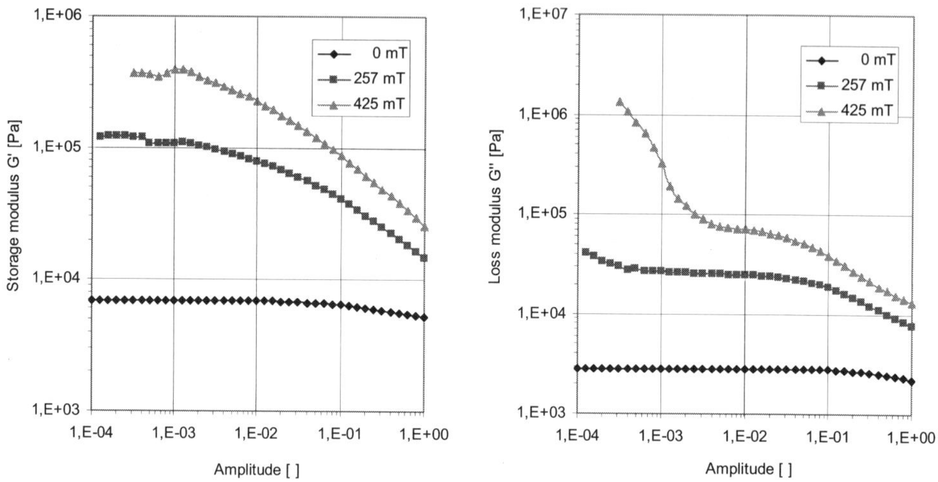

Figure 6. Dependence of storage modulus (left) and loss modulus (right) on deformation amplitude for samples with 20 % by volume iron particles and soft silicone at different magnetic flux densities (frequency 10 Hz)

3.5 Shape memory effect

MR elastomers also demonstrate another exciting phenomenon resembling shape-memory effect observed in other materials. Fig. 7 depicts an MR elastomer cube held between the pole shoes of an electromagnet. In the presence of a magnetic field, the cube is vertically elongated and adheres to both pole shoes of the magnet. Furthermore, the cube may also be deformed by external forces whereafter it remains in the deformed shape. However, the deformed body returns to its original cubic shape on cessasion of the field. Whereas the behavior of the MR elastomer in the field-free state is dominated by elastic forces of the matrix, magnetic forces become dominant under the influence of a field.

Such shape memory effects are well known with metallic and polymer materials. In most cases the control parameter is temperature, which leads to a relatively slow response. For MR elastomers, shape deformation is controlled by a magnetic field which can be switched on and off within a few milliseconds. It is conceivable that the MR elastomeric shape memory effect will find applications in a new class of actuators with large deformations and low forces.

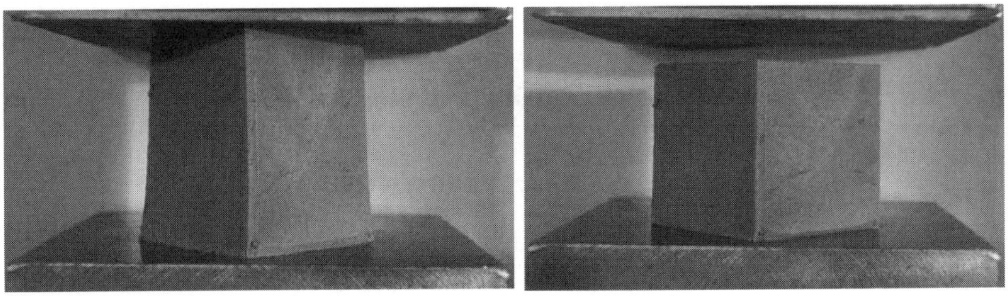

Figure 7. A cube-shaped MR elastomer sample between the pole shoes of an electromagnet with field on (left) and off (right)

4 Conclusions

The viscoelastic properties of magnetorheological (MR) elastomers composed of iron particles in silicone matrices have been outlined. In this study the influence of the particle concentration, the crosslink density and the application of a magnetic field during the curing process on the material properties have been systematically investigated. It has been shown that the storage and loss moduli of the composites can be drastically increased under the influence of a magnetic field. This effect was found to be even more pronounced for samples cured in a magnetic field. A new MR elastomeric shape memory effect, with potential for use in actuators, was observed.

5 Acknowledgements

The author wishes to thank C. Bayer, A. Hesler and A. Trendler for the preparation of samples, performance of measurements and general technical support during the investigations.

References

1. Jolly M. R., Carlson J. D., and Munoz B. C., A model of the behavior of magnetorheological materials. *Smart Mater. Struct.* **5** (1996) pp. 607–614.
2. Lokander M. and Stenberg B., Performance of isotropic magnetorheological rubber materials. *Polymer Testing* **22** (2003) pp. 245–251.
3. Lokander M. and Stenberg B., Improving the magnetorheological effect in isotropic magnetorheological rubber materials. *Polymer Testing* **22** (2003) pp. 677–680.
4. Ginder J. M., Nichols M. E., Elie L. D., and Tardiff J. L., Magnetorheological elastomers: properties and applications. *SPIE* **3675** (1999) pp. 131–138.
5. Bossis G., Abbo C., Cutillas S., Lacis S., and Métayer C., Electroactive and electrostructured elastomers. *Proc. 7th Int. Conf.* on Electrorheological Fluids and Magnetorheological Suspensions (R. Tao, Ed.), World Scientific, Singapore (2000) pp. 18–27.

THE STATIC YIELD STRESS OF FERROFLUID-BASED MAGNETORHEOLOGICAL SUSPENSIONS

Y. B. YANG, L. LI AND G. CHEN
School of Mechanical & Aerospace Engineering, Nanyang Technological University
50 Nanyang Avenue, Singapore, 639798
E-mail: ybyang@pmail.ntu.edu.sg

W. H. LI
School of Mechanical, Materials & Mechatronic Engineering University of Wollongong, Wollongong, NSW 2522, Australia

The Ferrofluid-based magnetorheological suspensions (F-MRS) have been newly developed, in which ferrofluids were used as the carrier media, and microsized iron particles as the magnetic material. The advantageous properties of the F-MRS were presented by comparing with commercial mineral oil based MR suspension (M-MRS). Furthermore, the static yield stresses τ_{sy} of four F-MRS samples were measured in fixed shear rate method and step-increased strain method. The different behaviors under the two methods were analyzed. And finally the dependence of the static yield stress on the magnetic induction B and the sample weight fractions ϕ were summarized.

1 Introduction

Magnetorheological (MR) suspensions consist of magnetic particles suspended in a liquid. They can respond to an applied magnetic field with a change in rheological behavior. Typically, this change is manifested by the development of a yield stress that monotonically increases with applied field [1].

Although MR suspensions have been utilized in many aspects [1-5], they are not very satisfying, because the micro-sized magnetic particles (bulk iron) in the suspension are easy to sediment because of gravity. To overcome this difficulty, one approach is to improve the viscosity of the carrier media. Wilson at al. [6] produced MR gel using polyurethane and silicone polymer as the media, the polymers were cross-linked so that the iron particles were entangled in the cross-linking. And they obtained comparatively stable MR gel. Alternatively, grease has also been tested as the carrier media [7]. Surely stable MR gel can be obtained using these similar methods, but normally the viscosity of the gel media is very large, more or less the MR effect will be discounted. Another approach is that smaller magnetic particles were used to replace the big micro-sized particles, so nano-MR suspensions were developed. Ferrofluids of nanosized Fe_3O_4 suspended in oil are typical representatives [8]. And small iron particles (26nm) also have been used for magnetic fluids [9, 10]. However, the yield stress of nano-MR suspensions is much smaller than that of common microsized iron based MR suspensions [11, 12].

Taking the advantages of both MR suspensions and nano magnetic suspensions, a type of newly MR suspensions was developed in this project. Where ferrofluids were used as the carrier media, and microsized iron particles were mixed into it additionally. This kind of hybrid magnetic fluids were called ferrofluids based MR suspensions (F-MRS). Actually, the magnetic components in F-MRS were the mixture of nanosized Fe_3O_4 ($\approx$ 12nm) of the media ferrofluids and microsized iron particles (4.5-5.2μm). The stability and viscoelastic properties were firstly investigated through the comparison with commercial mineral oil based MR suspensions (M-MRS). The static yield stresses of the F-MRS were then measured by two different methods: (1) fixed shear rate method, (2) step-increased strain method. The results of the two different methods were summarized, the dependence of the yield stress on the magnetic field and the weight fraction were analyzed.

2 Materials Preparation and Observation

In the experiment, ferrofluids (viscosity 300mPa, magnetic saturation 400G/emu, Ferrofluidics, USA) were used as the liquid carrier for our samples, carbonyl iron particles (particle size 4.5-5.2μm, density 4.4g/cm^3, Aldrich) as microsized particles. The samples with different weight fractions of ϕ = 20wt%, 40wt%, 60wt%, 80wt% were prepared by mixing the iron particles into ferrofluids, respectively. For comparison purpose, the commercial mineral oil based MR Suspension (M-MRS) MRF-132LD (80wt%, carbonyl iron particles, 4.5-5.2μm, LORD Corporation, USA) was diluted to produce a sample of weight fraction 60wt%.

We know that, in ferrofluids, the magnetic component is magnetite (Fe_3O_4) particles around 12 nm, however the microsized iron particle size is hundreds times more than that of Fe_3O_4. It's very difficult to observe both of them in transmission electron microscope (TEM) or scanning electron microscopy (SEM). For observation purpose, a sample of nanosized nickel particles (60-100nm) mixed with ferrofluids was prepared, because the surface property of nickel and iron are almost the same. Thus, the microstructure of the interaction between Fe_3O_4 and the big particles can be observed under TEM.

As the nanosized Fe_3O_4 particles are so small, they have high specific surface area and high surface energy [13]. When big particles were added, the smaller Fe_3O_4 particles may be absorbed onto the big ones. Fig. 1 showed the TEM micrographs of nanosized bulk nickel particles (60-100nm) mixed into ferrofluids. The TEM showed that the smaller Fe_3O_4 particles surrounds the big nickel ones. And a lot of Fe_3O_4 particles were attached onto the surface of the nickel particles. The strong repulsive force between Fe_3O_4 particles helps to reduce the agglomeration between big particles. Experimentally, we also found the whole mixture was easy to form gel-like structure with the iron weight fraction beyond 50wt%. And below this fraction, all the iron particles were also agglomerated like paste and floated in the media. While after mild stirring, the whole system can be re-dispersed and flow fluently.

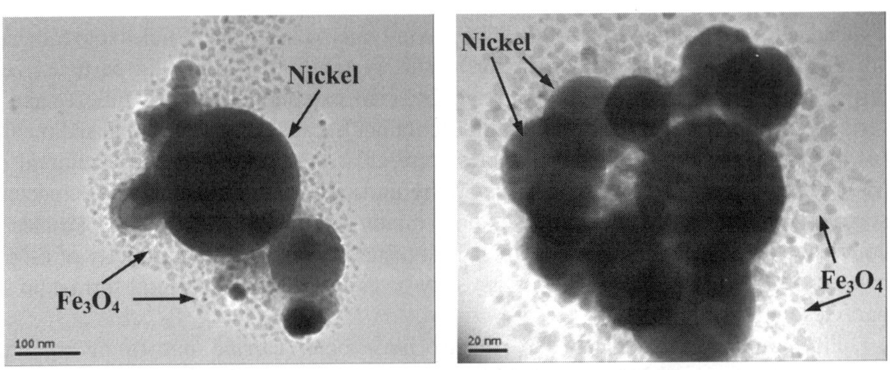

(a) (b)

Fig. 1 The TEM illustration of nanosized Fe_3O_4 particles attached on the big nickel particles.

3 The Measurement Methods of the Static Yield Stress

The yield stresses of F-MRS were measured on a shear rate controlled rheometer (Paar Physica, Germany) with plate-plate geometry. The copper coil was built inside which can provide magnetic field. To avoid sample slipping between the two plates, the surface of the plates were treated by crude sandpaper so that the surfaces became rough.

Two different methods were used to measure the yield stress as shown in Fig. 2 [14-17]. Fig. 2(a) showed the fixed shear rate method, a very small shear rate $\dot{\gamma}$ is applied to the sample, and the shear stress is taken. The static yield stress τ_{sy} is defined as the maximum shear stress in the whole time range. Fig. 2(b) illustrated the step-increased strain method, a logarithmically increasing shear strain γ is applied to the sample. Similarly, the static yield stress τ_{sy} is the maximum stress in the measuring strain range.

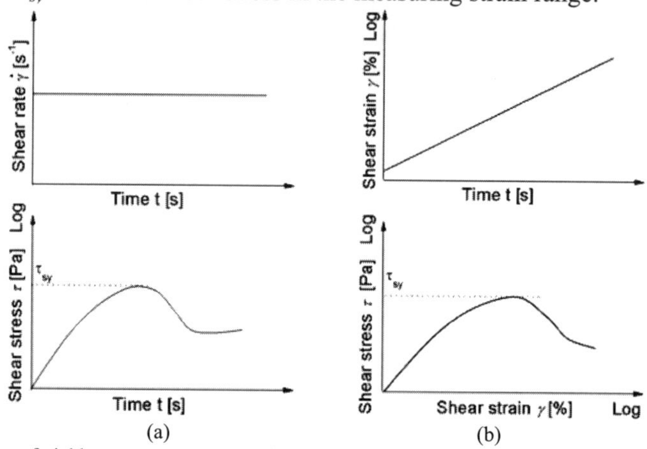

Fig. 2 The illustration of yield stress measurements for suspensions in two methods, (a) fixed shear rate method, (b) step increasing strain method.

4 Results and Discussion

4.1 Properties comparison between F-MRS and M-MRS

Typical MR suspensions are produced by mixing microsized iron particles into mineral oil, silicon oil or synthetic oil, with some surfactants as stabilizer. Because the particle size is so large, the particles are easy to sediment. Over an extended period of time, "cake-like" formation is observed at the bottom of the container because of gravitational sedimentation [6]. As shown in Fig. 3(a), the particles were clearly separated from the mineral oil in M-MRF, and it's not easy to re-disperse the particles in the mineral oil by mechanical stirring. While for F-MRS, there was only minor sedimentation, the iron particles were congregated with ferrofluids as a whole. That means the anti-settling property of F-MRS is better than M-MRS. This good property may be caused by the structural formation of the particles in the system.

Furthermore, rheological experiments have been carried out to investigate the difference between M-MRS and F-MRS. Fig. 4(a) shows the strain growth difference in stress sweeping between the two suspensions at $B = 97$mT for sample $\phi = 60$wt%. It could be seen that under the same magnetic induction and weight fraction, the strain of F-MRS increased continuously all along. While the strain of M-MRS jumped to a very high value when the strain approached to a certain value (yield stress). This phenomenon was also observed in the stress oscillation as shown in Fig. 4(b). In the figure, the stress was sweeping with the angular frequency $\omega = 2.0$ rad/s at $B = 145$mT. Both the elastic modulus G' and the loss modulus G'' for F-MRS were decreased continuously. And the G' and G'' were also jumped to another lever abruptly. These two experiments indicated that the continuity of the performance of F-MRS was better than M-MRS even a magnetic field is applied. This property is very important because in a real application, we hope that the behavior of the smart suspensions can be controlled continuously and smoothly.

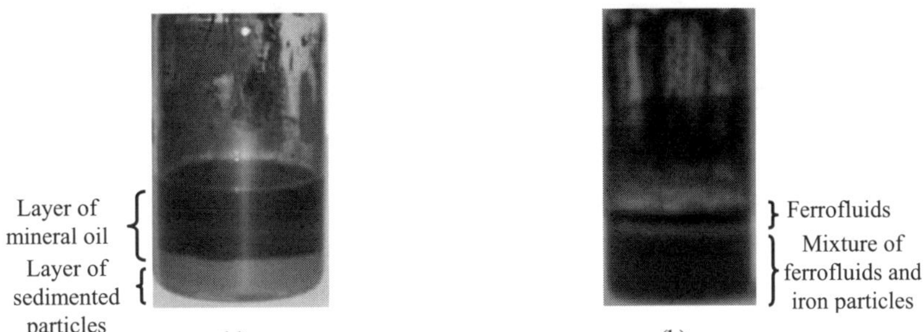

(a) (b)

Fig. 3 (a) "Cake-like" formation in diluted MRF132-LD after 7 days, (b) F-MRS with $\phi = 60$wt% in gel status.

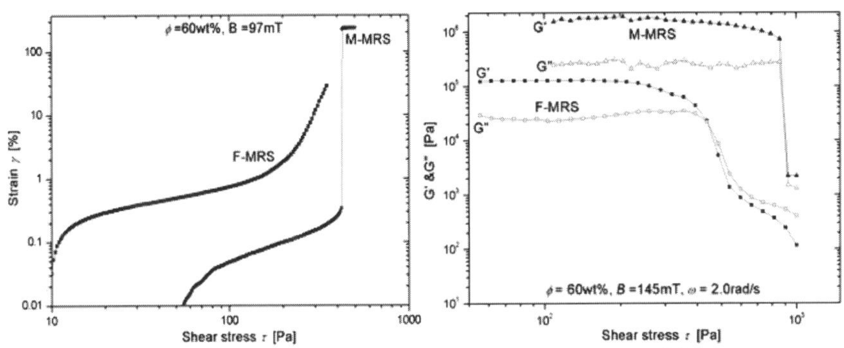

(a) (b)

Fig. 4 The stress controlled properties comparison between F-MRS and M-MRS both having $\phi = 60$wt%, (a) Strain increase at $B = 97$mT. (b) The dynamic moduli curves at $B = 145$mT with angular frequency $\omega = 2.0$rad/s.

4.2 Static Yield Stress Measurement for F-MRS

The yield stresses of the prepared four samples were measured in the two different methods as mentioned in section 3.

(a) Fixed shear rate mode

Fig. 5 shows the curves of yield stress vs time in a small fixed shear rates under different magnetic fields. In all the measurements, the shear rate was kept at a very small value, and the output of stress was taken per second. Fig. 5(a) displayed the yield stress for the F-MRS sample $\phi = 60$wt% at $\dot{\gamma} = 0.005s^{-1}$. The yield stress increases evidently with the external magnetic fields. And at the same shear rate, it costs more time for the suspension to reach the yield stress in higher magnetic flux. We can see that there is a small overshoot when the sample was yielding. Here we define the overshoot peak as the static yield stress τ_{sy}. After yielding, the shear stress remains in a horizontal line.

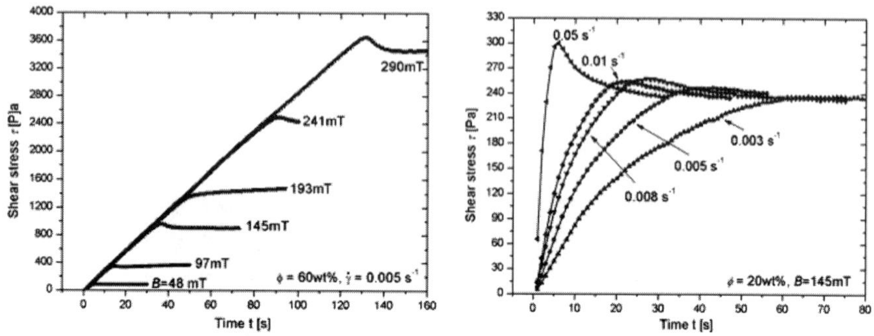

(a) (b)

Fig. 5 Yield stress measured in fixed shear rate method, (a) shear rate $\dot{\gamma} = 0.005s^{-1}$ for sample ϕ= 60wt% at different magnetic flux, (b) Stress curves at different constant shear rates at B = 145mT for sample ϕ= 20wt%.

Fig. 5(b) shows the stress curves for sample ϕ= 20wt% at B = 145mT in different fixed shear rates. As can be seen, at all the shear rates from $0.003s^{-1}$ to $0.05s^{-1}$, all the stresses increased to an overshoot point and then decreased slowly. And with the increasing shear rate, the yield stress was also increased a little bit. It's interesting to find that after the overshoot, all the stress curves approximately converged to a common line.

(b) Step-increased strain mode

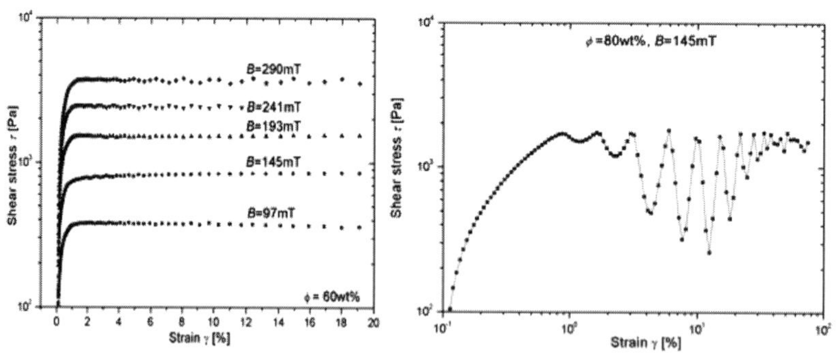

(a) (b)

Fig. 6 Yield stress measurement by strain increase method. (a) The stress vs. strain curves for sample ϕ= 60wt% at different magnetic flux, (b) The yield stress curve for sample ϕ= 80wt% at B = 145mT.

Fig. 6 shows the yield stress measured by strain increase method. Fig. 6(a) presented the yield stress for sample ϕ= 60wt% at different magnetic flux. In all the experiments, the strain γ was sweeping from 0.1% to 1000% logarithmically. In the linear plotting, the stress suddenly approached to its maximum value, and then kept at the maximum value with little decrease. Therefore, unlike the yield stress plotted in Fig. 5, the stress didn't overshoot at its maximum point.

Fig. 6(b) plotted the stress-strain relation for sample ϕ = 80wt% at B = 145mT. The stress first approached to the maximum value near stress γ = 1.0%. After the yield stress, the strain kept on increasing, but the stress began to "struggle" near the yield stress, forming a vibrating wave. The output of stress exhibited the chain clusters being destructed and re-constructed periodically. The interesting phenomena may be explained as follows. At very small shear strain before yielding point, the chain clusters were perturbed, but they were not broken down. Beyond the yield stress, some of the clusters cannot bear the deformation and broke down. So during the breaking, fewer clusters bear the same loading, and consequently more clusters will break. Therefore, the stress suddenly decreases during the broken time, and the stress valley in Fig. 6(a) demonstrated this part. However, because of the magnetic fields, the broken clusters will reconstruct. And at a certain time, all the clusters were completely re-formed, so the maximum loading they can endure should be equal to that at yielding moment. This process of destruction and construction was repeated, so the stress wave was formed, and the maximum stress is always the yield stress.

Table 1 Yield stresses measured by the two methods for the samples.

Weight fraction	Measurement mode of τ_{sy}	0.0mT	48mT	97mT	145 mT	193 mT	241 mT	290 mT
20wt%	Fixed shear rate (a)	2.3	22.1	115	208	400	630	860
	Strain controlled (b)	0	26	65	197	350	581	792
40wt%	Fixed shear rate (a)	2.1	19.5	391	650	1175	1625	2510
	Strain controlled (b)	0	20	211	475	810	1260	2380
60wt%	Fixed shear rate (a)	2.2	130	655	1120	2040	3000	4100
	Strain controlled (b)	8.7	23	420	746	1550	2010	3220
80wt%	Fixed shear rate (a)	1.2	460	820	2340	3840	5700	8500
	Strain controlled (b)	21.5	40	530	1430	3350	4400	6100

In a summary, the yield stresses for all the samples in different magnetic inductions were listed in Table 1. It could be seen that for one sample, the yield stress measured by fixed shear rate method were much bigger than that of step-increased strain method when magnetic induction B $\geq$ 97mT. We know that in fixed shear rate method, the strain was evenly increased very slowly. However, in step-increased strain method, the strain was increased slowly initially, while the step length was accelerated in larger strain. It's possible that the strain broke through the yield strain and caused the destruction of the chain structure, so the yield stress must be larger than that measure in this method. And under the same magnetic field, the yield stress was approximately doubled when the weight fraction was increased by 20wt% each time.

5 Conclusions

In this project, a type of newly magnetorheological suspensions (F-MRS) was developed by mixing microsized iron particles into ferrofluids. Comparing to mineral oil based MR suspension (M-MRS), the F-MRS were better in many aspects. They are slower in sedimentation and easier to re-disperse even after sedimenting. The performance of F-MRS was more continuous and smooth, which is significantly important for application.

The magnetorheological effect of F-MRS was also very evident, because the yield stress increased significantly when the magnetic induction increased. The static yield stress measured in fixed shear rate method was much bigger than that measured in step-strain creased method. For the prepared four samples with different weight fractions, two methods were used to measure the static yield stress under magnetic field. We found that the yield stresses measured by fixed shear rate method were much bigger than that of step-increased strain method. And under the same magnetic field, the yield stress was approximately doubled when the weight fraction was increased by 20wt% each time.

References

1. Jolly, M.R., J.W. Bender, and J.D. Carlson. *Properties and applications of commercial magnetorheological fluids.* in *Smart Structures and Materials 2002: Damping and Isolation, Mar 18-20 2002.*
2. Ahmadian, M., *Design and development of magneto rheological dampers for bicycle suspensions.* American Society of Mechanical Engineers, Dynamic Systems and Control Division (Publication), The ASME International Mechanical Engineering Congress and Exposition **67** (1999) pp. 737-741.
3. Anon, *New, cost effective braking, damping, and vibration control devices made with magnetorheological fluids.* Materials Technology **13**(1998) pp. 96-99.
4. Ashour, O., C.A. Rogers, and W. Kordonsky, *Magnetorheological fluids: materials, characterization, and devices.* J. Intell. Mater. Syst. Struct. **7**(1996) pp. 123-30.
5. Carlson, J.D., *Low-cost MR fluid sponge devices.* J. Intell. Mater. Syst. Struct.**10** (1999) pp. 589-94.
6. Wilson, M.J., A. Fuchs, and F. Gordaninejad, *Development and characterization of magnetorheological polymer gels.* J. Appl.Polym. Sci. **84** (2002) pp. 2733-42.
7. Rankin, P.J., A.T. Horvath, and D.J. Klingenberg, *Magnetorheology in viscoplastic media.* Rheol. Acta **38** (1999) pp. 471-7.
8. Odenbach, S., *Ferrofluids - Magnetically controlled suspensions.* Colloids and Surfaces A: Physicochemical and Engineering Aspects **217** (2003) pp. 171-178.
9. John, S., et al. *Magnetorheological fluids exploiting nanometer-sized particles.* in *Smart Structures and Materials 2002. Active Materials: Behavior and Mechanics, 18-21 March 2002, Proceedings of the SPIE - The International Society for Optical Engineering.* 2002. San Diego, CA, USA: SPIE-Int. Soc. Opt. Eng.
10. Poddar, P., et al., *Nanocomposite magneto-rheological fluids with uniformly dispersed Fe nanoparticles.* J.Nanosci.Nanotechnol.**4** (2004) pp. 192-6.
11. Phule, P.P. and J.M. Ginder. *Synthesis and properties of novel magnetorheological fluids having improved stability and redispersibility.* in *Proceedings of the 6th International Conference on Electro-Rheological Fluids, Magneto-Rheological Suspensions and their Applications, 22-25 July 1997.* 1998. Yonezawa, Japan: World Scientific.
12. Rosenfeld, N., et al. *Behavior of magnetorheological fluids utilizing nanopowder iron.* in *Eighth International Conference Electrorheological Fluids and Magnetorheological Suspensions, 9-13 July 2001.*
13. Kim, D.K., et al., *Starch-coated superparamagnetic nanoparticles as MR contrast agents.* Chem. Mater. **15** (2003) pp. 4343-51.
14. Caines, S.B., et al., *Yield stress and wall slip phenomena in colloidal silica gels.* J. Rheol. **47** (2003) pp. 847-868.

15. Stokes, J.R. and J.H. Telford. *Measuring the yield behaviour of structured fluids*. in *UWINNFM Conference on Rheometry, Journal of Non-Newtonian Fluid Mechanics*. 2004. Cardiff, UK: Elsevier.
16. Bonnecaze, R.T. and J.F. Brady, *Yield stresses in electrorheological fluids.* J. Rheol. **36** (1992) pp. 73-115.
17. Nguyen, Q.D.B., D. V., *Measuring the flow properties of yield stress fluids.* Annual Review of Fluid Mechanics **24** (1992) pp. 47-88.

ALTERNATIVE VIEW FOR YIELD STRESS OF ER/MR MATERIALS

F.E.FILISKO

Materials Science & Engineering, University of Michigan, Ann Arbor, Michigan 48109-2136

E-Mail: fef@umich.edu

Understanding ER behavior remains theoretically and mechanistically unsatisfactory. Although E induced particle aggregation is material dependent, shear yield stresses appear independent of materials, around 3-5 kPa . Recent studies of ER dispersions under compression and shear report higher yield values, 50 - 300 kPa. Some studies report ER fluids in preyield show shear moduli in the MPa range. ER/MR fluids of commercial concentration and at high field strengths at static conditions on microscope slides do not show single chains or columns but a dense continuous particulate network. However exposing certain dispersions to E/M and shear fields simultaneously aggregates particles into lamellar structures(walls) oriented in the shear direction. Such structures vs. columns require a much greater FORCE to shear which must be transmitted from the electrodes to the particulate structures by a " frictional force". Although non-slip conditions are assumed, this is mostly an unproven theoretical convenience. An alternate view suggests these " forces" may be inadequate to shear the particulate structures, the yield mechanism being slippage between electrodes and adjacent particle lamellae. Thus the strength limiting feature of ER/MR fluids is not particle lamellae shear strength, but " force of adhesion" between electrodes and lamellar surfaces. This is discussed and substantiated with experimental results.

1 Introduction

The term yield when applied to the response of a material to an unbalanced force is generally thought of as an irreversible change of shape. When applied to a ductile crystalline solid it is explained by motions of dislocations in preferred directions and along densest planes, whereas to a liquid flow is more appropriate. In the context of ER/MR fluids however, the term is applied to the shear stress necessary to cause the materials to flow at low shear rate. This yield stress is commonly associated with the strength of an ER/MR fluid, being in the range of 3-5 kPa for ERF and an order of magnitude higher for MRF. The mechanism for this yield is most commonly associated with the breaking of chains or columns of particles spanning the electrodes in response to a shear stress. This shearing between the electrodes/plates is a natural extension of a commonly assumed non-slip boundary condition assumed by theoreticians in solving fluid mechanics problems for homogenous fluids. However in the case of ER/MR fluids, under an E/M field the materials are no longer homogenous but consist of particulate structures induced to agglomerate into columns or lamellar structures under shear[3]. Although the non-slip condition does apply to the "oil" phase, it has never been shown to apply to ER/MR fluids under high fields regarding the particulate structure phase. A hypothesis is examined in this article that at normal ER/MR particle concentrations and fields, that the particulate structures may not adhere to the electrode/plate surfaces and may slip under shear. Some possible consequences of this hypothesis are that first, the particulate structures induced in ER/MR materials are much stronger than indicated by the measured "yield stresses" under shear; second, the yield stress is not a consequence of shearing within the fluid, i.e. between the electrodes/plates, but of slippage of the particulate structures along the plates; third, that to strengthen the fluids, efforts should be directed not solely to looking for better components but also to mechanisms of adhesion of the particulate structures to the electrodes/plates; and fourth, that ER/MR fluids can be made orders of magnitude "stronger" in shear by increasing the structure-electrode/plate adhesion.

For MR materials the mechanisms of adhesion are most likely magnetic interactions. However for ER the mechanisms of adhesion are totally unknown and in fact mechanisms such as electrode polarization, electrical double layers, and particles with non-zero zeta potential weaken such interactions. Attempts to increase these interactions by obvious means such roughening the surfaces, coating with dielectric particles, etc. have led to inconsistent results for various reasons including surface polarizations and charge concentrations at high points. It might be an interesting corollary that the difference in the apparent strengths between ER and MR materials are due to the greater and positive magnetic attraction of the particles in MR fluids to the walls.

In this article we will be examining the above hypothesis with evidence supporting it and then to check it in relation to the measure shear stress of ER fluids.

2 Results and Discussion

It is not uncommon for ER/MR activity to be illustrated by chains of particles bridging the electrodes/plates. When extrapolated to real fluids this can be very misleading for two reasons. The suspensions are very dilute and the situations are static. In reality, ER/MR dispersions are typically 20-30 volume % particles and are under shear. Thus under a field but in static conditions the particle structures are not single chains but form a continuous dense quasi-monolithic particulate structure between the electrodes as illustrated in figure 1. These are neither individual chains nor columns but a continuous particulate structure. Further under shear the particulate structures form into walls or lamellae oriented in the shear direction[3,9]. Such particulate structures require greater forces to shear which must be transmitted by the electrodes/plates, such forces being proportional to the cross sectional area of these particulate structures.

A B C

Figure 1. View of a 30% particle concentration ER fluid between wires 1 mm apart with voltages between the wires of A. 0 volts, B. 1 kvolt, and C. 2 kvolts under static conditions

The strength of ER/MR fluids are most commonly associated with the shear stress measured and extrapolated to 0 shear rate. However it is also known that the rheological response of an ER/MR fluid is not classically Bingham in nature but is more realistically described by a pre-yield region[1,2], i.e. a very low strain region of deformation where the response is linear, though dissipative, and a post-yield or flow region characterized by time dependent deformation. A measure of the strength of the particulate structures could thus be obtained in pre-yield region by appropriate methods, i.e. those involving very low strains so deformation of the particle structures will occur but not slip, and by compression studies under an E/M field where it has been shown that the particle structures do support static loads[4,5].

As illustrated in figure 2, substantial loads are supported by these fluids under field and this must be by the particulate structures of near MPa compressive stresses. The compressive strains to support these loads vary with field and gap (amount of strain) indicating that the yield stress of the particulate structures is a function of the E field. The normal stress calculated here is based upon the 25 mm diameter plate area, but if it is assumed that the particle structures carry all the forces then the normal stresses would actually be considerably higher.

It is possible to calculate a compressive modulus for the packed particle structures from figure 2 at the upper ends of the curves and these produce moduli in the range of MPa's. It is also possible to estimate shear modulus from dynamic studies from the portion of the curves where the data is (presumably) linear and these as well give moduli in the MPa range[1]. Finally, Oyadiji[6] obtained direct stiffness measurements on ER

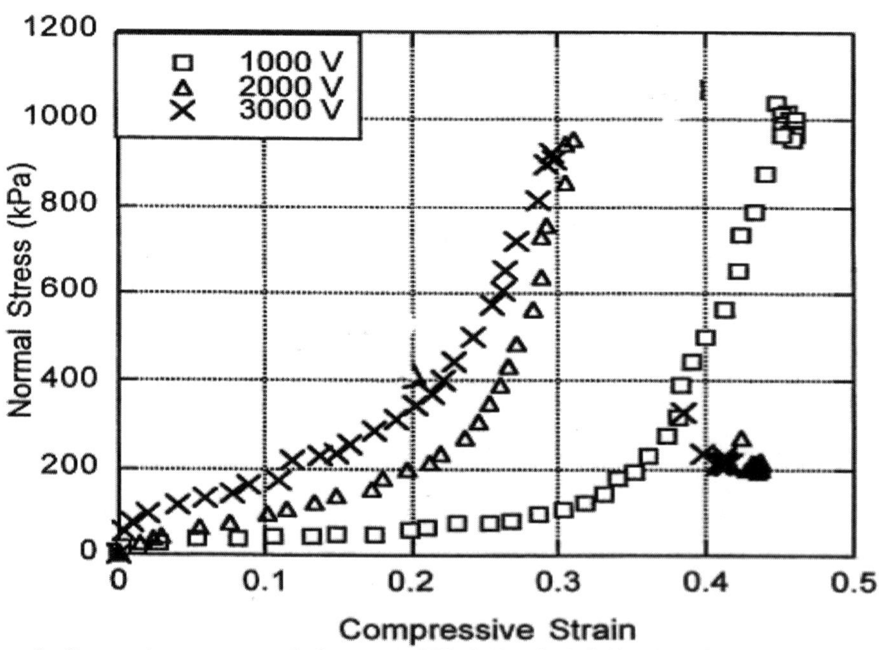

Figure 2. Compressive stress vs. strain for a general ER fluid under 3 fields. The points at 200 kPa and just above 0.4 are the residual pressures maintained after the 2000 and 3000 voltages were turned off.

fluids and recorded values for the modulus also in the MPa range.

These observations suggest that the ER/MR particulate structures are much stronger than the few kPa range and although it is understood that modulus is not a measure of yield strength, the large difference between shear determined strength and moduli is unusual considering that the particles are always of irregular shape and size and cannot pack into any regular pattern. Thus the mechanisms of yielding within the structures would be considerably higher than for regular packing of spheres for instance, which is unrealistic for a real ER/MR fluid[8].

Direct observation of the particulate structures between 25 mm parallel discs is illustrated in figure 3. With a field but no rotation, the upper figure shows a clear network of particulate structures, extending between the plates. These structures are thinner on the top than bottom but that is very typical. Upon rotation of 1 and 6 revolutions in the lower two figures there is a consolidation of the thinner structures into broader structures. But more importantly, upon careful examination, it can be seen that the structures after 6 revolutions are similar to those after 1 revolution, suggesting that the structures are slipping at the upper surface. That this is occurring is much clearer during direct observations of these tests. A word of caution however, the photographs view only the outer rim of the discs and this may not be representative of what is happening internally due in part to field fringing effects and shear rate variations across the disc's radius. Nonetheless these suggest that slippage of the structures at the upper surface is occurring.

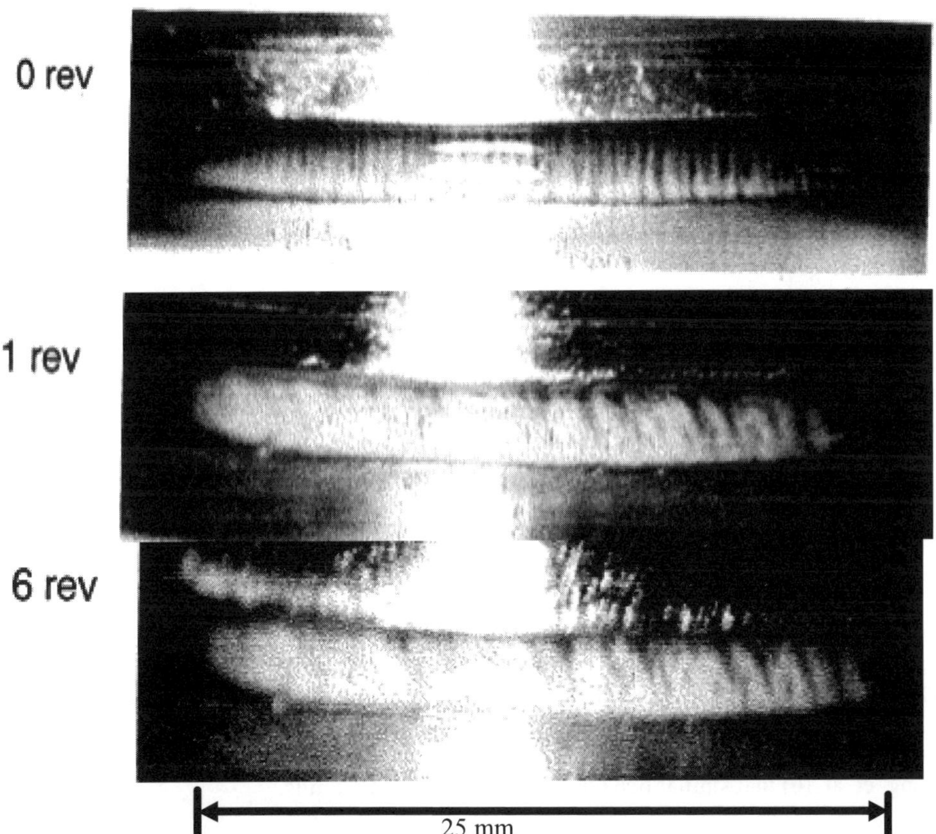

Figure 3. View between parallel discs 25 mm in diameter with a 1 mm gap of particle structures developed as a result of a 2kV voltage between the plates under static conditions at the top, after 1 revolution and after 6 revolutions. Gap is in the vertical direction.

From the above information it is not unreasonable to hypothesize that for real ER/MR fluids under high fields that the particulate structures are dense enough and strong enough such that the surface forces required between these structures and the electrodes/plates are insufficient to shear the structures and surface slippage occurs. Whereas a magnetic force can be estimated for MR materials, this cannot be done for ER materials because sources of interaction are complex and unknown. However, if a friction type mechanism is hypothesized, then one way of checking this is by increasing the normal force on the particulate structures. It is not intuitively obvious that such can be done, without the structures collapsing. However recent studies in fact show [4,5] and it is as well illustrated in figure 2 that substantial loads can be supported on these particulate structures.

Figure 4 illustrates the effect of increasing normal stresses on the fluids during shearing as a function of various voltages. The shear stresses are strong functions of the normal stresses, being nearly 100X greater than when the normal stresses are low. It is also interesting that the shear stresses increase nearly linearly with normal force as would

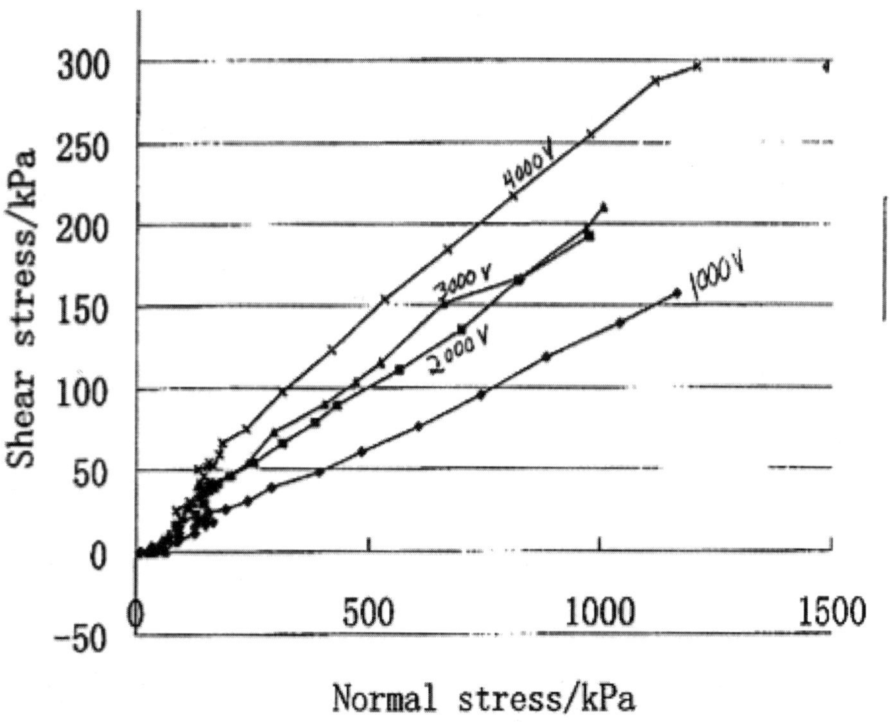

Figure 4. Shear stress vs. normal stress for a 28 volume % zeolite/paraffin oil ER fluid at various voltages.

be expected for a frictional type mechanism for surface adhesion. It has been shown by Tang, et. al. [6] that similar behavior is observed for MR fluids.

3 Conclusion

Various types of evidence have been presented to support the hypothesis that for ER/MR fluids of commercial concentrations and at high fields, the mechanism of yield, i.e. flow may not be shearing of the whole fluid as a homogenous fluid but by slippage of the particulate structures at the electrodes/plates. Thus it is proposed that the particulate structures may be orders of magnitude stronger that indicated by shear studies. Further it is suggested that (primarily for ER materials since MR fluids involve only carbonyl iron)attention should be directed more toward understanding the mechanisms of adhesion of the particulate structures to the electrodes and increasing this adhesion if the "apparent" strengths of the materials are to be increased.

References

1. Gamota D.R. and Filisko F.E., Dynamic mechanical studies of ER materials: moderate frequencies. J. Rheol. 35(3)(1991), pp. 399-427.

2. Gamota D.R., and Filisko F.E., Linear/non-linear properties of ER materials. In *Electrorheological fluids, Mechanisms, Properties, Structure,Technology, and Applications.* ed. by R. Tao(World Scientific, Singapore, 1991) pp. 246-263.

3. Henley S. and Filisko F.E., Flow profiles of ER suspensions, and alternative model for ER activity, J. Rheol. 43(5)(1999) pp. 1323-1336.

4. Lynch R., Meng Y., and Filisko F.E., Compression of dispersions to high stress under electric fields: effects of concentration and dispersing oil. J. Coll. Interf. Sci. 297(2006)pp.322-328.

5. Meng Y. and Filisko F.E., Unidirectional compression of ER fluids in electric fields. J. Appl. Phys 98(7)(2005)pp.21-27

6. Oyadiji S. O., Characterization of the complex shear modulus properties of ER fluids by the direct stiffness and master curve techniques. In *Electrorheological Fluids, Magnetorheological Suspensions, and Associated Technology* ed. by W.A.Bullough(World Scientific, Hong Kong, 1995) pp. 419-428.

7. Tang X., Zhang X., Rong Y., and Tao R., Structure enhanced yield stress of MR fluids. J. Appl. Phys. 87(5)(2000) pp. 2634-2638.

8. Tao R. and Sun T.M., Three dimensional structure of induced ER solid, Phys. Rev. Lett. 67(3)(1991) pp.398-401.

9. Vieira S.A., Nakano M., Henley S., and Filisko F.E., Transient behavior of the microstructure in ER fluids. In *Electrorheological Fluids and Magnetorheological Suspensions*. ed. by R. Tao(World Scientific, Hong Kong, 2000) pp. 152-161.

POLAR MOLECULE TYPE ELECTRORHEOLOGICAL FLUIDS

KUNQUAN LU[1*], RONG SHEN[1], XUEZHAO WANG[1], GONG SUN[1], WEIJIA WEN[1,2], JIXING LIU[1,3]

1. Beijing National Laboratory for Condensed Matter Physics, Institute of Physics, Chinese Academy of Sciences, Beijing 100080, China

2. Department of Physics, Hong Kong University of Science and Technology, Hong Kong, China

3. Institute of theoretical Physics, Chinese Academy of Sciences, Beijing 100080, China

** E-mail: lukq@aphy.iphy.ac.cn*

The static and dynamic shear stress of newly developed electrorheological (ER) fluids can reach more than 100 kPa and over 60 kPa at 3 kV/mm, respectively. The high yield stress of those ER fluids and its near linear dependence on the electric field are different from the conventional ER fluids and can not be explained with traditional dielectric theory. Experiment demonstrates that the polar molecules adsorbed on the particles play crucial role in those ER fluids, which can be named as polar molecule type electrorheological (PM-ER) fluids. To explain PM-ER effect a model is proposed based on the interaction of polar molecule-charge in between the particles, where the local electric field is much higher than the external one and can cause the polar molecules aligning. The main effective factors for achieving high-performance PM-ER fluids are discussed.

In ordinary Electrorheological (ER) fluids the field induced shear stress comes from the interaction between the polarized particles aligned in the electric field [1-3]. The physical mechanism of dielectric ER fluids has been well established [1,3,4] and the upper bound of the yield stress predicted by traditional dielectric theory is about 10kPa [5]. The low yield stress of the available ordinary ER fluids, usually few kPa, has blocked their applications in last half century.

By adding some polar molecules in the particles a series of high-performance ER fluids showing giant ER effect has been obtained in recent years [6-8]. Figure 1 shows the

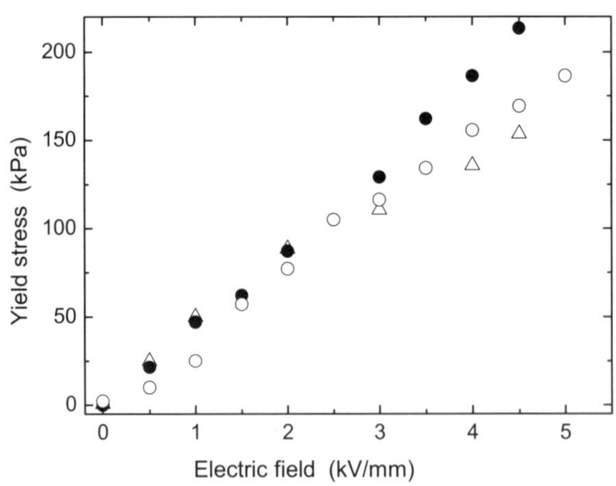

Figure 1. Yield stress versus electric field of some ER fluids. Dots: TiO2 nano-particles with (NH2)2CO contained; Circles: Ca-Ti-O nano-particles with C=O, O-H groups contained; Triangles: TiO2 nano-particles with C=O, O-H groups contained.

dependence of the yield stress on electric field for the ER fluids consisted of TiO_2 nano-particles with $(NH_2)_2CO$ molecules and C=O, O-H polar groups contained, as well as Ca-Ti-O nano-particles with C=O, O-H polar groups contained, respectively.

All the particles were prepared in spheral shape with diameter of 50-100 nm and were suspended in silicone oil with volume fractions about 35%. The yield stresses of the ER fluids were measured with a special designed flat shearing apparatus [9]. The yield stresses of those new ER fluids can reach up to 200 kPa, much higher than that of foregone ones, and behave a near linear dependence on electric field unlike the quadratic one in ordinary ER fluids. The current densities are less than 20 $\mu A/cm^2$ for TiO_2 nano-particle ER fluids and even less than 1 $\mu A/cm^2$ for Ca-Ti-O nano-particle ER fluid at 5 kV/mm.

The shear stress versus shear rate of an ER fluid consisted of TiO_2 nano-particles with $(NH_2)_2CO$ contained is shown in Fig. 2(a). For measuring the dynamic shear stress a cylindraceous rheometer has been established in which the ER fluid was sealed in order to avoid the suspension splashing (see Fig. 2(b)). The surfaces of the electrodes were coated with special layer for abating the slide of ER fluid on the electrodes [10]. A rather high shear stress at zero field is mainly due to the friction of the seal ring assembly of the rheometer. It can be seen that the dynamic shear stress is over 60 kPa at 3 kV/mm.

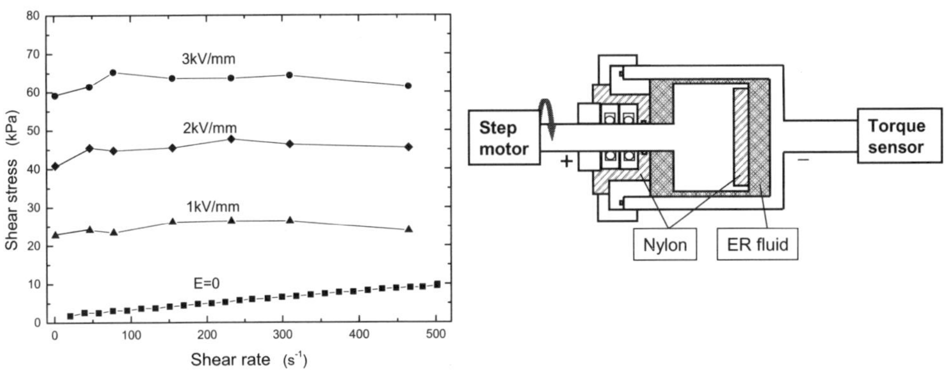

(a) (b)

Fig. 2 (a) Dynamic shear stress vs. shear rate at different field strength for the ER fluid of TiO2 with (NH2)2CO contained. **(b)** Scheme of the cylindraceous rheometer for measuring the dynamic shear stress.

The chemical analysis and infrared spectrum indicate the existence of polar molecules or polar groups contained in our prepared particles. By heating the particles to high temperature (500-800 °C) all the adsorbed polar groups are desorbed off. Then the ER fluids consisted of the sintered particles lose their character of high yield stress and become ordinary ER fluids. As an example, Fig.3(a) shows that the ER fluid consisted of Ca-Ti-O particles with C=O, O-H polar groups contained exhibits a high yield stress and that consisted of the sintered Ca-Ti-O particles loses its high yield stress after heating the particles at 500°C for 2h. Fig.3(b) is the infrared spectra of the Ca-Ti-O particles dried at 120°C and sintered at 500°C, respectively. The C=O, O-H polar groups are contained in Ca-Ti-O particles of dried at120 °C and all cleared away after heating the particles at

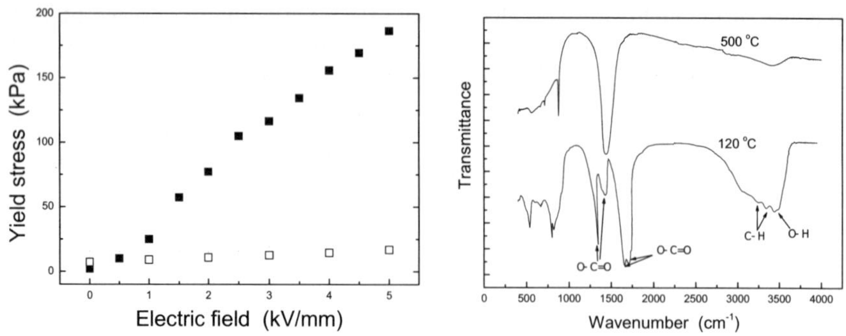

(a) (b)

Fig.3 (a) Yield stresses vs. electric field for ER fluid consisted of Ca-Ti-O particles dried at 120 °C and heated at 500 °C, respectively. **(b)** The infrared spectrum of the Ca-Ti-O particles dried at 120 °C and heated at 500 °C, respectively

500°C for 2h. These facts demonstrate that the polar molecules or polar groups contained in the particles play crucial role in the newly developed ER fluids. Therefore, those new ER fluids can be defined as polar molecule dominated ER (PM-ER) fluids.

The high yield stress and a near-linear dependence of yield stress on electric field in PM-ER fluids indicate that the mechanism of PM-ER fluids must be different from the conventional ER fluids and can not be explained by the traditional dielectric theory.

Although a model based on the polarization saturation layers in a urea coated particle system was proposed to explain the giant ER phenomenon [7], our further experiments indicate that this model should be improved. Firstly, the giant ER effect can be also observed if there are tiny polar molecules adsorbed on the particles instead of the whole layers coated. It means that a small quantity of the polar molecules on the surface of particles is the crucial factor for giant ER effect at all time. Secondly, the polar molecules in the thick coated layer are hardly to be aligned in the direction of applied electric field, because the layer is usually in a solid phase. Only the molecules adsorbed on the surface of the particles, or on the surface of the coated layer, are able to turn to the direction of the field in between the particles. Based on this consideration a more reasonable model for explaining the mechanism of PM-ER effect is proposed.

When an electric field E is applied the polarized particles in the ER fluid attract each other and close up along the field direction to form chains. Several authors calculated the local electric field E_{loc} in between the particles and indicated that E_{loc} must be much higher than E [11-14]. The value of E_{loc}/E depends on δ/r and $\varepsilon_p/\varepsilon_f$, where δ is inter-particle gap, r is the radius of particle, ε_p and ε_f are the dielectric constant of the particle and oil respectively. According to Davis's calculation [11] E_{loc}/E is about 10^3 when $\delta/r \approx 10^{-3}$ and $\varepsilon_p/\varepsilon_f \approx 100$. In our case the dielectric constant of the TiO_2 nano-particles ε_p can be about 10^3, much higher than that of the bulk materials [15,16]. Therefore the E_{loc} in between TiO_2 particles can be as high as about 10^9 V/m when $E \approx 10^6$ V/m, r =30 nm, $\delta \approx 0.2$ nm. It should be pointed that even if the local field is very high, the possibility for occurring the electrical breakdown is scarcely any in such narrow and small regions. Evidence is that an ordinary ER fluid can commonly bear more than 5×10^6 V/m without breakdown happened.

If there are polar molecules adsorbed on the surfaces with adsorption energy U_{ad} which is usually slightly larger than kT, they are relatively free for rotation and able to be oriented in the direction of a strong electric field [17,18]. Thus those polar molecules located in the gap between the particles are able to turn to the electric direction when the magnitude of μE_{loc} is comparable to U_{ad} or larger than about kT, where μ is the dipole moment of polar molecule. To align a molecule with μ=2 debye, for instance, the local field E_{loc} is needed to be higher than 6.4 x 10^8 V/m. This value can be achieved in the gap between particles when an external electric field 10^6 V/m is applied as described above.

Figure 4 illustrates the variation of orientation for the polar molecules on particle surfaces without and with field applied, respectively. In the case of E=0 the polar molecules on the surfaces are optionally distributed in an adsorbing geometry. As long as the field is increased to bring the particles approaching, together with E increasing, a high enough local field E_{loc} can be achieved and must cause the polar molecules in between two close particles aligning to field direction. It is obvious that the local field strengths are not same for various particles and the alignments of the molecules in every gap do not arise synchronously under a same external field due to the inhomogeneity of particle size and roughness. Unfortunately no proper technique is available to directly observe the rotation of the polar molecules in between the particles because the active molecules are in a very low ratio and in a so small space for such a complex system.

(a)

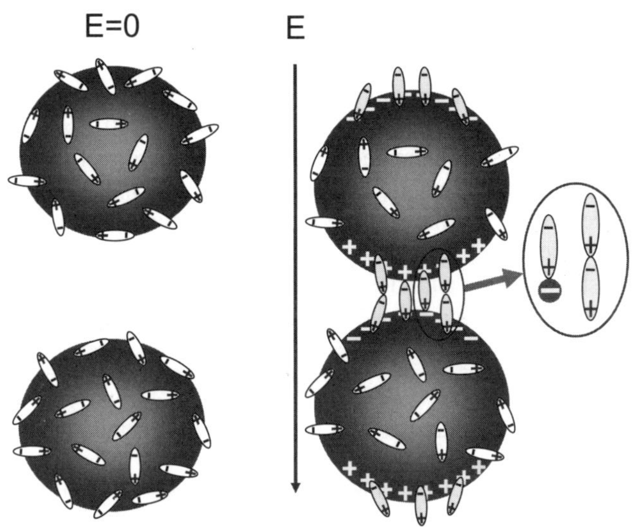

(b)

Fig. 4 The schematic illustration of the polar molecules aligned in the gap of the particles. **(a)** The polar molecules adsorbed on particle surface are optionally oriented when E=0. **(b)** The polar molecules in the gap of the particles are aligned along the filed when E is high enough. Two kinds of the interactions occur (in enlarged image): dipole-dipole interaction and dipole-charge interaction.

Once the polar molecules are aligned in between the particles, mainly two kinds of the interactions may occur. One is the attraction between polar molecules on a particle and that on the neighboring ones. In a point-dipole approximation, the interaction force of two same molecules lying in a head-versus-tail line, is $f_{m-m} = \dfrac{3\mu^2}{2\pi\varepsilon_0\varepsilon_f d^4{}_{m-m}}$ [19], where d_{m-m} is the center-center distance of two molecules. Another one is the attraction between polar molecule and the polarization charge on the opposite particles. The maximum attractive force between the polar molecule-charge is $f_{m-e} = \dfrac{Q\mu}{2\pi\varepsilon_0\varepsilon_f d^3{}_{m-e}}$ [19], where Q is the charge quantity, d_{m-e} is the distance from molecule center to charge. It is obvious that f_{m-e} is usually much larger than f_{m-m} because $d_{m-e} \approx \dfrac{d_{m-m}}{2}$ and then $\dfrac{f_{m-e}}{f_{m-m}} \approx \dfrac{16Qd_{m-e}}{3\mu}$. For instance, f_{m-e} =9.6x10^{-10} N when assume one polar molecule of μ=2 debye interacts with a unit charge (Q=e=1.6x10^{-19} C) in the case of d_{m-e} =0.2nm. Whereas f_{m-m}=3.8x10^{-11} N when taking d_{m-m}=0.4nm. Namely $f_{m-e}/f_{m-m}\approx 25$. So that, the origin of the PM-ER effect must come from the interactive force f_{m-e} of polar molecule-charge. When the electric field is turned off the polarized charges on the particles will vanish and the interaction between molecule-charge also disappears at the same time. As a result the aligned polar molecules come back to the original adsorbing state by the effect of thermal motion and $f_{m-e}\rightarrow 0$. In above calculations the influence of the induced polarization of the molecule is neglected, which enhances the interaction but much lower than that of inherent dipole moments.

On the other hand the interaction of two pure polarized particles can be calculated with traditional dielectric theory [11,12]. In a field of E=3x10^6 V/m for two neighboring TiO_2 particles with radius of 30nm and gap of 0.2 nm the attractive force f_{p-p} is 3.2x10^{-12} N. Therefore $f_{m-e}/f_{p-p}\approx 300$. Namely, even if there is one polar molecule active on the particles the interaction of polar molecule-charge is much higher than that of particle-particle in electric field.

The area in between the particles separated with $\delta=d_{m-e}$ is nearly $\pi rd/2$, where polar molecule-charge interactions may occur effectively. In this region the number of active molecule-charge interactions $N \propto \dfrac{\pi rd}{2}\rho_m n(E)$, where ρ_m is the density of the polar molecules adsorbed on the particles and the probability of the aligning polar molecules $n(E)\propto\mu E_{loc}\propto\mu E$. The number of the particle chains in a unit area is $\dfrac{3\phi}{2\pi r^2}$, where ϕ is the volume fraction of the particles in suspension. Therefore, the total attractive force can approximately be expressed as $F_{m-e} = \dfrac{3\phi}{2\pi r^2}Nf_{m-e} = A\dfrac{3\phi\rho_m e\mu^2 E}{\pi r\varepsilon_0\varepsilon_f d^2}$, where the parameter A is related to the properties of the particle and oil, geometry of the particles in the oil, as well as the adsorption energy U_{ad} et al. It is obvious that the F_{m-e} increases linearly with E due to $n(E)\propto\mu E$ and consequently the yield stress of PM-ER fluid exhibits a nearly linear dependence on applied electric field as shown in Fig. 1. $F_{m-e}\propto r^{-1}$ indicates that the yield stress of the PM-ER fluid consisted of smaller particles is higher as observed in the experiment [20]. This feature of PM-ER fluids is also different from that of the conventional ER fluid, in which the yield stress usually increases with particle size increasing [11, 12]. The nano-particles not only are favorable for obtaining high shear stress of PM-ER fluids, but also are advantageous for resisting the particle sedimentation in the fluids. The larger particles, however, are of benefit for getting higher local electric field. Using the particles with high dielectric constant can induce a higher local electric field in between particles, as a result the PM-ER response occurs at a lower applied field. For getting high local field in between particles the particles are always necessary to be prepared in spheral shape. $F_{m\text{-}e} \propto \mu^2 d^{-2}$ indicates that choosing the polar molecules with larger dipole moment and small size is required for getting higher yield stress of PM-ER fluids. Also the polar molecules with larger μ can be aligned at a lower E. To enhance the yield stress of PM-ER fluids a higher ρ_m of the adsorbed molecules should be pursued, as low ρ_m may cause the yield stress saturation at high E. It is obvious that the boundaries between the metal electrodes and the fluids are not well satisfied for the interaction of polar molecule-charge, where the attractive force must be reduced due to slide occuring. By using a designed method the intrinsic yield stress of PM-ER fluids can be measured [9]. Also the electrodes for applying the PM-ER fluids should be modified [10].

Using the expression of $F_{m\text{-}e}$ we can estimate the remarkable difference between the yield stresses of PM-ER fluid and the ordinary ER fluid. As an example, we consider the system of r=30 nm, δ=0.2 nm, $d_{m\text{-}e}$=0.2 nm, N=1, μ=2 debye, and ϕ=35%. Then $F_{m\text{-}e} \approx 18$ N/cm^2 in PM-ER fluid, corresponding to 180 kPa. Whereas the calculated attractive force in ordinary ER fluid consisted of pure TiO_2 particles under a field of E=3x10^6 V/m is $F_{p-p} = \frac{3\phi}{2\pi r^2} f_{p-p} \approx 0.06$ N/cm^2, i.e. the strength is 0.6 kPa. Although assuming there is only one polar molecule active in average on each particle in the calculation, $F_{m\text{-}e}$ is 3x10^2 higher than $F_{p\text{-}p}$. Namely the yield stress τ_y of PM-ER fluids can be 2-3 orders higher than that of conventional ER fluids, where $\tau_y \approx \frac{F_{m-e}}{3}$. The estimated results are on the whole consistent with the measured yield stresses of conventional ER fluids and PM-ER fluids, respectively. The more polar molecules are active on each particle, the higher yield stress of PM-ER fluids can be achieved.

By the same principle, if a medium containing polar molecules instead of that adsorbed on the particles one can be prepared, PM-ER effect is expected to occur. In this case the polar molecules in the liquid will orientate along the field in between the particles and interact with the polarized charges of the neighboring particles. In this situation the PM-ER effect is caused by the local freezing of the polar molecules and may be more effective.

The expression of $F_{m\text{-}e}$ clearly indicates how to achieve the PM-ER fluids with higher yield stress by optimizing all the effective factors. It is possible to obtain a PM-ER fluid with the yield stress τ_y up to MPa or more if one adopts μ and ρ_m larger, d and r smaller enough in material designing and manufacturing. For instance, assuming μ=2 debye, r=15 nm, d=0.3 nm, N=3, ϕ=35% and the particles being sufficiently dispersed in the fluid, then estimated τ_y can be about 1.7 MPa.

In PM-ER fluid the current density possesses a non-linear dependence on the electric field [8]. The $\ln j \propto E^{1/2}$ relation may come from Pool-Frenkel effect [7, 21].

It was found for a long time that a trace of water contained in the particle can consumedly increase the yield stress of ER fluids compared with that of ER fluids consisted of dry particles [22-24]. A ubiquitous opinion was that the 'water bridge' coming from the surface tension in between particles plays a role [25-27]. Based on the principle of PM-ER effect the role of the trace water can be immediately understood. It is the orientation and bonding of H_2O polar molecules in between particles causing the strong ER effect, whereas the water bridge is an improper explanation. Even so, the ER fluids with water contained have been forsaken in practice because of its easy evaporation.

In conclusion, the PM-ER effect comes from the interaction of polar molecules adsorbed on the particles instead of that of the particles themselves as in conventional ER fluid. When an electric field is applied and induces the particles polarizing and closing up to form chains, the local electric field in between the particles rapidly increases with the particles approaching and can be orders higher than the external electric field. As long as the local field strength in the gap between the particles is high enough, the polar molecules in the gap can turn to the field direction and bond with the polarized charges on the neighboring particle. This bonding effect of polar molecule-charge causes much larger strength than the interaction between pure polarized particles occurred in conventional ER fluids. That the increment of the active polar molecules is proportional to the electric field implies a near-linear dependence of the yield stress on the applied field. When the electric field shuts down the aligned polar molecules retour into the state of random distribution by thermal motion and polarized charges disappear at the same time. As a result, the bonding effect of polar molecule-charge and then ER effect vanish. The main influencing factors on the PM-ER effect have also been discussed. The PM-ER fluids with the yield stress higher than MPa can be expected by optimizing the factors in preparing PM-ER fluids.

Acknowledgement

The authors would thank Prof. M.Y.Hou, W.Y.Lai and Z.X.Cao for helpful discussion. This work was supported by the National Basic Research Program of China (No 2004CB619005), the Knowledge Innovation Project and Outstanding Overseas Chinese Scholars Fund of Chinese Academy of Sciences and NSFC.

References

1. T.C. Halsey and W. Toor, *Phys. Rev. Lett.* 65 (1990) pp.2820-2823.
2. T.C. Halsey, *Science* 258 (1992) pp.761-766.
3. R. Tao, J.M. Sun, *Phys. Rev. Lett.* 67 (1991) pp. 398-401.
4. Ma HR, Wen WJ, Tam WY, Sheng P, *Advances in Physics* 52 (2003) pp.343-383.
5. Ma HR, Wen WJ, Tam WY, et al., *Phys. Rev. Lett.*, 77 (1996) pp.2499-2502.
6. YL. Zhang, KQ. Lu, GH. Rao, Y. Tian, SH. Zhang, JK. Liang, *Appl. Phys. Lett.* 80 (2002) pp. 888-890;
7. W. Wen, X.Huang, S.Yang, K. Lu and P. Sheng. *Nature Materials*,2 (2003) pp.727-730.
8. Lu KQ, Shen R, Wang XZ, Sun G, Wen WJ,*International J. Modern Phys.*B19 (2005) pp.1065-1070; Shen R, Wang XZ, Wen WJ, Lu KQ, *International J. Modern Phys.* B19 (2005)pp.1104-1109; Wang XZ, Rong S, Wen WJ, Lu KQ, *International J. Modern Phys. B* 19 (2005) pp.1110-1113.
9. R. Shen, XZ. Wang , Y Lu, WJ. Wen, KQ. Lu to be published.
10. XZ. Wang ,R. Shen, Y Lu, WJ. Wen, G. Sun, KQ. Lu to be published.
11. L. C. Davis, *J. Appl. Phys.* 81 (1997) pp. 1985-1991.
12. R. Tao, Qi Jiang and H. K. Sim, *Phys. Rev. E* 52 (1995) pp. 2727-2735.

13. P. Gonon, J.-N. Foulc, P. Atten, and C. Boissy *J.Appl.Phys.*86 (1999) pp. 7160-7169.
14. Y. Chen, A.F. Sprecher, and H. Conrad, *J. Appl. Phys.* 70, (1991) pp. 6796-6803.
15. L. D. Zhang, *et al.*, *Physica Status Solidi A-Applied Research* **157**, (1996) 483.
16. A. Dey, S. De, A. De, S. K. De, *Nanotechnology* **15**(2004 1277).
17. Y. T. Shih， Y. Y. Liao And D. S. Chuu, *Physical Review B* 68(2003)075402.
18. Y.T. Shih, D.S. Chuu, And W.N. Mei, *Solid State Commun.* 99(1996) pp.819-824
19. J.N.Israelachvili, *Intermolecule and Surface Force*, Second Edition, Academic Press, 2000.
20. Wen WJ, Huang XX, Sheng P, *Applied Physics Letters*, 85(2004) pp.299-301.
21. J.Frenkel, *Phys. Rev.* 54(1938)647.
22. H. Uejima, *Jpn. J. Appl. Phys.* 11(1972) pp.319-326.
23. Yu.F. Deinega, G.V. Vinogradov, *Rheol. Acta* 23 (1984) 636.
24. A.V. Lykov, Z.P. Shulman, R.G. Gorodkin, A.D. Matsepuro, *J. Eng. Phys.* 18 (1970) 979.
25. J.E. Stangroom, *Phys. Techn.* 14 (1983)290.
26. H. See, U. Tamura and M. Doi, *J. Phys. D: Appl. Phys.* 26 (1993) pp.746-752.
27. H. Tamura, H. See and M. Doi, *J. Phys. D: Appl. Phys.* 26(1993) pp.1181-1187.

THE DYNAMIC BEHABIOR OF MAGNET ENCOMPASSED WITH MAGNETIC FLUID IN ALTERNATING MAGNETIC FIELDS

S. SUDO, T. YANO, T. KASHIWAGI AND H. HONMA

Akita Prefectural University, Ebinokuchi, Yurihonjo 015-0055, Japan
E-mail: sudo@akita-pu.ac.jp

This paper describes the dynamic behavior of permanent magnets encompassed with magnetic fluid subjected to the alternating magnetic fields. The dynamic behavior of NdFeB magnet encompassed with magnetic fluid in the magnetic fields is a fascinating subject that is attracting the attention of robot researchers and micromachine engineers. This subject is concerned with the development of the micro actuators driven by the wireless energy supply system. A permanent magnet adsorbs magnetic fluid, and it is suspended in magnetic fluid. The permanent magnet in the self-levitation phenomenon can be easily controlled by the external magnetic fields. Some basic responses of permanent magnets encompassed with magnetic fluid subjected to the alternating magnetic fields are studied with a high speed video camera system. Experimental results show the possibility of the control of the permanent magnet motions by the external magnetic field. According to the frequency of external alternating magnetic field, the responses of permanent magnet show reciprocating, rotary and metronomic motions. The effects of the magnetic fluid volume and magnetic field intensity on the dynamic behavior of permanent magnet are revealed experimentally.

1 Introduction

In the past few decades, the interest in the field of micro- and nanosystem greatly increased due to recent developments in Micro Electro Mechanical System (MEMS) technology, and physics of the micro world [1-3]. MEMS technology, which uses silicon micromachining techniques to create submillimeter mechanical structures, originated from IC technology. Moreover, with the critical development in micro processing technology and the availability of various functional materials and intelligent material systems, the importance of the developments of micro mechanisms and micro machines is well recognized.

To drive micro machines and micro systems, micro actuators are needed. Therefore, extensive investigations on micro actuators have been conducted and reported [1,2]. For example, thermo mechanical, piezoelectric, magnetostrictive, electrohydrodynamic, electrostatic, phase change, shape memory, electrorheological, and diamagnetism actuators have been proposed by a number of researchers. The wireless energy supply method is important for the drive of the micro machines. Authors also proposed new type of magnetic fluid actuator for driving micro machines in our previous paper [4]. In spite of many investigations, however, there still remains a wide unexplored domain. Research data for the development of the micro magnetic fluid actuator driven by the external magnetic fields are scanty, and there are some points which must be clarified.

In this paper, the dynamic behavior of NdFeB magnet encompassed with magnetic fluid in the alternating magnetic field is studied to develop the micro magnetic fluid actuator driven by the wireless energy supply system. Some basic responses of permanent magnet encompassed with kerosene-based magnetic fluid subjected to the external alternating magnetic fields are examined by a high speed video camera system. The reciprocating response and the rotational behavior of the magnet encompassed with magnetic fluid are observed. The effects of the magnetic fluid volume and magnetic field intensity on the dynamic behavior of NdFeB permanent magnet are revealed experimentally.

2 Element composition of examined actuator

In this paper, the dynamic behavior of the magnet encompassed with magnetic fluid as driving element of the micro magnetic fluid actuator is examined. Driving element of the micro actuator is composed of Nd2Fe14B permanent magnet and kerosene-based magnetic fluid. Table1 shows the magnetic properties of permanent magnet. In the experiment, the permanent magnet of the column shape of 4mm in the diameter and 2mm in height is used. The magnetic flux density of NdFeB magnet is 330mT on the surface of the magnet. Table2 shows the physical properties of magnetic fluid (ferricolloid HC-50) used in the experiment. The fundamental structure of magnet encompassed with magnetic fluid is shown in Fig.1.Magnetic fluid is adsorbed to the permanent magnet. The magnet encompassed with magnetic fluid is put on the acrylic plastic base with 3mm thickness. The iron plate with 1mm thickness is laid under the acrylic plastic base for the weak attraction of the magnet with magnetic fluid. The weak magnetic attraction keeps the magnet-magnetic fluid element at a constant state against a little inclination of the base.

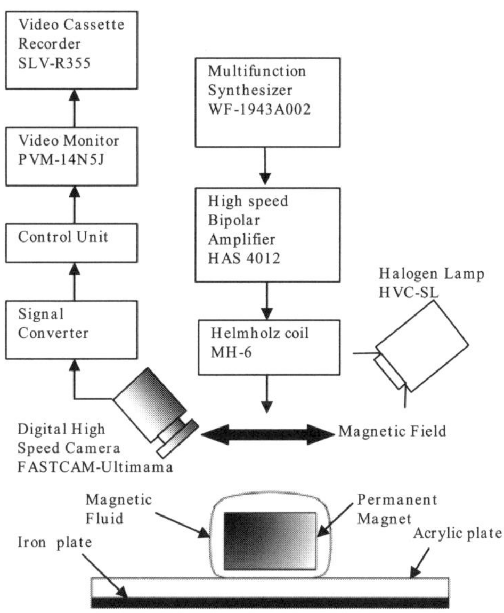

Fig. 1 Schematic diagram of experimental apparatus

Table 1. Magnetic properties of permanent magnet.

Residual magnetic flux density	Br(T)	1.62-1.3:
Coercive force	bHC(kA/m)	859-970
Coercive force	iHC(kA/m)	>955
Maximum energy product	(BH)max (kJ/m^3)	302-334

Table 2. Physical properties of magnetic fluid.

Magnetic fluid	Ferricolloid HC-5
Density	1250 kg/m^3
Viscosity	9 mPa·s
Saturation magnetization	32 kA/m
Surface tension	0.0277 N/m

The iron plate plays an important role as the stabilizer for the magnet-magnetic fluid element. The viscous effect of magnetic fluid can be neglected, because the viscosity of kerosene-based magnetic fluid is low.

3 Experimental apparatus and procedures

A schematic block diagram of the experimental apparatus to study the dynamic behavior of a permanent magnet encompassed with magnetic fluid in alternating magnetic fields is shown in Fig.1. The experimental apparatus is composed of a magnet-magnetic fluid element system, magnetic field generation system, and high-speed video camera system. The permanent magnet adsorbs magnetic fluid, and it is suspended in magnetic fluid. The permanent magnet in the self-levitation state can be easily controlled by the external magnetic fields. The alternating magnetic field is generated by applying alternating current voltage to the Helmholtz coil. The alternating current signal is supplied by the frequency synthesizer. The magnet-magnetic fluid element is located on a center part of the Helmholtz coil.

The direction of the magnetic field is parallel to the horizontal acrylic plastic base through all the experiments. In the experiment, a constant volume of magnetic fluid is adsorbed by the permanent magnet. Frequency responses of the magnet encompassed with magnetic fluid are studied with the high-speed video camera system.

In this experiment, the applied voltage E_c is given in the sinusoidal form as follows,

$$E_c = \frac{E}{2}\sin(2\pi f_0 t) \tag{1}$$

where $E/2$ is the amplitude of alternating voltage. The magnetic field is generated at a constant voltage E, where E is the total amplitude of alternating voltage. Frequency responses of the magnet encompassed with magnetic fluid are obtained for various applied voltages.

4 Experimental results and discussion

4.1 Driving characteristics of reciprocating motion

When the foregoing system is subjected to the alternating magnetic field with lower frequencies, the permanent magnet encompassed with magnetic fluid shows reciprocating motion for right and left along the direction of magnetic field [4]. In this experiment, the magnetic attraction between the permanent magnet and the iron plate is higher compared with experimental conditions in our previous paper [4], because the distance (thickness of acrylic plastic plate) is shorter. Figure 2 shows the response characteristics of the permanent magnet encompassed with magnetic fluid. In Fig. 2, S is the total amplitude of reciprocating motion, and V_m is the volume of magnetic fluid adsorbed to the magnet. It can be seen from Fig. 2 that the total amplitude S depends on applied voltage E and frequency f_0. The change of total amplitude S with magnetic fluid volume V_m for the condition $E = 8$ V is shown in Fig. 3. The total amplitude S increases with V_m at $f_0 = 1$ Hz. However the total amplitude S for $f_0 = 2$ Hz and $f_0 = 3$ Hz indicates the maximum value at $V_m = 7.5\text{x}10^{-8}$ m^3. The amplitude S for $f_0 = 4$ Hz does not show the change by V_m. At frequencies above $f_0 = 5$ Hz, the amplitudes S show very small values. The total amplitude S strongly depends on the duration of the magnetic field in the same direction.

In this system, magnetic fluid plays a lubrication role. Figure 4 shows the detail of

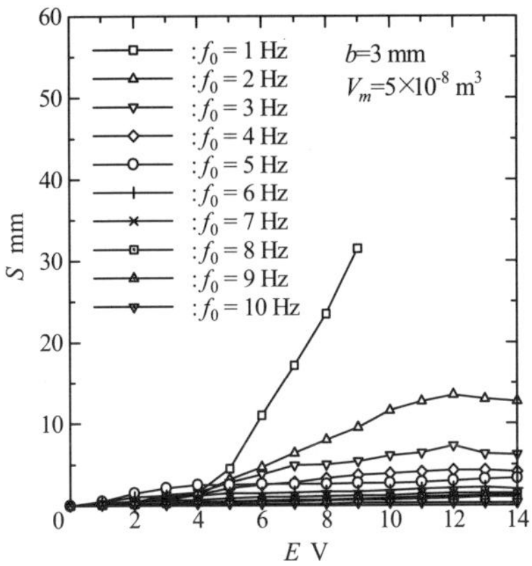

Fig. 2 Effect of applied voltage on the amplitude

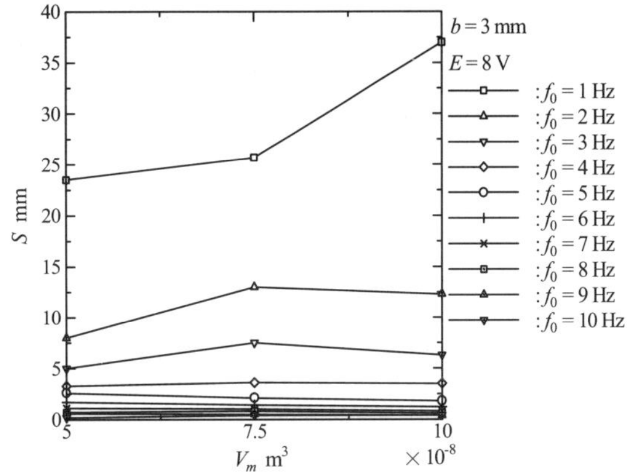

Fig. 3 Effect of magnetic fluid volume on the amplitude

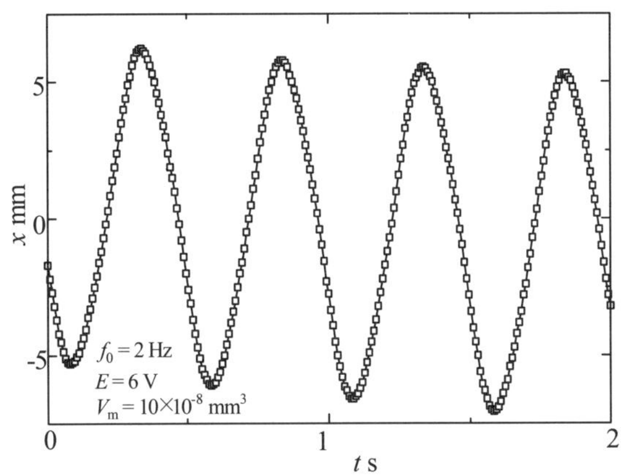

Fig.4 Reciprocating motion of magnet-magnetic fluid element

reciprocating motion of the permanent magnet encompassed with magnetic fluid. In Fig. 4, x is one-dimensional coordinate system parallel to the acrylic plate. The permanent magnet reciprocates regularly. The magnetic torque $\boldsymbol{T}$ acting on the permanent magnet with magnetic moment m in the external magnetic field $\boldsymbol{H}_e$ is given as follows;

$$\boldsymbol{T} = \boldsymbol{m} \times \boldsymbol{H}_e \tag{2}$$

In general, the permanent magnet in magnetic fluid is rotated by the magnetic torque described by Eq. (2). A slight rotation of the magnet on the iron plate brings ununiformity in the magnetic field. Ununiformity in the magnetic field means the gradient of the magnetic field. The permanent magnet encompassed with magnetic fluid is moved in the unidirection magnetic field of long duration at lower excitation frequencies.

4.2 Dynamic behavior of permanent magnet in higher magnetic fields

The response of the permanent magnet encompassed with magnetic fluid shows diversity according to the signal of the external magnetic field. At higher applied voltage E to the

Helmholtz coil, standing up and the rotational phenomena of the permanent magnet are observed. Figure 5 shows standing up and falling down of the permanent magnet encompassed with magnetic fluid at E = 16 V. In Fig. 5, the x-y two-dimensional coordinate is set on the surface of acrylic plate. The x coordinate axis is parallel to the acrylic plate, and the y coordinate axis is perpendicular to the plate. Figure 5 shows the movement trajectory of the fixed point on the magnet. The arrow indicates the direction of the movement, and S_t is the time interval of the data plot in Fig. 5. The upper figure shows the movement of the magnet to the right side, and lower figure shows the left side movement. First, the inclining magnet moves to the right, and stands up. Secondly, the magnet falls down and slides to left. Thirdly, the magnet stands up anticlockwise. This process is regularly repeated. Figure 6 shows the above-mentioned phenomenon for time. The upper figure shows the process on x axis, and lower figure shows the process on y axis. There are some photographs where states of the permanent magnet encompassed with magnetic fluid are shown. The magnet rapidly moves to the right and stands up. After word, the magnet returns gradually. The standing up phenomenon of the magnet is generated b the higher magnetic torque. The magnet rotates to one direction in higher magnetic torque area.

4.3 *Effects of volume of magnetic Fluid Systems Laboratory*

As mentioned above, the response of the permanent magnet encompassed with magnetic fluid shows reciprocating motion, standing up motion, and rotational motion according to the signal of the magnetic field. These response behaviors are different depending on the volume of magnetic fluid, V_m. Figure 7 shows the generation areas of each response behavior. The borderlines of each response decrease with the volume of magnetic fluid. The increase in the volume of magnetic fluid decreases the attraction force between the magnet and iron plate. The magnetic torque easily rotates the permanent magnet in larger volume of magnetic fluid.

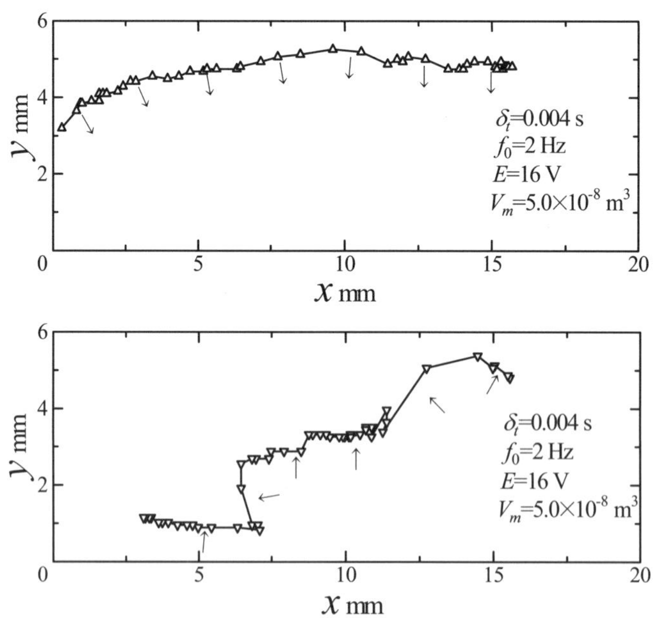

Fig.5 Dynamic behavior of magnet-magnetic fluid element

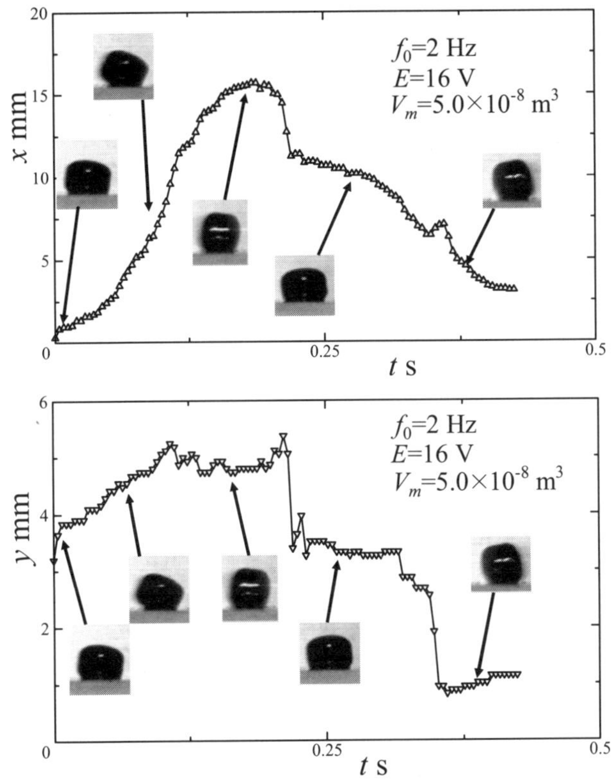

Fig.6 Dynamic response of magnet-magnetic fluid element

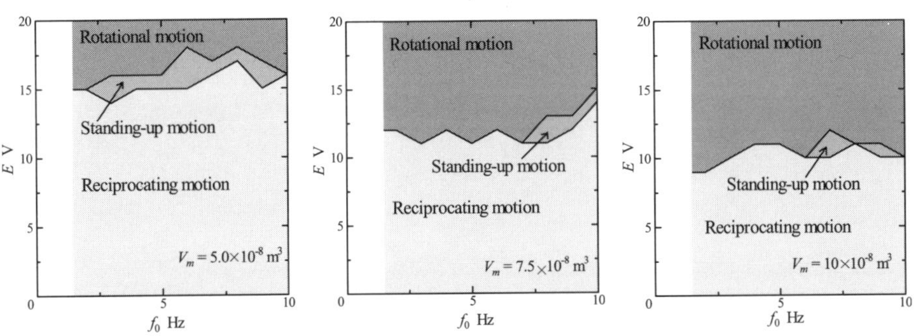

Fig.7 Each response region of magnet-magnetic fluid element

5 Conclusions

The dynamic behavior of permanent magnet encompassed with magnetic fluid in the alternating magnetic field was studied experimentally. The results obtained are summarized as follows.

(1) The response behavior of the permanent magnet encompassed with magnetic fluid in the alternating magnetic field depends on the applied voltage to the coil, the frequency, and the volume of magnetic fluid.
(2) The total amplitude of reciprocating motion decreases with the increase in the frequency of magnetic field.
(3) The response behavior of the permanent magnet at higher applied voltages shows standing up motion and rotational motion.
(4) The borderlines of response behavior of the magnet in E-f_0 chart decrease with the volume of magnetic fluid.

References

1. Esashi M., Micromachines (Sangyo Gijutsu Service Center, Tokyo, 2002) (in Japanese).
2. Higuchi T., Micromachine Technology (Sangyo Gijutsu Service Center, Tokyo, 2003) (in Japanese).
3. JSME, Proceedings of International Symposium on Micro-Mechanical Engineering, Tsuchiura and Tsukuba, (2003), pp.1-520.
4. Sudo S., Takaki Y., Hashiguchi Y. and Nishiyama H., Magnetic fluid devices for driving micro machines, *JSME Int. J.* 48(2005) pp.464-470.

A NEW MR FLUID-ELASTOMER (MRF-E) VIBRATION ISOLATOR

DAVID YORK, XIAOJIE WANG AND FARAMARZ GORDANINEJAD

Composite and Intelligent Materials Laboratory, Department of Mechanical Engineering, University of Nevada, Reno 89557, USA

E-mail: yorkd@unr.nevada.edu, xwang@scs.unr.edu, and faramarz@unr.edu

In this study, the performance of a new design concept utilizing a magnetorheological (MR) fluid composite material is examined through encapsulating a MR fluid into an elastomer. A prototype MR Fluid-Elastomer Vibration Isolator is built and its dynamic behavior is studied in oscillatory compressions for a wide range of frequencies under various applied magnetic fields. The experimental results show that both the stiffness and the damping capability of the MR Fluid-Elastomer Vibration Isolator is a function of the displacement amplitude and magnetic field strength, and only weakly dependent upon the frequency of excitation. This demonstrates that the new vibration isolator, whose mechanical properties can be controlled by an applied magnetic field, has potential in applications where tuning vibration characteristics are desired.

1 Introduction

The ability to tune the rheological properties of Magnetorheological (MR)/Electrorheological (ER) fluids has led to vast research opportunities into the field of mechanical vibration control. Such opportunities have directed researchers to explore such topics as semi-active or adaptive vibration control; a very promising and important application in the attenuation of vibrations.

The purpose of this research is to extend our previous studies by the designing and testing of a novel MR Fluid-Elastomer (MRF-E) Vibration Isolator for small amplitude and high frequency vibration control. Some general applications to this area of study are vibrations caused by rotary machinery and linear motion applications, such as controlled impact devices, and isolation tables [1,11]. A relatively new invention, the squeeze mode damper (SMD), has provided new opportunities for controlling such attenuations. The first non-conventional SMDs began to surface in the late 1980s and early 1990s and employed the use of Electrorheological (ER) fluid [6]. The ER SMDs were tested not only in linear applications but as rotary dampers as well [11]. These studies proved that not only did the ER SMD have the capability of generating larger forces in both compression and tension but were far easier to design and much more compact than ER dampers that work strictly in shear mode. Wang developed a new mixed mode ER damper that combined the capabilities of both shear and squeeze mode [8]. This damper was capable of generating much higher forces in compression and tension than ER dampers operating in shear mode. The MR SMD was first devised by Jolly and Carlson in 1996 [3]. The MR SMD produced much higher forces than the ER SMD due to the yield strength of MR fluid being much larger than the yield force of ER fluid [3,5,7].

Implementation of MR and ER SMDs into a vibrating mechanical system has many advantages. These dampers are capable of working in a wide range of frequencies (0Hz-85Hz) and small amplitudes (0.01mm-4mm) depending on the application [3]. The dampers are capable of generating larger forces with smaller strokes in compression and tension than conventional squeeze film and shear mode dampers. The dampers ability to dissipate energy is directly proportional to the displacement amplitude and field strength.

These dampers also have their disadvantages. The capability of the squeeze film damper (SFD) to dissipate energy decreases with increasing frequency and some dampers even develop a rheoplexic behavior as the number of oscillatory cycles increase [3,7,8]. The size of the fluid filled gap, between the two squeeze plates, has a direct effect on the dampers performance, and repeatable test results are often difficult to obtain due to the arduous task of refilling the fluid volume to the level it once was.

To overcome these disadvantages, which have been shown in most of MR/ER SMD devices, we proposed a new design concept utilizing a MR composite material through encapsulating MR fluids (or iron particles) into an elastomer [9,10]. This type of material is similar to other MR elastomer systems. However, the magnetizable particles are contained within a void in the elastomer casing and do not need alignment during the curing process before the matrices solidification, as with typical MR elastomers [2,4].

In order to evaluate the performance of the MRF-E Composite, a prototype MRF-E Vibration Isolator is designed and tested. A three-dimensional finite element analysis (FEA) commercial software is used to optimize the electromagnetic design that can generate a strong electromagnetic field strength inside the MRF-E Composite when fully activated. The dynamic behavior of the MRF-E Vibration Isolator is characterized using an Instron Model 8821S servo-hydraulic actuator.

2 MR Fluid-Elastomer Vibration Isolator

The schematic design of the MRF-E Vibration Isolator is shown in Figure 1. The device consists of several elements. Each element is optimized in order to obtain the most efficient electromagnetic device. The geometry of the core material placed above and below the MRF-E Composite had the most direct effect on the embedded MR fluid. The individual diameters and lengths of each part are adjusted (by halving or doubling) such that the model of encased MR fluid becomes saturated with magnetic flux. The optimization of each part is complete when no significant change (less than 10 percent) in the magnetic flux occurs when the dimensions of said part are altered further. The optimized electromagnetic design is realized using a three-dimensional FEA program. The amount of magnetic flux density inside the FE model of embedded MR fluid, of the fully optimized electromagnetic design, is shown in Figure 2. The magnetic flux density inside the activation area of the MRF-E Composite model increases with increasing current. Figure 3 shows the photograph of two MRF-E Composite prototypes which are used in the vibration isolation device.

3 Experimental Characterization

An Instron 8821S hydraulic dynamometer was used in the performance testing of the electromagnetic device under harmonic compressive motion. Figure 4 shows a photograph of the experimental setup. The upper end of the device is fixed while the lower end undergoes harmonic motion. The linear variable differential transformer (LVDT) (Model DCTH-100AG), which can measure displacements of +/- 2.5mm with resolution of 0.01mm, was used in the experiment to measure the deformation of the MRF-E Composite sample. The load cell used has a capacity of 500lb. The electromagnetic device performance was determined under various input currents (0.0, 1.0, 2.0, 3.0Amps) and different harmonic inputs.

The Instron is capable of producing the harmonic excitation needed for the performance testing at the required frequencies (0.1Hz-10Hz) and amplitudes (0.1mm-0.6mm). This performance test consists of a sinusoidal frequency sweep while maintaining amplitude at a constant displacement. Each specific frequency tested represents an element of the overall performance test. In each test, the input electric current applied to the MRF-E Vibration Isolator is kept at a constant level of 0.0, 1.0, 2.0, and 3.0Amps. Experimental data was collected using a data acquisition system at a sampling rate of 200 samples-per-second.

4 Results and Discussion

Figure 5a shows the typical force response results of the harmonic compressive deformation at a constant frequency and amplitude for different applied currents to the MRF-E Vibration Isolator. As can be seen in the off-state (0.0 Amp input), the force response curve of the MRF-E Composite is in-phase with the sinusoidal input trace, and that the MRF-E Composite behaves similar to a viscoelastic material. However, as the magnetic field is activated and the strength is increased the response force increases to a peak value of 470N at an input current of 3.0Amps compared to the peak value of 282N for the passive case without magnetic field, while the force response deviates from the in-phase sinusoidal response. These changes indicate that the MRF-E Vibration Isolator, through changing magnetic field strengths, has the ability to alter the dynamic behavior of the MRF-E Composite from linear viscoelastic to nonlinear viscoelastic, while also increasing its damping and stiffness capabilities. Figure 5b shows the typical experimental results of the hysteresis loop for the same constant frequency and amplitude as in Figure 5a. The areas of the loops enlarge with increasing current. The area inside each of the loops represents the energy dissipated-per-cycle and it increases with increasing current.

Figure 6 shows the relationship of the energy dissipated-per-cycle versus excitation frequency for the amplitude of 0.6mm. Figure 7 presents the energy dissipated-per-cycle for every experimentally tested frequency and amplitude (0.1-0.6mm). As can be seen, the energy dissipated-per-cycle increases with increasing amplitude and current. The closely clustered “dots” located at different measured amplitudes identify the energy dissipated-per-cycle at each tested frequency. Although there are larger variations for the measurements of energy dissipated-per-cycle at the smaller displacement amplitudes of (0.1mm-0.2mm) due to the noise from the signals, the energy dissipation mode of the MRF-E Composite displays a nearly constant behavior over the range of tested frequencies and for all the test displacement amplitudes. This may lead to the conclusion that the vibration isolator is only weakly dependent upon frequency, much like a friction damper. However, examination of the stiffness force, for the MRF-E Vibration Isolator, illustrates that it behaves similar to a controllable nonlinear spring with a stiffness coefficient that is a function of the vibration frequency. The equivalent viscous damping coefficient for the MRF-E Vibration Isolator can be obtained from the energy dissipated-per-cycle, W, as follows:

$$C_{eq} = \frac{W}{2\pi^2 f X_0^2} \tag{1}$$

where, f and X_0, are the input frequency and amplitude of the oscillatory motion. Figure 8 displays how the equivalent viscous damping coefficient, C_{eq}, changes with the displacement amplitude and input currents at a constant frequency of 8.0Hz. In order to clarify the changes in the values of C_{eq} for the range of tested frequencies and input currents, the calculated C_{eq} as a function of the applied current and excitation frequency for the amplitude of 0.6mm are presented in Table 1. As can be seen, the values of C_{eq}

decrease with increasing frequency, however, the values increase with increasing input current. This demonstrates the significant increase of the equivalent damping coefficient, for the MRF-E Vibration Isolator, that can be achieved for the tested frequency range (0.1-10.0Hz) at on-state (1.0-3.0Amp of applied current) compared to that of off-state (without applied current). Thus, it can be concluded that the MRF-E Vibration Isolator has the capability to control damping over a wide range of different frequencies and amplitudes.

5 Results and Discussion

From the results of the experimental investigation on the dynamic behavior of the MRF-E Vibration Isolator under different compressive oscillatory cycles, it can be concluded that through an applied magnetic field the MRF-E Composite has the capacity for controllable damping and dynamic stiffness over a wide range of frequencies and displacement amplitudes. It is also demonstrated, that unlike squeeze mode dampers, the MRF-E Composite's dynamic behavior is nearly independent of frequency. Therefore, the MRF-E Vibration Isolator's performance is clearly an improvement over current MR and ER squeeze film dampers.

6 Acknowledgements

This work is sponsored in part by the National Science Foundation. The authors are grateful for this support, and would like to specifically thank Dr. Chi Liu, Program Director.

References

1. Ahn, Y.K., Ha, J.Y., Yang, B.S., "A New Type Controllable Squeeze Film Damper Using An Electromagnet," *Journal of Vibration and Acoustics*, Vol. 126, pp. 380-383, 2004.
2. Ginder, J. M., Schlotter, W. F., and Nicholes, M. E. "Magnetorheological Elastomers in tunable vibration absorbers." *Proceedings of the SPIE - The International Society for Optical Engineering.* Vol. 4331, pp 103-110, 2001.
3. Jolly, M.R., Carlson, J.D., "Controllable Squeeze Film Damping using Magnetorheological Fluid," *5th International Conference on New Actuators*, Bremen, Germany, 26-28 June 1996.
4. Jolly, M. R., Carlson, J. D., Munoz, B. C., and Bullions, T. A., "The Magnetoviscoelastic Response of Elastomer Composites Consisting of Ferrous Particles Embedded in a Polymer Matrix," *Journal of Intelligent materials, Systems and Structures*, Vol. 7, 613-622, 1996.
5. Sims N.D., Stanway R., Johnson A.R., Mellor, P., "Design, testing and model validation of an MR squeeze-flow vibration damper," *Smart Structures and Materials 2001: Damping and Isolation Proceedings of SPIE*, Vol. 4331, pp. 111-120, 2001.
6. Stanway, R., Sproston, J.L., Prendergast, M.J., Case, J.R., Wilne, C.E., "ER fluids in the squeeze-flow mode: an application to vibration isolation," *Journal of Electrostatics*, Vol. 28, pp. 89-94, 1992.

7. Tang, X., Wang, X., Li, W.H., and Zhang, P.Q., "Testing and Modeling of an MR Damper in the Squeeze Flow Mode," *Proceedings of the 6th International Conference on Electro-Rheological Fluids, Magneto-Rheological Suspensions and Their Applications*, Ed. M. Nakana and K. Koyama, World Scientific, pp. 870-878, Singapore, 1998.
8. Wang, X., Zhang, P.Q., Tang, X, Tao, R., "Testing and Modeling a Cone-Shaped Squeeze-Film Mode Electrorheological Damper," *Journal of Intelligent Material Systems and Structures*, Vol. 10, pp. 748-752, September 1999.
9. Wang, X., Gordaninejad, F. and Hitchcock, G., "Dynamic Behaviors of Magnetorheological Fluid-Elastomer Composites Under Oscillatory Compression," *Damping and Isolation, Proceedings of SPIE Conference on Smart Materials and Structures*, Ed. Kon-Well Wang, Vol. 5386, pp. 250-258, 2004.
10. Wang, X., Gordaninejad, F. and Hitchcock, G., "A Magnetorheological Fluid-Elastomer Vibration Isolator," *Damping and Isolation, Proceedings of SPIE Conference on Smart Materials and Structures*, Ed. Kon-Well Wang, Vol. 5760, pp. 217-225, 2005.
11. Yao, G.Z., Qiu, Y., Meng, G., Fang, T., Fan, Y.B., "Vibration Control Of A Rotor System By Disk-type Electrorheological Damper," *Journal of Sound and Vibration*, Vol. 219, issue 1, pp. 175-188, 1999.

Table 1. Equivalent Viscous Damping Coefficient Values for a Constant Amplitude of 0.6mm.

C_{eq} [kN*s/m]				
Current [Amp] / **frequency [Hz]**	**0.0**	**1.0**	**2.0**	**3.0**
0.10	176.25	221.42	287.78	299.10
0.50	33.51	46.45	54.43	58.72
2.0	9.70	11.31	13.79	15.55
6.0	3.33	3.96	4.46	4.80
10.0	1.72	1.91	2.21	2.46

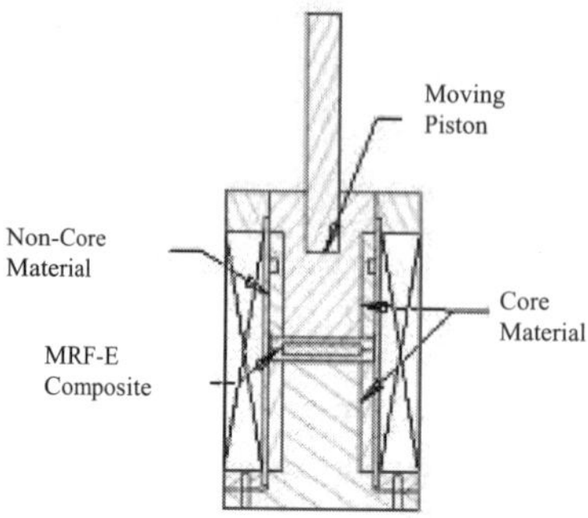

Figure 1. Schematic of MRF-Elastomer Vibration Isolator design.

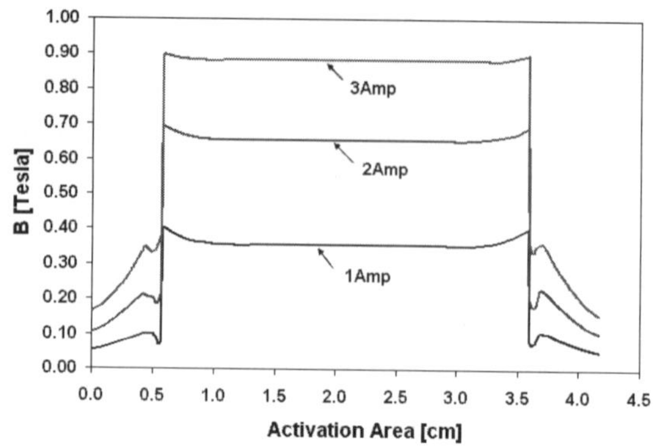

Figure 2. Results of Finite Element Analysis inside embedded MR Fluid.

Figure 3. Photograph of MR Fluid-Elastomer Composite structure prototypes. The MRF-E Composite on the left is a Poly-BD blend. The MRF-E Composite on the right is an RTV Silicon.

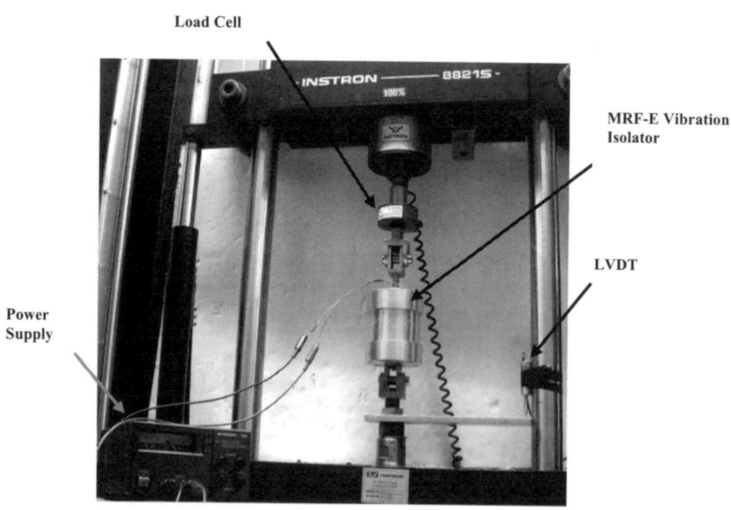

Figure 4. Experimental setup to measure the dynamic properties of MR Fluid-Elastomer Vibration Isolator.

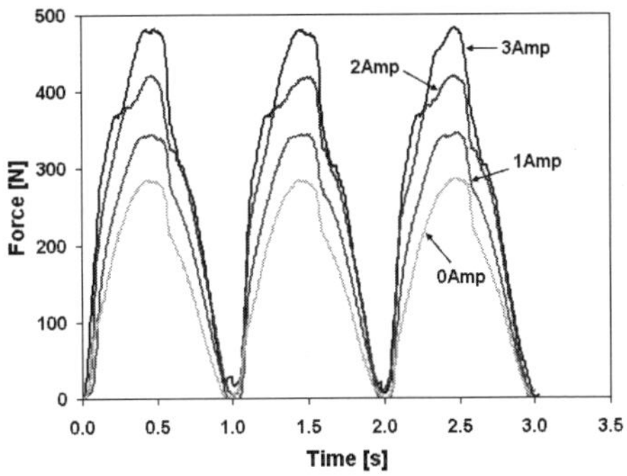

(5a)

Figure 5a. Typical force response curves to an amplitude of 0.4mm (peak-to-peak) and frequency of 1.0Hz with different input currents.

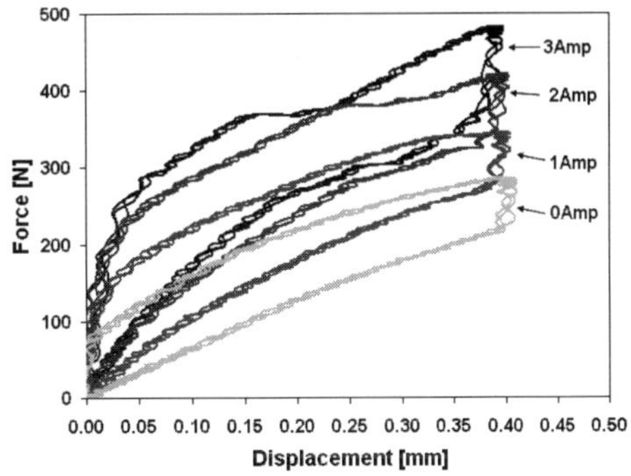

(5b)

Figure 5b. Hysteresis loop for the same inputs shown in Figure 5a.

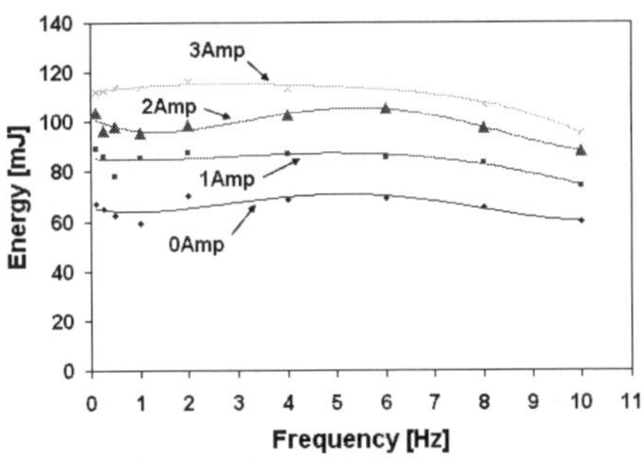

Figure 6. Energy dissipated-per-cycle for an amplitude of 0.6mm over all frequencies tested.

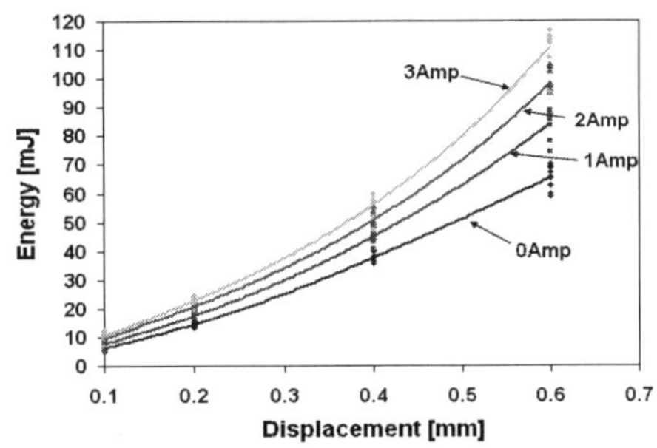

Figure 7. Energy dissipated-per-cycle for every experimentally tested frequency and amplitude.

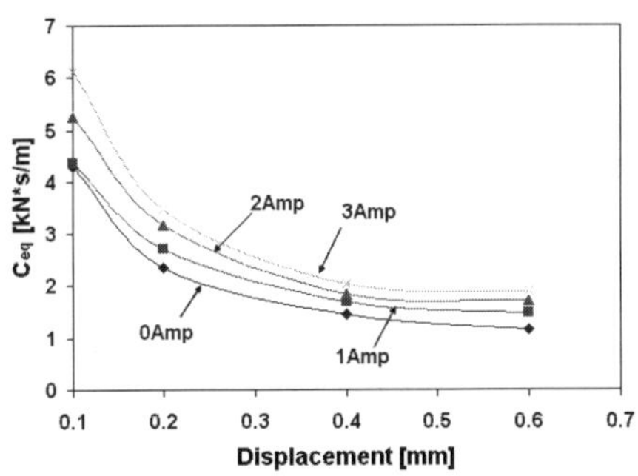

Figure 8. Illustrates the variability of the equivalent viscous damping coefficient at a constant 8.0Hz for all the tested amplitudes.

A NEW MASON NUMBER FOR MAGNETORHEOLOGY

DANIEL J. KLINGENBERG
Department of Chemical and Biological Engineering, University of Wisconsin, 1415 Engineering Dr., Madison, WI 53706, USA
E-mail: klingen@engr.wisc.edu

JOHN C. ULICNY AND MARK A. GOLDEN
General Motors R&D and Planning, 30500 Mound Rd, Warren, MI 48090-9055, USA
E-mail: john.ulicny@gm.com, mark.golden@gm.com

The electric field strength and shear rate dependence of the apparent shear viscosity of electrorheological (ER) suspensions can often be represented by a function of only the Mason number. A Mason number defined for magnetorheological (MR) suspensions by direct substitution of magnetostatic variables for electrostatic variables does not produce a similar collapse of shear viscosity data for MR suspensions. We show that a Mason number defined in terms of the suspension magnetization can be employed to produce a collapse of experimental data at various magnetic field strengths and shear rates. As for ER suspensions, the new Mason number can be calculated from experimentally measured quantities.

1 Introduction

Designing electro- and magnetorheological (ER and MR) suspensions and devices requires knowledge of the relationships between the suspension rheological properties and such variables as the deformation rate, applied field strength, and suspension composition. While predictive models have provided insight into the mechanisms governing suspension behavior, these models are not sufficiently quantitative for design purposes. Methods for correlating or reducing experimental data are thus desirable.

Marshall et al.[1] made significant progress toward developing methods for correlating experimental data for ER suspensions in shear flow. They reasoned that the primary forces governing the behavior of ER suspensions are electrostatic polarization forces induced by the applied electric field, and hydrodynamic forces caused by the particle motion relative to the continuous phase. Employing dimensional analysis, they concluded that suitably nondimensionalized suspension properties should depend on only the ratio of the magnitudes of these forces, defined as the Mason number. Writing the magnitude of the electrostatic polarization force as $F^E = 12\pi\epsilon_0\epsilon_c a^2\beta^2 E_0^2$, and the magnitude of the hydrodynamic force as $F^H = 6\pi\eta_c a^2\dot{\gamma}$, the Mason number may be defined

$$\mathrm{Mn} \equiv \frac{F^H}{F^E} = \frac{\eta_c\dot{\gamma}}{2\epsilon_0\epsilon_c\beta^2 E_0^2}, \tag{1}$$

where ϵ_0 is the permittivity of free space, ϵ_c is the relative dielectric constant of the continuous phase, a is the particle radius, β is the relative particle polarizability,[2] E_0 is the applied electric field strength, η_c is the viscosity of the continuous phase, and $\dot{\gamma}$ is the macroscopic shear rate.

The utility of this dimensional analysis is that the shear rate and field strength dependence of the suspension viscosity can be described in terms of the single independent variable Mn. This is illustrated in Fig. 1 where experimental data for a 20 wt% suspension of alumina particles in silicone oil are presented.[2] In Fig. 1(a), the apparent suspension

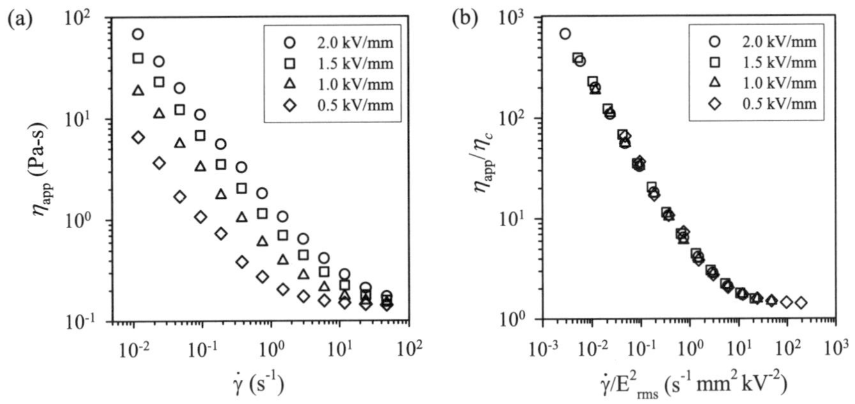

Figure 1. (a) Viscosity as a function of shear rate at several electric field strengths (rms field strength at 500 Hz) for a 20 wt% suspension of alumina particles in silicone oil. (b) Same data replotted as relative viscosity as a function of $\dot{\gamma}/E^2_{\mathrm{rms}} \propto$ Mn.

viscosity is plotted as a function of $\dot{\gamma}$ for several field strengths. In Fig. 1(b), the data are replotted as relative viscosity as a function of $\dot{\gamma}/E_0^2 \propto$ Mn. The data collapse on to a single curve, confirming the validity of the dimensional analysis. We expect this dimensional analysis to fail at large electric field strengths where nonlinear conduction phenomena become significant, [3,4] but Fig. 1(b) suggests that the Mason number dependence is appropriate for at least this range of field strengths.

One of the main benefits of the dimensional analysis described above is that relatively few experimental measurements are required to determine the behavior for a wide variety of combinations of $\dot{\gamma}$ and E_0. We wish to determine if a similar approach can be used to collapse shear viscosity data for MR suspensions.

2 Experimental

2.1 *Materials*

The data to be analyzed were extracted from Ulicny et al.,[5] where the torque was presented as a function of rotation rate and electromagnet current for MR suspensions in shear flow between concentric cylinders. Here we briefly describe the experiments.

The MR suspensions consisted of mixtures of two iron powders (all 50/50 mixtures by weight): a large particle iron (8 μm average diameter) and a small particle iron (2 micron average diameter). The particles were suspended in a hydrocarbon base liquid, with fumed silica added to reduce the sedimentation rate. The base liquid was a polyalphaolefin (PAO) of molecular weight of approximately 280. The fumed silica was an untreated material with a nominal average particle length of 0.2-0.3 μm. MR suspensions composed of iron volume fractions of $\phi = 0.20$, 0.30 and 0.45 were prepared.

The fumed silica imparted a weakly non-Newtonian character to the MR suspensions in the absence of an applied magnetic field. The zero-field rheological properties were measured and fit with a Bingham model. The yield stresses and plastic viscosities obtained

Table 1. Bingham parameters and base fluid viscosity for the three different suspensions in the absence of an applied magnetic field.

Iron Conc. (vol. %)	Silica Conc. (wt. %)	τ_0 (Pa)	η_{pl} (Pa·s)	η_c (Pa·s)
20	7.0	225.4 ± 0.8	0.132 ± 0.002	0.056
30	7.0	342.7 ± 0.9	0.200 ± 0.002	0.056
45	0.5	86.5 ± 1.4	0.195 ± 0.003	0.0089

from the fits are listed in Table 1. These yield stresses are much smaller than the field-induced stresses measured in the experiments described below. The base fluid viscosities η_c (PAO + fumed silica) are also listed in Table 1.

2.2 *Rheological measurements*

Rheometric experiments were performed with a custom-made concentric cylinder rheometer with an electromagnet coil imbedded in the outer cylinder. The MR fluid occupied the bottom half of the gap between the cylinders. The inner cylinder was rotated at a specified speed, and the torque transmitted to the outer cylinder was measured. Torque-rotation rate data are converted to average shear stress as a function of shear rate.

Samples were initially sheared at the desired rate with no current supplied to the coil. Current was then applied, producing a magnetic field within the MR suspension. The flux density B was measured for each current I employed, using a gauss probe placed in the upper half of the gap between the cylinders. Data were collected for 60 s. When the current was applied, the shear stress increased rapidly. For small currents, the shear stress then remained steady. For large currents, the shear stress subsequently increased slowly. This behavior was attributed to the formation of lamellar structures. In this study, we focus attention on the initial field-induced increase in the shear stress, measured 2 s after the magnetic field was applied.

For the MR fluid with $\phi = 0.45$, the suspension magnetization M was measured as a function of B using a vibrating sample magnetometer (EG&G PARC Model 155 VSM). This data permits evaluation of the magnetic field H, magnetization M, and permeability μ for any of the flux densities $B(I)$ measured during the rheometric experiments.

3 Results and Discussion

The rheological data are presented in Fig. 2(a) for $\phi = 0.45$, where the apparent suspension viscosity η_{app} divided by η_∞ is plotted as a function of the shear rate for different currents (and thus different magnetic field strengths). Here, η_∞ is defined as the high shear rate viscosity in the absence of an applied magnetic field. For this quantity we employ the plastic viscosity extracted from the steady shear stress-shear rate data measured in the absence of a magnetic field. The apparent viscosity data in Fig. 2(a) exhibit the typical shear thinning behavior of MR suspensions, with viscosity increasing with current (magnetic field strength), and decreasing with shear rate.

Following Marshall et al.,[1] we attempt to plot the experimental data as apparent viscosity (scaled by η_∞) as a function of the Mason number. By direct substitution of magne-

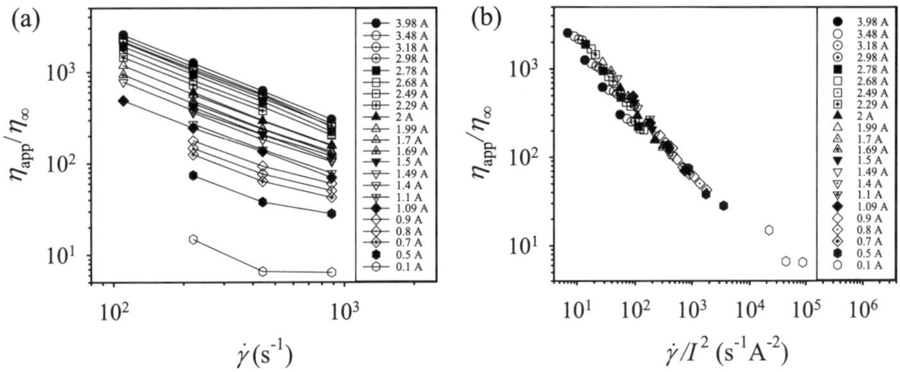

Figure 2. (a) Relative viscosity as a function of $\dot{\gamma}$ for different currents. (b) Relative viscosity as a function of $\dot{\gamma}/I^2$ ($\phi = 0.45$).

tostatic variables for electrostatic variables in the Mason number defined in Eq. 1, we write the Mason number for MR suspensions initially as

$$\text{Mn} = \frac{\eta_c \dot{\gamma}}{2\mu_0 \mu_c \beta_M^2 H^2}, \tag{2}$$

where μ_0 is the permeability of free space, μ_c is the relative permeability of the continuous phase, $\beta_M = (\mu_p - \mu_c)/(\mu_p + 2\mu_c)$ where μ_p is the relative permeability of the particle material, and H is the applied magnetic field strength.

As a first attempt, we assume $\beta_M \approx 1$ and $H \propto I$, and plot the relative viscosity as a function of $\dot{\gamma}/I^2$ in Fig. 2(b). The data clearly do not collapse on to a single function of $\dot{\gamma}/I^2$, with deviations most apparent at small $\dot{\gamma}/I^2$.

That the apparent viscosity cannot be represented as a function of $\dot{\gamma}/I^2$ is likely in part caused by nonlinear particle magnetization,[6] which results in the quantity β_M becoming magnetic field strength dependent. Thus as a next attempt to collapse the data, we relax the assumption of $\beta \approx 1$ and evaluate

$$\beta_M = \frac{\mu_p(H) - \mu_c}{\mu_p(H) + 2\mu_c}, \tag{3}$$

using $\mu_c = 1$ (here forward). The field-dependent particle permeability $\mu_p(H)$ is estimated from the Frohlich-Kennelly equation[6,7]

$$\mu_p(H) = 1 + \frac{(\mu^0 - 1)\, M_s/H}{(\mu^0 - 1) + M_s/H}, \tag{4}$$

with $\mu^0 = 10^3$ and $\mu_0 M_s = 2$ T, and with H equated with the applied magnetic field strength, determined from the measured flux density and suspension magnetic properties.

The scaled viscosity is plotted as a function of $\text{Mn}(H) \equiv \eta_c \dot{\gamma}/(2\mu_0 \mu_c \beta(H)^2 H^2)$ in Fig. 3(a). Again, the data do not collapse on to a single function of $\text{Mn}(H)$.

We believe that the lack of collapse of the data on to a single curve can still be attributed to nonlinear particle magnetization. The magnetostatic interparticle force is only proportional to H_0^2 at small field strengths. Including the field-strength dependence of β_M only partly accounts for the impact of nonlinear magnetization, as described below.

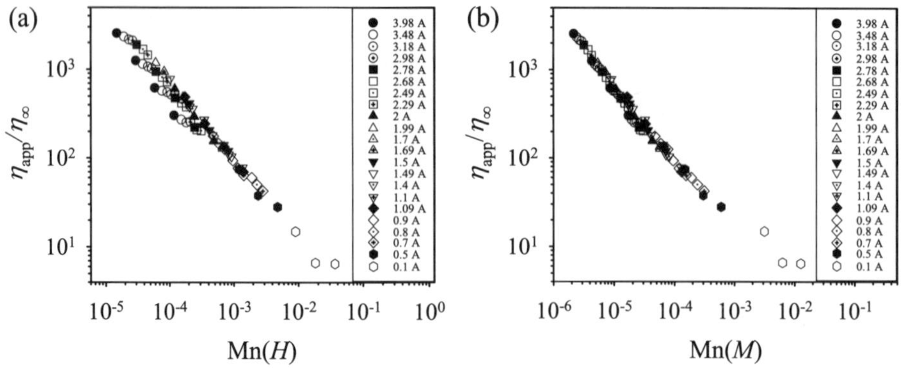

Figure 3. (a) Relative viscosity as a function of $\mathrm{Mn}(H)$. (b) Relative viscosity as a function of $\mathrm{Mn}(M)$.

Consider the point-dipole approximation to the magnetic force exerted on a magnetizable sphere at the origin of a spherical coordinate system caused by a second sphere at (R, θ) in a uniform magnetic field $\boldsymbol{H}_0 = H_0 \boldsymbol{e}_z$, written here in terms of the magnetic moment on each sphere $\boldsymbol{m} = m\boldsymbol{e}_z$,

$$\boldsymbol{F}^M = F_0 \left(\frac{\sigma}{R}\right)^4 \left[\left(3\cos^2\theta - 1\right)\boldsymbol{e}_r + \sin 2\theta \boldsymbol{e}_\theta\right], \tag{5}$$

where $\sigma = 2a$ is the sphere diameter, and

$$F_0 = \frac{3\mu_0 m^2}{4\pi\sigma^4}. \tag{6}$$

Consider the case where the applied magnetic field strength is sufficiently small such that $\mu_p =$ constant, and where $\beta_M \to 0$ or $R/\sigma \to \infty$. In such situations, the magnetic moment strength can be expressed

$$m = \pi\sigma^3 \beta_M H_0, \tag{7}$$

and we recover the commonly used form of the force magnitude, $F_0 \propto \mu_0 \sigma^2 \beta_M^2 H_0^2$.

For finite R/σ and nonvanishing β_M, Eq. 5 represents only the first term in a series expansion for $\boldsymbol{F}^M$.[8] In this case, F_0 still scales with H_0^2, but no longer scales with β_M^2.

Next consider sufficiently large magnetic field strengths where μ_p is field-strength dependent. For widely separated spheres, Eq. 7 is still valid, but β_M depends on the magnetic field strength,[7] and thus $F_0 \propto \beta_M^2 H_0^2$ is no longer proportional to H_0^2. For finite separations, the magnetic field strength, and thus the particle permeability and β_M, vary with position inside each sphere.[6] As a result, the force magnitude is no longer proportional to $\beta_M^2 H_0^2$ or H_0^2.

Instead of relating the sphere dipole moment to the applied magnetic field strength, one can simply relate the moment magnitude to the sphere magnetization, $\boldsymbol{M}_{\mathrm{sph}}$,

$$m = \frac{\pi}{6}\sigma^3 \langle M_{\mathrm{sph}} \rangle, \tag{8}$$

where $\langle M_{\mathrm{sph}} \rangle$ is the average magnitude of the sphere magnetization. For suspensions of magnetizable particles (e.g., iron particles) in a nonmagnetizable medium, the sphere magnetization is related to the suspension magnetization $\langle M \rangle$ via

$$\langle M_{\text{sph}} \rangle = \frac{1}{\phi} \langle M \rangle . \tag{9}$$

Thus Eq. 6 for the magnitude of the point-dipole force magnitude can be rewritten

$$F_0 = \frac{\pi}{48} \frac{\mu_0 \sigma^2 \langle M \rangle^2}{\phi^2} . \tag{10}$$

As before, the point-dipole approximation using Eq. 10 for F_0 is valid for small magnetic fields, when the magnetization is weak or the spheres are widely separated. However, the point-dipole approximation using Eq. 10 for F_0 is also valid in the limit $H_0 \to \infty$ as the particle magnetization saturates; in this limit the magnetization is constant throughout each sphere, and thus each sphere acts magnetically as only a dipole.

We therefore define a new Mason number using Eq. 10 for F_0,

$$\text{Mn}(M) \equiv \frac{F^H}{F_0} = \frac{9}{2} \frac{\eta_c \dot{\gamma} \phi^2}{\mu_0 \mu_c \langle M \rangle^2} , \tag{11}$$

where $F^H = 6\pi\eta_c a^2 \dot{\gamma}$ as discussed in the Introduction. We note that Eq. 11 for $\text{Mn}(M)$ reduces to Eq. 2 for small magnetic field strengths, and that every term in Eq. 11 can be measured experimentally. Similar Mason numbers employing the particle magnetization or magnetic moment have been defined by others. [9,10,11]

The ability of $\text{Mn}(M)$ defined by Eq. 11 to collapse experimental data is illustrated in Fig. 3(b), where the data presented in Fig. 2 are replotted as $\eta_{\text{app}}/\eta_\infty$ vs $\text{Mn}(M)$. For these experiments, the suspension magnetization was measured experimentally, and thus $\text{Mn}(M)$ could be evaluated directly. The data collapse very well on to a single curve, even at large magnetic field strengths (small $\text{Mn}(M)$) where the particle magnetization is significantly nonlinear. de Gans et al. [9,10] observed a similar collapse of viscosity data when plotted as a function of a similarly-defined Mason number, for suspensions of Brownian silica particles dispersed in ferrofluids.

Creating a plot of scaled viscosity vs $\text{Mn}(M)$ requires knowledge of the magnetic properties of the suspensions. This can be obtained from direct measurements as described above, or from models. For our suspensions with $\phi = 0.20$ and 0.30, we have not measured the magnetic properties, and therefore must rely on a magnetization model. We model the suspensions as chains of magnetizable spheres in axisymmetric cells, and solve for the magnetostatic field variables using the finite element method much like that reported by Ginder and Davis.[6,12] From this, the magnetization as a function of current for our systems can be obtained.

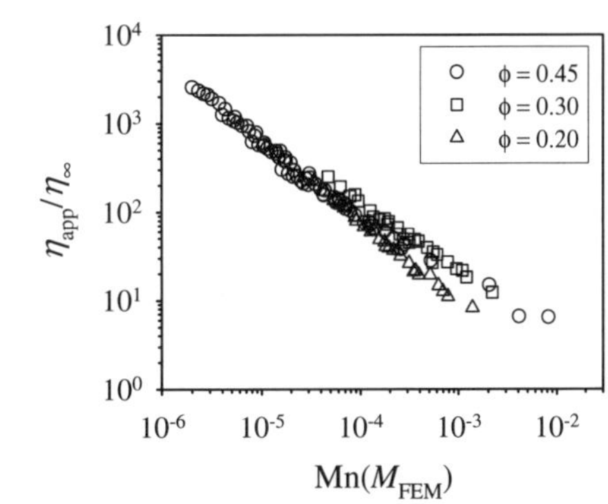

Figure 4. Relative viscosity as a function of $\text{Mn}(M)$ for $\phi = 0.2$, 0.3 and 0.45, where M is determined from predicted magnetic properties.

In Fig. 4, the relative viscosity is plotted as a function of $\text{Mn}(M)$ for $\phi = 0.2$,

0.3 and 0.45, using calculated magnetic properties to determine Mn(M). Comparison with Fig. 3(b) reveals that the data for $\phi = 0.45$ collapse on to a single function of Mn nearly as well as when the magnetic data are used to calculate Mn(M), and much better than the collapse depicted in Figs. 2(b) and 3(a). For $\phi = 0.20$ and 0.30, the data collapse very well. We note that the results for all three volume fractions collapse to nearly the same function of Mn(M).

4 Conclusion

The conventional Mason number, defined for ER suspensions as $\text{Mn} = \eta_c\dot{\gamma}/(2\epsilon_0\epsilon_c\beta^2 E_0^2)$, is useful for representing experimental data. The shear rate and field strength dependent apparent suspension viscosity reduces to a function of only the Mason number. Defining an MR Mason number by directly substituting magnetostatic variables for electrostatic variables in the ER Mason number is not as useful. Shear viscosity data at various shear rates and magnetic field strengths do not collapse to a single function of $\text{Mn}(H) \equiv \eta_c\dot{\gamma}/(2\mu_0\mu_c\beta_M^2 H_0^2)$, with deviations most apparent at large field strengths. The lack of collapse appears to be in part caused by nonlinear particle magnetization.

We find that a Mason number defined in terms of the suspension magnetization, $\text{Mn}(M) \equiv 9\eta_c\dot{\gamma}\phi^2/(2\mu_0\mu_c \langle M\rangle^2)$, is more useful. Apparent shear viscosity data at various shear rates and magnetic field strengths collapse to a single function of Mn(M). The primary benefit of this result is that relatively few experiments are required to describe the shear rate and field strength dependence of the apparent viscosity of an MR suspension over a wide range of conditions.

References

1. Marshall, L., C. F. Zukoski, and J. W. Goodwin, J. Chem. Soc. Faraday, Trans. I, **85**, 2785 (1989).
2. Parthasarathy, M., and D. J. Klingenberg, Mat. Sci. Eng. R, **R17**, 57 (1996).
3. Felici, N., J. N. Foulc, and P. Atten, in R. Tao and G. D. Roy (eds), *Electrorheological Fluids, Mechanisms, Properties, Technology, and Applications, Proceedings of the Fourth International Conference on Electrorheological Fluids, Feldkirch, Austria, July 20–23, 1993,* World Scientific, Singapore, 1994, p. 139.
4. Foulc, J. N., P. Atten, and N. Felici, J. Electrostatics, **33**, 103 (1994).
5. Ulicny, J. C., M. A. Golden, C. S. Namuduri, and D. J. Klingenberg, J. Rheol., **49**, 87 (2005).
6. Ginder, J. M. and L. C. Davis, **65**, 3410 (1994).
7. Shkel, Y. M., and D. J. Klingenberg, J. Rheol., **45**, 351 (2001).
8. Clercx, H. J. H., and G. Bossis, Phys. Rev. E, **48**, 2721 (1993).
9. de Gans, B. J., H. Hoekstra and J. Mellema, Faraday Discuss., **112**, 209 (1999).
10. de Gans, B. J., N. J. Duin, D. van den Ende, and J. Mellema, J. Chem. Phys., **113**, 2032 (2000).
11. Dominguez-Garcia, P., S. Melle, O. G. Calderon, and M. A. Rubio, Coll. Surf. A, **270-271**, 270 (2005).
12. Ulicny, J. C., and D. J. Klingenberg, in preparation (2006).

HIGH PERFORMANCE MAGNETORHEOLOGICAL FLUIDS TAILORED FOR A 700 NM AUTOMOTIVE 4-WHEEL-DRIVE CLUTCH

C. KIEBURG, G. OETTER, R. LOCHTMAN, C. GABRIEL, H. M. LAUN AND J. PFISTER

BASF Aktiengesellschaft, Carl-Bosch-Strasse, 67056 Ludwigshafen/Rhein, Germany
E-mail: christoffer.kieburg@basf.com

G. SCHOBER AND H. STEINWENDER

MagnaPowertrain AG & Co KG, Industriestrasse 35, 8502 Lennach, Austria
E-mail: gerald.schober@magnapowertrain.com

MRF performance requirements for a commercial 4WD clutch are outlined and appropriate characterization methods presented, a prerequisite to identify and overcome shortcomings of MRF formulations, popped up in a clutch test rig. These methods include the response to rapidly changing shear rates and switching magnetic flux densities as well as a characterization of durability and redispersibility. Parallel to the improvement of the MRF performance, design studies of the 4WD clutch with regard to the required torque, maximum allowable weight/dimensions, and benchmark to alternative technologies were conducted, making intensive use of computer aided design. Combining the expertise of the collaborating parties, a 12 kg 700 Nm 4WD clutch was realized and the expected performance based on CAD simulations verified, demonstrating the feasibility of a new technology.

1 Introduction

Automotive components make use of a broad variety of functional fluids. For magnetorheological devices, the functionalities of the component and of the fluid are closely intertwined. To provide an optimal solution, an intensive cooperation between device manufacturer and MRF producer is mandatory. For the project presented here, MagnaPowertrain, the leading supplier of drivetrain systems to the automotive industry, and BASF, the leading chemical company and largest producer of carbonyl iron powder – the dominant component of the MRF – have joined forces.

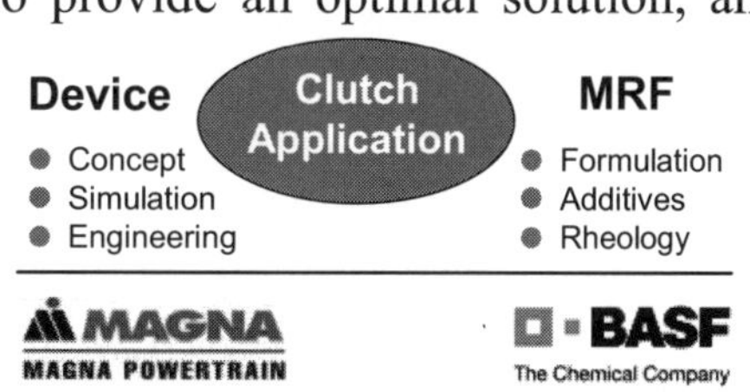

Key MRF performance needs for an automotive drivetrain clutch are:

- □ High yield stress in the activated state
- □ Low off-state viscosity - in particular at low temperatures
- □ Temperature range of operation -30 °C up to 150 °C
- □ Short response time to changes of magnetic flux density
- □ Long-term stability and durability - low in use thickening
- □ Easy redispersibility after centrifugation or long rest

2 Magnetorheology

Laboratory measurements on small-scale samples for simple conditions (isothermal flow, homogeneous magnetic field) covering a wide range of magnetic flux density, shear rate, and temperature, are required to both optimize MRF formulations and to provide data for the design of devices. A Physica MRD180 [1] magneto-cell attached to a Physica MCR501 rotational rheometer (both from Anton Paar GmbH) using plate-plate geometry (radius 10 mm, gap 0.3 mm, magnetic flux density B perpendicular to the plane of shear) was used. The introduction of an additional non-magnetic plate, acting as a Hall probe housing (**Fig. 1**), allows an online measurement of the true magnetic flux as well as a radial flux profile check with MRF. In addition, Maxwell 2D simulations of the magnetic field distribution (**Fig. 2**) verified the Hall probe results.

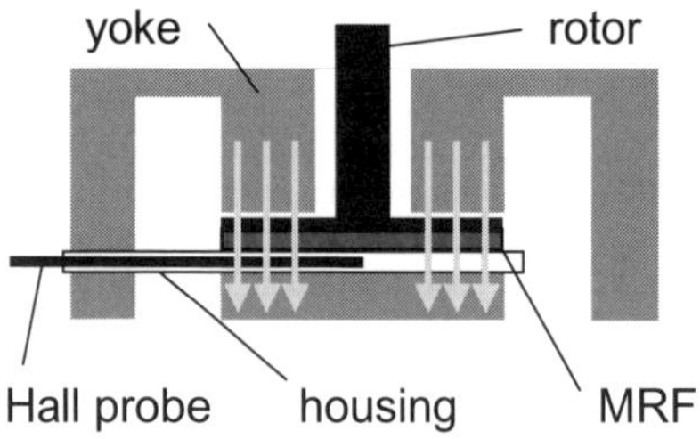

Figure 1: Online flux density measurement

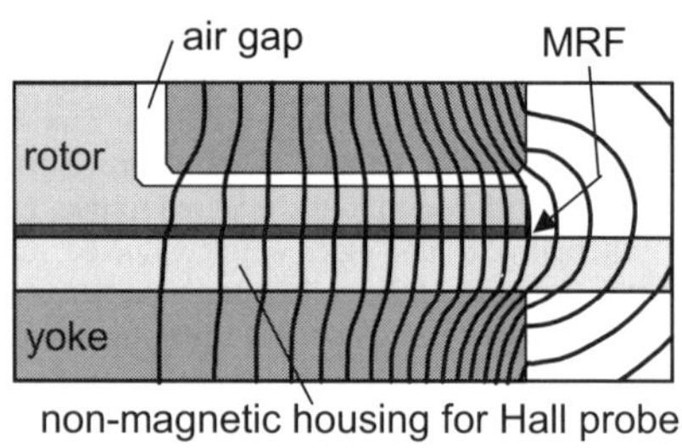

Figure 2: Maxwell 2D simulation of magnetic flux

2.1 Flow curves and characteristic

Fig. 3 shows flow curves of MRF A at 25°C up to shear rates of 3000 s^{-1} and flux densities up to 1.3 T (Note: The shear rate given is the shear rate r at the rim of the plate-plate gap, whereas the Newtonian shear stress is evaluated from the torque by assuming a Newtonian fluid in the gap). The true shear stress τ follows from the Newtonian shear stress τ_N as

$$\tau = \tau_N (3+n)/4 \quad \text{with} \quad n = \partial \log \tau_N / \partial \log r.$$

At high flux density, the shear stress is independent of the shear rate ($n = 0$), thus $\tau = \tau_N\, 3/4$. Noteworthy are the low shear stresses without magnetic field, which allow an active shear stress control by a factor of about 100 at the highest shear rates applied!

The MRF characteristic for 25°C in **Fig. 4** represents the shear stress vs. magnetic flux density, which hardly depends on the applied shear rate. The initial rise is less than quadratic in B and levels off when the MRF approaches its saturation magnetization. The on-state shear stress typically increases with decreasing temperature.

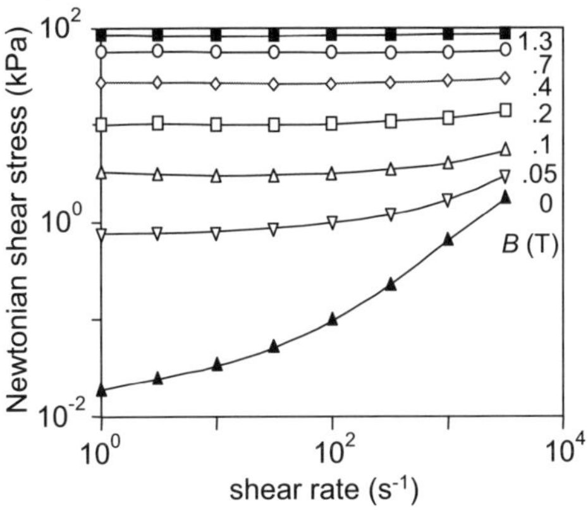

Figure 3: Flow curve of MRF A at 25°C

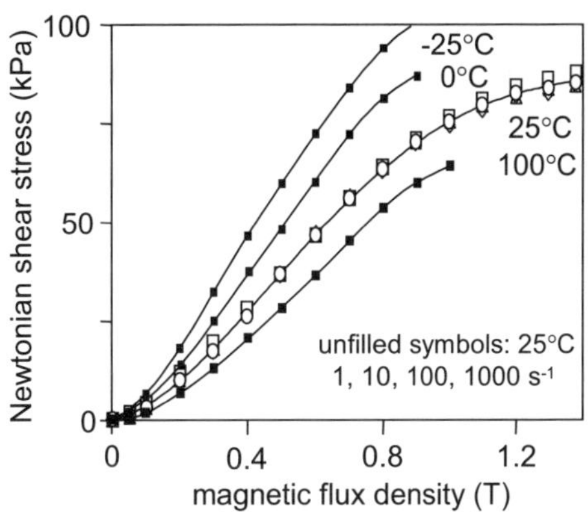

Figure 4: Characteristic of MRF A at 25°C

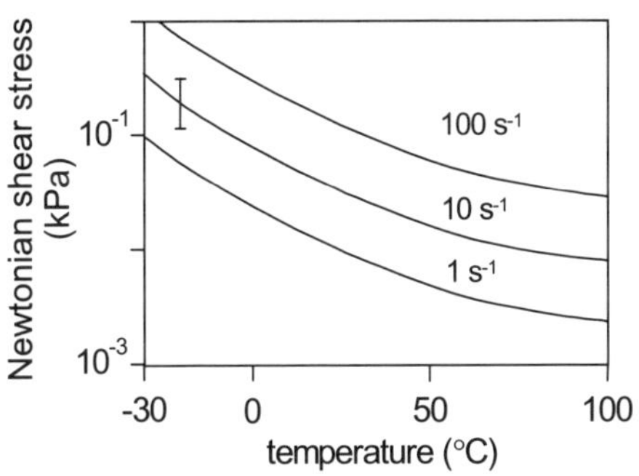

Figure 5: MRF A off-state shear stress versus temperature

The clutch requires a low off-state torque at the minimum temperature of operation. The main challenge here remains to achieve a low shear stress also at high shear rate **(Fig. 5)**, because the viscous behavior of the MRF dominates in the off-state regime.

2.2 Switching behavior

The MRF switching behavior is characterized by imposing a step in coil current by means of a power operational amplifier as depicted in **Fig. 6**. The transient signals of flux density (Hall probe) and torque were recorded at 0.1 ms sampling rate **(Fig. 7)**.

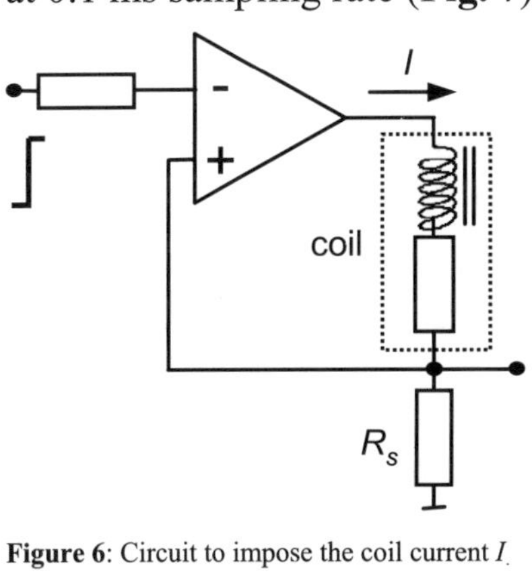

Figure 6: Circuit to impose the coil current I

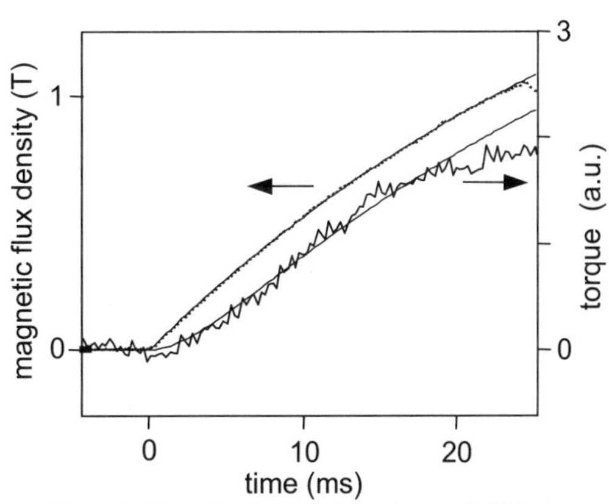

Figure 7: Flux density and shear stress of MRF A

A steep ramp of current causes a transient flux density, the initial part of which may be approximated by $B = B_o(1- e^{-t/\Lambda})$, where Λ stands for the time constant of the electromagnetic circuit, including eddy current effects. If one attributes a characteristic switching time λ to the MRF, the predicted shear stress response is $\tau = \tau_o\,(1-\lambda/(\lambda-\Lambda)\,e^{-t/\lambda} - \Lambda/(\Lambda-\lambda)\,e^{-t/\Lambda})$. For MRF A at 25°C switched from 0 to 1.05 T (Fig.7), Λ = 35 ms and a switching time of λ = 2.3 ±0.5 ms allow to fit the shear stress response over more than 2/3 of the total change [2].

2.3 Response to shear rate changes

Rapid changes of relative speed in the clutch should neither cause pronounced torque overshoots at the change from rest (static yield stress) to shear (dynamic yield stress) and vice versa, nor distinct torque fluctuations at the relevant shear rates. The MRF performance is characterized by a ramp test: The shear rate is linearly increased from 0 to 3000 s^{-1} within 120 s and subsequently redescended to zero. **Fig. 8** shows the desired flat profile of the monitored shear stress with acceptable overshoot and weak fluctuations.

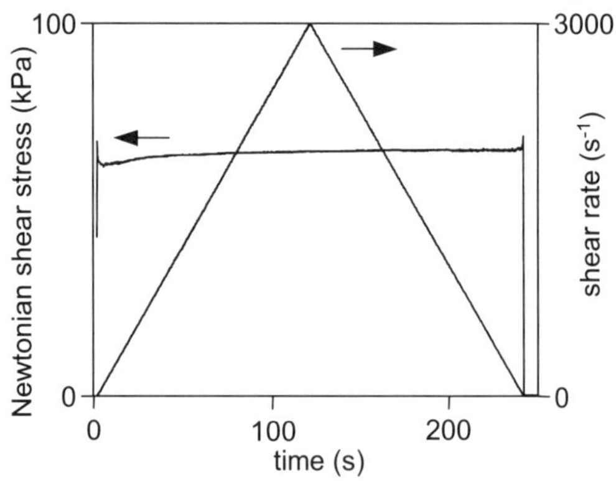

Figure 8: Ramp test on MRF A at 60°C and 0.8 T

2.4 In use thickening (IUT)

During clutch operation, high shear rate and shear stress levels may cause polishing/abrasion at the particle-wall and/or particle-particle interfaces. The effect is investigated in a sealed and cooled small disk clutch applying a narrow shear rate and flux density range to the whole MRF volume (**Fig. 9**).

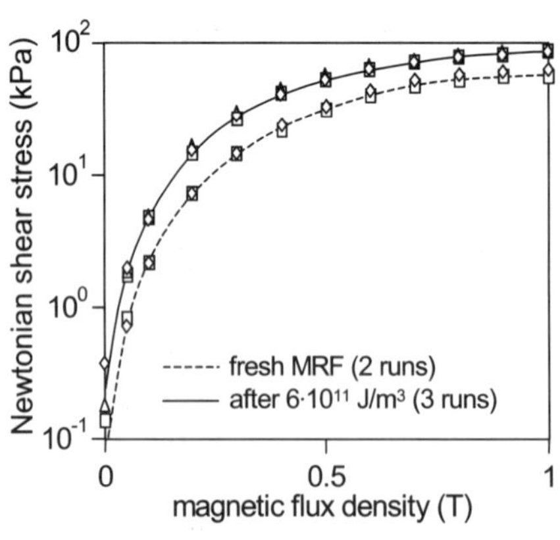

Figure 9: In use thickening of MRF B

The total mechanical work input during these durability tests was $6 \cdot 10^{11}$ Joule/m^3 at 25°C, flux density 0.5 T and shear rate 350 s^{-1}. The comparison of the characteristic of the fresh and sheared samples indicates that in use thickening (IUT) has a stronger effect on the off-state shear stress, whereas at high flux densities the characteristic is solely shifted upwards by a constant factor (here 1.5). As a consequence, the obtainable clutch torque level is increased, but the dynamic range slightly reduced.

2.5 Redispersibility

A quantitative measure for the MRF redispersibility after long-term at rest (sedimentation) or after centrifugation, is obtained by measuring the work to vertically penetrate a thin plate into the MRF at a constant speed (**Fig. 10**). This property reflects the ability of the additive system to prevent formation of strong and difficult to break aggregates. **Fig. 11** compares test results from 3 MRF formulations with different additive systems after 6 months rest in the test beaker at room temperature. The results correlate with a qualitative ranking obtained from manual spatula stirring tests.

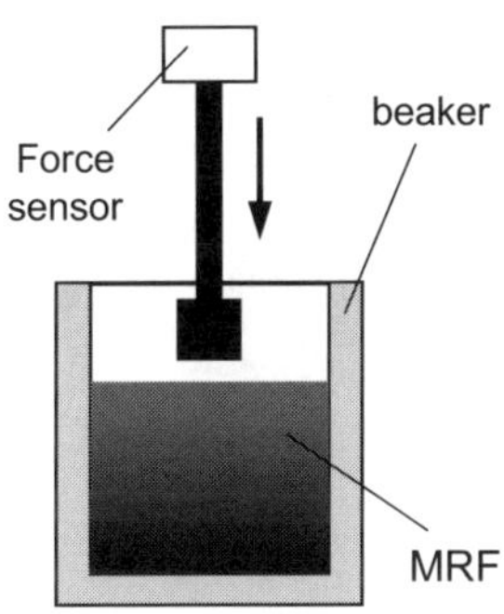

Figure 10: Penetrator test for redispersibility

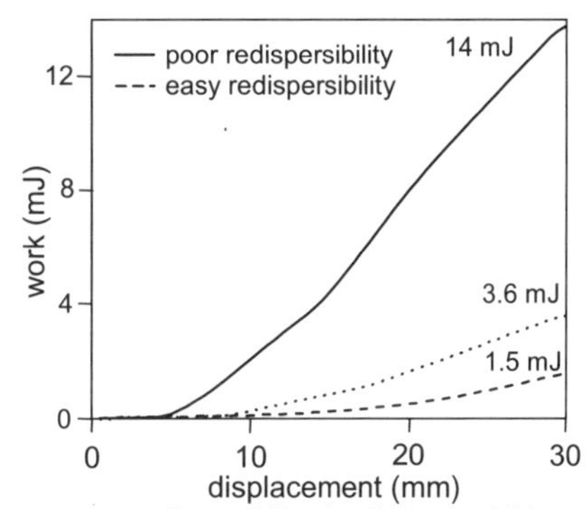

Figure 11: Redispersibility for different additive systems

3 Pilot clutch

3.1 Design and test rig

An important step in the project was the manufacture of a laboratory test rig (**Figs. 12 a,b**). The comparison of MRF characteristics measured by magnetorheology at BASF and of the test rig at MagnaPowertrain yielded valuable information for a reliable 3D-simulation of various possible clutch designs. Furthermore, the test rig results highlighted challenges for

MRF formulations in clutch applications and demonstrated the need to extend the range of conventional magnetorheological measurements.

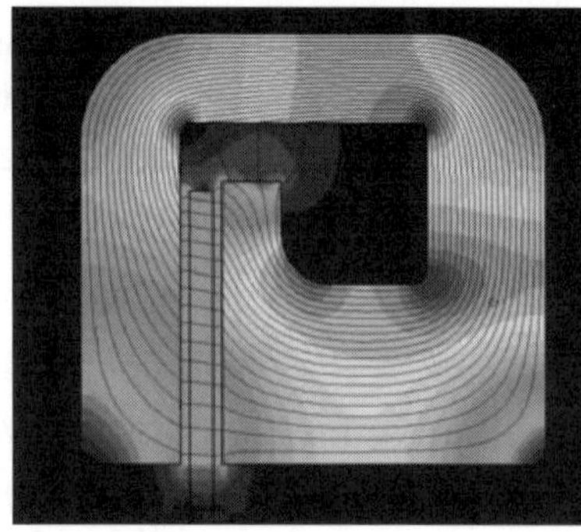

Figure 12: a) MRF clutch test rig b) Magnetic flux simulation of simple prototype clutch

Most important parameters in all evaluations were the calculated weight and device dimensions required to reach the targeted torque capacity of 700 Nm. Particularly in automotive industry, MRF devices have to be competitive against other technologies in these terms.

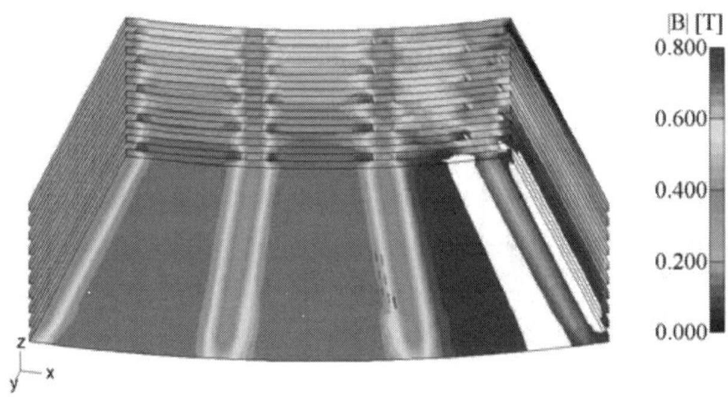

Figure 13: 3D-simulation of an evaluated clutch design filled with MRF

To meet the weight and torque targets, a lot of optimisation work was necessary, making extended use of the capabilities of 3D magnetic flux simulation. The MRF characteristic, combined with calculated magnetic flux densities in the fluid gaps of possible series designs, identified the toque capacities of several multi plate designs. An example of a segment of a multi plate disc design is shown in **Fig. 13**.

3.2 700 Nm MRF 4WD pilot clutch

The final clutch design (**Fig. 14**) matched the expectations in terms of torque capacity immediately. With its dimensions and functionality this MRF clutch can be used as a 4WD clutch in powertrain as well as in other automotive clutch applications. Nevertheless, further development and optimization is needed to raise the power weight ratio.

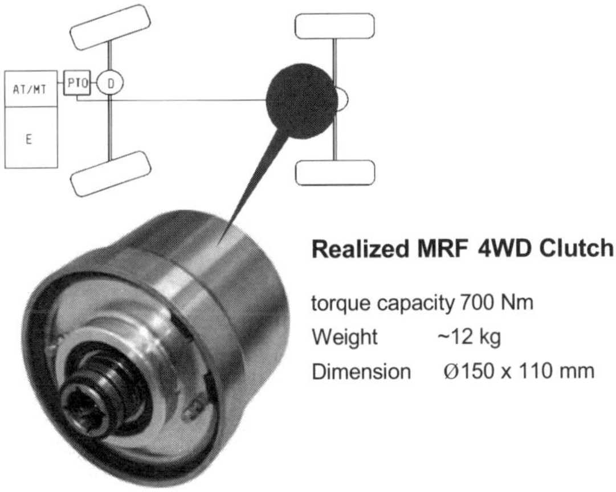

Figure 14: Realized 700 Nm 4WD clutch by Magna

4 Acknowledgements

We thank Dr. Huck of Anton Paar GmbH for providing details of the MRD180 magneto-cell for the Maxwell 2D simulations and a firmware allowing access to the MCR501 torque signal at 0.1 ms sampling rate. We are grateful to Dr. Renhart of TU Graz for 3D magnetic flux simulations of various clutch and yoke designs. P. Schuler is thanked for his help in designing laboratory test tools, G. Schmidt for the Maxwell2D simulations, and M. Bach for performing the magnetorheology measurements.

References

1. Laeuger L., Wollny K., Setting H. and Huck S., A new device for the full rheological characterization of magnetorheological fluids. In Electrorheological fluids and magnetorheological suspensions (ERMR2004) ed. by Lu K., Shen R. and Liu J. (World Scientific, 2005) pp. 370-376.
2. Laun H. M. and Gabriel C., Measurement modes of the response time of a magnetorheological fluid (MRF) for changing magnetic flux density, submitted to Rheol. Acta (2006).

YIELD STRESS IN FERROFLUIDS?

HAMID SHAHNAZIAN
Institute of Fluid Mechanics, Chair of Magnetofluiddynamics, George-Bähr-Str.3, 01062 Dresden, Germany
E-mail: hamid.shahnazian@tu-dresden.de

STEFAN ODENBACH
Institute of Fluid Mechanics, Chair of Magnetofluiddynamics, George-Bähr-Str.3, 01062 Dresden, Germany
E-mail: stefan.odenbach@tu-dresden.de

Recent experimental as well as theoretical investigations have shown that the formation of structures of magnetic nanoparticles has significant influence on the behaviour of ferrofluids. The dependence of this structure formation on the magnetic field strength and shear stress applied to the fluid leads to strong changes of the viscosity and to the appearance of viscoelastic effects in the fluids. The actual approaches for a description of the effects vary in the basic modeling of the fluid and its behaviour. Some models base on microscopic assumptions, other model the fluid on a mesoscale and even macroscopic descriptions abstaining from microscopic assumptions have been suggested. A point in which the predictions of the models differ is the question of an appearance of a magnetic field dependent yield stress in ferrofluids. For investigations concerning the appearance and field dependence of a yield stress a special stress controlled rheometer for ferrofluids has been designed. The preliminary results presented here, show a dependence of the yield stress on magnetic field strength for different kind of ferrofluids and magnetorheological fluids.

1 Introduction

Rheological investigations performed on ferrofluids have shown a strong change of viscosity caused by applied magnetic fields, the so-called magnetoviscous effect [1]. This effect can be explained with chain-like structures, which are formed due to interparticle interaction in the presence of a magnetic field. A shear flow, which causes viscous friction, yields a mechanical torque and leads to a disalignment of the chains from the magnetic field direction. The resulting magnetic torque countervails this misalignment and an increase of the viscosity can be observed. Further, the structures can break up under the influence of shear resulting in a reduction of the field dependent viscosity increase, an effect called shear thinning.

In order to explain the magnetoviscous effect different theoretical approaches have been developed. Classically such effects are explained on the basis of microscopic assumptions like the formation of chains [2, 3] or droplets [4] of magnetic particles. For numerical simulations recently a mesoscopic model has been used [5] and the macroscopic approach by Müller and Liu [6] – basing on the principles of irreversible thermodynamics – does not make use of any assumption concerning the microstructural make up of the fluid. A discrimination which of the models may give the best description of the rheological behaviour of ferrofluids relies on experimental data concerning effects for which the models given different predictions. One of these effects is the appearance of a yield stress in ferrofluids. With a shear rate controlled rheometer – as it has been used in former investigations [1] of the magnetoviscous behaviour of ferrofluids – the question whether a yield stress exists in ferrofluids can not be answered directly.

Extrapolating the measured flow curves to shear rate zero, a yield stress of 0.1 Pa could be estimated for the fluid APG513A and a magnetic field strength of 30 kA/m [1].

2 Experimental setup

The main topic of this work is to prove the existence of a yield stress for ferrofluids of different structural make up with a shear stress controlled rheometer. The magnetic field, which is necessary for the investigations, is produced by a cylindrical coil. A maximum field up to 80 kA/m with a homogeneity of 0.05 % in the region of the shear cell can be applied. The measuring cell is a cone-plate arrangement combined with a Couette region. With this combination the disturbing dewetting of the cone, which can arise by the appearance of fluid spikes due to the normal field instability, can be avoided [7].

fig. 1: The shear stress controlled rheometer for the investigations of the yield stress in ferrofluids. A detailed description is given in the text.

a)

b)

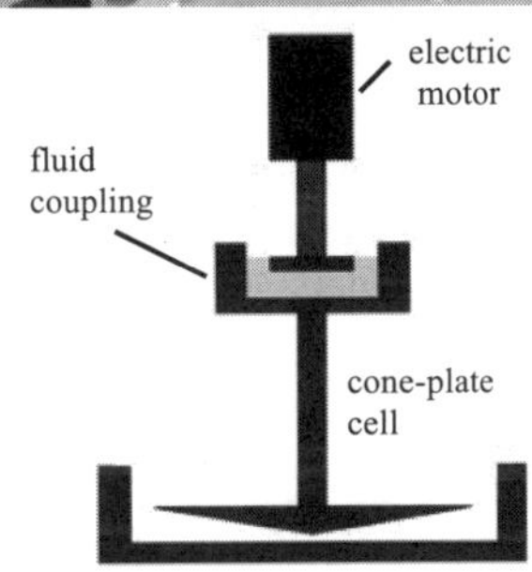

fig. 2: The driving unit consisting of an electric motor, a fluid coupling and a rotary encoder.
a) Photography of the experimental setup
b) Schematic sketch of the driving system

The cone is connected via an air bearing with an optical encoder. This rotary encoder is a versatile read head that is combined with a rotary disc. Its modular construction provides friction-free operation. To complete the system an interpolating decoder is used to increase the final resolution up to 2^{22} steps per revolution. With this device the frequency of rotation of the cone can be measured with a resolution of 10^{-7} Hz corresponding to a shear rate resolution of about 10^{-5} s^{-1}.

One of the most important features of the rheometer is the direct measurement of the yield stress as a function of the strength of an applied magnetic field. Shear stresses below the mentioned value of 0.1 Pa [1] can be applied with a conventional electric motor combined with a fluid coupling. The upper plate of the coupling (fig. 2b) is fixed to the electric motor. The torque is transmitted by the liquid to the second plate, which is fixed to the cone of the cone-plate cell providing a transmission of the torque to the fluid under investigation. The torque is proportional to the difference of the rotary frequencies of the coupling plates and the viscosity of the fluid in the coupling. The torque transmitted to the fluid in the shear cell is given by

$$M = (R^4/2h)\ \eta\ (2\pi f)\ .$$

The viscosity η in the fluid coupling is about 0.01 Pas. The upper plate of the coupling has a radius R of 0.5 cm and the gap h between the plates is 1 mm. With this geometry and a difference of rotary frequencies of about f = 0.1 Hz a minimum torque of 10^{-9} Nm can be realized, corresponding to a shear stress of 10^{-5} Pa in the given shear cell geometry.

Investigations of the yield stress were carried out in a shear stress range from 0 to 10 Pa for magnetorheological fluids and from 0 to 0.1 Pa for ferrofluids. Starting from stresses well below the estimated yield stress of 0.1 Pa [1] the shear stress is increased stepwise until a movement of the measuring fluid can be detected with the optical encoder. Repeating this for different magnetic field strengths information about the dependence of the yield stress on magnetic field strength can be obtained.

3 Experimental results

In the first experiments with the driving unit - electric motor combined with the fluid coupling - a magnetorheological (MR) fluid MRHBS4 (Liquids Research Limited) has been investigated as reference sample. As it is well known a much higher yield stress is expected for such fluids compared to ferrofluids containing nanosized particles. The results on magnetorheological fluids will give information whether the setup of the shear stress controlled rheometer allows a direct measurement of the yield stress as a function of an applied magnetic field.

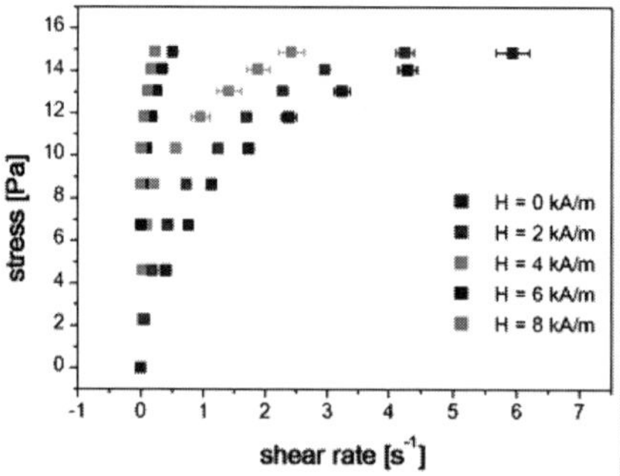

fig.3: The flow curve of the magnetorheological fluid for various magnetic field strengths.

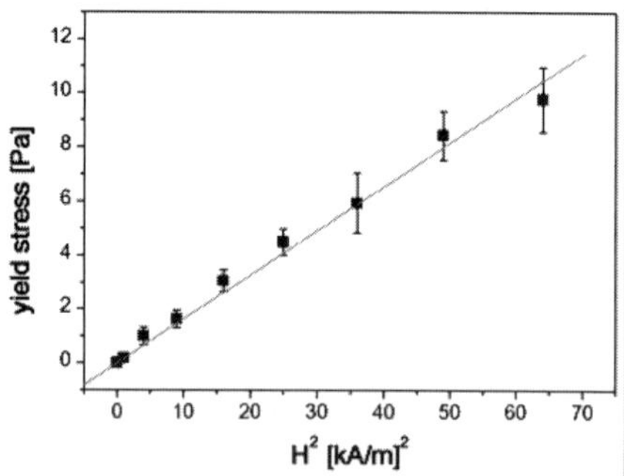

fig.4: The yield stress of the magnetorheo-logical fluid depends quadratically on the strength of the applied magnetic field.

Figure 3 shows the resulting flow curves for the MR-fluid for different magnetic field strengths. Since the rheometer has been designed for ferrofluids showing much weaker field dependent changes of the rheological behavior only weak magnetic fields have been used here to enable investigations in a shear stress range accessible with the setup.

First of all in presence of a magnetic field significant changes of the viscosity of the fluid are observed, which can be theoretically explained as an effect of structure formation in magnetorheological fluids [8 - 11]. Further a clear dependence of magnetoviscosity on shear rate is measured. As shear rate increases the magnetically induced increase of viscosity reduces significantly.

For the magnetorheological fluid the yield stress could be measured directly as a function of an applied magnetic field using the method described above. The effect grows quadratically with increasing magnetic field strength as it can be seen from the plot of the measured yield stress vs. H^2 in fig. 4. The found results confirm that the experimental setup as well as the procedure are suitable to enable a direct measurement of the yield stress and its dependence on magnetic field strength.

In a second step field dependent yield stress measurements for ferrofluids have been performed. Two different ferrofluids, Ferrotec APG513 (7.2 vol. % magnetite in synthetic ester, dynamic viscosity $\eta_{(T=20\,°C)} = 170$ mPas), and Co91 (3,25 vol. % cobalt in L9 oil, dynamic viscosity $\eta_{(T=20\,°C)} = 65$ mPas) prepared by N. Matoussevitch at Forschungszentrum Karlsruhe GmbH have been investigated.

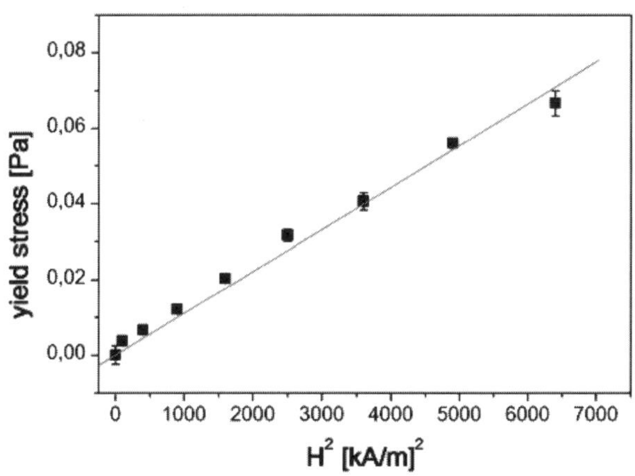

fig.5: The yield stress of the ferrofluid APG513A as a function of the square of the applied magnetic field.

In fig. 5 the yield stress for the fluid APG 513A is plotted versus the square of the magnetic field. The yield stress grows with increasing magnetic field strength and the expected proportionality to H^2 could also be confirmed. As the diagram shows the yield stress found for H = 30 kA/m equals $\tau_y = 0.012$ Pa which is well below the estimate taken from the previous extrapolations of flow curves obtained with the shear rate controlled rheometer. Nonetheless the values found here confirm the existence of a field dependent yield stress even in weakly interacting suspensions of nanosized magnetite particles.

The interparticle interaction – which has been proven earlier [1, 7, 12] to be the determining parameter for the appearance of magnetoviscous effects in ferrofluids – can be described by an interaction parameter λ given by [7]

$$\lambda = (\mu_0 M_0^2 V)/(24\,k\,T)$$

where M_0 denotes the spontaneous magnetization of the magnetic material, V the volume of the particles, k Boltzmann's constant, T the absolute temperature and μ_0 the vacuum permeability. Since λ is calculated for two magnetic particles in contact, it is obvious that for a ferrofluid with coated particles a modified interaction parameter

$$\lambda^* = \lambda\ (d/d + 2\ s)^3$$

has to be used [13]. For 16 nm magnetite particles, like they are usually assumed in the description of magnetoviscous effects in magnetite based ferrofluids one can easily calculate the $\lambda^* = 2.87$. For cobalt-based fluids with 10 nm particles a modified interaction parameter of $\lambda^* = 5.26$ can be obtained being about 2 times higher than the one for the magnetite-based ferrofluid. As a result one can expect the appearance of a significantly higher yield stress in the above mentioned fluid Co91.

Fig.6 shows the respective results for the cobalt based fluid again in a plot of τ_y as a function of H^2. For comparison the data obtained for APG513A has been plotted in the same diagram. It is clearly seen that the values for the yield stress of Co91 are significantly higher than for APG513A. For a field strength up to 30kA/m they differ by a factor of approximately 3.5.

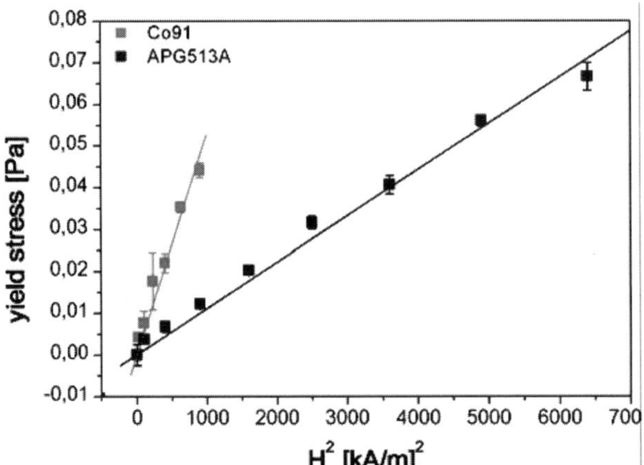

fig.6: The yield stress of the ferrofluids APG513A and Co91 as a function of the square of the strength of the applied magnetic field. A higher value of yield stress for the ferrofluid Co91 compared to ferrofluid APG513A is observed.

4 Discussion

The rheological investigations presented here have shown, that the newly designed stress controlled rheometer allows the determination of the yield stress and its field dependence even in weakly interacting suspensions of nanosized magnetite particles. It could be found that the effect is proportional to the square of magnetic field strength. Moreover its absolute value depends strongly on the interparticle interaction. This has been proven by investigations using ferrofluids containing magnetite and cobalt as magnetic component respectively.

Future investigations will focus on a successive variation of the interaction parameter by a further change of the size and material of the magnetic particles followed by a comparison of the experimental results with the different theoretical approaches. Moreover investigations concerning the specific interaction between the wall and the particles varying the material and roughness of the walls will have to be performed. Finally a variation of the macroscopic distance of the walls of the shear cell will provide information on the macroscopic size of the structures formed by the magnetic particles and its dependence on field strength as well as interparticle interaction.

5 Acknowledgements

Financial support by Deutsche Forschungsgemeinschaft (DFG) under grant Od18/8 providing the basis for our investigations is gratefully acknowledged.

References

1. S. Odenbach, Magnetoviscous effects in ferrofluids, LNPm71, Springer Verlag (2002)
2. S. Odenbach, H. Störk, Shear dependence of field induced contributions to the viscosity of magnetic fluids at low shear rates, Journal of magnetism and magnetic materials, 183 (1998) 188-194
3. A.Yu. Zubarev, S. Odenbach and J. Fleischer, Rheological properties of dense ferrofluids, Journal of Magnetism and Magnetic Materials, 252 (2002) 241
4. A.Yu. Zubarev, L. Iskakova, Yield stress in thin layers of ferrofluids, Physica A, 365 (2006) 265-281
5. P. Ilg, M. Kröger, S. Hess, Structure and rheology of model-ferrofluids under shear flow, Journal of Magnetism and Magnetic Materials, 289 (2005) 325-327
6. H. W. Müller, M. Liu, Structure of ferrofluid dynamics, Physical Review E, 64 061405 (2001)
7. Rosensweig, Ferrohydrodynamics. Cambridge University Press, Cambridge, New York (1985)
8. E. Lemaire, A. Meunier, G.Bossis, Influence of the particle size on the rheology of magnetorheological fluids, J. Rheol. 39(5), September/October 1995
9. E. Lemaire, G. Bossis, Y. Grasselli, Yield stress and structuration of magnetorheological suspensions, Journal of Magnetism and Magnetic Materials, 122 (1993) 51-52
10. Yoshiyuki Matsuno, Jun Itoh, Estimation of the yield stress of a magnetic fluid from flow curves, Journal of Magnetism and Magnetic Materials, 149(1995) 22-25
11. G. Bossis, S. Lacis, A. Meunier, O. Volkova, Magnetorheological fluids, Journal of Magnetism and Magnetic Materials, 252 (2002) 224-228
12. L. M. Pop, S. Odenbach, A. Wiedemann, N. Matoussevitch, H. Bönnemann, Microstructure and rheology of ferrofluids, Journal of Magnetism and Magnetic Materials, 289 (2005) 303-306
13. S. Thurm, S. Odenbach, Particle size distribution as key parameter for the flow behaviour of ferrofluids, Physics of Fluids 15 1658 (2003)

THE EFFECT OF DWELL TIME ON THE RHEOLOGICAL BEHAVIOR OF MR FLUIDS

MEHDI AHMADIAN
Center for Vehicle Systems & Safety, Virginia Tech, Blacksburg, VA 24061 USA

E-mail: ahmadian@vt.edu

FERNANDO D. GONCALVES
Engineering Specialist, Lord Corporation, Material Products Division, Cary, NC, 27511 USA

E-mail: Fernando_Goncalves@lord.com

This study intends to identify the behavior of MR fluid subject to high rates of shear and high flow velocities. A high shear rheometer is built which allows for the high velocity testing of MR fluids. The rheometer is capable of fluid velocities ranging from 1 m/s to 37 m/s, with corresponding shear rates ranging from $0.14\text{x}10^5$ s^{-1} to $2.5\text{x}10^5$ s^{-1}. Fluid behavior is characterized in both the off-state and the on-state. The MR fluid is run through the rheometer at various flow velocities and a number of magnetic field strengths. The term "dwell time" is introduced and defined as the amount of time the fluid spends in the presence of a magnetic field. Two active valve lengths were considered, which when coupled to the fluid velocities, generated dwell times ranging from 12 ms to 0.18 ms. The yield stress is found from the experimental measurements and the results indicate that the magnitude of the yield stress is sensitive to fluid dwell time. As fluid dwell times decrease, the yield stress developed in the fluid decreases. The results from the on-state testing clearly demonstrate a need to consider fluid dwell times in high velocity applications. Should the dwell time fall below the response time of the fluid, the yield stress developed in the fluid may only achieve a fraction of the expected value. These results imply that high velocity applications may be subject to diminished controllability for falling dwell times.

1 Introduction

MR fluid's success has been apparent in many fields of engineering and new applications continue to emerge. While much of the fluid's success has been realized in devices intended for automotive or civil engineering applications, recent studies have investigated the fluid's promise in impact or shock related applications. These emerging applications, and many current applications, subject the fluid to extreme flow conditions. Specifically, these flow environments or conditions include high shear and high velocity flow. The challenge in such devices becomes the lack of information regarding the behavior of MR fluid under these adverse operating conditions.

The great success of MR fluids continues to develop into new and emerging applications. Recently, a number of studies have considered the use of MR fluid dampers in impact or shock loading applications. Researchers have taken note of the success of MR dampers in conventional vibration suppression applications and are attempting to extend this success to impact or shock applications.

One such application is the use of MR dampers for active recoil of medium or large caliber gun systems. Consider the work by Ahmadian et al. [1] in which an MR damper was used to control the recoil dynamics of a 50-caliber rifle. The authors demonstrate that the MR damper is capable of controlling the recoil carriage displacement and velocity. The force transmitted was shown to be highly dependent on magnetic field strength. Several other studies have considered applications of MR dampers for similar high-speed applications [2-4].

El Wahed et al. [5-6] have studied the dynamic performance of MR and ER fluids when subjected to impulsive loads. The ER device used in this study operated in squeeze flow mode. A commercially available MR damper was also tested. A drop tower was used to simulate the impulsive loads. Norris [7] studied the behavior of MR dampers subject to impact and shock loading. Two types of MR dampers (mono-tube, double-ended) were tested in a drop tower with drop heights ranging from 0.305 m (12") to 2.43 m (96"), with corresponding impact velocities of 2.18 m/s (86 in/s) and 6.60 m/s (260 in/s). Norris claims that the results were dominated by fluid inertia and that the damper performance can be separated into two regions, controllable and uncontrollable. The results for varying field strength were indistinguishable until the piston velocity dropped below a certain value. Once the piston velocity dropped below a certain threshold value, the fluid better responded to various magnetic fields. A recent study by Browne et al. [8] has investigated the promise of MR fluid dampers in another form of impact–vehicular crashes. Other studies that may demand shear velocities higher than vehicle applications include those for aircraft landing gears and the U.S. Army's HMMWV.

The high shear rate demanded of MR fluid is not exclusive to only impact or shock applications. Consider the conditions found inside a MotionMaster™ damper in normal use where piston velocities are in the range of 5 cm/s – 20 cm/s. Corresponding shear rates range from $1\text{x}10^4$ s^{-1} – $4\text{x}10^4$ s^{-1}. In extreme conditions, this damper may experience speeds in excess of 1 m/s generating shear rates as high as $2\text{x}10^5$ s^{-1}. Similar conditions are observed for MR dampers used in automotive primary suspensions where shear rates may even reach 10^6 s^{-1} [9].

2 Test Apparatus

The principal purpose of this study is to investigate the high velocity and high shear behavior of MR fluid. Of key importance is the behavior of the fluid itself, not the behavior of the fluid operating in a particular device. The results must be general such that they are not specific to a particular device geometry. The simple geometry of the rheometer allows for the results to be extrapolated to any MR device geometry. To this end, the rheometer shown in Figure 1 was designed and built, which allowed for the high flow testing of MR fluid in both on and off states. As such, the rheometer was designed such that a magnetic field could be applied to the fluid.

The details of the rheometer, which works based on the Venturi concept, are described in details in [10]. The area ratio of the rheometer is greater than 1000. The rheometer was designed such that the parallel plate assumption could be made. If the aspect ratio is greater than or equal to 10, then there is no side wall effect [11]. As such, the dimensions of the rectangular channel were chosen such that the width of the channel is at least 10 times the height of the channel. The ratio of the length of the flow channel to the height of the slit is large enough that end effects can be neglected [12]. The MR valve is incorporated into the last 25.4 mm of the 101.6 mm channel. While two MR valve lengths were considered, 25.4 mm and 6.35 mm, the overall length of the flow channel remained at 101.6 mm. When testing with the 6.35 mm valve, a stainless steel blank was installed to make up the remainder of the length. It was important to maintain the overall length so that the results could be compared directly for both MR valve lengths. The best performance for an MR valve results from a rectangular geometry in which the fluid is in direct contact with the magnetic poles [13]. As such, the flow channel was made in a modular fashion; two stainless steel plates were sandwiched together, separated by a 1 mm stainless steel spacer. The valve portion of the rheometer is made from low carbon steel. Through this design we are able to modify the effective pole length of the valve by simply changing the length of the low carbon steel portion. Furthermore, the slit dimensions can be changed simply by replacing the stainless steel spacers.

A Hall probe was used to measure magnetic field strength. The probe itself was an "ultra-thin" (0.2 mm x 1.5 mm) transverse probe from F. W. Bell. The temperature of the fluid can contribute significantly to the behavior of the fluid. As such, a thermocouple was used to monitor the temperature of the fluid at the exit of the rheometer. The probe was used to read an initial temperature before the test began and also to record the rise in temperature due to shear heating of the fluid as it flows through the slit.

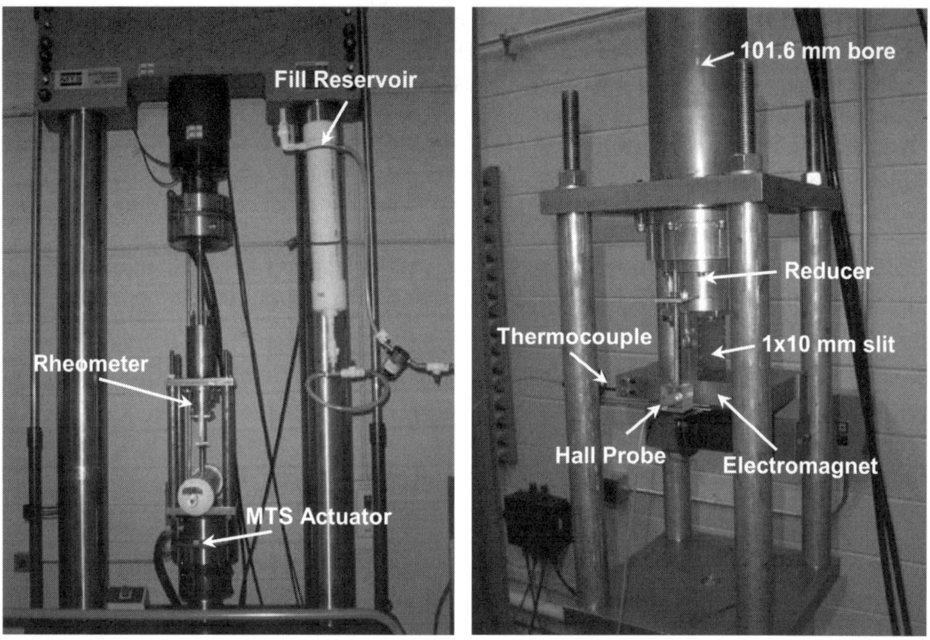

(a) (b)

Figure 1. (a) MR rheometer mounted in a hydraulic load frame; (b) Close-up of MR rheometer

Before the fluid was run through the rheometer, we precisely determined the friction force inherent in the system, through a series of experiments that included running the rheometer in a dry (without any fluid) condition. The results of these tests were used to model the rheometer friction force and account for it in the test results that will be described in the next section.

Piston velocities considered in this study ranged from 2 mm/s to 36 mm/s (~ 0.1 in/s – 1.4 in/s). With the 1x10 mm flow channel geometry, these piston velocities correspond to fluid velocities ranging from 2 m/s to 37 m/s (~ 6.7 ft/s – 121 ft/s). The maximum shear rate achieved is $2.5\text{x}10^5$ s^{-1}. For each velocity several magnetic field strengths were considered, ranging from 100 kA/m to 200 kA/m. For all velocities the fluid was also characterized in the absence of a magnetic field.

3 Test Results

Perhaps the most compelling feature of MR fluids is their ability to provide high dynamic yield stresses nearly instantaneously. While the response time of the fluid has yet to be thoroughly investigated, it is known that the fluid can respond to changes in magnetic field strength in a matter of milliseconds. This fluid response can be described as a change in the rheological behavior of the fluid, i.e., the development of a yield stress. The development of a yield stress upon the application of a magnetic field is commonly referred to as the MR effect. The degree of MR effect achieved by the fluid is a function of magnetic field strength. Since the fluid response time is finite, the ability of the fluid to realize the expected MR effect may be diminished if the fluid does not spend a sufficient amount of time in the presence of the magnetic field. We define this time as the dwell time, that is, the amount of time the fluid spends in the presence of a magnetic field.

The high velocity rheometer discussed in previous section is used to study the behavior of the MR fluid at high flow velocities with a magnetic field applied. The magnetic field is applied over a portion of the flow channel. Dwell time is calculated from the length of the MR valve and the fluid velocity through the valve. Two MR valve lengths are considered (25.4 mm and 6.35 mm). Fluid velocities through the valve range from 1 m/s to 37 m/s with corresponding dwell times ranging from 12.4 ms to 0.18 ms. Figure 2 shows the reduction in dwell time as fluid velocity increases, for the two valve lengths. With the 25.4 mm valve length, dwell times were between 12.4 ms and 0.69 ms, while for the 6.35 mm valve length, dwell times dropped to between 6.4 ms and 0.18 ms. For the 6.35 mm valve length, the piston velocities were chosen such that there was a large concentration of dwell times close to 1 ms. This was done in order to highlight the response of the fluid at dwell times close to 1 ms, which is estimated to be the response time of MR fluid.

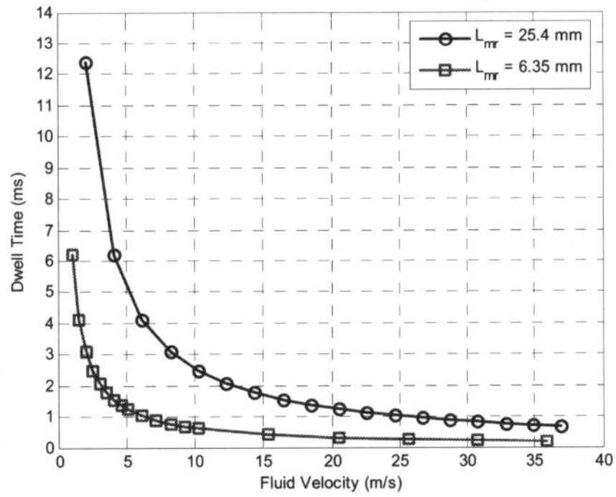

Figure 2. Dwell time of fluid through active valve length

As the fluid flows through the MR valve, it is subjected to a magnetic field in the active valve regions, and develops a yield stress that is expected to be proportional to the applied field. . Over this portion of the MR valve, the fluid develops its yield stress and allows for controllability in the force. As the fluid enters the active valve region, the transition from Newtonian-like flow to Bingham plastic flow occurs. Though it is known that MR fluid response to a magnetic field occurs in a matter of milliseconds, the degree of response achieved by the fluid has yet to be addressed.

Dwell time is defined as the amount of time the fluid spends in the presence of a magnetic field. In other words, dwell time is the time it takes the fluid to flow through the MR valve. Dwell time is calculated from

$$t_{dwell} = \frac{L_{mr}}{u_m} \tag{1}$$

where L_{mr} is the length of the MR valve and u_m is the mean fluid velocity through the flow channel. Consider, as an example, an extreme case in which the fluid is stationary in an active region of an MR valve. In such a case, the fluid will develop its full potential, or 100% of its expected yield stress with the application of a magnetic field. On the other hand, when the fluid is flowing through an MR valve, the time spent in the active region may influence the degree of response achieved by the fluid. In the absence of a magnetic field, the fluid exhibits Newtonian-like behavior; particle orientation is random and the velocity profile is parabolic. When the fluid is exposed to a magnetic field, the transition from Newtonian-like behavior to Bingham plastic behavior begins. During this transition, the particles begin to align along the flux path, increasing the yield stress of the fluid. Once the particles are fully aligned, the yield stress of the fluid reaches 100% of its expected value. If the dwell time exceeds the response time, then the fluid can achieve its full yield stress potential. As the fluid velocity increases, however, the dwell time decreases. Should this dwell time decrease below the time required to achieve the full yield stress, the fluid may only achieve a portion of the expected yield stress.

The yield stress was found from the measured force for each fluid velocity and evaluated as a function of the dwell time. A significant reduction in the yield stress is observed for dwell times less than 1 ms, as is shown in Figure 3. The 25.4 mm valve length was activated with field strengths of 100 kA/m and 200 kA/m at 18 piston velocities ranging from 2 mm/s to 36 mm/s. Piston velocities increased at an increment of 2 mm/s. Experiments with the 6.35 mm valve length were performed at three magnetic field strengths (100 kA/m, 150 kA/m, and 200 kA/m). Piston velocities for the 6.35 mm valve ranged from 1 mm/s to 35 mm/s. At this valve length, 10 piston velocities were tested between 1 mm/s and 5 mm/s, and another 5 velocities between 6 mm/s and 10 mm/s. This was done in order to highlight the response of the fluid at dwell times close to 1 ms. Above piston velocities of 10 mm/s, the velocity increment increased to 5 mm/s.

For the 25.4 mm MR valve, Figure 3a, a slight reduction in yield stress is observed as fluid velocity increases. Figure 3b shows the results for the 6.35 mm valve. With this valve configuration, a significant reduction in yield stress is observed as fluid velocity increases. The results in Figure 3b indicate that the yield stress at low velocities is nearly maintained. As the fluid velocity continues to increase, however, the yield stress falls off rather sharply. We also note that the differential between the yield stresses developed at low velocities for each magnetic field strength are significantly larger than those observed at high velocities. In other words, at high velocities, there is little change in the yield stress for varying magnetic field strengths. Though these results indicate some dependence on fluid velocity, the reduction in yield stress observed with both MR valves is more closely related to dwell time rather than fluid velocity.

Figure 4 shows the relationship between the yield stress developed in the MR valve and the dwell time. For the 25.4 mm valve length, Figure 4a, the yield stress remains relatively constant for dwell times greater than 2 ms. As the dwell time approaches 1 ms, however, we observe a roll-off in the yield stress. Similar results are observed for the 6.35 mm valve length, Figure 4b. For dwell times greater than 1 ms, the yield stress is relatively constant. Much like the 25.4 mm case, however, as the dwell time falls below 1 ms, the apparent yield stress drops considerably. The results for the 6.35 mm valve length clearly indicate that if the dwell time drops below 1 ms the fluid is unable to achieve its full potential. Figure 4b indicates that with a dwell time of 0.5 ms the yield stress developed in the gap is significantly less than the expected yield stress for a given magnetic field strength.

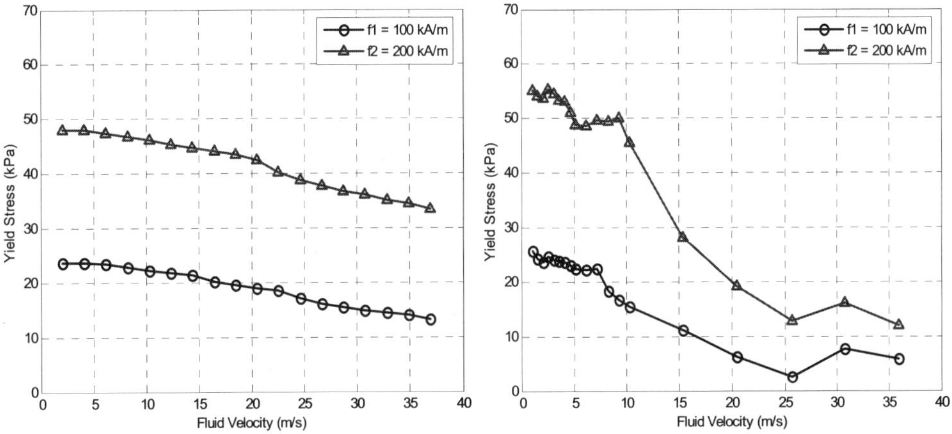

(a) (b)

Figure 3. Yield stress as a function of fluid velocity: (a) 25.4 mm MR valve; (b) 6.35 mm MR valve

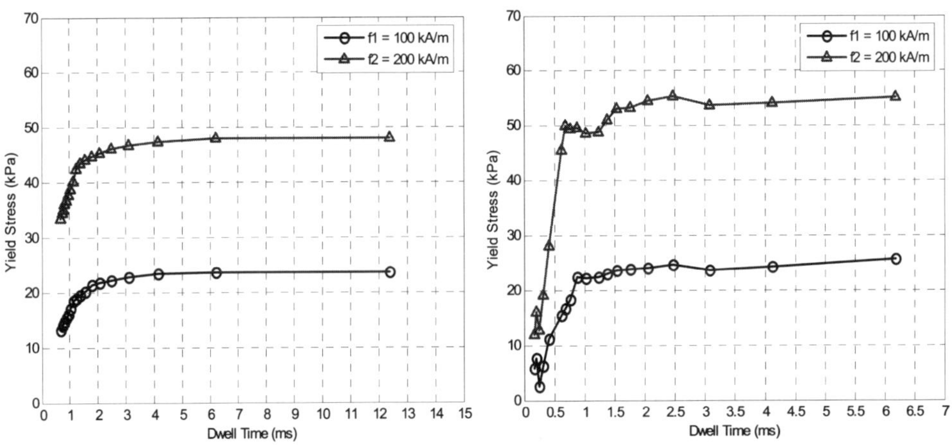

(a) (b)

Figure 4. Yield stress as a function of dwell time: (a) 25.4 mm MR valve; (b) 6.35 mm MR valve

4 Conclusion

The high velocity behavior of MR fluid subjected to a magnetic field was evaluated. Two active valve lengths were considered. Fluid dwell time was found for each valve length and fluid velocity. Experimental results indicate a strong correlation between dwell time and the yield stress developed in the MR valve. For the shorter valve—the 6.35 mm valve length—the ability of the fluid to reach its full yield stress was significantly diminished as the fluid velocity increased. The results of the study indicate that special care needs to be taken in design of MR devices to ensure that the valve design is suitable for the device application. The MR valve must be designed such that it allows the fluid sufficient amount of dwell time to reach its full yield stress, for the velocities at which the device is used for.

References

1. Ahmadian, M., Poynor, J. C., Gooch, J. M., "Application of Magneto Rheological Dampers for Controlling Shock Loading," *Proceedings of ASME Dynamic Systems and Control*, DSC-Vol. 67, p. 731-735, 1999.
2. Messina, N., "Multi-Role Armament & Ammunition System ATD-GDAS Program Scope and Approach," *Proceedings of 36th Annual Gun & Ammunition Symposium & Exhibition*, www.dtic.mil/ndia, presentation, April 9-12, 2004.
3. Chen, P. C., Wereley, N., "Magnetorheological Damper and Energy Dissipation Method," U.S. Patent 6,694,856 B1 2004.
4. Facey, W. B., Rosenfeld, N. C., Choi, Y. T., Wereley, N. M., "Design and Testing of a Compact Magnetorheological Damper for High Impulsive Loads," *Proceedings of the 9th International Conference on Electrorheological Fluids and Magnetorheological Suspensions*, Beijing, China, August 29 – September 3, 2004.
5. El Wahed, A. K., Sproston, J. L., Schleyer, G. K., "Electrorheological and Magnetorheological Fluids in Blast Resistant Design Applications," *Materials and Design*, 23, p. 391-404, 2002.
6. El Wahed, A. K., Sproston, J. L., Schleyer, G. K., "A Comparison Between Electrorheological and Magnetorheological Fluids Subjected to Impulsive Loads," *Proceedings of the 7th International Conference on Electrorheological Fluids and Magnetorheological Suspensions*, Honolulu, Hawaii, p. 401-410, 1998.
7. Norris, J. A., "Behavior of Magnetorheological Fluids Subject to Impact and Shock Loading," M.S. Thesis, Virginia Polytechnic Institute and State University, Blacksburg, VA, 2003.
8. Browne, A. L., McCleary, J. D., Namuduri, C. S., Webb, S. R., "Impact Performance of Magnetorheological Fluids," *Proceedings of IMECE*, 60542, Anaheim, CA, November, 13-20, 2004.
9. Carlson, J. D., "What Makes a Good MR Fluid? "*Journal of Intelligent Material Systems and Structures*, Vol. 13, No. 7, p. 431-435, 2002.
10. Goncalves F. D., "Characterizing the Behavior of Magnetorheological Fluids at High Velocities and High Shear Rates," PhD Dissertation, Virginia Polytechnic Institute and State University, Blacksburg, VA, 2004.
11. Baird, D. G., and Collias, D. I., Polymer Processing Principles and Design, Wiley Interscience, New York, NY, 1998.
12. Van Wazer, J. R., Lyons, J. W., Kim, K. Y., Colwell, R. E., Viscosity and Flow Measurement A Laboratory Handbook of Rheology, Wiley Interscience, New York, NY, 1963.
13. Lord Materials Division, "Magnetic Circuit Design," Engineering Note, 1999.

PREPARATION AND MAGNETORHEOLOGICAL CHARACTERIZATION OF POLYMER COATED-CARBONYL IRON SUSPENSIONS

H. J. CHOI, B. J. PARK, J. L. YOU AND M. S. CHO

Department of Polymer Science and Engineering, Inha University, Incheon 402-751, Korea
Email: hjchoi@inha.ac.kr

Despite its wide application as a suspended particle in magnetorheological (MR) fluids along with its extensive investigations on MR characteristics and dispersion stability, the pristine carbonyl iron (CI)-based MR fluid still possesses several drawbacks, including severe sedimentation of the CI particle due to its high density. To resolve these problems, we prepared poly(methyl methacrylate) (PMMA) coated-CI composite particles, possessing different shell thicknesses onto iron particles. The polymer coated composite magnetic particles (CIPMMA) were synthesized via an in-situ dispersion polymerization method using CI dispersions in methyl methacrylate monomer, and then adopted as dispersed phase of MR fluids. The CIPMMA was found to potentially resolve several problems of CI only based MR fluids such as severe sedimentation problem and poor dispersion stability. Flow and viscoelastic properties of the MR fluids were analyzed via a rotational rheometer equipped with a magnetic field suppler using measuring system of a parallel plate in both steady and oscillatory shear modes. Main focus was put on studying the effect of shell thickness in the internal structure formation and sedimentation characteristics. The MR characteristics including a yield stress were found to be affected by the CI contents in the composite particles.

1 Introduction

Magnetorheological (MR) fluid is a suspension of magnetically susceptible particles dispersed in nonmagnetic suspending medium. It is regarded as a smart material with controllable properties by an external magnetic field, which is abruptly transformed within milliseconds from a fluid-like to solid-like state under an applied magnetic field, by showing changes of its rheological properties such as yield stress and shear viscosity [1, 2]. In contrast to electrorheological (ER) fluids behaving similarly under an applied electric field [3-7], MR fluids generally show superior properties, including a high yield stress up to 10-100 kPa under a strong magnetic field [8-10]. Thereby MR fluids could be easily commercialized compared with the ER fluids since a magnetic field is more stable than the electric field. These MR fluids can be widely used in designing damping devices such as shock absorbers, clutch and position controllers [11-13]. The particles for MR fluids should be magnetized under an external magnetic field, suggesting that ferromagnetic, ferrimagnetic and paramagnetic materials can be used as dispersed phases of the MR fluids. Soft ferromagnetic materials such as magnetite and iron are generally used as MR materials [14]. However, most magnetic fluids including carbonyl iron (CI) based MR suspension have sedimentation and aggregation problems due to their high density and rigid surfaces. In order to enhance the dispersion stability of magnetic particles, several methods including polymer encapsulation have been reported [15]. Polymeric coating on inorganic particles provides better surface properties with chemical resistance and dispersion stability.

In this study, the core-shell structured composite particles were synthesized by a dispersion polymerization using methyl methacrylate (MMA) as a monomer and methacylid acid (MAA) as a surface modifier. The thickness of coating layer could be altered by adjusting the concentration of MMA and CI particles added to the systems. Rheological properties of the MR fluids under an applied magnetic field perpendicular to the shear flow direction were investigated by measuring flow behavior.

2 Experimental

2.1 Materials

Encapsulated magnetic particle based MR fluids were prepared by a relatively simple method, using carbonyl iron (CI, Aldrich, USA, Fe content is above 99.5%), whose average particle density and diameter of particles are 7.86 g/cm^3 and about 2.57 ㎛, respectively. Methyl methacrylate (MMA) was adopted as a monomer. 2.2'-azobis(isobutyronitrile) (AIBN) was used as an initiator in the dispersion polymerization after recrystallization from methanol. Deionized water, purified by distillation following by deionization using ion exchange resins was used.

2.2 Preparation of CIPMMA composite microspheres

The CI particles modified by the MAA were prepared with an aqueous solution in a round flask with vigorous stirring at room temperature. As for preparation of the CIPMMA particles, poly(vinyl pyrrolidone) (PVP) was dissolved in methanol with stirring. The mixture of MAA modified CI, MMA and AIBN was added to the solution phase, which was then polymerized for about 24h. The temperature was increased up to 55℃. A schematic diagram of the CIPMMA composite preparation step is provided in Fig. 1. Initially, the CI surface was modified by MAA to induce the dispersion polymerization reaction on the CI surface, and the modified CI was then dispersed in methanol containing PVP as a stabilizer. MMA monomer and a radical initiator of the AIBN were dissolved in the reactor. After the polymerization, synthesized CIPMMA particles were separated using a magnet and washed with water followed by methanol to remove PVP, unreacted monomer, and PMMA oligomers. The washed particles were finally dried at 60℃ oven. The fabricated particles are denoted as CI-A, CI-B and CI-C in which their ratios of CI and PMMA correspond to 2:1, 1:1 and 1:2. The magnetic microspheres were dispersed in mineral oil for preparing the MR fluid, respectively.

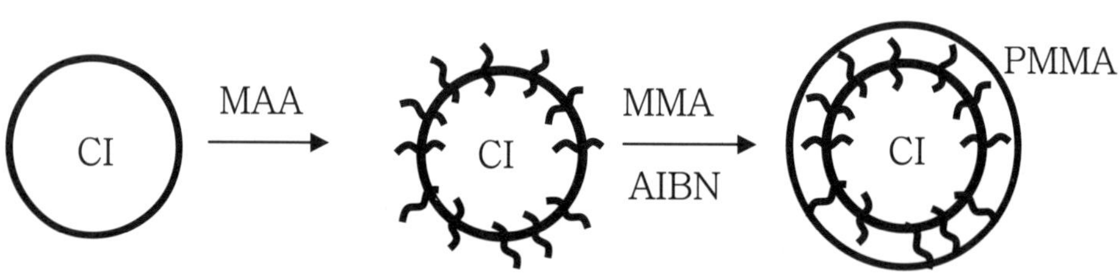

Figure 1. Schematic diagram of the synthesis of CIPMMA

2.3 Characterization

The diameter and surface characteristics of the CIPMMA particles were observed using scanning electron microscope (SEM, JSM-6700F, JEOL, Tokyo, Japan). MR characteristics were examined using a rotational rheometer (MCR 300, Physica, Stuttgart, Germany) with the magneto-rheological device at 25℃ for all tests. The coil current and magnetic field strength were controlled via a separate control unit and rheometer software (US200, Physica, Stuttgart, Germany). Homogeneous magnetic field was applied to the measuring system made of non-magnetic metal with a gap size of 1mm.

3 Results and discussion

Figure 2 shows characteristic morphologies of CIPMMA particles examined by the SEM, and the inset shows the SEM image of the pristine CI particles. The average particle size with spherical shape was observed to be about 4 ㎛, and the particle size increased due to the encapsulation of PMMA. With increased MMA concentration, the magnetic particle size was also increased. Furthermore, the surface of the synthesized CIPMMA particles became smoother as a result of the polymeric encapsulation.

Figure 2. SEM image of CIPMMA particles and the inset shows the pristine CI

Figure 3 represents X-ray photoelectron spectroscopy (XPS) which was used to analyze the surface sensitivity and its semi-quantitative specific chemical analysis. The CI spectrum contains a Fe 2p peak in the region from 711.7 to 720 eV and Fe 3p peak in 52.6 eV. The PMMA spectrum contains O 1s peak in 532.0 eV and 533.6 eV, while the lower binding energy C 1s component to the C-C contribution located at 285.1 eV. For PMMA, contributions of the carbon atoms directly bonded to oxygens are then located at a higher binding energy 286.7 eV. The CIPMMA magnetic particles have no Fe signals at 711.7, 720.0 and 52.6 eV. From these XPS analyses, the CI particles were confirmed to be finely encapsulated by the PMMA as a shell material.

Shear flow behaviors of the MR fluids were examined under a steady shear flow and a static magnetic field as represented in Fig. 4. It shows increased shear stress with increased magnetic field strength. As the magnetic field strength was increased, the yield shear stress of the MR fluid also increased, exhibiting typical MR behavior. Yield stress for each condition can be obtained by extrapolating the shear stress values at zero shear rate limit. This MR fluid shows a yield stress of approximately 14kPa at 343kA/m, which is a relatively high value compared with ER fluids and could be used for commercial applications. The yield stress decreased by coating polymers on the CI particle. The MR fluid of pure carbonyl iron particles showed larger yield stresses over the CIPMMA magnetic particles fluids due to the higher magnetic content.

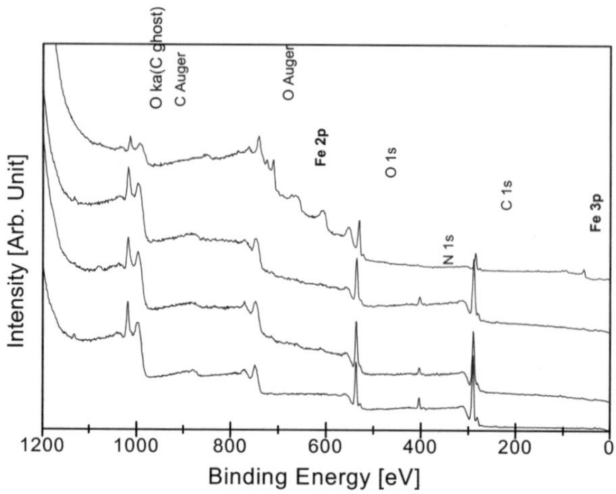

Figure 3. Surface characterization of CIPMMA from the XPS measurement (from top to bottom: CI, CI-C, CI-B and CI-A).

Different CIPMMA particles given in Fig. 4 show different shear stresses under magnetic fields because they contained different amount of the CI. When CIPMMA particles contain less magnetic CI particles in the case of CI-A than pristine CI particle, the shear stress increase of the CIPMMA particles was lower than that of pure carbonyl iron particles, demonstrating that the amount of CI in the MR fluids affects MR behavior.

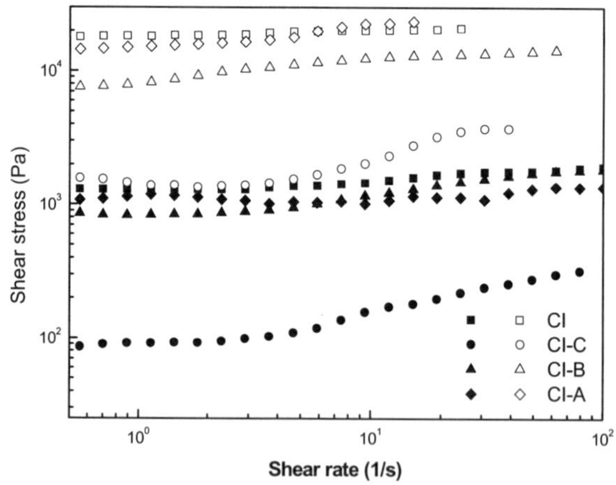

Figure 4. Shear stress versus shear rate of CIPMMA and CI particles (open: 343 kA/m, closed: 86 kA/m) .

MR fluid changes flow property from a liquid-like to a solid-like state under a magnetic field strength, in which the magnetic particles build up chain-like structure with an applied magnetic field due to dipole-dipole interaction, indirectly representing that the formed chain structure caused an increasing shear viscosity given in Fig. 5 [16, 17].

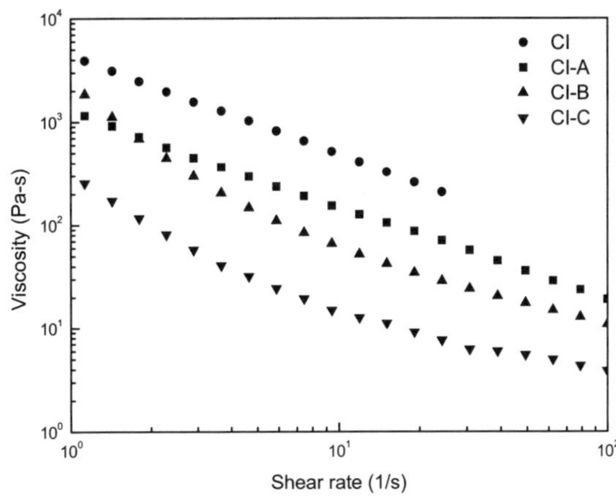

Figure 5. Shear viscosity versus shear rate of CIPMMA and CI particles based MR fluids under 86 kA/m.

Figure 5 gives shear viscosity of the MR fluids for different magnetic particles at a given magnetic field strength of 86 kA/m. The shear viscosity was increased with increased magnetic field strength and the ratio of CI to PMMA at a given shear rate. The pure carbonyl iron has higher shear viscosity than PMMA encapsulated materials [18, 19]. Yield behaviors which show constant shear stress with increasing shear rate were consistent for all magnetic field strengths tested. This result implied that particle cluster structures constructed within fluid under the applied magnetic field are continuously broken and then reformed until the shear rate of 100 sec^{-1} [20]. The shear thinning behaviors observed in the shear viscosity data stem from the changes of internal structures under shear deformation as well [21]. In other words, particle structures developed under external magnetic fields remained until the shear rate of 100 sec^{-1}.

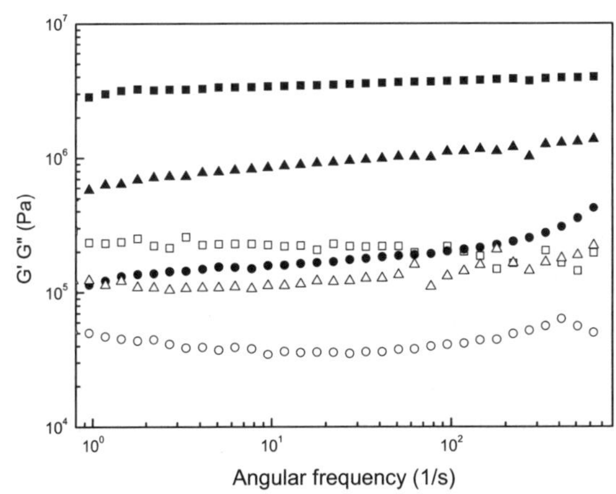

Figure 6. Storage modulus (a) and loss modulus (b) as a function of frequency for CIPMMA based MR fluid under 171 kA/m (■: CI-A, ▲: CI-B, ●: CI-C) (open : G", closed : G')

Figure 6 shows the oscillatory behavior of the MR fluids for different coating thickness under a magnetic field strength of 171kA/m. The storage modulus of MR fluid which is related with the elastic properties decreased with the coating thickness. When CIPMMA particles have more magnetic particles, loss modulus and storage modulus increase than CIPMMA particles with lower CI particle content. These results suggest that both G' and G" responses get reduced with increased polymer coating thickness. In addition, the G' value of storage modulus is obviously higher than that of loss modulus G", demonstrating larger elastic property (solid-like) rather than the viscous property under the magnetic field [22-27].

4 Conclusions

Core/shell structured CIPMMA composite particles were prepared by incorporating an in-situ dispersion polymerization with different thickness of coating layer. By increasing the MMA concentration, the magnetic particle size was also increased. The surface of the synthesized carbonyl iron particles became smooth as a result of the encapsulation. The shear stress was increased when increasing magnetic field. When magnetic particle concentration was increased, the yield stress and shear viscosity of the MR fluid was also increased.

5 Acknowledgements

This work was supported by the Advanced Environmentally-benign Structural Materials R&D Center, Korea (2005).

References

1. Bossis G. and Lemaire E., Yield stresses in magnetic suspension, *J. Rheol.* **35** (1991) pp.1345-1354.
2. Choi H. J., Cho M. S., Kim J. W., Kim C. A. and Jhon M. S., A yield stress scaling function for electrorheological fluids. *Appl. Phys. Lett.* **78** (2001) pp.3806-3808.
3. Zhao X. P. and Yin J. B., Advances in electrorheological fluids based on inorganic dielectric materials, *J. Ind. Eng. Chem.,* **12** (2006) pp. 184-198.
4. Quadrat O. and Stejskal J., Polyaniline in electrorheology, *J. Ind. Eng. Chem.,* **12** (2006) pp. 352-361.
5. Sung J. H., Park D. P., Park B. J. and Choi H. J., Universal yield stress characteristics of potato starch phosphate-based electrorheological fluid, *J. Ind. Eng. Chem.,* **12** (2006) pp. 301-305.
6. Pavlinek V., Saha P., Perez-Gonzalez J., de Vargas L., Stejskal J. and Quadrat O., Analysis of the yielding behavior of electrorheological suspensions by controlled shear stress experiments, *Appl. Rheol.,* **16** (2006) pp. 14-18.
7. See H., Sakurai R. and Saito T., Model of structure and conduction in an electrorheological fluid under flow, *Int. J. Mod. Phys. B* **10** (1996) pp.3267-3274.
8. Kuzhir P., Bossis G. and Bashtovoi V., Optimization of magnetorheological fluid valves, *Int. J. Mod. Phys. B,* **19** (2005) pp.1229-1235.

9. Facey W. B., Rosenfeld N. C., Choi Y. T., Wereley N. M., Choi S. B. and Chen P., Design and testing of a compact magnetorheological damper for high impulsive loads, *Int. J. Mod. Phys. B,* **19** (2005) pp.1549-1555.
10. Lim S. T., Cho M. S., Choi H. J. and Jhon M. S., Magnetorheological characterization of organoclay added carbonyl-iron suspensions, *Int. J. Mod. Phys. B,* **19** (2005) pp.1142-1148.
11. Carlson J. D., MR fluids and devices in the real world, *Int. J. Mod. Phys. B,* **19** (2005) pp. 1463-1470.
12. Lord Corporation [Online]. Available : http://www.rheonetic.com
13. Lita M., Popa N. C., Velescu C. and Vékás L. N., Investigations of a magnetorheological fluid damper, *IEEE Trans. Magn.* **40** (2004) pp. 469-472.
14. Cho M. S., Lim S. T., Jang I. B., Choi H. J. and Jhon M. S., Encapsulation of Spherical Iron-Particle With PMMA and Its Magnetorheological Particles, *IEEE Trans. Magn.* **40** (2004) 3036-3038.
15. Lee S. U. and Harris M. T., Surface modification of magnetic nanoparticles capped by oleic acids: characterization and colloidal stability in polar solvents. *J. Colloid Interf. Sci.* **293** (2006) pp. 401-408.
16. Jun J. B., Kim J. W., Lee J. W. and Suh K. D., Spherical polarization body: synthesis of monodisperse micron-sized polyaniline composite particles, *Macromol. Rapid Commun.* **22** (2001) pp.937-940.
17. Jang I. B., Kim H. B., Lee J. Y., You J. L., Choi H. J. and Jhon M. S.. Role of organic coating on carbonyl iron susepended particles in magnetorheological fluids. *J. Appl. Phys.* **97** (2005) pp. 10Q912.
18. Choi H. J., Kim C. A, Kwon T. M. and Jhon M. S., Viscosity of magnetic particle suspensions. *J. Magn. Magn. Mater.* **209** (2000) pp. 228-230.
19. Niraula B., King T. C. and Misran M., Rheology properties of dodecyl-β -D-maltoside stabilized mineral oil-in-water emulsions, *Colloid Surf. A-Physicochem. Eng. Asp.* **231** (2003) pp.159-172.
20. Choi H. J., Cho M. S. and Lee I. S., Electrorheology of monodisperse core/shell structured particle suspensions, *Int. J. Mod. Phys. B,* **19** (2005) pp.1077-1082.
21. Hong C. H. and Choi H. J., Synthesis and electrorheology of phosphoric acid doped polyaniline suspensions, *Int. J. Mod. Phys. B,* **19** (2005) pp.1149-1155.
22. Bombard A. J. F., Alcantara M. R., Knobel M. and Volpe P. L. O., Experimental study of MR suspensions of carbonyl iron powders with different particles sizes, *Int. J. Mod. Phys. B,* **19** (2005) pp.1332-1338.
23. Gordaninejad F., Kavlicoglu B. M. and Wang X., Friction factor of magneto-rheological fluid flow in grooved channels, *Int. J. Mod. Phys. B,* **19** (2005) pp.1297-1303.
24. Trendler A. M. and Bose H., Influence of particle size on the rheological properties of magnetorheological suspensions, *Int. J. Mod. Phys. B,* **19** (2005) pp.1416-1422.
25. Rankin P. J., Horvath A. T. and Klingenberg D. J., Magnetorheology in viscoplastic media, *Rheol. Acta,* **38** (1999) pp. 471-477.
26. Choi H. J., Jang I. B., Lee J. Y., Pich A., Bhattacharya S. And Alder H. J., Magnetorheology of synthesized core-shell structured nanoparticle, *IEEE Trans. Magn.* **41** (2005) 3448-3450.
27. Lim S. T., Cho M. S., Jang I. B., Choi H. J. and Jhon M. S., Magnetorheology of carbonyl-iron suspensions with submicron-sized filler, *IEEE Trans. Magn.* **40** (2004) 3033-3035.

SIMULATION OF MOTION OF MR- PARTICLES IN ROUGH MICRO TUBES CONTROLLED BY NON-UNIFORM MAGNETIC FIELD

SINAN OZCAN AND CAHIT EVRENSEL
Department of Mechanical Engineering, University of Nevada.Reno, USA
Email: cahit@unr.edu

Mark A. Pinsky
Department of Mathematics and Statistics, University of Nevada.Reno, USA
Email: pinsky@unr.edu

Modeling dynamic behavior of magnetorheolgocial (MR) fluids frequently based on a two-phase flow approach which homogenizes the particles' dynamics and ignores their individual motions. Particles motion in moving carriers was primarily studied in pure cylindrical tubes under the influence of inform magnetic field. The objective of this paper is to simulate flow of a MR-fluid consisting of a magnetically neutral carrier and ferrous particles in micro-tubes with rough wall under the influence of a non-uniform magnetic field developed by a permanent magnet. Carried in MATLAB simulations show that ferrous particles are concentrated around the magnets poles and bumps or dents of the tube wall which could clot the tube. A computationally efficient but simplified model developed in this study provides qualitative insights in behavior of MR-fluids while its quantitative accuracy declines if MR-particles form substantial clots that alter flow patterns. Analysis of these phenomena requires more accurate accounts of particles-fluid interaction and lead to more complex and computationally extensive models; the corresponding results are presented in a subsequent paper in this proceeding.

1. Introduction

Modeling of magnetorheological (MR) fluids is frequently based on a continuous model approach which homogenizes the effect produced by magnetic particles immersed in a magnetically neutral carrier, see, for example, monograph [1] and further references therein. Homogenization delivers computationally efficient models but ignores motion of individual particles and consequently obscures modeling of particles aggregation that determines key characteristics of MR-fluids and their flow patterns.

Analysis of structure formation in electroreological (ER) fluids received considerable attention in the past two decades. Klingenberg *et al.* [2-5] and K.C. Hass [6] pioneered with development of 2D and 3D models that simulate formation of the structures in ER suspensions. While the developed approach is based on several hypotheses that could limit its accuracy, the simulations were able to predict aggregation of the particles in experimentally observed structures.

Bonnecaze *et al* [7] developed a more detailed model which accounts for the far- and near- field effects for both the hydrodynamic interaction (by applying "Stokesian dynamics") and electrostatic interactions. However, this more accurate but computationally extensive model is able to simulate the motion of a small number of particles.

Experimental and theoretical studies of particle aggregation in MR-fluids have been reported in recent papers [8-10] which also include further references on this subject.

Application of field-controllable fluids to a biomedical field was considered in Pankhurst *et al.,* [11]. Utility of MR-fluids controlled by uniform magnetic field to target drag delivery was considered in Zachary *et al.* [12].

The first computational study on electrorheological suspensions with a non-uniform electric field was carried out by Kadaksham *et al.* [13]. The developed model is based on the method of distributed Lagrange multiplier and a point-dipole approximation for electrostatic forces; the hydrodynamic forces are neglected in this model. The simulations demonstrated that the magnetic particles accumulate in the regions where the magnitude of the electric field reaches local maximum or minimum values. In the same time a sufficiently large pressure gradient could sweep electrically polarizable particles away from the accumulation regions.

The objective of this paper is to simulate flow of MR-fluid consisting of magnetically neutral carrier and ferrous particles in micro-tubes with a rough wall controlled by a non-uniform magnetic field produced by a permanent magnet. Matlab simulations show that ferrous particles are concentrated around the magnet poles and bumps or dents of the tube wall which could clot the tube.

In the absence of flow (static fluid) the magnetic particles attract each other and align along the magnetic field lines in ray-like structure centered at the magnet pole where the strength of the magnetic field is maximal.

2. Simulation Model

In this study we assume that ferrous spherical particles are suspended in a resting or slowly moving magnetically neutral fluid. The governing equations are set up in a 2D-Lagrangian coordinate frame coupled with the resting tube. MATLAB code calculates the trajectories of each particle in the suspension subject to magnetic, hydrodynamic drag, gravitation, intra-particle attractive magnetic and repeling collision forces and particle-wall repeling collision forces. The particle trajectories are calculated by solving Newton's equations of motion for each spherical particle, a similar set of equations could be found in [14]. Figure 1 show the forces which are accounted by our model.

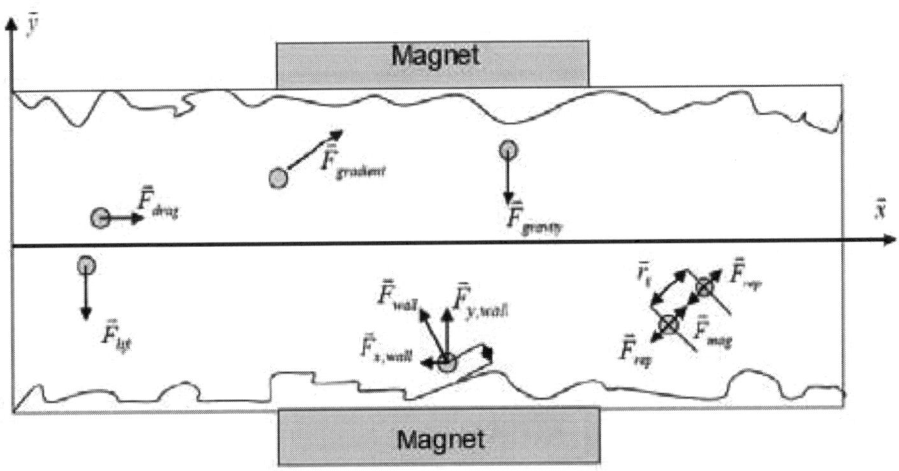

Figure 1 Schematic of the forces acting on each particle

The external magnetic field, applied to control particle motions, tends to align their dipole moments along its field lines. A non-uniform external magnetic field develops a force acting on each particle towards the regions of highest field gradient which are the poles of the magnet, see [16, 18, 17, 19 and 20] for more details. In this, as in other relevant studies, [2-7], the hydrodynamic drag force acting on a particle assumes Stokes' law. Note also that this model assumes that flow of magnetic particles does not affect flow of a magnetically neutral carrier. This assumption fosters computational efficiency, but its accuracy declines if the particles form substantial clots that alter flow patterns.

Analysis of these more complex phenomena require more accurate accounts of particle-fluid interaction and lead to more complex and computationally extensive models; the corresponding results are presented in a subsequent paper in this proceeding.

3. Simulation Results

Figure 2 plots the snapshots of the motion of magnetic particles that are aligned along the lines of external magnetic field. Figures 3 and 4 show snapshots of the particle motion in a tube with rough walls. The particle motions are affected by the pressure gradient, strength of magnetic field and particle size. Figure 3 shows that particles could accumulate around magnetic poles and dents in the tube wall. At the same time, a sufficient pressure gradient could push smaller-size particles through the tube, figure 4. However, particles tend to form chain-like structures while moving near wall imperfections in this case as well.

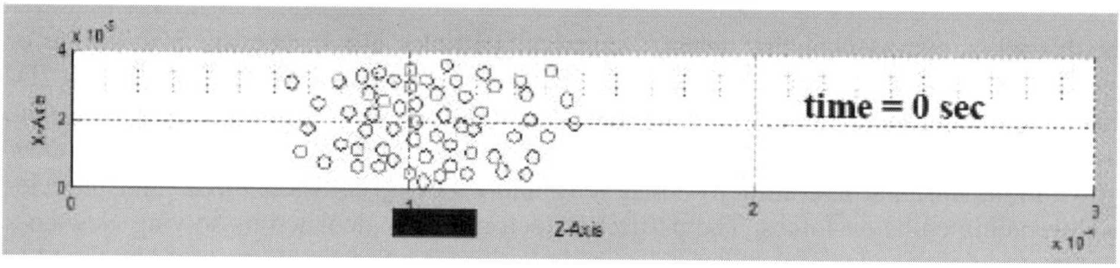

Figure 2 a. Static fluid. Initial positions of the particles.

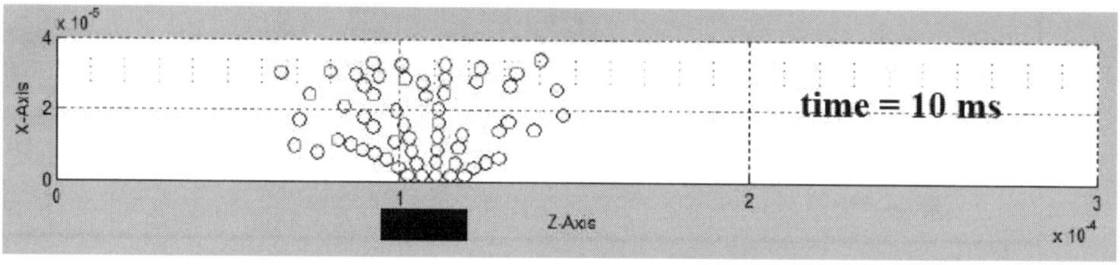

Figure 2b. Sixty particles in a static fluid; magnet strength is 0.2 T.

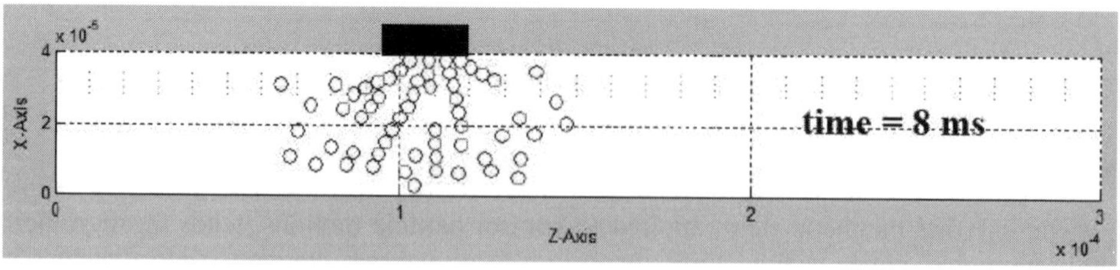

Figure 2c. Sixty particles in a static fluid; magnet strength is 0.2 T.

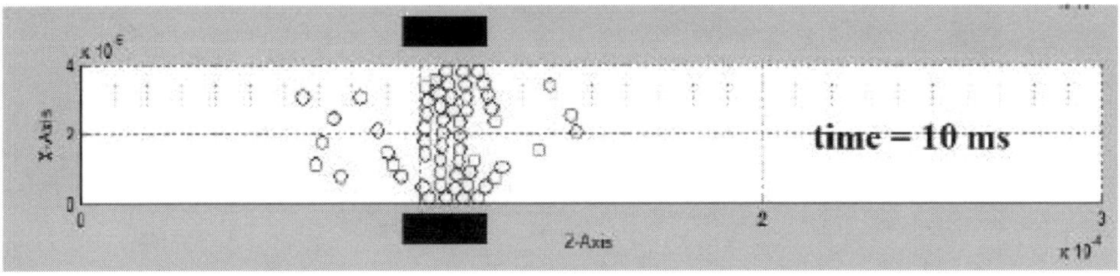

Figure 2d. Sixty particles in a static fluid with two equal magnets with strengths of 0.2 T

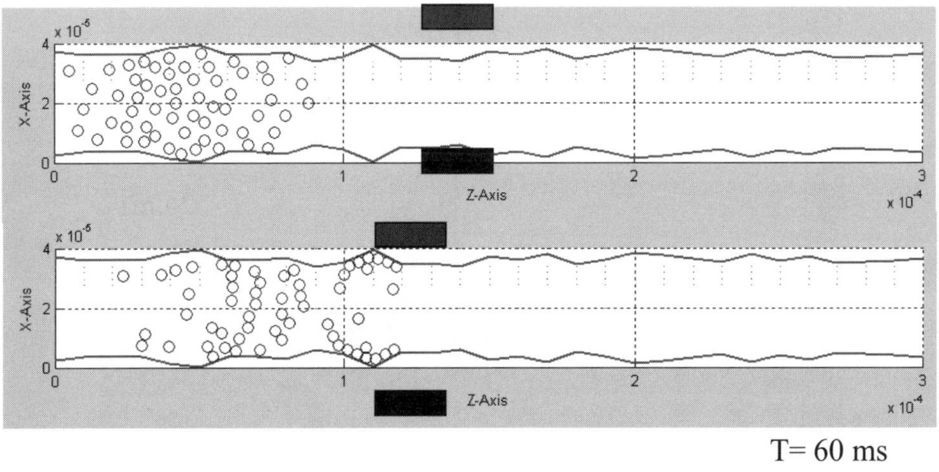

Figure 3. Flow of the particles in the tube with rough walls. Magnet strength is 0.7 T, and horizontal pressure gradient is 250 Pa.

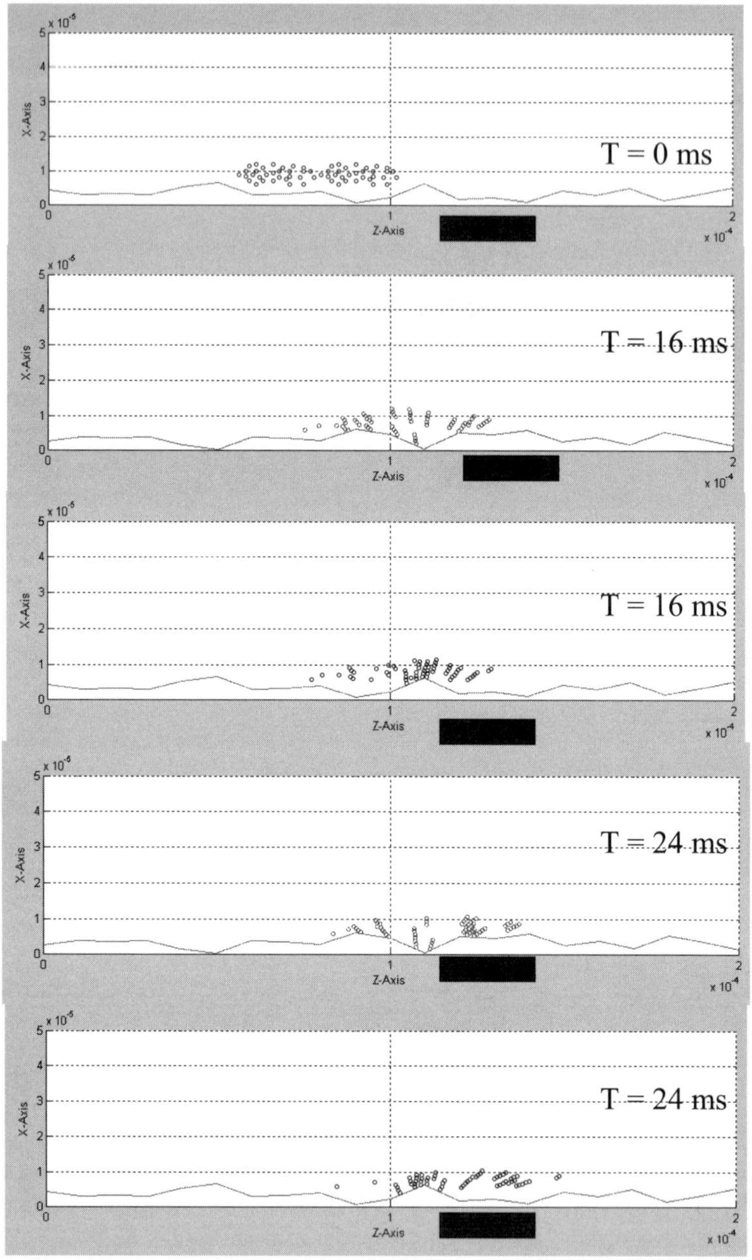

Figure 4. Flow of the particles through a tube with rough walls. Magnet strength is 0.2T and horizontal pressure gradient is 500 Pa.

4. Conclusion

This paper studies the dynamics of MR particles in a cylindrical tube with a rough wall which is controlled by a nonuniform magnetic field developed by a permanent magnet. We show that in case of static fluid the particles align along the magnetic field lines and form a multiple-chain structure centered at the magnet pole. In the moving carrier, particles tend to accumulate near magnetic poles and wall imperfections. They could aggregate and form chain-like structures which could clot the tube, however, sufficient pressure gradient could push particles through the tube.

References

1. R. E. Rosenbensweig, "Ferro hydrodynamics", Cambridge University Press, 1985
2. D. J. Klingenberg, Frank van Swol, and C. F. Zukoski, "Dynamic Simulation of Electrorheological Suspensions", Journal of Chemical Physics 91, No: 12, pp 7888, 1989.
3. D. J. Klingenberg, Frank van Swol, and C. F. Zukoski, "The Small Shear Rate Response of Electrorheological Suspensions", Journal of Chemical Physics 94, No: 9, pp 6160, 1991.
4. D. J. Klingenberg, Frank van Swol and C. F. Zukoski, "The Small Shear Rate Response of Electrorheological Suspensions II. Extension beyond the Point- Dipole Limit", Journal of Chemical Physics 94, pp 6170, 1991.
5. D. J. Klingenberg, C. F. Zukoski and J. C. Hill, "Kinetics of Structure Formation in Electrorheological Suspensions", Journal of Applied Physics 73, No: 9, pp 4644, 1993.
6. K. C. Hass, "Computer Simulations of Non-equilibrium Structure Formation in Electrorheological Fluids", Physical Review E 47, No: 5, pp 3362, 1993.
7. R. T. Bonnecaze and J. F. Brady, "Dynamic simulation of an Electrorheological Fluid", Journal of Chemical Physics 96, No: 3, pp 2183, 1992.
8. Mark Ivey, Jing Liu, Yun Zhu and Serge Cutillas, "Magnetic-Field-Induced Structural Transitions in a Ferrofluid Emulsion", *Physical Review E* **63**, 011403, 2000.
9. Howard See, "Constitutive Equation for Electrorheological Fluids Based on the Chain Model", *Journal of Physics: Applied Physics D* **33,** pp 1625, 2000.
10. Zuowei Wang, Haiping Fang, Zhifang Lin and Luwei Zhou, "Dynamic Simulation Studies of Structural Formation and Transition in Electro-Magneto- Rheological Fluids", *International Journal of Modern Physics B* **15**, No: 6&7, pp 842, 2001.
11. Q. A. Pankhurst, J. Connolly, S. K. Jones and J. Dobson, "Applications of Magnetic Nanoparticles in Biomedicine", Journal of Physics: Applied Physics D 36, pp 167, 2003.
12. Zachary G. Forbes, Benjamin B. Yellen, Kenneth A. Barbee and Gary Friedman, "An Approach to Targeted Drug Delivery Based on Uniform Magnetic Fields", IEEE Transactions on Magnetics 39, No: 5, pp 3372, 2003.
13. J. Kadaksham, P. Singh, and N. Aubry, "Dynamics of Electrorheological Suspensions Subjected to Spatially Nonuniform Electric Fields", Journal of Fluids Engineering 126, pp 170, 2004.
14. M. Mohebi, N. Jamasbi, and Jing Liu, "Simulation of the Formation of Non-equilibrium structures in Magnetorheological Fluids Subject to an External Magnetic Field", Physical Review E 54, No: 5, pp 5407, 1996.

15. Q. A. Pankhurst, J. Connolly, S. K. Jones and J. Dobson, “Applications of Magnetic Nanoparticles in Biomedicine”, Journal of Physics: Applied Physics D 36, pp 167, 2003.
16. George E. Mase, “Theory and Problems of Continuum Mechanics”, McGraw-Hill, 1970.
17. Zuowei Wang, Christian Holm and Hanns Walter Muller, “Molecular Dynamics Study on the Equilibrium Magnetization Properties and Structure of Ferrofluids”, Physical Review E 66, 021405, 2002.
18. Curtis M. Oldenburg, Sharon E. Borglin and George J. Moridis, “Numerical Simulation of Ferrofluid Flow for Subsurface Environmental Engineering Applications”, Transport in Porous Media 38, pp 319, 2000.
19. Malcolm McCaig, “Permanent Magnets in Theory and Practice”, John Wiley & Sons, New York, 1977.
20. Malcolm McCaig and Clegg A.G., “Permanent Magnets in Theory and Practice”, 2nd Edition, Halsted Press, Wiley, New York, 1987.

THE METHODS FOR MEASURING SHEAR STRESS OF POLAR MOLECULE DOMINATED ER FLUIDS

RONG SHEN[1], XUEZHAO WANG[1], YANG LU[1], GANG SUN[1], WEIJIA WEN[1,2] AND KUNQUAN LU[1]

1. Beijing National Laboratory for Condensed Matter Physics, Institute of Physics, Chinese Academy of Sciences, Beijing 100080, China
2. Department of Physics, The Hong Kong University of Science and Technology, Hong Kong, China
E-mail: lukq@aphy.iphy.ac.cn

A series of ER fluids materials with high shear stress have been developed recently, which named as polar molecule dominated electrorheological (PM-ER) fluids. Difficulties have been met in shear stress measurement process due to the slide of PM-ER fluids on the surface of metallic electrodes. In this paper, two shearing configurations have been developed to remove the interface effect. The intrinsic shear stress of ER fluids can be obtained by using the devices.

1. Introduction

Electrorheological (ER) fluids are a class of materials whose rheological property is controllable by applying an electric field. The mechanical properties of ER fluids may vary within a wide range (almost from a liquid to a solid), so how to measure and character the varieties is a key problem in the experiment. Rheometers with parallel-plate sensor system and inner-outer cylinder sensor system are widely used in the measurement of the rheological properties (viscosity, yield stress, etc.) of ER fluids. For traditional ER fluids whose shear stress is less than 10kPa, the interface state between electrode and ER fluids is similar to that inside of ER fluids according to dielectric theory. The parameters measured by using ordinary rheometers equipped with the metallic electrodes can reflect the rheological characteristics of ER Fluids.

A newly developed class of ER fluids, which we name as polar molecule dominated electrorheological (PM-ER) fluids [1], exhibits high yield stress [2-5]. The static yield stress of PM-ER fluids can be as high as hundreds of kPa or even higher, which is several orders higher than that of traditional ER fluids. The dynamic shear stress of the PM-ER fluids are also enhanced tremendously compared with that of ordinary ones. The principle of PM-ER fluids is quite different from the traditional dielectric theory, which we will report elsewhere [1]. The basic principle of PM-ER fluids is that the polar molecules adsorbed on the particles play crucial role and the high yield stress comes from the interaction of polar molecule-charge in between the particles, where the local electric field is much higher than the external one and can cause the polar molecules aligning.

At the interfaces of PM-ER fluid and the electrodes the boundary condition is not same as inside of the fluid. The fluid will slide at the surfaces of metallic electrodes, as a result, the measured shear stress must be much lower than the real value.

Several methods are applied to improve the reaction between PM-ER fluids and metallic electrodes, and excellent results have been gotten in our experiment. The detailed results will be reported in another paper [6]. Although the measured shear stresses of PM-ER fluids by using those modified electrodes can be several times higher than that measured with smooth metallic electrodes, the measured values still do not reflect the intrinsic shear stress of the fluids. It is necessary to find a way to measure the intrinsic rheological property for understanding the PM-ER fluids and for their applications.

In this paper, we will provide two methods for measuring the real yield stress of PM-ER fluids, with which the shortcomings of using the normal rheometer can be overcame in high shear stress measurement and the intrinsic shear stress of PM-ER fluids can be obtained. The solid character of the ER fluids at high electric field and the ordinary shear stress measurement applied for solid matter inspired us to use those methods.

2. Experimental

There are two designed shearing modes for measuring the shear stress of PM-ER fluids. One configuration designed for high shear stress measurement is sketched in Fig.1, which can be defined as unilateral shearing mode. Two metallic electrodes are embedded into two grooves of insulating blocks respectively. The materials of insulating block could be chosen from nylon, teflon, organic glass, etc. ER fluids are filled in the gap between the electrodes. A high DC voltage is applied to the electrodes, then step motor pushes one insulating block forward, meanwhile a force sensor attached to the block can detect the force F acted on the block by ER fluids. The data of F are collected by a computer connected to the force sensor. The shear stress τ of ER fluids can be calculated with $\tau = F/S$, where S is the cross section area of the ER fluids. In our measurement, the gap distance of two electrodes was about 3.3mm and the cross section area $S= h\times l = 7\times 2\text{mm}^2$, in which l is the length of the sample in moving direction and h is the thickness of the sample. The minimum moving velocity v of the upper block was 0.01mm/s.

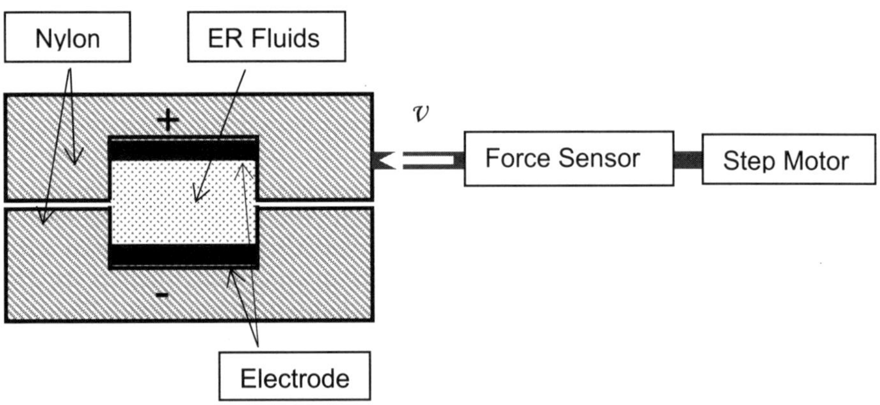

Figure 1. Sketch of unilateral shear configuration. The upper electrode (+) moves, and the lower electrode (-) is fixed.

Another configuration is named as slice shearing mode shown in Fig.2. Two metallic electrodes also are embedded in the grooves of insulating blocks just as in Fig.1. The gap between the electrodes is filled with PM-ER fluids, and the electrodes remain fixed. An insulating slice with 1mm thickness and 10mm height inserts the ER fluids under applied electric field. The movement of the insulating slice is pushed by a step motor; meanwhile the ER fluids act a force F to the slice. The calculation of shear stress of the ER fluids can be compared with double shear test in plastic metallic materials, that is $\tau = F / 2S$, where S is the cross section area of the ER fluids. In our measurement, the velocity of the insulation slice was 0.01mm/s, the gap width between two electrodes was 4 mm and the cross section area of the slice $S = 1 \times 10\ \mathrm{mm}^2$.

The dynamic shear stress of ER fluids could be measured conveniently by using the devices described above. The shear rate is controlled by step motor.

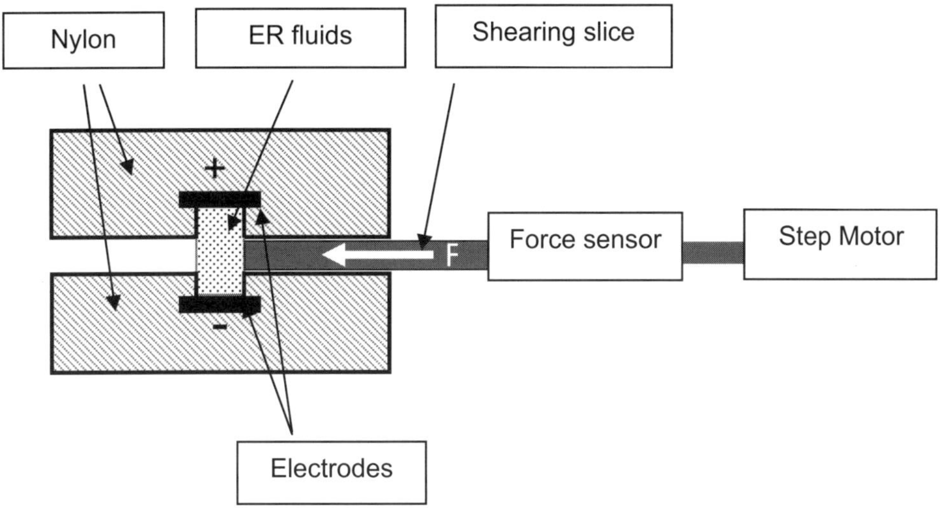

Figure 2. Sketch of slice shearing configuration. Two insulating blocks with electrodes are fixed, the slice moves to shear the PM-ER fluid.

3. Results and discussion

The rheological properties of nano-TiO_2/silicon oil ER fluid have been measured by using the experimental set-ups. The nano-TiO_2/silicon oil ER fluid is consisted of TiO_2 nano-particles, which contain polar molecules, with volume fraction about 35%. By measuring the yield stress with unilateral shearing mode under different electric fields respectively, the dependence of the yield stress on electric field can be obtained as shown in Fig.3. The values of the yield stress were taken from the curves of shear stress and displacement at different electric fields. As an example, the inset in Fig. 3 plots the relation of shear stress and shear displacement under 3 kV/mm, in which the yield stress corresponding to the maximum value is 150kPa at the displacement of 0.2mm. In Fig. 3 the dependence of the yield stress on electric field measured by using an ordinary rheometer consisted of parallel metallic electrodes is also shown for a comparison. It can be seen that the shear stress measured with unilateral shearing method is about five times of that measured with ordinary rheometer. The big reduction of the yield stress measured with ordinary rheometer indicates the slide of the PM-ER fluids arising on the metallic electrodes. Using the unilateral shearing method for the measuring the yield tress of PM-ER fluids is able to avoid such slide efficiently and obtain the intrinsic value of yield stress. The characters of the slide on metallic electrodes and the linear dependence of the measured yield stress on the electric field demonstrate that the PM-ER fluid is different from the traditional ER fluid and has been explained with a new mechanism [1].

The shear modules can be extracted from the data in inset of Fig.3. The shear modules $\kappa=\tau/\delta$, where τ is the shear stress and δ is the shear strain. The slope from start point to maximum of shear stress expresses the value of shear modules. Approximately, the shear strain δ is the ratio of the displacement and the half gap length between electrodes. It is calculated for the nano-TiO_2/silicon oil ER fluid that the shear modules κ=1.27 MPa at 3 kV/mm.

The same result can be obtained by using slice shearing mode. The variation of the yield stress measured with two methods is about 5%.

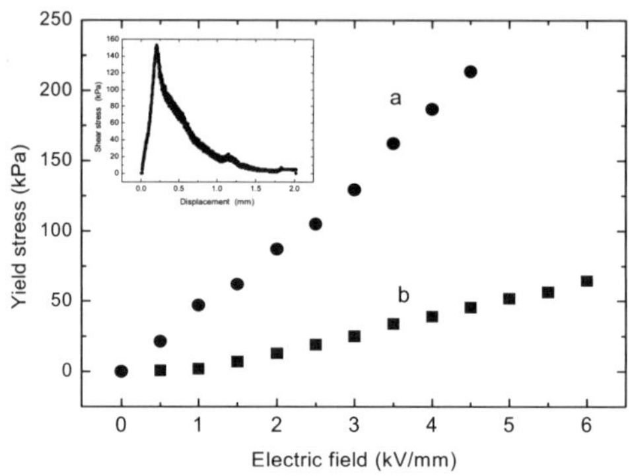

Figure 3. (a) Yield stress of nano-TiO_2/silicon oil ER fluids vs. electric field measured with unilateral shearing mode. (b) Yield stress of nano-TiO_2/silicon oil ER fluids vs. electric field measured with ordinary parallel metallic electrodes rheometer. The insert is the relation of shear stress and displacement measured using unilateral shear test under 3 kV/mm. The shearing velocity v was 0.01mm/s.

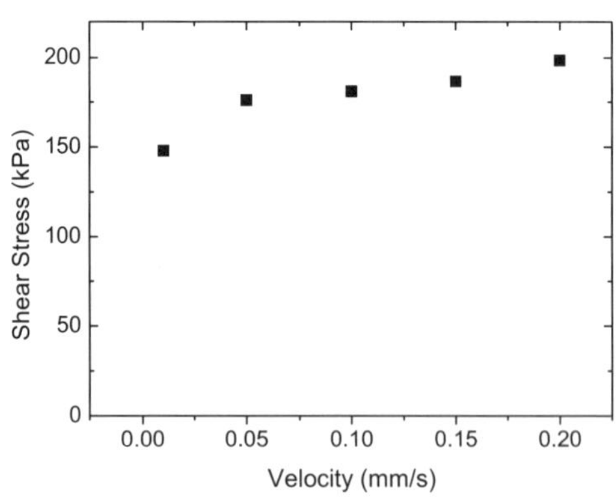

Figure 4. The shear stress vs. shearing velocity of CTO/silicon oil ER fluid measured by using unilateral shearing mode.

When the shearing velocity in the measurement increases the measured shear stress also increases as shown in Fig.4. The shear stress under higher shearing velocity is limited by data collection of our force sensor. However, in principle the dependence of shear stress on shear rate can be measured with those two shearing modes by using a high frequency data collection system.

The results obtained above clearly indicate that the shear stress of PM-ER fluid must be measured with new methods instead of using the ordinary rheometer as in the case of traditional ER fluids. Two devices with unilateral shearing mode and slice shearing mode can meet this requirement. They are able to measure the intrinsic the shear stress of PM-ER fluids and avoid the slide of the ER fluids on the electrode. In addition, both configurations are simple and easily operated in experiment.

For the applications of PM-ER fluids in practice the modified electrodes have been developed to reduce the influence of the slide on electrodes [7]. By using the modified electrodes the yield stress and the dynamic shear stress approaching the intrinsic ones of PM-ER fluids can be achieved.

4. Conclusion

Two kinds of devices, unilateral shearing mode and slice shearing mode, have been developed for measuring the rheological property of PM-ER fluids. The intrinsic characters of the fluids under electric field applied have been obtained and the interface effect of the electrodes can be removed efficiently.

5. Acknowledgements

This work was supported by the National Basic Research Program of China (No.2004CB619005)， the Knowledge Innovation Project and Outstanding Overseas Chinese Scholars Fund of Chinese Academy of Sciences and NSFC. We also thank Mr. Yongliang Zhai for his help in the experiment set-up.

References

1. K. Lu, R. Shen, X. Wang, G. Sun, W. Wen and J. Liu, *in the proceedings of this conference.*
2. K. Lu, R. Shen, X. Wang, G. Sun and W. Wen, The electrorheological fluids with high shear stress, *Intern. J. Mod. Phys. B*, 19 (2005) pp.1065-1070
3. R. Shen, X. Wang, Z. Wang, W. Wen, K. Lu, High yield stress TiO_2 based ER fluid, *Intern. J. Mod. Phys. B* , 18 (2005) pp.1104-1109
4. X. Wang, R. Shen, W. Wen, K. Lu, Calcium titanate nanopatical ER fluid, *Intern. J. Mod. Phys. B* , 18 (2005) pp.1110-1109
5. W. Wen, X. Huang, S. Yang, K. Lu, P. Sheng, The giant electrorheological effect in suspensions of nanoparticles, *Nature Material*, 2 (2003) pp.727-730
6. W. Wen, X. Huang, P. Sheng, Particle size scaling of the giant electrorheological effect, *Applied Physics Letters*, 85. (2004) pp.299-301
7. X. Wang et al, The Modified Electrodes for the applications of PM-ER Fluids, *in the proceedings of this conference*

NANOSTRUCTURED AND SURFACE POLYMERIZED MAGNETORHEOLOGICAL FLUIDS

BEN HU and ALAN FUCHS
University of Nevada, Reno, Department of Chemical Engineering, 1664 N. Virginia St, Reno, NV, 89501
Email: afuchs@unr.edu

FARAMARZ GORDANINEJAD AND CAHIT EVRENSEL
University of Nevada, Reno, Department of Mechanical Engineering, 1664 N. Virginia St, Reno, NV, 89501

A novel magnetorheological fluid, in which the surface of iron particles is coated with poly (butyl acrylate) by surface initiated atom transfer radical polymerization (ATRP), is investigated. The polymer coating procedure includes two steps, which are immobilization of initiator: 2-4(-chlorosulfonylphenyl)-ethytrichlorosilane (CTCS) on the iron particles surface and graft polymerization of butyl acrylate from the surface. The surface coating is characterized by FTIR and SEM. This magnetorheological fluid has controllable off-state viscosity and high shear yield stress. Coating polymer on the iron particles surface by ATRP can significantly reduce iron particles settling and improve stability of the MR fluid. Polymerization kinetics of bulk butyl acrylate are investigated using differential scanning calorimetry (DSC). Glass transition temperature is obtained using the step-scan DSC method. The molecular weight and conversion can be controlled by the molar ratio of monomer to initiator, reaction temperature and time. The reaction is first order determined by the plot of ln (M/M_0) against polymerization time. The overall activation energy is found to be 126kJ/mol by Kissinger's Method.

1 Introduction

Magnetorheological (MR) fluids are suspensions of magnetizable particles (e.g., iron particles) in a viscous or viscoelastic carrier fluid. They are field-controllable materials, whose rheological properties can be dramatically altered by applying a magnetic field. An MR fluid is in a free-flowing liquid state in the absence of a magnetic field. However, apparent viscosity can be increased by one or two orders of magnitude under a strong magnetic field in a very short time (milliseconds) and it exhibits solid-like characteristics.[1] The rheological behavior of MR fluids is typically described by the Bingham equation. A material with a yield stress in a magnetic field does not flow unless the applied stress exceeds the yield stress. This behavior is caused by the polarization induced in the suspended magnetic particles by application of an external field. The interaction between the resulting induced dipoles causes the particles to form columnar structure, parallel to the applied field. These chain-like structures restrict the motion of the fluid, thereby increasing the viscous behavior and causing the yield stress in the MR fluid. The change in viscosity is continuous and reversible, for example, after removing the magnetic field the MR fluid can revert to a free-flowing liquid.

A typical MR fluid contains 20-40% by volume of pure iron particles with nearly spherical shape in a non-magnetic carrier fluid. Iron particle sizes typically range form 10^{-7} to 10^{-5} m. [2]. The density of the particles is 7-8g/cm^3. The second component of a MR fluid is the carrier fluid, which serves as a continuous insulating medium. Carrier fluids are chosen based on their rheological properties and their temperature stability. Typically mineral oils, synthetic oils, water, polyethylene glycols, silicones and other fluids are used. MR fluids often contain additives to provide lubricating properties.

One of the most significant challenges in many applications is that iron particles are very dense and easily settle. One method for overcoming the problem is to disperse the particles in a fluid exhibiting a yield stress rather than in a Newtonian fluid. Some surfactant systems and greases are potential candidates. Another possibility is to add nano-size particles, such as fumed silica and organic clay, which can form a space particulate network. [3]. Both approaches can be used to prevent sedimentation. However, they significantly increase the off-state (no magnetic field) fluid viscosity and make them unusable in some applications.

Particle coating is another approach to improve the stability of particle containing fluids. By coating the particles one may stabilize the materials to prevent undesirable interactions with the environment, or alter optical, magnetic, conductive, adsorptive, dispersive, or surface properties of the dispersed particles. Methods in which fine particles are covered with polymers may involve either interaction of a preformed polymer with the inorganic cores, or direct polymerization of monomers adsorbed on the particles surfaces. Using the later method, a new generation of MRFs, known as magnetorheological polymer gels (MRPGs) used in vibration control and damping devices have been developed by Fuchs and coworkers.[4-6] These fluids contain partially crosslinked polymer gels, which are synthesized using non-stoichiometric amounts of monomers. Two MRPGs have been developed,[6] one is hydrocarbon polyol polyurethane MRPG, the other is silicone MRPG. Adjusting the ratio of resin to crosslinker and percentage of diluents the off-state rheology and the sedimentation behavior of the system can be controlled. These fluids have the advantage of controllable off-state viscosity as well as reducing the settling rate of the magnetic particles in the fluid. This behavior is possible because the polymer gel distributes itself in the carrier fluid and physically absorbs on the surface of magnetic particles, thus, preventing iron particles from approaching each other. Polymerization may take place before or after adding the magnetic particles. The latter case may result in precipitation of polymeric gel on the surface of the iron particles and improve the stability of the system.[4] However, a consideration relating to MRPG is that the interaction between polymer and iron particles' surfaces is physical absorption. Therefore the effect of inhibition of particles sedimentation and suspension stability has been limited.

There are several other approaches to prepare inorganic particles coated with polymers: grafting of polymers onto inorganic particles treated with silane coupling agents, [7, 8] surface-initiated ring-opening metathesis (ROMP) polymerization on particles, [9, 10] self-assembled mono-layer (SAM) adsorbed on particles [11, 12] and graphitic coating on particles.

A novel magnetorheological fluid, which is different from the previous generation of MRPGs, is investigated. In this system, to reduce sedimentation and agglomeration, the iron particle surface is covalently grafted with poly(butyl acrylate) via surface initiated atom transfer radical polymerization (ATRP). Grafting does not only improve the stability of the iron particles in suspension but also increases the compatibility of the particles with the carrier fluid system. Recently, controlled radical polymerization (CRP) has been successfully applied to surface-initiated graft polymerization in order to prepare a dense polymer layer with controlled structure on the surface of an inorganic substrate. [13] For the grafting process, the CRP technique is optimal, because this method affords control over the molecular weight, molecular weight distribution and structure of the resulting polymer. A major difference between conventional radical and controlled radical polymerization is the lifetime of the propagating radical during reaction. In conventional radical polymerization, radicals generated by decomposition of the initiator undergo propagation and termination reactions within a second, while the lifetime of the living radical can be extended to several hours in CRP which enables the preparation of polymers with predefined molar masses, narrow polydispersity, controlled compositions and functionality.[14] Metal catalyzed atom transfer radical polymerization (ATRP) is one important CRP method. ATRP has the following features: the polymerization can be performed at very mild conditions (room temperature), with high yield and on a broad range of monomers. The occurrence of transfer reactions (in solution) is negligible, because the radical species are always present at the end of the growing, surface tethered polymer chains. [15]

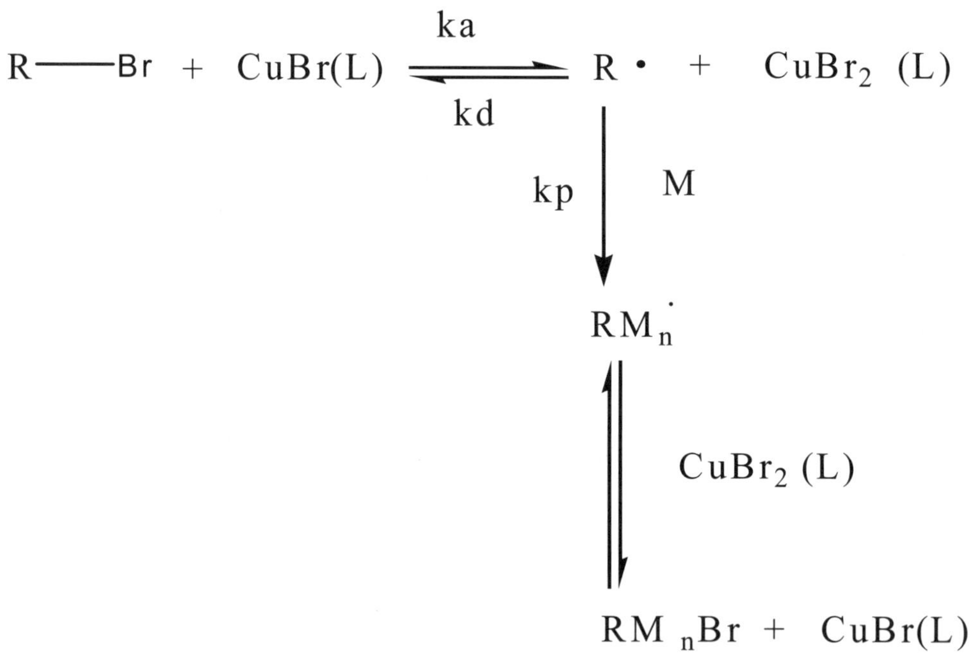

Figure 1. Schematic of mechanism of atom transfer radical polymerization.

Figure 1 describes the mechanism of ATRP, where L is a ligand that complexes with the cuprous salt and solubilizes it in the organic reaction system. k_a and k_d are rate constants for activation and deactivation of the halide initiator, with $K = k_a/k_d$. Activation of the initiator involves the CuBr metal center undergoing an electron transfer with simultaneous halogen atom abstraction and expansion of its coordination sphere. R· is the reactive radical that initiates polymerization. $CuBr_2(L)$ is called the deactivator and is the persistent radical that reduces the steady-state concentration of propagating radicals and minimizes normal termination of living polymers.

2 Experimental and Instrumentation

2.1 Materials

The MR fluid with the particles treated by ATRP includes carbonyl iron particles, carrier fluid, and polymer. Carbonyl iron particles (ISP Technologies Inc., Grade-R-2430) are 99.7% pure iron and are formed by thermal decomposition of iron pentacarbonyl $Fe(CO)_5$. The average diameter of the ferrous particles is about 3 to 4 μm. Each MR fluid sample contains 81 wt% carbonyl iron particles. The carrier fluid is an organic polar solvent typically chosen based on its viscosity, freezing and boiling points and vapor pressure. The solvent used is N-octyl-pyrrolidone (Aldrich Chemical Company, Inc.), which has a boiling point of 306°C, viscosity of 9cp at 20°C and a vapor pressure of less than 1.3Pa at 20°C. Butyl acrylate, copper bromide (CuBr, 99%), and L-(-)-sparteine (Sp) were obtained from Aldrich. Initiator 2-(4-chlorosulfonylphenyl)-ethyltrichlorosilane (CTCS), which dissolved in 50wt% toluene, was purchased from Gelest. All the reagents are used as received without further purification.

2.2. Immobilization of initiator and graft polymerization

Iron particles, 290g, with an average diameter 3-5um were dried in vacuum at 40C for 24 hours. The dried particles were then dispersed in 100g of non-polar solvent, toluene. CTCS with a chlorosulfonylphenyl group, which is an excellent initiating group for ATRP, was immobilized on the surface of iron particles by the self-assembled monolayer-deposition method. CTCS, 0.1g, in toluene solution was dripped into the iron particle, toluene suspension. The mixture reacted at room temperature and was kept in vacuum for 24 hours. The theoretically calculated initiators attached on the iron particle surface are about $3.5/nm^2$. The particles treated with initiator were then washed with tetrahydrofuran (THF) several times and dried in vacuum at room temperature for 24 hours.

The CTCS-functionalized iron particles were grafted with poly (butyl acrylate) using the ATRP approach. A degassed N-octyl-pyrrolidone (62g) solution containing 6.9g butyl acrylate, 0.04g CuBr, 0.06g Sp, 0.02g $CuBr_2$, and 290g initiator coated iron particles was sealed in a reactor. The mixture was reacted in vacuum at 70°C for 8 hours. Sp is the ligand coordinating with CuBr, providing a homogeneous ATRP system for butyl acrylate. After polymerization, the resultant suspension is the magetorheological fluid system.

3 Results and Discussion

3.1. Initiator immobilization and ATRP of butyl acrylate on iron particles.

The self-assembled monolayer deposition method to prepare iron particles carrying chlorosulfonyphenyl groups suitable for cooper complexes atom transfer radical polymerization is used to synthesize MR fluids. Chlorosilyl groups of CTCS react with hydroxyl groups on the surface of iron particles to yield a covalent bond Si-O-Fe, thus forming a monolayer of initiator on the surface. The formation of initiator monolayer on iron particles surface is confirmed by the FTIR spectrum. Figure 2 shows the FTIR spectra of pure iron particles and iron particles treated by CTCS (IP-CTCS). The spectrum of IP-CTCS has an absorption peak at 1371 cm^{-1}, which corresponds to the covalent bond S=O stretching of CTCS. The other characteristic band at 1171cm^{-1} is due to the Si-O stretch.

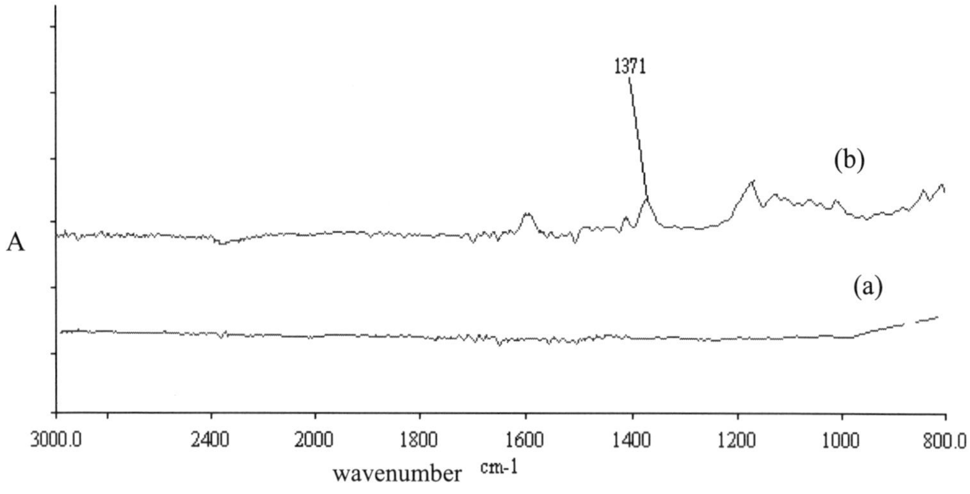

Figure 2. FTIR spectra of (a) pure iron particles and (b) iron particles coated with initiator monolayer

A fundamental concept of ATRP is the halogen exchange in the polymerizing system between the halogen-terminated growing polymer chain/ Cu(I)Br complex and macroradical /Cu(II) complex. Radicals are created by redox reaction between CTCS and CuBr/spartiene complex. In this reaction, Cu(I) is oxidized to Cu(II) by losing an electron while chloride atom in CTCS is reduced to Cl^{-1} by obtaining the electron from Cu(I). In this way, a free radical is produced.. In this free radical producing process, all the initiator decomposes at once or in a very short time period so that all propagating radicals grow for very close to the same time. Fast initiation is important, however, it is the fast equilibrium between the propagating radical and dormant species with an appropriate equilibrium constant K that determines the living characteristics of these reaction systems. The reactive radicals quickly initiate polymerization of butyl acrylate . The equilibrium constant must be low but not too low; that is, the concentration of propagating radical must be sufficient to achieve a reasonable propagation rate but not so high that normal bimolecular termination becomes predominant.

Cu(II) salt acts as a controlling or mediating agent because it is sufficiently reactive to couple rapidly with propagating chains to convert them reversibly into dormant, nonpropagating species. The overall result is that, with the introduction of the dormant state for living polymers, the bimolecular termination of living polymer is suppressed, and the average life time for living polymers is increased by at least 4 orders of magnitude. Better control over surface-initiated polymerization can be achieved by the addition of $Cu(II)Br_2$ to the monomer solution.

Comparing to the pure iron particles, there is an intense absorption peak at 1732cm-1 due to the C=O stretching of the carbonyl group of poly(butyl acrylate) in the spectrum of iron particles whose surface was treated with surface-initiated ATRP process. The observed poly(butyl acrylate) was grown from the surface of the iron particles because the polymerization could initiate only form the surface-bound CTCS and the treated iron particles were washed by THF followed by evaporation in the vacuum oven. The other characteristic absorption band is at 2955cm^{-1} is due to the C-H stretch in poly(butyl acrylate)

4 Conclusions

Surface initiated atom transfer radical polymerization (ATRP) was used to treat the surface of iron particles to stabilize the dispersion. The process includes two steps: immobilization of initiator CTCS onto iron particles surface, and atom transfer radical polymerization of poly(butyl acrylate). FTIR and SEM were used to characterize the surface of treated and untreated iron particles to confirm formation of the polymer layer coating on particles surface. The polymerization analysis and kinetics of ATRP of butyl acrylate are investigated. It was found that the molecular weight and conversion are controlled by the temperature, time, and concentration between monomer and initiator, and the overall activation energy is 126kJ/mol, with reaction order close to 1. The glass transition temperature of ATRP poly butyl acrylate is -47°C. The rheological properties of MR fluid synthesized by ATRP were investigated using a magnetic rheometer. Experiements confirmed that the rheological behavior can be described by the Bingham equation. Similar to a magnetorheological polymer gel, the off-state viscosity of ATRP MR fluid can also be controlled by the amount of polymer in the system. The stability of the MR fluid was greatly improved by surface polymerization using ATRP.

References:

1. Ginder J. M., Rheology Controlled by Magnetic Fields, Encyclopedia of Applied Physics, 1996; 16:487.
2. Carlson J. David, Jolly Mark R., MR fluid, Foam and Elastomer Devices, Mechatronics 2000; 10:555-569.
3. Klingenberg Daniel J., Magnetorheology: Applications and Challenges, AICHE Journal, February 2001; 47:246-249.
4. Wilson M., Fuchs A., Gordaninejad F., Development and characterization of magnetorheological polymer gels, J Appl Polym Sci 2002; 84:2733-2742.
5. Fuchs A., Gordaninejad, F., Blattman, D., Hamann, G., U.S pat. 2003; 6,527,972.

6. Fuchs A., Mei X., Gordaninejad, F., Xiaojie, W., Gregory H.H., Gecol H., Evrensel C., George K.; J Appl Polym Sci 2004; 92:1176-1182.
7. Yammoto M., Ohata M., New macromolecular silane coupling agents synthesized by living anionic polymerization; grafting of these polymer not inorganic particles and metals, Prog. Org Coat 1996; 27:277-285.
8. Ma M., Zhang Y., Yu W., Shen H.Y., Zhang H.Q., Gu N., Preparation and Characterization of magnetite nanoparticles coated by amino silane, Colloids and Surface A: Physicochem, Eng. Aspects, 2003; 212:219-226.
9. Marcus W., Jennifer J. Jackiw, Rossi R. R., Weiss P. S. and Grubbs R. H., Ring-opening Metathesis Polymerization from Surface, J. Am. Chem. Soc. 1999; 121:4088-4089.
10. Namyong Y. K., Jeo N. L., Choi I. S., Takami S., Harada Y., Krista R. Finnie, Gregory S. Girolami, Ralph G. Nuzzo, George M.Whitesides, and Paul E.Laibinis, Suface-Initiated Ring –Opening Metathesis Polymerization On Si/SiO2, Macromolecules 2000; 33:2793-2795.
11. Xing Y.F., O'Shea S.J., Li S.F.Y., Electron transfer kinetics across a dodecanethiol monolayer self assembled on gold, J. Electroanal. Chem. 2003; 542:7-11.
12. Huang X. H., Huang H.Z., Wu N. Z., Hu R. S., Zhu T., Liu Z. F., Investigation of structure and chemical states of self assembled Au nanoscale particles by angle-resolved X-ray photoelectron Spectroscopy, Surf Sci 2000; 459:183-190.
13. Eizo M., Shinpei Y., Tsedev N., Yoshinobu T., Takeshi F., Mikio T., Surface-initiated atom transfer radical polymerization of methyl methacrylate on magnetite nanopaticles, Polymer 2004; 45:2231-2235.
14. Pyun J., Matyjaszewski K., Synthesis of Nanocomposite Organic/Inorganic Hybrid Material Using Controlled / Living: Radical Polymerizaiton, Chem. Mater. 2001; 13:3436-3448.
15. Bontempo D., Tirelli N., Feldman K., Masci G., Crescenzi V., Hubbell J. A., Atom Transfer Radical Polymerizaiton as a Tool for Surface Functionalization, Adv. Mater. 2002; 14:1239-1241.

NONLINEAR ALTERNATING CURRENT RESPONSES IN ELECTRORHEOLOGICAL FLUIDS: DYNAMIC EFFECTS

K. W. YU

Department of Physics and Institute of Theoretical Physics, The Chinese University of Hong Kong, Shatin, New Territories, Hong Kong

W. J. TIAN AND J. P. HUANG

Department of Physcs, Fudan University, Shanghai 200433, China

kwyu@phy.cuhk.edu.hk

By using a perturbation approach, we investigate dynamic effects on nonlinear alternating current (ac) responses in electrorheological (ER) fluids under an ac or a direct current electric field. We show that the dynamic effect due to a shear flow plays a significant role in the responses. Our results can be well interpreted in the dielectric dispersion spectral representation, and they offer a convenient method to determine the relaxation time and rotation velocity of ER particles by measuring the nonlinear ac responses.

1. Introduction

An electrorheological (ER) fluid[1,2] contains polarizable particles suspended in a liquid of low dielectric constant. The rapid field-induced aggregation and the large anisotropy of ER fluids render these materials potentially important for applications. In a realistic situation, the fluid shear flow will set the particles in both translational and rotational motions[3,4,5]. Wan *et al.*[5,6] theoretically investigated such kind of rotation and pointed out that the rotation-induced displacement of the polarization surface charge reduces the interaction forces between the ER particles. Most recently, this was demonstrated in experiments[7].

When a suspension consisting of dielectric particles having nonlinear characteristics is subjected to an alternating current (ac) field with angular frequency ω, the electric response will generally consist of ac fields at frequencies of the higher-order harmonics[8,9,10,11].

In this work (see Ref. 12), we shall develop a perturbation approach[9,13] to investigate the dynamic effect on nonlinear ac responses in ER fluids under electric field.

2. Formalism

2.1 Without dispersion

We first consider a particle in the ER fluids under an applied external ac electric field $\mathrm{E}_{f0} = E_0\cos(\omega t)\hat{\mathbf{Z}} \equiv E_0\cos(\omega t)$, its induced dipole moment is given by: $\vec{p}_0 = p_0\,\hat{\mathbf{Z}}$. The shear flow in the ER suspension will give the particle a torque, which leads it to get a rotational motion about its center. We assume that the angular velocity caused by shear flow is $\omega_1\hat{\mathbf{Y}}$. The change of the dipole moment and the relaxation of various kinds have to be balanced against each other, resulting in a steady-state dipole moment $\vec{p}_0$, which deviates from the equilibrium dipole moment $\vec{p}_0$. Wan *et al.* gave the rate of change of the dipole moment in their work[5,6]:

$$\frac{d\vec{p}}{dt} = \vec{\omega}\times\vec{p} - \frac{1}{\tau}(\vec{p} - \vec{p}_0), \tag{1}$$

where τ is a relaxation time. When a steady state is reached, we obtain

$$\vec{p} = \frac{\vec{p}_0 + \tau(\vec{\omega}\times\vec{p}_0)}{1 + (\omega\tau)^2}. \tag{2}$$

And it can be shown that the angle between $\vec{p}_0$ and $\vec{p}$ is equal to ϕ and the magnitude of $\vec{p}$ is given by $|\vec{p}\,| = p_0\cos\phi$ where ϕ is given by tan $\phi = \omega\tau$ $= \omega\tau$.

Now, we try to solve the problem in another way. If we rotate with the particle, the particle will seem to be at rest, and it feels an electric field written as

$$\mathbf{E}_f = \frac{1}{2}e^{-it\omega_1}(e^{-it\omega} + e^{it\omega})\mathbf{E}_0. \tag{3}$$

While the suspended particles have a nonlinear dielectric constant, the nonlinear constructive relation between the electric displacement $\mathbf{D}_1$ and the local electric field $\mathbf{E}_1$ is defined as

$$\mathbf{D}_1 = \epsilon_1\mathbf{E}_1 + \chi|\mathbf{E}_1|^2\mathbf{E}_1 = \tilde{\epsilon}_1\mathbf{E}_1, \tag{4}$$

where □$_1$ denotes the linear dielectric constant of a suspended particle, and χ is the third-order nonlinear coefficient of the particle. It is worth noting that owing to the spherical symmetry under consideration, even-order nonlinearity disappears naturally. For convenience, we adopt the definition $|\mathbf{E}_1|^2 = \mathbf{E}_1 \cdot \mathbf{E}_1 \equiv E_1^2$. Throughout the work, only weak nonlinearity is considered. Next, the local electric field inside a particle is given by

$$\mathbf{E}_1 = \frac{3\epsilon_2}{\epsilon_1 + \chi E_1^2 + 2\epsilon_2}\mathbf{E}_f. \qquad (5)$$

Here □$_2$ represents the (linear) dielectric constant of the host medium (e.g., silicone oil). Owing to the small nonlinear coefficient χ; we can expand E_1 by taking χE_1^2 as a small perturbation. In the mean time, we shall ignore the smaller terms. Thus, we get

$$\mathbf{E}_1 = \frac{3e^{-it(\omega+\omega_1)}(1 + e^{2it\omega})\epsilon_2}{2(\epsilon_1 + 2\epsilon_2)}\mathbf{E}_0 - \frac{3e^{-it(\omega+\omega_1)}(1 + e^{2it\omega})\epsilon_2 \chi E_1^2}{2(\epsilon_1 + 2\epsilon_2)^2}\mathbf{E}_0. \qquad (6)$$

And the first term of the right part of Eq. (6) is the linear part.

On the other hand, the induced dipole moment

$$\mathbf{P} = \epsilon_2 \frac{\epsilon_1 + \chi E_1^2 - \epsilon_2}{\epsilon_1 + \chi E_1^2 + 2\epsilon_2} a^3 \mathbf{E}_f \qquad (7)$$

can be treated in the same way. In Eq. (7) a denotes the radius of the particle. Then we expand $\mathbf{P}(\omega \neq 0)$ by taking χE_1^2 as a perturbation and ignore the smaller terms. As a result, we obtain

$$\mathbf{P} = \frac{a^3 e^{-it(\omega+\omega_1)}(1 + e^{2it\omega})(\epsilon_1 - \epsilon_2)\epsilon_2}{2(\epsilon_1 + 2\epsilon_2)}\mathbf{E}_0 + \frac{3a^3 e^{-it(\omega+\omega_1)}(1 + e^{2it\omega})\epsilon_2^2 \chi E_1^2}{2(\epsilon_1 + 2\epsilon_2)^2}\mathbf{E}_0. \qquad (8)$$

Because the linear term (namely, the first term) of E_1 in the right-hand side of Eq. (6) is much larger than the second one (i.e., nonlinear part), we can introduce it as E_1 into Eq. (8). Then we obtain

$$\begin{aligned} P = {} & P_{\omega+\omega_1} e^{-it(\omega+\omega_1)} + P_{\omega-\omega_1} e^{it(\omega-\omega_1)} + P_{3\omega+3\omega_1} e^{-3it(\omega+\omega_1)} \\ & + P_{3\omega-3\omega_1} e^{3it(\omega-\omega_1)} + P_{\omega+3\omega_1} e^{-it(\omega+3\omega_1)} + P_{\omega-3\omega_1} e^{it(\omega-3\omega_1)}, \end{aligned} \qquad (9)$$

where the harmonic coefficients of various of frequencies are given by

$$
\begin{aligned}
P_{\omega\pm\omega_1} &= \frac{a^3(\epsilon_1 - \epsilon_2)\epsilon_2 E_0}{2(\epsilon_1 + 2\epsilon_2)}, \\
P_{3\omega\pm3\omega_1} &= \frac{27a^3\epsilon_2^4\chi E_0^3}{8(\epsilon_1 + 2\epsilon_2)^4}, \\
P_{\omega\pm3\omega_1} &= \frac{81a^3\epsilon_2^4\chi E_0^3}{8(\epsilon_1 + 2\epsilon_2)^4}.
\end{aligned} \tag{10}
$$

If ω = 0, we can easily sum the terms of the same frequency up. So we achieve

$$
p(\omega=0) = P_{\omega_1}e^{-it\omega_1} + P_{3\omega_1}e^{-3it\omega_1}, \tag{11}
$$

where

$$
\begin{aligned}
P_{\omega_1} &= \frac{a^3(\epsilon_1 - \epsilon_2)\epsilon_2 E_0}{\epsilon_1 + 2\epsilon_2}, \\
P_{3\omega_1} &= \frac{27a^3\epsilon_2^4\chi E_0^3}{(\epsilon_1 + 2\epsilon_2)^4}.
\end{aligned} \tag{12}
$$

Comparing Eq. (9) to Eq. (11), it is evident that harmonic signals can be detected at more frequencies due to the coupling of the frequencies ω and ω_1. Since we can detect each harmonic coefficient in experiment, it is easy to determine the angular velocity caused by shear flow with Eq. (10) as long as the external field frequency ω is known.

2.2 With dispersion

Until now, the dielectric constant □$_1$ and □$_2$ which have been used are both real numbers, that is, we have not taken into account the relaxation. To include the relaxation, one is allowed to see them as complex numbers[14.15], and they can be described as

$$
\begin{aligned}
\epsilon_1 &= \bar{\epsilon}_1 - i\bar{\sigma}_1/\omega_0, \\
\epsilon_2 &= \bar{\epsilon}_2 - i\bar{\sigma}_2/\omega_0,
\end{aligned} \tag{13}
$$

where both $\bar{\epsilon}$ and $\bar{\sigma}$ are real numbers, and they represent the real dielectric constant and the conductivity of the particle, respectively. In Eq. (13), ω_0 denotes the angular frequency whose details can be found below. In this case, the induced dipole moment should be[16]

$$\mathbf{P} = \bar{\epsilon}_2 \frac{\epsilon_1 + \chi E_1^2 - \epsilon_2}{\epsilon_1 + \chi E_1^2 + 2\epsilon_2} a^3 \mathbf{E}_f. \tag{14}$$

Because of $\overline{\sigma}_2 \ll \overline{\sigma}_1$ in real situation, we are allowed to see $\overline{\sigma}_2 = 0$ (i.e., $\epsilon_2 = \bar{\epsilon}_2$) in the following derivation as well as the numerical calculations presented in Section 3.2. And here, in the third-order harmonic terms, according to the mixing theory[17], the high-order harmonic output is combined with one input frequency and two mixing frequencies, by using of basic angular frequencies $(\omega+\omega_1)$ and $(\omega-\omega_1)$. Specifically, we can see $3\omega\pm3\omega_1 = 3(\omega\pm\omega_1)$ and $\omega \pm 3\omega_1 = 2(\omega \pm \omega_1) - (\omega \mp \omega_1)$. As a result, the angular velocityω_0 in Eq. (13) is replaced by harmonic angular velocities in the same way. So we can get the harmonic terms of $P(\omega\neq0)$. If $\omega = 0$, based on Eq. (12) we obtain

$$P_{\omega_1} = \frac{a^3(\bar{\epsilon}_1 - \bar{\epsilon}_2 - \frac{i\bar{\sigma}_1}{\omega_1})\bar{\epsilon}_2 E_0}{\bar{\epsilon}_1 - \frac{i\bar{\sigma}_1}{\omega_1} + 2\bar{\epsilon}_2},$$

$$P_{3\omega_1} = \frac{27a^3\bar{\epsilon}_2^4\chi E_0^3}{[\bar{\epsilon}_1 - \frac{i\bar{\sigma}_1}{\omega_1} + 2\bar{\epsilon}_2]^3[\bar{\epsilon}_1 - \frac{i\bar{\sigma}_1}{3\omega_1} + 2\bar{\epsilon}_2]}. \tag{15,16}$$

It is worth noting that in the actual use, only the real parts of the harmonic terms of induced dipole moment can be detected in experiments. Thus, we shall compute these real parts only in the following numerical calculations.

3. Numerical results

3.1 Without dispersion

We investigate the cases without dispersion. For an ac electric field with angular frequency ω and a rotating ER particle with ω_1, we have numerically calculated the harmonic responses of induced electric fields. (Figures have not been displayed here.) If dielectric constants $\square_1$ and $\square_2$ are given, higher nonlinear characteristics yield stronger harmonic responses. In other words, the magnitude of harmonic responses reflects the strength of nonlinear characteristics, as already experimentally reported by Klingenberg[8].

3.2 *With dispersion*

We calculate the harmonic responses of induced dipole moment versus the angular frequency ω_1 of the rotating particle, for the external electric field with different frequencies ω. (Figures have not been displayed here.) For $P_{\omega+\omega 1}$, $P_{3\omega+3\omega 1}$, and $P_{\omega+3\omega 1}$, increasing ω_1 yields decreasing $P_{\omega+\omega 1}$, but increasing $P_{3\omega+3\omega 1}$ and $P_{\omega+3\omega 1}$. Inverse results appear for $P_{\omega-\omega 1}$, $P_{3\omega-3\omega 1}$, and $P_{\omega-3\omega 1}$, due to the subtraction of the particle rotation frequency from the external field frequency. In case of a dc electric field, we can investigate the effect of !1 on the harmonic responses, see Fig. 1.

By analyzing Eq. (16), we find that, when

$$\omega_1 = 1.124 \frac{\bar{\sigma}_1}{\bar{\epsilon}_1 + 2\bar{\epsilon}_2}, \tag{17}$$

a maximum value should appear in Re($P_{3\omega 1}$), as displayed in Figs. 1(b) and 2 indeed. Here Re(...) means taking the real part of (...). In fact, this value could appear at the frequency $\omega_1 = \omega_1^*$ that satisfies the following known relation[18]

$$\tau\omega_1^* = 1. \tag{18}$$

According to Eq. (17), we obtain

$$\tau = \frac{\bar{\epsilon}_1 + 2\bar{\epsilon}_2}{1.124\bar{\sigma}_1} = \frac{\tau_{MW}}{1.124}. \tag{19}$$

Here the relaxation time τ is determined by the details of the relaxation process, and the Maxwell-Wagner relaxation time τ_{MW} is based on the Maxwell-Wagner theory of leaky dielectrics since the relaxation process is originated from a finite conductivity of the particle and host medium[18]. In Table 1 the τ was calculated according to Eq. (19), and ω_1^*, which was extracted from the curves of Fig. 2, is the frequency at which Re($P_{3\omega 1}$) reaches a maximum.

Fig. 1(b) shows that the nonlinear characteristic does not affect ω_1^*. Actually, ω_1^* can be

Table 1. List of τ and ω_1^*. Here τ was calculated according to Eq. (19) andω_1^* was extracted from the curves of Fig. 2. The two parameters satisfy $\tau\omega_1^* = 1$ [Eq. (18)].

	τ (s)	ω_1^* (rad/s)
$\bar{\sigma}_1 = 2\times 10^{-8.2}$ S/m	0.0156	64.1
2×10^{-8} S/m	0.0099	101.7
$2\times 10^{-7.8}$ S/m	0.0062	161.1
$\bar{\epsilon}_1 =$ 15□0	0.0078	127.1
20□0	0.0099	101.7
25□0	0.0118	84.7
$\bar{\epsilon}_2 =$ 2.2□0	0.0096	104.2
2.5□0	0.0099	101.7
2.8□0	0.0101	99.3

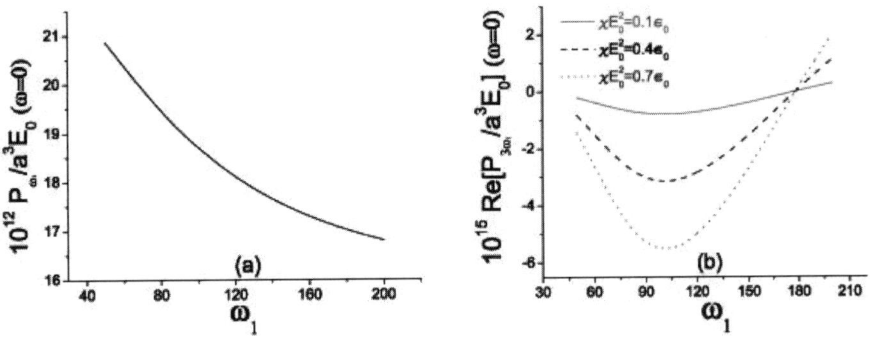

Figure 2. (Cited from Ref. 12) *With dispersion.* For an dc electric field ($\omega = 0$), the real part of harmonic terms of induced dipole moment P as a function of the angular velocity ω_1 of suspended rotating particles. And (b) for different nonlinear characteristicsχE_0^2. Parameters: □1 = 20□0, □2 = 2.54□0, $\sigma_1 = 2\times10^{-8}$S/m, $\sigma_2 = 1.0\times10^{-13}$S/m, and $\chi E_0^2 = 0.7\epsilon_0$.

affected by the linear responses of the particles only, as shown in Eq. (18) already. Thus, we investigate the effect of $\bar{\sigma}_1$, $\bar{\epsilon}_1$, and $\bar{\epsilon}_2$ on the peak, see Fig. 2. It is shown that all the quantities have an effect on the ω_1^*. Especially, decreasing $\bar{\sigma}_1$ or increasing $\bar{\epsilon}_1$ or $\bar{\epsilon}_2$ makes theω_1^* red-shifted (namely, located at lower frequencies). Consequently, in experiments, based on the observed peak frequencyω_1^*, we can conveniently determine the relaxation time $\tau(= 1/\omega_1^*)$. That is, our results offer an effective way to determine the relaxation time, which seems up to now a challenge. In addition, since we can detect each harmonic response at different ω_1 in experiment, the corresponding angular velocity $\omega 1$

of particles caused by shear flow is obtained directly.

Klingenberg experimentally studied the nonlinear ac responses (harmonics) of electrorheological fluids[8]. Very interestingly, he experimentally showed that the harmonics of the electric current is caused to increase while the external electric field increases, which is qualitatively in agreement with our numerical results obtained from the local electric field and the induced dipole moment.

4. Discussion and conclusion

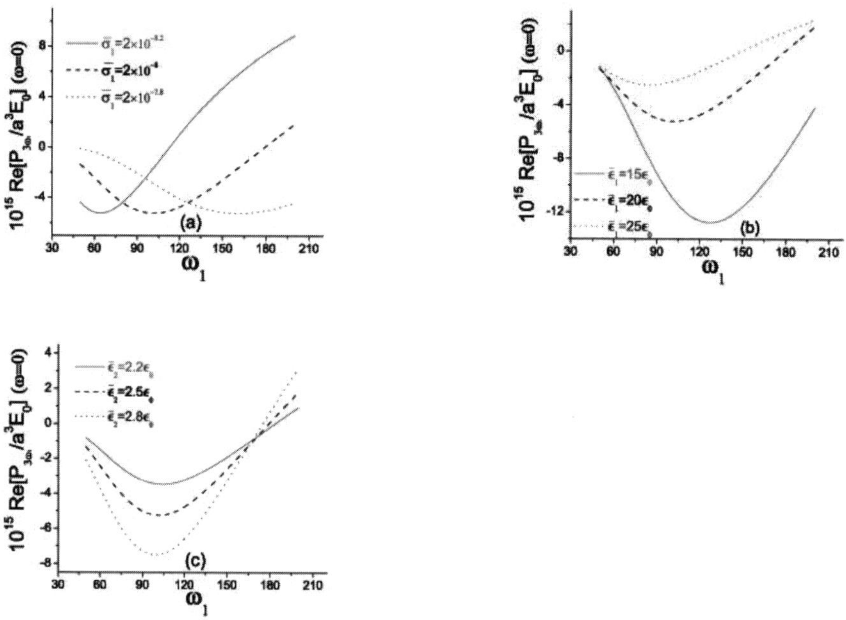

Figure 2. (Cited from Ref. 12) *With dispersion.* Same as Fig. 1, but for different (a) $\bar{\sigma}_1$ (b) $\bar{\epsilon}_1$, and (c) $\bar{\epsilon}_2$. Parameters: $\epsilon_1 = 20\epsilon_0$, $\epsilon_2 = 2.54\epsilon_0$, $\sigma_1 = 2\times10^{-8}$S/m, $\sigma_2 = 1.0\times10^{-13}$S/m, and $\chi E_0^2 = 0.7\epsilon_0$, if they are not the variables.

Throughout this work, only lower-order harmonics have been discussed. In fact, higher-order harmonics can also be studied as long as we keep more terms in Eqs. (6) and (8). However, the strength of the higher-order harmonics (e.g., fifth-order) is often several orders of magnitude smaller than the lower-order. To some extent, it is more attractive to detect the lower-order harmonics.

To sum up, we find that the dynamic effect due to a shear flow plays an important role in the nonlinear ac responses in ER fluids. Our results offer a convenient method to determine the relaxation time and the rotation velocity of ER particles, by measuring the nonlinear ac responses.

5. Acknowledgments

K.W.Y. acknowledges financial support from the RGC Earmarked Grant of the Hong Kong SAR Government. Two of us (W.J.T. and J.P.H.) acknowledge the financial support by the Shanghai Education Committee and the Shanghai Education Development Foundation ("Shu Guang" project) under Grant No. KBH1512203, by the National Natural Science Foundation of China under Grant Nos. 10321003, 1024402 and 10574027, and by Jiangsu Key Laboratory of Thin Films, Suzhou University, China.

References

1. T. C. Halsey, Science **258**, 761 (1992).
2. W. Wen, X. Huang, S. Yang, K. Lu, and P. Sheng, Nature Materials **2**, 727 (2003).
3. D. J. Klingenberg, F. van Swol, and C. F. Zukoski, J. Chem. Phys. 9461601991.
4. L. Lobry and E. Lemaire, J. Electrost. **47**, 61 (1999).
5. J. T. K. Wan, K. W. Yu, and G. Q. Gu, Phys. Rev. E **62**, 6846 (2000).
6. J. T. K. Wan, K. W. Yu, and G. Q. Gu, Phys. Rev. E **64**, 061501 (2001).
7. R. Tao and Y. C. Lan, Phys. Rev. E **72**, 041508 (2005).
8. D. J. Klingenberg, MRS Bull. **23**, 30 (1998).
9. G. Q. Gu, P. M. Hui, and K. W. Yu, Physica B **279**, 62 (2000).
10. J. P. Huang, J. T. K. Wan, C. K. Lo, and K. W. Yu, Phys. Rev. E **64**, 061505(R) (2001).
11. P. M. Hui, C. Xu, and D. Stroud, Phys. Rev. B **69**, 014203 (2004).
12. W. J. Tian, J. P. Huang, and K. W. Yu, Phys. Rev. E **73**, 031408 (2006).
13. G. Q. Gu and K. W. Yu, Phys. Rev. B **46**, 4502 (1992).
14. J. P. Huang, Chem. Phys. Lett. **390**, 380 (2004).
15. K. Kim, D. Stroud, X. Li, and D. J. Bergman, Phys. Rev. E **71**, 031503 (2005).

16. T. B. Jones, Electromechanics of Particles (Cambridge University Press, Cambridge, 1995).
17. Y. R. Shen, The Principles of Nonlinear Optics (J. Wiley and Sons, New York, 1984).
18. W. B. Russel, D. A. Saville, and W. R. Schowalter, Colloidal Dispersions (Cambridge University Press, Cambridge, England, 1989).
19. H. Ma, W. Wen, W. Y. Tam, and P. Sheng, Phys. Rev. Lett. **77**, 2499 (1996).

INVESTIGATION OF MAGNETEREHOLOGICAL PROPERTIES OF A MIXTURE OF CARBONYL IRON POWDERS SUSPENDED IN IONIC LIQUID

G. DODBIBA*, H.S. PARK, K. OKAYA, T. FUJITA

Department of Geosystem Engineering, Graduate School of Engineering,
The University of Tokyo, 7-3-1, Bunkyo-ku, Hongo, Tokyo, 113-8656, Japan
**E-mail: dodbiba@geosys.t.u-tokyo.ac.jp*

Properties of a magnetorheological (MR) fluid synthesized by dispersing a mixture of two carbonyl iron powders (CIP) in an ionic liquid have been investigated. At first, mixtures of CM (d_{50} = 7 μm) and HQ (d_{50} = 1.1 μm) CIP powders, prepared at the same solid weight fraction but at various weight fractions of large particles, were dispersed in the ionic liquid (N,N-Diethyl-N-methyl-N-(2-methoxyethyl) ammonium tetrafluoroborate), which is stable even at high temperature. Then, the magnetic clusters of the synthesized MR fluids were observed by using an optical microscope, whereas the magnetorheological properties were investigated by using a bi-cylindrical viscometer. Each apparatus was equipped with a magnetic field generator to create a uniform magnetic field. After finding the most suitable mixing ratio of powders, a new batch of MR fluid was synthesized by the addition of surfactant (ETC-7) to enhance its stability. Finally, the properties of the MR fluid with or without addition of surfactant were compared. The experimental results showed that the MR fluid with 60 wt% fraction of large particles exhibited the highest MR response.

1 Introduction

Generally speaking, magnetorheological (MR) fluids are dispersions of (1 to 10 μm) ferromagnetic particles in an organic or aqueous carrier liquid [1]. They must posses certain characteristics including stability over a broad temperature range, stability against settling, chemical stability, etc. [2-5]. Furthermore, MR fluids should exhibit a large field-inducted yield stress with small apparent viscosities in absent of an applied magnetic field, [6]. In this frame, the type of the carrier liquid and the effect of particle size of magnetic particles on rheology of MR fluid are significant and much attention has been given to this subject [7-14].

In an attempt to improve the performance of MR fluid the authors have investigated the properties of a MR fluid synthesized by dispersing a mixture of two carbonyl iron powders (i.e. BASF: soft grade CM and hard grade HQ) in an ionic liquid (N,N-Diethyl-N-methyl-N-(2-methoxyethyl) ammonium tetrafluoroborate). The primary reason for selecting an ionic fluid as carrier liquid is that they have a relatively low melting point, are nonvolatile, and do not evaporate [15]. Hence ionic liquid may be used in high-vacuum systems and eliminate containment problems.

At first, the random packing density of binary mixtures of spherical iron particles obeying a uniform distribution was calculated by using discrete element method (DEM), [16]. Then, mixtures of CM (d_{50} = 7μm) and HQ (d_{50} = 1.1μm) CIP powders, prepared at the same solid weight fraction but at various weight fraction of large particles, were dispersed in the ionic liquid. In this work, the magnetic clusters of the synthesized MR fluids were observed by using an optical microscope, whereas the magnetorheological properties were investigated by using a bi-cylindrical viscometer. Each apparatus was equipped with a magnetic field generator to create a uniform magnetic field. After finding the most suitable mixing ratio of powders, a new batch of MR fluid was synthesized by adding also a surfactant (ETC-7) to enhance its stability. Finally, the properties of the MR fluid with or without addition of surfactant have been compared.

2 Experimental

2.1 *Materials*

In this work an ionic liquid, namely N,N-Diethyl-N-methyl-N-(2-methoxyethyl) ammonium tetrafluoroborate, $C_8H_{20}ONBF_4$ (FW: 233.1), a commercial product of Kanto Chemical, Japan was used as carrier liquid. Figure 1 shows the structure of $C_8H_{20}ONBF_4$ (melting point: 9 °C; and specific gravity: 1.17), whereas Figure 2, on the other hand, shows the viscosity of $C_8H_{20}ONBF_4$ as a function of temperature, [17].

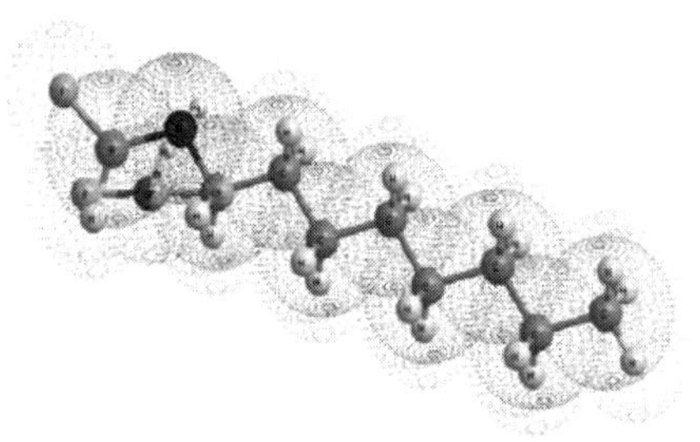

Figure 1. Structure of N,N-Diethyl-N-methyl-N-(2-methoxyethyl) ammonium tetrafluoroborate, $C_8H_{20}ONBF_4$

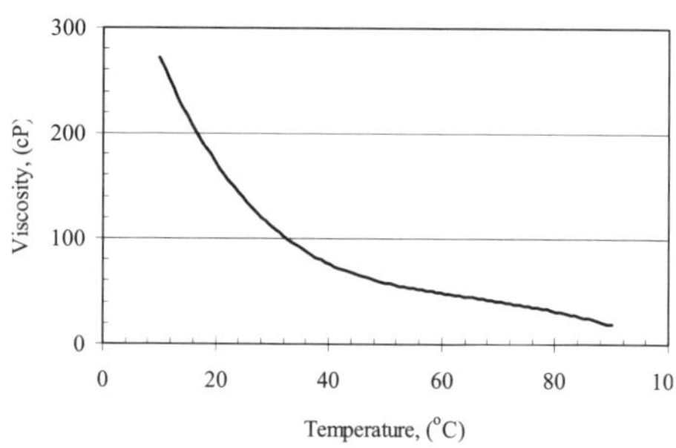

Figure 2. Viscosity of $C_8H_{20}ONBF_4$ as a function of temperature, [17].

Table 1. Composition of the MR fluids synthesized for the investigation

MR fluids; Sample No.	Magnetic particles					Ionic Liquid	Mass of MR fluid	Solid fraction
	soft grade CM (fraction of large Fe particles)		hard grade HQ (fraction of small Fe particles)		solid weight fraction			
	(gr)	(wt%)	(gr)	(wt%)	(gr)	(gr)	(gr)	(wt%)
1	100	100	0	0	100	24	124	80.6
2	80	80	20	20	100	24	124	80.6
3	70	70	30	30	100	24	124	80.6
4	60	60	40	40	100	24	124	80.6
5	50	50	50	50	100	24	124	80.6
6	20	20	80	80	100	24	124	80.6
7	0	0	100	100	100	24	124	80.6

Carbonyl iron powder (CIP), used throughout this work, were commercial products of BASF Corporation. Two types of CIP were selected for investigation: (1) CIP - soft grade CM (hereafter named also as fraction of large Fe particles), d_{50} = 7μm; and (2) CIP - hard grade HQ (hereafter named also as fraction of small Fe particles), d_{50} = 1.1μm. The mean diameter (d_{50}) ratio (λ) of large to small particles is ca. λ = 6.4.

Mixtures of CM and HQ powders, prepared at the same solid weight fraction ($\Phi = 80.6\ \text{wt}\%$) but at various weight fractions of large particles, were dispersed in the ionic liquid. Table 1 tabulates the composition of the MR fluids synthesized for the investigation.

Several batches of MR fluid were synthesized also by the addition of 0.25 wt% (i.e. 0.31 g) surfactant, trideceth-8 carboxylic acid (also know as ECT-7). It is important to note that when the surfactant was used in preparing the MR fluid, the amount of ionic liquid was reduced to 23.69 g in order to keep the solid fraction unchanged (i.e. $\Phi = 80.6\ \text{wt\%}$).

2.2 Measurements

The width of magnetic clusters of the synthesized fluids (see Table 1) was examined by using a digital microscope (Keyence VXH-100), equipped with an 18 million-pixel CCD camera. A set of two permanent magnets were also fixed to the stand of the microscope to create a uniform magnetic field.

Moreover, the magnetorheological properties of the synthesized fluids have been investigated by using a bi-cylindrical viscometer. The radius of inner cylinder and the radius of outer cylinder of viscometer were 1.35 and 1.50 cm respectively. The outer cylinder was filled with MR fluid at a depth of 7.5 cm. The viscometer was also equipped with a magnetic field generator to create a uniform magnetic field. The measurements were carried at room temperature.

3 Results and Discussions

3.1 DEM Simulation of Close Packing Density

The random packing density of binary mixtures of spherical iron particles obeying a uniform distribution was calculated after calculating the random packing, i.e. the volume filled by the particles. The volume filled by solid particles was calculated by using discrete element method (DEM), [16]. The mean diameters of the large and small components of the mixtures were 7 and 1.1 μm respectively. It should be noted that the random packing of the mixture was calculated for various volumes of large particles, while keeping constant the total number of particles and the diameter ratio. For random packing, the algorithm is composed of: (1) generation of random initial position of particles within a defined domain; (2) relocating particles to eliminate overlaps; and (3) "free settling" of particles.

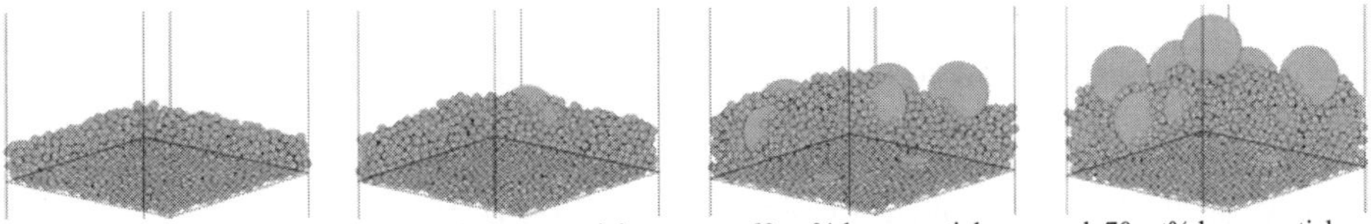

a. 0 wt% large particles b. 40 wt% large particles c. 60 wt% large particles d. 70 wt% large particles

Figure 3. Three-dimensional views of packing of a mixture of particles at various weight fractions of large particles. Total number of particles 3000. Diameter of large particles: 7 μm; Diameter of small particles: 1.1 μm.

Figure 3 shows three-dimensional views of random packing of a mixture at various weight fractions of large particles, whereas Figure 4 gives the numerical results of the simulation. Figure 4 shows that the packing density increases with increasing the fraction of large particles and approaches a maximum at ca. 75 wt% before starting to decrease with further increase of fraction of large particles (Fig. 4).

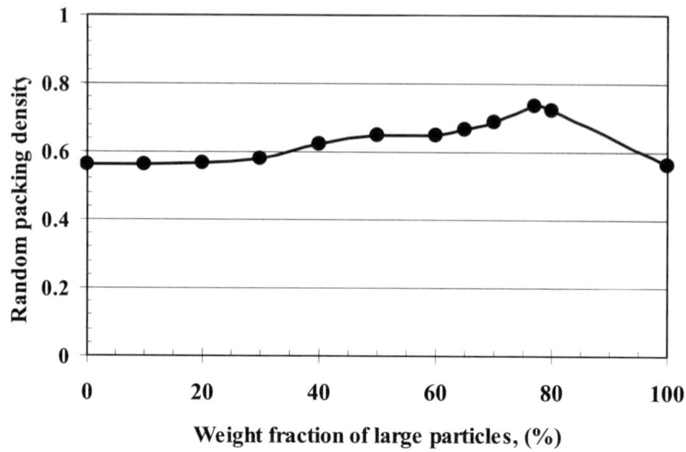

Figure 4. Random packing density as a function of weight fraction of large particles. Number of particles 3000.

3.2 *Observation of Magnetic Clusters: Physical Characterization*

The structure evolution of magnetic clusters was observed for successively increasing the weight fraction of large particles. The observation was carried out by employing a digital microscope, equipped with a CCD camera and two permanent magnets. The digital images were then processed by using an image processing software. The software (ImageJ Version 1.34s) was used to measure the width of magnetic clusters.

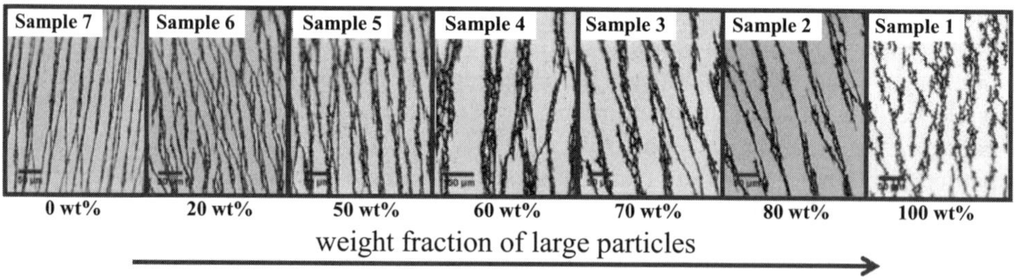

Figure 5. Structure evolution of magnetic clusters with increasing the weight fraction of large particles, at the magnetic flux density of 0.08 T. Observed by digital microscopy; magnification 600 x.

Figure 5 illustrates the structure evolution in ionic liquid, at the magnetic flux density of 0.08 T. Moreover, Figures 6 and 7 show the numerical results of the observation, i.e. the width of magnetic clusters as a function of magnetic flux density and weight fraction of large particles. As one can expect, the width of magnetic clusters increases with increasing magnetic flux density (Fig. 6). Figure 7, on the other hand, shows that the width of magnetic clusters increases not only with increasing magnetic flux density but also with increasing the weight fraction of large particles until reaching the peak at ca. 60 wt%. The same tendency was observed at all applied values of magnetic flux density. It should be noted that the peak in Figure 7 correspond to Sample 4 (see Table 1; Fig. 5).

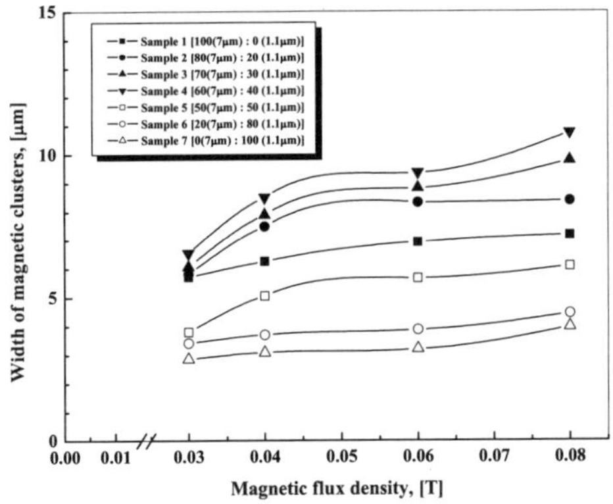

Figure 6. Width of magnetic clusters as a function of magnetic flux density.

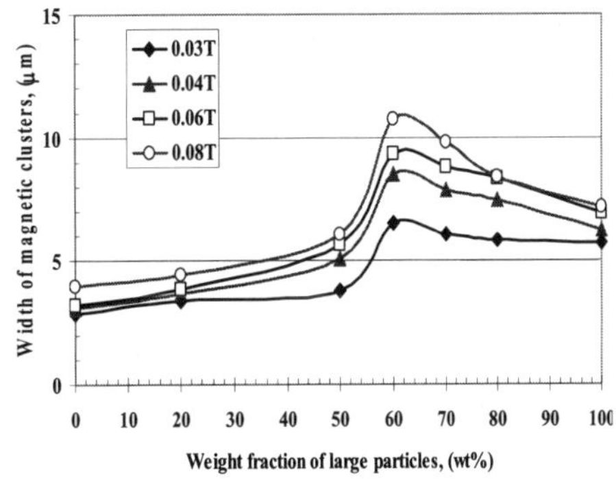

Figure 7. Width of magnetic clusters as a function of weight fraction of large particles

3.3 *The shear rate - shear stress curves*

Figures 8 shows the shear stress (τ) as a function shear rate (γ') for all samples (see Table 1) in absent of magnetic field (a), and at magnetic flux density of 0.08 T (b). It can be seen that all samples behave as Bingham Fluids. It can also be seen that the MR fluid prepared by dispersing only large magnetic particles (i.e. Sample 1; see Table 1) showed the highest shear stresses in absent of magnetic field (Fig. 8a); whereas the sample with 60 wt% weight fraction of large particles (i.e. Sample 4) showed the highest shear stresses when applying the magnetic field (Fig. 8b).

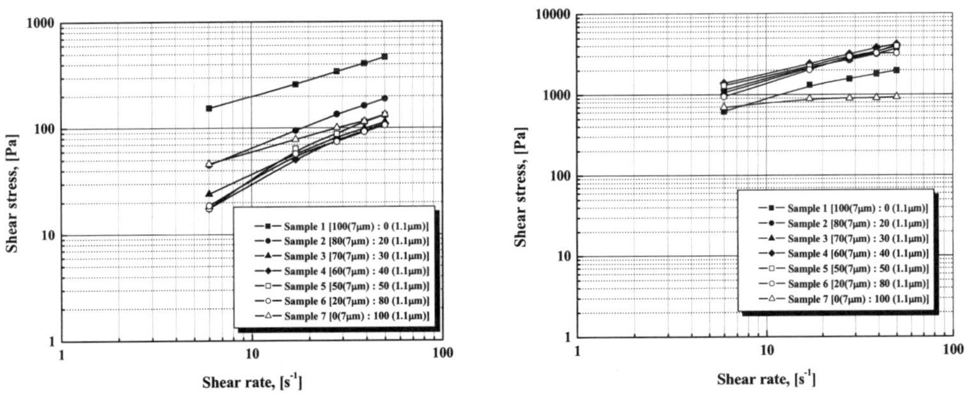

a. Magnetic flux density: 0 T

b. Magnetic flux density: 0.08 T

Figure 8. Variation of shear stress (Pa) with shear rate (s^{-1}) for different samples Magnetic flux density: (a) 0 T and (b) 0.08 T.

Figure 9 shows weight fraction of large particle as a function of the magnetorheological response ($\Delta\tau$) at various magnetic flux densities and shear rates. The quantity of $\Delta\tau$ is defined as the increase in the shear stress (τ) at a given magnetic flux density (B) from the one at $B = 0$ T, [6], thus:

$$\Delta\tau = \tau_{|B} - \tau_{|B=0T} \quad (1)$$

It can be seen that when the weight fraction of large particles is ca. 60 wt% the magnetorheological response reach its maximum (Fig. 9). The same tendency was observed for all values of magnetic flux density or shear rate applied. The fact that the suspension with 60 wt% weight fraction of large particles (Sample 4; see Table 1) exhibits the highest magnetorheological response can be explained with it high value of packing density. These results are in good agreement also with the simulation results, which suggested that packing density strongly depends upon the weight fraction of large particles.

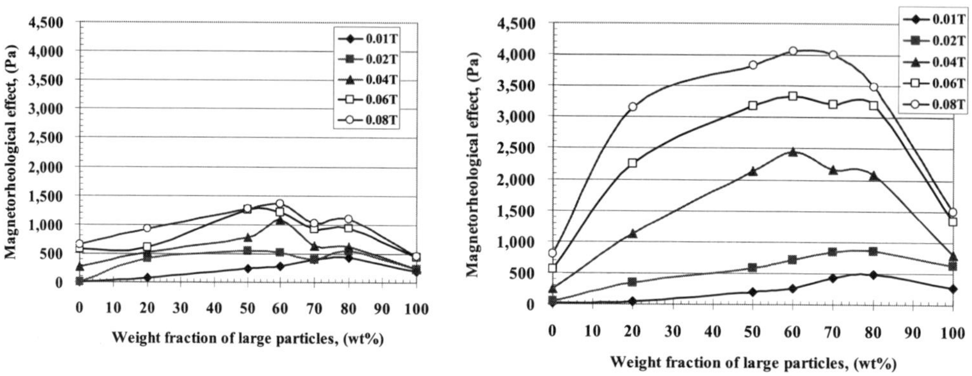

a. Shear rate: 5 s^{-1} b. Shear rate: 45 s^{-1}

Figure 9. Weight fraction of large particle as a function of the magnetorheological response ($\Delta\tau$) under various magnetic flux densities: (a) shear rates 5 s^{-1}; (b) shear rate: 45 s^{-1}.

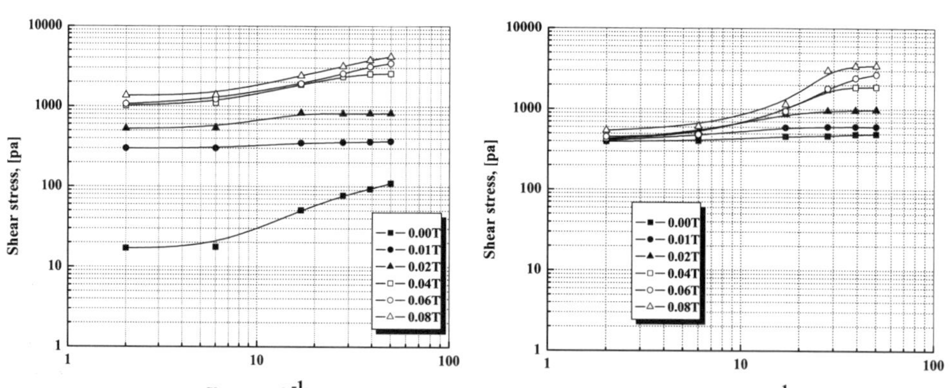

a. No surfactant b. With surfactant (ETC-7: 0.25 wt%)

Figure 10. Shear stress as a function of shear rate for samples with 60 wt% weight fraction of large particles (i.e. Sample 4) at various magnetic flux densities. The MR fluids were synthesized with (b) or without (a) the addition of 0.25 wt% surfactant (ETC-7).

Shear stress as a function of shear rate for the sample with the most suitable mixing ratio, i.e. 60 wt% weight fraction of large particles (i.e. Sample 4), are shown in Figure 10. The curves are drawn for several values of magnetic flux densities. Figure 10 also compares the properties of MR fluids synthesized with or without the addition of surfactant. When adding surfactant, the suspensions exhibited relatively low yield stress, when compared with those without surfactant, especially at high value of magnetic flux density. Moreover, when adding surfactant to the suspension, the overall stability of the MR fluid and its magnetorheological properties were improved remarkably.

4 Conclusion

In this paper properties of a magnetorheological (MR) fluid synthesized by dispersing a mixture of two carbonyl iron powders (CIP) in an ionic liquid that is stable even at elevated temperature, have been investigated. The simulation and experimental results pointed out that when the weight fraction of large particles is ca. 60 wt% the magnetorheological response reach its maximum. Moreover, it was also demonstrated that the synthesized MR fluid with a suitable mixing ratio can easily change its shear stress as a function of shear rate and the magnetic flux density, especially with addition of surfactant.

References

1. P. P. Phule, Smart Material Bulletin, 2001/2 (2001), pp. 7-10
2. J. D. Carlson, D. M. Catenzarite, K.A.St. Clair, Int. J. Modern Phys. B, 10 (1996) pp. 2857-2865.
3. G. Bossis, P. Khuzir, S.Lacis, O. Volkova, J. Magn. Magn. Mater., 456 (2003) pp. 258-259.
4. M. S. Cho, S. T. Lim, I. B. Jang, H. J. Choi, IEEE Trans. Magn. 40/4 (2004) pp. 3036-3038.
5. B. J. Park, I. B. Jamg, H. J. Choi, A. Pich, S. Bhattacharya, H. J. Adler, J. Magnetism and Magnetic Materials, 303 (2006) pp. 290-293.
6. P. J. Rankin, A. T. Horvath, D. J. Klingenberg, Rheol. Acta. 38/5 (1999) pp. 471-477.
7. Y. H. Shih and H. Conrad, Int. J. Mod. Phys. B, 8/20-21 (1994) pp 2835-2853.
8. Y. Asako, S. Ono, R. Aizawa, and T. Kawakami, Proceedings of the 5th International Conference on Electrorheological Fluids, Magnetorheological Suspensions and Associated Technology, edited by W. A. Bullough World Scientific, Singapore, 1996, p.342.
9. G. Bossis, S. Lacis, A. Meunier and O. Volkova, J. of Magnetism and Magnetic Materials, 252 (2002) pp. 224-228.
10. C.W. Wu and H. Conrad, J. Appl. Physics, 83, 7 (1998) pp. 3880-3884.
11. H. See, A. Kawai, F. Ikazaki. Rheol Acta, 41 (2002) pp 55 - 60.
12. D. He, N. N. Ekere, and L. Cai. Physical Review E, 60/6 (1999) pp. 7098-7104.
13. D. He, and N.N. Ekere, Rheol. Acta, 40 (2001) pp. 591-598.
14. A. J. F. Bombard, M. R. Alcantara, M. Knobel, P. L. O. Volpe, J. Modern Physics B, 19 (2005) pp. 1332-1338.
15. T. Welton, Chem. Rev., 99 (1999) pp. 2071-2083.
16. T. Inue, K. Okaya, Inter. J. of Mineral Processing, 44-45 (1996) pp. 425-435.

17. Kanto Chemicals, Ionic Liquids, Available from internet: http://www.kanto.co.jp/siyaku/ion.html (in Japanese).

INVESTIGATING THE TIME DEPENDENCE OF THE MR EFFECT

FERNANDO D. GONCALVES AND J. DAVID CARLSON
LORD Corporation, Global Research & Development, 406 Gregson Drive, Cary, NC 27511, USA
email: fernando.goncalves@lord.com, jdcarlson@lord.com

Magnetorheological fluids are known to respond in a matter of milliseconds to the application of a magnetic field. To date, however, very little work has been done to study the time dependence of the MR response. The purpose of this study is to investigate the response time of the fluid. Experiments were conducted on a high shear rate rheometer capable of fluid speeds in excess of 35 m/s. With an MR valve length of 6.35 mm, the resulting dwell times were as low as 0.18 ms. For each of three magnetic field strengths, a reduction in yield stress is observed as dwell time decreases. A model is proposed to represent the time response of the fluid to the application of the magnetic field. The experimental data and the proposed model are used to identify the response time of the fluid for each field strength. Results indicate that as the magnetic field increases, the response time of the MR fluid decreases. For the range of magnetic field strengths considered in this study the response time of the fluid ranged from 0.24 ms to 0.19 ms.

1 Introduction

Perhaps the most compelling feature of MR fluids is their ability to provide high dynamic yield stresses nearly instantaneously. While the response time of the fluid has yet to be thoroughly investigated, it is known that the fluid can respond to changes in magnetic field strength in a matter of milliseconds. In a previous work by the authors, however, it has been shown that the MR fluid may only achieve a fraction of the expected MR effect if the dwell time is not sufficiently large [1]. Dwell time is defined as the amount of time the fluid spends in the presence of a magnetic field. Consider the flow schematic shown in **Figure 1**, illustrating MR fluid flow through an MR valve. In the absence of a magnetic field the fluid exhibits Newtonian-like behavior; particle orientation is random and the velocity profile is parabolic. When the fluid reaches Section I, it is exposed to a magnetic field and the transition from Newtonian-like behavior to Bingham plastic behavior begins. When the fluid reaches Section II the particles have begun to align along the flux path. At this stage the yield stress is not yet fully developed, perhaps only reaching 70% of full potential. Only once the fluid reaches Section III has the yield stress reached 100% of the expected value. In the schematic, the dwell time exceeds the response time and the fluid can achieve its full potential. However, as the fluid velocity increases, the dwell time decreases. Should this dwell time decrease below the time required to achieve the full yield stress, the fluid may only achieve a portion of the expected yield stress. For example, if the active valve length were only half of that shown in **Figure 1**, the fluid would reach Section II and exit the valve only achieving 70% of full potential.

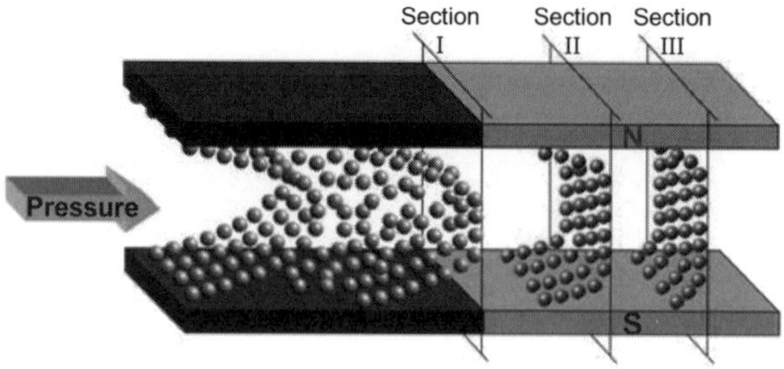

Figure 1. MR valve flow schematic illustrating fluid response to a magnetic field

The focus of this study is to examine the degree of response achieved by the fluid as a function of dwell time. The time response of the fluid is investigated and a model is proposed to characterize the response. From the proposed model, the response time of the fluid is found for three magnetic field strengths.

2 Approach

Figure 2 shows the high shear rate rheometer built for this study and the hydraulic actuator used to drive the fluid through the rheometer. The main tube has a bore diameter of 101 mm while the MR valve is a 1 mm x 10 mm rectangular channel. With an area ratio of greater than 1000, low piston velocities can be used to generate high fluid velocities through the valve. An electromagnet is used to activate the fluid in the valve. A Hall sensor is used to measure the field strength.

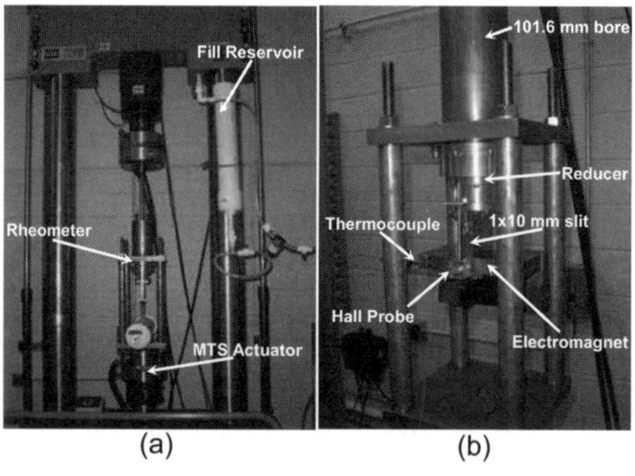

Figure 2. High shear rate rheometer

The rheometer was designed such that the parallel plate assumption could be made. With an aspect ratio greater than or equal to 10, the side wall effect can be neglected [2]. **Figure 3** shows the flow channel with the 1 mm x 10 mm channel. The ratio of the length of the flow channel to the height of the slit is large enough that end effects can be neglected [3]. The active length of the MR flow channel in this study was 6.35 mm. The best performance for an MR valve results from a rectangular geometry in which the fluid is in direct contact with the magnetic poles [4]. As such, the flow channel was made in a modular fashion; two stainless steel plates were sandwiched together, separated by a 1 mm stainless steel spacer. The valve portion of the rheometer is made from low carbon steel. An electromagnet is used to generate the magnetic field. Through this design we are able to modify the effective pole length of the valve by simply changing the length of the low carbon steel portion. Furthermore, the flow channel dimensions can be changed simply by replacing the stainless steel spacers.

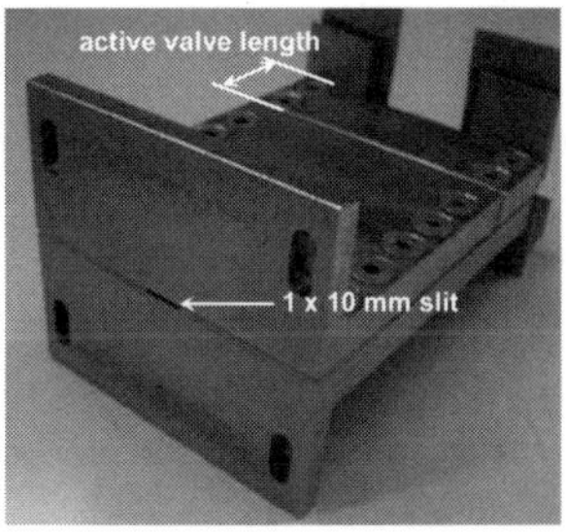

Figure 3. Flow channel showing MR active valve portion

Experiments were run by defining the piston velocity and measuring the corresponding force. The constant velocity input ensured that the force reached steady-state and that the flow was steady flow. In making the parallel plate approximation, the pressure developed in the valve can be written in the familiar non-dimensional form [5]

$$\mathcal{P}^3 - (1 + 3\mathcal{T})\mathcal{P}^2 + 4\mathcal{T}^3 = 0 \tag{1}$$

where $\mathcal{P}$ and $\mathcal{T}$, defined as

$$\mathcal{P} = -\frac{dp}{dx}\frac{h^2}{12u_m\eta} \tag{2}$$

$$\mathcal{T} = \frac{\tau_o h}{12u_m\eta} \tag{3}$$

are the non-dimensional pressure gradient and yield stress respectively. In expressions (2) and (3), dp/dx is the pressure gradient, h is the channel height, u_m is the mean fluid velocity, η is the fluid viscosity, and τ_o is the yield stress. In order to extract the yield stress from the measured force, we rewrite equation (1) as a cubic in $\mathcal{T}$, the non-dimensional yield stress.

$$\mathcal{T}^3 - \frac{3}{4}\mathcal{P}^2\mathcal{T} + \frac{1}{4}\left(\mathcal{P}^3 - \mathcal{P}^2\right) = 0 \tag{4}$$

In this form, we can solve for the roots of $\mathcal{T}$ in terms of $\mathcal{P}$, where $\mathcal{P}$ is found from the measured force, F, as

$$\mathcal{P} = \frac{F}{A_p}\frac{h^2}{12u_m\eta L} \tag{5}$$

where A_p is the piston cross sectional area and L is the length of the flow channel. The solution of equation (4) has a realizable root at [6]

$$\mathcal{T}(\mathcal{P}) = \mathcal{P}\cos\left(\frac{1}{3}\mathrm{acos}\left(\frac{1-\mathcal{P}}{\mathcal{P}}\right) + \frac{4\pi}{3}\right) \tag{6}$$

The yield stress is then found from equation (3) as

$$\tau_o = \frac{12u_m\eta}{h}\mathcal{T} \tag{7}$$

The yield stress calculated in equation (7) represents the average yield stress measured over the time interval defined by the valve length and fluid speed. In integral form the yield stress can be written as

$$\tau_{avg} = \int_0^{t_{dwell}} \frac{\tau(t)}{t_{dwell}}\,dt \tag{8}$$

where $\tau(t)$ represents the instantaneous yield stress and t_{dwell} is the dwell time. The instantaneous yield stress can be assumed to have the form

$$\tau(t) = \tau_H\left(1 - e^{-t/t_{mr}}\right) \tag{9}$$

where τ_H is the asymptotic yield stress at field strength H, and t_{mr} is the response time (one time constant) of the MR fluid. **Figure 4** shows both equation (8) and equation (9). Combining equation (8) and (9), the average yield stress, as illustrated in **Figure 4**, and can be written as

$$\tau_{avg} = \tau_H\left[1 - \frac{t_{mr}}{t_{dwell}}\left(1 - e^{-t_{dwell}/t_{mr}}\right)\right] \tag{10}$$

The experimentally determined yield stress can then be used to compute the response time, t_{mr}, of the MR fluid. Because equation (10) cannot be solved explicitly for t_{mr}, an optimization routine in MATLAB is used to find a t_{mr} that minimizes the error between the calculated τ_{avg} and the experimentally determined τ_o.

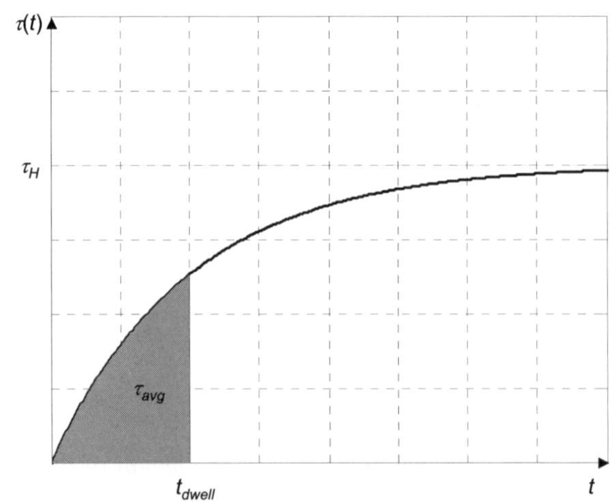

Figure 4. Instantaneous yield stress illustration

3 Results

Experiments were conducted at 100 kA/m, 150 kA/m, and 200 kA/m with piston velocities ranging from 1 mm/s to 35 mm/s. Corresponding fluid velocities ranged from 1 m/s to 36 m/s. **Figure 5** shows the measured pressure as a function of fluid velocity for each magnetic field strength. For low fluid velocities the pressure differential between field strengths is maintained. However, as the fluid velocity continues to grow the pressure differential decreases; pressures for the 100 kA/m, 150 kA/m, and 200 kA/m tests appear to converge towards the zero field case. This implies that, with a 6.35 mm active valve length, the yield stress being developed in the fluid is not being maintained as fluid velocity increases. The fluid is not spending enough time in the presence of a magnetic field to achieve the expected or full value of the yield stress.

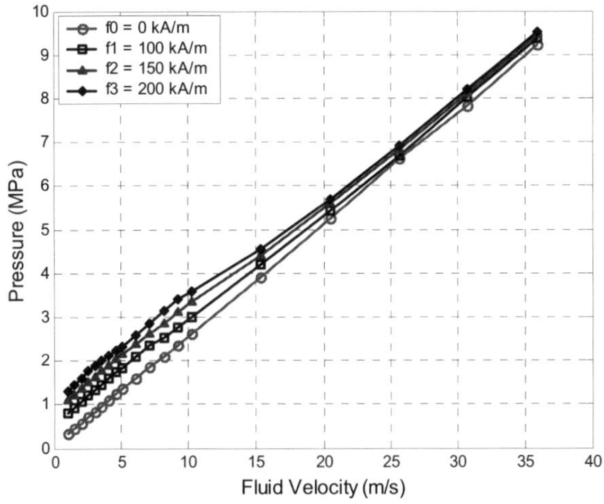

Figure 5. Pressure in flow channel as a function of fluid velocity

To further study the behavior in **Figure 5**, the yield stress is calculated. From the total pressure, the non-dimensional pressure, $\mathcal{P}$, developed in the MR valve can be extracted from equation (5). The non-dimensional yield stress, $\mathcal{T}$, is then found from equation (6). With the non-dimensional yield stress known, the apparent yield stress is found from equation (7). **Figure 6** shows the experimentally determined yield stress extracted from the experimental data. A significant reduction in yield stress is observed as fluid velocity increases. The result in **Figure 6** shows that the yield stress at low velocities is nearly maintained. However, as the fluid velocity continues to increase, the yield stress falls off rather sharply. Moreover, the rate of decay in the yield stress appears to be higher for larger magnetic field strengths. We also note that the differential between the yield stresses developed at low velocities for each magnetic field strength are significantly larger than those observed at high velocities. In other words, at high velocities, there is little change in the yield stress for varying magnetic field strengths. Though these results indicate some dependence on fluid velocity, the reduction in yield stress is more closely related to dwell time rather than fluid velocity [1].

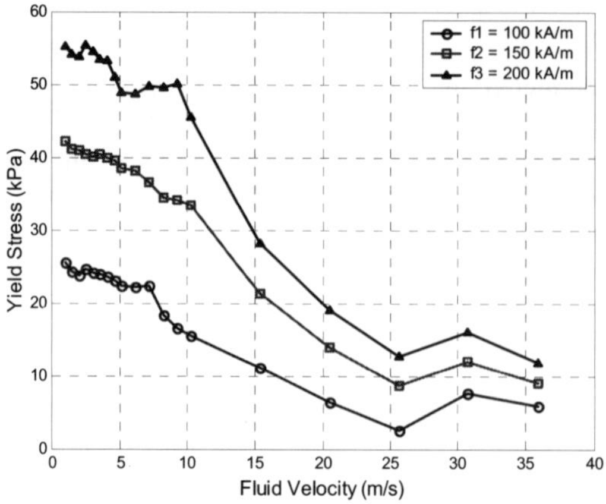

Figure 6. Yield stress as a function of fluid velocity

Expressing the results in **Figure 6** as a function of dwell time demonstrates the significance of dwell time on the achieved yield stress. The results for the 6.35 mm valve length clearly indicate that as the dwell time drops below 1 ms the fluid is unable to achieve full potential as shown in **Figure 7**. At a dwell time of 0.5 ms, for example, the yield stress developed in the gap is significantly less than the expected yield stress for a given magnetic field strength. Another significant result illustrated in **Figure 7** is that the sudden decay in the measured yield stress seems to occur sooner for lower field strengths; the yield stress for the 100 kA/m test begins to drop sooner than the 150 kA/m and 200 kA/m tests.

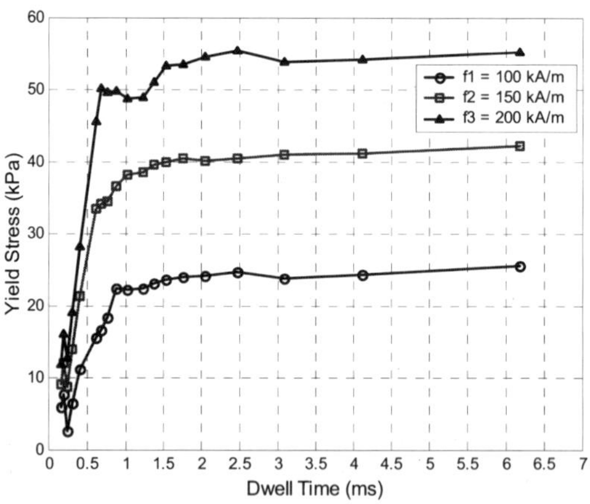

Figure 7. Yield stress as a function of dwell time

The proposed model given in equation (10) was fit to the data in **Figure 7**. Minimizing the error between the model and the experimental data, an optimization routine determined t_{mr} for each magnetic field strength. **Figure 8**a shows the experimental data along with the proposed model. The resulting response times for the 100 kA/m, 150 kA/m, and 200 kA/m tests are shown in **Figure 8**b and were determined to be 0.24 ms, 0.20 ms, and 0.19 ms respectively. As the results indicate, the response time of the fluid decreases as magnetic field strength increases. This finding supports the observation made in **Figure 7** regarding the initiation of the decay in yield stress; the decay in the yield stress begins sooner for lower field strengths because the response time of the fluid is higher. As the response time decreases, the decay in the yield stress is delayed and the fluid can withstand lower dwell times.

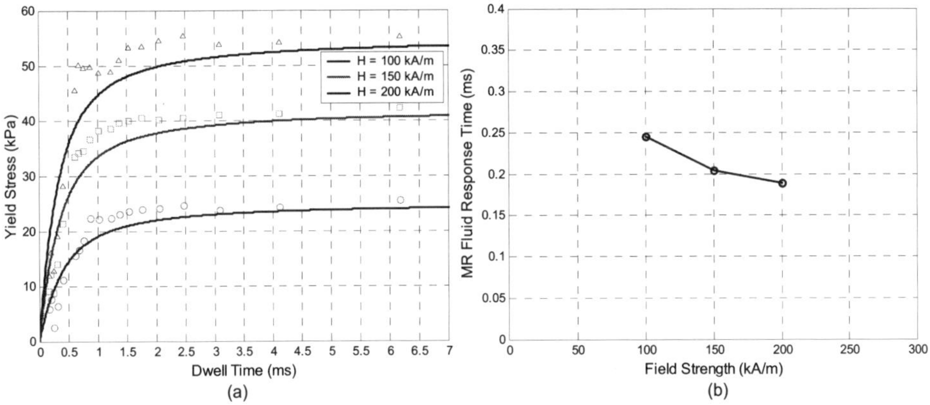

Figure 8. Model results: (a) experimental yield stress along with proposed model; (b) MR fluid response time as a function of magnetic field strength

4 Conclusion

The time response of MR fluid was investigated for three magnetic field strengths. A high shear rate rheometer was used to measure the yield stress of the fluid at fluid velocities ranging from 1 m/s to 36 m/s. Corresponding dwell times ranged from 6 ms to 0.18 ms. The yield stress was calculated for each magnetic field strength and a model was proposed to characterize the response. The experimental data and the proposed model were used to extract the response time for each magnetic field strength. Results indicate that the response time of the fluid decreases as magnetic field strength increases. Measured response times ranged from 0.24 ms to 0.19 ms.

References

1. Goncalves F. D., Ahmadian, M., Carlson, J. D., "Investigating the MR effect at high flow velocities," Smart Materials and Structures, 15, p. 75–85, 2006.
2. Baird, D. G., and Collias, D. I., Polymer Processing Principles and Design, Wiley Interscience, New York, NY, 1998.
3. Van Wazer, J. R., Lyons, J. W., Kim, K. Y., Colwell, R. E., Viscosity and Flow Measurement A Laboratory Handbook of Rheology, Wiley Interscience, New York, NY, 1963.
4. LORD Materials Division, "Magnetic Circuit Design," Engineering Note, 1999.
5. Phillips, R. W., "Engineering applications of fluids with a variable yield stress," Ph.D. Dissertation, University of California, Berkeley, 1969.

6. Gavin, H. P., "Electrorheological Dampers for Structural Vibration Suppression," Ph.D. Dissertation, University of Michigan, Ann Arbor, MI, 1994.

EFFECTS OF BODY FORCES ON THE STRUCTURE AND RHEOLOGY OF ER AND MR FLUIDS

DANIEL J. KLINGENBERG

Department of Chemical and Biological Engineering, University of Wisconsin, 1415 Engineering Dr., Madison, WI 53706, USA
E-mail: klingen@engr.wisc.edu

JOHN C. ULICNY AND ANTHONY L. SMITH

General Motors R&D and Planning, 30500 Mound Rd, Warren, MI 48090-9055, USA
E-mail: john.ulicny@gm.com, anthony.l.smith@gm.com

We employ particle-level simulations to show that body forces, such as gravity or centrifugal forces, can significantly influence the structure and rheology of ER and MR suspensions even when the magnitude of the body force acting on a particle is small compared to the field-induced force. We also report an experimental investigation of the effects of body forces on the structure of ER suspensions. Experimental results agree qualitatively with predictions.

1 Introduction

The role of hydrodynamic and interparticle forces in electrorheological (ER) and magnetorheological (MR) fluids has long been appreciated.[1,2,3] External body forces also act on particles within ER and MR fluids, but their effects have not been probed in any depth. Gravity causes particles to sediment in the absence of an applied field, but its effects in the presence of a field are typically ignored.[2,3] Other body forces include centrifugal forces in rotating devices, inertial forces from vibrations and translational accelerations, electrophoretic forces, and dielectrophoretic/magnetophoretic forces in nonuniform fields (e.g., near device edges).

In a previous article, we reported a study where we employed particle-level simulations to investigate the effects of a constant body force on the rheological properties of ER/MR suspensions in shear flow.[4] The main conclusions were that (1) body forces can reduce the field-induced shear stress when the body force acts normal to the shearing surfaces (see Fig. 1), even when the body force magnitude is small compared to the field-induced interparticle force, and (2) the effect of the body force depends on the system size (e.g., the ratio of the gap between the shearing surfaces and the particle diameter). In this report, we briefly summarize the simulation method and results for simple shear flow. Effects of body forces on the predicted suspension structure are also presented. We then describe an experimental investigation of the effects of a body force on the structure of ER suspensions. The experimental results agree qualitatively with the results predicted by the simulations.

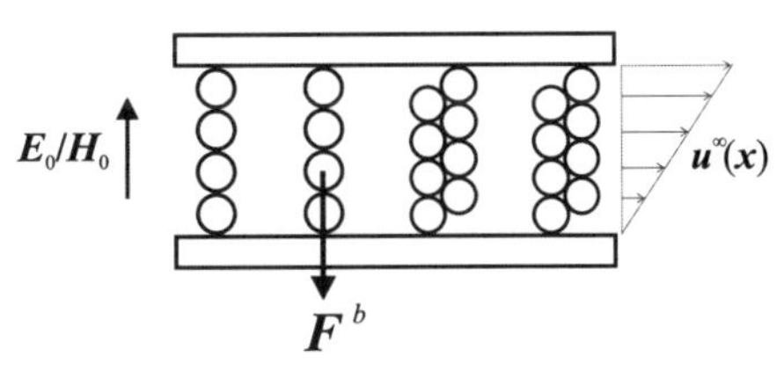

Figure 1. Schematic diagram of an ER/MR fluid in which the particles experience a body force directed normal to the shearing surfaces.

2 Methods

2.1 Simulations

The system to be simulated is depicted in Fig. 1, where a constant, uniform body force acts parallel or antiparallel to the applied field. This geometry corresponds to horizontal, planar sheared ER or MR suspensions in which the particles experience a gravitational body force normal to the shearing surfaces. This geometry also approximates shear of an ER or MR fluid in a concentric cylinder device (e.g., a typical geometry for an ER/MR clutch or brake), where for small gaps and similar rotation speeds of the inner and outer cylinders, the particles experience a nearly uniform outward centrifugal body force (in the frame of the cylinder). In either case, the body force may be written

$$\boldsymbol{F}^b = \frac{\pi}{6}\sigma^3 \Delta\rho a \boldsymbol{e}, \tag{1}$$

where σ is the particle diameter, $\Delta\rho = \rho_p - \rho_c$ is the density difference between the particles and continuous phase, and a is either the acceleration of gravity g or the centripetal acceleration, and $\boldsymbol{e}$ is the unit "downward" vector.

The simulation method employed has been described previously.[4] Here we briefly summarize the essential features. The suspension is modeled as a system of N spheres between bounding surfaces located at $z = \pm L_z/2$, with the electric or magnetic field ($\boldsymbol{E}_0$ or $\boldsymbol{H}_0$) applied in the z direction. Field-induced forces are described using the point-dipole approximation, with the force magnitude $F_0 = (3/16)\pi\epsilon_0\epsilon_c\sigma^2\beta^2E_0^2$ for linearly polarizable ER suspensions, $F_0 = (3/16)\pi\mu_0\mu_c\sigma^2\beta_M^2H_0^2$ for linearly magnetizable MR suspensions, or $F_0 = (1/48)\pi\mu_0\mu_c\sigma^2M_s^2$ for saturated MR suspensions. Here ϵ_0 is the permittivity of free space, ϵ_c is the relative permittivity of the continuous phase, β is the relative polarizability, μ_0 is the permeability of free space, μ_c the relative permeability of the continuous phase, β_M is the relative magnetizability, and M_s is the particle saturation magnetization. Short-range repulsive forces are employed to prevent particle overlap. Hydrodynamic resistance is treated as Stokes drag. The reader is referred to Ref. 4 for additional details.

The ratio of the body force to the electrostatic or magnetic force is written

$$G = \frac{\pi\sigma^3\Delta\rho a}{6F_0}, \tag{2}$$

and thus depends on the particular model chosen for F_0. Using the force magnitudes above, we get three ratios representing dimensionless body force magnitudes,

$$G = \begin{cases} \frac{8\sigma\Delta\rho a}{9\epsilon_0\epsilon_c\beta^2E_0^2} & \text{(linear polarization)}, \\ \frac{8\sigma\Delta\rho a}{9\mu_0\mu_c\beta_M^2H_0^2} & \text{(linear magnetization)}, \\ \frac{8\sigma\Delta\rho a}{\mu_0\mu_c M_s^2} & \text{(saturated magnetization)}. \end{cases} \tag{3}$$

Employing typical parameter values, one finds that G is typically small, $10^{-3} \lesssim G \lesssim 1$. However, we show below that significant effects of body forces are obtained for values of G in this range.

2.2 Experiments

Electrorheological suspensions were composed of barium titanate particles in mineral oil. Barium titanate particles were ground with a mortar and pestle, and sieved to obtain fractions with two average diameters (58 ± 5 μm and 77 ± 14 μm). The particles were dried at 110 °C for 24 hrs, then stored in a closed vial. The mineral oil (viscosity 0.15 Pa·s) was dried over molecular sieves. Suspensions with particle concentrations of 2 and 5 wt% were prepared.

Various cells were constructed as illustrated schematically in Fig. 2. The electrodes were aluminum, and the glass slides were fixed to the electrodes using epoxy. Three cells with different heights L_z and widths w were constructed [(L_z, w) = (2 mm, 3 mm), (3 mm, 3 mm) and (5 mm, 7 mm)]. All cells were 77 mm in length.

Well-shaken suspensions were loaded into a cell, and the open cell ends were sealed with nonconducting clay. Experiments were performed by turning the cell upside down to disperse the particles, and then rapidly applying a large electric field to create uniform chain-like structures. The electric field strength was then decreased in steps, and maintained at each level for 2 min. A video camera was employed to record snapshots of the suspension. When the electric field was decreased to a sufficiently small value, the chains would begin to collapse under the force of gravity. The field strength was decreased until no percolating chains remained. The suspension structure was characterized by the number of percolating chains present in the field of view.

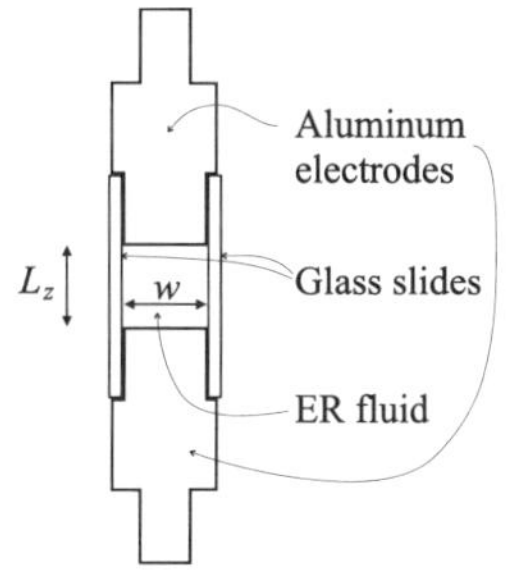

Figure 2. Schematic diagram of a cell.

3 Results

3.1 Simulations

Consider first suspensions in applied electric or magnetic fields with no ambient flow (shear rate $\dot{\gamma} = 0$). Various manufactured structures were created and allowed to relax by integrating the equations of motion with $G = 0$ until motion ceased. A dimensionless body force of magnitude G was then applied, and the equations of motion were again integrated to determine the evolution of the structure.

The behavior for a "double-chain" structure is illustrated in Fig. 3 where snapshots of the equilibrium structures are presented for various values of G. The initial structure is composed of two touching chains offset from one another in the z direction by $\sigma/2$. The structure illustrated has $N = 100$ spheres in a gap of $L_z = 50.5\sigma$, with $L_x = L_y = 20$. For small values of $G \lesssim 0.01$, the structure is essentially unaltered by the body force. As G is increased, spheres are forced downward. For $G \gtrsim 0.05$, the structure detaches from the top surface. As G is increased further, the structure collapses further.

Clusters were subjected to simple shear at a small shear rate [$\dot{\gamma} = 10^{-4}F_0/(3\pi\eta_c\sigma^2)$]. Shear stress is plotted as a function of shear strain in Fig. 4(a) for a double-chain cluster ($L_z = 50.5\sigma$) at several values of G. For each value of G, the stress initially increases with strain, and then fluctuates as the clusters rearrange. The maximum stress at low shear rates can be associated with a yield stress.[2,5] The maximum shear stress $\sigma_{xz,\max}$ is plotted as a function of G for both the double and single-chain clusters with different L_z in Fig. 4(b). For each data set, $\sigma_{xz,\max}$ decreases as G is increased.

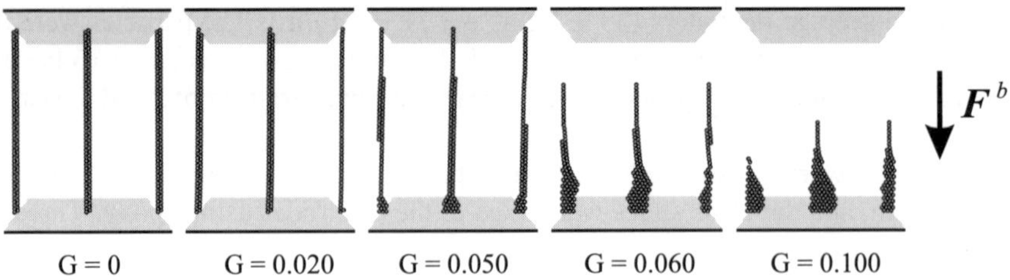

Figure 3. Simulated structures at equilibrium for various values of G ($L_z = 50.5\sigma$).

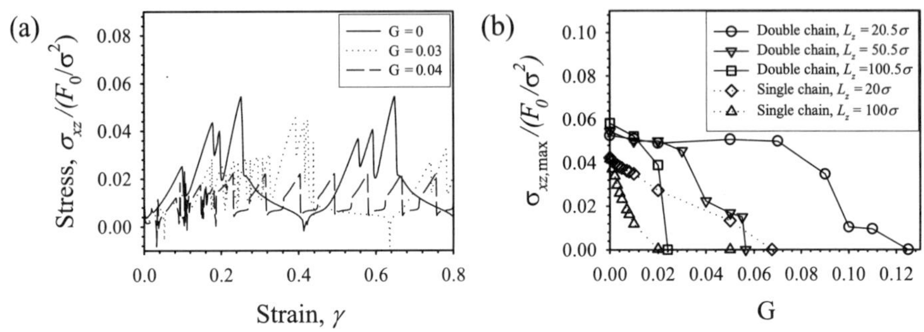

Figure 4. (a) Shear stress as a function of shear strain for the double-chain cluster with $L_z = 50.5\sigma$, for various values of the dimensionless body force magnitude G. (b) Maximum shear stress as a function of G for single and double-chain clusters for various L_z.

The value of G at the point where $\sigma_{xz,\max}$ goes to zero is defined as a "critical" value, G_c. Figure 4(b) illustrates that rheological properties can be reduced even for $G < G_c$. The mechanism is illustrated in Fig. 5, where snapshots of the double-chain cluster under shear ($L_z = 50.5\sigma$) are presented for $G = 0$ and 0.04. Even though $G = 0.04 < G_c \approx 0.565$, the structure evolution for $G = 0.04$ is significantly different than that for $G = 0$. For $G = 0$, the cluster is tilted and elongated under shear, and eventually ruptures. For $G = 0.04$, the cluster partly collapses, but does not rupture. Instead, the top sphere periodically gets pulled away from the top surface, allowing the cluster to remain inclined (at a smaller angle relative to the applied field than the cluster with $G = 0$ prior to its rupture).

The critical body force G_c is plotted as a function of L_z in Fig. 6 for percolating single and double-chain clusters. The data are well represented by the relationships

$$G_c = \frac{1.29}{L_z/\sigma} \text{ (single chains)}, \quad G_c = \frac{2.61}{L_z/\sigma} \text{ (double chains)}. \tag{4}$$

These results have several implications. First, the influence of body forces is significant for realistic system sizes even for small values of G; for 1μm diameter spheres in a 1 mm gap, $G_c \approx 10^{-3}$. Values of $G \geq 10^{-3}$ can easily occur in practical situations. Furthermore, the stress is reduced even for values of $G < G_c$. Second, assuming this scaling (Eq. 4) holds for smaller and larger gaps, simulations need not be performed for actual gap dimensions ($L_z/\sigma \sim 10^3$).

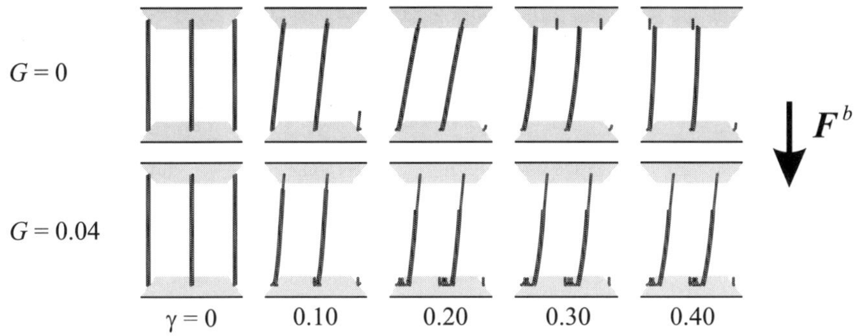

Figure 5. Simulated structures in shear for $G = 0$ and 0.40 ($L_z = 50.5\sigma$).

The scaling $G_c \propto (L_z/\sigma)^{-1}$ also suggests that the ability of a suspension to support a shear stress depends on both G and L_z/σ, and is characterized by the ratio

$$GL_z/\sigma = \frac{\pi L_z \Delta\rho a}{6F_0/\sigma^2}. \qquad (5)$$

For $GL_z/\sigma < (GL_z/\sigma)_c \sim O(1)$, the suspension can support a shear stress, otherwise it cannot. The parameter GL_z/σ is independent of particle size (i.e., $F_0 \propto \sigma^2$); this somewhat counterintuitive result arises because particles in clusters experience not only the body force acting on themselves, but also the body forces acting on all of the spheres above them transmitted via the short-range repulsive forces. This result also suggests that one must be careful when attempting to probe the effects of body forces experimentally. Experiments in which the σ is varied at fixed L_z are not predicted to show any variation, even when body forces are present and influence behavior; experiments must be performed at fixed L_z/σ to show a dependence on σ.

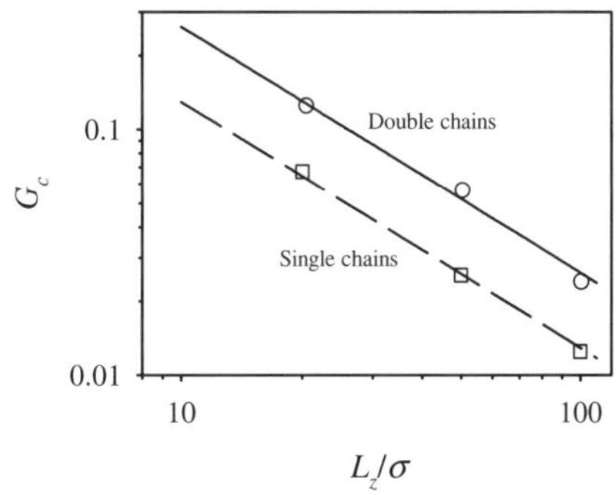

Figure 6. Critical dimensionless body force G_c as a function of L_z for single and double-chain clusters. The maximum uncertainty in the data is $\pm 4\%$. The curves are fits of each set of data to $G_c =$ constant $\times$ $(L_z/\sigma)^{-1}$.

3.2 Experiments

Snapshots of barium titanate suspensions ($E_0 = 469$ V/mm initially) at successively smaller electric field strengths are presented in Fig. 7. Figure 7(a) illustrates that the initial structure is relatively uniform. As the field strength is decreased, the chain-like aggregates eventually begin to collapse (Fig. 7(c)). Figure 7(d) illustrates that the chains completely collapse for sufficiently small but nonzero field strengths.

The fraction of percolating chains (relative to the number present initially) is plotted as a function of E_0 for different L_z in Fig. 8(a). As E_0 is decreased from its initial large

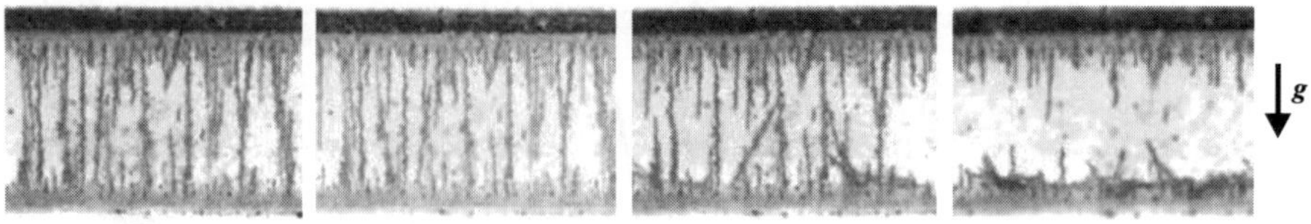

Figure 7. Snapshots of an ER suspension at various field strengths (2 wt%, 58 ± 5 μm diameter, $L_z = 3$ mm, 500 Hz).

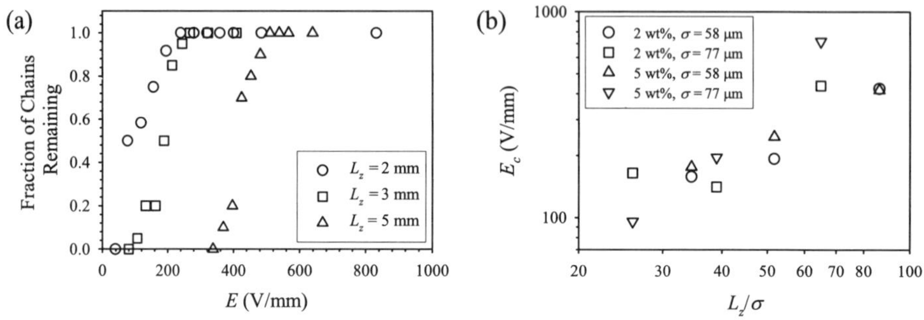

Figure 8. (a) Fraction of chains remaining as a function of field strength for various gaps (5 wt%, 77 ± 14 μm diameter). (b) Critical field strength as a function of L_z/σ for various concentrations and particle diameters.

value, the fraction of percolating chains initially remains constant. For sufficiently small E_0, the fraction begins to decrease as chains collapse. Again we note that the fraction of percolating chains decreases to zero at small but nonzero E_0.

We define the "critical" field strength E_c as the field strength where the fraction of percolating chains decreases to 0.5. In Fig. 8(b), E_c is plotted as a function of L_z/σ for various σ and concentrations. As observed in the simulations, larger gaps cause percolating chains to collapse at larger field strengths. Within experimental uncertainty, the G_c vs L_z/σ is independent of σ, consistent with the simulations.

In Fig. 9, $G_c = 8\sigma\Delta\rho g/(9\epsilon_0\epsilon_c\beta^2 E_c^2)$ is plotted as a function of L_z/σ for both the experiments and the simulations. In both cases, G_c decreases with increasing L_z/σ. The simulations yield $G_c \propto (L_z/\sigma)^{-1}$; fitting all the experimental data with $G_c \propto (L_z/\sigma)^{-p}$ yields $p = 2.56 \pm 0.82$, which constitutes a stronger gap dependence than observed in the simulations.

The values of G_c obtained from the simulations are significantly larger than those obtained experimentally. There are several possible explanations for quantitative differences. First, the critical "events" in the simulations and experiments are defined similarly, but not identically. Second the simulations employ the point-dipole approximation for the field-induced interparticle force. This form is known to underpredict the magnitude of the actual interparticle force. Finally, simulations conducted thus far have employed idealized, periodic structures, whereas those observed experimentally are more disordered. How each of these issues affects the results remains to be determined.

4 Conclusion

Particle-level simulations show that body forces, such as gravity or centrifugal forces, can significantly influence the structure and rheology of ER and MR suspensions even when the magnitude of the body force acting on a particle is small compared to the field-induced force. Experimental observations agree qualitatively with predictions.

The investigations described in this report considered only uniform body forces exerted normal to the flow in quiescent or weakly sheared systems. Numerous other situations can appear in devices, which deserve attention (e.g., body forces in the flow or vorticity directions).

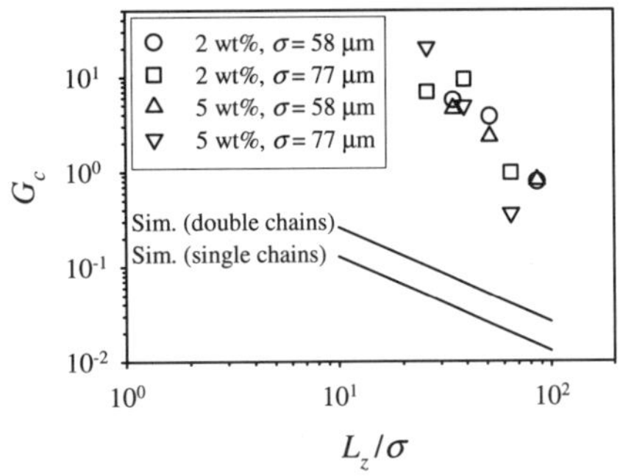

Figure 9. Critical dimensionless body force G_c as a function of L_z/σ for various concentrations and particle diameters.

Acknowledgments

The authors thank B. Droessler and D. Gloudemans for making the experimental measurements.

References

1. Gast, A. P., and C. F. Zukoski, Adv. Coll. Int. Sci., **30**, 153 (1989).
2. Parthasarathy, M., and D. J. Klingenberg, Mat. Sci. Eng. Rep., **R17**, 57 (1996).
3. Ginder, J. M., in *Encyclopedia of Applied Physics*, vol. 16, G. L. Trigg, ed., VCH, Weinheim, p. 487 (1996).
4. Klingenberg, D. J., J. C. Ulicny, and A. Smith, Appl. Phys. Lett., **86**, 104101 (2005).
5. Klingenberg, D. J., and C. F. Zukoski, Langmuir, **6**, 15 (1990).

CHARACTERIZATION OF RHEOLOGICAL PROPERTIES OF NOVEL MAGNETORHEOLOGICAL FLUIDS

S. MANTRIPRAGADA, X. WANG, F. GORDANINEJAD

Department of Mechanical Engineering, University of Nevada, Reno, Nevada 89557, USA
E-mail: faramarz@unr.edu

B. HU, AND A. FUCHS

Department of Chemical Engineering, University of Nevada, Reno, Nevada 89557, USA

The rheological properties of various novel MR fluids are characterized using a parallel plate MR shear rheometer. In these MR fluids the surface of iron particles is coated with a polymer. The rheological properties are measured and compared at various magnetic field strengths shear rates and strain amplitudes. .It has been shown that these MR fluids exhibit stable and desirable rheological properties such as, low viscosity and high yield stress which are comparable to or higher than the conventional MR fluids.

1 Introduction

Magnetorheological (MR) fluids are suspensions of micron-size ferrous particles in a viscous medium that display controllable and reversible changes in their rheological properties under externally applied magnetic fields. Each MR fluid is unique in its properties and responses to external stimuli. Factors such as the size and volume fraction of the iron particles and nature of the carrier fluids and additives used will affect these properties. Among these properties, the off state (i.e., no magnetic field applied) viscosity and the field dependent yield stress are major parameters which are considered in the design of MR devices [1]. Additionally, agglomeration and sedimentation of dense iron particles in a less dense carrier medium have significant effects on the performance of these devices in practice [2, 3]. The ability of particles to redisperse in the carrier medium is an indicator of the stability of the MR fluid [4]. Characterization of a MR fluid and examination of rheological properties, as well as, settling rate under various operating conditions is needed before its integration with various mechanical devices.

In this study, several novel MR fluids have been developed in which the surface of iron particles is coated with a polymer. These fluids provide high shear yield stress and adjustable off state viscosity, while achieving reduced particle settling rates and improved stability. The polymer coating is achieved by surface initiated atom transfer radical polymerization (ATRP) [4]. The focus of this study is to characterize the rheological properties of these fluids using a commercial MR fluid, MRF 132AD, as a benchmark. A series of oscillatory and rotational shear tests are performed to obtain a complete rheological behavior of these MR fluids. The effects of particle coating on the settling rate of the novel MR fluids are also investigated.

2 Material Background on UNR MR Fluids

Each fluid considered in this study is unique in composition and synthesis procedure. These fluids have been developed at the University of Nevada, Reno (UNR). The three novel MR fluids will be referred to as UNR1, UNR2 and UNR3. The compositions of the fluids, particularly the base fluid and the stabilizing polymers used, are as follows:

UNR1: Ferromagnetic particles dispersed in a carrier medium of PAO, with Polyurethane as a stabilizer.
UNR2: Carbonyl iron particles, surface grafted with butyl acrylate, in a carrier medium of N-octyl-pyrrolidone.
UNR3: Carbonyl iron particles, surface grafted with pentafluorostyrene, in a carrier medium of N-octyl-pyrrolidine.

MRF 132AD is the commercial benchmark fluid used in this study. Reference [4] provides the detailed synthesis of stable MR fluids in which the iron particles are covalently grated with a poly chain using atom transfer radical polymerization (ATRP). In this reaction system, an ATRP initiator, 2-(4- chlorosulfonylphenyl) - ethyltrichlorosilane (CTCS), is immobilized on the iron particle surface and then initiates radical polymerization. Then, the polymers are grafted from the surface of iron particles. Polymerization is carried out at a temperature of 70^0C. Figure 1 presents a schematic of the surface initiated ATRP approach. The polymer concentration in the fluid determines the off state viscosity and the particle settling rate. Scanning electron microscopy (SEM) is used to characterize the surfaces of both pure iron particles and surface initiated ATRP particles. SEM images of the carbonyl particles before and after the polymerization process are presented in Figure 2 and Figure 3, respectively.

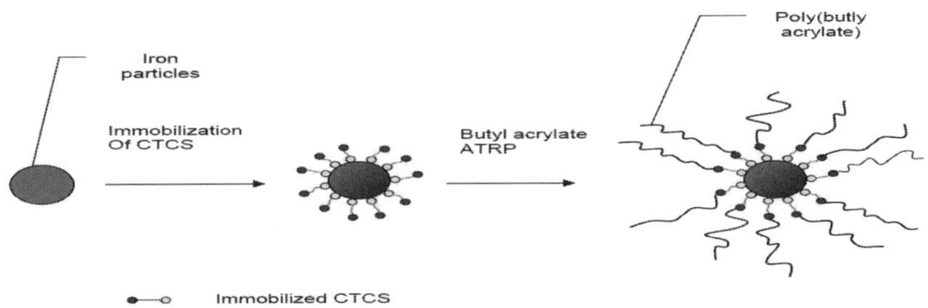

Figure 1. Schematic of ATRP structure.

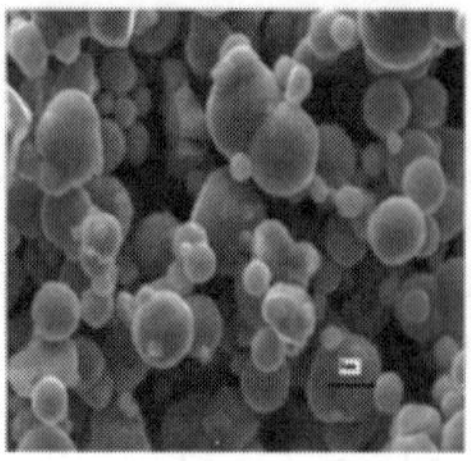

Figure 2. Pure Iron Particles

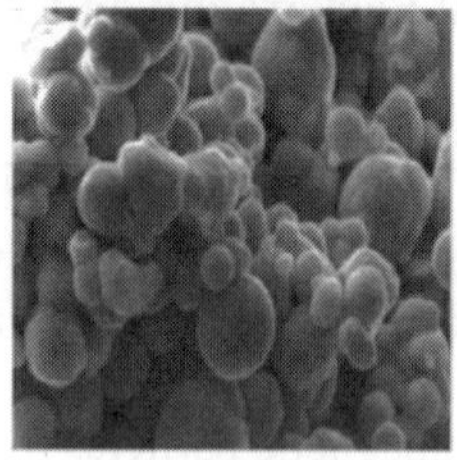

Figure 3. ATRP treated Iron Particles

3 Experimental Setup

A modular shear rheometer (Paar Physica MCR 300) having parallel plate geometry, with on board data acquisition and temperature control, is used for the tests. A fixed plate gap of 1mm and a fluid volume of 0.3 ml under the gap are used as standard experimental settings for all the samples tested. The rheological properties of the MR fluids are established under varying magnetic field intensities with temperature control. A maximum magnetic flux density of 0.5T can be achieved on the rheometer.

The following properties are considered for the characterization: off state viscosity, shear yield stress, ability of the fluids to maintain a high shear stress over a period of time, viscoelastic properties and settling rate.

4 Results and Discussion

4.1 Off State Viscosity

The off state viscosities of the fluids at room temperature are presented in Figure 4. Viscosity of the fluids decreases with increasing shear rates. Off state viscosities of UNR fluids are lower, and in the case of UNR3 only slightly higher, compared to the viscosity of the commercial benchmark fluid. At a shear rate of 400 1/s, UNR1, UNR2, UNR3 have off state viscosities of 0.1, 0.18, 0.23 Pa.s, respectively, while the benchmark fluid has a viscosity of 0.2 Pa.s. The off state viscosity in the polymer coated novel MR fluids can be controlled by the polymer fraction in the fluids, shown in Figure 5. Increase in the polymer concentration from 0.5% to 1.5% and 2%, shows a respective increase in the off state viscosity. Low off state viscosity is desirable in certain applications such as dampers.

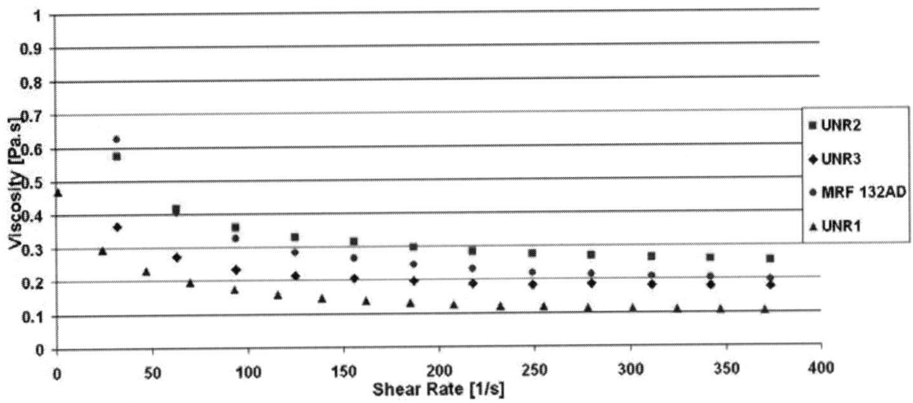

Figure 4. Off State viscosity as a function of shear rate at 25^{0}C.

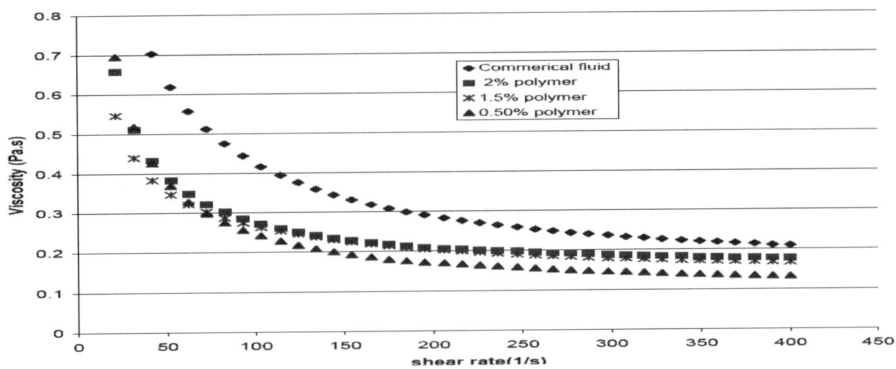

Figure 5. Effect of Polymer coating on the Viscosity of UNR MR fluids.

4.2 *Shear Yield Stress*

Shear yield stress is an important rheological property of MR fluids. Shear yield stress is obtained by extrapolating the shear stress versus shear rate curves to zero shear rate following the Bingham model. The field dependence of the yield stress for the four fluids is presented in Figure 6, where the yield stress is studied as a function of the magnetic flux density B. As expected the shear yield stresses of the fluids increase with increasing magnetic flux density. MR fluid UNR3 shows a higher increase in the yield stress by increasing the magnetic field strength compared to the other UNR and benchmark fluids. At 0.5T, UNR3 has a yield stress value of about 35KPa while that of UNR1, UNR2 and MRF 132AD is around 30KPa. Ability of the fluids to retain a high shear stress is another important criterion which is presented for UNR fluids in Figure 7, where the shear stress is shown as a function of time under a continuous shear rate of 9001/s and a flux density of 0.5T. The unstable nature of the curves is a direct consequence of the spillage of the fluid volume from the test region on the rheometer at high shear rates.

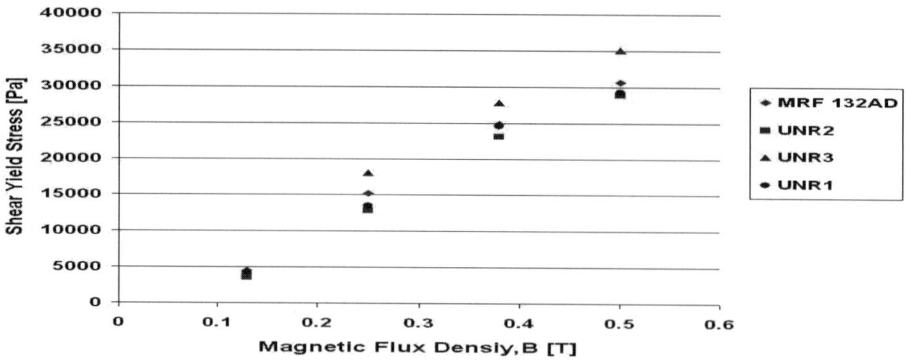

Figure 6. Shear yield Stress versus Magnetic Flux density at 25^{0}C.

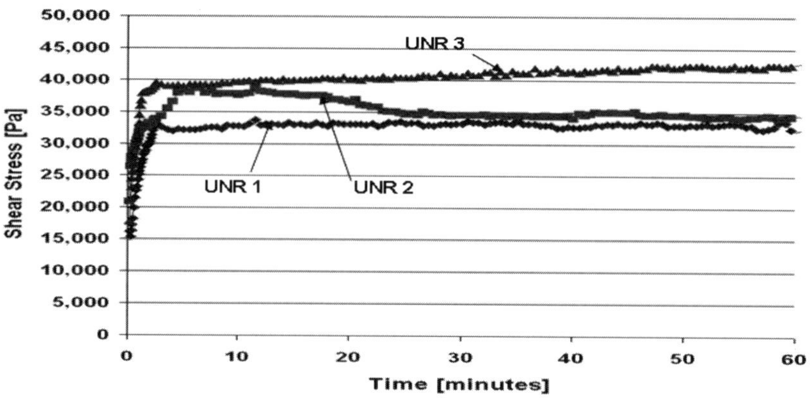

Figure7. Shear Stress versus Time at 900 1/s and 25^{0}C.

4.3 Viscoelastic Properties

All MR fluids exhibit viscoelastic properties under an applied magnetic field. Oscillatory testing is used to determine the viscoelastic behavior of the materials. Storage modulus (G') and loss modulus (G") are measured as functions of strain amplitude and frequency to completely understand these properties. An amplitude sweep test is conducted by linearly increasing the strain amplitude from 10^{-5} to 1 at a frequency of 0.5 rad/s to obtain the storage modulus (G') and loss modulus (G") versus the strain amplitude. These results are indicative of the structural changes taking place in the fluid under the applied field conditions. The fluid exhibits an elastic nature below a limiting strain value beyond which the viscous nature dominates, as clearly seen from curves in Figures 8 and 9. ATRP fluids UNR2 and UNR3 have higher critical strain values of around 0.008 compared to those of the UNR1 and MRF 132AD at about 0.005. The frequency sweep tests are performed at the constant strain amplitude of 0.05 %, which is in the pre yield region. The plateau in the storage and loss modulus curves in Figures 10 and 11, points to the stability of the fluids in this region. All results shown in Figures 8 to 11 are at 0.5T and 25^{0}C.

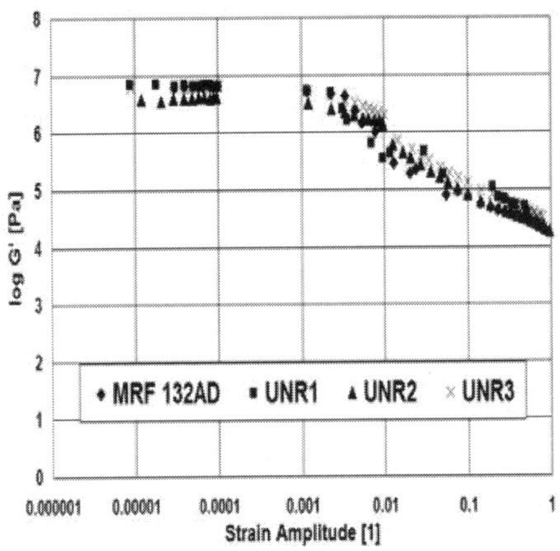

Figure 8. log G' versus Strain Amplitude at 0.5 rad/s. rad/s.

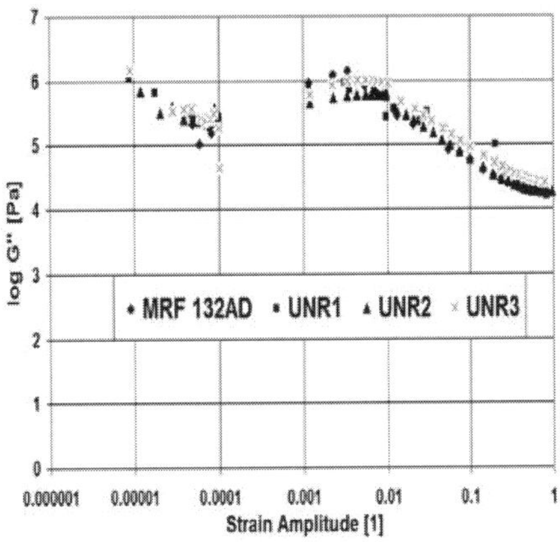

Figure 9. log G'' versus Strain Amplitude at 0.5

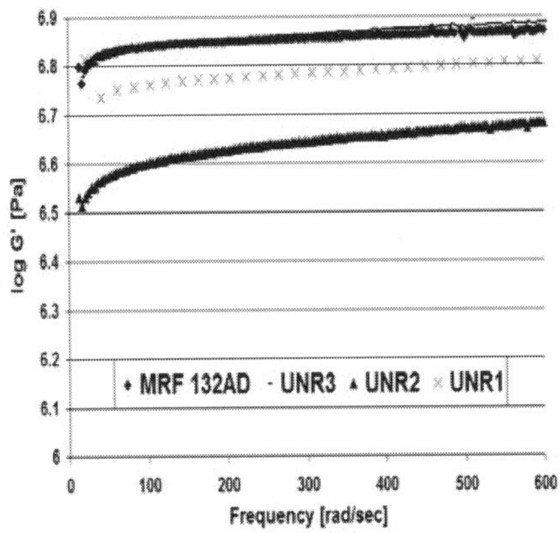

Figure 10. log G' versus Frequency at 0.05% strain. strain.

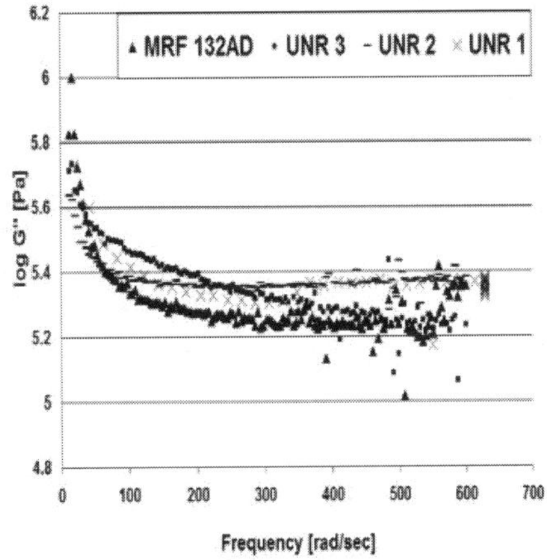

Figure 11. log G'' versus frequency at 0.05%

4.4 Particle Settling Rate

The settling rate of the MR fluids can be controlled by the percentage of polymer coating on the iron particle surface. Higher polymer fractions would result in lower yield stress. The settling rate for the MR fluid containing pure iron particles is significantly higher than the ATRP fluids. The particle settling rates of the novel fluids compared to the commercial benchmark fluid is presented in Figure 12. The stability of the MR fluids has been investigated using the gravity yield parameter [4]. The percentage of volume settled is measured against time to obtain the particle settling rate curve. Particle settling rate can be a good pointer towards the stability of MR fluids.

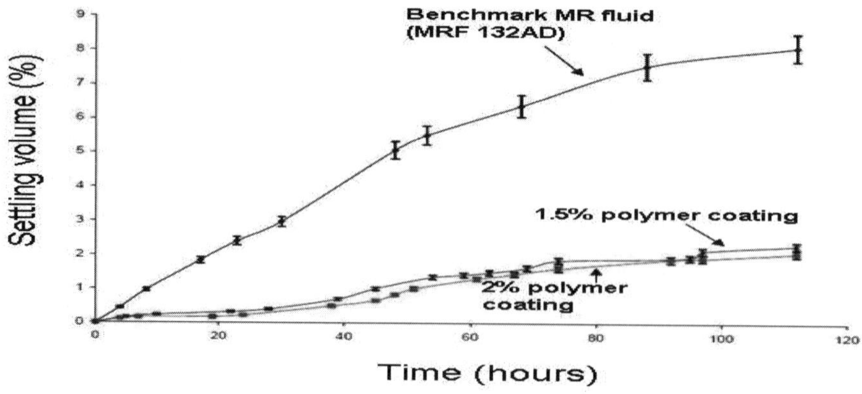

Figure 12. Particle settling results for various UNR and commercial MR fluids.

5 Conclusions

The rheological properties of the various novel MR fluids are characterized and compared with that of a commercial MR fluid. The new MR fluids exhibit stable and desirable rheological properties such as, low viscosity and high yield stress which are comparable to or higher than the benchmark fluid over a range of applied magnetic fields. The UNR fluids synthesized through the ATRP process however have higher yield strain limits, as can be seen from the viscoelastic properties. A low particle settling on the ATRP MR fluids is also achieved.

References

1. Ashour, O., Rogers, C. A., and Kordonsky, W., "Magnetorheological Fluids: Materials, Characterization, and Devices," *Journal of Intelligent Material Systems and Structures*, **Vol. 7**, pp. 123 – 130, March 1996.
2. Jolly, M. R., Bender, J. W., Carlson, J. D., "Properties and Applications of Commercial Magnetorheological Fluids," *Journal of Intelligent Material Systems and Structures,* **Vol. 10**, pp. 5 – 13, January 1999.
3. Park, J. H., Chin, B. D., Park, O. O., "Rheological Properties and Stabilization of Magnetorheological Fluids in a Water – in – Oil Emulsion," *Journal of Colloid and Interface Science **240***, pp. 349 – 354, 2001.
4. Hu, B., Fuchs, A., Sahin, H., Gordaninejad, F., and Evrensel, C., "Atom Transfer Radical Polymerized MR Fluids, *Proceedings of AIChE, #556 - Polymer Thin Films and Interfaces III (08A08),* November 2005).
5. Weiss, K. D., Carlson, J. D., Nixon, D. A., "Viscoelastic Properties of Magneto- and Electro-Rheological Fluids," *Journal of Intelligent Material Systems and Structures,* **Vol. 5**, pp. 772 – 775, November 1994.
6. Li, W. H., Chen, G., Yeo, S. H., "Viscoelastic Properties of MR fluids, *Smart Materials and Structures,*" **8,** pp. 460 – 468, 1999.

PHOSPHATE COATING ON THE SURFACE OF CARBONYL IRON POWDER AND ITS EFFECT IN MAGNETORHEOLOGICAL SUSPENSIONS

ANTONIO J. F. BOMBARD

UNIFEI – ICE - Av. BPS, 1303 – Itajubá – MG – CEP 37.500-903 – BRAZIL
E-mail: bombard@unifei.edu.br

MARCELO KNOBEL

Instituto de Física "Gleb Wataghin" – UNICAMP – Campinas – SP – BRAZIL
E-mail: knobel@ifi.unicamp.br

M. R. ALCÂNTARA

Instituto de Química – USP – Av. Prof. Lineu Prestes, 748 - São Paulo – SP – BRAZIL
E-mail: mralcant@iq.usp.br

Two types of carbonyl iron powders, (CIP's, BASF AG), the HS and HS-I (I = insulated, due a coating with phosphate), and two kinds of silica, one hydrophobic (Cab-O-Sil® TS610) and other hydrophilic (Cab-O-Sil® M5), were used to evaluate the influence of the surface treatment of the magnetic particle and the kind of fumed silica on the formulation of some magnetorheological suspensions (MRS). Oscillatory measurements at no field showed an evident difference between the silicas, but not a specific interaction with the phosphate coating on HSI. On the other hand, steady flow experiments also without magnetic field showed that the kind of silica and its specific interactions with the coating on iron powder drove the rheological behavior of the MRS on all region of the shear rate. Under magnetic field, the flow curves differences will be due to the iron particles and its magnetic properties, mainly on the region of higher shear rate.

1 Introduction

A Magnetorheological fluid consists, in general of micro particles (~ 1 – 10 μm) of a magnetic material dispersed in an organic or aqueous medium. The development of new magnetorheological formulations for technological applications is complicated by the lack of stability of these MR suspensions due to the natural sedimentation process caused by the large density mismatch between the magnetic particles and the dispersion phase.

Generally this problem was solved by the addition of electrosteric polymers [1], or thixotropic additives as organoclays [2], fumed silica [3], to the MR formulations in order to reduce and slowdown the settling process. Viscoplastic media (off – yield stress) was pointed out by *Klingenberg et al* [4] to prevent the sedimentation of MR suspensions. Other approaches to reduce settling are the use of nanoparticles, as chromium dioxide [5] and iron [6] or water-oil emulsions [7]. It should be stressed that there is a practical upper limit to increase the amount of gel forming additive without changing largely the off – plastic viscosity, which is not desirable [3]. However, studies addressing the surface properties of iron powders and their interaction with additives are sparse [8,9].

In this paper we studied the influence of the surface treatment of an iron powder with fumed silicas on the rheological behavior of the MRS. Two types of carbonyl iron powders, (CIP's), the HS and HSI (I = insulated), this second coated with phosphate, and two kinds of silica, one hydrophobic and other hydrophilic, were used.

2 Experimental

Both CIPs used, the HS and HSI was from BASF and were used without any treatment. Two kinds of fumed silica were used: Cab-O-Sil® M-5 that is the hydrophilic silica and Cab-O-Sil® TS-610 that is a hydrophobic one. Table 1 shows some manufacturer characteristics of the fumed silicas [6].

Table 1: Some characteristics of Cab-O-Sil® fumed silica M-5 and TS-610

Fumed silica	Surface area (m^2/g)	Carbon content (wt. %)	Chemical group at surface	Sp. gravity (g/cm^3)
M-5	200 ± 25	NA	Hydroxyl	2.2
TS-610	120 ± 20	0.85 ± 0.15	Methyl	2.2

The magnetization curves for all CIPs were obtained using a SQUID Magnetometer (Quantum Design MPMS XL). The particle size and particle size distribution were measured on the dry powders dispersed in water, by a Malvern Mastersizer Micro that operate using a laser light scattering technique.

The MR suspensions were prepared by pre-treating each CIP with 0,5% w/w of fumed silicas over the CIP mass. The final formulations were obtained by mixing handily and thoroughly the oil (Nujol®, Schering Plough, pharmaceutical grade) to the dry powders previously prepared to obtain the final composition. All MRS have 80% w/w of CIP. The nomenclature used defines firstly the CIP, followed by the kind of fumed silica, i.e., HSM5 is a MRS with CIP BASF HS and fumed silica Cab-O-Sil® M5.

The rheological behaviors of the MRS were measured in a controlled stress rheometer (Paar Physica MCR 300) equipped with the magnetorheological device of Physica (MRD-180) and plate-plate geometry (20 mm diameter, 1 mm gap). The temperature was kept constant at 25°C through a water bath circulator. The values of the magnetic field were measured on an empty gap that even if not exact, can be used to compare different MRS.

Before the oscillatory experiments the samples were pre-sheared during one minute using a shear rate of $100s^{-1}$ and left to rest for another 5 minutes to eliminate the rheologic history on each sample. The amplitude sweep was changed from 0,001 to 100 % with an angular frequency constant of $10s^{-1}$. For experiments with magnetic field sweep under strain constant it was used deformations of 0.01 %.

The flow curves were obtained at no magnetic field or under applied field sweeping the shear rate from 0.05 to $500s^{-1}$. The experiments of magnetic field sweep were performed at constant shear rate of $1\ s^{-1}$.

3 Results and Discussion

3.1 Magnetization and Particle size measurements of the pure CIP

The results of the magnetization curves for the dry powders HS and HSI are shown at Figure 1.

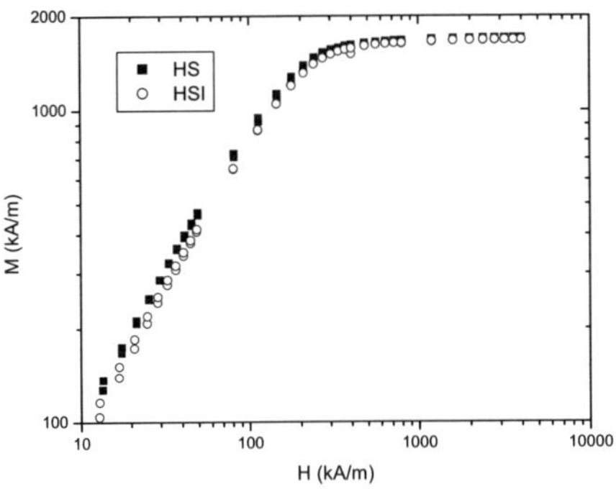

Figure 1: Magnetization M as function of the magnetic field strength H for the pure CIP HS and HSI.

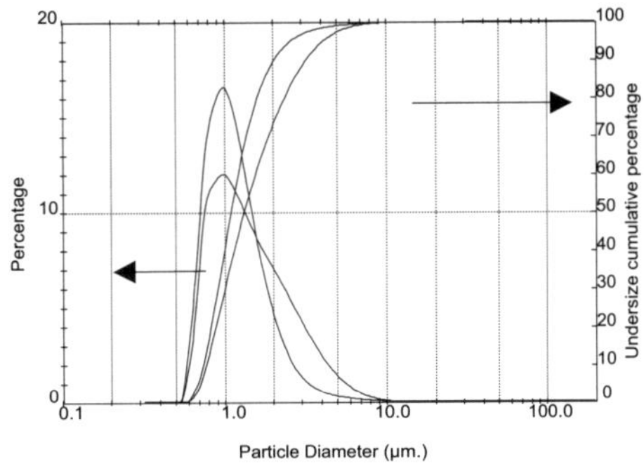

Figure 2: Particle size distribution (number) and undersize cumulative curves for CIP HS (dashed lines) and HSI (solid lines), measured by LALLS.

Figure 1 shows that both CIP almost has no hysteresis. Although the HS powder has a little more magnetization than HSI at low field strength, above 400 kA/m (~ 95% of magnetic saturation) both powders have practically the same magnetic saturation. Besides, the change on the initial magnetic susceptibility (below 100 kA/m) was more significant for HS than for HSI. In others words, the phosphate coating on the surface of HSI powder decreases only slightly the magnetic saturation.

Figure 2 shows the particle size distribution for both CIP's. It can be seen that the powders have almost the same size; with the HSI having particles slightly larger than HS. Perhaps, this size difference could be due particle clusters, because the interactions between the phosphate groups on the surface of the HSI particles.

Table 2 resume the magnetic properties and the percentile diameters obtained for both powders. It could be noted that the HSI shows particles with larger size diameters leading to higher values for d50% and d90%.

Table 2: Magnetic properties and distribution percentile diameters for HS and HSI (μm)

CIP	HS	HSI
$\chi_{initial}$	9.72	8.56
M_s (kA/m)	1704	1681
d10%	0.74	0.77
d50%	1.09	1.32
d90%	2.03	3.21

3.2 Rheological measurements

3.2.1 Oscillatory measurements

Firstly, the MRS samples were investigated through strain amplitude sweep without magnetic field, to verify the effect of the type of silica and their interactions with the CIP's used. The obtained results in Figure 3 showed clearly that the MRS based on hydrophobic fumed silica (squares) present a linear viscoelastic region (LVE) clearly defined with limit strain values of approximately 0.05 % for HSI610 and 0.2 % for HS610 systems. Maybe, for the MRS based on hydrophilic fumed silica (circles), the

LVE region (if it exist), is downward the minimum limit of strain achievable by the rheometer, and therefore was not observed.

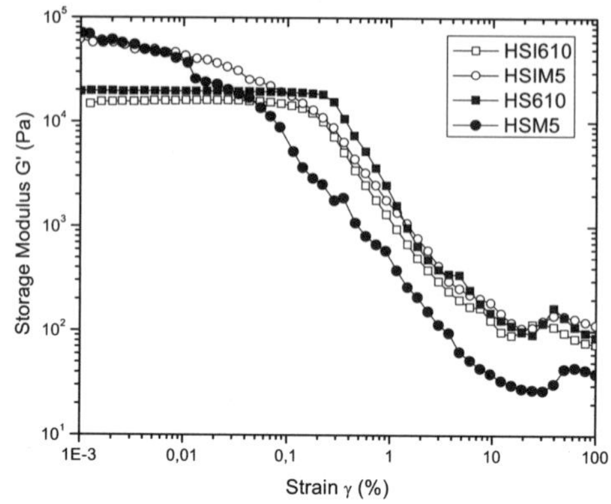

Figure 3: Storage modulus as function of the strain, without magnetic field, for the MRS.

It can be noted that for the HSI powder, the storage modulus obtained for hydrophilic silica (circles) is over twice the modulus obtained for the MRS with hydrophobic (squares) until c.a. 0.01% strain, decreasing until c.a. 0.1% and after that, changes practically to the same, independent of the kind of silica. For the HS powder, the initial storage modulus obtained for hydrophilic silica was larger than that of hydrophobic but on c.a. 0.03% of strain an inversion occurs and the hydrophobic system become to have a larger value. Measures of magnetic field sweep were also performed to evaluate the MR effect of the MRS. Figure 4 shows the results under constant strain $\gamma = 0.01\%$.

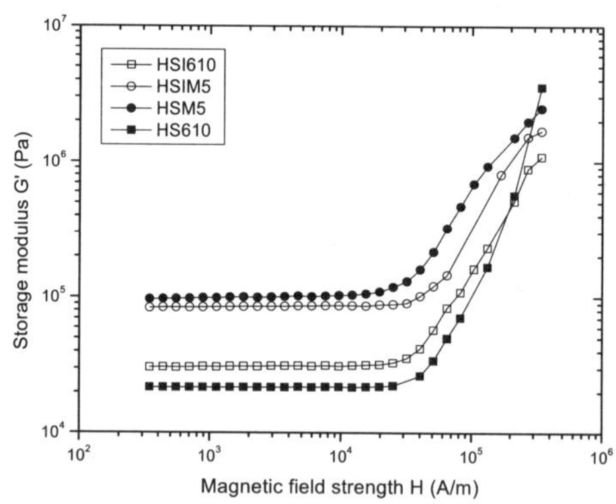

Figure 4: Storage modulus as function of the applied magnetic field strength, measured through a magneto-sweep with constant strain of 0.01%.

Oscillatory magneto sweep obtained under small deformations (0.01%) (Figure 4) showed that the systems using hydrophilic silica have larger elastic contribution ($G^* \sim 10^5$ Pa), than the hydrophobic one ($G^* \sim 2 - 3 \times 10^4$ Pa). Besides, considering the relative MR effect as the G' increase, the MRS of HSIM5 (phosphate – hydroxyl) showed the smaller effect, probably due a stronger network between hydrophilic silica and phosphate coating on HSI. Here, the rheological behavior was droved by the kind of silica used.

3.2.2 Steady Flow Measurements

The Figure 5 shows the flow curves for the MRS without magnetic field, while the Figure 6, the viscosity curves for the MRS with and without magnetic field. Open points under field strength = 340 kA/m.

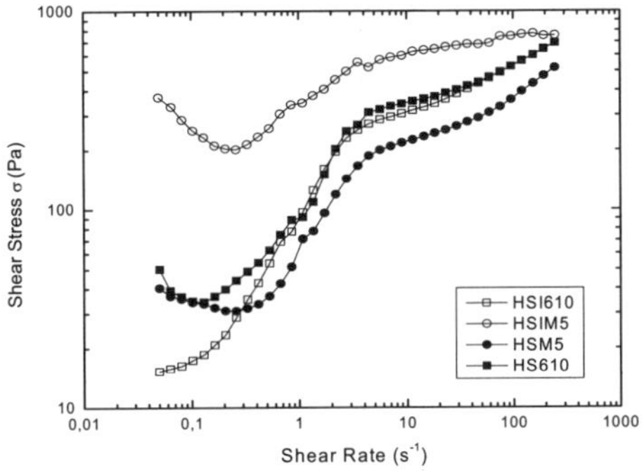

Figure 5: Flow curves for the MRS, without magnetic field.

It can be seen at Figure 5, that the curve of MRS HSIM5 have the highest shear stress values in all region tested, but especially at low shear region, around one order magnitude higher, when compared to the MRS HSI610. This is other evidence, which reinforces the idea of some specific interaction between phosphate and hydroxyl groups. These results could be explained by the presence of specific interactions and/or preferential adsorption of nanoparticles of silica on the iron powder surface.

From figure 3 is evident that there is a difference between the silicas, but no a specific interaction with the phosphate coating on HSI. The manufacturer of the fumed silicas reports for the hydrophilic grade the absence of network formation or at least a very weak interaction between silica nanoparticles, when groups responsible for hydrogen bonds are present (alcohols, carbonyl compounds, ethers) [10]. Khan and Zoeller confirm these behaviors for the silica Aerosil D150 (hydrophilic) dispersed on poly(propyleneglycol). They call "no bridging" the absence of hydrophilic silica network: "the solvent attaches itself to the surface silanol group of the fumed silica rendering it inactive for further network formation" [11]. Sazanova *et al*, studying the adsorption of tributyl-phosphate on silica gel, concluded that the first layer adsorbed on the particle is due to the hydrogen bonding between the protons of silanol groups at surface and the electron-donating oxygen atom of the adsorbate molecule [12]. So, it is reasonable that $FePO_4$ over the surface of HSI will changes the network between hydrophilic silica nanoparticles.

The Figure 6 showed that, for the entire region studied, the HSIM5 presented the highest "off – viscosity". Under the presence of magnetic field, the apparent viscosity was almost equal for all MRS. It become evident that the kind of silica and its interactions with the particles present on the formulation will be responsible by the apparent viscosity only without magnetic field.

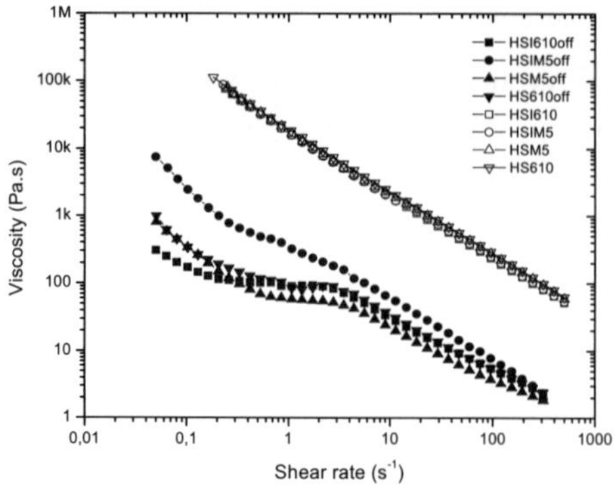

Figure 6: MR effect in terms of the viscosity increase. Full points without field; Open points under field strength = 340 kA/m.

3.2.3 Magneto-sweep under steady shear flow

The studies of magneto-sweep under steady shear flow were performed at relatively low shear rate: 1 s^{-1}. The results are shown at Figure 7.

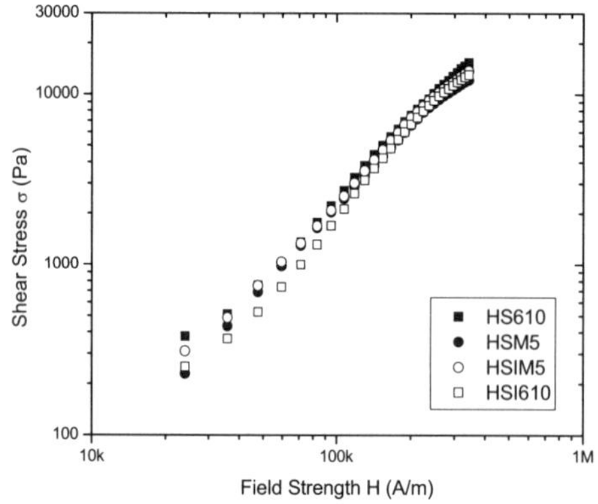

Figure 7: Shear stress, measured under constant shear rate (1/s) and sweeping the applied field.

In Figure 7, one can see that again, there is no significant difference between MRS under field. Besides, any difference is reduced at higher fields.

If we consider the entire region of applied magnetic field it is possible to observe that the shear stress increases with the magnetic field according a power law with n ≈ 1.5, in good agreement with the model proposed by Ginder *et al.* [13]. In their model, Ginder *et al* suggest the existence of three regions for the dependence of the yield stress with the magnetic field. A quadratic dependence was observed only at very low field region, below 30 kA/m.

For intermediate magnetic field region this dependence becomes very close to 3/2, in good agreement with Ginder's model. Then, a maximum value would be reached with the saturation of the magnetic particles. At this point, the yield stress becomes dependent only to the volume fraction of particles and to the square of the magnetization of saturation (M_s).

From the data presented at Figure 1 it is possible to choose three intervals of magnetic field: a low field region until 100 kA/m, an intermediate region comprising the values from 100 to 250 kA/m and a higher region above 250 kA/m. These regions were chosen taking into account the Ginder model. From the Figure 1 it can be seen that above 400 kA/m, both particles HS and HSI are very close to the saturation. In our case the study under higher magnetic fields is not possible due to the limitation on reaching, with MRD-180, values of magnetic field closer to the magnetic saturation.

The fitting parameters obtained for all the MRS's in Figure 7 are listed on Table 3. It can be seen that even for the low field region, the dependence of the shear stress with the square of the field was not observed, maybe because this dependence is valid only for very low fields. At higher fields, the shear stress increased linearly with the field strength.

Table 3: Fitting parameters of the data in Figure 7. Slope = 'n' and correlation R^2.

MRS	**Low Field** $30 < H < 100$ **kA/m**		**Intermediate Field** $100 < H < 250$ **kA/m**		**High Field** $H > 250$ **kA/m**	
	'n'	R^2	'n'	R^2	'n'	R^2
HS610	1.53	0,99970	1.54	0,99985	1.32	0,99947
HSM5	1.57	0,99952	1.55	0,99945	0.99	0,99924
HSIM5	1.49	0,99881	1.62	0,99961	1.10	0,99978
HSI610	1.62	0,99996	1.72	0,99989	1.12	0,99947

4 Conclusions

Without magnetic field, oscillatory measurements showed that apparently there is no interaction between the fumed silicas with phosphate coating of the HSI powder. On the other hand, under steady shear, the higher viscosity of the MRS made of the HSI with hydrophilic silica strongly suggest that the network in this system, it is stronger and a possible explanation would be the bridging between the phosphate groups on the iron powder with hydroxyl groups on nanoparticles of silica. It should be emphasized that the kind of silica and its interactions will be responsible by the apparent viscosity only at no magnetic field.

Under applied magnetic field, measurements of the storage modulus, showed that under very small strain ($0.01s^{-1}$) the difference between the two CIP's was a secondary factor and the rheological behavior was droved by the kind of silica used.

The flow experiments performed under magnetic field, showed that the interaction of the phosphate coating on CIP with hydroxyl groups on silica is secondary and the flow curves difference will be due to the iron particles and its magnetic properties, mainly on the region of higher shear rate.

The results obtained for the magneto-sweep under constant shear rate, showed that the extension as an approximation, of the dependence of Yield Stress with the magnetic field to the shear stress obtained under steady shear flow, as proposed by See [14] is valid only for the intermediate region of the magnetic field, with the adequate choice of the region limits.

5 Acknowledgements

AJFB acknowledges BASF AG for support to attend ERMR'06 as well as the cooperation agreement BASF AG – UNIFEI. MRA acknowledges the financial support from Fundação de Amparo à Pesquisa do Estado de S. Paulo (FAPESP).

References

1. Viota JL, de Vicente J, Duran JDG and Delgado A; "Stabilization of magnetorheological suspensions by polyacrylic acid polymers"; Journal of Colloid and Interface Science 284 (2): 527-541; 2005.
2. Lim ST, Choi HJ and Jhon MS; "Magnetorheological characterization of carbonyl iron-organoclay suspensions"; IEEE Trans. on Magnetics 41 (10): 3745-3747; 2005.
3. de Vicente J, Lopez-Lopez MT, Gonzalez-Caballero F and Duran JDG; "Rheological study of the stabilization of magnetizable colloidal suspensions by addition of silica nanoparticles"; J. Rheology; vol. 47(5); pp. 1093 – 1109; 2003.
4. Klingenberg DJ, Rankin PJ and Horvath AT; "Magnetorheology in viscoplastic media"; Rheol. Acta; vol. 38(5); pp. 471 – 477; 1999.
5. Park OO, Park JH and Kwon MH; "Rheological properties and stability of magnetorheological fluids using viscoelastic medium and nanoadditives"; Korean Journal of Chemical Engineering; vol. 18(5); pp. 580 – 585; 2001.
6. Wereley, N.M. *et alli*; "Bidisperse MR fluids using Fe particles at nanometer and micron scale"; J. Intelligent Material Systems & Structures; vol. 17: 393 – 401; 2006.
7. Park OO, Chin BD, Park JH and Kwon MH; "Rheological properties and dispersion stability of magnetorheological (MR) suspensions"; Rheol. Acta; vol. 40(3); pp. 211 – 219; 2001.
8. Johansson E and Nyborg L; "XPS study of surface-active organic compounds on fine ferrous powder"; Surface and Interface Analysis 30 (1): 333-336; 2000.
9. Heim L, Farshchi M, Morgeneyer M, et al.; "Adhesion of carbonyl iron powder particles studied by atomic force microscopy"; Journal of Adhesion Science and Technology 19 (3-5): 199-213; 2005.
10. Cabot Corporation, "Cab-O-Sil® Untreated fumed silica properties and functions"; Cab-O-Sil Division; Tuscola – USA; 1993.
11. Khan, S.A. and Zoeller, N.J.; "Dynamic rheological behavior of flocculated fumed silica suspensions," J. Rheology vol. 37; pp. 1225 – 1235, (1993).
12. Sazonova, V.F., Kojemyak, M.A. and Perlova, O.V.; "Adsorption of tributyl phosphate on silica gel," Adsorption Science & Technology vol. 19: 211 - 217 (2001).
13. Ginder JM, Davis LC and Elie LD; "Rheology of magnetorheological fluids: models and measurements"; Int. J. Mod. Physics B; vol. 10(23&24); pp. 3293 – 3303; 1996.
14. See, H., "Field dependence of the response of a magnetorheological suspension under steady shear flow and squeezing flow"; Rheologica Acta 42: 86 – 92; 2003.

MONITORING INTERPARTICLE DISTANCE IN MAGNETORHEOLOGICAL COMPOSITES

G. BOSSIS[1], E.COQUELLE[1], C.NOËL[2] F.GIULIERI[,2],A.M.CHAZE[2] ,

[1] *Laboratoire de Physique de la Matière Condensée, Université de Nice, Parc Valrose, 06108 Nice Cedex 2, France*

[2] *Chimie des Matériaux Organiques et Métalliquesx, Université de Nice, Parc Valrose, 06108 Nice Cedex 2, France*

We describe two different systems, the first one based on a magnetorheological elastomer and the second one on magnetic particles inside a liquid crystal. In both system we manage to have chain structures with particles that are not in contact. The effect of the gap on the viscoelastic properties. We show in particular how in magnetorheological elastomers, the energy dissipation is closely related to the creation and the motion of cavities in the gap between the particles. In liquid crystal chaining of particles can occur without applying a magnetic field. This happens if the anchoring of liquid crystal on the surface of the particles is homeotropic.We demonstrate how the combination of elastic defects and of a magnetic field allow to obtain microscopic springs made of a pair of magnetic spheresp.

1. Introduction

Filled elastomers are used in a wide range of industrial application, going from tires to dampers. In some cases, it is useful to get an active control on vibrations[1]; the magnetorheological elastomers (MRE) perfectly respond to this demand, because of their ability to rapidly and reversibly change their elastic and viscoelastic properties, depending on the intensity of an applied external magnetic field[2,3,4,5,6]. Those properties are given by the organization, during the curing of the elastomeric matrix, of micron-sized magnetic particles in a unidirectional chain like structure. Therefore, it is possible to induce by an external magnetic field an attractive force between neighbour particles that makes the chains acting like pseudo-fibres, strengthening the material. This makes it possible to adjust the natural frequency of structures (through an equivalent stiffness).

Under dynamic solicitations, filled rubbers are known for their non-linear strain response. It has been firstly studied by Payne in the 60's, and his name was given to this unexpected behaviour [7], which seems to be a constant of all filled rubbers [8,9,10] Indeed, Payne has observed a decrease of the conservation and the loss modulus with the strain, starting from an initial value E_0' (limit when $\varepsilon \to 0$), and falling down to a plateau E_∞' for $\varepsilon \to \infty$. Payne suggested that the dynamic deformations break some aggregates and the filler network, while the plateau results from strong links and hydrodynamic interactions in the elastomer [11]. The higher the strain is, the more the number of broken aggregates that cannot be reformed are [12]. This qualitative explanation is supported by some experiments which show that, if one takes a special care for the dispersion of the filler, the modulus decrease $\Delta E' = E_0' - E_\infty'$ is reduced [13].

In this paper we shall focus on the effect of the interparticle gap on the rheology of elastomers filled with particles having a chain like structure.

In section 2 we give a brief presentation of the preparation of MRE- especially concerning the coating used to increase adhesion between the polymeric matrix and the particles and also of the measurement of adhesion energy. In section 3 we describe the model we use to predict the stress-strain curve in the presence of detachment of elastomer and we compare the predictions of this model with the experiments done with two macroscopic spheres and for the composite. At last in section 5 we introduce a new system made of iron particles immersed in a liquid crystal and we show that this kind of system gives chain of aprticles with a well determined gap between particles.

2. Elastomer's preparation and characterization

The first step is a chemical clean-up of the iron particles, to remove their impurities and to get a reproducible surface. Using an ultrasound sonicator, one washes them successively with trichloroethylene, ethanol, pure water, ethanol and acetone, 10 minutes each. Depending on the chemical coating of the particles, we are able to study a weakly bonded system (untreated particles) and a strongly bonded one. To achieve this last configuration, the particles are coated with a silane coupling agent to improve adhesion to the elastomer: a 5% solution of aminosilane is prepared in pure water; the iron powder is then blended inside the solution still using the sonicator. The treated powder is at last washed using ethanol and then dried under vacuum.

Then, particles are carefully dispersed in the pre-polymer Rhodorsil RTV 1062S (initial viscosity η=45 Pa.s, initial Young modulus E_0=230kPa). The sample is degassed under vacuum, then poured in a cylindrical mould and put in rotation inside an electromagnet in order to avoid sedimentation during the structure creation. More details can be found elsewhere[14].

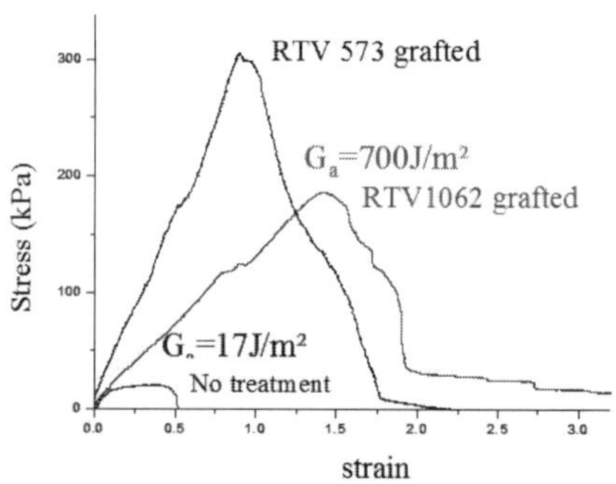

Fig. 1 Traction curves for coated and uncoated iron

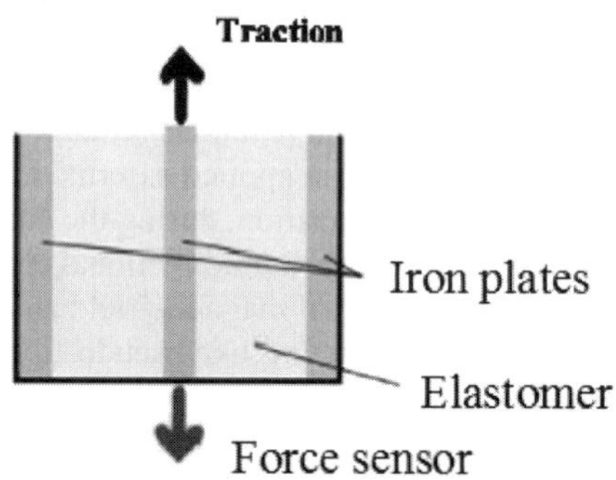

Fig.2 schematic view of adhesion measurement

To measure the adhesion energy between iron and the elastomer we have use the same treatment on iron plates as on iron particles. The traction curves are realized in the configuration schematized in fig.2 The traction curves for the same elastomer (RTV 1062) with and without coating are shown in fig.1 The efficiency of the coating is clearly demonstrated. The adhesion energy is obtained by identifying the elastic energy store at the rupture, to the surface energy : $G_a = e\int_0^{\gamma_c}\sigma d\gamma$, where e=5mm is the distance between two plates. It drops from 700J/m^2 with the coating to 17J/m^2 without coating.

3. Modelling of the two spheres system

Let's consider two close spheres, representing at the macroscopic scale two particles of a chain. Thus, during uniaxial tensions, we expect that the maximum local stress is located in the elastomer, at the vicinity of the poles of the spheres, leading to a detachment of the polymer matrix from the particles (provided the polymer-to-particle bonds are weaker than the polymer-to-polymer bonds). This process, called debonding, has been extensively studied in the 80's on static experiments by Gent [16]

It has been shown that, in the frame of linear elasticity, a two spheres system separated by a small gap, could be described by the same approximations as those used in lubrication theory [17]. In this approach the effective modulus E_{eff} of a cell including two half spheres of radius a and separated by a gap g is:

$$E_{eff} = \frac{1}{4} E_m \frac{(2a+g)}{a} \int_0^{\frac{\pi}{2}} \frac{\sin^3\theta \cos\theta}{\left(\frac{g}{2a} + 1 - \cos\theta\right)^3} d\theta \qquad (1)$$

where Em is the modulus of the elastomer. We have verified by comparing with the results obtained with a finite element code, that this approximation was justified.

Now, if the elastomer can debond from the surface, a cavity is initiated at the surface of the particles and growth up till it reaches their equator. The modelling involves three steps: a stress/strain relationship before the initiation of any damage that is based on Eq.(1), a criteria predicting the creation and the growth of a debonded area, and a stress/strain relationship for the 2 spheres system with a cavity inside the gap. As the strain is increased, the detachment of the elastomeric matrix from the spheres takes place as soon as the local elastic energy released by the matrix overcomes the energy needed to create a new surface, and hence to propagate debonding. From this condition the stress below the cavity is related to the angle θ_{cavity} defining the cavity (cf. fig.(3) by Griffith criterion[18]: $\sigma_c = \sqrt{\frac{8\pi E G_a}{6a\sin(2\theta_{cavity})}}$ On the other hand the stress below the cavity is obtained from Mooney equation for a stretch $\lambda(r)$:

$$\int_0^{a\sin\theta_c} \frac{E}{3}\left(\lambda(r) - \frac{1}{\lambda(r)^2}\right) \frac{2r}{(a\sin\theta_c)^2} dr = \sqrt{\frac{8\pi E G_a}{6a\sin(2\theta_c)}} \qquad (2)$$

Solving numerically Eq.(1) for θ_c allows to obtain the associated stress $\sigma_c(u)$ that is one member of the equality. This solution is obtained for a given displacement u that defines the local stretch $\lambda(r)$. The local stretch is obtained from the knowledge of the cavity shape: $f(r,\theta_c)$. From FEM simulation we have found[15] that the shape of the cavity is well approximated by $f(r,\theta_c) = -C(u, \theta_c)\, r^{2.2} - g/2$ with $C(u, \theta_c)$ satisfying to the condition on the surface: $f(a\sin\theta_c, \theta_c) = -g/2 - u - a(1-\cos\theta_c)$ (cf Fig. (3)).

The total stress is obtained by adding to the stress below the cavity the one of the area $\theta > \theta c$ that is modelled in the same lubrication approximation as Eq.(1):

$$\sigma = \sin^2\theta_{cavity}\int_0^{a\sin(\theta_{cavity})}\frac{E}{3}\left(\lambda(r)-\frac{1}{\lambda(r)^2}\right)\frac{2r}{\left(a\sin\left(\theta_{cavity}\right)\right)^2}dr$$
$$+\left(1-\sin^2\theta_{cavity}\right)\frac{1}{4}E_m\varepsilon\frac{(2a+g)}{a}\int_{\theta_{cavity}}^{\frac{\pi}{2}}\frac{\sin^3\theta\cos\theta}{(g/_{2a}+1-\cos\theta)^3}d\theta \qquad (3)$$

To check the validity of this model we have done experiments on two macroscopic iron spheres (radius 1cm) separated by a gap of 0.5mm. One pair of spheres was grafted as indicated in section 2. and the other one was untreated. The stress strain curve is linear for the grafted pair whereas it shows an abrupt transition for the other pair. The dotted lines are the predictions of Eq.(3). For the coated spheres there are no parameters at all- the modulus E is the one measured on a cylinder of the same elstomer at the same frequency- and the analytical model predicts perfectly well the behavior of this system. For the second system the only parameter is the critical debonding stress, or from Eq. (1) the initial defect size θ_i (in this example 1°), since the adhesion energy was independently measured. We see that the drop of stress and further increase is quite well predicted. The second test concerns the dissipated energy, that is to say tg(δ). The creation of a cavity can increase the dissipation of energy; by introducing a supplementary phase shift between the force and the strain but after one cycle the stress and the strain should come back in phase (if we neglect the dissipation in the elastomer). This is actually not the case because when the cavity becomes closed the elastomer readheres and still some energy is needed to debond the elastomer at the next traction. This behavior is well identified on the tow spheres system(fig.5) where the second traction curve is between the first traction and the retraction curves. From this experiment we have calculated a new adhesion energy G_2=0.45 G_a .Replacing G_a by G_2 in the determination of θ_c (Eq.1) and using Eq.(3) for the stress versus strain curve we can calculate the dissipated energy (area between the traction

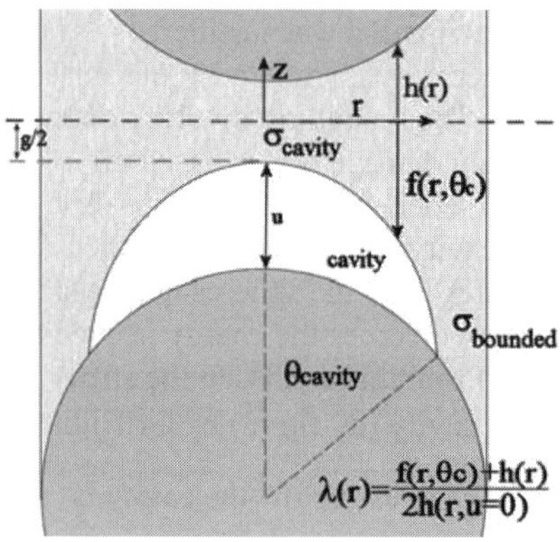

Fig.3 Definition of the local stretch λ(r);g is the initial gap between the particles. At r=0 the elastomer remains unstretched.

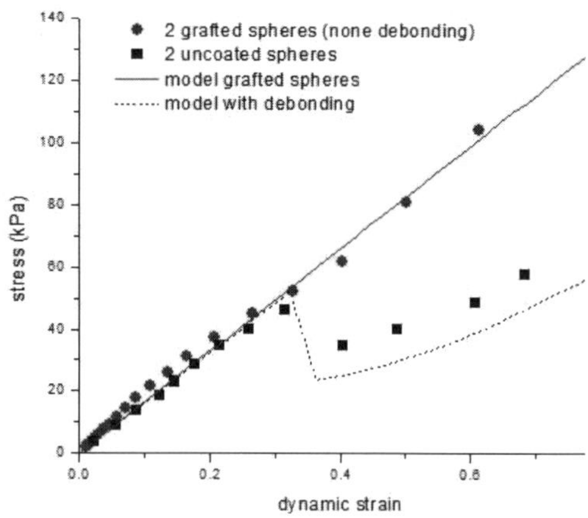

Fig 4 Stress versus amplitude of dynamic strain: u/g for two particles with initial gap g/a=0.05. Dots: grafted spheres: squares: :no treatment. Frequency 5Hz.

and retraction curve) for different strain amplitudes. The comparison with the experiment is made in fig.6. First we see that debonding of the uncoated spheres is associated with a jump in tg(δ), so this is an experimental proof of the increase of dissipation due to debonding. Secndly our model well capture the amplitude of this jump even if at higher strain the agreement is not so good.

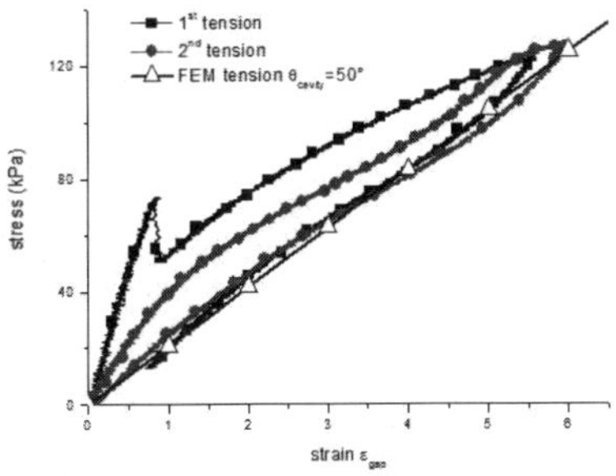

Fig.5 Hysteresis in the stress strain curve showing a readhesion during the second tension

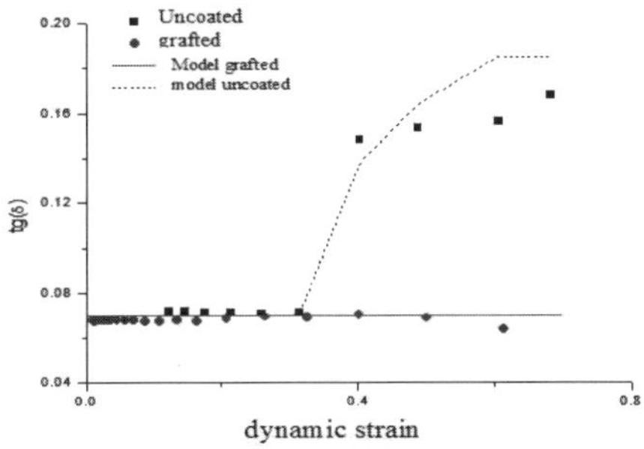

Fig.6 tg(δ) versus amplitude of dynamic strain: u/g for two particles with initial gap g/a=0.05. Dots: grafted spheres: squares: :no treatment. Frequency 5Hz.

The transposition of this two spheres model to the composite is possible on the basis of a chain of particles with a distribution of gaps that follows a Gaussian distribution. Then the solution of the equation of motion for N particles can be obtained on the same basis as for two particles. This model is able to predict the Payne effect that is to say the decrease of the modulus with the strain amplitude. More interesting would be the prediction of the effect of the magnetic field. This is not so simple since the magnetic forces between particles depend strongly on the interparticle gap. A rather striking feature is the fact that the dissipation increases in the presence of a magnetic field as can be seen in fig 7. for a structured composite with a volume fraction of 10% of iron particles. A possible mechanism which could explain this behavior is schematized in fig.8: If a cavity is created between two particles during the traction, then the gap between the two particles that include the cavity becomes larger. In this situation the magnetic force between these two particles is no longer balanced by the ones coming from the two others at extremity and a net magnetic force will be exerted that will increase the size of the cavity. This mechanism, coupled to the readhesion process could explain this increase of dissipation in the presence of a magnetic field.

We have seen that the existence of a small gap between the particles can determine both the storage and the dissipation modulus through debonding phenomena. We would

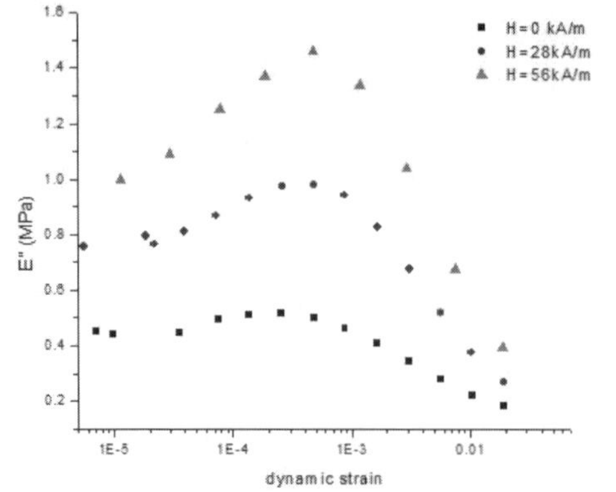

Fig.7 Increase of dissipated energy by application of the field; RTV1062, ϕ=10%

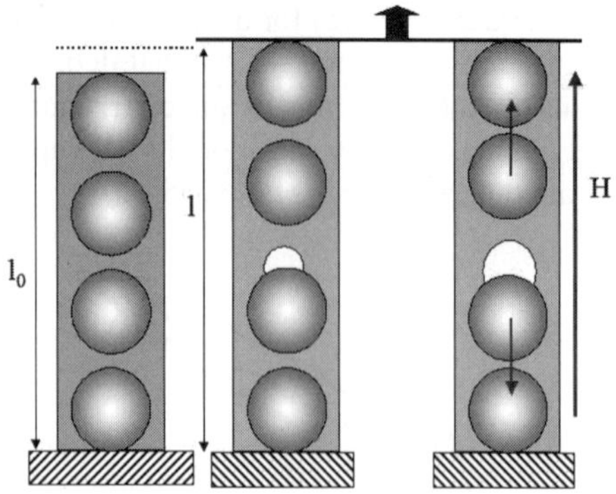

Fig.8 : Due to asymmetric gaps, the field increases the size of existing cavities

like to be able to master the formation of this gap but in practice it depends on the structuring process and on the strength of adhesion between the prepolymer and the iron surface. However we would like to show that, using a different dispersing medium, it is possible to obtain a well defined gap between iron particles.

4. Dispersion of iron particles in a liquid crystal

It is known that droplets or solid particles inserted in a nematic liquid crystal will create defects of orientation which cost some elastic energy to the system. In order to reduce this elastic energy the particles tend to aggregate. More interestingly, if the anchoring of the liquid crystal molecules is perpendicular to the surface the long range attractive force is of dipolar nature and lead to the formation of nice chains of droplets in the absence of any applied field[19]. An example is shown in fig.9 with iron particles of radius 2 microns covered with molecules favoring a strong perpendicular anchoring of the liquid crystal[20].

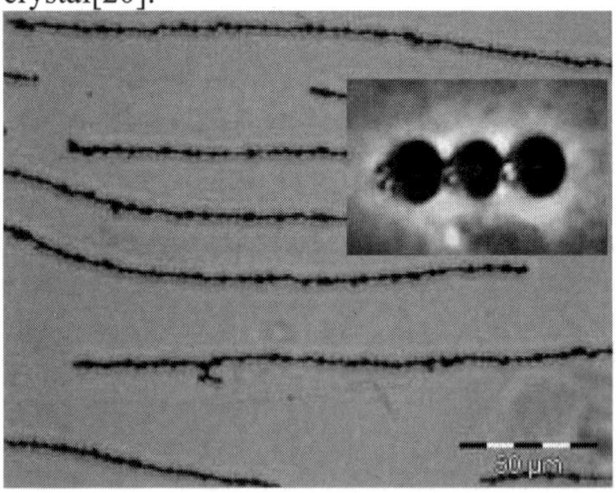

Fig.9 alignment of iron particles obtained in an E7 nematic liquid cystal. Upper insert: zoom on particles showing the existence of a well defined gap.

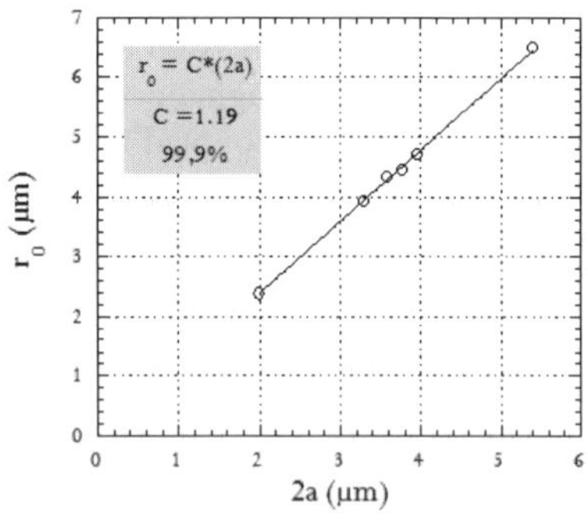

Fig.10 Average distance between two particles versus the average diameter of the pair.

The important point is that the particles do not come into contact because a defect remains blocked between the particles and causes the particles to repel each other at short distances. Theories based on a minimization of Franck energy have predicted that the equilibrium distance was 1.23 diameter for equal size particles[21]. Our measurement gave r_0= 1.19 ±0.05 <d> where d is the average diameter of the pair of particles. We still did not verify that this values holds also in long chain of spheres but it is very likely so.

Such a nice regular arrangement with a constant gap between particles could be kept by adding some polymer able to form a gel like structure. An other solution to keep this structure would be to use a nematic molecule able to polymerize after the arrangement of particles have been obtained. In this case we would have a soft magnetic composite offering very large magnetostrictive effect and certainly strong change of elasticity in the presence of a magnetic field. It is also worth noting that these alignments with aconstant gap betweebn particles can be obtained with other kinds of particles, as for instance silica particles, that could be actuated by an electric field instead of a magnetic field.

References

1. Dominguez A., Sedaghati R., Stiharu I., 2004, Smart Material. Struct. **13,** 1351-1361
2. Carlson J., Jolly M. 2000, Mechatronics **10**, 555-569
3. Ginder J., Nichols M., Elie. M, Tardiff J., 1999, SPIE *Con.f on Smart material Technologies.*, 131-138
4. Bellan C., Bossis G. ,2002, Int. J. Mod. Phys **16**, 2447-2454
5. Jolly M., Carlson J., Munoz B., 1996 ., Smart Material. Struct. **5,** 607-614
6. Zhou G., Li J., 2003, Smart Material. Struct. . **12,** 859-872
7. Payne A., 1962, J. Appl.Polymer Sci. **21**, 368-372
8. Brown J., Strenstein S., Yanyo L. 1997, Proceedings of the 4th Conference on Deformation and fractures of Composites, London, 49-52
9. Heinrich G., Kluppel M., Vilgis A. Current Opinion in Solid State and Material Science **6**, 195-203
10. Dutta N., Tripathy D., 1992, J. Appl. Polymer Sci. **44**, 1635-1648
11. Vieweg S., Unger R.? Heinrich G., Donth E. 1999 , J. Appl. Polymer Sci.**73**, 495-503
12. Payne A. 1963 , J. Appl. Polymer Sci,. **7** , 873
13. Payne A. 1999 , J. Appl. Polymer Sci., **73**, 495-503
14. Coquelle E. Bossis G. Giulieri F. 2006, Journal of Material Science, in press
15. Coquelle E. Bossis G. Giulieri F. 2006, Int. J. Solid Structure, in press
16. Gent A, 1980, Journal of Material Science **15,** 2884-2888.
17. Christensen R. 1979 “Mechanics of composite material”, Wiley,
18. Kendall K. J. 1994, J. of Adhesion Sci. and Technol. 11, 1271-1284
19. Poulin P., Stark H., Lubensky T. C., Weitz, D. A., 1997, Science **275**, 1770
20. Noel C. M., Bossis G., Chaze A. M., Giulieri F., Lacis S., 2006, Phys. Rev. Lett. 96, 217801
21. Fukuda J., Stark H., Yoneya M., Yokoyama H., 2004, Phys. Rev. E **69**, 041706

STUDY OF UTILIZABLE MAGNETORHEOLOGICAL ELASTOMERS

X. L. GONG, L. CHEN AND J. F. LI
CAS Key Laboratory of Mechanical Behavior and Design of Materials,
Department of Mechanics and Mechanical Engineering,
University of Science and Technology of China, Hefei 230027, China
E-mail:gongxl@ustc.edu.cn

This paper presents two kinds of magnetorheological elastomers (MREs). One is composed of appropriate silicon rubber, carbonyl iron particles and some other materials. It is cured under a strong magnetic field at a room temperature. Its relative modulus increment reaches 878%. Such high MR effect has not been reported until now. The other is composed by appropriate natural rubber, carbonyl iron particles and some other materials. After the compositions are mixed in a two-roll mill, they are cured under a strong magnetic field according to a temperature profile. Its relative modulus increment reaches 133%. Furthermore, it has enough hard strength to load bearing and absorb vibration energy. To measure their dynamic viscoelastic properties, DMA (Dynamic Mechanical Analyser) has been developed. A special magnetic field is fit to DMA. Its field intensity can reach 1.1T. This setup can measure the dynamic magnetoviscoelasticity of MREs under tension, compression, bending, and shear deformation modes. The shear mode results meet well with that measured by our previous self-developed measurement system. Electronic Pull Test Machine, Shaw's LX-A Rubber Hardness Gauge, Akron Abrasion Test Machine, Thermal Ageing Test Oven, Rubber Fatigue Test Machine, and Impact Elasticity Test Machine are also set up to measure their corresponding mechanical properties. All observed results show that the fabricated MREs are utilizable. They have successfully been utilized to adaptive tuned vibration absorbers, which will serve for vibration absorption of vehicles.

1 Introduction

Magnetorheological (MR) materials are a kind of intelligent materials whose performance can be controlled by an applied magnetic field. Since MR effect was discovered by Rabinow in 1948 [1], up to now, MR materials have already become a large family with MR fluids, MR foams and MR elastomers (MREs) etc. [2]. MREs are mainly composed of micrometer size magnetizable particles and elastomer matrix. After the particles and matrix material are mixed, the mixture is usually cured in the presence of a magnetic field. Some special structures, such as chains or columns of particles, remain in the matrix. When MREs are exposed to an external magnetic field, their storage moduli will change with the intensity of applied field.

Similar to MR fluids, MREs have also attracted considerable interest in both academic and industrial fields [3-13]. MREs based on soft silicone elastomers, hard natural rubber, poly (vinyl alcohol), gelatin and RTV polyurethane sealant have been prepared. From a manufacturer's point of view, recently isotropic MREs are also studied which are prepared without applying magnetic fields [14-17]. A number of models have been developed to describe MRE's performances [4,5,18,19]. MREs can be used in many applications that require variable stiffness components, such as adaptive tuned vibration absorbers (TVAs), stiffness tunable mounts and suspensions, and variable impedance surfaces [2,4,5]. The Ford Motor Company has patented an automotive bushing employing a magnetorheological elastomer [7,8]. Although there is currently little report on applications of MREs, it is still a kind of promising material that can be used in many applications with its controllable stiffness and other unique characteristics [2].

However currently developed MREs have not large enough MR effects, which have limited their wide industrial applications. And after adding the iron particles, their mechanical shouldn't become much weaker than the matrix. This paper aims to prepare MREs having good MR effect and good mechanical properties; a reliable evaluation system is also presented.

2 Experimental

2.1 Preparation of MRE Materials

Two kinds of MREs are prepared.

One is based on natural rubber. A conventional rubber-mixing technique, Two-Roll Mill, is used. Through the rolls uninterruptedly rotating, the molecular chains in natural rubber are breakdown, and the natural rubber losses its elastic and becomes viscous body gradually. Then crosslinkers, processing aids, carbonyl iron particles and plasticizers are added into the natural rubber. The used carbonyl iron particles are grade CN from BASF with the average sizes of 3.5μm, the natural rubber and other additives are provided by Hefei Wangyou Rubber Company. The resulting material is compressed into a mold. Then the mold is set in a self-developed magnet-heat coupled device (as shown in Fig.1) which is capable of applying the sample a magnetic field of 0 to 1.5Tesla and a temperature field in the range from 50℃ to 200℃. During the pre-cured stage, the heating system and the magnetic field are both turned on. The particles are magnetized and then form chains aligned along the magnetic field direction. 30 minutes later, the procedures of forming pre-configuration is finished. After shutting down the magnetic field, the temperature is raised to 153℃. At this condition, the sample is cured on sulfuration for 15min.

The other is based on silicone rubber. RTV silicon rubber and silicone oil are chosen as matrix. After the carbonyl iron particles are immersed in the silicone oil, they are mixed with RTV silicone rubbers. Then the mixture is put into a vacuum case to remove the air bubbles inside it. After cured about 24 hours at the room temperature under a magnetic field applied by above magnet-heat coupled device, MREs are prepared.

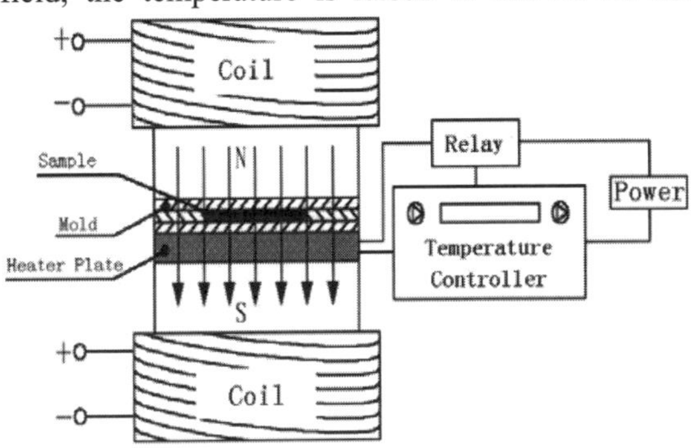

Fig.1 The sketch map of self-developed magnet-heat coupled device

2.2 Dynamic Testing System of MRE Performance

Dynamic Mechanical Analyser (DMA) is widely used to characterize materials' specific properties such as modulus and damping factor. DMA involves the application of a periodic stress to a sample and the measurement of the resultant strain. From these parameters a complex dynamic stiffness may be calculated, having two parameters, namely the in-phase component, or real stiffness and the out-of-phase component, or imaginary stiffness. The ratio of these two quantities defines the damping factor. A typical experiment consists of single or multiple frequencies applied to the test sample, which is subjected to a temperature ramp. Temperature profiles can be more complex, allowing for ramp and soak segments. The sample stiffness will depend upon its Modulus of Elasticity and its geometry or shape. The modulus measured will depend upon the choice of geometry, Young's (E*) for tension, compression and bending, Shear (G*) for torsion. For MRE, the most important properties are their viscoelastic properties under applied magnetic field. However, DMA cannot measure sample's properties under a magnetic field. So we have modified the DMA. A magnetic field ranging from 0mT to 1100mT is applied to Tritec 2000B of Triton Technology Ltd. (England). The system is shown as Fig.2. It can measure the shear modulus and damping factor at different frequency and different magnetic field. Using this system, the field dependence of shear modulus can be measured. The effects of strain amplitude and driving frequency as well as the temperature on viscoelastic properties of MREs can also be characterized.

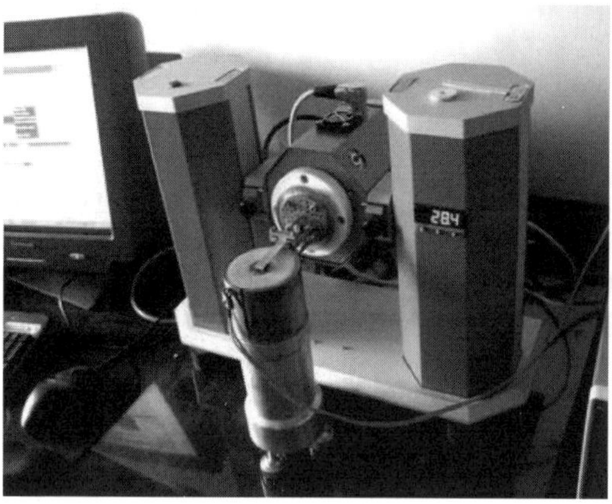

Fig.2 A magnet-mechanics coupled DMA

2.3 Mechanical Measurement System

The static properties can be observed by an Electronic Pull Test Machine, which is also fit by a self-developed magnetic field ranging from 0 to 400mT. In order to supply reference for properties of materials and applications, Shaw's LX-A Rubber Hardness Gauge, Akron Abrasion Test Machine, Thermal Ageing Test Oven, Rubber Fatigue Test Machine, and Impact Elasticity Test Machine are also set up to monitor their corresponding mechanical properties, respectively.

3 Results and Discussion

3.1 Natural Rubber Based MRE

The influences of fabrication condition and work condition on the natural rubber based MRE performances are experimentally investigated. Their mechanical properties are also monitored. It is noted that all percentages used in the context refer to weight percentages.

Fig.3 shows the field-dependence of shear modulus and damping factor of a group of MREs which contain same compositions (60% of iron particles, 20% of natural rubber and additives, and 20% of plasticizers) and pre-cured at same temperature of 80℃, but at different magnetic field of 0mT, 300mT, 600mT, 900mT, respectively. As can be observed from Fig.3 (a), shear modulus of each sample shows an increasing trend with magnetic field prior to the iron particle saturation, but strong magnetic field applied during curing leads to the high magneto-induced modulus. From Fig.3 (b), it can be observed that strong magnetic field applied during curing also leads to high loss factor.

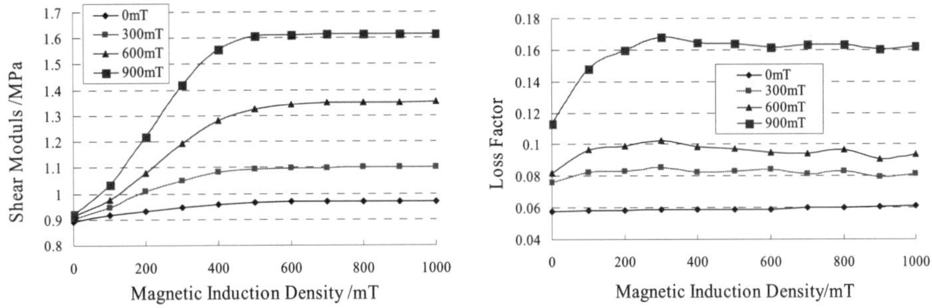

Fig.3 Magneto-induced modulus (a) and loss factor (b) *versus* applied magnetic strength for MREs cured under different magnetic field

Fig.4 shows the field-dependence of shear modulus of a group of MREs which contain same compositions as Fig.3 and pre-cured at same magnetic field (1T), but pre-cured at different temperature of 60℃, 80℃, 100℃, 120℃, respectively. As can be observed from Fig.4, the MRE pre-cured at 80℃ has the best MR effects in this group. 80 ℃ is the ideal temperature for the particles to move and form ordered pre-configuration.

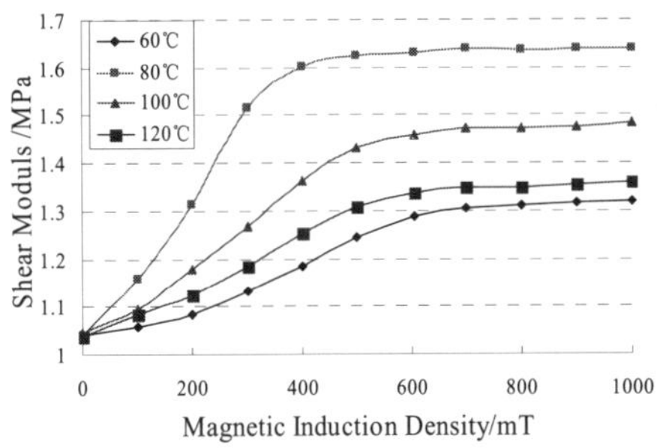

Fig.4 Shear modulus of MREs pre-cured at different temperature

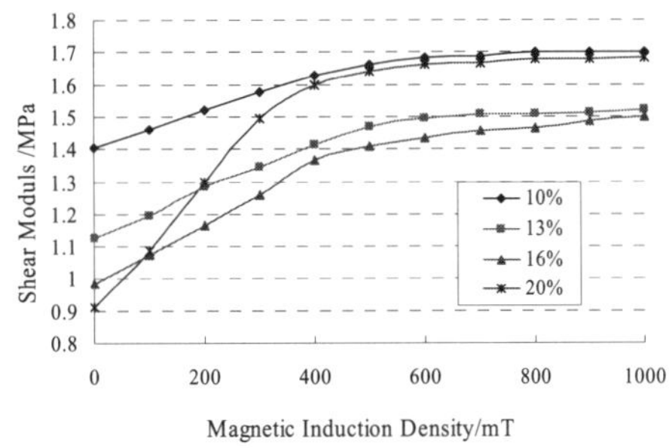

Fig.5 Shear modulus of MREs with different plasticizer

Fig.5 shows the field-dependence of shear modulus of a group of MREs which is pre-cured at same temperature of 80℃ and at same magnetic field (1T). They contain same content of iron particles (60%), but contain different content of plasticizers: 10%, 13%, 16% and 20%. From this figure, the zero-field moduli (G_0) of the samples with 10% and 20% of plasticizer are 1.4MPa and 0.9MPa, respectively. Also, their corresponding magneto-induced moduli are 0.2MPa and 0.7MPa. So the relative MR effects $\Delta G/G_0$ are 14% and 78%, respectively. Therefore, the amount of plasticizers in the matrix plays an important role in improving MR effects, especially the relative MR effects. Fig.6 shows the field-dependence of shear modulus of a group

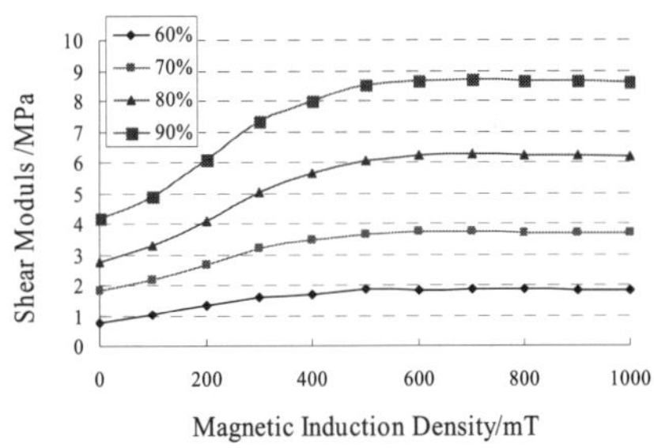

Fig.6 Shear modulus of MREs with different iron particles

of MREs which is pre-cured at same temperature of 80℃ and at same magnetic field (1T). But they contain different content of iron particles: 60%, 70%, 80% and 90%. The results show that the magneto-induced modulus increases dramatically with the particle content increasing. More the particles are, higher the magneto-induced modulus is. However, the increment of iron particles also enhances the zero-field modulus, which may decrease the relative MR effect. For example, the relative MR effect is reduced from 133% to 107% when particles content changed from 80% to 90%. Thus, it is not applicable to fabrication practical MRE by solely increasing particle contents.

The influence of work condition on MRE's properties is also investigated. From above results, sample with 80% iron particles, 10% natural rubber and 10% plasticizers is pre-cured at 80℃ and at the magnetic field of 1 T. Fig.7 shows the field dependence of modulus at various strain amplitudes, where the driving frequency is fixed as 5 Hz. The experimental results demonstrate the MREs behave as classical viscoelastic materials. In other words, the modulus of MREs shows a decreasing trend with strain amplitude.

Under the same dynamic stain amplitude of 0.7%, the frequency dependence of MRE's shear modulus is measured and shown in Fig.8. It is demonstrated that the exciting frequency has little influence on the magneto-induced modulus.

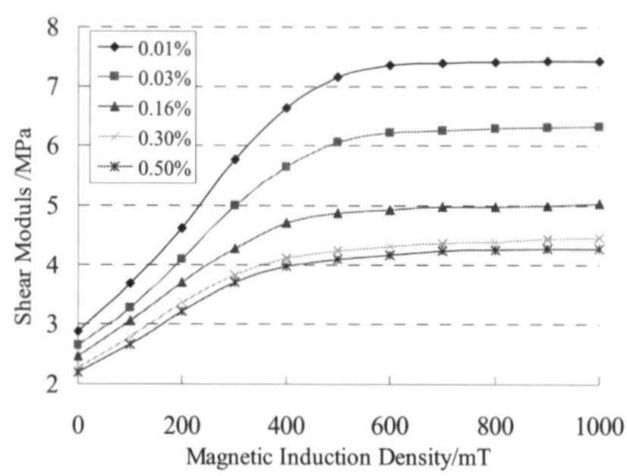

Fig.7 Shear modulus measured at different strain amplitudes

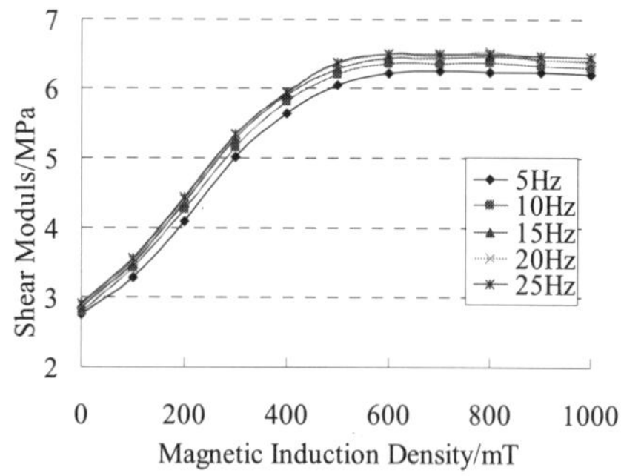

Fig.8 Shear modulus measured at different exciting frequencies

The mechanical properties of above MRE are monitored. Its tensile strength before thermal aging is 1.38MPa, after thermal aging at 70℃ for 2 days is 1.04MPa, angle tear strength is 5.5N/mm, resilience factor is 15%, and hardness is 65A, density is 2.9g/cm^3, abrasion decrement at press force of 28N after 50000 rounds is 1.1%. They are same order as usual natural rubber.

3.2 Silicone Rubber Based MRE

The influences of composition content and magnetic field strength during curing on the silicone rubber based MRE's performances are also experimentally investigated. The composition contents of samples are shown in Table 1. All percentages used in the context refer to weight percentages.

Table 1. Composition percentages of MREs based on silicone rubber

Sample No.	Carbonyl iron particle	Silicone rubber	Silicone oil
1	50%	25%	25%
2	60%	20%	20%
3	70%	15%	15%
4	80%	10%	10%

The field dependence of modulus of samples with different contents is shown in Fig.9. All samples are cured under a magnetic field of 1.5T. With the iron particles increase, the magneto-induced modulus increase quickly. When the MREs are composed by 80% carbonyl iron particles, 10% silicon rubber and 10% silicone oil, the shear modulus change from 0.34MPa at zero magnetic field to 3.34MPa at 1T magnetic field, the relative MR effect reaches 878%. The field dependence of modulus of samples cured under different magnetic field is shown in Fig.10. All samples have same composition as sample 4 in Table 1. Similar to MREs based on natured rubber, magnet field intensity in curing the mixture affects the MR effect greatly. Strong magnetic field applied during curing leads to high magneto-induced modulus. Their mechanical properties are also monitored. For sample 4 in Table 1, its tensile strength is 0.28MPa, angle tear strength is 1760N/mm, resilience factor is 9%, hardness is 28A, and density is 3.3g/cm^3. They are same order as usual silicone rubber.

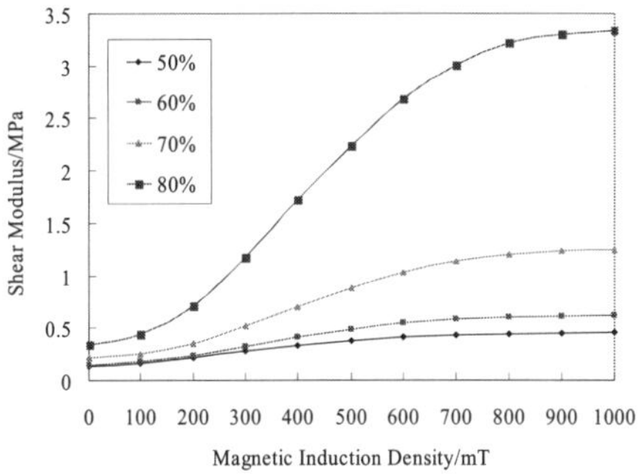

Fig.9 Shear modulus of samples with different contents

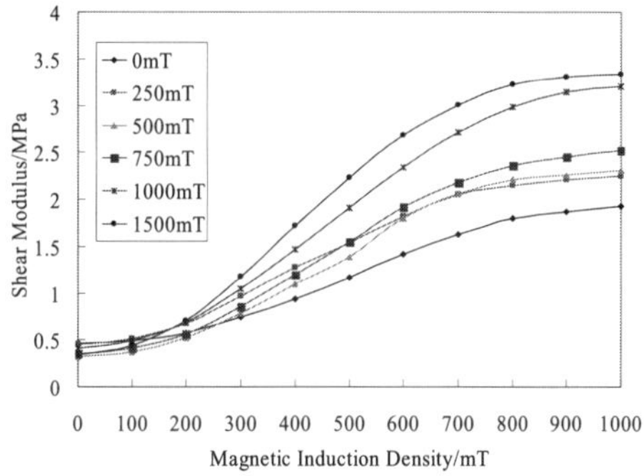

Fig .10 Shear modulus of samples cured under different magnetic field

4 Conclusions

1. A system of fabricating MREs is setup to mix the composition completely and cure the mixture at a coupled controllable magnetic field and temperature field.
2. A reliable evaluation system is setup to measure the dynamic properties and mechanical properties of MREs.
3. Prepared MREs based on natural rubber has good mechanical properties, its relative MR effect can also reach 133%. MREs based on silicone rubber has better MR effect, its best relative MR effect can reach 878%. These results have exceeded the best report in the literatures regarding MREs based on the same kind of matrix. They also have mechanical properties at same order as the matrix.

4. To fabricate utilizable MREs, the most important thing is to form good ordered-structure during curing. The following factors should be considered: a). suitable compositions and suitable contents; b). strong magnetic field during curing; c). suitable curing temperature profile for some kinds of MREs.

5 Acknowledgements

This study is granted by BRJH Project of Chinese Academy of Sciences and Specialized Research Fund for the Doctoral Program of Higher Education (No. 20050358010).

References

1. Rabinow J., The magnetic fluid clutch, *AIEE Transactions* **67** (1948) pp.1308-1315.
2. Carlson J.D., Jolly M.R., MR fluid, foam and elastomers devices, *Mechatronics* **10** (2000) pp.555-569.
3. Shiga T., Okada A., Kurauchi T., Magnetoviscoelasticity behavior of composite gels, *J. Appl. Polym. Sci.* **58** (1995) pp.787-792.
4. Jolly M. R., Carlson J. D., Munoz B. C. and Bullions T. A., The magnetoviscoelastic response of elastomer composites consisting of ferrous particles embedded in a polymer matrix, *Journal of Intelligent Material Systems and Structures* **7** (1996) pp.613-622.
5. Ginder J. M., Nichols M. E., Elie L. D. and Clark S. M. Controllable-stiffness components based on magnetorheological elastomers, *Proceedings of SPIE* **3985** (2000) pp.418-425.
6. Shen Y., Golnaraghi M. F. and Heppler G. R., Experimental research and modeling of magnetorheological elastomers, *Journal of Intelligent Material Systems and Structures* **15** (2004) pp.27-35.
7. Watson J. R., Method and apparatus for varying the stiffness of a suspension bushing, *US Patent* (1997) 5,609,353.
8. Stewart W. M., Ginder J. M., Elie L. D., Nichols M. E., Method and apparatus for reducing brake shudder, *US Patent* (1998) 5816587.
9. Bossis G., Electroactive and electrostructured Elastomers, *Int. J. Mod. Phys. B*, **15** (2001), pp.564-573.
10. Bednarek S., The giant magnetostriction in ferromagnetic composites within an elastomer matrix, *Applied Physics A*, **68** (1999), pp.63-67.
11. Demchuk S. A., Kuz'min V. A., Viscoelastic properties of magnetorheological elastomers in the regime of dynamic deformation, *J. Eng. Phys. Thermophys.* **75** (2002) pp.396-400.
12. Ginder J. M., Nichols M. E., Elie L. D., Tardiff J. L., Magnetorheological elastomers: properties and applications, *Proceedings of SPIE* **3675** (1999) pp.131-138.
13. Zhou G.Y., Shear properties of a magnetorheological elastomers, *Smart Materials and Structures* **12** (2003) pp.139-146.
14. Lokander M., Stenberg B., Improving the magnetorheological effect in isotropic magnetorheological rubber materials, *Polymer Testing* **22** (2003) pp.677-680.
15. Blom P., Kari L., Amplitude and frequency dependence of magneto-sensitive rubber in a wide frequency range, *Polymer Testing* **24** (2005) pp.656–662.

16. Gong X.L., Zhang X. Z., Zhang P. Q., Fabrication and characterization of isotropic magnetorheological elastomers, *Polymer Testing* **24** (2005) pp.669–676.
17. Wang Y.L., Hu Y., Deng H.X., Gong X. L., et al., Magnetorheological elastomers based on isobutylene-isoprene rubber, *Polymer Engineering & Science*, **46** (2006), pp.264-268.
18. Davis L. C., Model of magnetorheological elastomers, *J. Appl. Phys*, **85** (1999) pp.3348- 3351.
19. Dorfmann A., Brigadnov I. A., Constitutive modelling of magneto-sensitive Cauchy-elastic solids, *Computational Materials Science* **29** (2004) pp.270–282.

INCREASING ELECTRORHEOLOGICAL RESPONSE OF PARTICLES: THE EFFECT OF CONDUCTIVE POLYMER

A. LENGÁLOVÁ, V. PAVLÍNEK, Q. CHENG, P. SÁHA

Polymer Centre, Faculty of Technology, Tomas Bata University in Zlín, nám. TGM 275, 762 72 Zlín, Czech Republic

E-mail: lengalova@uni.utb.cz

The paper deals with the effect of two conductive polymers (polyaniline, PANI, and polypyrrole, PPy) in combination with inorganic particles on the response of silicone-oil suspensions. Conductive polymers cause a significant increase in electrorheological response. PANI combined with silica is much more efficient when in the form of a thin layer on the particle surface than if the two components are simply mixed. All combinations are followed from the viewpoint of viscosity and yield stress.

1 Introduction

ER fluids contain two basic phases: dispersed and continuous. The dispersed phase can be of various types, the two main being liquid and solid. In the group of solid dispersed phase a wide variety of particulate materials have been researched so far, among others alumina, ceramic materials, zeolites, metal oxides, silica and semiconductors (polypyrroles, PPy, and polyaniline, PANI) [1-5]. The last mentioned materials are one of those that enable design of anhydrous ER fluids.

The present paper deals with combined materials - inorganic particles and semiconducting polymers, which enables to join the effects of the two components and prepare materials with different properties. The effect of conductive polymers in several variations is researched: the difference in the behaviour of PANI being in the form of coating on and mixture with silica, and the behaviour of PPy in combination with silica; in this case the polymer is confined in the channels of mesoporous material.

2 Methods

2.1 Materials

The following combination of materials was used:

a) silica and polyaniline
b) mesoporous silica and polypyrrole.

2.1.1 Combination of silica and PANI

Silica and PANI were tested in two forms: as a mixture, and silica particles coated with PANI. The type of silica was Ultrasil VN3, Degussa-Hüls, Germany, average size of particles 18 μm.

Silica was coated with polyaniline through oxidative polymerization of aniline and subsequent deprotonation, as described in [8]. All PANI was deposited on the particles and no free polymer was found to accompany the coated particles. Depending on the amount of silica which was put in the polymerization bath, different amount of PANI was deposited on the particle surface. The thickness of the layer, taking the density of PANI as 1.24 $g.cm^{-3}$, was calculated to be from 0.1 to 2.2 μm.

The coated particles with different content of PANI (0, 5.7, 8.3, 15.3, 37.5, 64.3 and 100 wt.%) were properly mixed with silicone oil (Fluid 200, Dow Corning, UK, density 0.965 g/cm^3, viscosity 108 mPa.s) to create homogeneous 10% suspensions.
Pure PANI was prepared through the same process as described for coating. The size of particles was very close to that of silica and the conductivity was $6x10^{-11}$ $S.cm^{-1}$. PANI and silica particles were then mixed to get the same concentrations as in the previous case.

2.1.2 Combination of mesoporous silica and PPy

Two types of mesoporous silica were used. a) SBA-15 synthesised as described in [9]; b) MCM-41, which was prepared according to [10].

PPy was incorporated into pores in the following way: calcinated mesoporous silica was dried (SBA-15 at 250°C, for 6 h, MCM-41 at 200°C, for 4 h, both under vacuum), suspended above pyrrole monomer in a closed evacuated space for 24 h, then immersed in aqueous solution of $FeCl_3.6H_2O$ and stirred for another day. This was followed by repeated washing with deionized water and acetone, and drying under reduced pressure.
The incorporation of PPy in the pores was followed through the change in diameter of pores in mesoporous silica. The average pore size in SBA-15 was determined to be 8.7 nm before incorporation of PPy and 7.6 nm after it, which proves the incorporation of PPy in the channels. The amount of PPy in the channels of mesoporous silica was ca 38 wt.%. More details about the material characteristics can be found in [9].

In case of MCM-41 the portion of PPy incorporated in pores was 15 wt.%, the diameter of the pores changed from 3.3 nm to 2.5 mm. For both mesoporous silicas hexagonal pore structures remained after PPy incorporation. Neat PPy was prepared under the same conditions. Three types of 10 wt.% dispersions in silicone oil (the same type as in the case of PANI-containing suspensions) were prepared for SBA-15 measurements: mesoporous silica SBA-15, pure PPy, and SBA-15 with inside-pores incorporated PPy, all dried before suspension preparation. The other type of mesoporous silica – MCM-41 – was used in the same way.

2.2 Methods

The ER measurements were carried out using a coaxial-cylinder rotational viscometer (Bohlin Instruments, UK). The rotating inner cylinder of 14 mm in diameter and the stationary outer cylinder were separated by a 0.7 mm gap. They were connected to a DC power supply enabling application of the electric field strength $E = 0 – 3$ kV mm^{-1}. Both controlled shear rate and controlled shear stress were used, the former for the measurement of flow curves, the latter for the determination of static yield stress.
Before each measurement the suspension was properly stirred, and the DC voltage was applied for 3 min to ensure creation of equilibrium structure. Then ER properties were measured for each case, as indicated further.

3 Results and discussion

3.1 The effect of coating silica with PANI

The positive effect of coating with PANI on the ER response of the suspension can be seen in Fig. 1, which compares the relative increase of viscosity (η/η_0) in electric field (2 kV/mm) for pure silica particles and those coated with PANI. As can be seen, for the latter the ratio is significantly higher. This proves the positive effect of PANI on electrorheological properties of the tested suspension.

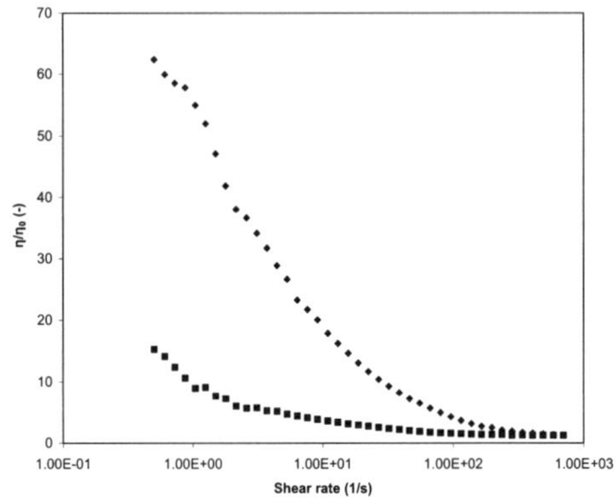

Figure 1 Increase in relative viscosity under electric field for pure silica and PANI coated particles. Electric field strength 2 kV/mm. Symbols: ■ pure silica, ♦ silica coated with PANI (8.3 wt.%).

3.2 The effect of the form of PANI

Following the suitability of PANI for electrorheological purposes, we further wanted to know whether the effect of the conductive polymer depends on its form. This was determined as differences in viscosities and yield stresses.

3.2.1 Viscosity

Although used in the same portion in the ER fluid, the form of PANI significantly influences the response of the fluid to electric field. Fig. 2 shows the comparison of the viscosities of suspensions of the same concentration of PANI (5.7 wt.%), in one case as coating on silica particles, in the other case in the form of simple mixture with silica. Two electric field strengths were chosen for comparison: 0.5 and 1.5 kV/mm. The figure shows that for coated particles the viscosity is much higher than for mixture (at low shear rate by one order of magnitude). Under the field of 0.5 kV/mm the former have the same viscosity as the latter at 1.5 kV/mm.

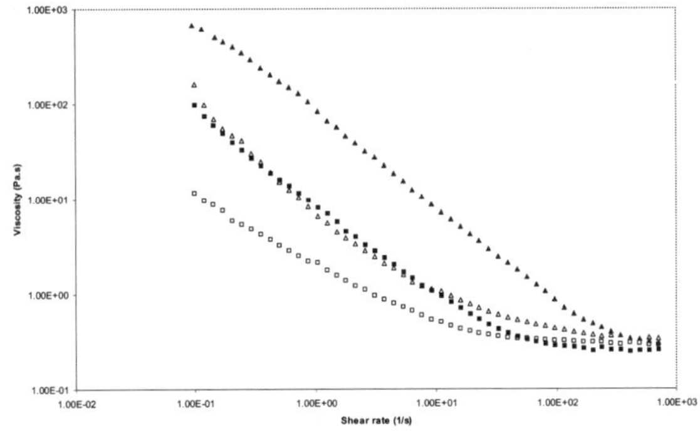

Figure 2 Behaviour of coated (full symbols) and mixed (empty symbols) particles under electric field: 1.5 kV/mm (triangles) and 0.5 kV/mm (squares).

The quantification of the differences is presented in Fig. 3 in the form of difference in viscosities vs. concentration of PANI. As can be seen, a weaker field causes smaller differences; under stronger fields the differences in viscosities are more pronounced.

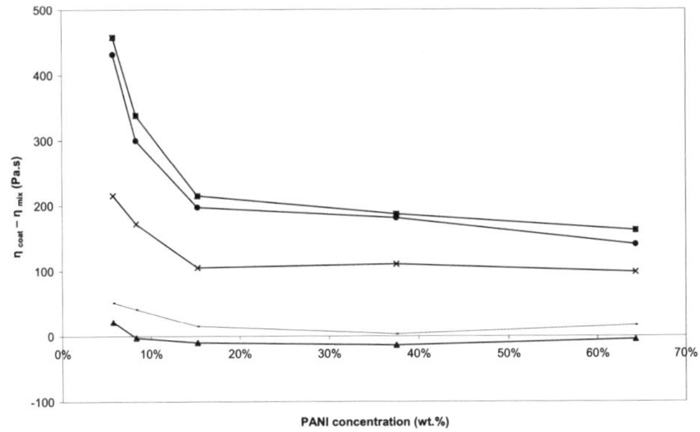

Figure 3 Difference in viscosities of coated and mixed particles suspensions at various concentrations of PANI. Electric field: ▲ 0, – 0.5, × 1, ● 1.5, ■ 2.0 kV/mm.

The figure also shows the effect of concentration. The clear trend here is – the lower the concentration, the higher the difference in ER response of the fluid. This is caused by the fact that if a small amount of PANI is in the form of a thin layer, its effect is much more important than if it is in bulk. It supports the idea that only the surface layer of conductive polymer is involved in ER effect [8]. Under zero field the suspensions, however, behave in a slightly different way; negative values at higher concentrations mean that the viscosity of the suspension of the mixture is higher than that of coated particles.

3.2.2 Yield stress

The relation between the structure rigidity, i.e. the yield stress and electric field strength is expressed in the form of exponential dependence $\tau_y \propto E^\alpha$ (the exponent depends on the response of particle polarizability to the field strength), thus in logarithmic coordinates, α will be the slope of the line. For fully developed structures the slope is 2.

Comparing the slopes for coated PANI particles and for mixtures (Figs. 4 and 5, respectively), one can see that the dependences of different concentrations are very close to each other, i.e. the suspensions behave in a similar way. For mixtures, however, they are much farther from each other, which means that for different concentrations the structures created in electric field are significantly different. This means that for the coated particles the ER behaviour only changes slightly with concentration.

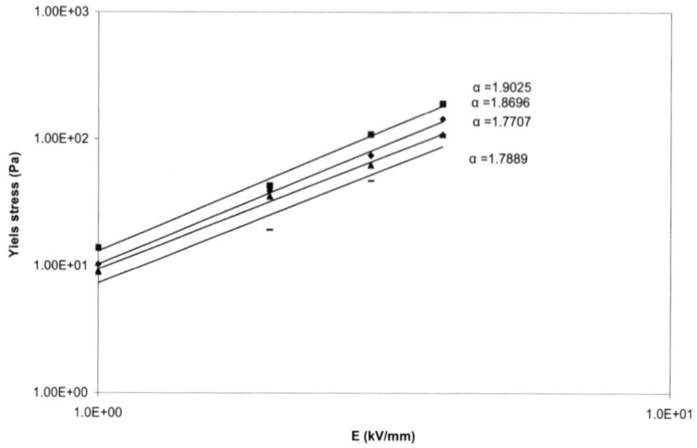

Figure 4 Dependence of yield stress on electric field for coated particles. PANI concentration: – 8.3, ▲ 15.3, ♦ 64.4, □ 100 wt.%.

An opposite is true for suspensions made of simple mixture of silica and PANI particles. Here, the concentration plays an important role in ER response.

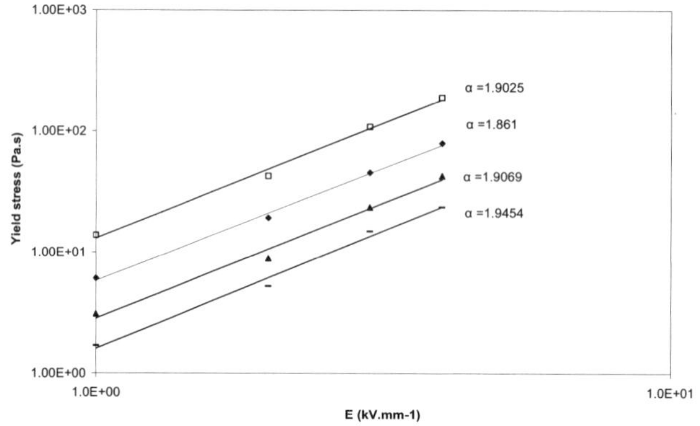

Figure 5 Dependence of yield stress on electric field for mixtures. PANI concentration: – 8.3, ▲ 15.3, ♦ 64.4, □ 100 wt.%.

3.3 Effect of PPy

3.3.1 Viscosity

Figure 6 presents the dependence of viscosity on shear rate for SBA-15 and MCM-41 doped with PPy. To avoid crowding of the graph, only field strengths of 0 kV/mm and 2 kV/mm are given here. As can be seen, the conductive polymer incorporation causes an increase in viscosity under electric field in orders of magnitude.

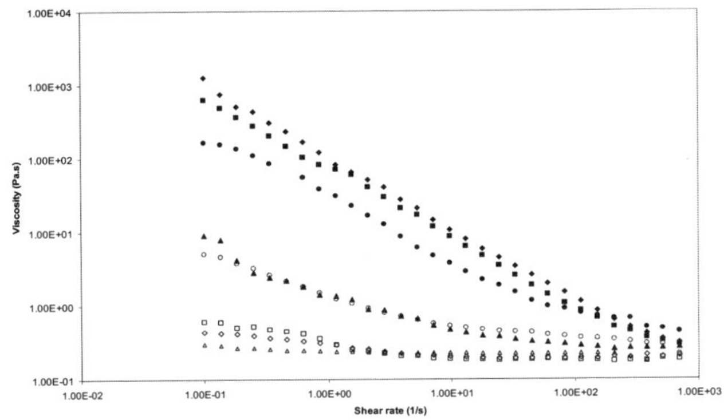

Figure 6 Change in viscosity under electric field. Symbols: ▲SBA-15, ● SBA-15/PPy, ■ MCM-41/PPy, ♦ PPy, electric field 0 kV/mm – empty symbols, 2 kV/mm – full symbols.

3.3.2 Yield stress

The same as for PANI combined with silica, also here the dependence of yield stress on the electric field was followed. The results are shown in Fig. 7. The slopes of lines for mesoporous silica doped with PPy clearly show how the polarizability of the materials increases from pure silica through the doped material to pure PPy. This is in agreement with previous findings because silica as dry inorganic material does not show high ER response, unlike semiconducting PPy.

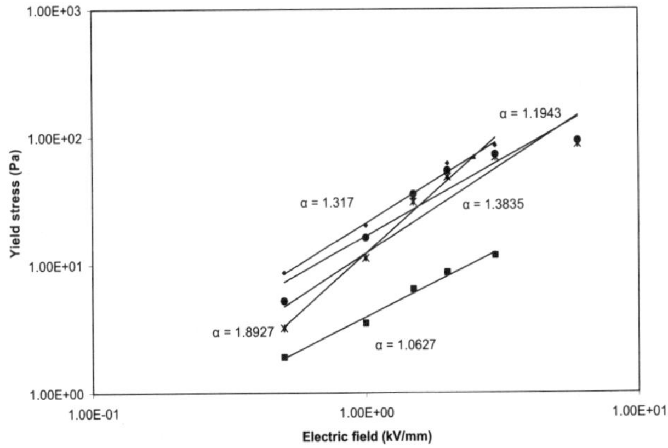

Figure 7 Dependence of yield stress on electric field strength for pure and PPy-doped mesoporous silica. Symbols: ■ SBA-15, ♦ SBA-15/PPy, ▲ PPy, ● MCM-41, * MCM-41/PPy.

4 Conclusion

The effect of semiconducting polymer on ER properties of inorganic silica particles was researched. The polymer was either on the surface, or incorporated in pores. In all cases the polymer increases the ER response of the particles suspended in silicone oil. The form of the polymer proved to be an important factor, as only the surface layer of the polymer is active in polarization. Thus, to reach a similar ER effect for simple mixture requires much higher concentration of the polymer than if it is in the form of coating.

5 Acknowledgements

The research was supported by the Ministry of Education, Youth and Sports of the Czech Republic in the frame of projects MSM 26500015, MSM 70 888352101 and KONTAKT ME 564, and by the Grant Agency of the Academy of Sciences of the CR (A 4050313).

References

1. Cho M.S., Choi H. J., Chin I.-J. and Ahn W.-S. Electrorheological characterization of zeolite suspensions *Microporous and Mesoporous Materials* **32** (1999) pp. 233-239.
2. Lengalova A., Pavlinek V., Saha P., Stejskal J., Kitano T. and Quadrat O., The Effect of Dielectric Properties on the Electrorheology of Suspensions of Silica Particles Coated with Polyaniline, *Physica A: Statistical Mechanics and its Applications* **321 (3-4)** (2003) pp. 411-424.
3. Lengalova A., Pavlinek V., Saha P., Quadrat O. and Stejskal J., The Effect of Dispersed Particle Size and Shape on the Electrorheological Behaviour of Suspensions. *Coloid and Surfaces A: Physicochem. Eng. Aspects* **227** (2003) pp. 1-8.
4. Kim Y. D., Park D. H., The effect of dispersed particle size and shape on the electrorheological behaviour of suspensions *Synthetic Metals*, **142** (2004) pp. 147-151.
5. Lengalova A., Pavlinek V., Saha P., Quadrat O., Kitano T. and Stejskal J., Influence of Particle Concentration on the Electrorheological Efficiency of Polyaniline Suspensions, *European Polymer Journal* (2003) **39**, pp. 641-645.
6. Lim Y. T., Park J. H. and Park O. O. Improved Electrorheological Effect in Polyaniline Nanocomposite Suspensions. *Journal of Colloid and Interface Science* **245** (2002) pp. 198-203.
7. Hirano S., Yogo T., Sakamoto W., Banno K. and Fukuzava, R. Improved Electrorheological Effect in Polyaniline Nanocomposite Suspensions. *Journal of European Ceramic Society* **24** (2004) p. 1911-1917.
8. Pavlinek V., Saha P., Kitano T., Stejskal J. and Quadrat O. The effect of polyaniline layer deposited on silica particles on electrorheological and dielectric properties of their silicone–oil suspensions. *Physica A* **353** (2005) pp. 21-28.
9. Zhao D., Huo Q., Feng J., Chmelka B. F., Stucky G. D., Nonionic Triblock and Star Diblock Copolymer and Oligomeric Surfactant Syntheses of Highly Ordered, Hydrothermally Stable, Mesoporous Silica Structures. *J. Am. Chem. Soc.* **120** (1998) pp. 6024-6036.
10. Cheng Q., Pavlinek V., Lengalova A., Li Ch. and Saha P. *Microporous and Mesoporous Materials* **xxx** (2006) pp. 263-269.

FLOW OF FIELD CONTROLLABLE FLUIDS WITH SUSPENDED FERROUS PARTICLES IN MICRO TUBES UNDER MAGNETIC FIELD

SINAN OZCAN

Department of Mechanical Engineering, University of Nevada, Reno, USA
Graduate Student

E-mail: ozcans@unr.nevada.edu

CAHIT A. EVRENSEL[1] , MARK A. PINSKY[2], ALAN FUCHS[3]

[1]Department of Mechanical Engineering, University of Nevada, Reno, USA
[2]Department of Mathematics, University of Nevada, Reno, USA
[3]Department of Chemical Engineering, University of Nevada, Reno, USA
[1,2,3]Professor
E-mail: cahit@unr.edu, pinsky@unr.edu, afuchs@unr.edu

This computational study focuses on the dynamics of individual ferrous particles and the flow of the incompressible Newtonian fluid under the effect of an externally applied magnetic field and pressure gradient in a micro tube with smooth wall. The particle dynamics is simulated as a discrete phase using MATHLAB code and the fluid flow is solved as a continuous phase using Computational Fluid Dynamics Software FLUENT. Interaction between the particle and fluid phases are included as hydrodynamic forces predicated by the fluid phase simulation and updated particle locations determined by the particle phase solution under non-uniform magnetic field. Non-uniform magnetic field forces the particles to move to poles of the magnet, and results in their accumulation. This causes drastic change on the continuous phase flow and pressure distribution, which in turn influences the particle motion. Predicted dynamics of the suspended ferrous particles under magnetic field and flow of the carrier fluid with pressure gradient is in reasonably well agreement with previous work. The results show that non-uniform magnetic field generated by externally placed magnets can be used to control the locations of the particles and flow of the fluid in a micro tube.

1 Introduction

Although there has been a considerable amount of research during the past twenty years there is still a growing interest on modeling MR and ER fluids. Existing studies mainly focus on the experimental investigation of the particle aggregation which is the cause of the viscosity increase and theoretical studies to predict the experimental results [1-3].

Klingenberg et al. [4-7] and K.C. Hass [8] are among the first researchers who developed 2D and 3D simulation methods to investigate the structure formation in ER suspensions. Their model is based on several assumptions which may limit the applicability of the results. First, they assumed that the electrostatic interaction between polarized particles is due to the isolated spheres via point-dipole approximation and neglected the multi-body interactions. However, the qualitative nature of the interaction forces was simulated successively. Second, they modeled the hydrodynamics on each sphere separately by Stokes' drag law. Thus, the multi-body Stokesian interactions were neglected. Their simulation results qualitatively predict the structures obtained experimentally.

Bonnecaze et al [9] developed a more detailed model which accounts for the far- and near- field effects for both the hydrodynamic interaction by applying "Stokesian dynamics" and electrostatic interactions. One disadvantage of this model is that it is restricted to a small number of particles due to the computational expenses when compared to the previous works discussed above.

Recently, many studies are focused on MR fluids' possible applications in the biomedical field. Pankhurst et al. [10] reviewed the basic physics of these applications of magnetic nanoparticles in detail. However, most of these applications are still at the early development stage and have not been marketed yet. Zachary et al. [11] demonstrated that it is possible to concentrate the magnetic microparticles at the specified locations in the vascular system by applying a uniform magnetic field. For this, they used magnetic particles anchored to the desired locations to stop the aggregating particles. They proposed their method as an inexpensive technique to block the catheter in order to kill the tumors and to seal off the arteriovenous malformations.

The first computational study on Electrorheological suspensions with a non-uniform electric field is carried out by Kadaksham et al. [12]. The model they developed is based on the distributed Lagrange multiplier method (DLM). They used the point-dipole approximation for electrostatic forces, and neglected the hydrodynamic forces. Their simulations demonstrated that the magnetic particles accumulate in the regions where the magnitude of the electric field is locally maximum or minimum and if the pressure gradient is high enough electrically polarizable particles are swept away from the accumulation region.

2 Mathematical Model

In this study, the spherical, particles suspended in a base fluid are simulated in a two dimensional Lagrangian frame of reference. The code developed in MATLAB for this purpose calculates the trajectories of each particle in the suspension by considering the magnetic forces, hydrodynamic drag forces and gravitational force. The trajectory for each particle is calculated by solving the differential Newton's equation of motion for each spherical particle [13]. For the trajectory along x-axis, the differential Newton's equation of motion for the particle i would be;

$$m\frac{d^2\vec{x}_i}{dt^2} = \vec{F}_{i,x}^{magnetic} + \vec{F}_{i,x}^{hydro} + \vec{F}_{i,x}^{grav} + \vec{F}_{i,x}^{rep} + \vec{F}_{i,x}^{wall} + \vec{F}_{i,x}^{gradient}$$

The detailed descriptions of forces acting on a particle are shown in Figure 1. The motion of the particles can be controlled by an externally applied magnetic field [14-17]. The magnetic field tends to align the dipole moments induced in the spherical particles along the direction of itself. The interaction energy between the two dipoles can be written as [15];

$$E_{dd} = \frac{1}{4\pi\mu_0}\left[\frac{\vec{m}_i\vec{m}_j}{\left|\vec{r}_{ij}\right|^3} - \frac{3}{\left|\vec{r}_{ij}\right|^5}(\vec{m}_i\vec{r}_{ij})(\vec{m}_j\vec{r}_{ij})\right]$$

The interaction force acting on dipoles due to this energy [15];

$$\vec{F}_{dd} = -\nabla E_{dd} = -\left[\frac{dE_{dd}}{dx_{ij}}\vec{i} + \frac{dE_{dd}}{dy_{ij}}\vec{j} + \frac{dE_{dd}}{dz_{ij}}\vec{k}\right]$$

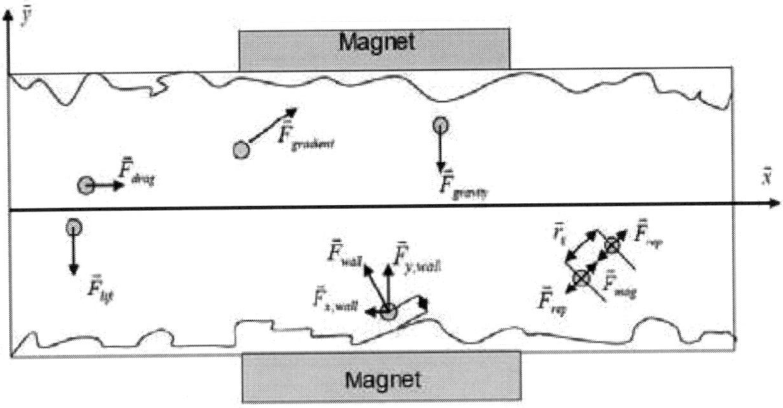

Figure 1. Schematic of the typical forces acting on each particle

If the externally applied magnetic field is not uniform, there will be a force on each particle due to this gradient. These forces attract the particles toward the region of highest field gradient which are the poles of the magnet [16, 18, 19 and 20]. The total magnetic force on each spherical particle in the suspension is governed by the equation given below;

$$\vec{F} = (\vec{\nabla} \cdot \vec{m})\vec{H}_o$$

The spherical particles experience a resistance as they move through the continuous phase fluid. The multibody hydrodynamic approach causes a computational complexity and an increase in computational time. As it was done in most of the computational studies [4-9] which were successful in capturing the mechanical physics in their simulations, the hydrodynamic drag force acting on the magnetic particle is given by Stokes' law in this study.

3 Procedure for the MATLAB-FLUENT Interface

Field controllable fluid modeling needs the modeling of particle phase and the fluid phase and their interaction. To account for this interaction between the particle and fluid phases, hydrodynamic lifting and drag forces acting on each particle in a domain where the liquid flows due to the existing differential pressure are simulated using the commercially available Computational Fluid Dynamics (CFD) code FLUENT.

The hydrodynamic forces acting on the particle due to the fluid flow is computed by assuming the particles in the domain as stationary walls in FLUENT to reduce the data transfer time between MATLAB and FLUENT. Also, hydrodynamic forces due to the motion of the particle in the domain in addition to the hydrodynamic forces calculated previously due to the fluid flow are calculated in MATLAB by assuming particles are in motion in a stationary fluid and by using the Stokes' drag law. This assumption brings some inaccuracy to the solution because in Stokes' law it neglects the other particles' effect on the other which may cause compression of the fluid in two phase flow. Since the relatively low particle velocities for the case studied in this paper this inaccuracy is expected to be small.

The first step of this procedure is the solution of the fluid phase in FLUENT for the defined initial positions of the particles to calculate hydrodynamic forces acting on each particle. The second step is the transferring the hydrodynamic forces from FLUENT to MATLAB in order to solve the particle phase in MATLAB per defined time step. The third step and the final step in the loop is to transfer the particle positions at the end of the time step from MATLAB to FLUENT and the solution of the fluid phase for the defined particle positions in the domain and finally the transfer of the calculated hydrodynamic forces to the MATLAB again.

In the solution of fluid phase, the flow was assumed as unsteady and a laminar due to small tube size and low flow velocities which results in low Reynolds number. Also, the drag and the lift forces were assumed to be linear function of the flow velocity.

4 Results

In this study, the pressure driven flow cases are simulated. The number of particles, sixty, their initial locations and the 0.5 kPa pressure gradient are kept the same in all simulations. Two different, 0.2 T and 0.1 T, magnetic field strengths are used in the simulations.

The main goal is to control the location of the magnetic particles and try to stop them in a pressure driven flow. In order to accomplish this objective, a non-uniform magnetic field definition was used. Due to their acquired dipoles, the particles interacted with each other. This caused the particles to align themselves along the magnetic field lines. Also, the non-uniformity on the environment created an attractive force on each particle towards the local maximum which was the pole of the magnet.

These results show that the magnetic particles can be stopped or slowed near the magnet pole, resulting in their accumulation near that region in a flow. This may in turn slow or block the flow. The simulation results obtained agree well qualitatively with the behaviors reported in the literature [12, 14 and 21].

The simulation results showed that the magnetic field strength has a great influence on the particle interaction. Lower field strength created weaker interaction forces. This resulted in more unstable formations and the flow can not be blocked. On the other hand, for higher field strengths the flow blockage may be achievable.

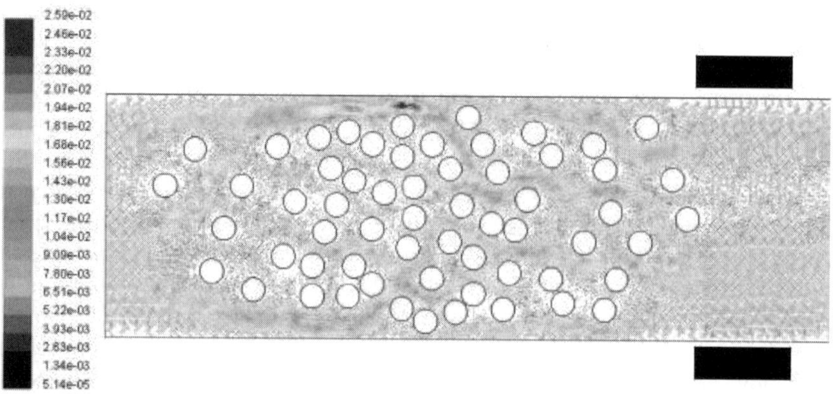

Figure 2. Initial Positions for the Cases Given Below

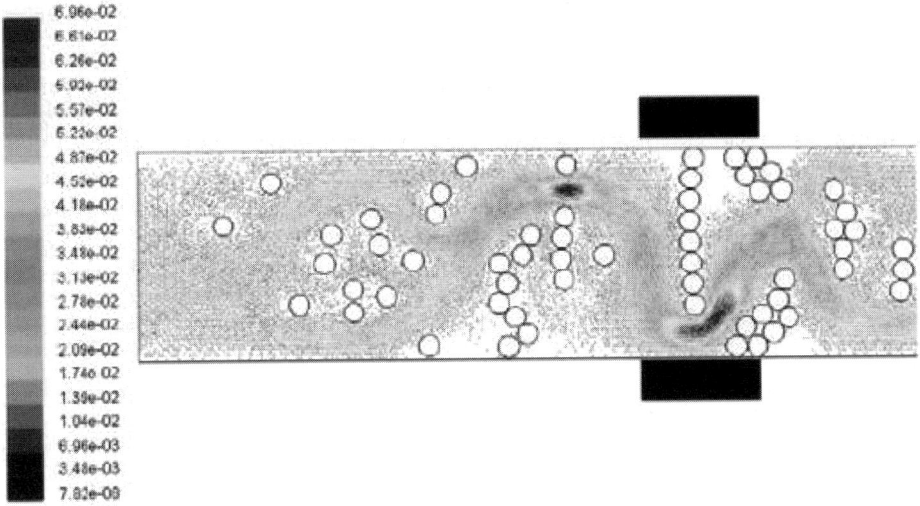

Figure 3. Velocity Vectors Colored by Velocity Magnitude (m/s) at time = 115 ms for 0.5kPa pressure gradient driven flow with 2 magnets with strengths of 0.1 T

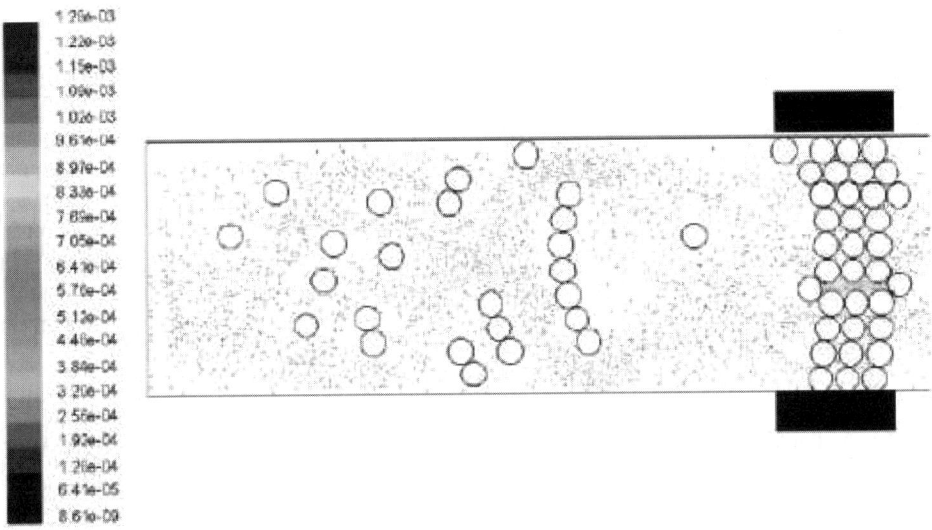

Figure 4. Velocity Vectors Colored by Velocity Magnitude (m/s) at time = 250 ms 0.5kPa pressure gradient driven flow with 2 magnets with strengths of 0.2 T

5 Conclusion

The model developed in this paper needs the solution of two phases; the particle dynamics as a discrete phase and the fluid flow as a continuous phase. Also, an interaction between the particle and fluid phases has to be defined in the model. Model used in this paper predicts that non-uniform magnetic field generated by externally placed magnets can be used to control the locations of the particles [21]. Particles can be stopped depending on the magnetic field strength and pressure gradient. Increasing the magnetic field strength results in the stronger interactions between the particles and increases the pressure gradient they can resist.

References.

1. Mark Ivey, Jing Liu, Yun Zhu and Serge Cutillas, Magnetic-Field-Induced Structural Transitions in a Ferrofluid Emulsion, Physical Review E **63** (2000), 011403.
2. Howard See, Constitutive Equation for Electrorheological Fluids Based on the Chain Model, Journal of Physics: Applied Physics D **33** (2000), pp. 1625.
3. Zuowei Wang, Haiping Fang, Zhifang Lin and Luwei Zhou, Dynamic Simulation Studies of Structural Formation and Transition in Electro-Magneto- Rheological Fluids, International Journal of Modern Physics B **15** (2001), No: 6&7, pp. 842.
4. D. J. Klingenberg, Frank van Swol, and C. F. Zukoski, Dynamic Simulation of Electrorheological Suspensions, Journal of Chemical Physics **91** (1989), No: 12, pp. 7888.
5. D. J. Klingenberg, Frank van Swol, and C. F. Zukoski, The Small Shear Rate Response of Electrorheological Suspensions, Journal of Chemical Physics **94** (1991), No: 9, pp. 6160.
6. D. J. Klingenberg, Frank van Swol and C. F. Zukoski, The Small Shear Rate Response of Electrorheological Suspensions II. Extension beyond the Point- Dipole Limit, Journal of Chemical Physics **94** (1991), pp. 6170.
7. D. J. Klingenberg, C. F. Zukoski and J. C. Hill, Kinetics of Structure Formation in Electrorheological Suspensions, Journal of Applied Physics **73** (1993), No: 9, pp. 4644.
8. K. C. Hass, Computer Simulations of Non-equilibrium Structure Formation in Electrorheological Fluids, Physical Review E **47** (1993), No: 5, pp. 3362.
9. R. T. Bonnecaze and J. F. Brady, Dynamic simulation of an Electrorheological Fluid, Journal of Chemical Physics **96** (1992), No: 3, pp. 2183.
10. Q. A. Pankhurst, J. Connolly, S. K. Jones and J. Dobson, Applications of Magnetic Nanoparticles in Biomedicine, Journal of Physics: Applied Physics D **36** (2003), pp. 167.
11. 12.Zachary G. Forbes, Benjamin B. Yellen, Kenneth A. Barbee and Gary Friedman, An Approach to Targeted Drug Delivery Based on Uniform Magnetic Fields, IEEE Transactions on Magnetics **39** (2003), No: 5, pp. 3372.
12. J. Kadaksham, P. Singh, and N. Aubry, Dynamics of Electrorheological Suspensions Subjected to Spatially Nonuniform Electric Fields, Journal of Fluids Engineering **126** (2004), pp. 170.
13. M. Mohebi, N. Jamasbi, and Jing Liu, Simulation of the Formation of Non-equilibrium structures in Magnetorheological Fluids Subject to an External Magnetic Field, Physical Review E **54** (1996), No: 5, pp. 5407.

14. Q. A. Pankhurst, J. Connolly, S. K. Jones and J. Dobson, Applications of Magnetic Nanoparticles in Biomedicine, Journal of Physics: Applied Physics D **36** (2003), pp. 167.
15. R. E. Rosenbensweig, Ferro hydrodynamics, Cambridge University Press, 1985.
16. George E. Mase, Theory and Problems of Continuum Mechanics, McGraw-Hill, 1970.
17. Zuowei Wang, Christian Holm and Hanns Walter Muller, Molecular Dynamics Study on the Equilibrium Magnetization Properties and Structure of Ferrofluids, Physical Review E **66** (2002), 021405.
18. Curtis M. Oldenburg, Sharon E. Borglin and George J. Moridis, Numerical Simulation of Ferrofluid Flow for Subsurface Environmental Engineering Applications, Transport in Porous Media **38** (2000), pp. 319.
19. Malcolm McCaig, Permanent Magnets in Theory and Practice, John Wiley & Sons, New York, 1977.
20. Malcolm McCaig and Clegg A.G., Permanent Magnets in Theory and Practice, 2nd Edition, Halsted Press, Wiley, New York, 1987.
21. Ranjan Ganguly, Amit P. Gaind, Swarnendu Sen, and Ishwar K. Puri, Analyzing Ferrofluid Transport for Magnetic Drug Targeting, Journal of Magnetism and Magnetic Materials, Vol. **289** (2005), pp. 331-334

ELECTROSENSITIVE LUBRICANTS

E. V. KOROBKO, Z. A. NOVIKOVA, N. A. BEDZIK, V. L. BASINUK, A. V. IVAHNIK, V. A. MANSUROV
A. V. Luikov Heat and Mass Transfer Institute, National Academy of Sciences of Belarus,
15 P. Brovka str., Minsk, 220072, Belarus
E-mail: eva@itmo.by

Electrosensitive compounding of lubricants were developed for use in the friction pairs in the transitive regimes of their work. The investigation of their viscoplastic properties and friction coefficients in the electric field was made.

1 Introduction

At interacting friction junctions in technological equipment, working nonstop at stationary regimes of loading in the conditions of liquid sliding friction, the most intensive wear of working surfaces of bearings and supports of sliding appears at stopping and launching the gear drive. In these periods of object functioning, the hydrodynamic component of the thickness of a lubricant layer decreases proportionally to relative sliding speed of the interacting elements. Friction of boundary lubrication and even in some cases dry friction appears in some parts that leads to a drastic increase of wear of friction junctions and to accelerated loss of efficiency of bearing sites and of the unit on the whole. This problem is of a great importance for expensive and unique equipment. One of the ways of solving it is to use lubricants with special antifriction reducers, and also lubricants possessing higher viscosity [1].

The increase of viscosity of the lubricant materials allows to shorten the duration of unfavorable for the bearing site period of interaction of friction junctions in the condition of friction of boundary lubrication but it also leads to big losses on friction and additional power inputs.

One of the ways to solve this problem is to use lubricant materials with governed, in particular, by the electric field rheological properties, that allows to provide viscosity needed for liquid friction adaptive to the conditions of the loading interaction of the contacting interfaces of bearings and sliding supports.

The aim of our investigation is to develop compositions of electrorheological fluids (ERF), that could be used as adaptive lubricants. Electric field changes significantly viscoplastic properties of such fluids, forms and heightens fluidity limit. After switching the electric field off the medium completely returns to the initial stage. The most important parameter which characterizes the plastic lubricants is the fluidity limit τ_0. This parameter corresponds to the yield stress and determines the thickness of a lubricant and divides the areas of elastic and plastic deformation and determines the conditions for launching a mechanism motion.

2 Experimental part

For developing compounding of plastic electrosensitive lubricants, investigation of ERF, containing high dispersive silicon dioxide — aerosil as a filler, that has an ability to form initial structural frame of the lubricant and quick recovery of the bonds between particles after their destruction at deformation was carried out. Aerosil — is a pure nonporous SiO_2, that is characterized by high dispersivity, homogeneity of the particle sizes and big specific surface — from 150 up to 450 m^2/g. The particle sizes are from 5 up to 40 nm (generally 10 – 15 nm). Except SiO_2 (99.8 %), it consists of aluminum, titan, boron, iron and other oxides. The activator was polyethylenepolyamine (PEPA), that has a much bigger sensitivity than the compounds according to the well known research [2]. Besides, this amine has not a big volatility which is important for practical application. As the dispersed medium a dried and purified transform oil, synthetic motor oil Mobil-1 SAE5W-50 , mineral motor oil Castrol JTX3 and PES-5 (polyethylsiloxane) fluid were used. For preventing sedimentation and heightening suspension fluidity a surfactant — monooleate glycerin was added.

The measuring method of the fluidity limit τ_0 (or yield stress) is based on the determining minimum torque, which enough to fix plastic deformation [3]. The measuring of the fluidity limit of electrorheological lubricants lies in the formation of a sample deformation, linearly changing in time, and in measuring the shear stress in the sample and its change in the conditions of electric field impact [3, 4]. The working site of the device represents a measuring system disk-disk, used for application of electric potential. Such geometry of a measuring unit allows to create a homogeneous electric field and to set accurately a value of the measuring slot which corresponds to the conditions of further mechanical lubricant property tests, as the value of the measuring slot could be thousand fraction of a millimeter. In this very site a plastic flow develops at certain deformation values and a certain intensity of electric field (Fig. 1).

The measuring was carried out on the Vibrorheometer (SDO IPCS, Moscow), used for the investigation of anomalous viscous and viscoelastic fluids at nonstop deformation at constant speed and cyclic deformation with different frequencies and amplitudes. In the regime of nonstop deformation the measuring was done at constant speed of change of deformation (0.1 rad/min, in the range of turning angles φ 0 – 0.06 rad, that corresponds to the range of changing of relative deformation ε, equal to 0 – 500 %) at temperature (25±1) C. The gap between disks was 0.2 mm. The range of changing electric voltage was 0 – 500 V, that corresponds to the intensity of electric field 0 – 2.5 kV/mm in the measuring unit.

13 samples of ERF were prepared and investigated, the dependencies of yield stress on the composition and the value of the voltage applied were determined. The compositions of ERF are submitted in Table 1.

Table 1. The composition of ER lubricants.

Component, %	Samples under investigation												
	1	2	3	4	5	6	7	8	9	10	11	12	13
Aerosil	30	30	35	40	45	30	30	35	30	30	30	35	35
PEPA	20	20	20	20	20	10	10	20	20	20	20	20	20
MOG	20	10	10	20	30	20	10	10	10	20	10	10	-
Oil	Transform oil							PES-5		Castrol GTX_3		Mobil-1 5W-50	

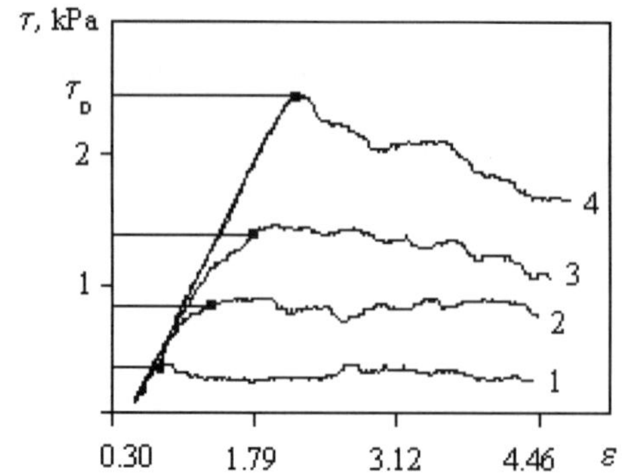

Figure 1. The dependence of shear stress on the deformation applied for a sample №3: 1 – U=0 V; 2 – 125; 3 – 250; 4 - 500.

In Fig. 1 a typical dependence of shear stress on the deformation applied for the sample 3 is shown. It can be conditionally divided into three areas: rectilinear — describes elastic deformation, then nonlinear — characterizes the developing of plastic deformation, the third area characterizes viscoplastic flow. From Fig. 1 it is seen that with an increase of intensity the nonlinear area broadens. The value τ_0 corresponds to the maximum value of the shear stress on the dependence $\tau(\varepsilon)$. The higher intensity of the applied electric field, the bigger value of τ_0 is.

The dependence of fluidity limit on the intensity of electric field is shown in Fig 2a,b. The biggest growth of τ_0 in the electric field (Fig. 2a) is seen for samples № 4 and 5, containing the biggest quantity of aerosil (40 and 45 wt.%) and activator – PEPA (20wt.,%). The smallest growth of the fluidity limit was marked for the samples № 6 и 7, with a minimum concentration of aerosil (30%) and PEPA (10%). The curves of changing fluidity limit for other samples are in the interval between these values , depending on the composition. In Fig. 2 b the change τ_0 in the electric field is shown for the samples that have PES-5, mineral oil Castrol GTX_3 and synthetic oil Mobil-1 5W-50 as dispersed medium. The data obtained for linear dependencies (Fig 2b and curves 4-8 Fig. 2a) could be approximated with an enough degree of accuracy by the function, containing two terms — a constant τ_{0n} and a power one: $\tau_0 = \tau_{0n} + kE^n$ and which is characterized by the inclination k (Fig. 2).

Coefficient k could be conditionally named as a "sensitivity coefficient of a plastic part of the flow", and coefficient τ_{0n} as a "thickness coefficient" at the zero value of electric field. In table 2 the values of these coefficients for different samples are present

Table 2 Sensitivity coefficients of the plastic part of the flow and thickness coefficient of electrorheological lubricants.

Sample Number	1	2	3	4	5	6	7	8	9	10	11	12	13
K	0.81	0.33	1.33	1.87	1.57	0.35	0.34	0.22	0.62	1.13	1.04	1.41	1.52
τ_{0n}	0.27	0.16	0.63	1.50	0.61	0.20	0.16	0.65	0.76	0.26	0.60	0.75	0.67

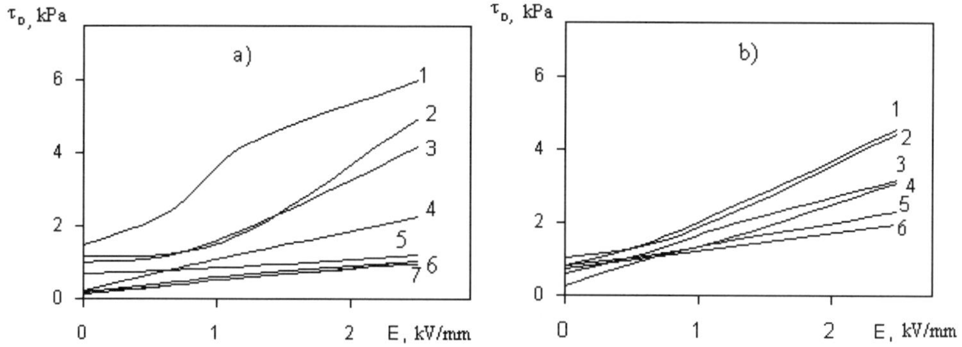

Figure 2. The dependence of yield stress growth on the value of the electric field applied for ER lubricants a) dispersed medium – transform oil: 1 – №4; 2 – № 5; 3 – № 3; 4 – № 1; 5 – № 6; 6 – № 2; 7 – № 7; б) dispersed medium – synthetic, mineral oils and PES-5: 1 – № 13; 2 – № 12; 3 – № 10; 4 – № 11; 5 – №9; 6 - №8.

Three samples of ER suspensions have been chosen to determine exploitation characteristics of the lubricants. In table 3 the results of ER lubricant tests are presented. The comparison of the data from the Table 3 with the main indices, characterizing plastic lubricants of the series production has shown that the compositions developed by us absolutely correspond to the requirements, made to the industrious lubricants. A loading curve has been received on the fourball friction machine (Fig. 3). At the identical loading the mean diameter of the wearing spot of the developed electrorheological lubricants is smaller, than for a lithol and pure oil Mobil-1.

Table 3 The results of the electrorheological lubricant tests

Sample number	11	12	13
Colloidal stability, % of the oil released	0.5	2.8	3.5
Drop falling temperature, °C	130	120	118
Penetration at 25 °C with mixing, mm^{-1}	165	215	230
Viscosity Pa·s at 20 °C and mean deformation speed gradient 10 s^{-1}	950	630	540
Mean diameter of a wearing spot, mm (determining critical loading (Pc) on the four ball friction machine), critical loading, H=1235	0.50	0.65	0.65
Mean diameter of a wearing spot, mm (determining welding loading (Pw) on the four ball friction machine), welding loading, H=1960	1.50	1.35	1.48
Fin index $I_f = \Sigma Q/n$ (determined on the four ball friction machine)	84.79	78.19	67.67

The main problem of the investigation was the experimental attempt to evaluate the possibilities of governing tribotechnological characteristics of friction junctions with oxide-ceramics coatings using electro manageable rheological fluid for their lubrication [5]. Cylindrical samples with a plain periphery working surface and a counter body in the form of a plate, made of deformed aluminum alloy D16 were used as for friction pair under investigation (Fig. 4). Oxide-ceramics coatings with the thickness of 90-120 μm,

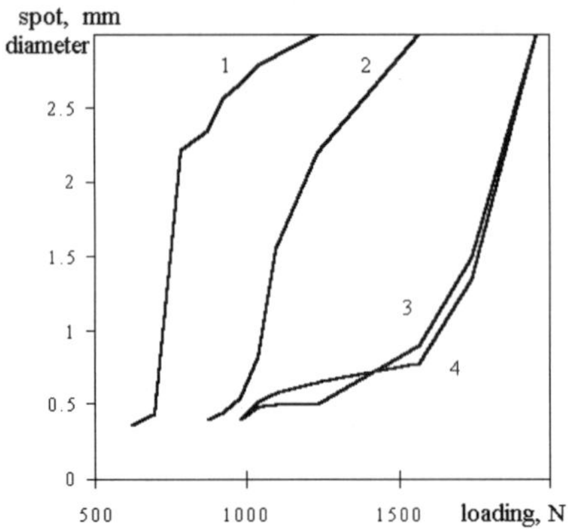

Figure 3. Loading curves for the samples of lubricants: 1 – litol-24; 2 – oil; 3 – ER-lubricant № 12; 4 - ER-lubricant № 13.

that have hardness changing in depth over the range of 12-16 GPa were formed on the interacting surfaces of the samples and counter body in the electrolyte of the alkaline type by the anode-cathode way.

Estimating friction coefficients was carried out in the modified friction machine with the plain parallel scheme of realization of interaction between the samples in the regime of sliding friction, staffed with microprocessed system of reading and preliminary processing of diagnostic information. As periphery sensor of friction forces, high sensitive tensormeter element was used, that allows fixing momentary values of a friction force.

During investigation the sample and the counter body were electroisolated from the friction machine and measurement equipment by means of special rigging The

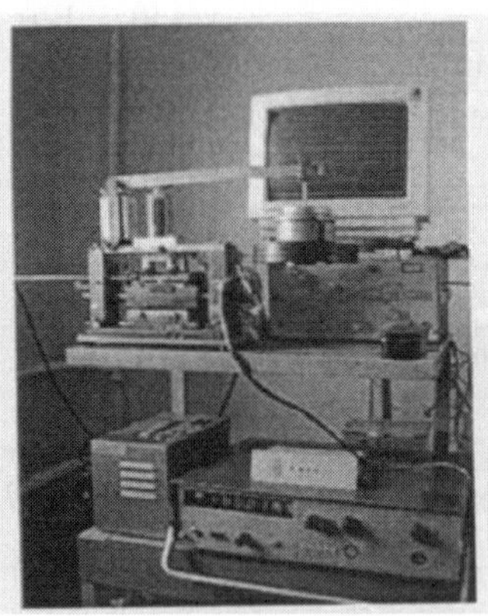

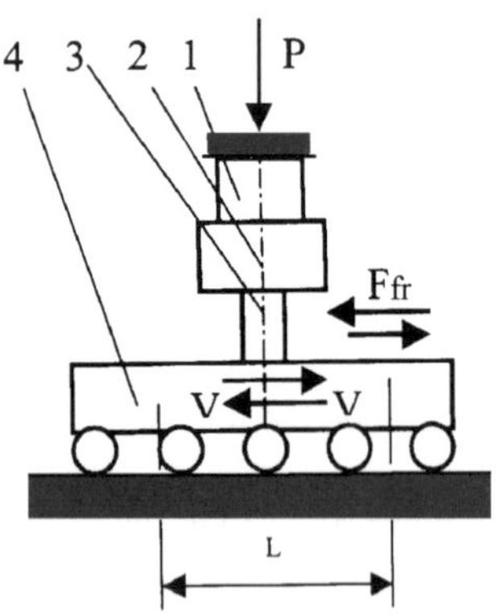

a b

Figure 4 Common vied of the device (a) and of the investigation scheme (б) for a research of tribotechnical characteristics of sample interaction in the regime of sliding friction: 1 – friction force sensor; 2 – a hold for samples investigated; 3 – samples investigated; 4 – a counterbody; Ffr – a friction force; L – the length of the sliding path at reciprocal motion of a counterbody; V – sliding speed.

investigation was carried out at room temperature and ambient humidity. Reading of information about momentary values of the friction coefficients was made in automotive regime.

In the scheme used the sliding speed of the object relative to the counter body increased smoothly according to the sinusoidal law from zero up to maximal value. Then it decreased smoothly to the full stop and the movement reversed; contact pressures after working out the given number of cycles increased stepwise from 0.5 to 3.2 MPa.

In the lubrication layer voltages changing from zero up to 1200 V were created. As early as 1180-1220 V substantial leaked current appeared which increased extensively at the following increase of the voltage. On the whole, at doing research leaked current was not higher than some milliamperes and power losses bounded with it were small, not more then some Wt. Estimations of the results received had comparative nature - the ration of friction coefficient in the lubricant layer with and without intensity was found experimentally out.

The change of the coefficients of the sliding friction and k_{fE} at the increase of the voltage of the electric field applied to the lubrication layer is shown in the Fig. 5 a, b.

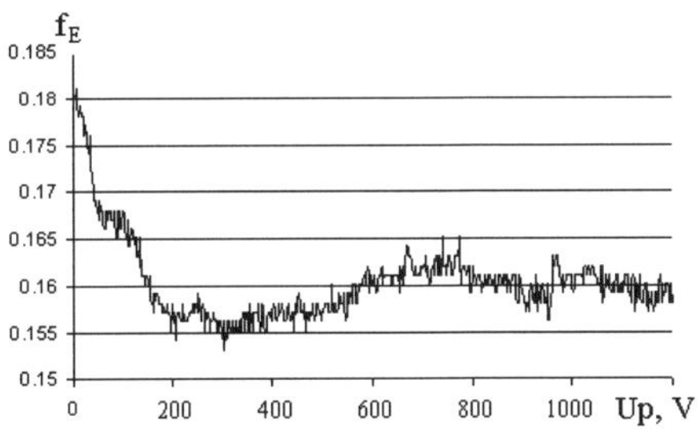

a

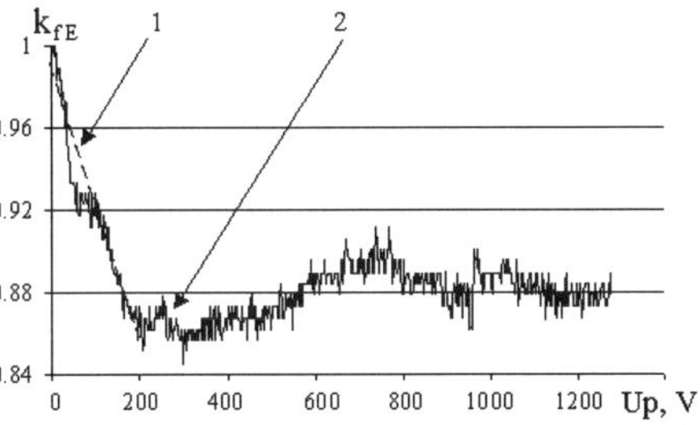

b

Figure 5. The change of experimentally received mean maximum values of the friction coefficients at the increase of the intensity of electric field (a) and the coefficient k_{fE} – ratio of the coefficient of the sliding friction with electric field in the lubrication layer to the coefficient of the sliding friction without electric field (b), calculated (1) and experimental (2)

The analysis of the results received has shown the following. With an increase of the intensity of electric field mean values of the coefficients of friction decrease to a certain level and then increase a little bit and stabilize in a certain range of values (Fig. 5). This can by explained be the fact that with an increase of the viscosity the thickness of the lubrication layer increases, and the friction of boundary lubrication converts into liquid friction, that leads to a change of the interaction nature of the friction junction and the decrease of friction coefficient.

Simultaneously with this the growth of the tractive resistance takes place, which is linked with the change of viscosity of ERF, that should lead to the increase of the friction coefficient; and the forces of disjoining pressure appear, typical to the electrorheological interaction of the contacting surfaces with oxide-ceramics coatings [5]. More over the increase of the thickness of the lubrication layer leads to a decrease of intensity of electric field applied to ERF. Thus, the influence of the increase of electric tension between contacting surfaces with oxide-ceramics coatings affects nonlinearly on the changes of the friction coefficients and the thicknesses of the lubrication layer. In this connection at its certain values the values of friction coefficients and thicknesses of lubrication layers establish on a certain level and further do not increase.

3 Conclusion

The use of ERF on the basis of mineral oil as the lubricant material for friction junction with oxide-ceramics coatings on the contacting surfaces leads to the reverse change of the coefficient of the sliding friction, that allows to provide effective governing of the tribotechnological characteristics of friction junctions due to controlled change of rheological properties of the lubricant in the elastic-hydrodynamic contact. At this the most rational range of electric voltages applied to the contacting surfaces exists, the smaller values of which do not provide the accelerated transition from the friction of boundary lubrication to the liquid friction in initial moment of motion, and the biggest values — practically do not influence the friction coefficients and thicknesses of the lubrication layer. Their values are determined by the properties of ERF and the roughness of the contacting surfaces. For the sample discussed this range is equal to 120-150 V.

The results obtained have shown, that the realization of the suggested way allows to create the rubbing junctions of expensive and unique objects with manageable tribological characteristics. Also it results in heightening their longevity.

Acknowledgement

We would like to thank Belarussian Foundation for Fundamental Research for the support of our research work (Grant T06-149).

References

1. Korobko E.V., Basinyuk V.L., Mardosevich E.I., Goncharova N.A., The managing of tribotechnical parameters of rubbing interfaces, *Friction and Wear* **24** (2004) Pp. 687-693.
2. Macepuro A.D., Novikova Z. A. The flow of electrorheological suspensions based on the aerosol at heightened temperatures, In Rheophysics (Minsk, HMTI AS BSSR, 1977) Pp. 52-54.
3. State Standards of USSR. Lubricants. **P. 2.** (M.: 1982).
4. DIN 53018 (web).
5. Korobko E.V. Electrostructured (electrorheological) fluids: the peculiarities of hydromechanics and the possibilities of application. Minks, 1996. – 189 p.

FORMATION OF POLARIZED CONTACT LAYERS AND THE GIANT ELECTRORHELOGICAL EFFECT

XIANXIANG HUANG, WEIJIA WEN, AND PING SHENG

Department of Physics and the Institute of Nano Science and Technology
Hong Kong University of Science and Technology
Clear Water Bay, Kowloon, Hong Kong

SHIHE YANG

Department of Chemistry and the Institute of Nano Science and Technology
Hong Kong University of Science and Technology
Clear Water Bay, Kowloon, Hong Kong

The discovery of the giant electrorheological (GER) effect poses a challenge to the theorists for an explanation of its physical underpinnings. This is particularly the case as the GER effect breaks the yield stress upper bound predicted on the basis of induced polarization mechanism. The GER effect was observed in suspensions of coated nanoparticles, each consisting of a core particle of barium titanyl oxalate (20-50 nm), coated with a 5-7 nm layer of urea. The suspending oil is also very important, as the "wrong" oil can mean no electrorheological effect at all. Our prior work has shown the GER data, on the field dependence of yield stress, to be quantitatively accountable by the formation of aligned molecular dipolar layers in the region of particle-particle contact. Here we justify this picture through microscopic statistical mechanic considerations. We present a model of hydrogen bonding which can give rise to both electrowetting between the particles and the oil, as well as inducing the formation of aligned surface molecular dipolar layers. Monte Carlo simulation is used to account for both the entropy and energetic effects. It is shown that the formation of aligned surface dipolar layers can indeed be a robust phenomenon

1. Introduction

Electrorheological fluids constitute a type of colloidal suspensions consisting of dielectric particles dispersed in an insulating oil [1-13]. The marvelous feature of an ER fluid is that it can solidify into a jelly-like state almost instantaneously (1-10 ms) when subject to an externally applied electric field with moderate strength (a few kV/mm), with a stiffness varying proportionally to the field strength. The liquid-solid transformation is reversible. Once the applied field is removed, the original flow state is recovered. Such characteristics enable ER fluids to be an electro-mechanical interface that, when coupled with motion sensors and electronic control systems so as to correlate external conditions with the ER characteristics, can mean controllability and responsiveness of mechanical devices. In particular, ER-based active dampers and clutches have recently been demonstrated [14-17]. However, ER fluids have yet to find widespread commercial applications, mainly due to the fact that the maximum yield stress achievable by most of the conventional ER fluids is less than 10 kPa, much below ~30 kPa required by many mechanical devices.

Recently, a new type of ER fluid with the giant electrorheological (GER) effect was discovered [18]. The GER fluids consist of coated nanoparticles ($BaTiO(C_2O_4)_2 + NH_2CONH_2$) suspended in silicone oil. In the core/shell structure (Inset of Fig. 1), urea coating serves as an ER promoter. The static yield stress curves for two volume fractions, measured under DC electric fields, are shown in Fig. 1.

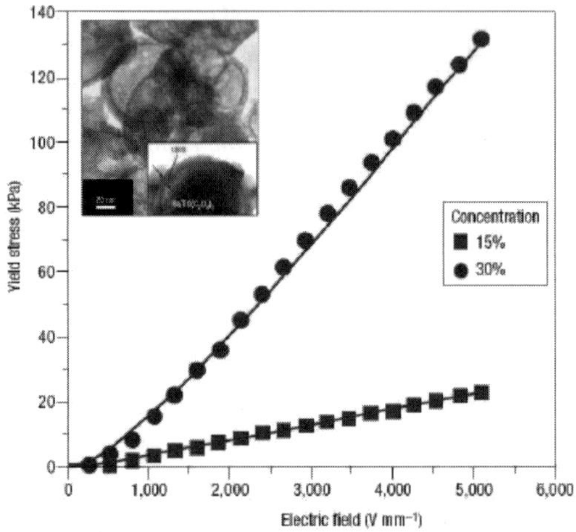

Figure 1: Static yield stress plotted as a function of applied electric field for two solid concentrations. Symbols denote experiment; solid lines are theory. Inset: TEM image of coated nanoparticles. Urea coatings are clearly seen.

The GER fluids behave differently from the conventional ER fluids. In the conventional ER fluids, the dispersed particles are polarized under applied electric fields, owing to the dielectric constant contrast to the fluid. In the heuristic dipole model for induced particle, the randomly dispersed particles would tend to form chains due to the dipole-dipole interaction, and the chains would coalesce into columns at higher field strength. An exact treatment which goes beyond dipole approximation is possible, and the local field effect and electrical conductivity effects can be also included as well [**Error! Reference source not found.**]. Based on the assumption that components of the ER fluids exhibit linear response to the electric field, theoretical upper bounds on the yield stress may be obtained in this so-called dielectric ER (DER) model. It varies as

$$1.38\sqrt{R/\delta} \tag{1}$$

in units of $\varepsilon_f E^2 / 8\pi$. Here δ denotes the separation between the surfaces of neighboring particles and ε_f is the dielectric constant of the liquid [13]. For the purpose of calculating the magnitude of the upper bound, it is generally taken to be 1-2 Angstroms. Several aspects of this expression should be noted. First, the yield stress varies as the square of the applied electric field. This follows from the fact that the yield stress is proportional to the electrostatic energy density $-\vec{P}\cdot\vec{E}$, where $\vec{P}$ is the polarization density. If the polarization is induced, then in the linear regime $\vec{P}$ is proportional to $\vec{E}$, and hence the square dependence on $\vec{E}$. Second, the upper bound scales as the square root of particle size. In other words, the induced polarization mechanism favors larger particles. This behavior has been experimentally verified [19]. Third, for a fixed electric field and particle size, there is a maximum value for the yield stress, given by Eq. (1). We shall see that the GER effect violates all three aspects of the induced polarization DER model. Hence it represents a new paradigm that requires new modeling considerations.

From Fig. 1, it is seen that the yield stress can be over 100 kPa. For the particle size of 50-80 nm, this value is more than one order of magnitude larger than the upper bound. The electric field variation is seen to be mostly linear instead of quadratic, indicating a constant (saturated) polarization. The particle size scaling behavior is also opposite to that predicted by Eq. 1 [20].

We propose the mechanism of saturation surface polarization, in the elastic contact region of the neighboring spheres, to be responsible for the observed GER effect [18,20]. In Fig. 2 we show a schematic picture of our model, consisting of two spherical coated particles whose centers are aligned along the direction of external electric field, with a (flat) deformed contact area. The parameters of the model are stated in the caption. Our prior work has shown that the measured data, on the field dependence of the GER yield stress, can be explained quantitatively by the formation of aligned molecular dipole layers in the region of particle-particle contact. Here the dipoles are urea's molecular dipoles (4.6 Debye each). However, a microscopic understanding of the aligned dipolar layers has been lacking.

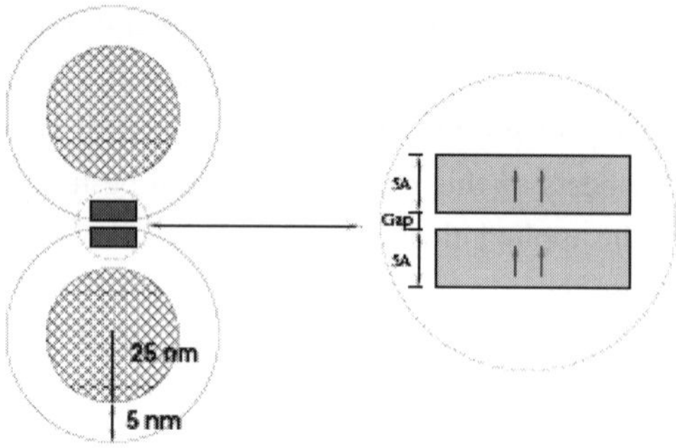

Figure 2: Surface polarization layer model. Radius of the core is 25 nm and the coating thickness is 5 nm. The gap between two flatten layers is 5–10*A* and the thickness of the saturation polarization layers is 5*A*. The overlap of two spheres is determined by the equilibrium of the electrostatic force and the elastic force.

The aim of this paper is to establish a microscopic model to account for the formation of aligned dipolar layers (in the contact region between two particles) in an externally applied electric field. The statistical mechanical underpinnings of the model is examined through Monte Carlo simulations. We propose electrowetting, with hydrogen bonding as a contributing element, in inducing the alignment layers by energetically favoring the aligned state in the form of lower interfacial energy between the particles and the suspending liquid.

2. Statistical mechanics of surface aligned layers

In order to investigate the formation of the surface dipolar layers from the statistical mechanic point of view, the contact area of the urea coating are modeled by two cubic lattices (8×8×8 each). A urea molecule is put on each lattice site and is simplified as a dipole with a permanent moment (4.6 Debye), same as that for bulk urea. The dipoles in the outer most layer, i.e., at the coating-core interface, are fixed in anti-parallel pairs, simulating the locked state in bulk urea. The dipoles in other layers can rotate freely as in the liquid state and the dipole-dipole interaction energy is given by

$$W_{ij} = \frac{\vec{p}_i \cdot \vec{p}_j}{r_{ij}^3} - \frac{3(\vec{p}_i \cdot \vec{r}_{ij})(\vec{p}_j \cdot \vec{r}_{ij})}{r_{ij}^5} \qquad (2)$$

At the inner-most layers, corresponding to the coating-oil interfaces, an additional "electrowetting" effect is proposed which gives rise to perpendicularly-aligned dipoles on the same layer. This is described below.

It has been observed in experiments that particle-oil wetting plays a crucial role in the GER effect [21]. Non-wetting can imply no GER effect. Thus a synergistic effect between the particles and the suspending oil exists and is responsible for the appearance of the GER effect. We propose that the electrowetting effect plays a role in the energetics of the aligned dipole layers, by lowering the energy of the aligned dipoles as compared to the usual anti-parallel alignment in the locked configuration (in bulk urea, for example). It is known that the electrowetting effect may be expressed by the Lippmann equation [22,23]:

$$\gamma = \gamma_0 - cE^2 \tag{3}$$

where γ is the surface tension and E is the magnitude of the local electric field. c is a small length scale measured in Å. The electro-wetting effect is generally small when the local electric field is weak. However, if the dipoles are aligned, the local field can be enormous (on the order of 10^7 V/cm), leading to a significant lowering of the surface energy. This electrowetting effect would mean that (1) the dipoles are not only favored to be aligned, but also to be perpendicular to the layer, as otherwise the electric field would be weak in the fluid region; (2) the effect remains the same if the dipoles are reversed in their orientations; and (3) the effect would counter the repulsive intra-layer interaction between parallel-aligned dipoles in the same layer, acting in conjunction with the attractive inter-layer interaction across the gap.

In mean-field form, the electrowetting effect may be expressed as

$$W_i = -J \sum_j p_i^z p_{j \neq i}^z \tag{4}$$

Here the superscript on p denotes the component of the dipole perpendicular to the layer, and J is a positive parameter whose microscopic origin will be explored below. Here we first consider the necessary magnitude of J to realize the aligned dipolar layers.

In order to account not only the energetics but also the entropy effects, we have carried out Monte Carlo simulations on the model. It was found that if J is assumed to be 0.4 eV, with the lattice constant being the same as that in bulk urea, i.e. $a = 5$ Å and the gap between two inner most layers being 5 Å, and at the room temperature and under a local electric field 7×10^6 V/cm, the surface aligned dipole layers can be stabilized, as seen in Fig. 3.

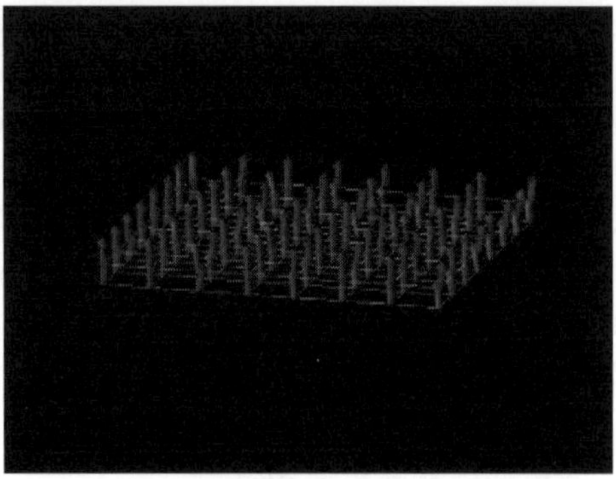

Figure 3: Monte Carlo simulation result shows that the aligned dipolar layers. Here we show mostly the inner layers in the gap region, i.e., the 8th and 9th layers

For a weaker electrowetting effect, i.e., $J < 0.4$ eV the aligned dipole layers are not stable. Here $J = 0.4$ eV corresponds to a value of $c = 16\,\mathrm{\mathring{A}}$ in Eq. 3, obtained by the difference of surface energy densities on the contact layers (with and without the electric field), divided by the square of local electric field (the threshold value $E_c = 0.7 \times 10^7$ V/cm, obtained numerically from a weak externally applied field $E_c^{(ext)} = 500$ V/mm through finite element simulation).

There can be two elements contributing to the electrowetting effect. The first element is the hydrogen bonding interaction. An illustration of the plausible hydrogen bonding network is shown in Fig. 4, where it is assumed that in the gap region there can be some urea

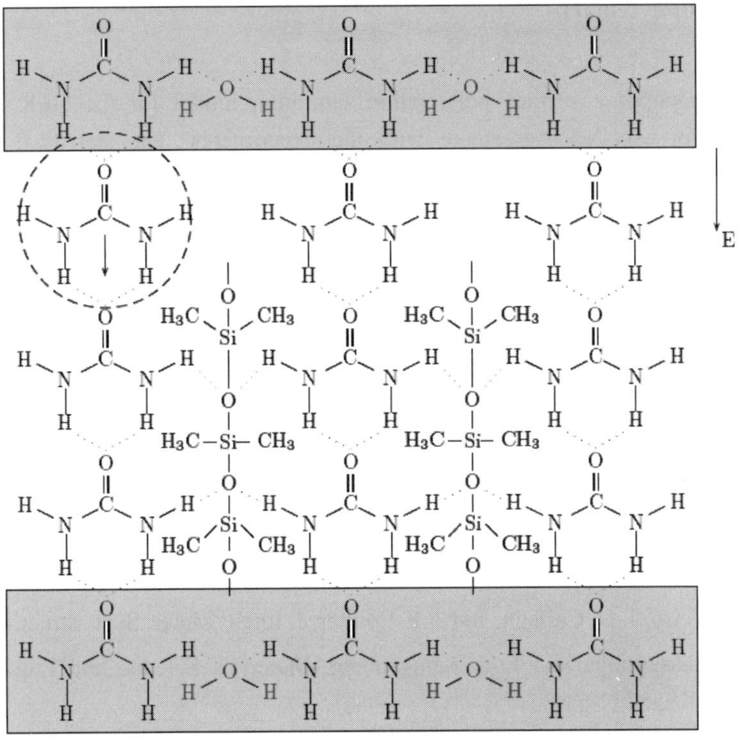

Figure 4: A schematic hydrogen bonding (indicated by the dotted lines) network favoring the formation of parallel alignment of urea molecular dipoles. The urea molecular dipole unit is indicated by dashed line, with direction as shown.

molecules, some water molecules, plus the silicone oil molecules. They network together to give rise to the favorable effect of parallel alignment of the urea molecular dipoles. As each hydrogen bond is on the order of 0.1 eV, it can be seen that the proposed network can contribute a considerable amount to the required $J = 0.4$ eV. The second element is the $-\alpha E^2$ interaction energy between the induced polarization of the molecules in the gap (α being the polarizability) and the local electric field. That energy can also be on the same order, provided the local electric field is on the order of 10^7 V/cm. It is seen that the silicone oil, with the presence of oxygen in its structure, can partly contribute to the hydrogen bonding network. It is conjectured that the presence of oxygen in silicone oil is a contributing factor to its synergistic wetting of the coated nanoparticles.

3. Concluding remarks

We have proposed a surface polarization saturation model for the GER effect, with predictions in excellent agreement with the experiments. The statistical mechanical underpinnings of the aligned dipolar layers was studied by using Monte Carlo simulations, with elements of hydrogen bonding contributing to the energetics favorable to the formation of such layers. We wish to acknowledge the support of grants from Hong Kong RGC (No. 604104).

References

1. H. Block and J. P. Kelly, J. Phys. D **21**, 1661 (1988).
2. A. P. Gast and C. F. Zukoski, Adv. Colloid Interface Sci. **30**, 153 (1989).
3. T. C. Jordan and M. Shaw, IEEE Trans. Electr. Insul. **24**, 849 (1989).
4. C. F. Zukoski, Annu. Rev. Mater. Sci. **23**, 45 (1993).
5. K. D. Weiss, J. D. Carlson, and J. P. Coulter, J. Intell. Mater. Syst. Struct.**4**, 13 (1993).
6. M. Parthasarathya and D. J. Klingenberg, Materials Science and Engineering: R: Reports **17**, 57 (1996).
7. T. C. Halsey and W. Toor, Phys. Rev. Lett. **65**, 2820 (1990).
8. R. Tao and J. M. Sun, Phys. Rev. Lett. **67**, 398 (1991).
9. H. J. H. Clercx and G. Bossis, Phys. Rev. E **48**, 2721 (1993).
10. R. A. Anderson, in *Proceedings of the International conference on Electrorheological fluids : mechanisms, properties, structure, technology and applications*, edited by R. Tao (World Scientific, Singapore, 1992), pp. 81–90.
11. N. Felici, J. N. Foulc, and P. Atten, in *Proceedings of the International conference on Electrorheological fluids : mechanisms, properties, technology and applications*, edited by R. Tao and G. D. Roy (World Scientific, Singapore, 1994), pp. 139–152.
12. L. C. Davis and J. M. Ginder, in *Proceedings of an American Chemical Society Symposium on Electrorheological (ER) Materials and Fluids: Progress in electrorheology : science and technology of electrorheological materials*, edited by K. O. Havelka and F. E. Filisko (Plenum Press, New York, 1995), pp. 107–114.
13. H. R. Ma, W. J. Wen, W. Y. Tam, and P. Sheng, Adv. Phys. **52**, 343 (2003).
14. K. P. Tan, R. Stanway, and W. A. Bullough, Mech. Adv. Mat. Struct. **13**, 1 (2006).
15. T. Nakamura and N. Saga, IEEE/ASME Trans. Mechatron. **10**, 154 (2005).
16. L. Y. Liu, X. X. Huang, C. Shen, Z. Y. Liu, J. Shi, W. J. Wen and P. Sheng, Appl. Phys. Lett. **87**, 104106 (2005).

17. N. M. Wereley, J. Lindler, N. Rosenfeld, and Y. Choi, Smart Mater. Struc. **13**, 743 (2004).

18. W. J. Wen, X. X. Huang, S. H. Yang, K. Q. Lu and P. Sheng, Nat. Mater. **2**, 727 (2003).

19. W. Y. Tam, G. H. Yi, W. J. Wen, H. R. Ma, M. M. T. Loy, and P. Sheng, Phys. Rev. Lett. **78**, 2987 (1997).

20. W. J. Wen, X. X. Huang, and P. Sheng, Appl. Phys. Lett. **85**, 299 (2004).

21. C. Shen, W. J. Wen, S. H. Yang and P. Sheng, J. Appl. Phys. **99**, 106104 (2006).

22. G. Lippmann, Ann. Phys. **149**, 546 (1873).

23. G. Beni and S. Hackwood, Appl. Phys. Lett. **38**, 207 (1981).

PROGRESS UPDATE OF MAGNETORHEOLOGICAL FINISHING (MRF®) TECHNOLOGY DEVELOPMENT

ARIC B. SHOREY, MARC TRICARD, WILLIAM KORDONSKI, AND DAVID KOLCZYNSKI

QED Technologies, Inc.

1040 University, Ave

Rochester, NY 14507

E-mail: shorey@qedmrf.com

Magnetorheological Finishing (MRF) is a precision finishing process that has become an industry standard tool for polishing high quality optical surfaces. This novel technology utilizes a magnetorheological (MR) fluid and a magnetic field to generate material removal for this sub-aperture polishing process. The properties of the MR fluid provide a significant number of advantages for generating a flexible and stable tool making it uniquely suitable for finishing a wide range of optics to extremely high precision. This includes diverse applications such as finishing flats/prisms, spheres, aspheres, and cylinders of materials that range from glass, single crystals, ceramics and some metals – all of which could be polished on the same platform. What is more, the unique attributes make it a valuable tool for lightweight optics and high-energy applications. The original MRF platform was developed to make routine work of finishing high precision optics (flats, spheres, aspheres etc.) ~20-200mm in size. Since then, the size of optic that can be finished with this technology has been scaled to both larger and smaller sizes over a range of less than 5 mm to more than 2000 mm. A number of technical challenges have been realized and overcome as the platform geometry has been modified. An overview of some of these challenges, and their solutions will be presented.

1 Introduction

New and more sophisticated tools are required as the needs in the optics fabrication community become more challenging and require higher precision. Figure 1 shows some examples of the types of challenging optical lenses and windows that are required for current and next generation systems. Optics are used in a wide range of applications, so, in addition to challenging shapes, a wide range of aperture sizes must be polished to high precision. In optics fabrication processes, the final finishing step is difficult, time consuming and often unpredictable. The traditional approach to polishing spherical and flat optics has been to use a full aperture pitch or pad covered tool formed to have the same shape of the part being polished. Pitch is a viscoelastic material, which means it flows as a result of the forces applied during polishing. This requires that the optician stop and check the part and lap shape frequently to avoid surface shape degradation over time. This iterative process is highly skilled, labor-intensive and non-deterministic. Tools that utilize a polyurethane pad instead of pitch have a similar problem in that the pad wears during polishing and the shape and removal characteristics change. Furthermore, it is extremely difficult or impossible to use a full aperture tool to polish complex shapes such as aspheres and conformal optics [1].

As the optic's shape becomes more complex (asphere, conformal or freeform), polishing to high precision requires the use of tools that are smaller than the full aperture size of the workpiece. Such sub-aperture tools made of pitch or pad have the problems described above. That is, the flow of pitch and wear of the pad give an inconsistent, non-deterministic removal rate. Furthermore, these polishing processes induce a small damaged layer on the order of tens of nanometers deep [2]. A number of approaches to generate sub-aperture polishing tools have been developed [3.4]. In addition to removal rate variability, many tools face significant challenges when polishing non-standard shapes in that the local curvature of the workpiece surface varies. This means that the contact zone between the polishing tool and workpiece changes. In order to have a predictable process, these changes should to be minimized and must be predictable so that they can be accounted for in the finishing process (e.g. in the software that controls the process).

Fig. 1: Range of design and size of optics used in various applications [5-8].

Magnetorheological Finishing (MRF) is a particularly effective tool with several unique properties that provide the stability and predictability required to achieve very high precision with few polishing iterations, as well as a compliant tool that allows for finishing complex shapes. MRF uses a sub-aperture lap created by a magnetorheological (MR) fluid inside a magnetic field [9]. By using careful characterization of material rate of removal and the figure error on the part to be polished, along with CNC control, MRF has been shown to be able to successfully polish a range of optics to better than 30 nm peak-to-valley (PV) [1]. This process gives microroughness of <1.0 nm rms, and furthermore has shown the ability to remove sub-surface damage (SSD) from previous processing steps and improve surface integrity [10-11]. The MR fluid lap in MRF gives advantages for use in a wide range of applications [12]. The compliant tool allows the user to polish flats, spheres, cylinders, aspheres and even freeform shapes all using the same polishing platform.

2 Magnetorhological Finishing

The unique properties of magnetorheological (MR) fluids are key to their successful implementation into a precision polishing process. As proposed by Kordonski, et al. [9], material removal is accomplished through hydrodynamic forces resulting from the shear flow of the MR fluid beneath the workpiece surface [9]. The MRF process is shown schematically in Fig. 2a. A pump delivers MR fluid through a nozzle onto a rotating wheel. This rotating wheel carries the fluid through the workzone (see inset in Fig. 2b) where it is stiffened by a magnetic field generated by an electromagnet of appropriate design. The applied magnetic field increases the apparent viscosity by several orders of magnitude. The flow of the magnetized fluid between the rotating wheel and the workpiece surface generate the shear forces required to remove material in the polishing regime. It is interesting to note the interaction of the MR fluid and workpiece shown in Fig. 2b. Notice that as the fluid flows beneath the workpiece, it perfectly conforms to the surface figure. This is an important attribute that enables the flexibility of MRF in finishing complex shapes. After the fluid flows beneath the workpiece surface, it is carried by the wheel to the suction device that removes the fluid from the wheel, and is subsequently pumped to the fluid reservoir, where the fluid properties are monitored and actively adjusted in situ to minimize variability.

The most important consideration when understanding the MRF process is the wheel/part/fluid interaction shown in Fig. 2b. This interaction is shown schematically in Fig. 3. The MR fluid flows from left to right, carried by the rotating wheel (lower boundary) into a zone of contact with the workpiece surface. The MR polishing fluids are comprised of micron-sized magnetic particles and non-magnetic polishing abrasives (typically cerium oxide or nano-diamond). Inside the magnetic field, the MR fluid is modeled as a Bingham fluid, whose yield stress is a function of the applied field [9]. Application of Bingham lubrication theory to the hydrodynamic flow between the wheel and workpiece provides insight into the material removal process. Given the nonzero yield stress of the MR fluid, this model predicts the existence of sheared and un-sheared regions in the flow depending on the local magnitude of the shear stress [9]. The gradient magnetic field is applied such that the magnitude is greater at the wheel surface than it is at the workpiece surface. This has the effect of drawing the magnetic particles towards the wheel, which results in the abrasives being concentrated at the workpiece surface. The magnetic particles provide a conformal lap, supporting the abrasives that remove material. In Fig. 3, the non-magnetic polishing abrasive particles are designated by squares. Darkened areas represent unsheared regions of magnetic fluid and the lighter areas denote MR fluid flowing through the gap. The model predicts the existence of a thin—but highly sheared—layer close to the part surface, where the high shear stresses enable the concentrated non-magnetic abrasives to polish the part. The removal function, or MRF spot, caused by this fluid flow is shown to the right in Fig. 3.

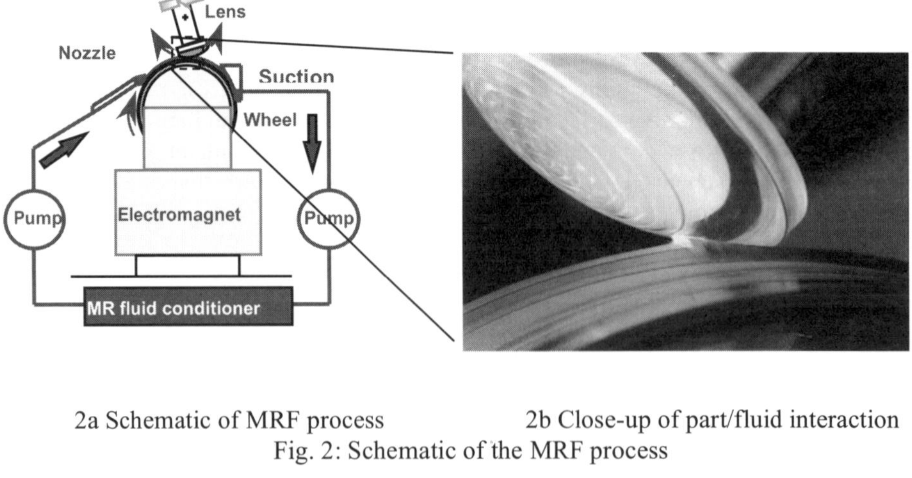

2a Schematic of MRF process 2b Close-up of part/fluid interaction

Fig. 2: Schematic of the MRF process

While a more complete understanding of the MR fluid flow is being sought, there is sufficient experimental evidence to support the model described above. The fluid viscosity, velocity, magnetic field, and workpiece positioning—all of which can be controlled to high precision—determine the shear stress distribution and magnitude. Since removal rate is proportional to shear stress, a stable removal process is achieved. The removal function can be adjusted by changing parameters such as wheel speed, lens positioning in the fluid, or magnetic-field strength. Relatively high, stable removal rates are achieved while creating a smooth, damage-free surface finish. Unlike conventional polishing where the particle's normal load can result in scratching, MRF creates relatively low normal load on the abrasive particles. This shear based mechanism of material removal, with low particle normal force, is what gives MRF advantages in removing SSD while polishing, as well as not generating SSD of its own.

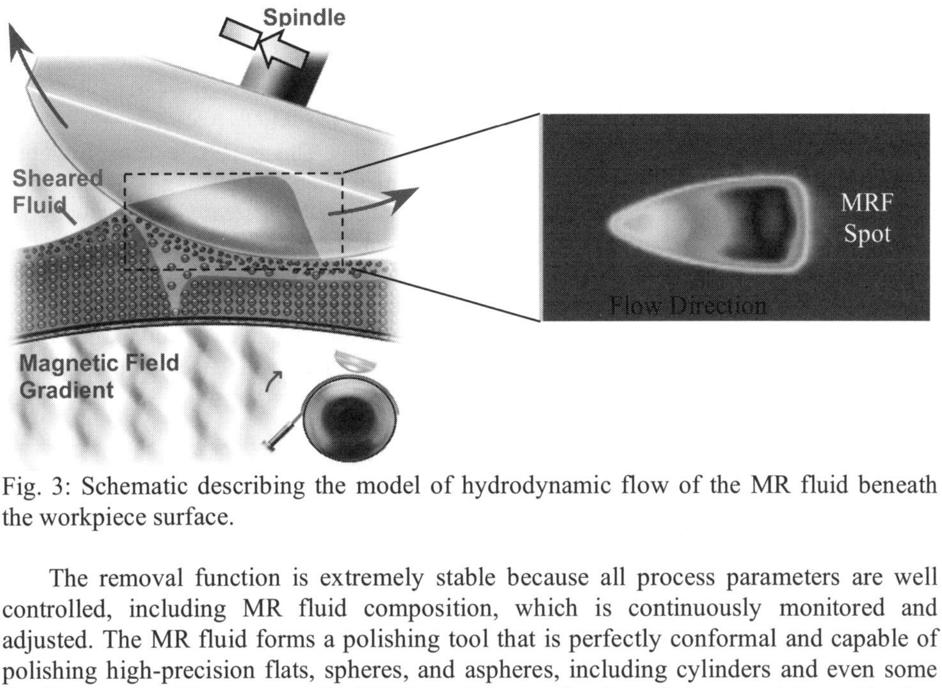

Fig. 3: Schematic describing the model of hydrodynamic flow of the MR fluid beneath the workpiece surface.

The removal function is extremely stable because all process parameters are well controlled, including MR fluid composition, which is continuously monitored and adjusted. The MR fluid forms a polishing tool that is perfectly conformal and capable of polishing high-precision flats, spheres, and aspheres, including cylinders and even some freeform optics. The success of MRF for volume production of ultra-precision optics is a result of its predictability. The process uses an empirical characterization of the figure error in the workpiece surface and de-convolves this with a material removal function to create dwell time instructions. Using computer numerical control (CNC), the part is moved through the polishing zone according to the dwell schedule.

3 Recent Advances in Scaling MRF technology

The first commercially available MRF platform, the Q22-X, polished optics from ~20 mm to 200 mm in size and was designed to polish rotationally symmetric parts (see Fig. 4a). In this design, the wheel was mounted on the platform below the workpiece and the electromagnet was designed to fit outside and underneath the wheel and workpiece. The workpiece is held by vacuum in the spindle, which moves the part through the work zone. This part used a rotational tool path and was best suited for round optics. The Q22-Y platform was introduced with the ability to polish using a raster motion and enabled precision finishing of rectangular, and other non-round parts.

There are two primary ways that the size of the contact zone in MRF can be adjusted. The first is by changing the depth that the part is inserted into the MR fluid ribbon. The deeper the part is placed into the fluid, the larger the contact zone and therefore the removal spot size. This allows for some adjustment of the removal function size, but as the size of the part to be finished increases to up to 1 meter, for example, it is desirable to increase the spot size more than can be accomplished by varying the depth the part is placed into the fluid. Furthermore, there is a limit to how small the spot size can be made with this method. For this, and other reasons, new MRF platforms have been developed for finishing large and small aperture sizes.

3.1 MRF Platforms for Large Apertures

There are a number of applications for large aperture precision optics, such as photolithography lenses, large/lightweight mirrors (e.g. astronomical telescopes) large windows (e.g. reconnaissance), etc. Development of the MRF platforms for larger apertures has provided new challenges to overcome. The part size capability has increased in stages; starting with the Q22-400X, which enabled finishing of rotationally symmetric apertures up to 400 mm in size. It has since been increased further with the introduction of the Q22-750P2 (750 mm x 1000 mm flat parts), the Q22-950F (950 x 1200 mm freeform parts), and continuing with development of multi-meter platforms. The size of the wheel used is typically increased as the size of the part to be polished increases. For example, the Q22-400X employs a 200 mm diameter wheel and the Q22-950F utilizes a 370 mm diameter wheel. In some novel applications, one of which will be discussed below, multiple sized wheels are used on the same platform.

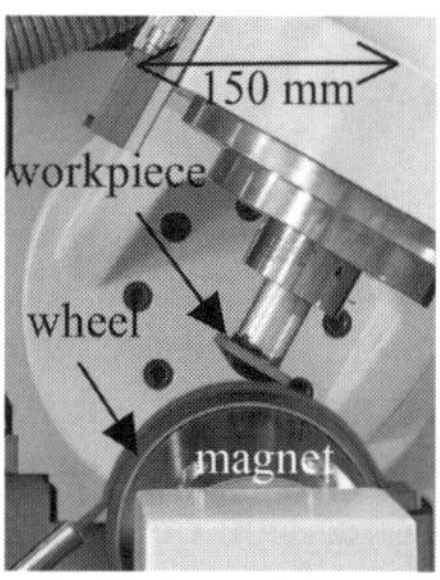

4a: Q22-X with 150 mm diameter wheel

4b: Q22-950F with 370 mm diameter wheel

4c: Q22-XE with 20 mm diameter wheel

Fig. 4: Images of various MRF polishing platforms

Among the challenges overcome was the relative orientation of the wheel and workpiece. When the workpiece size increases, it becomes more challenging (and higher risk) to hold the workpiece in a spindle with vacuum. Instead, these large aperture platforms are designed with the part safely held on a 2 or 3 axis stage (depending on the particular platform) below the "upside down" wheel, which carries the MR fluid ribbon from above, as shown in Fig. 4b. This design provided a number of challenges, particularly for the design of the applied magnetic field. A significant challenge associated with this design is ensuring that the fluid stays on the wheel as it flows through the wheel apex and back up to the device that removes the fluid from the wheel. This electromagnet *inside* the wheel is specifically designed to facilitate this process. Overcoming these technical challenges has enabled successful scale-up of the MRF technology to efficiently finish meter class optics to high precision.

3.2 MRF for Millimeter-Sized Features

In addition to increasing the size of optics that can be finished with MRF to include larger aperture sizes, wheels as small as 20 mm in diameter have been developed to address parts and features < 5 mm in size (see Fig. 4c). This, of course, led to a new set of challenges caused by the close proximity of the fluid delivery nozzle with the magnetic field as well as miniaturization of all of the components. A solution to a number of the challenges was accomplished by replacing the electromagnet with a suitable *permanent* magnet design. Implementation of this and other design changes have resulted in the ability to generate a polishing spot as small as 1 mm in size. The consequences of these changes are highlighted in the polishing results that are described in Section 4.

4 Results

MRF has demonstrated a great deal of success in precision finishing in a number of applications. These include such diverse applications as flats, prisms, spheres, cylinders and aspheres [1, 12]; optics for high energy applications [11]; optical molds [13]; lightweight mirrors and other large optics [14, 15]; domes [16]; and transmitted wavefront correction [17]; among others. Below we outline a novel application that was enabled by recent advancements in the MRF process.

4.1 Phase Plates

In some cases, the ideal surface after polishing is not perfectly flat, spherical, aspherical, etc. A significant advantage of MRF is that when generating the dwell map for the polishing process, the figure used as the desired final surface can be generated mathematically. The contour map in Fig. 5a shows the desired final surface for a part that was designed to precisely shape a laser beam according to a specific prescription. The meaning of this map is that the most material should be removed in the red regions and minimal material removal should occur in the purple regions. Given the sub-aperture nature of MRF, the flexibility of CNC control, and the ability to adjust a number of parameters to change removal spot size, MRF was able to precisely shape the surface according to this prescription. Figure 5b shows the final figure, which is the inverse of the desired prescribed removal shown in Fig. 5a. The machine that generated this map included both a 370 mm wheel and 50 mm wheel on the same MRF platform (Q22-750P2). The 370 mm diameter wheel was used to efficiently remove the large amounts of material over the 500 mm aperture, after which the 50 mm wheel was used to precisely print in the high frequency features seen in these figures [18, 19].

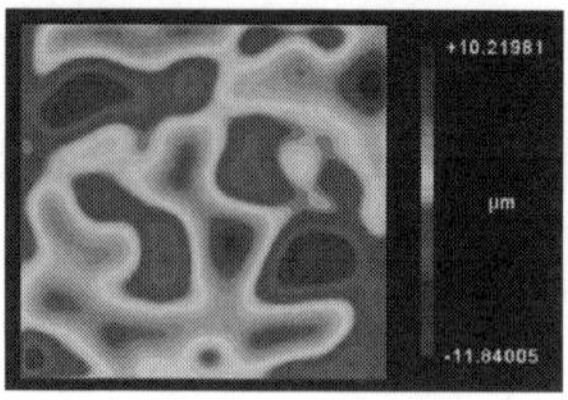

5a: Hit Map (desired removal)

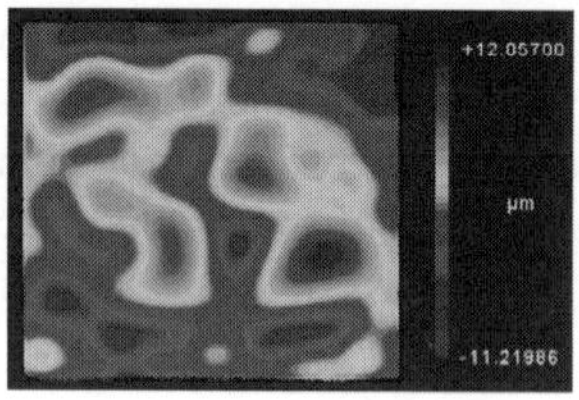

5b: Actual final figure

5c: Q22-750P2 outfitted with 370 mm and 50 mm wheels

5 Conclusion

Using a magnetorheological fluid and magnetic field, it is possible to generate a polishing lap with a number of unique attributes important for precision optics fabrication. The conformal nature of the MRF polishing lap enables precision finishing of a full range of current and next generation optics. The nature of the shear based removal mechanism improves surface integrity and removes SSD generated from previous processing steps. Finally, technology advancements have proven MRF to be scalable for final finishing of optics with sizes that range from < 5 to > 2000 mm in size.

6 References

1. D. Golini, W.I. Kordonski, P. Dumas, S. Hogan, "Magnetorheological Finishing (MRF) in Commercial Precision Optics Manufacturing", Proc. SPIE, **3782**: 80-91 (1999).
2. H. Yokota, H. Sakata, M. Nishibori and K. Kinosita, "Ellipsometric Study of Polished Glass Surfaces", Surface Science, **16**, pp. 265-274, (1969).
3. C.J. Evans, E. Paul, D. Dornfield, D.A. Lucca, G. Byrne, M. Tricard, F. Klocke, O. Dambon, and B.A. Mullany, "Material Removal Mechanisms in Lapping and Polishing", Annals of the CIRP, **52**/2:611-633.
4. S. D. Jacobs, "Innovations in Polishing of Precision Optics", International Progress on Advanced Optics and Sensors , Frontiers Science Series #40, edited by H. Ohmori and H. M. Shimizu, (Universal Academy Press, Tokyo, 2003) pp. 3-14.
5. Cell phone camera photo courtesy of Topcon Corp., Tokyo, Japan.
6. R. Shannon, J.P. Mills, H.M. Pollicove, P.A. Trotta and L.N Durvasula, "Optics That Fit", Photonics Spectra, April (2001).
7. J. Mulkens et al., "Optical lithography solutions for sub-65 nm semiconductor devices:, Proc. SPIE, **5040**: 753-762 (2003).
8. Images of Corona Spy Satellite and Hobby-Eberly Telescope courtesy of Eastman Kodak's Precision Optics Youth Forum, http://kodak.com/US/en/government/po/poyf
9. W.I. Kordonski, S.D. Jacobs, "Magnetorheological Finishing", International Journal of Modern Physics B, **10**: 2837-2848 (1996).
10. S.R. Arrasmith, S.D. Jacobs, J.C. Lambropoulos, A. Maltsev, D. Golini, and W.I. Kordonski, "The Use of Magnetorheological Finishing (MRF) to Relieve Residual

Stress and Subsurface Damage on Lapped Semiconductor Silicon Wafers", Proc. SPIE, **4451**: 286-294 (2001).

11. Menapace et al., "Combined advanced finishing and UV-laser conditioning for producing UV-damage-resistant fused-silica", Proc. SPIE, **4679**:56-68 (2002).
12. M. Tricard, P. Dumas and G. Forbes, "Sub-aperture approaches for asphere polishing and metrology", Photonics Asia 2004, Beijing, China, 8-11 November 2004.
13. C. Hall, M. Tricard, H. Murakoshi, Y. Yamamoto, K. Kuriyama, H. Yoko, "New mold manufacturing techniques", Proc. SPIE, **5868**: 264-273 (2005).
14. A.B. Shorey, W. Kordonski, M. Tricard, "Magnetorheological finishing and subaperture stitching interferometry of large and lightweight optics", Proc. SPIE, **5494**: 81-90 (2004).
15. M. Tricard, A. Shorey, B. Hallock, P. Murphy, "Cost-effective, subaperture approaches to finishing and testing astronomical optics", *to be published* Proc. SPIE, **6273** (2006).
16. A. Shorey, W. Kordonski, M. Tricard, "Deterministic precision finishing of domes and conformal optics", Proc. SPIE, **5786**: 310-318 (2005).
17. V. Bagnoud, M.J. Guardalben, J. Puth, J.D. Zuegel, T. Mooney, and P. Dumas, "High-energy, high-average-power laser with Nd:YLF rods corrected by magnetorheological finishing", Applied Optics, **44**, 282-288, 2005.
18. P. Dumas. et. al., "Flexible Polishing and Metrology Solutions for Free-form Optics," Proc. ASPE Winter Topical Meeting on Free-Form Optics: Design, Fabrication, Metrology, Assembly, Feb. 2004.
19. J.A. Menapace, S.N. Dixit, F.Y. Genin, W.F. Brocious, "Magnetorheological finishing for imprinting continuous-phase plate structures onto optical surfaces", Proc. SPIE, **5273**: 220-230 (2004)

PROPERTIES OF ULTRASONIC PROPAGATION IN FUNCTIONAL FLUIDS UNDER MAGNETIC FIELDS

M. MOTOZAWA AND T. SAWADA

Department of Mechanical Engineering, Keio University
3-14-1 Hiyoshi, Kohoku-ku, Yokohama 223-8522, Japan
E-mail: sawada@mech.keio.ac.jp

Y. MATSUMOTO

Canon Inc.,
3-30-2 Shimomaruko, Ohta-ku, Tokyo 146-8501, Japan

The properties of ultrasonic propagation velocity in water-based and kerosene-based magnetic fluids and an MR fluid are examined experimentally. The ultrasonic frequency used is 1 MHz and the measurement scheme is based on the pulse method. The external magnetic field intensity is varied from 0 mT to 550 mT and is swept at a constant rate dB / dt (sweep rate). The ultrasonic propagation velocity changes with applied magnetic field and hysteresis is observed. The properties of ultrasonic propagation seem to be highly influenced by the formation of chain-like clusters (in the magnetic fluids) and robust clusters (in the MR fluid).

1 Introduction

Magnetic fluid and Magnetorheological (MR) fluids are typical functional fluids, which change their rheological properties when an external magnetic field is applied. However, there are some characteristic differences in the rheological properties between magnetic fluids and MR fluids. These differences result from differences in the inner structures of each fluid (e.g. magnetic particle size, cluster formation, etc.). A magnetic fluid is a stable colloidal suspension of small surfactant-coated magnetic particles in a liquid carrier, such as water or kerosene, etc. The magnetic particles are about 10 nm in size. The preparation of a stable magnetic fluid involves a delicate balance between the attractive and repulsive forces on each particle. Under no external magnetic field, these particles are in Brownian motion and remain suspended[1]. When an external magnetic field is applied to a magnetic fluid, some of the colloidal particles coagulate and form chain-like clusters. It seems that these chain-like clusters and the Brownian motion of the magnetic particles have an influence on ultrasonic propagation velocity in the magnetic fluid. In our previous studies[2,3], we precisely measured ultrasonic propagation velocity in magnetic fluids under a magnetic field and discussed the influence of cluster formation on ultrasonic propagation velocity.

On the other hand, the composition of an MR fluid is similar to that of a magnetic fluid, however, the typical magnetic particle size is in a range of 100 nm to 10 μm. An MR fluid has controllable yield stress and large viscosity under a magnetic field because of its high magnetization[4, 5]. Various applications such as clutches, dampers and shock absorbers have been extensively developed using MR fluids. When an external magnetic field is applied to an MR fluid, the magnetic particles form robust clusters rather than the chain-like clusters found in magnetic fluids. It seems that these differences of cluster formation can be examined based on the properties of ultrasonic propagation velocity in magnetic and MR fluids.

Even though several theoretical[6] and experimental[7] studies have been carried out to clarify ultrasonic propagation properties in magnetic fluids, the mechanism is still not clear. Moreover, there have been very few studies concerned with ultrasonic propagation properties in an MR fluid. In this paper, we describe experiments involving precise measurements of ultrasonic propagation velocity in water-based and kerosene-based magnetic fluids and in an MR fluid. The experiments are performed under varying magnetic field intensity, sweep rate and angle ϕ between the direction of magnetic field and the direction of ultrasonic propagation. We also discuss the relationship between cluster formation and the properties of ultrasonic propagation.

2 Experiment

2.1 Experimental Apparatus

Figure 1 shows the experimental apparatus. The left figure is a block diagram of the experimental apparatus and the right figure shows the detailed structure of the test cell. The test cell has a concentric structure. The test fluid is in the inner rectangular vessel. The outer cylindrical container is filled with circulating water, which is provided by a temperature control unit to keep the test fluid at a constant temperature. The temperature of the test fluid is measured by a thermistor. Two ceramic oscillators are applied on opposite sides of the inner vessel as an emitter and a receiver, respectively. A burst wave is transmitted to one ceramic oscillator from the burst wave generator. The ultrasonic wave generates and propagates between these ceramic oscillators. The travelling time of the ultrasonic pulse through the test fluid is measured by the digital oscilloscope. The ultrasonic propagation velocity in the test fluid is calculated based on this time.

A magnetic field is applied to the test fluid by an electromagnet positioned outside of the test cell. The magnetic field intensity can be changed from 0 mT to 550 mT and swept at a constant rate dB / dt (sweep rate). This sweep rate can be adjusted freely. The angle ϕ between the direction of the magnetic field and the direction of ultrasonic propagation can be adjusted freely from 0° to 180°.

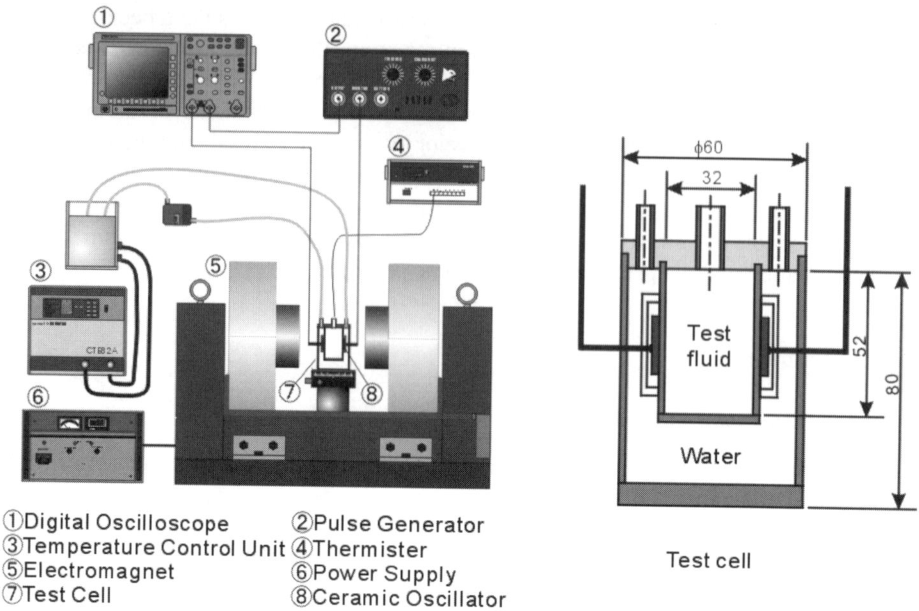

Figure 1. Experimental apparatus

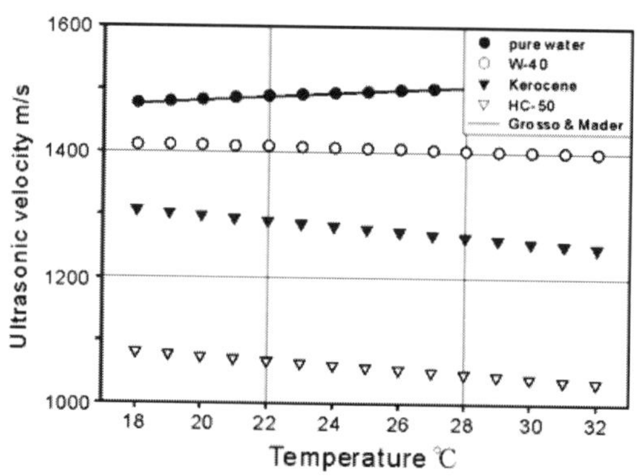

Figure 2. Ultrasonic propagation velocities in pure water, kerosene, W-40 and HC-50

Figure 2 shows the ultrasonic propagation velocities in pure water, kerosene, water-based (W-40) and kerosene-based (HC-50) magnetic fluids. These plotted points were obtained in our experiment, while the solid line is a measure of sound velocity in pure water obtained by Grosso and Mader[8]. The Grosso and Mader data is widely accepted as a standard for sound velocity in pure water. It can be seen that the ultrasonic propagation velocity in pure water obtained in our experiment matches very well with Grosso and Mader's result. Thus, our measuring system provides for high-accuracy results.

Table 1. Properties of test fluids

Test fluid	W-40	HC-50	MRF-132DG
Particle material	Magnetite	Magnetite	Iron
Particle size	About 10 nm	About 10 nm	1-3 μm
Carrier liquid	Water	Kerosene	Hydrocarbon
Weight concentration	40 wt%	50 wt%	80.98 wt%
Density (at 25 °C)	1.40×10^3 kg/m^3	1.39×10^3 kg/m^3	3.09×10^3 kg/m^3
Viscosity (at 25 °C)	25 mPa·s	15 mPa·s	90 mPa·s (at 40 °C)

2.2 *Test Fluid*

The test fluids are a water-based magnetic fluid (W-40), a kerosene-based magnetic fluid (HC-50) and an MR fluid (MRF-132DG). The properties of these fluids are indicated in Table 1[9, 10]. The magnetic fluids are produced by Taiho Industries Co. Ltd. and the MR fluid by Lord Co.

3 Result and Discussion

3.1 *Ultrasonic propagation velocities in Magnetic Fluids*

Figure 2 shows that the ultrasonic propagation velocity in both magnetic fluids is smaller than that in the carriers because the magnetic particles have an influence on the ultrasonic propagation velocity. As the temperature increases, the ultrasonic velocity in the magnetic fluids slightly decreases. This seems to be because the Brownian motion of the magnetic particles increases with the temperature. Therefore, when we take account of the properties of ultrasonic propagation in functional fluids, it is important to analyze the inner structure of fluids.

3.2 *Time Dependence of Ultrasonic Propagation Velocity*

Figure 3 shows the change of ultrasonic propagation velocity in magnetic fluids (a) and an MR fluid (b) versus the elapsed time of applying the magnetic field. The change of ultrasonic propagation velocity is expressed by $\Delta V/Vo$. Here, ΔV is defined by $\Delta V = V - Vo$ where V and Vo are ultrasonic propagation velocities with and without an external magnetic field, respectively.

In Figure 3(a) (in magnetic fluids), the magnetic field intensity is kept at 100 mT, and the angle ϕ is 0°. The ultrasonic propagation velocity in the magnetic fluids slightly increases over the time the magnetic field is applied. This increase seems to be caused by chain-like cluster formation. As the inner magnetic particles form chain-like clusters along the direction of the magnetic field, these clusters have an influence on

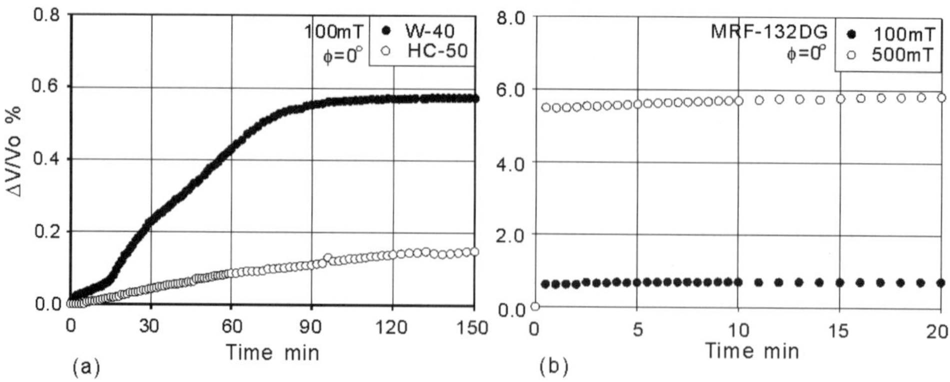

Figure 3. Time dependence of ultrasonic propagation velocity: (a) in magnetic fluids (100 mT), (b) in an MR fluid (100 mT and 500 mT)

ultrasonic propagation velocity in the magnetic fluid. As shown in Figure 3(a), the ultrasonic propagation velocities in both magnetic fluids continue to increase as the elapsed time of applying magnetic field increases, indicating growth of the chain-like cluster formations. The clusters continue to grow for about 1.5 hours in W-40, and 2 hours in HC-50. The ultrasonic propagation velocity change in HC-50 is smaller than that in W-40. This is apparently due to differences in the chain-like cluster formation between W-40 and HC-50. We suggest that the chain-like clusters in HC-50 are smaller, shorter and thinner than those in W-40 because of differences in the structure of surfactants. The magnetic particles in a water-based magnetic fluid are coated with a double layer of absorbed dispersant molecules, whereas, in a kerosene-based magnetic fluid, magnetic particles are coated with a single layer. Although detailed rheological properties have not been clarified, this different particle structure may be the reason for the difference in $\Delta V/Vo$ between the magnetic fluids.

In Figure 3(b) (in an MR fluid), the magnetic field intensity is held at either 100 mT or 500 mT, and the angle ϕ is 0°. As soon as the magnetic field is applied to the MR fluid, the ultrasonic propagation velocity in the MR fluid instantly increases. This indicates that the clusters form almost instantly with the applied magnetic field. This is likely because the particle size is larger and the magnetization is higher in comparison with the magnetic fluids. Moreover, the ultrasonic propagation velocity in an MR fluid is larger than that in the magnetic fluids. In an MR fluid, the clusters grow as robust clusters because of strong magnetization. The change rate in ultrasonic propagation velocity at 500 mT is larger than that at 100 mT. It is supposed that there are larger-scale clusters formed at 500 mT.

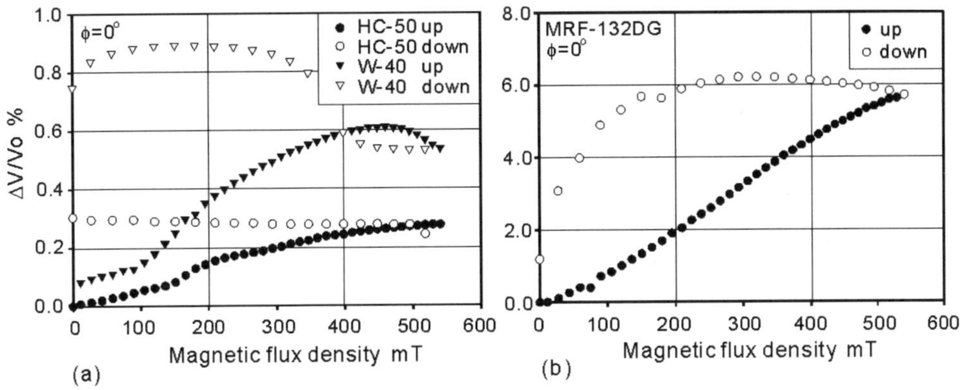

Figure 4. Hysteresis of ultrasonic propagation velocity: (a) in magnetic fluids, (b) in an MR fluid

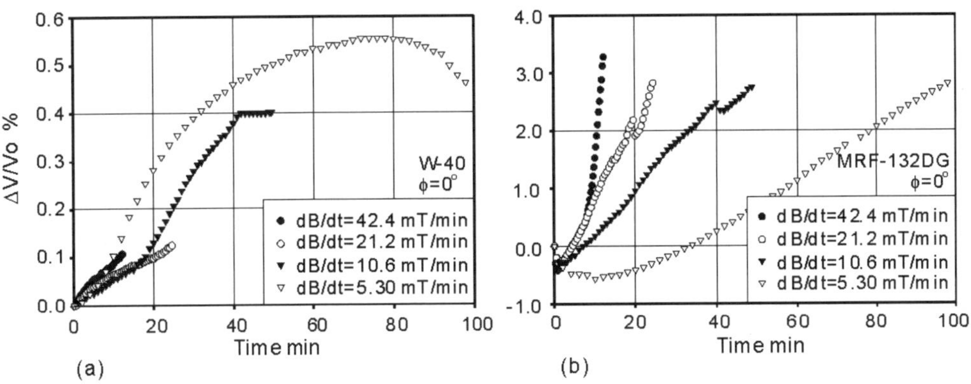

Figure 5. Ultrasonic propagation velocity with changing sweep rate: (a) in magnetic fluids, (b) in an MR fluid

3.3 Hysteresis of Ultrasonic Propagation Velocity

Figure 4 shows hysteresis of ultrasonic propagation velocities in (a) magnetic fluids and in (b) an MR fluid in relation to the magnetic field intensity for $\phi = 0°$. In these experiments, the magnetic field is applied using the following process: increase the magnetic field intensity by 12 mT every 2 minutes until reaching 550 mT. Thereafter, decrease the magnetic field intensity by 24 mT every 2 minutes. As shown in Figure 4, there is clear hysteresis in relation to the applied magnetic field.

When the magnetic field intensity increases, $\Delta V/Vo$ for W-40 and HC-50 also increases and their tendencies are similar. As describe above, the increase in ultrasonic propagation velocity is caused by chain-like cluster formation in proportion to increases in magnetic field intensity. Here again the change in ultrasonic propagation velocity in W-40 is larger than that in HC-50. When the magnetic field decreases, however, the ultrasonic propagation velocity remains at almost the same level through the process.

This seems to indicate that the chain-like clusters do not break easily in the decrease process.

Hysteresis of ultrasonic propagation velocity in an MR fluid is illustrated in Figure 4(b). The hysteresisis is similar to that in the magnetic fluids. In an MR fluid, $\Delta V/Vo$ also increases with magnetic field intensity. However, the change in ultrasonic propagation velocity in an MR fluid is about one order of magnitude larger than that in the magnetic fluids. As mentioned above, when the external magnetic field is applied to an MR fluid, the magnetic particles form robust clusters. In comparison with the chain-like clusters, the robust clusters have a stronger influence on the ultrasonic propagation velocity.

These complicated and interesting phenomena are believed to be derived from the change of bulk modulus according to the cluster formation. In order to understand these interesting phenomena, it is necessary to further investigate the structure of magnetic particles and the process of cluster formation.

3.4 Growth of Cluster with Sweep Rate

Figure 5 shows the ultrasonic propagation in (a) W-40 and in (b) an MR fluid under four sweep rates. In this experiment, the magnetic field is applied using the following sweep rates (rate of increasing magnetic field) until reaching 530 mT: dB/dt = 5.30, 10.6, 21.2 and 42.4 mT/min. For example, it takes 100 minutes to reach 530 mT in case of dB/dt = 5.30 mT/min. In Figure 5(a) (in W-40), when the magnetic field increases slowly (at lower sweep rate), the increase rate of ultrasonic propagation velocity is larger. The chain-like clusters seem to grow bigger and longer steadily when the increasing magnetic field is applied slowly. On the other hand, the chain-like clusters can not grow sufficiently when the magnetic field increases quickly. Thus, the ultrasonic propagation velocity is small at the higher sweep rate. As shown in Figure 5(b), this tendency was not obtained in an MR fluid. As mentioned in Sec. 3.2, the robust clusters grow instantly with applying magnetic field. Therefore, the ultrasonic propagation velocity in an MR fluid increases similarly with each sweep rate. Growth of cluster in an MR fluid is not related with these sweep rates.

4 Conclusion

Ultrasonic propagation velocities in water-based and kerosene-based magnetic fluids and in an MR fluid under uniform magnetic field are investigated experimentally. Using measurements of ultrasonic propagation velocity, the influence of the elapsed time of magnetic field application, the magnetic field intensity and the sweep rate on the ultrasonic propagation velocity are examined. The ultrasonic propagation velocity increases with the elapsed time and intensity of the magnetic field because of cluster formation. In addition, hysteresis is observed when reducing the magnetic field. This hysteresis indicates that clusters grow with applying magnetic field, but do not break down quickly as the magnetic field is reduced. These results are believed to be dependent on the characteristic rheological properties of magnetic fluids and MR fluids and the inner structure of the fluids when under a magnetic field. At the present stage, it is necessary to carry out more various and detailed experiments to clarify the mechanisms producing the clusters and these ultrasonic propagation properties.

Acknowledgements

This work was partially supported by a Grant-in-Aid for Scientific Research (C) of Japan Society for Promotion of Science.

References

1. Rosensweig R. E., *Ferrohydrodynamics*, (Cambridge University Press, Cambridge, 1985).
2. Motozawa, M., Matsumoto, Y. and Sawada, T., Effect of external magnetic field on ultrasonic propagation velocity in magnetic fluids, *JSME Int. J. B*, **48** (2005) pp. 471-477.
3. Motozawa, M. and Sawada, T., Influence of magnetic field on ultrasonic propagation velocity in magnetic fluids, *J. Mag. Magn. Mater.*, **289** (2005) pp.66-69.
4. Carlson J. D. and Jolly M. R., MR fluid: form and elastomer devices, *Mechatronics*, **10** (2000) pp. 555-569.
5. Zipser L., Richter L. and Lange U., Magnetorheologic fluids for actuators, *Sensors and Actuator A*, **92** (2001) pp. 218-325.
6. Taketomi, S., The anisotropy of the sound attenuation in magnetic fluid under an external magnetic field, *J. Phys. Soc. Jpn.*, **55** (1986) pp. 838-844.
7. Chung D. Y. and Isler W. E., Ultrasonic velocity anisotropy in ferrofluids under the influence of magnetic field, *J. Appl. Phys.*, **49** (1978) pp. 1809-1811.
8. Del Grosso V. A. and Mader C. W., *J. Acoust. Soc. Am.*, **52** (1972) pp. 1442-1446.
9. http://www.taihokogyo.co.jp/
10. http://www.mrfluid.com/

BASIC STUDY ON THE MICRO POLISHING OF METAL MOLD USING ELECTRO-RHEOLOGICAL GEL

NAOTA NAKAMURA, YASUHIRO KAKINUMA, TOJIRO AOYAMA

School of Integrated Design Engineering, Keio University, 3-14-1 Hiyoshi Kohoku-ku, Yokohama 223-8522, Japan
E-mail:kakinuma@sd.keio.ac.jp

KATSUTOSHI TANAKA, HIDENOBU ANZAI

Toshiba Machine CO., LTD,
FUJIKURAKASEI CO., LTD

In recent years, the automation of polishing and finishing technologies has become a requirement because polishing and finishing with high accuracy takes a long time and often depends on the experience and special skills of technical engineers. One of the most promising polishing technologies is Field-assisted Fine Finishing (FFF), which has attracted a lot of attention from many researchers. In FFF, an electric field or a magnetic field is applied for finishing, polishing, and so on. In this study, a new polishing technology using Electro-rheological Gel (ERG) is proposed. The ERG we have developed is a functional material whose induced shear force changes according to the applied electric field intensity. A thin ERG sheet with abrasive grains is applied to the polishing pad of a rotary polisher. The properties of the ERG polishing pad can be changed by adjusting the intensity of the electric field applied to the ERG pad. We conducted experiments to analyze the basic performance of the ERG polishing method. The results show that the polishing efficiency can be improved by applying an electric field

1 Introduction

As science and technology progresses, component manufacturers increasingly require high efficiency and high precision on the order of nanometers. For example, small optical lenses are efficiently produced by the injection molding process, and an ultra-precision mold with nanometer-order surface roughness is necessary. Fine polishing processes have been applied to the finishing process of the metal molds used to make these lenses. However, a long polishing time and the special skills of technical engineers are required to finish the surface with high accuracy. New polishing processes have been proposed to increase the polishing efficiency and accuracy. One of the most promising polishing technologies is Field-assisted Fine Finishing (FFF)[1, 2], which has attracted a lot of attention from many researchers. In FFF, an electric field or a magnetic field is applied to control the behavior of the abrasive grains in the polishing process.

We propose a new polishing technology using Electro-rheological Gel (ERG) [3]. ERG is a functional material whose induced frictional characteristics change according to the applied electric field intensity. A thin ERG sheet is applied to the polishing pad of a rotary polisher, and abrasive grains are spread on the surface of the ERG pad. The performance of the ERG polishing pad is variable in response to the change of the surface frictional property of the ERG pad. In this study, simple ERG polishing equipment is developed and the basic performance of the ERG polishing process is experimentally analyzed. The behavior of the abrasive grains under a high electrical field is also observed during the polishing process.

2 ERG assisted micro polishing

2.1 Basic structure of ERG

ERG is made from ERF (Electro-rheological Fluid)[4] through a gelling process [5, 6]. The structure of ERG is illustrated in Fig. 1. This functional material consists of primarily three components: gel, ER particles and silicone oil. The diameter of an ER particle is 16 μm on average. The ERG effect is obtained when an electrical field is applied to the ERG sheet using a pair of electrodes. The ER particles are suspended in the gel structure under no electric field, and a significant number of ER particles protrude from the gel component at the ERG surface (Fig. 1 (a)). In this condition, the upper electrode is supported by the slippery ER particles. When a high electric field is applied, the protruding ER particles retract into the gel component due to the dielectric polarization of the ER particles, and the upper electrode comes in contact with the sticky gel component (Fig. 1 (b)).

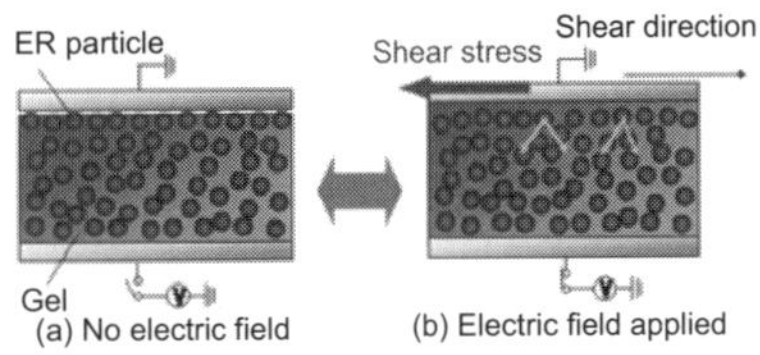

Figure 1 Structure and characteristic of the ERG

The behavior of the ERG surface under a high electrical field can be observed by using glass with an indium tin oxide (ITO) film. Fig. 2 shows the result of observing the surface of the ERG through the glass electrode. When a high electric field is applied, many black spots appear on the contact surface of the glass. These spots are the contact area of the gel component sticking to the glass electrode. The gel has a high adhesive property; thus, it sticks to the opposite electrode. The contact area of the gel component to the electrode is almost proportional to the electric field intensity, and the change of the frictional property of the ERG surface occurs in response to the applied electric field.

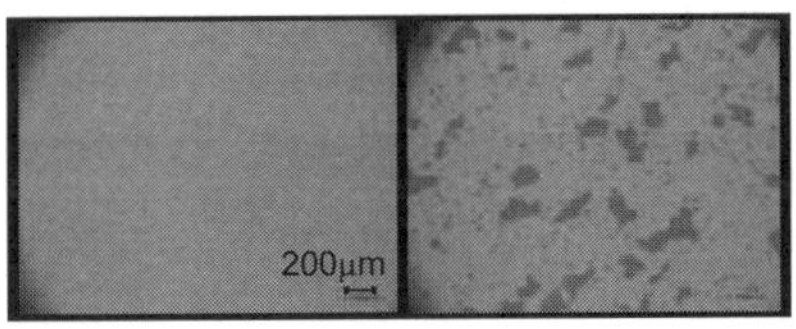

(a) No electric field (b) 2kV/mm applied

Figure 2 Observation of the ERG surface

2.2 Mechanism of the ERG polishing process

The mechanism of the proposed ERG polishing process with abrasive grains is illustrated in Fig. 3. Under no electric field, the abrasive grains are held between the polishing surface and slippery ER particles. In this case, the abrasive grains behave as loose grains, so finer polishing results are expected, although the polishing process is not very efficient.

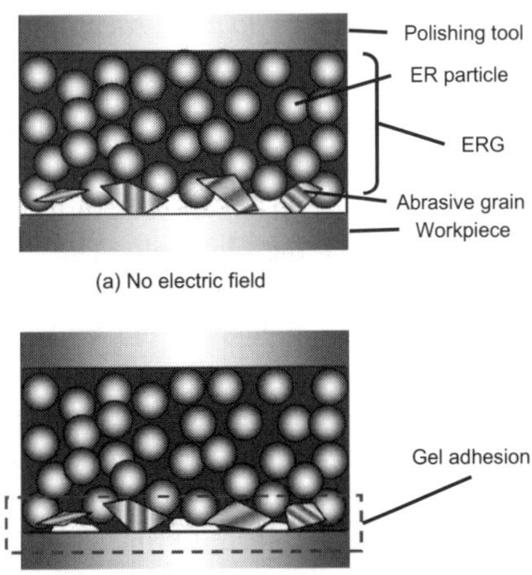

Figure 3 The mechanism of the ERG polishing process

When a high electric field is applied to the ERG polishing pad, the gel exudes to the surface due to the behavior of the ER particles described in Section 2.1, and the abrasive grains are surrounded and fixed by the gel. The abrasive grains perform as a bonded abrasive under the electric field, and the polishing speed increases.

Fig. 4 is a photograph of the ERG polishing pad with its surface covered with through the ITO glass. Some abrasive grains are surrounded by the gel under the high electric field. Even if the polishing grains are placed on the surface of the ERG polishing pad, the surface property changes from a slippery to a sticky state by applying the high electric field. Consequently, it is expected that the polishing performance is controlled dynamically by changing the electric field intensity.

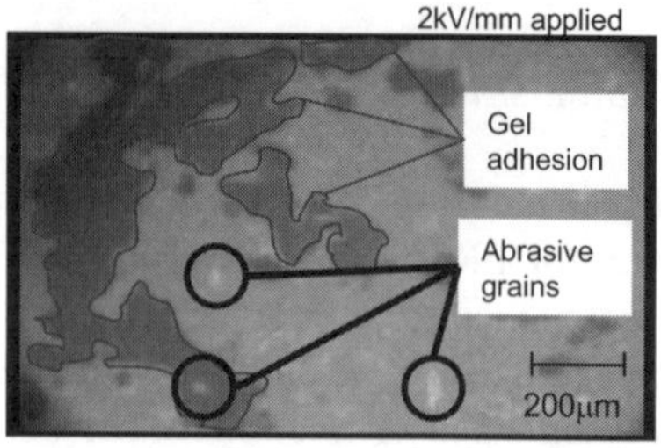

Figure 4 Observation of gel adhesion in a state of the ERG polishing pad surface covered with abrasive grains

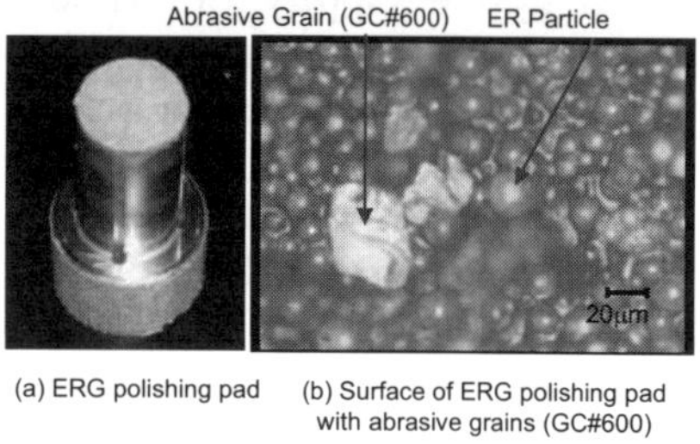

Figure 5 Pictures of ERG polishing pad

Fig. 5 shows photographs of the polishing tool and the ERG polishing pad with its surface covered with abrasive grains. The ERG polishing pad is molded at the top end of the polishing tool, and the abrasive grains are placed on the ERG surface. In this study, abrasive grains of Green carbon (GC) #600, which have a grain diameter of about 30 μm – 50 μm, are used.

3 Experimental analysis

3.1 Experimental setup and procedure

Fig. 6 shows a schematic diagram of the experimental setup for ERG polishing. A photo of the experimental setup is also shown in Fig. 7. The motor turns the polishing tool via the flat belt. The ERG polishing pad, with its surface covered with abrasive grains, is molded to the tip of the polishing tool by using a specially designed metal mold. The negative pole of the electric supply is connected to the polishing tool and the positive pole is connected to the screw that fixes the workpiece to the insulated housing. The cylindrical workpiece, which is 20 mm in height and 20 mm in diameter, is polished.

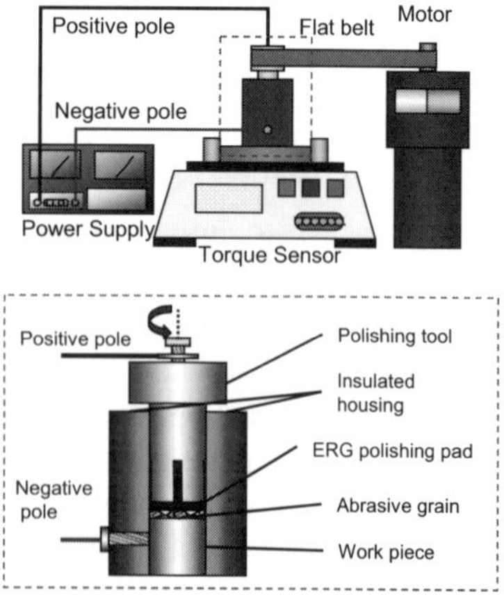

Figure 6 Experimental equipment for ERG polishing

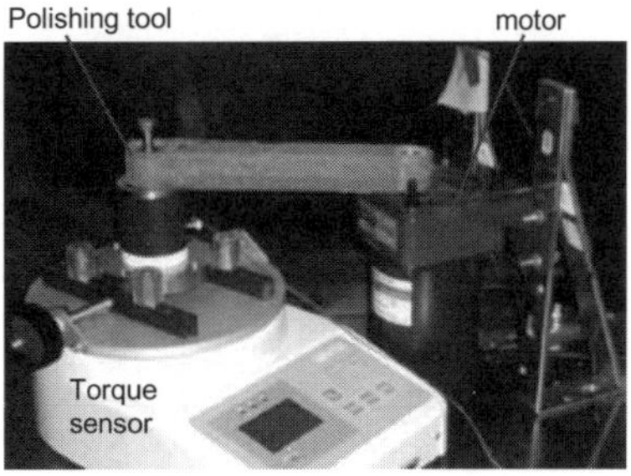

Figure 7 Picture of experimental setup

Table 1 Polishing conditions

Workpiece	**:stainless steel**
Polishing pad	**:ERG polishing pad (Thickness: 0.5mm)**
Abrasive grain	**:GC#600(Green Carbon)**
Rotational speed	**:200min-1**
Polishing time	**:140min**
Electric field intensity	**:0kV/mm, 2kV/mm**

In order to evaluate the effect of the ERG pad on the polishing process, polishing tests are performed under two different conditions. One of tests is carried out under a high electric field intensity of 2.0 kV/mm, and the other test is performed by eliminating the electric field. The polishing conditions are summarized in Table 1.

3.2 Experimental result

Fig. 8 shows the relationship between the electric field intensity and the polishing torque. When the electric field is applied to the ERG polishing pad, the polishing torque increases due to the ERG effect. It is supposed that the sticking of the gel to the polishing surface increases and the number of abrasive grains effective at polishing increases.

This result indicates the polishing pressure of the abrasive grains increases and the polishing efficiency increases, too.

Fig. 9 shows the relation between the average roughness (Ra) of the workpiece surface and the polishing time. The surface roughness improves as the polishing time increases. Furthermore, the rapid improvement of the surface roughness is obtained by applying the electric field to the ERG polishing pad. The same result is obtained for both the center (a) and the outer (b) parts of the workpiece surface.

The height topographies of the polished surface are shown in Fig. 10. The micro scratches on the surface disappear as the polishing process proceeds. The circular scratches at the center of the workpiece can be rapidly reduced by

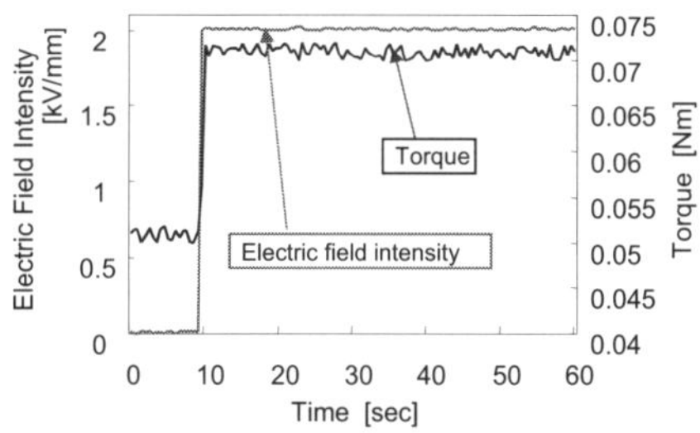

Figure 8 Relation between the electric field intensity and the torque

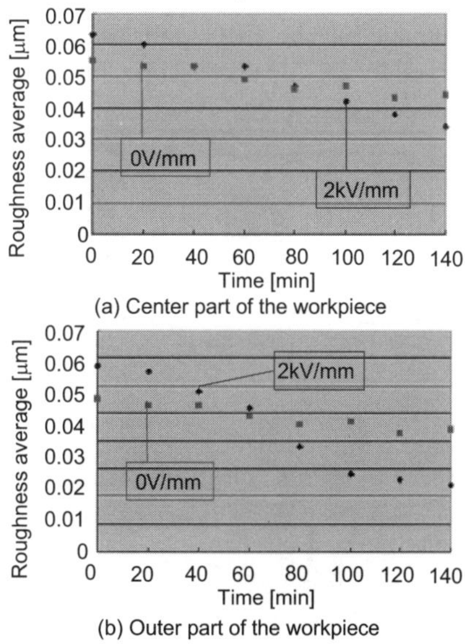

Figure 9 Relation between Ra of the workpiece surface and polishing time

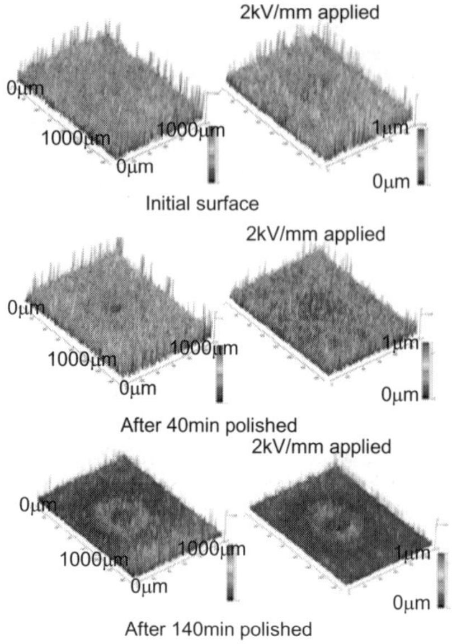

Figure 10 Height topography of the workpiece surface center

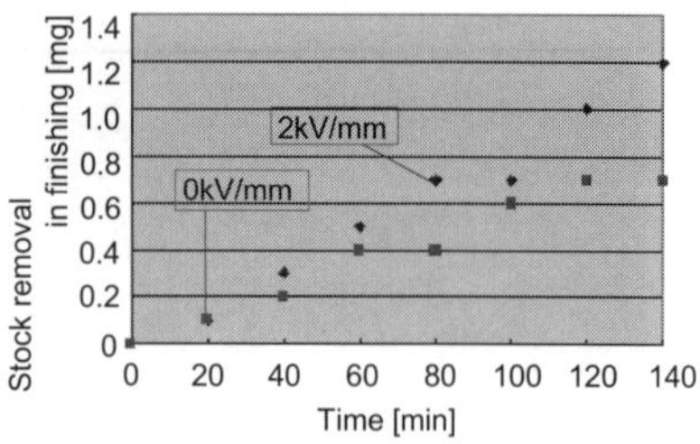

Figure 11 Relation between stock removal in finishing and polishing time

applying the proposed ERG polishing process.

The relation between the stock removal and the finishing and polishing time is shown in Fig. 11. The stock removal increases by applying an electric field to the polishing pad. This is thought to be caused by the change of the behavior of the abrasive grains from a loose abrasive to a bonded abrasive.

4 Observation of the behavior of the abrasive grains during polishing process

To analyze the mechanism of polishing using the ERG polishing pad, an observation of the ERG polishing process is carried out. Fig. 12 shows a diagram of the observation instrument for the ERG polishing process. The surface of the ERG polishing pad is observed by a CCD microscope through the ITO glass electrode. The negative pole of the

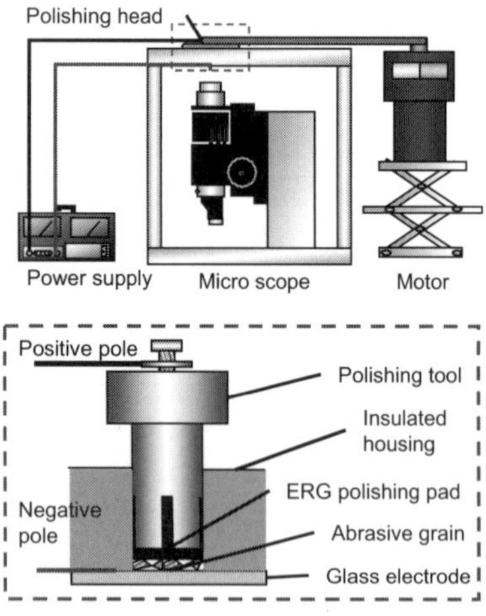

Figure 12 Outline drawing of the observation instrument

Table 2 Observation condition

Polishing pad	**:ERG polishing pad (Thickness: 0.5mm)**
Abrasive grain	**:GC#600**
Rotational speed	**:5min-1**
Electric field intensity	**:2kV/mm**

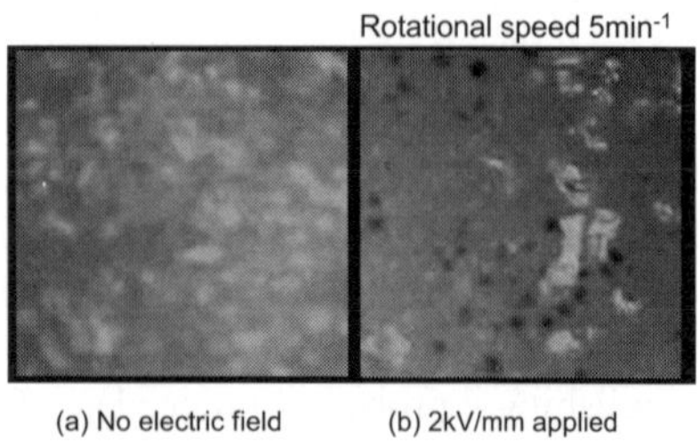

Figure 13 Polishing pressure increase between abrasive grains and glass electrode surface.

electric supply is connected to the polishing tool, while the positive pole is connected to the ITO glass electrode.

The observation conditions are shown in Table 2. The same abrasive grains (GC#600) used for t he polishing tests are selected. However, the rotational speed of the polishing tool is changed to 5 min^{-1} due to the specification limit of the CCD microscope.

The observation is shown in Fig. 13. It is also observed that, during the rotation of the polishing pad, the gel exudes to the surface of the glass electrode when a strong electric field is applied. With the gel sticking, it is recognized that the contact pressure between the abrasive grains and glass electrode surface is increased. The ERG effect is again expected during the polishing process.

5 Conclusion

The performance of Electro-rheological Gel (ERG) is applied to the micro polishing process of a metal mold for optical components. Test equipment for ERG polishing is developed and the effect of the proposed ERG polishing method is experimentally analyzed. The behavior of the ERG polishing pad during pad rotation is also observed through ITO glass. The results are summarized as follows:

(1) The proposed ERG polishing pad is effective for controlling the polishing rate of a metal mold.
(2) The polishing rate of the metal mold changes in response to the applied electric field, and rapid improvement of the surface roughness of the polished surface is obtained by activating the ERG polishing pad.
(3) It is estimated from the experimental results that the behavior of the abrasive grains changes from loose abrasive to bonded abrasive by applying the electric field to the ERG polishing pad.

6 Acknowledgements

This work was supported by a Grant-in-Aid for Scientific Research (B),Japan Society for the Promotion of Science

References

1. T. Kikuchi, J. Fujikawa, J. Furusho, A. Inoue, Polishing Using ER Slurry on One-sided Patterned Electrodes, Proceedings of the 9th International Conference of Electrorheological Fluids and Magnetorheological Suspensions (2004), 943-949.
2. T. Kuriyagawa, M.Saeki, K. Syoji, Electrorheological Fluid-assisted Ultra-precision Polishing for Small Three-dimensional Parts, Precision Engineering, 26 (2002) 370-380
3. Y. Kakinuma, T. Yakoh, T. Aoyama, H. Anzai, K. Ikebe, Development of Gel-structured Electro-rheological Fluids and their Application to Mechanical Damper Elements, the 8th IEEE International Workshop on Advanced Motion Control (AMC'04), Kawasaki, Japan, March 25 – 28, (2005), pp.541-545.

4. W. B. Kim, S. J. Lee, Y. J. Kim, E. S. Lee, The Electromechanical Principle of Electrorheological Fluid-assisted Polishing, International Journal of Manufacture 43 (2003) 81-88.
5. Y. Kakinuma, T. Aoyama, H. Anzai, A Study on the Application of Electro-rheological Gel to Torque Transfer Device, International Conference on Leading Edge Manufacturing in the 21st Century, Oct. 19-22, (2005), Nagoya, Japan.
6. Y. Kakinuma, T. Aoyama, H. Anzai, Basic performance of ER gel on one-sided structured electrodes, Proceedings of the JSPS international Symposium on Fluid Power, THUKUBA, Nov. 7-10, (2005).

HIGH VERSUS LOW FIELD VISCOMETRIC CHARACTERIZATION OF BIDISPERSE MR FLUIDS

G. T. NGATU AND N. M. WERELEY

Department of Aerospace Engineering, University of Maryland, College Park, MD 20742
e-mail: wereley@umd.edu

Our bidisperse magnetorheological fluids are suspensions of micron (2-10μm) and nanometer (~40n m) scale magnetic iron particles in silicone or hydraulic oil. Earlier studies were conducted to determine the yield stress of these fluids at low magnetic field induction. These studies have shown the absence of saturation yield stress implying the possibility of a higher yield stress by increasing the applied field. In this study, three different bidisperse MR fluids were investigated to determine the maximum available yield stress that can be obtained at or near saturation magnetic flux density. The iron loading in the fluids varied from 50% to 80% by weight. Two types of MR cells, a low field and a high field cells, were used for the investigation. Using a parallel disc rheometer alternatively equipped with one of the two MR cells, the flow curves of the MR fluids were obtained and their yield stress determined. The yield stress of the MR fluids as a function of applied magnetic field was identified using the Bingham-Plastic constitutive model. Results show that the high field cell (maximum 1 Tesla) was able to measure shear stress up to saturation, whereas the low field cell (maximum 0.26 Tesla) could not.

1 Introduction

Magnetorheological fluids typically consist of suspended micron sized magnetic particles, such as iron or cobalt, in a fluid medium, such as silicone oil and hydraulic oil. Their main advantage is that their rheological properties, like yield stress and viscosity, can be continuously varied and controlled by applying a varying magnetic field. In the presence of a magnetic field, magnetic dipoles of the suspended particles align themselves roughly parallel to the applied field and form chains which act as a flow barrier. A finite stress must be developed to yield these chained structures, thus giving rise to the so-called yield stress of MR fluids. MR fluids are gaining a wider application in active and semi-active systems, including dampers and clutches.

Bidisperse MR fluids employ nanometer scale particles in combination with micron scale ones in the suspension. These nanometer scale particles are introduced into the dispersion to alleviate the sedimentation problem encountered when using micron sized particles [1-3]. However, the addition of nanometer scale particles tends to decrease the yield stress of the MR fluid. Previous studies have been conducted in characterization of bidisperse MR fluids and the trade off between improved settling and reduced yield stress [1,2,4]. The goal of this paper is to investigate and determine the maximum measurable yield stress of bidisperse MR fluids using improved measuring equipment, which can help to optimize trade-offs when selecting an MR fluid for the desired application.

In this study, we investigated and compared the variation of yield stress for different iron loading bidisperse fluids at high and low fields using two types of MR cells. The low field test utilized TEK70 MR cell while the high field utilized new and improved MRD180 MR cell. Both cells were obtained from Anton Paar. Based on the input current, both cells are capable of continuously generating a magnetic field. The TEK70 input current varies from 0 to 2 A while it varies from 0 to 5 A for the MRD180 cell. Magnetic flux lines are oriented normal to the parallel discs. The tests were conducted using Anton-Paar Physica MCR300 Parallel Disc Rheometer (Fig 1). A standard gap of 1mm was maintained between the parallel discs. A sample of nominally 3 ml was injected into this gap for the test using a small diameter syringe. Rotational tests were carried out to determine the flow curves (shear stress vs shear rate). A Bingham-Plastic (BP) constitutive model [5] was fitted to the flow curve in a parameter identification procedure to determine yield stress.

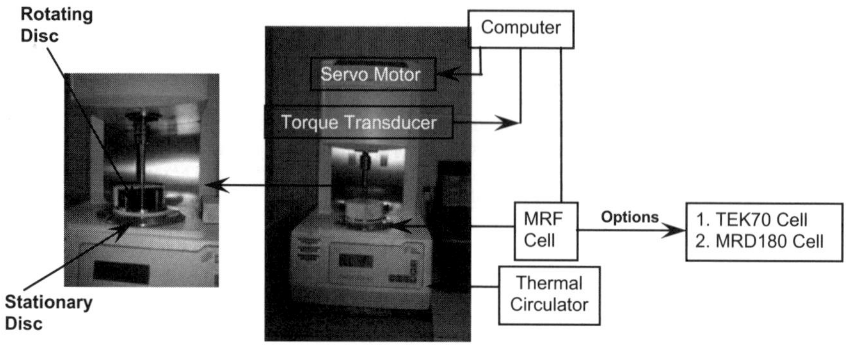

Fig. 1 Paar Physica MCR300 Rheometer

To cross-reference the input current to the rheometer with the magnetic flux density experienced at the gap, a Hall sensor was placed between the parallel plates while current was being applied. This calibration was conducted for all samples with both the high field MRD180 and the low field TEK70 MR cells. The results for a sample MR fluid are shown in Fig.1.

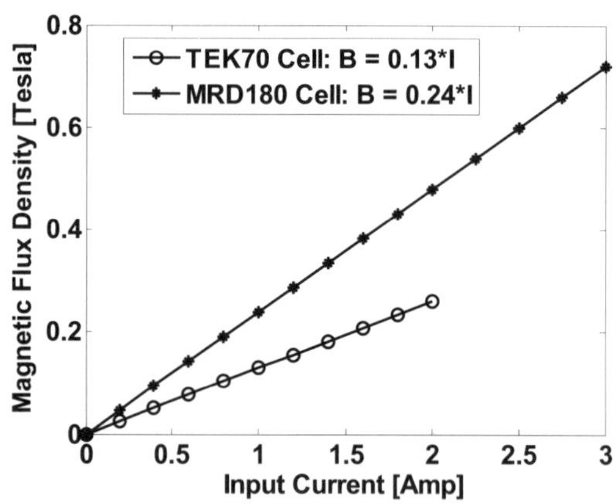

Fig.2 Calibration of Magnetic Flux Density

2 MR Fluid Synthesis and Samples Descriptions

A patented microwave-based process was developed by Material Modification Inc to synthesis nanometer scale powders [US Patent # 6,409,851]. In this microwave plasma synthesis technique, microwave energy was employed to generate plasma through ionization, disassociation and recombination of organometallic gas as a precursor. The high temperature vaporized the precursors, which promoted chemical reactions at the molecular level in the presence of microwaves. These vapors were rapidly cooled in an inert atmosphere to form powders. Nanometer scale iron powders were synthesized from iron carbonyl precursors according to the following chemical reaction:

$$\mathbf{Fe\,(CO)_5 \longrightarrow Fe\,(s) + 5CO\,(g)}$$

The average size of nanometer scale particles used in this study was 40 nm. Micron scale iron powder was commercially purchased with average particle size between 2 and 10 μm. Silicone oil was chosen as the carrier fluid to prepare stable MR fluids. Lecithin was used as a surfactant for producing these MR fluids. It was mixed in the oil using high speed emulsifier at nearly 11000 rpm. Iron powders were then added to the silicone oil and mixing continued until a uniformly dispersed mixture is obtained.

Table 1: MR Fluid Samples

Total Fe Loading	%wt	50	60	80
Micron scale particles	%	90	85	85
	%wt	45	51	68
Nanometer scale particles	%	10	15	15
	%wt	5	9	12

3 Rheological Characterization Method

MR fluids present non-linear behavior under the application of magnetic field, where their yield stress is strongly dependent on the magnetic field. MR fluid behavior can be categorized into pre-yield and post-yield regions. This behavior is the basis for the BP flow model characterizing MR fluids [4-6]. The BP constitutive model is below:

$$\tau = \tau_y + \mu\dot{\gamma}$$

where τ_y is the yield stress, μ is the post-yield viscosity and $\dot{\gamma}$ is the shear rate. The model predicts that in the pre-yield region, where the shear stress is below the yield stress (τ_y), the fluid exhibits rigid behavior. In the post-yield region, the fluid demonstrates a viscous behavior.

The BP model was fitted to the data using a weighted least squares error minimization procedure by selecting the parameters τ_y and $\dot{\gamma}$. Due to the use of measured shear rate as weights, high shear stress values obtained at high shear rates were chosen to better fit the model asymptotically and estimate the yield stress values of the MR fluids. The yield stress is given by the intercept of the high shear rate asymptote with the shear stress axis. A sample flow curve for a sample MR fluid from the rheological measurement fitted with the BP flow model is shown in Fig. 3(a) and 3(b).

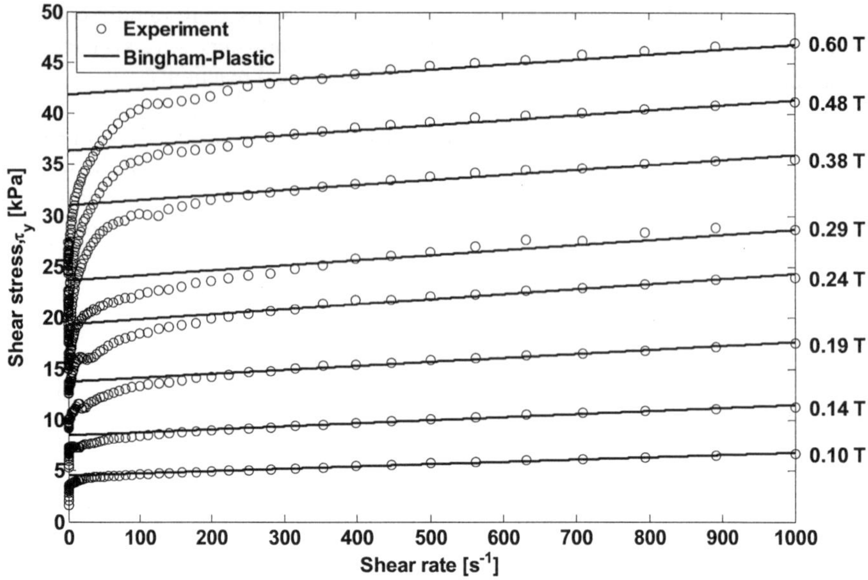

a. MRD180 MR Cell

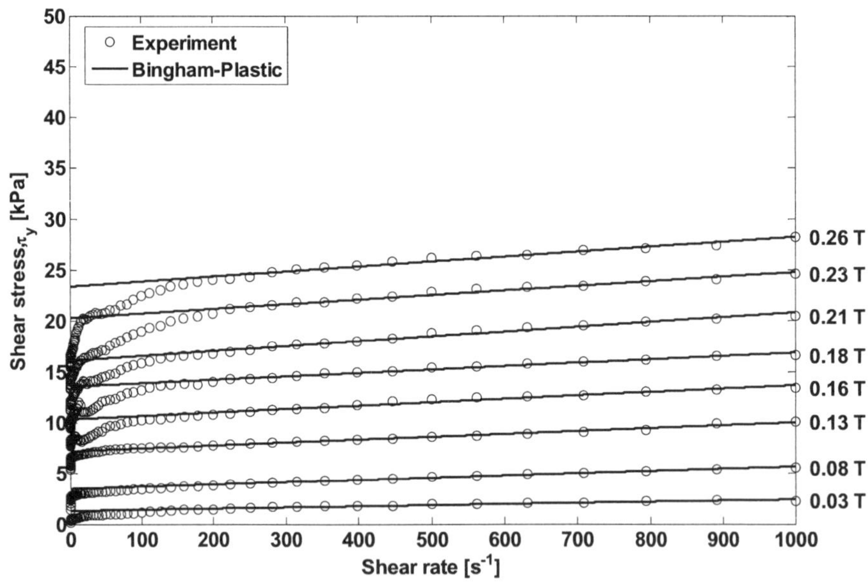

b. TEK70 MR Cell

Fig. 3 Flow Curve for MR Fluid with 80wt% of Iron Powder

The results of the yield stress estimation for all samples with both the high field (MRD180) and low field (TEK70) MR cells are shown in Fig 4. The plots clearly depict that the estimated yield stress increases smoothly for both MR cells. The predicted yield stress values are also in relatively good agreement for both cells until flux density of 0.26

Tesla, beyond which the low field TEK70 cell could not measure. However, for the high field MRD180 cell, we notice a higher yield stress corresponding to saturation or near saturation flux density, beyond which an increase in magnetic field will have little or no effect on the yield stress of the MR fluid. This is not manifested in the low field TEK70 cell, which showed a maximum flux density of 0.26 Tesla. Thus, with the TEK70 cell, yield stress at saturation is not captured and the estimated maximum yield stress value is well below the actual yield stress that the MR fluid can exhibit.

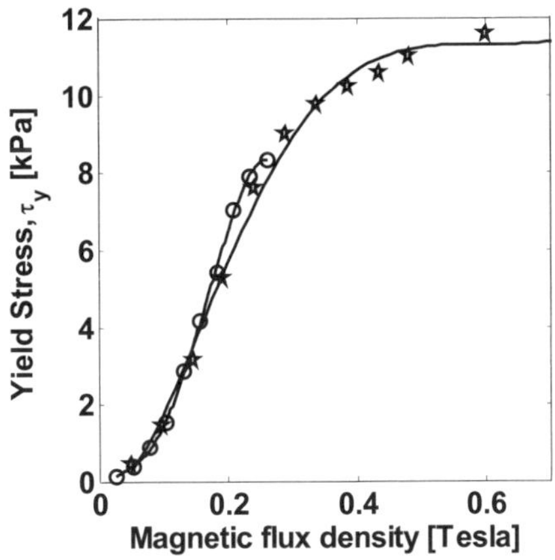

a. 50wt% Fe, 90% micro 10% nano

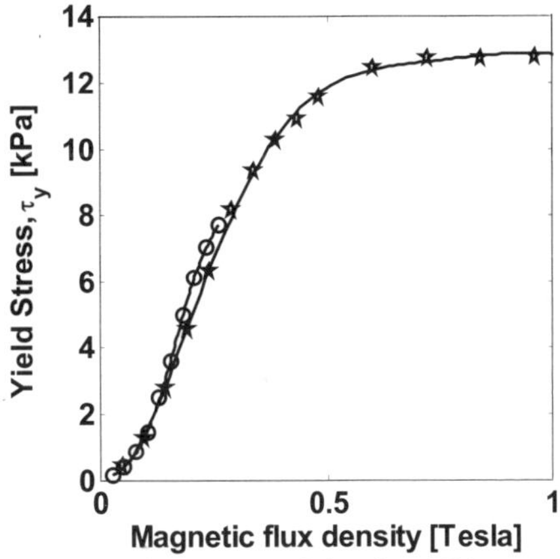

b. 60wt% Fe, 85% micro 15% nano

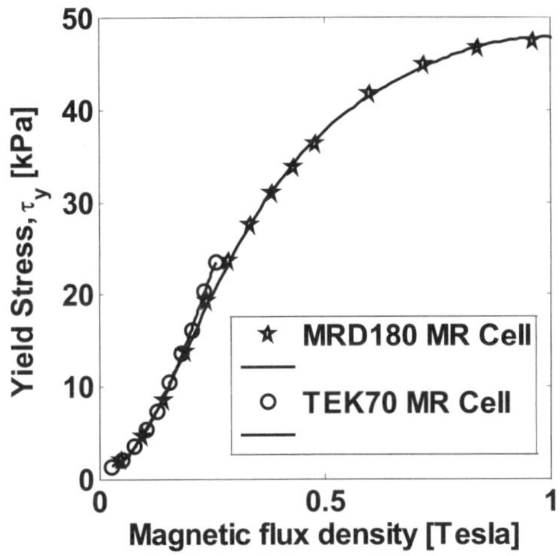

c. 80wt% Fe, 85% micro 15% nano

Fig. 4 Yield Stress vs Magnetic Field

Figure 5 shows the maximum measurable yield stress estimated using the two MR cells. The bar graph shows that there is a marked difference in the max measurable yield stress between the low field TEK70 cell and the high field MRD180 cell and we get a much better yield stress value with the MRD180 cell capable of generating a higher magnetic field. The difference in yield stress grows substantially as the iron loading in the MR fluid increases, with a maximum yield stress gain occurring at the 80wt% Iron loaded MR fluid.

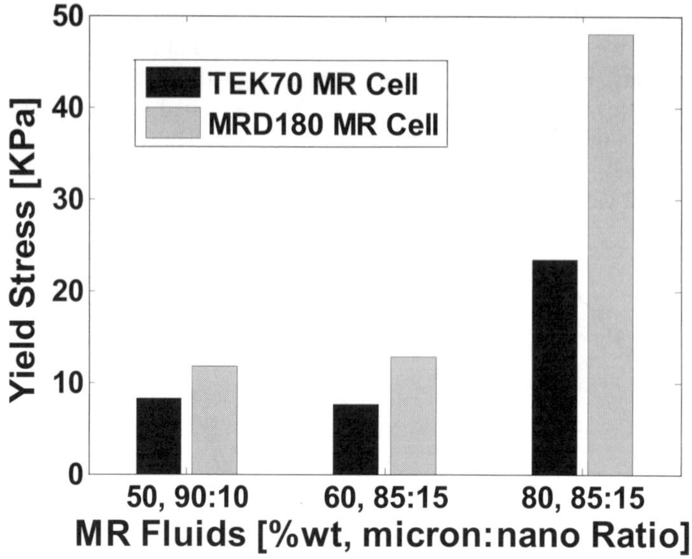

Fig. 5 Maximum Measurable Yield Stress Comparison

4 Conclusion

We investigated the variation in yield stress due to variation in applied magnetic field using two MR cells. One was a low field TEK70 MR cell, which was also employed in previous studies, and the other a high field MRD180 cell. Both MR cells were acquired from Anton-Paar. The key issues of the study were to determine if yield stress data from both the low and high field cells agree over the low field range and whether the saturation yield stress, thus the maximum yield stress, of the MR fluid samples can be captured with the high field cell. Previous studies conducted with the low field cell have shown the absence of saturation indicating the possibility of a higher yield stress if a higher field is applied. This study is also in agreement, showing the inadequacy in using the low field TEK70 cell in capturing the max measurable yield stress.

We investigated the variation in yield stress for 3 bidisperse MR fluids with varying iron loading. At low field, flow curves and yield stress curves for all the three samples show the same trend in both the low field TEK70 and high field MRD180 cells. However, the high field MRD180 cell provided a much better capability for measuring yield stress. The gain in yield stress when using the MRD180 cell was noticed to be considerably higher as the iron loading of the fluid increased from 50wt% to 80wt%. For instance, for the 80wt% iron loaded fluid, the maximum measurable yield stress was nearly doubled when the high field cell was utilized. In addition, the high field MRD180 MR cell can predict MR fluid behavior at or close to saturation. Thus, the saturation or near saturation dynamic yield stress, hence the maximum measurable yield stress, of MR fluids can be estimated with a high degree of confidence. The ability to predict the saturation yield stress of the MR fluid will help in optimization and selection of MR fluids based on their best performance for the desired application.

Acknowledgement

The authors would like to thank R. Radhakrishnan at MMI (Fairfax,VA) for providing the MR fluid samples.

References

1. A. Chaudhuri, G. Wang and N.M. Wereley, "Substitution of Micron by Nanometer Scale Powders in Magnetorheological Fluids," Int. J. Mod. Phys. B **19** (2005) pp. 1374.
2. N.M. Wereley, A. Chaudhuri, J. –H. Yoo, S. John, S. Kotha, A. Suggs, R. Radhakrishnan, B. Love and T. Sudarshan, "Bidisperse Magnetorheological Fluids Using Fe Particles at Nanometer and Micron Scale," J. Intel. Mat. Syst. Str. **17** (2006) pp. 393.
3. C. Kormann, H. Laun and H. Richter, "MR Fluids with Nano-Sized Magnetic Particles," Int. J. Mod. Phys. B **10** (1996) pp. 3167.
4. A. Chaudhuri, N.M. Wereley, R. Radhakrishnan and S.B. Choi, "Rheological Parameter Estimation for a Ferrous Nanoparticle-Based Magnetorheological Fluid using Genetic Algorithms," J. Intel. Mat. Syst. Str. **17**, pp. 261.
5. M. Jolly, J. Carlson and J. Bender, "Properties and Application of Commercial MR Fluids," J. Intel. Mat. Syst. Str. **10** (1999) pp. 1045.
6. X. Wang and F. Gordaninejad, "Dynamic Modeling of Semi-Active ER/MR Fluid Dampers," Proc. of the SPIE **4331** (2001) pp. 82.

STUDY ON CHARACTERISTICS OF AN ELECTRORHEOLOGICAL FLUID COUPLING

YONGGANG MENG, YU TIAN AND HONG JI

State Key Laboratory of Tribology, Tsinghua University, Qinghua Yuan, Beijing 100084, CHINA
E-mail: mengyg@tsinghua.edu.cn

A prototype of a cylindrical ERF coupling was fabricated, and its transmission capability in steady and unsteady running conditions was investigated. The torque transmitted by the coupling with viscous force in wide ranges of driving speed and field strength was measured. The relationship between the transmission torque, the zero field viscosity of ERF and the shear yield stress of ERF under field was obtained, and was compared with theoretical prediction. Moreover, by modulating the strength of the applied DC field, under a fixed load and input speed, the output speed of the coupling was controlled in the range of transmission ratio from 0 to 1. This provides a means to construct transmission systems in which the output speed can be conveniently adjusted and controlled. Further, square wave electric pulses were applied on the ERF coupling, with a fixed external load added on the output shaft, a uniform stepping rotation of the output shaft has been realized. Experiments showed that under certain conditions the rotation angle of the output axle can be precisely controlled.

1. Introduction

Electrorheological fluids (ERF) are smart materials whose apparent viscosity increases greatly as an external electric field is applied. The change is rapid, reversible and controllable with an electric voltage. The mechanism of ER effect is now widely accepted as a phase transition induced by the micro-structure changes under an electric field, and the extent of the change is determined by the mismatch of the conductivity and the dielectric constant between the solid and the liquid phases [1] as well as the strength of the applied electric field. ER fluids are usually Newtonian fluids at zero field condition and become Bingham fluids under external electric fields. This electric field controlled characteristic has attracted interests of researchers and engineers. Applications, such as dampers, clutches, engine mounts, and other smart structures and actuators, have been constructed based on this kind of intelligent materials [2].

Under zero electric field condition, the shear stress of ER fluids is described by,

$$\tau = \eta\dot{\gamma} \tag{1}$$

where τ is the shear stress of ER fluid, η is the viscosity of ER fluid, and $\dot{\gamma}$ the shear rate. The equation indicates that the shear stress is proportional to the shear rate and the viscosity of the fluid. When an external electric field is applied, the shear stress of ER fluids yields

$$\tau = \tau_E + \eta\dot{\gamma} \tag{2}$$

where τ_E is the field induced shear stress, an additional contribution to the shear stress of ER fluids under shear.

An original application suggested for ER fluids is clutch using multi parallel plates or multi concentric cylinders. The key parameters of the clutches are the adjustable ratios of torque transmission at the zero field to that under external electric fields. The zero field torque, M_o, and the torque under field, M_E, can be estimated based on equation (1) and equation (2) respectively. The greater the ratio of M_E to M_o, the better the performance of an ER clutch is. The ratio is determined by the slipping between the plates or cylinders, the zero field viscosity of ER fluid and the shear yield stress of ER fluid under electric field. The zero field torque transmission ability M_0, the field induced torque transmission ability M_E, and the total transmission torque M of a multi-concentric-cylinder clutch can be calculated as the following equations

$$M_0 = \frac{2\pi L \eta_0 \omega}{h}(R_1^3 + R_2^3 + ... + R_n^3) \quad (3)$$

$$M_E = 2\pi L \tau_E (R_1^2 + R_2^2 + ... + R_n^2) \quad (4)$$

$$M = M_0 + M_E \quad (5)$$

where ω is the relative angular velocity between the driving part and the driven part, R_i, i=1,..,n, represents the radius of the i-th cylinder, L the effective length of each cylinder, and h the gap width between two neighboring cylinders. Here L and h are assumed to be same for each cylinder for the sake of simplicity.

Torque transmission of ER fluids has been studied by several investigators (for instance [3], [9] and [10]). Stevens et al. designed and constructed an experimental facility of parallel plates to uncover the temperature effect on the torque transmission. The dynamics of a radial electrorheological clutch was also investigated recently [5]. An actuator to display force using ER fluid was fabricated and displayed a step response time of 3~5ms [6]. Also ER clutches were compared with MR clutches and concluded that the response characteristic of the ER clutch is better than that of the MR clutch [4].

As well known, the transmission ability and dynamic characteristics of ER clutches or couplings are determined by the dimension of the device and the properties of the ER fluids employed. In this study, a prototype of a concentric cylindrical ERF coupling was fabricated and a self-made zeolite based ERF with high strength was tested on the prototype ERF coupling. The characteristics of this kind of ER fluids with high shear strength have been reported by the authors [8]. Transmission characteristics of the coupling in steady running condition with constant electric field and under square wave electric filed are investigated in this study. For the steady running condition, a continual adjustment of the output speed is realized. Under square wave electric field applied, a uniform stepping rotation of the coupling output is obtained. Comparing with the dynamic characteristics of the ER fluid, the dynamic behaviors of the coupling are discussed.

2. Experiment apparatus

The cylindrical ERF coupling system is consisted of several parts as shown in figure 1. The stepping motor has a torque output ability of 1.3 Nm at rate of 300 rpm. The speed of the stepping motor is controlled by the signal frequency output from the signal source A. High electric field is provided by a high voltage power source and is controlled by an on/off switching circuit to apply to the cylinders through two brushes. The connection part between the two cylinders is made of nylon for insulating. Another nylon part is used to insulate the output axis from the base. The inner cylinder is connected to the anode and the outer cylinder is connected to the cathode. The rotating speed of the output axis is measured by a photoelectric diode. The load of the output axis is applied by suspending a dead weight on the output wheel through a rope as shown in figure 1.

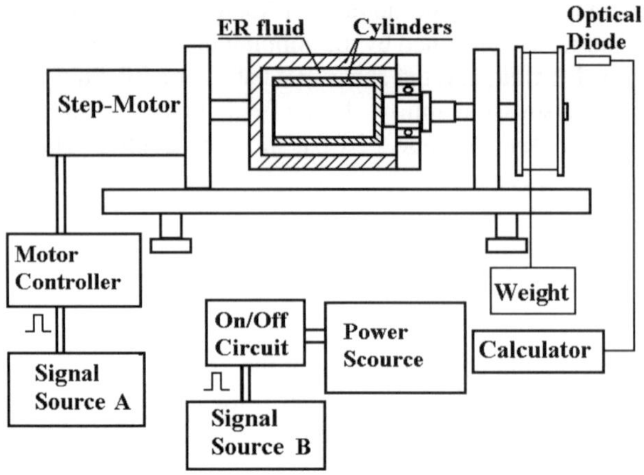

Fig.1 The sketch of the test system

In our clutch configuration, only the inner surface of the outer cylinder is effective to transport torque. The two end circular surfaces are insulated. The effective radius of the cylinder, R, is 14.5 mm, and the effective length of the cylinder, L, is 40 mm. The radius of the output wheel suspending weights is 25mm. All the experiments were done at room temperature (18°C).

3. Zero field torque transmission ability

At zero field condition, the ER fluid behaves as Newtonian fluid, the shear stress is proportional to the shear rate as described by equation (1) above. The theoretical prediction of torque transmission ability of the coupling can be expressed by

$$T = \eta \dot{\gamma} sr = \eta \frac{(\omega_{in} - \omega_{out})r}{h} sr \qquad (6)$$

where $s = 2\pi rL$ is the effective shearing area of the cylinder, and r is the radius of the inner cylinder. We tested the torque transmission ability of the pure viscous force by setting the output speed ω_0 to be approximate zero. That means the coupling can just lift the weight off the ground at a very slow speed and at this time the input speed is recorded. The viscosity of the ER fluid employed is measured on a HAAKE RV20 rheometer to be 13.3Pa.s. The prediction of T with the measured viscosity and the recorded speed is compared with experimental results as shown in figure 2. Good accordance between the experiment and prediction has been obtained. When the input speed of the motor was 120rpm, a weight of 1.08kg can be lifted off while keeping the output speed nearly zero.

4. Field induced torque transmission ability

When an electric field is applied on the ER coupling, the torque transmitted by ER fluid can be calculated by :

$$T = [\tau_E + \eta \frac{(\omega_{in} - \omega_{out})r}{h}] sr \qquad (7)$$

When ω_{in} is about 0.1 rad/s, the viscous force contributes only a little portion to the torque transmission, thus the load is mainly carried by the field induced shear stress. As the shear rate is about 3 s^{-1}, the field induced shear stress can be looked as the first approximation of the shear yield stress of the ER fluid. The test result is shown in figure 3. Also, the corresponding shear yield stress is shown in the figure.

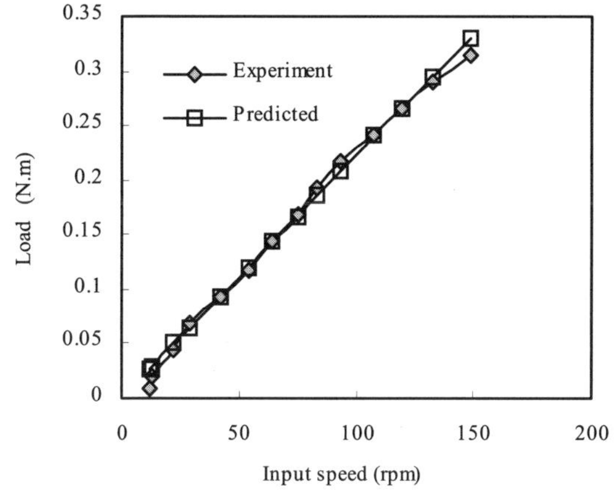

Fig. 2 Transmission ability by viscous force at zero field condition

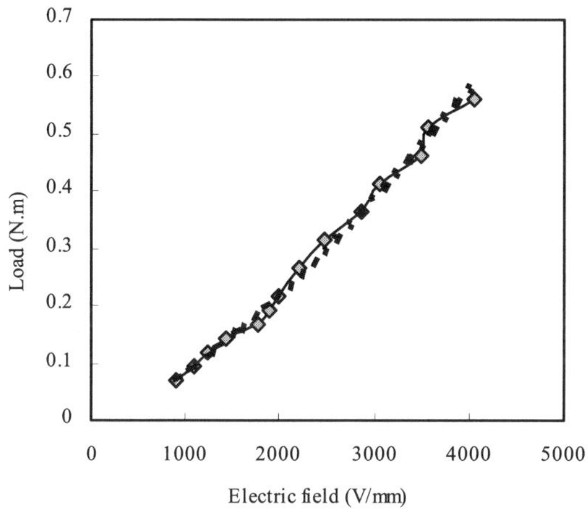

Fig. 3 The field induced torque transmission ability

The figure implies that the torque transmission ability of the coupling is proportional to the strength of the applied field with an exponent of 1.4 as described by equation (8).

$$T_E = 5\times10^{-6}E^{1.4} \qquad (8)$$

where T_E is in N.m, and E is in V/mm. The fitting curve of equation (8) is shown in the same figure as the dashed line. This is also the relationship between the shear yield stress and the electric field strength. In literatures, shear yield stress is proportional to the electric field with an exponent of 1.4-1.6 for zeolite silicone oil suspensions [7]. Our experimental torque transmission data is consistent with the relationship well.

5. Speed adjustment by external field

The above two experiments respectively showed the viscous torque transmission and the field induced torque transmission. In the following, the speed adjustment of the coupling by changing the external field will be shown.

A fixed load of 900g, corresponding to an effective torque resistance of 0.221Nm, is suspended on the output wheel. At a given input speed, the output speeds were recorded under various field strengths. Fig.4 shows the changes of the output speed with the applied field at three different input speeds, 29.3 rpm, 58.6 rpm and 117.2 rpm.

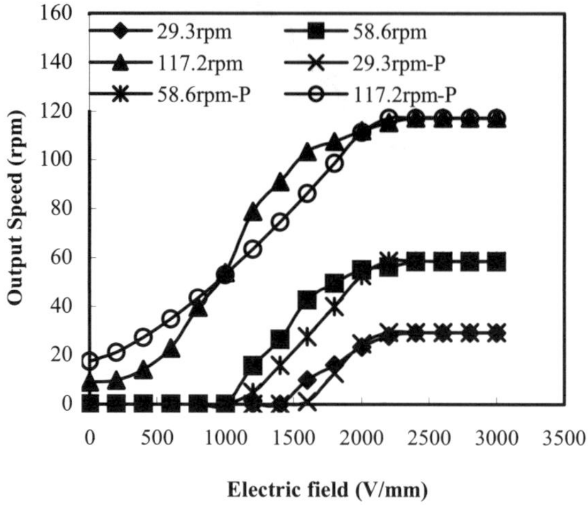

Fig. 4 The adjustment of output speed by electric field

When the input speed is as low as 29.3 rpm, the torque produced by viscous force is 0.069 N.m according to the previous experiment and not enough to provide a torque of 0.221 N.m to lift the load. When electric field is 1400 V/mm, according to the experimental result in figure 3, the field induced torque transmission ability was 0.143 N.m, the total transmission ability was 0.212 N.m and still can not lift the load. When electric field is 1600 V/mm, the field induced torque transmission ability reaches 0.153 N.m, the total ability exceeded the load and the output speed is a little higher than zero. When electric field of 2200 V/mm is applied, the output speed catches the input speed up. Non-slipping transmission is realized.

$$\tau_E sr + \eta sr(\omega_i - \omega_o) r / h = mgR \qquad (9)$$

According to figure 2, figure 3 and equation (9), the theoretical prediction of the output speeds at different input revolution speeds and electric field strengths, is also shown in figure 4. The predicted output speeds roughly agree with the experimental results. There always showed a little higher of field induced shear stress than tested in the former experiment but a little lower viscosity shear force. This may be due to the temperature rise during the test, especially when a great difference between the input speed and the output speed exists. The slipping between the two cylinders produced some thermal effects. It caused a rise of the temperature of the coupling, resulting in a slight enhancement of the ER effect. If these factors were taken into account in the theoretical calculation, the prediction would agree with the experimental results better.

From the experiment, we obtained that at a given load, until the electric field strength reached a critical value E_a, the load could not be driven. And when the field strength exceeded a critical value E_b, the output speed became the same as the input speed. Thus through the adjustment of the external electric field, the output speed can be adjusted in this whole speed range. If a feed back control circuit is embedded in the system, the output speed can be accurately controlled.

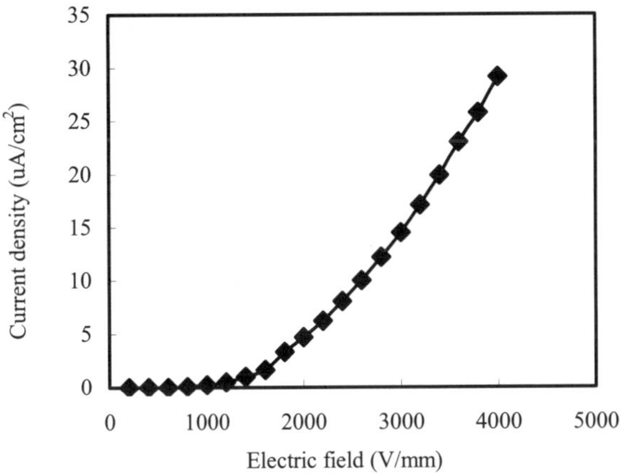

Fig. 5 The current density flow through the ER fluid

The current density passing through the ER fluid is shown in figure 5. When electric field is low, the current density is very small, in nA/cm^2 scale. With the increase of the external field, the current density increased rapidly. When 1000V/mm is applied, the current density needed is only 0.2uA/cm^2. But when 4000V/mm is applied, the current density needed is 30uA/cm^2. Much higher power is needed at a high electric field than at a low electric field.

If the speed is need to be adjusted from 0 to the input speed n (in the unit of rpm), and the load of the output axes is T_L, then the viscosity η, the maximum shear yield stress of the ER fluid τ_E and the dimensions of the cylinder coupling should satisfy the following equation (10). In equation (10), D is the diameter of the inner cylinder, L the effective length of the inner cylinder and r the radius of the inner cylinder. ΔT is a small positive value.

$$2\pi n(\pi DL)r\eta / h = T_L - \Delta T \leq \tau_E (\pi DL)r \qquad (10)$$

The equation described that the maximum shear yield stress of the ER fluid can promise the output speed to be the same as the input speed. Also to maintain a constant working condition of ER fluids to facilitate the speed control, thermal dissipation measures should better be taken in a practical system.

6. Response to square wave electric fields

The clutches based on ER fluids have great advantage of rapid response. It is much easier to be controlled than other mechanical driven clutches. The above experiments showed the transmission ability of ER coupling and demonstrated a speed adjustment approach.

Furthermore, by changing the loading method from the pulley form in the above experiment to a frictional one as shown in figure 6., the response of the coupling to square wave electric field was also investigated. In the modified testing setup, the rotation of the output axle was detected with an encoder, which has a resolution of 600 divisions per revolution.

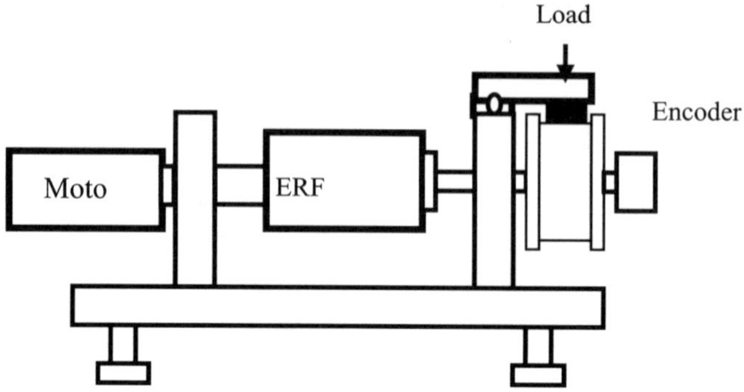

Fig.6 Modified test system for square wave field application

When the transmitted torques of the coupling is smaller than the load torque, the output wheel does not rotate. By increasing the input speed or imposing a voltage enough high, the load torque can be overcome, and the output wheel begins to move. As the output axle rotates one cycle, 600 square wave signals are acquired. Sampling the time interval, Δt, between two neighboring rising edge of the signals, the average speed ω_0 of the output axle during that interval can be written as

$$\omega_0 = \frac{2\pi}{600 * \Delta t} \qquad (11).$$

Taking the resistance torque T_L, the input speed ω_i, the moment of inertia of the output axle J, when electric field is turned on, the system can be described by:

$$\tau_E sr + \eta \frac{(\omega_i - \omega_0)r}{h} sr - T_L = J\dot{\omega}_0 \qquad (12)$$

When electric field is turned off, the system can be described by

$$\eta \frac{(\omega_i - \omega_0)r}{h} sr - T_L = J\dot{\omega}_0 \qquad (13).$$

In experiments, we set $\eta \frac{\omega_i r}{h} sr < T_L$ and $\tau_E sr > T_L$ so that $\omega_o = 0$ when electric field is off. After electric field is on, the output speed changes according to equation (12) and becomes synchronous with the input one after a short transient period. When the electric field is turned off, the output speed decreased according to equation (13) and finally decayed to zero. Therefore, in one cycle of the square wave electric field, the output axle forwards one small angular step.

In our experiment, the experiment parameters were set as following,

$$n_i = 17 rpm \qquad (14)$$

$$J = 6.22 \times 10^{-5}\ \text{kg·m}^2 \qquad (15)$$

The peak value of electric field strength in one square wave cycle was 1600 V/mm. According to the relationship described by equation (8), the corresponding field induced shear stress is τ_E =3kPa. The frictional torque load was estimated to be $T_L = 0.07Nm$ in the way described below. First applying a load on the friction wheel, then increasing the speed of the input axle gradually until the output axle starts to rotate. Based on the relative speed at that moment and the viscosity of the ER fluid, the viscous torque was calculated, which is equal to the frictional resistance T_L..

The measured movements of the output axle at 3 different frequencies, 0.48 Hz, 0.95 Hz, and 1.9 Hz, of the square wave electric fields are respectively shown in figure 7. The figures showed a fluctuation of about ±3rpm around 17rpm, which is the nominal value of the input speed. The fluctuation in output speed may be resulted from the unsteadiness of the driving stepping motor and the influence of vibration on the encoder. Anyway, the average output speed in the period of field-on is approximately 17rpm.

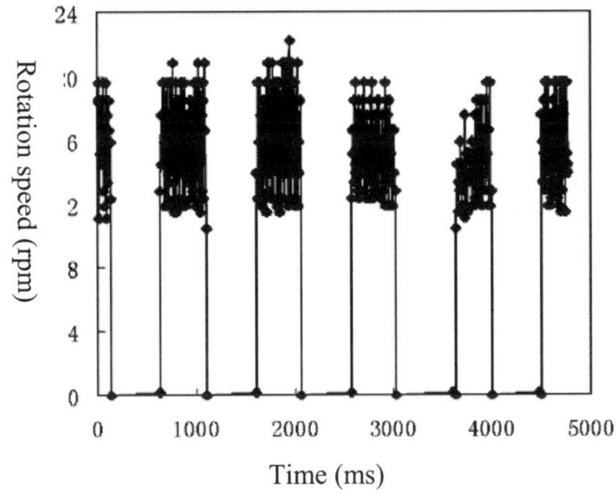

(a

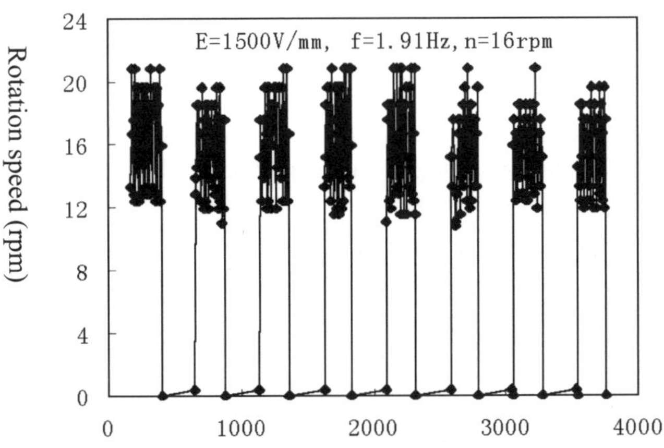

(b)

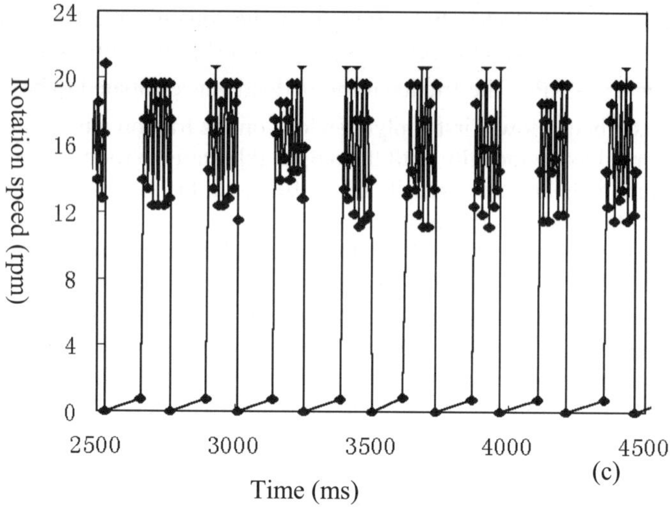

Fig. 7 Variation of output speed at different frequency square waves: (a)0.48Hz, (b)0.95Hz, (c)1.9Hz.

During each cycle of square wave field application, the ER fluid inside of the coupling experiences a hardening and a softening process. Meanwhile, the driven part of the test system accelerates from 0rpm to the running speed, about 17rpm, and then decelerates back to 0rpm correspondingly. According to an experimental investigation on the dynamic response of ER fluids, the shear stress of ER fluids changes with time, following the expression (16), as a pulse electric field is on [11].

$$\tau = \tau_E(1 - e^{\frac{t_0 - t}{T}}) \qquad (16)$$

The time needed for the hardening process is in the range of 1ms-7ms when the shear rate is in the range of 120/s-5/s [11]. From the acceleration curve obtained in this experiment, the time interval between the lower speed (0rpm) and the higher speed (about 17rpm) is less than 14ms. This means that the rotation system responds to the change in the applied field very fast. One the other hand, as the applied field is off, the shear strength of the ER fluid drops to zero almost linearly with time in less than 2ms [11]. The apparent deceleration time in this experiment is about 14ms too. Considering the limited time resolution due to the relative rough divisions of the encoder used in the experiment, the real deceleration period should be shorter than this time.

The above experiment demonstrated just a sequential way for realizing a stepping forward rotation of ER fluid couplings. If a feedback control is installed into the system, precise angular motion and positioning would be possible. Therefore, ER fluid couplings have potentials in application to automation production lines and robotics where variety and precision positioning control is required.

7. Conclusion

A prototype of a cylindrical ERF coupling was fabricated and studied in this paper. Its transmission capability under different field conditions was investigated. Relationship between the transmission torque, the zero field viscosity of ERF and the shear yield stress of ERF under field was obtained. Experiments also showed that, with proper parameters, by adjusting the external field strength, the output speed of the coupling can be easily adjusted in the range from zero to the given input speed. The response of the prototype coupling to square wave electric field was also investigated. The system showed a quick response to the variation of the electric field, and stepping rotation was realized. The experiment results indicated that the stepping motion of the coupling can be controlled in a precision better than 0.6° even with a simple sequential control.

8. Acknowledgements

This work is supported by the National Natural Science Foundation of China with grant No. 50525515 and the National Basic Research Program of China with grant No.2003CB716205..

References

1. Weiss K. D., Carlson J. D. and Coulter J. P., Material Aspects of Electrorheological Systems. *J. Intelligent Material Systems and Structures*. **4**(1993) pp.13-34.
2. Coulter J. P., Weiss K. D. and Carlson J. D., Engineering Applications of Electrorheological Materials, *J. of Intelligent Materials Systems and structures*. **4**(1993) pp. 248-259.
3. Stevens N.G., Sproston J. L. and Stanway R., An experiment study of electrorheological torque transmission, *J. Mechanism, Transmission, and Automation in Design, Transaction of The ASME*, **110**(1988) pp.182-188.
4. Choi S.B. et al. Comparison of Field-Controlled Characteristics Between ER and MR Clutches, *J. Intl. Mat. Sys. and Struc.* , **10**(1999) pp.615-619.
5. Whittle M., Aktin R.J. and Bullough W.A., Dynamics of a radial electrorheological clutch, *Intel. J. Mod. Phys. B.* **13**(1999) pp.2119-2126.
6. Furusho J. and Sakaguchi M., New actuators using ER fluid and their applications to force display devices in virtual reality and medical treatments, *Intel. J Mod. Phys. B* **13**(1999) pp.2151-2159.
7. Wu C. W. and Conrad H., A modified conduction model for the electrorheological effect, *J. Phys. D.* **29**(1996) pp.3147-3153.
8. Tian Y., Meng Y. and Wen S., Electrorheology of zeolite based ER fluids, *J. Appl. Phys.* , **90**(2001) pp.493-496.
9. Papadopoulos C.A., Brakes and clutches using ER fluids, *Mechatronics*. **8**(1998) pp. 719-726.
10. Brookfield D. J. and Dlodlo Z.B., Robot torque and position control using an electrorheological actuator, *Proceedings of the Institution of Mechanical Engineering Part 1-Journal of Systems and Control Engineering* . **212**(1998) pp.229-238.
11. Tian Y., Meng Y. and Wen S., Dynamic responses of zeolite-based ER fluid sheared between two concentric cylinders, *Journal of Intelligent Material Systems and Structures*, **15**(2004) pp. 621-626.

THE MODIFIED ELECTRODES FOR THE APPLICATION OF POLAR MOLECULE DOMINATED ELECTRORHEOLOGICAL (PM-ER) FLUIDS

XUEZHAO WANG[1] RONG SHEN[1] GANG SUN[1] WEIJIA WEN[1,2] KUNQUAN LU[1]

1. Institute of Physics, Chinese Academy of Sciences, Beijing, China
2. Department of Physics, The Hong Kong University of Science and Technology, Hong Kong, China
E-mail: lukq@aphy.iphy.ac.cn

For polar molecule dominated electrorheological (PM-ER) fluids [1], the surfaces of ordinary metallic electrodes cannot satisfy the condition of polar molecules' aligning and interaction. Slide must occur at the interface between PM-ER fluids and electrodes so that the measured shear stress is much lower than the true value. Several modified electrodes are developed and tested for increasing the adhesion of PM-ER fluids to the electrodes and weakening the slide in shearing. The modified electrodes can not only improve the measured shear stress to approach the intrinsic one, but also greatly reduce the current density of PM-ER fluids. Using the modified electrodes is certainly necessary for measuring and applying the PM-ER fluids.

1 Introduction

Electrorheological (ER) fluids are complex suspensions consisting of nanometer or micron sized solid particles mixed with insulating oil, whose shear stress and yield stress can be continuously tunable with an applied external electric field. While the electric field strength is high enough, suspensions can be changed into a solid-like state. The reversible transformation of soft to hard state can occur in about millisecond. Those characters of ER fluids open up the broad prospects for their application in industry and technology. However, the shear stress of the conventional dielectric ER fluids, which are based on the interaction of polarized particles [2-6], is quite low (usually less than 10kPa [2]). The low shear stress of available ER fluids has blocked their application widely.

A new type of ER fluids [7, 8] with high yield stress has been developed in recent years, which are defined as polar molecule dominated electrorheological (PM-ER) fluids [1]. The yield stress of PM-ER fluids can reach hundreds of kPa which is orders higher than that of the conventional ER fluids, that owes to the polar molecules adsorbed on the particles. A model based on the interaction of polar molecule-charge in between the particles is proposed to explain the PM-ER effect [1]. The polar molecules in the gap turn to the field direction due to the high local field and interact with the polarized charges of the neighboring particles. The attractive force of the polar molecule-charge can be orders higher than that of pure polarized particles in conventional ER fluids.

The essential difference between the PM-ER and conventional ER fluids both in principle and characters results in the difference on measuring method.

For conventional ER fluids, the interaction between the electrodes and ER fluids is the same as inside the ER fluid because of the image effect of the particles at the boundary of metallic electrodes. Therefore the ordinary metallic electrodes can be used for the measurement and application of the conventional ER fluids. For PM-ER fluids, however, because the condition at the surfaces of metallic electrodes is not the same as in the gaps between the particles, where the interaction is weak obviously, the slide of the fluids on the electrodes will occur when the interaction of polar molecule-charge is strong enough in particles' chains. Then the measured yield stress must be much lower than the exact value of PM-ER fluids. This problem will also restrain the application of PM-ER fluids. Although by roughing the surfaces of metallic electrodes the slide can be weaken, the break down is easy to occur on the surfaces at high electric field.

In this paper, several modified electrodes substituting for the ordinary metallic ones will be presented. The modified electrodes can basically avoid the slide by increasing the adhesion of material to electrodes. With the application of the modified electrodes not only the measured shear stress is close to the intrinsic value, but also the current density is reduced essentially for PM-ER fluids. The principle and those modified electrodes can be used in all kinds of devices of PM-ER fluids for their application.

2 Experimental

The calcium titanate (CTO) nanopatical PM-ER Fluid was used in the experiment. Spheral nano-CTO particles of 50 nm in diameter were synthesized by means of co-precipitation method without adding any special additives to the particles [8]. Infrared spectrum demonstrated that there were C=O, O-H polar groups contained in Ca-Ti-O nano-particles. The contents of C=O, O-H polar groups in the particles were regulated in the washing procedure during the material manufacture. The CTO particles were suspended in silicone oil with the volume fraction $\phi \approx 35\%$.

The yield stresses of CTO PM-ER fluid were measured with a parallel-plate rotating rheometer. For comparing the surface effect of the electrodes in the yield stress measurement five different kinds of electrodes were used. (A) Smooth metallic electrodes, of which the surfaces were polished carefully. (B) Rough metallic electrodes, of which the surfaces were ground with a sand paper. (C) The surfaces of the metallic electrodes were covered with a layer of double sticking tape 70 μm thick. (D) The surfaces of the metallic electrodes were coated with alumina. By using plasma spraying technique 1 μm sized alumina particles were sprayed on the surfaces of electrodes to form a rough insulating layer 20 μm thick. (E) The surfaces of the metallic electrodes were plated with diamond grains, of which the grain size was 100 μm and the layer thickness was 20 μm.

3 Results and Discussion

Because the surfaces of metallic electrodes can not well satisfy the condition of polar molecules' aligning and interaction in the PM-ER fluids, the slide at the interface between fluids and electrodes must occur inevitably and be a hindrance in the application and measurement for PM-ER fluids. By using the above 5 pairs of electrodes respectively to measure a CTO PM-ER fluid, we found that the surface effect of electrodes on the measurement of the yield stress is obvious as shown in the Fig. 1. In order to obtain the exact yield stress of PM-ER fluids an apparatus has been developed, which can measure the intrinsic rheological property and avoid the influence of the electrodes [9]. The measured data with this method are also shown in Fig. 1 for a comparison. Under the same electric field strength 5000V/mm, the measured yield stress with smooth metallic electrodes (data A) is less than 30kPa, which is only 1/6 of real value of yield stress (data F). By using rough metallic electrodes the measured yield stress is 60 kPa (data B) which is double of that measured with smooth metallic electrodes. With other surface modified electrodes the measured yield stresses can be much improved. Especially, by using the electrodes coated with diamond and alumina grains the measured yield stresses is about 120 kPa (data D and E), which are about 4 times of that measured with smooth metallic electrodes and can reach 2/3 of the real one at 5000V/mm. At the accidented surfaces

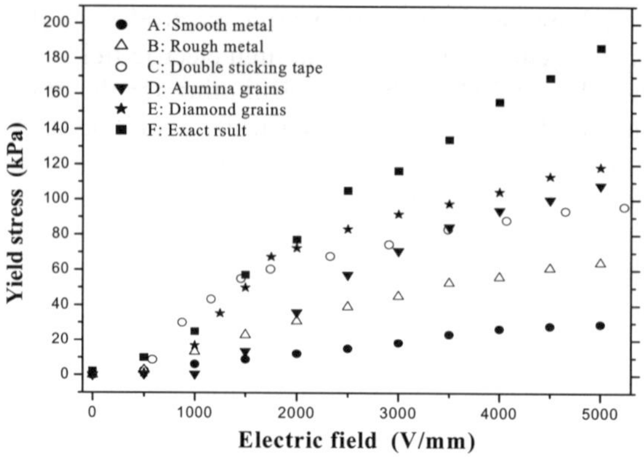

Fig.1 The variation of yield stress vs. electric field measured with different electrodes.

consisted of diamond and alumina grains the nano-CTO particles can fill into the clearances between the grains. The salients of the grains resist the fluid flowing to create a shear stress, which is much higher than that caused by a flat surface and can be estimated to be about 2/3 of the real shear stress. It can be seen that the values of the measured yield stress in data A-E of Fig. 1 deviate from the exact one (data F) more seriously at high electric field. This phenomenon may come from the mechanic property of the ER fluid at different electric field. The ER fluid is sticky at low electric field and is solid at high electric field. The slide of the solid ER fluid on the electrodes can easier occur. In the dynamic shear stress measurement by using the modified electrodes coated with diamond grains, the dynamic shear stress at shear rate of 250 s^{-1} can reach about 70 kPa as the electric field is 3 kV/mm. A detailed description will appear elsewhere [1]. However the dynamic shear stress is unable to be measured successfully when the electrodes are made of smooth metal due to the slide of the ER fluid on the electrodes.

Besides avoiding slide, the modified electrodes can also reduce the electric

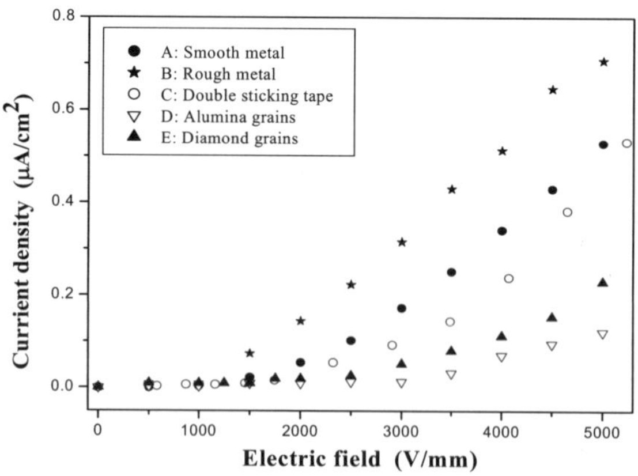

Fig.2. The variation of current density vs. electric field measured with different electrodes.

current through the sample greatly as shown in Fig. 2. The current density of our prepared ER fluid is very low, less than 1μA/cm^2. The current density with rough metallic electrodes is the largest one. It is easy to cause breaking down by using rough metallic electrodes, although applying a file-like metallic electrode can also greatly increase measured yield stress. The current densities are reduced greatly, even less than 0.2μA/cm^2, in using the modified electrodes with coating alumina or diamond grains because of their insulating property.

According to the above results, electrodes play an important role in the measurement of PM-ER fluids and their application as well. Actually the ordinary metallic electrodes are unsuited to be used in the measurement and application of PM-ER fluids. In principle, if there are the particles with polar molecules contained on the electrodes, the boundary condition at the surface of electrodes can be satisfied for the PM-ER effect. Therefore the most convenient way is to glue those particles on the electrodes. Using the double sticking tapes is an example. However those electrodes can only be used for few times because of their infirmness. Although double sticking tape is not strong enough to be used in applied devices for a long time, it's convenient to be used for testing the material. Diamond and alumina are both durable, insulated and stable in the range of working temperature -50~200℃, which are suitable for modifying the electrodes in the measurement and application of PM-ER fluids.

4 Conclusions

PM-ER fluids based on the interaction of polar molecule-charge in between the particles are newly developed [1]. The slide of the fluids inevitably exists at the surfaces of metallic electrodes because the interfaces cannot satisfy the boundary condition of polar molecules' aligning and interaction. Comparing the influence of different electrodes on the ER effect and electric current density of PM-ER fluids, we find that the electrodes are one of the critical factors in their measurement. The modified electrodes coated with diamond and alumina grains not only avoid the slide effectively to improve the measured stress, but also reduce the current density greatly, which are favorable for measuring and applying the PM-ER fluids.

5 Acknowledgement

The authors would thank Prof. S.Z.Yang and B.R.Zhao for technical help in the modification of electrodes. This work was supported by the National Basic Research Program of China (No 2004CB619005), The Knowledge Innovation Project and Outstanding Overseas Chinese Scholars Fund of Chinese Academy of Sciences and NSFC.

References

1. K. Q. Lu, R. Shen, X. Z. Wang, G. Sun, W. J. Wen and J. X. Liu, Polar molecule dominated electrorheological effect, to be published.
2. H. R. Ma, W. J. Wen, W. Y. Tam and P. Shen, Frequency dependent electrorheological properties: Origin and Bounds, *Phys. Rev. Lett.* **77** (1996) pp. 2499-2502
3. T. C. Halsey, Electrorheological fluids, *Science* **258** (1992) pp. 761-766
4. R. Tao and J. M. Sun, Three-dimensional structure of induced electrorheological

solid, *Phys. Rev. Lett.* **67** (1991) pp. 398-401
5. Y. Chen, A.F. Sprecher, H. Conrad, Electrostatic particle-partile interactions in electrorheological fluids, *J. Appl. Phys.* **70** (1991) pp. 6796-6803
6. L. C. Davis, Time-dependent and nonlinear effects in electrorheological fluids, *J. Appl. Phys.* **81** (1997) pp.1985-1991.
7. R. Shen, X. Z. Wang, W. J. Wen, K. Q. Lu, TiO_2 based electrorheological fluid with high yield stress, *International J. Modern Phys. B* **19** (2005) pp. 1104-1109.
8. X. Z. Wang, R. Shen, W. J. Wen, K. Q. Lu, High performance calcium titanate nanopartical ER fluids. *International J. Modern Phys. B* **19** (2005) pp. 1110-1104
9. R. Shen et al, The methods for Measuring Shear Stress of Polar Molecule Dominated Electrorheological (PM-ER) Fluids, in this proceedings

DYNAMIC BEHAVIOR OF FIELD-RESPONSIVE PARTICULATE SUSPENSIONS

H. SEE, C. JOUNG AND C. EKWEBELAM

School of Chemical and Bio-molecular Engineering, The University of Sydney, Sydney, NSW 2006, Australia

E-mail: howards@chem.eng.usyd.edu.au

We have examined the small strain response of an inverse ferrofluid system, consisting of micron-sized inert particles dispersed in a ferrofluid, which is a magnetisable liquid consisting of single domain magnetite nanoparticles. Under a magnetic field the inert particles will form elongated aggregates in the field direction, analogous to a magnetorheological fluid. It was found that the fluid appeared to have a Bingham fluid-like yield stress when analysed using the flow curve. However careful study of the behavior at very low shear rates revealed an ever decreasing shear stress. In addition, the behavior of conventional magnetorheological fluids at large strains under steady shear flow and constant magnetic field was also studied, and the results compared to particle-level computer simulations.

1 Introduction

Recent years have witnessed continued strong interest in magneto-rheological fluids (MRFs). These are "smart fluids" which show a dramatic but reversible increase in flow resistance when placed under an externally applied magnetic field. There has been considerable research activity worldwide in both industry and academia with the aim of improving our understanding of the fundamental mechanisms behind the behavior of MRF (see for example the reviews in [1,2]).

There is still a need for a general constitutive framework for this material. Towards this end, careful rheometrical measurement of these systems under small strain deformations will improve our understanding of the relationship between aggregate microstructure and the rheological response. At the other extreme of strain dependent behavior, the response at large strains is also of considerable practical interest – for example the question as to whether the shear stress remains at the same level under constant field and continuous steady shearing is an important one from the viewpoint of device design, particularly if the device is meant to be operated in an activated (field-on) state under prolonged deformations (for example in some vibration damping systems [3]).

An experimental technique which has succeeded in illuminating key aspects of the field-induced response of MRF is based on a physically related system, the "inverse ferrofluid". This typically consists of magnetically inert particles (e.g. polystyrene or silica particles) dispersed in a magnetizable ferrofluid. The ferrofluid itself is comprised of 10nm-sized magnetite particles dispersed in a mineral oil. Each magnetite nanoparticle is a magnetic mono-domain and hence possesses a moment. Thus, in the absence of an external field, thermal motion causes these moments to be randomly oriented and there is no net magnetic moment in the ferrofluid. If a field is applied there is a tendency for the magnetite particles to align in the field direction, leading to a net magnetic moment in the ferrofluid. Since the micron-sized inert particles are much larger than the nanoparticles in the ferrofluid, the latter can be effectively considered as a continuum liquid which polarises under a magnetic field. This system is called an “inverse ferrofluid”.

Application of an external magnetic field will induce magnetostatic interactions between the inert particles analogous to those in MRF, and the particles will form elongated aggregates in the field direction (Fig. 1), with an accompanying increase in flow resistance. In general the stress levels are somewhat lower than conventional MRF, and indeed it is not proposed to use the inverse ferrofluid as an alternative to MRF in engineering applications. However, they serve as a useful "model" system [4]. Recent work using this approach has been reported by de Gans and co-workers [5], who used silica particles somewhat smaller (53nm to 189nm) than those typically used in MRF and examined the effects of particle size.

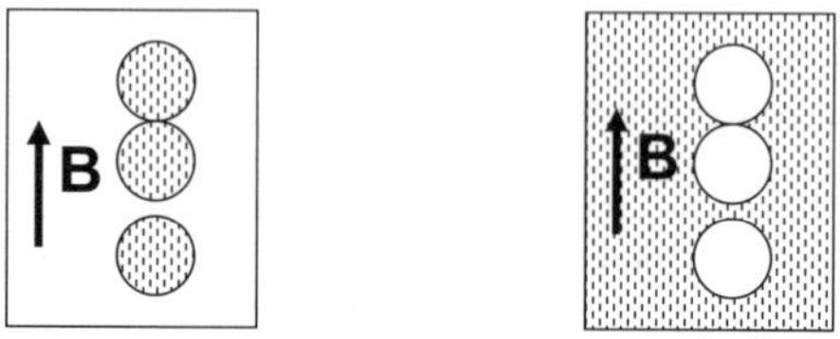

Figure 1. Schematic diagram of a conventional magneto-rheological fluid (left) and an inverse ferrofluid (right). In the latter, neutral particles are dispersed in a magnetizable ferrofluid carrier. In both systems, upon application of a magnetic field, the particles will move to form elongated aggregates in the field direction.

This paper will report on the small strain behavior under a constant magnetic field using the inverse ferrofluid system, with particular focus on characterizing the yielding behavior. The long strain behavior of a conventional MRF has also been explored - this choice of sample was made because the results would then be more readily applicable to the long time performance of actual engineering devices. Further, recent work using a particle-level computer simulation predicts a decrease in shear stress as strain is increased (under constant shear deformation and magnetic field) – the comparison with the experimental results will provide a useful perspective on the direction of MRF modeling.

2 Experimental

2.1 Materials and sample preparation

Two sample types were used.

<u>Inverse ferrofluid:</u> To produce this, magnetically inert polystyrene particles (4.5μm average size) were dispersed at a volume fraction of 30% in a commercially available ferrofluid (Ferro-fluidics Corp., USA).

<u>Conventional magneto-rheological fluid (MRF):</u> This was a suspension of carbonyl iron particles (ISP Technologies INC Iron Micropowder Grade S-3700, average particle size 3.5μm) dispersed in Dow Corning 200 silicone oil (0.1 Pas) made up to a volume concentration of 30%.

2.2 Rheometrical measurements

The rheological measurements were carried out on an Anton Paar Physica MCR300 rheometer, operating in the controlled strain mode with parallel plates (diameter 20mm, gap 1mm). A magneto-rheological test cell (Physica TEK 70-MR) was used. This consisted of a unit under the bottom plate housing the lower half of a magnetic circuit constructed from iron elements. This unit contained a coil of 495 windings through which a dc electric current was passed to generate the uniform magnetic field perpendicular to the plates. The magnetic flux density across the plates was controlled by adjusting the electric current through the coil. The maximum magnetic flux density across the plates achievable with this system was 0.54 T. A detailed description of this system has been presented in other publications (eg [6]).

The rheometrical tests were carried out as follows:

1. Before all tests, the samples were pre-sheared for 30s at a constant shear rate of $50s^{-1}$, to enhance dispersion of the particles. The magnetic field was then applied to the quiescent suspension for 3min prior to the rheological test, to allow for the structures to build up.
2. Test 1 – Small strain tests using the inverse ferrofluid: First, a constant shear rate (e.g. $0.01s^{-1}$) was applied to the initially stationary sample, and the shear stress recorded as a function of shear strain. The plateau value at long shear strains was noted. This test was repeated for a range of applied shear rates, and the plateau values of the shear stress were taken to represent the steady state value of the shear stress at that particular shear rate. On a separate plot, these plateau values of the shear stress were plotted as a function of shear rate – this is essentially the "flow curve" of the material obtained by slowly incrementing the shear rate.
3. Test 2 – Large strain tests using conventional MRF: The change in shear stress with time (or equivalently strain) under a constant shear rate was monitored over a long time. This was compared with computer simulations we have recently carried out (details of these will be described in the next section).

For all tests, the temperature was maintained at 25°C. Further it should be noted that at the end of each test, the samples were inspected and no noticeable clumping or sedimentation of particles was observed.

3 Results and Discussion

3.1 Test 1 – Small strain tests using inverse ferrofluid

Fig. 2 shows a plot of shear stress as a function of strain for constant shear rate of 0.01 s^{-1} and magnetic field strength of 0.54 T. Three different data sets are shown to indicate the spread of the data. The data shows initial elastic-like behavior at very small strains (ie the shear stress is linear in strain at the start-up), but the curves then plateau at quite low strain values (typically strain ~ 0.002). In the case shown, the plateau value is approx. 60Pa. We can take 0.002 to be a characteristic "strain to yielding" - this is somewhat low compared to other materials, but it is known that electro- and magneto-rheological fluids show very low limits of linear mechanical response (e.g. 10^{-3} is reported in [7]). Further, from this test, it is possible to define a measure of the yield stress by noting the value of the stress when the curves begin to plateau under increasing strain – in the case of Fig.2, this would appear to be in the vicinity of 50Pa (although the curvature of the plot makes this hard to judge unambiguously).

Fig. 3 shows a plot of shear stress as a function of shear rate, obtained from the plateau values (as seen in Fig 2 for example). We see that the stress levels are raised as the magnetic field is applied (B =0.54T), compared to no field at all. Thus this inverse ferrofluid does indeed show a field-induced increase in flow resistance, and is therefore appropriate as a model of MRF. Further, we note that both axes in Fig. 3 are linear. With this type of plot, the "yield stress" can be found by invoking the Bingham fluid model: that is, we extrapolate to zero shear rate and find for the 0.54T field case that the intercept on the stress axis is about 90 Pa (or possibly closer to 100Pa if we extrapolate using the flatter parts of the plot at shear rates $1s^{-1}$ or above). This represents the "yield stress" obtained from this flow curve method.

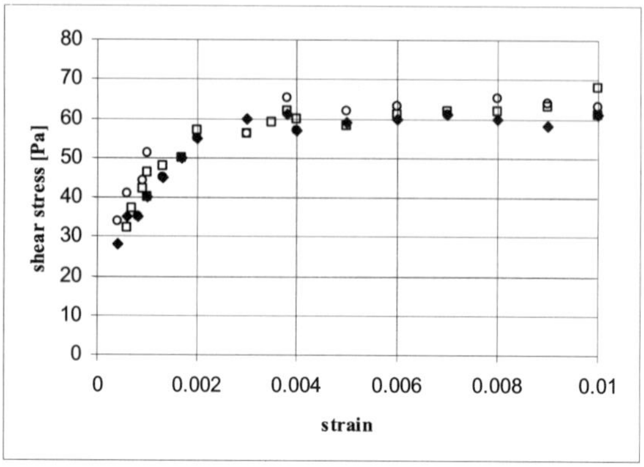

Figure 2. Shear start-up curve under a magnetic flux density of 0.54T under a constant shear rate of $0.01s^{-1}$.

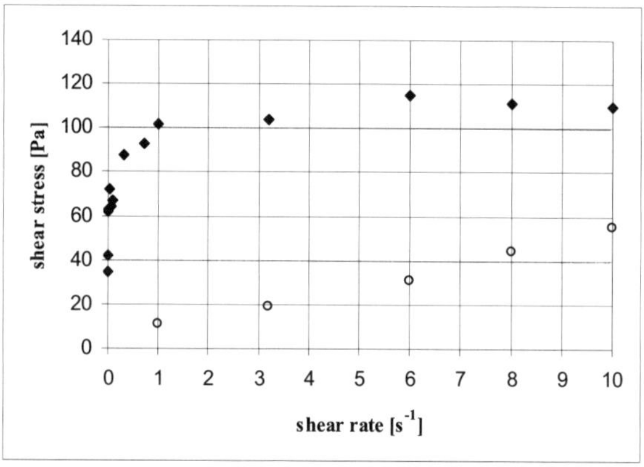

Figure 3. Shear stress as a function of shear rate under a magnetic flux density of 0.54 T. The zero-field data is also plotted (open circles).

Since the shear rate axis in Fig. 3 is linear, it highlights the larger shear rate values. If we now employ a logarithmic axis, starting at a value of 0.001 s^{-1}, and plot the same data of shear stress versus strain we obtain Fig 4. We now see that the shear stress takes a value of only 35 Pa at the slowest shear rate of 0.001 s^{-1}, and that there is a monotonic increase in shear stress with shear rate.

The implication from Fig. 4 is that the "yield stress" deduced from Fig 3 (the "flow curve" method using linear axes of shear stress and strain rate) is an over-estimate. This has occurred because the extrapolation of the data using the Bingham fluid model uses linear axes, and does not capture the finer resolution yielding behavior occurring at very low shear rates. Indeed, similar ideas have been discussed by Barnes and Walters [8,9] for a range of materials (not just electro- or magneto-rheological materials) which purport to show a yield stress but will actually flow with smaller shear stresses at very low shear rates. It would seem that the inverse ferrofluid systems also display this behavior, and so the concept of yield stress needs to be used with some caution. However, this is not to dismiss the use of yield stress *per se* – the concept of "yield stress" remains a useful engineering description of the material – but to point out that, when analyzed more closely, a complex behavior of yielding can observed under low shear rates. Particle level simulations are currently under way to shed light on the microstructural mechanisms responsible for this behavior.

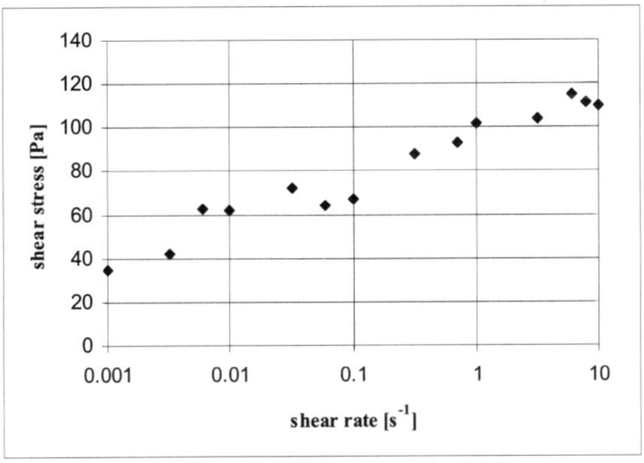

Figure 4. Shear stress as a function of shear rate, using a logarithmic axis. The magnetic flux density is 0.54 T.

3.2 *Test 2 – Large strain tests using conventional MRF*

Fig 5 shows the change in shear stress for the case of a constant shear rate of 100 s^{-1} and constant magnetic field of 0.54T. We observe that there is a slight increase in shear stress over the time period studied. Similar behavior was observed for the other conditions of shear rate and magnetic field studied.

At this stage, it is worth noting that this behavior is in contrast to particle-level computer simulations we have been performing. These simulations are a reduced version of the Stokesian Dynamics method [10] with modifications to model the magneto-rheological response. MRF particles are modeled as rigid magnetizable spheres suspended in a Newtonian fluid. The Rotne-Prager Yamakawa tensor is used in the construction of the resistance matrix accounting for interparticle hydrodynamics, and particle-wall hydrodynamics are included with a formulation for sphere-wall interactions [11]. MR forces are incorporated using superimposed sphere-pair dipole interaction forces.

We studied an initially randomly placed monolayer suspension in a magnetic field sheared between infinite parallel walls (infinite periodic in the shear direction). Typical suspensions of order 10% volume fraction initially developed the expected chain-like microstructure seen in ERF, but showed substantial qualitative divergence over longer strains. The shear stress increased as the particles formed the elongated aggregates in the field direction. However, interestingly, the simulation found that the shear stress decreased with large strain. The microstructures were correspondingly observed to become large, rather rounded clusters in the middle of the flow field.

The cause is believed to be due to a combination of the periodic boundary domain, limited particle numbers used in simulation, and the lack of wall image dipoles peculiar to MR systems. While the discrete particle method is now a well established and accepted approach for general suspension modelling, under these specific conditions, a fundamental deficiency is highlighted here which calls for some attention when applied to MR fluids. This simulation work has recently been reported [12].

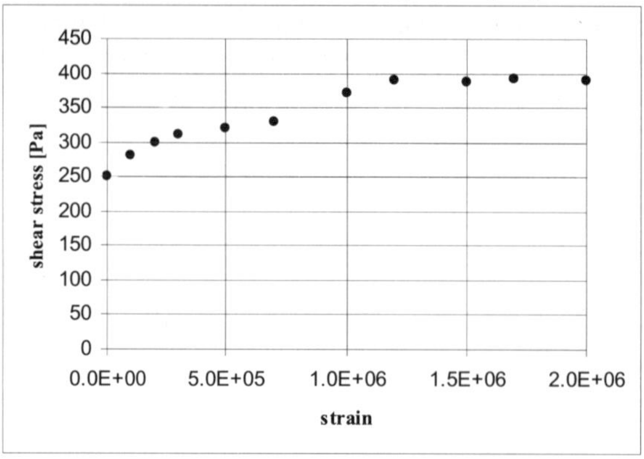

Figure 5 Shear stress as a function of strain, under a constant shear rate of $100s^{-1}$ and magnetic flux density of 0.54T.

It is worth noting that a feature of this present simulation which distinguishes it from most others is the fact that there are no attraction forces to the walls (there are only hydrodynamic/repulsion force interactions between particles and the wall). This is because, in the magnetic case, there is no "image dipole"-type force keeping particles anchored to the wall surface [13], in contrast to the situation for electro-rheological fluids [14].

4 Concluding remarks

We have seen that the "inverse ferrofluid" provides a useful way to probe the physical mechanisms governing magnetorheological behaviour. In particular, small strain rheometry has given us many insights into the complex yielding behaviour of these systems. Currently the response of these inverse systems to a small amplitude oscillatory shear flow is being investigated experimentally to give insights into the dynamic mechanisms, and the results will be reported in the near future.

We have also revealed a limitation in the particle simulation method when applied specifically to MR fluids bound by walls. The modeled microstructure and rheological response is shown gradually to diverge from the observed behavior towards a clearly artificial final state. This becomes increasingly apparent at larger strains. This finding shows that the modelling of MRF by discrete particle models with the inclusion of wall effects requires further refinement in order to correctly capture the observed microstructural evolution and stress response.

5 Acknowledgements

We gratefully acknowledge the support of the Australian Research Council. We thank Christopher Ekwebelam for his assistance with the experiments.

References

1. Ginder J.M., Behavior of magnetorheological fluids. *Materials Research Society Bulletin* (August 1998) pp. 26-29.
2. Klingenberg D.J., Magnteorheology: Applications and Challenges. *AIChE Journal* **47** (2001) pp. 246-249.
3. Wang J. and Meng G., Magnetorheological fluid devices: principles, characteristics and applications in mechanical engineering. *Proceedings of the I MECH E Part L Journal of Materials:Design and Applications* **215** (2001) pp. 165-174.
4. Popplewell J., Rosenweig R.E. and Siller J.K., Magnetorheology of ferrofluid composites, *J. Magnetism and Magnetic Materials* **149** (1995) pp. 53-56.
5. de Gans B.J., Duin N.J., van den Ende D. and Mellema J., The influence of particle size on the magnetorheological properties of an inverse ferrofluid. *J. Chem. Phys.* **113** (2000) pp. 2032-2042.
6. See H. and Chen R., The behaviour of field-responsive fluids during shear start-up. *Rheol. Acta* **43** (2004) pp. 175-179.
7. Parthasarathy M. and Klingenberg D.J., A microstructural investigation of the nonlinear response of electrorheological suspensions. II. Oscillatory shear flow. *Rheol. Acta* **34** (1995) pp. 430-439.
8. Barnes H.A. and Walters K.,The yield stress myth? *Rheol. Acta* **24**(1985) pp.323-326
9. Barnes H.A., The yield stress – a review or 'παντα ρει' – everything flows? *J. Non-Newtonian Fluid Mech.* **81** (1999) pp. 133-178.
10. Durkofsky L., Brady J.F., and Bossis G., Dynamic simulation of hydrodynamically interacting particles. *J. Fluid Mech.* **180** (1987) pp. 21-49.
11. Binous H. and Phillips R., The effect of sphere-wall interactions on particle motion in a viscoelastic suspension of FENE dumbbells. *J. Non-Newtonian Fluid Mech.* **85** (1999) pp. 63-92.
12. Joung C. and See H., Simulation of Magneto-Rheological Fluids incorporating hydrodynamic effects. In Y.S. LUO, Q.H. RAO, Y.Z. XU (eds) *Proc. 4th Pacific Rim Conference on Rheology - Advances in Rheology and its Applicatons* ed. by Luo Y.S., Rao Q.H. and Xu Y.Z. (Shanghai, 7-11 Aug. 2005) pp. 658-661.
13. Martin J.E., Field-induced rheology in uniaxial and biaxial fields. In R. TAO (ed) *Proc. 7th International Conference on Electro-rheological Fluids and Magneto-rheological Suspensions* ed. by Tao R. (Honolulu, 19-23 July 1999) pp. 28-49.
14. Klingenberg D.J., van Swol F. and Zukoski C.F., The small shear rate response of electrorheological suspensions. I. Simulation in the point-dipole limit. *J. Chem. Phys.* **94** (1991) pp. 6160-6169.

INFLUENCE OF COMPOSITION AND MICROSTRUCTURE ON ELECTRORHEOLOGICAL PROPERTY OF M-DOPED (M=Na, Zr) NANO-SIZED TiO_2 PARTICLE MATERIALS

YAN-LI SHANG, YUN-LING JIA AND JUN-RAN LI*

State Key Laboratory of Rare Earth Materials Chemistry and Applications , Peking University, Beijing 100871, China.

E-mail: lijunran@pku.edu.cn; Fax: 861062751708; Tel: 861062753517

YAN-LI SHANG, YUN-LING JIA AND MING-XIU LI

Department of Chemistry, Hebei Normal University, Shijiazhuang, 050091, China.

Nano-sized TiO_2 particle materials doped with Na_2O or ZrO_2 have been synthesized. The electrorheological (ER) properties of the materials have shown that both Na-doping and Zr-doping can effectively enhance the ER performance of TiO_2 materials with suitable degree of doping. A material, the suspension of which has a high relative shear stress (τ_r=84.0), has been obtained. Such high ER activity is advantageous in its application as an ER fluid. The composition, especially the composition in the surface layer of particles, plays a dominant role in influencing the ER performance of the particle materials.

1. Introduction

Electrorheological (ER) fluids have attracted much interest for their potential applications in various mechanical devices, such as clutches, valves, damping devices, etc [1, 2]. However, ER fluids have not been widely utilized because their ER effect is not strong enough to meet the requirements of most of these applications. The mechanism of the ER effect has been intensively discussed in many articles in an effort to obtain better ER materials. Hao and his co-workers [3,4] have demonstrated that interfacial polarization would contribute to the ER effect of the materials. The material, which has a large dielectric loss, could give a large interfacial polarization once it is dispersed into a liquid, namely, a large dielectric loss is required for a good ER material. In the particle polarization process under an electric field, the particle dielectric constant is dominant [3]. It is well known that the composition and microstructure of the particles, which would influence the physical and chemical natures including surface properties of a material, are essential to the dielectric and polarization properties of the material. Therefore, it is possible to modify a material's dielectric and polarization properties by adjusting the synthetic procedure such that an ER material having a desired microstructure and composition can be produced. ER materials composed of composite oxide have been prepared and studied [5-6], showing that doping La_2O_3, CeO_2 or Cr_2O_3 in TiO_2 can enhance the ER performance of TiO_2 particle materials.

In order to gain a better understanding of the ER mechanism, and obtain suitable ER materials, it is critical to investigate the effect of composition, especially the composition in the surface layer of particles, and microstructure on the ER performance of a particle material.

2. Experimental

Nano-sized TiO_2 materials doped with Na_2O or ZrO_2 have been synthesized by sol-gel methods. First, the ethanol solution of the metal nitrate (M=Zr) or acetate (M=Na) was added into ethanol solutions containing tetrabutyl titanate with a required M/Ti molar ratio. After the transparent gels were formed at room temperature, they were then dried at 85-90℃, and finally roasted for 8 hours at 550℃. Samples 2-7 were thus obtained. Pure TiO_2 material (sample 1) was prepared by a procedure similar to the one mentioned above. The compositions of these samples are listed in Table 1.

Table 1 Composition and characteristic of materials

Sample No.	Composition (M/Ti, molar ratio)	τ_0(Pa)	τ_E(Pa)	τ_r (τ_E/τ_0)	Surface area (m^2/g)	Pore volume (cm^3/g)	Grain size (nm)	Composition in surface layer (molar ratio)	
								M/Ti	O_w/Ti
1	TiO_2	17.5	172	9.79	5.41	0.028	15.5		
2	Na_2O-TiO_2(0.05)	12.0	337	28.1	23.7	0.081	15.8	0.314	0.118
3	Na_2O-TiO_2 (0.1)	19.1	1606	84.0	37.7	0.14	13.6	0.409	0.154
4	Na_2O-TiO_2 (0.2)	20.3	1104	54.3	33.5	0.21	17.3	0.546	0.201
5	ZrO_2-TiO_2(0.05)	16.4	573	34.9	71.6	0.12	9.7	0.0756	
6	ZrO_2-TiO_2 (0.1)	20.8	856	41.1	50.0	0.082	11.3	0.135	
7	ZrO_2-TiO_2 (0.2)	15.3	152	9.90	70.2	0.10	6.4	0.289	

All particle materials were mixed with dimethyl silicone oil under stirring and ultrasonically dispersing for five minutes to produce all the ER suspensions (25wt%). The shear stresses and apparent viscosities of the ER suspensions were determined in electric fields (dc field) of various strengths (E) and at a shear rate (γ) range of 0 to 300s^{-1} and 25 ℃ on a German Rotary Viscometer (Type HAAKE CV20).

3. Results and discussion

In order to obtain a clear comparison between the ER performances of these materials, three types of shear stresses were employed: the shear stresses of the suspension with and without applied electric field (τ_E and τ_0), and relative shear stress (τ_r), which is defined in our work as the ratio of the shear stress at an electric field to zero-field shear stress ($\tau_r=\tau_E/\tau_0$). Relative shear stress was also used to represent the magnitude of the ER activity. The corresponding results at E = 4.2 kv/mm and γ = 200 s^{-1} are listed in Table 1.

Figure 1 illustrates the shear stresses as a function of electric field strength at shear rate γ = 200 s^{-1} for pure TiO_2 and M-doped TiO_2 materials, respectively. From Figure 1, the shear stresses of the seven samples are similar in the absence of an electric field. Whereas in an electric field, the shear stresses of samples 2-6 are higher than that of TiO_2 (Sample 1); sample 7 shows slightly lower shear stresses than that of sample 1; and sample 3 has the highest shear stress among all the samples. From Table 1, we can see that the τ_r values of the M-doped samples are larger than that of pure TiO_2 material except for sample 7, and that of sample 3 is the largest among samples 1-7. This indicates that both Na-doping and Zr-doping can enhance the ER performance of TiO_2 materials when the doping degree is suitable. The ER activity of sample 3 (Na/Ti = 0.1, molar ratio) is the highest among these samples. The ER effect of sample 2 (Na/Ti = 0.05, molar ratio) is lower than that of sample 4 (Na/Ti = 0.2, molar ratio), namely, an optimal degree of Na_2O doping exists, which can be close to a Na/Ti ratio of 0.1 (molar ratio). For the materials doped with ZrO_2, the τ_r value of sample 5 (Zr/Ti = 0.05, molar ratio) is larger than that of sample 7 (Zr/Ti = 0.2, molar ratio), and sample 6 (Zr/Ti = 0.1, molar ratio) has the largest τ_r value; therefore, the optimal degree of ZrO_2 doping should also be close to a Zr/Ti ratio of 0.1 (molar ratio). Comparing samples 3 and 6, both have the same doping degree (M/Ti = 0.1, molar ratio); however, the τ_r value of sample 3 (84.0) is two times that of sample 6 (41.1). Therefore, an appropriate Na-doping can improve more effectively the ER performance of a TiO_2 material than Zr-doping. The high τ_r value of sample 3 is very advantageous in its application as an ER fluid. The ER behaviors of the materials mentioned above are related to their composition, microstructure and dielectric property.

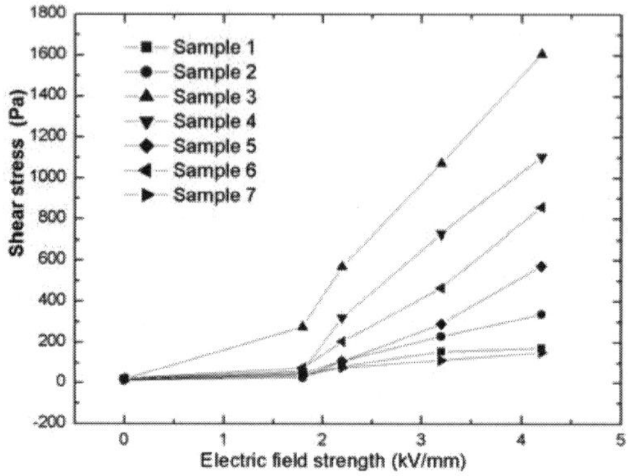

Figure 1 Shear stresses of suspensions of materials under different electric field strength

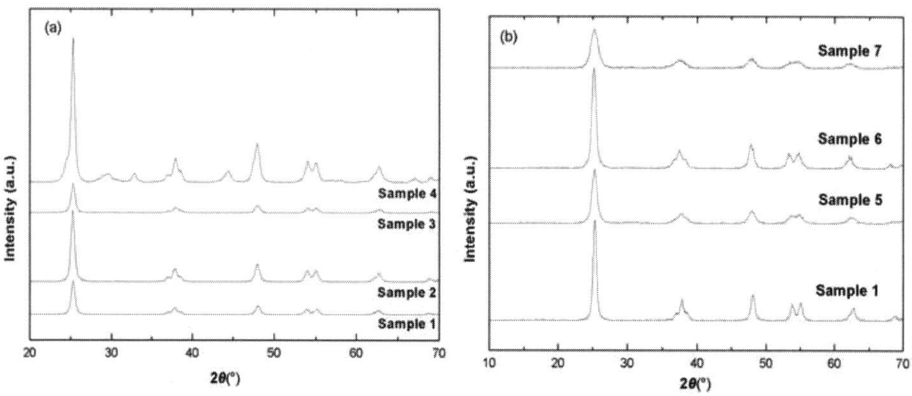

Figure 2 XRD patterns of Na-doped (a) and Zr-doped (b) materials

The XRD analyses of the materials were carried out on a Bruker D8 ADVANCE X-ray diffractometer with Cu-Kα radiation at a wavelength of 1.5406 nm in a range of 5–70°. Figure 2 illustrates the XRD patterns of samples 1-7. From the XRD patterns of the materials doped with Na_2O [see Figure 2 (a)], we can find that the crystal structures of sample 2 and 3 are the same as that of pure TiO_2 (sample 1), they all belong to the body-centered tetragonal crystal system and space group 141/amd (anatase titania) (JCPDS No. 21-1272); whereas, the XRD patterns of sample 4 show not only the characteristic peaks of anatase titania, but also the characteristic peaks of sodium titanate. These results show that Na^+ ions have entered into the TiO_2 crystal lattice with doping degrees of Na/Ti = 0.05 and 0.1 for samples 2 and 3, respectively. For sample 4, the presence of characteristic peaks of sodium titanate implies that part of the doped Na^+ ions can not enter into the TiO_2 crystal lattice, such that the sodium titanate powder from the reaction between TiO_2 and Na_2O would exist on the surface of the Na_2O-TiO_2 crystal particles as an impurity, which may cause the dielectric constant and dielectric loss to decrease, thus reducing the ER activity of the material. Considering the ER property of samples 2-4, an elevation of the ER activity from sample 2 to sample 3 can rise from the larger Na^+ ions (radius 0.102 nm) substituting for Ti^{4+} ions (radius 0.0605 nm), causing the distortion degree of the TiO_2 crystal lattice to increase with increasing doping degree. However, the superabundant doping, which causes the sodium titanate powder to emerge on the surface of the Na_2O-TiO_2 particles, results in sample 4 having a lower ER effect than that of sample 3.

The XRD patterns from Figure 2 (b) illustrate that the pure and the three materials doped with ZrO_2 exhibit only the characteristic peaks of the anatase phase, with no independent phase from the doped ZrO_2 or any other phase observable, indicating that the ZrO_2 has entered into the TiO_2 crystal lattice. This shows that Zr-doping is different from Na-doping, in that the doped ZrO_2 can still exist in the crystal lattice of TiO_2 when the doping degree is larger, as in the case of sample 7 (Zr/Ti = 0.2, molar ratio). Here, it should be noted that the ER activity of sample 7 is much lower than that of sample 6, even though a superfluous ZrO_2 phase, which can influence the interface polarization of the ER suspension, has not been detected. This unusual result can be related to the dielectric property of these materials (see Section 3.3).

Hao et al [3] have suggested that a large interfacial polarization would facilitate the particle to attain a large amount of charges on the surface, and lead to the turn of the particle along the direction of an electric field to form a fibrillation structure. They have also indicated that if the particle surface charge density is d, its surface area is S, and the particle is uniform, then one single particle bounds charge $q_s = Sd = k_1 E$, where E is the electric field strength, k_1 is a constant, $k_1 = S\varepsilon_s/4\pi$, and ε_s is the static dielectric constant. Therefore, the magnitude of the q_s value should be related to the ER effect. According to the equation, q_s would increase with the surface area S when both ε_s and E are kept constant. The surface area containing the surface area of pores and pore volume of the particle materials, measured on an ASAP 2010 Accelerated Surface Area & Porosimetry made by the US Micromeritics Company, are listed in table 1. Comparing sample 2-7 with sample 1, their surface area and pore volume are larger than those of sample 1. Based on the observed ER performance mentioned above, we can speculate that larger surface area and pore volume are advantageous in enhancing the ER activity of a material. However, the ER activity of sample 7 is similar to that of sample 1, despite the fact that its surface area and pore volume are larger, and its dielectric constant and dielectric loss are also higher than that of sample 1 (see Section 3.3 and Figure 3 and 4). Such a result should be ascribed to the grain size (6.4nm) of sample 7 being too small.

The average grain sizes of the particle materials are estimated by applying the XRD line broadening analysis using the Scherer formula [7], revealing values ranging from 6 to 18 nm (see Table 1); therefore, these materials are nano-sized particle materials, and doping ZrO_2 results in an evident decrease in the grain sizes of the TiO_2 materials. At the same time, the Zr-doped materials have much larger surface area than the Na-doped materials. The fact that the ER effect of the Zr-doped materials is lower, as mentioned above, shows that when the grain size of a material is too small its ER activity will drop even if it has a large surface area. It is well known that the Brownian motion tends to compete with particle fibrillation [2], and as a result, large grain sizes should be more beneficial in enhancing the ER performance of nano-sized particle materials. Among the Zr-doped materials, sample 6 has the largest grain size, which can be a significant contributing factor to it having the highest ER activity; on the other hand, the grain size of sample 7 being the smallest may be the reason why it shows the lowest ER activity.

To compare the dielectric property of these materials, the dried powder (0.3g) was dry-pressed at 6 MPa for 5 minutes to pellets of 8mm×2mm in diameter and thickness. The capacitance (C) and dielectric loss tangent (tanδ) of the pellets were obtained on a HP4274A Multi-frequency LCR Meter at room temperature. The dielectric constant (ε) of the materials were derived from the measured C according to the conventional relation, $\varepsilon = C(4\pi d)/S$, where d and S are the thickness and area of the pellet, respectively.

Figures 3 and 4 illustrate the frequency (f) dependence of ε and tanδ in the range of 200 Hz-40 kHz for samples 1-7. These experimental results have shown that sample 1 (pure TiO_2), the ER effect of which is the lowest, has the lowest ε value and tanδ value among all samples. The correlation between the dielectric property and the ER effect is consistent with the empirical criterion obtained by T. Hao et al [8].

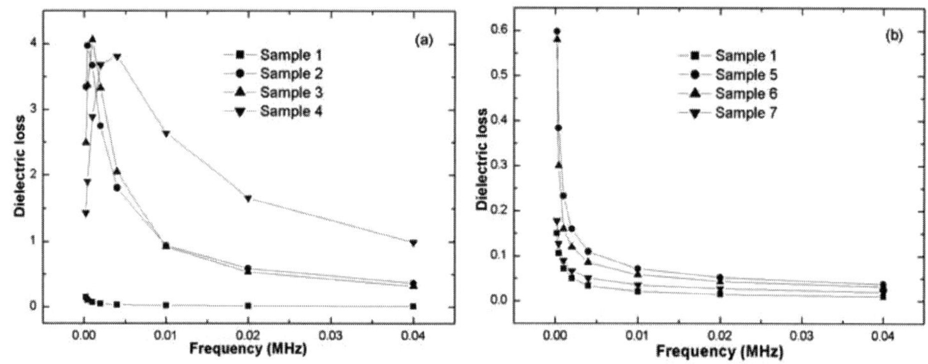

Figure 3 Frequency dependence of dielectric loss tangent for Na-doped (a) and Zr-doped (b) materials

From the tanδ values of the Na-doped materials under different frequencies [see Figures 3(a)], we can see that the dielectric loss peaks appear at 400 Hz, 1 kHz and 4 kHz for samples 2, 3 and 4, respectively, the tanδ value of sample 3 is the largest at 1 kHz, sample 2 has the largest tanδ values at f < 1 kHz, and at f > 1 kHz, the tanδ values of sample 4 are the largest. Based on the observed ER property of these materials, a good ER material should have a high tanδ value at 1 kHz, and the weaker ER effect of sample 2 may be owing to its lower tanδ values in the range of 1-4 kHz. Among the three materials doped with Na_2O, the change in tanδ with frequency is notable for sample 4. Its tanδ values are the lowest in the range of 200 Hz-1 kHz, and highest thereafter. Although the sodium titanate powder exists as an impurity on the surface of the Na_2O-TiO_2 particles, this high tanδ may come from the higher polarity of the sodium salt. The special dielectric behavior of sample 4 may be responsible for its ER activity, which is lower than that of sample 3 and higher than that of sample 2.

The frequency dependency of the tanδ values for the TiO_2 materials doped with ZrO_2 [see Figure 3(b)] is different from that of the materials doped with Na_2O. In Figure 3(b), no dielectric loss peak is present. Sample 7, the ER effect of which is the lowest, reveals the lowest tanδ values in the range of 200 Hz-40 kHz among all doped materials. The tanδ values of sample 6 are slightly lower than those of sample 5, while the τ_r is slightly larger for sample 6 than for sample 5, that is, sample 6 has better ER performance than sample 5. This unusual result indicates that for two materials with different compositions, the relative magnitudes in their ER activity cannot be confirmed according to their dielectric properties, especially when their dielectric properties are similar, where the impact of the grain size is dominant in influencing the ER property of these nano-sized particle materials (see section 3.2).

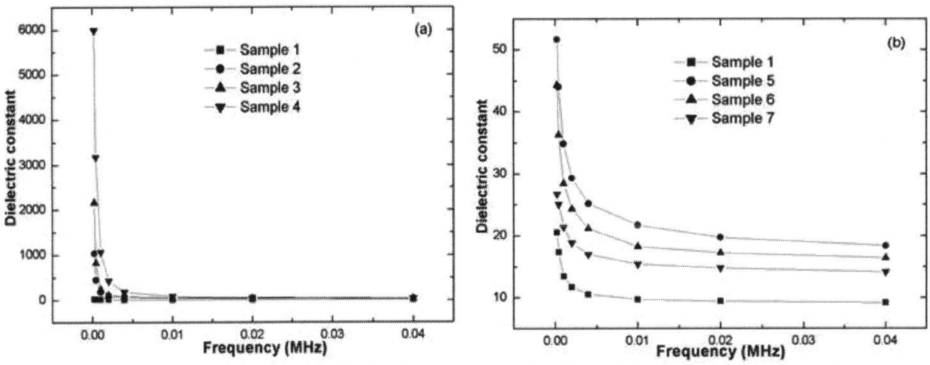

Figure 4 Frequency dependence of dielectric constant for Na-doped (a) and Zr-doped (b) materials

When comparing the ε (see Figure 4) and tanδ values (see Figure 3) of the Na- and Zr-doped materials, we can see that the ε and tanδ values are larger for the TiO_2 doped with Na_2O than the TiO_2 doped with ZrO_2 when the doping degree and frequency for both are the same. This can be an important reason why the Na-doped materials have better ER performance than the Zr-doped materials, except for sample 2. The ER activity of sample 2 is lower, even though its ε and tanδ values are larger, than that of sample 5. The reason may lay in the fact that the surface area ($23.65m^2/g$) of the particles of sample 2 is much smaller than that ($71.57m^2/g$) of sample 5, which may influence their interface polarization in a suspension. In addition, there is another obvious difference between the Na- and Zr-doped materials in the variation of their ε values, that is, the ε values increase for the Na-doped and decrease for the Zr-doped materials with increasing doping degree. These experimental phenomena can be explained by the surface composition of these particle materials.

In order to understand the relationship between the dielectric property and surface composition of the doped TiO_2 materials, the XPS analyses of samples 2-7 were performed on an AXIS-Ultra instrument from Kratos Analytical using monochromatc Al Kα radiation (225 W, 15 mA, 15 kV) and low–energy electron flooding for charge compensation. The values of M/Ti (molar ratio, M=Na, Zr) and O_w/Ti (molar ratio, O_w is oxygen atom in H_2O molecule) in the surface layer of the particles are listed in Table 1. The experimental results show that there is trace water on the surface layer of the particles for the Na-doped, and none for the Zr-doped materials, even though the conditions for the preparation and testing of these two types of materials are entirely the same. Such result can be attributed to the ability of Na_2O to strongly absorb water. The trace water could be an important factor in increasing the ε and tanδ values of the Na-doped materials. The ε value has been shown to increase with increasing water content. The values of M/Ti in the surface layers of the particles (see Table 1) are much larger for the Na-doped than the Zr-doped materials. This can be explained by the fact that the radius (0.102 nm) of the Na^+ ion is much larger than (0.0605 nm) that of the Ti^{4+} ion, whereas the radius (0.072 nm) of the Zr^{4+} ion is very close to that of the Ti^{4+} ion, namely, the match between Zr^{4+} and Ti^{4+} is better in terms of charge and radius, and ZrO_2 can thus enter the TiO_2 crystal lattice more easily than Na_2O. In addition, as commonly known, the polarity of the ZrO_2 molecule is much weaker than that of the Na_2O molecule. Comparing to the Zr-doped materials, the Na-doped materials have characteristics such as larger ε and tanδ and consequently better interface polarization and ER effect, and the ε value increases with increasing Na/Ti ratio; these characteristics of the Na-doped materials can be attributed to their larger Na/Ti values in the surface layers of the particles and the stronger polarity of the Na_2O molecule. Therefore, the dopant property including the charge and radius of the ion and the polarity of the molecule, namely the composition of the surface layer of a particle material, is also an important factor in influencing the dielectric property and ER performance of the material. A proper match between the dopant property and the doping degree should not be neglected.

4. Conclusion

Both Na-doping and Zr-doping can effectively enhance the ER effect of a TiO_2 material when the doping degree is suitable. The microstructure and composition of the surface layer of particles are important factors in influencing the dielectric property and ER effect of a material. The impact of the grain size on the ER property can be dominant for the nano-sized particle materials.

5. Acknowledgements

This Project is supported by the National Natural Science Foundation of China (20023005, 29831010) and the National Key Project for Fundamental Research (G1998061305).

References

6. Halsey T. C., Electrorheological fluids, *Science, New Series*, **258** (1992), pp. 761-766.
7. Hao T., Electrorheological Fluids, *Advanced Materials*, **13** (2001) pp. 1847-1857.
8. Hao T., Kawai A.and Ikazaki F., Mechanism of the Electrorheological Effect: Evidence from the Conductive, Dielectric, and Surface Characteristics of Water-Free Electrorheological Fluids, *Langmuir*, **14** (1998) pp 1256-1262.
9. Hao T., Electrorheological suspensions, *Advances in Colloid and Interface Science* **97** (2002) pp 1-35.
10. Zhao X.P. and Yin J.B., Preparation and Electrorheological Characteristics of Rare-Earth-Doped TiO_2 Suspensions, *Chemistry of Materials* **14** (2002) pp 2258-2263.
11. Yin J. B.and Zhao X.P., Preparation and Enhanced Electrorheological Activity of TiO_2 Doped with Chromium Ion, *Chem. Mater*. **16** (2004) pp 321-328.
12. Guiner A., Theorie et Technique de la Radiocrystallographie, Third Ed., Dunod, Paris, France, (1964) pp 482.
13. Hao T. Xu Z. and Xu Y., Correlation of the dielectric properties of dispersed particles with the electrorheological effect, *J. Colloid Interface Sci.* **190** (1997) pp 334-340.

YIELD STRESS PREDICTIONS FOR MAGNETORHEOLOGICAL FLUIDS FROM A KINETIC THEORY BASED PARTICLE PAIR MODEL

C. CIOCANEL, G. LIPSCOMB AND N.G. NAGANATHAN

University of Toledo, MIME Dept., 2801 W. Bancroft St., Toledo, OH 43606 USA
E-mail: cciocane@eng.utoledo.edu

The paper introduces a kinetic theory based particle pair model for magnetorheological fluid characterization. Fluid constitutive equations are derived from an analysis of the motion of a pair of particles subject to hydrodynamic and magnetic field forces. The effect of neighboring particles from the same chain on the magnetic force is included. The model can predict fluid behavior when particles separate or remain in contact during flow. For yield stress predictions, the case when particles separate is considered. Predictions for the yield stress are compared with experimental data and with other theoretical models. Predictions from the proposed model are in good agreement with experimental data.

1 Introduction

While magnetorheological (MR) fluids are used in commercially available products like shock absorbers [1], engine/powertrain mounts [2], breaks [3], the physics of fluid behavior are poorly understood. The microstructural mechanisms governing the fluid properties, i.e. yield stress, viscosity, normal stress, etc., are dependent on the magnetization force acting between particles in a chain, the interaction force between particles in adjacent chains, and the hydrodynamic forces applied by the fluid on the suspended particles/clusters of particles. To improve our understanding, these mechanisms are continuously investigated both on Earth and in space [4].

Publications from the last decade contain a significant number of experimental and theoretical studies on MR fluids. Some of these studies, relevant for the topic of this paper, are summarized in the following paragraphs. Lemaire and Bossis [5] proposed a methodology to calculate the static yield stress for a MR fluid based on the force required to break the number of chains per unit area between two parallel disks. They have also investigated the effect of the boundaries on the yield stress magnitude, concluding that boundaries made of low magnetic permeability materials are detrimental for the fluid. The model did not account for the magnetization induced by the neighboring particles.

Volkova *et al.*[6] used a chain model to analyze the experimental shear stress *vs.* shear rate curves of MR fluids. They considered a chain of particles to be stable when the hydrodynamic force at the center of the chain was lower than the attractive force between two adjacent particles. The model is not applicable at high shear rates, when aggregates disintegrate. Bechtel *et al.* [7] developed a set of constitutive equations based on the dumbbell model assuming hydrodynamic, Brownian and magnetic forces to act on the beads. The model did not consider short or long range particle hydrodynamic interactions, and the expressions for the Brownian and magnetic forces were selected to match the kinetic model predictions to experimental shear data.

Melle and Martin [8] acknowledged the limitations of the polarization model and added the effect of the dipoles from a long chain, on the dipole interaction between two contacting particles from the same chain.

The microstructural model described in this paper allows predictions of the fluid macroscopic properties from the microstructural mechanics, i.e. the hydrodynamic and the magnetic forces acting on the particles. The model is capable to simulate a broad range of particle concentrations and flow through complex geometries. A constitutive equation for MR fluids is derived, based on an analysis of the motion of pairs of magnetic particles in the suspending fluid in the presence of a magnetic field. The analysis is similar to that of bead-spring models of polymeric liquids with replacement of the elastic connector force by a magnetic force. Yield stress predictions for simple shear flow are compared with data acquired experimentally and reported in literature.

2 Microstructural model development

The proposed model is developed considering a pair of magnetizable particles of radius *a*, connected by a center-to-center vector $\overline{Q}$ that describes the overall orientation and the internal configuration of the two particles. The instantaneous location in space of the center of mass of the particle pair, relative to the origin of a coordinate system, is given by a vector $\overline{R}$, while their position relative to the origin is given by position vectors $\overline{R}_1$ and $\overline{R}_2$. The particle pair is exposed to a magnetic field $\overline{H}$ and a velocity field v.

Application of the kinetic theory for model development requires the establishment of an evolution equation for the particle pair orientation and the development of an expression for the stress tensor. The evolution equation is derived by combining the equations of motion of the particles with an equation of continuity describing conservation of system points in configuration space [11]. The equation of motion for each particle immersed in solvent is derived considering that the particle diameter is small enough that the inertia of both particles is negligible. Each particle was assumed to experience a hydrodynamic drag force (Stokes drag) as it moves through the solution, and an external magnetic force. A thermal force was not included in the equation of motion. According to Calderon and Melle [12] and Carletto and Bossis [13], when particle size is larger than few micrometers, the magnetic force dominates the thermal forces, and therefore the latter may be neglected.

2.1 Forces expressions

The hydrodynamic force was calculated as:

$$F^h = \frac{1}{2}\zeta\left(\dot{\overline{Q}} - \omega\cdot\overline{Q} - \lambda\left(D - D : \frac{\overline{Q}\overline{Q}I}{4a^2}\right)\cdot\overline{Q}\right), \tag{1}$$

where $2\omega = (\nabla v)^T - \nabla v$ and $2D = (\nabla v)^T + \nabla v$ are the vorticity tensor and the rate of deformation tensor, respectively. The motion is a function of particle aspect ratio through the material constant λ as discussed in the literature. The friction coefficient ζ was approximated with Stokes drag as $\zeta = 6\pi\eta_s a$, where η_s represents the solvent viscosity.

The magnetic force was calculated as the particles are inside a long chain of particles. Consequently, the force expression, derived from the point dipole approximation, was modified [9] to account for the effect of the neighboring particles from the chain. The derived equation is:

$$\overline{F}^m = \frac{4\mu_0\pi a^6\chi_{eff}^{\;2}}{3}\left[\kappa_2\frac{H^2}{Q^5}\overline{Q} - (3\kappa_1 + 2\kappa_3)\frac{(\overline{H}\cdot\overline{Q})^2}{Q^7}\overline{Q} + 2\kappa_3\frac{(\overline{H}\cdot\overline{Q})}{Q^5}\overline{H}\right] \tag{2}$$

In Equation 2, $\mu_0 = 4\pi \times 10^{-7}$ Tm/A is the permeability of vacuum and χ_{eff} is the effective magnetic susceptibility of the particles. According to Furst and Gast [14] the effective magnetic susceptibility can be calculated as $\chi_{eff} = 3\beta$, where $\beta = (\mu_p - \mu_s)/(\mu_p + 2\mu_s)$ is the permeability contrast factor, and μ_p and μ_s are the permeability of the particles and of the solvent, respectively. The $\kappa_{i\ ,i=1\div3}$ coefficients are functions of $\xi = \beta\zeta(3)$, where $\zeta(3) \approx 1.202$ is the Riemann zeta function. These coefficients are calculated as $\kappa_1 = (1+\xi^2/8)/(1-\xi/4-\xi^2/8)^2$, $\kappa_2 = 1/(1+\xi/4)^2$, and $\kappa_3 = 1/(1-\xi/4-\xi^2/8)$.

To account for both far from contact and near contact conditions, the magnetic force has been modified using the Heaviside step function as follows

$$\overline{F}^m = \frac{4\mu_0 \pi a^6 \chi_{eff}^{\ 2}}{3}\left[\hbar\kappa_2 \frac{H^2}{Q^5}\overline{Q} - \hbar(3\kappa_1 + 2\kappa_3)\frac{(\overline{H}\cdot\overline{Q})^2}{Q^7}\overline{Q} + \right. \\ \left. + 2\kappa_3 \frac{(\overline{H}\cdot\overline{Q})}{Q^5}\overline{H} - 2(1-\hbar)\kappa_3 \frac{(\overline{H}\cdot\overline{Q})^2}{Q^7}\overline{Q}\right] \tag{3}$$

where the Heaviside function $\hbar$ is defined as

$$\hbar\left(tr\langle\overline{Q}\overline{Q}\rangle - 4a^2\right) = \begin{cases} =1 \ for \ tr\langle\overline{Q}\overline{Q}\rangle - 4a^2 > 0 \\ =0 \ for \ tr\langle\overline{Q}\overline{Q}\rangle - 4a^2 \le 0 \end{cases}, \tag{4}$$

with $tr\langle QQ\rangle = Q_1Q_1 + Q_2Q_2 + Q_3Q_3$. As introduced in Equations 3 and 4, the Heaviside function prevents particle overlap. The attractive magnetic force increases considerably when the particles get closer due to the divergence of the magnetostatic field between particle surfaces. The overlap is prevented by removing that component of the magnetic force acting in the $\hat{Q}$ direction when the particles are in contact; this being equivalent to introducing an opposing contact force.

2.2 The stress tensor

In order to predict fluid yield stress, the expressions of the forces were considered for the case when the particles forming a pair can separate. This requires that the coefficient λ from Equation 1 to be set to zero, and the Heaviside function to be set equal to one ($\hbar = 1$) in Equation 3.

With the new expressions of the particle forces, the kinetic theory is involved to determine the time evolution equation for the moments of the particle distribution function and the stress tensor. The step by step procedure involved in calculating the stress tensor components is provided in [15].

The second order moment of the distribution function $\langle\overline{Q}\overline{Q}\rangle$, required to calculate the particle hydrodynamic and magnetic contributions to the stress, determined from the Fokker Planck equation, is evaluated by numerically integrating the following equation

$$\frac{\partial}{\partial t}\langle\overline{Q}\overline{Q}\rangle = (\omega + \lambda\cdot D)\langle\overline{Q}\overline{Q}\rangle - \langle\overline{Q}\overline{Q}\rangle(\omega - \lambda\cdot D) -$$

$$-\frac{\lambda D}{2a^2}:\langle\overline{QQ}\rangle\langle\overline{QQ}\rangle+\frac{4}{\zeta}\frac{\Delta H^2}{Q_0^5}\hbar k_2\langle\overline{QQ}\rangle+ \tag{5}$$

$$+\frac{4}{\zeta}\frac{\Delta}{Q_0^7}(3k_1+2k_3)(\hbar-2)\overline{HH}:\langle\overline{QQ}\rangle\langle\overline{QQ}\rangle+\frac{4}{\zeta}\frac{\Delta}{Q_0^5}k_3\left(\langle\overline{QQ}\rangle\cdot\overline{HH}+\overline{HH}\cdot\langle\overline{QQ}\rangle\right)$$

where $\Delta = 4\mu_0\pi a^6\chi_{eff}^{\ 2}/3$ and $Q_0 = 2a$.

The particle hydrodynamic ($\tau^{p,h}$) and magnetic ($\tau^{p,m}$) contributions to the stress are then calculated as:

$$\tau^{p,h}=\phi\cdot\left[\mu_1\frac{\langle\overline{QQ}\rangle}{4a^2}+\mu_2 D:\frac{\langle\overline{QQ}\rangle\langle\overline{QQ}\rangle}{16a^4}+2\mu_3\left(D\cdot\frac{\langle\overline{QQ}\rangle}{4a^2}+\frac{\langle\overline{QQ}\rangle}{4a^2}\cdot D\right)\right] \tag{6}$$

$$\tau^{p,m}=\phi\cdot\left[\frac{\alpha\hbar k_2 H^2}{Q_0^5}\langle\overline{QQ}\rangle+\frac{\alpha(\hbar-2)(3k_1+2k_3)}{Q_0^7}\overline{HH}:\langle\overline{QQ}\rangle\langle\overline{QQ}\rangle+\frac{2\alpha k_3}{Q_0^5}\overline{HH}\cdot\langle\overline{QQ}\rangle\right] \tag{7}$$

where μ_1, μ_2, and μ_3 are material constants [16], ϕ is the particle volume fraction, and $\alpha = \mu_0\,\chi_{eff}^{\ 2}a^3/2$.

Algebraic summation of the solvent and particle contributions to the stress yields the total stress tensor

$$\tau = \tau^s + \tau^{p,h} + \tau^{p,m} \tag{8}$$

where $\tau^s = 2\eta_s D$ is the solvent contribution to the stress and η_s is the solvent viscosity.

Model parameters are either available in the literature or have been evaluated experimentally.

3 Model predictions

3.1 Particles orientation

The balance between particle forces determines particle pair behavior under steady shearing conditions. An oscillatory behavior is observed at low magnetic fields. In this case, the particle pairs tend to align with the magnetic field, but the hydrodynamic force arising from fluid shearing is stronger than the magnetic force and particle pairs rotate with the fluid (see Figure 1.a). The frequency of oscillation increases with the increase in shear rate while the

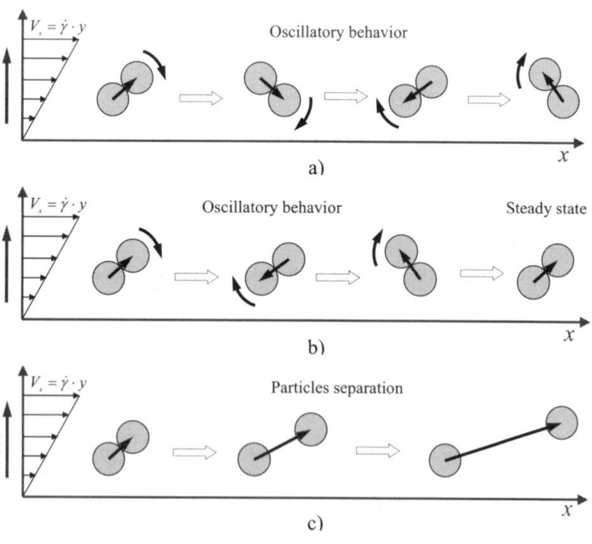

Figure 1. a) Oscillatory behavior of the particle pairs. b) Transition to steady state. c) Particle separation under flow.

applied magnetic field remains the same. At moderate magnetic fields, the oscillatory behavior ceases and shortly the fluid reaches a steady state. If both the shear rate and the field are large the steady state establishes soon after the flow begins. The particle-pair equilibrium position is reached somewhere between the direction of the flow and the direction of the field (see Figure 1.b). Figure 1.c shows the case when particles separate under shear flow. This case was considered in developing the procedure for yield stress predictions using the proposed model. This is discussed in detail in the following section.

3.2 Yield Stress Calculation

In this study, it is assumed that the yield stress is determined by a localized phenomenon, i.e. the breakage of a chain spanning the gap between fluid boundaries, as shown in Figure 2. In addition, it is assumed that the breakage occurs somewhere within the gap spanning the fluid boundaries, and the upper part of the chain will keep moving with the velocity of the upper boundary, while the other part will stick to the lower boundary (the fixed boundary). This breakage mechanism implies that a local shear rate, $\dot{\gamma}_{local}$, is responsible for the particle separation, and accordingly this must be identified through simulation.

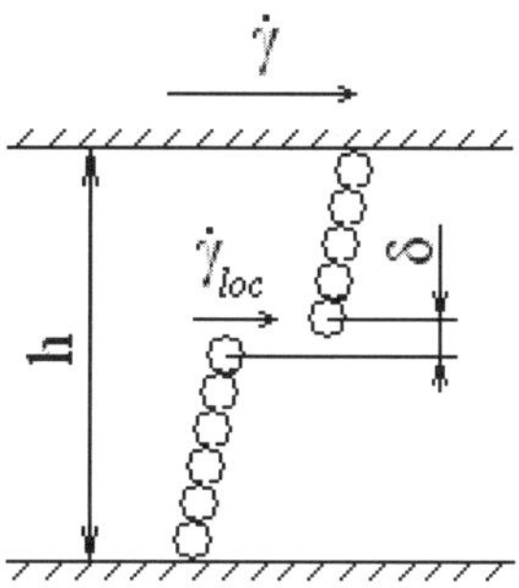

Figure 2. Idealized chain breakage mechanism.

To calculate $\dot{\gamma}_{local}$, it is assumed that the particles in a pair are initially in contact. The simulation is started under a constant magnetic field and a given shear rate. Once the flow is onset, the particles are allowed to separate, by switching the Heaviside function from $\hbar=0$ to $\hbar=1$. Under this condition, the evolution of the component of the connecting vector in the flow direction is monitored to identify the value of the shear rate that yields a value greater than one for it. This shear rate is considered to be the local shear rate that causes chain breakage and therefore it is further used to calculate the fluid yield stress. However, due to the localized nature of this analysis, the value of the local shear rate causing the breakage cannot be used immediately for calculations of the fluid yield stress that would be readily comparable with experimental data. That is because the proposed approach does not account for the actual gap between the fluid's boundaries, denoted in Figure 2 by *h*. The local shear rate needs to be translated into an apparent shear rate, $\dot{\gamma}_{app}$, accounting for the actual geometry used in the experimental setup.

To calculate the relationship between the local shear rate, the apparent shear rate and the parallel plate geometry, the apparent shear rate is first calculated as (see Figure 2)

$$\dot{\gamma}_{app} = \frac{V_{plate} - 0}{h} \tag{9}$$

and then, the local shear rate is determined as

$$\dot{\gamma}_{loc} = \frac{V_{plate} - 0}{a} = \dot{\gamma}_{app} \cdot \left(\frac{h}{a}\right) \tag{10}$$

The apparent shear rate, $\dot{\gamma}_{app}$, is plugged back into the simulation, along with the corresponding value of the magnetic field that yielded $\dot{\gamma}_{local}$, and calculate the magnetic and hydrodynamic contribution to the macroscopic yield stress. These two components of the yield stress, as given by the simulation, must be added to the Newtonian component of the stress calculated as

$$\tau^{s} = 6\pi\eta_{s}\phi\dot{\gamma}_{app}\frac{h}{a} \tag{11}$$

in order to get a value for the yield stress that is readily comparable with the experimental yield stress.

Yield stress calculated using the procedure described above have been compared with experimentally measured yield stress_ for the MRF-132LD fluid at shear rates of 5, 50 and 500 1/s and magnetic fields of 20, 40, 60, 80 and 160kA/m. In addition, model predictions have been compared with fluid data provided by Lord Corporation [16]. A good agreement between these sets of data is indicated by the cumulative plot from Figure 3.

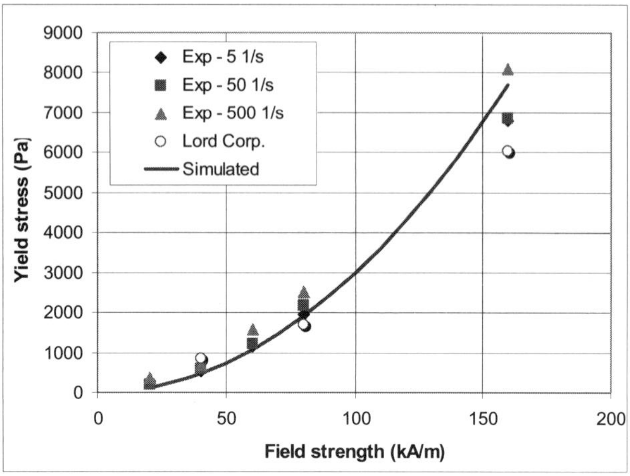

Figure 3. Comparison between simulated and experimental values of the yield stress for the MRF-132LD manufactured by Lord Corp.

4 Conclusions

A set of constitutive equations for magnetorheological fluids has been developed. The equations can monitor the evolution of the configuration of particle pairs in the presence of a complex flow and a magnetic field. The equations account for the particle volume fraction, hydrodynamic drag and magnetic forces, including the effect of the neighboring dipoles on the attractive magnetic force between the particles in a pair.

A particular case that allows calculation of the fluid yield stress was discussed. Model predictions of the yield stress, for low to moderate magnetic fields, are in good agreement with the experimental measurements and data reported in literature. The difference between model predictions and experimental data at higher magnetic fields are attributed to: a) changes in the assumed chain breakage mechanism that may occur at higher fields; b) neglect of the effects of the particles from the neighboring chains on the particle pair.

References

1. Farhan G., Wang K.W. and Xia L., Magnetorheological fluid damper feedback linearization control for helicopter rotor application. *Smart Mater. Struct.* **10** (2001) pp. 96–103
2. Ahn Y. K., Yang B.-S., Ahmadian M. and Morishita S., A small sized variable-damping mount using magneto-rheological fluid. *J. Int. Mat. Sys. and Struct.* **16** (2005) pp. 127-133.
3. Ye S.and Williams K.A., MR fluid brake for control of torsional vibration. *Proc. 2005 SAE World Congress* Paper no. 2005-01-1503 (2005) pp. 1-7.

4. http://www.spaceref.com/news/viewsr.html?pid=20951, *NASA Space Station On-Orbit Status*. 9 June 2006.
5. Lemaire E. and Bossis G., Yield stress and wall effects in magnetic colloidal suspensions. *J. Phys. D: Appl. Phys.* **24** (1991) pp. 1473-1477.
6. Volkova O., Bossis G., Carletto P. and Cebers A., Shear Banded Structures and Nematic to Isotropic Transition in MR Fluids. *Proc. 7th Int. Conf. on ER/MR Susp.* (Ed. R. Tao, World Scientific Singapore) (2000) pp. 358-365.
7. Bechtel S., Washington G., Ahmadkhanlou F. and Wang Y., Microstructural analysis and control of magneto-rheological fluid. *Proc. IMECE2004* Paper no. 61693 (2004) pp. 1-6.
8. Melle S. and Martin J.E., Chain model of a magnetorheological suspension in a rotating field. J. Chem. Phys. 118 (21) (2003) pp. 9875-9881.
9. Choi Y.T., Cho J.U., Choi S.B. and Wereley N.M., Constitutive models of electrorheological and magnetorheological fluids using viscometers. *Smart Mater. Struct.* **14** (2005) pp. 1025–1036.
10. Bird R.B., Curtiss F.C., Armstrong R.C. and Hassager O., Dynamics of polymeric liquids. **1&2**, 2nd Ed. (John Wiley & Sons, New York) (1987).
11. Calderón O. G. and Melle S., Dynamics of simple magnetorheological suspensions under rotating magnetic fields with modulated mason number. *J. Phys. D: Appl. Phys.* **35** (2002) pp. 2492-2498.
12. Carletto P. and Bossis G., Field-induced structures and rheology of a magnetorheological suspension confined between two walls. *J. Phys. Cond. Matter* **15** (2003) pp. S1437-S1449.
13. Furst E.M. and Gast A.P., Dynamics and lateral interactions of dipolar chains. *Phys. Rev. E* **62** (5) (2000) pp. 6916-6925.
14. Ciocanel C., Simulation of magnetorheological fluids under dynamic conditions. *Ph.D. Dissertation* The University of Toledo (2006).
15. Lipscomb II G.G., Denn M.M., Hur D.U. and Boger D.V., The flow of fiber suspensions in complex geometries. *J. of Non-Newtonian Fluid Mechanics* **26** (1988) pp. 297-325.
16. Lord Corporation (2002) – MRF-132LD Product Bulletin.

ELECTRORHEOLOGICAL BEHAVIOR OF NANOMETRIC POWDERS SUSPENSIONS

ELŻBIETA ŻERO, IRENEUSZ WIELGUS AND JANUSZ PŁOCHARSKI

Department of Chemistry, Warsaw University of Technology
Noakowskiego3, 00-664 Warszawa, Poland
E-mail: ela@ch.pw.edu.pl, janusz@ch.pw.edu.pl

The sol-gel method was used to prepare amorphous TiO_2 powders with sub-micrometric grain size. These powders together with commercial nanometric TiO_2 were characterized with FT-IR spectroscopy, X-ray diffraction, scanning electron microscopy, and a light scattering method. Then electrorheological suspensions were composed out of these powders and their flow curves under electric field were recorded. The ER activity of the synthesized powders was relatively high which probably resulted from residual amounts of polar organic matter present in the prepared titania. It was found, however, that the studied suspensions underwent a very well pronounced agglomeration process clearly observed in suspensions of higher concentrations and particularly in those fluids which were exposed to electric field for some time. Therefore, the studied ERFs which were originally prepared out of nanometric powders contain in fact aggregated particles of considerably bigger size. The agglomeration process is hardly reversible and strongly influences all basic properties of the studied electrorheological suspensions.

1 Introduction

Electrorheological fluids (ERFs) consisting of solid particles dispersed in an insulating liquid can reversibly increase their viscosity and transform into a solid upon application of electric field [1]. This phenomenon can be exploited in many technical applications [2-5], however, the insufficient usable parameters of the available ER fluids significantly limit the scope of these applications.

A new and good information for the development of the ER technology came with the series of publication on the so called "Giant Electrorheological Effect" abbreviated sometimes as GER [6-9]. The authors of these publications were able to prepare fluids whose yield stress value was more than 20 times higher than of any other ERF before. These exceptionally strong fluids consisted of highly concentrated silicone oil dispersions of oxygen compounds of titanium, barium, strontium and some other elements modified with various organic and inorganic additives. The other common and unusual feature of these new ERFs was that the grain size of the powders used to prepare the dispersions was relatively low – reaching nanometric size in a few cases [6-8] and significantly below 1 μm in all others. This stays in a considerable contrast to the earlier practice since most of ERFs comprised grains of at least a few micrometer size as well as with some ER effect theoretical models proving that the magnitude of the ER effect should decrease with decreasing size of dispersed particles [10]. In addition a few other questions related to reversibility and stability of the ER effect and zero field viscosity have been raised.

Taking the above into account we have decided to perform a series of experiments on model ER fluids comprising well characterized submicrometric grains. Titania based powders have been chosen because of the high ER activity of this kind of material. We have paid a particular attention to the particle agglomeration process since we believe it can be responsible for the different behavior of the GER-type fluids compared to the classic ones.

2 Experimental

2.1 Preparation of nanometric powders and ER suspensions

Nanometric powders of titanium dioxide were prepared using the sol-gel method [11]. Titanium tetrabutoxide $Ti(OBu)_4$, water and hydrochloric acid as hydrolysis inhibitor were used as the substrates. The alkoxide was slowly added into acidified water and magnetically stirred to complete the reaction. In order to obtain different granulations of the powders two different concentrations of water were used - 1:2 and 1:6 volume ratio of $Ti(OBu)_4$ to water. The obtained gels were filtrated and then dried in vacuum for 10 hours at 80°C. The Titanium Dioxide P25 from Degussa was used for comparison. According to the manufacturer's information it has a grain size of about 21 nm.

The ER fluids were prepared from the above powders by dispersing them in a silicone oil (polydimethylsiloxane) of 0.1 Pas viscosity using a mortar with a pestle. Concentrations of the studied dispersions were 10, 20, 30 and 40% by weight.

2.2 Sample characterization and rheological measurements

X-ray diffraction (XRD) patterns were recorded with the HZG-4 diffractometer (Seifert GmbH) using CuK_α radiation. FT-IR spectra were recorded on the Perkin Elmer System 2000 spectrometer at resolution of 1 cm^{-1}. The KBr pellet technique was used. Morphological studies of the powders were carried out on the Quanta 200 SEM (FEI company). The grain size distribution within the ER suspensions was measured using the Zetasizer Nano-2S apparatus (Malvern Instruments Ltd.).

Electrorheological properties of the suspensions were measured with the Bohlin Visco 88 viscometer equipped with a homemade measuring unit of two concentric cylinders geometry. The flow curves were recorded within a shear rate range of 6-530 s^{-1} and at DC field strength of 0-3 kV/mm. All measurements were taken at ambient temperature.

3 Results and Discussion

3.1 Structure characterization

The X-ray diffraction patterns of the synthesized powders as well as the P25 sample shown that the prepared materials were amorphous while the P25 sample was a crystalline mixture of anatase and rutile (about 80:20 confirmed also by the manufacturer's information).

The results of the FT-IR analysis were shown in Fig.1. The bands observed at 2960-2860 cm^{-1} and 1480-1350 cm^{-1} are characteristic of stretching and deformation vibrations of a C-H bond, respectively. The IR spectrum of butyl alcohol is shown for comparison. It is clearly seen that the synthesized titania powders contain some organic residues being most probably adsorbed butanole – the product of the hydrolysis reaction. The crystalline P25 sample contained no traces of organic matter. The weak band around 3400 cm^{-1} could be attributed either to residual water or hydroxyl groups present on the titania particles surface.

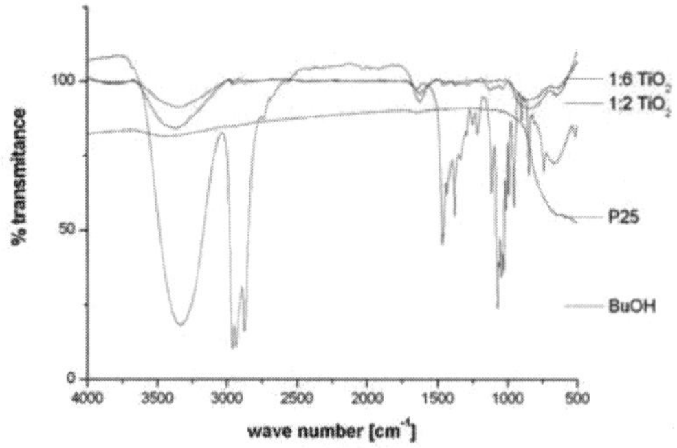

Figure 1. FT-IR spectra of P25, TiO_2 1:2, TiO_2 1:6 and BuOH

3.2 Morphological analysis

The morphology of the studied powders was shown in Fig.2. The prepared titania powders formed agglomerated grains of irregular shape. Their average size can be estimated as about 500 and 200 nm for TiO_2 1:2 and 1:6, respectively, however due to the large level of aggregation these estimations are not precise. The P25 material took the form of aggregates of 50-100 nm size.

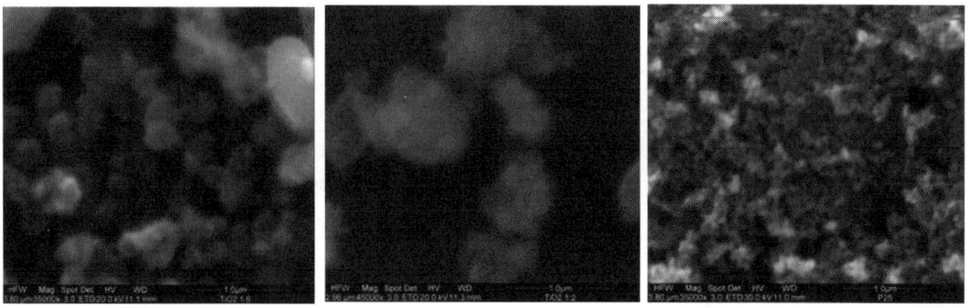

(a) (b) (c)

Figure 2. SEM images of (a) TiO_2 1:6, (b) TiO_2 1:2, (c) P25

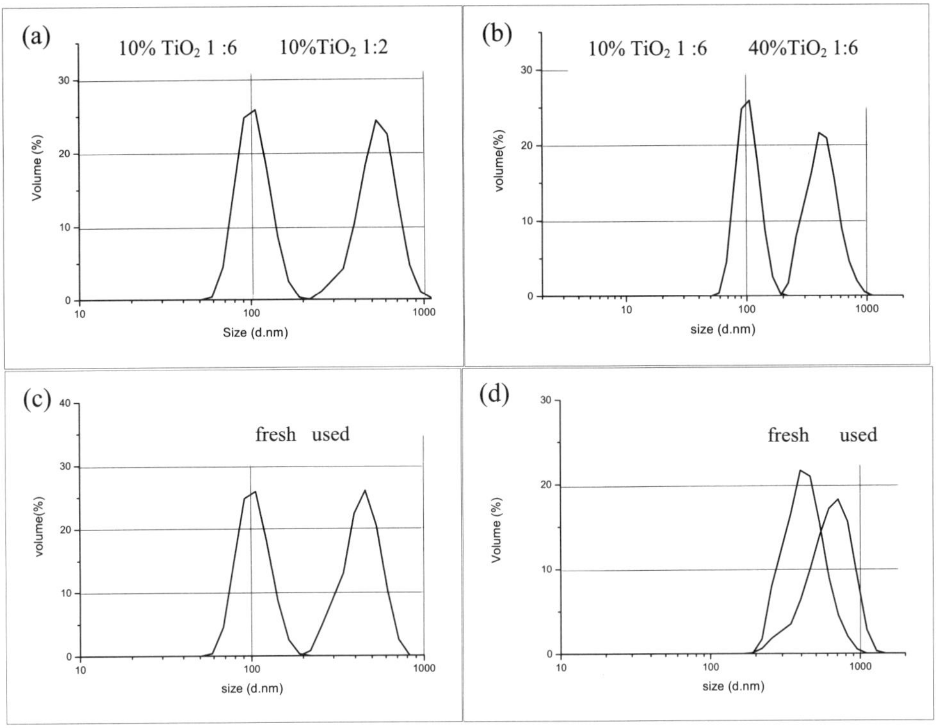

Figure 3. Size distribution of particles in ER suspensions: (a) 10% (m/m) TiO_2 1:2 and TiO_2 1:6; (b) 10 and 40% (m/m) of TiO_2 1:6; (c) 10% TiO_2 1:6 before and after field application; (d) 40% 1:6 TiO_2 before and after field application.

Fig.3. shows the results of direct measurements of particle size performed within the ER suspensions by the light scattering method. Fig.3a presents the particle size distribution in two 10% TiO_2 suspensions. The mean size of TiO_2 1:2 particles was 430 nm and 147 nm for TiO_2 1:6. These results are in a very reasonable agreement with the SEM image data and confirm that the particles are well dispersed and not agglomerated in the freshly prepared suspension. However, when the suspension concentration goes up a significant increase of the average particle size increases (as shown in Fig.3b) what can be explained as the rise of aggregates. We have also checked the particle size distribution after application of electric field. The ER suspensions were introduced into the rheometer and the full set of rheological measurements as described in p. *3.3* was performed. Then the suspensions were poured out from the rheometer and the particle size measurement was taken. Figs.3c and d show the results that clearly prove that significant and irreversible agglomeration takes place upon application of the electric field. In the case of lower concentration the four-times size increase was observed. In the case of higher concentration this increase was twofold only but this suspension was already agglomerated before the experiment. The irreversible agglomeration can be also observed when looking on the sedimentation rate of freshly prepared suspensions and comparing it with suspensions taken out from the rheometer after the electrified experiment. The settling rate of a fresh fluid is very low while used fluids settle surprisingly fast irrespectively of their nominally small particle size.

3.3 *Rheological properties*

Fig.4 presents a set of flow curves recorded for the 10% P25 suspension. The ER effect in this suspension is relatively low with the yield stress value reaching 140 Pa and the leaking current density about 14 μA/cm^2 at E=3 kV/mm. The zero field curve also exhibits some yield stress which results from high viscosity of this suspension and its paste-like nature. This was also the reason why P25 suspensions of higher concentrations were not studied.

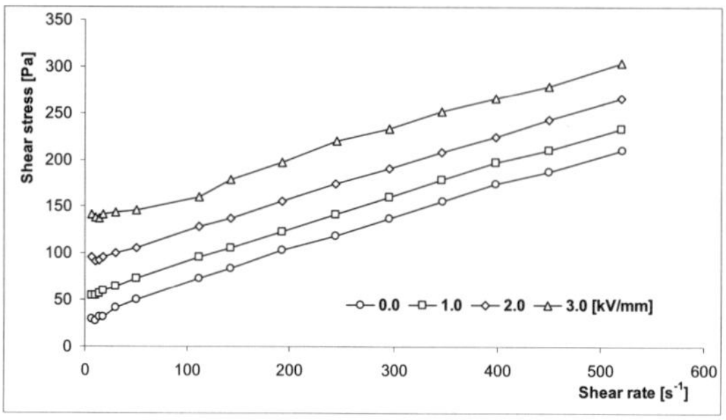

Figure 4. Flow curves of 10% (m/m) suspension of P25

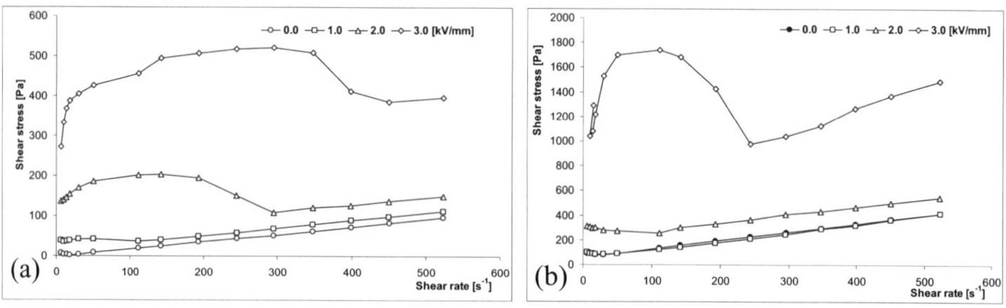

Figure 5. Flow curves of (a) 10% (m/m) and (b) suspension of TiO_2 1:6

Figs.5a and b show the flow curves of 10 and 40% suspensions of TiO_2 1:6. It could be seen that in this case at least two times higher values of τ_0 were obtained (270 and 1050 Pa at 3kV/mm for 10% and 40%, respectively) as well as the zero field viscosity was significantly lower. Therefore, preparation of the highly concentrated suspension was possible. The leaking current density was below 0.1 μA/cm^2 for 10% and about 0.35 μA/cm^2 for 40% at E=3kV/mm. However, the recorded flow curves, particularly when taken at higher voltages, were not as smooth as in the case of P25. In addition a series of repeated measurements showed that the particular traces of the flow curves were poorly reproducible. Similar character of curves at higher electric field was observed by Yin and Zhao [7] and Wen [8]. We believe that this phenomenon is related to the irreversible agglomeration process documented in our experiments.

Figs. 6a and b show the flow curves of 10% and 40% suspensions of TiO_2 1:2 powders, respectively. The yield stress for the 10% suspension was about 490 Pa with the leaking current about 2 μA/cm^2 at E= 3kV/mm. These values for the 40% suspension were about 1050 Pa and 2.8 μA/cm^2 at 3kV/mm. It means that the yield stress of the more coarse-grained TiO_2 is higher in the case of 10% suspensions and just the same in the case of the more concentrated fluid. This result is clear when we assume that 10% suspensions are less agglomerated and real grain size in the case of TiO_2 1:2 is bigger.

So, the bigger grain, the higher effect. In the case of the 40% suspensions, however, the real grain size is just the same for both titania powders and thus the yield stress is the same. The difference is in the leaking current density. It seems that the aggregates consisting of bigger number of small individual grains (the case of TiO_2 1:6) conduct worse.

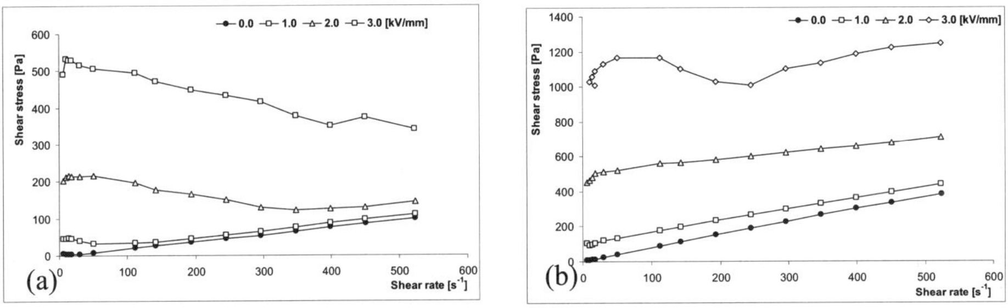

Figure 6. Flow curves of (a) 10% (m/m) and (b) 40% suspension of TiO_2 1:2

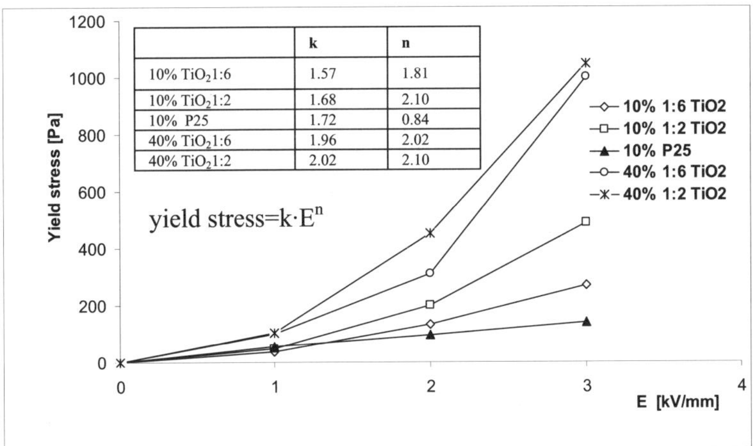

	k	n
10% $TiO_2$1:6	1.57	1.81
10% $TiO_2$1:2	1.68	2.10
10% P25	1.72	0.84
40% $TiO_2$1:6	1.96	2.02
40% $TiO_2$1:2	2.02	2.10

Figure 7. Yield stress dependence on electric field strength

A typical feature of many ERFs is the quadratic relation between yield stress and electric field strength. In order to check whether this applies also for our suspensions we have fitted our experimental data to the equation $\tau_0 = k \cdot E^n$, where τ_0 is the yield stress, E is the electric field strength, k and n are fitting parameters. The yield stress dependence on electric field strength for our ERFs is presented in Fig.7 that contains also the calculated parameters. It can be clearly seen that all relations are almost precisely quadratic with the exception of the 10% P25 suspension showing the linear dependence. The reasons of this difference are not clear at the moment.

Finally we have performed a test for time stability of the observed ER effect. The suspensions were sheared at $24 s^{-1}$ for 5 minutes at zero field, then with field on for 15 minutes (at 3kV/mm) and finally with field off for 5 minutes. The results for the 30 and 40% suspensions of TiO_2 1:6 are shown in Fig.8. It can be seen that the development and stabilization of the shear stress take time in the range of minutes. The decay of the formed ER structures is also not immediate. The observed slow response of the studied ERFs is particularly well pronounced in the case of highly concentrated suspensions and again the reasons of this behavior can be attributed to the slow and hardly reversible agglomeration process.

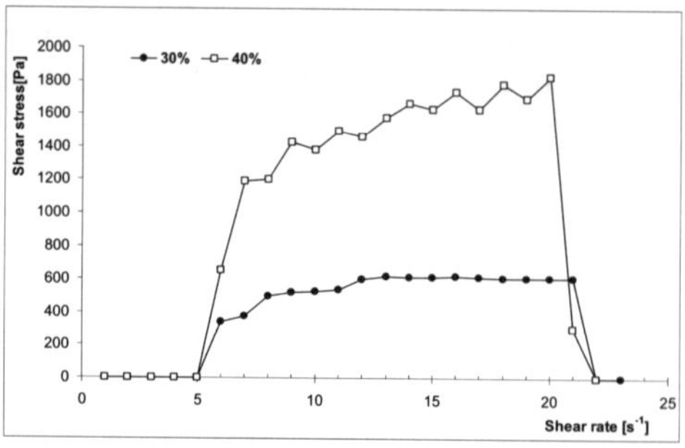

Figure 8. Time dependence of shear stress for 30% and 40% suspensions of TiO_2 1:6. Shear rate 24 s^{-1}; field strength 3 kV/mm.

4 Conclusions

A series of sub-micrometric modified titania powders were synthesized, characterized and then used to prepare ER suspensions. The grain size distribution within the dispersions was directly measured by the light scattering method. These results were correlated with rheological measurements showing that particles in the dispersions mostly occurred in a form of relatively stable aggregates whose dimensions were considerably higher then the initial size of the powder grains. It seems that the high ER effect in such ERFs as well as the reported problems with stability and reversibility are due to this aggregation process.

5 Acknowledgements

This work was partially financed by Ministry of Science within the PW-004 Long-term Program.

References

1. Winslow, W.M., J. Appl. Phys. 20 (1949) 1137.
2. S.P. Couter, K.D. Weiss, J.D.Carlson, J. Intell. Mater. Syst. Struct. 4 (1993) 248.
3. T. Hao, Adv. Mater. 13 (2001) 1847.
4. J.J. Fan, X.P. Zhao, X.M. Gao, C.N. Cao, J. Phys. D 35 (2002) 88.
5. X.P. Zhao, Q. Zhao, X.M. Gao J. Appl. Phys. 93 (2003) 4309.
6. W.Wen, X. Huang., S.Yang, K.Lu, and P.Sheng, Nat. Mater. 2 (2003) 727-730
7. J.B. Yin, X.P. Zhao, Chem. Phys. Lett. 398 (2004) 393-399
8. W. Wen, X. Huang, P. Sheng, Appl. Phys. Lett. 85 (2004) 299-301
9. Y. Zhang, K. Lu, G. Rao, Y. Tian, S. Zhang, J. Liang, Appl. Phys. Lett. 80 (2002) 888-890
10. M. Parthasarathy, D.J. Klingenberg, Mat. Sci. Eng R17 (1996) 57-103
11. B. Li, X. Wang, M. Yan, L. Li, Mater. Chem. and Phys.78 (2002) 184-188.

POLY(ETHYLENE GLYCOL)-COATED CARBONYL IRON PARTICLES: PREPARATION AND MAGNETORHEOLOGICAL BEHAVIOR

JIAN GUO GUAN AND LI CHUN LI

State Key Laboratory of Advanced Technology for Materials Synthesis and Processing, Wuhan University of Technology, Wuhan 430070, China
E-mail: guanjg@mail.whut.edu.cn

A novel water-based magnetorheological (MR) suspension is based on the poly(ethylene glycol)-coated carbonyl iron particles, which is prepared by the co-sol-gel reaction using carbonyl iron particles as raw materials, and water without any other additive. The microstructure, anti-oxidation properties, magnetic and magnetorheological properties of the core-shell particles are characterized by scanning electronic microscopy, transmission electronic microscopy, Fourier transform infrared spectroscopy, thermogravimetric-differential thermal analysis, vibrating sample magnetometer, and ARES2000 rheometer, respectively. The results show that the as prepared particles have core-shell structure with a polymer shell of less than 100 nm and its anti-oxidation property is enhanced due to the compound coating. Both the core-shell particles and its based MR suspension exhibit excellent soft magnetic performance. The based MR suspension shows low zero viscosity and high MR effects, especially greatly improved anti-settlement and redispersibility. The phenomena are attributed to the hydrophilic thin shell of the core-shell particles.

1 Introduction

Magnetorheological (MR) suspensions, invented by Rabinow in the late 1940s[1], undergo rapid, nearly completely reversible, significant changes in their viscosity upon application of an external magnetic field. MR suspensions take advantages of fast response time, controllable yield stress, low plastic viscosity, broad operational temperature range and low power requirement. Therefore they have tremendous possibilities in providing simple and effective interfaces between electronic controls and mechanical systems, such as controllable dampers, clutches, brakes, valves, high resolution polishing, engine mounts, virtual reality and so on [2~4].

In the field of MR polishing [5], water is regularly used as the carrier fluid to obtain water-based MR fluids for some mechanical and chemical reasons. As up to now very little study has been reported on water-based MR suspensions [6], the current problems of particles settling, agglomeration and poor anti-oxidation are limiting the wide application, even though several approaches of to add surfactants, to disperse the particles in a non-Newtonian liquid of grease or elastomer or to utilize different size of particles (e.g., by adding nonmagnetic or magnetic nanoparticles) are developed. For example, most surfactants used for MR suspensions are organophosphorus or organosulfur compounds and may cause pollution to the environment. MR grease and elastomer are only feasible for MR dampers. The nanoparticles can effectively reduce the settlement rate of MR fluid but it will usually reduce the MR effect. On the other hand, study on core-shell particles, which endow the pristine particles with novel properties, for MR fluid has evoke extensive interests in recent years [7~11].

It is considered that hydrophilic poly (ethylene glycol) is environment friendly (even used in medicament) and low-cost, thus has the probability of improving the suspension stability of water-based MR suspensions if it is chemically coated in the surface of the magnetic suspended particle. The poly (ethylene glycol)/carbonyl iron core-shell particles and its water-based MR fluid have been prepared and characterized at the first time in the present work. It is found that the as prepared core-shell particle of carbonyl iron with both even, compact transition layer of SiO_2 and comb-coil poly (ethylene glycol) brushes on the surface can obviously reduce the settlement and improve the redispersibility and anti-oxidation property of the water-based magnetorheological suspensions.

2 Materials

Micron sized carbonyl iron powder (ranging from 1 μm to 3 μm in diameter) is purchased from Jiangsu Tianyi superfine metal powder Ltd Co.. α-methoxy-ω-triethoxy-silicon propyl urethano polyethylene glycol with poly(ethylene glycol) chain length of 1900 was self-made in our lab and abbreviated to MTEPAMPEG. Deionized water with a resistivity of 10 MΩ.cm is produced by Aquapro ultrapure water system. Other reagents such as tetraethyl orthosilicate (TEOS) and ammonia water etc. were used without further treatment.

3 Experimental procedure

3.1 Preparation of the core-shell particles

In a 1000 mL three-necked flask. 56.0 g of carbonyl iron powder, 500 mL of DMF, 1.2 mL of TEOS, 50.0 ml of DMF solution containing 0.04 $mol.L^{-1}$ MTEPAMPEG was added in sequence and then dispersed with an electric stirrer for 30 min at a high speed, next 10 mL of ammonia water was slowly added (1.0 $mol.L^{-1}$) via a constant pressure funnel. The reaction lasted for 48 h under a stirring rate of 500 r/min. The resulting magnetic particles were separated from the mixture by a strong magnet and then ultrasonic washed with deionized water and ethanol, finally vacuum dried for 8 h. The dried powder was later carefully grinded using an agate mortar and sealed in a plastic bag for further usage.

3.2 Preparation of the water based MR fluid

40.0 g of the core-shell particles and 10.0 g of deionized water were mixed and grinded in the agate mortar to form a fine dispersion. The water-based MR fluid with a volume fraction of about 30% was then produced and sealed in a plastic bottle.

3.3 Characterization of the core-shell particles and its water based MR fluid

Scan electron micrograph (SEM) was taken by JSM-5610 LV scan electron microscope with an accelerating voltage of 20 kV. Transmission electric micrograph (TEM) was taken on Model H-8100 transmission electric microscope made by Hitachi of Japan. Diffuse reflection FTIR of the samples were characterized by Thermo Nicolet USP AVATAR300 of USA, the sample was pressed with KBr of reagent's purity. The thermal stability of the core-shell particles was taken on a PYRIS1 DSC TALAB thermal analyzer with a heating rate of 5℃/min under air and Argon atmosphere respectively. Magnetic property was performed at room temperature on ADE Model 4HF VSM whose maximal magnetic field can be set to 1.75 T and precision can reach 0.01 Oe. A modified ARES2000 Magnetorheometer with parallel plate geometry was exploited to measure the magnetorheological behavior. The magnetic field can be controlled by software MRF, the maximal field can be set to 0.6 Tesla without any leakage of the magnet space in the outside.

4 Results and discussion

4.1 Characteristics of the poly(ethylene glycol)-coated carbonyl iron particles

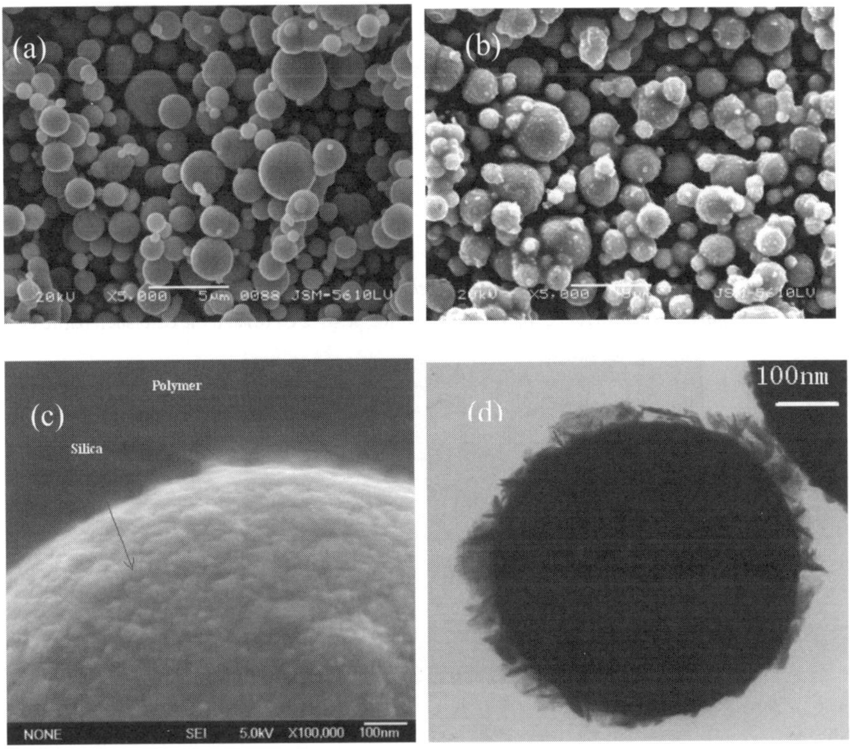

Figure 1. Electronic microscopic image of the pristine carboxylic iron particles (a) and the core-shell particles(b-d)

Fig.1(a) shows that the pure carbonyl iron has a statistically mean particle size of about 1~2 microns, all the carbonyl iron particles are spherical with very smooth surface. Fig.1(b) shows that the surface of the core-shell particles is obviously wrapped by a lot of flocky substance, quite different from that of pure carbonyl iron particles. According to the preparing process, through the after-treatment such as being thoroughly ultrasonic washed and vacuum dried, all the reactant that wasn't grafted on the surface of the carbonyl iron can be totally removed from the mixture, thus the resulting shell of the compound core-shell particles must be SiOx and PEG definitely. A more obvious core-shell structure was observed in Fig.1(c) and Fig.1(d), which indicated that the coating shell had a thickness less than 100 nm with its branches stretched in the water solution. This is the typical hydrophilic behavior of the poly (ethylene oxide) chain, hence the shell of the particles was conformed to be PEG. Combined with the reflection FTIR spectra (Fig.2), the core-shell particles presented two new wide absorption bands within 600~700 cm^{-1} and 1000~1100 cm^{-1}, which were ascribed respectively to $-CH_2-$ pendulous vibration band in PEG, C-O-C flexible vibration band in PEG and Si-O vibration absorbing band in SiO_x. This provided us the strong evidence for judgment that the PEG and SiO_x have been grafted onto the surface of carbonyl iron particles.

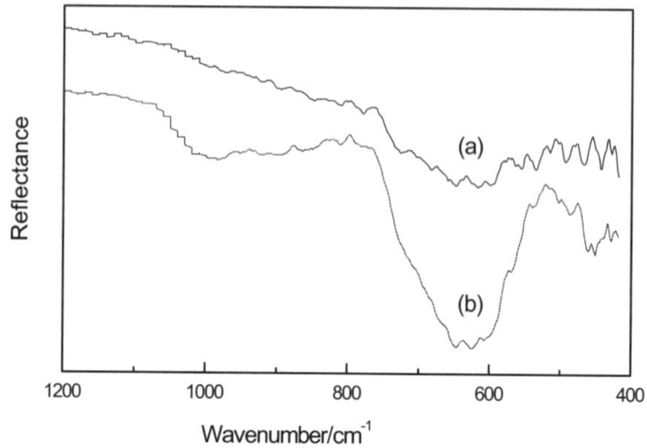

Figure 2. Diffuse reflection IR spectra of the pristine CI (a) and the core-shell particles (b)

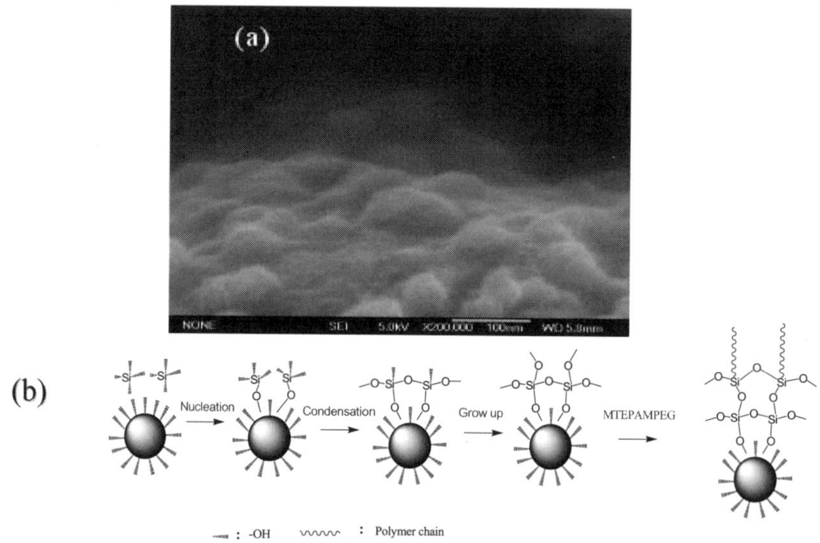

Figure 3. SEM(a) of particle surface and the mechanism of the formation of the core-shell particles(b)

Fig.3(b) presented the possible mechanism for the formation of the core-shell particles. Under the catalysis of ammonia water, the silanol groups of TEOS and MTEPAMPEG were hydrolyzed to high active silicon-hydroxyls, which easily bonded with the hydroxyls at the surface of carbonyl iron powder to form anchor points. Once the anchor points were formed, the processes of nucleation, condensation and growth-up will occur at the surface of the carbonyl iron particles, which can be seen clearly in Fig.3(a). Finally a Carbonyl iron-core / PEG polymer-shell structure with an even, compact transition layer of SiO_x was formed; As the ether oxygen bonds -O- in the long chain of PEG are apt to bend, leaving the oxygen atoms on the outside of hydrophobic chain -C-C-, this kind of ether oxygen atoms have an excellent affinity of forming hydrogen bonds with water molecules. Consequently, the core-shell particles exhibit excellent hydrophilic behaviors. At the same time, the stretched long chain of the shell in the water can effectively prevent the hard settlement of the water-based MR fluid as well as enhance the stability and redispersibility of the suspension system to a certain extent.

4.2 Thermal properties

TG curves indicate that the core shell particles show a much more slowly increasing in weight than that of the pure carbonyl iron particles. The DSC of the core-shell particles has two separate peaks at 338.7°C and 488.9°C, indicating the formation of FeO and Fe_3O_4 in sequence, higher than that of pure carbonyl iron at about 321°C and 417°C, which was not obvious. All these differences can be attributed to both the PEG polymer brushes and especially the even, compact transition layer of SiO_2, which protect the core (iron) from oxidation.

4.3 Magnetic properties

The purchased carbonyl iron powders show high *Ms* of 176.7 emu/g, low *Mr* of 0.13 emu/g and *Hc* of 2.76 Oe. The *M-H* curves of the pristine carbonyl iron particles, the core-shell particles and the water-based MR fluid have the same shape with squareness of zero. There is no obvious hysteresis, indicating that they are perfect soft magnetic characteristic. Compared to the pristine carbonyl iron particles, *Ms* of the core-shell particles decreases by 0.8% while *Hc* and *Mr* increase a little, in good agreement with the micro-sized magnetic iron/nano-nonmagnetic polymer core-shell structure. The nanometer-coating technique hasn't much influence on the magnetization characteristic of the core-shell particles. The MR fluid, in which water takes up a mass fraction of about 20%, show a decrease of *Ms* by 25% but still exhibits good soft magnetic characteristic, assuring that very little energy can be lost in appliance. The saturation field for both the core-shell particles and the MR fluid is about 0.6 T, this can be a guideline for us to design such smart equipments.

4.4 Magnetorheological properties of the water-based MR fluid

Two kinds of water based MR suspensions with a same volume fraction (30%) are prepared with the pristine carbonyl iron and the core-shell particles. The settlement behavior is found that the MR fluid based on the pristine carbonyl iron appears obvious two layers in one day while that based on the core-shell particles shows no observable settlement in two weeks, confirming the significant role of the hydrophilic PEG chain in improving the anti-settlement performance of the MR fluid.

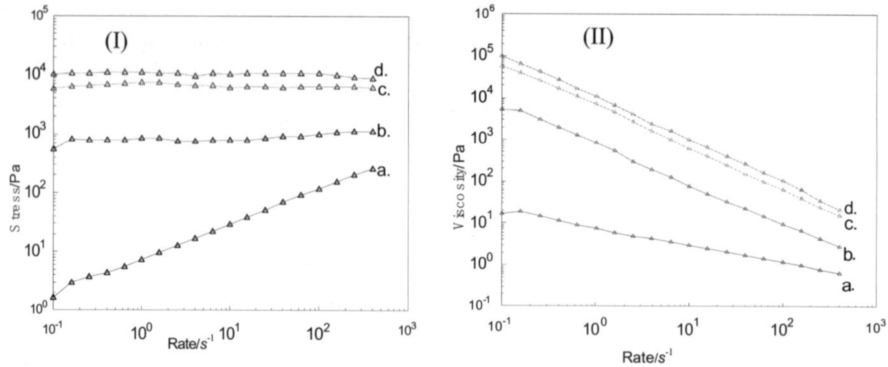

Figure 4. Viscosity(I) and shear stress(II) vs shear rates for the water-based MR suspension under different magnetic fields of (a.) 0;(b.),0.05T; (c.) 0.15T and (d.) 0.20T

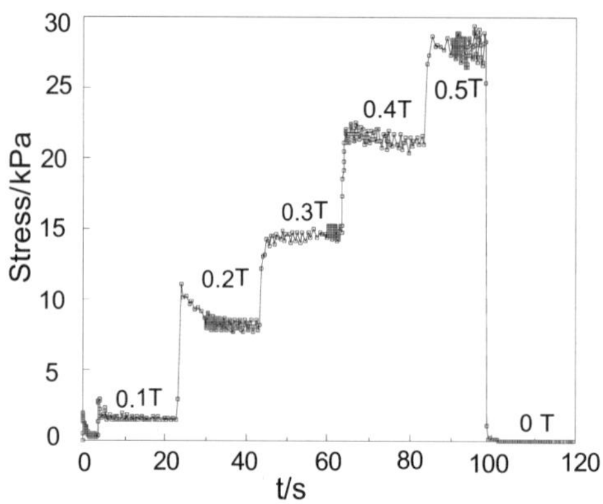

Figure 5. Shear stress for the water-based MR suspension under stepwise magnetic field and shear rate of 100 s^{-1}

The viscosity and shear stress vs. shear rates for different magnetic fields as well as the response characteristic to stepping field from 0 to 0.5 T were investigated on ARES2000 at room temperature of 25℃. Both Fig.4 and Fig.5 present that the zero viscosity of the MR fluid is below 3.0 Pa.s when the shear rate exceeds 10 s^{-1}, lower than that of LORD Company's commercial products water-based MRF-214ES and MRF-240BS [http://www.lord.com, 2004]. For different magnetic fields, the apparent visc -osity of the MR fluid tends to decrease with increasing shear rates; this is the typical shear thinning behavior of MR suspensions. Shear stress rises linearly with shear rate at zero field while upon application of an external field there is no such relation between the shear stress and shear rate. Shear stress increases with the magnetic field. At shear rate of 100 s^{-1}, shear stress of 28.5 kPa@0.5T is achieved 20 Pa@0T, demonstrating a remarkable MR effect. Meanwhile, the significant switching effect with the abrupt jump of the magnetic field is noticed, suggesting its practical application in controllable MR polishing and smart MR damping systems.

5 Conclusions

Utilizing carbonyl iron powder as raw materials and DMF as carriers, the poly (ethylene glycol)-coated carbonyl iron core-shell particles are prepared by hydrolysis and co-sol-gel reaction of TEOS and PEG. The core-shell structure has been confirmed by TEM, SEM and FTIR of the particles as well as the anti-settlement performance of its based MR fluid. The core-shell particles remain good soft magnetic performance and improved anti-oxidation. The based MR fluid exhibits low zero viscosity and high MR effects, especially greatly improved anti-settlement and redispersibility due to the thin hydrophilic polymer shell of the core-shell particles.

6 Acknowledgements

The project was financially supported by the MOE Program of New Century Excellent Talents in Universities (NCET-05-0660) and 863 Hi-tech. Program (2001AA339020) of China.

References

1. Rabinow J., The magnetic fluid clutch, *AIEE Trans.*, 67(1948) pp. 1308-1315
2. Bullough W. A. (ed) *Electro-Rheological Fluids, Magneto-Rheological Suspensions and Associated Technology* (Singapore: World Scientific, 1996)
3. Carlson J. D. and Jolly M. R., MR fluid, foam and elastomer devices, *Mechatronics*, 10(2000) pp. 555-569
4. Yang J. N., Agrawal A. K., Samali B., et al., Benchmark problem for response control of wind-exicted tall buildings, *J. Eng. Mech. ASCE*, 130 (2004) pp. 437-446
5. Shorey A. B., Jacobs S. D., Kordonski W. I., et al. Experiments and observations regarding the mechanisms of glass removal in magnetorheological finishing, *Applied Optics*, 40 (2001) pp. 20-33
6. Viota J. L., de Vicente J., Duran J. D. G., et al. Stabilization of magneto- rheological suspensions by polyacrylic acid polymers, *J. Colloid Interf. Sci.*, 284 (2005) pp. 527-541
7. Guan J. G., Wang W., Gong R. Z., et al., One step synthesis of cobalt-phthalocyanine/iron nanocomposite particles with high magnetic susceptibility, *Langmuir*, 18(2002) pp. 4198-4204
8. Cho M. S., Lim S. T., Jang I. B., et al., Encapsulation of spherical iron-particle with PMMA and its magnetorheological particles, *IEEE Trans. Mag.,* 40 (2004) pp. 3036-3038
9. Jun J. B., Uhm S. Y., Ryu J. H., et al., Synthesis and characterization of monodisperse magnetic composite particles for magnetorheological fluid materials, *Colloids Surf. A-Phys. Eng.*, 260 (2005) pp. 157-164
10. Choi H. J., Jang I. B., Lee J. Y., et al., Magnetorheology of synthesized core-shell structured nanoparticle, *IEEE Trans. Mag.*, 41 (2005) pp. 3448-3450
11. Pu H. T., Jiang F. J., Yang Z. L., Studies on preparation and chemical stability of reduced iron particles encapsulated with polysiloxane nano-films, *Materials Letters*, 60 (2006) pp. 94-97

SELF-ASSEMBLY MORPHOLOGIES AND ELECTRORHEOLOGICAL PROPERTIES OF POLYANILINE-POLY(ETHYLENE GLYCOL)-POLYANILINE TRIBLOCK COPOLYMERS

HUI-RU MA[1,2], JIAN-GUO GUAN[1] AND RUN-ZHANG YUAN[1]
1. State Key Laboratory of Advanced Technology for Materials Synthesis and Processing, Wuhan University of Technology, Wuhan 430070, China
E-mail: guanjg@mail.whut.edu.cn
2.School of Science, Wuhan University of Technology, Wuhan 430070, China

The self-assembly properties and ER effects of the Polyaniline-Poly(ethylene glycol)-Polyaniline (PAn-PEG-PAn) triblock copolymers were studied in this paper. The results indicate that with the increase of solubility parameter of the solvent, PAn-PEG-PAn copolymers form into different morphologies of spheriods, vesicles and rods. PAn-PEG-PAn copolymers with vesicles morphology show the highest polarization strength, while those with rods have the most rapid polarization rate. Among the PAn-PEG-PAn copolymers of different morphologies, the PAn-PEG-PAn copolymer vesicles show the strongest ER effect.

1 Introduction

The electrorheological(ER) fluids are suspensions of polarizable particles in insulated oil. The ER behavior is closely related to dielectric phenomena[1, 2] and the suspending particles polarizations are regarded as the fundamental factors for producing electrorheological activity. The particle shape and size was reported to have significant influence on the ER effectiveness[3, 4]. According to the polarization model, the dielectric constant ε' is related to the polarization strength and the dielectric loss ε'' related to the polarization rate and particle chains stability. At present, it is well accepted that the slow polarization rate, especially the interfacial polarization, is mainly responsible for the electrorheological effect [5, 6].

Among various anhydrous ER materials, polyaniline and its derivatives were considered to be one of the most promising materials for their relatively lower density, better thermal stability and controllability of permittivity and conductivity by changing pH [7, 8]. To improve ER properties of polyaniline particles, some organic and inorganic substances were successfully used as coatings of the particles [4, 9, 10, 11]. In this paper, the polyaniline core-shell particles with different morphologies were obtained directly by treatment of PAn-PEG-PAn copolymer in different solvents. The relationships between the self-assembly behaviors of the triblock polymer and ER properties are studies. The results indicate that both the PEG shell and the morphologies of the particles have important roles in the interfacial polarization and ER properties. Different self-assembly morphologies of the triblock polymer particles have different dielectric and electrorheological properties. The PAn-PEG-PAn copolymer with vesicles morphology shows the strongest ER effect.

2 Experimental section

2.1 Preparation of PAn-PEG-PAn triblock copolymer micelles

2.0 g PAn-PEG-PAn copolymer with conductivity of 6.97×10^{-9} S/cm, which was made in our laboratory and well characterized by FTIR and element analysis, was weighted and put into 200 ml selective solvents of dimethylfomamide, dichlomethane or xylene, respectively. The copolymer micelle particles with different morphologies were obtained by stirring the mixture for 24 h at room temperature or 60 , filtering, washing with water and drying overnight in vacuum at 65 .

2.2 The preparation of ER fluids

The dedoped PAn-PEG-PAn particles with different morphologies were weighted, ground in a mortar and well dispersed into silicone oil to obtain the ER fluids. In order to compare the ER effect, the weight fractions of all the ER fluids here are fixed to be 30 %.

2.3 Structure characterization and property measurement

The particles morphologies were observed by transmission electron microscopy (TEM, HITACHI H-600, Japan). Dielectric properties of the prepared ER fluids were measured using an Agilent 4284A precision impedance analyzer, which links to ARES rheometer fitted with parallel plate. The diameter of the plate is 25.0 mm and the electrode gap is 0.6 mm. The frequency of AC electric fields was varied from 100 Hz to100 kHz and the strain was chosen to be 5 %. Rheological properties of the suspensions were investigated using a modified controlled-strain fluid ARES rheometer with a pair of parallel plates (diameter of 14.0 mm). The electric field for ER measurement was applied using a high voltage power supply and electrode gap was 0.8 mm.

3 Results and Discussion

Fig.1 shows the TEM images of the PAn-PEG-PAn copolymer aggregates formed in different selective solvents. It shows that in dimethylfomamide, the uniformed rods with the lengths about 150~200 nm and the width about 60 nm are available. In dichlomethane, the PAn-PEG-PAn copolymers firstly form into separate vesicles, then aggregate each other as the vesicles have large specific surface areas and are unstable. In xylene, Spherical micelles with obvious core-shell structure are observed. The particle size is about 110 nm and the ratio of core to shell is about 5:1. This can be attributed to the fact that the PEG chains are much easier to soluble in xylene than the PAn chains. Fig.1 also suggests that with increasing the solubility parameter of the solvent increasing (xylene is $18.0(J/cm^3)^{1/2}$, dichlomethane is $19.8(J/cm^3)^{1/2}$, dimethylfomamide is $24.8(J/cm^3)^{1/2}$), the self-assembly morphologies change from sphere, vesicles, to rods. Different solubility parameter of solvents leads to different solubility or swelling ability to core and shell chains, which influence the repulsive actions between the core and shell. As a result, the aggregates exist as different morphologies.

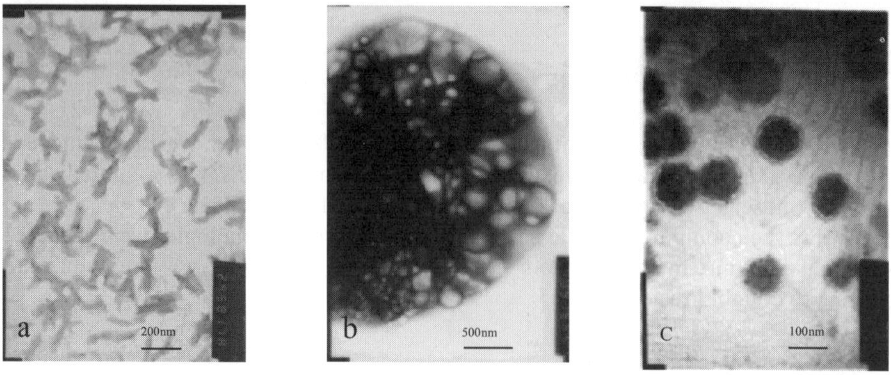

Fig.1 TEM micrographs of PAn-PEG-PAn aggregates from different selective solvents
(a) DMF; (b) dichlomethane; (c) xylene

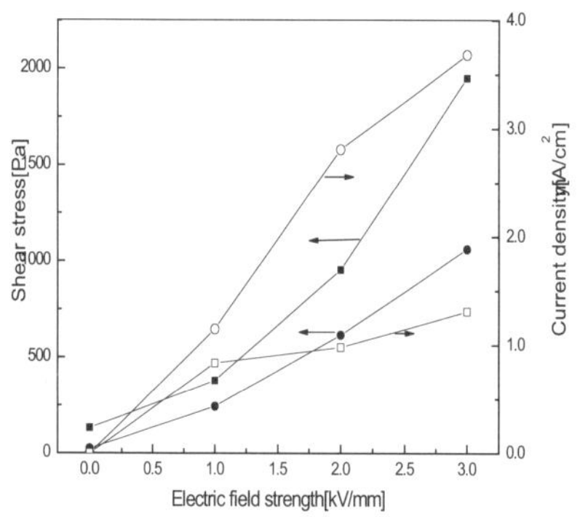

Fig.2 Shear stress (filled signs) and current density (filled signs) vs DC electric field strength for the ER fluids based on the PAn (●, ○) and PAn-PEG-PAn with spherical particles (■, □) at shear rate $0.1s^{-1}$

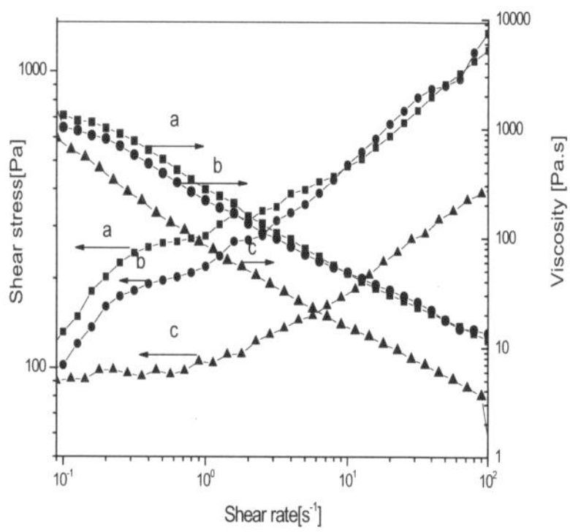

Fig.3 The rheological curves of the ER fluids based on the PAn-PEG-PAn particle with different morphologies of (a) vesicles, (b) spheres, (c) rods in absent of electric field

Fig.2 shows the ER behaviors of the ER fluids respectively based on PAn-PEG-PAn or polyaniline particles in different electric field. It is obvious that the ER fluid of PAn-PEG-PAn shows better ER properties than that of Polyaniline. This can be reasonably attributed to the intrusion of PEG insulated shell. The PEG shell mainly plays two key roles. One is to effectively hamper electronic hop among different PAn cores which helps to lower power consumptions of ER fluids, and the other is to increase the interfacial polarization ability of the ER particles, resulting to enhance the shear stress.

Fig.3 is the rheological curves of the ER fluids based on PAn-PEG-PAn particles of different morphologies at zero electric field. All 3 ER fluids behave like non-newtonian fluids and the shear-thinning phenomena are clearly observed. The particles of different morphologies show different rheological bahaviors. At lower shear rate, the ER fluids of vesicles bases have higher shear stress and zero viscosity than that of ER fluids of spheres bases, while at higher shear rate, their flow curves almost superpose together. This can be explained that the particles with vesicle structure for large specific surface areas have good compatibility with silicone oil, leading to increase the volume fraction of ER fluid and subsequently enhance its viscosity. With the increase of shear rate, the compatibility becomes worse owing to centrifugal effect. Therefore, shear stress and zero-field viscosity of the ER fluids gradually have the same values as that of ER fluids based on spherical particles. At the whole shear region, the ER fluid based on rods shows the lowest values of shear stress and viscosity, indicating the weakest resist ability to shear field due to the easy orientation tendency of rods along external force.

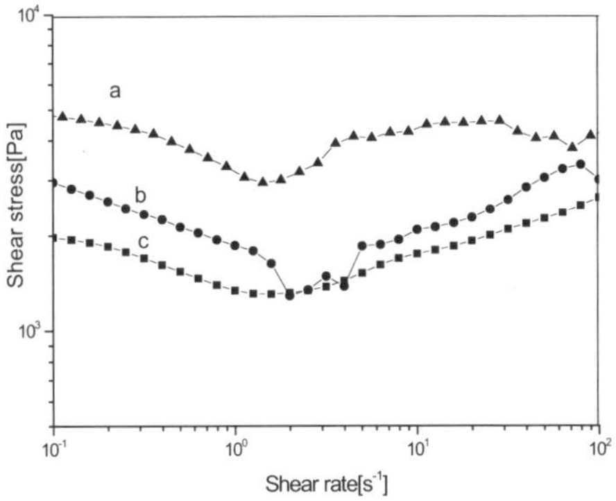

Fig.4 Shear stress as a function of shear rate for ER fluids of PAn-PEG-PAn particle with different morphologies of (a) vesicles, (b) rods, (c) spheres at 3 kV/mm

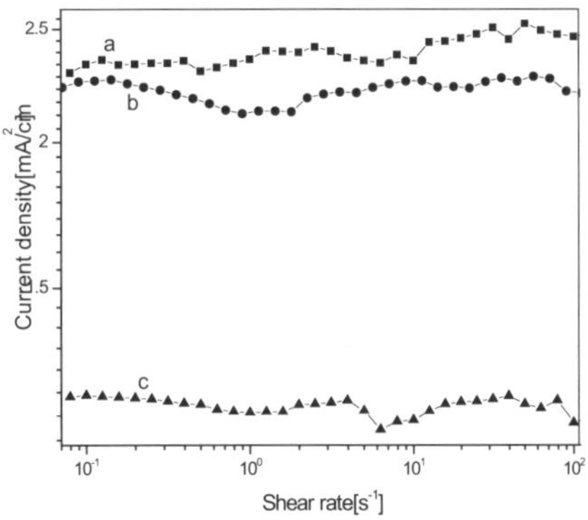

Fig.5 Current density as a function of shear rate for the ER fluids of PAn-PEG-PAn particle with different morphologies of (a) vesicles, (b) rods, (c) spheres at 3 kV/mm

Fig.4 gives shear stress-shear rate curves of the ER fluids of PAn-PEG-PAn particles with different morphologies at electric field strength of 3kV/mm. When the electric field was applied to these suspensions, the shear stress rapidly increased with a yield behavior. The important effect of particle morphologies on ER properties is obvious. The ER fluid based on vesicle particles has the strongest ER effect and its shear stress reaches about 5.0 kPa at 3kV/mm. The particles of rod show better ER effect than that of spheroid. This is because that rod particles of larger length-diameter ratio have a greater induced dipole moment when they are oriented along the electric field, while ER fluid based on spherical particles still has the larger stress than that of PAn for the interfacial polarization. Over the whole shear range, the flow instability appears at all 3 ER fluids. Shear stress gradually decreases with shear rate until the lowest value appears. This phenomenon has been found in many ER fluids, which is related to the breakage rate of the chain structures caused by the applied shear and is related to the rate of structure reformation induced by an external electric field. But it is also found from Fig.4 that the lowest value of shear stress for the ER fluid of rods appears at the highest shear rate. This suggests that the particle chains composed of rods can keep the most stable under the electric field.

Fig.5 is the current density-shear rate curves of ER fluids of PAn-PEG-PAn particles with different morphologies at the electric field strength of 3kV/mm. it can be seen from

Fig.5 that all the ER suspensions show low current density. This is useful to reduce power consumptions for ER devices.

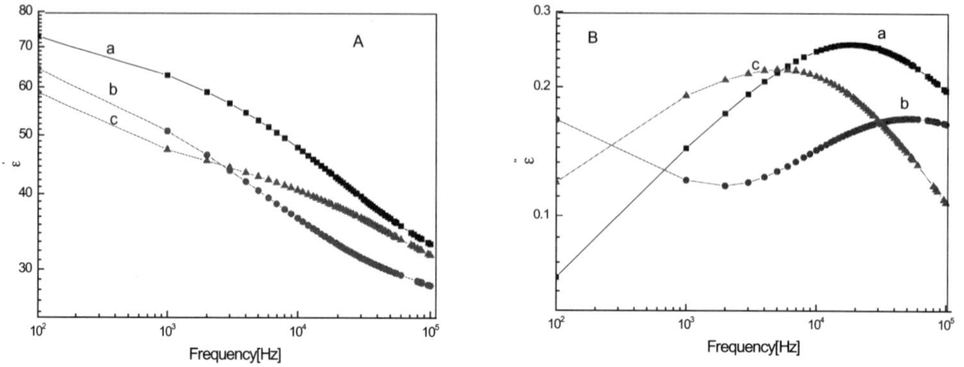

Fig.6 Dielectric spectra as a function of electric field frequency for the ER fluids of PAn PEG-PAn particles with different morphologies of (a) vesicles, (b) rods, (c) spheres

The effects of particle morphologies on ER properties are also analyzed via dielectric spectra shown in Fig.6. Block et al considered that both the difference in dielectric constant $\triangle\varepsilon'$ at 10^2 and 10^5 Hz and the local maximum dielectric loss factor between 10^2 and 10^5 Hz are response for a "good" ER properties. The $\triangle\varepsilon'$ is related to the polarization strength of ER fluids and the local maximum of dielectric loss factor is related to the proper polarization rate. Fig.6A manifests that relative permittivity for the ER fluid of vesicles and rods decreases steeply with frequency, indicating that they have strong polarization strength. ER fluids composed of rods have the maximum dielectric loss at highest frequencies. This corresponds to short relaxation time in electric field and to a fast polarization rate, which consist with its stable chain structures showed in Fig.4.

References

1. Tian Hao., Electrorheological suspensions. *Advances in colloid and interface science* **97** (2002) pp.1–35.
2. Harry Block, Kelly J. P., Qin A. and Watson T., **M**aterials and mechanisms in electrorheology. *Langmuir* **6** (1990) pp. 6–14.
3. Cho, M. S., Cho, Y. H., Choi, H. J., et al, Polyaniline-coated PMMA microsphere for electrorheological fluids. *Synthetic Metals* **135–136** (2003) pp. 15–16.
4. Yong Taik Lim, Jong Hyeok Park, O Ok Park1, Improved Electrorheological effect in polyaniline nanocomposite suspensions. *Journal of Colloid and Interface Science* **245** (2002) pp. 198–203.
5. Yin Jianbo and Zhao Xiaopeng, Giant electrorheological activity of high surface area mesoporous cerium-doped titania templated by block copolymer. *Chemical Physics Letter* **398** (2004) pp. 393–399.
6. Anežka Lengálová, Vladimír Pavlínek and Petr Sáha, Electrorheology of polyaniline-coated inorganic particles in silicone oil. *Journal of Colloid and Interface Science* **258** (2003) pp. 174–178.
7. Vladimír Pavlínek, Petr Sáha, Takeshi Kitano, et al, The effect of polyaniline layer deposited on silica particles on electrorheological and dielectric properties of their

silicone oil suspensions. *Physica A: Statistical Mechanics and its Applications* **353** (2005) pp. 21–28.

8. Lengálová, A., Pavl´ınek, V., Sáha a, P., et al. The effect of dispersed particle size and shape on the electrorheological behaviour of suspensions. *Colloids and Surfaces A: Physicochem. Eng. Aspects* **227** (2003) pp. 1–8.
9. Jan Trlica , Petr Sáha , Otakar Quadrat et al, Electrorheological activity of polyphenylenediamine suspensions in silicone oil. *Physica A* **283** (2000) pp. 337–348.
10. Guan Jianguo, Xie Hongquan and Guo Junshi, Preparation and properties of highly active electrorheological fluid based on polyaniline. *Chemical Journal of Chinese Universities-Chinese* **17** (1996) pp. 965–967.
11. Choi, H.J., Kim, J.W. and K.To, Synthesis and electrorheological behavior of semiconducting poly(aniline-co-o-ethoxyaniline). *Synthetic metals* **101** (1999) pp. 697–698.

ON THE CONTROL OF A MR TORQUE TRANSFER DEVICE

M. H. ELAHINIA, C. CIOCANEL, K, MOLYET, AND N. NAGANATHAN

Dynamic and smart Systems Laboratory, University of Toledo, 2801 West Bancroft St., MS 312, Toledo, Ohio, USA

E-mail:mohammad.elahinia@utoledo.edu

In this paper, a magnetorheological (MR) torque transfer device is designed, modeled, and controlled. MR fluids possess the unique ability to undergo dramatic and nearly completely reversible changes in their rheological properties under the application of a magnetic field. These controllable fluids can serve as quiet, rapid interfaces between electronic controls and mechanical systems. One area of application is to use these fluids as actuators. The MR torque transfer device can function as either a clutch or a brake. This coupling device was designed and built using a parallel plates configuration, and uses a stationary electromagnetic coil to activate the fluid. A PID controller is designed and experimentally evaluated. In the experimental control setup, the output variables are the position, velocity, and torque at the output shaft and the control input is the electromagnet current. Angular position of the output shaft and the transferred torque are measured using an encoder and a torque transducer, respectively. A dSpace control system was used to experimentally implement the control algorithms. The closed loop performance of system was studied for both torque regulation as well as torque tracking.

1 Introduction

Magnetorheological (MR) fluids typically consist of micron-sized polarizable particles, such as iron, dispersed in a polar or nonpolar liquid, along with surfactants to prevent sedimentation. Applying a magnetic field to these fluids causes reversible changes in their rheological properties within milliseconds. These changes are related to the increase in their strength by developing a yield stress. MR fluids behave as a Bingham solid; that is, they behave like a solid until the applied shear stress becomes equal to the yield stress, which marks the onset of flow. The electric analogs of MR fluids are electrorheological (ER) fluids, which are qualitatively similar. Because of their rapid response, these controllable fluids can serve as quiet, rapid interfaces between electronic controls and mechanical systems. Some areas of applications include shock absorbers, clutches, brakes, engine mounts, and active vibration control.

Several authors have discussed the performance of various ER clutches, such as Whittle et al. [1], Johnson et al. [2], Tan et al., [3] and Nakamura et al. [4]. Choi et al. designed and built ER and MR disk type clutches to compare their performance [5]. They derived a nondimensional model to predict the torque developed assuming each fluid behaved as a Bingham plastic. They also demonstrated the long term torque controllability of the clutches using a PID controller. Lampe et al. used the Bingham plastic model to derive equations for the transmitted torque in MR clutches of two different geometries, parallel disks, and concentric cylinders [6]. They also designed and tested a MR clutch using a dihedral disk shape. Lee et al. studied a concentric cylinders MR clutch by taking into consideration nonuniform magnetic fields and complicated fluid flows [7].

This paper presents the design, modeling, and control of a parallel disk torque-transfer device, which can function as a clutch or brake. This devise was built and the output torque measured and compared to predicted values based on the results of the fluid testing.

2 Modeling and Design

The torque generated by shearing MR or ER fluids in the gap between two parallel disks can be predicted by using the well-known Bingham plastic model [5-6]:

$$\tau = \tau_y + \eta\dot{\gamma} \tag{1}$$

where τ is the shear stress, τ_y is the field dependent yield stress which marks the onset of flow, η is the viscosity, and $\dot{\gamma}$ is the shear rate. Assuming a linear velocity distribution in the fluid between the disks, the shear strain rate can be written as:

$$\dot{\gamma} = \frac{dV}{dz} = \frac{(\omega_2 - \omega_1)r}{s} \tag{2}$$

where ω_1 and ω_2 are the angular velocities of the disks, r is the radial distance from the center, and s is the distance between the disks. Substituting Eq. (2) into Eq. (1) results in:

$$\tau = \tau_y + \eta\left[\frac{(\omega_2 - \omega_1)r}{s}\right] \tag{3}$$

The torque transmitted by a differential area of the disks can be written as:

$$dT = 2\pi\tau r^2 dr \tag{4}$$

Substituting Eq. (3) into Eq. (4) and integrating from R_i to R_o, the inside and outside radius, respectively, results in:

$$T = \frac{2}{3}\pi\tau_y\left(R_o^3 - R_i^3\right) + \frac{\pi\eta}{2}\left[\frac{(\omega_2 - \omega_1)}{s}\right]\left(R_o^4 - R_i^4\right) \tag{5}$$

Equation (5) can be used to predict the output torque in a MR torque transfer device of parallel disk configuration, the design of which will be discussed next.

A parallel disk MR torque-transfer device that can function as a clutch or brake was designed and built. The device contains an electromagnet that can apply a uniform magnetic field adjustable up to a value of 0.4 T. To test the device, an experimental setup was built to vary the input rotational speed, to measure the input power, and to measure the input and output speed and output torque of the device [8]. As shown in Fig. 1, the clutch consisted of an outer disk/housing assembly made of two halves, which were bolted together. A rubber o-ring was placed in the groove between the two halves to form a seal against leakage of the MR fluid. The 60 mm diameter inner disk fit inside the outer disk/housing assembly with a 1 mm gap on either side for the MR fluid. An oil seal was placed in the outer housing to prevent fluid leakage along the inner disk shaft, and a sealed radial bearing to allow relative motion between the inner and outer disks. The outer disk/housing is the input side and the inner disk is the output side. Also shown in Fig. 1, a stationary electromagnetic coil was placed around the outside of the clutch with a steel shell enclosure to direct the magnetic field to the clutch. Gaps of 1.59 mm separated the steel shell from the clutch on either side. The inner disk and a portion of the outer disk/housing were made of AISI 1018 steel while the rest of the outer disk/housing was made of aluminum, which will direct the magnetic flux from the coil enclosure to the steel portions of the clutch, maximizing the magnetic field strength applied to the MR fluid inside. The magnetic field is concentrated on the outer section of the inner disk, which is the active portion of the clutch. This section has an outer radius of 30 mm and an inner radius of 14.3 mm.

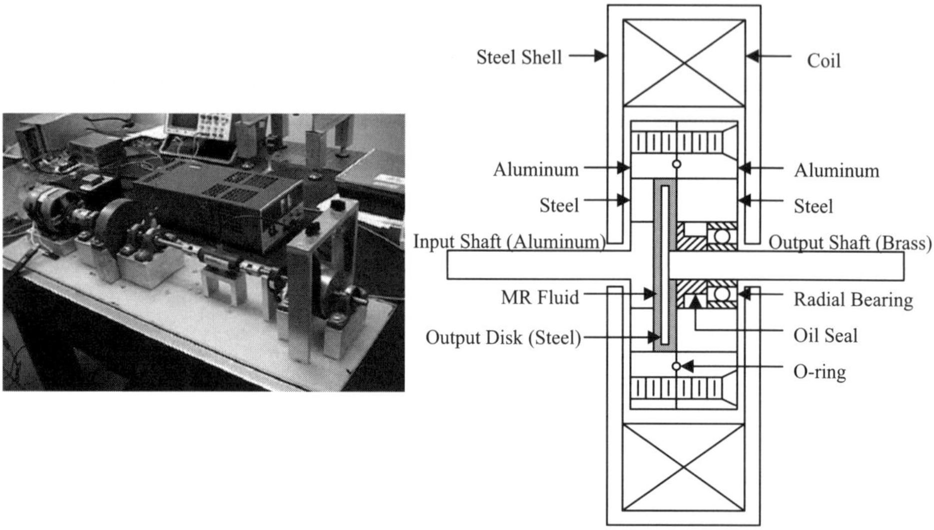

Figure 1. Clutch Experimental clutch setup and clutch assembly with electromagnetic coil arrangement.

The electromagnetic coil was made with 480 turns of 18 gauge copper wire, and fit inside a shell made of AISI 1018 steel. Finite element analysis was used to design the coil and clutch assembly and to determine the magnetic field strength [8]. The radial magnetic field distribution in each of the MR fluid gaps is uniform along the active portions of the disk, with a value of about 0.384 T for the fluid gap on the input side of the clutch, and a constant 0.4 T for the fluid gap on the output side of the clutch. The resistance in the wire was measured to be 4.3 Ω, and the power consumed at 2.82 A is calculated to be 34.2 watts. Figure 1 also shows the overall experimental setup. The clutch is placed inside the electromagnetic coil, and each shaft of the clutch assembly is supported by pillow block bearings, which are mounted on aluminum blocks. A small variable speed DC motor is used to drive the clutch. A DC power supply, with a voltage capacity of 35 V and a current capacity of 15 A, supplies power to the electromagnetic coil. A rotating torque transducer from Cooper Instruments & Systems is connected to the output shaft to measure the torque developed by the clutch. A dSPACE hardware-in-the-loop system is used to study the closed-loop torque-transfer behavior of the device. A digital encoder from the US Digital Corp. is connected to the output shaft to measure the rotational speed of the output shaft. A variable prony brake, which is a simple friction device, is used to simulate loading of the device. The prony brake consists of a 101.6 mm diameter aluminum disk with an adjustable leather belt rubbing against it to provide friction. The torque transfer device was tested by using it in brake mode, with the prony brake tightened to prevent motion of the output shaft. MRF-132LD was used in the device. Experiments were run at various constant speeds between 250-1500 rpm and magnetic fields of 0.1, 0.2 and 0.3 T. At the beginning of each test, the motor was turned on and set to a constant speed, and then the power supply to the electromagnetic coil was turned on, with the current preset to provide a given level of magnetic field. Torque readings were measured and recorded once per second for one minute, and an average was taken. These results are plotted in Fig. 2, along with the predicted output torque calculated by applying Eq. (5) to each of the two MR fluid-filled gaps inside the device. An effective outer radius of 27 mm instead of the actual outer radius of 30 mm (1.18 in.) was used in Eq. (5). The percent differences between the measured and predicted torque values for 0.2 T and 0.3 T were under 10%, while for 0.1 T there was about a 30% difference.

3 Control Experiments

In order to demonstrate the controllability of the MR torque-transfer device, a PID controller was designed to regulate the output shaft torque. To this end, the measured torque is compared to the desired value in order to adjust the applied electric voltage to the electromagnet. The torque error is defined as the difference between the desired torque and the measured torque:

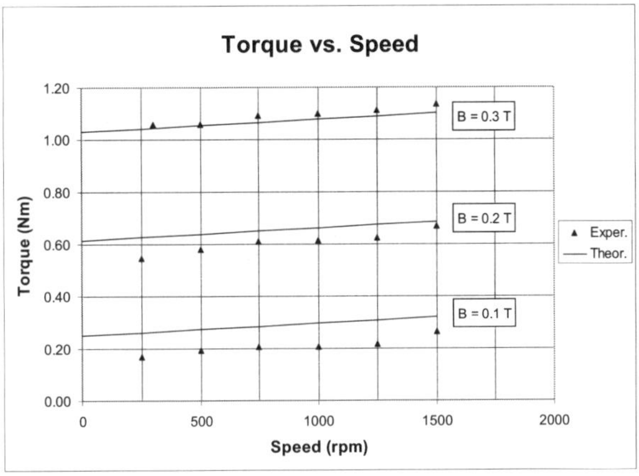

Figure 2. Comparison of predicted and measured output torques.

$$e = \tau - \tau_d \tag{6}$$

where τ and τ_d are the measured and desired torques, respectively. The PID controller adjusts the voltage to the electromagnet as:

$$U = [K_P + K_D s + \frac{K_I}{s}]e \tag{7}$$

where K_P, K_D, and K_I are the proportional, derivative, and integral control gains, respectively. This section presents the results from the control experiments with both constant as well as variable desired torques. The controller was developed in Simulink and was implemented with the dSPACE control solution.

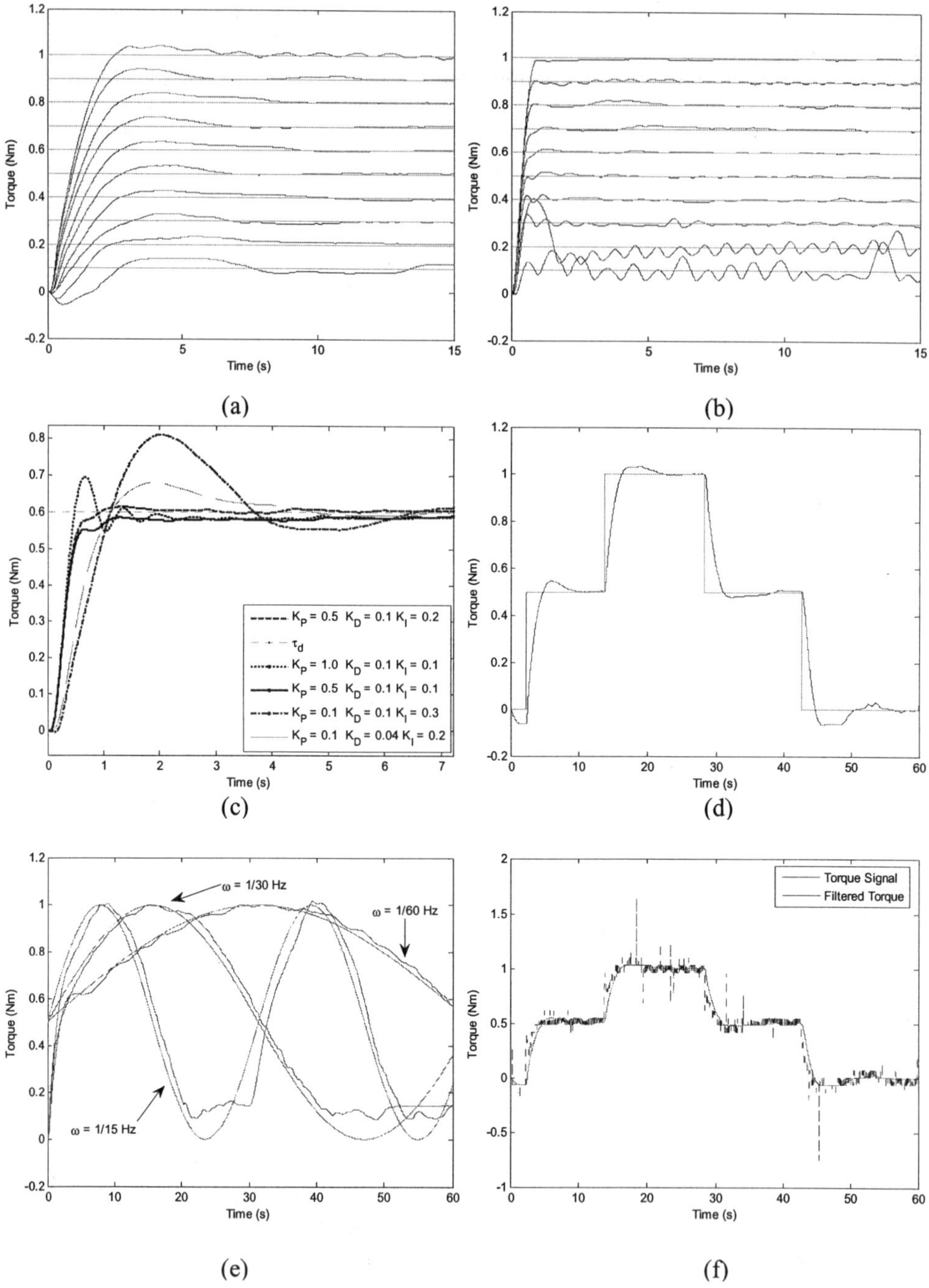

Figure 3. Close-loop performance of the MR clutch.

Figure 3 illustrates the closed-loop torque transfer performance of the device. In the experiments, shown in Fig. 3(a) and (b), the desired torque varies from 0.1 Nm to 1 Nm. The controller gains for these experiments are shown in Table 1. It can be seen that the performance of the system depends on both the desired torque as well as the controller gains. In other words, the regulation performance of the PID controller is acceptable for certain set-points. However, the performance deteriorates for other desired torques. Figure 3(c) further illustrates the effect of controller gains on the regulation performance of the device. In this experiment, the desired torque is maintained at 0.6 Nm.

Table 1. PID controller gain for the regulation torque-transfer experiments.

Controller	K_P	K_D	K_I
Fig. 3 (a)	0.1	0.1	0.1
Fig. 3 (b)	0.5	0.1	0.2

Figure 3 (d) represents a tracking control experiment. In this experiment, the torque transfer device is to follow a user-defined desired torque profile. Similar results are shown in Fig. 3 (e), where the desired torque profile is a sinusoidal function. It can be seen that the tracking performance of the device is frequency dependent. That is, the tracking error is larger for higher frequencies.

For all the experiments, the measured signal of the torque transducer was filtered for high frequency noise. It is essential to remove the high-frequency noise in order to improve the quality of the control results. High frequency noise, for example, is amplified through the derivate element of the PID controller, which leads to large high-frequency control input (to the electromagnet). To this end an analog second order low pass Bessel filter was used. The pass-band edge-frequency of the filter is 20 Hz. The sampling time of 0.05 seconds was chosen for all the experiments. Figure 3(e) illustrates the measured torque and the filtered signal for a tracking control experiment.

4 Conclusion

Torque transfer devices are essential part of variety of electromechanical systems. It was shown in this paper that MR fluid-based torque transfer devices can be analyzed and designed based on material behavior models. A model for predicting the transferred torque for a MR clutch was developed. This model was experimentally verified and acceptable correlation between the model prediction and the clutch performance was established. Furthermore, controllability of a MR clutch was experimentally demonstrated. A PID controller was designed to regulate the transferred torque. Both regulation as well as tracking torque control was successfully achieved with this controller. Based on the experimental results, however, it is evident that there is a need for more advanced control methods. These controllers will provide robust torque transfer behavior over a wide range of working conditions and for a range of desired torques.

References

1. Whittle, M., Atkin, R.J. and Bullough, W.A., 1995, "Fluid Dynamic Limitations on the Performance of an Electrorheological Clutch", Journal of Non-Newtonian Fluid Mechanics, 57, pp. 61-81.
2. Johnson, A.R., Bullough, W.A. and Makin, J., October 1999, "Dynamic Simulation and Performance of an Electro-Rheological Clutch Based Reciprocating Mechanism", Smart Materials and Structures, 8(5), pp. 591-600.

3. Tan, K.P., Bullough, W.A., Stanway, R., Sims, N., Johnson, A.R. and Tozer, R.C., 2002, "A Simple One Dimensional Robot Joint Based on the ER Linear Reversing Mechanism", Proceedings of the 8th International Conference on Electrorheological Fluids and Magnetorheological Suspensions, Ed. G. Bossis, World Scientific, Singapore, pp. 323-328.
4. Nakamura, T., Saga, N. and Nakazawa, M., 2002, "Impedance Control of a One Shaft-Type Clutch Using Homogeneous Electrorheological Fluid", Proceedings of the 8th International Conference on Electrorheological Fluids and Magnetorheological Suspensions, Ed. G. Bossis, World Scientific, Singapore, pp. 153-159.
5. Choi, S.B., Hong, S.R., Park, D.W., Cheong, C.C. and Park, Y.K., 2000, "Comparison of Field-Controlled Characteristics Between ER and MR Clutches", Proceedings of the 7th International Conference on Electro-Rheological Fluids and Magneto-Rheological Suspensions, Ed. R. Tao, World Scientific, Singapore, pp. 603-610.
6. Lampe, D., Thess, A. and Dotzauer, C., June 17-19, 1998, "MRF Clutch Design Considerations and Performance", 6th International Conference on New Actuators, Bremen, Germany, pp. 449-453.
7. Lee, U., Kim, D., Hur, N. and Jeon, D., 2000, "Design Analysis and Experimental Evaluation of an MR Fluid Clutch", Proceedings of the 7th International Conference on Electro-Rheological Fluids and Magneto-Rheological Suspensions, Ed. R. Tao, World Scientific, Singapore, pp. 674-681.
8. Molyet, K., Ciocanel, C., Yamamoto, H., Naganathan, N., "Design and Performance of a MR Torque Transfer Device", Proceeding of 2005 ASME International Mechanical Engineering Congress and Exposition November 5-11, 2005, Orlando, Florida USA.

DEVELOPMENT OF SELF-SENSING MAGNETORHEOLOGICAL DAMPERS FOR STRUCTURAL VIBRATION CONTROL

S. W. OR
Department of Applied Physics
The Hong Kong Polytechnic University, Hung Hom, Kowloon, Hong Kong

Y. F. DUAN, Y. Q. NI* AND T. K. CHAN
Department of Civil and Structural Engineering
The Hong Kong Polytechnic University, Hung Hom, Kowloon, Hong Kong
**E-mail: ceyqni@polyu.edu.hk (corresponding author)*

Magnetorheological (MR) dampers have been successfully used in vibration control of civil and mechanical structures. Application examples include vibration control of stay cables in cable-stayed bridges, vibration damping of automotive seats and suspension systems, and vibration isolation of automation and precision equipment/machines. While usefulness, the existing MR dampers are incapable of self-sensing structural vibration for implementing real-time, closed-loop vibration control. This paper describes the development of a kind of novel MR dampers possessing the attractive sensing-while-damping function with a high degree of sensor-damper collocation. Lead zirconate titanate (PZT) piezoceramic sensors and PZT/polymer piezocomposite sensors have been embedded, respectively, with actuation-only MR dampers to form two self-sensing MR dampers. The sensors have specifically designed geometry and electrode pattern to provide an optimal sensibility. Performance tests have been conducted by mounting the self-sensing MR dampers on a material test system operating in various displacement-controlled excitations, including sinusoidal and ramp excitations with different combinations of frequency and amplitude. Good agreements between the sensor outputs and the force inputs have been observed. The damper embedded with piezocomposite sensor has shown a higher sensitivity and a better reliability than the one integrated with piezoceramic sensor. The self-sensing MR dampers have great potential to be used in real-time close-loop vibration control of civil and mechanical structures.

1 Introduction

Semi-active MR dampers have been increasingly applied for vibration control of civil and mechanical structures [1-7] While the force of MR dampers is adjustable and controllable through altering the applied magnetic field, the existing MR dampers are incapable of sensing structural vibration and damper force for implementing real-time closed-loop vibration control. Implementation of closed-loop control using MR dampers requires an inverse dynamic model for the dampers so as to determine appropriate current input for facilitating the desired control force. It is still a challenge to develop an accurate inverse dynamic model for MR dampers with high nonlinearity. Alternatively, a force feedback control loop is often deployed to eschew the deficiency of requiring inverse dynamic modeling. This in turn calls for the installation of force sensors between the vibrating structure and the dampers at the expense of reducing the control effectiveness and increasing both the implementation difficulty and engineering cost. Therefore, it is highly desirable to develop smart MR dampers with embedded force sensors for providing real-time force feedback control in a cost-effective and reliable manner.

The piezoelectric materials, such as piezoelectric ceramics, polymers and composites, are promising for this function due to their effectiveness over a large frequency range, simplicity, reliability, compactness, and lightweight [8]. Piezoelectric ceramics, polymers and composites can convert changes in mechanical energy into electrical signals and also can exhibit the effects conversely. By integrating the piezoelectric materials into the traditional MR dampers, the damping force can be sensed and thus smart MR dampers with sensing-while-damping function are developed. In this study, PZT piezoceramic embedded MR dampers and PZT piezocomposite embedded MR dampers are designed, fabricated and tested.

2 Configuration of Self-Sensing MR Dampers

A conventional MR damper as shown in Figure 1(a) generally includes a cylinder housing MR fluid (13), electromagnet (14), diaphragm (15), accumulator (16), a piston (17), and a connection support (18). The bearing and seal (12) direct the movement of the piston and prevent leakage of the MR fluid (13). By applying different current/voltage inputs to the electromagnet through the wires (11), the MR fluid (13) may reversibly change from a free flowing, linear viscous fluid to a semi-solid with adjustable and controllable yield strength such that the damper force can be changed accordingly.

Figure 1(b) illustrates a self-sensing MR damper of the present study. Similar to the conventional MR damper, the self-sensing MR damper also includes a cylinder housing MR fluid (13), an electromagnet (14), a diaphragm (15), an accumulator (16), and a piston (17). It further includes a piezoelectric sensor (28) embedded into the connection support for the measurement of damper force. The piezoelectric sensor (28) senses the variation in the external force exerted on the damper and generates electrical signals in accordance with the variation in the external force imposed on the electrode surfaces. As the piezoelectric sensor is capable of sensing real-time variation in the external force, real-time control of the damper force can be achieved according to a devised control strategy.

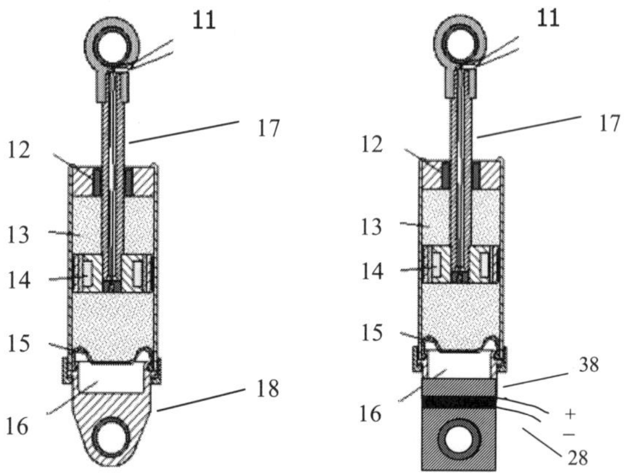

(a) Conventional MR damper (b) Self-sensing MR damper

Figure 1. Configuration of conventional MR damper and self-sensing MR damper

As shown in Figure 2, sensory components in the devised self-sensing MR damper include two wafer electrodes (31) and (32) mounted on either side of a piezoelectric wafer (30). An insulating wafer (33) is mounted between the electrode (31) and the neighboring surface (40) of the upper part (38) of the connection support. These sensory components are sandwiched centrally in a stack (35) under the mechanical pressure by using a threaded shaft (41) penetrating through the lower part (42) of the connection support and a nut (43) concave in the upper part (38) of the connection support. The preloading pressure is large enough so that the piezoelectric wafer (30) remains in compression during operation. The threaded shaft (41) should be insulated from the electrodes and piezoelectric wafers. Electric wires are connected to the electrodes (31) and (32) for the measurement of voltage developed across the wafer piezoelectric (30) through a signal conditioning unit (24) and a data acquisition unit (25). The results can be recorded and processed using a personal computer (26) and displayed on a monitor (27), enabling the damper force to be measured during the operation of the damper.

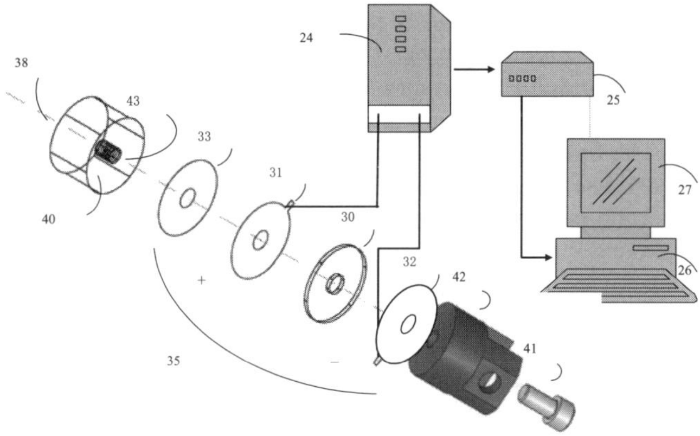

Figure 2. Exploded isometric view of sensory components in self-sensing MR damper

The piezoelectric wafer (30) can be any suitable piezoelectric material including piezoelectric ceramics, polymers, and composites. A piezoelectric ceramic and four piezoelectric ceramic/polymer composites with different configurations (0-3, 1-3, radial 2-2, and parallel 2-2), as illustrated in Figure 3, are explored in the present study. Usually, piezoelectric ceramic sensors have sharp resonances and high sensitivity within a narrow bandwidth. Signals with frequencies within their resonances will be greatly amplified and artifacts may be created. As ceramics are hard and brittle, it is difficult to produce ceramic sensors with large element size and complex shape and damage caused by mechanical shock or vibration is more serious. Piezoelectric polymer sensors, however, have wider bandwidth, and all signals will be received with more or less equal sensitivity over a wide range of frequency. They can be fabricated into complex shapes and are more resilient to mechanical stress as they are more flexible. Their major drawbacks are lower sensitivity and less temperature stability. Piezoelectric composite sensors, on the other hand, can be tailored to combine the desired properties of ceramics and polymers and are the most suitable choice for this sensor.

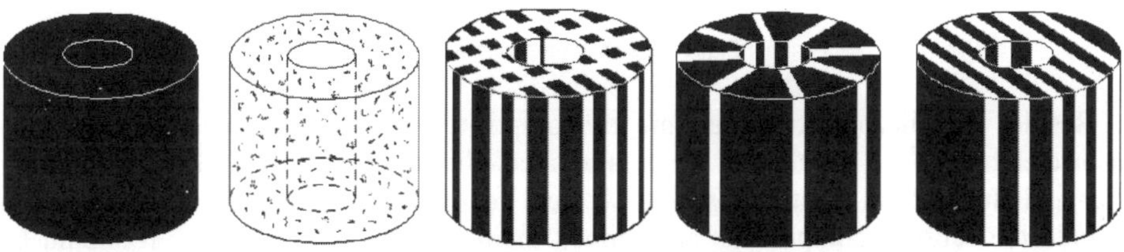

(a) Ceramic (b) Composite (0-3) (c) Composite (1-3) (d) Composite (radial 2-2) (e) Composite (parallel 2-2)

Figure 3. Piezoelectric wafers made of different piezoelectric materials

3 Fabrication of Self-Sensing MR Dampers

A ring-shaped lead zirconate titanate (PZT) piezoceramic sensor and a ring-shaped PZT/ epoxy piezocomposite sensor have been fabricated and integrated with traditional MR dampers to form two dampers possessing the attractive sensing-while-damping function with a high degree of sensor-damper collocation.

The piezoceramic sensor and piezocomposite sensor have similar inner diameters, outer diameters and thicknesses of about 13 mm, 38 mm and 2 mm, respectively and are electroded on their two major surfaces perpendicular to their thicknesses along which polarization is induced. Figure 4 shows the piezoceramic (100% ceramic) wafer. Dice-and-fill technique was employed to fabricate the piezocomposite wafer. A Disc DAD321 automatic dicing saw equipped with a flange type diamond saw blade was used in the dicing operation. The first set of parallel grooves with equal space was diced into the PZT ring. Then the second set of grooves was diced perpendicular to the first set. As a result, grooves with a certain pattern were produced as shown in Figure 5. Volume fraction of piezoceramic is around 90% by applying 21 cuts in each direction. Epoxy CIBA-GEIGY and Ciba Hardener were mixed in a ratio of 100:51. The mixture was filled into the grooves to form the piezocomposite wafer of the second sensor.

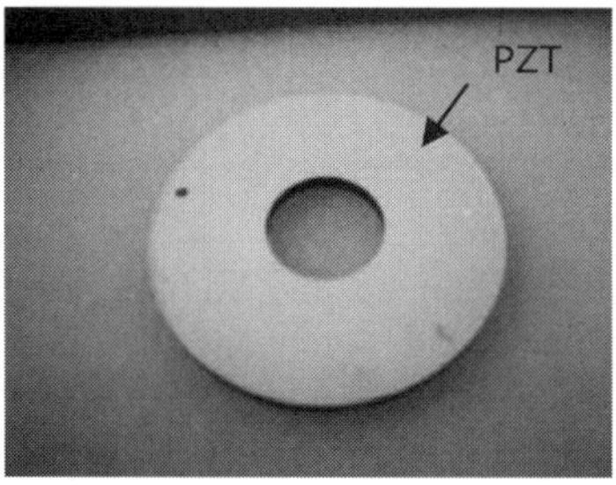

Figure 4. Piezoceramic wafer

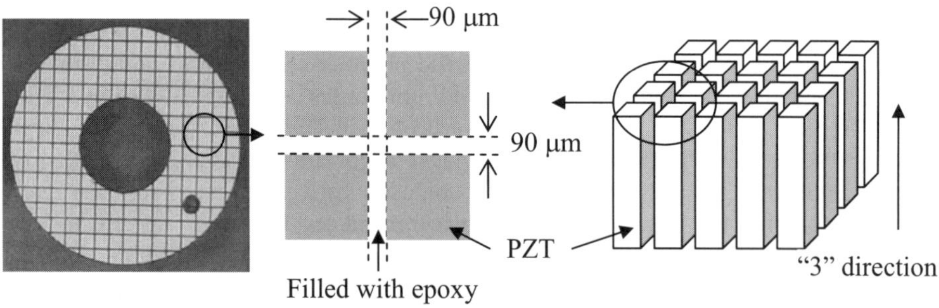

Figure 5. Piezocomposite wafer

Besides the piezoelectric (ceramic or composite) wafers, an insulating ring, two sensor electrodes, and an adapter were fabricated. Figure 6 shows the plastic insulating wafer and BeCu alloy electrode wafers. A steel adapter was designed and fabricated to integrate the sensory components and embed them into the MR dampers. It should be noted that a good surface finishing is essential for achieving highly reliable and sensitive sensors. Figure 7 illustrates an insulation band which is deployed on the surface of the threaded shaft to insulate it from the electrodes and piezoelectric wafers.

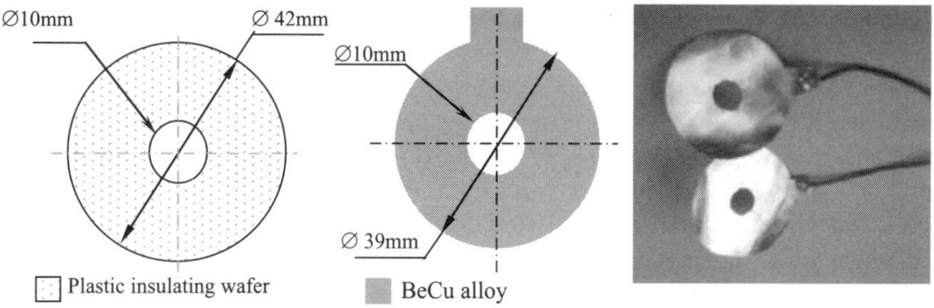

Figure 6. Plastic insulating wafer and BeCu alloy electrode wafer

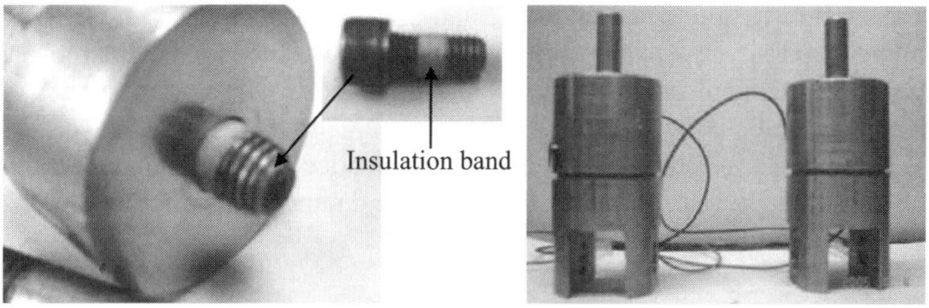

Figure 7. Bolt tied with insulation band and integration of sensory components

Piezoelectric sensors should be pre-loaded to a certain level to avoid free conditions in their operation because they work only under compressed conditions. Level of preload depends on the magnitude of the load to be measured and pressure capacity of the wafer. Output signal from the load cell was recorded by KYOWA Data logger UCAM-10B. After calibration, a torque of 40 N·m by the torque driver can produce 20 kN compression force, equivalent to stress of about 20 MPa on piezoelectric sensors. Using torque driver with load capacity set as 40 N·m, torque was applied slowly to the bolt until torque driver is overloaded. As a result, the piezoelectric sensor was preloaded to 20 MPa. During the preloading, positive and negative electrodes should be connected in short circuit to avoid accumulation of charges on the surfaces of the piezoelectric wafers. Figure 8 shows the preloading of sensory components and a fabricated self-sensing MR damper.

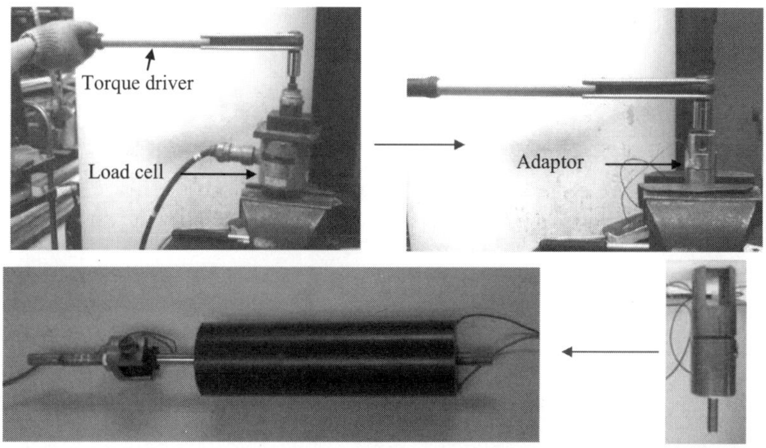

Figure 8. Preloading of sensory components and integration of self-sensing MR damper

4 Performance Tests of Self-Sensing MR Dampers

The performance of the fabricated self-sensing MR dampers was tested by using a Material Test System (MTS 810). The displacement-controlled tests were carried out in different waveforms (Sine and ramp, respectively) with different displacement amplitudes (2.5 and 5.0 mm, respectively) and different frequencies (0.2, 1.0, 5.0 and 10.0 Hz, respectively), while different voltages (0.0, 5.0 and 10.0 V, respectively) were applied to the self-sensing MR dampers.

Due to the limited space, only partial test results are presented in this paper. Figure 9 shows a comparison of the damper force measured by the embedded sensor and by the MTS for the MR damper embedded with piezoceramic sensor. The results were obtained under the excitation frequency equal to 0.2 and 5.0 Hz, respectively. Figure 10 shows the corresponding experimental results for the MR damper embedded with piezocomposite sensor. A good agreement between the results obtained by the embedded sensor and by the MTS is observed for both dampers. The agreement is better in the case of higher frequency. It is also found that the MR damper embedded with piezocomposite sensor has higher sensitivity and better reliability than the MR damper embedded with piezoceramic sensor.

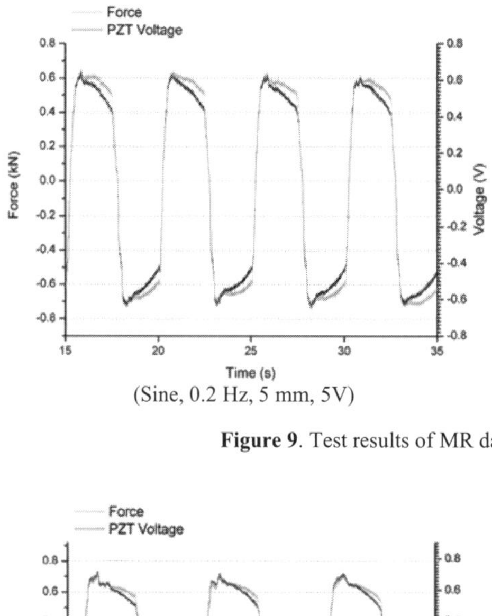

(Sine, 0.2 Hz, 5 mm, 5V)

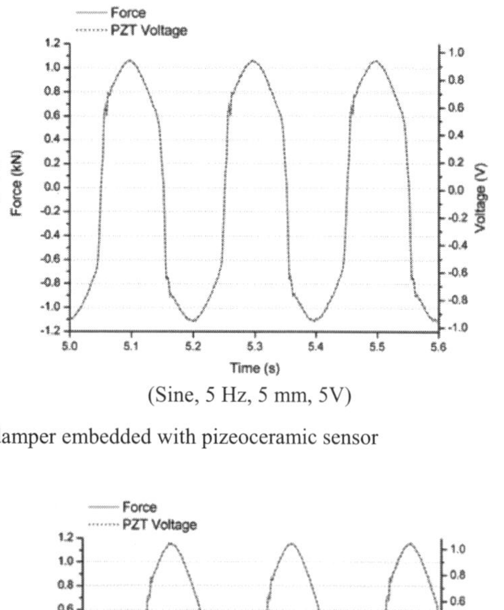

(Sine, 5 Hz, 5 mm, 5V)

Figure 9. Test results of MR damper embedded with pizeoceramic sensor

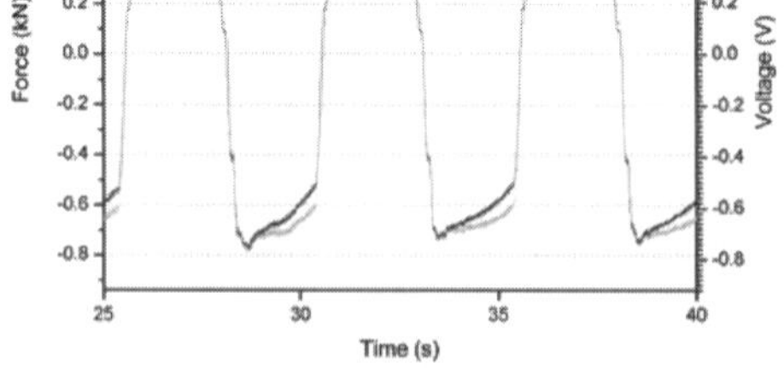

(Sine, 0.2 Hz, 5 mm, 5V)

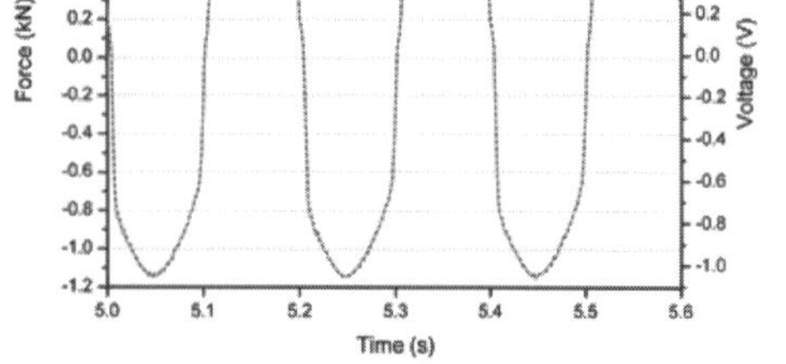

(Sine, 5 Hz, 5 mm, 5V)

Figure 10. Test results of MR damper embedded with pizeocomposite sensor

5 Conclusions

In the present study, self-sensing magnetorheological (MR) dampers have been formed by ingeniously integrating piezoceramic/piezocomposite sensors into the conventional MR dampers. Prototype MR dampers with the attractive sensing-while-damping function have been designed, fabricated and tested. The experimental results show a good agreement between the damper forces measured by the embedded sensor and by the MTS for both MR dampers embedded with a ring-shaped lead zirconate titanate (PZT) piezoceramic sensor and a ring-shaped PZT/epoxy piezocomposite sensor, respectively. It is also found that the MR damper embedded with piezocomposite sensor has higher sensitivity and better reliability than the MR damper embedded with piezoceramic sensor.

The self-sensing MR dampers with embedded piezoelectric force sensors have great potential for application to real-time closed-loop vibration control of civil and mechanical structures. The further research will include the establishment of mathematical models for this kind of novel MR dampers and the development of force feedback control algorithms specific for implementing the devised self-sensing MR dampers for real-time semi-active vibration control of structures.

6 Acknowledgements

This study was supported by an earmarked interdisciplinary research project (G-YD85) from The Hong Kong Polytechnic University. The support is gratefully acknowledged.

References

1. Breese D.G. and Gordaninejad F., "Semi-active, fail-safe magneto-rheological fluid dampers for mountain bicycles". *International Journal of Vehicle Design* **33** (2003) pp. 128-138.
2. Carlson J. D. and Jolly M. R., "MR fluid, foam and elastomer devices". *Mechatronics* **10** (2000) pp. 555-569.
3. Choi S. B., Nam M. H. and Lee B. K., "Vibration control of a MR seat damper for commercial vehicles". *Journal of Intelligent Material Systems and Structures* **11** (2000) pp. 936-944.
4. Duan Y. F., Ni Y. Q. and Ko J. M., "Cable vibration control using magneto-rheological dampers". *Journal of Intelligent Material Systems and Structures* **17** (2006) pp. 321-325.
5. Dyke S. J., Spencer B. F., Sain M. K. and Carlson J. D., "Modeling and control of magnetorheological dampers for seismic response reduction". *Smart Materials and Structures* **5** (1996) pp. 565-575.
6. Jung H.-J., Spencer Jr. B. F., Ni Y. Q. and Lee I.-W., "State-of-the-art of semiactive control systems using MR fluid dampers in civil engineering applications", *Structural Engineering and Mechanics* **17** (2004) pp. 493-526.
7. Kamath G. M., Wereley N. M. and Jolly M. R., "Characterization of magneto-rheological helicopter lag dampers". *Journal of the American Helicopter Society* **44** (1999) pp. 234-248.
8. Or S. W., Chan H. L. W. and Choy C. L., "P (VDF-TrFE) copolymer acoustic emission sensors". *Sensors and Actuators A*: *Physical* **80** (2000) pp. 237-241.

MAGNETORHEOLOGICAL (MR) JET™ FINISHING TECHNOLOGY

W. KORDONSKI AND A. SHOREY

QED Technologies Inc, 1040 University Avenue, Rochester NY 14607, USA
E-mail: kordonski@qedmrf.com

One suitable way to polish optics of complex shapes is by using a jet of abrasive fluid. In doing so, the energy required for polishing is supplied by the radial spread of the jet, which impinges upon a surface to be polished. Generally, the jet instability results in a non-deterministic polishing process. A method of jet stabilization has been proposed, developed and demonstrated whereby the round jet of MR fluid is magnetized by an axial magnetic field as it flows out of the nozzle. It has been experimentally shown that in this case a stable and reproducible material removal function can be achieved at a distance of several tens of centimeters from the nozzle. At the same time, the interferometrically derived distribution of material removal for the MR Jet coincides well with distribution of the fluid power density calculated using CFD modeling. Polishing results support the assertion that the MR Jet finishing process may produce high precision surfaces on glasses and single crystals.

1 Introduction

Conformal (or freeform) and steep concave optics are important classes of optics that are difficult to finish using conventional techniques due to mechanical interferences and steep local slopes. A missile aerodynamic nose cone, which is also an optical window, is one example. One suitable way to polish these classes of optics is by using a jet of abrasive/fluid mixture. A fundamental property of a fluid jet is that it begins to lose its coherence as the jet exits a nozzle. This results in an unstable flow over the impact zone and consequently polishing spot instability. The destabilizing disturbance is dramatically increased with jet velocity. This causes the high-speed liquid jet, which is the prime interest for finishing, to break into droplets and progressively spread out. For this reason, the diameter of a water jet used for cutting has to be very small in order to provide precision machining and high unit pressure. Also, a nozzle is situated as close as it is practically possible to the work piece to be cut. Typically, such cutting jets contain abrasive particles with a high kinetic energy capable of sputtering (chipping) material away from the surface in the impact zone. Although a flow regime where the jet can polish rather than cut requires lower jet velocity, the problem of jet stability still persists especially when applied to deterministic, high precision finishing. A reduction of jet velocity to obtain a coherent jet is impractical because it results in low impact energy, and therefore, low material removal rate. Increasing jet stability with fluid viscosity proportionally increases the resistance to fluid flow in the delivery system and consequently increases the pumping power required to deliver the fluid to the nozzle. At the same time, previous work has shown an attempt to use the water jet for polishing such materials as glass, diamond, ceramics, stainless steel and alloys [1]. An appropriate adjustment of the abrasive water jet process parameters, such as jet velocity, abrasive size and concentration makes reduction of surface roughness on glass to Ra = 1.2nm possible [2]. To be utilized in deterministic high precision finishing, a stable, relatively high-speed, low-viscosity fluid jet, which remains collimated and coherent before it impinges upon the surface is required. Such a unique tool may also resolve the challenging problem of high precision finishing of steep concave surfaces and cavities.

2 MR Jet polishing

2.1 Jet Stabilization

In contrast to known jet polishing methods [1, 2] where material removal relies on the kinetic energy of impacting abrasive particles, the technique discussed in this paper is based on the assumption that the energy required for the abrasive to provide polishing may be supplied by the radial spread of a liquid jet when it impinges upon the surface to be polished [3]. Such shear flow of abrasive slurry may provide sufficient surface stress and result in material removal, which is characteristic to polishing. A method of jet stabilization has been proposed, developed and demonstrated whereby the round jet of magnetorheological fluid, containing abrasive particles, is magnetized by an axial magnetic field when it flows out of the nozzle [3, 4, 5]. A local magnetic field induces longitudinal fibrillation and high effective viscosity within the MR fluid [6]. This results in the suppression of the most dangerous initial disturbances. As a result, the MR fluid ejected from the nozzle creates a highly collimated, coherent jet. The stabilizing structure induced by the magnetic field within the jet gradually begins to decay while the jet passes beyond the field. However, the remnant structure still suppresses disturbances and, the consequent stabilization of the MR Jet can persist for a sufficient time in which the jet may travel up to several meters (depending on the jet diameter) without significant spreading and loss of structure. The foregoing is illustrated in Fig. 1.

Figure 1. Jet snapshot images (velocity = 30 m/s, nozzle diameter = 2 mm).

In the case of water, the jet remains stable only for ~ 2 nozzle diameters (transparent section of the jet at the outlet). MR fluid has a higher viscosity than water therefore the coherent portion of the jet extends to ~7 - 8 diameters. Initial disturbances (visible in the form of ripples on the surface of the coherent part of the jet) eventually result in the jet breakdown and rapid spreading. When magnetized at the outlet, the jet of MR fluid remains coherent for more than 200 diameters. The MR fluid jets shown in Fig.1 have the same fluid viscosity and jet velocity. This means that with the magnet off, the viscosity is too low and the velocity is too high to provide a stable jet of fluid. With the magnet on, the same low-viscosity, high-velocity jet of MR fluid is stabilized by the application of the magnetic field.

2.2 *Removal Function Stability*

Polishing spot stability is the most important characteristic required for high precision deterministic finishing. The stability of the removal function generated by magnetically stabilized MR Jet was evaluated using the experimental setup described in [3]. The result is demonstrated in Fig. 2, where the maps of the material removal distribution (polishing spots) are subtracted.

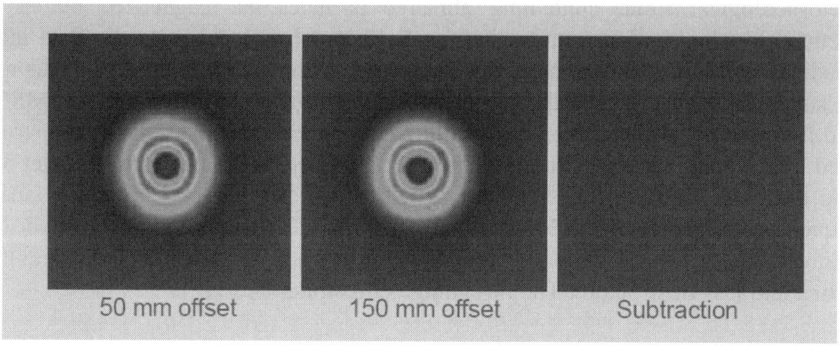

Figure 2. Subtraction of the full material removal maps (spots) showing insensitivity to stand-off distances.

The maximum variation in these removal maps is only 6.5%, even though the separation between the part and the nozzle changes by 100 mm. This quality is particularly important when considering precision finishing of steep concave (or freeform) optics.

2.3 *Polishing Performance*

The MR Jet finishing machine [3] was used to demonstrate the ability of achieving precision figure correction on concave shapes, namely, on the concave surface of a fused silica glass dome. The dome is shown in Fig. 3a.

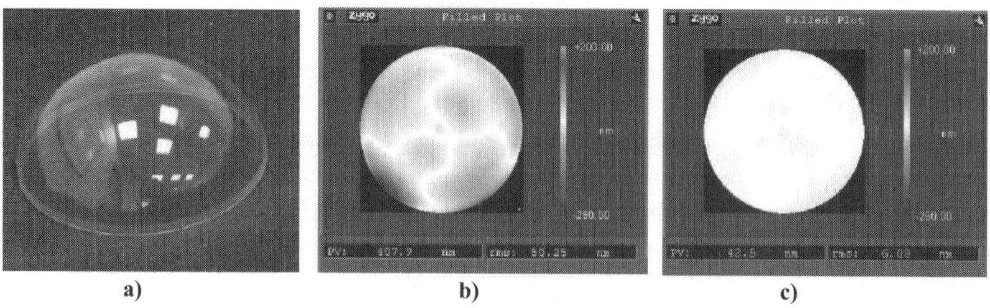

a) **b)** **c)**

Figure 3. Spherical dome polishing: (a) spherical dome; (b) the initial figure error of 408 nm P-V and 50.3 rms: (c) the final figure error of 42.5 nm P-V and 6nm rms after polishing with MR Jet.

The figure error of the concave surface was characterized and then polished using MR Jet. Figure 3b shows the initial surface error, and Fig. 3c shows the error after polishing. The figure was improved almost 10x - from 408 nm PV (50.3 nm rms) to 42.5 nm PV (6.1 nm rms). The surface roughness was < 1nm rms. This demonstrates the effectiveness of MR Jet to correct the figure of the concave surface of hemispheric domes to extremely high precision.

3 Modeling

3.1 Theoretical considerations

The proposed model of jet polishing assumes that the energy for abrasive particles/surface mechanical interaction is provided by the radial laminar flow, which occurs as a result of impingement of the coherent liquid column upon the surface. Also, the process of material removal is considered in the context of the fundamental Preston's statement that "the rate of polishing of the glass (that is, the rate at which material is removed) is proportional to the rate at which work is done on each unit area of the glass" [7]. To put it another way, the rate of material removal at each point of the surface is proportional to the local power density:

$$\dot{R} = Kw \qquad (1)$$

where $\dot{R}$ is material removal rate, K is the process/material defined constant and w is the power density (Watt/m^2). In turn, the power density can be defined as

$$w = \frac{\dot{W}}{S} \qquad (2)$$

where $\dot{W}$ is rate of work done at the surface S.
Generally, the rate of work done by the fluid at the control surface is [8]:

$$\dot{W} = -\int_S \vec{F} \cdot \vec{V} dA \qquad (3)$$

where is F the vector fluid stress applied to the element dA of the control surface and $\vec{V}$ is the vector fluid velocity. The fluid stress can be divided into components perpendicular to dA and in the plane of dA

$$\vec{F} = \vec{s}_n + \vec{\tau}_x \qquad (4)$$

where $\vec{s}_n$ is the normal stress and $\vec{\tau}_x$ is the shear stress. These considerations imply that $\dot{W}$ can be expanded as follows:

$$\dot{W} = -\int_S \vec{\tau}_x \cdot \vec{V} dA - \int_S \vec{s}_n \cdot \vec{V} dA \qquad (5)$$

The fluid shear stress can arise only from viscosity, whereas the normal stress can result from both pressure and viscosity. However, in the case under consideration when there is no normal fluid velocity component at the surface, s_n will be dominated by the compressive stress of fluid pressure:

$$\vec{s}_n = -p\hat{n} \qquad \text{but} \qquad \vec{s}_n \cdot \vec{V} = 0$$

where p is the fluid hydrostatic pressure and n is the normal to the surface. The rate of work done at the surface then becomes:

$$\dot{W} = -\int_S \vec{\tau}_x \cdot \vec{V} dA \qquad (6)$$

and the power density can be written as:

$$w = |\vec{\tau}_x \cdot \vec{V}\ | \qquad (7)$$

From Eq. (1) and Eq. (7) follows that the removal rate is proportional to the product of the shear stress and the fluid velocity, which is the power density.

3.2 *Model verification*

An effort was made to find the correlation between the experimental removal function and the theoretical model presented above. An experimental removal rate distribution was compared with the distribution of the power density calculated using a commercially available CFD package [9]. Due to the fact that the MR fluid is not affected by the magnetic field at the impingement zone, the assumption was made that it can be considered as a Newtonian fluid. The 3-D solutions for the laminar jet flow were obtained using the free surface volume of fluid (VOF) method. The distribution of the power density over the jet impingement zone was calculated as follows:

$$w = \left|\vec{\tau}_x \cdot \vec{V}\right| = \left|\tau_{yx} V_x\right| + \left|\tau_{yz} V_z\right| \qquad (8)$$

In this paper only the normal jet impingement will be discussed. One of the results is shown in Fig. 4.

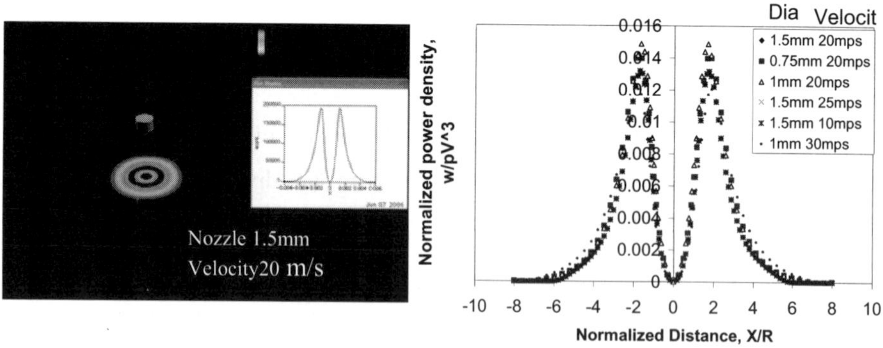

a) b)

Figure 4. Computed power density distribution at normal jet impingement: (a) Iso-surface, (b) Profile of generalized distribution of power density.

Qualitatively, by calculating with Eq. 8, distribution of the power density over the plane in Fig. 4a is similar to the distribution of removal rate in the polishing spot shown in Fig. 2. An attempt was made to generalize all obtained results with a dynamic fluid power density as applied to the surface power density and jet radius for the radial distance. The fluid dynamic power density can be presented as a product of fluid dynamic pressure $0.5\rho V^2$ and jet velocity V. Here ρ is fluid density. The result of such generalization is shown in Fig. 4b. The diametric profiles of power density distribution for a wide range of jet velocities, as well as jet diameters, are generalized by one universal profile. The same approach was used to process experimental data for jet material removal rate. The result of the generalization is shown in Fig. 5a. Again, one universal profile, in this case for the removal rate distribution, can be attributed to all jet flow parameters involved in the experiment. Finally, both average dimensionless profiles were plotted together as it shown in Fig. 5b.

a)

b)

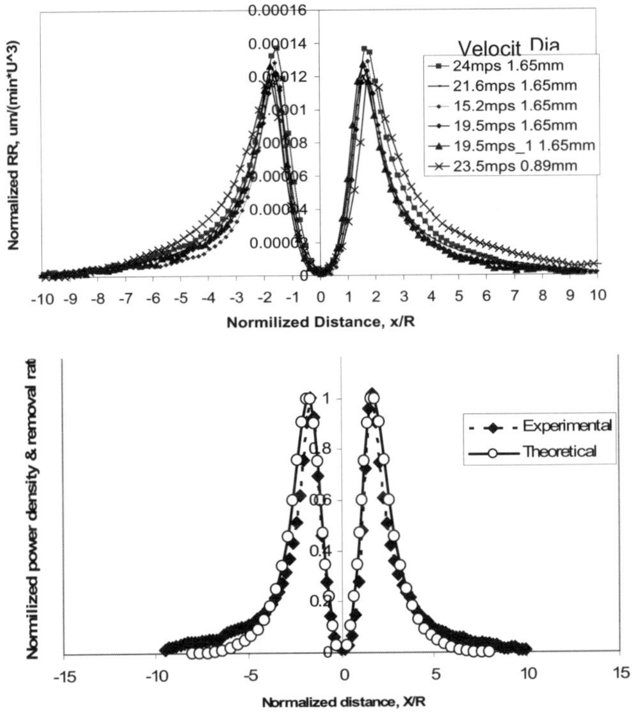

Figure 5. (a) Experimental MR Jet generalized removal rate profile; (b) Normalized profiles of the generalized removal rate and surface power density

A very good correlation is observed between both the generalized removal rate profile and the distribution of the fluid power density at the surface which provides strong support for the model presented above.

4 Summary

It has been demonstrated that impingement of a magnetically stabilized, collimated jet of MR fluid induces radial surface flow, which results in the generation of a stable polishing spot. The MR Jet finishing can produce high precision surfaces on the order of tens of nanometers p-v with roughness < 1 nm rms. Due tothe insensitivity of the offset distance, this technique may be valuable in finishing complex shapes, especially those with steep concaves and parts with a variety of cavities. In numerous experiments with different process parameters (jet velocities, nozzle diameters, and fluid viscosity) it was shown that material removal in the polishing spot closely correlates with the computed fluid surface power density. In practical terms, this means that the CFD modeling can be used to generate synthetic polishing spots for a variety of surfaces.

5 Acknowledgements

Authors appreciate support from Don Golini, Marc Tricard, Arpad Sekeres, Justin Tracy and other colleagues at QED Technologies. Portions of the research reported in this presentation were performed in connection with contract W911QX-05-C-0047 with the U.S. Army Research Laboratory and contract No. N68936-06-C-0015 from Naval Air Warfare Center Weapons Division. The views and conclusions contained in this document are those of the authors and should not be interpreted as presenting the official policies or position, either expressed or implied, of the U.S. ARL, the NAWCW Division, or the U.S. Government unless so designated by other authorized documents. Citation of manufacturer's or trade names does not constitute an official endorsement or approval of the use thereof. The U.S. Government is authorized to reproduce and distribute reprints for Government purposes notwithstanding any copyright notation hereon.

6 References

1. Momber, A. and Kovacevic R. Principles of Abrasive Water Jet Machining (Springer, New York, 1998)
2. Booij, S., Hedser van Brug, Braat J. and Fahnle O., Nanometer deep shaping with fluid jet polishing. *Opt. Eng.* **41(8),** (2002) pp.1926-1931
3. Kordonski W., Shorey A. and Tricard M. Jet-induced high-precision finishing of challenging optics. *Proc. .SPIE* **5869** (2005) pp. 44-51.
4. Kordonski, W., Golini, D., Hogan, S., and Sekeres, A. System for abrasive jet shaping and polishing of surface using magnetorheological fluid. US Patent # 5,971,835 (1999).
5. Kordonski W., Apparatus and method for abrasive jet finishing of deeply concave surfaces using magnetorheological fluid. US Patent # 6,561, 874 (2003).
6. Kordonski W., Elements and Devices Based on Magnetorheological Effect. *Journal of Intelligent Material Systems and Structures* **4:1** (1993) pp. 65-69.
7. Preston, F., W.,The theory and design of plate glass polishing machines. *J. Soc. Glass Tech* **11** (1927) pp. 214-256.
8. Blevins, R. D., Applied Fluid Dynamics Handbook. (Van Nostrand Reinhold Co., Inc. 1984).
9. Storm/CFD 2000 (Adaptive Research,, Alhambra, CA

FUZZY FINITE-ELEMENT ANALYSIS OF MAGNETORHEOLOGICAL DAMPERS

S. A. GHAZAVI AND N. YASREBI

Department of Mechanical Engineering, College of Engineering
University of Tehran, P.O. Box 11365 / 4563, Tehran, Iran
E-mail: aghazavi@ut.ac.ir

A fuzzy finite element (FFE) based method is used for modeling Magnetorheological(MR) Dampers performance with vague or imprecise uncertainties. The successful performance of MR devices or dampers requires them to be tested under imprecisely defined conditions. However, physical prototyping and testing under vague conditions are expensive and time consuming. The limitations associated with physical prototype testing can be circumvented through numerical studies.

In this work, fuzzy sets are used to represent the uncertainties present in the MR fluid, piston, housing properties (magnetic properties) and material property of magnet wire (electric resistance). The first vertex fuzzy analysis technique is used to compute the possibility distributions of a specific function that reflects some characteristics of an MR damper (electrical power and time constant). The finite element analysis of electromagnetic circuit is done by ANSYS Parametric Design Language (APDL) to obtain the magnetic flux density and to perform the optimization process. Next a new method is proposed for reducing uncertainties effects. The results show that the FFE method and the proposed method found to be useful and suitable for use in the design of MR dampers.

1 Introduction

Jacob Rainbow, an inventor at the US National Bureau of Standards, developed the first magnetorheological (MR) fluids in the late 1940s. MR fluids are suspensions of magnetizable particles that are on the order of tens of microns (20-50 microns) in diameter, in a carrier fluid. Magneto-rheological fluids are activated by the application of a magnetic field across the fluid. The induced field causes alignment of the iron particles in column-type structures parallel to the field lines. The resulting structure resists fluid flow and therefore causing a change in fluid apparent viscosity.

MR fluids can be used in many controllable devices. By applying a controllable magnetic field which subsequently causes a change in the viscosity of the fluid used in an MR device, we can control the response of a MR device. External magnetic field is applied by passing a varying current through a wired part. Thus, by changing the current we can control the applied magnetic field and consequently the performance of the MR device.

Several methods are employed for modeling MR fluids and equipments [1-5]. In all of these methods, the input parameters including the material and geometric properties of the MR device are characterized as deterministic variables. In reality, these variables have many uncertainties. For example, the properties and related graphs presented in the material data sheets are the typical properties which have some deviations from the actual properties of the material that is being used. Also many material properties change by changing performance conditions. For instance metals electric resistance changes by changing environment temperature. All these uncertainties have direct effects on the response and performance of the MR device.

Fuzzy theory is a suitable framework to model and analyze uncertainty in structures and for calculating the possibility distribution of the MR device response. The merits of a fuzzy approach are twofold. First a better prediction of the parameter is achieved by quantifying the range of the response parameter. Second, the resulting response bounds can be compared with the allowable design limits which subsequently make for an economic and safer design of smart structures. [9] Modeling of the uncertainties in structural systems, by Fuzzy theory in conjunction with Finite Element Method is presented by several researchers [6-9].

In this study we have utilized Fuzzy Finite Element Method to determine the effect of uncertainties on the performance of MR damper. Finite element calculation was performed using ANSYS Parametric Design Language (APDL) for calculating the By using this approach, a new method for designing MR damper is proposed which reduces the effect of uncertainties.

2 Fuzzy Finite Element (FFE) Method

In this section an overview of fuzzy modeling and computation is presented. Also the basic steps of using FFE are described in detail.

2.1 Modeling Uncertainties Using Fuzzy Variable

In fuzzy logic instead of using deterministic and crisp variables, fuzzy variables are used. Every fuzzy variable is associated with a fuzzy set. A fuzzy set is an imprecisely defined set that dose not have crisp boundary. It means that in a fuzzy set transition from 'belonging' to 'not belonging' to a set is a gradual transition. 'Belonging' of members of a fuzzy set is assessed using a function, which is called membership function. Membership function of a fuzzy set has values in [0,1] interval, unlike a crisp set that there are only two cases: one thing belongs to a crisp set and its membership function value is equal to 1 or it dose not belong to that crisp set and its membership function value is equal to 0.

The process of defining a fuzzy set and its related membership function for a variable with uncertainty is called fuzzification. Also determining the membership function value for a specific member is known as fuzzifying. Many different membership functions can be used, but the simplest membership function is a triangular membership function (Fig. 1), which is used in this study. Fuzzification of input variables is the first step in fuzzy finite element analysis.

2.2 Computation of Fuzzy Response [9]

The next step is computing the fuzzy response quantity. For computing the fuzzy response, 'extension principle' is used. Two different numerical methods can be used for applying 'extension principle'. These are the discretization and vertex methods .In this study vertex method is used which is described in following paragraph.

The first step in the implementation of the vertex method is the α-level representation of the fuzzy input variables. An α-level representation of a fuzzy variable x_i, denoted by $[x^{\alpha}_{iL}, x^{\alpha}_{iR}]$, is the interval in which the possibility of x_i is at least equal to α. This is illustrated in figure 1, where x^{α}_{iL} and x^{α}_{iR} are the extreme left and right values, respectively, of the fuzzy variable, x_i, at level α. Thus, at the α-level, each fuzzy variable is discretized into two crisp values, x^{α}_{iL} and x^{α}_{iR}. Once the fuzzy variables are discretized all the binary combinations of the extreme left and right values of all the fuzzy variables are formed. The binary combinations are then fed sequentially into a numerical model (finite element model) and the required response quantity, y, for each combination is obtained. Assuming there are N fuzzy variables and denoting the binary combinations at the α level by $C_{\alpha j}$ (j = 1, 2, . . . $N_{c/\alpha}$), where $N_{c/\alpha}$ is the total number of binary combinations, then the fuzzy response corresponding to $C_{\alpha,j}$ can be written in the abbreviated form $y(C_{\alpha,j})$. For example, for N fuzzy variables, the total number of binary combinations at the α-level, $N_{c/\alpha}$, is 2^N. The extreme values of the fuzzy response at the α-level are then given by

$$[y_L^{\alpha}, y_R^{\alpha}] = [\min|_{\lambda,j}\,(y(C_{\lambda,j})),\ \max|_{\lambda,j}\,(y(C_{\lambda,j}))]$$
$$\lambda \geq \alpha\,;\ j = 1, 2, \ldots N_{c/\alpha} \qquad (1)$$

where y_L^{α}, and y_R^{α}, are the minimum (left) and maximum (right) values of the response quantity, respectively. Equation (1) is an approximate implementation of the extension principle, through the vertex method, and must be repeated for all the α-levels of interest ranging from zero to one. The results obtained from equation (1), for a sweep of α-levels, are used to construct the possibility distribution, $\mu(y)$ of the response. The accuracy of the possibility distribution of the response quantity depends on the number of α-levels that is used.

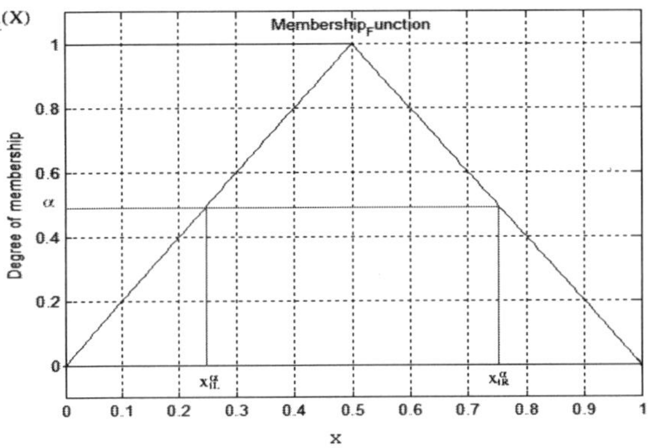

Figure 1. A simple Triangular membership function one.

3 Fuzzy Response of a MR Damper

In this section at first an optimization is done to calculate the best dimensions of an MR damper for a special case and then the fuzzy response of the optimization objective for these dimensions is calculated. Next another optimization problem is introduced in order to reduce the sensitivity of the optimization objective in the presence of uncertainty.

3.1 Optimized Design of a MR Damper

The force of a double-ended MR damper can be calculated by [10]

$$F = \Delta P \frac{(D_p + t_g)^2 - D_r^2}{4} \tag{2}$$

Where D_p is the piston diameter, t_g is the fluid gap, and ΔP is the pressure drop across the piston length which can be calculated by

$$\Delta p \approx 2\tau_y \frac{l_p}{t_g} \tag{3}$$

Where l_p is the pole length, τ_y is the yield stress of the MR fluid which depends on the exerted magnetic flux density and can be approximately calculated by [10]

$$\tau_y = \alpha B \tag{4}$$

Where B is the magnetic flux density and

$$\alpha \approx 80 \frac{KPa}{T} \tag{5}$$

The objective of optimization is assumed to be

$$J = T(= \frac{L}{R}) + \beta i^2 R \tag{6}$$

Where T is the time constant, L is the coil inductance, R is the coil resistance, and β is the weighting factor. And by this mean we can minimize time constant and electric power consummation of the MR damper [10].

L and R can be calculated from following equations:

$$L = \frac{N\Phi_B}{i} \qquad R = rN\pi D_c \tag{7}$$

Where N is the number of wire turns, r is the resistance per unit length of magnet wire, Φ_B is the magnetic flux which is calculated by finite element software, and D_c is the coil diameter.

t_g, l_p, N, i, D_c are considered as optimization parameters.

The following constrains are applied: F>1000 N, i<5 Amp, and for the finite element analysis, the magnetic flux density for the coil core is assumed B_c<1.5 T, and for the wall B_w <1.5 T.

In the next section, the calculation of the fuzzy response of the optimization objective with calculated optimized parameters is presented.

3.2 Fuzzy Response of the Optimization Objective

By using parameters values obtained in section 3.1 and utilizing the process described in section 2.2, fuzzy response of the optimization objective is calculated by assuming 10% deviation for the following variables:

- Magnetic property of MR fluid, piston and housing. 10% deviation from the following graph (Fig. 2) is considered by multiplying all B's values by a constant, γ, which varies in [0.9,1] and $B = \gamma f(H)$ where H is the magnetic field strength.

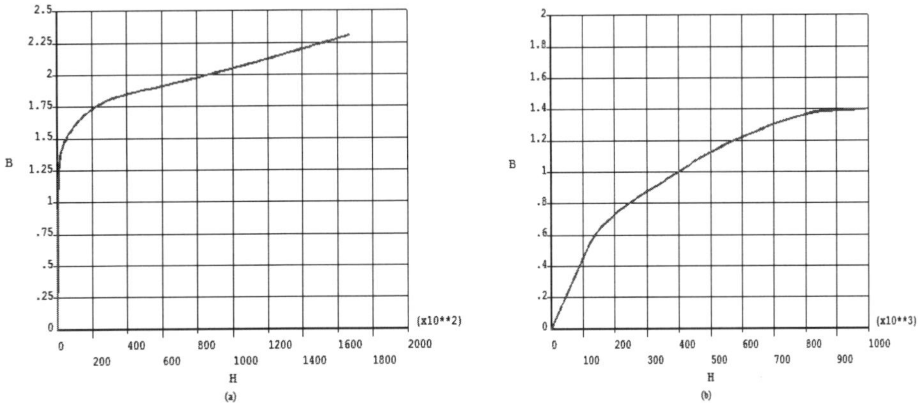

Figure 2. B-H Curve. a) Steel, b) MRF.

- Electric resistance of magnetic wire. 10% change for r (the resistance per unit length) is assumed.

3.3 Sensitivity Reduction

It is possible to reduce the range of objective function variations by changing design parameters (material properties or geometric dimensions). In other word the effects of uncertainties on the objective function can be decreased. The difference between the maximum and minimum values of the objective function ($J_m = J_{0max} - J_{0min}$) in zero-level ($\alpha=0$) is a criterion for the effects of uncertainties on the objective function. Therefore by means of reduction of this interval, a robust design is achievable. This is done by defining an optimization problem whose objective is equal to J_m. For this problem l_p, N, i, D_c are considered to be the optimization parameters and the following constrains are applied; F>1000 N, i<5 Amp, T<0.01, B_c<1.5, and B_w <1.5 T.

4 Discussion of the Results

4.1 Fuzzy Response of the Optimization Objective

The following possibility distribution of the optimization objective is obtained (Fig. 3). Average electric power is equal to 23.5W and average time constant equals to 515x 10^{-3}s and the objective optimization variation is 9.1% in zero-level ($\alpha=0$).

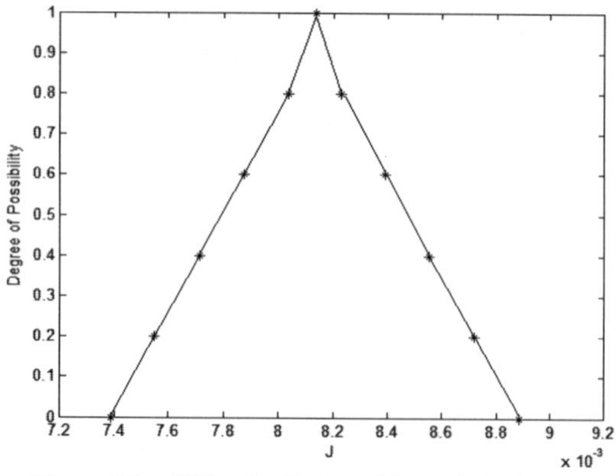

Figure 3 Possibility distribution of the optimization objective (P=23.5W, T=515e-3s)

4.2 Sensitivity reduction

The possibility distribution of the objective function J, by using the proposed method in section 3.3, is obtained (Fig. 4). The average electric power is equal to 28.75W and the average time constant equals to 427×10^{-3}s and the objective optimization variation is 4.8% in zero-level (α=0). By comparing Fig. 4 with Fig. 3 it is clear that the proposed method for reducing the sensitivity, in spite of increasing the electrical power, decreases the time constant and the objective optimization variation.

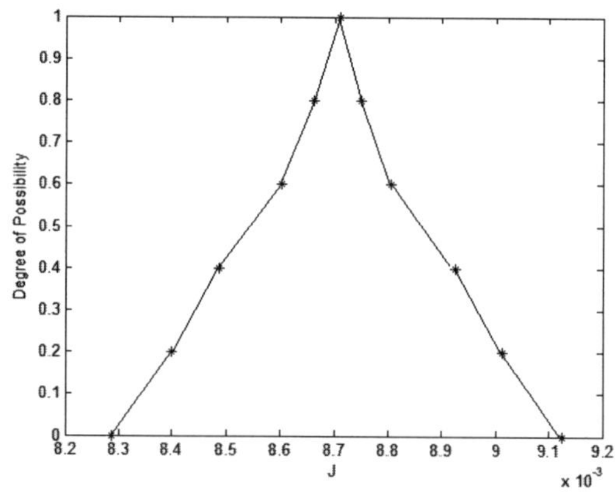

Figure 4 Possibility distribution of the optimization objective (P=28.75W, T=427e-3s)

5 Conclusions

A fuzzy finite element analysis of MR damper is performed and for a specific function J, that reflects some characteristics of an MR damper, the fuzzy response is calculated. The same process and calculation can be done for other characteristics and specification of a

MR damper and other different inputs may be considered as fuzzy variables.
Also a new method is introduced and successfully applied in order to reduce the effects of uncertainties on the objective functions. The computation cost of this method approximately is equal to the fuzzy response calculations. Therefore the proposed method may be used as a numerical design method to design any structures especially smart structures in which accuracy plays an important role.

References

1. Ericksen, E. O and Gordaninejad, F., A Magneto-Rheological Fluid Shock Absorber for an Off-Road Motorcycle, *International J. Vehicle Design* **33** (2003) pp. 138-152.
2. Norman M Wereleyy and Li Pang, Nondimensional analysis of semi-active electrorheological and magnetorheological dampers using approximate parallel plate models, *Smart Mater. Struct.***7** (1998) pp. 732–743.
3. -GuangqiangYang, Large-Scale Magnetorheological Fluid Damper for Vibration Mitigation: Modeling, Testing and Control, University of Notre Dame,PHD thesis,2000
4. Wojcieh Szelag,Finite element analysis of coupled phenomena in magnetorheological fluid devices,*the international journal for computation and mathematics in Electrical and Electronic Engineering* **23** (2004) pp.813-824.
5. MANAS DAS, CFD analysis of magnetorheological polishing fluid (MRPF) flow in magnetorheological abrasive flowfinishing process, *ROLL* **-Y210517** (2000) pp.30-137.
6. Rao S S and Sawyer J P ,A fuzzy finite element approach for the analysis of imprecisely-defined systems, *AIAA J.* **33** (1995) pp.264–370.
7. Chao R-J and Ayyub B , Finite element analysis with fuzzy variables, *Building an International Community of Structural Engineers (Proc. Structures Congress XIV)*ed S K Ghosh and R Mohammadi, pp.643–50.
8. Chen L and Rao S S , Fuzzy finite element approach for the vibration analysis of imprecisely-defined systems, *J. Finite Elements Anal. Design* **27** (1997) pp.69–83
9. U.O.Akpan, T.S. Koko, I.R. Orisamolu and B.K. Gallant, Fuzzy finite-element analysis of smart structures, *Smart Mater. Struct.* **10** (2001) pp.273–284
10. Henri GAVIN, Jesse HOAGG and Mark DOBOSSY, OPTIMAL DESIGN OF MR DAMPERS,*Proc. U.S.-Japan Workshop on Smart Structures for Improved Seismic Performance in Urban Regions Seattle WA*, ed. K. Kawashima, B.F. Spencer, and Y. Suzuki, (2001) pp.225-236.

QUALITATIVE ANALYSIS OF MAGNETO-RHEOLOGICAL TUNED VIBRATION ABSORBERS: EXPERIMENTAL APPROACH

JEONG-HOI KOO

Department of Mechanical and Manufacturing Eng., Miami University, Oxford, OH 45056 USA
E-mail: koo@muohio.edu

MEHDI AHMADIAN

Department of Mechanical Engineering, Virginia Tech, Blacksburg, VA 24061 USA
E-mail: ahmadian@vt.edu

The primary purpose of this study is to provide an experimental evaluation of some of the characteristics and benefits of magneto-rheological tuned vibration absorbers (MR TVAs). Tuned vibration absorbers (TVAs) often provide a simple and elegant solution to many vibration problems. To extend the concept of passive TVAs to MR TVAs, we discuss four different control methods, called velocity-based on-off groundhook control, velocity-based continuous groundhook control, displacement-based on-off groundhook control, and displacement-based continuous groundhook control. Using the test apparatus, a series of tests were conducted to investigate the dynamics of the MR TVA with each control policy. The performances of each of the cases were then analyzed along with the equivalent passive TVA. The performance index was the transmissibility between the input and the output displacement of the structure. The experimental results indicate that the most suitable control method for MR TVAs is the displacement-based groundhook control.

1 Introduction

Tuned Vibration Absorbers (TVAs) have been effective engineering devices to suppress vibrations of machines and structures since their invention in the early 1900s. A TVA is a vibratory subsystem attached to a primary system. It normally consists of a mass, a spring, and a damper. Mounted on the primary system, a TVA counteracts the motions of the primary system, "absorbing" the primary structure's vibrations. However, a conventional passive TVA is only effective when it is tuned properly; otherwise, it can magnify the vibration levels of the primary system, hence, the name "tuned" vibration absorber. In many practical applications, the off-tuning of a TVA often occurs because of system parameter changes over time. To cope with these problems, extensive studies over the past two decades have been performed to develop new designs and concepts for passive TVAs. These designs include adaptive, semi-active, and active TVAs. A comprehensive review of configurations, developments, and applications of TVAs can be found in the papers by Sun et al. [1], Housner et al. [2], and Soong and Spencer [3].

As an alterative to passive TVAs, this paper considers a semi-active TVA with a Magneto-Rheological (MR) damper. The semi-active MR TVA model (Figure 1b) replaces a passive damping element with a controllable damping element, which distinguishes it from the conventional passive system (Figure 1a). In the passive TVA model, the primary structure is coupled with a damped TVA, and the mass of the structure and absorber are defined by m_1 and m_2, with their corresponding displacements as x_1 and x_2, respectively. The absorber's spring (k_2) and damper (c_2) are mounted on the structure. The stiffness and damping of the structure are represented by k_1 and c_1, respectively. A controllable damping element (i.e. MR damper) is the key element for the semi-active MR TVA.

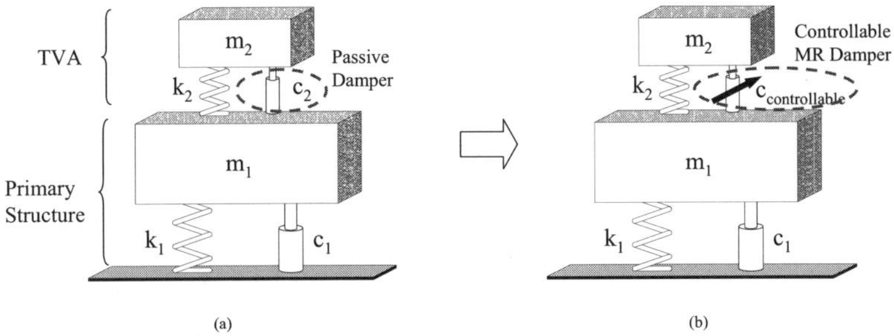

Figure 1. (a) Conventional passive TVA (b) Semi-active MR TVA

2 Semi-active Controllers

This section briefly introduces four semi-active control policies examined for the MR-TVA. Details of theses control algorithms and physics behind them can be found in reference [4]. The four strategies are:

- velocity based on-off groundhook control (On-off VBG)
- displacement based on-off groundhook control (On-off DBG)
- velocity based continuous groundhook control (Continuous VBG)
- displacement based continuous groundhook control (Continuous DBG)

Table 1. Semi-Active Damping Control Policies

On-ff VBG	$v_1 v_{12} \geq 0 \quad c_{controllable} = c_{on}$ $v_1 v_{12} < 0 \quad c_{controllable} = c_{off}$	(1)
On-off DBG	$x_1 v_{12} \geq 0 \quad c_{controllable} = c_{on}$ $x_1 v_{12} < 0 \quad c_{controllable} = c_{off}$	(2)
Continuous VBG	$c_{controllable} = \left\| \frac{G v_1}{v_{12}} \right\|$	(3)
Continuous DBG	$c_{controllable} = \left\| \frac{G x_1}{v_{12}} \right\|$	(4)

In the above equations, v_1 is the velocity of the structure and v_{12} represents the relative velocity across the damper. The displacement of the structure mass is represented as x_1. The parameter G represents a constant gain that can be pre-adjusted to control the overall damping of the TVA.

3 Test Setup

A 2-DOF test rig (see Figure 2) was designed to experimentally evaluate an MR damper for a semi-active TVA. For the damping element of the semi-active TVA, an "MR sponge damper" was used. For the input to the test rig, a chirp signal was used. The chirp sweeps frequency ranges 0.5 Hz to 10 Hz in 68 seconds. A detailed description of the experimental setup including the test procedures can be found in reference [5].

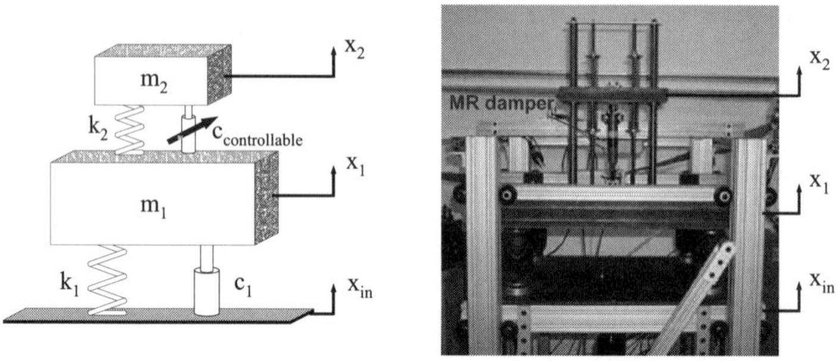

(a) (b)

Figure 2. (a) Mathematical Model of Semi-Active TVA (b) Physical Model of MR TVA

4 Dynamic Analysis of MR TVA in Passive Mode

This section presents the dynamic behavior of the passive TVA, which will then be compared with those of the semi-active TVA. For passive TVA testing, the MR damper was used in a passive mode. In other words, the current to the MR damper was constant throughout the test, making the damper behave like a passive damper.

4.1 Tuning the Passive TVA

Typically, three parameters need to be determined to design a TVA; its mass, stiffness, and damping ratio. This experimental study uses a fixed TVA mass (m_2), which weighs 32 lb (the TVA plate mass + fixture mass). The passive TVA was tuned by fine tuning the parameters experimentally. Table 2 summarizes the tuned parameters.

Table 2. Parameters for Passive TVA

TVA Mass (m_2)	32 lb
Mass Ratio (μ)	5.56 %
Each spring Stiffness (k_{cs})	26 lb/in
Current (i)	0.12 Amps

4.2 Dynamic Analysis of the Passive TVA

The dynamic analysis is performed through the use of the frequency domain response. Figure 3 shows the transmissibility and phase plots of the passive TVA testing. Transmissibility, which is the ratio between output displacement of the structure and the input, was used as a performance index. As the transmissibility increased, the magnitude of vibrations in the primary structure also increased. Thus, the purpose of using a TVA was to reduce transmissibility. Before adding a TVA to the primary system, there was one resonant peak with a maximum transmissibility of fifteen as shown in Figure 3a, indicating a single-degree-of-freedom system (SDOF). The SDOF resonant peak indicates that the primary system's natural frequency was 5.1 Hz. Mounting a TVA on the structure reduced the transmissibility. When the current was 0.0 Amps, there were two resonant peaks, indicating that the TVA mass added one more degree of freedom to the system. Increasing the current from 0.0 Amps to 0.1 Amps reduced the two resonant peaks. However, this occurred at the expense of increasing the transmissibility at the valley. As the damping was further increased (by increasing current), the two resonant peaks merged into one peak, and the peak grew. This indicates that the TVA mass and the primary mass were linked, disabling the TVA and magnifying the vibrations. At a high current (i.e. 0.6 Amps in this experiment), the coupled system acted much like a SDOF system, showing one distinct resonant peak (see Figure 3a). However, the frequency of the new SDOF peak was lower than that of the no-TVA-attached SDOF peak. This was because the effective mass of the coupled system became the sum of the TVA mass and the structure mass and reduced the natural frequency.

Figure 3b further shows the dynamics of the passive system with phase angles. A phase angle is the angle between the TVA and the structure. When the current was 0.0 Amps or 0.1 Amp, the phase angle was close to 90 degrees around the resonant frequency of the structure. This indicates that the TVA mass and the structure mass were 90 degrees out of phase, and the TVA counteracted the motions of the structure. Consequently, the transmissibility decreased, implying a reduction of structure vibrations. However, when the current was 0.3 Amps, the phase angle dropped below 90 degrees. With a current of 0.6 Amps, the phase angle became about 10 degrees, implying that the TVA and the structure were highly coupled. Thus, the benefits of using the TVA were negated at this high current.

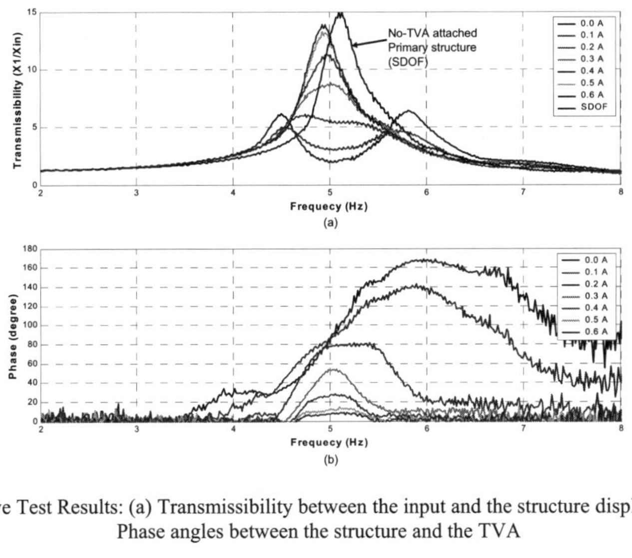

Figure 3. Passive Test Results: (a) Transmissibility between the input and the structure displacement and (b) Phase angles between the structure and the TVA

5 Dynamic Analysis of MR TVA with Semi-Active Control

The subsequent sections study the dynamics of semi-active TVAs by performing parametric studies. The primary purpose of the parametric studies is to understand the dynamics of each semi-active TVA. The parametric studies look into the effects of on-state/off-state currents as well as control gains.

5.1 *Design of Semi-active TVAs (Tuning Semi-active TVAs)*

Semi-active TVAs were tuned in the same manner as the passive TVA case. Simulation results were used to estimate the initial coil spring rate (k_{cs}), on-state and off-state currents. Table 3 summarizes the parameters for each semi-active TVA after fine tuning.

Table 3. Parameters for Semi-active TVAs

Parameters	On-off VBG	Continuous VBG	On-off DBG	Continuous DBG
Coil spring rate	25 lb/in	25 lb/in	25 lb/in	25 lb/in
On-state current	0.2 Amps	0.6 Amps	0.6 Amps	0.6 Amps
Off-state current	0 Amps	0 Amps	0 Amps	0 Amps
Gain	N/A	0.1	N/A	1000

5.2 *On-Off Groundhook Controlled TVAs*

Dynamic performance of MR TVAs with on-off groundhook controllers are evaluated by changing the on/off-state currents. To study the effect of on-state currents, the on-state currents were discretely set at 0.0, 0.1, 0.3, and 0.6 Amps, while the off-state current was set to 0.0 Amps. The lowest off-state current (0.0 Amps) ensured the MR damper's lower boundary current limit and offered a wider range of force envelopes.

Figure 4a shows the transmissibility plot of the on-off VBG controlled semi-active TVA. With an on-state current above 0.3 Amps, the semi-active TVA experiences coupling of the two masses, which is similar to the passive TVA dynamics shown in Figure 3a. However, the rate of transmissibility growth at the valley occurs slowly when compared with that of the passive TVA. This indicates that the coupling of the two masses in the on-off VBG case is less than in the passive case.

Figure 4b shows the results of the on-state currents changes of the on-off DBG TVA. As the on-state current increased, the two resonant peaks decreased. Unlike the passive system and the on-off VBG case, the reduction in the magnitudes of two resonant peaks occurred without the expense of raising the isolation valley. In other words, the isolation valley maintained the minimum transmissibility around the resonance of the structure, keeping the two masses decoupled at high on-state currents. This observation highlights one of the relative merits of the on-off DBG controlled, semi-active system.

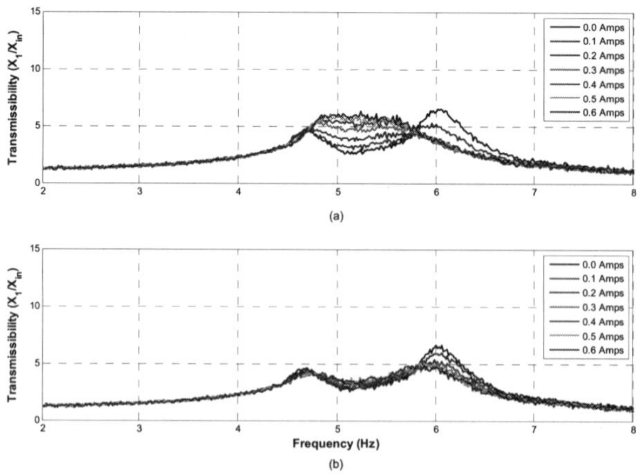

Figure 4. Effect of On-State Current of the Semi-Active TVA: (a) On-off VBG (b) On-off DBG

To further evaluate the dynamics of the on/off-groundhoook TVAs, the effect of changing the off-state currents was investigated. Figure 5a shows that as the off-state current increased, the two resonant peaks came together and the isolation valley grew. This indicates that the dynamics of the on-off VBG system with high off-state currents slowly approached those of the passive system. Figure 5b shows that as the off-state damping (current) increased, the amplitudes of the two resonant peaks also increased and the isolation valley grew. The results in Figure 5 indicate that the dynamics of the semi-active system with high off-state currents slowly approached those of the passive system. These suggest that the semi-active TVA should keep the off-state current as low as possible (if not optimal) to obtain its performance gains.

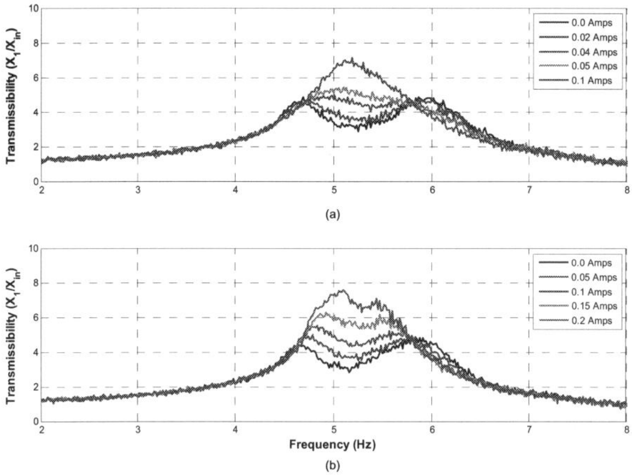

Figure 5. Effect of Off-State Current of the Semi-Active TVA: (a) On-off VBG (b) On-off DBG

5.3 Continuous Groundhook Controlled TVAs

This section presents the dynamic performance of the continuous semi-active TVAs. For the Continuous VBG case, the on-state current was set at the maximum operation current (i.e., 0.6 Amps) and the off-state current was set at 0.0 Amps (minimum damping boundary). To change the dynamics of the TVA, the gain was adjusted. The gain, the constant parameter 'G' in equation 3, varied discretely from 0 to 1000. This gain determined the distribution of damping (current) levels between the two boundaries. A lower gain indicates that damping levels are more closely distributed in the lower boundary (or off-state); conversely, a higher gain indicates that the damping levels are more distributed in the upper boundary. As the gain increased, the continuous groundhook control followed the on-off damping control. This is because the large gain pushed the damping distribution up to the maximum (or on-state) damping limit. Figure 6a shows the results of continuous VBG control. When the gain increased, the two peaks came together and the peak grew, exhibiting similar dynamics to those of passive and on-off VBG control. The continuous DBG control case was studied the same manner as the continuous VBG case. Figure 6b shows the results of the continuous DBG controlled semi-active TVA. As the gain increases, the two peaks come down, but the valley rises slightly. For the most part, the dynamics of continuous DBG are the same as on-off DBG. Both displacement-based semi-active controls outperform the passive and velocity based controls in reducing the resonant vibrations of the primary structure.

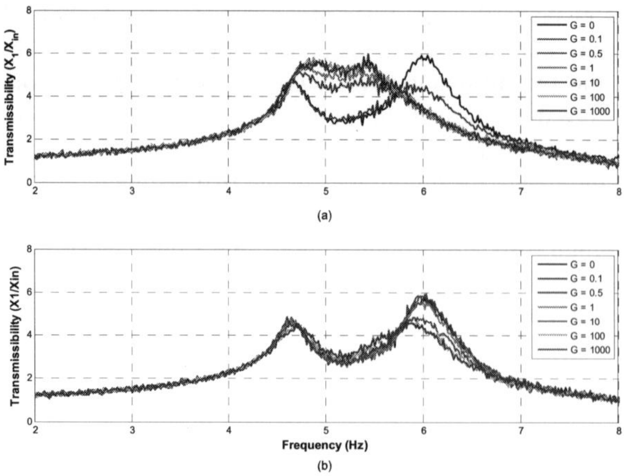

Figure 6. Effect of gains of Semi-Active TVA: (a) Continuous VBG and (b) Continuous DBG

6 Conclusion

This paper presented an experimental evaluation of the dynamic performance of magneto-rheological tuned vibration absorbers (MR TVAs). After introducing four different control methods used to control damping in the MR TVA, a 2-DOF MR TVA test rig was briefly described. Using the test rig, a series of tests were conducted to investigate the dynamics of the MR TVA with each control policy. The performances of each of the cases were then analyzed along with the equivalent passive TVA. The results suggest that the semi-active TVAs should keep the off-state current as low as possible (if not optimal) to obtain their performance gains. The results further indicate that the most suitable control method for MR TVAs is the displacement-based groundhook control.

References

1. Sun, J. Q., Jolly, M. R., and Norris, M. A., “ Passive, Adaptive and Active Tuned Vibration Absorbers- A survey,” Transactions of the ASME, 50th Anniversary of the Design Engineering Division, Vol. 117, pp. 234-242, 1995.
2. Housner, G. W., Bergman, L. A., Caughey, T. K., Chassiakos, A. G., Claus, R. O., Masri, S. F., Skelton, S. E., Soong, T. T., Spencer, B. F., and Yao, J. T. P., “Structural Control: Past, Present, and Future,” Journal of Engineering Mechanics, 123(9), pp. 897-971, 1997.
3. Soong, T. T. and Spencer, B. F. Jr., “Supplemental Energy Dissipation: state-of-the-art and state-of-the-practice,” Journal of Engineering Structures, 24, pp. 243-259, 2002.
4. Koo, J-H., Ahmadian, M., Setareh, M., and Murray, T., “In Search of Suitable Control Methods for Semi-Active Tuned Vibration Absorbers,” Journal of Vibration and Control, Vol. 10, No. 2, pp 163-174 February, 2004.
5. Koo, J-H., Ahmadian, and M., Setareh, M, “Experimental Robustness Analysis of Magneto-Rheological Tuend Vibration Absorbers Subjected to Mass Off-Tuning,” Journal of Vibration and Acoustics, Vol. 128, pp 126-131 February, 2006.

ANALYSIS AND DEVELOPMENT OF A MAGNETORHEOLOGICAL FLUID - BASED TRANSPARENT HAPTIC SYSTEM

FARZAD AHMADKHANLOU, GREGORY N. WASHINGTON, AND STEPHEN E. BECHTEL

The Ohio State University, 650 Ackerman Rd, Suit 255, Columbus, OH 43202, USA

Email: Ahmadkhanlou.1@osu.edu, Washington.88@osu.edu, Bechtel.3@osu.edu

The performance of teleoperation systems can be improved if force feedback is employed. When the operator controlling the master has the feeling of direct interaction with the remote environment, the system is called "transparent". The goal of this paper is to combine novel magnetorheological (MR) fluid-based haptic systems with microstructural analysis and modern control techniques to design and develop a transparent haptic system. For this purpose, a MR fluid-based joystick is designed. The work in this paper overcomes the sticky wall phenomena, adds restoring force feeling to the user by using a variable compliance, and increases the overall transparency of the system.

1 Introduction

Recently force feedback techniques have been used to increase the effectiveness of human-machine interfaces. The applications range from biomedicine to computer games and virtual reality devices. A teleoperation system consists of a pair of manipulators: master and slave. The operator handling the master directs control signals to the slave which acts on an external object. The performance of teleoperation systems can be improved by the addition of force feedback. When the operator controlling the master has the feeling of direct interaction with the remote environment, the system is called "transparent". Most traditional force feedback systems employ DC motors or hydraulic actuators in the master to simulate the feedback forces. Slave inertias are in general considerably higher than their corresponding master inertias, which leads to the feedback of large signals emanating from impacts between the slave and its external object. These abrupt signals must be significantly attenuated to prevent instability, but this attenuation produces nontransparent systems that have a characteristically mushy feel. The comparative advantage of using magnetorheological (MR) fluids in haptic devices lies in the increase in feedback signals that can be gained without instability. The goal of this paper is to combine novel MR fluid-based haptic systems with microstructural analysis and modern control techniques to design and develop a transparent haptic system.

The main objective of any force feedback system is to reproduce at the user's (master) end the forces encountered by the actual (slave) or virtual system. This can be seen as a tracking problem where the goal is to track the force at the 'virtual end' and regenerate it at the user end. The authors have developed a MR fluid-based force feedback joystick used in a tracking system [1, 2]. This system provides the user with tactile information in an environment. Although the results show significant improvement by adding force feedback to visual-only feedback, the tracking system is found to be lacking in two specific areas: (i) The most notable negative is the "sticky wall" phenomena (the user senses an unphysical force when leaving a surface). (ii) The system is unable to provide restoring forces to the user, so that the user cannot sense the stiffness of the external object. The work in this paper overcomes these two limitations through the addition of variable compliance to the tracking system. Variable compliance is incorporated by the addition of torsional springs driven by DC motors, either in parallel or series with the dampers.

This paper is organized as follows: Kinetic theory-based microstructural modeling of MR fluids is used in section 2 to derive the constitutive equation relating stress, strain rate, and magnetic field. Section 3 describes the preparation of MR fluids with a range of particle sizes and concentration, and carrier fluids, and characterization of these MR fluids to identify the optimal fluid. Section 4 describes the design and control of a force feedback system utilizing the MR fluid. Section 5 describes how the haptic device together with a DC motor and a force sensor can be combined to create a transparent system (Figure 1). The paper is concluded by presentation of experimental results and conclusions in sections 6 and 7.

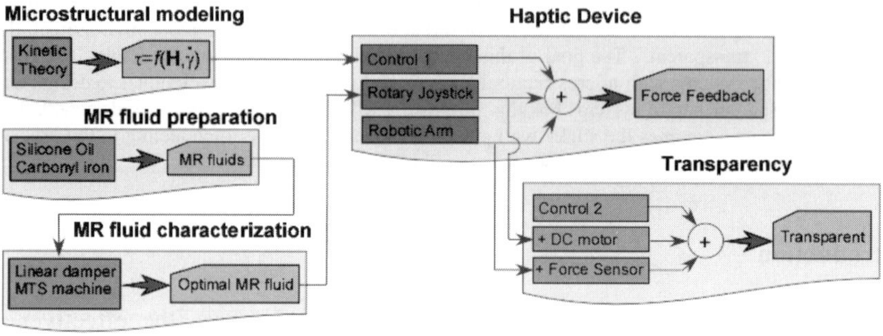

Figure 1. MR fluid based transparent haptic system

2 The kinetic theory-based elastic dumbbell model of MR fluids

A magnetorheological fluid consists of micron sized magnetizable particles in a carrier fluid. In kinetic theory-based models, the suspended particle is considered as either an elastic dumbbell or a rigid dumbbell. The distributed force from the carrier fluid on the particle is idealized as being localized on the beads at the ends of the particle, with a rigid or elastic connector between the beads. Macroscale constitutive equations relating flow, stress, and particle orientation are produced by integrating the coupled equations governing forces, flow, and orientation over a representative volume of particles and carried fluid. In the elastic dumbbell model of the iron particles, two beads of mass m are joined by a spring. The connector **q** between two beads is representative of the orientation of the iron particle in the carrier fluid. The evolution equation for particle orientation is deduced from the Smoluchowski (kinetic) equation that describes the rate of change of orientation **q** with time:

$$\dot{q}_i = \underbrace{\left(L_{ij} - \mu D_{ij}\right) q_j}_{\textbf{effect of carrier fluid flow}} + \underbrace{a_{ij} \frac{2kT}{\psi} \frac{\partial}{\partial q_k} \left(\xi_{kj}^{-1} \psi\right)}_{\textbf{effect of Brownian motion}} + \underbrace{2a_{ij} f_j^{(c)}}_{\textbf{effect of intraparticle force}} + \underbrace{a_{ij} \left(f_{1j}^{(m)} - f_{2j}^{(m)}\right)}_{\textbf{effect of magnetic field}}$$

The left hand side of this equation is the Lagrangian time derivative of particle orientation, and the terms on the right hand side arise from the four physical effects that can contribute to this change of orientation. Definitions of all parameters and coefficients, and details of the derivation, are in [3]. We consider the particles to be elastic with modulus β, so that the internal particle force is $f_i^{(c)} = \beta q_i$ and introduce $f_{1i}^{(m)} - f_{2i}^{(m)} = \frac{c\dot{\gamma}}{(1+\chi)|\mathbf{H}|} q_i$. The 3-D mesoscale equations for the MR fluid is obtained by integrating the microscale Smoluchowski equation and a constitutive for stress over a representative volume element containing many particles. To evaluate the modeling choices of $f_i^{(c)}$ and $f_{1i}^{(m)} - f_{2i}^{(m)}$ against experiments, we specialize the 3-D theory to the 1-D application of steady simple shear in the presence of a transverse magnetic field by inserting the special velocity and magnetic fields $\mathbf{v} = [\dot{\gamma}x_2 \quad 0 \quad 0]^T$, $\mathbf{H} = [0 \quad H \quad 0]^T$, with positive constant shear rate $\dot{\gamma}$ and magnetic field strength H. This reduces the above constitutive equations for the 3-D stress tensor to a 1-D constitutive equation for shear stress in simple shear

$$\tau_{12} = \eta_s \dot{\gamma} + \frac{nkT\,\zeta}{4\left(\beta + \dfrac{c\dot{\gamma}}{2(1+\chi)H}\right)}\dot{\gamma}$$

Explicitly, the scalar τ_{12} in this equation is the component of the stress tensor accompanying the velocity and magnetic fields. From experimental results we observe that this expression replaces the strict yield stress of the Bingham/Herschel-Buckley-type visco-plastic models with a steep gradient, which is in better agreement with experimental observation.

3 Preparation and characterization of MR fluids

Thirty one different MR fluids were employed in the investigation of this paper, consisting of carbonyl iron powder in two different particle size ranges (2-5μm and 4-7μm) mixed with silicone oils with five different viscosities (η=0.00935, 0.01906, 0.34055, 0.974, and 12.1875 Pa-s). Weight ratios of carbonyl iron powder to silicone oil ranged from 2 to 6 percent.

These MR fluid samples were utilized in a linear damper designed and constructed by the authors [5, 6]. Conventional MR fluid-based dampers have a reservoir of MR fluid sealed by an o-ring or gasket; these mechanical seals create sizable off-state resistant shear viscosity. The authors' damper saturates an open-celled polyurethane sponge with MR fluid, confining the fluid and eliminating the need for seals, thereby reducing the off-state viscosity.

The damper was used to characterize the MR fluid behavior: Solenoid was subjected to a controlled current ranging from 0 to 4 amps, generating a transverse magnetic field across the foam saturated with MR fluid. An MTS machine was used to control the axial motion of the piston in the damper and measure the required axial force.

The MR fluid used by Rabinow in his investigation [4] consisted of 9 parts by weight of carbonyl iron to one part of silicone oil; we investigated MR fluids with a range of particle sizes, carrier fluids, and weight ratios, to measure the dynamic range (the ratio of maximum viscosity in on-state to minimum viscosity in off-state) as a function of the MR fluid ingredients. Among all the 31 different samples, the 12 samples made with reported 1000 and 12500 cSt silicone oils were too viscous to saturate the foam. All other samples were tested by the MTS machine. The resulting force vs. displacement measurements of the damper for varieties of MR fluids were used to characterize the behavior of each MR fluid/damper system.

The MR fluid suitable for a haptic device should have the following characteristics: very low off-state viscosity, large dynamic range, and low particle settlement. The MR fluid sample made of 5 parts by weight of 4-to-7μm carbonyl iron powder to one part of 10 cSt Silicone oil was determined to be the one most appropriate to be used in the MR fluid-based haptic system in the next section.

4 MR fluid-based force feedback rotary system

In a previously designed single degree of freedom (SDOF) rotary system [2], a user operates the rotary MR damper joystick to command the motion of a rotary actuator (Figure 2). The end effector (EE) for the force feedback system is an SRV-02, a Quanser supplied robotic arm. The rotation of the probe (robotic arm), which is held at the center by two springs, is controlled by the joystick. The position of the joystick is sensed by a digital encoder. The encoder signal is sent to a data acquisition system that, in turn, sends a voltage to the robotic arm. When the probe contacts a stiff or compliant object, the probe will pivot and the angular change is sensed by an encoder that is coupled to a spring. A flexible force sensor, FlexiForce®, is placed on the robotic arm. This coupling allows one to produce a simulated stiffness from force-displacement data.

When the EE comes in contact with an external object, depending upon the position of the arm and the sensed force, the DC power supply sends a current to the solenoid within the MR damper joystick. As the angular change or force increases, the current sent to the MR joystick increases and the user feels the force from MR joystick proportional to the force encountered by the springs or the force sensor.

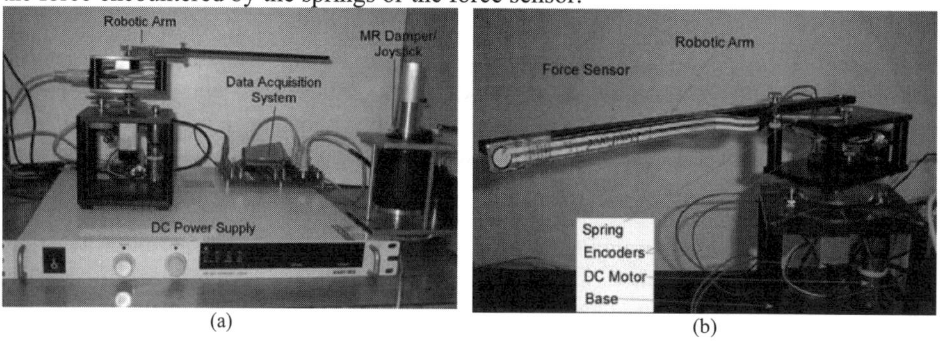

Figure 2: SDOF system: (a) Rotary force feedback system setup (b) Force Sensor

The state space form of the equation of motion for the SDOF rotary system with explicit connection to the microstructural modeling can be written as [1, 7]

$$\begin{bmatrix}\dot{\theta}\\ \ddot{\theta}\end{bmatrix}=\begin{bmatrix}0 & 1\\ \dfrac{-K_D}{J_D} & -\dfrac{4\pi R^3 L\eta}{hJ_D}\end{bmatrix}\begin{bmatrix}\theta\\ \dot{\theta}\end{bmatrix}+\begin{bmatrix}0 & 0\\ \dfrac{1}{J_D} & -\dfrac{2\pi R^2 LnkT\,\zeta B_1 B_2}{\chi B_{\max} J_D}\end{bmatrix}\mathbf{u}\,;\,\mathbf{u}=\begin{bmatrix}T_D\\ V_{in}\end{bmatrix}$$

Definitions of all parameters and coefficients, and details of the derivation, are in [1, 7]. This equation relates the inputs to the master, T_D and V_{in}, to the output position θ which is sent to the slave. Full state feedback control is implemented to control the motion of the robotic arm. Through proportional – derivative (PD) control, a matrix gain is used to give an output voltage to the arm to move the base to the desired position. A 4th order Bessel function is used to place the gain values of gain such that the system has the quickest response possible while eliminating overshoot [2].

5 Transparency

When the operator controlling the master has the feeling of direct interaction with the remote environment, the system is called "transparent" (Figure 3.a). It physically means that the impedance of the environment (external object) is equal to the transferred impedance from slave to the master or operator's hand [8-10]. In this case for the same force we will have the same motion. The transparency of the system will be discussed further in two subsections.

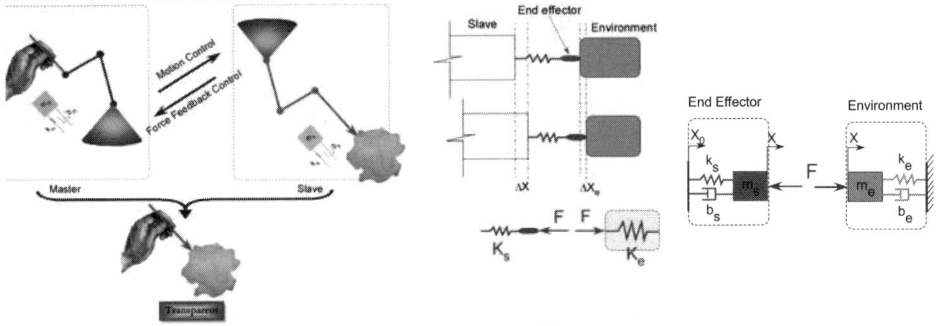

(a) (b) (c)

Figure 3: (a) Master-Slave bilateral teleoperator (b) Steady state/quasi-static (c) Transient/dynamic

5.1 *Steady State/Quasi-Static*

Figure 3.b shows an EE at the moment of touching the environment. In this system the EE is connected to the slave through a linear spring. Assume a mass-less force sensor is attached to the EE, where it comes in touch with the external object. As the operator commands the slave to move further, a compression force F is exerted to the EE spring and force sensor. The following equations can be written for the EE and environment

$$\Delta X_M = \Delta X + \Delta X_{Err}\,;\ \Delta X = \Delta X_s + \Delta X_e = \frac{F}{K_s} + \frac{F}{K_e}$$

Where ΔX_M, ΔX, ΔX_{Err}, ΔX_s, ΔX_e, K_s, K_e are the commanded displacement by master, the actual displacement of slave, displacement error, the compression in EE spring, compressed deflection in external object, stiffness of EE spring and stiffness of environment. Therefore the stiffness of the environment can be evaluated by

$$K_e = \frac{F}{\Delta X - F/K_s}$$

5.2 *Transient/Dynamic*

An EE and environment are shown in Figure 3.c. The equation of motion (EOM) for the EE is given by

$$m_s\ddot{x} + b_s\left(\dot{x} - \dot{x}_0\right) + k_s\left(x - x_0\right) = -F$$

All the parameters in this differential equation are known so we can solve for x. The EOM for the environment is given by

$$m_e\ddot{x} + b_e\dot{x} + k_e x = F$$

By defining the impedance of the environment as $Z_e = \frac{F}{\dot{x}}$, one can compute it by the received signals from a force and displacement sensors. In case of a rotary system, an encoder will give the required information for the deflections in EE spring and the environment. The impedance of the master is defined by

$$Z_m(s) = M_m s + B_m + \frac{K_m}{s}$$

Therefore there are two variables to change in order to match the impedance of master to the impedance of slave: damping (B_m) and stiffness/compliance (K_m). Figure 4 shows a proposed system with variable damping and stiffness. The damping coefficient is a function of shear stress developed in the MR damper $B_m = B_m(\tau)$ and the stiffness/compliance is a function of MR damper spring $K_m = K_m(\theta_s)$ and the current applied to the DC motor. The torsion spring is used to simulate the restoring force and the DC motor is used to adjust the initial angle of rotation in the spring.

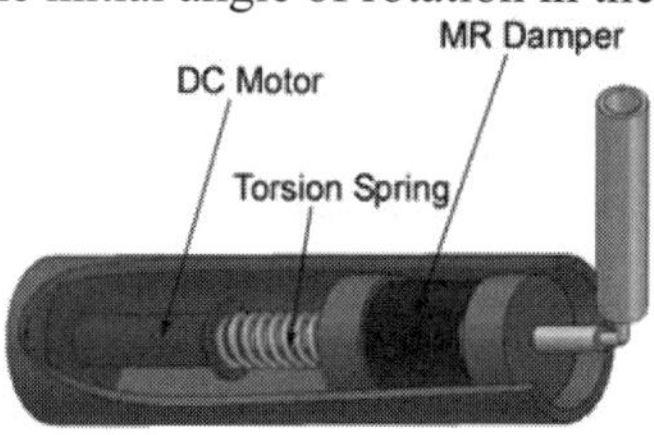

Figure 4: A proposed variable compliant MR damper system

6 Experimental Results

The aforementioned SDOF rotary system was used to test five different objects with different stiffness/compliance. Three different reference input were applied to the system: a reference input by hand and two automated sinusoidal reference inputs with low and high frequencies, 0.1 and 1 Hz, to model the quasi-static and dynamic cases. The force-deflection graphs of these objects were investigated. The deflections of stiffer objects were less than that of compliant ones. The sensed forces were also more for stiffer objects. Using the collected data on displacement and forces the stiffness of different objects were computed and simulated in the master (Figure 5). The results show by using the force and deflection sensor in the EE and a DC motor in the master one can simulate the stiffness/compliance and restoring force of the environment.

7 Conclusion

By coupling a force sensor with an encoder in a rotary slave (end effector) along with a DC motor and torsional spring in a SDOF MR based force feedback master, the restoring force exerted by the environment to the end effector is simulated and generated in the joystick. Consequently, it adds the sense of external object stiffness to the user by the addition of variable compliance to the system. The work in this paper overcomes sticky wall phenomena as well. This research is under progress for a two degree of freedom system.

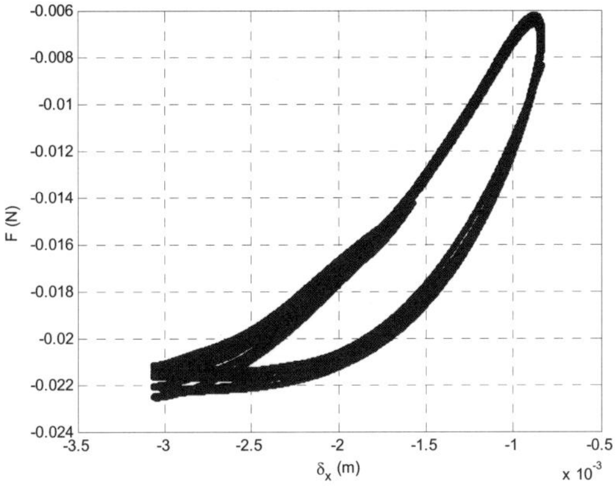

Figure 5: Force-deflection curves for an external object in the environment

8 References

1. Ahmadkhanlou, F., Washington, G.N., Wang, Y., and Bechtel, S.E. *The Development of Variably Compliant Haptic Systems using Magnetorheological Fluids*. in *12th SPIE International Symposium*. 2005. San Diego, CA.
2. Roy, A., Neelakantan, V.A., and G.N., W. *Design and Development of Passive and Active Force Feedback Systems using Magnetorheological Fluids*. in *ASME International Mechanical Engineering Congress and Exposition*. 2003.
3. Bechtel, S., Washington, G., Ahmadkhanlou, F., and Wang, Y., *Microstructural Analysis and Control of a Magnetorheological Fluid.* ASME International Mechanical Engineering Congress and Exposition, 2004.
4. Carlson, J.D. *What Makes a Good MR Fluid?* in *8th International Conference on Electrorheological (ER) Fluids and Magneto-rheological (MR) Suspensions*. 2001. Nice, France.
5. Neelakantan, V.A., Washington, G.N., and Wolfe, R. *Force Feedback System using Magneto Rheological Fluids for Telerobotic Surgery*. in *SPIE Int. Conference on Smart Structures and Materials: Industrial and Commercial Applications of Smart Materials*. 2002. San Diego, CA.
6. Neelakantan, V.A., *Modeling, Design, Testing, Control and Applications of Magnetorheological Fluid Devices*. 2003, Ohio State University.
7. Ahmadkhanlou, F., Washington, G., Bechtel, S., and Wang, Y. *Magnetorheological Fluid Based Automotive Steer-By-Wire Systems*. in *13th SPIE International Symposium*. 2006. San Diego, CA.
8. Hannaford, B., *A Design Framework for Teleoperators with Kinesthetic Feedback.* IEEE Transactions on Robotics and Automation, 1989. **5**(6): p. 426-434.
9. Lawrence, D.A., *Stability and Transparency in Bilateral Teleoperation.* IEEE Transactions on Robotics and Automation, 1993. **9**(5): p. 624-637.
10. Lawrence, D.A., Pao, L.Y., Salada, M.A., and Dougherty, A.M. Quantitative experimental analysis of transparency and stability in haptic interfaces. in in Proceedings of the Fifth Annual Symposium on Haptic Interfaces for Virtual Environment and Teleoperator Systems. 1996.

MAGNETORHEOLOGICAL-FLUID DAMPER WITH INTEGRAL STEP-AND-REPEAT ACTUATOR

ERIC H. ANDERSON, MICHAEL F. CASH, MICHAEL E. EVERT AND SCOTT C. PENDLETON

CSA Engineering, Mountain View, CA, 94043, USA
Email: eric.anderson@csaengineering.com

Magnetorheological (MR) fluids provide a means for controlling properties of dampers to achieve various performance objectives. Some applications that require dampers may benefit from an additional actuation capability. While MR fluid alone is not sufficient for actuation, it can be used in combination with other components to create an actuator while adjustable damping properties are retained. A concept for a device that serves as both a controllable damper and a linear actuator is described. Such a device might be useful in a suspension or leveling system, or a repositioning system incorporated in a recoil damper. Basic design considerations are reviewed and the particular step-and-repeat architecture, in which two MR damper subassemblies are used as variable holding clamps, is presented. Two candidate actuator cores are considered. A voice coil core was used in a prototype unit, and a magnetostrictive or piezoelectric core is suggested for a higher-force device. Prototype design features and results of testing are described. A test unit achieved speeds up to 10 mm/s. Several non-ideal effects were noted, and must be addressed to make the technology practical.

1 Introduction

Magnetorheological (MR) fluid dampers are sometimes referred to as "actuators," but they are generally not capable of actuation. Rather, they can be used in various configurations to manage energy at the device or system level. The properties of MR fluids, principally the yield stress, can be controlled in certain ways so devices achieve some, but not all, of the same results possible with components that incorporate actuators. An MR fluid based damper with true actuation capacity must incorporate another physical effect for energy transformation to actual mechanical power output.

In the best case, this would be done without significant compromise to damper performance. The starting point for the present development is the premise that certain applications that require dampers also require or could benefit from actuation. Normally the damping (with MR fluids or not) and actuation functions are split into two devices (Figure 1), but perhaps an electronic infrastructure in place to drive and control MR effects might also support actuation.

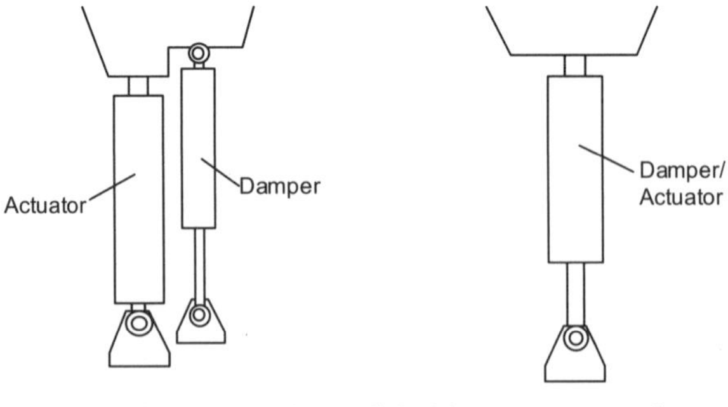

Figure 1. When applications require a damper in parallel with an actuator, an alternate configuration would combine the two functions in one device

MR-based suspensions are in use in vehicles. In some instances, the ability to raise and lower the vehicle for leveling or adjustment of ride height would prove useful. Aircraft landing gear use dampers and deployment/retraction actuators in parallel. An integrated device might serve both functions. Recoil damping for larger guns and artillery can benefit from the MR fluid augmentation and perhaps positioning for reloading or munitions transfer. Finally, several applications in human rehabilitation are considering MR devices and may benefit from a positioning control. The proposed class of devices is not suggested as a general replacement for modular actuators and damper struts.

The basic idea explored here for damper augmentation is to begin with a relatively typical MR damper configuration, and then use multiple fluid filled regions as controllable clamps, in combination with an interior actuation stage, to form a step-and-repeat actuator.

2 Requirements and Basic Device Architecture

A primary characteristic of MR devices is the ratio of a particular quantity when measured with a magnetic field applied or not applied (zero field). In dampers, this characteristic is the ratio of the damper force capacity in the ON vs. OFF positions for typical rates. Maximizing this ratio, while maintaining mass- and volume-efficient devices, is a fundamental design consideration. To this design driver we will now add a series of constraints related to actuation performance.

A critical assumption or guideline is the overlap between the controllable damping properties and the actuation properties. In general there are two cases to consider: a) actuation is required simultaneously with the full range of controllable damper properties, or, b) actuation is required only at times when the full range of controllable damper properties is not needed. This second situation is much more feasible, and it is also consistent with many of the motivating applications. Damper properties we would like to maintain include total travel or stroke, rates, damping force capacity, equivalent viscous damping in an OFF state, heat generation, performance with varying temperature, and degree of controllability including any desired directionality in properties. Physical packaging constraints on rods, housings, seals, magnets, and fluids are also important.

Among the key actuation properties are force, stroke, velocity or bandwidth, linearity, power input, heat generation, reliability and electromagnetic compatibility. This discussion has a bias towards linear or translational actuation and damping, but similar requirements apply in rotation.

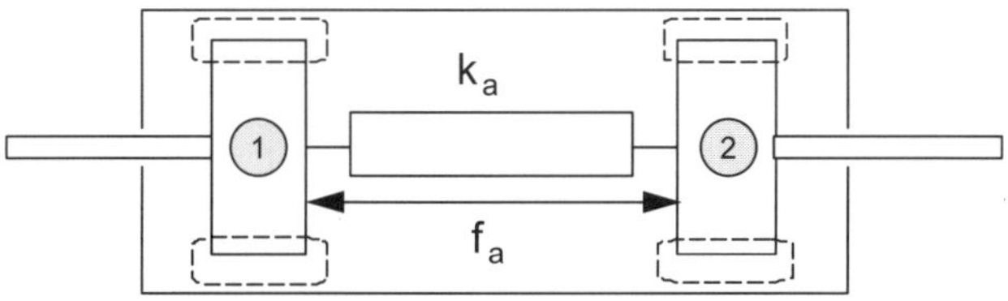

Figure 2. Basic concept shows an MR damper split into two parts, 1 and 2, with an actuator in the middle

The configuration shown schematically in Figure 2 begins with a typical MR damper piston configuration, splits it into two sections, and puts an actuator in between. With proper design, damper functions should be relatively unaffected, replicating those of a unit without actuation. The interior actuator is a device with force, stroke and stiffness characteristics matched to different applications.

The essence of actuator operation is that the two parts of the damper function as controllable clamps. By proper sequencing of these clamps, in a motion that mimics an inchworm, the rod can be moved in one direction or the other. Figure 3 shows schematically how the clamps and the expansion actuator are operated and sequenced. At a minimum, the clamping or holding force should exceed the actuation force.

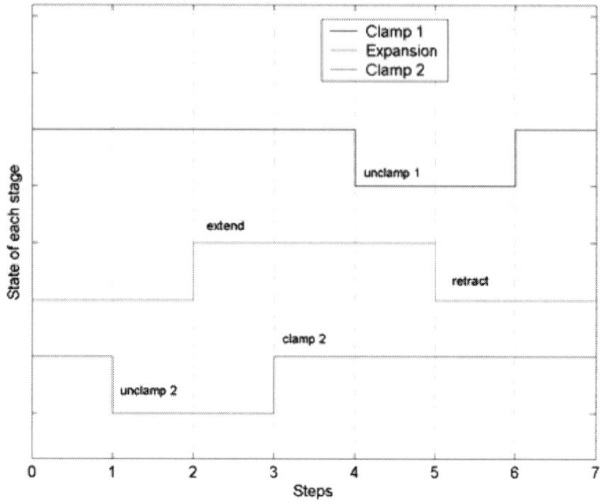

Figure 3. Operation of the step-and-repeat actuator showing repeated 6-step cycle. Clamps are either open or closed, and expansion stage is either expanded or retracted.

Device-level step size depends on the stroke of the extension actuator, and in practice, on the speed of operation. Step size is likely to be constant up to some frequency, and then will roll off at higher frequencies. Thus, the maximum device velocity will likely be achieved at some intermediate frequency.

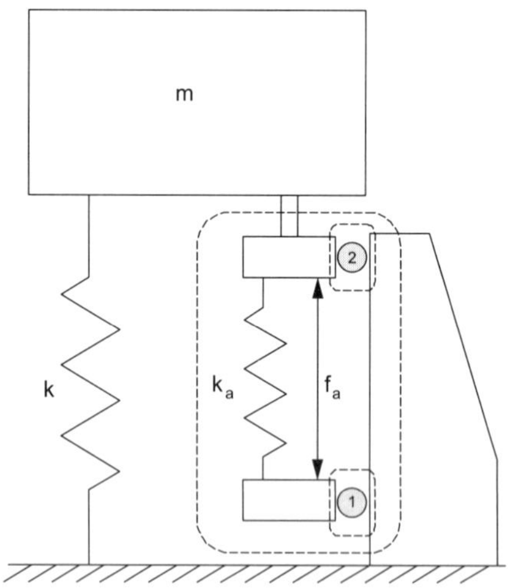

Figure 4. The basic damper-actuator architecture on the right with damping/clamps at 1 and 2 shown with a parallel spring that supports the main payload weight

A more critical aspect of the design is the force capacity of the actuator. Consider the case of a vehicle suspension, shown schematically in Figure 4. A spring, k, supports the weight of the payload (e.g. a quarter vehicle or equivalent). If the damper-actuator on the right is initiated with the spring, k, in a compressed state and carrying the full weight, then whether or not clamps 1 and 2 are locked, it will not carry load. Now assume that the payload is to be raised in the presence of gravity. Clamp 1 would be locked and clamp 2 would be open when the actuator stage expands. Clamp 2 would then be locked, the actuator retracted, and clamp 1 would move up before being locked out again. The redistribution of load would continue to the point where the actuator force requirement would need to match the weight of the payload, and further, where the actuator would have to lift the payload and counter the restoring force of the spring, k. For this class of application, the clamps must be capable of holding a large force, and the actuator must be capable of producing one.

For a given physical volume and mass, large force capacity is likely to result in reduced velocity. Properties of several candidate actuation approaches that trade force and velocity are summarized in Table 1.

Table 1. Qualitative comparison of options for the core actuator component

Type	Force	Stroke	Velocity	Packaging
Motor-driven screw	Moderate-high	High	Moderate	Several options
Linear motor	Moderate	Moderate	Moderate	Compact
Voice coil	Low	Moderate	Moderate	Compact
Magnetostrictive	High	Low	Low	Magnetic circuit
Piezoelectric	High	Low	Low	Compact

Four of the five approaches listed are electromagnetic and would employ some type of magnetic circuit, although it is unlikely that one circuit will support the clamp and extension stages. Use of a motor-driven screw would not take full advantage of that actuation type. The concept of operation does not require a long actuator stroke. Rather, total device stroke would accumulate from a series of small steps. A linear motor or a voice coil actuator may be preferable, and are likely to have approximately the same performance for a given package size and mass. They are not likely to match the force capacity of a rotary motor. Among short-stroke actuators, magnetostrictive and piezoelectric materials are nearly direct replacements for one another, although a piezoelectric stack is likely to be more compact.

Mechanical power output, expressed as a product of force and velocity, is the ultimate measure of actuator performance. With similar power outputs for all options, the choice of actuation will depend on whether large force is required, as in the suspension application, and preferences on packaging and other secondary considerations such as ease of electronic drive and heat generation. Two actuator types were selected for this concept development: a voice coil unit that was built and tested, and a second unit, not yet built, will use a piezoelectric or magnetostrictive.

3 Concept Design

Several design considerations are independent of the actuation type, and others are specific to the voice coil approach. The device concept began with an MR damper having two separate regions of controllable yield stress. While the design requires independent control for the actuation mode, in operation as a damper there is not much value in controlling the two regions separately.

The goal of this effort was proof of concept, with an emphasis on the actuation portion of the device. The damping portion was designed to exploit proven geometries and methods. Goals for the prototype properties are summarized in Table 2. The major property of interest for the prototype is the velocity. The projected product of force and velocity is increased in the future piezoelectric or magnetostrictive device. This is believed to be possible because the velocity is limited by the ability to open and close MR clamps and not just by the inherent actuator performance. Thus, the high force actuator core is expected to allow the device to deliver more overall mechanical power.

Table 2. Target properties for the prototype device and future high-force limit

Quantity	Prototype	Future	Notes
Diameter	< 100 mm	< 100 mm	Future devices could be smaller
Length	< 250 mm	< 250 mm	Excludes external rods
Damping force	5000 N	5000 N	Sets maximum gap width
ON vs. OFF force	10:1	10:1	OFF force depends on rate
Total stroke	±25 mm	±25 mm	Packaging issue only
Velocity	10 mm/s	1 mm/s	Both directions
Force output	40 N	2000 N	Low force with voice coil

During the design of the voice coil device it became apparent that the relatively low actuator stiffness would likely result in a low frequency axial mode of the actuator subassembly. This was seen as an opportunity to exploit the expansion stage resonance to increase motion, perhaps achieving a 2-4 times dynamic amplification, although operation at the resonance frequency may be difficult in practice. The negative aspect of the resonance is its effect on the device dynamics in the damper mode.

The electronic drive and control system was required to drive three loads, independently and simultaneously: the expander stage and the two MR clamp valves in a sequence shown in Figure 3. Relative timing between the three drives, including on/off, overlap, overall clock rates, were required. A position sensor was baselined to close the loop around device position.

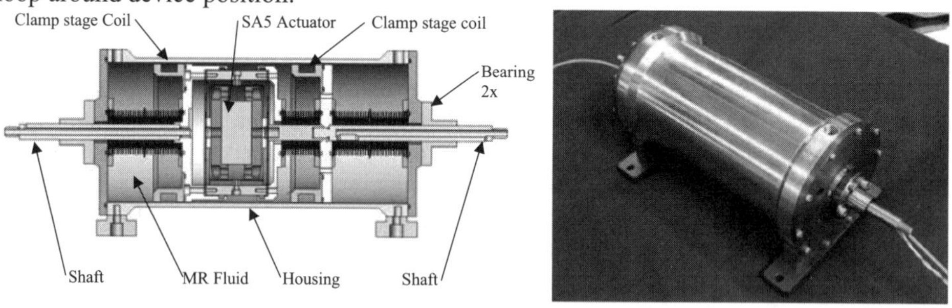

Figure 5. Cross-section and photograph of the prototype damper-actuator show central voice coil actuator and damper/clamp components on either side

A cross-section and photograph of the prototype device are shown in Figure 5. The main components are centrally located in the outer housing and end caps, and the device is nominally left-right symmetric nearly axisymmetric about the central rod. Shafts protrude from either end of the housing end caps and travel on low friction linear bearings. Each shaft is attached to a clamp stage coil within the housing, and is threaded on the exposed end to attach to measurement sensors or other hardware components. Bellows integral to the system provide a flexible seal from the MR fluid (Lord MRF-

132AD) and can extend and compress as the actuator translates. Thermal expansion is addressed with an external volume. Two custom aluminum end caps attach to the actuator and seal it from the MR fluid. At locations where o-rings cannot be used, self sealing fasteners are used to contain an airtight seal.

The drive and control system was designed to supply up to 7 A at 24 V to each of the three coils in a full H-bridge arrangement and used a Texas Instruments C2000 DSP. Cycle time, delays, current magnitudes, etc. are modifiable in software. The significant high frequency noise from the switching drives was partly filtered by the electrical loads, magnetic circuit and mechanical system.

4 Test Results

This section summarizes actuator performance, with emphasis on measured motion and velocity as a function of operating frequency. The frequency proved to be limited by electrical and mechanical/fluid effects, but the concept was demonstrated as intended. The internal electromagnetic actuator was tested for basic function prior to the addition of the MR fluid, and exhibited a resonance at about 20 Hz. . The actuator was tested at several combinations of current levels and overall cycle speeds with a high-precision mechanical roller slide and encoder combination that provided 100 nm resolution. A crosslink was used in some device tests to study the effect of a fluid shunt path.

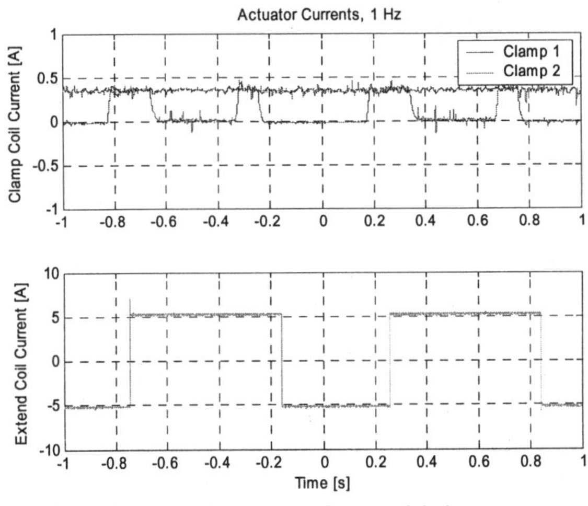

Figure 6. Actuator current waveforms at 1Hz; step waveforms and timing are apparent

Measured actuator currents are shown in Figure 6, where the basic sequencing is apparent. Note that the core actuator was driven with 5 A, but the clamp components were driven with only 0.35 A. The shape of the clamp current waveforms degraded in tests at higher frequencies (20 Hz and above). These limitations resulted from prototype drive electronics and are not considered fundamental limitations to the concept.

Example position versus time results are shown in Figure 7 for 20 Hz drive, with motion from the center near to one end of the 65 mm total stroke. On closer inspection, each step appears to be progressively smaller. At the end of the stroke the step size must be zero, and the actuator support flexures and bellows act to increase the steps on one side of center and decrease them on the other. There may also be an effect involving the fluid transport that is not well understood.

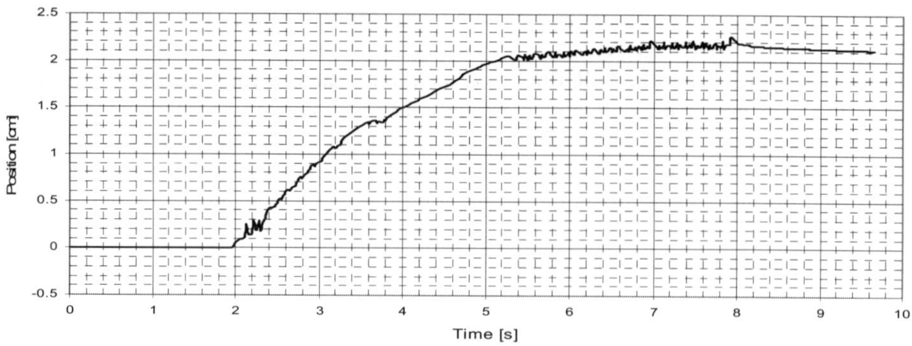

Figure 7. Positioning performance at 20 Hz drive reached 9.4 mm/s, with individual steps barely discernable

At a higher drive frequency (20 Hz in Figure 7), the velocity is higher and step sizes are smaller. The internal actuator resonance helps. The smaller step size could be due to inadequate release of the clamps at the start of the expansion stage action. Velocity decreases toward the end of stroke. Table 3 summarizes the velocity data, which shows the basic increase in velocity proportional to frequency, the effect of the actuator resonance, the limits imposed by incomplete current drive, incomplete clamping and other yet to be defined fluid effects.

Table 3. Maximum velocity under zero load

Drive frequency (Hz)	Velocity (mm/s)
1	1.5
8	2.8
20	9.4
30	0.2

5 Conclusions and Recommendations

This paper reports on early concept development for a device that retains properties of an MR damper while adding the ability to position the damper output rod and move it axially. The applications of interest do not necessarily require the damping and actuation functions to operate simultaneously. An initial concept device demonstrated basic movement over 65 mm stroke at speeds up to about 10 mm/s. It was not tested comprehensively to determine if damping functions were at all compromised.

The next steps in this development include:

1. Upgrade the electronic drive system to remove constraints on current delivery.
2. Test the unit against various loads to quantify force output capacity and develop force-speed curves.
3. Measure force-displacement and force-velocity responses to characterize the device in ways normally used to understand MR dampers.
4. Explore different timing strategies and clamp current levels.
5. Enhance device Simulink models.
6. Design, build and test a device using either a piezoelectric actuator.
7. Review the device performance under prolonged periods of load support.

MAGNETORHEOLOGICAL ISOLATORS USING MULTIPLE FLUID MODES

MIKEL BRIGLEY, YOUNG-TAI CHOI AND NORMAN M. WERELEY

Smart Structures Laboratory, Department of Aerospace Engineering, University of Maryland, College Park, MD 20742 U.S.A.

This study presents the experimental and theoretical evaluation of an MR (magnetorheological) isolator using multiple fluid modes including shear, flow and squeeze. For doing so, a novel type of multi-mode MR isolator against multi-degree-of-freedom excitations is proposed and fabricated. The experimental testing of the proposed MR isolator is conducted by an MTS machine and its damper characteristics are experimentally evaluated by equivalent damping and complex stiffness methods. To construct a theoretical model of the MR isolator, its dynamic equation is derived and important model parameters are identified by the force averaging method using the force-displacement or the force-velocity plots. Using the theoretical model, the damper characteristics of the MR isolator are also predicted and compared with those using the experimental data.

1 Introduction

Magnetorheological (MR) isolators have been designed based on three working fluid modes of damper operation: shear mode, flow mode, and squeeze mode. The shear mode occurs when one of two magnetic poles translates or rotates relative to the other. In the shear mode, the fluid is directly sheared by relative motion of these surfaces. The flow mode is characterized by pressurized flow between two magnetic poles that form the opposite walls of a rectangular or annular duct. The squeeze mode occurs when the magnetic poles move toward each other, squeezing out the fluid. However, the shear mode isolators require large magnetic pole surface area contacted with the fluid in order to achieve high isolator force, and the squeeze mode isolators can effectively work in the very small vibration situation. The flow mode isolators have unwanted lock-up state between the magnetic pole gap against relatively high frequency and small amplitude of excitations. In the lock-up state, the flow motion of the MR fluid in the gap does not take place. Therefore, in this study, an MR isolator using multiple fluid modes such as the shear, flow and squeeze is developed to lessen drawbacks of single fluid mode MR isolator. For doing so, a novel type of multi-mode MR isolator against multi-degree-of-freedom excitations is proposed and fabricated. The experimental test of the proposed MR isolator is conducted by an MTS machine and its damper characteristics are experimentally evaluated by equivalent damping and complex stiffness methods. To construct a theoretical model of the MR isolator, its dynamic equation is derived and important model parameters are identified by the force averaging method using the force-displacement or the force-velocity plots. Using the theoretical model, the damper characteristics of the MR isolator are also predicted and compared with those using the experimental data.

2 Multi-Mode MR Isolator

A schematic diagram of the multi-mode MR isolator proposed in this study is presented in Figure 1(a). The electromagnetic coil is wrapped around the bobbin and the outer cylinder serves as a magnetic flux return. The bobbin is attached to the top elastomer through the connecting screw. As a result, the elastomer motion results in the bobbin motion. Axial motion of the elastomer—and consequently the bobbin—imposes radial flow of the fluid in the gap between the bottom of the bobbin and the outer cylinder, thus creating squeeze mode damping. The radial flow also causes a flow along the sides of the bobbin. In this area, shear and flow mode damping is present. The photo of the multi-mode MR isolator fabricated in this study is shown in Figure 1(b). The height and the outer diameter of the MR isolator are 69 mm and 52 mm, respectively. The bobbin and the outer cylinder were made from 1018 carbon steel and 24 gage copper wire was wound around the bobbin. There were 188 turns of the wire upon completion of winding. Note that, in single flow mode MR isolators, a lock-up status where any flow does not allow in a gap can be taken places. As a result, in the lock-up status, single flow mode MR isolators produce no damping and works as passive isolators. However, in the multi-mode MR damper, since the elastomer is connected to the bobbin, the lock-up status does not occur.

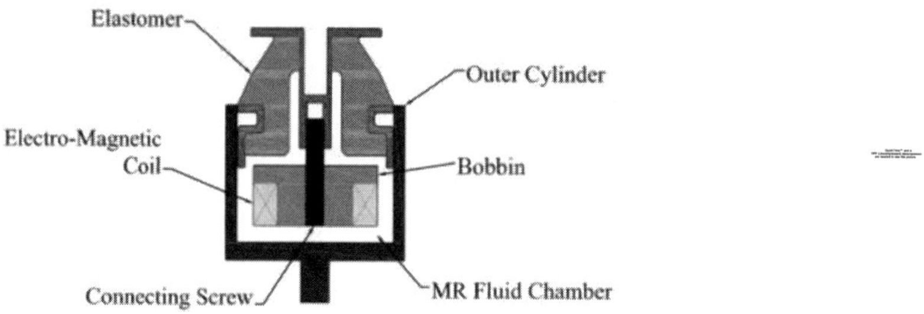

(a) schematic diagram (b) fabricated isolator

Figure 1. The multi-mode MR isolator using shear, flow and squeeze modes

3 Damper Characterization

For damper characterization, the experimental testing of the multi-mode MR isolator is conducted using a hydraulically powered MTS machine. In this case, the sinusoidal displacements with constant amplitudes are used as excitation inputs. The excitation frequencies are varied from 5-20Hz in increments of 1Hz as well as at 1Hz and 2.5Hz. The displacement amplitudes consist of 0.1mm, 0.25mm, 0.5mm, and 0.75mm. The applied currents are ranged from 0-2A in increments of 0.5A.

Figure 2 presents the experimental responses of the multi-mode MR isolator under the sinusoidal displacement input with 5Hz frequency and ±0.5mm amplitude. As shown in figure 2(a), as the current input increases, the area enclosed by the force-displacement plot increases because of an increase in the damping of the MR isolator. In addition, as the current input increases, the force-displacement plot is rotated counter clockwise because of an increase in the stiffness of the MR isolator. On the other hand, when the current input is not applied to the MR isolator (zero field), the hysteresis loop in the force-velocity plot in figure 2(b) is elliptical in shape, typical of a purely viscous damper. However, upon application of the current input, the hysteresis plots deviate from the elliptical behavior because of a yield force due to the MR fluid.

Two approaches, equivalent viscous damping and complex stiffness, are used to characterize the damper behavior of the MR isolator.

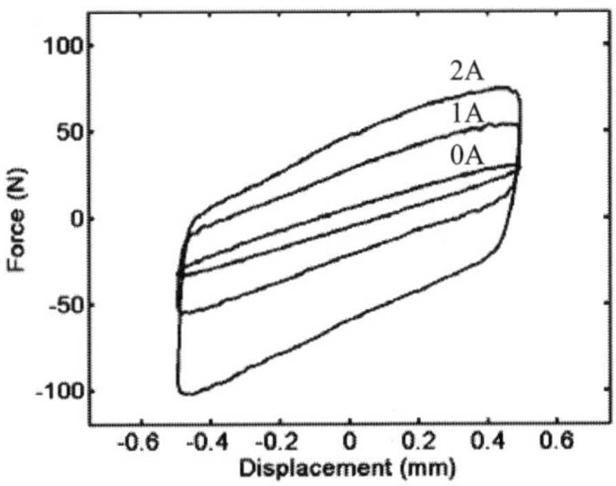

(a) force vs. displacement

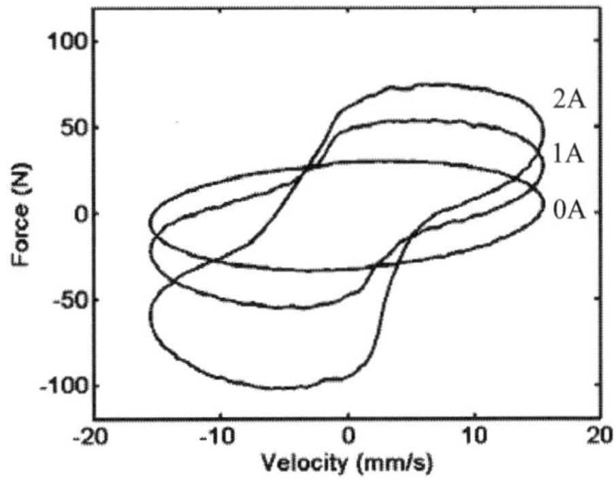

(b) force vs. velocity

Figure 2. Experimental responses of the multi-mode MR isolator under the sinusoidal displacement input (5Hz and ±0.5mm)

3.1 Equivalent Viscous Damping

The first characterization technique is the evaluation of the equivalent viscous damping of the MR isolator. This is a standard linearization technique that can be applied to a nonlinear damper to find an equivalent viscous damping force proportional to velocity.

The equivalent viscous damping is computed by equating the energy dissipated by the MR isolator over a single cycle, E, at a given frequency, ω to that of the viscous damping. The energy dissipated by the MR isolator is given by

$$E = \int_{cycle} F dx = \int_{0}^{2\pi/\omega} F x dt \tag{1}$$

For a viscous damper, the dissipated energy is given by

$$E = C_{eq} \pi \omega X^2 \tag{2}$$

Equating energies gives an equation for the equivalent damping, C_{eq}, by

$$C_{eq} = \frac{E}{\pi \omega X^2} \tag{3}$$

Figure 3 shows the experimental equivalent damping for the multi-mode MR isolator. The current input shows big increment in the equivalent damping. However, as the displacement amplitude increases (see figure 3(a)), the equivalent damping for the MR isolator decreases. At no current input, the equivalent damping varies only slightly with the displacement amplitude. On the other hand, as the frequency increases (see figure 3(b)), the equivalent damping for the MR isolator decreases. In addition, at no current input, the equivalent damping does not vary much with the frequency.

3.2 Complex Stiffness

The second characterization technique is the complex stiffness approach. This is a linearization technique in the frequency domain, which replaces the nonlinear hysteresis loop with an equivalent ellipse. In this approach the complex damper stiffness, K^*, is characterized as an in-phase or storage stiffness, K', which is a measure of the energy stored over a cycle and a loss stiffness, K'', which is a measure of the energy dissipated over a period. The complex stiffness can be represented as a Kelvin chain (a spring and viscous damper in parallel). The complex stiffness is expressed as

$$K^* = K' + jK'' = K'(1+\eta) \tag{4}$$

where η is the loss factor.

In the complex stiffness, the damper force is reconstructed as a Fourier series by

$$f(t) = F_c \cos(\omega t) + F_s \sin(\omega t) = K'x(t) + \frac{K''}{\omega} x(t) \tag{5}$$

where F_c and F_s are the cosine and sine Fourier coefficients of $f(t)$ at frequency ω.

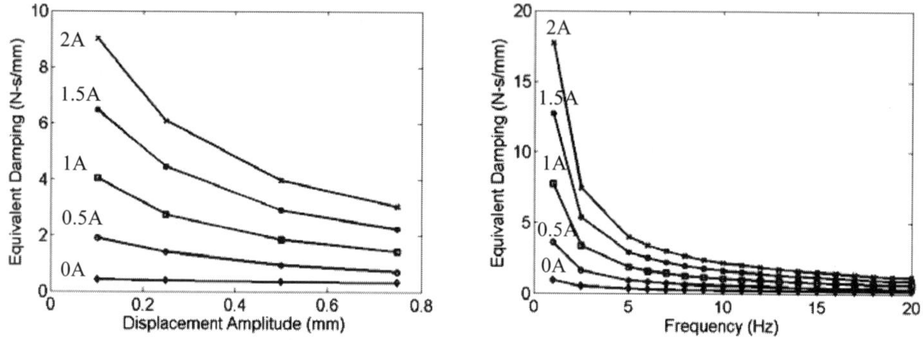

(a) at 5Hz frequency (b) at ±0.5mm amplitude

Figure 3. Experimental equivalent damping for the multi-mode MR isolator

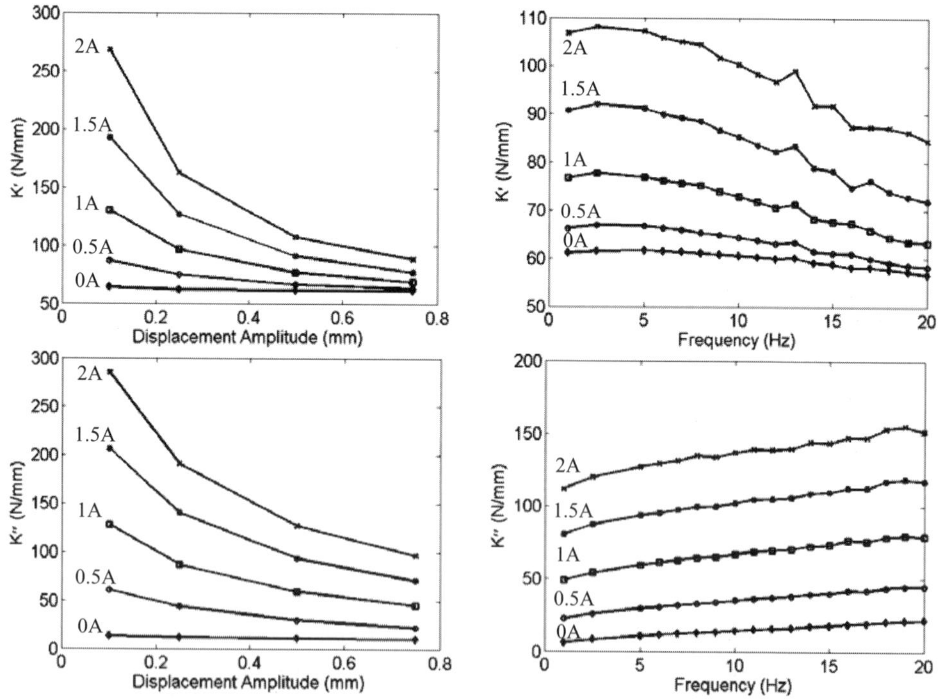

(a) at 5Hz frequency (b) at ±0.5mm amplitude

Figure 4. Experimental complex stiffness for the multi-mode MR isolator

The sinusoidal displacement, $x(t)$, is also reconstructed by

$$x(t) = X_c \cos \omega t + X_s \sin \omega t \tag{6}$$

where X_c and X_s are the cosine and sine Fourier coefficients of $x(t)$ at frequency ω.

Substituting Eq.(6) into Eq.(5) and equating the sine and cosine terms yields expressions for the storage stiffness as well as the loss stiffness

$$K' = \frac{F_c X_c + F_s X_s}{X_c^2 + X_s^2} \quad \text{and} \quad K'' = \frac{F_c X_s + F_s X_c}{X_c^2 + X_s^2} \tag{7}$$

Figure 4 presents the experimental complex stiffness for the multi-mode MR isolator. As seen in figure 4, the storage and loss stiffness of the MR isolator is greatly increased as the increased current is applied. From figure 4(a), it is observed that the storage and loss stiffness decreases as the displacement amplitude increases. However, as the frequency increases, the storage stiffness decreases and the loss stiffness increases.

4 Theoretical Modeling

The total force of the multi-mode MR isolator can be obtained by a superposition of the force developed by each operating mode of the fluid as well as the force from the elastomer. The elastomer is modeled as a spring-damper, its force is given by

$$F_{el} = K_{el} x + C_{el} x \tag{8}$$

The force of the MR isolator due to the shear flow is given as [1-3]

$$F_{shear} = \frac{\mu b L}{dx} x + b L_a \tau_{y,sh} \operatorname{sgn}(x) \tag{9}$$

The force of the MR isolator due to the flow mode is given by [4,5]

$$F_{flow} = \frac{12 A_p^2 \mu L}{b d^3} x + \frac{2 A_P L_a \tau_{y,fl}}{d} \operatorname{sgn}(x) \tag{10}$$

The force due to the squeeze flow is given as [6]

$$F_{squeeze} = \frac{3\pi \mu r_p^4}{2(x_0 + x)^3} x + \frac{4\pi \tau_{y,sq} r_a^3}{3(x_0 + x)} \operatorname{sgn}(x) \tag{11}$$

Then, the total force of the multi-mode MR isolator is obtained as follows:

$$F_d = F_{el} + F_{shear} + F_{flow} + F_{squeeze} = K_{el} x + \left(C_{el} + C_{fl}\right) x + F_y \operatorname{sgn}(x) \tag{12}$$

where C_{fl} and F_y are the total fluid damping and total yield force due to each fluid mode, respectively. In this study, it is assumed to be $\tau_{y,sh} = \tau_{y,fl} = \tau_{y,sq} = \tau_y$ for simplicity.

To implement the isolator model, the knowledge of the fluid properties, μ and τ_y, as well as the damping and stiffness properties of the elastomer, C_{el} and K_{el} are needed. Since the equivalent damping at zero field input is almost constant over wide ranges of displacement and frequency, the fluid viscosity and the damping of the elastomer are chosen to be constant: μ=0.09 Pa·s, C_{el}=0.3 N/mm. However, K_{el} and τ_y are dependent on the frequency, the displacement amplitude and the current input. In this study, the force averaging method, using the force-displacement plot or the force-velocity plot, is used to identify the model parameters of K_{el} and τ_y.

If averaging the force for a given value of displacement in the force-displacement plot, then only the stiffness force is left in the force-displacement plot. The slope of the stiffness force around the y-axis in the force-displacement plot yields the elastomer stiffness, K_{el}. On the other hand, the averaging force for a given value of velocity in the force-velocity plot extracts the viscous and yield force from the total force. The extracted viscous and yield force, $F_{d,ext}$, at the maximum and minimum velocity should be equal to the viscous and yield force predicted by the model as follows:

$$F_{d,ext}(\dot{x}_{\max}) = (C_{el} + C_{fl})\dot{x}_{\max} + F_y \,\mathrm{sgn}(\dot{x}_{\max})$$
$$F_{d,ext}(\dot{x}_{\min}) = (C_{el} + C_{fl})\dot{x}_{\min} + F_y \,\mathrm{sgn}(\dot{x}_{\min}) \qquad (13)$$

From the average yield force, F_y, of Eq.(13), the yield stress, τ_y, is obtained.

Figure 5 presents the comparison between experimental and predicted responses of the multi-mode MR isolator. Since the theoretical model is based on a Bingham model, the hysteresis loop in the force-velocity plot is not described well by the model. However, overall behavior of the MR isolator is favorably captured by the model.

Figure 6 presents the comparison between experimental and predicted complex stiffness of the multi-mode MR isolator. The predicted loss stiffness is in good agreement with the experimental results. However, the predicted storage stiffness is

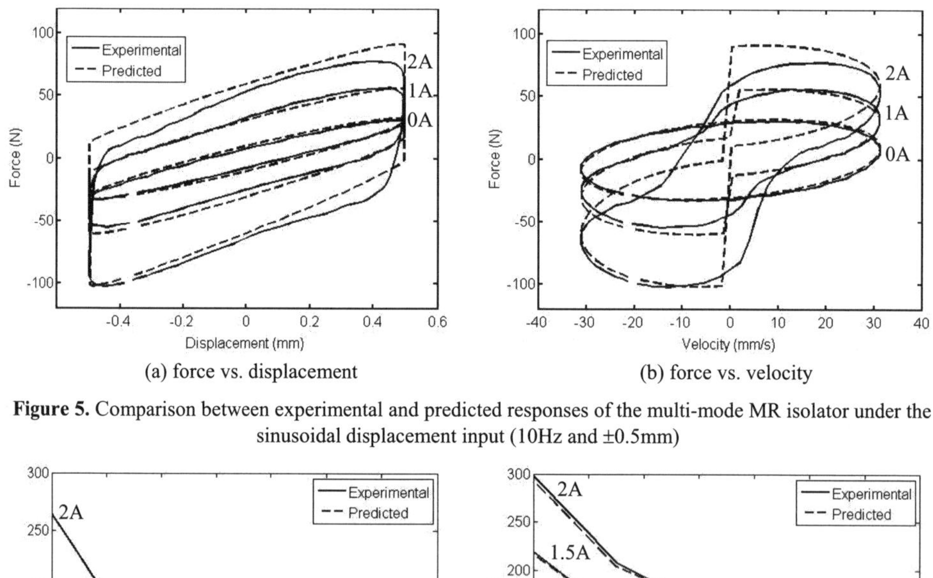

(a) force vs. displacement (b) force vs. velocity

Figure 5. Comparison between experimental and predicted responses of the multi-mode MR isolator under the sinusoidal displacement input (10Hz and ±0.5mm)

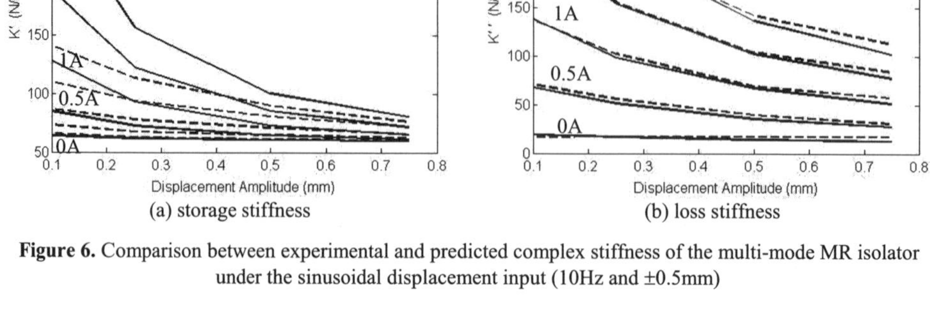

(a) storage stiffness (b) loss stiffness

Figure 6. Comparison between experimental and predicted complex stiffness of the multi-mode MR isolator under the sinusoidal displacement input (10Hz and ±0.5mm) smaller than the experimental results. The difference between the experimental and predicted storage stiffness results from the fact the theoretical model does not attempt to account for the low speed hysteresis loop in the force-velocity plot. In the complex stiffness approach, the hysteresis in the force-velocity plot is described by the additional storage stiffness so that the experimental storage stiffness is apparently larger than the predicted storage stiffness.

5 Conclusions

In this study, the damper characteristics of the multi-mode MR isolator were evaluated experimentally and theoretically. For doing so, a novel type of MR isolator using shear, flow and squeeze modes was proposed and fabricated. For damper characterization, the experimental damper testing was conducted by an MTS machine and its damping capability was evaluated by equivalent damping and complex stiffness methods. It is observed that, as the displacement amplitude and frequency of excitations decreases, the equivalent damping increases. However, at zero-field input, the equivalent damping does not vary much with the amplitude and frequency. The storage and loss stiffness decreases, as the displacement amplitude increases. However, as the frequency increases, the storage stiffness decreases and the loss stiffness increases. A theoretical model of the multi-mode MR isolator was constructed based on a Bingham model and the important model parameters such as the storage stiffness and the yield stress were identified by the force average method using the force-displacement or force-velocity plots. Comparison on experimental and theoretical behavior of the MR isolator was conducted. It is shown that the predicted equivalent damping of the MR isolator is in a good agreement with the experimental results. In addition, the predicted loss stiffness shows a good match with the experimental results. However, the predicted storage stiffness is smaller than the experimental results since the theoretical model does not consider the low speed hysteresis loop or compression loop in the force-velocity plot.

References

1. Choi, S.B. and Choi, Y.T., "Sliding mode control of shear-mode type ER engine mount," *KSME International Journal*, Vol. 13, No.1, pp. 26-33, 1999.
2. Choi, S.B., Choi, Y.T., Cheong, C.C., and Jeon, Y.S., "Performance evaluation of a mixed mode ER engine mount via hardware-in-the-loop simulation," *Journal of Intelligent Material Systems and Structures*, Vol. 10, pp. 671-677, 1999.
3. Lou, Z., Ervin, R.D., Filisko, F.E., and Winkler, C.B., "An electrorheologically controlled semi-active landing gear," *SAE Technical Paper Series* 931403, 1993.
4. Wereley, N.M. and Pang, L., "Nondimensional analysis of semi-active electrorheological and magnetorheological dampers using approximate parallel plate," *Smart Materials and Structures*, Vol.7, No.5, pp. 732-743, 1998.
5. Lindler, J. and Wereley, N.M., "Analysis and testing of electorheological bypass dampers," *Journal of Intelligent Material Systems and Structures,* Vol. 10, pp. 363-376, 1999.
6. Jolly, M.R. and Carlson, J.D., "Controllable squeeze damping using magnetorheological fluid," 5th *International Conference on New Actuators*, pp. 333-336, 1996.

FORCE FEEDBACK CONTROL OF A MEDICAL HAPTIC MASTER USING AN ELECTRORHEOLOGICAL FLUID

Y. M. HAN, P. S. KANG, K. G. SUNG AND S. B. CHOI

Smart Structures and Systems Laboratory, Department of Mechanical Engineering, Inha University, Incheon 402-751, Korea

E-mail: seungbok@inha.ac.kr

This paper presents force feedback control performance of a spherical haptic device featuring an electrorheological (ER) fluid that can be used for minimally invasive surgery (MIS). As a first step, a spherical ER joint composed of rotational and stationary electrodes is designed and optimized based on the mathematical torque modeling. The active force produced in MIS is generally small, even though the passive force is large. In order to meet this agreement, both clutch and brake mechanism are adopted for the ER joint. In this operation, the active (small) force feedback by the rotational electrodes and/or semi-active (large) force feedback are achieved by the stationary electrode. Subsequently, the master device is manufactured by integration the spherical ER joint with AC motor. In order to achieve desired force trajectories, a sliding mode controller, which is robust to uncertainty, is formulated by considering mechanical friction and hysteretic behavior of the ER fluid as uncertainty. The controller is then experimentally realized. Tracking control performances for various force trajectories are presented in time domain.

1 Introduction

In the modern medical treatment, it is very popular to adopt robotic technology. Typically, minimally invasive surgery (MIS) needs a support of robotic master-slave system [1]. MIS requires the surgeon to perform operation through usage of long surgical tools that are inserted through incision points. Robotic engineers need to design a better environment for performing these surgeries. The commercially available medical master-slave systems are the da VinciTM system and the Zeus surgical system [2,3]. However, these robotic systems are large, expensive and heavy. Only visual information is sent back to the master system, and there is no information about viscosity and stiffness of the touched tissue, bone or organ. Therefore, the surgeon that remotely controls the robot is not able to feel what is happening inside the patient

Among several smart materials, the ER fluid is actively researched as an actuating fluid for vehicle applications and robotic system[4]. ER fluids have the ability to change their rheological properties such as yield stress by the intensity of applied electric field. This is an ideal capability for haptic systems in which it is important to simulate various different strengths of forces. The controllable rheological property by electric field gives the possibility for ER haptic devices to accurately mirror remote or virtual compliance. This has led several research works. Sakaguchi and Furusho [5] developed an ER fluid-based planar force-feedback manipulator system. Mavroidis et al. [6] developed force feedback devices called MEMICA (MEchanical MIrroring using Controlled stiffness and Actuators). Böse et al. [7] developed an ER fuid-based force-feedback joystick. However, these researches have difficulties in realization of multi-DOF force-feedback. Furthermore, ER haptic device is simple and effective candidate because force produced in surgery is mostly passive, but it is impossible to generate active force.

Consequently, the main contribution of the present work is to provide force tracking control of a haptic device featuring ER fluid and spherical geometry by realizing the

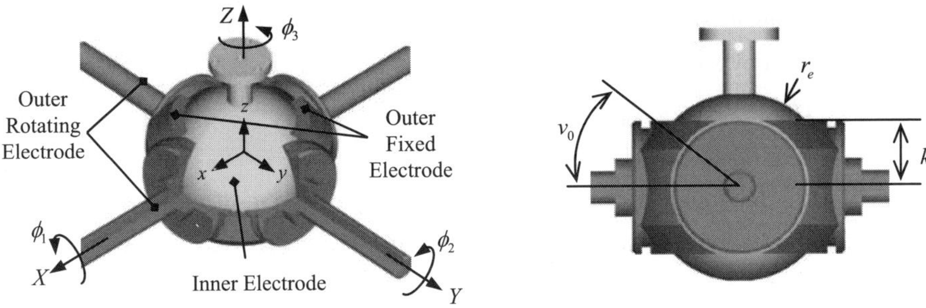

(a) view point (1,1,1) (b) view point (1,0,0)

Figure 1. Geometric feature of the spherical joint.

sliding mode controller (SMC). The proposed spherical joint is adopting both of clutch and brake mechanism in order to achieve active (small) force feedback by the rotational electrodes and semi-active (large) force feedback by the stationary electrode. The master device is manufactured by integrating the spherical ER joint with AC motor. In order to achieve desired force trajectories of the haptic device, a SMC is formulated and experimentally realized. Tracking control performances for various trajectories are presented in time domain.

2 Spherical ER Joint

A spherical ER joint is designed and manufactured for haptic system. The employed ER fluid is characterized by high yield stress up to $6kPa$. In general, ER fluid behaves as the Bingham fluid [8] whose constitutive equation is given by

$$\tau = \eta\dot{\gamma} + \tau_y(E), \qquad \tau_y(E) = \alpha E^{\beta} \tag{1}$$

where, τ is the shear stress, η is the dynamic viscosity, $\dot{\gamma}$ is the shear rate and $\tau_y(E)$ is the dynamic yield stress of the ER fluid. It has been proved that $\tau_y(E)$ is a function of the electric field E and exponentially increases with respect to the electric field. The parameters of α and β are characteristic values of the ER fluid to be determined by experiment. The measured yield stress of the employed ER fluid is $1177.6E^{1.77}\,Pa$ at room temperature. This experimentally obtained Bingham model is to be incorporated with the torque model of the proposed ER joint.

The ER joint proposed in this study is shown in Fig. 1 in which bi-directional ER clutch actuators [9] are employed in X and Y directions. Each bi-directional ER clutch actuator is composed of inner sphere electrode and two outer 1/4 sphere electrode driven in the opposite direction by using two AC driving motors. The gab between inner and outer electrodes is fully filled with the ER fluid. Therefore, one can readily obtain positive or negative torque from the ER clutch actuator by applying the electric field selectively to each part. In this study, the two bi-directional ER clutch actuators hold the inner sphere electrode in common in x and y directions. Therefore, the proposed ER joint can give 2- DOF active force reflection. It is noted that the proposed system can operates in two different mode; semi-active mode for passive and large force, and active mode for active and small force. The torque transmission T_t of the ER clutch actuator consists of three elements. Those are controllable field-dependent torque T_c, viscous friction torque

T_η, and mechanical friction torque T_f, By considering the Bingham model, T_c and T_η can be expressed by surface integral as follows:

$$T_c = \iint_S \tau_y \, r_c dA, \qquad T_\eta = \iint_S \eta\dot{\gamma} \, r_c dA \tag{2}$$

where, S is surface area of the spherical electrode, r_c is the distance of the point on electrode from rotational axis, and which is defined by

$$r_c = r_e \cos v \tag{3}$$

were, r_e is effective radius of electrode. In this study, the electrode has spherical shape, and x and y axes are identical. Therefore, spherical coordinate system is employed to mathematically describe the produced torque in x axis. Consequently, Eq. (2) and (3) are rewritten in spherical coordinate system as follows [10]:

$$\begin{aligned} T_{t,active} &= \int_{\frac{\pi}{2}-v_0}^{\frac{\pi}{2}} \int_0^{2\pi} (\tau_{y,cw} - \tau_{y,ccw}) r_e^{\,3} \cos^2 v \, du dv + \int_{\frac{\pi}{2}-v_0}^{\frac{\pi}{2}} \int_0^{2\pi} \eta \frac{(\omega - \dot{\theta})}{h} r_e^{\,4} \cos^3 v \, du dv \\ T_{t,semi} &= \int_{\frac{\pi}{2}}^{\frac{\pi}{2}} \int_0^{2\pi} \tau_{y,ER} \frac{2}{\pi} \sin^{-1}\left(\frac{k_0}{r_e \cos v}\right) r_e^{\,3} \cos^2 v \, du dv + \int_{\frac{\pi}{2}}^{\frac{\pi}{2}} \int_0^{2\pi} \eta_{ER} \frac{\dot{\theta}}{h} \frac{2}{\pi} \sin^{-1}\left(\frac{k_0}{r_e \cos v}\right) r_e^{\,3} \cos^2 v \, du dv \end{aligned} \tag{4}$$

where, $\tau_{y,cw}$ (or $\tau_{y,ccw}$) represents the yield stress of the ER fluid when the electric field is applied to the clockwise (or counterclockwise) rotating part of the ER clutch actuator. In addition, h is the gap size. ω and $\dot{\theta}$ are the angular velocity of the motor and the inner sphere, respectively. It is noted that the angular velocity of the inner sphere is assumed to be zero, and the friction term in Eq. (2) is considered as uncertainty. From the Eq. (2) and (5), the total produced torque can be rewritten as follows:

$$\begin{aligned} T_{t,active} &= 6314.6 r_e^{\,3} \{2v_0 - \sin(\pi - 2v_0)\} + \frac{0.4163}{h} r_e^{\,4} \left\{ 8 - 9\sin(\frac{\pi}{2} - v_0) - \sin(\frac{3}{2}\pi - 3v_0) \right\} \\ T_{t,semi} &= 12629.2 r_e^{\,3} \{2v_0 + \sin(2v_0)\} + \frac{0.5551}{h} r_e^{\,4} \left\{ 9\sin(\frac{\pi}{2} - v_0) + \sin(3v_0) \right\} \end{aligned} \tag{5}$$

where, the yield stress of the employed ER fluid is 4.02*kPa* at 2*kV/mm*, and fluid viscosity is 0.53*Pas*. The angular velocity of the geared AC motor is 1.5*rad/s*. A desired active force level is generally 1*N*, which is much smaller than passive force (up to 10*N*). Before deciding design parameter values, absolute and relative control torque is investigated for two design parameters where the electrode gap size is assumed to be 1*mm*. The obtained design parameter values are $r_e = 30\,mm$ and $v_0 = 0.72\,rad$ considering geometric interference between axes. Consequently, v_0 is decided so as to have maximum value considering spatial condition for manufacture, and then r_e is decided so as to achieve the required force level, which is set by 1*N* in this work. Fig. 2 shows the manufactured ER haptic joint. The inner electrode is hollowed to decrease inertia effect, and it is supported by spacer form the body of haptic device. The 4 outer electrodes are directly driven by 4 geared AC motors. Fig. 3 shows measured force response. The produced force is 1.08*N* at 2*kV/mm* field intensity, which is well matched with torque modeling result.

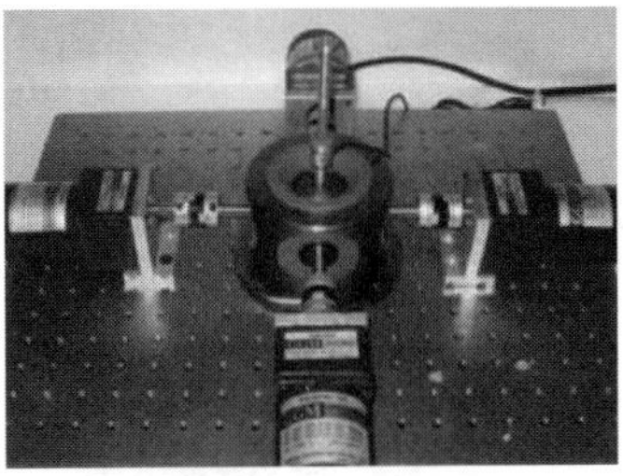

Figure 2. Photograph of the ER haptic system.

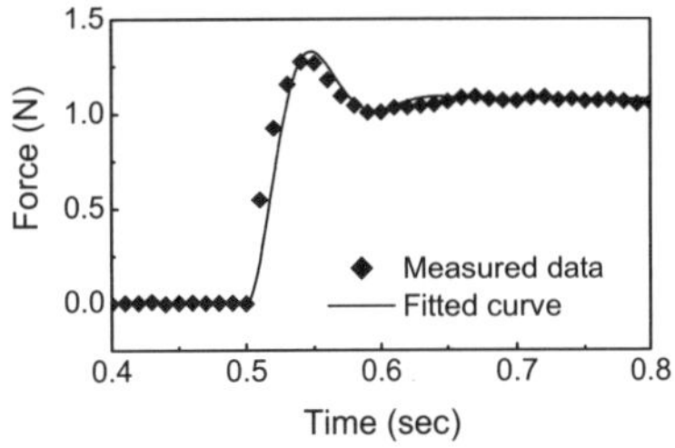

Figure 3. Measured torque response.

3 Sliding mode controller

The inherent dynamic characteristics of the proposed haptic device should be identified to achieve successful control implementation. This is normally accomplished by investigating step-input response of the ER joint system. Fig. 3 presents the measured step response of the ER joint. It is clearly seen that the proposed system behaves like the 2nd order dynamic model. Thus the system dynamics can be expressed by

$$\ddot{F}_i + 2\varsigma\omega_n \dot{F}_i + \omega_n^2 F_i = \omega_n^2 I_i \cdot u_{\theta i}, \qquad i = 1,2 \tag{6}$$

where, u_θ is generalized force, F is the produced force at the operation link, and I is the influence coefficient which is determined by produced torque and length of link. However, the control input direction is not same with the generalized force direction. As shown in Fig. 1, the clutch actuators are located in the fixed reference coordinate system *XYZ,* and the inner electrode with operation link is attached to the rotating coordinate system *xyz*. Therefore, the rotation angles of the inner electrode with respect to the fixed reference coordinate system *XYZ* are described by the Euler angle. It is noted that the Euler angle (θ) should not be confused with the rotation angle (ϕ) of ER clutch driven by the motor. Considering this point, the velocities of the actuators can then be expressed as a function of the time derivatives of the Euler angles:

$$\begin{bmatrix} \dot{\phi}_1 \\ \dot{\phi}_2 \\ \dot{\phi}_3 \end{bmatrix} = \begin{bmatrix} \cos\theta_2 \cos\theta_3 & -\sin\theta_3 & 0 \\ \cos\theta_2 \sin\theta_3 & \cos\theta_3 & 0 \\ -\sin\theta_2 & 0 & 1 \end{bmatrix} \begin{bmatrix} \dot{\theta}_1 \\ \dot{\theta}_2 \\ \dot{\theta}_3 \end{bmatrix} \tag{7}$$

Combining Eq. (7) with relationship between u_ϕ and u_θ by virtual work principle yields the following governing equation of motion:

$$\begin{bmatrix} \ddot{F}_1 + 2\varsigma\omega_n \dot{F}_1 + \omega_n^2 F_1 \\ \ddot{F}_2 + 2\varsigma\omega_n \dot{F}_2 + \omega_n^2 F_2 \end{bmatrix} = \begin{bmatrix} \cos\theta_2 \cos\theta_3 & \cos\theta_2 \sin\theta_3 \\ -\sin\theta_3 & \cos\theta_3 \end{bmatrix} \begin{bmatrix} \omega_n^2 I_1 \cdot u_{\theta 1} \\ \omega_n^2 I_2 \cdot u_{\theta 2} \end{bmatrix} \tag{8}$$

The impending control issue of this work is to achieve an accurate force reflection. As well-known, ER fluids exhibit hysteretic behaviors on their responses, and mechanical friction exists in the joint. Thus, there exist parameter uncertainties in the control model such as influence coefficient (*I*). This requires to use robust control schemes such as sliding mode controller (SMC) which is very effective for nonlinear and parameter uncertain systems. In general, the control robustness to parameter uncertainties is not

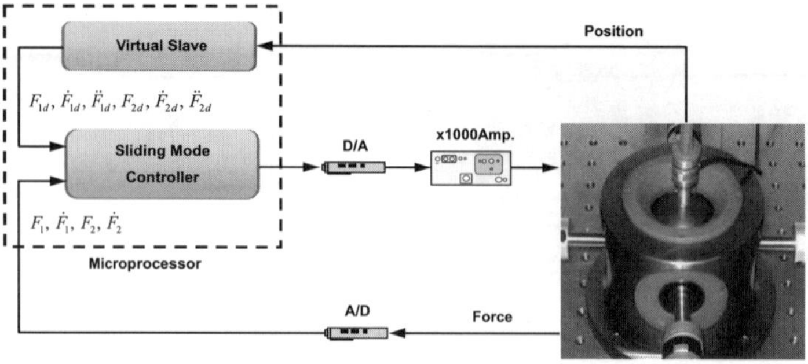

Figure 4. The experimental configuration of the force feedback control.

guaranteed by adopting conventional PID or LQG controller. The first step to formulate the SMC is to define a sliding surface which guarantees stable sliding mode motion on it. Since there are two control inputs in equation (10), the following two sliding surfaces are defined which are asymptotically stable:

$$\begin{aligned} S_i &= \dot{e}_i + c_i e_i = 0 \\ e_i &= F_{id} - F_i \end{aligned}, \qquad c_i > 0, \qquad i = 1,2 \tag{9}$$

Then, a sliding mode exists on the sliding surface whenever the following sliding mode condition is satisfied. Prior to formulating the SMC, we consider parameter variations which exist in practice. Consequently, the following parameter variation is imposed:

$$I = I_0 + \delta I, \qquad |\delta I| < a I_0 \tag{10}$$

where, I_0 is the nominal influence coefficient under the condition of all system parameters known. The δI is corresponding possible variation. It is noted that the variation limit of δI needs to be known in the synthesis of the controller by the weighting factor (a). By incorporating Eq. (10) into the governing equation (6) and applying the sliding mode condition, the SMC is as follows:

$$u_{\theta i} = \frac{1}{\omega_n^2 I_0}\left\{c_i \dot{e}_i + \ddot{F}_{id} + 2\varsigma\omega_n \dot{F}_i + \omega_n^2 F_i + k_i \,\mathrm{sgn}(S_i)\right\}, \qquad i = 1,2 \tag{11}$$

In practice, it is not desirable to use the discontinuous control law in equation (11) due to the chattering. Therefore, the discontinuous control law is normally approximated by a continuous one inside boundary layer width. This is accomplished by replacing $\mathrm{sgn}(S_i)$ by a saturation function. Once the control input is determined, the input field to be applied to each ER clutch actuator on the fixed reference system is obtained as follows:

$$E_1 = \frac{\left[u_{\theta 1}\cos\theta_3 - u_{\theta 2}\cos\theta_2 \sin\theta_3\right]^{\frac{1}{\beta}}}{\alpha\cos\theta_2}, \quad E_2 = \frac{\left[u_{\theta 1}\sin\theta_3 + u_{\theta 2}\cos\theta_2\cos\theta_3\right]^{\frac{1}{\beta}}}{\alpha\cos\theta_2} \tag{12}$$

4 Results and Discussions

In order to evaluate control performance of the proposed system, an experimental setup is established as shown in Fig. 4. It is noted that the algorithm of the virtual slave part is not

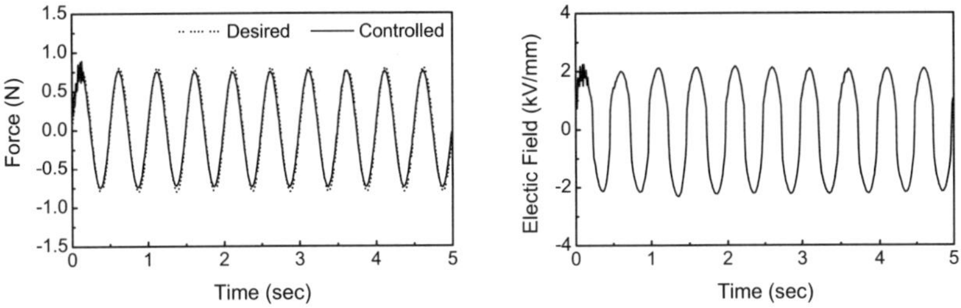

Figure 5. Active Force tracking control response.

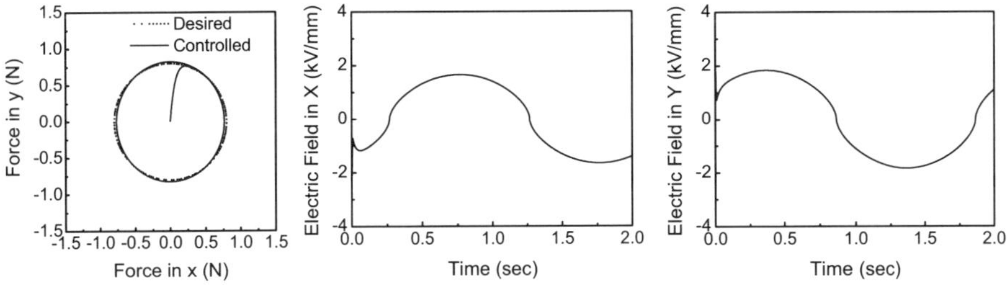

Figure 6. 2-DOF force tracking control responses.

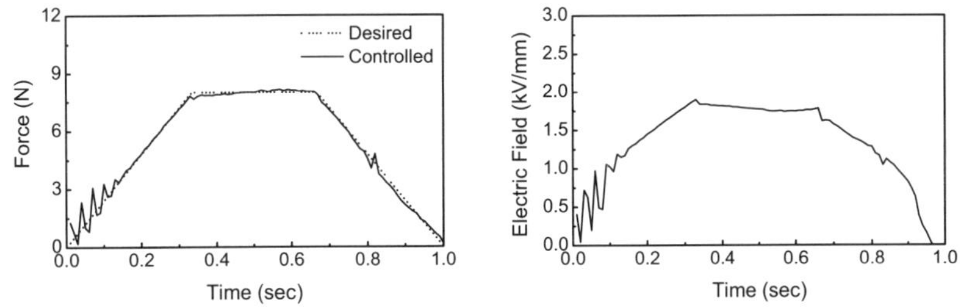

Figure 7. Semi-active Force tracking control responses.

considered in this study. The proposed spherical joint with two bi-directional ER actuators is connected with 4 AC motors. The operation link connected to inner electrode is fixed at a certain angle position by fixture. The produced force is measured by 6 axis force/torque sensor (ATI, Nano17) at the middle of the link. Then it is converted to digital signal via a 16bits A/D converter, and fed back to the microprocessor. The SMC given by Eq. (11) and (12) is then activated in the microprocessor to reflect the desired force from virtual slave. In this study, gradient matrix of sliding surface and discontinuous gain matrix are determined by $[38, 38]$ and $[4800, 4800]$, respectively.

As a first step, the active force tracking control responses are investigated by activating one bi-directional type of ER actuator in *X* axis. The operation link is fixed at origin. Fig. 5 presents the measured control responses for the sinusoidal trajectories (magnitude: 2*Hz*, amplitude: 0.8*N*) in *Y* axis. It is clearly seen that the desired force trajectories are well achieved, but input magnitude of field intensity is a little higher than that in simulation because of the effect of the parameter variation; influence coefficient due to hysteresis and mechanical friction. The maximum tracking error is evaluated by 0.09*N*. Fig. 6 presents the simulated control responses for 2-DOF force-feedback. The operation link is fixed at a certain angle ($\theta_2 = \pi / 12rad$, $\theta_3 = \pi / 6rad$), and weighting factor (a) of uncertainty variation is imposed by 0.8 and 1.2 for x and y directions, respectively. Two different desired trajectories are imposed to *x* and *y* directions, whose amplitude is 0.8*N*, and frequency is 0.5*Hz*. In this case, the two bi-directional ER actuators are activated simultaneously. It is clear seen from the results that the each trajectory in *x* and *y* directions well follows the desired one without exhibiting significant tracking errors. Fig. 7 shows the semi-active force tracking control responses, in which a voluntary human operator moves the operation link. The desired trapezoidal trajectory is favorably achieved with much higher force level than active mode.

5 Conclusion

In this work, a new type of ER haptic device featuring spherical geometry was proposed for reflection of active and passive force produced in minimally invasive surgery (MIS). An ER master has been designed and manufactured with the fixed electrodes, the rotating electrodes and the 4 geared AC motors, in which shape of electorde is spherical. In order to achieve desired force trajectories, a sliding mode controller (SMC) is formulated by considering friction and hysteresis as uncertainties, and experimentally realized. Subsequently, it has been demonstrated that the desired force trajectories are well achieved in semi-active and active modes by adopting the spherical ER joint and SMC.

6 Acknowledgement

This work was supported by the Korea Research Foundation Grant (KRF-2004-041-D00050). This financial support is gratefully acknowledged.

References

1. Abovitz, R., Digital surgery: the future of medicine and human-robot symbiotic interaction. *Industrial Robot: An International Journal*, **28(5)** (2001) pp. 401-405.
2. Intuitive Surgical, http://www.intuitivesurgical.com.
3. Computermotion, http://www.computermotion.com.
4. Stangroom, J. E., Carlson, J. D. and Duclos, T. G., ER fluid clutches and brakes -the synopsis. *ER Fluid Developments Ltd.* (1992).
5. Sakaguchi, M. and Furusho, J., Force display system using particle-type electrorheological fluids. *Proceedings of the 1998 IEEE International Conference on Robotics and Automation* (Leuven, Belgium, 1998) pp. 2586-2590.
6. Mavroidis, C., Pfeiffer, C., Celestino, J. and Bar-Cohen, Y., Design and modeling of an electro-rheological fluid based haptic interface. *Proceedings of the 2000 ASME Mechanisms and Robotics Conference* (Baltimore MD 2000).
7. Böse, H., Berkemeier, J. and Trendler, A., Haptic system based on electrorheological fluid. *Proceedings of the ACTUATOR 2000 Conference* (Bremen, Germany, 2000).
8. Shames, I. H. and Cozzarelli, F. A., Elastic and inelastic stress analysis. *Prentice Hall* (Englewood Cliffs, New Jersey, 1992).
9. Saito, T. and Sugimoto, N., A study on electro-rheological motion control using an antagonistic rotary actuator. *Proc. of Int. Conf. ER Fluids* (1997) pp. 54-65.
10. Han, Y. M., Design and control of haptic interface utilizing smart materials. *Ph. D. Dissertation* (Inha University 2005).

MR FLUID TECHNOLOGY - COMMERCIAL STATUS IN 2006

J. DAVID CARLSON
Lord Corporation, 406 Gregson Drive, Cary, NC 27511, USA
Email: jdcarlson@lord.com

Large-scale, commercial use of magnetorheological (MR) fluids to provide semi-active control in mechanical systems is now a reality. This is due primarily to their application in controllable, automobile suspension systems. Controllable MR fluid dampers are presently available on a number of General Motors vehicles as well as the new Ferrari Fiorano and Audi TT Coupe. Commercial MR fluid production is presently on the order of 100,000 liters per year. The number of MR devices in use worldwide is estimated to be in excess of 250,000.

1 Introduction

MR fluid technology has progressed to the point where it is routinely used on a commercial scale to provide semi-active control in a variety of automotive and industrial applications. MR fluids have been used commercially since the mid-1990s. Today, the greatest driving force behind MR fluid technology is primary automotive suspension systems. Other, commercially significant, application areas include steer-by-wire (haptic) systems and secondary vehicle suspension systems in heavy-duty vehicles. Additional developing areas that are poised to become commercially significant include engine mounts, fan clutches, all-wheel drive clutches, position locking and detent devices, crash protection systems, civil engineering infrastructure dampers, military vehicle suspensions, and prosthetic devices.

2 Automotive Primary Suspensions

In January of 2002, General Motors Corporation introduced the Cadillac Seville STS automobile with a MagneRide™ suspension system having real-time controllable MR fluid shock absorbers and struts as standard equipment [1]. The Magneride™ shock absorbers are made by Delphi Corporation with the MR fluid being made by Lord Corporation. Controllable MR fluid suspension systems are currently available on six General Motors' vehicles. These include the Chevrolet Corvette, Cadillac SRX Roadster, Cadillac XLR sport utility vehicle, Cadillac STS sedan [2] and, last year, the Cadillac DTS sedan [3] and Buick Lucerne [4].

In February 2006 Ferrari introduced the 599 GTB Fiorano [5] and in April 2006 Audi introduced the TT Coupe [6]. These are the first non General Motors production automobiles to feature a real-time controllable MR fluid-based suspension systems as standard equipment. The application history of MR fluids in primary automotive suspension systems is illustrated in Figure 1.

All commercial MR fluid controlled primary suspension systems are based on a monotube shock absorber that utilizes a single-stage, axi-symmetric MR valve contained in the piston. Such a MR fluid piston/valve design efficiently provides high on-state and low off-state force in a very compact and robust package. The controllers for these systems process inputs from relative position sensors at each wheel. In addition, inputs from a lateral accelerometer yaw rate sensor, steering angle sensor and speed sensor all feed by way of a CAN BUS into the controller. The control algorithms are quite complex and seek to simultaneously optimize a wide range of factors that influence overall handling and ride comfort. System response time is under 10 ms. Since their introduction, MR fluid-based, real-time controlled suspension systems have garnered virtually universal praise from automotive reviewers [7,8].

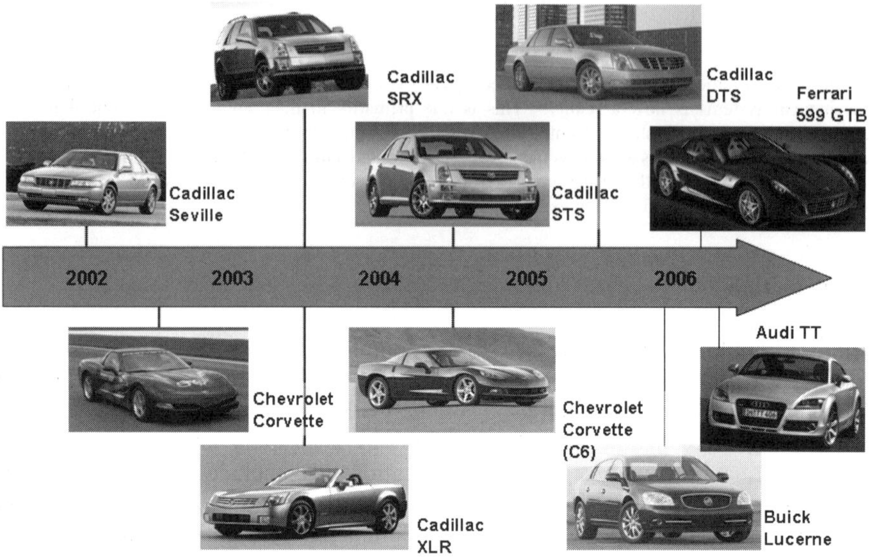

Figure 1. Application history of MR fluid based primary automotive suspensions.

Common to virtually all commercial MR fluid shock absorbers is single-coil, axi-symmetric MR fluid valve having an annular flow path as shown in Figure 2. This simple design has proven to be very reliable and is amenable to efficient, mass production. The straight flow path without any bends or changes in flow direction enables very low off-state forces without compromising the on-state. Automotive MR fluid shock absorbers have high durability with an operational life in excess of 150,000 km in a vehicle. Commercial MR fluid dampers are available in a variety of configurations including: coil-over shock, air-lift shock or Macpherson strut.

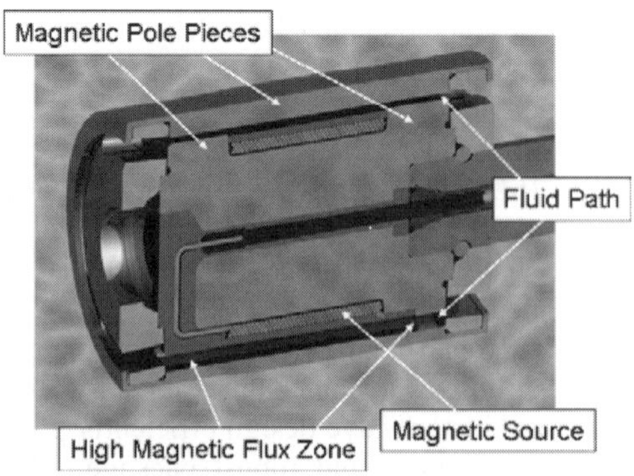

Figure 2. Typical damper piston with axi-symmetric MR fluid valve.

3 Steer-By-Wire

The trend in vehicle industries toward control-by-wire (steer-by-wire, shift-by-wire, throttle-by-wire, brake-by-wire, etc.) has created a need for highly controllable, rugged, cost-effective haptic devices to provide realistic force-feedback, sensations to the operator, whether the manual device is a wheel, a joystick, a pedal, or a lever. Commercialized products for forklift vehicles use a system that includes a MR brake, position sensor and control electronics in an integrated package to sense steering wheel position, control over-steer when the vehicle is moving quickly, and simulate the end-stop limit of driving wheel turn. The system even provides the necessary support and bearings for the steering wheel. Recently, such systems have also been applied to marine, i.e. boat, steering systems. Key benefits of MR technology in steer-by-wire applications include:

- Smooth, realistic force feedback with high fidelity feel
- High torque density - 10 times greater than DC motors
- Compact size
- Inherent system stability
- Low power consumption (typically 12V, 1 Amp max current)

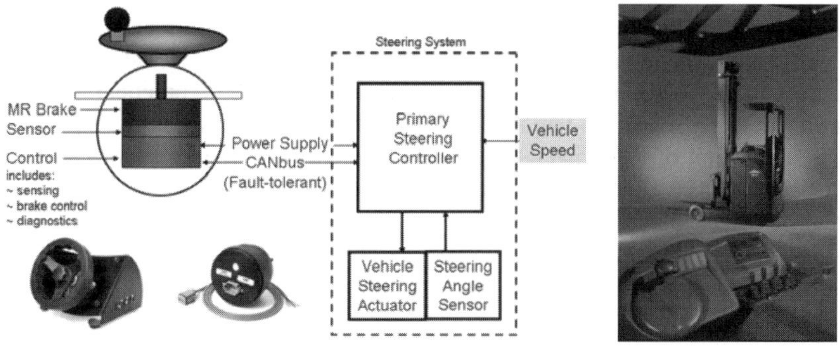

Figure 3. Steer-by-wire systems with integrated sensor, brake and controls.

4 Secondary Vehicle Suspension Systems

Secondary suspension systems include controlled semi-active damper systems for heavy-duty vehicle seating and cab suspension units for agricultural and construction vehicles. MR fluid damper seat suspensions based on the Lord RD-1005-03 damper have been in use since 1998. There are presently about 30,000 systems in use with more than 10^9 kilometers logged in truck, bus and agricultural operation. Many academic researchers are familiar with these dampers because of their ready availability as their convenient size for laboratory studies. These dampers have proven to be very robust and routinely outlive the seats on which they are installed. They meet or exceed all of the pertinent quality standards including SAE J1385, SAE J1386, ISO 7096 and ISO 5007.

MR fluid cab suspension units are integrated devices that include damper, air spring, sensors and controls in single package as shown in Figure 4. MR fluid control enables the use of low-stiffness springs without compromise between ride and cab stability. They eliminate the need for a roll bar and allow control of vertical, pitch and cabin roll motion.

Figure 4. Integrated MR fluid cab suspension unit.

5 Common Questions, Issues and Concerns

5.1 MR Fluid Response Time

The speed of an MR fluid device is largely determined by factors extrinsic to the MR fluid, particularly the inductance of the MR device and the characteristics of the current source (amplifier). Recently, experimental time-response data on practical MR fluids has become available. Goncalves has made measurements of the hydraulic response of MR fluids as a function of fluid dwell-time in a well-defined MR fluid valve [9,10]. From such measurements a fluid response time in valve-mode operation is estimated to be on the order of 0.2 ms.

5.2 Settling

As with any micron scale particulate suspension in which a density mismatch exists between the particle and fluid, long-term settling must be controlled. The specific application dictates to what extent and how the settling will be controlled. Most automotive applications are highly dynamic such that complete suspension stability is not required. Rather, it is entirely sufficient to ensure that, while the MR fluid suspension may slightly stratify upon standing, it does not form hard sediment and is easily remixed by the normal motion of the device.

The rapid and reliable redispersability of commercial MR fluids in shock absorbers has been extensively studied and validated by Delphi [11]. Shock absorbers were stored horizontally or vertically and thermally cycled for multiple days to simulate settling for periods up to the equivalent of one year or were subjected to real-time tests for up to one year. At the end of the settling time, the shock absorbers were carefully loaded into a test frame and the forces from the very first cycle of use were recorded. Figure 5 shows typical results for a test time of one year. While the first cycle shows an initial increase in force due to particulate settling, by the second cycle, the damper produces forces identical with the original performance of the damper when it was new. Thus, it takes only one stroke of the damper to redisperse the MR fluid, even after 1 year of settling.

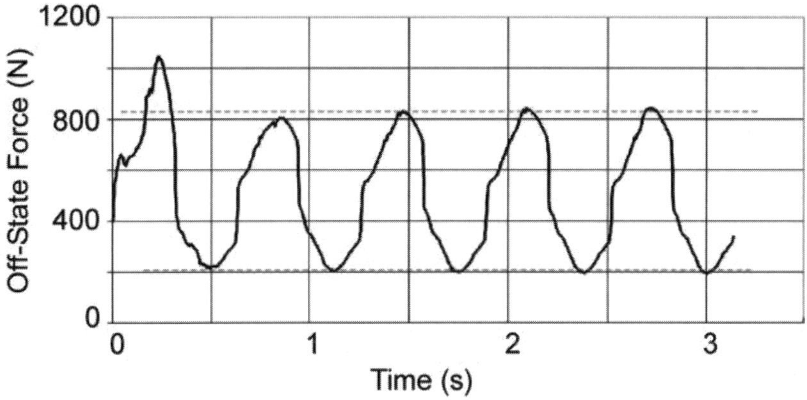

Figure 5. MR damper force during cycling immediately after one year of settling [11].

5.3 Component Wear and Seal Life

As with most dynamic hydraulic devices, seal integrity is crucial. While early MR fluids were indeed abrasive and could cause rapid seal wear and eventual catastrophic loss of MR fluid, today's commercial MR fluids and devices have been designed as a system to deliver long life. For automotive shock absorbers this means no MR fluid leakage for the guaranteed life of the vehicle of 100,000 miles or 160,000 km. Factors crucial to achieving such long life include seal geometry, fineness of shaft finish relative to iron particle size, and the addition (or omission) of certain specific additives to the fluid formulation.

The ultimate test is actual use. In addition to accelerated and track testing, Delphi has tested MR dampers that have been in actual use for over two years [11]. See Fig. 6.

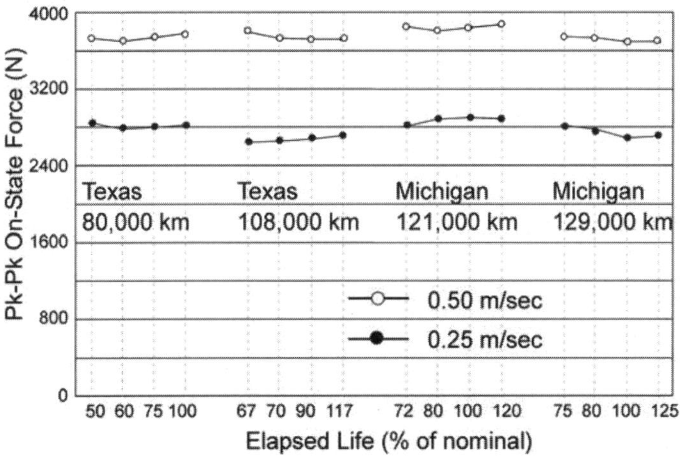

Figure 6. On-state shock performance from field life to 160,000+ km [11].

Magneride shock absorbers harvested from vehicles from Texas and Michigan, with between 80,000 and 129,000 km, were subjected to performance tests and further laboratory durability testing equivalent to the nominal life (or more) of the dampers

(160,000 km = 100% Field life), to verify the long-term behavior of both the MR fluid and shock absorber. After harvesting, all of the dampers, from both northern and southern climates, performed within original, "as-manufactured" specifications. During subsequent, additional laboratory testing the on-state damping force changes were nominal, varying only 3-5%, even after an additional equivalent 80,000 km of laboratory durability were added to the actual mileage on the shocks. Similarly, off-state forces and seal friction with and without side-load showed only nominal change.

5.4 MR Fluid Life/Durability

Depending on usage conditions, all MR fluids will eventually show some degree of deterioration. Such deterioration is usually manifested as a thickening of the fluid that is apparent as an increase in the off-state viscosity or off-state damper force. The amount of deterioration depends on shear rate, temperature and time. Silicone oil based fluids, are particularly prone to cross-linking if exposed to high temperatures for extended periods, particularly when used in an environment rich with an iron.

An ad hoc measure that has proven useful in estimating the expected life of MR fluid in a particular application is the lifetime dissipated energy or LDE [12] defined as:

$$LDE = \frac{1}{V}\int_{0}^{life} P \cdot dt$$

P is the instantaneous mechanical power converted to heat in the MR device. V is the volume of MR fluid in the device. LDE is simply the total mechanical energy converted to heat per unit volume of MR fluid over the life of a device. The best MR fluids today can sustain a LDE in excess of 10^8 J/cm^3 before they thicken to the point where device performance is compromised. Poor MR fluids, on the other hand, may become unusable with LDE's as low as 10^5 J/cm3. Current commercial MR fluids will last the equivalent of hundreds of thousands of km in an automotive shock absorber application with no degradation in performance. Long life MR fluids suitable for high-energy rotary applications such as automotive clutches and brakes are far more challenging due to the much smaller volume of MR fluid that is generally involved.

6 MR Fluid Status Today

Current production of MR fluid is on the order of 100,000 liters per year. It is estimated that more than a quarter million MR devices are presently in use worldwide. A ten-fold increase in MR fluid production over the next decade is anticipated. MR fluids are routinely produced in accordance with ISO 9001 quality standard and the even more stringent ISO/TS-16949 requirements demanded by the automotive industry. The cost of MR fluid varies depending on the specifics of the formulation and the annual quantity that is purchased. While their price in small quantity, i.e. ~ liter, is about \$600/liter, the price can fall to well below \$100/liter for large volumes of some MR fluid formulations [13].

While several "standard" MR fluids are commercially available today, MR fluids for significant applications are tailored to meet the specific requirements of the application. The liquid type, viscosity, particle size, particle shape and morphology, surface chemistry, volume fraction and additive package are all chosen to optimize the MR fluid for the specific conditions of the application. MR fluids can be formulated to have maximum yield strength in the range of 30 to 90 kPa. Proper choice of additives can insure that the MR fluid is as stable against particle settling, as the application requires. Finally, all MR fluid formulation additives must be evaluated with respect to fluid life and durability. The interdependency of all of these factors is very complex and important to understand in optimizing the performance of these fluids within an MR device and control system for a particular application, as trade-offs must constantly be made in the design process.

References

1. Delphi Corporation, www.delphi.com/pdf/ppd /chsteer/magneride.pdf (2005).
2. General Motors Corp. (2002, 2003 & 2005).
3. W. J. Keegan, Jr., AutoBlog, www.*autoblog. com/entry/1234000637030854/* (2005).
4. *www.nctd.com/sneakpreview.cfm? Vehicle=2006_Buick_Lucerne&ReviewID=80,* (2005).
5. www.mph-online.com/web/prtranslated/00581 (2006).
6. www.media.audiusa.com/article_di splay.cfm? article_i d=9732 (2006).
7. H. Halverson, 2003, *The Idaho Corvette Page*, CV World Internet Pub., *www.idavette.net/_magride/* (2003).
8. www.popsci.com/popsci /bown/article/0,16106,388074,00.html (2002).
9. F.D. Goncalves, *Characterizing the Behavior of Magnetorheological Fluids at High Velocities and High Shear Rates*, Ph.D. Thesis, Virginia Polytechnic Institute and State University, Blacksburg (2005).
10. F. D. Goncalves, M. Ahmadian and J. D. Carlson, Smart Materials and Structures, **15**, 75-85 (2005).
11. L.C. Yanyo, *IIR Suspension and Damping Conference*, Germany (2004).
12. J. D. Carlson, *J. of Intelligent Material Systems and Structures*, **13**, 431-435 (2003).
13. J. R. Toscano, Lord Corporation (2006).

COMPARISON OF ERF CLUTCH DESIGNS

D. J. ELLAM[1], W. A. BULLOUGH[2] AND S. M. CHEN[3]

[1,2] *Smmart Unit , Department of Mechanical Engineering, The University of Sheffield, Sheffield S1 3JD, UK, Email: W.A.Bullough@sheffield.ac.uk*

[3] College of Mechanical Engineering, Fuzhou University, Fujian Province, 350002, China

Email: smch2002@vip.sina.com

A comparison of the mechanical performances of radial plate and concentric clutches is made on the basis of: torque/size, acceleration, control ratio and maximum casing temperature – for both a single clutch pair and multi plate models, all with double acting surfaces. The various embodiments examined herein assume a constant rotor and case thickness, negligible inertial effects of the fluid and the clutches are, in the main, always slipping steadily. Electrical heating is not considered.

CFD is used for the computations and is verified against analytical solutions and experimental data where possible. The fluid, which may be MRS or ERF is taken to be a Bingham plastic with temperature invariant plastic viscosity and yield stress. Where acceleration is calculated the initial value (when full excitation is just applied to an already slipping clutch) is given. Input and output shafts are not included in the comparisons though they may have some effect on acceleration and heat transfer: Casing temperatures are taken to be uniform, as they approximately will be in the steady state, as highly conductive materials of construction being used. Through these simplified models some idea of this relative overall performances of the two clutch forms can be gleamed. Perhaps more importantly the CFD method is shown to be useful for detailed investigation of more specific situations. FLUENT [1] was chosen as the operating programme with sub-routines to handle the non Newtonian nature of the fluid. No regulation of the input drive speed due to loading is accounted for.

1 Torque and Acceleration Comparisons

Assuming a uniform casing and rotor thickness (z), it is possible to derive approximate equations for torque (N) and inertia (I) based upon thc number of internal rotors (n).

1.1 Multi channel concentric clutch

With reference to Figure 1, the following applies to a multi-channel concentric clutch having each rotor of effective radius (r_c), electrode length (L_e), outer casing length and radius L_c and R_c respectively [1]

$$N_{cn} \approx \sum_n 2\left\{2\pi r_{cn}{}^2 L_e\left(\tau_y + \frac{\mu_e r_{cn}\Omega}{h}\right)\right\} \quad \text{and} \quad I_{cn} \approx \sum_n 2\pi\rho z r_{cn}{}^3 L_e \tag{1}$$

where τ_y is yield stress in the electrode gap h. ρ is the metal density and μ_e the plastic viscosity of the fluid.. Ω is the relative (slipping) angular speed and z is the rotor thickness.

$$when\ n = 1 \qquad \frac{N_{c1}}{I_{c1}} = \frac{2}{\rho z}\left(\frac{\tau_y}{r_c} + \frac{\mu_e\Omega}{h}\right), \tag{2}$$

$$\mathrm{R_c} = (\mathrm{r_c}) + (\mathrm{h}+1.5\mathrm{z}) \text{ and } L_c = L_e + 2(z+h) \quad \text{(the outer casing dimensions)} \quad (3)$$

1.2 Multi channel radial clutch

With reference to Fig 2 the equivalent equations for a radial clutch with outer casing length and radius L_r and R_r respectively and where $\mathrm{r_r}$ is the effective radius are;

$$N_r = \sum_n 2\left\{2\pi r_r^{\,3}\left(\frac{\tau_y}{3} + \frac{\mu_e r_r \Omega}{4h}\right)\right\},\; I_r = \frac{1}{2}\rho n \pi z r_r^{\,4} \quad (4)$$

$$\frac{N_c}{I_r} = \frac{2}{\rho z}\left(\frac{4}{3}\frac{\tau_y}{r_r} + \frac{\mu_e \Omega}{h}\right) \quad R_r = r_r + (h+z) \quad L_r = 2n(z+h) + z \quad (5), (6)$$

The radial clutch torque is dependent upon $\mathrm{r_r}^3$, the concentric clutch torque on $r_c^{\,2}$, but L_e also plays a role. For the same torque {eqn(1) = eqn(4)}, when $N_o << N_e$, (suffixes e and 0 here are for field alone and zero field respectively) and $L_e = r_c$ then:

$$\mathrm{r_c} = 0.69\mathrm{r_r}, \quad \mathrm{L_r} = 3\mathrm{z} + 2\mathrm{h} \text{ and } \mathrm{L_c} = \mathrm{r_r} + 2\mathrm{z} + 2\mathrm{h} \quad \text{here } (z \approx h) << \mathrm{r_r} \quad (7)$$

Hence the concentric device is 31% smaller in the radial direction but is significantly larger in the axial direction. The concentric design may therefore be more suited to a radially confined geometry but will always be much longer than the radial clutch design i.e. for the same torque output etc. In addition, the ratio of N_e/N_o in the radial design is 25% greater than in the concentric design. It can be noted that for both devices a smaller radius gives greater acceleration. On an acceleration basis, for a single rotor/dual channel (n=1) and when N_o is small, the radial clutch is more favourable by 33%. However, adding more channels to the concentric design will slightly improve the acceleration (α) as the effective radius decreases.

2 Heat Transfer Comparison (Single Rotor Clutches)

CFD Model

CFD was used to model the rate of convective heat transfer from the clutch casing surface: it was deemed uneconomical to model the three-dimensional problem at this stage. The verified CFD procedure [2] was used most significantly for checking the major assumption of a uniform surface temperature. When setting up the CFD model four regions had to be specified. The two fluid regions were the smart fluid and the surrounding air, the two solid material regions included the inner output rotor and outer casing. The outer casing was surrounded with a 10 mm thickness of air which itself was closed off using a pressure boundary condition at temperature T_∞. The inner rotor was set to rotate with speed Ω. All solution parameters were left at default values.

For the parameters, r = 30 mm, h = 0.5 mm, z = 1.0 mm and T_∞ = 25 C, clutch torque and acceleration results for both designs were compared: Table 1, Fig 3 and Fig 4, indicate that for the same surface temperature T_s, the radial clutch offers more acceleration. Also for the same T_s, the overall size and consequent surface area of the concentric design is larger than in the radial. This allows it to convect more heat, and thus the generation of slightly more torque for the same surface temperature.

The relative performances of the clutch designs can be expressed as a ratio. For a variety of fluid properties and speeds, this ratio does not change significantly: see Fig 4 which clearly shows that for the same nominal radius and surface temperature, the radial clutch gives approximately 10% less torque but 30% more acceleration than the concentric clutch. In addition the radial clutch length is little more than half that of the concentric clutch. Fig 5 shows how T_s increases with speed. The plastic-viscosity is found to play a noticeable but relatively insignificant role, due to it having a minimal effect on N_{eo}. The CFD results for torque and generated heat are within 1-3% of the analytical approximations (Table 1). The CFD values include rotor end effects, which are not accounted for in the semi-empirical solution. In the CFD model, the case surface temperatures are found to be much higher than in the semi-empirical solutions; this is because the CFD convective model is two-dimensional, when it should really be three-dimensional. However, since the smart fluid properties are not set to be temperature dependent the resulting conductive model inside the clutch is still quite valid and allows the distribution and differences in temperature to be analysed. In both designs, only a very small variation in temperature occurs within the clutch itself and thus the uniform surface temperature assumption is quite valid. The high thermal conductivity of the outer casing evens out the surface temperature. In the radial design, the temperature is seen to be greater at higher radii due to increased shear rates there.

3 Multi-Plate Clutches

The single rotor study shows that the radial clutch type is often a more favourable design – for a smaller size the radial device gives a similar torque output but greater acceleration. For the radial design, increasing the casing length has no direct effect on α. However, this is not true for the concentric form should the unused space inside of the first output rotor can be utilised. The multi-plate clutch device was therefore studied next.

Practical sizes of casing were chosen (L and R). This dictated the number (n) of double channel segments that could be accommodated in both design configurations. Using a similar method to that described in section 3, for a particular speed and fluid properties, the amount of torque, viscous heating and surface temperature could be calculated for both devices. In practice, the smart fluid properties would be a function of temperature. Since the shear rates should be relatively low, the plastic-viscosity plays no significant role, in which case it is only the assumption that the yield-stress was the same for all temperatures that is in contention. The validity of the constant surface temperature assumption and the accuracy of the magnitudes of torque and viscous heating were successfully checked out in CFD studies.

3.1 Summary of the Models

The input parameters were as follows; h = 0.5 mm, z = 1.0 mm, $\tau_y = 1.0$ kPa, μ_e = 50mPa.s, Ω = 250 rpm and T_∞ = 25 C. Several practical configurations of outer casing geometry were investigated (R and L). The input speed and yield-stress have a proportional effect on the magnitude of viscous heating in constant speed operation but it was discovered that even for relatively low loadings that the temperature rose to well above the range in which air properties were linearised. To reduce viscous heating, an additional criterion had to be added. This was that the clutch would be in shear for only 10% of the total operating cycle. This would be a more realistic state of operation: The clutches were modelled as a steady problem, but with only 10% of the maximum input power so as to reduce the average torque and viscous heating to correspond with the analytical model a viscosity of 10% of μ_e (previous) was set to accomplish this.

3.2 Results

Looking at the ratios of relative performances (Fig 6) there is now virtually no difference between the two designs. For the same nominal casing dimensions, torque, acceleration and surface temperatures are very similar.
Table 2 illustrates how acceleration is greatly enhanced by a reduction of the outer radius of the clutches and that increasing the length of a clutch has no effect on acceleration. The CFD results for torque and viscous heating are seen to agree well with empirically modified theory (taken from [1]).

It appears that for a multi-plate clutch, both radial and concentric designs fare equally well. The CFD results do however illustrate one other major consideration, namely how well the heat generated in the annular gaps can be conducted into the outer casing. To maximise acceleration of the output rotor it is tempting to construct it from a lightweight material (expected to have a low thermal conductivity). This would however reduce heat transfer rates. The thermal conductivity of the inner rotor was set to equal that of nylon with the result being a hindrance to heat transfer and increased temperature. This is more of a problem for the concentric clutch, but by only a few degrees. For example, with dimensions (R = 60 mm, L = 60 mm), the difference between maximum and minimum temperatures within the clutch is 2.3 C for the concentric design and 1.5 C in the radial design.

4 Conclusions

It is evident from comparisons of the experimental and theoretical data for a viscous clutch device set-up, that the correct magnitudes of heat transfer can be effectively predicted. For a more complex set-up, such as a multi-plate clutch in an unsteady operation, CFD promises to be a simple and important tool for heat transfer evaluation.

In previous [2] experimental studies, the viscous medium was Newtonian Shell oil. Heat transfer in non-Newtonian fluids is a complex subject. However, it is proposed that modelling heat transfer in smart fluids need not be that complicated due to the following reasons: All smart fluid devices operate within a very narrow channel, for which the distribution of temperature across the gap width is likely to be small. Using CFD, the most important variable (yield-stress) can be entered as a function of temperature, excitation and shear rate. Furthermore, correctly determined non-Newtonian velocity profiles will give the rate of heat generation on a cell-by-cell basis.

In general, there is little to separate the performances of the two clutch types. Electrically speaking the centre portion in a radial clutch can be masked off to advantage.

Further work is required with respect to the similarity of flows in radial and concentric clutches and the more general problem of non-isotropic fluid behaviour. Better thermal properties of fluids are needed. For a further description of the methods employed see [1] and[3].

I would seem that smart fluid clutches would be best employed in fast engagement/disengagement short slip duration duties.This would probably be achievable with a cylindrical model where the outer driving plate could be fitted with cooling fins.

Some lack of precision will apply to the temperature predictions ,.Also , these will vary with the properties of a real ER fluid and clutch material:,the above results assume aluminium clutch parts and a nominal fluid specific gravity.

Figures

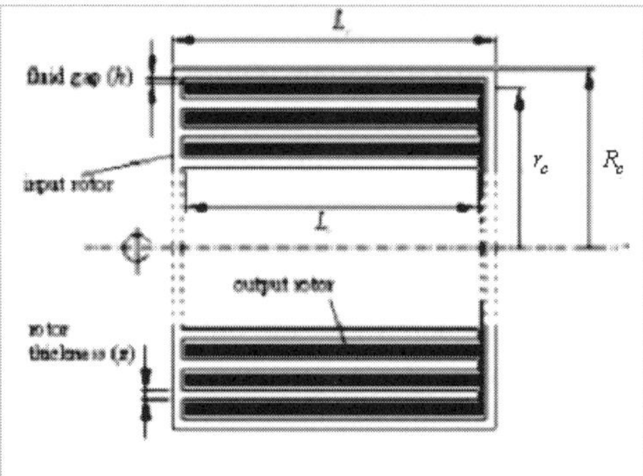

Fig. 1 Concentric clutch geometry for a multi channel device

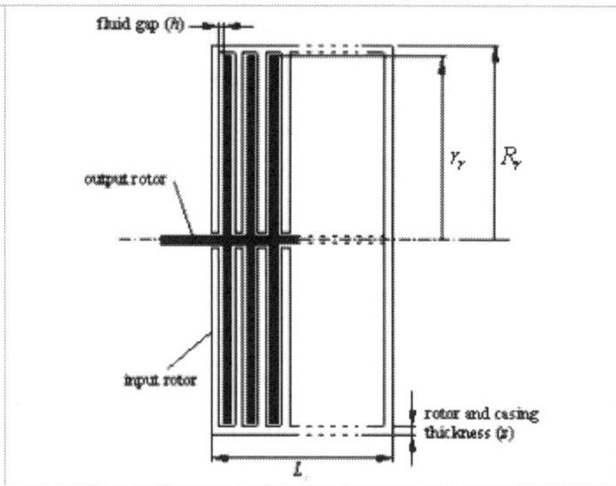

Fig. 2 Radial clutch geometry for a multi channel device

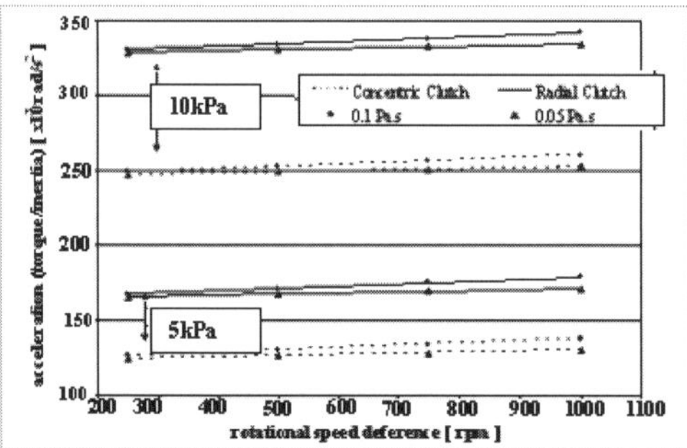

Fig. 3 Single rotor dual chancel concentric and radial clutch design comparison. Acceleration is calculated for the same clutch surface temperatures

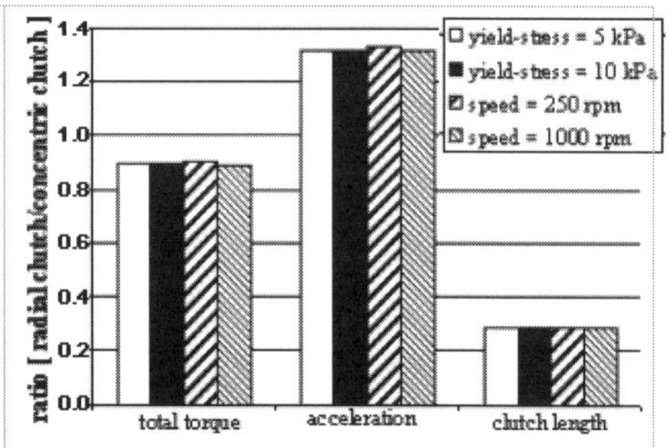

Fig. 4 Single rotor dual chancel concentric and radial clutch design comparison. Ratio of torque, acceleration, and clutch length for the same clutch surface temperature.

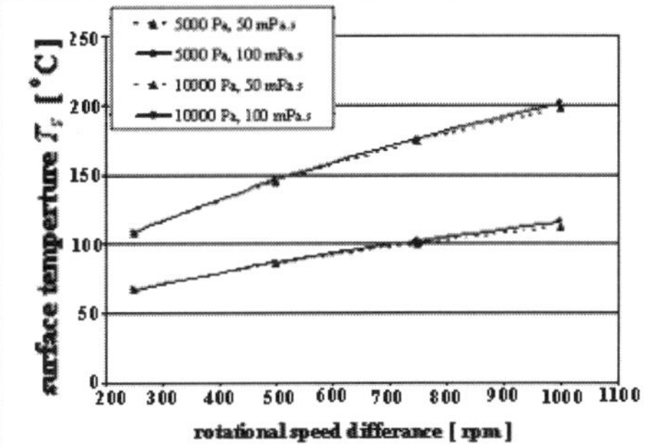

Fig. 5 Single dual chancel concentric and radial clutch design comparison. Ratio of torque, acceleration, and clutch length for the same clutch outer surface temperature

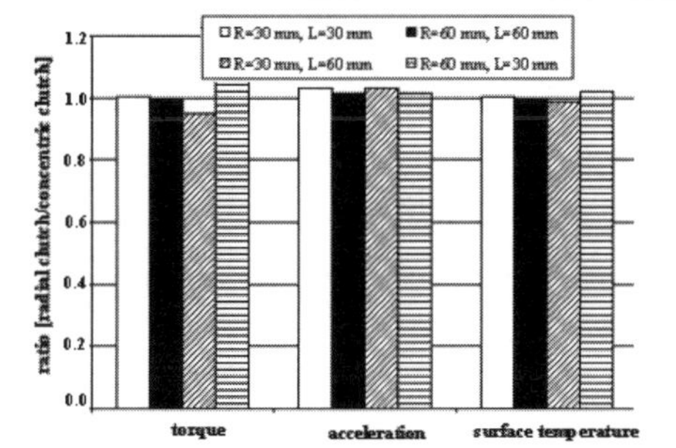

Fig.6 Multi chancel concentric and radial clutch design comparison. Ratio of torque, acceleration, and clutch length for the same clutch yield-stress value

yield-stress		τ_y	kPa		5				5				10				10			
plastic-viscosity		μ_e	mPa.s		50				100				50				100			
speed		Ω	rpm		250	500	750	1000	250	500	750	1000	250	500	750	1000	250	500	750	1000
radial design	total torque	N_{eo}	analytical	Nm	0.572	0.579	0.585	0.592	0.579	0.592	0.605	0.619	1.138	1.144	1.151	1.158	1.144	1.158	1.171	1.184
			CFD		0.581	0.589	0.596	0.604	0.588	0.603	0.617	0.632	1.154	1.163	1.171	1.179	1.162	1.177	1.192	1.207
			difference	%	1.5	1.7	1.8	1.9	1.6	1.8	1.9	2.1	1.4	1.6	1.7	1.8	1.5	1.6	1.8	1.9
	viscous heating	P_{in}	analytical	Watts	15.0	30.3	46.0	62.0	15.2	31.0	47.6	64.8	29.8	59.9	90.4	121.2	30.0	60.6	92.0	124.0
			CFD		15.2	30.7	46.7	63.1	15.4	31.5	48.4	65.9	30.1	60.8	92.0	122.9	30.4	61.7	93.7	126.7
			difference	%	1.5	1.3	1.5	1.7	1.6	1.6	1.8	1.7	1.1	1.5	1.7	1.4	1.5	1.8	1.9	2.1
	casing length	L_r	mm		4.0	4.0	4.0	4.0	4.0	4.0	4.0	4.0	4.0	4.0	4.0	4.0	4.0	4.0	4.0	4.0
	casing radius	R_r	mm		31.5	31.5	31.5	31.5	31.5	31.5	31.5	31.5	31.5	31.5	31.5	31.5	31.5	31.5	31.5	31.5
	acceleration	α	x10^3 rad/s^2		165	167	169	171	167	171	175	179	329	331	333	335	331	335	338	342
concentric design	total torque	N_{eo}	analytical	Nm	0.636	0.647	0.657	0.666	0.643	0.661	0.678	0.694	1.264	1.277	1.289	1.299	1.271	1.292	1.310	1.328
			CFD		0.650	0.663	0.675	0.685	0.658	0.678	0.697	0.715	1.282	1.302	1.314	1.325	1.290	1.318	1.338	1.357
			difference	%	2.2	2.4	2.7	2.8	2.3	2.5	2.7	2.9	1.4	1.9	1.9	2.0	1.5	2.0	2.1	2.2
	viscous heating	P_{in}	analytical	Watts	16.6	33.9	51.6	69.7	16.8	34.6	53.3	72.7	33.1	66.9	101.2	136.0	33.3	67.6	102.9	139.0
			CFD		16.9	34.4	52.6	71.5	17.1	35.2	54.4	74.6	33.5	67.9	102.9	138.5	33.7	68.7	105.1	142.1
			difference	%	1.5	1.5	2.0	2.5	1.6	1.6	2.1	2.5	1.2	1.5	1.6	1.8	1.3	1.6	2.1	2.2
	casing length	L_c	mm		14.07	14.09	14.09	14.07	14.03	14.00	13.96	13.91	14.09	14.12	14.13	14.13	14.07	14.07	14.06	14.04
	casing radius	R_c	mm		32.0	32.0	32.0	32.0	32.0	32.0	32.0	32.0	32.0	32.0	32.0	32.0	32.0	32.0	32.0	32.0
	acceleration	α	x10^3 rad/s^2		125	126	128	130	126	130	134	138	247	249	251	253	249	253	257	261
surface temperature		T_s	^{0}K		340.6	358.8	373.2	385.7	341.1	360.2	375.7	389.6	381.9	417.9	446.0	470.3	382.4	419.3	448.6	474.3

Table 1: Comparison between measured experimental temperatures and theoretical predictions

Design	Quantity	Symbol	Method	Unit	Case 1	Case 2	Case 3	Case 4
	yield-stress	τ_y		Pa	1000	1000	1000	1000
	plastic-viscosity	μ_e		mPa.s	50	50	50	50
	speed	Ω		rpm	250	250	250	250
	casing length	L		mm	30	60	60	30
	casing radius	R		mm	30	60	30	60
radial design	total torque	N_{eo}	analytical	Nm	1.02	18.70	2.048	9.35
radial design	total torque	N_{eo}	CFD	Nm	1.01	18.32	1.980	9.12
radial design	total torque	N_{eo}	difference	%	-1.4	-2.1	-3.4	-2.5
radial design	viscous heating	P_{in}	analytical	Watts	2.68	49.0	5.36	24.5
radial design	viscous heating	P_{in}	CFD	Watts	2.64	48.2	5.29	23.8
radial design	viscous heating	P_{in}	difference	%	-1.5	-1.6	-1.3	-2.8
radial design	acceleration	α		rad/s^2	3634	18691	36338	18691
radial design	surface temperature	T_s	analytical	°C	32.0	55.5	36.6	42.3
concentric design	total torque	N_{eo}	analytical	Nm	1.021	18.71	2.155	8.86
concentric design	total torque	N_{eo}	CFD	Nm	1.000	17.83	2.054	8.48
concentric design	total torque	N_{eo}	difference	%	-2.1	-4.9	-4.9	-4.5
concentric design	viscous heating	P_{in}	analytical	Watts	2.67	49.0	5.64	23.2
concentric design	viscous heating	P_{in}	CFD	Watts	2.61	48.1	5.52	22.6
concentric design	viscous heating	P_{in}	difference	%	-2.4	-1.8	-2.2	-2.7
concentric design	acceleration	α		rad/s^2	35245	18419	35245	18419
concentric design	surface temperature	T_s	analytical	°C	31.9	55.5	37.2	41.4

Table 2: Comparison between measured experimental temperatures and theoretical predictions

References

1. D J Ellam PhD Thesis, University of Sheffield 2003, Modelling Smart Fluid Devices using CFD.
2. D J Ellam, W A Bullough and V Oravsky, Heat Transfer From an ESF Radial Plate Clutch Surface, Proc 8th Int. Conf. ER/MR Fluids held Nice July 2001. P 104 – 110. World Scientific Pub.
3. R Smyth, K H Tan and W A Bullough, Heat Transfer Modelling of a Cylindrical ER Clutch, Int Jnl. Mod Phys B.8, p 3033 – 3050 (1994) World Scientific Pub.

ANALYSIS AND VALIDATION OF SEMI-ACTIVE MR RECOIL DAMPER

J.-H. YOO*, J.-W. PARK†, AND N. M. WERELEY*

*Department of Aerospace Engineering, University of Maryland, College Park, MD 20742, USA
E-mail:wereley@eng.umd.edu

Daewoo Precision Industries Co., LTD, Busan, South Korea, 619-873

We present the design and evaluation of a controllable, semi-active magnetorheological (MR) fluid shock absorber. This study experimentally validates asymmetric quasi-steady MR damper model, using an idealized Bingham plastic shear flow mechanism, for the flow mode damper configuration. Essentially, the MR valve within the piston of the damper comprises a magnetic circuit and annular orifice, where the inlet and outlet geometry of the MR valve can be the same (symmetric) or different (asymmetric). In the asymmetric model, geometric properties of active valve length and gap, as well as the magnetization are different at each end of the MR valve, so that the field dependent yield stress is also different. Two dampers were tested using a MTS machine for sinusoidal stroke of 30mm, over a range of velocities below 0.4m/s. The damping force and damping effectiveness vs. velocity diagram was experimentally validated using the data over a range of applied magnetic field.

1 Introduction

MR dampers have recently been utilized in vehicle primary suspensions [1], recoil damper systems [2-3], as well as a growing number of applications. Recently, research has focused on the shock isolation damper design and impulse force mitigation strategies. A key issue is the control of energy flows in a gun recoil system. The unique controllable energy dissipation of MR dampers enables control of these energy flows, as well as alleviation of vibration and shock loads for recoil forces. Facey *et al.* [2] designed a compact MR damper for impulsive loads with a bifold MR valve configuration. Ahmadian *et al* [3] evaluated MR dampers for a gun recoil system, and demonstrated enhanced performance. The MR valve within the piston of the damper comprises a magnetic circuit and annular orifice, where the inlet and outlet geometry of the MR valve can be the same (symmetric) or different (asymmetric). Most studies have assumed that the MR valve presents a symmetric configuration. However, in the asymmetric MR valve, geometric properties of active valve length and gap, as well as the magnetization, are different at each end of the MR valve, so that the field dependent yield stress is also different. Often due to packaging and machining considerations, asymmetries arise in the MR valve design. In this study, we develop the asymmetric MR valve model to account for different valve length, gap, and yield stress at each end of the MR valve. Characterization of two dampers is experimentally performed at speeds up to 0.4 m/s with 30 mm sinusoidal stroke on a servohydraulic MTS testing machine. Analysis of the damping force vs. velocity diagram was validated using the experimental data over a range of applied magnetic field. From these results, we deduced that a small improvement in the magnetic flux performance, accompanied by a minor change in the valve design, substantially enhanced damping force.

2 Flow Mode MR Damper

A schematic of the flow mode damper is shown in Figure 1. Figure 2 shows the schematic and dimensions of the MR valve in the damper. The MR valve controls the flow rate and pressure drop through the gap, and so controls damping level. To evaluate the performance of this asymmetrical MR valve, a generalized MR valve model is developed. Essentially, the problem is one of two different MR valves in series.

In Figure 2, different active lengths, L_1 and L_2, and gaps, d_1 and d_2, are assumed. Also, the cross-sectional duct areas are defined as A_{d1} and A_{d2}. For a Newtonian fluid fluid (approximating the off-state or absence of field), the pressure loss across each active valve is given as [4]:

$$Q_1 = \frac{A_{d1}d_1^2}{12\mu L_1}(P_1 - P_0) \qquad Q_2 = \frac{A_{d2}d_2^2}{12\mu L_2}(P_0 - P_2) \tag{1}$$

Here μ is the viscosity of the MR fluid. Force can be expressed as a function of velocity using continuity, $Q_1=Q_2$, and the relationship between pressure and force of the damper shaft, $\Delta P = F_d / A_P$, in a valve as:

$$F_d = 12\mu A_p^2 \frac{A_{d1}d_1^2 L_2 + A_{d2}d_2^2 L_1}{A_{d1}d_1^2 A_{d2}d_2^2} v_0 = C_N v_0 \tag{2}$$

Here: F_d is the damping force, v_0 is the piston or damper shaft velocity, and A_p is the piston area. From this equation, we deduce the Newtonian damping coefficient with viscosity of fluid and dimensions of the MR damper.

The flow rate in the on-state (presence of field) through each MR valve in Figure 2 can be expressed using the Bingham-plastic model as [4]:

$$Q_1 = \frac{A_{d1}d_1^2}{12\mu L_1}(1-\bar{\delta}_1)^2(1+\bar{\delta}_1/2)(P_1 - P_0) \qquad Q_2 = \frac{A_{d2}d_2^2}{12\mu L_2}(1-\bar{\delta}_2)^2(1+\bar{\delta}_2/2)(P_0 - P_2) \tag{3}$$

Force can be expressed as function of velocity using continuity and the definition of pressure, $\Delta P = F_d / A_P$, as:

$$F_d = \frac{12\mu A_p^2\left[A_{d1}d_1^2 L_2(1-\bar{\delta}_1)^2(1+\bar{\delta}_1/2) + A_{d2}d_2^2 L_1(1-\bar{\delta}_2)^2(1+\bar{\delta}_2/2)\right]}{A_{d1}d_1^2(1-\bar{\delta}_1)^2(1+\bar{\delta}_1/2)A_{d2}d_2^2(1-\bar{\delta}_2)^2(1+\bar{\delta}_2/2)} v_o = C_{eq} v_0 \tag{4}$$

From equations (2) and (4), we can define the damping coefficient or dynamic range as the ratio of C_{eq} to C_N as:

$$\frac{C_{eq}}{C_N} = \frac{A_{d1}d_1^2 L_2(1-\bar{\delta}_1)^2(1+\bar{\delta}_1/2) + A_{d2}d_2^2 L_1(1-\bar{\delta}_2)^2(1+\bar{\delta}_2/2)}{(A_{d1}d_1^2 L_2 + A_{d2}d_2^2 L_1)(1-\bar{\delta}_1)^2(1+\bar{\delta}_1/2)(1-\bar{\delta}_2)^2(1+\bar{\delta}_2/2)} \tag{5}$$

The damper shaft volume inside the damper varies as the piston moves. In order to simplify the analysis of the MR damper, it is assumed that the MR fluid is incompressible. The volumetric change of the damper produces a spring force from the pneumatic reservoir compliance, as shown in Figure 1. By assuming that the gas inside the reservoir is adiabatic, the compliance of the gas reservoir is given by

$$C_g = \frac{V_0}{P_0 \kappa} \tag{6}$$

Here V_0 and P_0 are initial volume and pressure of the pneumatic reservoir, respectively, Also, κ is the specific heat ratio of the gas. Now, the total force for MR damper can be written as follows:

$$F = \frac{A_r^2}{C_g} x_0 + F_d$$

Here A_r is the piston rod area and x_0 is the piston displacement.

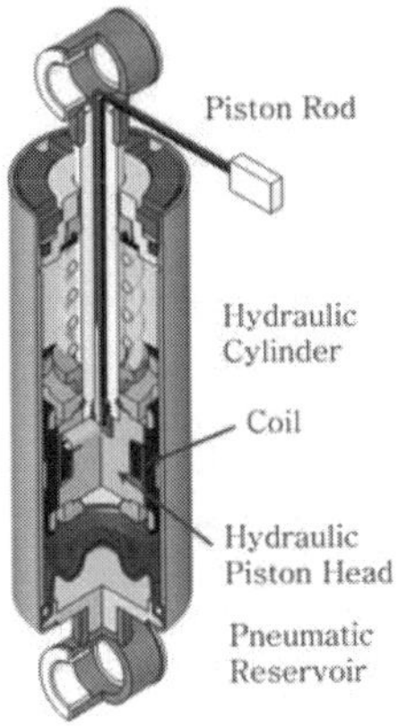

Figure 1. Schematic of MR damper.

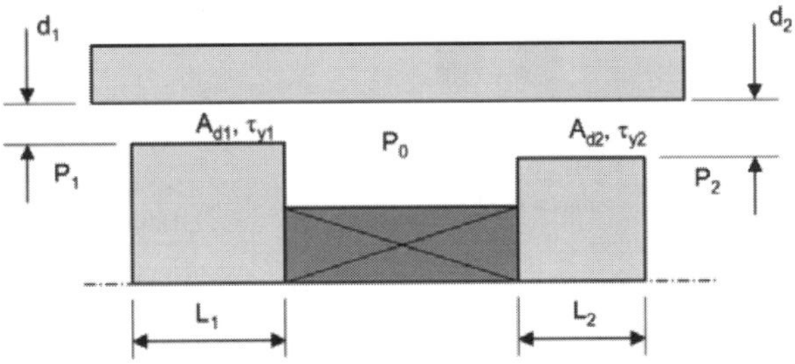

Figure 2. MR valve configuration and dimensions.

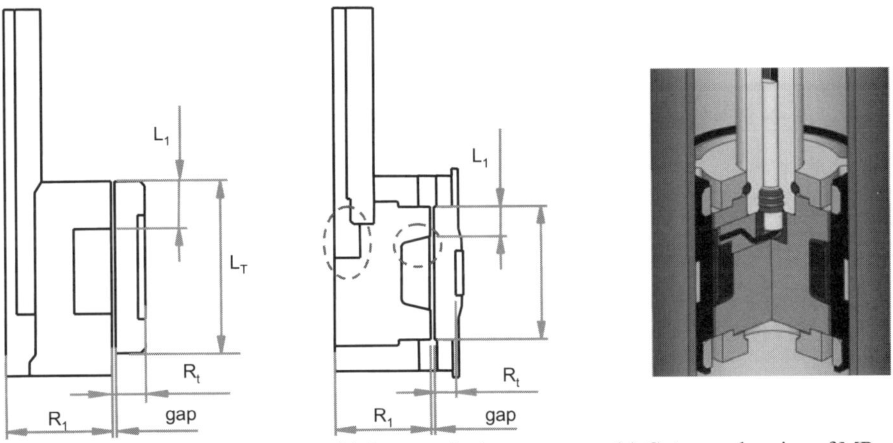

(a) Asymmetric damper (b) Symmetric damper (c) Cutaway drawing of MR valve

Figure 3. MR valve configurations

3 Magnetic Field Analysis

In Figure 3, schematics of the (a) asymmetric, and (b) symmetric valves are shown. The MR valve consists of solenoid core, flux ring and magnetic coil. For the symmetric model case, we made a little improvement on the magnetic circuit at the solenoid core part as indicated in the figure. Detailed dimensions are listed in Table 1. Based on these dimensions, we conducted non-linear magnetic field analysis with Flux-2D software to verify the magnetic flux density at the gap. Figure 4 (a) and (b) show the B-H curves for solenoid core and flux ring, respectively. These non-linear B-H data are used for the magnetic field analysis. In Figure 5, the averaged magnetic flux densities at each gap are shown for each damper model. The magnetic field density distribution on the FE model is depicted in Figure 6 (a) for an example and Figure 6 (b) shows the magnetic flux density results along the valve gap center according to the magnetic current input. As can be seen in Figure 5 and Figure 6 (b), the field intensity at each gap does not have the same value even in the symmetric damper configuration.

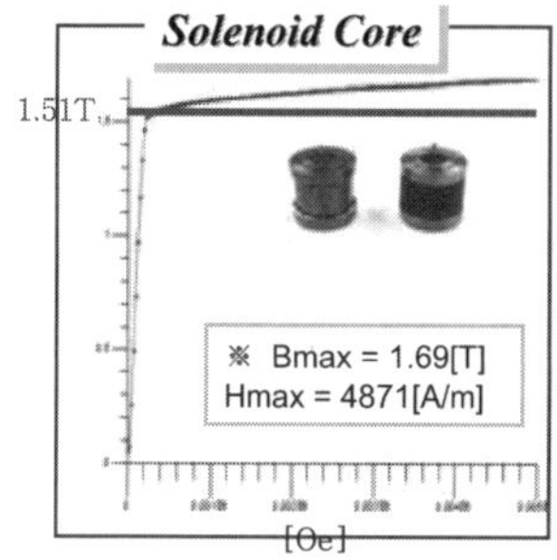

(a) Solenoid Core

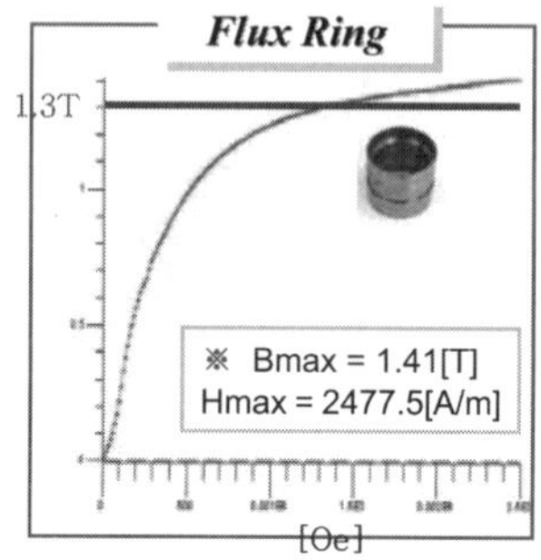

(b) Flux Ring

Figure 4. B-H curve for each component in magnetic circuit of the MR valve

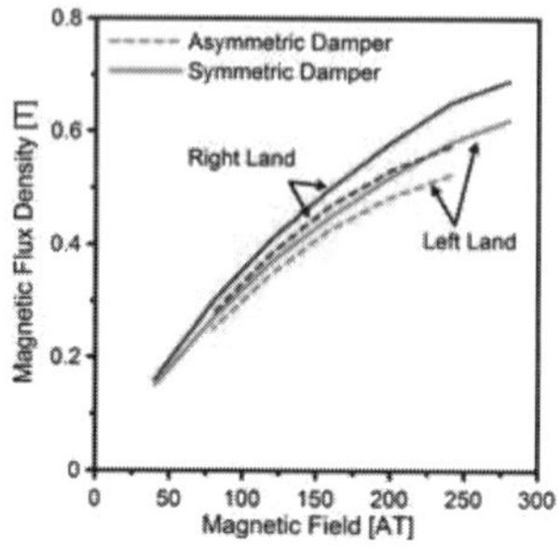

Figure 5. Magnetic performance results for each damper

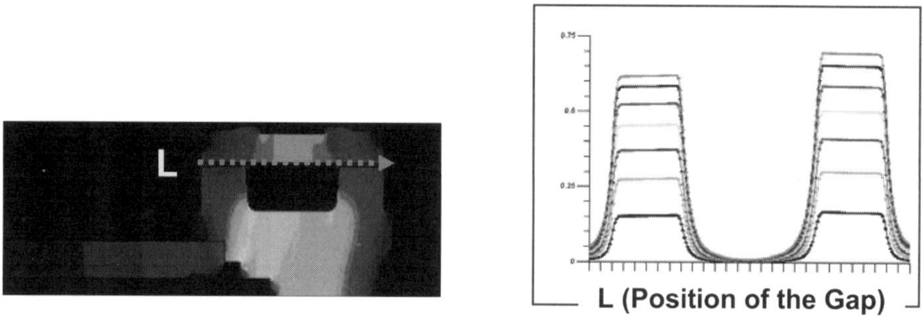

(a) Magnetic Density Distribution (b) Magnetic Density along the Gap

Figure 6. Magnetic field analysis result for symmetric damper.

Table 1. Specifications of MR valves design

	L_1	L_2	R_1	Gap	R_t	C_N
Asymmetric	5.94 mm	7.72 mm	14.25 mm	0.5mm	3.125mm	973 Ns/m
Symmetric	5.80 mm	5.80 mm	15.00 mm	0.5mm	3.300mm	1019 Ns/m

4 Experiment

4.1 Experimental Setup

For the experimental validation of the damper analyses equations, force measurements from sinusoidal displacement cycles were recorded on a MTS dynamic test machine at DPI (Daewoo Precision Industrial Co.). The test setup, shown in Figure 7 with the MR damper mounted in the clevises, sinusoidally oscillated the shaft of the damper, measured the shaft displacement using an LVDT and measured the resulting load using a 5000 lb load cell. A matrix of tests for 30 mm of stroke with difference experimental conditions was collected for experimental validation of the quasi-steady flow mode damper analysis. Oscillation velocity ranged from 0.01 to 0.4 m/s, in increments of nominally 0.1 m/s. The applied current ranged from 0 to 1.2A in increments of 0.2 A. This translated to a range of magnetic field of 0 to 240 A-T in increments 40 A-T because the number of turn is about 200 turns. The magnetic field range was based on the magnetic field FE analysis in former section.

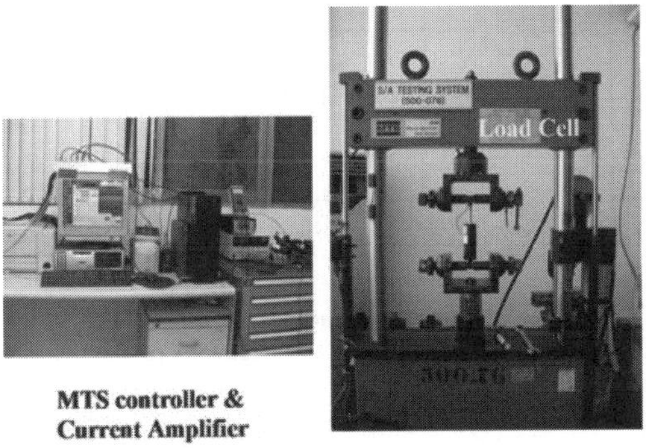

MTS controller & Current Amplifier

MTS Machine

Figure 7. Experimental setup

4.2 Experimental Results

In Figure 8, the experimental force vs. displacement hysteresis cycles are compared with simulation results using the asymmetric MR valve model. In Figure 8 (a) and (b), these force vs. displacement hysteresis cycles are plotted as magnetic field increases in value for 0.2 m/s and 0.4 m/s shaft velocity, respectively. As the magnetic field increases, the area within the hysteresis cycle also increases, indicating that the damping or energy dissipated by the damper is increasing. In Figure 8, the solid curves refer to modeling results presented in former section.

In Figure 9, the force versus velocity diagram is plotted for 30 mm stroke, and magnetic field input values of 0, 80, 120, 160, 200, and 240 A-T. These data are shown for a excitation frequency of 2 and 4 Hz in Figures 8(a) and 8(b), respectively. The Bingham plastic analysis with asymmetric MR valve configuration is used to model the force versus displacement and force versus velocity hysteresis cycle behavior in Figures 8 and 9. In all cases, the analysis correlates well with the data for all values of applied magnetic field. *The preyield compression loop observed in the force versus velocity response was not modeled in the current study.* Figure 10 depicts the maximum force vs. maximum velocity diagrams for the asymmetric and symmetric dampers. Here, the Bingham plastic damper model based asymmetric valve model well represents the experimental data. The asymmetric damper presented a controllable force of 1803 N, which was smaller that the symmetric damper controllable force of 2308N. Thus, from these test data, we could deduce that a small improvement in the magnetic flux performance, accompanied by a minor change in the design will give substantial enhancement in damping force.

We calculated equivalent viscous damping (C_{eq}) and the damping coefficient (C_{eq}/C_N) using the measured force vs. displacement hysteresis data and the results are shown in Figure 11 for each damper configuration. Note that the solid curves refer to modeling results presented in a former section. These results give an idea of the control effectiveness of the dampers. The simulation results agree well with the experimental result over this velocity range.

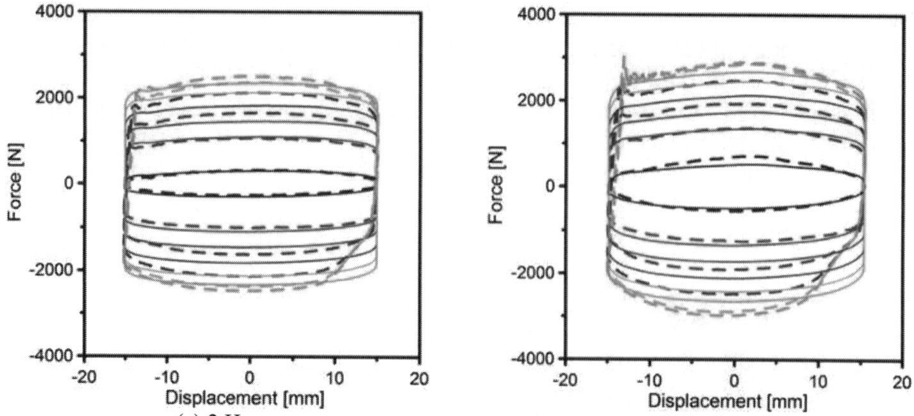

(a) 2 Hz (b) 4 Hz

Figure 8. Typical force vs. displacement hysteresis cycle data. (The dotted lines refer to experimental data and the solid curves designate modeling results using the asymmetric MR valve).

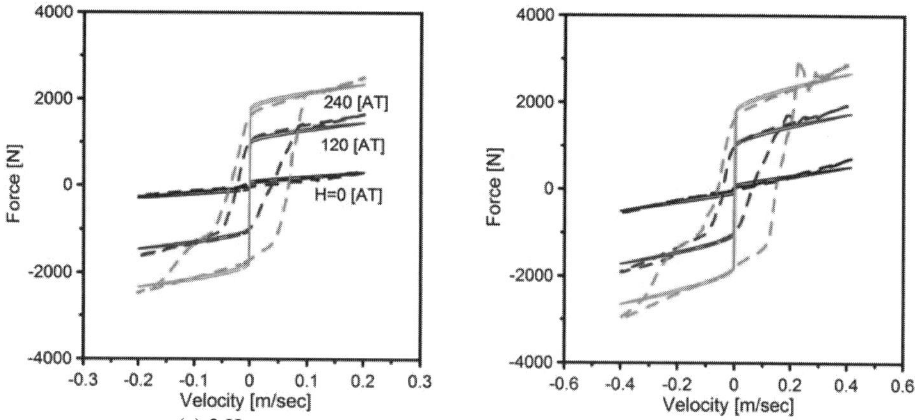

(a) 2 Hz (b) 4 Hz

Figure 9. Typical force vs. velocity hysteresis cycle data. (The dotted lines refer to experimental data and the solid curves designate modeling results using the asymmetric MR valve)

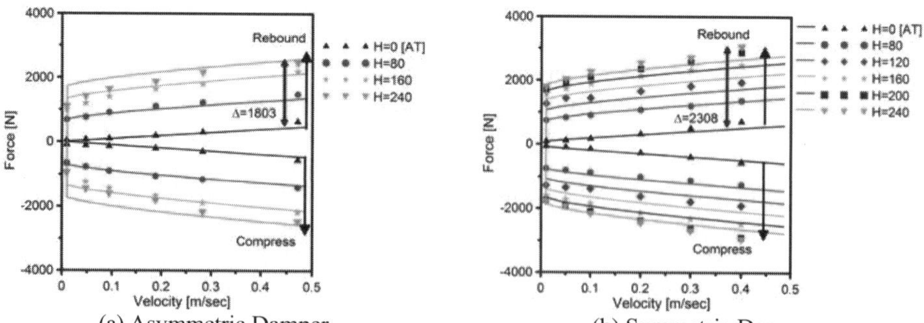

(a) Asymmetric Damper (b) Symmetric Damper

Figure 10. The maximum force versus maximum velocity response (The symbols refer to experimental data and the solid curves designate modeling results using the asymmetric MR valve).

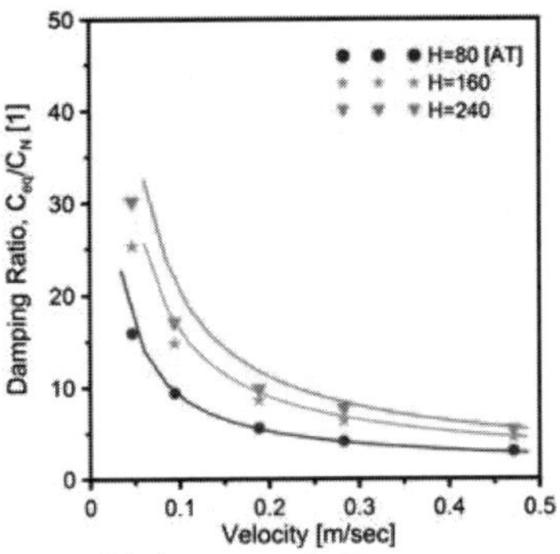

(a) Asymmetric Damper

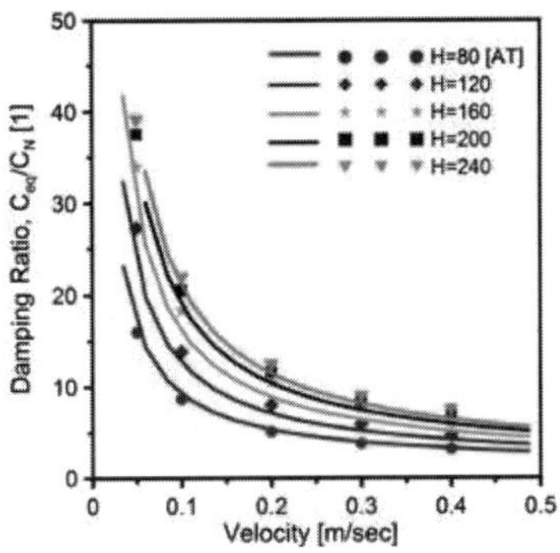

(b) Symmetric Damper

Figure 11. Effectiveness of the MR damper (The symbols refer to experimental data and the solid curves designate modeling results using the asymmetric MR valve).

5 Summary and Conclusion

The objective of this study was to compare asymmetric and symmetric valve designs. To evaluate the damper performance, we introduced an asymmetric damper model and derived damper effectiveness (C_{eq}/C_N) based on Bingham plastic shear flow mechanism for the flow mode damper configuration. Two dampers were tested using a MTS machine for sinusoidal stroke of 30 mm. The damping force vs. velocity and damping coefficient vs. velocity diagrams were experimentally validated using data taken over a range of applied magnetic field. Based on this study, the following conclusions can be drawn:

1. The asymmetric MR valve equation successfully predicted damping and energy dissipation in MR fluid dampers with asymmetric valve geometry and magnetic performance.
2. The slight improvement in the magnetic flux performance with minor change in the design in this design example gave substantial enhancement in damping force.

Acknowledgments

This research supported by DAEWOO Precision Inc. under the Dual Use Technology Program in South Korea (Dr. E.-J. Choi, technical monitor).

References

1. J.E. Lindler, Y.-T. Choi, and N.M. Wereley, "Double Adjustable Shock Absorbers Utilizing Electrorheological and Magnetorheological Fluids," *International Journal of Vehicle Design*, **33**(1-3), (2003), pp.189-206.
2. W.B. Facey, N.C. Rosenfeld, Y.-T. Choi, N. M. Wereley, S.-B. Choi and P. Chen, "*Design and Testing of a Compact Magnetorheological Damper for High Impulsive Loads,*" International Journal of Modern Physics B, **19**(7-9), (2005), pp.1549-1555.
3. M. Ahmadian, and J.C. Poynor, "An Evaluation of Magnetorheological Dampers for Controlling Gun Recoil Dynamics," Shock and Vibration, **8**(3-4), (2001), pp. 147-155.
4. Wereley, N.M. and Pang, L., "Nondimensional analysis of semi-active electrorheological and magnetorheological dampers using an approximate parallel plate model," *Smart Materials and Structures* **17** (1998) pp. 732-743.

MAGNETO-RHEOLOGICAL FLUID DAMPER FOR A FRONT-LOADING WASHING MACHINE: DESIGN, FABRICATION AND CHARACTERIZATION

GOKHAN AYDAR, C. A. EVRENSEL, F. GORDANINEJAD

Mechanical Engineering Department University of Nevada, Reno, Nevada 89557, USA
E-mail: aydarg@unr.nevada.edu,cahit@unr.edu, faramarz@unr.edu

ALAN FUCHS

Chemical Engineering Department University of Nevada, Reno, Reno, NV, USA
E-mail: afuchs@unr.edu

This study focuses on the theoretical analysis, design, fabrication and characterization of a magneto-rheological (MR) fluid damper for a horizontal axis, front-loading washing machine. In such washing machines, there are two main cycles; washing and spin cycles. During acceleration from washing cycle to high-speed spin, the tub passes through its resonant speed requiring relatively high damping. On the other hand, high damping results in increased force transmission to the housing and noise at high-speed spin cycle. Controllability of the MR fluid damper allows adjustment of damping requirements for different cycles and helps reduce the noise at high speed spin cycle while limiting the tub motion at resonance. A new MR valve is used for the design of the MR fluid damper for washing machine to satisfy the requirements of the application. Fabricated prototype damper is characterized using harmonic displacement input. Test results are in good agreement with the theoretical predictions.

1 Introduction

One of the major problems associated with high-speed front-loading washing machines is high levels of vibration and noise. The tub of a front-loading washing machine is supported by a pair of springs and a pair of dampers. While the dampers help limit the amplitude of vibrations at resonance, they result in higher force transmitted to the housing and increased noise at spin cycle.

The goal of this study is to minimize the amplitude of oscillations at resonance without resulting in relatively high force transmitted to the housing at high-speed spin cycle. This can be achieved using a controllable damper, such as magneto-rheological fluid (MRF) dampers which are classified as semi-active devices.

MR fluids are suspensions of micron-sized intensely polarized particles in a carrier fluid [1]. These particles' diameters are usually in the range of 1-10 μm [2]. The first invention and development of MR fluids and devices can be accredited to Rabinow at the US National Bureaus of Standards in the late 1940's [3, 4]. Recent developments led to MR fluids with desirable properties such as low viscosity, high yield strength, and stable hysteretic characteristic over a wide temperature scale [1]. Since MR fluid dampers do not add energy to the system but dissipate, they do not have the destabilizing potential active devices may have. Their power requirement is much lower compared to active devices. The force controllability of MRF dampers makes them very attractive in vibration isolation applications [5]. Better understanding of the characteristics and benefits of MRFs at the end of the 1980's and beginning of 1990's resulted in greatly increased interest in this technology [1]. A list of devices that utilize MR fluids is continually expanding and includes vibration isolators, shock absorbers, brakes clutches, etc., can be found in Ref. [2].

This study aims at the design, fabrication and characterization of a MRF damper with possible application in front-loading washing machines. In order to limit the tub displacement near resonance and minimize the noise at high-speed spin cycle the damper is desired to be active with the maximum damping force up to about 1.4 times the resonant tub rotational speed. The MRF damper should be on the active state with minimum damping at higher speeds. To achieve this, the following criteria are used for the design of the controllable MRF damper:

- Maximum activated damping force should be at least 150 N at resonance.
- Passive (no-current) damping force should be as low as possible.

2 Design of the MRF Damper

The geometry of the design is selected so that different type of materials can be used for the parts of the damper while satisfying the required magnetic field strength and distribution. A patented geometry of a MR valve developed at the University of Nevada, Reno is used as the basis of this new design to satisfy the requirements of the application [6]. Electromagnetic finite element analysis software Ansoft and fluid mechanics models are utilized to minimize the off-state viscous force while providing the necessary force in the on-state. A disk-type MR valve is used. The theoretical study indicated that the required magnetic field flux density inside the MR damper valving area should be over 0.4T in order to meet the design criteria for the maximum force. The result of finite element analysis of the selected damper design is presented in Figure 1. As can be seen in this Figure, the maximum magnetic field is about 0.47T in the valving area ($0<r<17.5$mm) for input current of 3A. Also, in the design process, minimum force at 0 Amp is estimated to be about 60N.

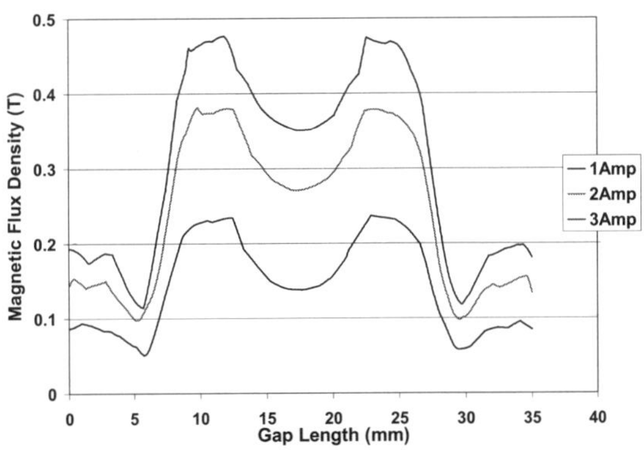

Figure 1. Three-dimensional finite element analysis result for the magnetic field through the MR gap.

For the theoretical force calculations of the damper; pressure drops due to viscous losses, minor losses and MR force are determined separately and added together to find the total damper force. To obtain the viscous pressure drop, the fluid is assumed to be an incompressible Newtonian fluid, and the flow to be steady and laminar. In this design, there are two main sources for the viscous pressure drops:

- Fluid flow between parallel disks, identified as MR channel, or gap.
- Fluid flow through the channel inside the core.

Pressure drop for the radial, laminar, steady flow of a Newtonian fluid through the gap between two parallel disks is given as [7]:

$$\Delta P_{vis_{gap}} = \frac{6\mu Q}{\pi h^3} \ln(\frac{r_2}{r_1}) \tag{1}$$

where Q is the volumetric flow rate, h is the gap height, $r_1$is the outer radius of the disk (in this design, rod radius), r_2 is the radius of the inner radius of the disk, and μ is viscosity of the fluid. Minor losses are one of the major pressure drop sources of this design and they are given in the general form of [7]:

$$\Delta P_{\min or} = \frac{K_L \rho V^2}{2} \tag{2}$$

where K_L is the overall minor pressure drop coefficient, ρ is the fluid density and V is the average velocity of the fluid. The pressure drop for the MR effect for a single-stage disk type MR valve is given as [7]

$$\Delta P_{disk\ mr} = 2.85 \frac{\tau_y}{h}(r_2 - r_1) \tag{3}$$

where τ_y is the yield stress of the fluid due to the applied magnetic field, h is the gap height . The yield stress of the fluid can be calculated as [8]:

$$\tau_y(P) = 1.03 \times 10^5 B(T) \tag{4}$$

where B(T) is the magnetic flux density in Tesla.

A theoretical prediction of the MR damper performance for harmonic input with 3Hz frequency and 2 cm peak-to-peak amplitude is given in Figure 2. It shows that it meets the maximum force criteria at 3 ampere. Also, the damper fabricated based on this design is shown in Figure 3.

3 MRF Damper Characterization

The characterization tests of the MRF damper are performed on the Instron 8821s hydraulic shaker. Figure 3 shows the test equipment and the damper. A power supply is used to provide the required current of 3Amps for the damper. For the low frequency tests, 2cm peak-to-peak displacement is applied while this displacement is reduced to 0.6cm or 0.4cm at higher frequencies not to exceed the limits of the hydraulic shaker used.

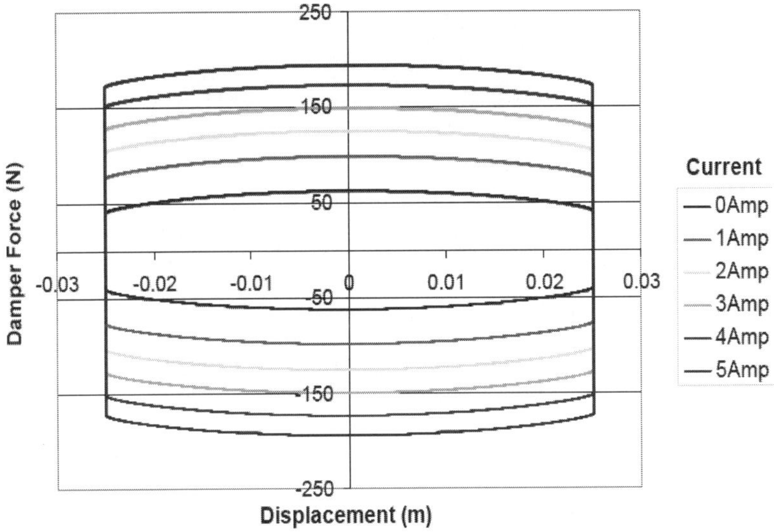

Figure 2. Predicted total damper force as a function of applied displacement, harmonic input with amplitude of 2.5 cm and frequency of 3 Hz.

Figure 3. Designed MRF Damper.

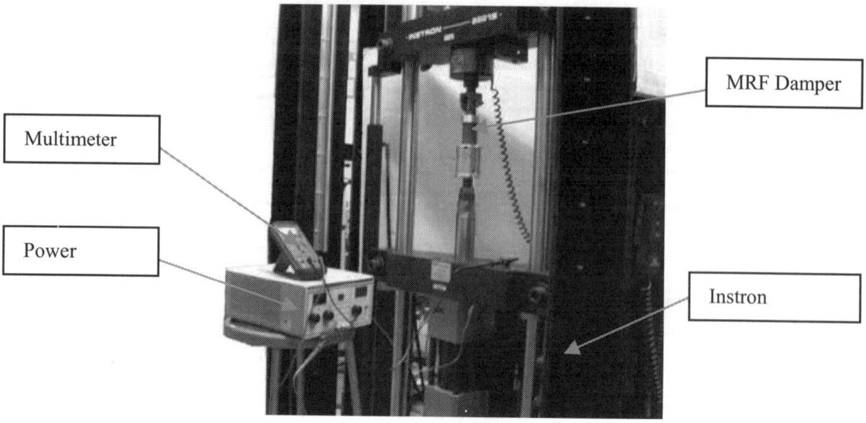

Figure 4. Test equipment for the MRF damper.

Results are presented in the form of the force–displacement and force–velocity behavior of the MRF damper (Figures 5-10). They are given for 0.5Hz and for 7.5 Hz frequencies and various input currents. The force requirement 150 N can be achieved with a 2.5A applied current. The condition representing the resonance case of the system is examined by a harmonic displacement input of 7.5Hz frequency and 2cm peak-to-peak amplitude at different currents.

As it can be seen from the graphs, the increase in the off-state force is low (around 10-15N) between 0.5Hz and 15Hz frequencies. However, the experimental results, especially force-velocity, have some deviation from the theoretical results. This may be explained as follows: (1) In high frequencies, the air in the system can form a non-linear spring effect, (2) It is noticed that the seal friction is changing, this is due to penetration of particles between the seals and the shafts, and (3) The testing equipment has some limitations, and for the high frequency tests, the test conditions were very close to test limits, and these can cause some errors in the test results.

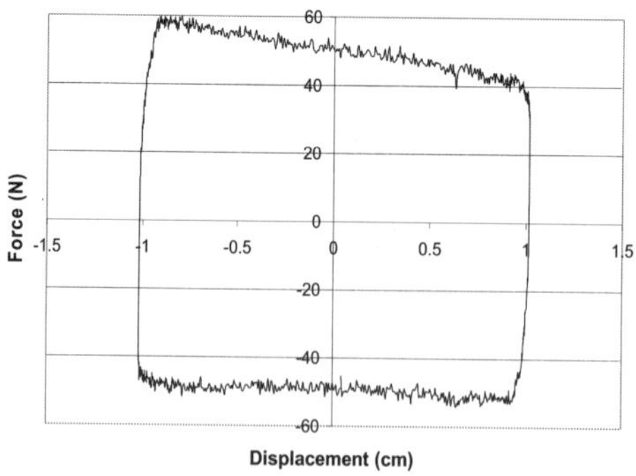

Figure 5. Experimental force-displacement loops of MRF damper with 0.5 Hz and 2.0cm peak-to-peak amplitude sinusoidal input.

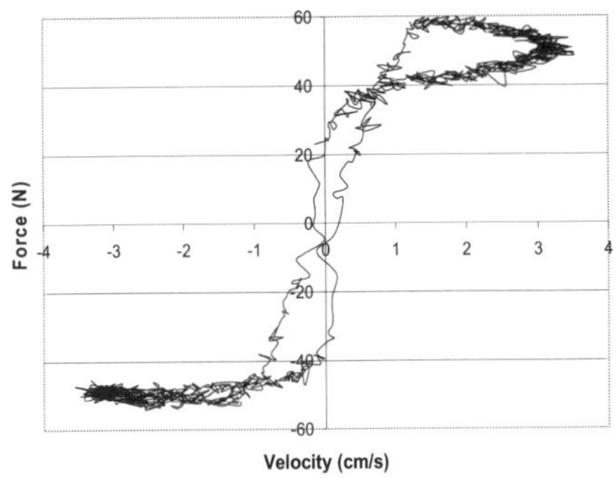

Figure 6. Experimental force-velocity loops of MRF damper with 0.5Hz and 2.0cm peak-to-peak amplitude sinusoidal input.

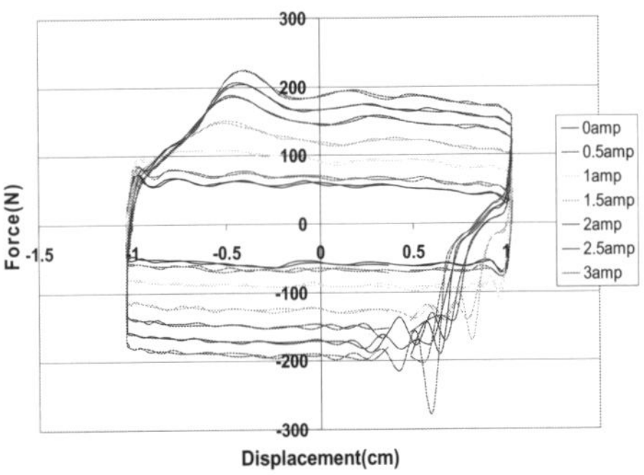

Figure 7. Experimental force-displacement loops of MRF damper with 7.5 Hz and 2.0 cm peak-to-peak amplitude sinusoidal input and for different currents.

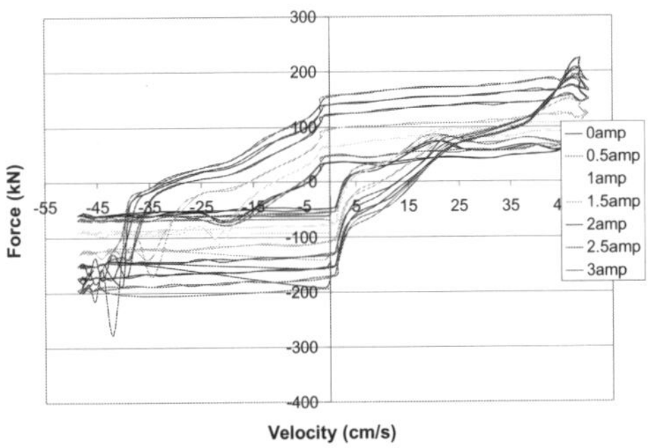

Figure 8. Experimental force-velocity loops of MRF damper with 7.5 Hz and 2.0 cm peak-to-peak amplitude sinusoidal input and for different currents.

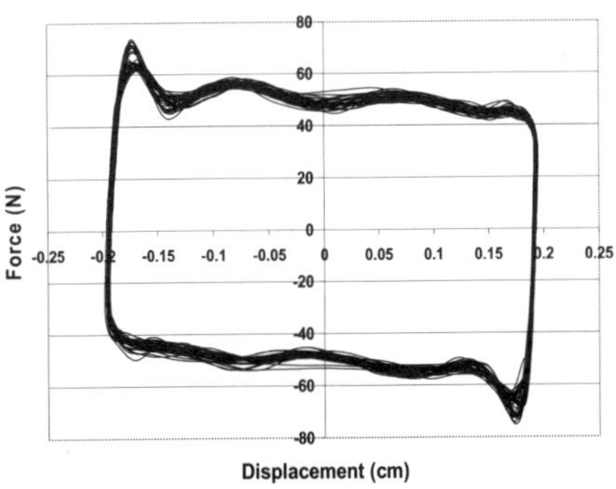

Figure 9. Experimental force-displacement loops of MRF damper with 15 Hz and 0.4 cm peak-to-peak amplitude sinusoidal input.

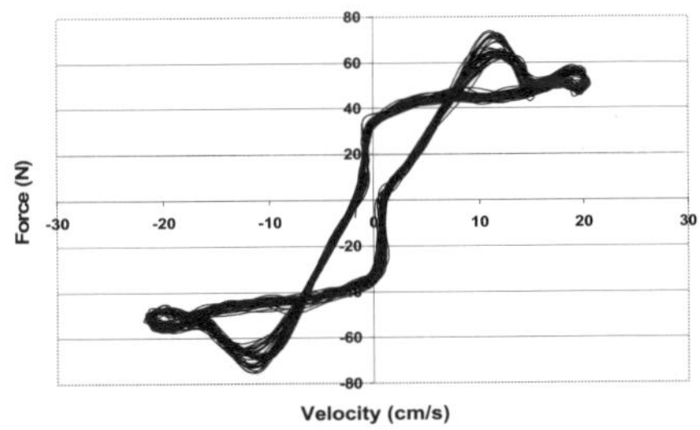

Figure 10. Experimental force-velocity loop of MRF damper with 15 Hz and 0.4 cm peak-to-peak amplitude sinusoidal input.

4 Summary and Conclusions

In this study, a novel design of a damper with design criteria of 150N maximum force, low off-state force is presented. Potential application of this damper is in a front-loading washing machine. Design geometry and dimensions are selected to fulfill the requirements of the application by using fluid mechanics principles and electro-magnetic finite element analysis. Test results of the fabricated damper are in good agreement with the theoretical predictions. They also show that design objectives are satisfied.

5 Acknowledgements

Authors thank Huseyin Sahin, Yanming Liu and Xiaojie Wang for their contribution to this study.

References

1. Sassi, S., Cherif, K., "An Innovative Magnetorheological Damper for Automotive Suspension: From Design to Experimental Characterization," *Smart Materials and Structures,* 14, pp 811-822, 2005.
2. Kavlicoglu, B. M., Gordaninejad, F., Evrensel, C., Fuchs, A., and Korol, G., "A Semi-Active Magnetorheological Fluid Limited Slip Differential Clutch," ASME *Journal of Vibration and Acoustics*, Vol. 128, No. 5, pp. 604-610, 2006.
3. Qing, S., Ke, Z., "New Type Magnetorheological Damper and Its Application on Semi-Active Control of Structural Seismic Response." *Progress in Safety Science and Tech.,* 3, pp 376-382, 2002.
4. Rabinow, J., "The magnetic Fluid Clutch," *Transactions of the American Institute of electrical Engineers,* 67, pp 1307-1315, 1948.
5. Symans, M.D., Constantinou, M.C., "Experimental Testing and Analytical Modeling of Semi-Active Fluid Dampers for Seismic Protection," *Journal of Intelligent Materials, Systems and Structures,* Vol. 8, pp.644-656, 1997.
6. Gordaninejad, F. and Breese, D. G., "Magneto-Rheological Fluid Dampers," U.S. patent No. 6,019,201, 2000.
7. Dogruer, U., Gordaninejad, F., and Evrensel, C. A., "A Magneto-Rheological Fluid Damper for High-Mobility Multi-purpose Wheeled Vehicle (HMMWV)," *Damping and Isolation, Proceedings of SPIE Conference on Smart Materials and Structures*, Ed. Kon-Well Wang, Vol. 5386, pp. 195-203, 2004.
8. Gordaninejad, F. and Kelso, S. P., "Fail-Safe Magneto-Rheological Fluid Dampers for Off-Highway, High-Payload Vehicles," *Journal of Intelligent Material Systems and Structures*, Vol. 11, No. 5, pp. 395-406, 2000.

APPLICATION OF ERG WITH ONE-SIDED PATTERN ELECTRODES TO A MICRO ALIGNMENT AND FIXTURE DEVICE

Y. KAKINUMA AND T. AOYAMA

Faculty of Science and Technology, Keio University, 3-14-1 Hiyoshi, Kouhoku-ku, Yokohama 223-8522, Japan
E-mail: kakinuma@sd.keio.ac.jp

H. ANZAI

Fujikura Kasei Co., Ltd, 2-6-15 Shibakouen, Minato-ku, Tokyo, Japan

Electro-rheological gel (ERG) is a functional material whose induced shear force is changed according to the intensity of the applied electric field. One-sided patterned electrodes have been proposed in a recent study of ERF, in order to simplify the structure of the wiring. In this study, one-sided patterned electrodes were applied to the ERG and the ER effect was experimentally confirmed. The frictional characteristics of the ERG surface vary in response to the electric field applied to the one-sided patterned electrodes, and a very small change in the ERG thickness occurs due to contraction of the ERG. Performance of the ERG in a micro alignment and fixture device was experimentally investigated. The results show that ER effects are obtained not only for metallic materials, but also for nonmetallic materials, such as glass, bakelite and so on, as sliding objects on the ERG surface. Based on these results, a new fixture device with micro positioning for micro machining of glass is proposed and its performance is experimentally analyzed.

1 Introduction

A machining process, which is a direct patterning method and environmental friendly process, is one potential approach for manufacturing micro textures on thin metal plates [1], and forming micro channels on a lab-on-a-chip for micro total analysis systems (μ-TAS) [2,3]. Such an approach allows for easier design changes and is much faster than clean room photolithographic methods. The photolithography process is not environmentally friendly, because it uses a large amount of energy and involves hazardous chemicals. In addition, photolithographic methods have restrictions in the small number of materials that can be used, or the shapes that can be formed. A machining process also has an advantage in that manufacturing of 3D shapes is possible. Several studies for micro grooving using a micro milling cutter have been performed [4,5]. However, fixing a thin workpiece without distortion, and adjusting the alignment of the work surface is a time consuming process. Therefore, application of a developed electro-rheological gel (ERG) [6,7] to a new fixture device for micro machining has been proposed.

The ERG, which is composed of electro-rheological (ER) particles and silicone gel, varies its surface friction according to the applied electric field. This characteristic is called the ER effect of the ERG. In addition, ERG undergoes micro shrinkage when an electric field is applied [7].

As shown in Fig. 1(a), the ERG is sandwiched between a pair of parallel plane electrodes and the ER effect is obtained under an electric field, because the parallel lines of electric force pass through the ERG. In a recent study, it was confirmed that the ER effect could be obtained by applying the structure of a one-sided electrode. Figure 1(b) shows the concept of ER effect generation using one-sided electrodes. Change in the frictional properties of the ERG, due to the ER effect, is obtained when using one-sided electrodes, even if an insulating material is used as the sliding plate. By utilizing these characteristics of the ERG, a new fixture device can be developed which can easily fix and detach a workpiece, and the surface alignment can be adjusted by application of a voltage. It is considered that the workpiece undergoes minimal distortion, because the ERG fixes the workpiece not by adding an external force, such as a mechanical clamp, but by the frictional force generated by the ER effect. In this study, a prototype ERG fixture device with a one-sided electrode configuration was developed and its performance was experimentally evaluated.

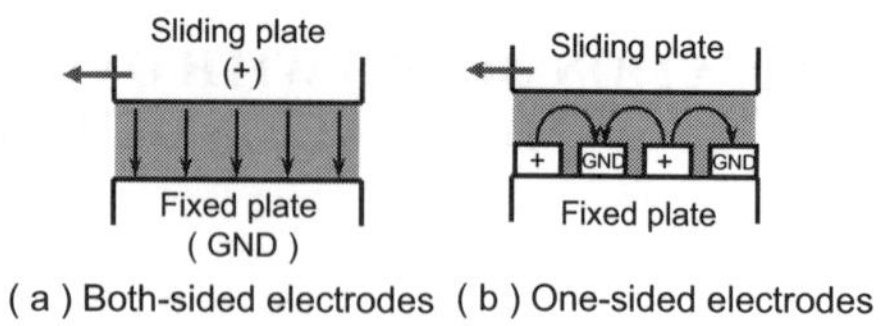

Figure 1. Concept of both-sided electrodes and one-sided electrodes

2 Development of the ERG fixture device

The structure of the prototype ERG fixture device is illustrated in Fig. 2. The one-sided electrode for the ERG fixture device is divided into 4 blocks. An ERG sheet of 0.5 mm thickness is formed on each block of the one-sided electrodes. It is possible to apply different voltages to each block of electrodes. Therefore, by adjusting the applied voltages to the 4 blocks of the electrodes, the alignment of the work surface is possible, due to the different shrinkage of the ERG in the vertical direction according to the voltages applied to the one-sided electrodes. The effective area of the ERG fixture is 20×70 mm (W×L). The electrode width and gap in each block of one-sided electrodes is 1 mm. To investigate the ability for fixing of metal and insulating materials, an aluminum plate (0.2 mm thick) and borosilicate glass (0.8 mm thick) were prepared as workpieces.

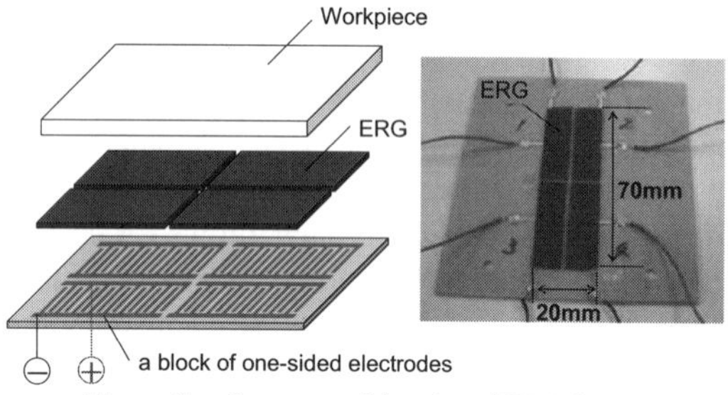

Figure 2. Structure of developed ERG fixture

3 Basic performance of the ERG fixture

3.1 Procedure for measurement of frictional force and alignment

The maximum frictional force which is generated from the ER effect depends on the intensity of the electric field at the surface of the ERG sheet. When a metal plate is used as the workpeice, a both-sided electrode mechanism is more suitable to effectively fix the plate. However, it is necessary to apply the one-sided electrode mechanism for a glass plate workpiece. To fix both metal and glass materials with the same ERG on one-sided electrodes, it is proposed that different electrode arrangements be applied according to the material. Figure 3 shows the fixing method for both materials used by ERG fixture. In the case of the metal plate, all blocks of the one-sided electrodes serve as the anode and the metal plate is earthed as the cathode. As shown in Fig. 3(a), this electrode arrangement realizes a both-sided mechanism. For the glass workpeice, a one-sided electrode mechanism is applied, as shown in Fig. 3(b).

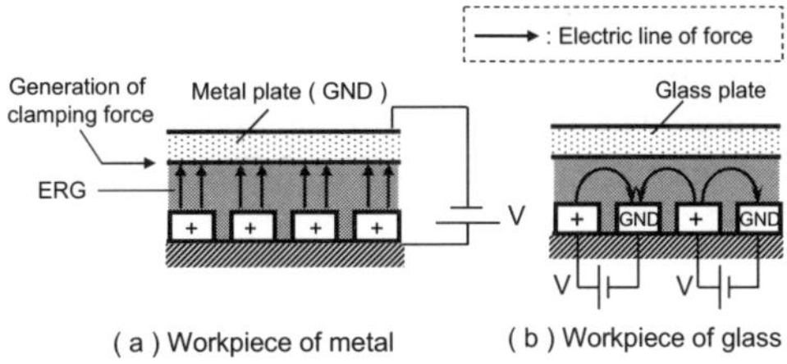

Figure 3. Method of fixing workpieces

The basic performance of the ERG fixture under an electric field was evaluated for frictional force and alignment of the workpiece surface. A schematic of the experimental setup for measurement of the maximum frictional force is given in Fig. 4. A workpiece is set on the ERG fixture and various voltages are applied to the one-sided electrodes (the aluminum plate serves as the cathode for the aluminum workpiece). Shear motion is applied to the workpiece by a slider using motor-driven equipment. The frictional force is measured by a strain gauge load cell attached to the slider. In order to investigate the change in the surface alignment of the workpiece under an electric field, vertical displacement was measured at three points on the workpiece surface. Figure 5 shows a schematic diagram of the equipment used for measuring the vertical displacement. A steel plate is put on each measurement point of the workpiece surface (points A, B and C) and the displacement at each point is measured with an eddy current displacement sensor.

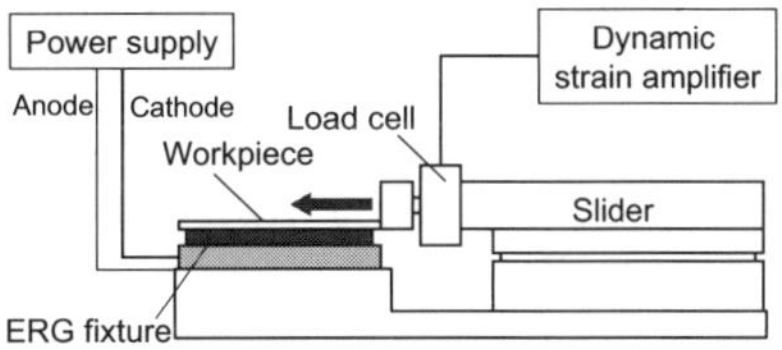

Figure 4. Experimental setup for measurement of maximum friction force

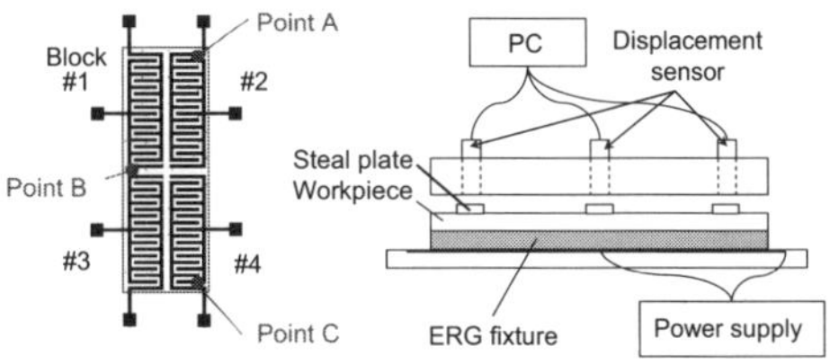

Figure 5. Experimental setup for measurement of vertical displacement

3.2 Relation between frictional force and applied voltage

The relation between the maximum frictional force and the voltage applied to the one-sided electrodes is summarized in Fig. 6. In both workpieces, the maximum frictional force increases as the applied voltage increases. In addition, the generated frictional force is increased by increasing the weight applied to the workpiece. For the aluminum and glass workpieces, the maximum applied voltage is different, because the generated electric field in the ERG changes according to the electrode arrangement shown in Fig. 3. While the maximum applied voltage between the one-sided electrodes and the aluminum plate is 750 V for the aluminum plate workpiece, the voltage between the one-sided electrodes can be applied up to 4 kV for the glass plate workpiece. The frictional force generated by the applied voltage is different between both workpieces. The frictional force increases widely in proportion to the square of the applied voltage in the aluminum plate. On the contrary, the frictional force increases only slightly in proportion to the voltage in the glass plate.

These results indicate that the fixing force obtained using the ERG fixture is a resultant force of the static frictional force under no electric field, and the force due to the ER effect. The equations for the fixing force of the ERG fixture for the aluminum and glass workpieces are defined as follows.

$$F_{al} = \mu_0 N + (\alpha + m\beta) V^2 \quad (1)$$

$$F_{gl} = \mu_0 N + \gamma V \quad (2)$$

where $\boldsymbol{F_{al}}$ [N] and $\boldsymbol{F_{gl}}$ [N] are the maximum fixing force for the aluminum plate and glass plate, respectively, μ_0 is the coefficient of static friction, $\boldsymbol{N}$ [N] is the normal component of reaction for the workpiece, $\boldsymbol{m}$ [g] is the mass of the workpiece, $\boldsymbol{\alpha}$ [N/kV^2] $= 4.4$, $\boldsymbol{\beta}$ [N/g kV^2] $= 0.016$ and $\boldsymbol{\gamma}$ [N/kV] $= 0.16$ are constants, and $\boldsymbol{V}$ [kV] is the applied voltage. The coefficients for static friction between the ERG and the aluminum plate and between the ERG and the glass plate are obtained as 0.42 and 0.48, respectively. The maximum fixing force calculated from Eq.(1) and Eq.(2) are illustrated in Fig. 6 as the dashed line. It was confirmed that the experimental value and the converted value are in close correspondence.

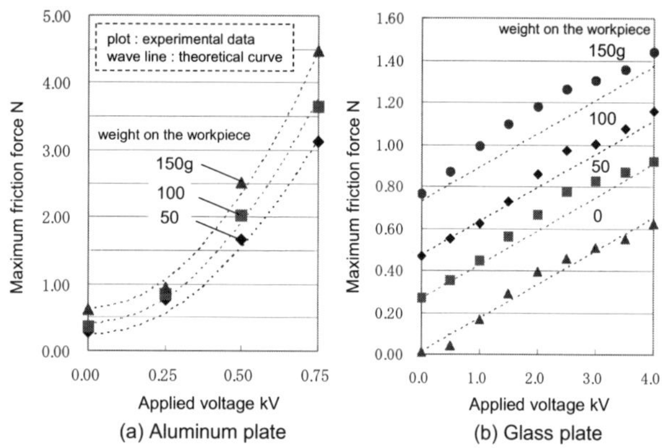

Figure 6. Relation between maximum friction force and applied voltage

3.3 *Relation between the surface alignment and applied voltage*

An aluminum plate was used as the workpiece and various voltages were applied only between block #1 of the one-sided electrodes and the aluminum plate, and the change of surface alignment was investigated. The relation between surface alignment and the

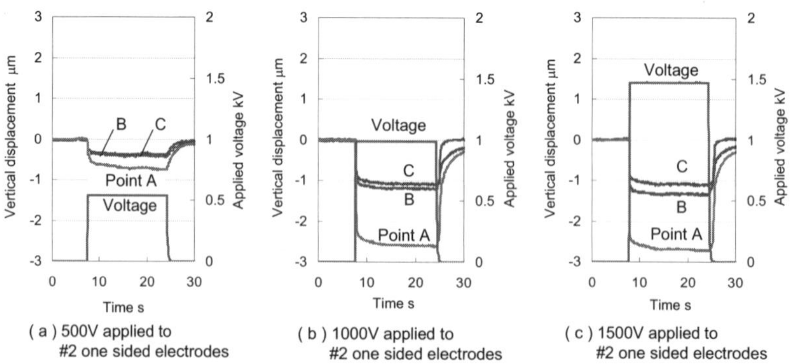

Figure 7. Relation between surface alignment and applied voltage in aluminum plate

applied voltage is shown in Fig. 7. By applying the voltage to block # 2 of the one-sided electrodes, point A on the aluminum plate sinks and the inclination of the surface changes. Moreover, the displacement in the vertical direction is changed in proportion to the applied voltage. Point A is displaced by approximately 2.7 μm when the applied voltage between the one-sided electrode and aluminum plate was 750 V. It is clear that the alignment of the workpiece surface can be adjusted by the voltage applied to the ERG fixture.

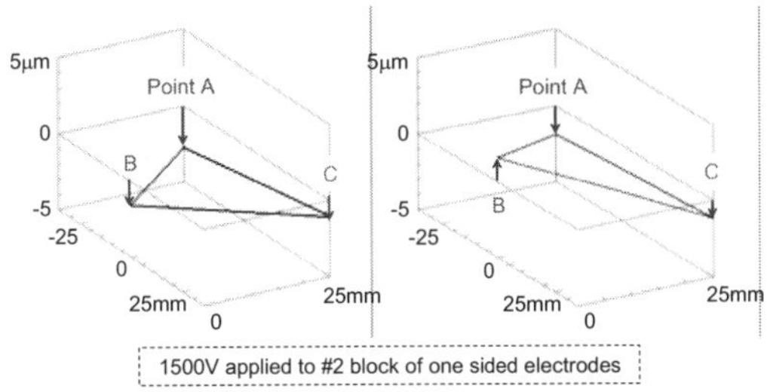

(a) Metal plate (b) Glass plate

Figure 8. Difference of alignment between metal plate and glass plate

When the glass plate is used as the workpiece, the variation of the surface alignment due to application of voltage differs from that for the aluminum plate. As shown in Fig. 8, each point in the metal plate sinks deeper than that in the glass plate. This is because the generated electric field in the ERG is different between the metal plate and the glass plate.

4 Evaluation of fixture characteristics by the cutting test

4.1 Experimental setup and cutting conditions

To evaluate whether the ERG fixture can be practically used as the fixture element of a micro machining tool, micro groove cutting tests were performed on thin aluminum and glass plates set on the ERG fixture, using a machining center (Mori Seiki, NVD-1500). The experimental setup is shown in Fig. 9. For the aluminum plate, a maximum voltage of 750 V is applied between the ERG and all blocks of the one-sided electrodes. For the glass plate, 4 kV is applied to all blocks of the one-sided electrodes. The cutting forces were measured using a piezo-electric dynamometer, which was placed between the ERG fixture and the stage of the machining center. Table. 1 summarizes the cutting conditions for the aluminum and glass workpieces. The rotational speed of the spindle was set at 20000 min^{-1} for cutting of both workpieces. From past studies of glass cutting, it was confirmed that the depth of the cut per tooth in ductile mode was limited to 12 nm [5]. Therefore, the feed rate was set at 0.48 mm/min for cutting of the glass in ductile mode. Water, which is suitable for glass cutting, was used as the cutting fluid [5]. The optimum tools selected for cutting of

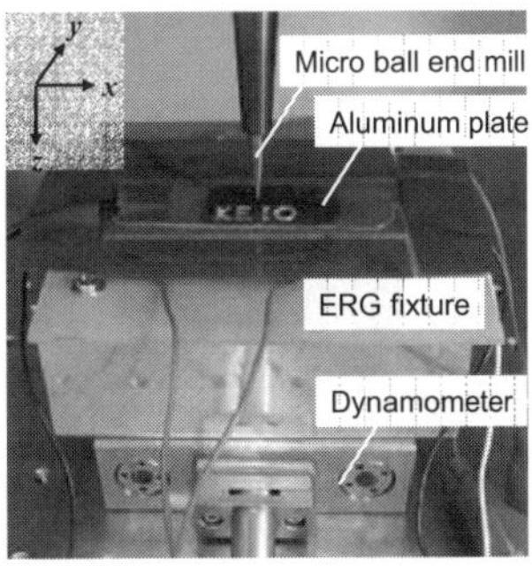

Figure 9. Experimental setup for cutting test

Table.1 Cutting condition for cutting of aluminum and glass

Workpiece	Aluminum	Borosilicate glass
Rotational speed	20000 min^{-1}	20000 min^{-1}
Feed rate	100 mm/min	0.48 mm/min
Tool	Carbide ball end mill	Diamond coat ball end mill
Radius of tool	0.5 mm	0.1 mm
Cutting fluid	Dry	Water

aluminum and glass were carbide and diamond coated ball end mills, respectively.

4.2 Results of the cutting test

Figures 10 and 11 show the 3 spatial components of cutting force during machining of aluminum and glass, respectively. Although the magnitude of the cutting force is different for aluminum and glass workpieces, the absolute value of each component is 0.4 N or less in both workpieces. These results indicate that the fixturing force of the ERG fixture is sufficient to tolerate the cutting force during micro machining. The appearance of both machined workpieces after the cutting test is shown in Fig. 12. It was confirmed that the characters "KEIO" were clearly processed on the surfaces of the workpieces, because both workpieces were sufficiently fixed by the ERG fixture during machining. It is considered that application of an ERG fixture as a new mechanism for fixing workpieces has high potential.

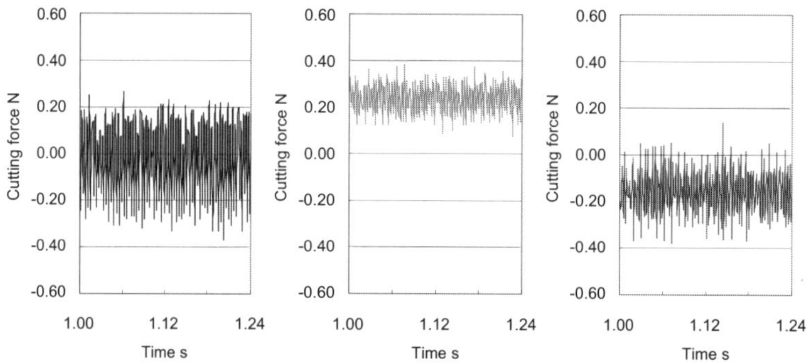

(a) x component (b) y component (c) z component

Figure 10. Cutting force using workpiece of Aluminum

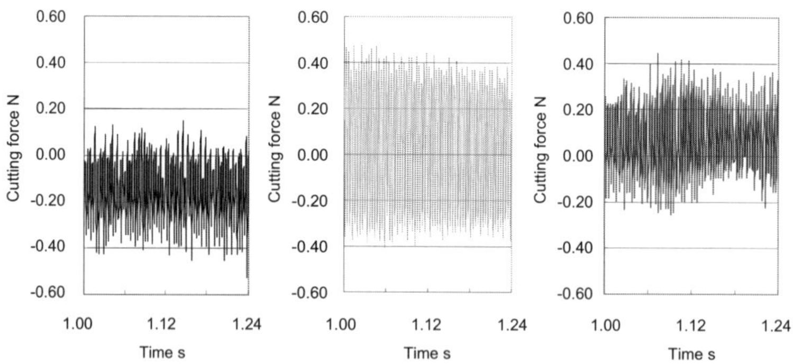

(a) x component (b) y component (c) z component

Figure 11. Cutting force using workpiece of glass

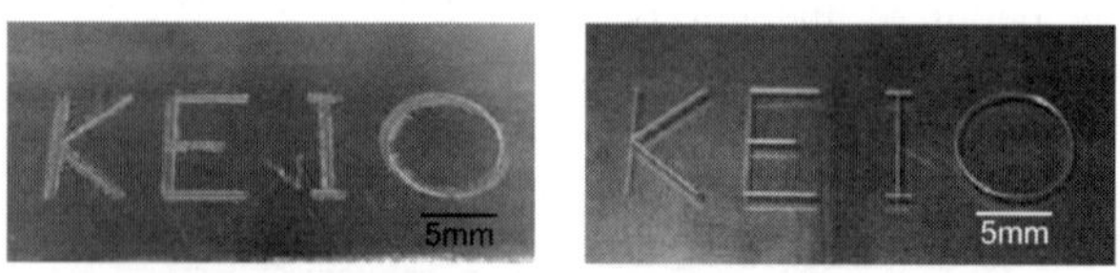

(a) Aluminum plate (b) Glass plate

Figure12. Results of cutting test

5 Conclusion

In this study, ERG, which changes its surface friction widely according to an applied electric field, is used as a fixture device for micro machining. The developed ERG fixture is composed of an ERG sheet and four blocks of one-sided electrodes, to which a voltage can be independently applied. The basic performance of the ERG fixture was experimentally analyzed during the fixing test and measurement of the surface alignment. The results indicate that the fixing force changes widely according to the applied voltage to the ERG fixture. It was also confirmed that the surface alignment of the workpiece can be minutely changed by the applied voltages. This performance suggests that the surface alignment can be controlled by adjusting the applied voltages to 4 blocks of one-sided electrodes. Based on the basic performance analysis, a micro groove cutting test using the ERG fixture was carried out. The ERG fixture exhibits sufficient fixing force during micro machining. The present study demonstrates a high potential for application of an ERG fixture, and the ERG fixture mechanism also has functionality for ease of workpiece alignment and detachment.

6 Acknowledgements

This work was partially supported by a Grant-in-Aid for Young Scientists (B), The Ministry of Education, Culture, Sports, Science and Technology. The authors also wish to express their sincere thanks to MTTRF for providing the machine tools for micro cutting test.

References

1. Moriwaki N. and Furukawa Y., Frictional Properties of the Micro-Textured Surface of Anisotropically Etched Silicon, Annals of the CIRP, 52/1, (2003) pp. 471-474.
2. Gracias A., Xu B. and Castracane J., Fabrication of Three Dimensional Micro-Channels in SU8, *Proceedings of 9th International Conference of Miniaturized Systems for Chemistry and Life Sciences*, (2005) pp. 663-665.
3. Bundgaard F., Perozziello G. and Geschke O., Rapid Prototyping Methods for All-COC/Topas Waveguids and Microfluidic Systems, *Proceedings of 9th International Conference of Miniaturized Systems for Chemistry and Life Sciences*, (2005) pp. 1200-1202.
4. Moriwaki T., Shamoto E. and Inoue K., Ultra-precision Ductile Cutting of Glass by Applying Ultrasonic Vibration, Annals of CIRP, 41/1, (1992) pp. 141-144
5. Matsumura T. and Ono T., Glass Machining with Ball End Mill, *Transactions of the North American Manufacturing Research Institution of SME*, 33, (2005) pp. 319-326.
6. Kakinuma Y., Aoyama T., Anzai H., Sakurai K., Isobe K. and Tanaka K., Basic Properties of Gel-structured Electro-rheological Fluids, *International Journal of Modern Physics B*, **19** (2005) pp. 1339-1345.
7. Kakinuma Y., Aoyama T., Anzai H., Sakurai K., Isobe K. and Tanaka K., Development of Gel-Structured Electro-rheological Fluids and their Applications to the Control of Positioning, *Proceedings of 8th International Research/Expert Conference*, (2004) pp. 459-462.

ISOKINETIC EXERCISE MACHINE USING ER BRAKE AND ITS VELOCITY CONTROL AT HIGH-SPEED TRAINING

TAKEHITO KIKUCHI, JUNJI FURUSHO

Osaka University, 2-1 Yamadaoka, Suita, Osaka, Japan
E-mail: kikuchi@mech.eng.osaka-u.ac.jp

KUNIHIKO ODA

Osaka Electro-Communication University, 1130-70 kiyotaki, shijyo-nawate, Osaka, Japan

Strengthening muscle force by training, e.g., an isokinetic exercise is effective for not only healthy persons but also those who need rehabilitation. The isokinetic exercise is a training method under the restriction of constant rotational speed of a joint during the exercise. This enables dynamic trainings and at the same time, trainings at maximum output force over the whole range of motion. In many conventional isokinetic exercise systems, a high-power servomotor is used for an exact speed control. These systems are useful in clinical and investigative researches. However, these systems can be dangerous due to system trouble. Therefore, safety countermeasures are extremely important but lead to system complexity and higher costs. In this study, we developed an isokinetic exercise system using an "ER brake". This brake includes particle-type ER fluid as the working fluid. Because of the rapid response of the ER fluid, the ER brake shows superior response performance compared with conventional brakes. In other words, a simple and safe training system with accurate rotational speed can be realized with this brake. In addition, a high-speed (800 deg/sec at an elbow joint) control was realized with this system.

1 Introduction

Strengthening muscle force by training, e.g., isokinetic exercises, is effective for not only healthy persons but also those who need rehabilitation. Training methods can be classified into an "isometric exercise", an "isotonic exercise" and an "isokinetic exercise" depending on the contraction forms of a skeletal muscle. The isometric exercise is a training in which a muscle can be contracted without moving the joint. It allows training at a maximum output force, but does not allow recovery of the dynamic function of the muscle. The isotonic exercise is a training using a constant load. It enables a dynamic training to be realized, but not at maximum output force over the entire range of motion. The third type is the isokinetic exercise, or training under the restriction of constant rotational speed of a joint during the exercise. This enables dynamic training and at the same time, the training at maximum output force over the whole range of motion, too. This exercise can be considered a training method offering the benefits of both isometric and isotonic exercise.

It is known that the relation called a force-velocity curve is established between the output force of a muscle and a joint velocity [1]. The model drawing of the force-velocity curve is shown in Fig.1. The case of the zero joint velocity is called an "isometric contraction." In the isometric contraction, the joint is fixed but inside the muscle, muscle fibers are shortened and elastic sites such as a tendon are extended. The case of the positive velocity of the joint while generating the force, that is to say, the state of the joint movement toward the same direction with the direction of the output force is called a "concentric contraction." This corresponds to the positive zone of the velocity in Fig.1. In the case of the concentric contraction, the output force decreases with the increase of the joint velocity and exceeding certain joint velocity leads to the zero output force. This curve under the concentric contraction is formulated by A.V.Hill (1938) [2] as:

$$V = b\,(P_0 - P) / (P + a)$$

where 'P' is the force, 'V' is the velocity of a contraction of a muscle, 'P_0' is the force at the isometric contraction, and 'a' and 'b' are parameters depending on each muscle. This is called "Hill's Curve." As such a relationship holds between the output force of the muscle and the joint velocity, a quantitative evaluation might be possible if the maximum output force of the muscle at each joint velocity could be measured under the constant velocity of the motion.

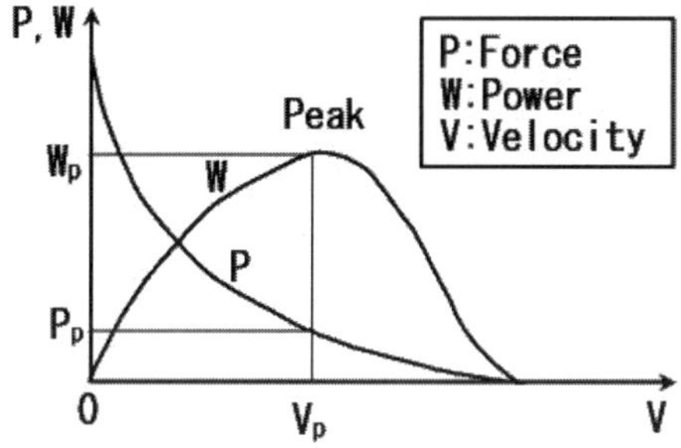

Figure 1. Hill's curve

In many conventional isokinetic exercise systems, a high-power servomotor is used for an exact speed control. These systems are useful in clinical and investigative researches because they allow realization of the isokinetic exercise and a quantitative evaluation of a muscle is possible by measuring the generated muscle force and the movement velocity during training. However, these systems can be dangerous due to system trouble if actually used for training. Therefore, safety countermeasures are extremely important but lead to system complexity and higher costs. The person undergoing the training generates the acceleration force, therefore any actuator is not needed to realize the isokinetic exercise for concentric contraction mechanically; brakes alone are sufficient. Because no active action occurs in a system using only brakes, it is safer than actuator-type systems. Although some brake-type systems have already been developed, they cannot control the rotational speed accurately because of the response delay of conventional electromagnetic brakes.

We developed an isokinetic exercise system using an "ER brake" [3], [4]. This brake includes particle-type ER fluid as the working fluid. Because the response characteristics of the particle-type ER fluid are of very high speed, the response of the ER brake shows superior performance compared with any other conventional brakes. By using this brake, a simple and safe training system with accurate rotational speed can be realized with this brake. In this paper, control method and its result of a high-speed (800 deg/sec at an elbow joint) control during training with the developed system are described.

2 Methods

2.1 ER Brake

Figure 2 shows the basic structure of the ER brake consisting of a fixed cylinder and a rotating cylinder with the ER fluid between them. The two cylinders also serve as a pair of electrodes. The rotating cylinder is fitted with an output shaft and is driven by an external force through this shaft. When a voltage is applied between a pair of cylinders, or electrodes, the electric field is produced in the ER fluid between them, increasing the viscosity of the fluid. Thus the rotating cylinder is forced to slow down, providing the rotating output shaft with resistance (braking torque). Figure 3 is a photo of the developed ER brake. The specifications of this brake are shown in Table 1. Unlike the conventional friction brake, this brake can generate an accurately controlled dynamic braking torque. And at the same time, this responds faster than any other brakes. The braking torque of it can rise just in several milliseconds [3], [4].

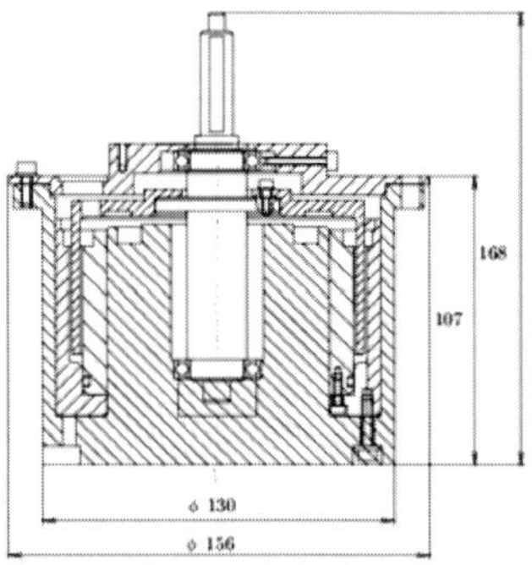

Figure 2. Sectional view of ER Brake

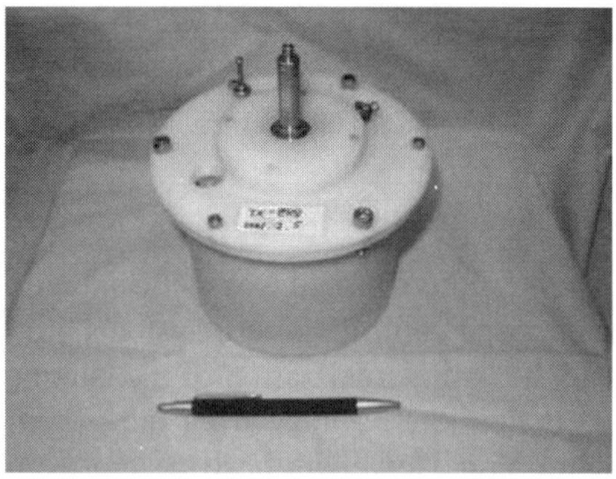

Figure 3. ER Brake

Table 1. Specifications of ER Brake

Moment of inertia (kgm^2)	$6.14 * 10^{-4}$
Diameter of rotational cylinder (mm)	108
Effective height of rotational cylinder (mm)	40
Thickness of rotational cylinder (mm)	3.0
Electrode-to-electrode distance(mm)	1.0
Effective electrode surface area (cm^2)	264
Maximum brake torque (Nm)	4.0

2.2 *Isokinetic Exercise System Using ER Brake*

As shown in table 1, the possible output torque generated from single brake is limited to 4.0 [Nm]. In other words, the output torque that is enough for training cannot be obtained by only itself. Therefore, we determined to install a speed reducer between the brake and an operation section. The speed reducer was designed using a belt-pulley mechanism.

Reduction ratio has to be determined with considering possible maximum output force by human. In this time, we mainly considered an elbow joint exercise. Referring to data books of human engineering, it was designed for an average of men with age of 19 as target users. Because the training system uses only brake, only concentric contraction training is targeted.

The maximum output torque of the system will need to exceeds 90 [Nm] according to the data books with any margins. The maximum output torque of the ER brake is about 4 [Nm], but taking care of safety aspect, it was estimated to be 3 [Nm]. From this consideration, it is judged that 1/30 of the reduction ratio is necessary. But with increase of the reduction ratio, the rotational speeds of the electrode inside the ER brake increases. At the same time, the shear rate of the fluid increases and so, the effect of a base viscosity of the fluid cannot be ignored.

Based on discussions like above, 1/2.00, 1/3.11, 1/6.85, 1/10.67, 1/19.28 and 1/30.00 were adopted as the reduction ratio in this system. The structure was designed to be able to change the reduction ratio by changing the pulleys. System design specification was determined as table 2.

Table 2. Specifications of training system for flexion and extension of elbow joints

Maximum training speed (deg/s) [in the case of 1/2 of the reduction ratio]	800
Maximum torque (Nm) [in the case of 1/30 of the reduction ratio]	90

The assembly drawing and the appearance of the developed training system are shown in Fig.4 and Fig.5, respectively. This equipment is the training system with which the isokinetic exercise for flexion and extension of an elbow can be realized. Since the beginning of training, a trainer will try to rotate the handle with the force as large as possible. Then, elbow joint will be rotated with constant velocity by the velocity control of the ER brake. In this system, training speed can be set to arbitrary value. By setting acceleration time, deceleration time, angle at the beginning of training and the maximum angle, the velocity curve during the training is calculated automatically. The rotation angle, the velocity and the output torque of the elbow are recorded during the exercise by a computer. The angle was measured using a rotary encoder (Resolution: $1.94*10^{-5}$ [rad]) and the velocity was approximated by difference calculation. Operation torque was measured from torsion of the final axis using strain gauges.

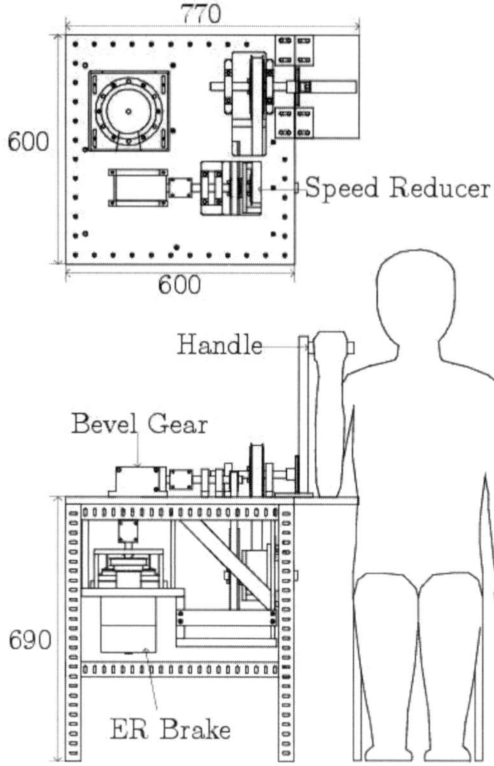

Figure 4. Isokinetic exercise system using ER brake

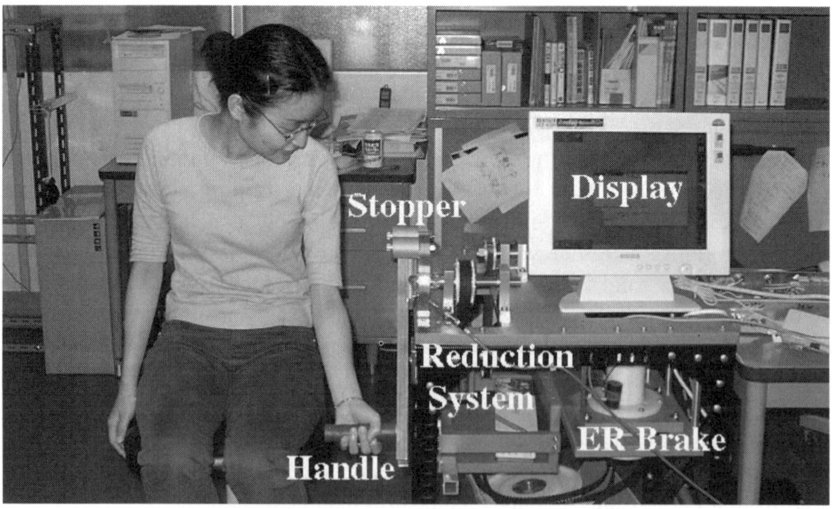

Figure 5. Appearance of developed training system

3 Results

The isokinetic exercise with the developed training system was carried out. As the control method, IP controller shown in Fig.6 was used, where 'ω_{ref}' is reference velocity, 'ω' is measured velocity, 'T_B' is braking torque, 'T_O' is operating torque, 'G(s)' is a transfer function of ER brake. Sampling time of the control was 0.5[ms].

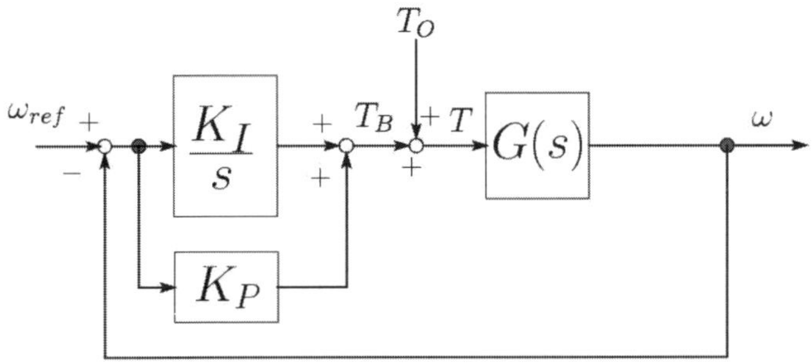

Figure 6. PI controller for velocity control of ER brake

We set the reduction ratio 1/2.00, and carried out the high speed training at 800[deg/s]. The result is shown in Fig.8. It was too rapid to carry out a training over the angular velocity of 400[deg/s] using conventional machines. However, this system can realize training over 800[deg/s], safety and easily.

Using this system, an exact evaluation of exercise was carried out. And safe training was performed at the same time. From above result, this system was proved to be effective also for a quantitative evaluation of the training in high-speed range. Dynamic behavior of ER fluid under high shear rate is not figured out. Therefore, there is room for improvement in velocity control.

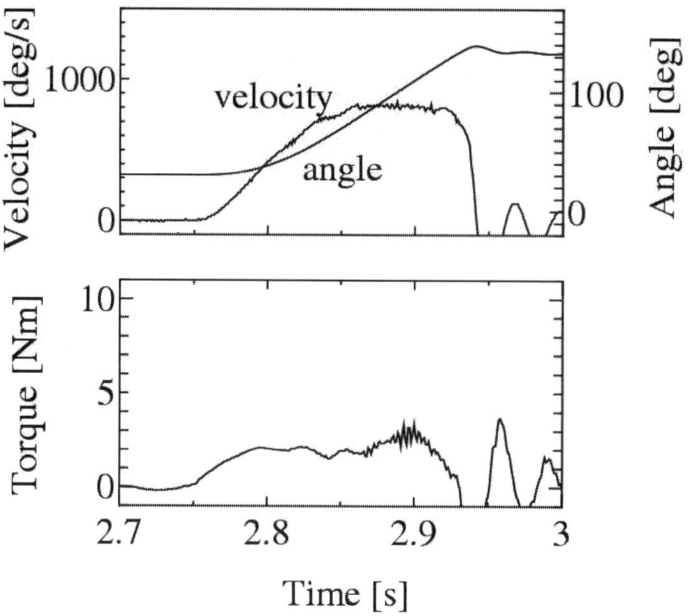

Figure 7. Experimental result of High-speed (800[deg/sec]) velocity control for isokinetic exercise with developed machine

4 CONCLUSIONS

(1) We developed the ER brake using the particle-type ER fluid. This device has quick and stable response characteristics.
(2) We developed an exact and the safe isokinetic training system using the ER brake.
(3) The high-speed training over 800[deg/s] with the developed training system was carried out. Any other machines cannot realize such a high-speed training.

References

1. D. A. Winter, Biomechanics and Motor Control of Human Movement, Second Edition, John Wiley & Sons, (1990).
2. Hill A.V., The mechaics of active muscle, Proceedings of the Royal Society of London. Series B, Boiological sciences, Vol.141, pp104-117 (1953).
3. Takehito KIKUCHI, Junji FURUSHO, Velocity Control of Brake Using Particle-Type ER Fluid and It's Application, the Sixth International Conference on Motion and Vibration Control 2002, pp 837-842.
4. Takehito KIKUCHI, Junji FURUSHO and Kunihiko ODA, Development of Isokinetic Exercise Machine Using ER Brake, Proceedings of 2003 IEEE International Conference on Robotics and Automation(ICRA2003, CD-ROM), pp214-219(2003.8).

OPTIMAL CONTROL OF A HALF-CAR VEHICLE MODEL WITH A VARIABLE DAMPING SEMI-ACTIVE SUSPENSION

Y. ERCAN

TOBB Economics and Technology University, Sogutozu Caddesi No. 43, Sogutozu, 06560 Ankara, Turkey

E-mail: yercan@etu.edu.tr

A half-car vehicle model with variable dampers placed between the sprung mass and the unsprung masses was considered in the study. A suitable set of state variables were defined and state equations of the system were obtained. An integral performance index involving a weighted combination of the average squared heave accelerations of the sprung mass, and squares of suspension deflections was defined. The closed loop optimal control law which gives the optimal value of the damping coefficients of the variable dampers was obtained by using Pontryagin principle. The performance of a vehicle with an optimally controlled semi-active suspension was calculated for measured road profile inputs, and trade-off curves were obtained between the average sprung mass acceleration and suspension deflections. Performance of the vehicle with semi-active system was compared to the corresponding performance of a vehicle with a passive suspension system. Then, a time delay was introduced to the response of the damper and its effect on the performance of the semi-active system was investigated. The results of the study have shown that performance of the vehicle with variable damping semi-active suspension is much better than that of the vehicle with a passive suspension.

1 Introduction

A set of suitable parameters, normally the coefficients of the dampers between the sprung and the unsprung masses are changed in semi-active suspensions to optimize the system performance. Semi-active suspensions are much simpler compared to active suspensions and may be implemented easily by using magneto-rheological dampers. They can be activated by use of negligibly small amounts of enegy. Dampers with variable coefficients are commercially available, and semi-active suspensions are now applied to some commercial vehicles [1,2].

Semi-active suspensions were first proposed more than 30 years ago by Karnopp and Crosby [3,4]. Many studies have been carried out since then, including linear quadratic Gaussian (LQG) control of ¼- and ½-vehicle models, control schemes with and without preview, systems which employ active as well as semi-active elements [5-13]. There have been also some studies specifically on suspensions with magneto-rheological dampers [14,15]. A review of these studies may be found in Hrovat [16].

The optimal control of suspension systems is usually derived from the solution of the linear quadratic Gaussian (LQG) problem which assumes white noise disturbance inputs to the system, and unbounded control inputs. However, real road profiles are not white noise; control inputs (damping coefficients) are contrained, and control commands are executed with a time delay. As a result, performance of a system using the optimal control solution of the LQG problem will be adversely effected by all of these. The purpose of this study to obtain the optimal control algorithm for a half-car model with a variable damping semi-active suspension, and evaluate its performance when the vehicle is subjected to measured road inputs, and the control inputs are constrained and delayed.

2 System Model

A half-car vehicle model as shown in Figure 1 was considered in this study. Here b_{21} and b_{22} are the variable dampers, the coefficients of which may be adjusted to control the system behavior.

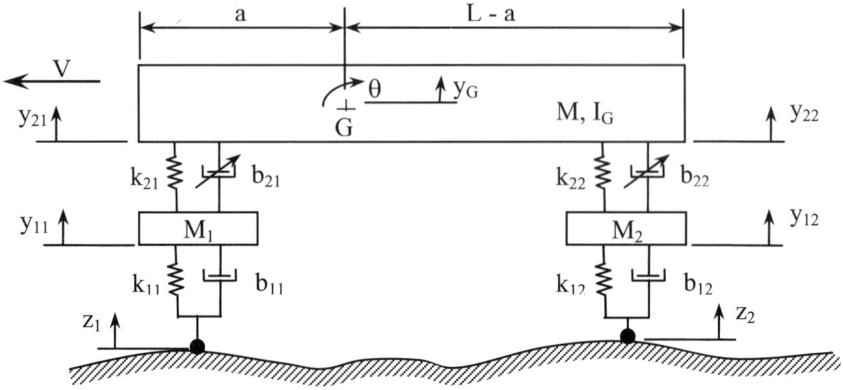

Figure 1. System Model.

The dynamic suspension deflections, tire deflections and their rates of change were defined as the state variables of the system as $x_1 = y_{21} - y_{11}$, $x_2 = \dot{y}_{21} - \dot{y}_{11}$, $x_3 = y_{11}$, $x_4 = \dot{y}_{11}$, $x_5 = y_{22} - y_{12}$, $x_6 = \dot{y}_{22} - \dot{y}_{12}$, $x_7 = y_{12}$, $x_8 = \dot{y}_{12}$. Control variables were defined as $u_1 = b_{21}$ and $u_2 = b_{22}$. Then the state equations of the system were obtained as follows:

$$
\begin{aligned}
\dot{x}_1 &= x_2 \\
\dot{x}_2 &= -A_1x_1 - A_2x_2u_1 + A_3x_3 + A_4x_4 + A_5x_5 + A_6x_6u_2 - A_3z_1 - A_4\dot{z}_1 \\
\dot{x}_3 &= x_4 \\
\dot{x}_4 &= B_1x_1 + B_2x_2u_1 - A_3x_3 - A_4x_4 + A_3z_1 + A_4\dot{z}_1 \\
\dot{x}_5 &= x_6 \\
\dot{x}_6 &= C_1x_1 + A_6x_2u_1 - C_5x_5 - C_6x_6u_2 + C_7x_7 + C_8x_8 - C_7z_2 - C_8\dot{z}_2 \\
\dot{x}_7 &= x_8 \\
\dot{x}_8 &= D_5x_5 + D_6x_6u_2 - C_7x_7 - C_8x_8 + C_7z_2 + C_8\dot{z}_2
\end{aligned} \tag{1}
$$

Here the coefficients are defined accordingly in terms of the system parameters, and the control inputs are constrained as $0 < U_{min} < u_i < U_{max}$ $(i = 1, 2)$. The state equations given above involve terms in which state and control variables are multiplied by each other. Such systems are called bilinear. Solutions of the state regulator problem and the quadratic Gaussian problem for linear systems are the same and may be obtained by the use of Riccati equation, which yields a closed loop control. It can be shown that the solution of the state regulator problem for bilinear systems may also be obtained from Riccati equation. In the present study, state regulator problem was solved first for the system described by equations (1), then performance of the resulting optimally controlled system was evaluated for constrained control inputs and measured road profile inputs.

The heave acceleration of a point on the sprung mass which is located at a distance x from the center of gravity G may be expressed is terms of the heave acceleration $\ddot{y}_G$ of the center of gravity and the pitch acceleration $\ddot{\theta}$ as

$$\ddot{y}(x) = \ddot{y}_G - \ddot{\theta}x \tag{2}$$

Then, the average squared heave accelerations of the points on the sprung mass can be obtained as

$$(\ddot{y}^2)_{av} = \frac{1}{L}\int_{-a}^{L-a} (\ddot{y}_G - \ddot{\theta}x)^2 dx = \ddot{y}_G{}^2 - \ddot{y}_G\ddot{\theta}(L-2a) + \ddot{\theta}^2(\frac{L^2}{3} - La + a^2) \tag{3}$$

A quadratic performance index was assumed in this study as follows, involving the average squared heave accelerations of the sprung mass, squares of suspension deflections and tire deflections.

$$P = \int_0^{t_f} [(\ddot{y}^2)_{av} + \rho_1(x_1{}^2 + x_5{}^2) + \rho_2(x_3{}^2 + x_7{}^2)]dt \tag{4}$$

The performance index was first expressed in terms of the state and control variables. Then, the Pontryagin function was constructed as $H = L(\mathbf{x},\mathbf{u}) + \boldsymbol{\lambda}^T\mathbf{f}$, where L represents the integrand of P, elements of **f** are the right-hand sides of equations (1), and elements of λ are the co-state variables. Then partial derivatives of H with respect to the control variables u_1 and u_2 were equated to zero, optimal u_1 and u_2 were solved in terms of the state and co-state variables and subtstituted into the expression for H to obtain the optimal Pontryagin function H_o. Then, writing co-state and state equations yielded two equations of the form $\dot{\boldsymbol{\lambda}} = \mathbf{T}_1\mathbf{x} + \mathbf{T}_2\boldsymbol{\lambda}$ and $\dot{\mathbf{x}} = \mathbf{H}_1\mathbf{x} + \mathbf{H}_2\boldsymbol{\lambda}$ where $\mathbf{T}_1$, $\mathbf{T}_2$, $\mathbf{H}_1$ and $\mathbf{H}_2$ are coefficient matrices of dimensions 8×8 involving the weighting coefficients of the performance index and the system parameters. If one assumes $\boldsymbol{\lambda} = \mathbf{R}(t)\mathbf{x}$ where $\mathbf{R}(t)$ is a positive definite symmetric matrix, these equations yield the Riccati equation,

$$\dot{\mathbf{R}} + \mathbf{R}\mathbf{H}_2\mathbf{R} + \mathbf{R}\mathbf{H}_1 - \mathbf{T}_2\mathbf{R} - \mathbf{T}_1 = 0 \tag{5}$$

with the boundary condition $\mathbf{R}(t_f) = \mathbf{0}$. If $t_f = \infty$, then $\mathbf{R}(t)$ becomes a constant matrix and an algebraic equation is obtained. Once the matrix **R** is solved, the elements of $\boldsymbol{\lambda}$ as well as u_1 and u_2 can be expressed in terms of the state variables, giving closed loop control. During numerical simulations the restrictions on control inputs were further imposed.

3 System Parameters and Method of Solution

Parameters of the physical system were assumed to be as follows:

$L = 7$ m	$a = 4.5$ m	$M = 3500$ kg
$k_{22} = 225000$ N/m	$k_{21} = 125000$ N/m	$k_{11} = 625000$ N/m
$M_1 = 150$ kg	$M_2 = 300$ kg	$k_{12} = 1125000$ N/m
$b_{11} = 7000$ N·s/m	$b_{12} = 13000$ N·s/m	$I_G = 13000$ kg·m^2
$U_{min} = 2000$ N·s/m	$U_{max} = 20000$ N·s/m	

$t_f = \infty$ was assumed in this study. However, rather than solving the algebraic Riccati equation, differential Riccati equation (5) was integrated in the negative time direction starting from the final condition $\mathbf{R} = \mathbf{0}$ until the steady state solution is reached. Calculations are carried out by using MATLAB.

A measured road profile was used to excite the system. For this purpose, the left track of a randomly selected road profile (No. TX90030602) from The University of Michigan Transportation Research Institute (UMTRI) World Wide Web site was used [17].

4 Performances of the Passive and Optimal Semi-Active Suspensions

4.1. *Passive Suspensions*

Performance of the passive system was examined first by performing a parametric optimization. Runs were made on MATLAB. The performance parameters corresponding to the minimum rms value of the average heave acceleration of the sprung mass were found to be $b_{21} = 7000$ N·s/m, $b_{22} = 13000$ N·s/m, $(\ddot{y}_{av})_{rms} = 0.2283$ m/s^2, $(y_{21}\text{-}y_{11})_{rms} = 1.446$ mm, $(y_{22}\text{-}y_{12})_{rms} = 1.507$ mm, $(y_{11} - z_1)_{rms} = 6.897$ mm, $(y_{12}\text{-}z_2)_{rms} = 6.857$ mm.

4.2. *Semi-Active Suspensions*

Performances of the semi-active suspensions were determined by using MATLAB programs developed for this purpose. First, the weighting coefficient ρ_2 is assumed to be zero, thus assigning no penalty to the value of tire deflections, and the value of ρ_1 was changed. The results corresponding to the minimum rms value of the average heave acceleration of the sprung mass were as follows: $\rho_1 = 100$, $\rho_2 = 0$, $(\ddot{y}_{av})_{rms} = 0.1916$ m/s^2, $(y_{21}\text{-}y_{11})_{rms} = 1.624$ mm, $(y_{22}\text{-}y_{12})_{rms} = 1.662$ mm, $(y_{11} - z_1)_{rms} = 6.892$ mm, $(y_{12}\text{-}z_2)_{rms} = 6.853$ mm.

Figures 2 and 3 show the trade of curves between rms average heave acceleration of the sprung mass and suspension deflections as the value of ρ_1 changes. Comparisons of these results with those of the passive system show that $(\ddot{y}_{av})_{rms}$ is reduced from 0.2283 m/s^2 to 0.1916 m/s^2 as a result of using semi-active suspensions, representing a reduction of 16%.

Runs were also made by increasing the value of ρ_2, which penalizes the tire deflections. However, the results show that controling the coefficients of dampers located between the sprung and unsprung masses is ineffective in reducing the tire deflections.

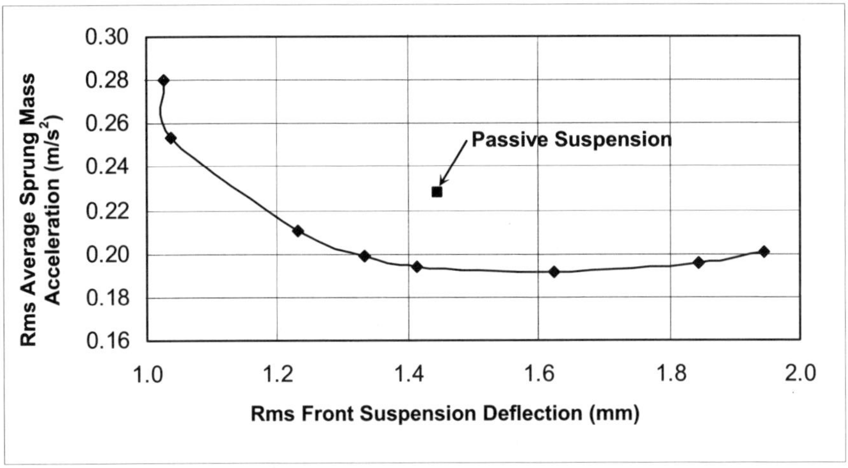

Figure 2. Sprung Mass Acceleration versus Front Suspension Deflection Trade-Off Curve for Changing ρ_1.

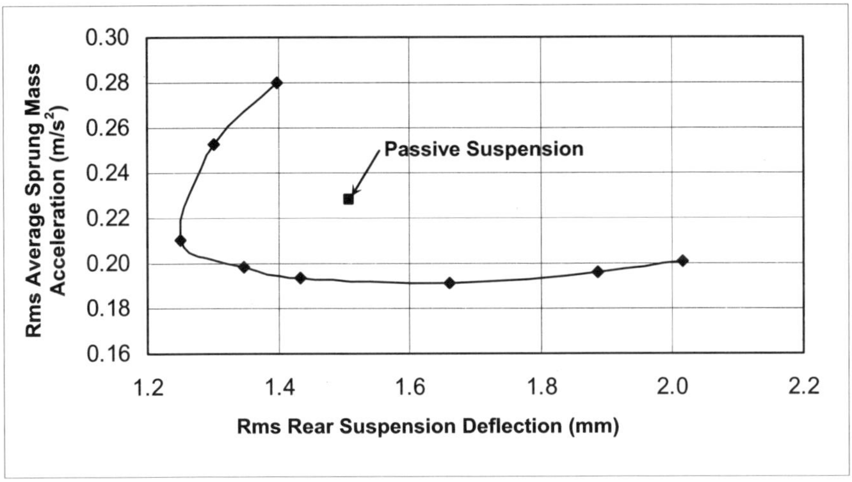

Figure 3. Sprung Mass Acceleration versus Rear Suspension Deflection Trade-Off Curve for Changing ρ_1.

4.3. *Effect of Delayed Control Input on the System Performance*

The dynamic response characteristics of the variable dampers and the system which drives these dampers have not been considered in the above analysis. If magneto-rheological dampers are used, a time delay (τ_d) of about 30 ms occurs due to digital processing and the response time of the dampers. In order to determine the effect of this time delay, a first order variable damper model was introduced into the model, and the control algorithm found in section 2 was applied to the resulting system. The investigation was carried out for time constants ranging from 10 ms to 30 ms. The resulting values of the minimum rms accelerations and the corresponding suspension deflections are depicted in Figure 4. The results show that rms sprung mass acceleration increases by approximately 8% as a consequence of the delayed control inputs, compared to the system without delays.

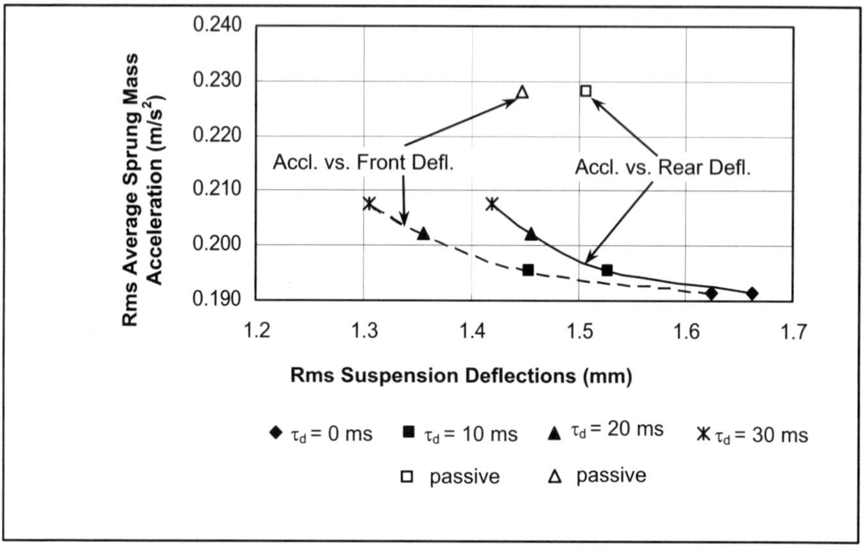

Figure 4. Average Sprung Mass Acceleration Versus Suspension Deflections for Delayed Control Inputs.

5 Conclusions

This study shows that heave accelerations of the sprung mass can be reduced effectively by using semi-active suspensions as compared to those of the passive system while keeping the suspension deflections at comparable and acceptable levels. For the vehicle parameters and the road input profile considered in this study, a reduction of 16% in rms average heave acceleration of the sprung mass has been achieved. A time delay of 30 ms in applying the control inputs increases the rms average sprung mass heave acceleration of the optimal system only by 8% while reducing the suspension deflections, which still represents considerable improvement compared to the passive suspensions.

6 Acknowledgements

Support of TOBB Economics and Technology University is gratefully acknowledged.

References

1. Lord Corporation World Wide Web site, www.lord.com, *Lord Corporation, Cary, NC, USA.*

2. Cadillac World Wide Web site, www.cadillac.com, *General Motors Corporation, Detroit, MI, USA.*
3. Karnopp, D.C., Crosby, M.J. and Harwood, R.A., Vibration Control Using Semi-Active Force Generators, *ASME Journal Of Engineering for Industry,* **96** (1974) pp. 619-626.
4. Hrovat, D., Margolis, D.L. and Hubbard, M., An Approach Toward the Optimal Semi-Active Suspension, *ASME Journal Of Dynamic Systems, Measurement and Control*, **110** (1988) pp 288-296.
5. Hac, A. and Youn, I., Optimal Design of Active and Semi-Active Suspensions Including Time Delays and Preview, *ASME Journal Of Vibration and Acoustics*, **115** (1993) pp. 498-508.
6. Hac, A. and Youn, I., Optimal Semi-Active Suspension with Preview Based on a Quarter Car Model, *Journal Of Vibration, Acoustics Stress and Reliability in Design*, **114** (1992) pp. 84-92.
7. ElMadany, M.M. and Abduljabbar, Z., Alternative Control Laws for Active and Semi-Active Automotive Suspensions - A Comparative Study, *Computers and Structures*, **39** (1991) pp. 623-629.
8. Hwang, Y.R. and Wang, S.C., Computer Control Algorithm for Semi-Active Suspensions Based on a Two-Wheel Vehicle Model, *International Journal of Computer Applications in Technology*, **17** (2003) pp. 37-46.
9. Tseng, H.E. and Hedrick, J.K., Semi-Active Control Laws – Optimal and Sub-Optimal, *Vehicle System Dynamics*, **23** (1994) pp. 545-569.
10. Karnopp, D., Design Principles for Vibration Control Systems Using Semi-Active Dampers, *ASME Journal Of Dynamic Systems, Measurement and Control,* **112** (1990) pp. 448-455.
11. Elbeheiry, E.M., A Method for Preview Vibration Control of Systems Having Forcing Inputs and Rapidly Switched Dampers, *Journal Of Sound and Vibration*, **214** (1998) pp.269-283.
12. Elbeheiry, E.M., Suboptimal Bilinear Control Methods Applied to Suppressing Car Vibrations, *Journal Of Vibration and Control*, **7** (2001) pp. 279-306.
13. Ercan, Y., Theoretical Investigation of an Optimally Controlled Semi-Active Suspension with Variable Damping, Proceedings of 11th National Machine Theory Symposium – Turkey, **2** (2005) pp. 355-366.
14. Karkoub, M.A. and Mohamed, Z., Optimal Control of Suspension Systems Using Smart Actuators, *IEEE Symposium on Emerging Technologies and Factory Automation, ETFA*, **1** (2001) pp. 87-93.
15. Gordaninejad, F., Evrensel, C.A., Yeo, M.S., Liu and Y., Karakaş, E.S., Study of a Quarter Model HMMWV Suspension System Using a Magneto-Rheological Fluid Damper, *Proceedings of SPIE - The International Society for Optical Engineering*, **5056** (2003) pp. 506-513.
16. Hrovat, D., Survey of Advanced Suspension Developments and Related Optimal Control Applications, *Automatica*, **33** (1997) pp. 1781-1817.
17. The University of Michigan Transportation Research Institute (UMTRI) World Wide Web site, www.umtri.umich.edu, *University of Michigan, Ann Arbor, MI, USA.*

HYDRO-MECHANICAL ANALYSIS OF A MAGNETORHEOLOGICAL BYPASS DAMPER

S.R. HONG[1], GANG WANG[1], WEI HU[1], N. M. WERELEY[1], AND S.B. CHOI[2]

[1]*Smart Structures Laboratory, Alfred Gessow Rotorcraft Center*
Department of Aerospace Engineering, University of Maryland, College Park MD 20742
E-mail: srhong@umd.edu, gwang@umd.edu, weihu@umd.edu, wereley@umd.edu
[2]*Department of Mechanical Engineering, Inha University, Incheon, South Korea*
E-mail: seungbok@inha.ac.kr

Damper behavior of a magnetorheological (MR) bypass damper is analyzed using hydro-mechanical modeling approach. An MR bypass damper consists of piston and double end rods in a hydraulic cylinder, and a bypass comprising cylindrical tubing and an MR valve. Damping forces are developed in the annular bypass via Poiseuille flow. A hydro-mechanical model for the MR bypass damper is derived by considering lumped hydraulic parameters, which are compliance of MR fluids inside the cylinder and flow resistance through the MR bypass valve. Numerical simulations are conducted to predict the controllable damping force due to variable yield shear stress of MR fluid, the post-yield damping due to viscosity of MR fluid, and hysteresis behavior in the low velocity domain due to the compliance effect of the fluid. A laboratory-scale MR bypass damper will be fabricated and tested in order to validate the hydro-mechanical MR damper model.

1 INTRODUCTION

Magnetorheological (MR) fluids undergo reversible and rapid changes in material characteristics when subjected to a magnetic field [1]. This change is primarily observed as a significant increase of the yield shear stress of MR fluids and can be continuously controlled by tuning the magnetic field intensity. Thus, MR fluids can be effectively utilized in shock absorbers for control of automotive primary suspension [2-5], dampers for seismic mitigations of building and structure [6-7], impact dampers for recoil system [8], and mounts for control of flexible structural vibration [9]. Most MR fluid based dampers or shock absorbers are designed with single rod, and have an accumulator that is used to prevent cavitation on the low pressure side of the piston and to compensate for the volume change in the MR fluid induced by the rod motion. On the other hand, double rod dampers whose rod diameters are same have also been studied [7, 10]. Because the volume change of MR fluid inside those dampers is designed to be small to avoid spring effect due to fluid compressibility, the spring effect of the dampers are so small to be negligible. Thus, those studies on MR devices have been focused on the controllable damping mechanisms.

In order to achieve desirable control performance, it is necessary to have an accurate damping force model that can capture the inherent hysteresis behavior of an MR damper. Several types of damper models have been developed to describe the damping force behavior of MR dampers [11-17]. However, many damper models used phenomenally determined mechanisms to capture post-yield and hysteretic pre-yield force of MR dampers. Those model parameters cannot be derived from damper geometric and material properties. Comparatively, the fluid compressibility was considered in the MR damper models by the introduction of the fluid bulk modulus [18]. Using known damper properties, this fluid mechanics based MR damper model can accurately predict hysteretic damper behavior of ER/MR damper. Similarly, hydro-mechanical model based on lumped hydraulic parameters of flow resistance and chamber compliances was established to investigate hysteretic damping force behavior of electrorheological (ER) dampers and the damping force predictions were compared to other four existing ER damper model [19]. In this paper, we adopt same hydro-mechanical damper model and apply it to a by-pass MR damper configuration. The bypass MR damper consists of a double rod hydraulic cylinder and MR bypass valve. The hydro-mechanical model of the bypass MR damper is derived by considering lumped hydraulic parameters such as compliances of fluids inside the cylinder and flow resistances through the MR bypass valve. The force characteristics of the MR damper are analyzed and will be compared to the experimental data using the theoretical model.

2 MAGNETORHEOLOGICAL BYPASS DAMPER

The prototypical configuration of the bypass MR damper proposed in this study is presented in **Figure 1**. The damper consists of a piston, two rods of same diameters, a high-pressure MR fluid filled hydraulic cylinder, and a bypass between one chamber of the cylinder and the chamber on the opposite side of the piston. The MR fluid flows from one chamber, through the bypass, and onto the chamber on the opposite side of the piston as a result of piston motion. The bypass is serially connected by tubing and flow mode MR valve. Thus, the total flow resistance in the bypass is the sum of flow resistances in tubing and MR valve. Therefore, the damping force of the bypass MR damper can be controlled by adjusting the flow resistance of the MR valve.

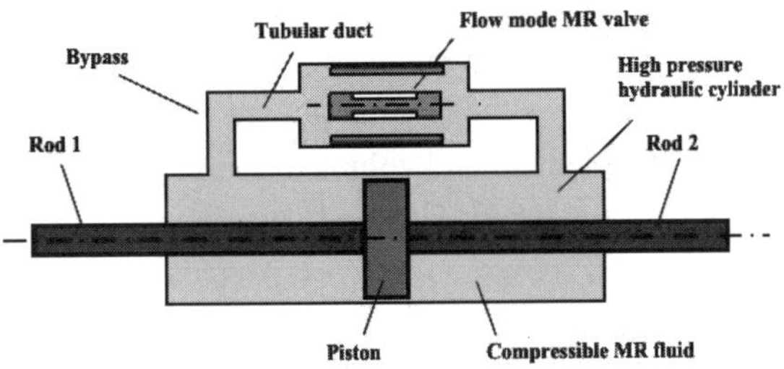

Figure 1: Schematic of Bypass MR Damper

3 HYDRO-MECHANICAL MODELING

Fluid compressibility was accounted for in the damper can be expressed by damping force model of the MR damper by the inclusion of the bulk modulus [18]. This fluid mechanics-based damping force model accurately predicted hysteretic damping force behavior owing to fluid compressibility of an ER/MR damper. An alternative approach using a hydro-mechanical model based on lumped hydraulic parameters of flow resistance and chamber compliances was

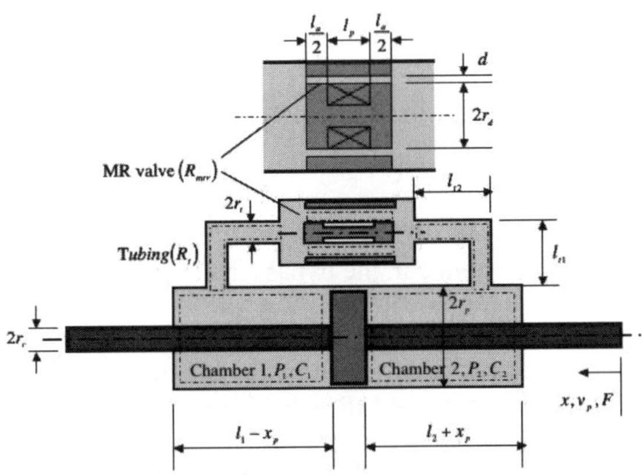

Figure 2: Hydro-Mechanical Model of Bypass MR Damper

established to investigate the hysteretic damping force behavior of ER dampers [19]. In this study, the hydro-mechanical modeling approach was adopted to derive a model that can account for the compressible fluid spring and variable damping characteristics of a bypass MR damper.

Figure 2 shows the schematic of a bypass MR damper. The lumped parameter hydraulic model was developed in order to capture the damper behavior by considering the appropriate control volumes as well as the geometry of the damper. Because the fluid volume in the bypass is small relative to the volumes of chambers 1 and 2, the compressibility of the MR fluid in the bypass is not considered. Volumetric continuity for chambers 1 and 2 yields,

$$C_1\dot{P}_1 = A_p v_p - Q_f \tag{1}$$

$$C_2\dot{P}_2 = -A_p v_p + Q_f \tag{2}$$

In the above, P_1 and P_2 are the fluid pressures of the chambers 1 and 2, C_1 and C_2 are the compliances of the chambers 1 and 2, A_p is the piston area, v_p is the rod/piston velocity, and Q_f is the volume flow rate through the bypass.

The compliances and volume flow rates of Eqs. (1, 2) can be expressed below:

$$C_1 = V_1/E = A_p(l_1 - x_p)/E \tag{3}$$

$$C_2 = V_2/E = A_p(l_2 + x_p)/E \tag{4}$$

$$Q_f = A_t v_t = A_d v_d \tag{5}$$

where

$$A_p = \pi(r_p^2 - r_r^2),\ A_t = \pi r_t^2,\ A_d = bd = 2\pi(r_d + d/2)d \tag{6}$$

In the above, V_1 and V_2 are the fluid volumes of the chambers 1 and 2, E is the effective bulk modulus of the MR fluid, l_1 and l_2 are the lengths of each chamber when load is not applied to the strut, x_p is the displacement of the rod/piston, r_p is the radius of piston head, r_r is the radius of rod, A_t and A_d are the cross-sectional areas of the tubing and MR valve annulus, r_t is the radius of the circular tubing, b, d, and r_d are circumference of the annulus, gap of the annulus, inner radius of the annulus of MR valve, respectively. v_t and v_d are average velocity of fluid flow through the tubing and MR valve. The pressure drop due to MR fluid passing through the bypass is given by

$$P_1 - P_2 = \Delta P_{mrv} + \Delta P_t = R_f Q_f \tag{7}$$

where

$$R_f = R_{mrv} + R_t \tag{8}$$

In the above, ΔP_t and ΔP_{mrv} are the pressure drops in the tubing and MR valve. R_f is the fluid resistance in the bypass, and represented as the sum of fluid resistances R_{mrv} and R_t of the MR valve and tubing.

By considering pressure drops in the active and passive domains of the MR valve, the flow resistance in the MR valve is expressed as [11]

$$R_{mrv} = \frac{\Delta P_{mrv}}{Q_f} = \frac{\Delta P_p}{Q_f} + \frac{\Delta P_a}{Q_f} = \frac{12\mu l_p}{bd^3} + \frac{12\mu l_a}{bd^3}\frac{1}{(1-\bar{\delta})^2(1+\bar{\delta}/2)} \tag{9}$$

In the above, μ is the post-yield viscosity, and l_p and l_a are the gap lengths in the passive and active domains of the MR valve. $\bar{\delta}$ is the non-dimensional plug thickness defined by the ratio of the plug thickness δ and gap size d. And the nondimensional plug thickness $\bar{\delta}$ can be obtained by considering following polynomial equation:

$$\frac{6\bar{\delta}}{(1-\bar{\delta})^2(1+\bar{\delta}/2)} = \mathrm{Bi},\ 0 \le \bar{\delta} < 1 \tag{10}$$

where

$$\mathrm{Bi} = \frac{\tau_y d}{\mu v_d} \tag{11}$$

In the above, the Bingham number, Bi, is the ratio of dynamic shear stress to Newtonian shear stress [20], and is determined by the yield shear stress, τ_y, the viscosity, μ, and the average velocity of the MR fluid flowing through the MR valve, v_d. On the other hand, the flow resistance of in the cylindrical tubing is given by [21]

$$R_t = \frac{\Delta P_t}{Q_f} = \frac{8\mu l_t}{\pi r_t^4} \tag{12}$$

In the above, $l_t = 2l_{t1} + 2l_{t2}$ is the total length of the tubing. The total flow resistance in the bypass is obtained from Eqs. (9,12) and expressed as

$$R_f = \frac{12\mu l_p}{bd^3} + \frac{12\mu l_a}{bd^3}\frac{1}{(1-\bar{\delta})^2(1+\bar{\delta}/2)} + \frac{8\mu l_t}{\pi r_t^4} \tag{13}$$

As it can be seen from Eq. (13), the flow resistance or pressure drop in the bypass is tuned by the nondimensional plug thickness which is a function of the yield shear stress of MR fluid.

The force of the bypass MR damper exerted by rod/piston motion is given by

$$F = A_p(P_1 - P_2) \tag{14}$$

where

$$P_1 = \frac{A_p}{C_1}x_p - \frac{A_d}{C_1}x_d \tag{15}$$

$$P_2 = -\frac{A_p}{C_2}x_p + \frac{A_d}{C_2}x_d \tag{16}$$

In the above, x_d is the displacement of the lumped fluid in the annulus of the MR valve. The Eqs. (15,16) for chamber pressures P_1 and P_2 were obtained by integrating Eqs (1,2). By substituting the variables P_1 and P_2 of Eqs. (15,16) into Eq. (7), the first order differential equation relating the lumped fluid displacement in the MR valve x_d and rod/piston displacement x_p is obtained as follows

$$v_d = \dot{x}_d = \left(-\left(\frac{1}{C_1} + \frac{1}{C_2}\right)A_d x_d + \left(\frac{1}{C_1} + \frac{1}{C_2}\right)A_p x_p\right)\Big/ A_d R_f \tag{17}$$

The variable x_d can be obtained by solving Eq. (17), and is to calculate the chamber pressures P_1 and P_2 of Eqs. (15,16).

4 MODELING RESULTS

The bypass MR damper designed in this study consists of three main parts: a commercial high-pressure hydraulic cylinder, rods/piston, a flow bypass including industrial fittings and MR valve. As shown in **Figure 2**, the geometry of the bypass MR damper is summarized in **Table 1**.

Table 1: Bypass MR Damper Geometry

Bore diameter ($2r_0$)	38.1mm
Fluid space lengths l_1 and l_2	152mm
Rod diameter ($2\,r_r$)	15.875mm
MR valve annular gap (d)	0.508mm
MR valve inner radius of annular gap(r_d)	14.25mm
MR valve active length (l_a)	5.33mm
MR valve passive length (l_p)	10.92mm

In this study, silicone oil based MR fluid (MRF-336AG, Lord Corporation) diluted with 10cSt silicone oil is chosen to fill the MR damper. By adding silicone oil to the MR fluid, the particle volume fraction is reduced from 36% to 28%. The bulk modulus of the MR fluid is predetermined to be 1.1 GPa, and the post yield viscosity of the fluid is assumed to be 1.0Pa·s. The tubing diameter $2r_t$ and total length l_t of the are 9mm and 340mm, respectively.

Based on above MR damper geometric parameters and MR fluid properties, we can predict the hysteresis behavior of the MR damper using proposed hydro-mechanical damper model. **Figure 3** shows the damper force versus rod displacement and velocity relationship under a sinusoidal displacement excitation with amplitude of 10 mm and excitation frequency of 1 Hz. The energy dissipated by the damper over one vibration cycle is the measure of its damping capacity, which is given by the area enclosed with the force vs. displacement cycle. As shown in **Figure 3**(a), when the applied field (yield stress) is high, the area within the force vs. displacement cycle becomes large, which indicates the increase of the damping or energy dissipated by the damper. From **Figure 3** (b), we can see that the hydro-mechanical model accurately represents the force vs. velocity behavior for pre-yield and post-yield hysteresis. **Figure 4** shows the predicted damper behavior as the amplitude of the applied displacement becomes 20 mm.

5 CONLUSIONS

By utilizing the compressibility and variable yield stress of MR fluid, a hydro-mechanical model was developed to characterize MR dampers. A bypass MR damper was proposed to evaluate the hydro-mechanical model. The bypass consisting of fittings and MR valve was adopted for the damping augmentation. The damper behavior was analytically investigated by considering lumped hydraulic parameters which are damper parameters and MR fluid properties. The hydro-mechanical model can be used to predict both force vs. displacement and force vs. velocity hysteresis behavior of the MR damper. Since the hydro-mechanical model consists of physically motivated hydro-mechanical lumped parameters, the compliance effects manifested in the pre-yield hysteresis behavior of the MR damper are successfully captured. A bypass damper will be fabricated and tested in the future in order to validate the hydro-mechanical model in predicting the hysteresis behavior.

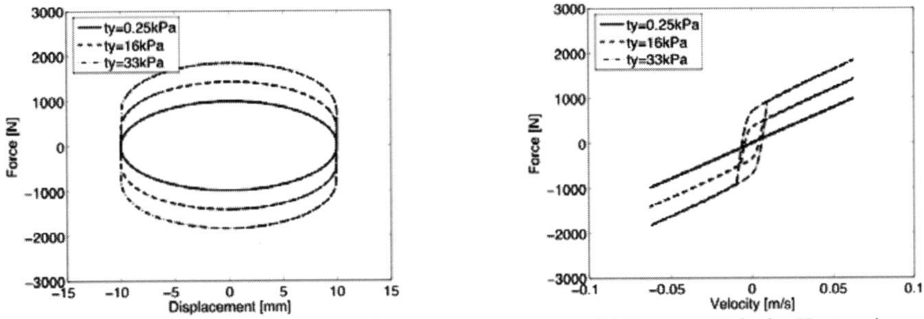

(a) Force vs. Displacement Hysteresis (b) Force vs. Velocity Hysteresis

Figure 3: Hysteresis Damper behavior using Hydro-Mechanical Model (10 mm Amplitude & 1 Hz)

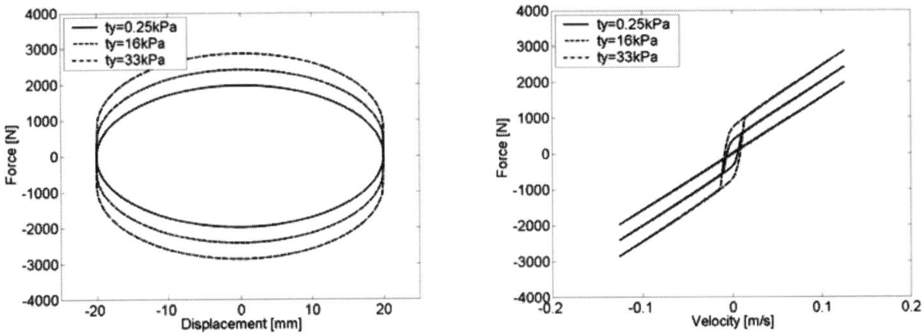

(a) Force vs. Displacement Hysteresis (b) Force vs. Velocity Hysteresis

Figure 4: Hysteresis damper behavior using Hydro-Mechanical Model (20 mm Amplitude & 1 Hz)

Reference

1. G. Bossis, S. Lacis, A. Meunier, and O. Volkova, "Magnetorheological Fluids," *Journal of Magnetism and Magnetic Materials*, 252, pp.224~228, 2002.
2. H. S. Lee and S. B. Choi, "Control and Response Characteristics of a Magneto-Rheological Fluid Damper for Passenger Vehicles," *Journal of Intelligent Material Systems and Structures*, 11, pp.80~87, 2000.
3. A. H. Lam and W. H. Liao, "Semi-Active Control of Automotive Suspension Systems with Magneto-Rheological Dampers," *International Journal of Vehicle Design*, 33(1-3), pp.50~55, 2003.
4. E. S. Karakas, F. Gordaninejad, C. A. Evrensel, M. S. Yeo, Y. Liu, and H. Sahin, "Control of a Quarter HMMWV Suspension System Using a Magnetorheological Fluid Damper," *Proceedings of SPIE*, 5386, pp.204~213, 2004.

5. N. D. Sims, N. J. Holmes, and R. Stanway, "A Unified Modelling and Model Updating Procedure for Electrorheological and Magnetorheological Vibration Dampers," *Smart Materials and Structures*, 13, pp.100~121 , 2004.
6. F. Yi, S. J. Dyke, J. M. Caicedo, and J. D. Carlson, "Experimental Verification of Multi-input Seismic Control Strategies for Smart Dampers," *ASCE Journal of Engineering Mechanics*, 127, pp.1152~ 1164, 2001.
7. H. Sodeyama, K. Suzuki, and K. Sunakoda, "Development of Large Capacity Semi-Active Seismic Damper Using Magneto-Rheological Fluid," *Journal of Pressure Vessel Technology*, 126, pp.105~ 109, 2004.
8. W. Facey, N. C. Rosenfeld, and N. M. Wereley, "Design and Characterization of a Compact Impact Damper for High Impulsive Loads," *44th AIAA/ASME/ASCE/AHS Structures, Structural Dynamics, and Materials Conference*, AIAA2003-1645, 2003.
9. S. R. Hong and S. B. Choi, "Vibration Control of a Structural System Using Magneto-Rheological Fluid Mount," *14th International Conference on Adaptive Structures and Technologies*, pp.182~191 , 2003.
10. J. Lindler and N. M. Wereley, "Analysis and Testing of Electrorheological Bypass Dampers," *Journal of Intelligent Systems and Structures*, 10, 363~ 376, 1999.
11. N.M. Wereley, and L. Pang, "Nondimensional Analysis of Semi-active Electrorheological and Magnetorheological Dampers Using Approximate Parallel Plate Models," *Smart Materials and Structures*, Vol. 7, No. 5, 732 (1998).
12. R.W. Phillips, *Engineering Application of Fluids with a Variable Yield Stress,* Ph.D. Thesis, Mechanical Engineering, University of California at Berkeley (1969).
13. R. Stanway, J.L. Sproston, and A.K. El-Wahed, "Application of Electrorheological Fluids in Vibration Control: A Survey," *Smart Materials and Structures*, Vol. 6, No. 3, 351 (1997).
14. L. Pang, G.M. Kamath, and N.M. Wereley, "Analysis and Testing of A Linear Stroke Magnetorheological Damper," *AIAA/ASME Adaptive Structures Forum*, Long Beach, CA (1997).
15. G.M. Kamath, N.M. Wereley, and M.R. Jolly, "Characterization of Magnetorheological Helicopter Lag Dampers," *Journal of the American Helicopter Society*, Vol. 44, No. 3, 234 (1999).
16. D.R. Gamota, and F.E. Filisko, "Dynamic Mechanical Studies of Electrorheological Materials: Moderate Frequencies," *Journal of Rheology*, Vol.35, No. 3, 399 (1991).
17. B.F. Spencer(Jr), S.J. Dyke, M.K. Sain and J.D. Carlson, "Phenomenological Model of a Magnetorheological Damper," *Journal of Engineering Mechanics*, Vol. 123, No. 3, 230 (1997).
18. X. Wang and F. Gordaninejad, "Dynamic Modeling of Semi-Active ER/MR Fluid Dampers," *Proceedings of SPIE*, 4331, pp.82~ 91, 2001.
19. S. R. Hong, S. B. Choi, Y. T. Choi, and N. M. Wereley, "Comparison and Damping Force Models for an Electrorheological Fluid Damper," *International Journal of Vehicle Design*, 33(1-3), pp.17~ 35, 2003.
20. W. Prager, *Introduction to Mechanics of Continua*, Ginn and Company, New York, 1961.
21. F. M. White, *Fluid Mechanics*, McGraw-Hill, New York, 1998.

NONDIMENSIONAL ANALYSIS OF SHEAR-MODE TYPE ROTARY DRUM DAMPERS USING MAGNETORHEOLOGICAL FLUIDS

NORMAN M. WERELEY AND YOUNG-TAI CHOI

Smart Structures Laboratory, Department of Aerospace Engineering,
University of Maryland, College Park, MD 20742, U.S.A.
Email: wereley@umd.edu and nicechoi@wam.umd.edu

SEUNG-BOK CHOI

Smart Structures and Systems Laboratory, Department of Mechanical Engineering,
Inha University, Incheon 402-751, Korea
Email: seungbok@inha.ac.kr

In this study, nondimensional analysis of shear-mode type rotary drum dampers using MR (magnetorheological) fluids is theoretically conducted. A shear-mode type rotary drum MR damper is constructed and its governing equation of motion is derived based on three different constitutive flow models: Bingham-plastic, biviscous, and Herschel-Bulkley. The damping performance of the rotary drum MR dampers is characterized in terms of the damping coefficient, which is the ratio of the equivalent viscous damping at field-on status to the damping at field-off status. To this end, the damping coefficient for the rotary drum MR damper is theoretically obtained using a key nondimensional parameter, the Bingham number. The performance characteristics of the rotary drum MR dampers are evaluated with respect to different constitutive flow models.

1 Introduction

Generally, rotary drum dampers have been used on rotating machinery to inhibit the occurrences of unwanted oscillations. These oscillations adversely affect machine performance and can result in permanent damage to the machine itself. Thus, the function of rotary drum dampers is to dampen undesired rotation of an object by providing a retarding torque to a shaft. Conventional passive rotary drum dampers work on the principle of viscous fluid shearing, i.e., energy is dissipated by the shear motion of a viscous fluid between two different wall surfaces. As a result, the damper force (or torque) of the passive rotary drum dampers is highly dependent on the angular velocity. Recently, the rotary drum dampers using magnetorheological (MR) fluids are spotlighted because of a controllable damper force (or torque) and fast response time. Rotary drum MR dampers in the absence of a magnetic field are similar to conventional passive rotary drum dampers. However, rotary drum MR dampers can produce additional damper force (or torque) resulting from the yield stress of MR fluids in response to the magnetic field.

In this study, nondimensional analysis of shear-mode type rotary drum MR dampers is theoretically conducted. A shear-mode type rotary drum MR damper is constructed and its governing equation of motion is derived based on three different constitutive flow models: Bingham-plastic, biviscous, and Herschel-Bulkley. The damping performance of the rotary drum MR dampers is characterized in terms of the damping coefficient, which is the ratio of the equivalent viscous damping at field-on status to the damping at field-off status. To this end, the damping coefficient for the rotary drum MR damper is theoretically obtained using a key nondimensional parameter, the Bingham number, which is the ratio of the yield stress to the viscous stress. The performance characteristics of the rotary drum MR dampers are evaluated with respect to different constitutive flow models.

2 Rotary drum dampers

A typical rotary drum damper is schematically shown in Figure 1. It comprises an inner cylinder of length L and radius r_b, and an outer cylinder of radius r_c. When the inner cylinder rotates with an angular velocity Ω, the applied torque T is obtained by [1,2]

$$T = 2\pi r_b^2 L \bar{\tau}_b \tau_y \tag{1}$$

where $\bar{\tau}_b$ is the nondimensional shear stress, $\bar{\tau}_b = \tau_b / \tau_y$, τ_b is the shear stress at the inner cylinder, and τ_y is the yield stress of the MR fluid. On the other hand, the angular velocity can be described by

$$\Omega = \int_{r_b}^{r_c} \frac{\dot{\gamma}}{r} dr = \frac{1}{2} \int_{\tau_c}^{\tau_b} \frac{\dot{\gamma}}{\tau_{r\theta}} d\tau_{r\theta} \tag{2}$$

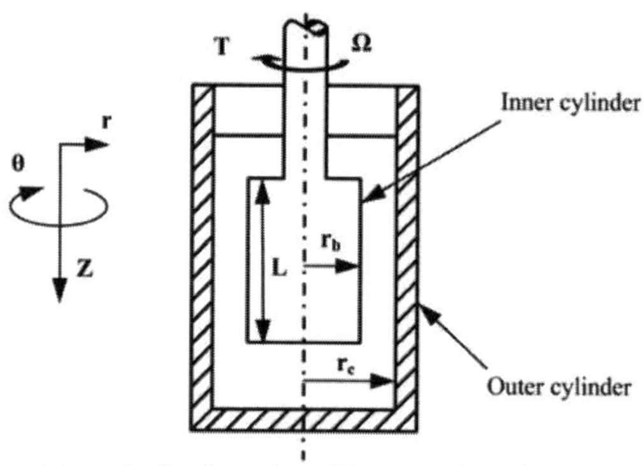

Figure 1. Configuration of the rotary drum damper.

2.1 *Bingham-plastic model*

For the Bingham-plastic model, two distinct conditions may occur in the gap of the rotary drum damper: (i) $\tau_c < \tau_y < \tau_b$; (ii) $\tau_c > \tau_y$. Therefore, the shear stress is obtained by solving two distinct equations as follows [2]:

$$\frac{2(\psi - 1)}{\mathrm{Bi}} = \begin{cases} (\bar{\tau}_b - 1) - \ln \bar{\tau}_b & 1 < \bar{\tau}_b < \psi^2 \quad \text{(3a)} \\ \bar{\tau}_b \left(1 - \dfrac{1}{\psi^2}\right) - \ln \psi^2 & \bar{\tau}_b > \psi^2 \quad \text{(3b)} \end{cases}$$

$$\text{with} \qquad \psi = \frac{r_c}{r_b} \qquad \text{and} \qquad \mathrm{Bi} = \frac{\tau_y (\psi - 1)}{\mu \Omega} \tag{4}$$

The critical Bingham number, $\mathrm{Bi}_{\mathrm{crit}}$, is defined as the Bingham number when $\bar{\tau}_b = \psi^2$ as follows:

$$\mathrm{Bi}_{\mathrm{crit}} = (\psi - 1) \left[\frac{(\psi^2 - 1)}{2} - \ln \psi \right]^{-1} \tag{5}$$

If $\mathrm{Bi} > \mathrm{Bi}_{\mathrm{crit}}$, the rotary drum damper stays at the condition of $1 < \bar{\tau}_b < \psi^2$ governed by

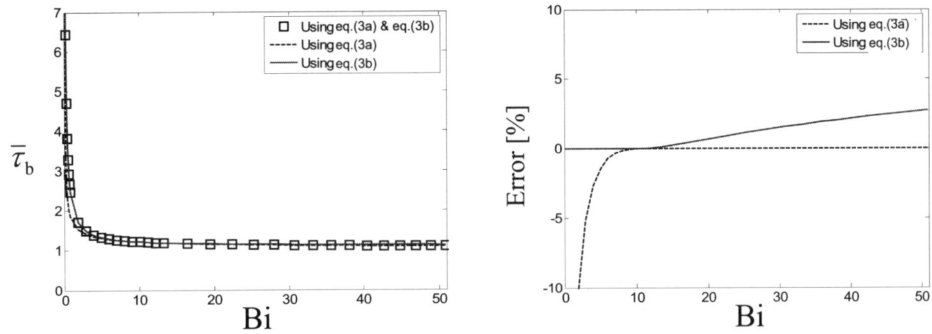

(a) nondimensional shear stress (b) approximation error

Figure 2. The nondimensional shear stress, $\overline{\tau}_b$, for the Bingham-plastic model ($\psi = 1.1$).

Eq. (3a). This condition physically implies a high yield stress due to a high current input and/or small viscous stress due to small rotational speed and small viscosity. If $\text{Bi} < \text{Bi}_{\text{crit}}$, the condition of $\overline{\tau}_b > \psi^2$ governed by Eq. (3b) is only dominant.

Figure 2 shows the nondimensional shear stress vs. the Bingham number. As you can see this figure, Eq. (3b) can approximate the behavior using Eqs .(3a) and (3b). When we limit the Bingham number to the range of $\text{Bi} \le 5\text{Bi}_{\text{crit}}$, the approximation error using Eq. (3b) in the Bingham-plastic model is less than 5%. Because the gap for most MR applications is generally less than 2 mm, the nondimensional radius is assumed to be limited to the range of $1 < \psi \le 1.2$ in this study.

The nondimensional shear stress for the Bingham-plastic model can be obtained from Eq. (3b) as follows:

$$\overline{\tau}_b = 2\left[\frac{(\psi - 1)}{\text{Bi}} + \ln\psi\right]\left(\frac{\psi^2}{\psi^2 - 1}\right) \tag{6}$$

Substituting Eqs. (6) into (1) yields

$$T = \frac{4\pi r_b^2 \mu \psi^2 L}{(\psi^2 - 1)}\left[1 + \frac{\ln\psi}{(\psi - 1)}\text{Bi}\right]\Omega \tag{7}$$

The equivalent damping C_{eq} for the Bingham-plastic model is defined as

$$C_{eq} = \frac{T}{\Omega} = \frac{4\pi r_b^2 \mu \psi^2 L}{(\psi^2 - 1)}\left[1 + \frac{\ln\psi}{(\psi - 1)}\text{Bi}\right] \tag{8}$$

Then, the damping coefficient for the Bingham-plastic model is obtained by

$$\frac{C_{eq}}{C} = 1 + \frac{\ln\psi}{(\psi - 1)}\text{Bi} \tag{9}$$

where C is damping at field-off status and is determined by setting $\text{Bi} = 0$ in Eq. (8).

2.2 *Biviscous model*

For the biviscous model, three conditions may occur in the gap of the rotary drum damper: (i) $\tau_b \le \tau_{ys}$; (ii) $\tau_c < \tau_{ys} < \tau_b$; (iii) $\tau_c > \tau_{ys}$. From a previous study [2], the nondimensional shear stress for the biviscous model is given by

$$\frac{2(\psi-1)}{\mathrm{Bi}}=\begin{cases}\bar{\tau}_b\left(1-\dfrac{\bar{\mu}}{\psi^2}\right)-(1-\bar{\mu})(1+\ln\bar{\tau}_b) & 1<\bar{\tau}_b<\psi^2 \quad (10a)\\ \bar{\tau}_b\left(1-\dfrac{1}{\psi^2}\right)-(1-\bar{\mu})\ln\psi^2 & \bar{\tau}_b>\psi^2 \quad (10b)\\ \bar{\tau}_b\left(1-\dfrac{1}{\psi^2}\right)\bar{\mu} & \bar{\tau}_b\le 1 \quad (10c)\end{cases}$$

$$\text{with}\quad \bar{\tau}_b=\frac{\tau_b}{\tau_{ys}}=\frac{\tau_b(1-\bar{\mu})}{\tau_{yd}},\quad \bar{\mu}=\frac{\mu_{po}}{\mu_{pr}}\quad\text{and}\quad \mathrm{Bi}=\frac{\tau_{ys}(\psi-1)}{\mu_{po}\Omega} \qquad (11)$$

The nondimensional shear stress for the biviscous model vs. the Bingham number is presented in Figure 3. The approximation error using Eq. (10b) in the biviscous model is less than 11.5% when the Bingham number is limited to the range of $\mathrm{Bi}\le 5\mathrm{Bi}_{\mathrm{crit}}$.

Approximating and rearranging Eq. (10) yields the nondimensional shear stress for the biviscous model as follows:

$$\bar{\tau}_b=\begin{cases}2\left[\dfrac{(\psi-1)}{\mathrm{Bi}}+(1-\bar{\mu})\ln\psi\right]\left(\dfrac{\psi^2}{\psi^2-1}\right) & \bar{\tau}_b>1\\ \dfrac{2\psi^2}{\bar{\mu}\mathrm{Bi}(\psi+1)} & \bar{\tau}_b\le 1\end{cases} \qquad (12)$$

Substituting Eqs. (12) into (1) yields the torque

$$T=\begin{cases}\dfrac{4\pi r_b^2\mu\psi^2L}{(\psi^2-1)}\left[1+\dfrac{(1-\bar{\mu})\ln\psi}{(\psi-1)}\mathrm{Bi}\right]\Omega & \bar{\tau}_b>1\\ \dfrac{4\pi r_b^2\mu\psi^2L}{(\psi^2-1)}\left(\dfrac{1}{\bar{\mu}}\right)\Omega & \bar{\tau}_b\le 1\end{cases} \qquad (13)$$

Then, the damping coefficient for the biviscous model is

$$\frac{C_{eq}}{C}=\begin{cases}1+\dfrac{(1-\bar{\mu})\ln\psi}{(\psi-1)}\mathrm{Bi} & \bar{\tau}_b>1\\ \dfrac{1}{\bar{\mu}} & \bar{\tau}_b\le 1\end{cases} \qquad (14)$$

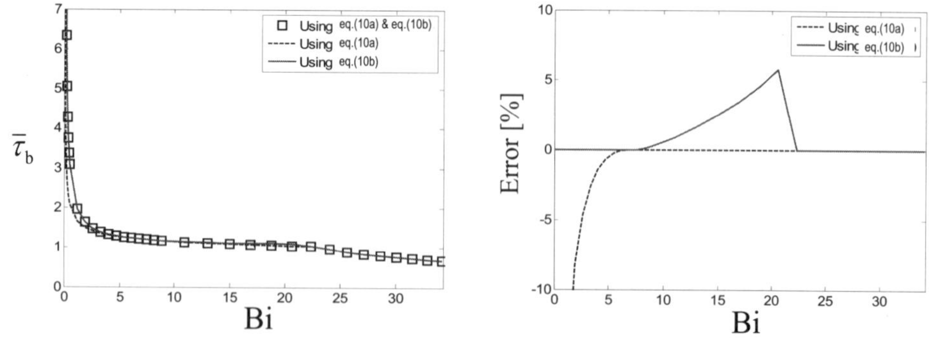

Figure 3. The nondimensional shear stress, $\bar{\tau}_{\mathrm{b}}$, for the biviscous model (ψ =1.1 and $\bar{\mu}$ =0.05).

From Eq. (10), the critical Bingham number for the biviscous model is given by

$$\mathrm{Bi}_{\mathrm{crit}} = (\psi - 1)\left[\frac{(\psi^2 - 1)}{2} - (1 - \bar{\mu})\ln\psi\right]^{-1} \tag{15}$$

On the other hand, from Eq. (14), the preyield Bingham number, $\mathrm{Bi}_{\mathrm{preyield}}$, to ensure that the rotary drum damper is in the preyield region is

$$\mathrm{Bi}_{\mathrm{preyield}} = (\psi - 1)(\bar{\mu}\ln\psi)^{-1} \tag{16}$$

If $\mathrm{Bi} \geq \mathrm{Bi}_{\mathrm{preyield}}$, then the rotary drum MR damper presents the preyield damping behavior.

2.3 *Herschel-Bulkley model*

Like the Bingham-plastic model, two conditions may occur in the gap of the rotary drum damper: (i) $\tau_c < \tau_y < \tau_b$, (ii) $\tau_c > \tau_y$. The nondimensional shear stress for the Herschel-Bulkley model is given by

$$\frac{2^n(\psi - 1)^n}{\mathrm{Bi}} = \frac{1}{\alpha^n}(\bar{\tau}_b - 1)^{\alpha n}\left[{}_2F_1(\alpha,1;\beta;1-\bar{\tau}_b)\right]^n, \qquad 1 < \bar{\tau}_b < \psi^2 \tag{17a}$$

$$\frac{2^n(\psi - 1)^n}{\mathrm{Bi}} = \frac{1}{\alpha^n}\left[(\bar{\tau}_b - 1)^{\alpha}\,{}_2F_1(\alpha,1;\beta;1-\bar{\tau}_b) - \left(\frac{\bar{\tau}_b}{\psi^2} - 1\right)^{\alpha}{}_2F_1\left(\alpha,1;\beta;1-\frac{\bar{\tau}_b}{\psi^2}\right)\right]^n, \qquad \bar{\tau}_b > \psi^2 \tag{17b}$$

$$\text{with } \alpha = \frac{(1+n)}{n},\ \beta = \frac{(1+2n)}{n},\ \mathrm{Bi} = \frac{\tau_y(\psi - 1)}{\mu_{eq}\Omega},\ \text{and } \mu_{eq} = K\left(\frac{\Omega}{\psi - 1}\right)^{n-1} \tag{18}$$

where the exponent, n, is the flow behavior index, K is the consistency parameter, and ${}_2F_1(\alpha,1;\beta;x)$ is the hypergeometric function [2].

The nondimensional shear stress for the Herschel-Bulkley model is shown in Figure 4. The approximation error using Eq. (17b) in the Herschel-Bulkley model is less than 5% in the range of $\mathrm{Bi} \leq 5\mathrm{Bi}_{\mathrm{crit}}$ and the flow index of $0.5 \leq n \leq 2$.

Now we assume the solution of Eq. (17b) as follows:

$$\bar{\tau}_b = 2\lambda\left[\frac{(\psi - 1)}{\mathrm{Bi}} + \ln\psi\right]\left(\frac{\psi^2}{\psi^2 - 1}\right) \tag{19}$$

where λ is a correction factor multiplied by $\bar{\tau}_b$ for the Bingham-plastic model to be determined. Then, the torque for the Herschel-Bulkley model from Eqs. (1) and (19) is

$$T = \frac{4\lambda\pi r_b^2\mu_{eq}\psi^2 L}{(\psi^2 - 1)}\left[1 + \frac{\ln\psi}{(\psi - 1)}\mathrm{Bi}\right]\Omega \tag{20}$$

Then, the damping coefficient for the Herschel-Bulkley model is given by

$$\frac{C_{eq}}{C} = \left[1 + \frac{\ln\psi}{(\psi - 1)}\mathrm{Bi}\right] \tag{21}$$

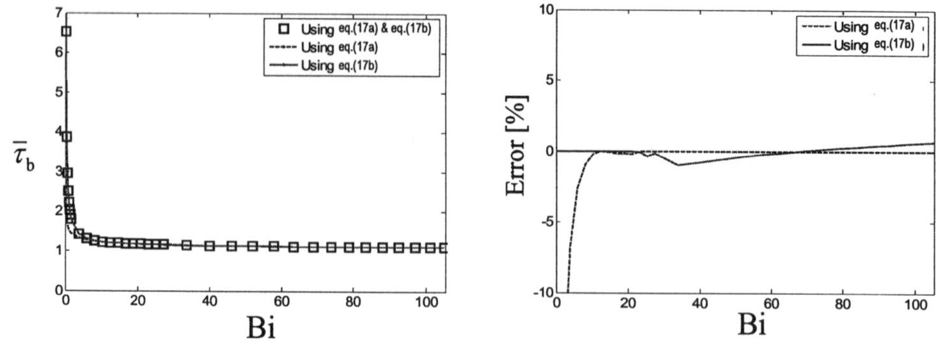

Figure 4. The nondimensional shear stress, $\bar{\tau}_b$, for the Herschel-Bulkley model ($\psi = 1.1$ and $n = 2$).

The critical Bingham number for the Herschel-Bulkley model is given by

$$\mathrm{Bi}_{\mathrm{crit}} = (\psi - 1)\left[\frac{(\psi^2 - 1)}{2\lambda} - \ln\psi\right]^{-1} \tag{22}$$

Substituting the shear rate for the Herschel-Bulkley model into Eq. (2) and integrating, we obtain the shear stress in the absence of field as follows:

$$\tau_b = \frac{2^n K\Omega^n}{n^n\left(1 - \frac{1}{\psi^{2/n}}\right)^n} \tag{23}$$

Substituting Eqs. (23) into (1) also yields the torque of the rotary drum damper in the absence of field for the Herschel-Bulkley model

$$T = \frac{2^{n+1}\pi r_b^2 K\psi^2 L}{n^n\left(\psi^{2/n} - 1\right)^n}\Omega^n \tag{24}$$

By setting $\mathrm{Bi} = 0$ in Eq. (20) and equating it to Eq. (24), the correction factor, λ, is obtained as follows:

$$\lambda = \frac{2^{n-1}(\psi^2 - 1)(\psi - 1)^{n-1}}{n^n\left(\psi^{2/n} - 1\right)^n} \tag{25}$$

3 Nondimensional analysis

Damping coefficient of the rotary drum MR damper for the Bingham-plastic, biviscous and the Herschel-Bulkley models is presented in Figure 5. Note that damping coefficient for the Bingham-plastic model is similar to the result for the Herschel-Bulkley model using $n = 1$. In Figure 5(a), as the nondimensional radius, ψ, increases, the damping coefficient of the rotary drum damper decreases. This implies that the increasing gap in the rotary drum MR damper decreases the damping coefficient. Under a constant gap condition, the rotary drum MR damper with a small cylinder radius produces lower damping coefficient. In addition, we observe that the damping coefficient of the rotary drum MR damper for the Herschel-Bulkley model is not dependent on the exponent n.

Damping coefficient of the rotary drum MR damper for the biviscous model is presented in Figure 5(b). In this case, the nondimensional radius, ψ, is selected to be constant $(\psi = 1.1)$. As observed in this figure, as the nondimensional viscosity, $\bar{\mu}$, decreases, the damping coefficient for the biviscous model becomes close to the trends of the Bingham-plastic model. In the mean while, as the nondimensional viscosity, $\bar{\mu}$, increases, the damping coefficient for the biviscous model becomes close to the behavior of the Newtonian flow. In addition, we can observe from Figure 5(b) that the damping coefficient lines for the biviscous model are piecewise continuous. This is the reason why the biviscous constitutive model has two distinct regions such as preyield and postyield. For sufficiently high Bingham number greater than the preyield Bingham number, $\mathrm{Bi}_{\mathrm{preyield}}$, the rotary drum MR damper dominantly shows the preyield damping behavior.

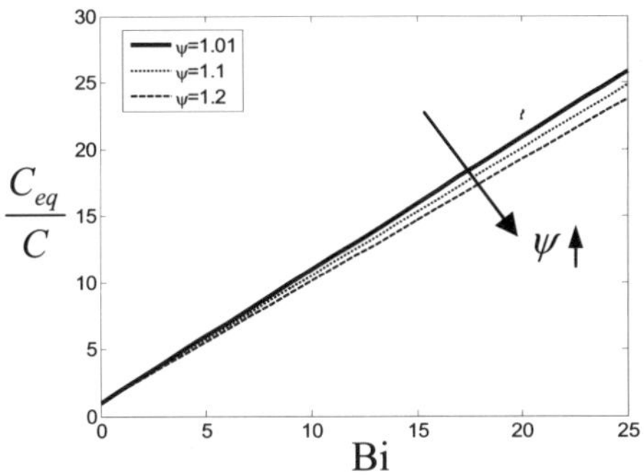

(a) Bingham-plastic and Herschel-Bulkley models

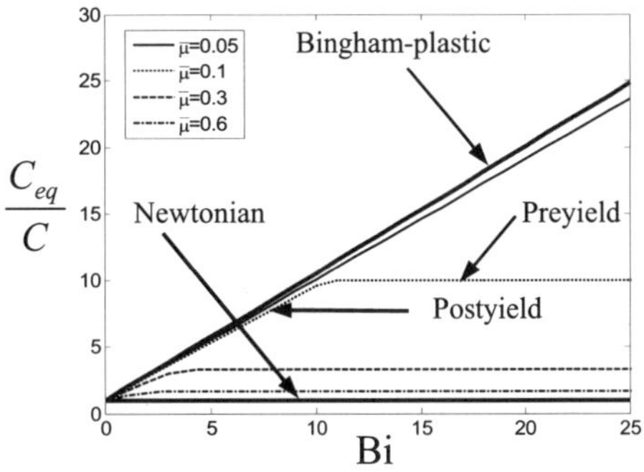

(b) biviscous model

Figure 5. The damping coefficient, C_{eq}/C, with respect to different constitutive models.

4 Conclusions

The damping performance of a shear-mode type rotary drum MR damper was theoretically analyzed based on three different constitutive flow models: Bingham-plastic, biviscous, and Herschel-Bulkley. It is shown that the damping coefficient of the rotary drum MR damper is proportional to the Bingham number as follows:

$$\frac{C_{eq}}{C} = 1 + \sigma \mathrm{Bi} \tag{26}$$

where

$$\sigma = \begin{cases} \ln\psi/(\psi-1), & \text{for Bingham - plastic and Herschel - Bulkley} \\ (1-\bar{\mu})\ln\psi/(\psi-1), & \text{for biviscous model} \end{cases} \tag{27}$$

References

1. Van Wazer J R, Lyons, Lyons J W, Kim K Y, and Colwell R E 1963 *Viscosity and flow measurement* Interscience Publishers Chapter 2
2. Choi Y T, Cho J U, Choi S B and Wereley N M 2005 Constitutive model of electrorheological and magnetorheological fluids using viscometers *Smart Materials and Structures* **14** 1-12

FULL-SCALE MAGNETO-RHEOLOGICAL FLUID DAMPERS FOR HEAVY VEHICLE ROLLOVER

H. SAHIN, Y. LIU, X. WANG

Composite and Intelligent Materials Laboratory, University of Nevada Reno, Reno, Nevada 89557, USA

E-mail: ciml@unr.edu

F. GORDANINEJAD, C. EVRENSEL

Department of Mechanical Engineering, University of Nevada Reno, Reno, Nevada 89557, USA

E-mail: faramarz@unr.edu , evrensel@unr.edu

A.FUCHS

Department of Chemical Engineering, University of Nevada Reno, Reno, Nevada 89557, USA

E-mail: afuchs@unr.edu

A unique MRF by-pass damper for heavy vehicle controllable suspension systems is designed, manufactured and tested. The damper can generate nearly 8000 N which meets the maximum force requirement for preventing the heavy vehicle rollover under certain crucial circumstance. A dynamic simulation of the roll over performance of a heavy vehicle incorporated with four MR dampers is carried out using vehicle dynamic software. Emergency maneuver and roll over scenario are simulated. The results show that the controllable MRF dampers could achieve better performance for protection from the vehicle rollover, which the roll angle can be reduced by 45% compared to the regular OEM dampers.

1 Introduction

Ride quality and handling performance of ground vehicles depend highly on the performance of vehicle's primary suspension systems [2] A semi-active controllable suspension system consists of a spring and a controllable damper with adjustable damping capacity. Magnetorheological fluid (MRF) semi-active dampers can be used in a controllable suspension system since their damping force can be altered by an applied magnetic field using an electromagnet.

As a key component of a controllable suspension system, a MRF damper has advantages over other controllable dampers, such as, electro-rheological fluid dampers, controllable friction dampers and variable fluid dampers. MRF dampers are robust in design and have a wide range of capabilities, such as, the absence of mechanical valving (for flow control) in the damper, insensitivity to impurity penetration, fluid stability, and long operational life. Most importantly, MRF dampers can be considered "fail-safe" devices, i. e., when there is an electrical malfunction in the control system; the damper behaves as a passive device with similar damping capacity to that of OEM. [1]

The purpose of this research is to develop full-scale MRF dampers for a controllable suspension system of heavy vehicles such as HMMWV, track vehicles, Stryker, buses, trailers, etc. The MRF damper is implemented by retrofitting an OEM passive damper of a heavy vehicle with a designed by-pass MR valve. The MRF by-pass damper performance is examined with series of dynamic tests under harmonic motions for different amplitude, frequency and electrical current input. Performance of full-scale

MRF dampers for heavy vehicle rollover protection is also carried out with a dynamic vehicle simulation method.

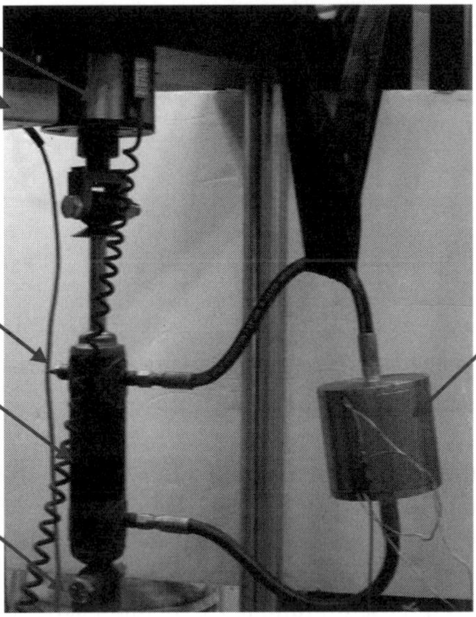

Figure 1. Experimental setup of the MRF by-pass damper.

2 Dynamic Characterization of The MRF By Pass Damper

A two-stage, disk- type, by-pass MR valve is designed and tested to obtain a controllable pressure of over 3.5 MPa. The MRF valve is then used to retrofit a commercial OEM damper of a heavy vehicle (more than 5440 kg in weight) as a controllable MRF bypass damper [3]. Experimental setup for characterizations of the MRF by-pass damper is presented in Figure 1. Both sinusoidal and triangular displacement inputs with various frequencies and electric currents were applied. The damping forces of the MRF damper were measured by a force transducer. A linear variable differential transducer (LVDT) is mounted on the Instron to measure the velocity and displacement.

Typical results of the force-displacement and force-velocity hysteresis loops for the sinusoidal excitation at frequency 1.0Hz subjected to different applied current are presented in Figure 2. As can be seen, the damping forces of the MRF by-pass damper changes significantly with the applied currents.

The maximum damping force versus the peak piston velocity of the sinusoidal excitations is shown in Figure 3. The solid lines are the theoretic predictions for the performance of the MRF damper at maximum input current 1.3A. The dot-dash symbols represent the experimental results of the MRF damper as well as the OEM damper from the dynamic tests. As can be seen, when the MRF damper is fully activated, it can generate about 8000N damping force comparing with 3000N for the off-state (zero current input) and 4200N for the passive OEM damper at piston velocity of 19cm/s.

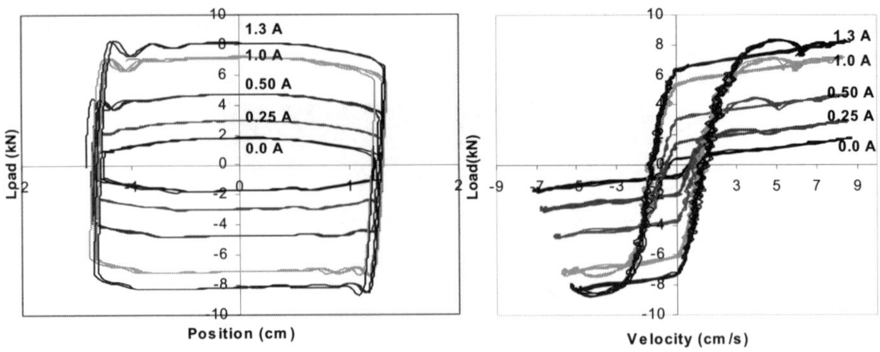

(a) (b)

Figure 2. (a) Force-displacement (b) Force-velocity loops for MRF by-pass damper at 1.0 Hz 2.54 cm peak-to-peak displacement for various electrical current inputs.

In the application, the required maximum controllable force for the heavy vehicle suspension system is 6000 N, the MRF by-pass damper can meet the design criteria and the experimental results are in good agreement with the theoretical prediction. The damper can be considered as a fail-safe device which must have a certain viscous damping performance. From these preliminary dynamic characterization results, it shows that in the absence of the electric current to the damper, the MRF damper can still provide sufficient damping force as that of the OEM damper.

3 Model and Controls

Based on the tests results shown in Figure 3, the model of the developed MRF damper can be expressed as following equations, which is used in the control application.

$$F_{off} = \begin{cases} 9.8258\text{v} - 924.7 & v < -0.8 \\ 122.87\text{v} & -0.8 \le v \le 0.8 \\ 12.563\text{v} + 850.72 & v > 0.8 \end{cases}$$

$$F_{on} = \begin{cases} F_{off} - 6000 & v < 0 \\ F_{off} + 6000 & v \ge 0 \end{cases}$$

where F_{off} (N) and F_{on} (N) are the passive damper force and fully activated damper force, v (mm/s) is the piston velocity of the MRF damper.

To demonstrate the effectiveness of the controlled MRF dampers for protecting the rollover of the heavy vehicle, the lateral acceleration based control system is constructed as Figure 4, and two control methods are applied as follows:

Controlled_MRF4: Fully activate the 4 MRF dampers when absolute value of lateral acceleration is large than 0.1g.

Controlled_MRF2: Fully activate the 2 MRF dampers on the left side when the lateral acceleration is large than 0.1g, and activate 2 MRF dampers on the right side when the lateral acceleration is small than –0.1g.

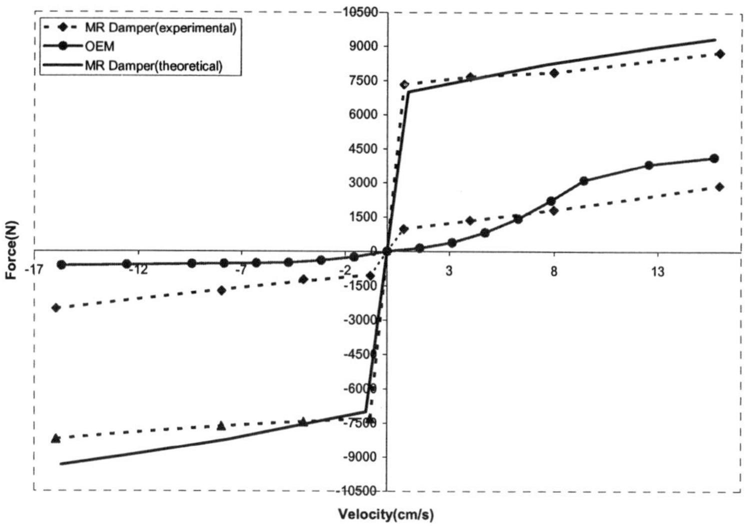

Figure 3. Comparisons among the theoretical results, and the experimental data for the MRF and OEM dampers

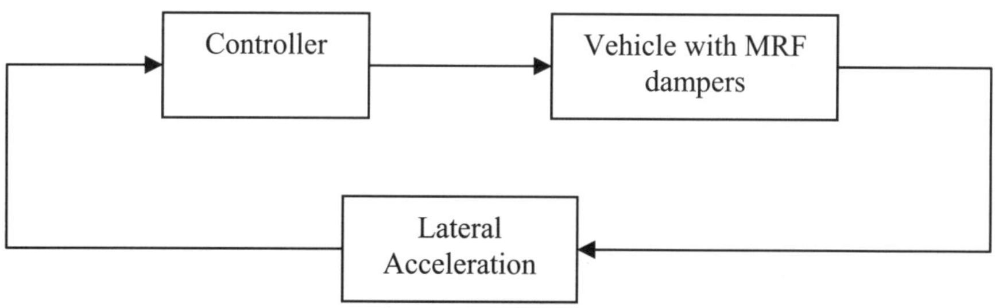

Figure 4. Control schematic of the MRF dampers

4 Simulations

Simulink programs of control methods are developed to integrate to TruckSim software for roll stability simulation. A typical road profile shown in Figure 5 is used in the simulation, which presenting the emergency lane change.

Figure 6 and Figure 7 show roll angle simulation results of the heavy vehicle during the emergency lane change at different vehicle speed. "Baseline" and "MRF" are the simulation with OEM damper and passive MRF damper. "Controlled_MRF4" and "Controlled_MRF2" are the simulation with controlled MRF dampers.

We can see that both "Controlled_MRF4" and "Controlled_MRF2" have better stability performance than un-controlled MRF and OEM dampers by reducing the roll angles. But one of the important finding is that "Controlled_MRF2" has best protection in all cases, which the roll angle of the vehicle reduce more than 45% compared to OEM damper for a range of vehicle speed (from 40mile/h to 70mile/h).

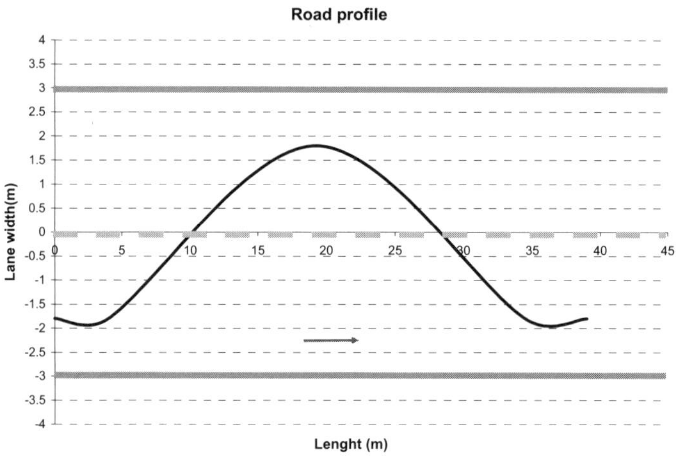

Figure 5. Road profile for simulation

5 Conclusions

Dynamic simulations of the roll over performance of a heavy vehicle incorporated with controlled full-scale MR dampers are carried out using vehicle commercial software by integrating the Simulink control program. Two control methods, "Controlled_MRF4" and "Controlled_MRF2", are examined in the simulations and compared to un-controlled MRF dampers and OEM dampers. Typical emergency lane change situation is selected for the road profile. The results clearly demonstrated that the controlled MR fluid dampers have high capability to prevent roll over during the emergency lane change than un-controlled and OEM dampers, and the control method plays important rule. As conclusion, this study demonstrates the potential safety and handling benefits of application of the controlled MRF damper technology for commercial vehicles.

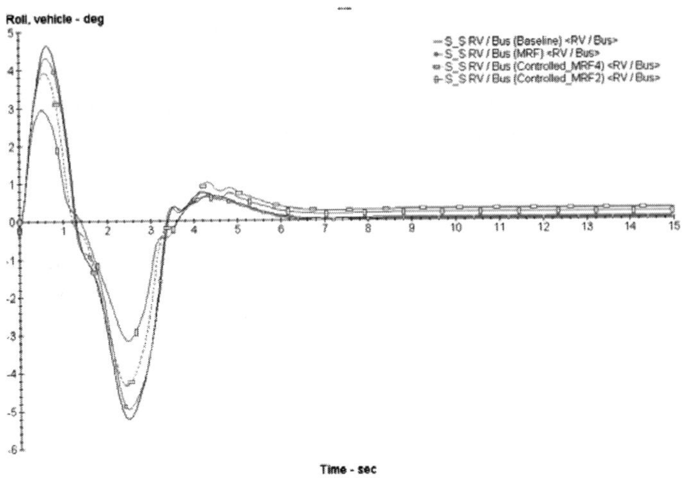

Figure 6. Double lane at 40mile/h

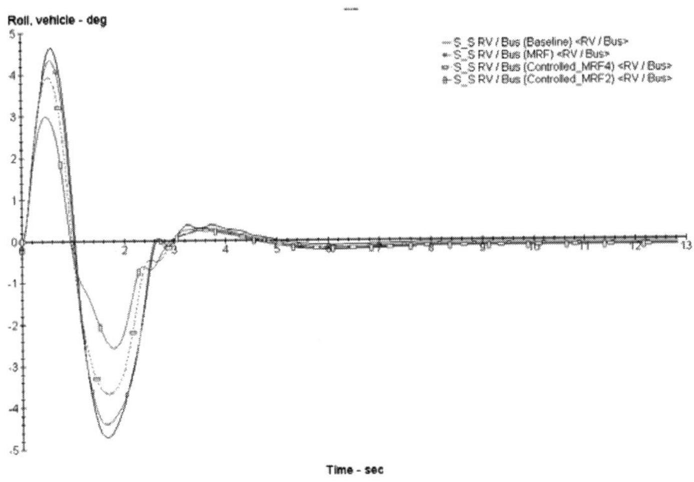

Figure 7. Simulation results at 70mile/h

References

1. Dogruer, U., Gordaninejad, F., and Evrensel, C. A., "A New Magneto-Rheological Fluid Damper for High-Mobility Multi-purpose Wheeled Vehicle (HMMWV)," *Proceedings SPIE Smart Structures and Materials Conference, Damping and Isolation, Ed. By A. S. Agnes and K-W Wang,* Vol. 5052, 2003, pp.198-206.
2. Ivers, D. E., and Miller, L. R.. "Semi-Active Suspension Technology: An Evolutionary View,"
3. Sahin, H., Gordaninejad, F., Wang, X., Evrensel, C., Fuchs, A., Hu, B., "A New Magneto- Rheological Fluid Damper for Semi-Active Suspension of Heavy Vehicles," *ANCRISST Workshop and KISTEC Seminar,* 2005, July 21 – 24.

CHARACTERISTICS OF A TUNED LIQUID DAMPER USING A MAGNETIC FLUID

K. OHNO, M. SHIMODA AND T. SAWADA

Department of Mechanical Engineering, Keio University
3-14-1 Hiyoshi, Kohoku-ku, Yokohama 223-8522, Japan
E-mail: sawada@mech.keio.ac.jp

S. HORIE

JGC Corporation, 2-3-1 Minatomirai, Nishi-ku, Yokohama 220-6001, Japan

The characteristics of a tuned magnetic fluid damper (TMFD) are examined experimentally. Using a tuned mass damper analogy, the concept of effective and non-effective masses is introduced and the qualitative relation among fluid mass ratio, the natural frequency and magnetic field intensity are demonstrated. Effective damping is realized by using a suitable application of magnetic fields based on a frequency response diagram.

1 Introduction

A tuned liquid damper (TLD) is a passive mechanical damper which is based on exciting a liquid to make a sloshing motion and by tuning the sloshing frequency to the system frequency. For example, TLDs have been used in the reduction of the vibration of tall buildings, of long span bridges, and of offshore structures subjected to wind and earthquakes[1].

Modi and Munshi[2] performed a parametric study focused on enhancing the energy dissipation efficiency of a rectangular TLD through introduction of a two-dimensional obstacle. They obtained a significant increase in energy dissipation when using the obstacle. Sun et al.[3] measured liquid motion in a shallow TLD, including rectangular, circular and annular tanks, subject to harmonic base excitation. Using a tuned mass damper (TMD) analogy, they calibrated the TLD parameters from experimental results.

These types of TLDs, however, can not control the vibrations actively. To improve the performance of TLDs, the use of functional fluids has been introduced so that electric or magnetic forces can be used as additional forces. Truong and Semercigil[4] used an electro-rheological fluid as the working fluid of a sloshing absorber and performed simple numerical simulations. Abé, Fujino and Kimura[5] proposed a new active TLD using a magnetic fluid and the performance of the proposed active TLD was verified experimentally using a two-story building model. Ohira, Houda and Sawada[6] developed an analytical model for the tuned magnetic fluid damper (TMFD), expanding on the study of Abé et al. The mechanism of the TMFD is shown in Figure 1. The TMFD dampens oscillations by means of a magnetic force in the horizontal direction, and an intensity of an applied magnetic field can also be changed according to the phase of the oscillating structure. Ohira et al. demonstrated the potential of a TMFD using a cylindrical container and four electromagnets. They pointed out that the performance of the TMFD was improved by switching the applied magnetic field at a point near the resonant frequency of the undamped structure.

In these TMFDs, the damping effect is considered to be caused by the magnetic force. We have measured the magnetic force directly in order to examine the effect of the magnetic force on TMFD. However, we found that the order of the magnetic force is considerably smaller than that of the inertia force of the structure. It is believed that the magnetic force does not play an important role in damping directly. In the present paper, we introduce a new concept, suggesting that the liquid mass is divided into two parts, i.e. an effective mass and a non-effective mass as shown in Figure 2. In particular, we focus on a change of natural frequencies owing to the change in ratio of effective and non-effective masses. When the fluid mass ratio is changed by the magnetic field, a natural frequency of a TMFD is also changed. This means that a natural frequency of the magnetic fluid sloshing can be varied by a vertical magnetic field.

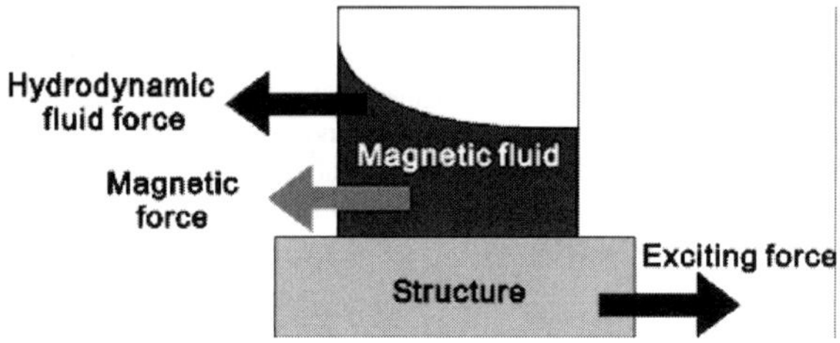

Figure 1. Original concept of a TMFD

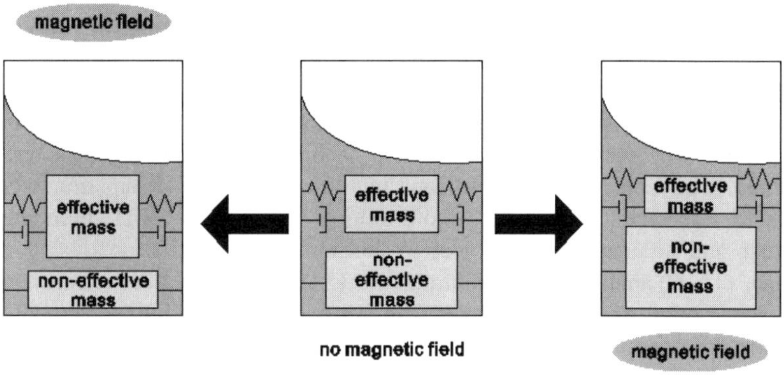

Figure 2. Effective and non-effective masses of a TMFD

2 Sloshing of Magnetic fluids

Consider a cylindrical container partially filled with a magnetic fluid subjected to sinusoidal lateral excitation. The unsteady Bernoulli equation applied to a magnetic field[7] is given by

$$\rho\frac{\partial\phi}{\partial t}+\frac{1}{2}\rho|v|^2+p+\rho gz-\mu_0\int_0^H MdH=\rho X_0\omega^2 r\sin\omega t \tag{1}$$

where ρ is the density, ϕ is the velocity potential, z is the distance from a free surface, μ_0 is the magnetic permeability of vacuum, M is the magnetization, H is the magnetic field, and ω is the angular frequency. If the magnetization and magnetic field are assumed to be parallel and a susceptibility of the magnetic fluid is constant, the magnetic term of Eq.(1) becomes

$$\mu_0 \int_0^H MdH = \frac{\mu_0}{2} \chi H(z)^2 = \frac{\mu_0}{2} \chi H_0^{\ 2} e^{-2\alpha(z+h)} \tag{2}$$

where χ is the susceptibility, h is the depth of the magnetic fluid, H_0 is the magnetic field intensity at the bottom of the container, $z = -h$, and α is a constant determined from experimental measurements of the magnetic field. The natural frequency of the magnetic fluid calculated from Eq. (1) and a continuity equation is written in the form

$$f_{mn} = \frac{1}{2\pi} \sqrt{\frac{\varepsilon_{mn} g^*}{R} \tanh \frac{\varepsilon_{mn} h}{R}} \tag{3}$$

where R is the radius of the container and the suffixes m and n indicate oscillation modes. g^* is an apparent gravitational acceleration and is given by

$$g^* = g + \frac{\alpha \mu_0 \chi H_0^{\ 2} e^{-2\alpha h}}{\rho} \tag{4}$$

The sign of α corresponds to the direction of the magnetic field. For example, when the magnetic field is applied from the bottom of the container, g^* increases because $\alpha > 0$.

3 Analytical Model of TMFD

Figure 3 shows a simple analytical model of a TLD based on an analogy with the TMD. Here X_0 and X_s are displacements of the structure and the vibrating table, respectively. m_{fe} is the effective mass and m_{fie} is the non-effective mass (which is regarded as a part of the structure). The ratio of m_{fe} to m_{fie} can be changed according to the applied magnetic field. Thus the total mass of the liquid is $m_f = m_{fe} + m_{fie}$ and the total mass of the structure model becomes $m_s + m_{fie}$ where m_s is the net mass of the structure model including the container for the liquid and other components. k_s and c_s are stiffness and damping in relation to the structure model, respectively. Also, x_f, k_f, c_f are the apparent displacement, stiffness, damping of the effective mass, respectively.

In the case of a TMFD, the effective mass may be changed by changing the magnetic field, and the natural frequency of the fluid sloshing will also change. Equations of motion for the TMFD are given by

$$(m_s + m_{fie})\ddot{x}_s = -k_s(x_s - X_0 e^{i\omega t}) - c_s(\dot{x}_s - i\omega X_0 e^{i\omega t}) + k_f(x_f - x_s) + c_f(\dot{x}_f - \dot{x}_s) \tag{5}$$

$$m_{fe}\ddot{x}_f = -k_f(x_f - x_s) - c_f(\dot{x}_f - \dot{x}_s) \tag{6}$$

The best damping can be obtained by changing an exciting frequency.

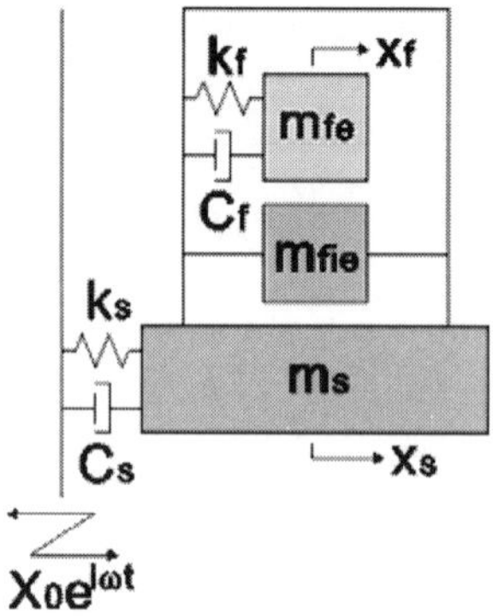

Figure 3. Analytical model of TLD

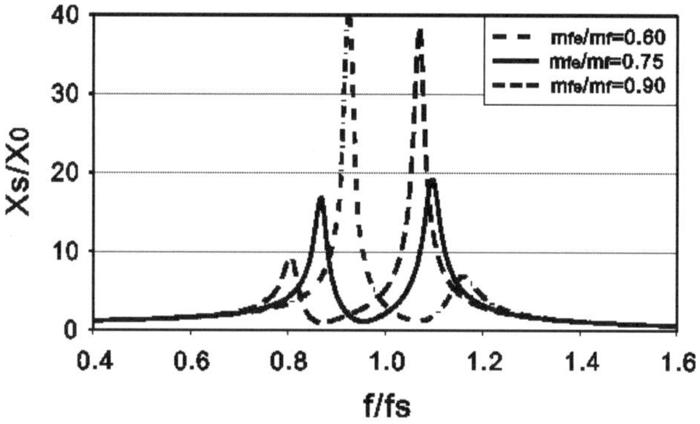

Figure 4. Example of frequency response based on numerical simulation

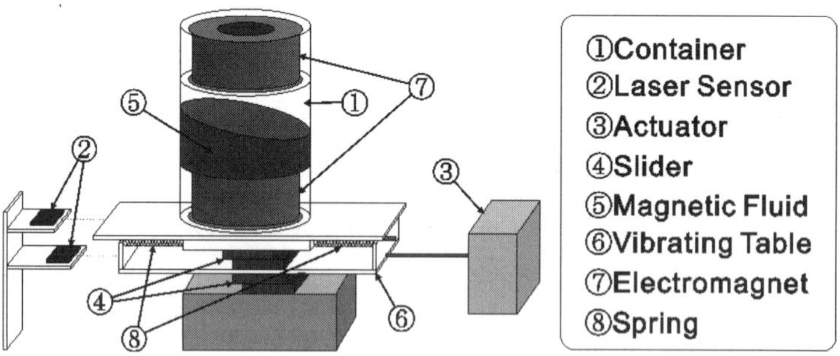

Figure 5. Schematic diagram

Using this analytical model, frequency responses of a TMFD can be obtained by numerical simulation. Figure 4 shows an example of a simulation showing the results of changing the fluid mass ratio. The values of parameters used for the calculation are as follows: m_f / m_s = 0.0745, $\sqrt{(k_f/m_f)/(k_s/m_s)}$ = 0.817, $c_s / \sqrt{m_s k_s}$ = 0, $c_f / \sqrt{m_f k_f}$ = 0.048. From Figure 4 it is clear that even a small change in the fluid mass ratio m_{fe} / m_f produces a significant change on frequency response.

4 Excitation Experiment

4.1 Experimental Apparatus

Figure 5 shows a schematic diagram of the experimental apparatus. The TMFD is a cylindrical container with an inner diameter of 144 mm, which contains a water-based magnetic fluid. Its depth and mass are 37 mm and 0.76 kg, respectively. Two electromagnets are placed on the bottom and top walls and a magnetic field can be applied to the magnetic fluid from topside or downside. The TMFD is placed on a roller table supported on a vibrating table. Displacements of the roller table (considered as the structure in the model) and the vibrating table are measured by laser displacement transducers. The total mass of the structure with two electromagnets is 16.0 kg.

4.2 Natural Frequency and Frequency Response

Excitation experiments were carried out with an excitation amplitude of X_0 = 0.15 mm. The current supplied to the electromagnets is varied from 0.0 A to 5.0 A measuring every 0.5A and the exciting frequency is varied from 1.0 Hz to 3.4 Hz. The maximum magnetic field intensity at the barycentric position of the magnetic fluid is 61 mT and is -10 mT in the case of the bottom and the upper electromagnets, respectively. Here, a minus magnetic field corresponds to a magnetic filed applied from the upper electromagnet. The natural frequency with no magnetic field is f_s = 2.13 Hz. Figure 6 shows the change in the natural frequency with applied magnetic field. Because the apparent gravity increases with the intensity of the magnetic field as indicated in Eqs. (3) and (4), the natural frequency also increases with magnetic field intensity.

The frequency responses under three kinds of magnetic field are shown in Figure 7. With increasing magnetic field intensity, the best damping point moves to a higher frequency region and the right peak decreases. This result shows a similar tendency to the results obtained by the numerical simulation as shown in Figure 4. That is, a decrease of m_{fe} / m_f corresponds to an increase of the magnetic field intensity.

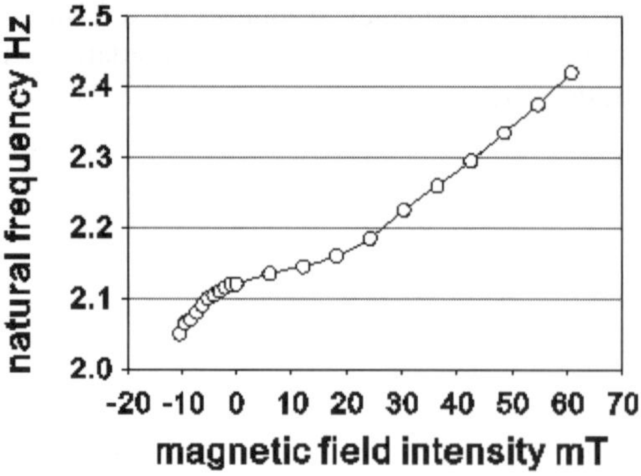

Figure 6. Natural frequency with magnetic field

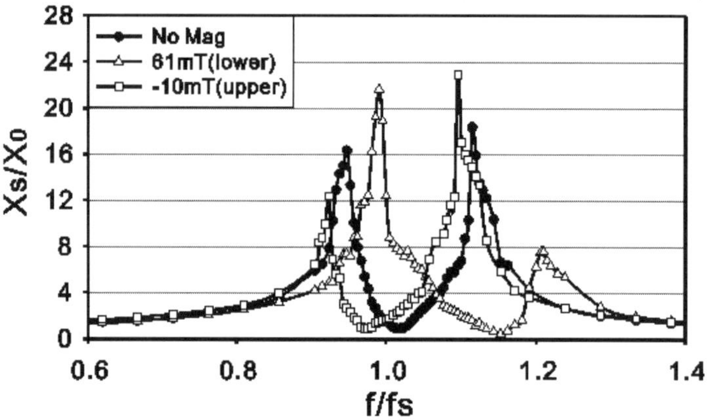

Figure 7. Frequency response

5 Active Control

Based on the experimental results shown in Figure 7, in order to reduce larger vibration amplitudes X_s / X_0, we attempted to use several kinds of magnetic field intensity as shown in Table 1. Here again, the minus supplied current/magnetic field corresponds to a magnetic field caused by the upper electromagnet. In the region of $0.976 \leq f/f_s \leq 1.143$, the trial equations based on the experimental results are used while constant magnetic fields are used in the other regions.

Figure 8 shows the frequency responses both when under active control and with no magnetic field. It is clear that effective damping is obtained by controlling the magnetic field such that the vibrating peaks are decreased. In the low frequency region, even the lowest reduction rate achieves a good reduction of about 35 %. In the wide range of $0.96 \leq f/f_s \leq 1.15$, a remarkable damping effect is observed. This effect can not be produced by a normal TLD.

Table 1. Magnetic field conditions for an active TMFD

f/f_s	Current A	Magnetic field mT
$f/f_s \le 0.875$	0.0	No Mag
$0.875 < f/f_s \le 0.929$	5.0	61
$0.929 < f/f_s < 0.976$	-5.0	-10
$0.976 \le f/f_s \le 1.143$	$342.165 f_s^3 - 2380.62 f_s^2 + 5529.56 f_s - 4284$	Control
$1.143 < f/f_s < 1.190$	5.0	61
$f/f_s \ge 1.190$	0.0	No Mag

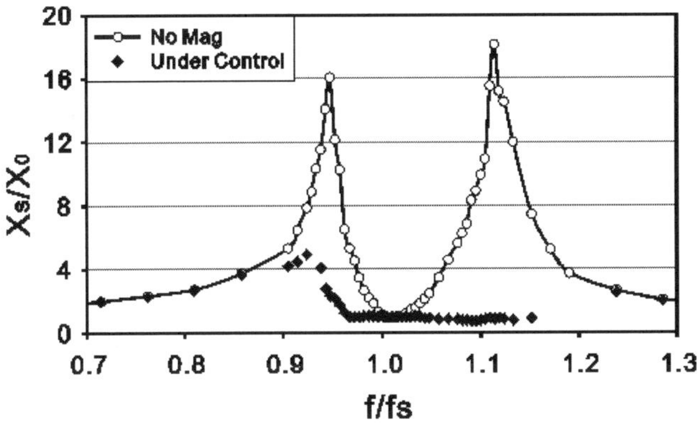

Figure 8. Frequency responses under the active control

6 Concluding Remarks

The damping characteristics of a TMFD are examined experimentally and it is determined that the characteristics are influenced by a change in the natural frequency of the magnetic fluid sloshing. The best damping points are obtained over a wide range of exciting frequencies. It is shown that active control of a TMFD can be realized by using suitable magnetic field conditions.

References

1. Ibrahim R. A., *Liquid Sloshing Dynamics*, (Cambridge University Press, Cambridge, 2005) Chap.10.
2. Modi V. J. and Munshi S. R., An efficient liquid sloshing damper for vibration control, *J. Fluids Struct.*, **12** (1998) pp.1055-1071.

3. Sun L. M. *et al.*, The properties of tuned liquid dampers using a TMD analogy, *Earthquake Eng. Struct. Dyn.*, **24** (1995) pp.967-976.
4. Truong T. D. and Semercigil S. E., A Variable damping tuned absorber with electro-rheological fluid for transient resonance of light structures, *J. Sound Vib.*, **239** (2001) pp.891-905.
5. Abé, M., Fujino, Y. and Kimura, S., Active tuned liquid damper (TLD) with magnetic fluid, *Proc. SPIE*, **3329** (1998) pp. 620-623.
6. Ohira, Y., Houda, H. and Sawada, T., Effect of magnetic field on a tuned liquid damper using a magnetic fluid, Int. J. Appl. Electromag. Mech., 13 (2001/2002) pp.71-78.
7. Rosensweig R. E., *Ferrohydrodynamics*, (Cambridge University Press, Cambridge, 1985).

SEMI-ACTIVE CONTROL OF SUSPENSION SYSTEM USING A SERIAL MULTIELECTRODE ELECTRORHEOLOGICAL DAMPER

W. H. KUO
Department of Automation Engineering, Tung Nan Institute of Technology, Taipei, Taiwan 22211
E-Mail:whkuo@mail.tnit.edu.tw

T. N. WU* J. GUO* M. H. CHIANG** Y. N. CHEN* Y. P. WANG
**Department of Engineering Science and Ocean Engineering, National Taiwan University, Taipei, Taiwan 10617*
*** Institute of Automaton and Control, National Taiwan University of Science and Technology, Taiwan 10617*
Department of Automation Engineering, Tung Nan Institute of Technology, Taipei, Taiwan 22211

This study is to explore the controllable ratio of damping characteristics based on a serial multielectrode electrorheological(SMER) flow mode damper. The SMER damper is applied to the two-degree-of-freedom model of a suspension system. Semi-active suspension control system equipped with the SMER damper and the fuzzy controller is designed for the evaluation of the performance of the SMER damper. The SMER damper is designed, fabricated and tested. The design method for ER dampers based on the requirements of the system damper such as controllable damping force, controllable ratio and the ratio of rebound damping force and compression damping force is described. The experimental results of the semi-active control system have demonstrated that the displacement and the acceleration power spectral density are significantly reduced by applying the semi-active control on the SMER dampers. The simulation and experimental time and frequency response of the displacement and the acceleration indicate the effectiveness of the design of the SMER damper.

1 Introduction

Electrorheological (ER) fluids can change their mechanical properties, such as viscosity and yielding shear stress, through applied electric field. Such a change in properties is fast, field-dependent, and reversible[1]. Many new mechanical and hydraulic devices in the control technology, such as clutch, brake, engine mount, hydraulic control valve, damper, structure vibration control, and electro-mechanical coupling devices [2,3,4], can be designed with specific system characteristics adjustable by using different electric fields to meet performance requirements.

The ER damper is a device of controllable damping force or damping coefficient using ER fluid, which can be applied to control vibration system or semi-active suspension system. For instances, Wong et al. [5] and Wu et al. [6] used a parallel multielectrode ER damper as well as various composite control strategies for improving both ride comfort and road holding. Weyenberg et al. [7] applied ER dampers to the semi-active suspension systems of vehicles controlled by modified sky-hook algorithm, and directly installed accelerometers and position sensors in the vehicle to measure the feedback signals of the vehicles. The results showed that the level of vehicle body acceleration and the tire load variation were improved. Choi et al. [8] proposed a cylindrical ER damper for a passenger car and showed that a continuously variable ER damper can be satisfactorily used for the feedback control of desired damping forces. Its controllability of damping force was improved by implementing a skyhook controller. Suh et al. [9] used a sky-hook control algorithm to improve the riding quality through computer simulation, and then the ER damper was installed in the passenger car to test. The results of experiment showed that the RMS acceleration of control mode is generally better than that of passive mode. Choi et al. [10] adopt the hardware-in-the-loop simulation method and used a sliding mode controller. The simulated results showed that the full-car ER suspension system is very effective for vibration isolation. Choi et al. [11] evaluated the performance characteristics of a semi-active ER suspension system associated with skyhook controller and four ER shock absorbers to a full car. The results of experiment showed that the semi-active system can be effectively employed to the passenger vehicle improving both riding comfort and steering stability.

From these previous studies, it is obvious that the most of the ER dampers are long-stroke (long-electrode) dampers, parallel multi-electrode flow mode ER dampers and multi-electrode shear mode ER dampers. The ER damper developed in this paper is a SMER damper[12], and is suitable to short stroke damper with high damping coefficient. The SMER damper consists of a set of concentric cylindrical electrodes. All electrodes are electrically parallel and the flow paths are arranged in hydraulically series. Hence the developed ER damper has short stroke and better ER effect. This paper demonstrates the effectiveness of the controlled vibration suppression using the SMER damper through experiment. The SMER damper is applied to the two-degree-of-freedom model of vibration system for evaluating the performance of the suppressive vibration system using fuzzy controller. The SMER damper is shown to be effective to control the vibration motion experimentally.

2 The Characteristics of the SMER Damper

The 3-layer SMER damper with the electrode length $L_e = 100mm$ and the electrode gap $h_e = 0.6mm$ has been designed and manufactured. The SMER damper is redesigned for the requirements of the system damper such as controllable damping force, controllable ratio and the ratio of rebound damping force and compression damping force, as shown in Figure 2.1(a). Simulated and experimentally measured characteristics of the SMER damper are shown in Figure 2.1(b).

In order to change the relative amounts of compression and rebound damping force, the piston of the SMER damper makes the damping force of compression stroke less than that of rebound stroke. The check valve mounted on the piston results in free flow from the compression chamber to rebound chamber, and the diameter of check valve can adjust the relative amounts of compression and rebound damping force in SMER dampers, as shown in Figure 2.2(a). The term, F_{c0} / F_{r0}, indicates the damping force at compression and rebound stroke without electric field.

ER fluids are suspension mixed with fine particles and non-conducting liquids. The particles are randomly suspended in ER fluids without electric field. As an electric field is applied to ER fluids, particles chained between electrodes are generated to resist the flow. The formulating and releasing processes of the particle chains with electric field take some time for the movement of the particles, such that the response time is delayed. Thus, the time delay has to be taken into consideration in designing the control vibration system for real time control. The response time of ER fluid is required to be tested firstly for controlling vibration systems.

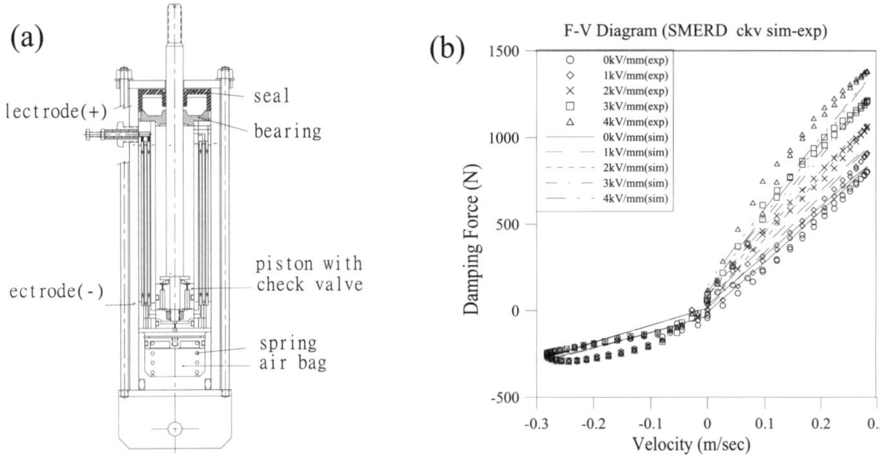

Figure 2.1 (a)The 3-layer SMER damper with check valves. (b) Simulated and experimentally measured characteristics of the SMER damper

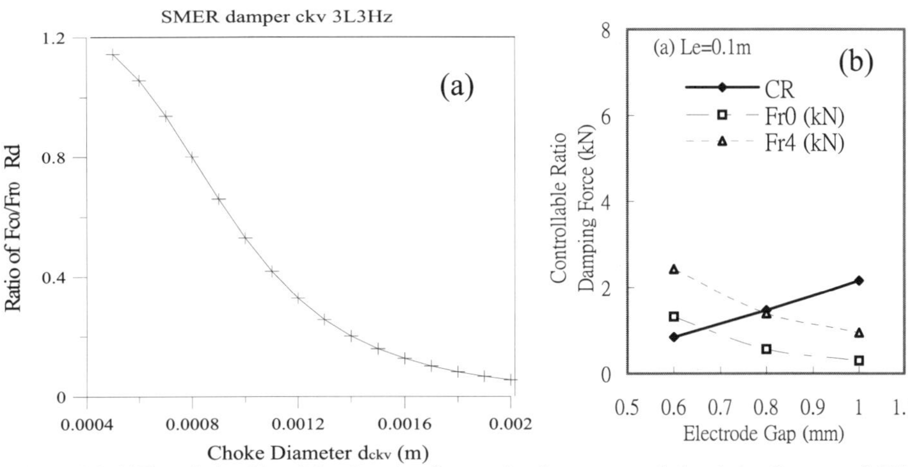

Figure 2.2 (a)The relationship of the damping force ratio of F_{c0}/F_{r0} and the choke diameter of SMER damper with the check valves. (b)The relationship of controllable ratio(CR), damping force and electrode gap of SMER damper with the check valve, Le=electrode length. $CR=(F_{r4}-F_{r0})/F_{r0}$

The SMER damper experiment is performed by the excitation of triangular displacement waves, which develop constant velocity for testing step response. The input electric field is from 0 to 4 kV/mm of square wave, and a peak-to-peak displacement is 0.015 m. The test results are shown in Figure 2.3.

The time response characteristics can be evaluated with respect to the electric field and the excited motion. The motion of the SMER damper is excited by the triangular wave with the displacement of ±7.5 mm and 1 Hz frequency, as shown in Figure 2.3(a) Therefore, the constant damping force at a constant velocity 0.03 m/sec, as shown in Figure 2.3(b), can be generated. The 5 Hz square waves from 0 to 4 kV/mm are used for the control electric field, as shown in Figure 2.3(c). The time response characteristics of the SMER damper to square wave control electric field is shown in Figure 2.3(b). From Figure 2.3(b), the response time about 65ms is evaluated. It depends on not only the response characteristics of the ER fluid itself but also the size of the ER damper.

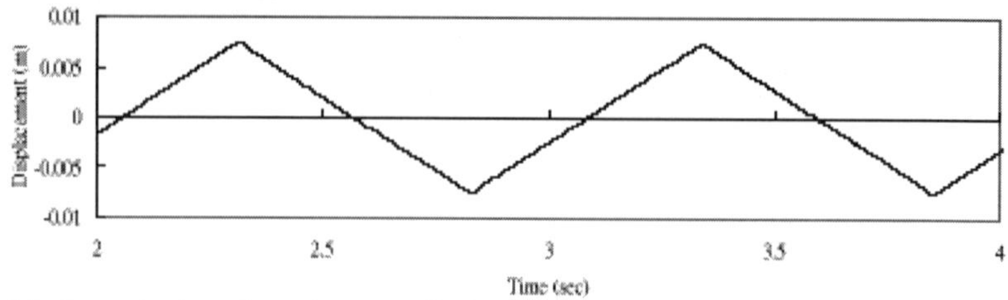

(a) The excitation triangular wave displacement ±7.5 mm at 1 Hz frequency

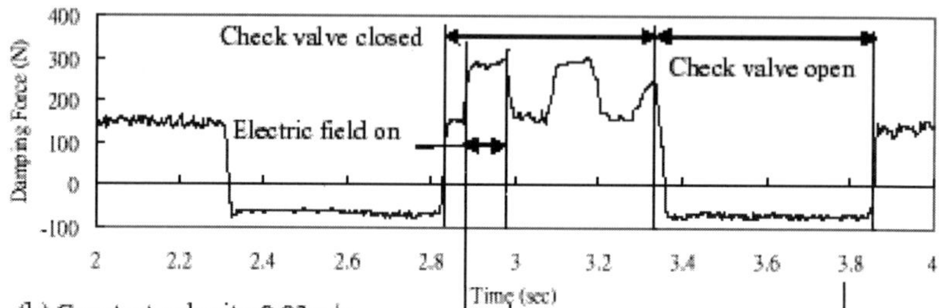

(b) Constant velocity 0.03m/s

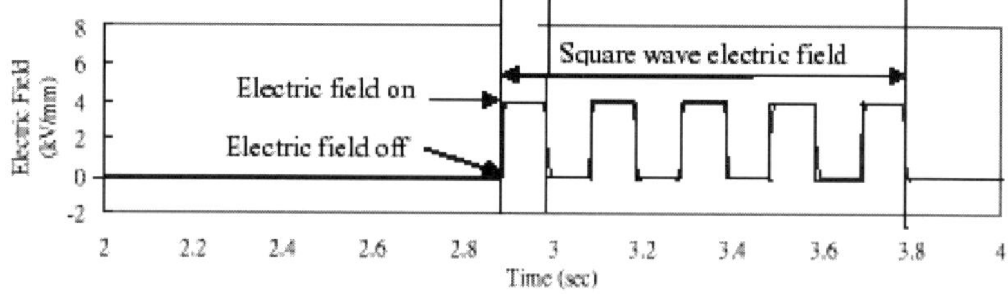

(c) The square wave control electric field 0 to 4 kV/mm at 5 Hz frequencies

Figure 2.3 Time response characteristics of the SMER damper to square wave control electric field

3. Two-degree-of-freedom vibration system

A semi-active suspension system with appropriate control strategy should provide improved ride comfort and road holding for a given suspension. Ride comfort may be evaluated by the response of the sprung mass to the excitation from the ground. Usually, the transmissibility ratio is used as a basis for assessing the vibration isolation characteristics. Road holding is related to the variation of the normal force between the tire and the road during vibration. It can be represented by the dynamic tire deflection or the displacement of the tire centre relative to the road profile. A typical schematic of semi-active suspension system as shown in Figure 3.1 represents the two-degree-of freedom (two-DOF) model of vibration system that described the quarter part of a vehicle. The system consists of a two-DOF model, input signal, and a controller. The SMER damper and the suspension spring (k_2) are mounted between the sprung mass and the unsprung mass. The suspension spring is considered as a linear spring and the SMER damper with check valve is used. The tire is lumped to be a linear spring element (k_1). The input signals with sine waves serve to excite the sprung mass through the tire and suspension system. The velocity $\dot{x}_2$ of sprung mass and relative velocity $\dot{x}_2 - \dot{x}_1$

between sprung and unsprung mass are assumed measurable, and x_0 is the given vertical displacement. The sprung and unsprung masses are considered as rigid body. Therefore, the equations of motion for this two-DOF model of vibration system can be derived as.

$$m_1\ddot{x}_1 - C(E)(\dot{x}_2 - \dot{x}_1) - k_2(x_2 - x_1) + k_1(x_1 - x_0) = 0 \tag{3.1}$$

$$m_2\ddot{x}_2 + C(E)(\dot{x}_2 - \dot{x}_1) + k_2(x_2 - x_1) = 0 \tag{3.2}$$

where m_1 indicates the sprung mass; m_2 is the unsprung mass; k_1 indicates the equivalent spring stiffness of the tire; k_2 is the suspension spring stiffness. $C(E)$ which is a function of the electric field is the damping coefficient of the SMFER damper. Noted that the static equilibrium positions of the sprung and unsprung masses are consider as the origins for the co-ordinates.

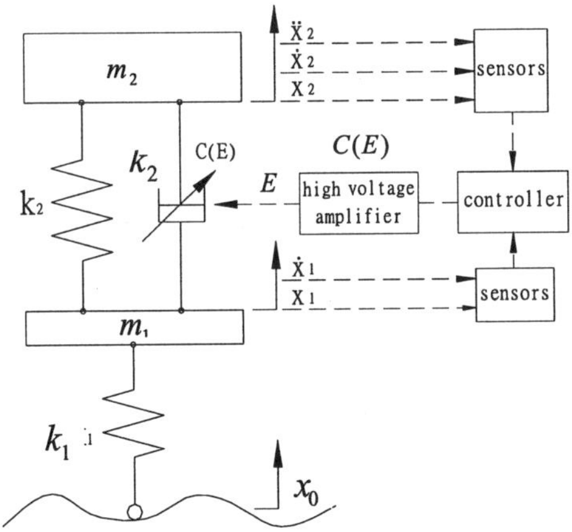

Figure 3.1 Two-degree-of-freedom model of vibration system

3.1 Control Strategy

The control strategy is developed based on the "sky-hook" concept. The objective is to reduce the sprung mass acceleration by modulating the damping force in accordance with operating conditions. This control strategy can be described as follows:

If $\dot{x}_2(\dot{x}_2 - \dot{x}_1) > 0$, then the maximum damping is required, and

If $\dot{x}_2(\dot{x}_2 - \dot{x}_1) < 0$, then the minimum damping is required.

This strategy indicates that the maximum damping force should be applied to reduce the sprung mass acceleration if the relative velocity of the sprung mass with respect to the unsprung mass is in the same direction as that of the sprung mass velocity. On the other hand, if the two velocities are in the opposite direction, the damping force should be at a minimum for minimizing the acceleration of the sprung mass. This control strategy requires the measurement of the absolute velocity of the sprung mass. Although the accurate measurement of the absolute vibration velocity of the sprung mass on a moving vehicle is very difficult to achieve, it is easy to be realized in the laboratory experiment in order to evaluate the performance of damper and controller.

The "sky-hook" concept is an on-off control strategy. In order to determine the continuous control input voltage, the fuzzy control algorithm is adopted. The following two fuzzy variables: the vertical velocity of sprung mass $\dot{x}_2 = SP$ and relative velocity between sprung and unsprung mass $\dot{x}_2 - \dot{x}_1 = SR$ are chosen. The following conditional statements are established.

If SP is negative large and SR is negative large, then e is large.
If SP is negative large and SR is positive large, then e is small.
If SP is positive large and SR is negative large, then e is small.
If SP is positive large and SR is positive large, then e is large.
....

(3.3)

Table 3.1 Fuzzy control rule

SR \ SP	NL	NM	NS	ZO	PS	PM	PL
NL	PL	PL	PL	PL	PM	PS	ZO
NM	PL	PL	PL	PM	PS	ZO	NS
NS	PL	PL	PM	PS	ZO	NS	NM
ZO	PM	PM	PS	ZO	NS	NM	NM
PS	PM	PS	ZO	NS	NM	NL	NL
PM	PS	ZO	NS	NM	NL	NL	NL
PL	ZO	NS	NM	NL	NL	NL	NL

Now, the linguistic input variables are determined to describe SP and SR as follows:

$$\widetilde{SP} = \{NL, NM, NS, ZO, PS, PM, PL\}$$
$$\widetilde{SR} = \{NL, NM, NS, ZO, PS, PM, PL\} \tag{3.4}$$

where NL =negative large, NM =negative medium, NS =negative small, ZO =zero, PS=positive small, PM=positive medium, PL=positive large. Similarly, a linguistic output variable is defined to describe the weighting parameter as follows:

$$\tilde{e} = \{NL, NS, ZO, PS, PL\} \tag{3.5}$$

where NL, NS, ZO, PS, PL are fuzzy variables of $\tilde{e}$. Control rules of Equation (3.3) are represented in Table 3.1 in terms of a 'look up' table, and these rules can be inferred from the center-of-gravity method.

The basic configuration of the fuzzy logic consists of fuzzification interface, decision-making logic and defuzzification interface, as shown in Figure 3.2. In the fuzzification interface, SP and SR are estimated, and these are modified into the linguistic values $\widetilde{SP}$ and $\widetilde{SR}$. In the defuzzification interface, fuzzy output e decided for the decision-making logic is changed into a numerical value for the purpose using real input. For the computer simulation in a subsequent section, the membership functions of the fuzzy input variables for SP and SR, and output parameter shown in Figure 3.3 are used. The two-DOF model of vibration system can be expressed in block diagram as shown in Figure 3.4. Figure 3.5 shows the Simulink block diagram of fuzzy system. These block diagram can be solved using MATLAB/Simulink simulation software [13]. The simulation of control responses with and without the fuzzy controller is shown in Figure 3.6.

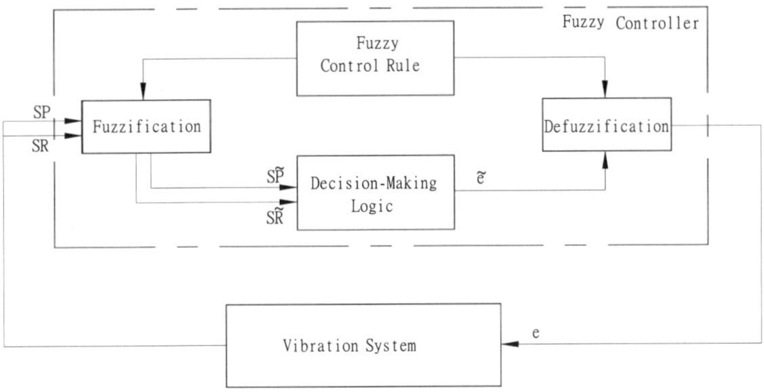

Figure 3.2 Block diagram of fuzzy control system

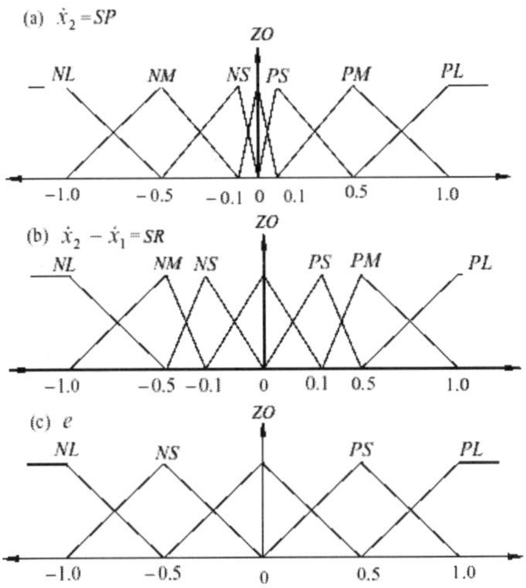

Figure 3.3 Fuzzy variables and membership functions of the fuzzy controller

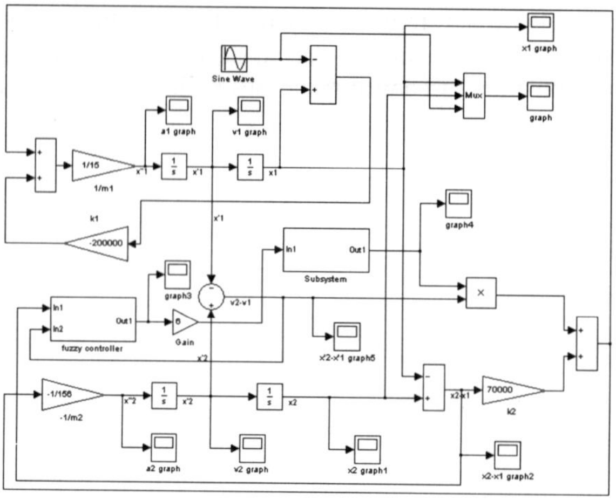

Figure 3.4 Simulink block diagram of two-DOF vibration system

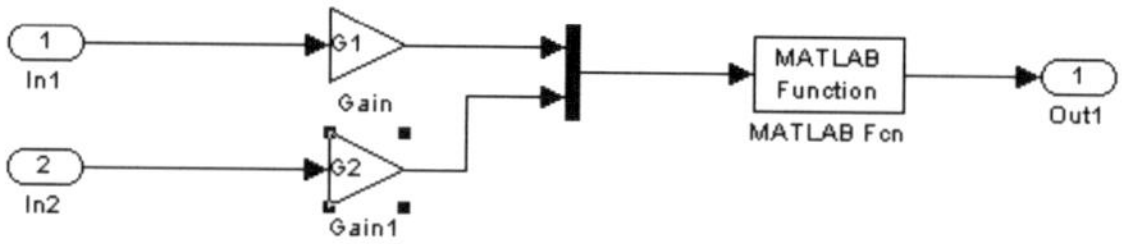

Figure 3.5 Simulink block diagram of the fuzzy controller

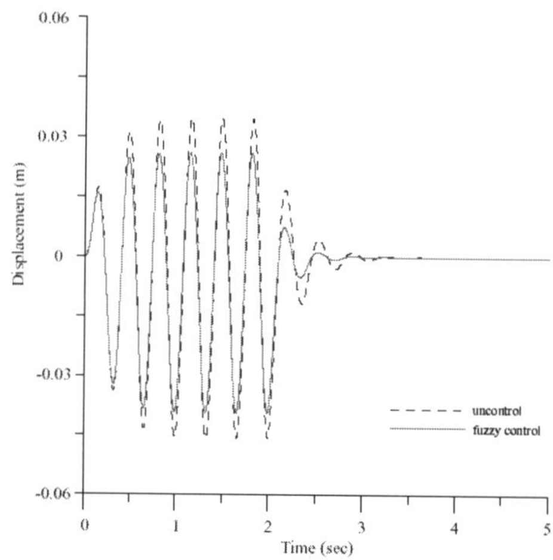

Figure 3.6 Simulation of control responses with and without the fuzzy controller

3.2 Experimental Setup

The schematic diagram of the two-DOF vibration control testing system is shown in Figure 3.7. The parameter values of the spring constant is $k_1 = 2\times10^5\ N/m$ and the unsprung mass is $m_1 = 15\ kg$. The parameter values of two-DOF model are shown in Table 3.2. The photograph of the testing system is shown in Figure 3.8.

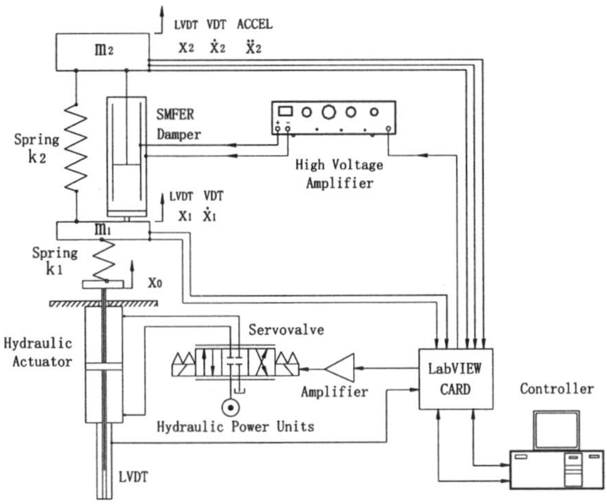

Figure 3.7 The schematic diagram of the two-DOF vibration control testing system

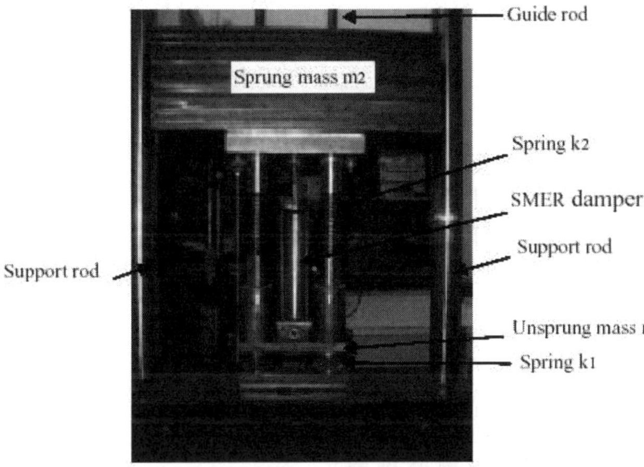

Figure 3.8 The photograph of two-DOF vibration control testing system

Table 3.2 Parameter values of the two-DOF of vibration system model

Parameters	Symbol	Unit	Numerical values
Sprung mass	m_2	kg	156
Unsprung mass	m_1	kg	15
Spring constant of suspension	k_2	N/m	70000
Spring constant	k_1	N/m	200000
Damping coefficients	$C_r(E)$	$N\cdot s/m$	2724~4710
(E=0~4 kV/mm)	C_c	$N\cdot s/m$	1030

$C_r(E)$: the rebound stroke, C_c: the compression stroke.

3.3 Experimental Results and Discussions

Experimental results with and without the fuzzy controller are shown in Figure 3.9 where the exciting amplitude xo=30 mm. Figure 3.9(a) shows the amplitude response of exciting frequency 3Hz with amplitude 30mm. Figure 3.9(b) shows the acceleration response of exciting frequency 3Hz with amplitude 30mm. Figure 3.9(c) shows the electric field input to the SMER damper. The output electric field is set within the range 0 to 3 kV/mm in order to avoid the dielectric breakdown. It was verified experimentally data that the proposed sky-hook fuzzy controller provides satisfactory reduction in vibration amplitude in both of the two cases.

Figure 3.10(a) shows the frequency responses obtained from power spectral density (PSD) of the vertical displacement of the sprung mass and Figure 3.10(b) shows the frequency responses obtained from power spectral density (PSD) of the vertical acceleration of the sprung mass. It is clearly observed that the acceleration PSD are substantially reduced over than 50% by applying the fuzzy controller, but the vertical displacement PSD are reduced about 35%.

In order to improve the experimental system setup in the future, the Coulomb frictional force exerting on guide rods by the sprung mass must be reduced in this two-DOF vibration control testing system. The Coulomb frictional force of the current system is over 120N due to the poor parallelism of two guide rods.

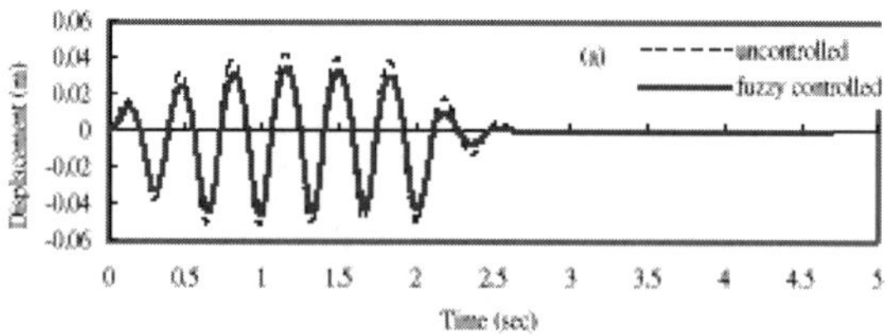

(a) Amplitude responses

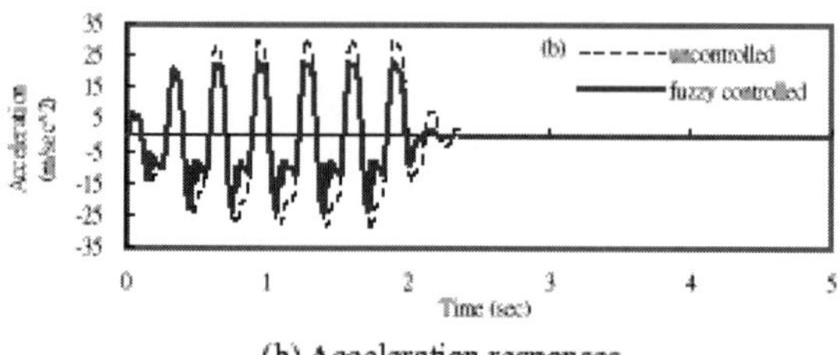

(b) Acceleration responses

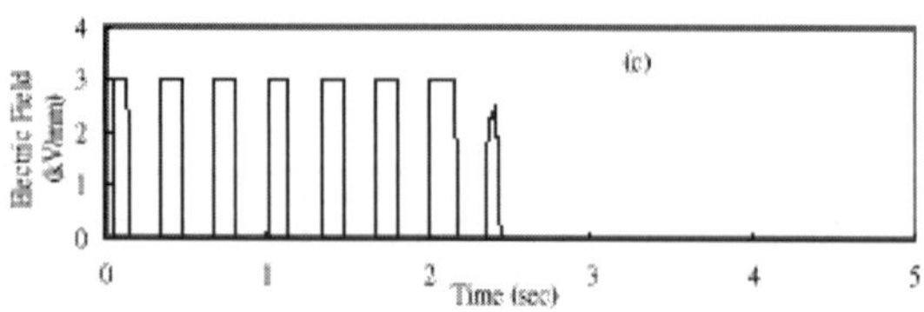

(c) The electric field input to the SMFER damper

Figure 3.9 Experimental results with and without the fuzzy controller (exciting frequency 3 Hz and exciting amplitude xo=30 mm)

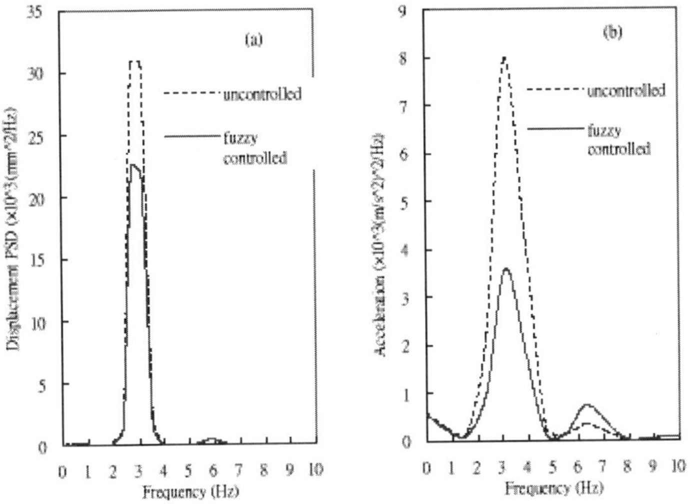

Figure 3.10 Frequency responses of the vibration control system
(a)Vertical displacement PSD with amplitude 30 mm.
(b)Vertical acceleration PSD with amplitude 30 mm.

4 Summary

This paper focuses on the performance evaluation of the SMER damper applied to the two-DOF vibration system. The control algorithm of the vibration system is developed using a fuzzy controller for two-DOF vibration system. The simulation and experimental time and frequency responses of the displacement and the acceleration indicate the effectiveness of the developed SMER damper. It is verified that the developed SMER damper with changeable number of electrode and electrode gap can produce a relatively high damping force with a short-stroke damper as well as low ER effect of ER fluid.

Experimental results of a vibration control system using a 3-layer SMER damper and controllers demonstrate that the displacement and acceleration power spectral density are substantially reduced to about 50% under input sine wave excitations. The design and applications as well as the optimal control algorithms of SMER dampers for road vehicle suspension system will be the future research topics.

References

1. Block H. and Kelly ,J. P., Electro-rheology, Journal of Physics D: *Apply Physics* 21, (1988) pp. 1661-1677.
2. Weiss K. D. , Carlson J. D. and Coulter J. P., Material Aspects of Electrorheological Systems, *Journal of Intelligent Material Systems and Structures*, Vol.4, (1993) pp. 13-34.
3. Brooks D. A., Devices Using Electro-Rheological Fluids, Proceedings of the 2nd International Conference on ER Fluids, Raleigh, NC, (1989) pp. 371-401.
4. Tanaka Y. and Gofuku A., Development and Analysis of an ER Fluid Pressure Control Valve, *Mechatronics*, Vol.7, No.4, (1977) pp. 317-335.
5. Wong J. Y. , Wu X. M., Sturk M. and Bortolotto C. On the Application of ER Fluid to the Development of Semi-active Suspension Systems for Ground Vehicles, *Transactions of the CSME*, Vol.17,No.4B, (1993)pp. 789-800.

6. Wu X. M., Wong. J. Y., Sturk M. and Russell D. L., Simulation and Experimental Study of a Semi-active Suspension with an ER Damper, *International J. of Modern Physics B*, Vol.8, Nos.20 & 21, (1994) pp. 2987-3003.
7. Weyenberg T. R., Pialet J. W. and Petek N. K., The Development of ER Fluids for an Automotive Semiactive Suspension System, *International J. of Modern Physics B*, Vol.10, Nos.23 & 24, (1996) pp. 3201-3209.
8. Choi S. B., Choi Y. T., Chang E. G., Han S. J. and Kim C. S., Control Characteristics of a Continuously Variable ER Damper, *Mechatronics* 8, (1998) pp. 143-161.
9. Suh M. S. and Yeo M. S., Development of Semi-active Suspension Systems Using ER Fluids for the Wheeled Vehicle, *J. of Intelligent Material Systems and Structures*,Vol.10, (1999) pp. 743-747.
10. Choi S. B., Choi Y. T. and Park D. W., A Sliding Mode Control of a Full-Car Electrorheological Suspension System via Hardware in-the-Loop Simulation, *Transactions of the ASME* 122, (2000) pp. 114-121.
11. Choi S. B., Lee H. K. and Chang E. G., Field Test Results of a Semiactive ER Suspension System Associated with Skyhook Controller , *Mechatronics* 11, (2001) pp. 345-353.
12. Kuo W. H., Wu T. N., Guo J., Chiang M. H. and Chen Y. N., Design and Performance Evaluation of a Serial Multielectrode Electrorheological Damper, *Journal of Sound and Vibration 292*, (2006) pp. 694-709.
13. Cavallo A., Setola R. and Vasca F., Using MATLAB, SIMULINK and Control System Toolbox: A Practical Approach, *Prentice Hall* (1997).

MAGNETORHEOLOGICAL FLUID FAN DRIVE FOR TRUCKS

ANTHONY L. SMITH AND JOHN C. ULICNY

General Motors R&D and Planning, 30500 Mound Rd, Warren, Michigan 48090 USA
E-mail: anthony.l.smith@gm.com, john.ulicny@gm.com

LAWRENCE C. KENNEDY

Behr America, Inc., 2700 Daley Dr, Troy, Michigan 48083 USA
E-mail: larry.kennedy@us.behrgroup.com

A magnetorheological fluid (MRF) fan drive prototype for automotive truck application was designed and tested for both performance and durability. A dual concentric gap with drum rotor design was chosen to meet the required torque capacity, packaging, and mass constraints. Finite element magnetostatic modeling was performed to size the electromagnetic circuit and achieve the desired flux density levels in the MR fluid gaps. The clutch was filled with a custom-formulated MR fluid. Performance testing showed excellent speed control and response. The required 40 N-m torque capacity was achieved along with low drag speed, which is a key design characteristic of the fan drive. The clutch successfully passed several 500-hour durability tests in a test cell environment. Performance testing indicated that the MRF clutch maintained its required torque capacity with very little increase in drag speed over the duration of the tests indicating no fluid thickening issues. The total dissipated energy in these tests is about 3.8 GJ. The total specific dissipated energy for these tests is more than seven times higher than the 10^7 J/cm^3 upper limit previously suggested in the literature.

1 Introduction

Numerous magnetorheological fluid (MRF) based clutches have been designed and tested since the idea was first proposed by Rabinow.[1] However, few investigations have been conducted to test the durability of such devices[2] and no MRF clutches for automotive use have been successfully demonstrated which meet both performance and long term durability requirements. Recent literature[3,4] has cast doubt on the possibility of high power versions of MRF clutches by suggesting that there is a total energy input limit which an MRF can tolerate.

This paper describes the performance and durability testing of a MRF fan clutch designed for automotive use. This clutch, combined with a custom-formulated MRF, was found to meet the performance and durability targets for a 40 N-m fan drive application typically used in trucks and SUVs. In practical terms, this also means that the clutch must withstand high internal temperatures (on the order of hundreds of degrees Celsius) and slip energies on the order of kilowatts for hundreds of hours.

2 MRF Fan Drive

The prototype MRF fan drive (clutch)[5,6,7,8] used for this testing is a dual concentric gap with drum rotor design as shown in Fig. 1. This design was chosen to meet the required torque capacity, packaging and mass constraints. The rotor lies in an annular groove between the stator and the coil housing, forming 0.9 mm inner and outer gaps, and is driven by the input shaft at a ratio of the engine speed. The MR fluid[9] is contained mostly in the inner and outer gaps; the composition of the MR fluid is shown in Table 1. The fan drive was filled with 51.2 cm^3 of MR fluid.

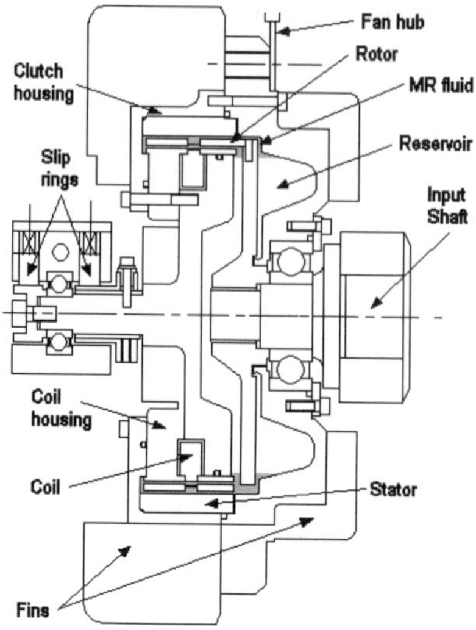

Figure 1 - MRF Fan Drive

Table 1. MR Fluid Composition

Component	Value
Base fluid	Polyalphaolefin
Volume fraction	42% carbonyl iron
Density	3.75 g/cm^3
Additive package	Proprietary formulation

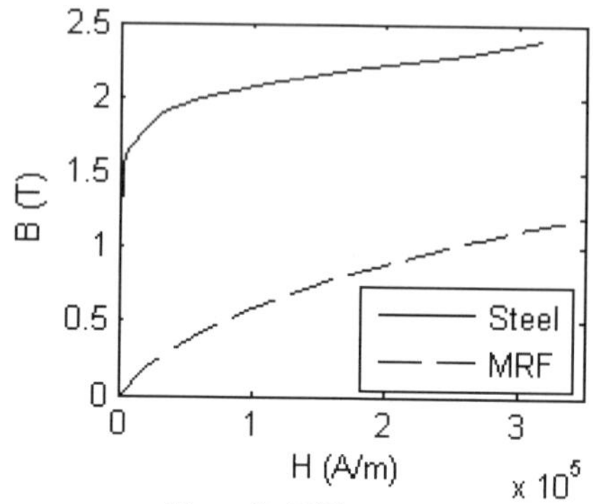

Figure 2. B(H) curves

The electromagnetic circuit is formed by the 110-turn electromagnetic coil, coil housing, rotor, MR fluid and stator. The coil housing, rotor and stator were fabricated from low carbon steel. Webs in the center of the rotor and top of the coil housing prevent shunting of the electromagnetic circuit. Finite element magnetostatic modeling was used to size the electromagnetic circuit such that the maximum flux density in the steel portions of the circuit was 1.7 Tesla at the maximum coil current of 5A. The steel and MR fluid B-H curves used are shown in Fig. 2. The flux density in the magnetic circuit at a current of 5A is shown if Fig. 3. Fig. 4 shows the flux density in the MR fluid at the center of the gaps for various current levels. The fan drive was designed to reach the maximum fan speed of 3600 rpm with less than 5 A of current at beginning of life so that as the MR fluid degraded over time the fan drive would still be able to reach 3600 rpm. Slip rings were used to provide current to the coil.

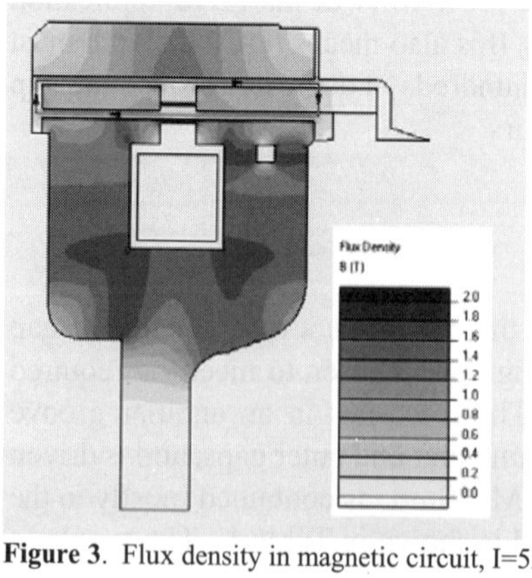

Figure 3. Flux density in magnetic circuit, I=5A

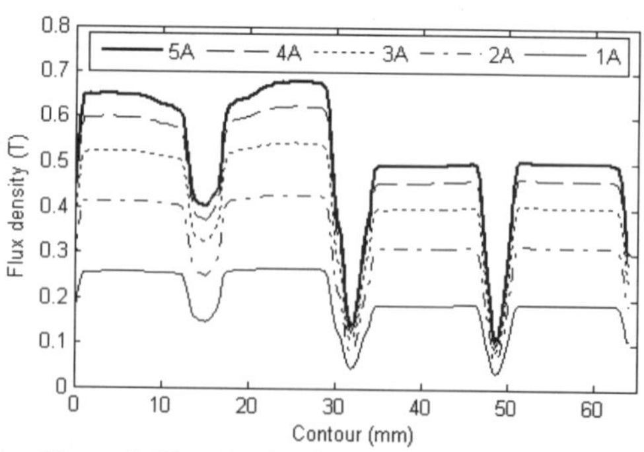

Figure 4. Flux density along the contour shown in Fig. 3 (0 mm is bottom right hand corner of contour)

The clutch housing was made of aluminum with fins on the front and back surfaces sized to reject a maximum 5kW of slip power. Behind the rotor hub, the reservoir had an angled wall with a step to reduce the amount of MR fluid required to fill the clutch. A sealed single-row deep-groove ball bearing supported the clutch housing assembly on the input shaft. The hub of the axial fan was bolted to the clutch housing. The fan torque is proportional to the fan speed squared, $T = k_{fan}\omega^2$, with $k_{fan} = 2.814 \times 10^{-4}$ Nm-s^2.

3 Experimental Setup

Durability and performance testing of the fan drive was performed in a test cell maintained at 150°F (65.6°C). The input shaft of the fan drive was attached to an electric motor driven spindle. The MRF fan drive was computer controlled at a 20 Hz rate using a feed forward + PID controller. The controller commanded an H-bridge current amplifier which provided the current to the coil. The voltage input to the H-bridge current amplifier was limited to 14V. The skin temperature was measured with an infra-red temperature sensor monitoring the fins radially outboard of the stator.

The input speed and fan speed profiles for the durability test cycle are shown in Fig. 5. The cycle profile is 2 hours long and was repeated 250 times. The input speed profile is 3000 rpm for the 1st hour except for a 2 minute period at 5000 rpm centered about 30 minutes; the input speed for the 2nd hour is 5000 rpm. The fan speed profile consisted of four commanded speed values (600, 1600, 2000, 2700 rpm) with ramps at 30 rpm/sec to transition between each value. At an input speed of 5000 rpm the fan speed command was limited to 2475 rpm to maintain slip power [10] at 5 kW. The lowest commanded fan speed, 600 rpm, was set below the drag speed of the clutch in order to monitor the off-state drag speed over the course of the durability test. The shear rate in the center of the outer gap is also plotted in Fig 5. The maximum value is 2.50×10^4 s^{-1} and the minimum value is 1.91×10^3 s^{-1}. In the center of the inner gap, the maximum value is 2.34×10^4 s^{-1} and the minimum value is 1.80×10^3 s^{-1}.

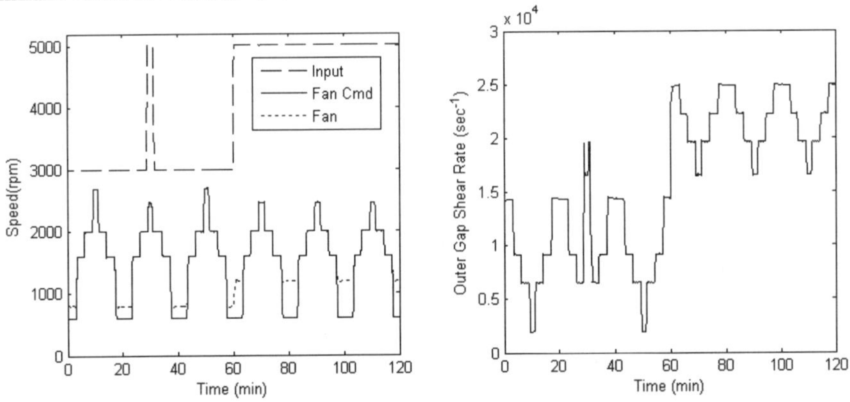

Figure 5 - Durability test profile and corresponding shear rate in the outer gap.

The clutch was performance tested three times during the course of the durability test: at the beginning, near the middle, and at the end of the test. The performance test consisted of an input speed of 4000 rpm and with the fan drive commanded at values from 600 rpm to 3600 rpm and then back to 600 rpm in 600 rpm increments. Each step was maintained for three minutes with the command ramped at 30 rpm/sec between command steps.

4 Results

4.1 Durability Test

The results of the durability test are shown in Figs. 6 and 7. Fig. 6 shows the first cycle of the durability test. Subplots of speed, current, slip power and skin temperature are shown. Since the fan drive performance was consistent, Fig. 6 is representative of the clutch performance for the entire durability test.

The fan speed tracked the commanded fan speed very well. When the commanded fan speed was constant and the current commanded was non-zero the mean absolute error was 3.5 rpm. When the commanded fan speed was being ramped the mean absolute error was 45.0 rpm. When the commanded fan speed is 600 rpm the off-state drag speed can be seen. The current required to achieve the commanded speed is shown; in general, the current increased with increasing commanded fan speed. The slip power for the cycle is shown; note that the maximum slip power is 5 kW. The total slip energy dissipated in one test cycle was 14.9 MJ. The skin temperature trend closely follows the slip power; the peak temperature is at $t \approx 52$ minutes and at $t \approx 72$ minutes for input speeds of 3000 rpm and 5000 rpm, respectively.

To illustrate the behavior of the fan drive for the entire 500 hour durability test, Fig. 7 shows how maximum and minimum values during the two hour test cycle vary over the course of the 500 hours. The current and temperature subplots were de-noised [11] for clarity.

The fan speed subplot shows the maximum fan speeds when the input speed is 3000 and 5000 rpm. For 3000 rpm and 5000 rpm input speed, the fan speed always reached the 2700 rpm and 2475 rpm commanded fan speed, respectively. The minimum fan speed at 3000 rpm and 5000 rpm input speed show how the off-state drag speed varied during the 500 hours. At 3000 rpm input, the drag speed started at 771 rpm and increased to 798 rpm at the end of 500 hours; this corresponds to a 0.1 Nm change in drag torque. At 5000 rpm input, the drag speed started at 1178 rpm and increased to 1228 rpm at the end of 500 hours; this corresponds to a 0.4 Nm change in drag torque. The drag torque on the fan drive is a combination of off-state viscous drag from the MR fluid and opposing drag torques from the two bearings.

The current subplot shows the maximum currents required to reach the commanded fan speed during each 20 minute cycle of the fan speed when the input speed is 3000 rpm and 5000 rpm. For 3000 rpm, the current started at 2.8 A to reach 2700 rpm and reached 3.0 A after 500 hours. For 5000 rpm, the current started at 2.8 A to reach 2475 rpm and reached 3.6 A after 500 hours. Note that the current decreases initially, implying a break-in period. The 3000 rpm plot reaches a minimum of 2.6 A at 70 hours, the 5000 rpm plot reaches a minimum of 2.5 A at 83 hours and once again at 107 hours. After those minimum values are reached, the maximum currents at both 3000 rpm and 5000 rpm both increase over time.

The temperature subplot shows that the maximum skin temperature at input speeds of 3000 rpm and 5000 rpm and the test cell temperature remained relatively constant. The mean maximum temperatures were 138.4 °C and 88.7 °C for when the input speed was 5000 rpm and 3000 rpm, respectively. The mean room temperature was 66.9 °C.

The total slip energy dissipated during the 500 hour test was 3.73 GJ. Most of the energy was dissipated by the MR fluid since the bearing drag torques are small and oppose each other. Therefore, the total specific dissipated energy for the MR fluid is approximately $7.3\text{x}10^7$ J/cm^3 which is more than seven times higher than the 10^7 J/cm^3 upper limit previously suggested in the literature.[3,4]

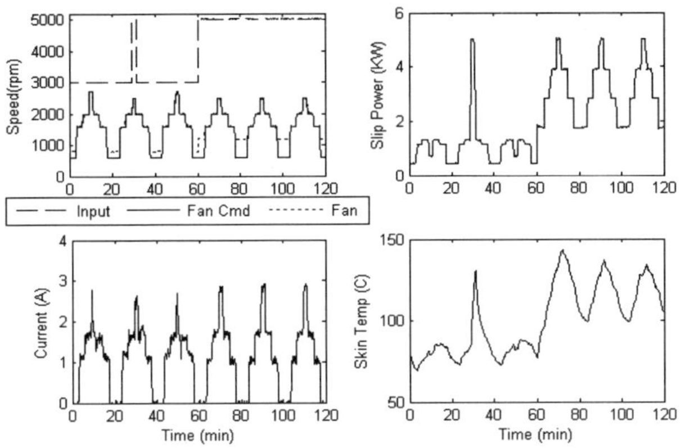

Figure 6 - Durability test results (0-2 hours)

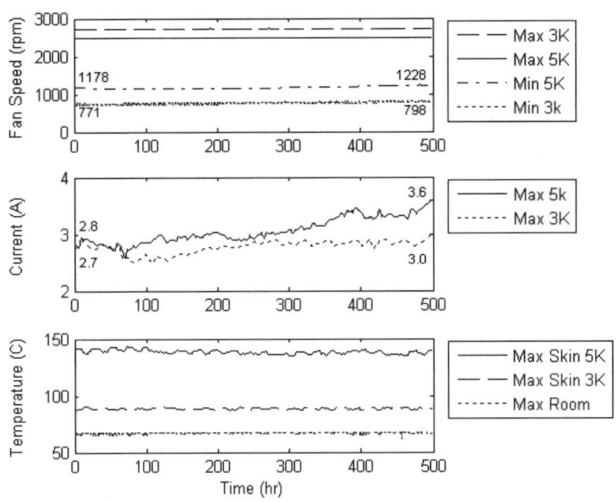

Figure 7 - Durability test results (0-500 hours)

4.2 Performance Test

Fig. 8 shows the performance testing results obtained at times of 0, 244, and 500 hours of durability testing. There are subplots of fan speed (with commanded fan speed), current, and skin temperature. The current and temperature subplots were de-noised[11] for clarity.

The fan drive tracked the commanded fan speed well for all three tests. The mean absolute error (MAE) at commanded fan speeds of 1200 rpm and above for the tests were 7.3, 8.5 and 7.9 rpm at 0, 244, and 500 hours, respectively. At 500 hours, when the fan drive was commanded at 3600 rpm the MAE was 27.3 rpm due to a 0.1Hz oscillation in the fan speed. Adjustment of the PID gains in subsequent testing eliminated this phenomenon.

The current subplot shows a generally increasing trend with time in current required to reach the commanded fan speeds with the exception of 3600 rpm. This increase in current with time is indicative of MR fluid oxidation. The clutch sometimes required less current to operate at 3600 rpm than it did to operate at 3000 rpm. This was attributed to several factors: lower slip power levels would tend to reduce the fluid temperature thus increasing torque capacity, body force effects, and particle migration.

The skin temperature subplot shows similar temperature for the 244 and 500 hour tests. The 0 hour test had a slightly different profile. This may be due to the 244 and 500 hour tests both having many hours of durability testing prior to the performance test whereas the 0 hour test only had a short warm up period.

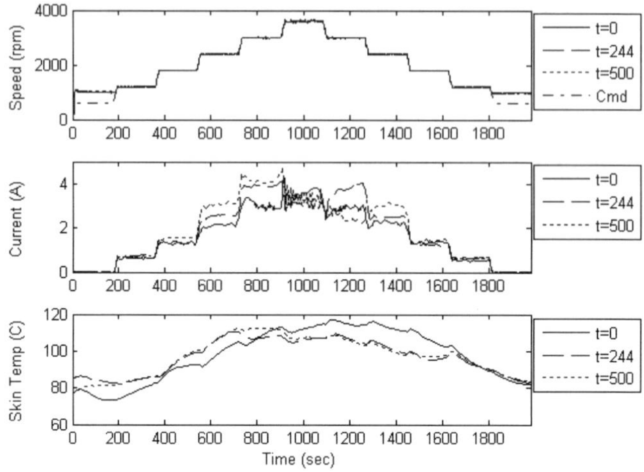

Figure 8 - Performance test results (0, 244, 500 hours)

4.3 *Fluid Analysis*

After the final performance test, the MRF fan drive was disassembled and the MR fluid was sampled from various locations in the fan drive. The iron volume fraction (V_f) was measured using thermogravimetric analysis and corrected for the uptake of oxygen during the test. The oxygen content of the iron was measured using the inert gas fusion technique.

Table 2. MR Fluid Analysis Results (Viscosity and Yield Stress @ T = 40 °C)

Sample	Iron V_f (%) Raw Value	Iron V_f (%) O_2 Corrected	O_2 in Fe (%)	Viscosity [Pa-s]	Yield Stress [Pa]
Annular Gap	42.8	42.2	2.4	0.188	71
Inside of rotor	55.7	55.2	2.2		
Coil housing face	55.6	55.2	1.7		
Outside of rotor	43.1	42.5	2.4		
Reservoir	47.5	47.2	2.2		
MRF as made	43.0	42.9	0.4	0.121	81

Table 2 shows the results for all of the sample locations. The MR fluid in the gaps had a slightly decreased raw V_f. Some fluid packing on the inside surface of the rotor and the coil housing face was observed as noted by their higher V_f values. The oxygen content of all of the samples was higher than the initial "as made" sample. The increase in current required to maintain a particular fan speed over time in both the durability and performance tests is likely due, in part, to oxidation of the MR fluid since iron oxides have a much lower magnetization than elemental iron. Table 2 also shows MR fluid viscosity and yield stress results before and after the 500 hour test. The increase in MR fluid viscosity is consistent with the increase in off-state drag speed seen over the course of the durability and performance tests. This small measured change in viscosity, along with the minimal increases of off-state drag speed, indicate that there were no problems with in-use thickening of the MR fluid. The decrease in off-state yield stress over 500 hours is small compared to the change in viscosity.

5 Conclusions

We have demonstrated an MRF fan drive for automotive use that is capable of maintaining a 40 N-m torque capacity and meeting all speed performance requirements, including a low drag torque, before and after rigorous durability and performance tests. Furthermore, the clutch and the fluid were able to withstand a total specific dissipated energy many times greater than was previously thought possible. This performance sets a new milestone for MRF clutch performance and demonstrates a truly practical capability for this technology in a most demanding application.

Acknowledgements

The authors thank Mark Golden (GM) for providing the MR fluid; Brian McDermott (Behr America) for clutch buildup/teardown and running the performance and durability tests; Keith Snavely (GM) for performing thermogravimetric fluid analysis, and Noel Potter (GM, retired) for performing the oxygen analyses.

References

1. Rabinow, J. "The Magnetic Fluid Clutch." AIEE Transactions 67(1948): 1308-1315.
2. Jones, W.P. "Magnetic Mixtures for Clutch Application." AIEE Transactions 72(1953): 88-92.
3. "FAQs - Attributes of MR Materials." Lord Magneto-Rheological (MR) Technology. Lord Corp. 10 May 2006 <http://www.lord .com/Default.aspx? tabid=2038&pid=3>.
4. Carlson, J.D. "What Makes a Good MR Fluid? " 8th International Conference on ER Fluids and MR suspensions. July 9-13, 2001.
5. Moser, Sommer, Usoro, and Smith, US6,032,772, *Viscous Clutch Assembly*. March 7, 2000.
6. Moser, Sommer, Usoro, and Smith, US6,102,177, *Viscous Clutch Assembly*. August 15, 2000.
7. Moser, Sommer, Usoro, and Smith, US6,173,823, *Viscous Clutch Assembly*. January 16, 2001.
8. Usoro, Smith, Kao, Moser, and Sommer, US6,318,531, *Magnetorheological Fluid Clutch*. November 20, 2001.
9. Ulicny, Smith, Golden, and Hayden, US6,932,917, *Magnetorheological Fluids with Stearate and Thiophosphate Additives*. April 19, 2005.

10. Smith, and Usoro, US,648,115, *Method for Slip Power Management of a Controllable Viscous Fan Drive*. November 18, 2003.
11. Misiti, Misiti, Oppenheim, and Poggi. Wavelet Toolbox User' s Guide, Version 3 Natick: The MathWorks, 2005.

FUZZY-SLIDING MODE CONTROL OF SEMI-ACTIVE SUSPENSION SYSTEMS WITH MR DAMPERS

L. ZHENG AND Y. N. LI
State Key Laboratory of Mechanical Transmission, Chongqing University, Chongqing 400044, P.R.China
E-mail : zling@cqu.edu.cn

A. BAZ
Department of Mechanical Engineering, University of Maryland College Park, MD 20742, USA
E-mail : baz@eng.umd.edu

A fuzzy-sliding mode controller is presented to control the dynamics of semi-active suspension systems of vehicles using magneto-rheological (MR) fluid dampers. A full-car model is used to design and evaluate the performance of the proposed semi-active controlled suspension system. Four mixed mode MR dampers are designed, manufactured, and integrated with four independent sliding mode controllers. In order to overcome the chattering of the sliding mode controllers, a fuzzy logic control strategy is merged into the sliding mode controller. The proposed fuzzy-sliding mode controller is designed and fabricated. The performance of the semi-active suspensions is evaluated in both the time and frequency domains. The obtained results demonstrate that the proposed fuzzy-sliding mode controller can effectively suppress the vibration of vehicles and improve their ride comfort. Furthermore, it is shown that the "chattering" of sliding mode control is smoothed when it is integrated with a fuzzy logic control strategy.

1 Introduction

Recently, suppression of the vibration of vehicles has been a very active research area which aims primarily at attenuating the disturbances resulting from various road excitations. This is normally accomplished by employing a wide variety of vehicle suspension system. Semi-active suspension systems, which consist of conventional springs and controllable dampers, are conceived to be a practical solution for enhancing the performance of suspension systems and improving the ride comfort. Most of these systems rely in their operation on MagnetoRheological (MR) dampers because of their potential use as semi-active control devices [1-2]. The feasibility of these devices has been successfully demonstrated in numerous automobile engineering applications [3-5].

However, integration of the MR dampers into suspension systems poses serious challenges as the MR dampers are highly nonlinear and the parameters that govern the vehicle dynamics are often uncertain due to changes in the loading conditions such as the number of passengers and payload. Such challenges can be overcome by using robust control algorithms. For example, Choi *et al.* [6] used a sliding mode controller (SMC) to attenuate the vibration of a semi-active seat suspension with an ElectroRheological (ER) fluid damper. Choi *et al.* [7] extended his work to a full-car suspension system featuring ER dampers and found it to equally effective. Yokoyama *et al.* [8] proposed an adaptive model following SMC for semi-active suspension systems with MR damper.

The sliding mode control theory has been widely studied as a robust control strategy for many years. However, the main drawback of SMC is the "chattering" phenomenon which can excite undesirable high-frequency dynamics. Several methods of chattering reduction have been reported such as a boundary layer around the switching surface and max–min type control by a unit vector function. However, these approaches provide no guarantee of convergence of the sliding mode and involve a tradeoff between chattering and robustness.

Fuzzy logic control (FLC) has been considered as one of the most efficient methods in addressing the inherent control problems related to nonlinear systems that contain uncertainties. In this paper, the smooth control feature of the FLC is used to overcome the "chattering" of SMC for a full car semi-active suspension with MR dampers. A novel fuzzy-sliding mode controller is designed and formulated. The control characteristics of a full-car suspension system featuring MR dampers for a passenger vehicle integrated with a FSMC (fuzzy-sliding mode control) controller is evaluated in both the time and frequency domains by considering bump and random excitation conditions.

2 MR Damper

In this study, mixed mode MR dampers are designed and manufactured. The schematic drawing of the MR damper is shown in Fig.1. This is a twin tube MR damper with double class magnetic circuits for a small-size passenger vehicle.

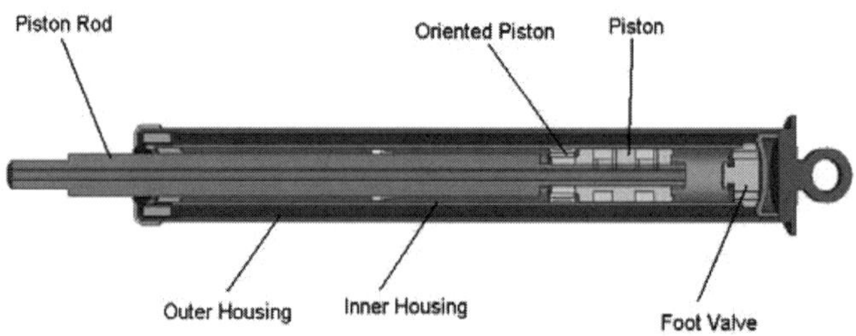

Figure 1 - The schematic drawing of the MR damper

For the MR damper with double class magnetic circuits in the mixed mode, the damping force can be derived by regarding the mixed mode as a combination of the flow and shear modes [9]:

$$F_{DAMPER} = \left[\frac{18\eta l A_p^{\ 2}}{\pi R_1 (R_2 - R_1)^3} + \frac{6\pi R_1 l \eta}{R_2 - R_1} \right] \dot{x}_p + \left[\frac{9 l A_p}{R_2 - R_1} + 6\pi R_1 l \right] \tau_y \tag{1}$$

where A_p is the working area of the piston, η is viscosity of the MR fluid without the magnetic field applied, l is the length of each duct, R_2 is inner radius of the inner housing, R_1 is the outer radius of the piston. In additional, the duct is divided into three parts by double magnetic circuits, we assume that each part has the same length.

In order to express the damping force in a simple way and apply it to the full-car model, Eq. (1) can be rewritten as follows:

$$F_{DAMPER} = C_s \dot{x}_p + F_{MR} sign(\dot{x}_p)$$

$$\text{where} \quad C_s = \frac{18\eta l A_p^{\ 2}}{\pi R_1 (R_2 - R_1)^3} + \frac{6\pi R_1 l \eta}{R_2 - R_1}, \text{ and } F_{MR} = \left[\frac{9 l A_p}{R_2 - R_1} + 6\pi R_1 l\right] K H^{\beta} \tag{2}$$

such that C_s is the equivalent viscous damping coefficient of the MR damper and F_{MR} is the controllable damping force of the MR damper. Also, K and β are the characteristic constants of the MR fluid used in the MR damper with K=0.0618 and β=1.25 for the damper under consideration. The main geometrical and physical parameters for MR damper are listed in Table 1.

Table 1 - Geometrical and physical parameters of the MR damper

η	5 Pa.s	R2	1.55×10^{-4} m
K	0.0618	R1	1.40×10^{-4} m
β	1.25	N	450
l	12×10^{-3} m	Dp	2.8×10^{-4} m
Ap	6.1575×10^{-4} m^2		

Fig.2 shows a comparison between the theoretical and experimental force-displacement characteristics of the MR damper at different control currents. The displayed characteristics are obtained for a sinusoidal excitation frequency of 2Hz with an excitation amplitude of 25mm. It can be seen that the theoretical predictions are in close agreement with the experimental results. Hence, Eq.(2) can be accurately used to design the full-car semi-active suspension system with MR dampers.

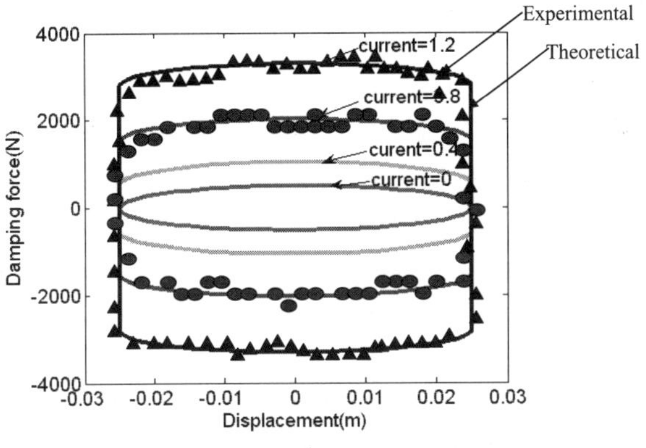

Figure 2 - Force-displacement loop for MR damper (input current = 0, 0.4, 0.8, 1.2A)

3 The Sliding Mode Controller

3.1 Dynamic model

A dynamic model for a full-car semi-active suspension system via MR dampers is constructed in Fig. 3. The governing equations of motion can be derived as follows:

$$M\ddot{h} + f_{ks1} + f_{ks2} + f_{ks3} + f_{ks4} + f_{cs1} + f_{cs2} + f_{cs3} + f_{cs4} = F_{MR1} + F_{MR2} + F_{MR3} + F_{MR4} \quad ,$$

$$I_{xx}\ddot{r}+(f_{ks1}+f_{cs1})l_{ylf}-(f_{ks2}+f_{cs2})l_{ylf}+(f_{ks3}+f_{cs3})l_{lyr}-(f_{ks4}+f_{cs4})l_{ylr}=F_{MR1}l_{ylf}$$
$$-F_{MR2}l_{ylf}+F_{MR3}l_{ylr}-F_{MR4}l_{ylr},$$

$$I_{yy}\ddot{p}-(f_{ks1}+f_{cs1})l_{xf}-(f_{ks2}+f_{cs2})l_{xf}+(f_{ks3}+f_{cs3})l_{xr}+(f_{ks4}+f_{cs4})l_{xr}=-F_{MR1}l_{xf}$$
$$-F_{MR2}l_{xf}+F_{MR3}l_{xr}+F_{MR4}l_{xr},$$

$$M_{w1}\ddot{z}_{w1}-f_{ks1}-f_{cs1}+f_{kt1}=-F_{MR1}, \qquad M_{w2}\ddot{z}_{w2}-f_{ks2}-f_{cs2}+f_{kt2}=-F_{MR2},$$

$$M_{w3}\ddot{z}_{w3}-f_{ks3}-f_{cs3}+f_{kt3}=-F_{MR3}, \text{ and } M_{w4}\ddot{z}_{w4}-f_{ks4}-f_{cs4}+f_{kt4}=-F_{MR4}, \tag{3}$$

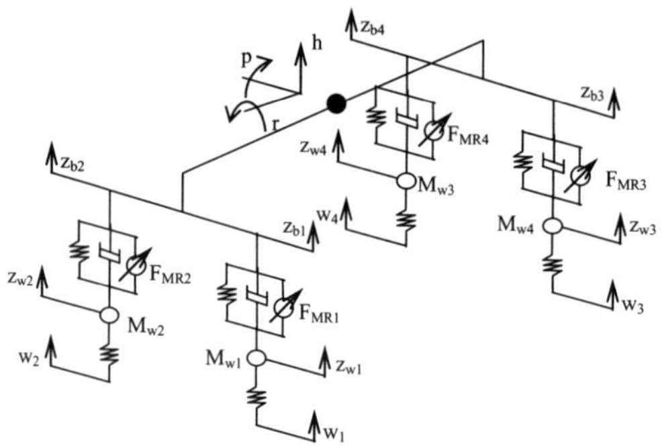

Figure 3 - Dynamic model for a full-car semi-active suspension system via MR dampers

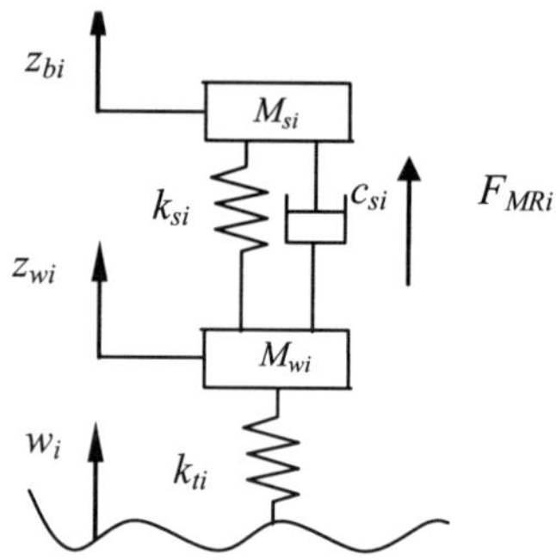

Figure 4 - A quarter-car model

where $f_{ksi}=k_{si}(z_{bi}-z_{wi})$, $f_{csi}=c_{si}(\dot{z}_{bi}-\dot{z}_{wi})$, $f_{kti}=k_{ti}(z_{wi}-w_i)$ for $i=1,2,3,4$. M is the sprung mass and M_{wi} ($i=1,2,3,4$) is the unsprung mass. I_{xx} and I_{yy} are the roll and pitch mass moment of inertia . k_{si} ($i=1,2,3,4$) is the stiffness coefficient of the suspension. c_{si} ($i=1,2,3,4$) is the damping coefficient of the suspension, and it is equal to C_s in Eq. (2). k_{ti} is the stiffness coefficient of the tire. Also, h , r and p are the vertical displacement, roll and pitch angular displacement. l_{xf} and l_{xr} are the distances between the front suspension and center of gravity of sprung mass (C.G.) and the rear suspension and C.G. l_{ylf} and l_{ylr} are half of front and rear wheel tracks. Furthermore, z_{bi} , z_{wi} and w_i ($i=1,2,3,4$) are the displacements of the sprung and unsprung masses, and the disturbance applied to the i th tire.

3.2 The Sliding Mode Controller

In order to simplify the control system of the full-car semi-active suspension with MR dampers, four independent sliding mode controllers are designed. Each controller is used to control one quarter of the car model as shown in Fig.3. In this way, the computational burden for the whole control system can be reduced. This is essential to obtaining a rapid response for various road conditions. Hence, a quarter car-model is considered to design these sliding mode controllers as shown in Fig. 4.

For the quarter car model, the governing equations of motion can be expressed as follows:

$$M_{si}\ddot{z}_{bi} + f_{ksi} + f_{csi} = F_{MRi} \quad , \quad M_{wi}\ddot{z}_{wi} - f_{ksi} - f_{csi} + f_{kti} = -F_{MRi} \tag{4}$$

where M_{si} is sprung mass.

The state vector is defined as

$$X_i = [X_{i1} \quad X_{i2} \quad X_{i3} \quad X_{i4}]^T = [z_{bi} - z_{wi} \quad \dot{z}_{bi} \quad z_{wi} - w_i \quad \dot{z}_{wi}]^T$$

So the state space equation is obtained:

$$\dot{X}_i = A_i X_i + B_i F_{MRi} + E_i \dot{w}_i \tag{5}$$

where

$$A_i = \begin{bmatrix} 0 & 1 & 0 & -1 \\ -\frac{k_{si}}{M_{si}} & -\frac{c_{si}}{M_{si}} & 0 & \frac{c_{si}}{M_{si}} \\ 0 & 0 & 0 & 1 \\ \frac{k_{si}}{M_{wi}} & \frac{c_{si}}{M_{wi}} & -\frac{k_{ti}}{M_{wi}} & -\frac{c_{si}}{M_{wi}} \end{bmatrix}, \quad B_i = \begin{bmatrix} 0 \\ \frac{1}{M_{si}} \\ 0 \\ -\frac{1}{M_{wi}} \end{bmatrix}, and\ E_i = \begin{bmatrix} 0 \\ 0 \\ -1 \\ 0 \end{bmatrix}.$$

In the present study, the sliding mode controller will be formulated based on an LQR control strategy. The total control force is divided into two parts. The first part is the linear part while the other part is the nonlinear part. The linear control force is determined by the LQR control and the nonlinear part is determined by the sign function as follows:

$$F_{MRi} = F_{MRlinear} + F_{MRnonlinear} \tag{6}$$

In order to determine a linear control force, LQR control for a quarter-car model is used. The cost function for LQR control can be expressed as follows:

$$J = \int_0^\infty (y_i^T Q_i y_i + rF_{MRi}^2)dt \tag{7}$$

where, $y_i = [\ddot{z}_{bi} \quad z_{bi} - z_{wi} \quad z_{wi} - w_i]^T =$ the output which can be written as:

$$y_i = cX_i + dF_{MRi} \tag{8}$$

Hence, the optimal gain matrix K is given by:

$$K = R^{-1}(B^T P + N^T) \tag{9}$$

Let $\quad Q = c^T Q_i c\ , \quad N = c^T Q_i d\ , R = r + d^T Q_i d$

Then, P is the solution of the associated Riccati equation:

$$A^T P + PA - (PB + N)R^{-1}(B^T P + N^T) + Q = 0 \tag{10}$$

and the linear control force is obtained:

$$F_{MRlinear} = -KX = -R^{-1}(B^T P + N^T)X \tag{11}$$

For nonlinear part, first a sliding surface is defined:

$$S = CX = R^{-1}(B^T P + N^T)X \tag{12}$$

In this way, the nonlinear control force can be given by sliding mode control theory:

$$F_{MRnonlinear} = -F^*(CB)^{-1} sign(S) \tag{13}$$

Substituting Eq. (13), Eq. (12), and Eq.(11) into Eq.(5), the stability condition is identified as follows:

$$S\dot{S} = X^T(PB + N)R^{-1}R^{-1}(B^T P + N^T)(AX + BF_{MRi} + E\dot{w})$$

$$= X^T(PB + N)R^{-1}R^{-1}(B^T P + N^T)AX - R^{-1}(B^T P + N^T)BS^2 - F^*|S| + SR^{-1}(B^T P + N^T)E\dot{w}$$

$$= X^T[(PB + N)(B^T P + N^T)A + A^T(PB + N)(B^T P + N^T)]X/2R^2 - (B^T P + N^T)BS^2/R$$

$$-F^*|S| + SR^{-1}(B^T P + N^T)E\dot{w} \tag{14}$$

Since $(PB+N)(B^T P + N^T)$ is positive semi-definite, and A is stable, thus, the matrix $[(PB+N)(B^T P+N^T)A + A^T(PB+N)(B^T P+N^T)]$ is negative semi-definite. Also, the scalar $(B^T P+N^T)BS^2/R$ is positive. Therefore, the first two terms in inequality (14) is negative. In this case, the stability condition $S\dot{S} < 0$ can be satisfied when

$$-F^*|S| + SR^{-1}(B^T P + N^T)E\dot{w} < 0 \tag{15}$$

We assume $|\dot{w}| \le \dot{w}_{\max}$, inequality (15) is modified into:

$$F^* \ge |(B^T P + N^T)E/R|\dot{w}_{\max} \tag{16}$$

4 The Fuzzy-Sliding Mode Controller

In this section, a fuzzy-logic control is introduced to design a fuzzy-sliding mode controller. The rules of fuzzy-sliding mode controller can be described as follows [10]:

Rule1: If S is NB, then $F_{MRnonlinear}$ is PB

Rule2: If S is NM, then $F_{MRnonlinear}$ is PM

Rule3: If S is ZE, then $F_{MRnonlinear}$ is ZE

Rule4: If S is PM, then $F_{MRnonlinear}$ is NM

Rule5: If S is PB, then $F_{MRnonlinear}$ is NB

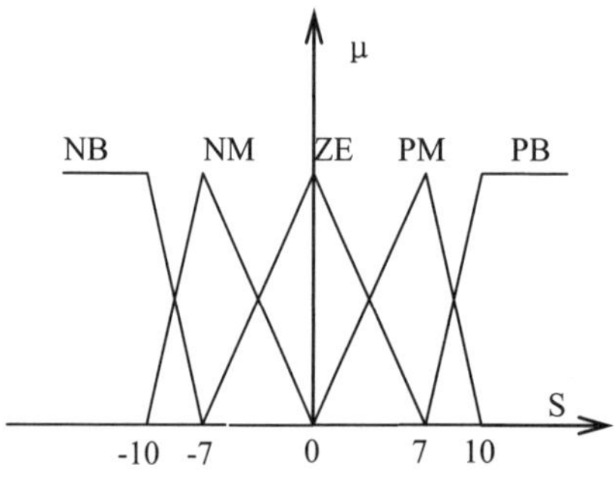

Figure 5 - The membership functions of S

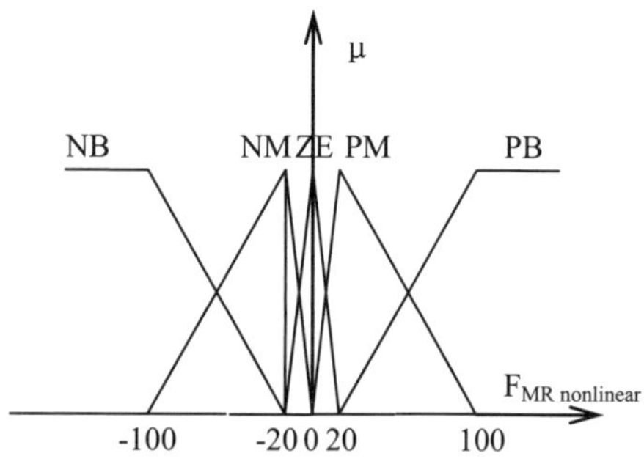

Figure 6 - The membership functions of $F_{MRnonlinear}$

where NB, NM, ZE, PM, PB are linguistic terms of antecedent fuzzy set, they mean negative big, negative medium, zero, positive medium, and positive big. Fig.5 and Fig.6 display the membership functions of S and $F_{MRnonlinear}$. In order to obtain non-fuzzy value of control, a method of defuzzification called "center of gravity method" is used here:

$$F_{MRnonlinear} = \frac{\sum_{i=1}^{5} F_{MRnonlinear} \cdot \mu_{Bi}(F_{MRnonlinear})}{\sum_{i=1}^{5} \mu_{Bi}(F_{MRnonlinear})} \tag{17}$$

where $\mu_{Bi}(F_{MRnonlinear})$ is the corresponding membership function .

The control force F_{MRi} given by Eq.(6) is calculated in an active actuating manner. However, the MR damper is a semi-active actuator. Thus, the control damping force should be modified to meet semi-active constrained condition:

$$F_{MRi} = \begin{cases} F_{MRi} & for F_{MRi}(\dot{z}_{bi} - \dot{z}_{wi}) > 0 \\ 0 & for F_{MRi}(\dot{z}_{bi} - \dot{z}_{wi}) \leq 0 \end{cases} \quad (i = 1,2,3,4) \tag{18}$$

5 Performance Evaluation and Discuss

The control characteristics for the full-car semi-active suspension system with MR dampers are evaluated under two types of road excitations. The first excitation is a single rectangular bump having the height of 0.01*m* and the width of 0.5*m*. It is used to reveal the transient response characteristics. In this bump excitation, the vehicle travels the bump with the velocity of 3 *km/h*. The second type of road excitation is the random road profiles which are used to evaluate the stable state response characteristic. As a typical random process, the power spectral density (PSD) of the road velocity excitation can be expressed as follows [11]:

$$G_{\dot{w}}(f) = (2\pi f)^2 G_w(f) = 4\pi^2 G_0 n_0^2 v \tag{19}$$

where n_0 is the reference spatial frequency $n_0 = 0.1\,(1/m)$. G_0 denotes the road roughness coefficient . The road disturbance can be obtained by integrating the white-noise signal produced by Eq. (19). The vehicle is assumed to travel at a constant speed of 20 *m/s* (72*Km/hr*).

The system parameters of the full-car suspension with MR dampers are listed in Table2.

Table2 Parameters of the full-car model with MR dampers

Parameter	Unit	Value
M_s	kg	1583
I_{xx} / I_{yy}	kg m2	531/2555
$M_{w1} / M_{w2} / M_{w3} / M_{w4}$	kg	48/48/74/74
$K_{s1} / K_{s2} / K_{s3} / K_{s4}$	N/m	35000/35000/34000/34000
$C_{s1} / C_{s2} / C_{s3} / C_{s4}$	Ns/m	600/600/600/600
$K_{t1} / K_{t2} / K_{t3} / K_{t4}$	N/m	220000/220000/220000/220000
I_{xf} / I_{xr}	m	1.116/1.438
l_{ylf} / l_{ylr}	m	0.77/0.765

Fig.7 presents the controlled time responses of the full-car semi-active suspension with MR dampers for the bump excitation. It can be observed that the vertical and pitch acceleration of sprung mass are substantially reduced by the sliding mode controller, the chattering of the SMC is smoothed by the proposed fuzzy-sliding mode controller (FMSC). The robust character of the proposed controller is studied by considering the parameter perturbations: $\Delta M = 791.5kg$, $\Delta I_{xx} = 265.5$ kg m^2, $\Delta I_{yy} = 1277.5$ kg m^2. It can be seen that the FSMC displays good robustness in the presence of parameter perturbation.

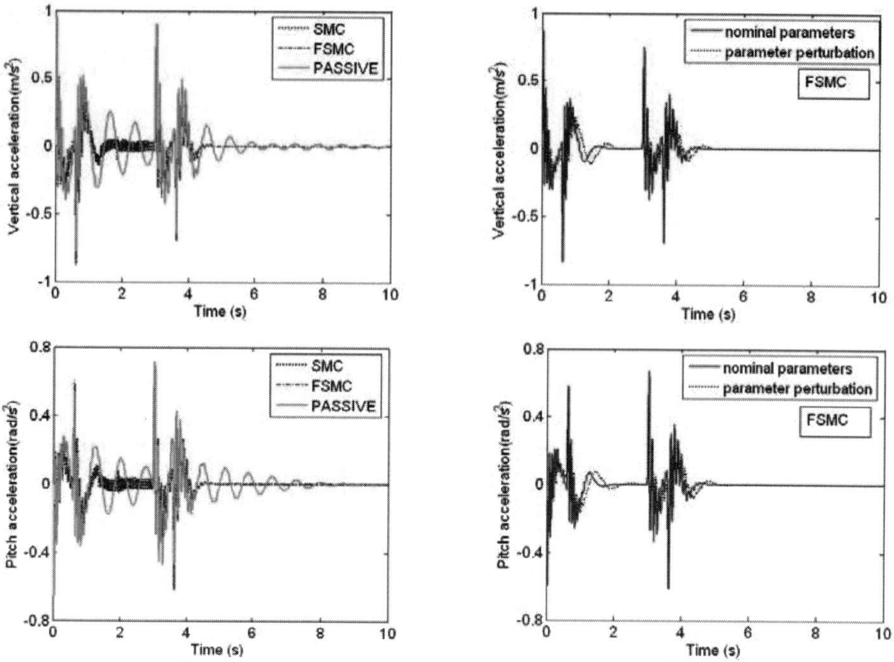

(a) (b)

Figure7. Time responses of the full-car suspension with MR dampers for the bump excitation: (a) using different controller under nominal parameters and (b) comparison between nominal parameters and parameter perturbations with FSMC controller

Fig.8 presents the controlled frequency response of semi-active suspension with MR dampers for random excitation. It can be seen that the vertical, pitch acceleration can be substantially reduced by FSMC controllers, especially effective in body resonance frequency.

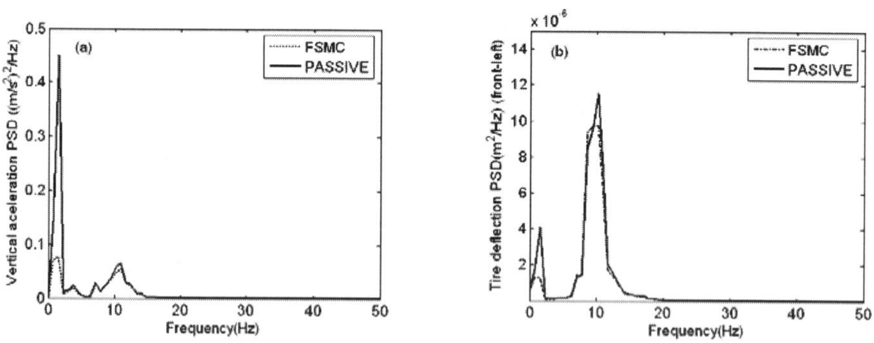

Figure8. Frequency responses of the full-car suspension with MR dampers for random excitation

6 Conclusion

A robust control system is developed for semi-active suspension systems of vehicles using magneto-rheological (MR) fluid dampers. MR dampers with the mixed mode are designed and manufactured, and their field-dependent damping forces are experimentally evaluated. A sliding mode controller (SMC) is designed and formulated and a fuzzy-logic control strategy is integrated with it. The performance of the resulting FSMC is evaluated. The obtained results show that when the fuzzy control is combined with the sliding mode controller, the "chattering" of the SMC is reduced considerably. It is also shown that the performance of the proposed fuzzy-sliding mode controller is stable and exhibits good robustness in the presence of parameter uncertainty.

7 Acknowledgements

This study was supported by a grant from the Chinese Nature Science Foundation (Grant No. 50475064)

References

1. Sean P. K., Experimental characterization of commercially practical magnetorheological fluid damper technology. SPIE conference on smart structure and materials, Dan Diego, CA, March 2001.
2. Dyke S. J. Spencer Jr B. F., Sain M. K., Carson J. D., Modeling and control of magnetorheological dampers for seismic response reduction, *Smart Mater. Struct.* **5** (1996) pp. 565–575
3. Ericksen E. O. and Gordaninejad F., A magneto-rheological fluid shock absorber for an off-road Motorcycle," *International J. Vehicle Design*, **33**(1/2/3)(2003) pp.139-152
4. Gordaninejad F. and Kelso S. P., Magneto-rheological fluid shock absorbers for HMMWV, *Proceedings of SPIE Conference on Smart Structures and Materials,* Newport Beach, CA, March, 2000.
5. Gordaninejad, F. and Kelso, S. P., Fail-safe magneto-rheological fluid dampers for off- highway, high-payload vehicles, *Journal of Intelligent Material System and Structures*, **11**(5)(2000) pp. 395-406
6. Choi S. B., Choi.H. J., Lee.Y. S., Han. M. S., Vibration control of an ER seat suspension for a commercial vehicle, *Transactions of the ASME*, *Journal of Dynamic Systems, Measurement, and Control,* **125**(2003)pp. 60-68
7. Choi S. B., Choi.Y. T., Park. D. W., A sliding mode control of a full-car electrorheological suspension system via hardware in-the-loop simulation. *Transactions of the ASME, Journal of Dynamic Systems, Measurement, and Control,* **122**(2000)pp. 114-121
8. Yokoyama M., Hedrick J.H., Toyama.S., A model following sliding mode controller for semi-active suspension systemswith MR dampers, Proceedings of the American Control Conference,Arlington, VA June 25-27, 2001
9. LORD Materials Business Unit, Designing with MR fluids, *Engineering Note*, December, 1999
10. Kim S.W., and Lee J. J., Design of a fuzzy controller with fuzzy sliding surface, *Fuzzy Sets and Systems*, **71**(1995) pp.359-367
11. Wang J. Y., *Theory of Ground Vehicles*, *3rd* ed. New York: John Wiley, 2001

FIELD TEST ON VIBRATION CONTROL OF VEHICLE SUSPENSION SYSTEM FEATURING ER SHOCK ABSORBERS

S. B. CHOI, Y. M. HAN, H. J. SONG AND J. W. SOHN

Smart Structures and Systems Laboratory, Department of Mechanical Engineering, Inha University, Incheon 402-751, Korea
E-mail: seungbok@inha.ac.kr

H. J. CHOI

Department of Polymer Science and Engineering, Inha University, Incheon 402-751, Korea
E-mail: hjchoi@ inha.ac.kr

This paper presents vibration control responses of a passenger vehicle equipped with continuously controllable electrorheological (ER) shock absorbers. As a first step, four ER shock absorbers (two for front parts and two for rear parts) are designed and manufactured based on the damping force levels and mechanical dimensions required for the middle-sized passenger vehicle. After experimentally evaluating the field-dependent damping force property and controllability, the manufactured ER shock absorbers are incorporated with the suspension of a car to be tested. Subsequently, the skyhook control algorithm, which is simple, but very effective in rear field, is formulated by considering vertical, pitch and roll motions. The controller is then realized by integrating feedback sensors such as accelerometer and the field test is undertaken under bump and random road conditions. The control responses for the ride quality and steering stability are evaluated in both time and frequency domains. In addition, the performance comparison between conventional suspension and ER suspension is carried out.

1 Introduction

One of salient properties of ER (electro-rheologocal) fluid is that it has fast response characteristic to electric field and hence wide control bandwidth. This inherent feature has triggered tremendous research activities in the development of various engineering applications including shock absorbers. Sturk *et al.* proposed a high voltage supply unit with ER shock absorber (or damper) and proved their effectiveness via quarter car suspension system [1]. Nakano constructed a quarter car suspension model using ER damper and proposed proportional control algorithm in order to isolate vibration [2]. Petek *et al.* constructed a semi-active full suspension system installed with four ER dampers and evaluated suspension performance through the implementation of a skyhook control algorithm which considers heave, pitch and roll motions of the car body [3]. Peel *et al.* proposed a long-stroke ER damper for effective vibration control [4]. Wereley and Pang proposed a nondimensional model for the analysis of ER damper operated in various working modes, and proved the validity of the model via experimental work [5]. Choi *et al.* proposed a cylindrical ER damper for passenger car and its controllability of damping force was proved by implementing a skyhook controller [6]. As is evident from previous recent works, most of research works have been focused only on damping force evaluation of ER damper itself or/and vibration isolation evaluation of a quarter suspension system installed with one ER damper. However, the field test of a full car installed with four ER dampers is considerably rare. This, of course, is due to that the real implementation of ER dampers to the full car requires a multi-disciplinary effort including ER fluids, ER devices, high voltage amplifiers, sensors, controllers as well as the test vehicle.

This paper presents performance evaluation of the ER suspension system through vehicle field test. Four cylindrical ER shock absorbers adaptable to a middle-sized passenger car are manufactured and then field-dependent damping characteristics are evaluated. A skyhook controller considering the vertical motions of the car body and wheel is formulated and realized to the test vehicle installed various sensors such as accelerometer and gyroscope. Control responses are demonstrated under various road conditions in time and frequency domain.

2 Design of ER Shock Absorber

The schematic configuration of the cylinder type ER shock absorber proposed on this work is shown in figure 1(a). The ER shock absorber is divided into the upper and lower chambers by the piston, and it is fully filled with the ER fluid. By the motion of the piston, the ER fluid flows through the duct between inner and outer cylinders from one chamber to the other. The gas chamber located outside of the lower chamber acts as accumulator of the ER fluid induced by the motion of the piston. By applying the electric field to the ER shock absorber, the damping force can be obtained as follows;

$$F = \frac{A_r^2}{C_g} X_p + (A_p - A_r)^2 \left[I_e \dot{V}_p + R_e V_p \right] + (A_p - A_r) P_{ER} + F_r \tag{1}$$

where, A_r is area of the piston rod, C_g is compliance of the gas chamber, X_p is piston displacement, A_p is area of the piston, I_e is fluid inertia, V_p is piston velocity, R_e is fluid resistance from the viscosity, P_{ER} is pressure drop due to the yield stress of the ER fluid and F_r is frictional force, respectively.

We clearly see from the above equation that the damping force of the ER shock absorber comprises five components. Among them, the fourth term is controllable by the intensity of the electric fluid. The size of ER shock absorber and the level of the required damping force are chosen on the basis of the conventional passive oil damper used for a middle-sized passenger car. The damping force model is analyzed in order to determine design parameters. The designed parameters are given in the table 1. ER shock absorber was manufactured with designed parameters as shown in the figure 1(b). Figure 2 presents the measured damping characteristics of the front ER shock absorber. PANI/T_1O_2 composite particle based ER fluid was employed. As shown in the figure 2(a), the damping force increases as the electric field increases. For instance, the damping force is increased up to 1000*N* at the piston velocity of 0.375*m/s* by applying the electric field of 3*kV/mm*. It is here remarked that the level of damping force of the conventional passive oil damper is almost same as the one at 0*kV/mm*. It is also observed from figure 2(b) that the enclosed area increases as the electric field increases. This directly implies that the energy dissipation is increased during the cycle by increasing the field intensity.

Table 1. Design parameters of ER shock absorbers.

	Electrode Gap	Electrode Length	Maximum Stroke	Area of Piston Head	Area of Piston Rod	Gas Pressure
Front	0.88*mm*	292.4*mm*	± 164*mm*	804.25mm^2	380.13mm^2	10 *bar*
Rear	0.88*mm*	280.4*mm*	± 154*mm*	572.56mm^2	132.73mm^2	10 *bar*

3 Performance Evaluation via Vehicle Field Test

The vehicle for the field test is 2000cc middle-sized passenger car donated by GM-DAEWOO. It has a total length of 4770*mm*, total width of 1815*mm*, total height of 1440*mm*, wheelbase of 2700*mm*, front track of 1550*mm,* rear track of 1535*mm* and weight of 1395*kg*. Four ER shock absorbers are prepared for vehicle field test and installed as shown in the figure 3. PANI/T_1O_2 composite particle based ER fluid is adopted and poured into ER suspension systems 400cc for front and 300cc for rear, respectively. For this field test, high voltage amplifier is developed in small size in order to install in the car as shown in the figure 4. It is designed to amplify 700 times of input voltage and its response time is 10*ms*. The electric power of amplifier can be supplied from car battery in 12*V*. The photograph of the test vehicle is shown in the figure 5. ER suspension system is installed to control the vehicle vibration excited from the road during driving. Measured data from the sensors are sent to real time data acquisition system through A/D channel. Vehicle movements are analyzed based on the measured data every 1*ms* by computer. Desired damping forces for the improvement of ride quality are calculated by computer based on the analyzed vehicle movement data. Control input is determined by the control algorithm implemented in the computer and input signal is sent to the high voltage amplifier through the D/A channel of data acquisition system and the amplified input electric field is applied to the ER shock absorber.

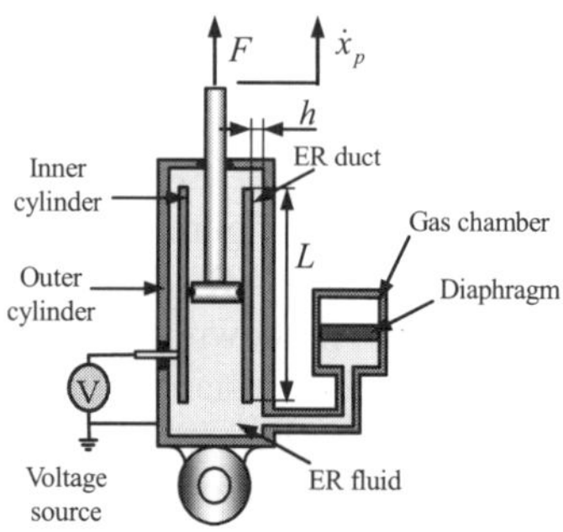

(a) schematic diagram

(b) photograph

Figure 1. Proposed ER shock absorber

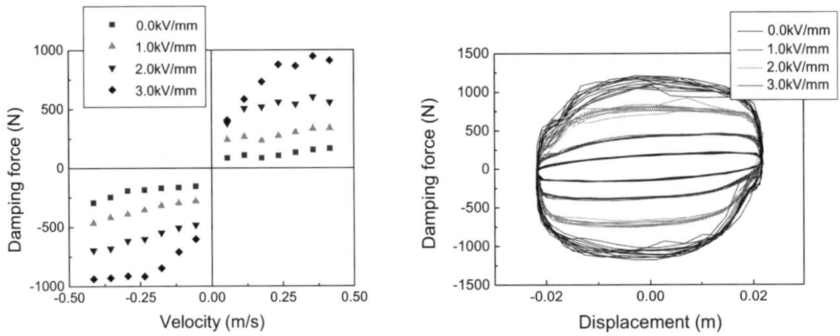

(a) damping force vs. piston velocity (b) damping force vs. displacement

Figure 2. Measured damping characteristics of the front ER shock absorber.

In order to get various vibration information such as pitch, roll, yaw and acceleration of the centroid of the vehicle during driving, accelerometer and gyroscope are located at the center of the vehicle and 3-axis accelerometer is located on the frame under the driver's seat. Wire type LVDT and accelerometer sensors are used to measure vertical vibration of each ER shock absorbers where LVDT is fixed at the ER shock absorber and the wire is hooked at the vehicle body.

For the implementation of skyhook control algorithm which is very effective for real vehicle environment, vertical velocity of the body is calculated by numerical integration of the measured data by accelerometer located at the suspension. Relative displacements between body and wheel are measured by using LVDT located at the suspension. Trigger sensor is adopted in which aluminum tape which can reflect the light is adhered to start and stop points of the road. Data acquisition and vibration control can be started after generated voltage is transferred to the data acquisition system.

In order to evaluate vibration control performance of the ER suspension system, two kinds of test environment are prepared such as bump and random road. For the bump test, bump is prepared with the 80*cm* width and 7*cm* height as shown in figure 6. Bump response test is carried out with 30*km/h* vehicle speed. For the evaluation of random road responses, various road conditions are prepared such as paved noise road, unpaved noise road and long waved road as shown in figure 7. Road test was undertaken with 72*km/h* vehicle speed. Vehicle speed is measured by GPS and regarded as a real speed instead of speed on speedometer. Professional test driver is selected in order to ensure the safety and drive with constant vehicle velocity.

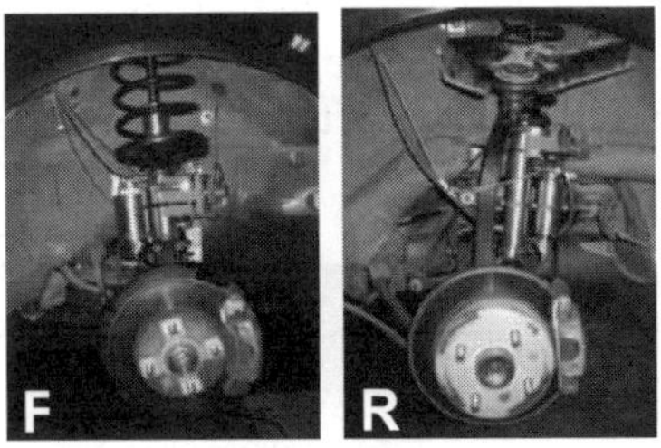

Figure 3. Installed front and rear ER suspension system.

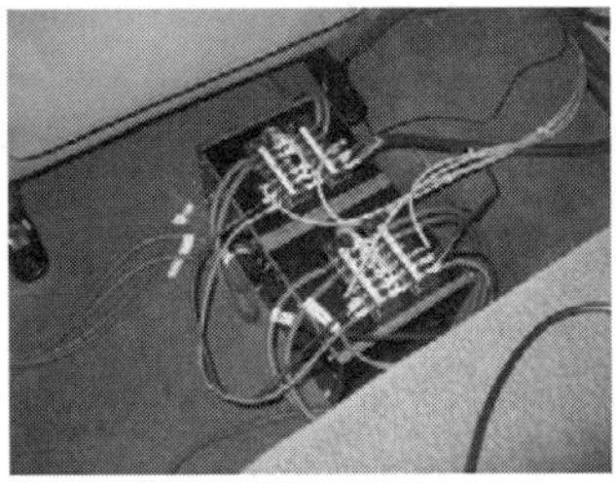

Figure 4. Installed high voltage amplifier.

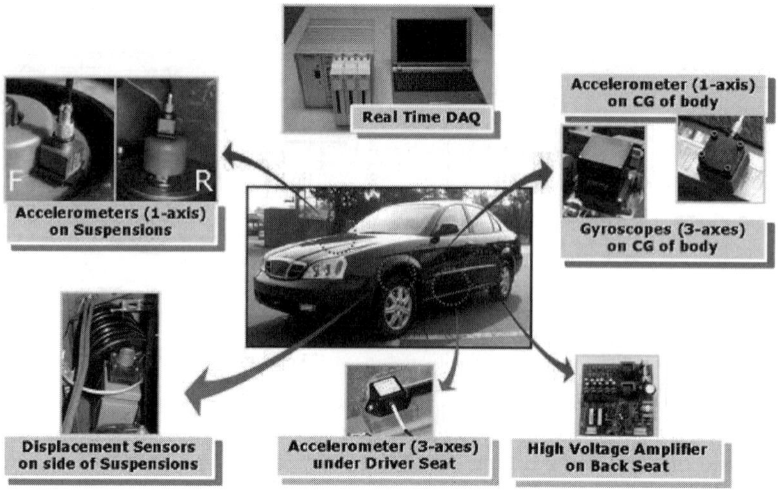

Figure 5. Photograph of sensor configuration

4 Field Test Results and Discussion

For the performance evaluation of the vibration reduction with ER suspension system, bump test is undertaken with 30*km/h* vehicle speed. Test results are presented in figure 8 by using pitch angular displacement, vertical acceleration of the vehicle body at C.G. and travel of the front and real suspension. By adopting ER suspension system with control algorithm, the vibration of vehicle is rapidly reduced after passing the bump. It is clearly observed that vertical acceleration of the vehicle body and pitch angular displacements are reduced up to 28% as shown in the figure. And also, travels of the front wheel and real wheel in the driver's side are effectively reduced. This means that driving stability is increased with control algorithm which is designed with focusing on the enhancement of the ride quality. In order to evaluate the improvement of the ride performance in driving, road tests are undertaken under various kinds of road conditions. Figure 9 shows the vehicle responses under long waved road. The vibration of vehicle was successfully attenuated up to 42% by activating the ER shock absorbers in the point of vertical acceleration and pitch angular displacements of the vehicle body, and travels of the front and real wheel. In figures 10 and 11, power spectral densities (PSD) are presented in the frequency domain for paved road and unpaved road, respectively. It is clearly seen that vibration energy of the vehicle was considerably reduced up to 50% at the resonance frequency area.

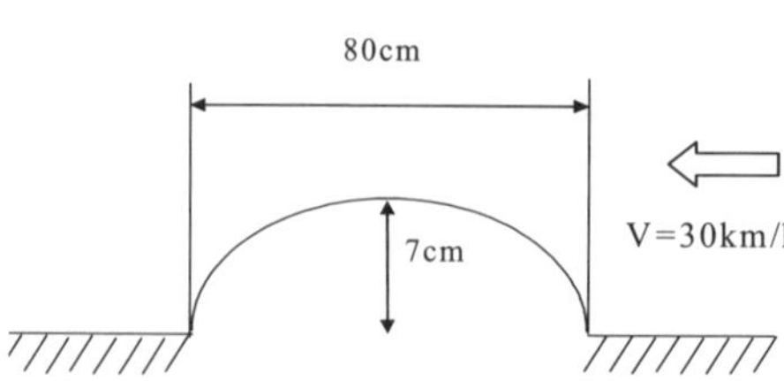

Figure 6. Bump excitation profile.

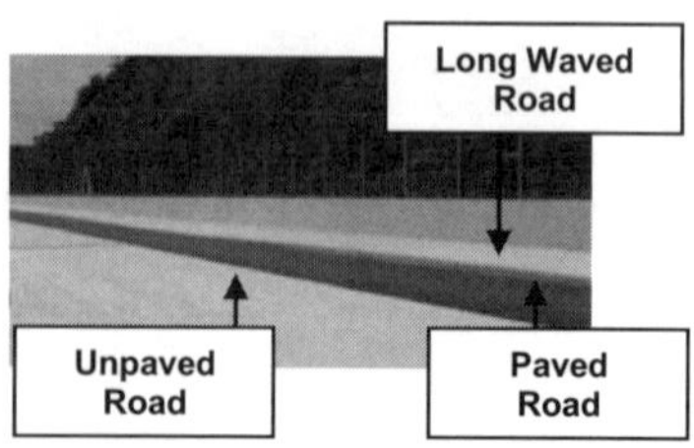

Figure 7. Various road conditions for road test.

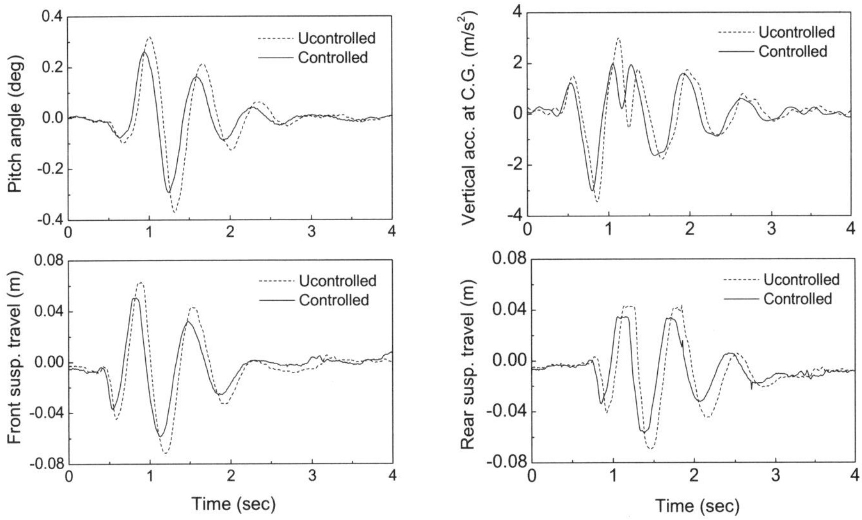

Figure 8. Bump responses with 30*km/h* vehicle speed.

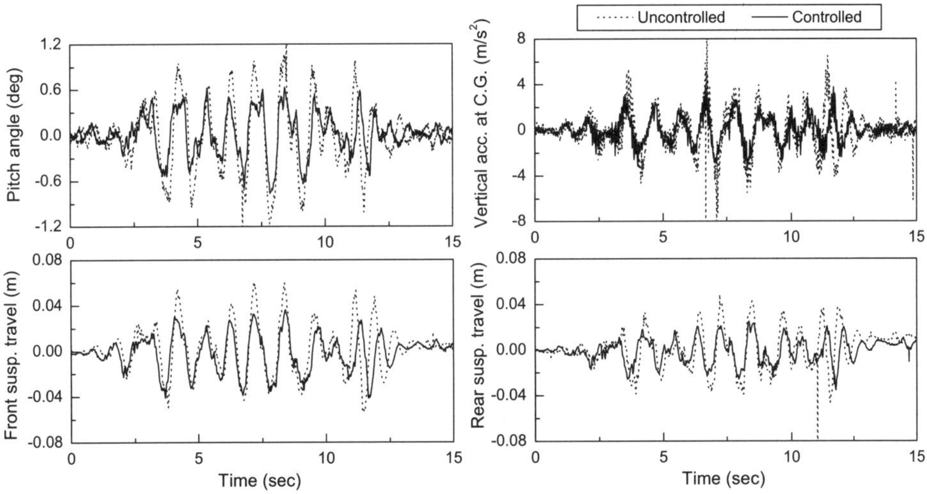

Figure 9. Random road responses under long waved road.

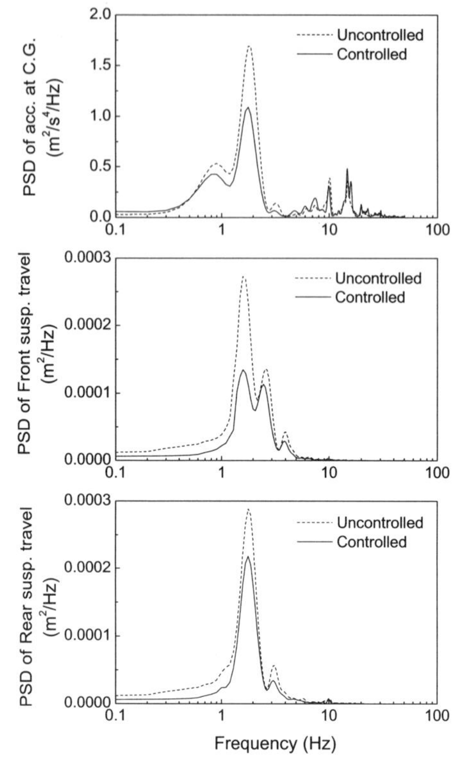

Figure 10. PSD of random road responses under paved road.

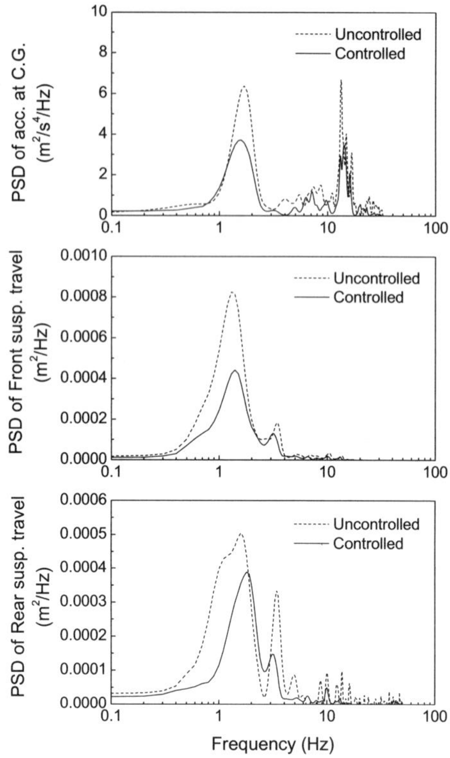

Figure 11. PSD of random road responses under unpaved road.

5 Conclusion

Vehicle field test was undertaken in order to evaluate performance characteristics of a semi-active ER suspension system associated with skyhook controller. As a first step, four ER shock absorbers were designed and their field-dependant damping characteristics were experimentally evaluated. After assembling the demonstration vehicle with ER suspension system, field tests were performed under various road conditions. It has been demonstrated through bump and driving road tests that ride comfort of the vehicle can be much improved by controlling the semi-active ER suspension system. In addition, the improvement of the steering stability was also demonstrated by showing the reduction of the pitch motions. The results presented in this work are quite self-explanatory justifying that the semi-active ER suspension system can be effectively employed to the passenger vehicle with improved both ride comfort and steering stability. It is finally remarked that field tests using various control strategies to much improve the performance, and control durability test will be undertaken in the near future as a second phase of this study.

6 Acknowledgement

This work was supported by the National Research Laboratory (NRL) program. This financial support is gratefully acknowledged.

References

1. Sturk, M., Wu, X. M. and Wong, J. Y., Development and evaluation of a high voltage supply unit electro-rheological fluid dampers. *Vehicle System Dynamics*, **24** (1995) pp. 101-121.
2. Nakano, M. and Novel, A., Semi-actave control of automotive suspension using an electrorheological shock absorber. *Proceedings of the 5th International Conference on ER fluids, MR Suspensions and Associated Technology* (Singapore, 1996) pp. 645-653.
3. Petek, N. K., Romstadt, D. L., Lizell, M. B. and Weyenberg, T. R., Demonstration of an automotive semi-active suspension using electro-rheological fluid. *SAE Technical Paper Series* 950586 (1995).
4. Peel, D. J., Stanway, R. and Bullough, W. A., Modeling and control of an ER long-stroke vibration damper. *Proceedings of SPIE on Smart Structures and Materials* (San Diego, U.S.A., 1998) pp. 416-427.
5. Wereley, N. M. and Pang, L., Nondimensional analysis of semi-active electrorheological and magnetorheological dampers using approximate parallel plate models. *Smart Materials and Structures*, **7** (1998) pp. 732-743.
6. Choi, S. B., Choi, Y. T., Chang, E. G. and Han, S. J., Control characteristics of a continuously variable ER damper. *Mechatronics*, **8** (1998) pp. 143-161.

EXAMINATION OF THROUGHFLOW IN A RADIAL ERF CLUTCH*

S. M. Chen [1], W. A. Bullough[2] and D J Ellam[3]

[1] *College of Mechanical Engineering, Fuzhou University, Fujian Province, 350002, China*

Email: smch2002@vip.sina.com

[2,3] *Department of Mechanical Engineering, The University of Sheffield, Sheffield S1 3JD, UK*

Email: W.A.Bullough@sheffield.ac.uk

A prototype clutch having horizontal rotational axis and a narrow gap between its two parallel radial disks is examined using CFD. The narrow gap between the discs is filled with an unexcited electro structured fluid (ESF); the through flow of {electrorheological (ERF) or magnetorheological (MRF)} fluid that would arise is contemplated as a means of contributing towards the cooling of the clutch. The throughflow, caused by body /centrifugal force action, is examined firstly when both plates rotate together at speed Ω . The aim is to estimate the flow rates arising under conditions of different rotational velocities, clutch geometries and fluid properties. Following this the cooling of the clutch is examined for the case of one plate fixed and the other rotating at a constant speed throughout. The inlet and outlet port pressures are made the same and the fluid is taken to be unexcited / Newtonian with viscosity μ .

1 Introduction

Previous works have considered 2d flow in radial and circumferential ER clutch geometries with a view to examining the improvement of temperature control-by pumping fluid through the gap between excited bounding stationary and moving electrode plates as per the hydraulic principle. Some limitation on the torque transmission capability is consequential since the yield stress is distributed between the two directions {1}. The problem of heat balance {2} in clutch designs based on both of these arrangements was studied in order to assess the magnitude of the cooling problem and to verify CFD techniques for future use. Now follows an examination of the usefulness of throughflow control on the maximum temperature generated.

The present paper considers the possibilities for self induced flow in a radial clutch. Initially, the maximal mass flow rate ($\dot{m}$) is assessed by calculating the flow rate that occurs when the excitation is switched of from a locked together clutch pair i.e., when the fluid is Newtonian and both plates rotate at the driver speed. Fluid inertia effects are neglected but all other contributing parameters except time are otherwise investigated. Secondly the steady slipping of a clutch is examined (for a selected case) to see if the induced flow has much effect on the temperature of the fluid leaving the clutch. In this case the flow field is assumed to be laminar and constant viscosity / adiabatic: the flow is unrestricted in the two situations compared.

---

* **The Author Gratefully Acknowledges the Support of K.C.Wong Education foundation, Hong Kong.**

2 Clutch and CFD modelling

2.1 Configuration

A virtual two plate radial clutch with horizontal rotational axis and narrow gap between the two perpendicular radial disks is shown in figure 1(a). The narrow gap between the discs is filled with electro-structured liquid; The ESF circulates freely from inlet to outlet.

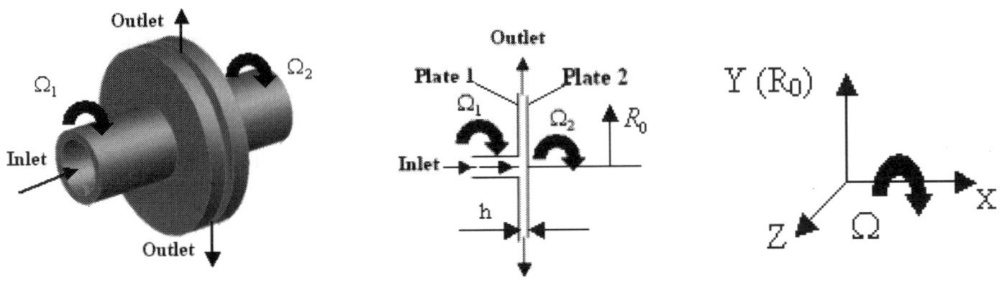

(a) Outline figure **(b)** Simple model **(c)** Coordinates

Figure.1 Model of radial clutch

The equations of motion are described in [3]. For reasons of future needs CFD is used to solve the fluid field.

2.2 CFD modeling procedure

Laminar flow was set in Fluent. Firstly, for set angular velocities $\Omega = 500$ to $1000(rpm)$ with outlet radial size R_0=40~80(mm), inlet radius size R_i =3 to 8 (mm), h=0.5 to1 (mm), and the invariant viscosity 100~300 (mPa.s). Table 1 summarizes the values of variables used in this problem. Reynolds number ($\mathrm{Re} = \rho\Omega R_0^2 / \mu$) = 800-2100.

Table 1. Variable of used in modeling

No	Contents	Unit	Variable
1	Plate angular velocity (Ω)	r.p.m.	500, 800, and 1000
2	Radial (y-direction) size (R_0)	mm	40, 50, 60, 70 and 80
3	Inlet aperture radius (R_i)	mm	3, 5, and 8
4	Gap size h	mm	0.5, 0.6, 0.7,0.8, 0.9and 1
5	Inlet pressure	Pa	0~1000, interval 100
6	Inlet dynamic viscosity (μ)	Pa.s	0.1, 0.15, 0.2, 0.25, and 0.3
7	Inlet density (ρ)	Kg/m^3	900,1000,1200, and 1500

The solution strategies involve 2D axisymmetric swirling flows and with both initial inlet and outlet pressures set at zero when both plates rotate together fluid flows 2D throughout due to centrifugal and viscous forces. The difficulties associated with solving swirling and rotating flows are a result of the high degree of coupling between the momentum equations, which is introduced when the influence of the rotational terms is large. A high level of rotation introduces a large radial pressure gradient, which drives the flow in both axial and radial directions. In turn, this determines the distribution of the swirl or rotation in the field. This coupling may lead to instabilities in the solution process, which may require special solution techniques in order to obtain a converged solution. Because of this the PRESTO scheme was used in the Pressure list for Discretization in the Solution Controls panel, which is well suited for the steep pressure gradients involved in swirling flows and will ensure the processing stability and convergence. Table 2 details the CFD model used in this problem. All numerical solutions are converged to criterion 0.0001. Only slight programmers adjustments are necessary for the slipping clutch computations.

Table 2 CFD Modeling

Software	Fluent 5/6, finite volume, segregated solver	
Physical Model	2D, steady, laminar, incompressible, isothermal flow	
Fluid Properties	Newtonian fluid, density constant	
Equations: Flow, swirl velocity	interpolation scheme	2st order for all equations
	pressure-velocity coupling	SIMPLE
Solution strategies	Discretization Pressure	PROESTO
	grid ($\boldsymbol{x}$, $\boldsymbol{y}$) [$\boldsymbol{y}$ = radial direction]	60 (axial) x 200(radial)
Boundary Conditions	Plate 1, wall-1	The working situation see table 1
	Plate 2, wall-2	The working situation see table 1
	Inlet	Pressure inlet
	Outlet	Pressure outlet

3 Results and Discussions

Figure 2 shows typical values of the clutch instantaneous mass flow rate ($\dot{m}$) magnitude for a range of practical situations. The standard conditions are plate separations h=0.5 (mm), outlet radius R_0 = 50(mm), density ρ =1000 (kg/m^3), Newtonian viscosity $\mu = 150$ (mPa.s), inlet radius R_i=5(mm), and inlet and outlet initial pressure as zero. The flow rate was calculated for steady conditions and is that induced when both plates are rotating at the same speed Ω and is approximately that situation which would occur when two locked plates are just de-energised.

The fluid inertial effects are likely to be negligible [3]; subsequently the driven plate deceleration would depend on the inertia of the load i.e. in the absence of a parallel operating brake. Thus the amount of fluid flowing out over time interval Δt would be have been approximately $\dot{m}\Delta t/2$, i.e. if the deceleration was linear. The cooling rate would depend on the amount of fluid in the circuit, the convection capability of the outer case, whether or not an appreciable proportion of viscous loading on the slipping take up of torque segment of the machine cycle is involved and the slip in the deceleration part of the cycle. [4]

For the present calculation the torque $N = 0$ since both plates do not slip in the period for which the flow rate is calculated. The flow is influenced over a range of radii R_0 =40-80 (mm) by factor of 4 indicating proportional R_0^2 and doubling of speed from 500-1000 (rpm) factor four plus, suggesting the $R^2 \times \Omega^2$ of the centrifugal term. Thus a higher $R \times \Omega$ produces an increase in flow of approximately some 20 times over the range shown in Fig 2. Of course the increase in the speed of the outer radius, would also increase the shearing load, which depends on μ Ω R/h and the electrical load approximately proportional to πR^2 if the full clutch surface is energized.

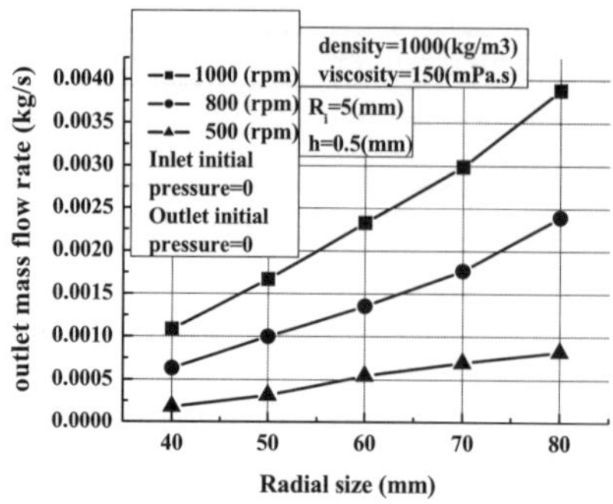

Fig.2 Effect radial size of plate on outlet mass rate when both plates rotational equal speed

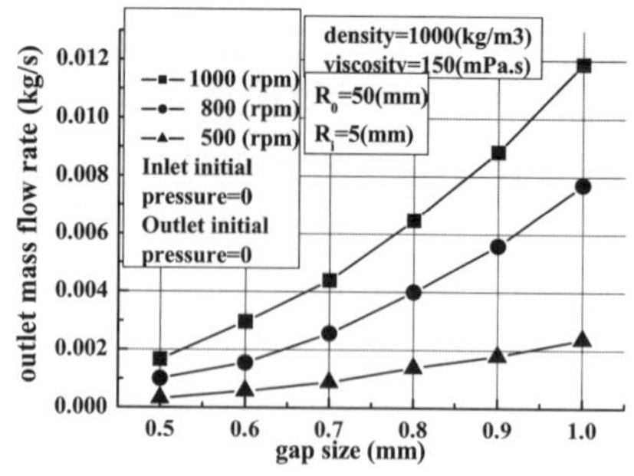

Fig.3 Effect gap size of plate on outlet mass rate when both plates rotational equal speed

Other factors – somewhat outside control if the engineer are h which shows approximately a $6\times$ increase in flowrate for a doubling the gap which also will double the electrical potential and heating were the clutch excited but decrease the viscous load, see fig.3.

The viscosity of the fluid has a very significant effect on flow rate (Fig 4), by a factor of approximately 4 over the 100 to 300 (mPa.s) practical range. Density has less effect, Fig.5 for a range of variation.

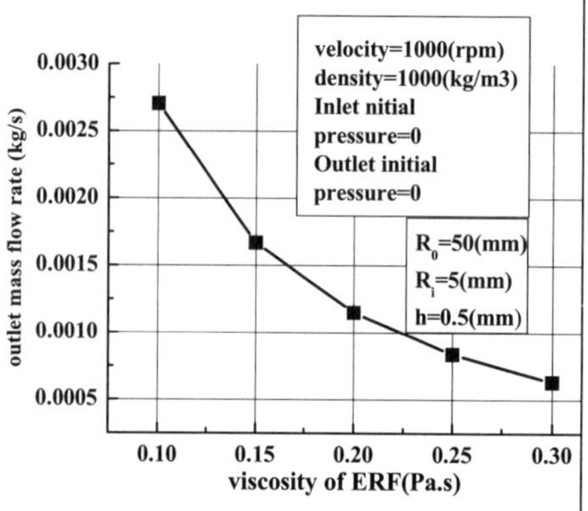

Fig.4 Effect viscosity on mass rate

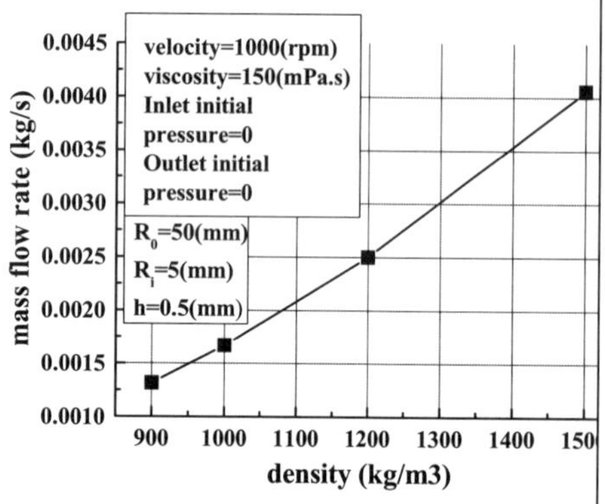

Fig.5 Effect density on mass rate

Other data extracted whilst the programme was available were the effect of throttling of the inlet (figure 6) and in figure 7 shows the slight effect of the amount of pressure drop in the cooling/circulation loop. Throughout the calculations no heat transfer is involved and any change due to temperature is due only to the changes implied in ρ and μ.

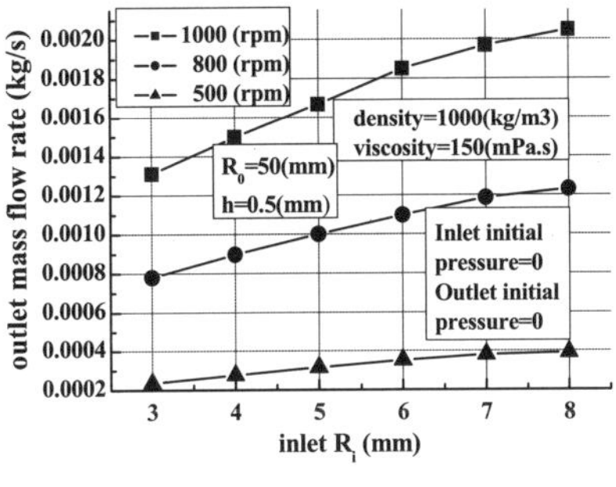

Fig.6 Effect inlet size on outlet mass rate

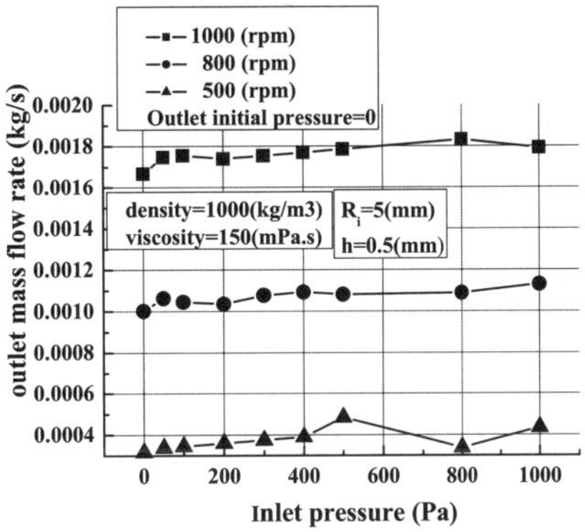

Fig.7 Effect inlet pressure on outlet mass rate

Figures 8 and 9 show the outcome of a similar situation to Figs **2** and **3** but for one plate stationary. In the slipping mode there will be a significant viscous shear load with heating of the fluid as it flows from inlet to outlet (tip radius), consequently thermodynamic considerations should apply. The magnitudes of these are difficult to estimate, see [5] for the conduction/convention case. For the present it is only required to assess the concept of throughflow cooling so, the simplest case of (impractical) clutch form, adiabatic boundaries and non-temperature dependant viscosity are chosen as boundary conditions i.e. in addition to the laminar, Newtonian flow approach already stated. Slipping mode and two rotating plate mode results differ only in the magnitude of the mass flow rates: compare figures 8 and 9 with **2** and **3**, where there is no viscous load arising save in the radial direction. For this reason and the approximate nature of the solution only the extreme (high speed) conditions are investigated.

Using the simplified heat balance equation $\Omega \cdot N = \dot{m} \cdot Cp \cdot \Delta T$ for a control volume over the clutch, calculating N by $2\pi R^4 \mu\Omega/4h$ where for $\Omega = 1000\text{rpm}$, Cp=4182J/Kg.^{0} C the temperature rise for h = 0.5mm R = 50mm tends to 3 to 4 °C. Allowing for the dual sided nature of the clutch in [5] and a possible lower value of Cp this order of temperature is handlable. It also goes some way to vindicating, the adiabatic and constant viscosity, and assumptions.

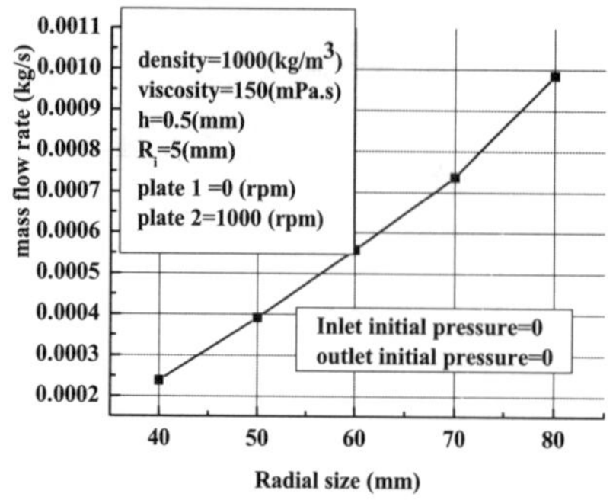

Fig.8 Effect radial size of plate on outlet mass rate when plate 1 is stationary

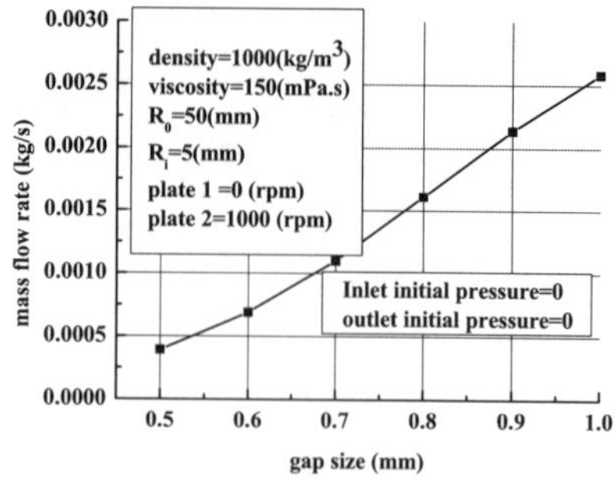

Fig.9 Effect gap size of plate on outlet mass rate when plate 1 is stationary

4. Conclusion

The results produced by The CFD exercise are not a solution to the clutch-cooling problem. Rather they throw light on what magnitudes of flow rates may be possible for on-off flow control (by excitation switching) or the thermodynamics of a stationary rotor case. Assumptions made are generally sensible but need consideration in more detail and refinement before CFD is applied to any real clutch circuit and operational sequence. No verification is available but in the past CFD has proved reasonable precise when applied to similar situations. Whether significant cooling can be achieved in any practical situation where the methods considered here can be applied is not known at this stage. However, a method of assessment is shown to be available and flow rates are likely to be significant in a removal of heat sense (given the high density and specific heat of ERF) compared to convection from the clutch surface [5].

It may be summarized that through flow cooling via the self induced flow of the radial type clutch is proved to the extent that a more rigorous analysis (CFD) of the concept, one incorporating convection from the case of a practical clutch form with conductance and non Newtonian fluid is called for: there is a serious prospect of success.

The difference between torques on input and output shafts stands in need of further investigation- a wide discrepancy could limit clutch speed and mass flow rate. Since no electrical conductance has been included in the considerations to date, the result applies to MR clutches (also). The effect of any casing, which is used to house the fluid, needs to be included in the model for a real clutch unit.

References

1. D J Ellam, RJ Atkin, and W A Bullough. Appraisal of an ESF Radial Plate Clutch with cooling flow using a NNF CFD Solver. FED-Vol.252, Rheology and Fluid Mechanics of Non linear Materials ASME 2000 ,117-125.

2. DJ Ellam,RJ Atkin and WA Bullough. Analysis of Smart Fluid Clutches with Cooling Flow using Two Dimensional Bingham plastic Analysis and Computational Fluid Dynamics. Proc. IMechE vol.219 Part A: J. Power and Energy. 639-652
3. D J Ellam. Modeling Smart Fluid Devices Using Computational Fluid Dynamics. PhD thesis 2004. University of Sheffield.
4. Makin J et al. Methodology Performances and Problems in ER Clutch Based Positioning Mechanisms. Proc. 6th Int. Conf. ER/MR Fluids, Yonezawa, Japan, July 1997.
5. D J Ellam, W A Bullough and S M Chen. Comparison of ERF Clutch Designs. 10th Int. Conf. on ER/MR Fluids,June 18-22,2006 ,Lake Tahoe, USA.

INVESTIGATION OF DISSIPATIVITY FOR SEMIACTIVE CONTROL OF THE BASE ISOLATED BENCHMARK BUILDING WITH MR DAMPERS

BARIS ERKUS AND ERIK A. JOHNSON

Civil and Environmental Engineering, University of Southern California, Los Angeles, CA 90089-2531, USA
E-mail: JohnsonE@usc.edu

This paper investigates the effects of dissipativity on the seismic performance of the base isolated benchmark building controlled by magnetorheological (MR) fluid dampers. In the past decade, MR dampers have received special attention in the structural control community due to the advantages over other control strategies from a seismic mitigation perspective. However, due to their nonlinear behavior, which is mainly characterized by their dissipative nature, development of control strategies for MR dampers may be challenging. In general, dissipativity is never taken into account in the design process; therefore, control of the damper may not be efficient in some cases. Recently, several dissipativity measures are introduced to aid the selection of control design for smart dampers. While, these indices have proved to be useful for simple structures and idealized dampers, they have not yet been investigated for realistic building and damper models. The recently introduced base isolated benchmark building and the 20-ton MR damper model are realistic and very suitable for the investigation of dissipativity. In this study, the linear quadratic regulator (LQR) is selected as the control strategy to command several MR dampers to control the base isolated benchmark building. It is shown that control designs with higher dissipativity characteristics are more suitable for MR dampers than the ones that have lower dissipativity.

1 Introduction

In the past decade, semiactive control of structures using magnetorheological (MR) dampers has received special attention in the structural control community [1, 7]. Particularly, robustness, performance and reliability characteristics of MR dampers make them a sound alternative to other control technologies in seismic protection of structures.

One important characteristic of an MR damper is its dissipative behavior; *i.e.*, an MR damper can only exert forces that dissipate energy from the structure. In practice, there is no control strategy or theory that is specifically tailored for the dissipative behavior of an MR damper; generally, common semiactive control strategies use linear control theories assuming that the damper is a fully active device. One of these strategies is clipped optimal control [1], where a primary controller uses a linear quadratic optimal control and a secondary controller commands the damper to produce the primary control force as efficiently as possible (Figure 1). While this approach is shown to be very effective for many structural systems, there are some cases where the semiactive control approach fails to be efficient [2,4].

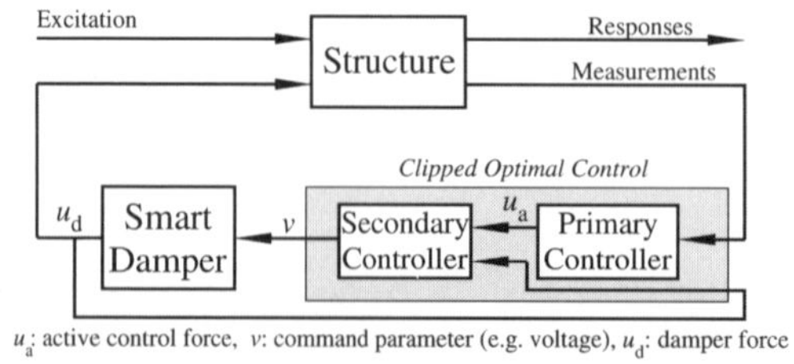

Figure 1. Clipped optimal control

Previously, the authors have proposed *dissipativity indices* to quantify the dissipative nature of a controller in the semiactive control design for an MR damper [5]. Dissipativity indices are not only useful to identify control designs that can and cannot be idealized by an MR damper efficiently, but also they can be used in the development of new control theories that will result in more efficient deployment of an MR damper. Numerical investigations of the semiactive control of two simple two-degree-of-freedom structures showed that control designs that result in mostly dissipative primary forces, which are identified by the dissipativity indices, are more suitable for a damper in the semiactive control [5]. However, these results were based on simple structure models. It would also be useful to investigate a realistic control problem to exploit further the applicability of the dissipativity indices. A recently introduced base-isolated benchmark problem [3, 6] is very suitable for this purpose as it includes a realistic structure model and the challenges that are common in a real-life control design.

This paper investigates the effects of dissipativity on the seismic performance of the base isolated benchmark building controlled by magnetorheological fluid (MR) dampers using recently introduced dissipativity indices. A linear quadratic regulator (LQR) is selected as the primary controller in a clipped optimal control strategy to command several MR dampers to control the base isolated benchmark building. A Kalman filter is used to estimate the states, and a Kanai-Tajimi filter is used to model the ground motion. Two control designs are considered. The first design is selected such that it yields high-dissipativity indices while the second design has low dissipativity. Extensive numerical studies show that dissipativity indices are very useful to identify control designs that are suitable for MR dampers for this realistic control problem.

2 Dissipativity and Dissipativity Indices

Consider an external force $f(t)$ applied to a system. Let the velocity of a point on the system be $v(t)$. A strictly dissipative force is defined by

$$f(t)v(t) \leq 0, \quad \text{for all } t \geq 0 \Leftrightarrow f(t) \text{ is strictly dissipative}$$

Based on this definition, a simple deterministic dissipativity index for a primary control force u_{a} in clipped optimal control can easily be defined as

$$D_{\%} = 1 - \frac{1}{T}\int_{0}^{T} H[u_{\mathrm{a}} v_{\mathrm{d}}]dt$$

where v_{d} is the damper velocity. In addition to $D_{\%}$, two stochastic indices are defined as

$$D_{\mathrm{e}} = E[u_{\mathrm{a}} v_{\mathrm{d}}] \quad \text{and} \quad D_{\mathrm{ne}} = \frac{E[u_{\mathrm{a}} v_{\mathrm{d}}]}{\sqrt{E[u_{\mathrm{a}}^2]}\sqrt{E[v_{\mathrm{d}}^2]}}$$

where D_{e} is called the mean energy flow rate and D_{ne} is called the normalized mean energy flow rate. The indices given above are discussed in detail in [5].

Another deterministic index, which is defined in the benchmark problem [6], is

$$D_{\mathrm{EDC}} = \frac{E_{\mathrm{CTR}}}{E_{\mathrm{EXC}}}$$

where E_{EXC} is the energy input to the system by the excitation, and E_{CTR} is the energy dissipated by the controllers.

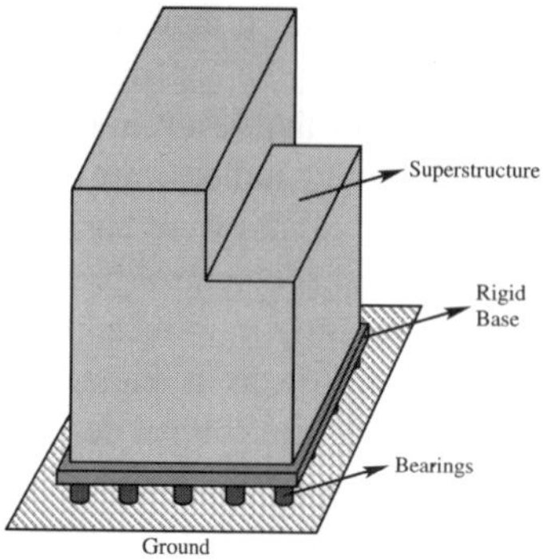

Figure 2. Benchmark structure

The indices given above are defined for one controller. For systems with more than one controller, a cumulative index is defined by

$$D^c = \sum_{i=1}^{N} w_i D_i \quad \text{and} \quad w_i = \frac{\text{RMS}(u_i)}{\sum_{j=1}^{N} \text{RMS}(u_j)}$$

where D^c is the cumulative dissipativity index, N is the number of control devices, D_i is the dissipativity index of the ith device, and u_i is the ith force.

In the following, dissipativity indices are used to analyze the benchmark structure.

Table 1. Performance indices

Index	Definition
J_1	Peak Base Shear
J_2	Peak First Floor Shear
J_3	Peak Base Drift
J_4	Peak Interstory Drift
J_5	Peak Absolute Floor Acceleration
J_6	Peak Controller Force
J_7	RMS Base Drift
J_8	RMS Absolute Floor Acceleration
J_9	Energy Absorbed by the Control Devices
J_{10}	Peak Controller Force
J_{11}	RMS Floor Drifts

3 Semiactive Control of Base-isolated Benchmark Building with MR Dampers

The base-isolated benchmark problem was introduced by the ASCE Technical Committee on Structural Control to allow engineers and researchers to implement their control approaches for a realistic base-isolated structure [6]. The benchmark structure is an eight-story frame building with linear beams, columns, bracing elements and rigid slabs (Figure 2) and represents a typical building in the Los Angeles area. The superstructure rests on a rigid concrete base, which is isolated from the ground by an isolation layer. To evaluate the control strategies, several performance indices are introduced (Table 1). The original benchmark problem uses seven historical earthquake ground acceleration data for numerical analyses. These earthquakes are: 1994 Northridge (Newhall, Sylmar, Rinaldi), 1940 El Centro, 1995, Kobe, 1999 Jiji and 1992 Erzincan earthquakes. Further details of the benchmark structure can be found in the definition paper [6].

In this research, a linear quadratic regulator is selected as the primary controller and defined as follows. Consider a linear time-invariant system:

$$\dot{\mathbf{q}} = \mathbf{A}\mathbf{q} + \mathbf{B}\mathbf{u} + \mathbf{E}\mathbf{w}$$
$$\mathbf{z} = \mathbf{C}_z\mathbf{q} + \mathbf{D}_z\mathbf{u} + \mathbf{F}_z\mathbf{w}$$

where $\mathbf{q}$ is the state vector, $\mathbf{u}$ is vector of control forces, $\mathbf{w}$ is a stationary white noise stochastic disturbance with unit intensity, and $\mathbf{z}$ is the vector of outputs to be minimized. In this paper, $\mathbf{z}$ is selected to be the drifts of the base corners and absolute floor accelerations. The LQR problem is to find a control gain that satisfies the optimization

$$\min \quad E[\mathbf{z}^{\mathrm{T}}\tilde{\mathbf{Q}}\mathbf{z} + \mathbf{u}^{\mathrm{T}}\tilde{\mathbf{R}}\mathbf{u} + \mathbf{z}^{\mathrm{T}}\tilde{\mathbf{N}}\mathbf{u} + \mathbf{u}^{\mathrm{T}}\tilde{\mathbf{N}}^{\mathrm{T}}\mathbf{z}]$$
$$\text{subject to} \quad \dot{\mathbf{q}} = \mathbf{A}\mathbf{q} + \mathbf{B}\mathbf{u} + \mathbf{E}\mathbf{w} \quad \text{and} \quad \mathbf{u} = -\mathbf{K}\mathbf{q}$$

Table 2. Conditions on the indices for the control designs

J	DES1	DES2
J_3	min (J_3)	$J_3 < 1.0$
J_4	$J_4 < 1.0$	
J_5	$J_5 < 1.0$	min (J_5)
J_{10}	$J_{10} < 0.15$	

Table 3. Classification of dissip. indices

	Dissipativity Index
GROUP 1	Normalized mean energy dissipation rate, D_{ne}
	Percentage of dissipative forces, $D_{\%}$
GROUP 2	Mean energy dissipation rate, D_e
	Energy dissipated by the controller, D_{EDC}

where $\tilde{\mathbf{Q}} \geq 0$ and $\tilde{\mathbf{R}} > 0$ are symmetric weighting matrices, and $\mathbf{K}$ is a constant feedback gain matrix. In this research, $\tilde{\mathbf{Q}}$ is selected as

$$\tilde{\mathbf{Q}} = b\begin{bmatrix} a\mathbf{I} & \\ & \mathbf{I}\omega^{-4} \end{bmatrix}$$

where a is the weight on the base corner drifts, ab is the weight on the absolute floor accelerations, and ω is a frequency term that is used to have a $\tilde{\mathbf{Q}}$ with consistent units. Further, a Kalman filter is used to estimate the states, where the measurements are selected as the floor and ground absolute accelerations and controller drifts. To model the earthquake ground motion, a Kanai-Tajimi filter is appended to the structure. A 20-ton MR damper model [8] is used in the analyses. Twelve dampers are located at the isolation layer: two dampers (x and y directions) for five corners and two dampers in the center of mass of the base.

4 Numerical Simulations

In this section, the dissipativity characteristics of the benchmark structure controlled with clipped optimal control and MR dampers are investigated. For this purpose, two control designs, DES1 and DES2, are defined, where the control parameters are found by averaging the design parameters that are selected for each of the earthquake data such that the conditions on the performance indices given in Table 2 are satisfied.

The control design of the benchmark building is a very complex trade-off problem. To illustrate the challenge in the control design and the general appearance of some of the performance indices for a fully active system, the designs DES1 and DES2 are graphed along with some of the performance indices for the Newhall earthquake in Figure 3. (Although not presented here, the indices for the other earthquakes are very similar to the indices given for the Newhall earthquake.)

From the results of the fully active system, it can be observed that design of an efficient semiactive controller is a very challenging task for a realistic control problem. For example, each earthquake requires different control designs for optimality, and these designs may be very different from each other as shown in Figure 3. The usual procedure in semiactive control is to select the best active controller that suits the design goals and use that controller as a primary controller in the clipped optimal control strategy. In this study, these designs are selected as shown in Figure 3. Clearly, J_3 and J_5 are two competing indices that must be traded off; it is not possible to have a control design that simultaneously yields very low J_3 and J_5.

To provide further insight into the benchmark problem, the dissipativity indices for the fully active system are given: Figure 4 shows the cumulative stochastic dissipativity indices, and Figure 5 shows the deterministic dissipativity indices for the Newhall earthquake. A very important observation in the dissipativity indices is the similarity between (D_{ne}, $D_{\%}$) and (D_e, D_{EDC}) pairs in Figures 4 and 5. This is a very crucial result for

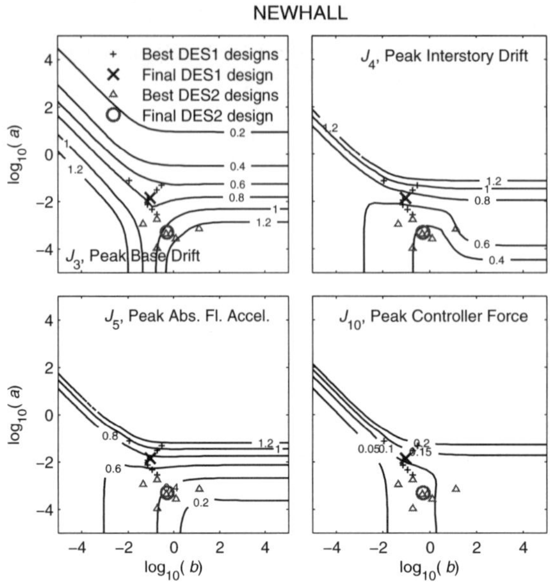

Figure 3. Indices J_3, J_4, J_5 and J_{10} for Newhall earthquake

the semiactive design because stochastic indices provide the dissipativity characteristics very efficiently via solution of a Lyapunov equation without making any numerical simulation, which is a very time-inefficient task and excitation dependent. Based on this result, the dissipativity indices can be classified as shown in Table 3.

Another very important observation that is derived from the performance index and dissipativity plots is that DES1 is a high dissipativity control design, while DES2 is a low dissipativity control design by nature. This result shows that it is more likely that the semiactive system will be more efficient in realizing DES1 than DES2 due to the dissipativity characteristics of the controllers. In fact, with the aid of the dissipativity indices, it is now very easy to comment on any other selected active control design, whether or not it is suitable for semiactive control with MR dampers.

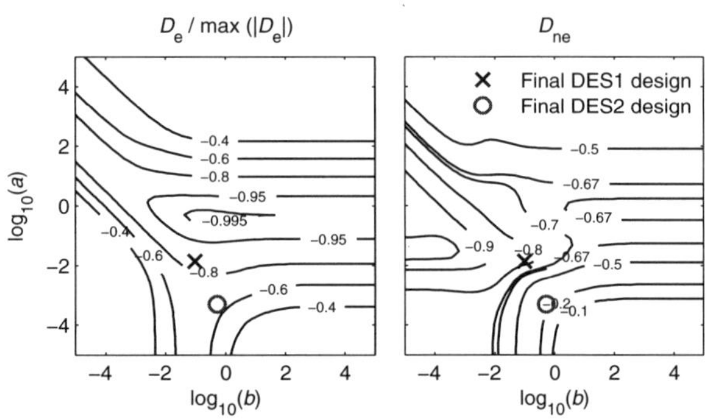

Figure 3. Stochastic dissipativity indices D_e and D_{ne}

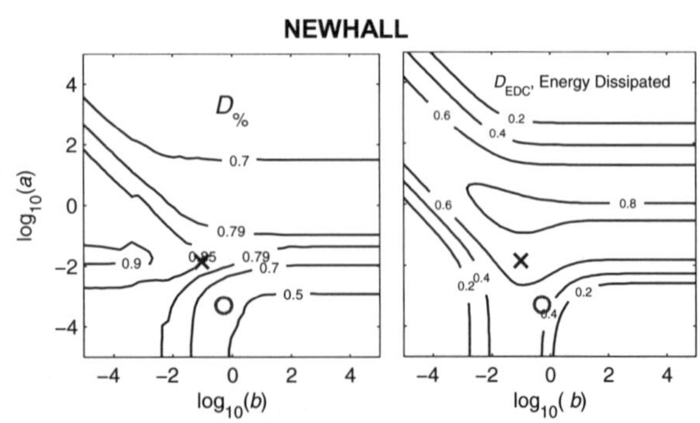

Figure 4. Deterministic dissipativity indices $D_\%$ and D_{EDC}

It is also observed that control designs with high dissipativity indices may not necessarily be the best control designs for semiactive control with MR dampers. For example, a control design with the parameters around $a = 1$ and $b = 1$ yield very high dissipativity indices (D_e, D_{EDC}). Although the J_3 index is very low for this design, other performance indices are very high; thus, this is not an acceptable active controller design in overall sense. In the next step, the semiactive simulations are evaluated for seven earthquakes. The performance indices for DES1 and DES2 are provided in Tables 4 and 5, respectively.

The results given in Tables 4 and 5 provide very

Table 5. Performance indices for DES1

Performance Indices		Newhall		Sylmar		El Centro		Rinaldi		Kobe		Jiji		Erzincan	
		ACT	SACT	ACT	SACT	ACT	SACT	ACT	SACT	ACT	SACT	ACT	SACT	ACT	SACT
J_1	Peak Base Shear	0.581	1.058	0.543	1.046	0.515	1.066	0.715	1.317	0.636	1.166	0.597	0.945	0.557	1.132
J_2	Peak First Floor Shear	0.611	1.24	0.583	1.183	0.514	1.039	0.71	1.442	0.644	1.149	0.607	0.917	0.558	0.998
J_3	Peak Base Drift	1.047	0.972	0.93	0.917	0.889	1.057	1.169	0.951	0.979	0.789	1.364	0.615	0.833	0.73
J_4	Peak Interstory Drift	0.379	1.123	0.695	1.297	0.375	0.962	0.693	1.277	0.597	1.461	0.64	0.911	0.612	0.916
J_5	Peak Absolute Floor Acceleration	0.372	1.157	0.699	1.545	0.292	1.12	0.653	1.625	0.504	1.438	0.636	1.102	0.527	1.071
J_6	Peak Controller Force	0.488	0.473	0.531	0.504	0.426	0.423	0.535	0.624	0.417	0.677	1.024	0.328	0.427	0.379
J_7	RMS Base Drift	1.034	1.025	0.755	0.925	1.444	1.068	0.969	0.977	1.06	0.885	1.674	0.639	0.704	0.767
J_8	RMS Absolute Floor Acceleration	0.423	1.066	0.421	0.999	0.322	0.911	0.553	1.165	0.411	1.157	0.636	0.877	0.416	0.877
J_9	Energy Absorbed by the Control	0.477	0.651	0.504	0.649	0.521	0.678	0.548	0.65	0.416	0.656	0.605	0.649	0.541	0.655
J_{10}	Peak Controller Force	0.087	0.084	0.126	0.12	0.035	0.035	0.129	0.15	0.056	0.091	0.389	0.125	0.094	0.083
J_{11}	RMS Floor Drifts	0.502	1.003	0.514	0.882	0.327	0.769	0.594	1.06	0.502	1.092	0.58	0.837	0.503	0.86

Table 4. Performance indices for DES2

Performance Indices		Newhall		Sylmar		El Centro		Rinaldi		Kobe		Jiji		Erzincan	
		ACT	SACT	ACT	SACT	ACT	SACT	ACT	SACT	ACT	SACT	ACT	SACT	ACT	SACT
J_1	Peak Base Shear	0.969	0.979	0.929	1.019	0.856	0.943	1.203	1.295	1.133	1.150	0.742	0.806	1.024	0.983
J_2	Peak First Floor Shear	1.071	1.188	1.005	1.087	0.914	0.965	1.210	1.302	1.183	1.160	0.766	0.802	1.028	0.994
J_3	Peak Base Drift	0.764	0.854	0.656	0.795	0.862	0.852	0.809	0.834	0.620	0.645	0.549	0.531	0.552	0.618
J_4	Peak Interstory Drift	0.689	1.103	1.169	1.338	0.725	0.930	1.204	1.284	1.139	1.568	0.821	0.876	1.006	1.032
J_5	Peak Absolute Floor Acceleration	0.750	1.087	1.309	1.580	0.594	1.122	1.226	1.507	1.012	1.559	0.861	0.965	0.984	1.021
J_6	Peak Controller Force	0.624	0.552	0.634	0.599	0.527	0.587	0.805	0.740	0.732	0.769	0.408	0.328	0.525	0.546
J_7	RMS Base Drift	0.774	0.848	0.606	0.747	0.904	0.854	0.692	0.777	0.748	0.709	0.617	0.563	0.519	0.606
J_8	RMS Absolute Floor Acceleration	0.804	1.110	0.799	0.996	0.564	0.929	1.069	1.240	0.830	1.232	0.743	0.820	0.768	0.840
J_9	Energy Absorbed by the Control	0.701	0.718	0.740	0.737	0.713	0.750	0.735	0.732	0.675	0.723	0.740	0.720	0.758	0.747
J_{10}	Peak Controller Force	0.111	0.099	0.150	0.142	0.044	0.049	0.194	0.178	0.099	0.104	0.155	0.124	0.115	0.120
J_{11}	RMS Floor Drifts	0.876	1.026	0.813	0.869	0.583	0.790	1.078	1.128	0.909	1.155	0.733	0.786	0.771	0.817

useful information about the semiactive performance of designs DES1 and DES2. It is easily observed that DES1 reduces J_3 successfully, while DES2 is not successful in reducing J_5. Considering the trade-off problem between DES1 and DES2, it is easily observed that a semiactive system is more suitable for DES1 than for DES2.

Table 6. Overall dissipativity indices

	$D_{\%}$		D_{EDC}		D_e	D_{ne}
	ACT	SACT	ACT	SACT	ACT	ACT
DES1	0.859	0.844	0.701	0.718	-1324	-0.828
DES2	0.588	0.688	0.477	0.651	-1012	-0.225

To observe the overall dissipativity characteristics of both active and semiactive systems, the cumulative indices are tabulated in Table 6. Considering the dissipativity analysis, the results of the semiactive simulations and the overall dissipativity indices, it can be concluded that dissipativity analysis provides very useful information in the semiactive design without making any computationally intensive simulations.

5 Conclusions

In this research, dissipativity indices are utilized to investigate the performance characteristics of the benchmark structure controlled with clipped optimal control and MR dampers, which represents a realistic structural control problem. It is found that the proposed stochastic indices efficiently provide the dissipativity characteristics of a control design without making computationally intensive simulations. It is also shown that control designs with high dissipativity indices are more suitable for semiactive control with MR dampers than control designs with low dissipativity indices.

6 Acknowledgements

The authors gratefully acknowledge the partial support of this research by the National Science Foundation under CAREER Grant CMS 00-94030, and by the US Department of Transportation through Grant 03-17 from the National Center for Metropolitan Transportation Research (METRANS).

References

1. Dyke, S. J., Spencer, B. F., Jr., Sain, M. K., and Carlson J. D. (1996). "Modeling and control of magnetorheological dampers for seismic response reduction." *Smart Materials & Structures*, **5**(5), 565–575.
2. Erkus, B., Abé, M., and Fujino, Y. (2002). "Investigation of semi-active control for seismic protection of elevated highway bridges." *Eng. Structures*, **24**(3), 281–293.
3. Erkus, B., and Johnson, E. A. (2006). "Smart base isolated benchmark building part III: a sample controller for bilinear isolation." *Journal of Structural Control and Health Monitoring*, **13**(2-3), 605-625..
4. Inaudi, J. A. (2000). "Performance of variable-damping systems: Theoretical analysis and simulation." *Proceedings of 3rd International Workshop on Structural Control*, Paris, France, 301– 316.
5. Johnson, E. A., and Erkus B. (2006). "Dissipativity and performance analysis of smart dampers via LMI synthesis," *Journal of Structural Control and Health Monitoring, published online at SCHM web page.*
6. Narasimhan, S., Nagarajaiah S., Gavin, H. P., and Johnson, E. A. (2006). "Smart base isolated building part I: problem definition." *Journal of Structural Control and Health Monitoring*, **13**(2–3), 573–588.
7. Patten, W. N., Kuo, C. C., He, Q., Liu, L., and Sack, R. L. (1994a). "Seismic structural control via hydraulic semiactive vibration dampers (SAVD)." *Proc. of the 1st World Conference on Structural Control*, Los Angeles, California, **FA2**, 83–92.
8. Yang, G., Spencer, B. F., Jr., Carlson, J. D., and Sain, M. K. (2002). "Large-scale MR fluid dampers: modeling and dynamic performance considerations." *Engineering Structures*, **24**(3), 309– 323.

A STUDY ON TIME-DELAY OF AN ER FLUID'S RESPONSE WHICH EFFECTS ON CONTROL PERFORMANCE OF SERVO-SYSTEMS

KEN' ICHI KOYANAGI

*Graduate School of Engineering, Osaka University**

Particle type ER fluids commonly have about 5-mili seconds response times. This delay needs to be taken into consideration when electric fields to energize ER fluids are controlled in real time. Especially, servo systems with ER fluid devices have a sampling time for the control loop that is about 1-mili second and so is shorter than the response time of ER fluids. Though the delay in this case has a great possibility of having effects on the performance or stability of servo systems, most research in this area has not mentioned the problem. We previously developed a particle ER fluid damper for fast and precise positioning of a direct-drive motor; its viscosity was controlled continuously in real time. In this paper, we investigated how the time delay affected damping performance and stability of this system.

1 Introduction

Many applications that use particle type ER fluids have been suggested, such as dampers, clutches, and brakes. Fluids of this type commonly have a response time of about 5-mili seconds. This delay is not a serious issue when constant electric fields are impressed like in the case of damping structures. However, the delay needs to be taken into consideration when the electric fields are controlled in real time. Especially, servo systems introduced with ER fluid devices have a sampling time for their control loop that is about 1-mili second and so is shorter than the response time of ER fluids. Though the delay in this case has a great possibility of having an effect on the performance or stability of these systems, most reported research in this area has not mentioned the problem.

In direct-drive (DD) motor systems [1] without reduction gears, disturbance directly influences motor dynamics and control accuracy because of the small damping [2]. In an earlier study, we introduced a particle type ER fluid damper to a DD motor system for fast and precise positioning. Viscosity of the damper was controlled continuously in real time to improve the performance of the DD system through dynamic viscosity changing [3]. However, even this study failed to discuss the problem of the ER fluid response time delay.

There have been many studies to clarify the mechanism of the ER effect [4], but few have studied the speed of the response itself. From concerns about experimental solutions, dead time or first-order lag from mechanical and/or control factors of the experimental system could be accumulated in results of experiments, and so it may be impossible to accurately measure just the delay of the fluid. However, just factors on the delay of the ER fluid can be extracted if the simulation results under various conditions are compared with the experimental ones. This paper investigates how time delay affects damping performance and stability.

---

* The author is now with the Department of Intelligent Systems Design Engineering, Faculty of Engineering, Toyama Prefectural University; Kurokawa 5180, Imizu, Toyama, 939-0398, Japan; E-mail: koyanagi@pu-toyama.ac.jp

Figure 1. ER damper and DD motor

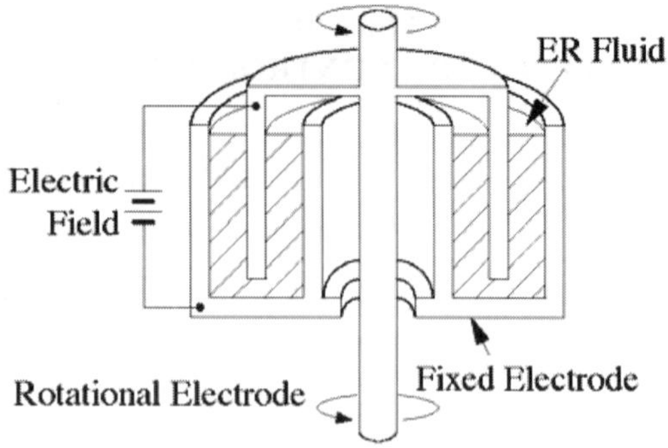

Figure 2. Model of ER damper

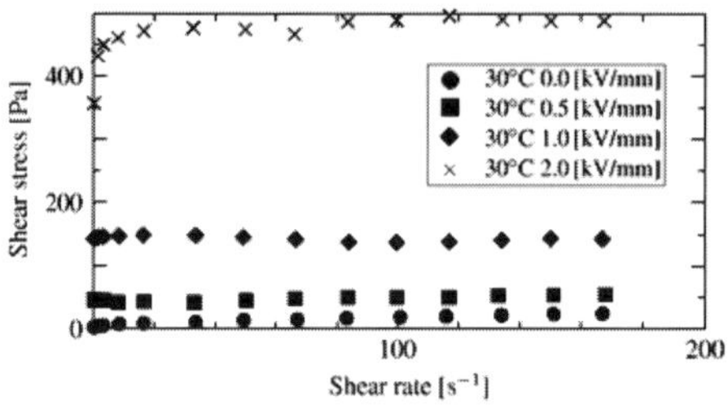

Figure 3. Electric field strength of ARP05

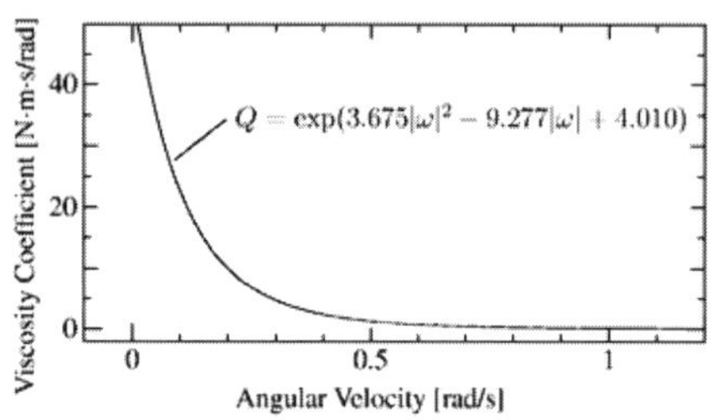

Figure 4. Relation between Q and ω

2 Experimental System

This study used the ER damper and DD motor system shown in Fig. 1 [2, 3]. A model of an ER damper is shown in Fig. 2. It uses particle type ER fluid, and consists of a rotational electrode connecting to the motor shaft and an electrode fixed to the outer circumstance of the motor housing. The gap between the electrodes is 0.5 mm. The maximum torque of the DD motor is 30 N·m. A resolver included in the DD motor has a resolution of 491,520 ppr. The sampling time for the control is set to 0.5 ms. Moment of inertia of the DD motor and the ER damper is $J = 0.075$ kg·m^2.

The particle type ER fluid used in this study was "ARP05" supplied by Asahi-Kasai Co. for research purpose. The characteristics of its share rate and share stress are shown in Fig. 3. The coefficient of viscosity of the damper, Q N·m·s/rad varied according to the angular velocity of the motor as in Fig. 4 [3].

3 Draft models of the delay of ER fluids

A microscopic or transient response model of particle type ER fluids has not been clarified. Common mathematical models of the response delay are written in a time-lag of first order or a dead-time style. In this paper, the response of the ER damper is captured macroscopically, and then the model is experientially considered as being approximated with a time-lag of first order or a dead-time.

First, the case of a first-order lag is presented. Here, the basic viscosity is Q_0, the variance of the viscosity is Q_1, and the time constant is T_Q. The transient response of the viscosity, $Q(s)$, can then be written as follows:

$$Q(s) = \frac{Q_1}{T_Q s + 1} + Q_0 \tag{1}$$

Second, the case of a dead-time is presented. Where, L_Q is the dead-time, and $Q(s)$ can be written as follows:

$$Q(s) = Q_1 e^{-L_Q s} + Q_0 \tag{2}$$

4 Theoretical verification of stability)

The transfer function from the input torque, T, to the output angle, θ, of the DD motor and ER damper system is written as follows.

$$G(s) = \frac{1}{Js^2 + Q(s)s} \cdot e^{-Ls} \tag{3}$$

The transfer function from the reference angle, θ_r, to the output angle, θ, of the model with the PD control shown in Fig. 5 is written as Eq. (4). In this chapter, the stability of the system is theoretically verified using this equation.

$$\frac{\theta}{\theta_r} = \frac{K_P e^{-Ls}}{Js^2 + \{Q(s) + K_D e^{-Ls}\}s + K_P e^{-Ls}} \tag{4}$$

4.1 The dead time in Q

In this section, the dead time lag style is considered with variations of the coefficient of viscous friction, Q; meaning that the variation of Q calculated with the equation in Fig. 4 is reflected in the system after the time lag, L_Q.

The closed loop transfer function of Eq. (4) becomes as follows using Eq. (2).

$$\frac{\theta}{\theta_r} = \frac{K_P e^{-Ls}}{Js^2 + (Q_0 + Qe^{-L_Q s} + K_D e^{-Ls})s + K_P e^{-Ls}} \tag{5}$$

The stability is analyzed by a parameter plane method [5] because of e^{-Ls} in Eq. (5). Substitution of $s = j\omega$ and $e^{-j\omega L} = \cos(\omega L) - j\sin(\omega L)$ to the characteristic equation gives the following stability limitation of K_P and K_D.

$$\begin{cases} K_P = (J\omega^2 - Q_1\omega \sin\omega L_Q)\cos\omega L + \omega(Q_0 + Q_1\cos\omega L_Q)\sin\omega L \\ K_D = (J\omega - Q_1\sin\omega L_Q)\sin\omega L - (Q_0 + Q_1\cos\omega L_Q)\cos\omega L \end{cases} \tag{6}$$

4.2 The 1st-order delay in Q

The closed loop function of Eq. (4) becomes the following equation using Eq. (1).

$$\frac{\theta}{\theta_r} = \frac{(T_Q s+1)K_P e^{-Ls}}{JT_Q s^3 + (J + Q_0 T_Q + T_Q K_D e^{-Ls})s^2 + \{(Q_1 + Q_0) + (T_Q K_P + K_D)e^{-Ls}\}s + K_P e^{-Ls}} \quad (7)$$

The same as in the previous section, the substitution of $s = j\omega$ and $e^{-j\omega L} = \cos(\omega L) - j\sin(\omega L)$ to the characteristic equation gives the following equations.

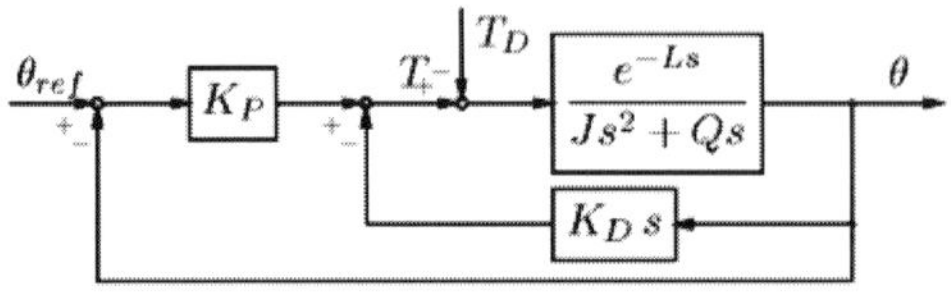

Figure 5. Block diagram of P-D control system

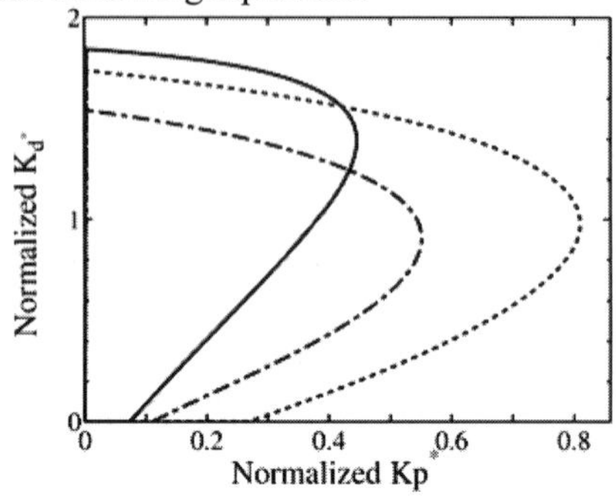

Figure 6. Stable area of P-D control

$$\begin{cases} K_P = \dfrac{1}{1+T_Q{}^2\omega^2}\{\omega^2(J - Q_1 T_Q + JT_Q{}^2\omega^2)\cos\omega L + \omega(Q_0 + Q_1 + Q_0 T_Q{}^2\omega^2)\sin\omega L\} \\ K_D = \dfrac{1}{1+T_Q{}^2\omega^2}\{\omega(J - Q_1 T_Q + JT_Q{}^2\omega^2)\sin\omega L - (Q_0 + Q_1 + Q_0 T_Q{}^2\omega^2)\cos\omega L\} \end{cases} \quad (8)$$

4.3 Comparison of stable regions

Changing the parameter ω in Eq. (6) and Eq. (8) gives stable regions for the models according to K_P and K_D. Figure 6 shows stable areas that are inside of curves; dotted-line: no delay, solid line: the dead time model, and dot-dashed line: the 1st-orderd delay model. Normalization is applied as $K_P=K_p^*J/L^2$ and $K_D=K_d^*J/L$.

For stability, transfer functions require that all coefficients of their denominator are plus. Therefore, stable regions are limited in the first quadrant of K_p^*–K_d^* plane as Fig. 6. Moreover, areas inside curves drawn by the parameter ω are the stable regions. As shown in the figure, if Q is variable and has a time delay, stable regions become smaller.

However, in this study, controlling gains for experiments and simulations in consideration of stability and performance are set at (K_p^*, K_d^*)=(0.0800, 0.464). The point of (K_p^*, K_d^*) is somewhat inside the stable regions.

5 Simulation experiments

5.1 Introduction of friction model

In this chapter, transient responses of systems with a time delay from an ER damper are investigated by simulating the step response. Different from the case of real experiments, a simulation model without disturbance hardly causes any vibration or other disturbing behavior; therefore, the differences of responses by a variety of lagged models are hard to confirm. Therefore, a friction model was introduced to the simulation model. A time delay for the DD motor system was also included in the simulation, so that the model as closely as possible resembled a real system.

The friction model had the relationship between angular velocity and command torque derived experimentally as shown in Fig. 7. The simulation result, including that of the friction model (solid lines) and the experimental results (dashed line) are shown in Fig. 8. The tendencies of the responses, such as the vibration cycle, were almost the same.

5.2 *Dead time in the coefficient of viscous friction Q*

First, Q is assumed to have dead time in its variance. Figure 9 shows the simulation results; a solid line indicates that the lag time L_Q was 5 ms, a dotted line that $L_Q = 0$ ms, and a dashed line is the experiment.

In this case, the behavior did not differ from the case where there was no delay in the simulation. It was also the same as the experimental result. However, a larger time delay of $L_Q = 50$ ms gave a different response, as in Fig. 10. A larger time delay clearly causes some deterioration to the damping property.

5.3 *Dead time in an applied electric field E*

This section shows the case of the dead time inserted after the command of E was outputted to the damper. That is, the electric field that accords to the command value of the viscosity at the sampling time is reflected in the system model after the time lag. In view of the damping property, the new Q is affected by the angular velocity when the new E has been inflicted on the system, so that Q may differ from the value of the Q desired when calculating the command for the viscosity.

Figure 11 shows the simulation result for 5 ms of dead time by the solid line, that of 0 ms dead time by the dotted line, and the experimental result by the dashed line. When the model had 0ms of dead time, the result was the same as when Q had dead time and was near to the experimental result. On the other hand, when the model had 5 ms of dead time, the response showed some difference and showed somewhat vibrational behavior, as shown in the figure.

5.4 *Time-lag of the first order in Q*

Where the time-lag of the first order showed lagged model of $Q(s)$ is treated. The simulation results of $T_Q = 5$ ms are shown in Fig. 12 by the solid line. The response shows little difference from that of the dead-time model of Q shown in Fig. 9. Therefore, the response of the system can be considered as being hardly affected, even if the system has a first ordered delay, in respect to the variance of Q.

5.5 *First ordered delay in E*

The simulation results when the time-lag of the first order is inserted after the command of E are shown in Fig. 13 by the solid line. The time constant was 5 ms.

This response is not similar to the case of the 1st-order delay in Q as in Fig. 12. The phenomenon can be explained as follows. When the motor got close to the reference angle and became slower, the desired Q increased, and then the desired E became bigger in the same way. However, the slow variance of E resulted in an insufficient magnitude of the electric field evening the vicinity of the reference. Therefore, the coefficient of viscosity was not large enough, and made $K_d^* + q^*$ smaller than that of the theoretical value; which in turn brought about an inadequate damping property and so a large overshooting. On the other hand, when the motor recovered from the overshooting, the electric field became large enough. Therefore, a large frictional torque was invoked, and so the angle was unable to return to the reference by PD control.

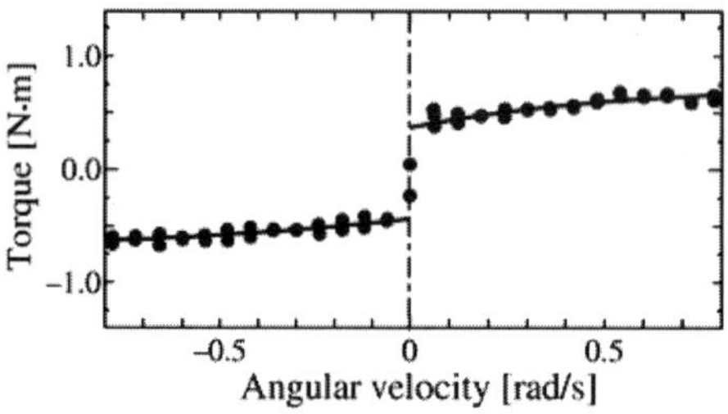

Figure 7. Friction model

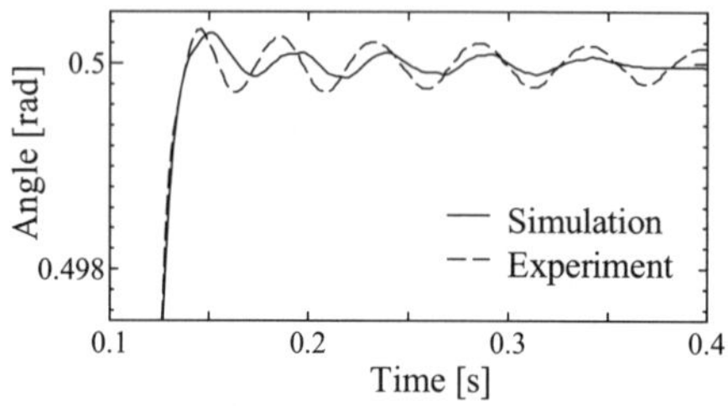

Figure 8. Verification of friction model

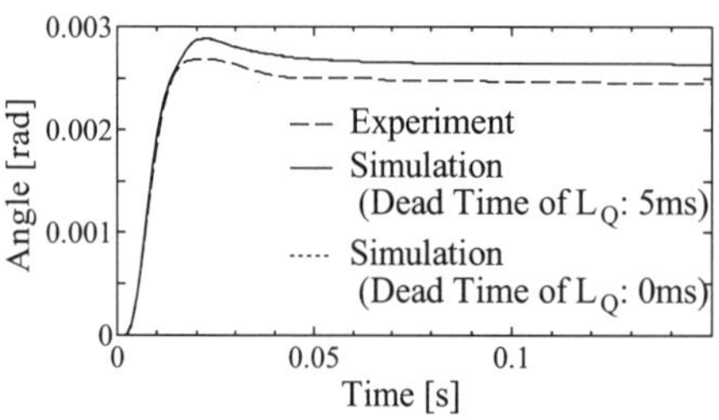

Figure 9. Result of dead time model of Q

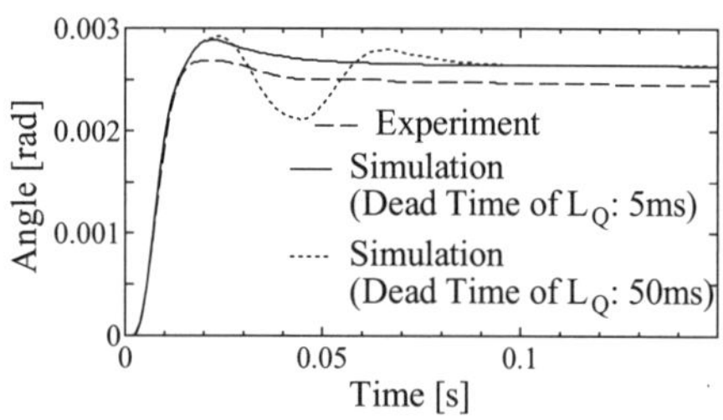

Figure 10. The case of large LQ

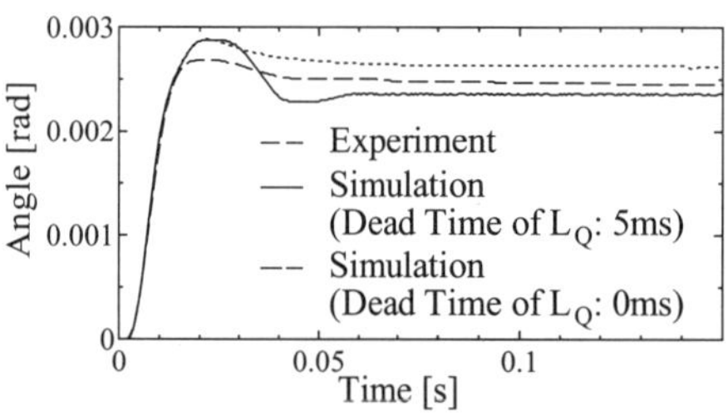

Figure 11. Result of dead time model of E

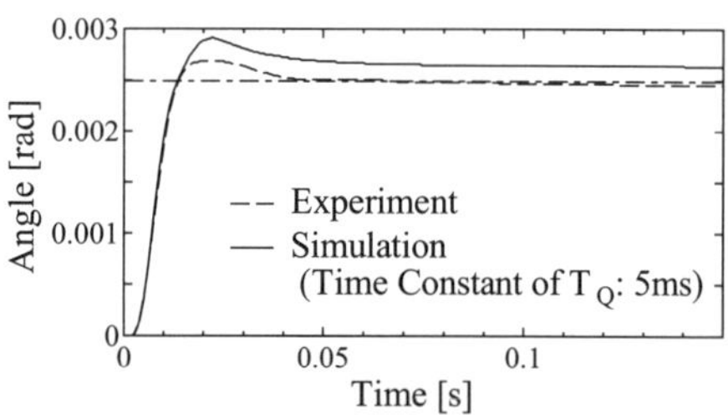

Figure 12. Result of 1st-order lag model of Q

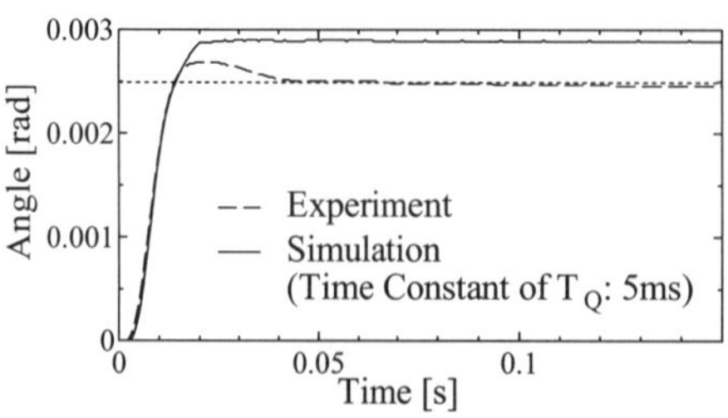

Figure 13. Result of 1st-order lag model of E

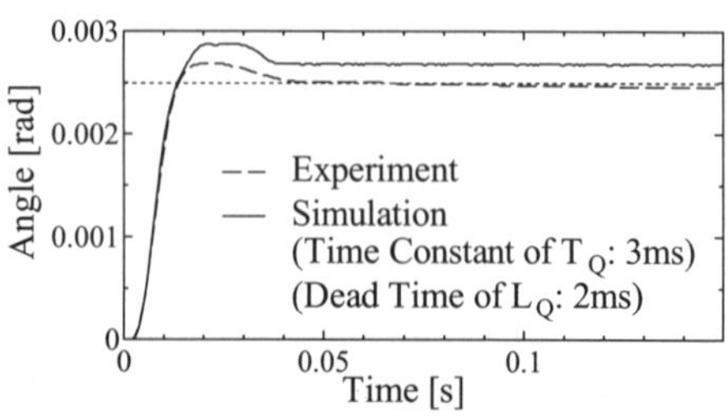

Figure14. Result of $L_Q = 2$ ms, and $T_Q = 3$ ms

5.6 *Model of both dead time and first ordered lag in E*

For further investigation, another simulation was carried out wherein the model had a delay model of E that included both dead-time and 1st-ordered delay; $L_Q = 2$ ms, and $T_Q = 3$ ms. As shown in Fig. 14, the results show a similar tendency to the experimental ones, in spite of there being a little difference from the other simulation noted above. The torque data of Fig. 14 also implied the same responses for the experiment and the simulation.

6 Discussions and Conclusion

In this paper, I investigated the effects of time delays from the viscosity of the damper on control performance and stability in a high speed and precision position control system with a variable viscous damper using particle type ER fluid.

First, two basic mathematical models presenting the delay of viscous changes were introduced: a dead-time model, and a time-lag model of the first order. Then, when the coefficient of viscosity was variable and included delay in its variation, the stability of a control system was verified by the parameter plane method. The results showed that stable regions of the parameters got smaller from the time delay.

Second, responses in a simulation, which had various kinds of time delay models with viscous changing of the damper, were compared with those of experiments. The simulation model was approximated to real equipment by including a delay of the DD motor system itself and a friction model so that the response was affected only by differences in the way the viscosity manifested. Considering the mechanism of viscosity, the case where the 1st-ordered delay was inserted before applying the electric fields appeared to be nearest to real equipment. However, the simulation results showed the largest deviation from the experimental ones. Whereas, if the delay model of the electric field had both of dead-time and 1st-ordered delay, this simulation results were approximated to the experimental response.

Above all, this study demonstrates the possibility that the transit response model of particle type ER fluid can be approximated to a composite function of a dead-time model and a time-lag model of the first order. For designing the controller phase, using a model that explicitly includes a delay for the ER dampers will cause stability analysis or parameter design to be more precise.

7 Acknowledgements

This work has been supported by grants from the Kurata Memorial Hitachi Science and Technology Foundation and the Electro-Mechanic Technology Advancing Foundation. I wish to thank Prof. J. Furusho of Osaka University and Dr. A. Inoue of ERtec Co. for useful advice.

References

1. Y. Ono, A direct-drive motor control system and its features, *Advanced Robotics* **6(2)** (1992) 243.

2. N. Takesue *et al.*, Motion Control of Direct-Drive Motor by a Homogeneous ER Fluid, *J. of Intelligent Material Systems and Structures* **10(9)** (1999) 728.
3. K. Koyanagi and J. Furusho, Direct-Drive Motor System with Particle-Type ER Fluid Damper, *Int. J. of Modern Physics B* **19(7-9)** (2005) 1661.
4. K. Lu, R. Shen and J. Liu, *editor*, *Proc. of the Ninth Int. Conf. on Electrorheological Fluids and Magnetorheological Suspensions* (World Scientific) (2005).
5. D.D. Ši ljak, Analysis and Synthesis of Feedback Control Systems in the Parameter Plane, I–Linear continuous systems, *IEEE Trans. Applications and Industry* **83(AI-75)** (1964) 449.

DEVELOPMENT OF 2-DOF MR JOYSTICKS FOR VIRTUAL REALITY APPLICATIONS

W. H. LI, B. LIU, P. B. KOSASIH AND X. Z. ZHANG

School of Mechanical, Materials and Mechatronic Engineering, University of Wollongong, Wollongong, Northfields Avenue, NSW 2522, Australia

E-mail: weihuali@uow.edu.au

The paper presents the design and development of a 2-DOF magnetorheological fluid (MRF) based haptic joystick and studies its applications in virtual reality. The developed system consists of three main parts: MR joystick, control hardware, and software. The MR joystick is composed of two actuators with disc shape positioned perpendicularly with a gimbal structure. The dimensions of the actuators were optimized using finite element analysis and their steady-state performances of the actuator were measured. The kinetics of the joystick in terms of working space and resistance force were discussed, where a subhysteresis model and a compensation technique were employed. The applications of the MR joystick in virtual reality were demonstrated using two interface examples.

1 Introduction

Very recently, virtual reality goes beyond the flat monitor that shows information to the users, and tries to immerse the user in a computer-based visual world. The users are able to feel as though they were in a location other than their physical place by a combination of virtual environment and certain haptic devices [1]. A haptic system provides the users an integrated perception for the people in entertainment, virtual manufacturing and medical training areas.

A variety of haptic devices bloomed for different applications in the last decade. The early one is a Programmable Keyboard [2], which consists of a piano-like keyboard providing computer-driven force feedback for each of its sixteen keys. The latest and the most successful haptic device is Phantom system. There are several variants but generally a stylus is grasped, or a thimble braces the user' s finger. There are six actuated degrees of freedom and six sensed orientations [3].

It is noted that conventional haptic devices are based on motor control to provide force and torque. Since the motor control is not a force control, but a position control, it is hard for traditional device to simulate accurate force and viscosity behavior of the virtual environment. This shortcoming is expected to be overcome by developing MR fluid based haptic devices. Carlson developed a portable hand and wrist rehabilitation MR device [4]. Some other researchers attempted to use MR fluids in constructing tactile displays to replicate perceived biological tissue compliance [5]. This project aims to develop a 2-DOF MR joystick and investigate their applications in virtual reality.

2 System Development

2.1 System Overview

The schematic MR joystick haptic system is shown in Figure 1. It consists of four main components: an MR haptic joystick, a server computer, a data acquisition (DAQ) board and a power amplifier. The central force performance part in this system is the MR haptic joystick, which ensures the user to feel different resistant forces according to a virtual environment. The joystick position is sent back to the computer by the DAQ board. The Computer performs as a control and display unit in this system. Firstly, the user designs a virtual environment by 2D or 3D picture modeling and control tool in the computer. Then the computer generates control signals using the 2D or 3D picture information, which correspond to the virtual environment and the position of the joystick handle detected by the position sensors. Meanwhile, the computer shows the updated picture to the user. The DAQ board is the information exchange platform between the computer and the joystick. It transfers the position signal, detected by the position sensor on the joystick, to the computer for force calculation and then sends the force control signal to the amplifier to drive the MR joystick. A power amplifier is used to amplify the coil current, since the computer and the DAQ board cannot provide current strong enough to drive the joystick directly. Consequently the operator can feel the different resistant forces when the operator moves the cursor, controlled by the joystick, through the images shown on the screen of the computer.

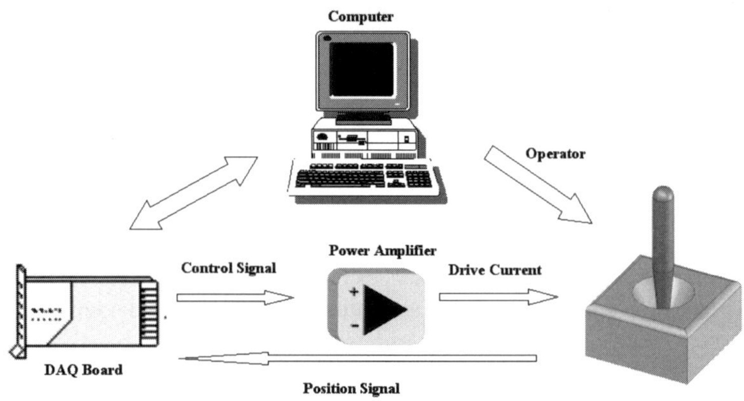

Figure 1. Schematic of the 2-DOF MR haptic system.

2.2 MR Joystick

The MR joystick is used to achieve virtual force simulation. The main components of the joystick are two MR disc-shape actuators, a gimbal structure, a handle and frames of the joystick, as shown in Figure 2. The two MR actuators are the torque generation elements, which are set orthogonally and connected to the shafts of the gimbal mechanism that are fixed in a square frame by bearings. Each actuator is used to detect one-dimensional position of the handle by a linear potentiometer attached to the shaft of the MR actuator. The actuator also generates resistance in the same direction according to the control current generated by the computer. The gimbal structure, made of standard steel, enables the user to feel resisting forces in a two-dimensional area by the handle of the joystick. Such structure converts the two orthogonal torques, generated by the MR actuators, into the resistant force that can be felt in 2D working area. All of the movable elements of the joystick are limited and linked by the frame, which is made of aluminium in the consideration of the weight and the capability and strength requirements.

MR actuators are key elements of the MR joystick. The MR actuator is in disc shape. The MR fluid is contained in gaps formed by the rotary plate and the two MILD steel casings. When the plate rotates in the MR fluid, resistant force is generated in these two gaps by the increased of viscous resistance to flow. The torque transmissibility of the MR actuator depends largely on the dynamic yield stress of the MR fluids, which is a function of a magnetic field excited by the coil. The design, manufacturing, testing and modeling of the MR actuator was detailed in another paper [6].

Figure 2. This is a caption for figure one This is a caption for figure one This is a caption for figure one and one.

2.3 MR Joystick Kinematics Analysis

The real working surface of the joystick is a part of a spherical surface, which is calculated by MATLAB and plotted in Figure. The length of the handle is 10cm. The top point in the figure shows the position of the end point of the joystick handle in a vertical state. The four corners of the figure represent the handle positions in the limiting cases: the two rotations get their maximum (+120°) or minimum (-120°) values respectively.

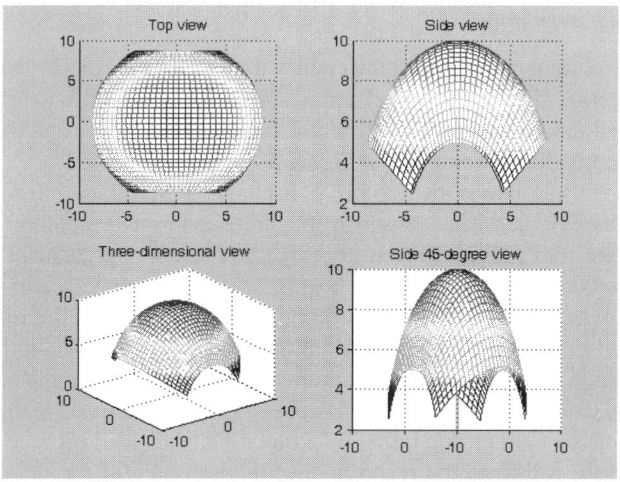

Figure 3. The working surface of the joystick.

The virtual environment in the computer is usually built up in rectangular coordinates rather than spherical coordinates, so a certain mapping transformation needs to be defined between the two environments. This spherical working surface is defined by the longitude lines and the latitude lines, which means by the corresponding rotary degrees of the two MR actuators. Each of them runs from −120° to +120°. The force users can feel is always tangent to the sphere surface. To make this working sphere surface simulate the virtual flat surface in the virtual environment, the spherical surface is deformed into a square surface by changing the angle units into length units as shown in Figure 4. This mapping method was based on human perception and reduces the calculation of the surface deformation by ignoring the coupling of the two rotations.

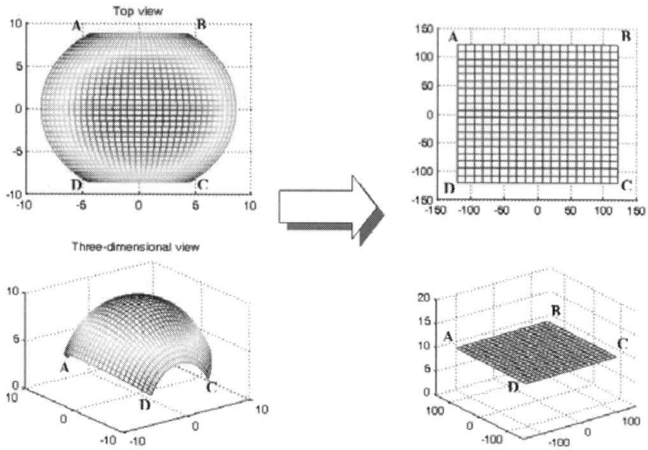

(a) Before deformation (b) After deformation

Figure 4. Working surface deformation.

3 Interface Demonstrations

3.1 System Implementation

This system uses a computer to act as the control module and the display module. The MR joystick performs as the human interface module, through which people can feel the corresponding virtual resistance and generate reactions to the virtual environment. A picture of this haptic system is shown in Figure 5.

The spirit of this system is the LabVIEW package, which is used to control the whole system and to display the visual interface for the user through a computer screen. This graphical software tool provides the user with an efficient approach for designing test, measurement, and control systems. It can also be seamlessly connected to equipment such as the PCI6221 DAQ board and the SCB-68 board that are used as the interface between the computer and the MR actuator in this haptic system. This software is composed of a block diagram panel for graphical programming and a front panel for the interface display. All the DAQ and control processes are in a while loop frame designed in the block diagram panel, the frequency of which can be set by a timer. Data acquisition is critical in a haptic application because it affects the response and the accuracy of the whole system. This project adopts DAQmx strategies to improve the DAQ performance instead of traditional NI-DAQ.

Figure 5. The MR haptic joystick system.

3.2 Interface demonstrations

The two different interfaces are shown in Figures 6 and 7. Each of them has a press switch, a slide switch, a simulation knob and a display unit. The press switch is the main switch of the program, through which the whole program can be turned on or off. Two different control modules are adopted in the two application interfaces.

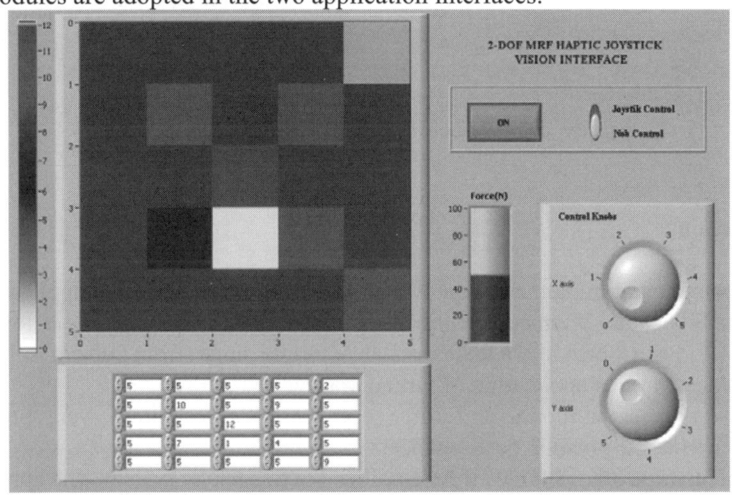

Figure 6. Demonstration of 2-DOF joystick vision interface.

In the joystick control module shown in Figure 6, the cursor displayed in the screen is controlled by the MR joystick. In the knob control module, the cursor movement can be demonstrated by the control knobs. The slide switch is used to change the working module between the joystick control and the knob control, which is used to simulate the case in the absence of the Joystick. Users can feel different forces when the cursor is moved into different color square areas in the screen. The matrix at the left bottom of Figure 6 is used to set the values of the resistant forces corresponding to different colors. It can be seen from this example that users are convenient to partition screen segments by changing the matrix in terms of dimensions and values.

The second demonstration is a virtual surgery interface as shown in Figure 7. The window on the left shows a 2D x-ray picture of a finger and the window on the middle shows a 3D model built up based on the 2D x-ray picture. The cursor position can be controlled by three control knobs, shown in lower right part of the figure. The resistance force is displayed by a vertical bar in the right of the screen. When the cursor (+ symbol) moves downward from the position marked in the figure, it will pass through three typical materials: air, tissue, and bone. The user will then feel different forces through the handle of the joystick.

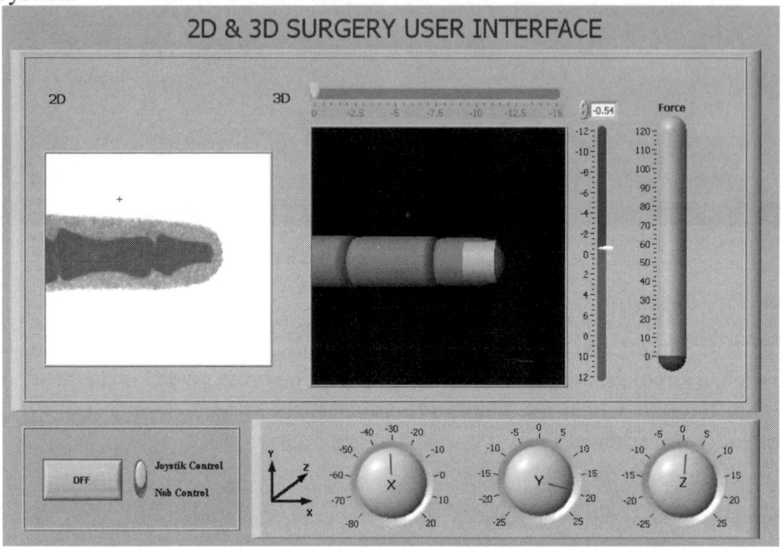

Figure 7. Demonstration of 2D & 3D surgery training interface.

4 Discussion

The development and implementation of 2-DOF MR joystick in virtual applications were presented in this paper. Main contributions or innovations are summarized below:

This MR joystick provides a wide resistance torque range (from 50Ncm to 1000Ncm), which can meet various application of virtual reality.

A sub-hysteresises model and force compensation technique were proposed to accurately predict the joystick performances.

Virtual instrument techniques were employed in this haptic system. The application of virtual instrument has the following advantages: (a) Modularized system easy for maintenance; (b) A user-friendly interface for the people who are not familiar with computer controls.

Both 2D and 3D interfaces were demonstrated to show basic applications of the developed system.

Finally, despite of the progress made thus far continuing research is still needed to further improve the performance of the joystick by adding active elements and reducing the volume and the weight of the joystick to meet more demanding industrial applications.

References

1. Salisbury K., Brock D., Massie T., Swarup N. and Zilles C., Haptic rendering: programming touch interaction with virtual objects. *Proc. Symposium on Interactive 3D Graphics*, ACM. 1995 pp.123-130.
2. Cadoz C. and Ramstein C., Capture, representation, and composition of the instrumental gesture. *Proc. International Computer Music Conference,* 1990 pp. 53-56.
3. Sensable Techonologies: Introduction to Phantom system, http://www.sensable.com/, retrieved Oct. 5, 2005.
4. Carlson J. D., Portable hand and wrist rhabilitation device, US Patent No., 6117093.
5. Liu Y. J., Davidson R. I., Taylor P. M., Ngu J. D. and Zarraga J. M. C., Single cell magnetorheological fluid based tactile display. *Displays*, **26** (2005) 29–35.
6. Liu B., Li, W. H., Kosasih, P. B. and Zhang X. Z., Performance testing and hysteresis modeling of a position-feedback MR actuator. *10th Int. Conf. on ER/MR Fluids*, Reno, Nevada, 18 – 22 June, 2006, paper number: P38.

RESEARCH AND DEVELOPMENT OF THE INTELLIGENT PROSTHETIC ANKLE JOINT WITH A MR LINEAR BRAKE

C. LI AND J. FURUSHO

Osaka University, Yamadaoka2-1, Suita, Osaka, Japan,
E-mail: furusho@mech.eng.osaka-u.ac.jp

S. MORIMOTO, M. TOKUDA AND Y. HASHIMOTO

Osaka Electro-Communication University, Osaka, Japan
E-mail: morimoto@sky.plala.co.jp

The purpose of this study is to develop a prosthetic ankle joint intelligently controlled by the specially designed linear MR brake of flow-type. MR fluid changes its rheology depending on the intension of an applied magnetic field. The brake using MR fluid has a simple structure and can control its braking force with a low voltage. Prosthesis users often tumble because they don't have dorsiflexon during swing phase, so that we propose putting a MR brake to keep dorsiflexion in order not to tumble. We hope that prosthesis users can walk more naturally and smoothly by use of developed intelligent prosthetic ankle joint with MR brake. We developed a prototype of the intelligent prosthetic ankle joint with the MR brake and evaluated it in walking experiments and obtained good results.

1 Introduction

All prosthesis users need "foot" part. This part moves so frequently and widely that it needs to be as light as possible. It also must be silent and strong because it is used in daily life. If stiffness or spring characteristics of prosthesises can be changed according to timing of walk and dorsiflection can be kept adequately, it will be much easier for users to walk and run [1,2].

Now, shapes of prosthesises' sockets has been improved and the range of actions of prosthesis users has varied. And also a lot of intelligent prosthetic knees have been developed and are sold in the market. But none of prosthesis that can control the ankle is in the market

Fig.1 is an example of a prosthesis's "foot" part. It is composed of a prosthetic ankle joint and a foot-ankle unit. Energy of movement stored at the elastic part (rubber) of the prosthetic ankle joint is relieved when kicking, and produces driving force. Also this rubber can absorb shocks when a heel hits on the ground, and change the ankle angle smoothly.

After kicking back, a leg leaves from the ground and swings forward during swing phase. Then the rubber goes back to its balanced midpoint, and so instantly the ankle turns from the dorsal direction (deflecting a tiptoe upward , as shown in Fig.1 (a)) to the plantar direction (swaging a tiptoe underneath, as shown in Fig.1 (c)).

Since there is a problem that the tiptoe of the prosthesis tends to collide on the ground, in order not to tumble, users have to walk paying attention to lifting legs. This is why users have to walk unnaturally and leads to using unnecessary energy.

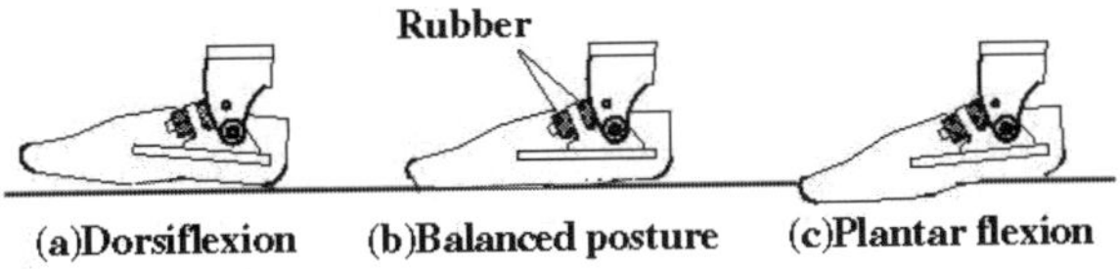

Figure 1. Prosthetic foot

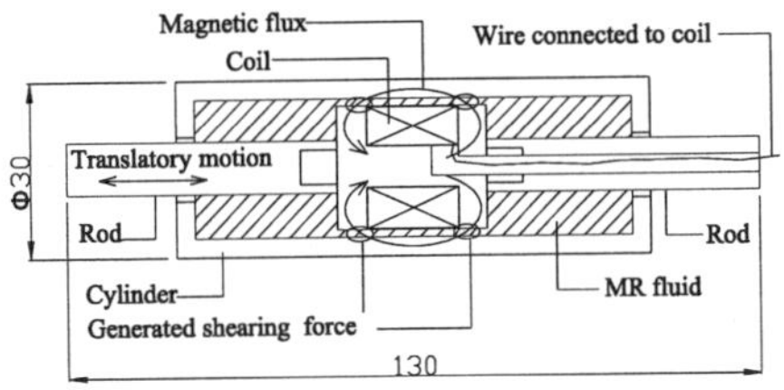

Figure 2. Construction of MRLB

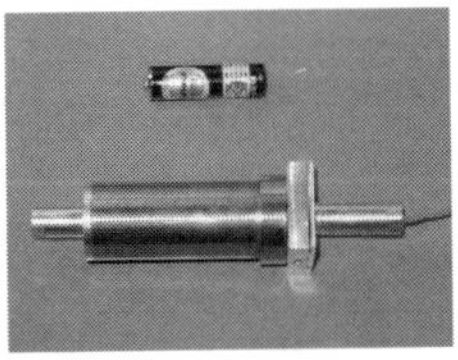

Figure 3. Picture of the MR brake

Table 1. Specifications of the MR Brake

Pole Gap	0.3	[mm]
Diameter	30	[mm]
Length	137	[mm]
Max Current	1.0	[A]
Max Electric Power consumption	1.6	[W]
Diameter of copper wire	0.4	[mm]
Coil Turn	148	[turn]
Amount of MR Fluid	15.1	[ml]
Weight	330	[g]

Table 2. Response time of braking force under constant piston speed

Coil Current	63.2%	90%
0.2 → 0.3[A]	11[ms]	57[ms]
0.2 → 0.1[A]	5[ms]	32[ms]

To solve this problem, we suggest setting a linear brake at "foot" part.

2 Development of A Linear MR Brake

2.1 Construction of The Linear MR Brake

A linear brake using the MR fluid (we call it MRLB) has a simple structure and can control its braking force depending on the electric current applied in the coil[5]. In addition, it is made of fluid so doesn't make mechanical noises. Furthermore it is compact, light and can produce strong force with less electric power.

Fig.2 is a schematic of MRLB, in which there are a piston composed of two rods and a bobbin sandwiched by the rods on both sides and MR fluid inside of a cylinder. If electric current is applied in the coil which rolls the bobbin, magnetic field as sort of loop is generated as follows; Bobbin → MR fluid → Cylinder → MR fluid → Bobbin.

When the piston is pushed in this situation, since a magnetic field is applied in the MR fluid between both these magnetic poles, its shear stress increases. As a

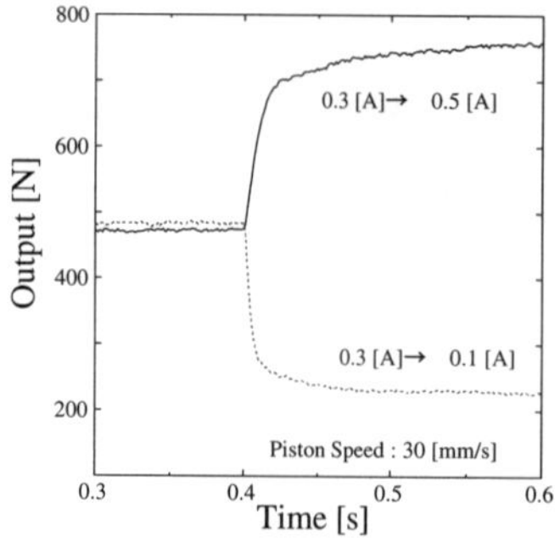

Figure 4. Step responses of braking force

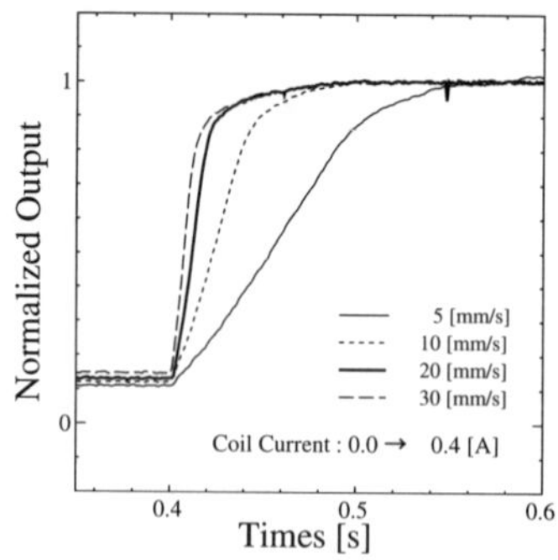

Figure 5. Response time of braking force under constant piston speed

result, it prevents the piston from moving. This force is the braking force of MRLB.

Fig.3 is the developed MRLB's photo and a AA battery. Specifications of the MR brake are shown in Table 1.

2.2 Response of The Linear MR Brake

To investigate dynamic specifications of MRLB, we carried out experiments of step response to the current. While the piston of the MRLB was being moved in controlled speed 30 [mm/s], the current changed with a step-shaped 0.4 [s] after starting the experiment.

The results are shown in Fig.4. Response time of settling for 63.2% and 90% are shown in Table 2. The experimental results revealed that response time in falling phase was faster two times than the one in rising phase.

Next, when the current in the coil change as step-shaped ($0 \rightarrow 0.4$ [A]) in different speed of piston, output responses are shown in Fig.5. The experimental results revealed faster speed of the piston makes step response faster.

3 Development of An Intelligent Prosthetic Ankle Joint and Its Experimental Evaluation

3.1 The Prototype of An Intelligent Prosthetic Ankle Joint

Fig.6 is a layout of a prototype of the intelligent prosthetic ankle joint with MRLB. The main body and the axis of MRLB are both fixed at the pin joint. A potentiometer is put on the rotation axis of the prosthetic ankle joint, so that we can measure the angle of the ankle joint directly. MRLB is controlled by this angle value. We estimate MRLB energy output on the basis of Fig.7, which shows relation between the current and brake force. It has been revealed that the relation between ankle joint angle θ [deg] and the torque T [N·m] as Eq. $T = 4.54\theta$.

MRLB stops the reaction force of rubber stored by the dorsal angle. Maximum dorsal angle is not more than 10[deg] during walking [1]. Therefore, the maximum torque of ankle joint is about 45.4[N·m]. And so we need maximum force of about

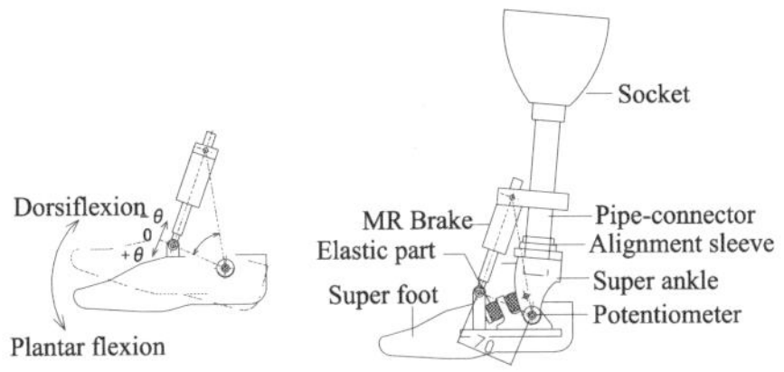

Figure 6. Layout of the intelligent prosthetic ankle joint

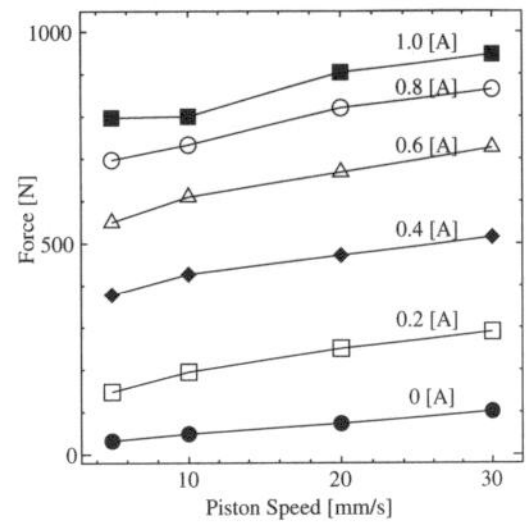

Figure 7. Current v.s. brake force

Table 3. Profile of subject

Gender	Male
Prosthesis	Left Below–knee Prosthesis
Statue	163[cm]
Age	56
Weight	81[kg]

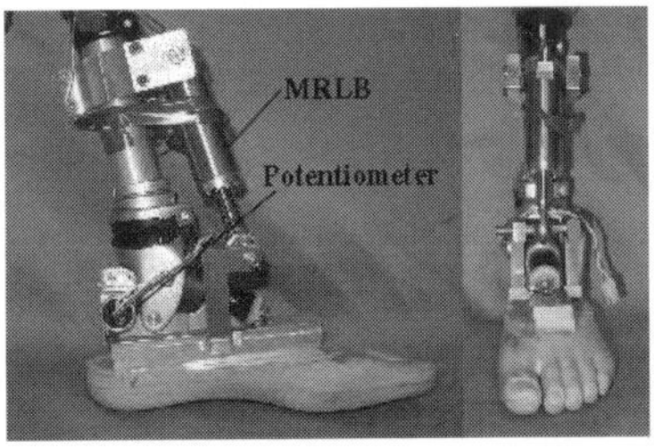

Figure 8. The prototype of an intelligent prosthetic ankle joint

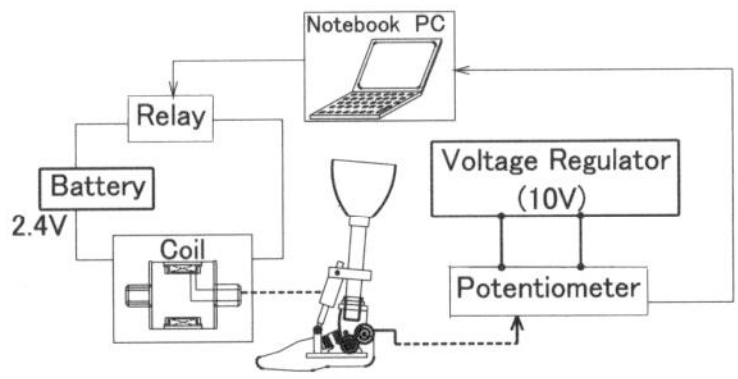

Figure 9. Control circuit of the prototype

649 [N] to keep dorsiflexion against the maximum torque during swing phase, because the MRLB is put in 0.07 [m] position from the rotation axis (Fig.6).

The developed prototype photo is shown in Fig.8. Weight of the prototype only is about 2.8[kg] and the electric power consumption including sensors is about 2.1[W].

3.2 Control Circuit

The construction of the control circuit of the prototype of the intelligent prosthetic ankle joint is shown in Fig.9. There are the MR brake part, the sensor part including a potentiometer and a notebook type computer.

If the ankle angle measured by the potentiometer is recognized as maximum by controller on the notebook type PC, the relay is turned on. Thus the relay's "on"

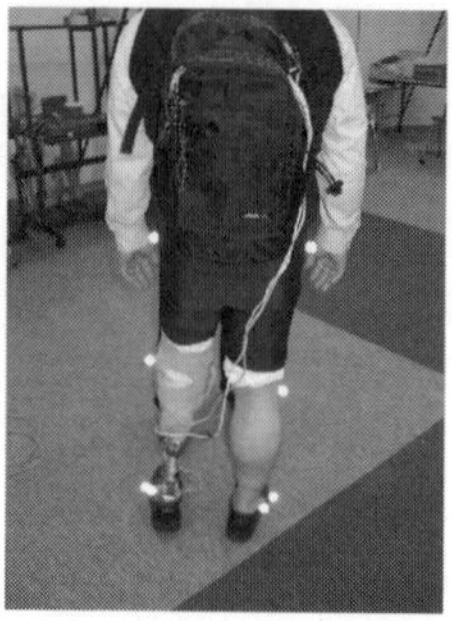

Figure 10. Appearance of walking experiments

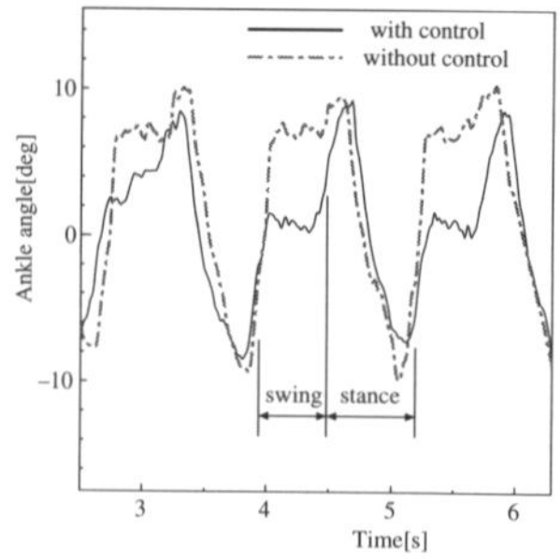

Figure 11. The change of the ankle angle

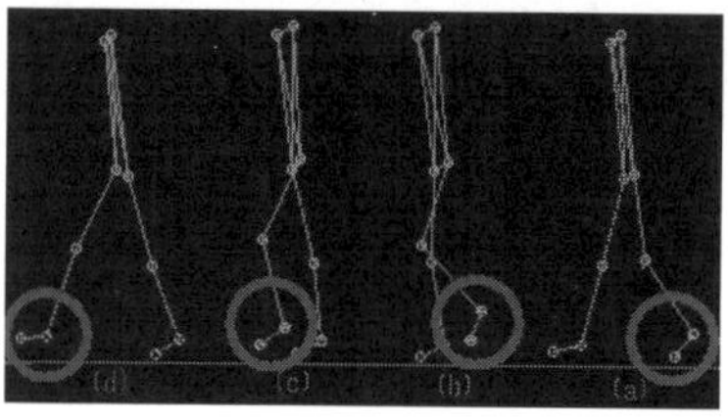

Figure 12. Walking positions with brake control

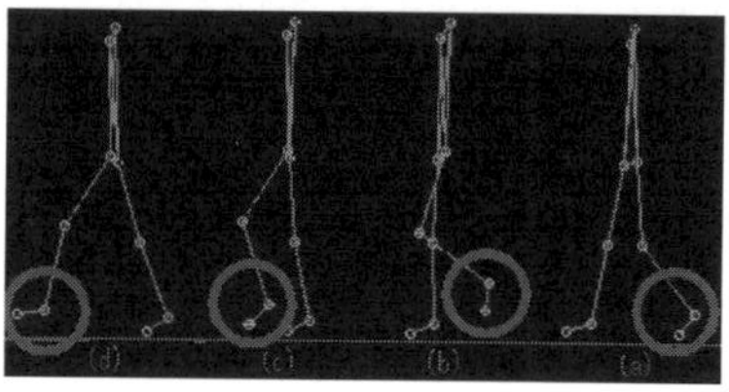

Figure 13. Walking positions without brake control

or "off" controls the MR brake.

The batteries for this control circuit is also divided into the two parts. The battery for the sensor part is two Cadnica N-6PT 7.2[V]/110[mAh] in series. The battery for the MR brake part is two Cadnica N-1CV 1.2[V]/2200[mAh] in series.

The control circuit and batteries are settled in a controller box (12.5[cm] $\times$ 19.5[cm] $\times$ 5.0[cm]) which is about 1.0[kg].

The test subject can move freely with rucksack packing the PC and the controller box (Fig. 10), so that we can carry out experiments in various situations. The total weight of the rucksack is about 3.5[kg].

3.3 Object and Methods of walking experiments

Walking experiments with the developed prototype were carried out. An optical motion capture system VICON was used to measured three dimensional coordinates, which is set in whole span of a laboratory about 8[m] $\times$ 8[m]. Reflecting markers were put on some positions of the subject and five high-speed cameras catch the three dimensional coordinates of the markers in the sampling time of 120[Hz].

Walking experiments were carried out several times for some hours modifying patterns. The patterns of walking experiments were as below:

- (A) : **Without brake control, normal speed, paying attention to lifting the knee** (The subject walks lifting the knee as usual in order not to tumble).

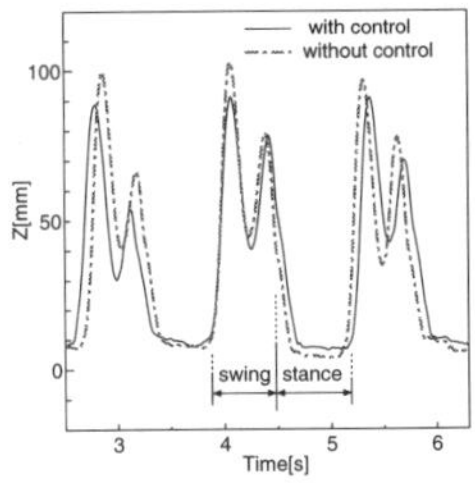

Figure 14. Variations of height of the tiptoe while walking

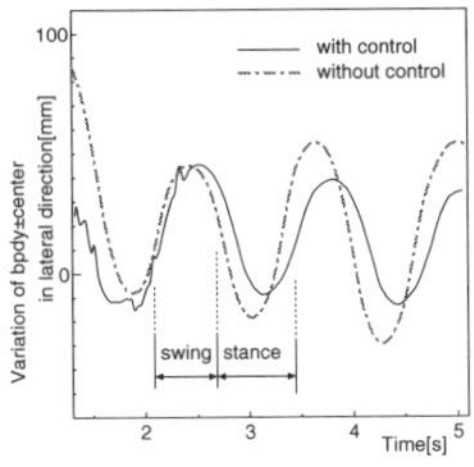

Figure 15. Variations of body–center in lateral direction with respect to time

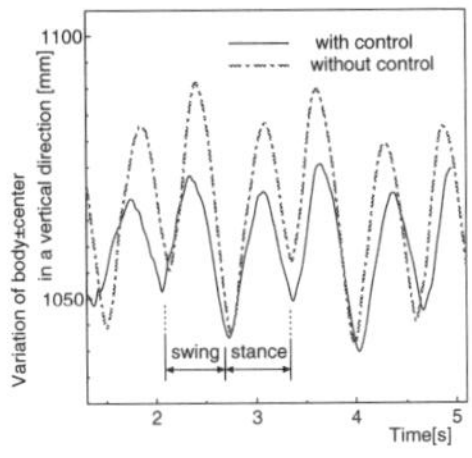

Figure 16. Variations of body–center in vertical direction with respect to time

- (B) : **With brake control, normal speed, paying attention to not lifting the knee** (The subject walks aiming regular walking).

3.4 Results and Discussion

Coordinate values of VICON data were translated to the angle of the ankle, the clearance of the tip toe from the ground and the body's gravity point of the subject. The dashed line in Fig.11 and Fig.14- 16 indicates the result of the case (A). And the full line in these figures indicates the result of the case (B).

Fig.11 shows a comparison of the ankle angle. The dorsal angle is kept from starting MR brake to last half of medium-term swing phase. Generally, in medium-term swing phase, "foot" part is most nearest to the ground. By applying MR brake during swing phase, keeping dorsifexion enables the tiptoe of the subject not to collision on the ground. As a result, it enables the subject to walk without stumble.

Fig.12 and Fig.13 are the series of static images extracted from moving images of the walking experiments. The ankle arounded with circles in these figures show the prosthetic side of the test subject(a swinging phase). (a), (b), (c) and (d) in Fig.13 shows the process of swinging without brake control. All the angles in Fig.13 are all angle of the midpoint. (a),(b),(c) and (d) in Fig.12 shows the process of swinging with brake control.

(a) in Fig.12 is the image just before the prosthetic leg left the ground the dosal angle which seemed to be maximum and was same as (a) in Fig.13. On the other hand, (b) is the image of swing phase's start and the ankle angle in Fig.12 (b) is apparently smaller than the angle in Fig.13 (b). It is because in Fig.13 the ankle angle was not controlled by MR brake and then the ankle angle returned to the neutral position, while (b)in Fig.12,the ankle angle was controlled by MR brake and so the dorsal angle was being kept. Next (c) in Fig.12 is the image in process of medium-term swing phase and the dorsal angle which is the same as (b) in Fig.12 is still being kept. Last (d) in Fig.12 is the image of later stage of swing phase and the ankle angle is larger than the angle in (c), because current in the coil is turned off and the angle is going to return to the neutral position.

Fig.14 shows the variations of the height of tiptoe to ground while walking. In this figure,the of the height in case(B) is smaller than the variations of case (A).This indicates that it is unnecessary to lift the knee while walking because of being kept dorsiflexion during swing phase.

Fig.15 shows the variations of body–center in lateral direction with respect to time Fig.16 shows the variations of body–center in vetical direction with respect to time. In these two figures, the variations of case(B)is smaller than the change of case (A). In short, though the distance of the body's gravity increases when the subject walks lifting the knee in order not to tumble, keeping the dorsal angle saved him from tumbling and decreased the unnecessary movement of the gravity.

Walking experiments were carried out for several hours.

4 Conclusion

In this study, we have originally developed a MR brake that has a simple structure and can control its braking force with a low voltage. we also proposed and developed a prototype of the intelligent prosthetic ankle joint using the MR brake and confirmed that keeping dorsiflexion during swing phase by applying MR brake was effective for users' walk.

References

1. S.Morimoto et al, "Ankle dynamics during gait: Effect of walkway inclination on ankle and moment in amputess and non-amputess," Annual Report of ROSAI Rehabilitation Engineering Center, 1-5, 1996
2. A.Nakagawa, "Present situation of lower limb external prosthesis in view of maehatronic technology application," Systems, Control and Information, vol.41, no.11, pp.478-483, 1997
3. "Magneto Rheological Fluid MRF132-LD Product Bulletin," Lord Corporation, 1999
4. J.Carlson et al, "Commercial Magneto-Rheological Fluid Devices," Proc. of 5th Int. Conf. on Electro-Rheological Fluids, Magneto-Rheological Suspensions and Associated Technology, World Scientific, pp20-28, 1995
5. J.Furusho et al, "Development of Intelligent Prosthetic Ankle Joint (1st Report, Development of Linear-Type MR-Fluid Brake)," Transactions of the Japan Society of Mechanical Engineers (C), vol.70, no.695, pp.275-282, 2004
6. C.Li et al, "Developement of Intelligent Prosthetic Ankle Joint (2st Report, Development of the 1st Prototype with Intelligent Prosthetic Ankle Joint)," Transactions of the Japan Society of Mechanical Engineers (C),vol.72, no.714, pp.493-498, 2005

ANALYSIS OF THE STRUCTURAL CHARACTERISTICS AND MAGNETIC FIELD INTENSITY OF MR DAMPER

W. K. LAN, L. ZHENG, Y. N. LI, S. S. LI

State Key Laboratory of Mechanical Transmission, Chongqing University, China 400044
Email :zling@cqu.edu.cn

A. BAZ

Department of Mechanical Engineering, University of Maryland, College Park, MD 20742, USA
E-mail : baz@eng.umd.edu

This study focuses on the design and characterization of magneto-rheological (MR) fluid dampers for a mini-bus based on mixed operational mode. The theoretical fluid mechanics-based model is developed and used to predict the performance of these MR dampers. The arrangement of coils in the piston and winding orientation are discussed and identified by electromagnetic finite element analysis. As a verification of the theoretical prediction, two prototypes of MR dampers are designed, manufactured and characterized. The theoretical results are compared with the experimental results for different input frequency and electric current. The theoretical predictions are shown to be in excellent agreement with the experimental data. The obtained results indicate that the performance of the MR damper with two coils is better than a damper with a single coil of the same activation length. Furthermore, the developed dynamic damping force is found to be controllable over a wide range suggesting that these proposed MR dampers can be used as a good actuator for the semi-active suspension system of the mini-bus Changan 6350.

1 Introduction

Recent renewed interest in MR fluids has been apparent in automobile suspension technologies. Lord Corporation pioneered the use of the fluid in the 1990s, introducing MR dampers for truck seat suspension called the "Motion Master." Shortly after, Lord Corporation developed a rotary MR brake for treadmill applications, as well as dampers for automobile use. As a field-dependent semi-active dissipating device, the MR dampers demonstrated their feasibility for applications in automobile vibration control. Several innovative prototypes MR dampers have been designed, developed and manufactured at the University of Nevada, Reno (UNR) such as MR dampers for mountain bicycles [1], motorcycles [2] and vehicles [3]. Delphi demonstrated its semi-active suspension system in vehicles [4], and Carrera Shocks, Inc. has developed its own MR dampers with the associated control systems.

As a component of semi-active suspension system, the role of the MR damper is to produce variable damping force in response to input road excitation in order to meet the ride comfort requirements of vehicles. An MR damper can be easily controlled by varying the input current enabling it to be integrated into the semi-active electric control suspension system. Compared with traditional variable orifice dampers, the operational speed of MR damper is much faster and its power requirement is minimal. More importantly, the MR dampers can be considered as "fail-safe" devices.

In this study, two prototypes of mixed mode MR dampers for a mini-bus are designed, manufactured and characterized. A theoretical fluid mechanics-based model is developed and used to predict the performance of these MR dampers. The arrangement of the coils in a piston and winding orientation are investigated using electromagnetic finite element analysis (EFEA). The geometric and material parameters of the dampers are evaluated and optimized by the EFEA. The theoretical predictions are compared with the experimental results for different input frequency and electric current. The performance characteristics of the prototype MR dampers are determined experimentally and significant conclusions are drawn out.

2 Theoretical Modeling of the MR damper

Fig.1 shows the basic geometrical parameters of the MR damper and Fig.2 displays the associated paths of the magnetic filed.

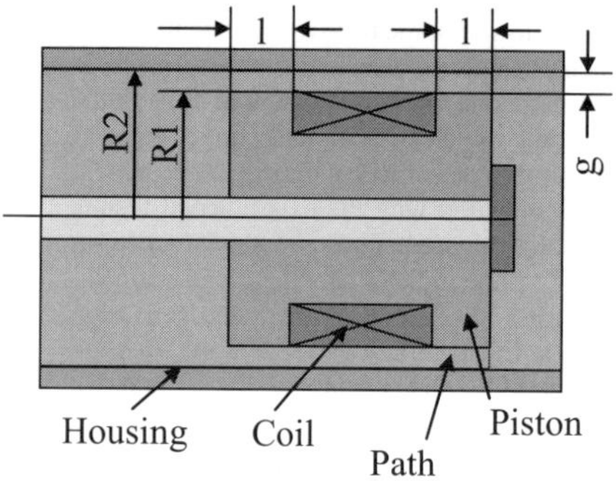

Figure 1 The mixed operational mode

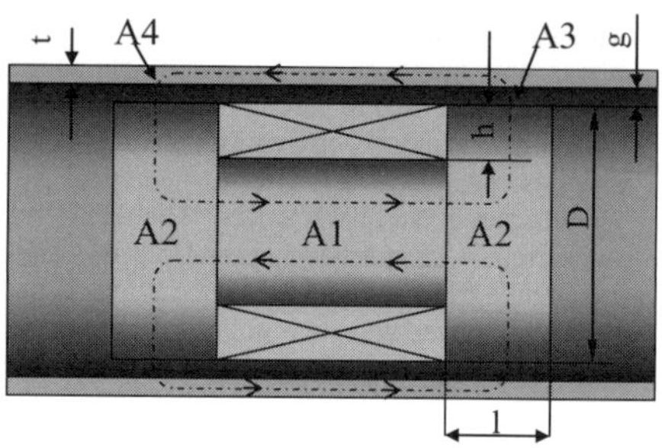

Figure 2 Magnetic field actuation areas

A MR fluid consists of micron-sized ferrous particles within a base carrier fluid, such as silicon or mineral oil, and additives. Under the influence of a magnetic field the ferrous particles polarize and attract each other resulting in the formation of chains or columns of particles parallel to the orientation of the magnetic flux lines. These chain-like structures restrict the motion of the fluid and thereby increasing the viscous characteristics of the suspensions. The mechanical energy needed to yield the microstructure increases as the applied field increases resulting in a field-dependent yield stress. Thus the behavior of the fluid is often represented as a Bingham plastic having variable yield stress. In this model, the constitutive equation is derived:

$$\tau = \tau_y + \eta\dot{\gamma} \tag{1}$$

where τ is fluid yield stress, τ_y is the field dependent yield stress, $\dot{\gamma}$ is the fluid shear rate, and η is the plastic viscosity.

Most devices that use MR fluids can be classified as having either fixed poles (pressure driven flow mode) or relatively moveable poles (direct-shear mode). These two basic operational modes are widely used in the MR controllable fluid devices.

The pressure drop developed in a device based on pressure driven flow mode is assumed to result from the sum of a viscous component P_η and a field dependent coulomb component P_τ. This pressure drop may be approximated by [5]:

$$\Delta P = \Delta P_\eta + \Delta P_\tau = \frac{12\eta QL}{g^3 w} + \frac{c\tau_y L}{g} \tag{2}$$

where L, g and w are the length, gap and width of the flow channel between the fixed poles, Q is the volumetric flow rate, η is the fluid viscosity with no applied field and τ_y is the field dependent yield stress. The parameter c has a value ranging from a minimum value of 2 to a maximum value of 3.

In a similar manner, the force developed by a direct-shear device is:

$$F = F_\eta + F_\tau = \frac{\eta S L w}{g} + \tau_y L w \tag{3}$$

where S is the relative pole velocity and $A = Lw$ is the shear area.

In this study, the MR damper for a small vehicle is designed based on the combination of the previous two basic operational modes, resulting in what is called as the mixed operational mode as shown in Fig.1

The damping force of a mixed mode MR damper can be written as:

$$F_{DAMPER} = P_\eta \cdot A_P + P_\tau \cdot A_P + F_\eta + F_\tau \tag{4}$$

Substituting Eqs. (2), (3) into Eq.(4), the total damping force become:

$$F_{DAMPER} = \frac{12\eta Q L}{g^3 w} A_p + \frac{c\tau_y L}{g} A_p + \frac{\eta S L w}{g} + \tau_y L w \tag{5}$$

Since $Q = A_p S, w = 2\pi R_1, g = R_2 - R_1, L = 2l$, c is given a value of 3, Eq.(5) reduces to:

$$F_{DAMPER} = \left[\frac{12\eta l A_p^2}{\pi R_1 (R_2 - R_1)^3} + \frac{4\pi R_1 l \eta}{R_2 - R_1} \right] S + \left[\frac{6 l A_p}{R_2 - R_1} + 4\pi R_1 l \right] \tau_y \tag{6}$$

The ratio of the viscous shear mode damping force to the flow mode damping force can be written as:

$$\frac{F_\eta}{P_\eta} = \frac{4\pi R_1 l \eta}{R_2 - R_1} \left[\frac{12\eta l A_p^2}{\pi R_1 (R_2 - R_1)^3} \right]^{-1} = \frac{\pi R_1^2 (R_2 - R_1)^2}{3 A_p^2} \tag{7}$$

Defining the ratio of piston rod area to piston area as α, then $A_p = (1-\alpha)\pi R_1^2$, Eq.(7) become:

$$\frac{F_\eta}{P_\eta} = \frac{1}{3(1-\alpha)} \left[\frac{R_2 - R_1}{R_1} \right]^2 \tag{8}$$

In a similar manner, the ratio of the coulomb flow mode damping force to shear mode damping force may be obtained as :

$$\frac{F_\tau}{P_\tau} = \frac{2}{3(1-\alpha)} \frac{R_2 - R_1}{R_1} \tag{9}$$

Since $R_2 - R_1 << R_1$, note that $F_\eta << P_\eta$, $F_\tau << P_\tau$ from Eqs.(8) and (9). If the damping force of shear mode is neglected, then Eq.(6) may be simplified to:

$$F_{DAMPER} = \frac{12\eta l A_p^2}{\pi R_1 (R_2 - R_1)^3} S + \frac{6 l A_p}{R_2 - R_1} \tau_y \tag{10}$$

where the field-dependent fluid yield stress can be written as follows:

$$\tau_y = K H^\beta \tag{11}$$

with K and β denoting the characteristic constants of the MR fluid used, which can be determined by magnetic property experiment for MR fluid. Here, K *and* β are 0.0618, and 1.25. Also, H is the magnetic field intensity in flow gap given by:

$$H = \frac{NI}{2(R_2 - R_1)} \quad (12)$$

where N represents the number of turns of the coil. Substituting Eq.(11) and (12) into Eq.(10) yields:

$$F_{DAMPER} = \frac{12\eta l A_p{}^2}{\pi R_1 (R_2 - R_1)^3} S + \frac{6 l A_p K}{R_2 - R_1} \left[\frac{NI}{2(R_2 - R_1)} \right]^\beta \quad (13)$$

Eq.(13) can be used to predict the theoretical damping force .

3 Electromagnetic Finite Element Analysis

3.1 Theoretical Analysis

The goal of the electromagnetic field design is to channel the field lines in a manner that makes them cross the fluid gap in a perpendicular fashion, as well as making sure that the fluid gap is the point of highest reluctance in the path of the field lines. Therefore, it is necessary to compute electromagnetic field distribution by using electromagnetic finite element analysis when all geometric parameters have been approximately determined according to the design equations discussed in Section 2. In order to compute electromagnetic field distribution, the areas perpendicular to the field lines can be divided into four parts as shown in Fig.2. These areas are *A1, A2, A3* and *A4*.

$$\mathrm{A1} = \frac{\pi}{4} \cdot (D - 2h)^2, \ \mathrm{A2} = \mathrm{A3} = \pi DL, \text{ and } \mathrm{A4} = \pi (D + g + \frac{t}{2}) \cdot t \quad (14)$$

Defining the total magnetic flux Φ :

$$\Phi = B_f \cdot A_f \quad (15)$$

where B_f represents the magnetic flux density of the MR fluid and A_f is the effective pole area of MR fluid . A_f can be regarded as A3 approximately.

Using the Principle of Continuity of Magnetic Flux:

$$\Phi = \Phi_{piston(A1)} = \Phi_{piston(A_2)} = \Phi_{fluid(A3)} = \Phi_{hou\sin g(A4)} \quad (16)$$

For the different areas perpendicular to the magnetic lines, the magnetic fluxes can be computed as follows:

$$\Phi_{piston(A1)} = B_{Iron} \cdot A1, \ \Phi_{piston(A2)} = B_{Iron} \cdot A2$$
$$\Phi_{fluid(A3)} = B_f \cdot A3, \Phi_{hou\sin g(A4)} = B_{Steel} \cdot A4 \quad (17)$$

In this study, iron and steel are used as the materials for piston and housings respectively.

Once magnetic flux density in different regions are computed by Eqs. (16) and (17), the magnetic field intensity H in different regions can be obtained by *B-H* curves of different materials. Finally, the number of turns of the coil can be determined from:

$$NI = \sum_{i=1}^{4} L_i H_i \tag{18}$$

One benefit of using MR fluid for varying damping force in lieu of a mechanical method is its response time is far less than the time a mechanical system requires to vary orifice size. Note that the MR fluid response time is defined as the time required to change the fluid apparent viscosity. Typically, this time response is limited by the time response of the electromagnetic circuit that creates the required flux density within the MR flow gap. The design challenge exists in creating a DC electromagnetic circuit that can generate sufficient flux across the MR flow gap in a minimum amount of time.

The time response of an electromagnetic circuit can be modeled as a function of inductance and electrical resistance. This can be expressed as [6]:

$$\tau = \frac{L}{R} \tag{19}$$

where τ is the response time constant, L is inductance, and R is resistance. Minimizing the L/R ratio reduces the time constant of the electromagnetic circuit. However, generating sufficient flux in the MR flow gap requires the existence of a minimum number of coil windings. This gives an estimate of the required electrical resistance of the winding. The remaining parameter which can be manipulated in order to reduce the time response is the inductance that can be represented as [7]:

$$L = \frac{4\pi^2 N^2 r^2 K_n}{l} \cdot 10^{-6}, \text{ with } K_n = \frac{1}{1 + 0.9\frac{r}{l} + 0.32\frac{t}{r} + 0.84\frac{t}{l}} \tag{20}$$

where r is the mean radius, t is thickness, and l is length of the coil along the centerline, K_n is geometry term. Decreasing the inductance can be achieved by manipulating these area terms.

Consider the magnetic arrangements displayed in Fig.3. Fig.3 (a) shows an arrangement of a single coil while Fig.3 (b) shows three smaller coils whose total cross-sectional area is equal to the area of the single coil.

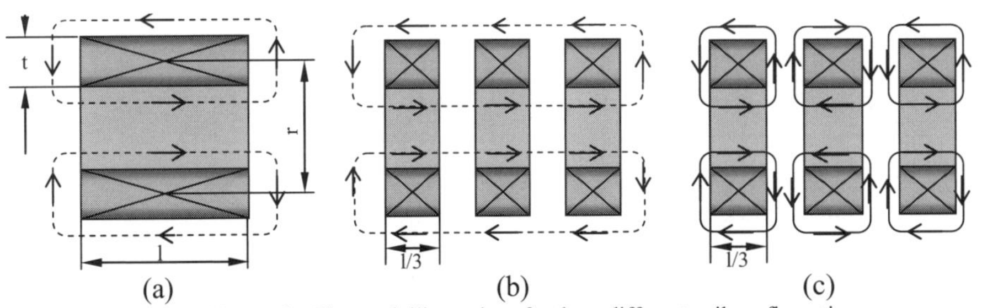

Figure 3 Flux path illustrations for three different coil configurations.
(a) flux lines around the single coil., (b) flux lines around three coils which have same winding orientations
(c) flux lines around three coils which have alternate winding orientations

It is assumed that the same number of coils and type of magnet wires are used for both cases, hence equivalent values for electrical resistance. The inductance of the electromagnetic circuit in Fig.3 (a) is calculated directly from Eq. (20). Inductance for the series electromagnetic circuit of Fig.3(b) must account for the effects of mutual inductance in addition to the series addition of three individual coil inductances [8], i.e.:

$$L = L_1 + L_2 + L_3 \pm 2(M_{12} + M_{23} + M_{13}) \tag{21}$$

here L is the summed inductance, L_1 is inductance of the first coil, L_2 the second, and L_3 the third. M_{12} is mutual inductance for the interaction between the first and second coil, M_{23} is mutual inductance for the interaction between the second and third coil, and M_{13} is mutual inductance of the external two coils. The plus/minus sign before the mutual inductance terms indicates that coupling is either additive or subtractive, depending on the connection polarity.

Fig.3 illustrates the difference in connection polarity. Fig.3 (a) has flux lines that flow around the single coil. Fig.3 (b) and Fig.3 (c) represent two three-coil arrangements with similar characteristics except for polarity. The main difference between Fig.3 (b) and Fig.3 (c) is the length of the flux path. The benefit of the coil arrangement of Fig.3 (c) is that the overall inductance of the circuit is much lower than the previous two cases; the mutual inductances are subtracted from overall inductance. This results in a reduced time response. The design of magnetic path in this research takes this arrangement into account.

3.2 Electromagnetic Finite Element Analysis

Electromagnetic Finite Element Analysis is used to optimize the design of the magnetic path in the MR damper. Consider the cases shown in Fig.4. It is noted that the difference between case1 and case 2 is only the length of the coil spool. Case 1 and case 2 have a single coil. Case 3 has two coils with alternate winding orientations. The geometric parameters of the three cases are listed in Table 1.

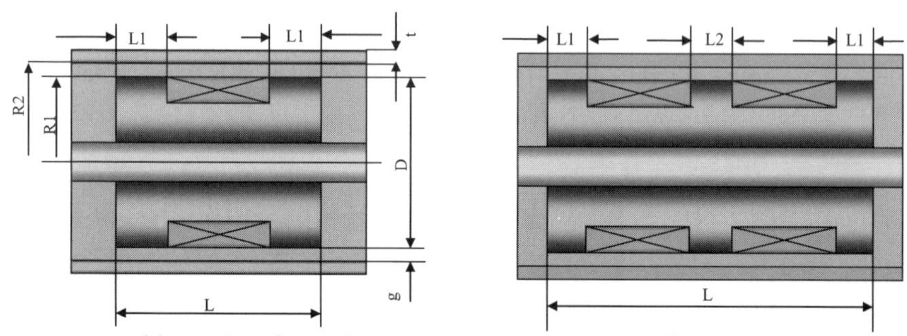

(a) case1 and case 2 (b) case 3

Figure 4 MR dampers with single coil (a) and for two coils (b)

Table 1 Geometric parameters

<table>
<tr><td rowspan="2">Case 1</td><td>D(mm)</td><td>L(mm)</td><td colspan="2">L1(mm)</td><td>g(mm)</td><td>t(mm)</td><td colspan="2">n</td></tr>
<tr><td>28</td><td>42</td><td colspan="2">11</td><td>1.5</td><td>1.5</td><td colspan="2">300</td></tr>
<tr><td rowspan="2">Case 2</td><td>D(mm)</td><td>L(mm)</td><td colspan="2">L1(mm)</td><td>g(mm)</td><td>t(mm)</td><td colspan="2">n</td></tr>
<tr><td>28</td><td>52</td><td colspan="2">11</td><td>1.5</td><td>1.5</td><td colspan="2">300</td></tr>
<tr><td rowspan="2">Case 3</td><td>D(mm)</td><td>L(mm)</td><td>L1(mm)</td><td>L2(mm)</td><td>g(mm)</td><td>t(mm)</td><td>n1</td><td>n2</td></tr>
<tr><td>28</td><td>52</td><td>7.5</td><td>7.5</td><td>1.5</td><td>1.5</td><td>150</td><td>150</td></tr>
</table>

The magnetic flux densities of cases 1 and 2 are shown in Fig.5 whereas the magnetic flux density of case 3 is shown in Fig.6. It can be seen that the magnetic flux density of case 1 is more concentrated than that of case 2. This will result in rapid magnetic field saturation for case1. Comparing Fig.6 (a) and Fig.6(b), it can be noted that the alternate winding orientation results in a relatively uniform magnetic flux density distribution along the MR fluid flow path. In this paper, a prototype MR damper is designed and manufactured based on this coil arrangement.

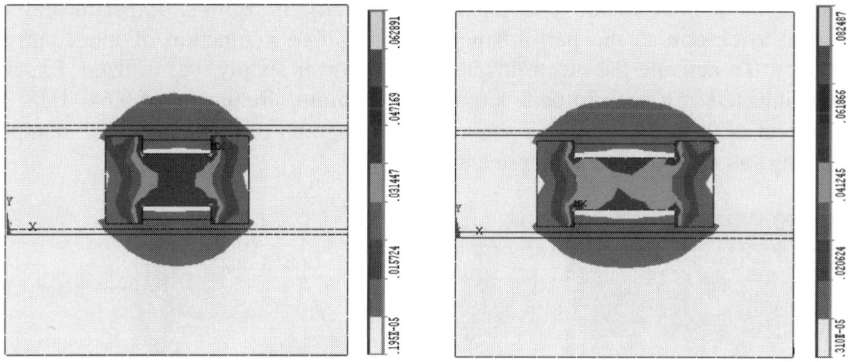

(a) case1 (b) case2

Figure 5 Magnetic Flux densities of cases 1 and 2

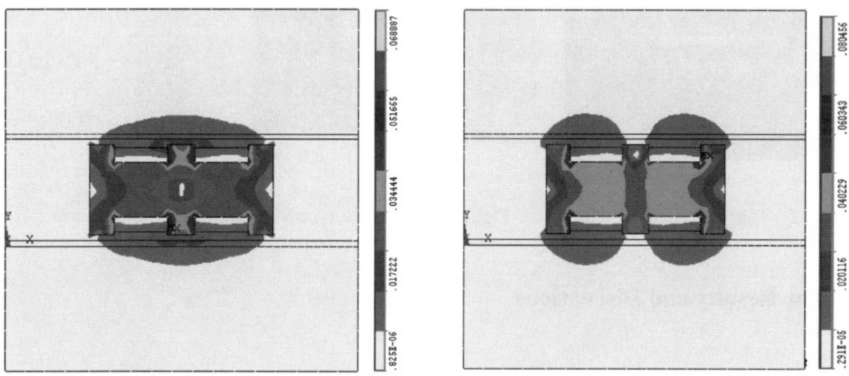

(a) alternative winding orientation (b) same winding orientation

Figure 6 Magnetic Flux density of case 3 with different winding orientations

4 Experimental Study

The geometric parameters of the prototypes of the MR dampers considered in this study are designed based on original equipment manufacturers (OEM) of passive viscous dampers. The design of magnetic path is optimized by using a three-dimensional electromagnetic finite element analysis (EFEA). This ensures that the prototypes of the MR dampers can be installed in the front suspension of the mini-bus Changan 6350. Two prototypes MR dampers are manufactured and characterized. These two MR dampers are of the twin-tube type. The only difference between the prototypes is the number of coils, one has a single coil while the other has two coils in series with alternative winding orientations. In this study, pure iron is used as the material of the piston to increase the magnetic flux density as much as possible. The MR fluid used in the two prototypes is supplied by Chongqing Instrument Material Research Institute.

The prototypes are tested using a servo-hydraulic shock testing dynamometer shown in Fig.7. It has a maximum stroke of 12.5cm and is equipped with a 50KN load cell that measures the damping forces of the shock absorber and a displacement transducer that measures its displacement. The maximum velocity is 2.5m/s. Experimental date is collected to determine the performance of the unit as a function of input current and frequency. To activate the electromagnet, a DC power supply was utilized. Experiments were conducted at a peak-to-peak amplitude of 25mm, frequency of 0.64, 1.28, 1.92Hz and current of 0, 0.5, 1.0A. Fig.8 shows is a photograph of the actual MRF damper used for testing and its associated components.

Figure 7 The experimental setup

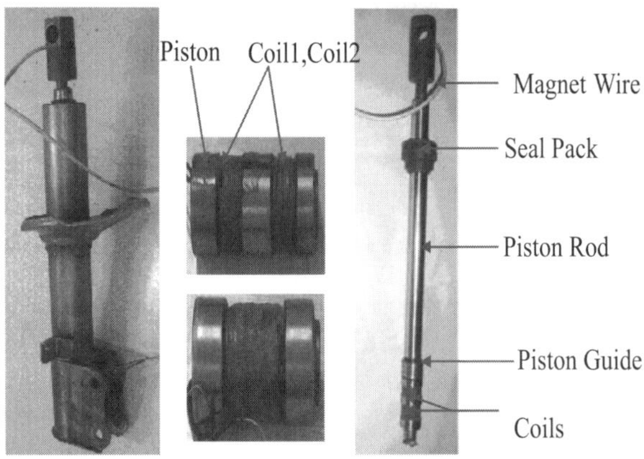

Figure 8 The prototype MR damper and components

5 The Results and Discussions

Fig.9 shows that comparisons between the theoretical and experimental force versus displacement characteristics for different input electric currents at input motion amplitude of 25mm and input frequency of 1.92Hz for the single coil MR damper. Fig.10 shows the corresponding characteristics for the MR damper with two coils. Excellent agreements are evident between the theoretical predictions and experimental results. Comparing Fig.9 with Fig.10, it can be seen that the theoretical and experimental force for the MR damper with two coils is obviously higher that that for MR damper with single coil at the same input current. This ascertains the validity of the theoretical analysis based on the electromagnetic theory. Hence, when a single coil is divided into series of coils with alternate winding orientations, the damping force of MR damper will be increased considerably for the same input current.

Fig.11 shows the experimental force-displacement characteristics for the MR damper with a single coil at different input frequencies at a constant displacement of 25 mm. In each of the force-displacement results shown in this work, the positive forces and displacements correspond to the rebound strokes and the negative forces and displacements correspond to the compression strokes. It can be seen that the compression stroke force increase with input frequency is the same with the rebound force increase. This means that the design of valve in piston still needs to be improved. The same results can be observed in Fig.12 when the MR damper is provided with two coils.

6 Conclusions

Two prototypes of MR dampers for a mini-bus are designed, manufactured and characterized. The difference between these two prototypes is the arrangement of coils. The theoretical fluid mechanics-based model is developed and used to predict the performance of these MR dampers. The arrangement of the coils in the piston and winding orientation is discussed and investigated by using electromagnetic finite element analysis. The theoretical predictions are compared with experimental results for different input frequency and electric current. The theoretical predictions are shown to be in excellent agreement with the experimental data. The experimental results show that the performance of the MR damper with two coils is better than that of the MR damper with only single coil of the same length.

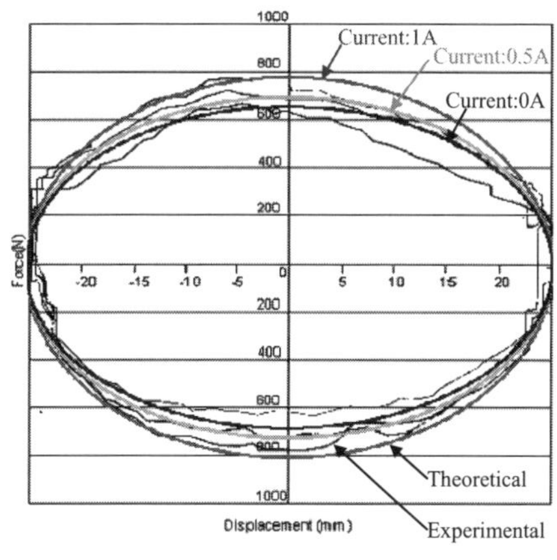

Figure 9 MR damper with single coil in 1.92 Hz

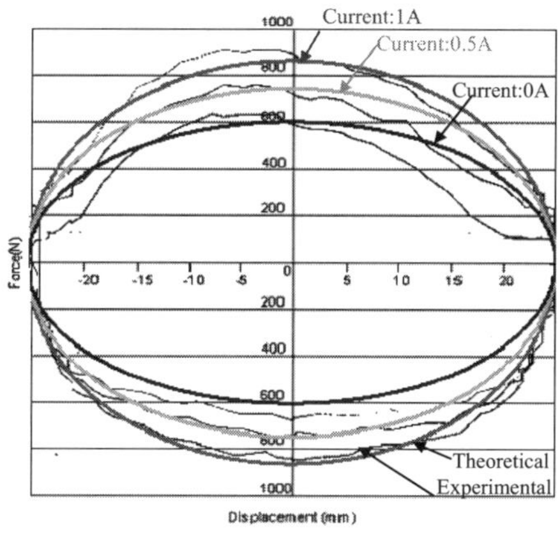

Figure 10 MR damper with two coils in 1.92 Hz

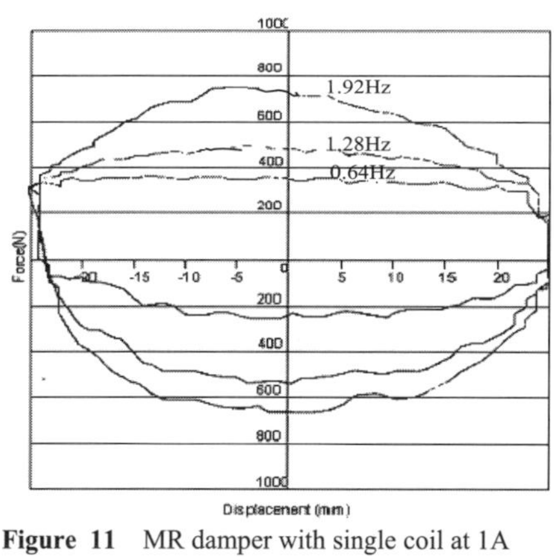

Figure 11 MR damper with single coil at 1A

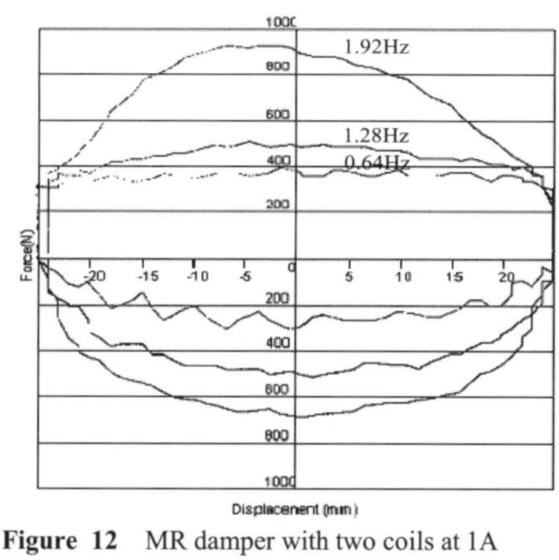

Figure 12 MR damper with two coils at 1A

7 Acknowledgements

This study was supported by a grant from the Chinese Nature Science Foundation (Grant No. 50475064)

References

1. Breese D. G., and Gordaninejad, F., Semi-Active, Fail-Safe Magneto-Rheological Fluid Dampers for Mountain Bicycles, *International J. Vehicle Design*, **33**(1/2/3) (2003) pp.128-138.
2. Ericksen E. O. and Gordaninejad F., A magneto-rheological fluid shock absorber for an off-road Motorcycle," *International J. Vehicle Design*, **33**(1/2/3) (2003) pp.139-152.
3. Gordaninejad F. and Kelso S. P., Magneto-rheological fluid shock absorbers for HMMWV, *Proceedings of SPIE Conference on Smart Structures and Materials on Damping and Isolation,* Paper No. 3989-21, Vol. 3989, Ed. by T. Tupper Hyde Newport Beach, CA, March, 2000.
4. www.delphi.com/pdf/e/magneride
5. LORD Materials Business Unit, Designing with MR fluids, *Engineering Note*, December, 1999
6. Kelso S.P., Experimental characterization of commerciallypractical magneto-rheological fluid damper technology, *SPIE Conference on Smart Structures and Materials on Industrial and Commercial Applications of Smart Structures Technologies,* paper# 4332-34, Vol. 4332, Ed. by A.-M. R. McGowan, and E. H. Anderson, San Diego, CA, March 2001.
7. V.G. Wellsby, *The Theory and Design of Inductance Coils,* , Macdonald & Co., Ltd.,1952
8. T.A. Bowland, *Electrical Circuit Formulae,* Bowest Pty Ltd, 1998.

TWO-LAYERED MAGNETIC FLUID SLOSHING IN A RECTANGULAR CONTAINER

S. YOSHIDA, A. BENVENUTI, K. TOMITA AND T. SAWADA

Department of Mechanical Engineering, Keio University
3-14-1 Hiyoshi, Kohoku-ku, Yokohama 223-8522, Japan
E-mail: sawada@mech.keio.ac.jp

The dynamic behavior of a two-layered fluid in a laterally oscillated rectangular container under a vertical non-uniform magnetic field was studied experimentally. The two-layered fluid was formed of water based magnetic fluid and silicone oil. The dynamic behaviour was studied by measuring the velocity profile of the two-layered fluid using an ultrasound velocity profile measuring technique while varying the vertical magnetic field. The displacement of the free surface was measured using a laser sensor. The effect of the magnetic field on the sloshing of the two-layered fluid was considered.

1 Introduction

A magnetic fluid is a stable colloidal solution of many surfactant-coated ferromagnetic particles in a liquid carrier such as water or kerosene. The particles are about 10 nm in size. The preparation of a stable magnetic fluid involves a delicate balance between the attractive and repulsive forces on each particle. Under no external magnetic field, these particles are in Brownian motion and remain suspended. When a magnetic field is applied to a magnetic fluid, several interesting characteristics are observed. These interesting characteristics seem to be caused by changes in magnetic body force.

When magnetic fluids are used as a work fluid in a dynamic fluid mechanical system undergoing a phenomenon such as sloshing, it is necessary to measure not only the displacement of the free surface but also the internal velocity profiles of the fluid. However, we cannot measure internal velocity profiles using optical methods such as laser Doppler anemometry because a magnetic fluid is opaque.

A technique called the Ultrasound Velocity Profile (UVP) method, which was developed by Takeda[1, 2], is a method of measuring a velocity profile along a line with respect to the velocity component along the ultrasonic beam. This UVP technique has two main advantages in comparison with conventional methods like LDV or PIV. First, it can be applied to an opaque fluid like a magnetic fluid. Second, flow mapping is practical because a line measurement is used. As such, the UVP technique has recently been tested as a useful method for a velocity measurement of a magnetic fluid for measuring flow[3, 4].

Several studies have been performed to investigate sloshing[5]. In particular, Handa and Tajima[6] researched two-layered fluid sloshing in a rectangular container, both experimentally and theoretically. Sawada et al.[4] analyzed sloshing using a magnetic fluid in a rectangular container subject to non-uniform magnetic fields and measured inner velocity profiles using the UVP method.

Although these various theoretical and experimental studies have been carried out to better understand sloshing, the mechanisms involved in two-layered fluid sloshing using a magnetic fluid remain unclear and inner velocity profiles of such a system have not been measured as yet. In the present paper, we experimentally measured the displacement of the free surface and the inner velocity profiles of a two-layered fluid sloshing under a non-uniform magnetic field. The test fluid included layers of a water-based magnetic fluid and a silicone oil. The influence of magnetic field on resonant frequency and inner velocity profiles are presented.

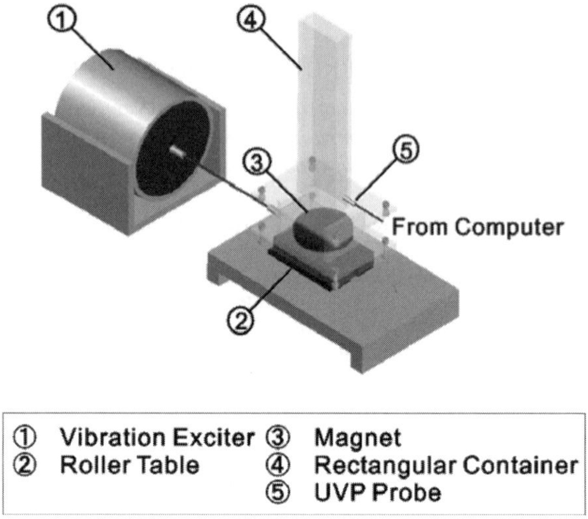

Figure 1. Experimental apparatus

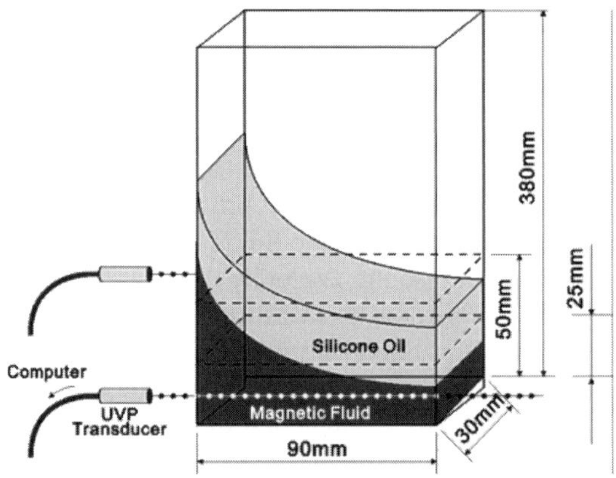

Figure 2. Test container

2 Experiment

Figure 1 shows the experimental apparatus and Figure 2 shows the test rectangular container. The rectangular container measures 90 mm × 30 mm × 380 mm and is made of acrylic resin. The oscillator has a frequency range from 0.8 Hz to 3.5 Hz. The amplitude of oscillation is 1.0 mm for all experiments.

A water based magnetic fluid (W-40) and a silicone oil (KF-56) are used. W-40 has a 40 % weight concentration of fine magnetic particles and is produced by Taiho Industries Co., Ltd., Japan. Its sound velocity, viscosity and density are c = 1440 m/s, $\mu = 2.5 \times 10^{-2}$ Pa·s and $\rho = 1.40 \times 10^3$ kg/m^3 at 25 °C. A porous SiO_2 powder (MSF-10M, Liquidgas, Co., Ltd.) having grains with an average diameter of about 0.9 μm is added to provide reflectors for the ultrasonic waves. KF-56 is produced by Shin-Etsu Chemical, Co., Ltd., Japan. Its sound velocity, kinematic viscosity and density are c = 1250 m/s, $\mu = 1.5 \times 10^{-2}$ Pa·s and $\rho = 1.0 \times 10^3$ kg/m^3 at 25 °C. A nylon powder produced by Ono Science Co. Ltd., Japan is added in order to provide reflectors for the ultrasonic waves for KF-56. The average diameter of grains is about 4 μm.

Three permanent magnets are used. Their averaged magnetic flux densities are as follows: 16.9 mT, 24.3 mT and 34.6 mT. An ultrasonic transducer is fixed on the side wall of the container in order to measure the horizontal velocity profile. The ultrasonic transducer has an active diameter of 5 mm and the measuring volume is a thin-disc shape element ϕ5 mm × 0.71 mm. The basic frequency is 4 MHz. LabVIEW is used in order to transmit and to receive the ultrasonic pulse wave.

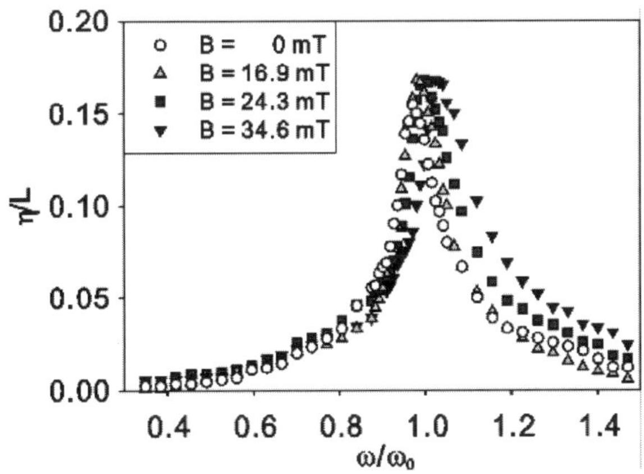

Figure 3. Frequency response of surface elevation of the magnetic fluid

3 Results and Discussions

3.1 Magnetic Fluid Sloching

The frequency response of the free surface of magnetic fluid sloshing is shown in Figure 3. The fluid height is 50 mm. Here B is the surface average magnetic flux density, η is the maximum free surface elevation at the side wall, L is the width of the container (L = 45 mm), ω is the forcing angular frequency and ω_0 is the resonant angular frequency obtained experimentally for B = 0 mT. This frequency is f = 2.775 Hz. As the forcing frequency increases, the surface elevation also increases until the free surface is intensively shaken near the resonant frequency. After the resonant frequency, the surface shaking is suppressed. As the magnetic field intensity increases, the first resonant frequency shifts to a higher frequency. For B = 34.6 mT, the first resonant frequency is 2.925 Hz.

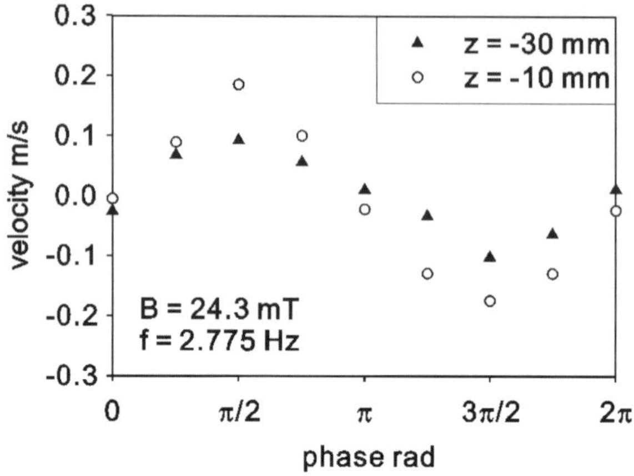

Figure 4. Time dependent velocity profiles of the magnetic fluid at the middle point

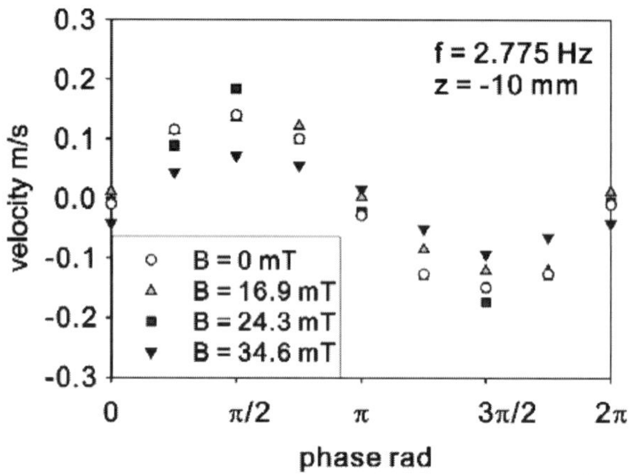

Figure 5. Magnetic field dependent velocity profiles of the magnetic fluid at the middle point

UVP measurements were carried out at the first resonant frequency. Figure 4 shows the time dependent velocity profiles for B = 24.3 mT. Here the start phase is the minimum surface elevation time and z is the position of the UVP transducer from the static free surface of a magnetic fluid. As the measuring point approaches the level of the free surface, the velocity increases. Fluid motion of the lower part appears to be suppressed because of the viscosity. Figure 5 shows the magnetic field dependent velocity profiles. When the magnetic field increases, the velocity decreases because of the magnetic force. The velocity for B = 34.6 mT is about half of that for B = 0 mT. These results show that we can indirectly control magnetic fluid sloshing by changing the magnetic field.

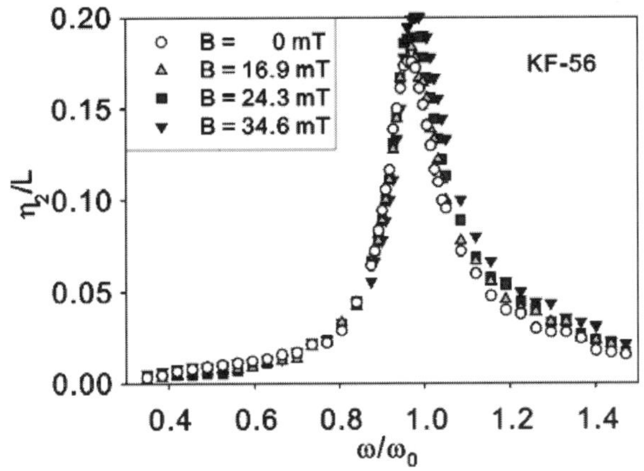

Figure 6. Frequency response of surface elevation for two-layered fluid sloshing

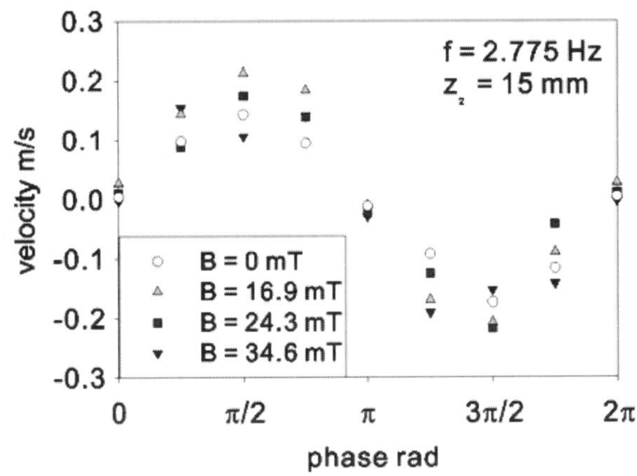

Figure 7. Magnetic field dependent velocity profiles for two-layered fluid sloshing

3.2 Two-layered Fluid Sloshing

The frequency response of the free surface of the two-layered fluid sloshing is shown in Figure 6. The fluid height of each fluid is 25 mm. Here η_2 is the maximum free surface elevation of a silicon oil at the side wall. As the forcing frequency increases, the surface elevation also increases until the free surface is intensively shaken near the resonant frequency. This tendency is similar to the result for single magnetic fluid sloshing. After the resonant frequency, the surface shaking is reduced. Similar to the above, the first resonant frequency shifts to a higher frequency as the magnetic field intensity is increased. For B = 34.6 mT, the first resonant frequency is 2.825 Hz. In this case, the rate of change of the resonant frequency and increasing rate of the surface elevation are smaller than those for single magnetic fluid sloshing because the magnetic force is relatively small.

Figure 7 shows the magnetic field dependent velocity profiles of two-layered fluid sloshing. Here z_2 is the position of the UVP probe from the interface between the two layers. Comparing Figures 5 and 7, there do not appear to be significant differences. The rate of change of the velocity in the two-layered sloshing is slightly smaller. The expected reduction of the inner velocities according to the magnetic field does not occur. In the present experiment, we have used a constant fluid volume ratio, i.e. 1:1. In the future, it will be useful to perform more detailed experiments while changing the arrangement/combination of the two layers.

4 Summary

Lateral sloshing of a magnetic fluid and a two-layered fluid in a rectangular container in a vertically-applied non-uniform magnetic fields has been investigated. In magnetic fluid sloshing, frequency response of the free surface and the inner velocity profile can be controlled using a magnetic field. In two-layered fluid sloshing, frequency response of the free surface and the inner velocity profile are generally similar to those in a single magnetic fluid. However, there appears to be the possibility for reducing the disturbance of the upper fluid by using magnetic fluid for the lower layer and applying a magnetic field.

Acknowledgements

This work was partially supported by the Keio Gijuku Fukuzawa Memorial Fund for the Advancement of Education and Research.

References

1. Takeda, Y., Velocity profile measurement by ultrasound Doppler shift method, *Int. J. Heat Fluid Flow*, **7** (1986) pp. 313-318.
2. Takeda, Y., Velocity profile measurement by ultrasonic Doppler method, *Exp. Therm. Fluid Sci.*, **10** (1995) pp. 444-453.
3. Kikura, H., Takeda, Y. and Durst, F., Velocity profile measurement of the Taylor vortex flow of a magnetic fluid using the ultrasonic Doppler method, *Expt. Fluids*, **26** (1999) pp. 208-214.
4. Sawada, T., Kikura, H. and Tanahashi, T., Kinematic characteristics of magnetic fluid sloshing in a rectangular container subject to non-uniform magnetic fields, *Expt. Fluids*, **26** (1999) pp. 215-221.
5. Ibrahim R. A., *Liquid Sloshing Dynamics*, (Cambridge University Press, Cambridge, 2005).
6. Handa, K. and Tajima, K., Sloshing of two superposed liquid layers in a rectangular tank, *Trans. Jpn. Soc. Mech. Eng. B*, **45** (1979) pp.1450-1457.

DEVELOPMENT OF NORMALLY CLOSED TYPE OF MAGNETORHEOLOGICAL CLUTCH AND ITS APPLICATION TO SAFE TORQUE CONTROL SYSTEM OF HUMAN-COLLABORATIVE ROBOT

T. SAITO AND H. IKEDA

Japan National Institute of Occupational Safety and Health (JNIOSH), 1-4-6 Umezono, Kiyose, Tokyo, Japan

E-mail: saitot@s.jniosh.go.jp, ikeda@s.jniosh.go.jp

In this study, a normally closed (NC) type clutch using magnetorheological (MR) suspension and a rare-earth permanent magnet is proposed as a safety device for human-collaborative robots. This clutch can transmit an effective torque under various conditions and function as a holding brake and a changeable torque limiter. To design this clutch to achieve both good controllable characteristics and safety performance, the normally closed structure with a permanent magnet was realized by placing two facing magnetic poles asymmetrically for efficient generation of the magnetic field, and by adopting the multi-sheet rotor for the improvement of the torque output. As the analysis results show, it was confirmed that the three-sheet rotor obtained about 1.5 times torque output as much as the single-sheet rotor. When the exciting current was 0.8A, the transmitted torque marked the minimum; when the exiting current was 0A, the torque output of approx. 41Nm was generated with the time constant of 0.15s.

1 Introduction

Recently, a new type robot sharing its working space with human working space has been under research and development for use in carrying heavy objects not only in factories but also at outdoor working sites and in welfare facilities. Since the general purpose of this robot is to substitute for conventional carriage supporting equipment, a working mode in which a human and a robot carry a heavy object together in a coordinated manner is being widely studied.

Since it is impossible to control the torque output of the load-lifting actuator not to exceed the human tolerance, a safety interlock system for monitoring the actuator torque output is essential to ensure the safe use of robots, especially those of vertical articulated type. That is, it is indispensable to equip a safety interlock that always monitors the actuator torque output and stop the torque output immediately in case of excessive output occurrence. In addition, the robot should be capable of maintaining its posture with no supply of energy in case of emergency or muting the interlock [1]. In order to fulfill these requirements, a torque-monitoring interlock system, composed of an antagonistic rotary actuator by means of magnetorheological (MR) suspension and a safety controller, was operated on a trial basis [2]. Although it had the normally closed (NC) type brake function, the antagonistic rotary actuator was not practicable due to improper size and weight.

Aiming to be a clutch device for use in joint axes of human-collaborative robots, this paper proposes a normally closed (NC) type clutch using MR suspension (hereinafter referred simply as “NC type MR clutch”) to control the torque output variably and secure a sufficient holding torque in case of emergency stop. Furthermore, this clutch is designed by using simulation to obtain efficient torque output characteristics and evaluated on the practical performance as a safety device.

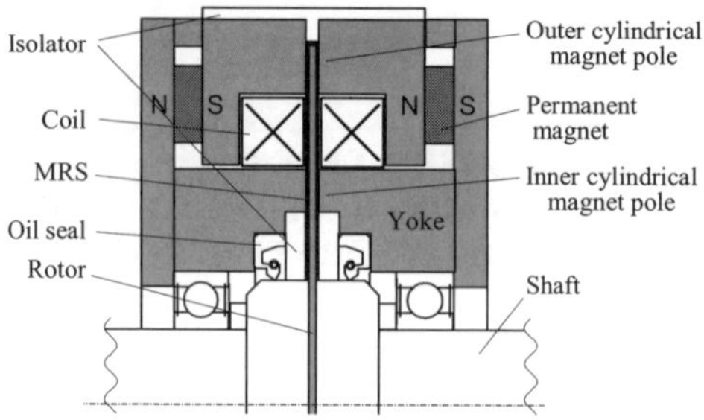

Figure 1. Schematic structure of NC type of MRC

Photo 1. Component parts of NC type MRC

2 Design concept of NC type MR clutch

The structure of the NC type MR clutch is shown in Fig. 1 and Photo 1. This clutch is so structured that magnetic circuits, each of which being composed of a permanent magnet (Sm_2Co_{17}) and an electromagnetic coil (0.5mm diameter enameled wire with 600 turns), are connected to each other via yoke and are arranged so that they can face each other with a rotor (diameter 158mm, thickness 2mm) connected to an output shaft between them. The clearance of 0.5mm wide formed between the inner and outer cylindrical magnetic poles and the rotor is filled with MR suspension.

The external power is applied to the rotating, left and right, mechanically connected magnetic circuits. When the coil is not excited, the filled MR suspension is solidified by the magnetic fields generated by the permanent magnets, and the torque is transmitted to the rotor. When the coil is excited, the magnetic fields generated by the permanent magnets are negated by the magnetic field generated by the excitation of the coil, and the MR suspension regains the suspensibility and makes the clutch open. According to this mechanism, when the exciting current flows in the direction in which it overlaps the magnetic field generated by the permanent magnet, the torque larger than that in the unexcited state can be transmitted.

The magnetic circuit layout shown in Fig. 1 works to negate the magnetic attraction force produced between the rotor and the magnetic pole and thereby prevent both the rotor and the magnetic pole from contacting each other. The reason is that the difference of the two magnetic poles facing each other differs in the convergence of the magnetic flux through the MR suspension, though the direction of the magnetic field does not influence the magnetic attraction force. According to the result of the magnetic simulation shown in Fig. 2, the magnetic flux density in the MR suspension filled up between facing magnetic poles in the opposite direction (asymmetric pole placement) is higher than that in the same direction (symmetric pole placement). In the case of symmetric pole placement, the magnetic flux flows in the rotor in the radial direction, thereby magnetic saturation appears in the rotor, and as a result, the transmitted torque is restrained.

According to the above-mentioned discussion, it was decided to place the facing magnetic poles asymmetrically. Moreover, it was aimed to increase the surface area of the rotor to obtain larger torque output, particularly larger holding torque when the NC type clutch was used. However, since increasing the rotor diameter required change to the magnetic pole shape, a multi-sheet rotor was adopted this time, which was composed of

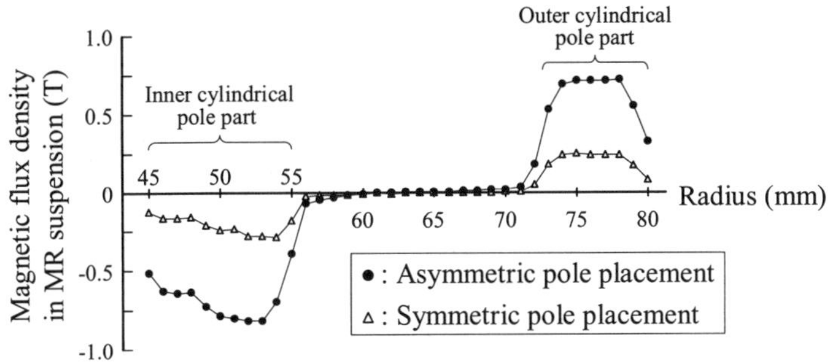

Figure 2. Simulation result of magnetic flux density caused by different magnetic circuit layouts

plural rotor sheets of the same shape as shown in Fig.3. This is because, in the asymmetric pole placement, magnetic resistance increases but the drop in magnetic flux penetrating through plural rotors sheets is presumed to be small, and therefore, the increase in the transmitted torque can be expected, even if the magnetic circuit is not changed. In addition, since this NC type MR clutch is assumed to be built in a robot joint axis, increasing the width is less restricted than that of increasing the outside diameter.

A composition with two ring-shaped magnetic substance spacers (thickness 2mm) placed with the gap of 0.5mm between them among three rotor sheets as shown in Fig. 3(b) was adopted. With this composition, the transmitted torque was estimated and compared with that of single rotor sheet shown in Fig. 3(a). Here, MRF-132DG (LORD Co.) was used as a test fluid, and the magnetic field H (kA/m) to be applied to the MR suspension variable according to the applied coil current was calculated by using the finite element method (FEM), and the yield stress τ_{MR} (Pa) of the MR suspension per radius variable according to H was estimated based on Eq. (1):

$$\tau_{MR} = 0.172\, H^3 - 0.709\, H^2 + 0.361\, H \tag{1}$$

The approximation formula delivered from the yield stress characteristics on the data sheet. Then, the transmitted torque T is calculated by using Eq. (2):

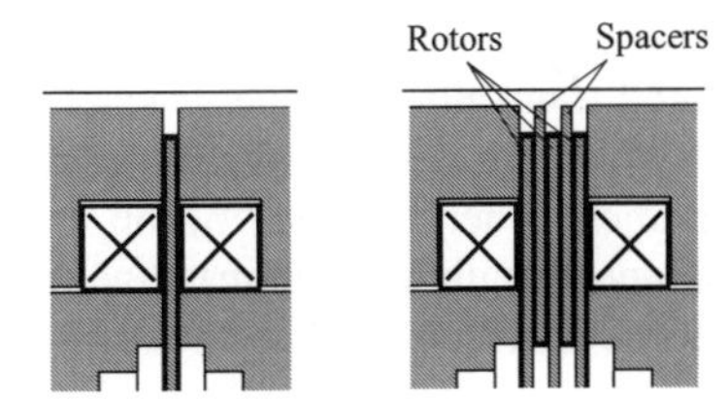

Figure 3. Difference of magnetic circuit parts caused by the number of rotors

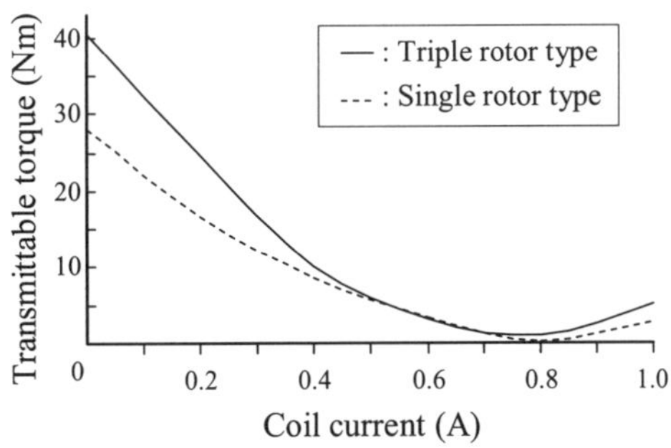

Figure 4. Transmitted torque calculated by equations in case of two type of rotor

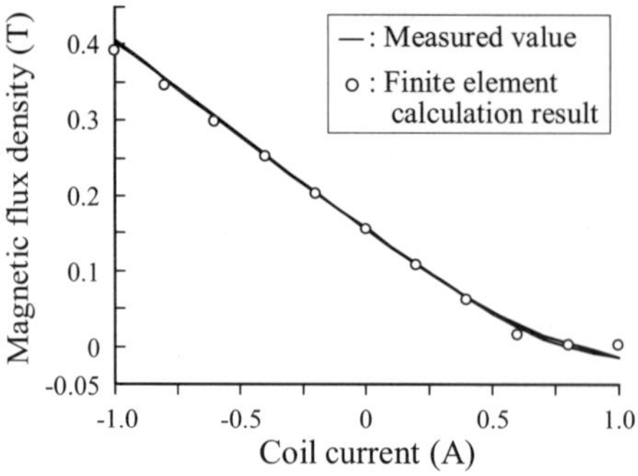

(a) Single rotor type

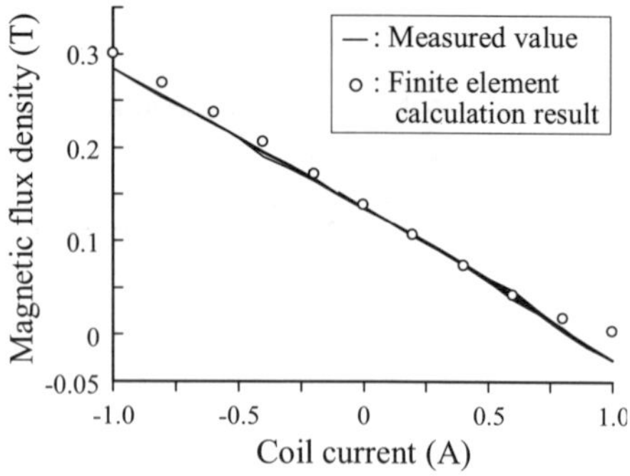

(b) Triple rotor type

Figure 5. Comparison of magnetic flux density between calculated value and measured value

$$T = \sum_{n=1}^{2n} \left(\int_{r_i}^{r_o} \tau_{MR} \; 2\pi r^2 \; dr \right) \tag{2}$$

where r_i is the radius on inner edge in the inner cylindrical magnetic pole, r_o is the radius on outer edge in the outer cylindrical magnetic pole, and n is the number of rotor sheets. Fig. 4 shows the calculation result, from which it can be expected that the transmitted torque in the three-sheet rotor in the unexcited state will be about 1.5 times as much as that with the single-sheet rotor.

3 Result of experiment on the transmitted torque characteristics

The NC type MR clutch, composed as shown in Fig. 1, was examined whether it had the characteristics obtained through the above simulation as to the single-sheet rotor and the three-sheet rotor. The magnetic field was measured at 1mm from the inner edge of the outer cylindrical magnetic pole with no filling of the MR suspension, and the variation in the magnetic field against the exciting current was compared with the simulation result (See Fig. 5). As a result, it was confirmed that the simulated values and measured values of both the rotor types were approximately consistent with each other excepting a case where the coil electric current was 1A and that the designed characteristics had been achieved.

Then, the transmitted torque estimated in Fig. 4 was measured with the experimental device shown in Photo 2 to verify whether the estimated transmitted torque was actually generated. In this experiment, the shaft was rotated while the designated coil exciting current was being applied to, and the torque at the moment when the shaft began to slip was measured with a load cell by pushing an arm (moment length of 200mm) attached to the outside of the external magnetic circuit. This torque, which was regarded as the positional holding performance of the NC type MR clutch, was recorded as the peak value of the measured value. However, it should be noted here that the shaft was rotated after 10min from when the exciting current was applied to the shaft in this experiment. The reason for this time lag will be described later.

Fig. 6 shows the measurement result. It was confirmed, as the analytical result showed, that in the unexcited status, the torque obtained with the three-sheet rotor was

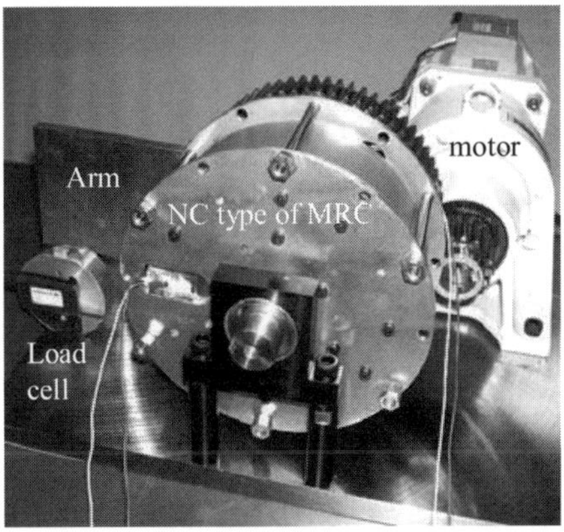

Photo 2. Experimental system to measure transmitted torque

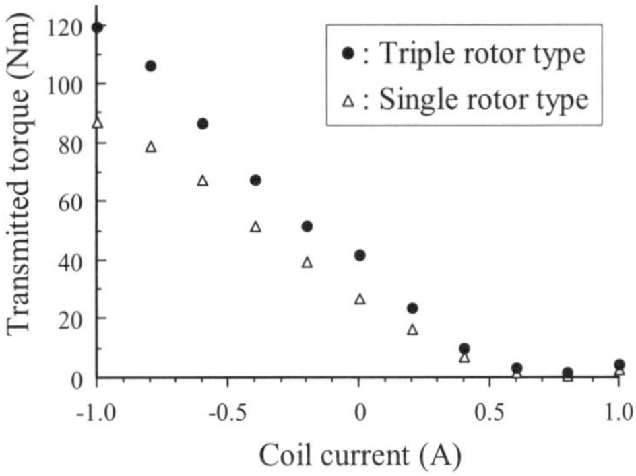

Figure 6. Relationship between exciting current of coil and measured transmitted torque

about 1.5 times as much as that obtained with the single-sheet rotor, and that when the exciting current was 0.8A, the transmitted torque was the smallest and the state in which the clutch was disengaged was obtained. In general, however, the torque value measured at each exciting current level, excepting 0.8A, was slightly smaller than the estimated value shown in Fig. 4. The torque values when the exciting current level was varied from 0A to -1A are also shown in Fig. 6. From this figure, it is understood that the torque measured at 0A is about three times as much as that obtained at the current of 1.0A.

4 Verification of safety function of the NC type clutch

Since the NC type MR clutch, which is a safety device built in a robot joint axis and functions as a holding brake to precisely hold the robot posture in collaboration with a variable torque limiter, the following characteristics are required:

1) The torque limit value designated by the coil exciting current shall not change even though the current is constant.
2) Also, the holding torque generated by the permanent magnet in the unexcited state shall not change until the coil is re-excited.
3) When the coil exciting current is shut off, the holding torque shall be generated promptly.

To verify that the NC type MR clutch satisfied the above requirements, the response and temporal variation in the transmitted torque were examined.

As to the requirements 1) and 2), Fig. 7 shows the result of measurement of temporal torque variation at a constant coil exciting current level. Due to the secondary growth of cluster well known as one of the characteristics of the MR suspension, the torque increased gradually until 10min passed after the change in the coil exciting current. Therefore, to have the MR suspension serve as a torque limiter, the steady state value after the lapse of 10min (torque value shown in Fig. 6) should be regarded as the limit value. On the other hand, to use the MR suspension for the holding brake, the small torque value before obtaining the steady state, which is about 80% of the settling value, should be considered.

Next, to examine the primary response characteristic of the torque, the step response of the transmitted torque when the coil exciting current became unexcited with the shaft rotating to be constant at 30 rpm with the AC servo motor was measured by using the

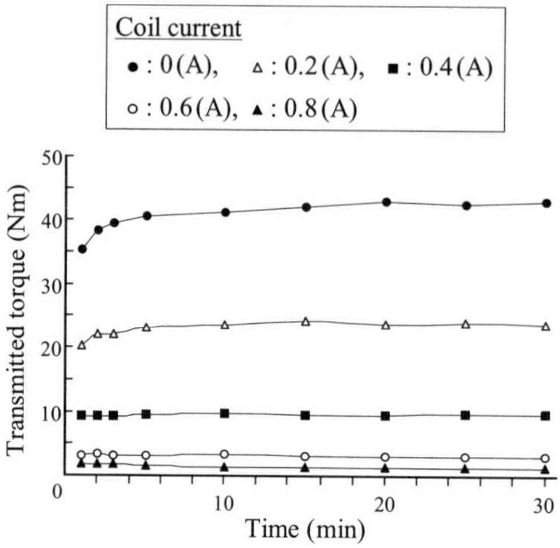

Figure 7. Transmitted torque change with passage of time under constant exciting currents

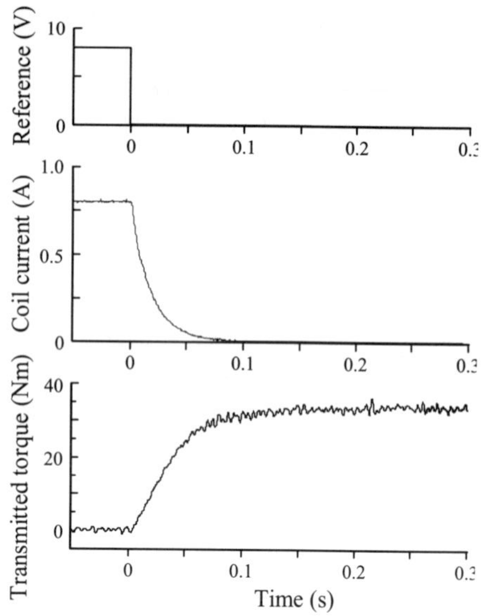

Figure 8. Step response of transmitted torque (holding torque)

above-mentioned experimental device. To evaluate the worst response, the torque value when the exciting current was changed from 0.8A to 0A of was measured. The result is shown in Fig. 8. From the response waveform of the exciting current, it is understood that the response delay of the current amplifier can be roughly approximated as a time-lag of the first order system of time constant 18ms. The transmitted torque followed to the change in this exciting current with little delay. The time constant of the torque as the first order system remained almost the same at approx. 0.15s even when the applied coil exciting current was changed.

5 Conclusions

To realize a safety device capable of controlling the torque output of the robot joint axis and securing the holding torque in case of emergency stop, a normally closed type of clutch by using the MR suspension (NC type of MR clutch) was developed and evaluated. This NC type MR clutch is well applicable as a safety mechanism for robots. From now on, the authors plan to promote the improvement of the magnetic circuit efficiency and the mass-to-output ratio.

References

1. ISO 10218-1, Robots for industrial environments – Safety requirements – Part 1:Robot (2006) p.17
2. T. Saito, H. Ikeda, Safe torque control system of human-collaborative material handling robot for crush prevention, Symposium digest of ISR2005, (2005) p.57

A MAGNETORHEOLOGICAL DAMPER WITH BIFOLD VALVES FOR SHOCK AND VIBRATION MITIGATION

MIN MAO, WEI HU, YOUNG-TAI CHOI, AND NORMAN M. WERELEY

Alfred Gessow Rotorcraft Center, Department of Aerospace Engineering, University of Maryland, College Park, MD 20742 USA

wereley@umd.edu

This paper presents the design and fabrication of a flow-mode bifold magnetorheological (MR) damper for shock and vibration mitigation for high piston velocity (15mph or 6.71 m/s) as well as an evaluation of its performance at low speed. Based on a Bingham-plastic model, as well as a Bingham-plastic model coupled with a low speed hysteresis model, two theoretical MR damper models for flow-mode MR dampers are constructed. Using the design strategy associated with the Bingham-model based damper model, two MR damper designs for achieving the performance requirement with a limited space are considered: first, the conventional MR damper that has an MR valve inside the piston head and second, the bifold MR damper that has MR valves at each end of the damper. After numerically comparing the damping performances of two MR damper designs, the bifold MR damper was chosen because its dynamic range was better at high speed. The bifold MR damper was tested at relatively low piston velocity using an MTS testing machine under sinusoidal loading. Experimental data are compared well with the results predicted by the theoretical models.

1 Introduction

Over the past decade, magnetorheological (MR) fluids and applications have experienced rapid growth. Many theoretical and/or analytical methods[1-3] were established and various kinds of MR devices were built and employed as efficient semi-active control components in numerous practical applications from seismic mitigation to helicopter lag mode control[4-5]. Recently, as MR dampers have been highlighted as a promising candidate for crashworthiness systems, the development of MR devices to cope with crash conditions where a high shock load is applied during a short time period is of great interest. Browne et al[6] experimentally conducted impact tests of an MR damper over high piston velocity range of from 1.0 to 10 m/s and showed that the MR damper force could be adjusted by applying magnetic field. Facey et al[7] developed a compact MR damper for the protection of sensitive devices against high impulsive loads and experimentally tested the damping capability of the damper at a peak piston velocity of 0.88 m/s. However, automotive crashworthiness system must operate at peak velocities of as much as 15 mph (6.71 m/s) or higher. The design and configuration of MR dampers for crashworthiness is not well explored in the literature. Therefore, this motivates our study, which focuses on the theoretical analysis, design and laboratory implementation of a compact high force MR damper for shock and vibration mitigation.

In this study, the design and fabrication of a flow-mode bifold MR damper for shock and vibration mitigation under high piston velocity (15 mph or 6.71 m/s) is presented. Based on Bingham-plastic (BP) model and a BP model coupled with a low speed hysteresis model (BPH), two theoretical MR damper models of flow-mode MR dampers are constructed. Using the design strategy[8] associated with the BP damper model, two MR damper designs for achieving the performance requirement with a limited volume (implying constraints on diameter and stroke) are considered: (1) a conventional MR damper having an MR valve inside the piston head and (2) a bifold MR damper that has MR valves at each end of the damper. After theoretically evaluating and comparing the damping performance of the two MR damper designs, the bifold MR damper was selected for fabrication. The damping performance of the bifold MR damper was conducted at relatively low piston velocity using an MTS testing machine. The experimental testing results were compared with the results predicted using both BP and BPH models.

2 Nonlinear Damper Modeling

The models used in this study are outlined below. For the BP model, the force, F of the MR damper is given by the nonlinear flow behavior as follows[8]:

$$F = F_\eta + (F_\tau + F_f)\cdot \mathrm{sgn}(V_p) \tag{1}$$

where

$$F_\eta = f\frac{A_p \rho L \bar{A}^2 V_p^2}{2D_h} \text{ and } F_\tau = \frac{2A_p L\tau_y}{d} \tag{2}$$

Here, F_τ is the yield force due to the MR effect, and F_f is the friction force due to shaft seals and can be identified by experiments. F_η is the Newtonian viscous force and V_p is the piston velocity. A_p is the effective piston area, and $A_p = \pi(D_p^2 - D_r^2)/4$, where D_p is the diameter of the piston, D_r is the diameter of the piston rod. $\bar{A}$ is the area ratio between the effective piston area and the cross sectional valve area and $\bar{A} = A_p / A_d$ with $A_d = d\cdot \pi D_b$, where D_b is the effective gap diameter. τ_y is the yield stress of an MR fluid, d is the gap of the MR valve and D_h is the hydraulic diameter, which is used for non-circular valve. L is the total length of the active area. f is the Darcy friction factor[8] and is chosen depending on the magnitude of the Reynolds number of the fluid flow in the valve. If the annular duct in the MR valve is approximated as a rectangular duct or parallel plates, the hydraulic diameter D_h is given by[8]

$$D_h = 2d \tag{3}$$

Note that the viscous force, F_η, for a laminar flow is reduced to the most familiar form as follows:

$$F_\eta = \frac{12\eta L A_p^{\,2} V_p}{\pi D_b d^3} \tag{4}$$

where η is the viscosity of the MR fluid.

For simplicity, the friction force, F_f is neglected during the design stage. Then the desired dynamic force range, D of the MR damper is represented as[8]:

$$D = \frac{F}{F_\eta} = 1 + \frac{16}{f}\left(\frac{Bi}{\mathrm{Re}_{D_h}}\right) \tag{5}$$

where

$$\mathrm{Re}_{D_h} = \frac{2\rho\bar{A}V_p d}{\eta} \quad \text{and} \quad Bi = \frac{\tau_y d}{\eta \bar{A} V_p} = \left(\frac{2\rho d^2 \tau_y}{\eta^2}\right)\frac{1}{\mathrm{Re}_{D_h}} \tag{6}$$

The Bingham-model based nonlinear damper model given by Eq.(1) does not capture a hysteresis loop in the force-velocity plot since the yield force shows a discontinuous around zero piston velocity. To more precisely describe practical hysteretic characteristics of MR dampers in the force-velocity plot, a hysteresis model[9] is also adopted in this study. In the BPH model, the force of the MR damper is given by

$$F = F_\eta + (F_\tau + F_f)\tanh\left[(V_p + \lambda_1 X_p)\lambda_2\right] \tag{7}$$

Here X_p is the piston displacement, tanh[·] is the hypertangent function, and λ_1, λ_2 are parameters to capture the shape of the hysteresis loop and are identified from experimental data.

3 Damper Design

For shock and vibration mitigation, MR dampers must be designed to produce sufficient controllable damper force at high piston velocity. In this case, the desired maximum piston velocity was 6.71 m/s (15 mph). In addition, the zero-field damper force was kept below 2 kN below the maximum piston velocity, and the dynamic force range must be greater than $D = 2$. Such a dynamic range implies that the damping force for maximum field should be greater than 4 kN at the peak piston velocity of 6.71 m/sec. While satisfying these specifications, the outer diameter of the MR damper must not exceed 2 inches.

First, an MR damper (Figure 1(a)) that has a conventional MR valve in the piston head was considered for the damper design. When a particular MR fluid (η=0.4 Pa ·s and $\tau_{y,\max}$=80 kPa) is used, the simulated damping performance of the conventional MR damper is showed in Figure 2. In this case, the dimensions were chosen using the design strategy[8] associated with the BP damper model: $D_p = 47.625, D_r = 11.125, D_b = 37.885$, $L = 6.0, d = 1.2$ (all the units are mm). From Figure 2, the conventional MR damper can generate a total damping force of 4 kN. However, the field-off damper force is higher than 2 kN at 15 mph, violating our specification, so that the dynamic range is less than the desired value of $D = 2$. Thus, the conventional MR damper cannot meet the design goal.

Therefore, a bifold MR damper was proposed for this application. The schematic diagram of the bifold MR damper is shown in Figure 1(b). The design parameters were also obtained using our BP model-based design strategy[8]: $D_p = 25.4, D_r = 11.125$, $D_b = 32.389, L = 64, d = 1.2$ (all units are mm). As seen in Figure 2, the bifold MR damper satisfies both the total damping force and zero field force requirements. In addition, up to the maximum piston velocity of 15 mph, the desired dynamic range is greater than 2. Therefore, the bifold MR damper design was chosen and fabricated in this study.

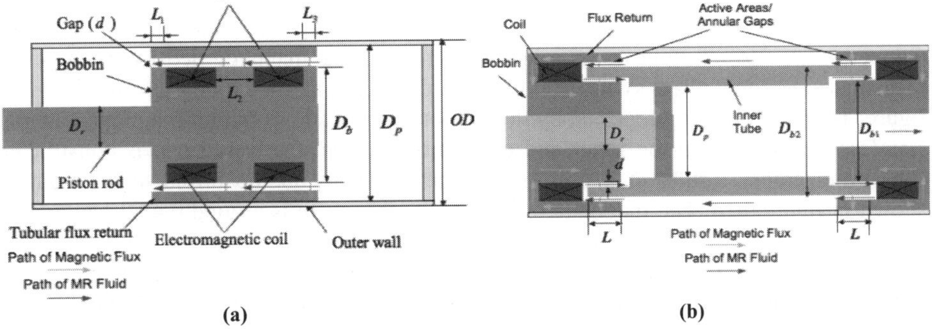

Figure 1. Schematic diagrams of MR dampers for shock and vibration mitigation. (a) conventional MR damper design, (b) bifold MR damper design.

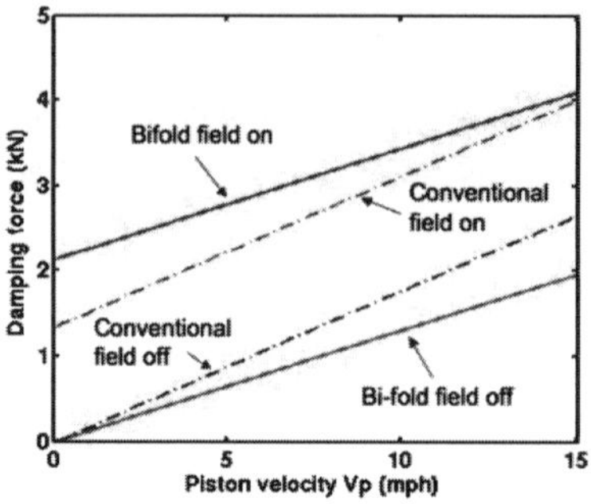

(a) Damping force vs. piston velocity

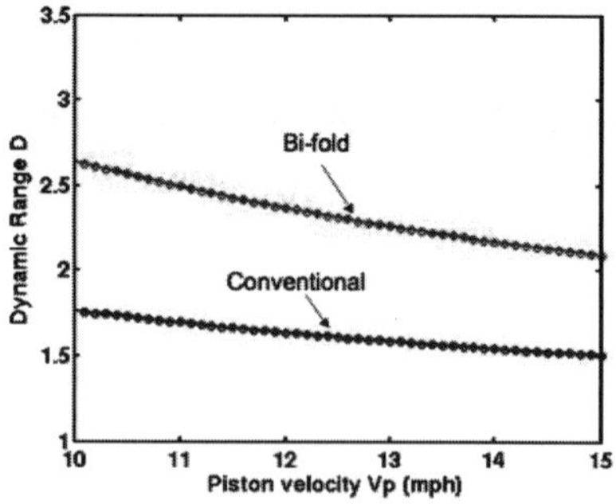

(b) Dynamic range vs. piston velocity

Figure 2. Comparison of simulated damping performance between conventional and bifold MR damper.

4 Damper Fabrication

Figure 3 presents the cross-sectional view of the bifold MR damper fabricated in this study. For designing the magnetic circuit, the use of high permeability material such as Hiperco alloy (saturates at 2.3 Tesla) enables high magnetic circuit performance (reduced eddy current losses and reduced magnetic circuit volume), but the cost will be high[7]. Thus, in this study, the material AISI 12L14 was used to reduce cost. The AISI 12L14 is low carbon steel (nominal carbon content is 0.14%) and has high magnetic saturation. For better magnetic property, the AISI 12L14 was annealing at 1600°F for 1 hour and was then slowly cooled to room temperature before the material was machined.

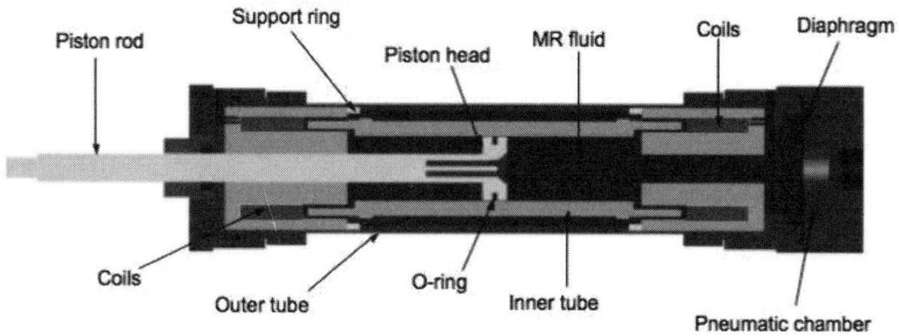

Figure 3. Cross sectional view of the fabricated bifold MR damper.

Figure 4 presents an FEM analysis of the magnetic characteristics of the bifold MR damper at the current input of 1.25 A. In Figure 4(a), the magnetic density in the gap is presented. The view is axi-symmetric with the damper centerline along left margin of the figure. The inner tube is supported by the support ring (made from non-magnetic material, stainless steel) between the inner and outer tubes so that the bifold gap is constructed. This arrangement results in a uniform magnetic density in the bifold annular gap as shown in figure 4(b). The magnetic density at the inner gap is around 0.48 Tesla and the magnetic density at the outer gap is around 0.40 Tesla. The average magnetic density in the gaps is 0.44 Tesla, which corresponds to the maximum yield stress of nominally 49 kPa, which produces the simulated damping performance showed above in Figure 2.

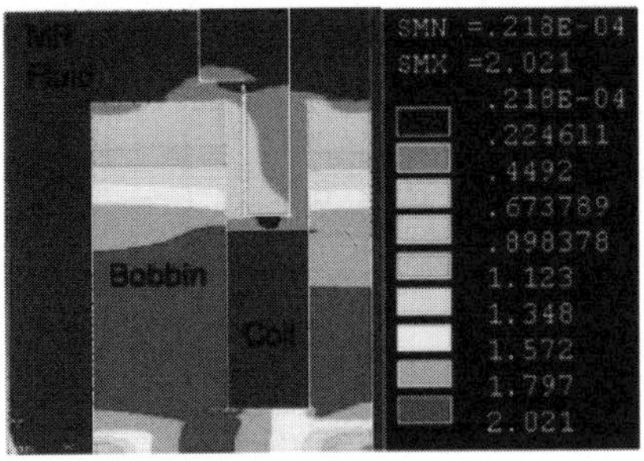

(a) Magnetic density in the gap

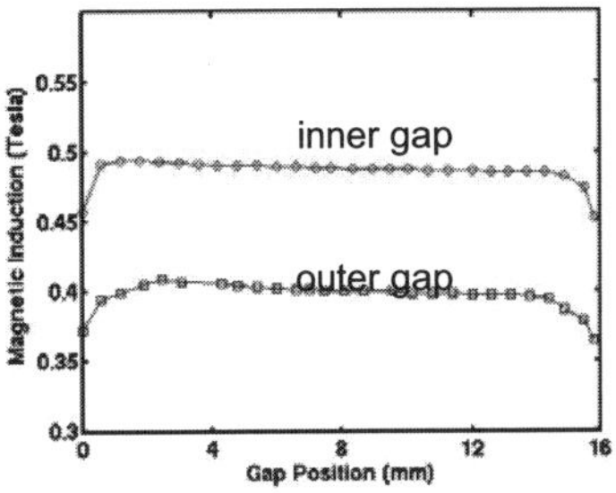

(b) Magnetic density vs. gap position

Figure 4. FEM analysis on the magnetic characteristics of the bifold MR damper at the current input of 1.25 A. (a) magnetic density in the gap, (b) magnetic density vs. gap position.

5 Damper Testing and Characterization

To evaluate the damping performance and validate the effectiveness of the design, the damper force testing at relatively low velocity was conducted on a hydraulically powered MTS testing machine using Lord MRF-132AD fluid (η=0.18 Pa ·s and $\tau_{y,\max}$=40 kPa). The tests were run at 0.5 to 5 Hz sinusoidal signals in 0.5 Hz increments. For each frequency, the applied currents vary from 0 to 1.75 A with 0.25 A increment, and the commanded damper stroke was ½ inch. The gas chamber was pressurized to 1000 psi for all testing.

In Figure 5, the measured force (solid line) vs. sinusoidal displacement and velocity are compared to the results predicted (dash-dot line) using the BP model in Eq.(1) and the BPH model from Eq.(7). In Figure 5, the excitation frequency is 2.5 Hz and the applied currents are 0 A and 1.25 A. It is observed that the damper force of the MR damper is greatly increased by the applied current input. At the peak piston velocity of 0.1 m/s, the dynamic range approaches $D \approx 7$. Test data compared well with results predicted by both theoretical damper models. However, the force versus velocity behavior predicted by the BP model does not account for the low speed hysteresis or compression loop in the force versus velocity diagrams. In contrast, the BPH model accurately accounts for this low speed hysteresis loop.

The equivalent (viscous) damping, C_{eq}, is computed by equating energy dissipated by the MR damper over a single cycle, E, at a given frequency, ω, to that of a viscous damper:

$$C_{eq} = \frac{E}{\pi\omega X_{p0}^{2}} \qquad \text{where } E = \int_{cycle} F dX_p = \int_0^{2\pi/\omega} F V_p dt \tag{8}$$

Here, X_{p0} is the amplitude of the excitation displacement. In figure 6, the measured equivalent damping of the bifold MR damper is evaluated and compared with the results predicted by both damper models. Here, the equivalent damping predicted by both models favorably matches the measured results for both damper models. However, the BPH model presents a better prediction of the equivalent viscous damping C_{eq} than the BP model over the piston velocity range tested in this study.

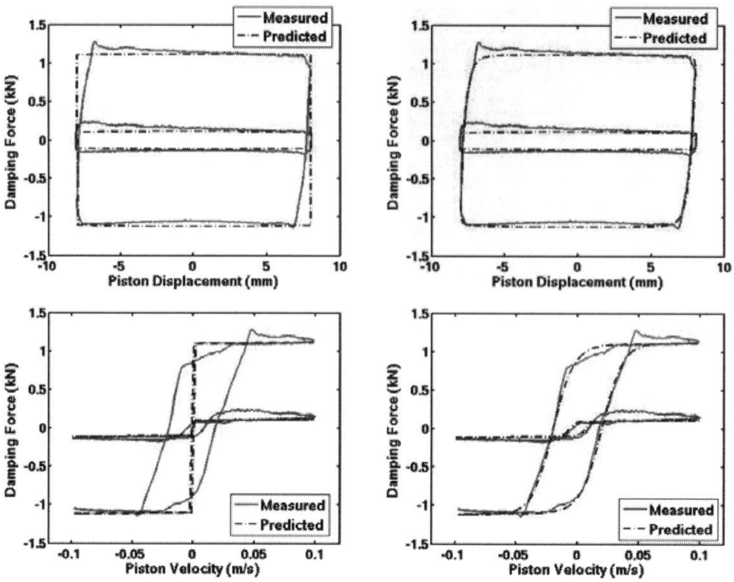

(a) BP model **(b)** BPH model

Figure 5. Comparison between the measured (solid line) and predicted (dash-dot line) damper forces of the bifold MR damper at 2.5 Hz: a) BP model, b) BPH model.

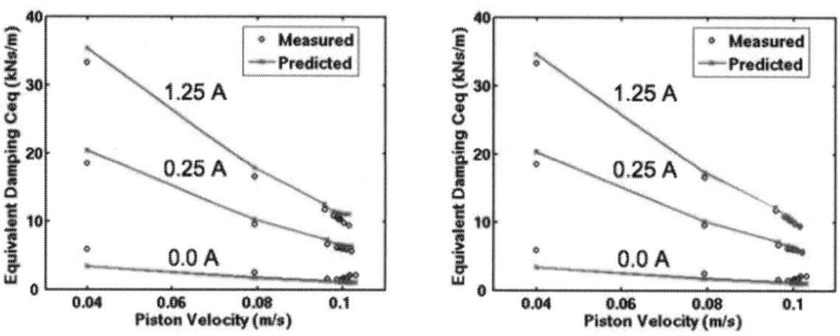

(a) BP Model **(b)** BPH Model

Figure 6. Comparison between the measured (discrete circles) and predicted (cross-marker lines) equivalent viscous damping vs. piston velocity at various currents.

6 Conclusions

The design and fabrication of a flow-mode bifold MR damper for shock and vibration mitigation under high piston velocity, as well as its performance at low speed, was presented. Design specifications included: maximum speed of 15 mph or 6.71 m/s, off state damping no greater than 2 kN and a maximum on state damping of at least 4 kN, implying a dynamic range of 2 at maximum speed. Two theoretical models were developed to aid in the design and evaluation process: Bingham-plastic model (BP) alone, as well as a Bingham-plastic model coupled with a low speed hysteresis model (BPH). To achieve the desired performance specifications under both a length and diameter constraint, two MR damper designs were considered: (1) a conventional annular orifice damper with an annular MR valve inside the piston head, and (2) a bifold damper having MR bifold valves at each end of the damper. Both configurations were able to meet the high speed maximum on-state damping requirement. However, it was observed that the conventional MR damper design could not meet the off-state damping requirement over this speed range, whereas the bifold MR damper could. Based on these numerical results, the bifold MR damper design was selected for fabrication. In this study, the low speed damping performance of the bifold damper was assessed at relatively low piston velocity (around 0.1 m/s or 0.22 mph) using an MTS testing machine. The experimental force vs. sinusoidal displacement testing results compared well with the results predicted by both theoretical damper models. However, the force versus velocity behavior predicted by the BP model lacked fidelity because the model does not account for the low speed hysteresis or compression loop. The BPH model has a representation of the low speed force versus velocity hysteresis loop, so that the force predictions more accurately matched experimental data. As a result, the equivalent damping predicted by the BPH model was also more accurate than that predicted by the BP model. Tests intended to characterize the behavior of this damper design over its entire speed range ($0 \le V_p \le 15$ mph) will be carried out in the near future.

References

1. D. J. Peel, R. Stanway and W. A. Bullough, "Dynamic Modeling of an ER Vibration Damper for Vehicle Suspension Applications," *Smart Mater. Struct.* **5**: 591-606, 1996.
2. N. M. Wereley and L. Pang, "Nondimensional Analysis of Semi-active Electrorheological and Magnetorheological Dampers Using Approximate Parallel Plate Models," *Smart Mater. Struct.* **7**: 732-743, 1998.
3. X. Wang and F. Gordaninejad, "Flow Analysis of Field-controllable, Electro- and Magneto-rheological Fluids Using Herschel-Bulkley Model," *J. Intelligent Mater. Systems and Structures*, **10**(8): 601-08, Aug, 2000.
4. G. Yang, B. F. Spencer, H. J. Jung, J. D. Carlson, "Dynamic Modeling of Large-Scale Magnetorheological Damper Systems for Civil Engineering Applications," *J. Engineering Mechanics*, **130**(9): 1107-1114, Sept. 2004.
5. W. Hu and N. M. Wereley, "Magnetorheological Fluid and Elastomeric Lag Damper for Helicopter Stability Augmentation," *Inter. J. of Modern Physics B*, **19**(7-9): 1471-1477, 2005.
6. A. L. Browne, J. D. McCleary, C. S. Namuduri, S. R. Webb, "Impact Performance of Magnetorheological Fluids," ASME International Mechanical Engineering Congress and Exposition, Anaheim, CA, USA, Nov. 2004.
7. W.B. Facey, N.C. Rosenfeld, Y.T. Choi, and N.M. Wereley, S.B. Choi, P.C. Chen, "Design and Testing of a Compact Magnetorheological Damper for High Impulsive Loads," *J. Modern Physics B*, **19**(7-9):1549-1555, 2005.
8. M. Mao, Y.T. Choi and N. M. Wereley, "Effective design strategy for a magnetorheological damper using a nonlinear flow model," SPIE Symposium on Smart Materials and Structures, **5760**: 446-455, 2005.
9. Y. T. Choi and N. M. Wereley, "Semi-Active Vibration Isolation Using Magnetorheological Isolators," *Journal of Aircraft*, **43**(4): 1244-1251, 2005.

SQUEEZE-FLOW MODE MAGNETORHEOLOGICAL FLUID MOUNT

C. CIOCANEL, T. NGUYEN, M. ELAHINIA AND N.G. NAGANATHAN

Dynamic and smart Systems Laboratory, University of Toledo, 2801 West Bancroft St., MS 312, Toledo, Ohio, USA

E-mail: cciocane@eng.utoledo.edu

The paper presents the design and of a magnetorheological (MR) fluid based damper that operates in squeeze and flow modes. The damper has been designed to reduce vertical vibrations, namely displacement/force transmissibility, from a pneumatic hammer to the operator body. The mount governing equations have been derived and the effect of system parameters on its performance was analyzed. Particular attention was given to the effect of the operator mass and the magnitude of the magnetic field on the mount response. Results analysis indicated that the resonant peak decreases as the applied magnetic field increases. Simultaneously, the resonance frequency decreases.

1 Introduction

Mechanical vibrations are a common presence in our day to day activities, either when riding a car, bus or train to work, or when we operate machineries at the workplace. Human perception and the impact on human's health of such vibrations highly depend on the frequency level and amplitude at which they occur. To mitigate or eliminate structure's vibrations one may involve passive, semiactive or active vibration isolation devices. Traditional mounts for vibration isolation, made of elastomers, are passive and tuned for a narrow range of frequencies and each design fits a certain application. These drawbacks encouraged the development of semiactive and active vibration mitigation mounts that can operate at a wider range of frequencies and that can be easily controllable. Lately, magnetorheological (MR) and electrorheological (ER) fluids have been used to develop semiactive and active mounts and dampers mostly for automotive applications [1,2,3]. The characteristics of these fluids, i.e. viscosity and yield stress, change instantaneously and reversibly upon the application of a magnetic/electric field. This specific behavior makes them very suitable for vibration isolation applications.

Hong *et al.* [4] developed a dynamic model for a flow mode type ER mount. Based on model predictions, parameters have been selected to manufacture a mount capable to support a 200 kg mass. Particular attention was given to the force and displacement transmissibility characteristics of the mount. The proposed dynamic model can be readily adopted for magnetorheological mounts. In 2002, Hong *et al.* [5] investigated the effect of an ER mount working in squeeze mode on the dynamic response of a flexible beam. The governing equations for the mount have been developed and an optimal control strategy was implemented. They confirmed experimentally that an ER mount with the appropriate control can efficiently mitigate structures vibrations.

Ha *et al.* investigated experimentally the damping characteristics of a squeeze mode MR mount. They observed that an increase in the applied magnetic field caused the resonant frequency to increase, contrary to what is observed in the response of a one degree of freedom system when the damping increases. In addition, the MR fluid was proven to be more effective at low frequencies than at high frequencies.

In 2006, Hong et al. [6] compared a squeeze mode and a flow mode ER mount concluding that the squeeze flow mount performed better at low levels of excitation. They also noted that the transmissibility of the mounts can be improved by adopting a skyhook control strategy.

This paper presents the design of an MR mount for vibrations mitigation of a pneumatic drill (jackhammer) in an attempt to reduce the vibration experienced by the operator of the tool. Extensive exposure to vibrations generated by the drill can lead to blood-circulation failures in the operator's fingers. The mount design is presented and the governing equations of motion are derived. The effect of the operator weight and of the applied magnetic field on mount displacement transmissibility is evaluated.

2 Design

The mount consists of a main chamber and a reservoir connected through six channels. In operation, the MR fluid can flow between the main chamber and the reservoir. The fluid flow is regulated by the magnetic field generated by a pair of circular coils installed in a Helmholtz configuration. This coils configuration was selected to maximize the applied magnetic field generated by a certain electric current passing through them. When the applied field is low, the fluid flows between chambers and it is squeezed simultaneously. The air gaps created around the reservoirs accommodates the change in volume of the elastomer. The metallic housing holding the elastomer has holes in it at the air gaps location to allow the free deformation of the elastomer.

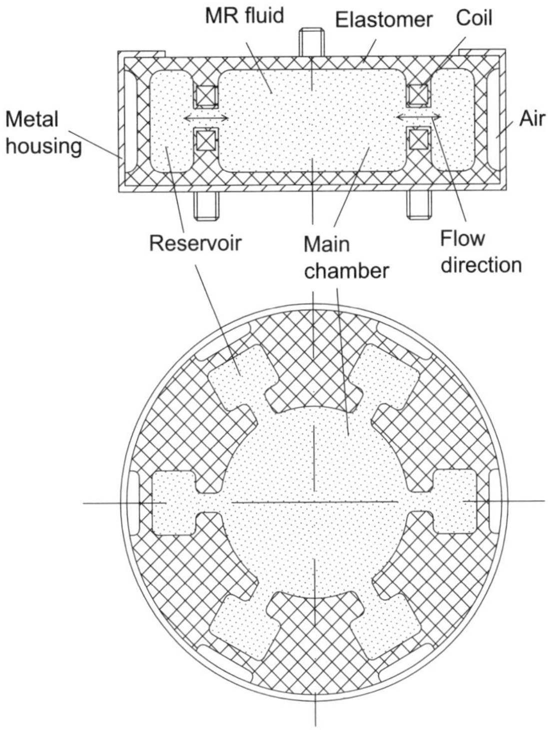

Figure 1. Cross sections through the proposed MR mount. a) side view. b) top view.

At high magnetic fields, the fluid flow is almost obstructed through the connecting channels and the operating mode is considered to be squeeze only. This is because the magnetic field generated by the coils cannot activate the whole fluid volume. The fluid located at the center of the main chamber would be affected much less than the fluid closer to the coils.

The following section introduces the mathematical model of the mount and the equations characterizing the mount.

3 Mathematical model

In deriving the equations characterizing the MR mount, it was assumed that the fluid is incompressible, exposed to an uniform pressure in each of the control volumes and that the mount exhibits no lateral motion. Based on the physical model illustrated in Figure 2 and the procedure proposed by Hong *et al.* [6], the equations characterizing the mount are derived as detailed below.

The equation of motion of the supported mass M is given by:

$$M\ddot{x} + k_e(x - y) + c_e(\dot{x} - \dot{y}) = A_p P_1 \quad (1)$$

where k_e and c_e are the stiffness and damping of the elastomer, A_p is the area of the main chamber, P_1 is the pressure in the main chamber, x is the mass displacement and y is the input displacement.

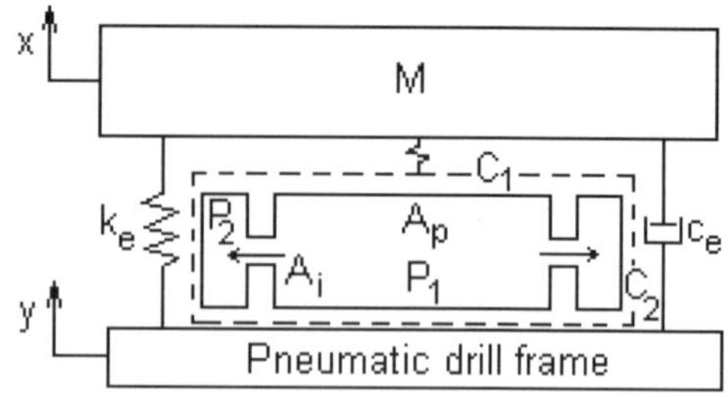

Figure 2. – Physical model of the MR mount.

The pressure drop due to the flow of the MR fluid through the inertia track of the mount is then evaluated from the linear momentum equation as:

$$P_1 - P_2 = I_i \dot{Q}_i + R_i Q_i + \Delta P_{MR} \quad (2)$$

where P_2 is the pressure in secondary chamber of the mount, I_i is the fluid inertia, R_i is the fluid drag at zero magnetic field, Q_i is the fluid flowrate through the inertia track, and ΔP_{MR} is the pressure drop due to the yield stress of the MR fluid.

In addition, the generalized flow continuity equations for the system [7] yield:

$$\dot{P}_1 = \frac{A_p}{C_1}(\dot{x} - \dot{y}) - \frac{Q_i}{C_1} \quad (3)$$

$$\dot{P}_2 = \frac{Q_i}{C_2} \quad (4)$$

where C_1 and C_2 are the compliance of the main chamber and reservoir [8], respectively.

Taking the derivatives of Equations 1 and 2 and then substituting $\dot{P}_1$ and $\dot{P}_2$ given by Equations 3 and 4 in the new equations yields:

$$M\dddot{x} + k_e(\dot{x} - \dot{y}) + c_e(\ddot{x} - \ddot{y}) = \frac{A^2{}_p}{C_1}(\dot{x} - \dot{y}) - \frac{A_p A_i \dot{x}_i}{C_1} \quad (5)$$

$$\frac{A_p}{C_1}(\dot{x} - \dot{y}) - \frac{A_i \dot{x}_i}{C_1} - \frac{A_i \dot{x}_i}{C_2} = I_i A_i \dddot{x}_i + R_i A_i \ddot{x}_i + \Delta \dot{P}_{MR} \quad (6)$$

where $A_i \dot{x}_i$, $A_i \ddot{x}_i$ and $A_i \dddot{x}_i$ were substituted for Q_i, $\dot{Q}_i$ and $\ddot{Q}_i$.

Equation 6 is then multiplied by the area of the main chamber, A_p, and then both Equations 5 and 6 are integrated once. The integration yields the following coupled equations of motion for the mount:

$$M\ddot{x} + c_e \dot{x} + \left(k_e - \frac{A_p^2}{C_1}\right)x + \frac{A_i A_p}{C_1} x_i = c_e \dot{y} + \left(k_e - \frac{A_p^2}{C_1}\right)y \quad (7)$$

$$I_i A_i A_p \ddot{x}_i + R_i A_i A_p \dot{x}_i + A_i A_p\left(\frac{1}{C_1} + \frac{1}{C_2}\right)x_i = \frac{A^2{}_p}{C_1}(x - y) + \Delta P_{MR} A_p \quad (8)$$

The last term of Equation 8 represents the force due to the MR fluid flow through the inertia track, and can be substituted by [9]:

$$\Delta P_{MR} A_p = 3\frac{L}{h} A_i \tau_y(H) + \frac{12 A_i^2 \eta(H) \dot{x}_i L}{bh^3} \quad (9)$$

The first term in Equation 9 is the fluid force contribution due to the yield stress. The second term is the fluid's viscous force developed at the flow through the inertia track. The parameters from Equation 9 represent: L – the length inside the inertia track channel over which the magnetic field is applied, h – the distance between the magnetic poles, which is equal to the height of the inertia track channel, b – the width of the channel, A_i – the cross-sectional area of the inertia track, $\tau_y(H)$ – the MR fluid yield stress that which magnetic field (H) dependent, and $\eta(H)$ – the MR fluid viscosity, which is field and shearing rate dependent.

To evaluate the effect of the MR mount on displacement transmissibility, Equations 7 and 8 are transferred into Laplace domain, under the assumptions that the initial conditions are zero, as follows:

$$m_i s^2 X_i(s) + c_{11} s X_i(s) + k_{11} X_i(s) + k_{12} X(s) = k_{11y} Y(s) + k_{12y} \frac{1}{s} \tag{10}$$

$$m s^2 X(s) + c_{22} s X(s) + k_{22} X(s) + k_{21} X_i(s) = k_{21y} s Y(s) + k_{22y} Y(s) \tag{11}$$

The coefficients from Equations 10 and 11 are defined as: $m_i = I_i A_i A_p$, $m = M$, $k_{11} = A_i A_p \left(\frac{1}{C_1} + \frac{1}{C_2} \right)$, $k_{12} = -\frac{A^2{}_p}{C_1}$, $k_{21} = \frac{A_i A_p}{C_1}$, $k_{22} = k_e - \frac{A_p^2}{C_1}$, $k_{11y} = -\frac{A^2{}_p}{C_1}$, $k_{12y} = 3\frac{L}{h} A_i \tau_y$, $k_{21y} = c_e$, $k_{22y} = k_e - \frac{A_p^2}{C_1}$, $c_{11} = R_i A_i A_p - \frac{12 A_i^2 \eta L}{bh^3}$ and $c_{22} = c_e$. Expressing $X_i(s)$ as a function of $X(s)$ and $Y(s)$ from Equation 11, substituting it into Equation 10 and then dividing it by $Y(s)$ yields the following expression for the transfer function of the system:

$$H(s) = \frac{X(s)}{Y(s)} = \frac{\left[(k_{21}s + k_{22})(m_i s^2 + b_{11}s + k_{11}) - k_{11y}k_{21} \right] Y_0 \omega s - k_{21}k_{12y}(s^2 + \omega^2)}{Y_0 \omega s \left[(ms^2 + b_{22}s + k_{22})(m_i s^2 + b_{11}s + k_{11}) - k_{12}k_{21} \right]} \tag{12}$$

Replacing s by $j\omega$ in Equation 12 and separating the real and imaginary parts yields the frequency response of the system as:

$$H(j\omega) = \frac{X(j\omega)}{Y(j\omega)} = \frac{AC - BD}{C^2 + D^2} + j\frac{AD + BC}{C^2 + D^2} \tag{13}$$

where

$A = k_{11}k_{22} - k_{11y}k_{21} - k_{21}b_{11}\omega^2 - k_{22}m_i\omega^2$

$B = k_{11}k_{21}\omega + k_{22}b_{11}\omega - k_{21}m_i\omega^3$

$C = mm_i\omega^4 - mk_{11}\omega^2 - b_{11}b_{22}\omega^2 - m_i k_{22}\omega^2 + k_{11}k_{22} - k_{12}k_{21}$

$D = (k_{22} - m\omega^2)b_{11}\omega + (k_{11} - m_i\omega^2)b_{22}\omega$.

The MRF 132AD was chosen for this application. The fluid yield stress dependence on the magnetic field was determined, through curve fitting, from data provided by Lord Corporation [10] (see Figure 3). To determine the best fit equation, the curve shown in Figure 3 was extrapolated to origin and then a best fit equation was determined. The resulting Equation 14 was then derived under the assumption that at zero field the yield stress is zero. Equation 15 was determined through experiments carried out by the authors. The viscosity dependence on the shear rate has not been included.

$$\tau_y = -8 \cdot 10^{-7} H^2 + 0.3662H \tag{14}$$

$$\eta = -2 \cdot 10^{-14} \cdot H^3 + 8 \cdot 10^{-9} H^2 - 48 \cdot 10^{-6} H + 0.03 \tag{15}$$

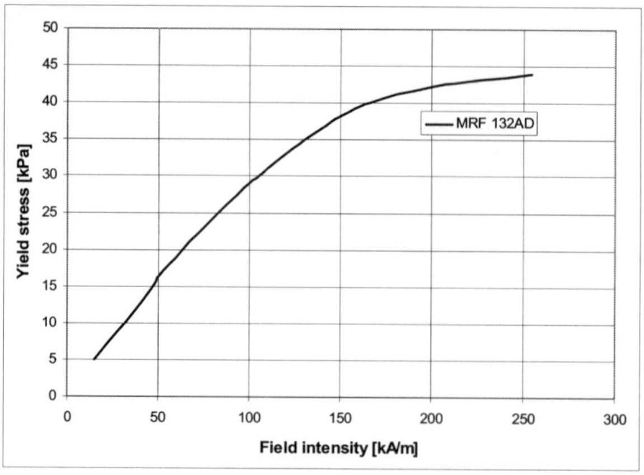

Figure 3. Yield stress variation with magnetic field intensity [10]

4 Results

The dynamic behavior of the mount has been simulated using the developed characteristic equations.

Particular attention has been given to the effect of the operator mass, magnetic field, cross sectional area of the inertia track and area of the main chamber on the mount response. Figure 4 and 5 indicate that the presence of the magnetic field causes a significant reduction in transmissibility of the mount accompanied by a small decrease in its resonant frequency. In addition, Figure 4 shows that an increase in the supported mass causes a drop in resonant frequency.

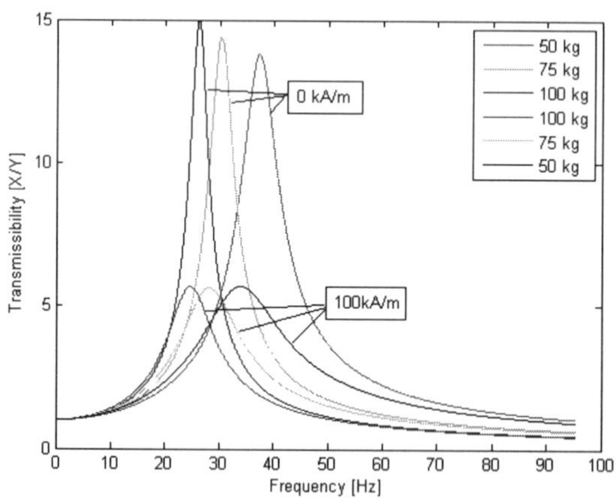

Figure 4. – Mount response for different masses, with and without field.

The effect of the mount geometrical characteristics is summarized in Figures 6 and 7. The larger the area of the main chamber, the smaller the transmissibility peak and the higher the resonant frequency of the mount (see Figure 6). Also, the smaller the area of the main chamber, the higher the reduction in peak transmissibility of the mount. The inertia track cross sectional area effect is shown in Figure 7. The smaller the inertia track for a given area of the main chamber, the smaller the resonant frequency of the mount and the displacement transmissibility peak.

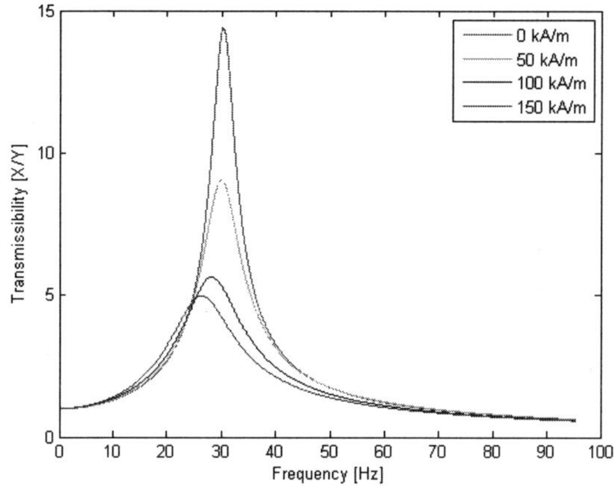

Figure 5. Field impact on transmissibility for a mass of 75 kg.

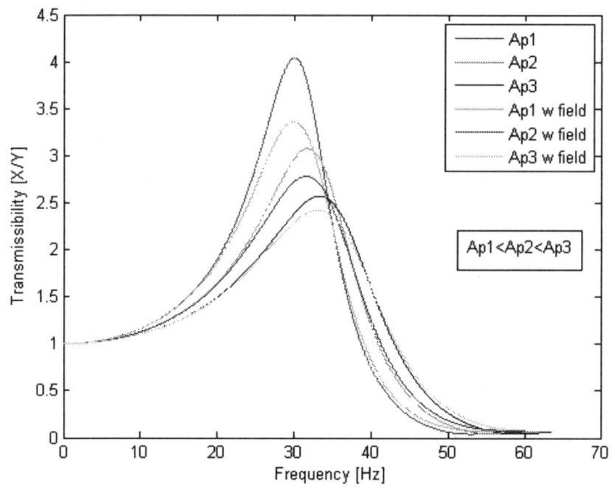

Figure 6. Main chamber area effect on the mount response.

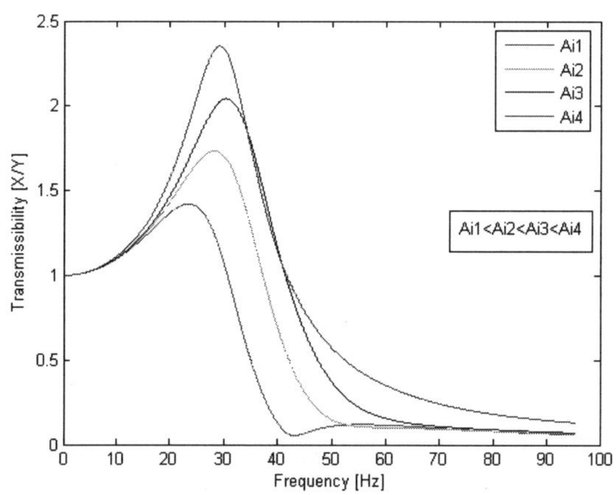

Figure 7. The effect of the inertia track area on the mount response.

If an increase in resonance frequency of the mount is desired, the inertia track may be increased. However, the increase over a certain value of the inertia track may cause a decrease in resonance frequency, as shown in Figure 7, the case of area A_{i4}.

5 Conclusions

A squeeze-flow mode MR mount was designed to mitigate vibration transmissibility to the operator of a pneumatic hammer. The dependence on the magnetic field of the fluid yield stress and fluid viscosity has been included in the equations characterizing the mount dynamic behavior. Simulated results indicate that an increase in magnitude of the applied magnetic field yields a decrease in transmissibility and a decrease in resonance frequency. At high magnetic fields, a continuous increase in magnetic field doesn't yield a significant decrease in transmissibility because of the onset of the fluid "lock-up" [6] state. The overall response of the mount fits its purpose. Experimental validation of the simulated results along with the implementation of a robust control strategy is being pursued.

References

1. Ahn Y., Ahmadian M. and Morishita S., On the design and development of a magneto-rheological mount, *J. Vehicle System Dynamics* (32) (1999) pp. 199-216.
2. Matsuoka H., Mikasa T. and Nemoto H., NV Countermeasure Technology for a Cylinder-On-Demand Engine Development of Active Control Engine Mount, *Proceedings of the SAE World Congress*, (2004).
3. Stelzer J.G., Schulz M.J., Kim J. and Allemang R.J., A magnetorheological semi-active isolator to reduce noise and vibration transmissibility in automobiles. *J. of Int. Mat. Sys. and Struct.* **14** (12) (2003) pp. 743-765.
4. Hong S.R., Choi S.B., Jung W.J., Ham I.B. and Kim D.K., Vibration control of an ER mount subjected to high static loads. *J. Sound and Vibration* **242** (4) (2001) pp. 740-748.
5. Hong S.R., Choi S.B., Jung W.J., and Jeong W.B., Vibration isolation of structural systems using squeeze mode ER mounts. *J. of Int. Mat. Sys. and Struct.* **13** (2002) pp. 421-424.
6. Hong S.R., Choi S.B., and Lee D.Y., Comparison of vibration control performance between flow and squeeze mode ER mounts: Experimental work. *J. Sound and Vibration* **291** (2006) pp. 740-748.
7. Woods R.I. and Lawrence K.L. Modeling and Simulation of Dynamic Systems (Prentice Hall, London, 1997).
8. Choi S.B., Cheong C.C., Jung J.M. and Choi Y.T., Position control of an ER valve-cylinder system via neural network controller. *Mechatronics* **7** (1) (1997) pp. 37-52.
9. Srinivasan A. V. and Michael McFarland D., Smart Structures: Analysis and Design (Cambridge University Press, New York, 2001).
10. Lord Corporation, MRF 132AD Product Bulletin (2003).

A MAGNETOELECTRIC RELATIVE DISPLACEMENT SELF-SENSING MAGNETORHEOLOGICAL DAMPER

D. H. WANG

Key Laboratory of Optoelectronic Technology and Systems of the Ministry of Education of China, Chongqing University, Chongqing, China 40004
E-mail: dhwang@cqu.edu.cn
Precision and Intelligence Laboratory, Department of Optoelectronic Engineering, Chongqing University, Chongqing, China 400044
E-mail: dklai@hotmail.com

D. K. LAI

Precision and Intelligence Laboratory, Department of Optoelectronic Engineering, Chongqing University, Chongqing, China 400044
E-mail: dklai@hotmail.com

In this paper, the innovative principle of a magnetoelectric relative displacement self-sensing magnetorheological damper (MSMRD), through integrating an integrated relative displacement sensor (IRDS) into a commercially available MR fluid damper, is realized and experimentally tested. The prototype of MSMRD and the testing setup with the rapid control prototyping (RCP) technique are established using 849 Shock Absorber Test System from MTS and the DS1103 PPC controller board from dSPACE GmbH under MATLAB/SIMULINK/RTW from MathWorks. Finally, the experimental testing on the first assembled prototype of MSMRD and the corresponding results are presented and analyzed.

1 Introduction

Recently, the academic and industrial research and development have placed much emphasis on how to improve the performance of magnetorheological (MR) dampers and many significant improvements in this field have been achieved. In order to let MR dampers be mass produced and applied in industry, the cost and reliability of the semi-active system based on MR dampers in system level are also very important. How to decrease the system cost and improve the reliability of the semi-active systems based on MR dampers should be paid much attention to when the performance of MR dampers are improved significantly.

According to Wang and Liao [1], the semi-active controller of the semi-active control system based on an MR damper consists of a system controller and a damper controller. In order to realize the feedback control of the semi-active system based on an MR damper, the dynamic responses, which include the relative displacement, the relative velocity, or the acceleration of the plant or across the MR damper, should be monitored and fed back to the system controller to generate the desired damping force. In this case, the monitoring of the dynamic response of the plant or across the MR damper is very important to compose a closed loop semi-active system based on an MR damper.

Based on the above description, we can image that it is a good ideal to integrate the dynamic response sensor into MR dampers to compose MR dampers with self-sensing ability, which can not only decrease the cost but also improve the reliability with eliminating the separate sensors and its connectors from the conventional systems. Through integrating integrated relative displacement sensors (IRDS) into MR dampers, the configuration of the semi-active suspension systems based on MR dampers can be simplified because the separate sensors for the dynamic response monitoring and the connectors are not needed again. It should be noted that the performance of the MR dampers and the semiactive system based on MR dampers can not be influenced by integrating the IRDS.

In 2005, Lai and Wang proposed a relative displacement self-sensing approach to make MR dampers self-sensing and the preliminary research results were presented [2]. In this paper, the innovative principle of a magnetoelectric relative displacement self-sensing magnetorheological damper (MSMRD) is realized and experimentally tested. The prototype of MSMRD and the testing setup with the rapid control prototyping (RCP) technique are established using 849 Shock Absorber Test System from MTS and the DS1103 PPC Controller Board from dSPACE GmbH under MATLAB/SIMULINK/RTW. Finally, the experimental testing on the prototype of MSMRD and the corresponding results are presented and analyzed.

2 Magnetoelectric Relative Displacement Self-Sensing Magnetorheological Damper

2.1 Principle and Prototype of MSMRD

The prototype of MSMRD is realized through integrating an IRDS into a commercially available MR damper. The principle and the first assembled prototype of MSMRD are graphically shown in Figs. 1 and 2. Per Fig. 1, the primary flux path is formulated along the piston head, the fluid gap, the cylinder, the cylinder cover, the upper cover of the cylinder, and the piston rod. In order to formulate the primary flux path as shown in Fig. 1, the slightly modification on the piston shape of commercially available MR dampers is needed. In addition, the piston rod, the piston head, the upper cover of the cylinder, and the cylinder cover must be made from soft steel materials with high magnetic permeability whereas the coil cover, the cylinder, the slide piston and the down lid of the cylinder, which are not shown in Fig. 1, are made from nonmagnetic stainless steel materials. When changing the energizing current to the exciting coil, which is usually wound on the piston head of the MR damper and also acts as the exciting coil for the IRDS, the yield stress of MR fluids within the MR fluid gap will be adapted correspondingly.

The IRDS based on electromagnetic induction of the prototype of MSMRD mainly comprises an exciting coil and an induction coil. The exciting coil is usually wound on the piston head of the MR damper and also acts as the exciting coil for the MR damper to adapt the yield stress of MR fluids as above mentioned. The winding methods of the induction coil of the IRDS determine the different integration methods and/or configurations of MSMRD, which includes (a) the rod sensing mode and (b) the cylinder sensing mode as shown in Fig. 1. For the rod sensing mode of the IRDS, the induction coil is wound uniformly in the piston rod by a constant coil density. On the contrary, the induction coil is wound on the cylinder uniformly by a constant coil density for the cylinder sensing mode of the

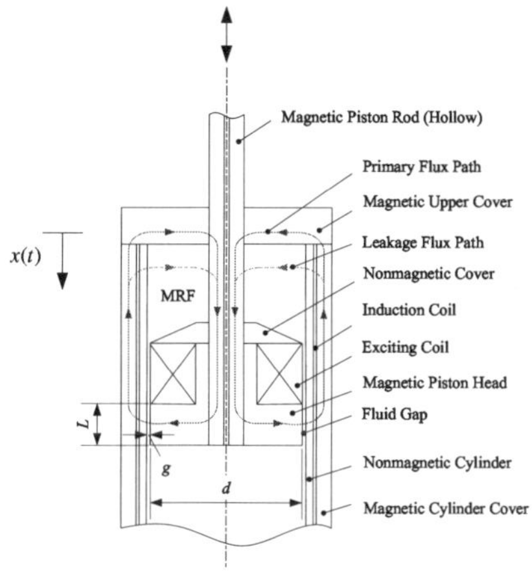

Figure 1. Principle of MSMRD.

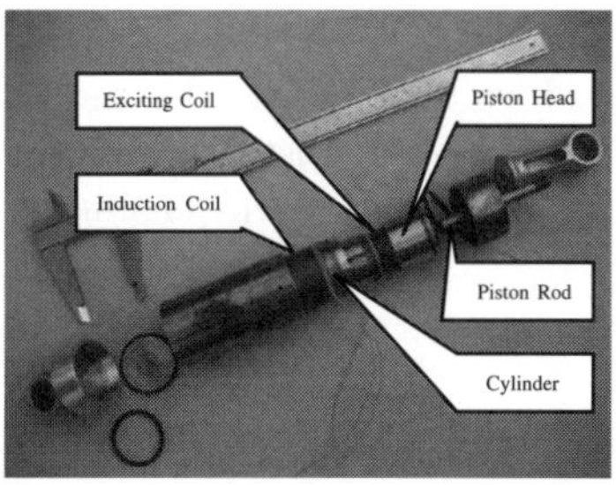

(a)

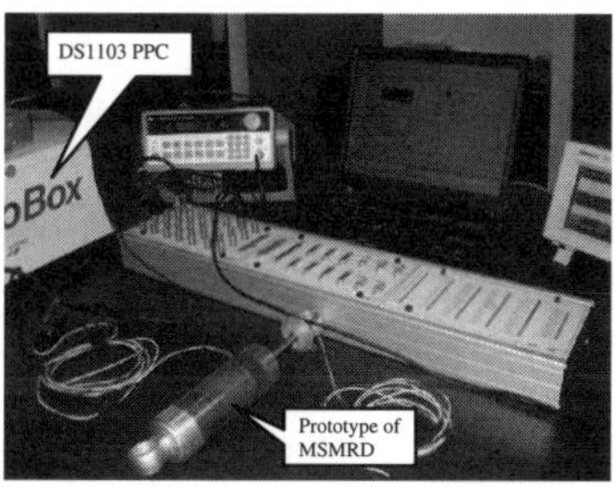

(b)

Figure 2. Prototype of MSMRD: (a) exploded view and (b) first assembled prototype.

IRDS. There are more leakage flux and more serious nonlinearity on the primary flux using the rod sensing mode of the IRDS than that using the cylinder sensing mode, which has been validated by the preliminary experiments and would evidently affect the performance of the IRDS. Furthermore, more strict manufacturing process is needed when using the rod sensing mode. Regarding the cylinder sensing mode, it is more practical and easy to realize the concept of MSMRD on commercially available MR dampers because the induction coil is mounted on a nonmagnetic cylinder.

For developing the prototype of MSMRD using the cylinder sensing mode, the maximum damping force of the MR damper with 5000.00N under an DC of 2.00A and the maximum induced voltage of the IRDS no less than 0.50V are designated. The exploded view and the first assembled prototype of MSMRD are shown in Fig. 2(a) and 2(b), respectively.

2.2 Mathematical Modeling of IRDS

Observing Fig. 1, the fluid gap between the cylinder and the piston head, which is used to let MR fluids pass in the primary flux path, is independent with the positions of the piston. In this way, the magnetic reluctance due to the fluid gap is constant when the MR damper works. In addition, the magnetic permeability of the soft steel elements is much larger than that of the fluid gap, so the total reluctance of the primary flux path is dominated by the fluid gap reluctance and is nearly independent of the positions of the piston. In this case, the instantaneous voltage $e(t)$ in the induction coil can be given by

$$e(t) \approx 2\pi f k \phi_m x(t) \cos(2\pi f t) \tag{1}$$

where k is the coil density, $x(t)$ is the relative displacement of the piston across the damper, f is the frequency of the sinusoidal carrier, ϕ_m is the peak value of the primary flux in the primary flux path. f, k, and ϕ_m are assumed to be constants.

According to Eq. (1), the virtual voltage value E across the induction coil is given by

$$E = \frac{2\pi}{\sqrt{2}} f \phi_m k x(t) \tag{2}$$

Eqs. (1) and (2) show that the induced voltage across the induction coil is directly proportional to the relative displacement of the piston across the prototype of MSMRD, which can be used to make the MR damper self-sensing.

3 Experimental Validation

After the first prototype of MSMRD is assembled, a series of experimental testing are carried out. In this section, the experimental setup and the testing results about the IRDS of the prototype of MSMRD are presented and analyzed.

3.1 Experimental Setup

Fig. 3 represents the experimental setup used to test the prototype of MSMRD. Observing Fig. 3, the experimental setup consists of the 849 Shock Absorber Test System from MTS, the first assembled prototype of MSMRD, the standard LVDT, and the data acquisition system, which is not shown in Fig. 3. The prototype of MSMRD is installed on the MTS machine in series with its hydraulic actuator and the LVDT is parallel installed with the prototype of MSMRD. The DS1103 PPC controller board from dSPACE GmbH under the MATLAB/SIMULINK/RTW from MathWorks is used as the data acquisition system. The software packages including the MATLAB/SIMULINK/RTW package provided by MathWorks, the ControlDesk version 2.0, the Real Time Interface to Simulink (RTI1103 version 3.0), the dSPACE ControlDesk to Simulink Interface (DSSIMULINK version 1.00), and the Matlab-dSPACE Interface Libraries (MLIB/MTRACE version 4.2) provided by dSPACE benefit the prototyping of MSMRD with RCP technique.

The energizing voltage to the exciting coil and the induced voltage from the induction coil are generated and interrogated by the DS1103 PPC controller board

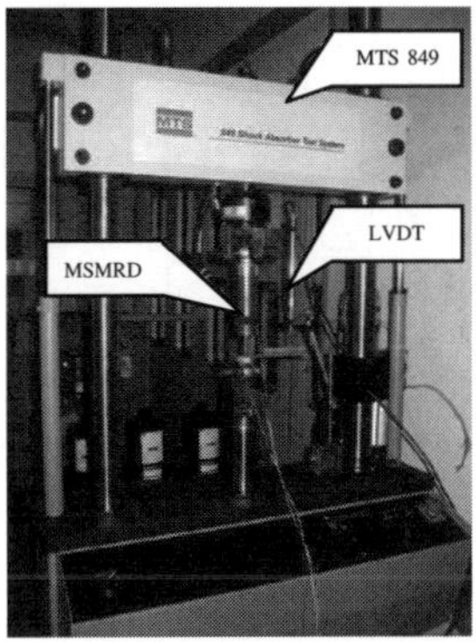

Figure 3. Photo of the experimental setup.

Table 1. Test groups and cases on the prototype of MSMRD.

Group	Case	Original Position	Displacement excitation		Carrier	
			Amplitude	Frequency	Amplitude	Frequency
		mm	mm	Hz	V (AC)	Hz
1	1	30.00	20.00	3.00	2.00	**100.00**
	2	30.00	20.00	3.00	2.00	**200.00**
2	3	30.00	20.00	**0.50**	2.00	100.00
	4	30.00	20.00	**1.00**	2.00	100.00
3	5	**15.00**	10.00	3.00	2.00	100.00
	6	**30.00**	10.00	3.00	2.00	100.00

under the MATLAB/SIMULINK/RTW. Via the MATLAB/SIMULINK/RTW, a series of signal processing units are built to realize the processing of the induced signal from the IRDS of the prototype of MSMRD, which include the signal amplifying, the modulation, the demodulation, the A/D conversion, and the displaying.

3.2 Testing Results under Sinusoidal Excitations

Six cases, which can be categorized into 3 groups, are tested to evaluate the characteristics of the IRDS of the prototype of MSMRD. The tested groups and cases are listed in Table 1 and the results are shown in Fig. 4.

The first group, which is composed of Cases 1 and 2 in Table 1, is conducted to evaluate the effect of the carrier frequency on the induced voltage from the IRDS of the prototype of MSMRD. When the exciting coil of the prototype of MSMRD is energized by carriers with an amplitude of 2.00V and frequencies of 100.00Hz and 200.00Hz, the test results of the IRDS of the prototype of MSMRD under the sinusoidal displacement excitation with a frequency of 3.00Hz and an amplitude of 20.00mm are shown in Figs. 4(a) and 4(b), respectively. It is worthy to note

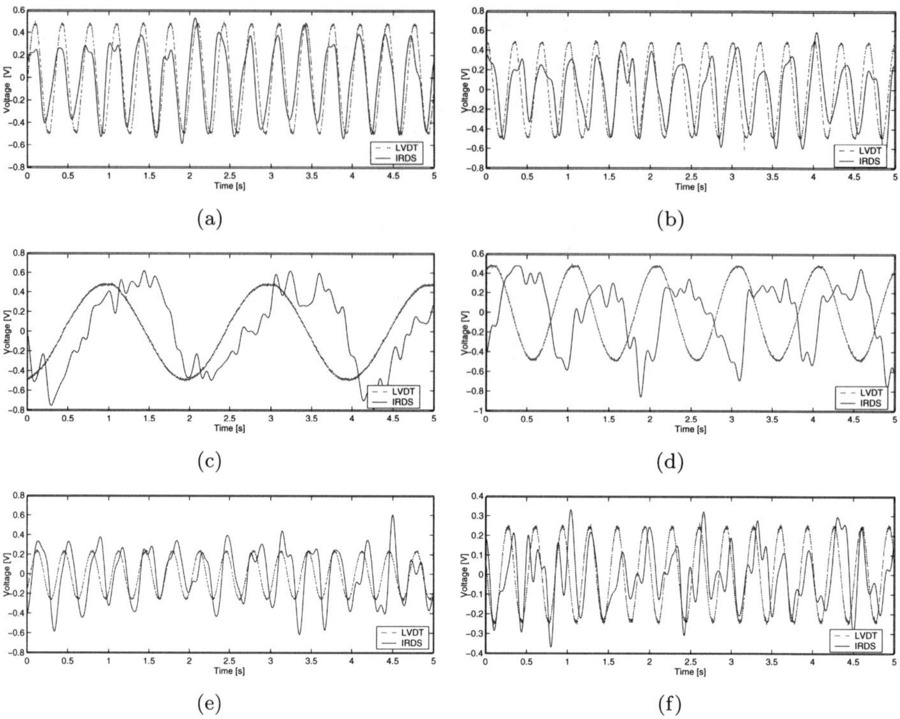

Figure 4. Comparison between the output of the IRDS of the prototype of MSMRD and the output of the LVDT under different conditions: (a) carrier frequency of 100.00Hz, (b) carrier frequency of 200.00Hz, (c) displacement excitation frequency of 0.50Hz, (d) displacement excitation frequency of 1.00Hz, (e) original position of 15.00mm, and (f) original position of 30.00mm.

that the original position of 30.00mm of the piston is assured when testing the first group. Observing Figs. 4(a) and 4(b), the lower the carrier frequency is, the more the behavior of the IRDS is close to that of the standard LVDT.

The second group with Cases 3 and 4 in Table 1 is carried out to evaluate the effect of the frequency of the sinusoidal displacement excitation on the IRDS of the prototype of MSMRD and the testing results are shown in Figs. 4(c) and 4(d), respectively. Observing Figs. 4(c) and 4(d), the lower the frequencies of the displacement excitation are, the more the behavior of the IRDS is close to that of the standard LVDT.

The third group with Cases 5 and 6 in Table 1 is carried out to evaluate the linearity behavior of the IRDS with different original position of the piston of the MR damper and the testing results are shown in Figs. 4(e) and 4(f), respectively. It should be noted that the induced voltages from the IRDS of the prototype of MSMRD illustrates the nonlinearity of the IRDS, which partially results from the increased flux leakage when the piston moves toward the bottom lid of the cylinder.

Observing Fig. 4, the induced voltage from the IRDS of the prototype of MSMRD basically accords with that of the standard LVDT under sinusoidal excitations. Because of the output of the standard LVDT directly indicates the vibration generated by the MTS machine, the preliminary conclusion can be achieved that the induced voltage of the IRDS of the prototype of MSMRD is directly proportional to the relative displacement of the piston across the prototype of MSMRD.

4 Conclusion and Discussion

In this paper, a novel magnetoelectric relative displacement self-sensing magnetorheological damper is developed and the experimental testing on the prototype of MSMRD is carried out. Only the testing results about the IRDS of the prototype of MSMRD are presented in this paper due to the page limit. Through the preliminary testing, the IRDS of the prototype of MSMRD functions well and meets a satisfying performance. The IRDS of the prototype of MSMRD can produce an inducted voltage with the maximum amplitude of 0.50V within the whole travel stroke of the piston in the damper.

Acknowledgments

The authors wish to acknowledge the financial supports by a Foundation for the Author of National Excellent Doctoral Dissertation of PR China (Grant No. 200132) and a Program for New Century Excellent Talents in University (Grant No. NCET-05-0765). The authors also wish to acknowledge the help from the Predetection Center of Zongshen Motorcycle Group R&D.

References

1. D. H. Wang and W. H. Liao, Journal of Intelligent Material Systems and Structures **16**, 983 (2005).
2. D. K. Lai and D. H. Wang, in *Proceedings of SPIE Conference on Smart Structures and Materials 2005: Smart Structures and Integrated Systems*, ed. A. B. Flatau (Proceedings of SPIE – Volume 5764, 2005).

VIBRATION SUPPRESSION OF ROTATING FLEXIBLE BEAMS USING ER FLUIDS: EXPERIMENTAL WORK

KEXIANG WEI[1, 2], GUANG MENG[1*] AND SHUO ZHOU[1]

[1] *State Key Laboratory of Vibration, Shock & Noise, Shanghai Jiao Tong University, Shanghai 200240, P. R. China*

[2] *Department of Mechanical Engineering, Hunan Institute of Engineering, Xiangtan 411101, P. R. China*

[*] *E-mail: gmeng@sjtu.edu.cn*

The main objective of this study is to experimentally quantify the feasibility of using ER fluids for suppressing vibrations of the rotating beams. An ER sandwich beam specimen, in which an ER fluids layer is sandwiched between two elastic surface layers, is constructed. An experiment set-up is established to investigate the effect of the ER fluids on vibration suppression of the rotating beams with point-to-point motion. The vibration response performances of the beam subjected to different electric field intensity, rotating speed, and acceleration are demonstrated and evaluated. The experimental results obtained indicate that significant vibration attenuation is achieved at different operating conditions by applying an electric field to the rotating beam.

1 Introduction

ER fluids are a kind of smart materials, physical properties of which such as viscosity and shear modulus can undergo instantaneous and reversible changes when subjected to different electric field. These unusual properties enable ER fluids to be employed in numerous potential engineering applications, such as shock absorbers [1], clutch/brake systems [2], fluid control [3] and intelligent structures [4].

One of the most commonly researched of ER fluids structures is the ER fluids treated sandwich beam, in which the ER fluids completely fills the gap between two containing surface layers [5]. These sandwich structures have adaptive control capabilities of varying the damping and stiffness of beam by changing the intensity of applied electric field. Since Gandhi et al [6] first proposed the application of ER fluids to adaptive structures, many achievements have been obtained in vibration control of beam structure. Choi et al [7] investigated the vibration properties of hollow cantilevered beams filled with an ER fluid and derived an empirical model for predicting field-dependent vibration characteristics. Yalcintas and Coulter [8] developed a theoretical model for adaptive beam structures with various boundary conditions. The semi-active vibration control capabilities of ER adaptive beam were theoretically and experimentally illustrated [9]. Gong and King [6] experimentally investigated the vibration properties of a sandwiched beam clamped at both ends and fully or partially treated with an ER fluid layer. Rahn and Joshi [10] applied the visco-elasticity and sandwich theory to model ER sandwich beam. A feedback control system was designed to control the vibration of cantilevered ER beam. Young et al [11] investigated the flexural vibration of laminated composite beams sandwiched by two ER fluids layers. Phani and Venkatraman [12] experimentally studied the behavior of a sandwich beam with ER fluid acting as the core material. However, most ER adaptive beams under investigation are considered to be non-motion. Rare studies on the vibration suppression of rotating beam using ER fluids can be found in the literature.

In our previous work, the dynamic characteristics of rotating sandwich beam filled with ER fluid were numerically analyzed [13]. In this paper, we experimentally present and discuss the feasibility of using ER fluid in controlling the vibration of rotating beams. An ER sandwich beam specimen, in which an ER fluids layer is sandwiched between two elastic surface layers, is constructed. An experiment set-up is established to investigate the effect of the ER fluid on vibration suppression of the rotating beams. Vibration control responses subjected to different electric field intensity, rotating speed, and acceleration are demonstrated and evaluated.

2 Experimental set-up

The ER beam specimen, as shown in Fig. 1, is treated with an ER fluid layer which confined by a 2mm silicon rubber dam on the edges sandwiched between two aluminum face-plates. The aluminum face-plates also serve as the electrodes for the applied electric field. At one end of the beam, a glass/epoxy pad with a length 20mm and width 35mm is used as the mid-layer to provide the rigidity for the clamping force.

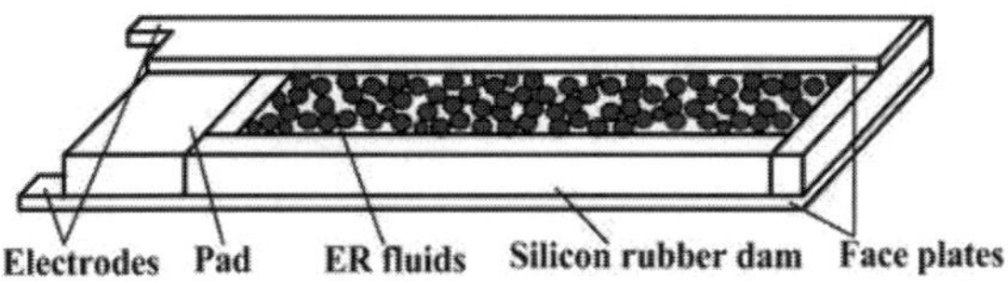

Fig.1 Configuration of a sandwich beam filled with ER fluids

The ER fluid used in this study consists of corn starch suspensions in silicone oil. The carrier fluid has a viscosity of 50cS, and the volume fraction of the suspension is 15%. The geometrical and material parameters of the ER sandwich beam specimen are listed in Table 1.

Table 1 Main properties of the ER sandwich beam

	Surface layer	ER fluids layer
Length (mm)	320	300
Width (mm)	35	35
Thickness (mm)	0.4	2
Density ($kg \cdot m^{-3}$)	2683.9	1338
Modulus (MPa)	70	

The experimental apparatus for investigating the vibration response of the rotating ER beam at different electric strength and motion conditions is shown in Fig. 2. The test beam is clamped at a rotating platform driven by a servo motor in a cantilevered configuration. An eddy current probe (model CWY-DO-20T08), which is used to measure the vibration displacement, is also anchored on the rotating platform, and can motion in company with the rotating beam. In order to avoid the deflection displacement of the measuring point surpassing its measuring range, the eddy current probe was located at 17mm far from the beam root. By the vibration theory, beam is a continuous system. Regardless of measuring point whether is located at the tip end of the beam, the deflection response curve should contain the vibration information of beam, like mode, frequency and so on. A dynamic signal analyzer, produced by Data Physics Corporation with model number SignalCalc430, is used to process the acquired analog signals from the eddy current probe. A high-voltage power supply connected with the aluminum face-plates provides the required electric field during the test.

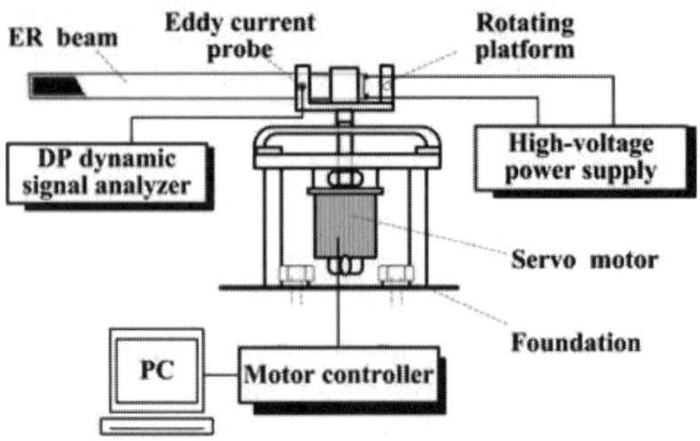

(a) Schematic diagram

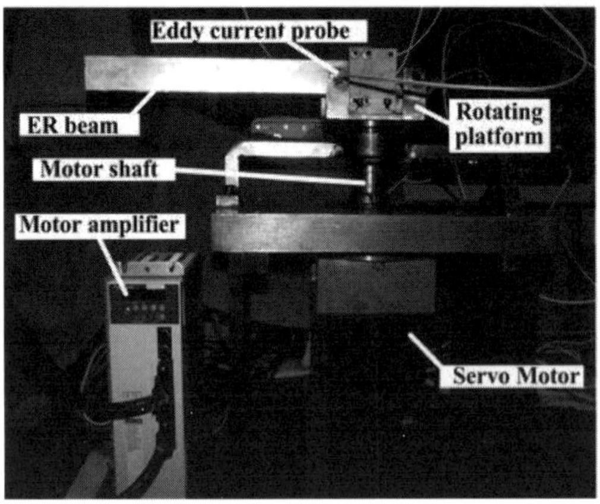

(b) Photograph

Fig. 2 Experimental set-up for the rotating ER beam

3 Results and discussions

In this test, the ER beam is operated at point-to-point motion. The procedure is to rotate the beam from $\theta=0$ to a fix angle $\theta=\pi$. The rotating angle acceleration is chosen as $20\text{rad}\cdot\text{s}^{-2}$, $40\text{rad}\cdot\text{s}^{-2}$, and $80\text{rad}\cdot\text{s}^{-2}$. Considering the specimen is sealed by silicon rubber, the maximum of rotating speed is limited to 120rpm. Four levels of electric field strengths (0kV/mm, 0.5kV/mm, 1kV/mm, and 1.5kV/mm) are applied to the rotating ER beam.

The vibration control responses of the rotating ER beam at different electric field strength, rotating speed and acceleration are shown in Figs. 3-6, respectively. Figure 3 presents the response curves at different intensity of applied electric field when angle acceleration $\ddot{\theta}=20\text{rad}\cdot\text{s}^{-2}$ and maximal rotating speed $\dot{\theta}=30\text{rpm}$. It is shown that the vibration decay time is shorten by applying the electric field. The beam vibration is eliminated in less than 3s with applied electric field intensity E=0.5kV/mm, while the vibration lasts for more than 5s without applied electric field. Increasing the intensity of electric field, the decay time further shorten. For example, when the electric field strength is increased from 0.5kV/mm to 1.5kV/mm, the vibration decay time is shorten about 1s. From Fig. 3, we also can see that the vibration frequency is not varied with the electric field strength. The number of vibration cycle is three in each second at different intensity of electric field. This demonstrates that the ER fluid mainly tunes the damping of system to suppress the vibration of beam by applying an electric field.

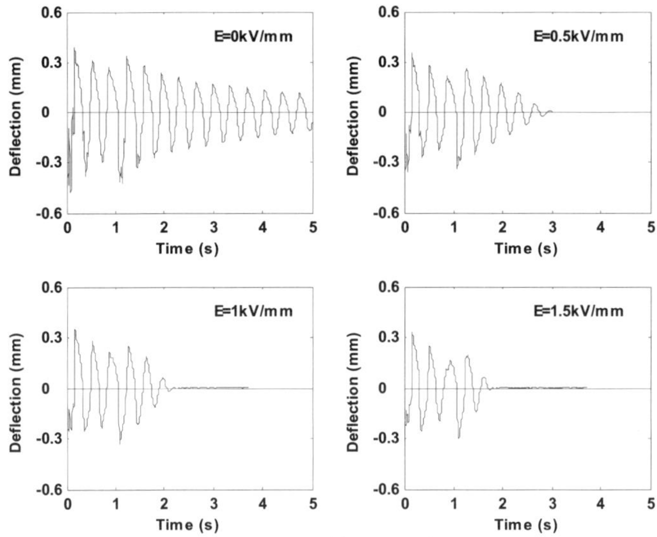

Fig. 3 Vibration responses at angle acceleration $\ddot{\theta}$ =20rad·s^{-2} and maximal rotating speed $\dot{\theta}$ =30rpm.

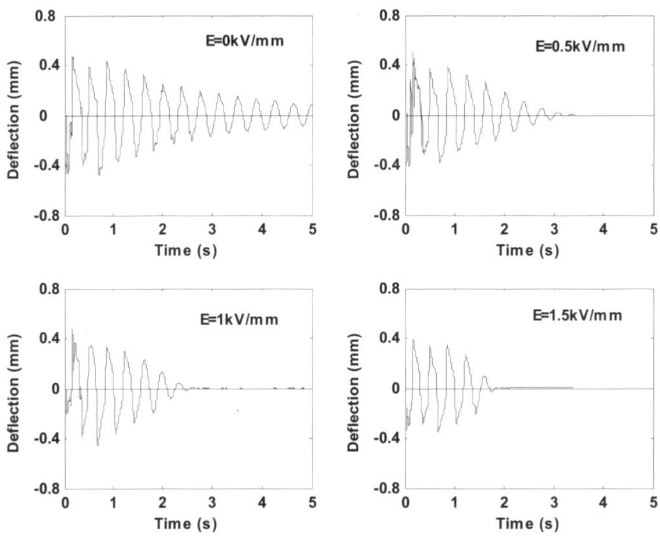

Fig. 4 Vibration responses at angle acceleration $\ddot{\theta}$ =40rad·s^{-2} and maximal rotating speed $\dot{\theta}$ =60rpm.

Figure 4 is the vibration response of the rotating beam under angle acceleration $\ddot{\theta}$=40rad·s^{-2} and maximal rotating speed $\dot{\theta}$=60rpm. It can be seen that the vibration suppression capability of rotating ER beam under applied electric field is similar to that when angle acceleration $\ddot{\theta}$=20rad·s^{-2} and maximal rotating speed $\dot{\theta}$=30rpm. Only the maximal vibration amplitude is increased. This is due to increasing rotating speed and acceleration will cause a larger centrifugal force applied on the beam, and intensify the vibration of beam.

Figure 5 illustrates the measurement results under angle acceleration $\ddot{\theta}$=80rad·s^{-2} and maximal rotating speed $\dot{\theta}$=90rpm at different electric field strength. In this case, the effectiveness of vibration control of ER fluid is most obvious. The decay time of beam vibration is less than 1.5s under applied electric field strength E=1.5kV/mm, while the decay time is about 2s in Fig. 3 and Fig. 4, respectively. This means that the rotating beam filled with ER fluid might have an optimum vibration control effect for one rotating speed by applying electric field. In the next study, this need to be further investigated.

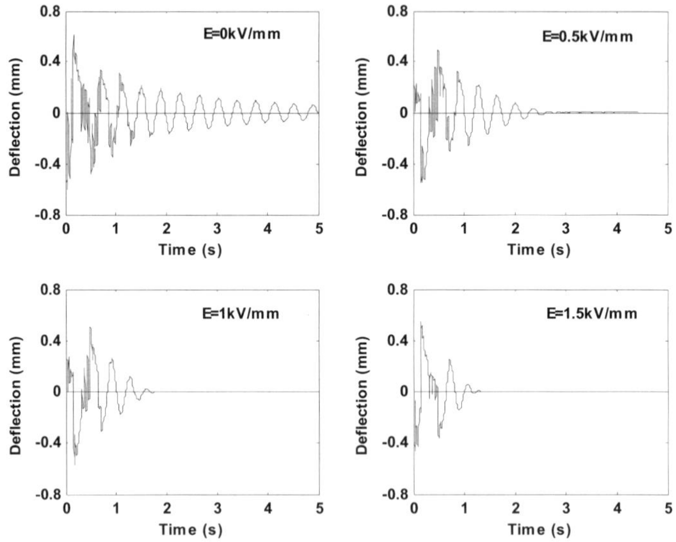

Fig. 5 Vibration responses at angle acceleration $\ddot{\theta}$ =80rad·s^{-2} and maximal rotating speed $\dot{\theta}$ =90rpm.

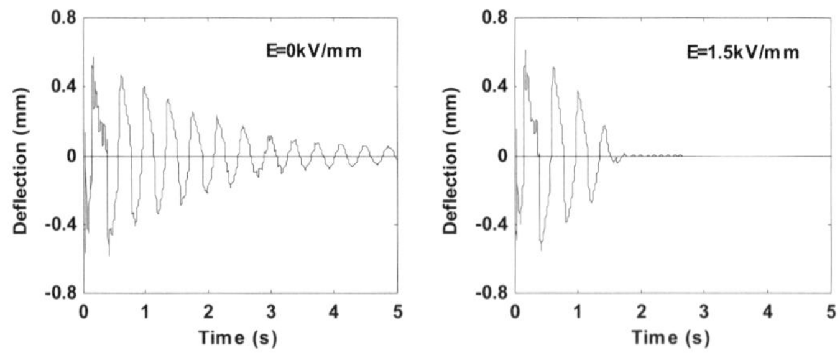

Fig. 6 Vibration responses at angle acceleration $\ddot{\theta}$ =80rad·s^{-2} and maximal rotating speed $\dot{\theta}$ =120rpm.

When the maximal rotating speed further increases, the vibration suppression capability of ER fluids attenuates. This can be seen from comparing Fig. 6 with Fig. 3 and Fig. 4. It is because high rotating speed will intensify the vibration of system and weaken the ER effect. However, the vibration control effect is very obvious at high speed.

4 Conclusion

The vibration of a rotating beam with point-to-point motion has been controlled by using sandwich beam filled with ER fluids. The vibration response performance of the rotating ER beam was experimentally evaluated with respect to the intensity of the electric field, the rotating speed, and acceleration. The results obtained have indicated that the residual vibration of the beam caused by base motion at different rotating speed and acceleration can be quickly suppressed by controlling the intensity of the electric field to the ER beam. These demonstrate the feasibility of ER fluids in attenuating the vibration of flexible robot arms.

5 Acknowledgments

This work was supported by the National Natural Science Foundation of China (NSFC) under the Grant No. 50390063, and the Foundation of National Key Lab. of Defense Technology under the Grant No. 51463040403JW0301.

Reference

1. Lindler J. E. and Wereley N. M., Double adjustable shock absorbers using electrorheological fluid, *J. Intell. Material Syst. Struct.* **10** (2000) pp. 652-657.
2. Nakamura T., Saga N. and Nakazawa M., Impedance control of a single shaft-type clutch using homogeneous electrorheological fluid, *J. Intell. Material Syst. Struct.* **13** (2002) pp. 465-469.
3. Wei K. X., Meng G. and Zhu S. S., Fluid power control unit using electrorheological fluids, *Int. J. Fluid Power* **5** (2004) pp. 49-54.
4. Li Q., Research into an integrated intelligent structure - A new actuator combining piezoelectric ceramic and electrorheological fluid, *J. Acoust. Soc. Am.* **111** (2002) pp. 856-860.
5. Gong H. and Lim M. K., Vibration characteristics of sandwich beams partially and fully treated with electro-rheological fluid, *J. Intell. Material Syst. Struct.* **8** (1997) pp. 401-413.
6. Gandhi M. V., Thompson B. S. and Choi S. B., A New Generation of Innovative Ultra-Advanced Intelligent Composite Materials Featuring Electro-Rheological Fluids: An Experimental Investigation, *J. Compos Mater.* **23** (1989) pp. 1232-1255.
7. Choi S. B., Park Y. K. and Kim J. D., Vibration characteristics of hollow cantilevered beams containing an electro-rheological fluid, *Int. J. Mech. Sci.* **35** (1993) pp. 757-768.
8. Yalcintas M. and Coulter J. P., Electrorheological material based adaptive beams subjected to various boundary conditions, *J. Intell. Material Syst. Struct.* **6** (1995) pp. 700-717.
9. Yalcintas M. and Coulter J. P., Analytical modeling of electrorheological material based adaptive beams, *J. Intell. Material Syst. Struct.* **6** (1995) pp. 488-497.
10. Rahn C. D. and Joshi S., Modeling and control of an electrorheological sandwich beam, *J. Vib. Acoust.* **120** (1998) pp. 221-227.
11. Young K. K., Kim J. and Choi S. B., Passive and active damping characteristics of smart electro-rheological composite beams, *Smart Mater. Struct.* **10** (2001) pp. 724-729.

12. Phani A. S. and Venkatraman K., Vibration control of sandwich beams using electro-rheological fluids, *Mech. Syst. Signal Processing* **17** (2003) pp. 1083-1095.
13. Wei K. X., Meng G. and Lu H. Q., Dynamic analysis of a rotating electrorheological sandwich beam, *Int. J. Mod. Phys. B* **19** (2005) pp. 1236-1242.

MICROFLUIDIC MANIPULATION IN LAB-CHIPS USING ELECTRORHEOLOGICAL FLUID

X.Z. NIU, L.Y. LIU, W.J. WEN and P. SHENG

Institute of Nano Science and Technology, Department of Physics, The Hong Kong University of Science and Technology, Clear Water Bay, Kowloon, Hong Kong, China

E-mail:phwen@ust.hk

Microfluidic manipulation is one of the main concerns in lab-chip society. This paper presents the design, fabrication and testing methodologies of electrorheological (ER) fluid-controlled active microfluidics chips. All of the fundamental building blocks, such as microvalves, micropumps and micromixers, are fabricated with similar processes in PDMS-based chips for easy integration. Such building blocks can be digitally controlled by on-off switching of electric fields applied on the parallel electrodes along the ER fluid channels. Experimental results show reliable functionalities of these devices. The ER active control scheme supplies additional flexibilities over the passive control of microfluidics. A hybrid approach active mixers, combining the ER active actuation with the passive baffle structures, was applied to achieve more efficient chaotic mixing than the passive mixers available.

1 Introduction

Microfluidic components play important roles in micro total analysis system or lab-on-a-chip systems that are widely used in DNA sequencing, drug screening and delivery, in situ monitoring and the other biotechnology and environment areas. Micro valves, mixers and pumps are fundamental building blocks to manipulate micro flows. Various kinds of active schemes were developed in recent years [1-3]. In most of the designs with moving parts, diaphragms or cantilevers were adopted to control the flows. Polydimethylsiloxane (PDMS) based microdevices were intensively developed due to their flexibility, easy fabrication using soft lithography and bio-compatibility. These devices are pneumatically controlled, mechanically controlled or hydrogel actuated. But an integrable and electronically addressable active control with fast response in milliseconds is still needed in micro total analysis systems.

Electrorheological (ER) fluid, as a well-known smart material, has been studied for many years, and was used in many engineering applications such as clutch, vibration damper or valves. A kind of giant electrorheological (GER) fluid reported in ref [4] can reach a yield stress as high as 200 kPa. The GER fluid consists of urea-coated 20nm-diameter nanoparticles (barium titanyl oxalate) suspended in insulating oil. The transformation from liquid-like to solid-like behavior is relatively quick, on the order of 1-10 milliseconds, and reversible. Such electric-field controlled rheological property implies a promising application for controlling micro flows.

This paper reports a new methodology of microfluidic control with ER fluid, as well as the design, fabrication and testing of ER active chips. Micro valves, pumps and mixing channels can be integrated in a single microchip, making possible an ER controlled lab-chip.

2 Design methodology and experiment of ER active microfluidic elements

ER fluid, as a kind of colloidal suspension, can sustain high shear stress through the application of an external electric field. Such property was used in the designing of ER controlled microfluidic chip. Figure 1 shows the side view of an ER controlled chip and the associated ER circle system. The micro chip was made of PDMS by soft lithography [5]. Two kinds of liquid channels: ER fluid channel and experimental fluid channel were arranged on different layers and be separated by thin diaphragms at the intersection areas. The ER fluid was circulated through the chip with a pump (Masterflex C/L© used in the experiment) as labeled P in Fig. 1. To stabilize the pressure difference between the inlet and outlet of the ER fluid channel, two reservoirs were connected on the both side of the pump, R_H dictates the reservoir with high pressure, and R_L dictates the reservoir with low pressure.

The basic working principle of the ER-fluid-control is a push-and-pull actuation caused by the flexible diaphragm sandwiched between the ER channel and flow channel when an on-off voltage is applied to ER fluid. Two pairs of parallel electrodes are arranged along each ER channel forming an upstream valve and a downstream valve for ER fluid itself. A control electric circuit was designed to output square wave DC voltage signals to the electrodes. If an adequate DC electric field is subjected to any pair of electrodes, the flow of ER fluid can be slowed or fully stopped due to the ER effect. The pressure in the ER channel between the two valves can therefore be modulated if the two valves are alternately opened and closed. Such a pressure change within ER channel will eventually result in the deformation of the flexible diaphragm with a vertical pull-and-push movement. Therefore, the liquid flow in the flow channel will be controlled with the pressure change in ER channel. This push-and-pull mechanism has the advantage of faster response time compared with valves with only a push source, because the PDMS diaphragm has a low Young's modulus and cannot resume its original position very quickly by its own elasticity. Therefore the movement of the diaphragm can be digitally controlled on its amplitude, frequency and phase by electric signals.

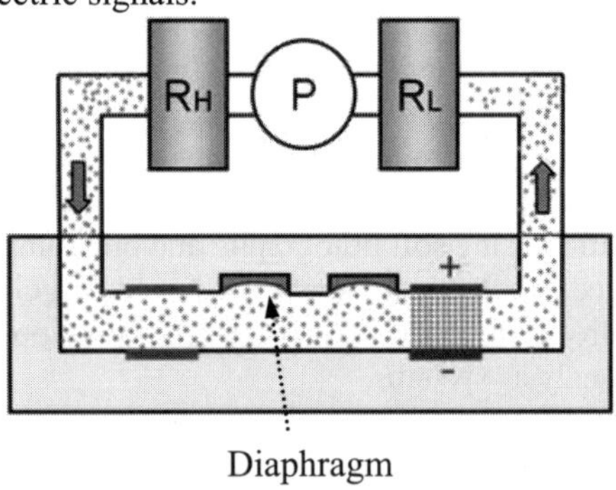

Figure 1. Schematic diagram of the electrorheologically controlled microfluidic chips and the ER circulation system; the diaphragm between the ER fluid channel and the experimental fluid channel can be pushed up and pull down.

ER active valves, pumps and mixers can be achieved in the similar working principles as shown in Fig. 1. In Fig. 2 (a), a micro valve is formed by simply overlapping perpendicular the ER fluid channel layer on a micro fluidic channel layer [6]. The flow channel was molded out with rounded photoresist mold to assure the flow can be fully closed [2]. It is tested that the response time of on-off switching of the flow can be less than

10 ms and with sound reliability, suitable for most microfluidic applications. It should be noted that the diaphragm area can not be too big in the valve chips to avoid fluctuation of flow speed. But if the diaphragm area is enlarged intentionally, large amount of liquid can be repelled and imbibed, therefore the diaphragms can be used as pulsating pumps, for example, in active pumps or mixers.

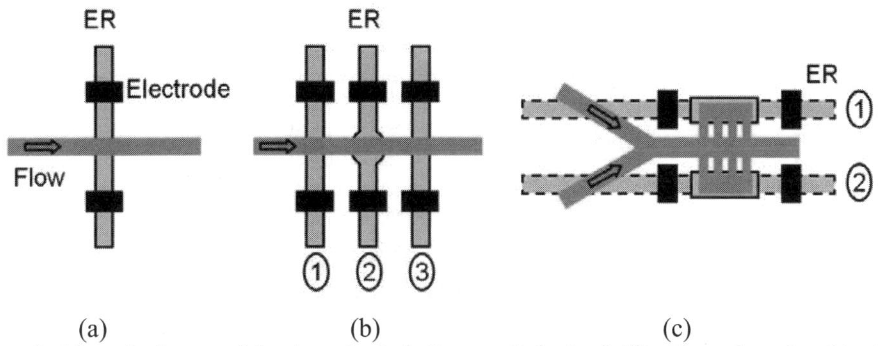

Figure 2. Schematic diagram of the electrorheologically controlled microfluidic chips; microvalve; (b) micro pump; (c) micro mixer

By arranging three pulsating pumps serially on a micro flow channel, an ER micro pump was fabricated as show in Fig 2 (b) [7]. The co-operative movements of the diaphragms actuated by the ER fluid valves are the direct driving mechanism for pumping, As the electrodes can be independently controlled, we are able to digitalize the signals so that the voltages are expressed and managed with binary codes. Besides the tunable voltages, digitized programmable control offers flexibility (e.g., reversing the pumping flow direction) and adjustable individual action steps. It was tested that the pumped flux reaches a peak of 1.4 μl/s at ~35 Hz (in a testing range of 0 to 100 Hz) when all the electric signals have the strength of 2kV/mm.

Microfluidic mixer is not only a critical component in lab-on-a-chip but also of interest in fundamental physics because in microscale systems only creeping flows can exist due to the small Reynolds number. It was numerically shown that chaotic mixing can occur in the orthogonal-channel pulsating mixers [8]. However an active chaotic mixer chip integrable with the other component is not an easy task due to the difficulties in the fabrication and operation in microscale. Using an ER controlled mixer [9] as shown in Fig. 2 (c), we have experimentally demonstrated that fully chaotic mixing can really be achieved. In the mixer chip, the two pulsating pumps were set to operate in a contrast phase. Amplitude and frequency were adjusted to achieve optimal mixing depending on the flow speed at the inlet.

3 Hybrid approached of ER active mixers

The digitally addressable ER active control offers more flexibility over the passive ways of microfluidic manipulations. If these two approaches can be combined together, may induce better or new functionalities. For example, in the active mixers proposed in ref [10], five active elements are needed to finish one process of mixing/reaction. By using the active mixer in part 2, a continuous mixing directly in the fluid channel can be achieved with only two active elements. But during the optimization of the control parameters, it was found that the actuation amplitude of the pulsating pumps should be increased proportionally with the increase of flow speed at the inlet, and there exist a optimal frequency range, out of which the mixing is poor and most of the ER pumping energy was wasted in the driving of the unmixed liquid blobs back and forth in the orthogonal side channels.

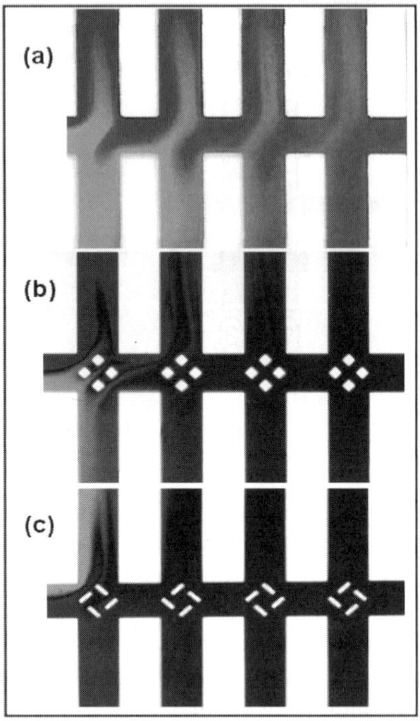

Figure 3. Hybrid approach on chaotic mixing. Compare of the three kinds of mixing channels in the same control parameters. (a) In the channel without baffles, regular blobs [8] exist; (b) In the channel with symmetric baffles, regular blobs are cut into thin layers that chaotic mixing can be achieved; (c) In the channel with asymmetric baffles, KAM blobs disappear more rapidly and stronger chaotic mixing can be achieved.

Based on a Poincaré section analysis and its physical understanding, different passive flow baffles were installed in the main microfluidic channel to enhance mixing at high frequencies [11]. Results shown in Fig. 3 indicate that the combined hybrid approach enables effective chaotic mixing at enhanced frequency and reduced passage distance in two-dimensional flows. Under the ER control parameters but using asymmetric baffles, regular blobs shown in Fig. 3 (a) was found to disappear in Fig. 3 (b) and (c). Therefore the hybrid approach do improve the mixing efficiency than the simple model in ref [8]

A continuous micro mixer chip, including three valves and one mixer, was further proposed as shown in Fig. 4 (a). The valves operate in the same manner as in ref [6]. But the mixer was further improved to avoid dead end for easy operation. Therefore, liquids can be stretched, folded, splitted and recombined in the channels. Figure 4 (b) shows that fully mixing can be achieved even after four pairs of side channels. By switching the microvalves, different streams of liquid can be mixed optionally. Such a chip and the future large scale integration may find applications in microfluidic display, chemical reactions and so on.

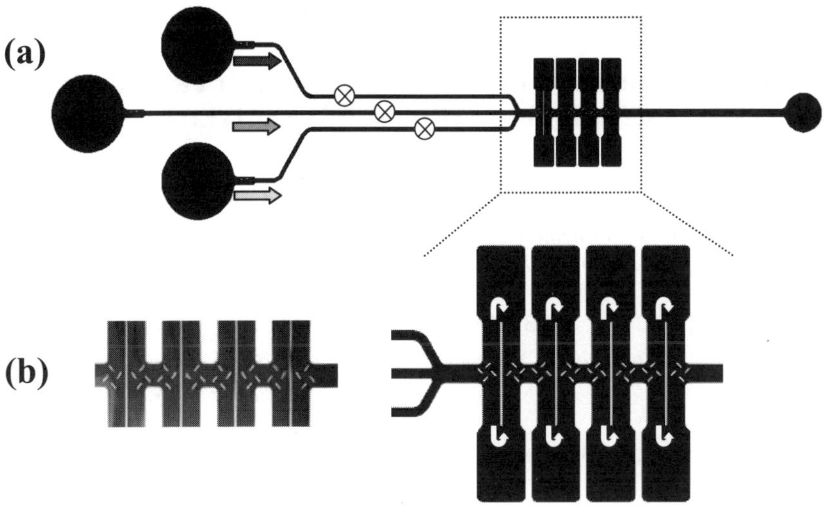

Figure 4. A continuous micromixer that can mix different streams of liquids, in a mechanism of splitting and recombination of the flows as well as chaotic mixing.

Acknowledgements

We acknowledge the support of Hong Kong RGC grants No. 604205 and 604104.

References

1 Z. Yang, H. Goto, M. Matsumoto, R. Maeda, *Electrophoresis*, 21, 116 (2000)
2 M. A. Unger, H-P Chou, T. Thorsen, A. Scherer and S. R. Quake, *Science*, 288, 113 (2000)
3 D.T. Eddington and D.J. Beebe *J. Microelectromech. S.* 13 (4): 586 (2004)
4 W. Wen, X. Huang, S. Yang, K. Lu, P. Sheng, *Nat. Mater.* 2 (11), 727 (2003)
5 Y.N. Xia and G.M.Whitesides, *Annu. Rev. Mater. S.* 28: 153 (1998)
6 X.Z. Niu, W.J. Wen, Y.K. Lee, *Appl. Phys. Lett.* 87, 243501, (2005)
7 L.Y. Liu, X.Q. Chen, X.Z. Niu, W.J. Wen and P. Sheng, *Appl. Phys. Lett.* (2006), in press
8 X. Z. Niu and Y. K. Lee, *J. Micromech. Microeng.* 13, 454 (2003)
9 X.Z. Niu, L.Y Liu, W.J Wen and P. Sheng, *Appl Phys. Lett.*, 88, 153508, (2006)
10 J.W. Hong, V. Studer, G. Hang, W.F.Anderson and S.R. Quake, *Nat. Boitechnol.* 22 435 (2004)
11 X.Z. Niu, L.Y. Liu, W.J. Wen and P. Sheng, *Phys. Rev. Lett.* (2006), in press

MR DAMPER GENERATING RESPONSE-DEPENDENT FRICTIONAL FORCE FOR CONTROLLING BUILDING STRUCTURES EXCITED BY SEISMIC LOAD

S.H. LEE, K.W. MIN, K.J. YOUN AND L.CHUNG

Department of Architectural Engineering, Dankook University,San-8, Han-nam dong, Young-San Gu, Seoul, KOREA, E-mail: lshyun00@dankook.ac.kr

S.B. CHOI, AND K.G. SUNG

Department of Mechanic Engineering, Inha University

H.G. LEE

Department of Mechanic Engineering, Daeduk University

In this study, a response-dependent MR damper of which frictional force is determined according to the assumed shape function composed of displacement and velocity of the damper piston itself, is designed and its performance is compared to the passive one and semi-active one using linear quadratic regulator. Numerical analysis results from SDOF system indicate that passive MR damper designed to have appropriate frictional force of which magnitude is less than about 30% of the base shear force, provides the smallest displacement response spectrum over all structural periods and response-dependent MR damper show better control performance in reducing absolute acceleration spectrum with increasing frictional force for the long period structure. Also, seismic analysis of a three story structure with the MR damper shows that well designed response-dependent MR damper provides performance equivalent to or better than the semi-active MR damper operated by LQR controller in spite of using only the structural response transferred to the damper.

1 Introduction

Since magneto-rheological (MR) damper can achieve large damping force by changing magnitude of the applied magnetic field dependent on the input current, extensive researches have been analytically and experimentally to verify validity of the MR dampers for controlling large-scale civil structures. The MR damper is said to be operated semi-actively when the magenetic field is modulated according to the structural response or excitation characteristics based on specific control logic while it is called as passive one when the constant input current is given regardless of the structural response. It is known that the optimally designed semi-active MR dampers generally show more excellent control performance than passive control. Various control algorithms such as Lyapunov, decentralized bang-bang, maximum energy dissipation, clipped optimal control and variable friction control algorithms [1], are adopted for designing the semi-active MR dampers. Although, semi active MR dampers do not need large supplemental power supply which is essential for active control devices, it requires additional signal processing systems for the measurement of the structural responses or excitation forces, calculation of the optimal control forces, and sending out the input voltage signals, which may be practically difficult to be installed at real scale civil structures. This study was motivated by considering following questions; 1) Does semi-active MR damper always show better control performance than passive one? Evenwhen the performance of the passive MR damper is slightly worse than the semi-active ones, should the MR damper be operated semi-actively? 2) Should the semi-active MRdamper use global information in addition to the damper-installed floor response for obtaining optimal performance? 3) Can the magnetic field be modulated without using any sensor and computer?

In order to answer these questions, in this study, a response-dependent MR damper of which frictional force is determined according to the assumed shape function composed of displacement and velocity of the damper piston itself, is designed and its performance is compared to the passive one and semi-active one using linear quadratic regulator which requires global structural information. The restriction of the information determining the variable frictional force in the response-dependent MR damper was placed in order to develop a MR damper of which magnetic field is automatically modulated according to the displacement or velocity transferred to the damper without using any sensing and computing systems in upcoming study. First, the performance difference of various models for describing the hysteretic characteristics of the MR damper are investigated using a numerical example of a three story building structure and a proper model is selected. Second, in order to determine appropriate range of the variable frictional force for controlling seismic response, numerical analysis is performed for a single-degree-of freedom system (SDOF) system. Third, control algorithm for designing response-dependent MR damper is presented. Finally its performance is compared to the passive one and semi-active one for SDOF and multi-degree-of freedom (MDOF) systems.

2 MR damper Model Comparison for Building Response Control

Many models such as Bingham, biviscous, hysteretic biviscous, simple Bouc-Wen, Bouc-Wen with mass element, and phenomenon models have been developed for capturing the force characteristics of the MR dampers. Yang compared those models and obtained the values of the parameters for the last four models though harmonic test using 0.5Hz sine displacement control for 20-ton MR damper [2]. In this study, the performance difference of those models in controlling seismic response is investigated. Figure 1 shows the force-velocity relationships of above 6 models when a sine function is given as displacement. It is observed that the hysteretic characteristics of each model are different from each other while the maximum value of damping force is almost equivalent. Particularly, Bingham and biviscous model have disadvantage that they cannot describe the hysteretic behavior. Figure 2 shows time histories of the third floor displacement and absolute acceleration responses of the three-story shear building structure with a MR damper at first floor, and the peak values of interstory displacement and absolute acceleration are listed in table 1. The considered building structure has identical story properties. Story mass, stiffness, and damping are, respectively, 100 metric ton, 98000kN/m, and 140.7kN·sec/m, and El Centro (1942, NS component) earthquake is used as input load. It is shown from Figure 2 and table 1 that structural responses are significantly reduced by the MR damper. The fact that the difference between controlled structural responses by each model cannot be identified in Figure 2, indicates that the essential parameter is not the shape of hysteretic curve but the magnitude of maximum control force. Bouc-Wen model requires integration process of nonlinear differential equation in calculating control force, and thus it takes much time effort. Accordingly, in this study, Bingham model is considered because it is mathematically the simplest one composed of friction and viscosity.

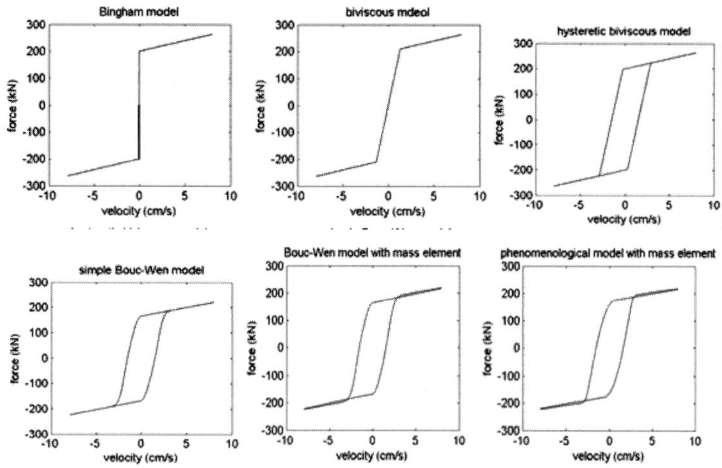

Figure 1. Force-velocity relationship of 6 models for MR damper

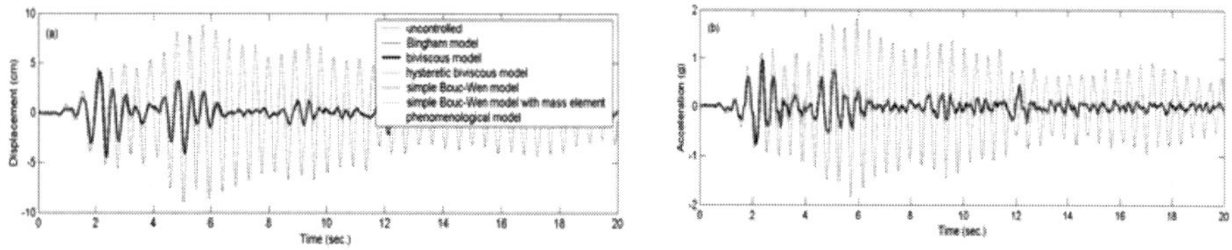

Figure 2. Time histories of displacement and acceleration responses of 3 story building

Table 1. Peak responses of 3 story building

	floor	uncontrolled	(1)	(2)	(3)	(4)	(5)	(6)
Max. Inter-story Disp (cm)	1	4.2	1.8	1.8	1.8	1.6	1.7	1.7
	2	3.2	1.5	1.5	1.5	1.4	1.4	1.5
	3	1.9	0.9	0.9	0.8	0.8	0.8	0.8
Max. Acc (cm/s^2)	1	975	461	468	495	461	499	522
	2	1477	672	662	665	618	649	661
	3	1814	880	872	831	769	806	811

(1) Bingham (2) biviscous (3) hysteretic biviscous (4) simple Bouc-Wen
(5) simple Bouc-Wen with mass element (6) phenomenological model

3 Response-dependent MR damper

3.1 The range of friction force of the MR damper

A SDOF system with the MR damper described by Bingham model is gven as

$$m_s\ddot{x} + c_s x + k_s x = -m_s\ddot{x}_g - f_f(H)\text{sgn}(\dot{x}) - C_o\dot{x} \tag{1}$$

where m_s, c_s, and k_s denote structural mass, damping, and stiffness, respectively. f_f and C_o are, respectively, frictional force dependent on the magnetic field H and viscosity.

The friction force of the MR damper is normalized by the peak shear force of the uncontrolled structure f_s.

$$\rho = f_f / f_s \tag{2}$$

Figure 3 shows the results from SDOF system with structural period 1.0 second and 5% damping ratio. Figure 3(a) and (b), respectively, are the peak displacement R_d and base shear R_f normalized by uncontrolled responses, and Figure 3(c) illustrates the earthquake energy R_e dissipated not by the MR damper but by the structure itself. SR denotes the stiffness ratio of the brace system for the installation of the MR damper to the structural stiffness. In Figure 3, it is observed that the reasonable value of ρ is lower than 0.3 because increasing ρ higher than 0.3 causes R_f and R_e to increase considerably and R_d to increase or change little except for *SR*=50. This indicates that the reasonable frictional force of the passive MR damper is less than 30% of f_s.

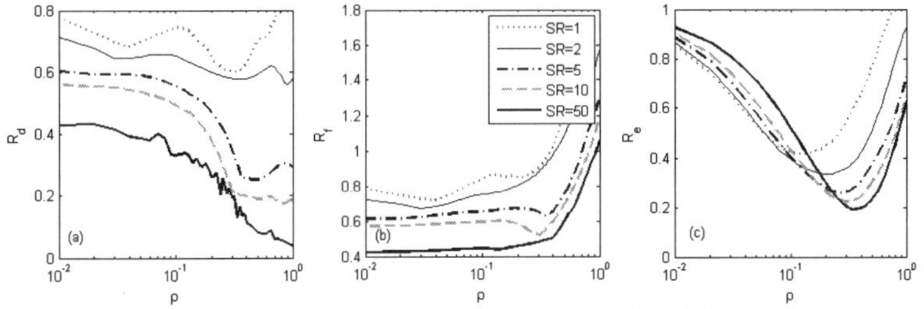

Figure 3. Response variation with respect to ρ (T_n = 1.0 second)

3.2 *Control algorithm*

The yield stress τ of the MR fluid, which is linearly proportional to f_f, is known to be nonlinearly proportional to the magnetic field.

$$\tau = \alpha H^{\beta} \tag{3}$$

Cho et al. developed so called smart passive MR damper in which the magnitude of the magnetic field automatically varies according the velocity of the damper piston by electro magnetic induction (EMI) system which doesn't need any sensing and control systems, showed that the smart passive MR damper performed better than passive-one and almost equivalently to existing semi-active MR damper [3]. However, the study by Cho et al. did not consider the effects of the frictional force magnitude normalized to the base shear and the fundamental period which are critical in the seismic analysis. In this study, a response-dependent MR damper is designed and its control performance is investigated by varying the range of the frictional force and structural period. The frictional force of the response-dependent MR damper is determined based on following equations (4) and (5) which are a function of the displacement x_d and velocity $\dot{x}_d$ of the MR damper piston itself.

$$f_f = \begin{cases} f_{\min} & V < V_1 \\ \dfrac{f_{\max} - f_{\min}}{V_2 - V_1}\left(V(x_d, \dot{x}_d) - V_1\right) + f_{\min} & V_1 \le V \le V_2 \\ f_{\max} & V > V_2 \end{cases} \tag{4}$$

$$V(x_d, \dot{x}_d) = \alpha_D \,|\, x_d \,|^n + \alpha_V \,|\, \dot{x}_d \,|^m \tag{5}$$

where $f_{\max}$ and $f_{\min}$, are, respectively, maximum and minimum frictional force, and V_1 and V_2 are important design factors defining the range of V corresponding to the transition of the friction force. α_D and α_V determine relative weightings between the relative displacement and velocity, and n and m determine the shape of friction force variation. If an MR damper of which magnetic field varies according to the movement of the damper piston can be developed, its frictional force would have the form of equation (4) when considering the relation between τ and H in equation (3). This is a reason why the response-dependent MR damper has the variable frictional force in equation (4).

4 Numerical Example

4.1 SDOF system

For performance evaluation of the response-dependent MR dampers, nonlinear time history analyses of the mass normalized SDOF systems with 5% damping ratio are conducted. The results are averaged over 20 earthquakes measure in rock-site [4]. The maximum ρ is increased from 0.1 to 0.3 by 0.1 and minimum ρ is 0.01. Three response-dependent MR dampers denoted by Semiactive-1, Semiactive-2 and Semiactive-3 have (α_D, α_V) equal to (0, $1/\omega_n$), (1, 0) and (1, $1/\omega_n$), respectively, where ω_n is the natural circular frequency of the structure. Also, m=1, n=1, V_1=0.05V_2, and V_2=$\alpha_D(\sigma_D)^n$+$\alpha_V(\sigma_V)^m$, where σ_D and σ_V are the RMS displacement and velocity without the MR damper, are used.

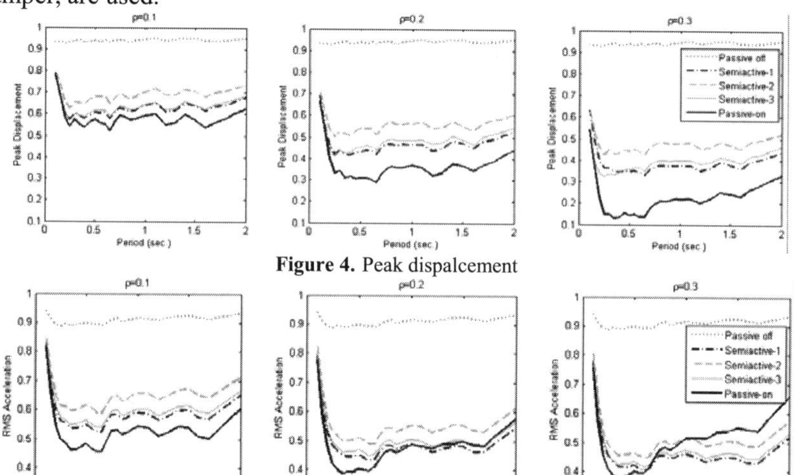

Figure 4. Peak dispalcement

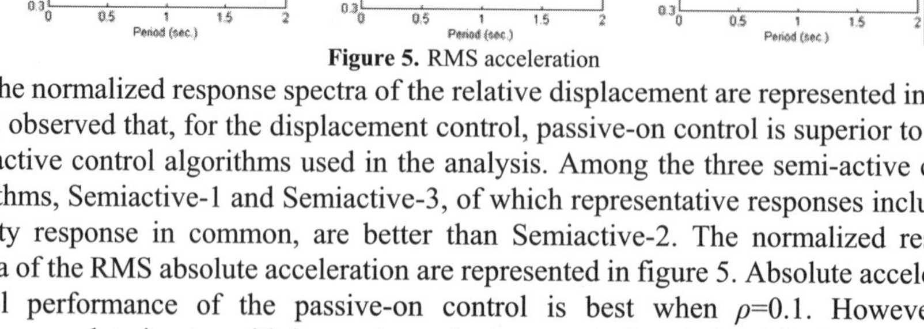

Figure 5. RMS acceleration

The normalized response spectra of the relative displacement are represented in figure 4. It is observed that, for the displacement control, passive-on control is superior to all the semi-active control algorithms used in the analysis. Among the three semi-active control algorithms, Semiactive-1 and Semiactive-3, of which representative responses include the velocity response in common, are better than Semiactive-2. The normalized response spectra of the RMS absolute acceleration are represented in figure 5. Absolute acceleration control performance of the passive-on control is best when ρ=0.1. However, the performance deteriorates with increasing ρ for long natural periods while Semiactive-1, 2 and 3 show the regular control performance regardless of the structural period.

4.2 A three story structure

General active control algorithms such as linear quadratic regulator (LQR), bang-bang, clipped optimal control and Lyapunov controller can be applied to the design of semi-active MR damper for controlling MDOF systems [5]. In this section, the numerical analysis is performed for a three story structure used by Dyke et al., in which the MR damper is connected between the ground and the first floor, and the performance of the respondent-dependent MR damper compared to that of passive one and the semi-active one of which control force is calculated based on LQR to minimize following performance index.

$$J = \int_0^{\infty} \left(z^T Q z + u^T R u \right) dt \tag{6}$$

where z denotes state vector composed of inter-story displacement and relative velocity to ground, and Q and R are weighting matrices for state and control, respectively. In this example, $R=10^{-3}$ is used and Q has diagonal elements are $Q_{11}=1.37\times10^7$, $Q_{22}=6.84\times10^5$, $Q_{33}=6.84\times10^5$, Q_{44}=98.3, Q_{55}=98.3 and Q_{66}=98.3, and the other elements are zero.

$$f_{MR}(t) = \begin{cases} \text{passive-off} & f_{LQR}(t) < f_{\min} \text{ or } f_{LQR}(t)\cdot f_{MR}(t) \le 0 \\ f_{LQR}(t) & f_{\min} \le f_{LQR}(t) \le f_{\max} \\ \text{passive-on} & f_{LQR}(t) > f_{\max} \end{cases} \tag{7}$$

where f_{LQR} is a control force computed by LQR control logic and f_{MR} is a force which can be realized by the MR damper. El Centro earthquake with peak ground acceleration scaled to 0.3g is used and $f_{\min}$ and $f_{\max}$ are determined with $\rho=0.01$ and $\rho=0.05$, respectively. Semiactive-1 in section 4.1 is used as response-dependent MR damper.

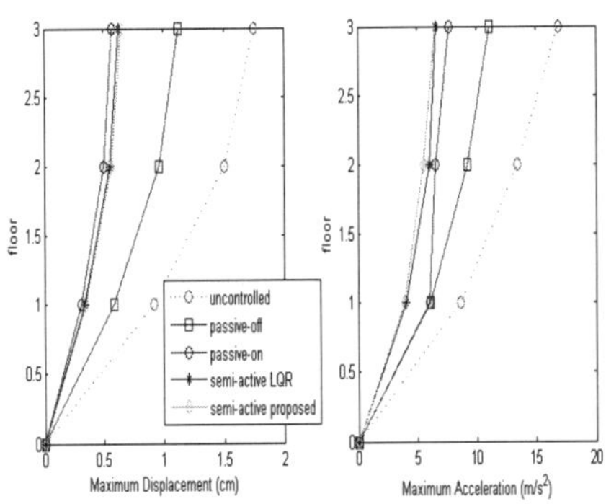

Figure 6. Structural response of the three story structure

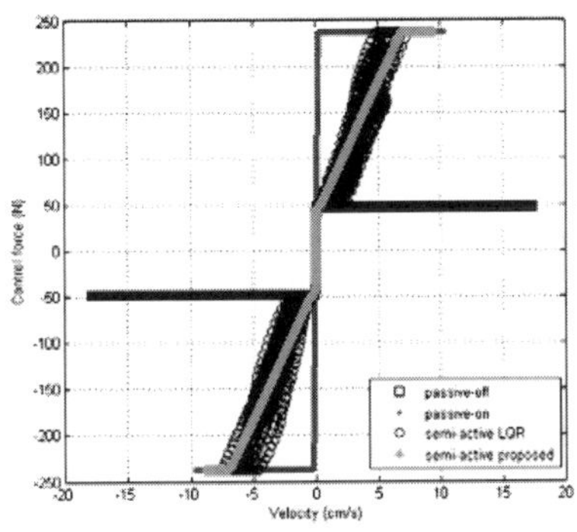

Figure 7. Force-velocity curve

Figure 6(a) and (b) show the peak displacement and absolute acceleration distribution, respectively. It is observed that semi-active cases provide slightly larger displacement than the passive-on MR, but they reduce absolute acceleration most significantly. It is also shown in figure 6(b) that the response-dependent MR damper proposed in this study shows slightly better control performance than LQR. The response-dependent MR damper is more effective than LQR considering that it only uses the velocity of the first floor in calculating the force of the MR damper. Figure 7 shows that the forces by the semi-active MR dampers vary according to the velocity while those by the passive ones are constant. It is also observed that the response-dependent MR damper generates obviously linearly varying forces between f_{min} and f_{max} while the MR damper operated by LQR algorithm doesn't do although LQR has positive correlation between the force and the velocity. This phenomenon is so natural considering that LQR calculates the control forces by using the information from all the floors.

5 Conclusion

Followings are findings in this paper; 1) Passive-on MR damper with appropriate magnitude of frictional force less than 30% of the base shear of uncontrolled structure show the smallest displacement spectra of SDOF system over all periods; 2) The response-dependent MR damper of which frictional force varies according to the displacement or velocity of the damper piston itself, show better acceleration control performance with increasing frictional force for longer period SDOF systems than passive cases; 3) Response-dependent MR can be designed to provide performance better than LQR in spite of using only the structural response directly transferred to the damper.

Other semi-active algorithms can be designed to show superior performance than passive or the response-dependent MR dampers. Especially, this would be more possible when the magnitude of the frictional force increases so that the rigidity of the MR damper has undesirable effects such as the response increase of the other stories where the MR damper is not installed. However, considering that passive and the response-dependent MR dampers do not need any sensing or computing system, slight performance deterioration can be accepted in the practical application. Efforts are currently underway to develop the real mechanism of the response-dependent MR damper.

6 Acknowledgements

The work presented in this paper was partially supported by the National Research Laboratory Program (Project No. M1-0203-00-0068) from the Ministry of Science and Technology in Korea and the Second Stage of BK21 through the Ministry of Education & Human Resources Development through in Korea.

References

1. Dyke S. J., Spencer Jr. B. F., Sain M. K. and Carlson J. D., Modeling and control of magnetorheological dampers for seismic response reduction. *Smart Materials and Structures*, **5** (1996) pp. 565 - 575.
2. Yang G., Large - scale magnetorheological Fluid Damper for Vibration Mitigation: Modeling, Testing and Control (University of Notre Dame, Indiana, 2001)
3. Cho S. W., Cho J. S., Kim C. H. and Lee I. W., Smart Passive System Based on MR Damper., *Earthquake Engineering Society of Korea,* **9** (2005) pp.51-59.
4. Lee S. H, Min K. W., and Lee Y. C, Modified sliding mode control using target derivative of Lyapunov function", *Engineering Structures*, 27 (2005) pp.49-59.
5. Jansen, L. M. and Dyke S. J., Semiactive control strategies for MR dampers: comparative study. *Journal of Engineering Mechanics*, ASCE, **126** (2000) pp. 795-803.

FUZZY LOGIC CONTROL FOR MAGNETORHEOLOGICAL DAMPER

C.A.M. PINHEIRO, J.J. LIMA, A.J.F. BOMBARD AND J.C. DIAS
Federal University of Itajubá, CP 50, CEP 37500-903, Itajubá - MG, Brazil
E-mail: pinheiro@unifei.edu.br

J.B. MACHADO
State University of Campinas - FEEC, CP 6101, CEP 13083-852, Campinas - SP, Brazil
E-mail: jeremias@dca.fee.unicamp.br

This paper proposes a control strategy for systems of automotive active suspension using fuzzy logic and shows how to generate the necessary rules. Conventional control algorithms for active or semi-active suspension, such as on/off and continuously variable will be shown. The results of the control laws are compared with the fuzzy strategy through numeric simulations and experimental tests. Analyses of results show the advantages and disadvantages of the used methods.

1 Introduction

The conventional or passive automotive suspension systems are used since many decades in vehicles and other devices, with the purpose of absorbing vibrations, due to their robustness and low price. The basic models studied in suspension systems use the representation of one vehicle wheel, the well-known "quarter-car" model. The fig. 1 illustrates this model [1].

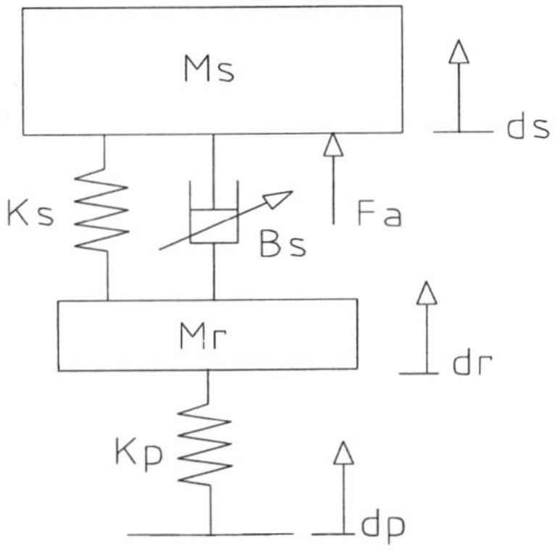

M_s – Sprung mass
M_r – Unsprung mass
B_s – Damper rate
K_s – Spring constant of suspension
K_p - Spring constant of tire
d_r – Vertical road displacement
d_u – Vertical unsprung mass displacement
d_s – Vertical sprung mass displacement
F_a – Action force

Figure 1. Basic model of a "quarter-car" automotive suspension

In the last two decades were developed and improved active (or semi-active) suspension systems, where modifications in suspension components, like pneumatic springs or controlled dampers, resulted in comfort and safety improvements. One example of such devices are magnetorheological dampers, whose damping forces are controlled through a magnetic field that changes the rheology of the fluid, the so called magnetorheological effect [2].

The spring and the passive damper in structures of conventional suspension are represented by its coefficients K_f and B_a, respectively. The parameter M_s corresponds to the vehicle suspended mass (body), sustained by the suspension system. The constant M_r models the set wheel more tire mass, and K_p represents the tire coefficient elasticity.

The displacements d_p, d_r and d_s are tire, wheel and body vertical actions respectively. The force F_a can represent the action exercised by a mechanism aiming to impose some dynamic characteristics in the suspension. The device can be an active damper that replaces a passive damper. The adopted basic modeling in this work is usual in the literature [1,3]. Other models more elaborated that consider longitudinal, vertical and lateral dynamic of a vehicle also are treated in the specialized literature [4].

The mathematical modeling of these systems can be made through differential equations, state variables or transfer functions. Considering the forces on the suspension elements, can be written the equations (1) and (2). Parameters example of a suspension system are [5]: $M_s = 400$ kg; $M_r = 50$ kg; $B_a = 1000$ Ns/m; $K_f = 20000$ N/m; K_p=250000 N/m. Generally the parameters are chosen to define a natural frequency around 1 Hz in d_s and of 10 Hz in d_r.

$$M_S \ddot{d}_S + B_a(\dot{d}_S - \dot{d}_r) + K_f(d_S - d_r) - F_a = 0 \qquad (1)$$

$$M_r \ddot{d}_r + B_a(\dot{d}_r - \dot{d}_S) + K_f(d_r - d_S) + K_p(d_p - d_r) - F_a = 0 \qquad (2)$$

1.1 Control Strategies Simulation

In order to obtain mathematical models appropriated to analyze systems of semi-active suspension in different control strategies, the modeling equations (1) and (2) will be converted in state variables. In this work will be adopted the substitutions of variable given below, defining the state equations (3).

$$x_1 = d_s \quad x_2 = \dot{d}_s = \dot{x}_1 \quad x_3 = d_r \quad x_4 = \dot{d}_r = \dot{x}_3$$
$$\dot{x}_4 = \ddot{d}_r \quad \dot{x}_2 = \ddot{d}_s \quad u_1 = d_p \quad u_2 = F_a$$

$$\begin{aligned}
\dot{x}_1 &= x_2 \\
\dot{x}_2 &= -\frac{K_f}{M_S}x_1 - \frac{B_a}{M_S}x_2 + \frac{K_f}{M_S}x_3 + \frac{B_a}{M_S}x_4 + \frac{1}{M_S}u_2 \\
\dot{x}_3 &= x_4 \\
\dot{x}_4 &= \frac{K_f}{M_r}x_1 + \frac{B_a}{M_r}x_2 - \frac{K_f+K_p}{M_r}x_3 - \frac{B_a}{M_r}x_4 + \frac{K_p}{M_r}u_1 - \frac{1}{M_r}u_2
\end{aligned} \qquad (3)$$

The variable x_1 represents the vertical displacement of the suspended mass, x_2 the speed of the same, and the derivative of this last the corresponding acceleration. The variable x_3 represents the wheel vertical displacement, x_4 its speed, and the derivative of this the resultant acceleration. The variable u_1 expresses a perturbation in the suspension, the tire vertical action. The variable u_2 represents the compensation force of an active suspension system (for example, of an active damper). This input with value always null ($F_a = 0$) represents a passive suspension system. In many structures of active suspension (or semi-active), an active damper can replace a passive damper as energy dissipation element of the suspension system. Thus, the value of the parameter B_a in the modeling can be null, because a variable force represented by Fa substitutes the passive damper.

Perturbations or usual disturbances in analyses of suspensions systems are the abrupt actions, as large displacements in the tire of the vehicle. The figure 2 illustrates the suspended mass acceleration (vehicle body), for a passive suspension with a displacement ($u_1 = d_p$) vertical (sudden) of 0.05 m in the tire.

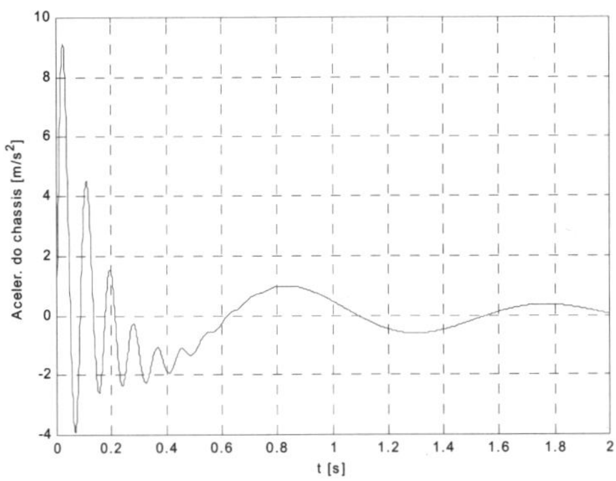

Figure 2. Acceleration of the body for passive suspension.

1.1.1 On/Off and Continuously Variable Control

Control strategies for active suspension systems have being proposed by several years. The more usual are on/off and continuously variable [5]. The basic strategy on/off is represented by (4), where F_a expresses the force exercised by the active damper, being C_{on} and C_{off} the attrition coefficients values of the same (by fluid flow in the damper or changing its magnetic properties). V_{abs} the absolute velocity of the suspended mass, and V_{rel} the relative velocity between suspended mass and the wheel set mass (respectively, x_2 and $x_2 - x_4$ in the state variables). If the absolute velocity and the relative velocity own the same algebraic signal, then an intense state of damping is selected through a coefficient C_{on}. Now, if the signals are opposite, a low state of damping is selected through a coefficient C_{off}. Positive relative speed is defined with the masses M_S and M_{r} (fig. 1) moving away itself. Positive absolute speed is defined with the action upwards of the suspended mass (M_S).

The equation (5) represents the usual strategy of continuously variable control. The basic difference regarding the strategy (4) it is just that the force produced (when the absolute speed and the relative speed own the same algebraic signal) is proportional the absolute speed as well. Usual values for the parameters are in C_{off} = 500 Ns/m and C_{on} = 1400 Ns/m.

$$F_a = \begin{cases} C_{on} V_{rel} & \text{if} \quad \mathrm{V}_{\mathrm{abs}} V_{rel} \geq 0, \\ C_{off} V_{rel} & \text{if} \quad \mathrm{V}_{\mathrm{abs}} V_{rel} < 0. \end{cases} \tag{4}$$

$$F_a = \begin{cases} C_{on}V_{abs} + \mathrm{C_{off}V_{rel}} & \text{if} \quad \mathrm{V_{abs}}V_{rel} \geq 0, \\ C_{off}V_{rel} & \text{if} \quad \mathrm{V_{abs}}V_{rel} < 0. \end{cases} \tag{5}$$

1.1.2 Fuzzy Strategy

In this work, fuzzy logic will be used as a control strategy for semi-active suspension, and we will show how to generate the necessary rules. The utilization of fuzzy logic was already considered in other papers about active suspension [7,8]. However, these control fuzzy strategies worked together with other control methods (poles allocation, optimal control, etc.), or it used rules obtained empirically without a basement that allows to guide improvements or modifications. One of the advantages of the fuzzy logic in applications in several field, is its characteristic of incorporate linguistics information in the decision strategies [9]. We can obtain fuzzy rules through the context analysis of the wished application, coding linguistically information of the controlled system, or using learning methods, clustering, and other to extract the necessary information [10]. In terms of modeling, fuzzy systems are universal approximation of functions.

In this work the fuzzy rules are obtained through the knowledge or of the basic information of an active suspension system previously established. Concepts of the continuously variable strategy will be used due its good characteristics, and because is frequently made in the practice. More details about how the fuzzy logic was implemented can be found in reference [11].

The fig. 3 illustrates the suspended mass acceleration curves using several strategies. The numbers identify the control strategy: 1 = Passive; 2 = On/Off; 3 = Continuously Variable and 4 = Fuzzy.

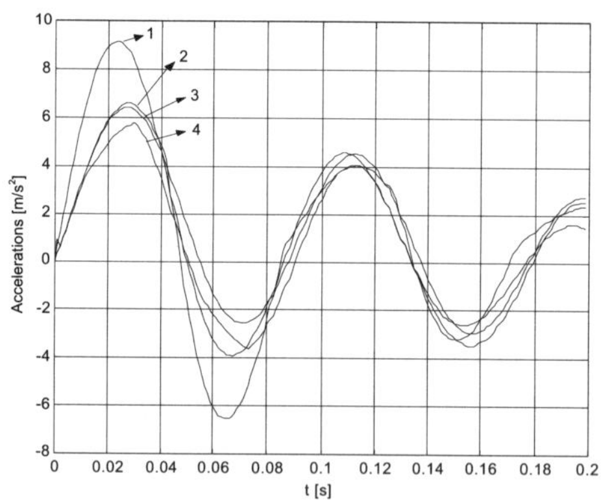

Figure 3. Acceleration damping for several strategies as simulated using MatLab. 1 = Passive; 2 = On/Off; 3 = Continuously Variable; 4 = Fuzzy.

In order to compare characteristics of the several strategies, it will be shown the graphs of the damping forces and displacements of the suspended mass (fig. 4 and 5). The table 1 shows the RMS values of the main physical variables of the control strategies. The table 2 shows the maximum values and the accommodation times (for ± 2% of the RMS values) of the variables.

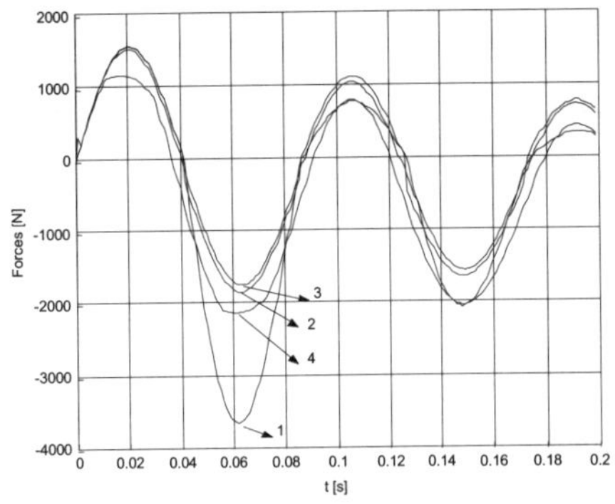

Figure 4. Force damping for several strategies as simulated using MatLab. 1 = passive; 2 = On/Off; 3 = Continuously Variable; 4 = Fuzzy.

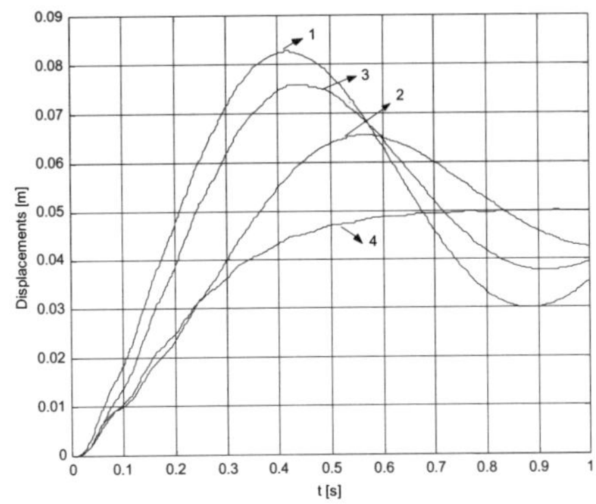

Figure 5. Displacement damping for several strategies as simulated using MatLab. 1 = Passive; 2 = On/Off; 3 = Continuously Variable; 4= Fuzzy.

Table 1. RMS Values of the main physical variables of the control strategies.

Logic	Passive	On / Off	Cont. Variable	Fuzzy
Acceleration [m/sec^2]	1.42	1.22	1.27	1.10
Damping Force [N]	457	497	418	470
Displacement [mm]	53	49	51	46
Velocity [m/sec]	0.125	0.074	0.102	0.054

Table 2. Maximum values / Accommodation times

Logic	Passive	On/Off	Cont. Var.	Fuzzy
Acceleration [m/sec^2].	9.0 / 1.85 s	6.6 / 2.1 s	6.5 / 2.0 s	5.7 / 1.6 s
Damping Force [kN]	2.86 / > 2 s	3.62 / 1.8 s	1.80 / 1.8 s	2.07 / 1.6 s
Displacement [mm]	82 / 2 s	65 / 1.95 s	75 / 1.95 s	0.0 / 0.76 s
Velocity [m/sec]	0.33 / 2.1 s	0.21 / 1.9 s	0.30 / 2.1 s	0.20 / 1.3 s

Analyzing the graphs and information tables can be verified that the passive suspension is that presents the biggest acceleration, displacement and body speed. The maximum active force damping value or the effective value is smaller for the strategy that uses Cont. Variable control, but the accommodation times are larger for the other physical variables. The on-off strategy is the easiest of implement, but it can introduce jerks problems and undesirable audible noises. The continually variable strategy is more advantageous than the on/off, and is relatively simple to be implemented. The proposed strategy with fuzzy logic was that introduced the best dynamic characteristics in general. The fuzzy strategy rules are obtained from the concepts of continually variable politics and the best performance of the fuzzy strategy is explained due to fuzzy systems are universal approximations of functions. The adequate choice of the membership functions, allows to minimize abrupt variations in the force of the active damper, reducing thus jerk problems and noise problems.

2 Methods

Objectifying to obtain real results, it was assembled a prototype (in reduced scale) to accomplish experimental tests as showed in the fig. 6.

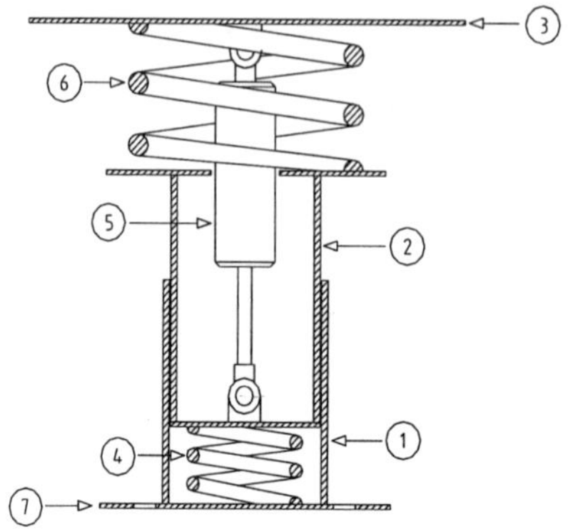

where:

1 - Suspension Guide;
2 - Joint for fixation of the springs, the damper, and the block representing the wheel;
3 - Base for support the suspended mass;
4 - Spring representing the tire;
5 - MR damper LORD RD-1005-3
6 - Spring of the shock-absorber system;
7 - Fixation base.

Figure 6. Experimental prototype drawing.

A microcomputer (PC) implements (by software) the control algorithms, and a data acquisition board accomplishes the interface with the sensors and the amplifier. With the information of acquired accelerations (a_{abs} and a_{rel}), the information of the velocities (V_{abs} and V_{rel}) was calculated through integration operations. The information in the control algorithm generates the activation command for the active damper amplifier. The response for an excitation of the suspended mass for passive suspension under different control strategies were measured through an accelerometer, and collected through the acceleration sensor conditioner.

3 Results

The experimental results were grouped in a same graph, in order to compare the strategies. The fig. 7 shows the results of real tests. The numbers identify the control strategy: 1 = Passive or conventional; 2 = On/Off; 3 = Continuously Variable; 4 = Fuzzy Control.

It can be observed that the control strategies performances were coherent with the previous simulations, even with parameters in reduced scale. Besides, the Fuzzy Control shows the best performance.

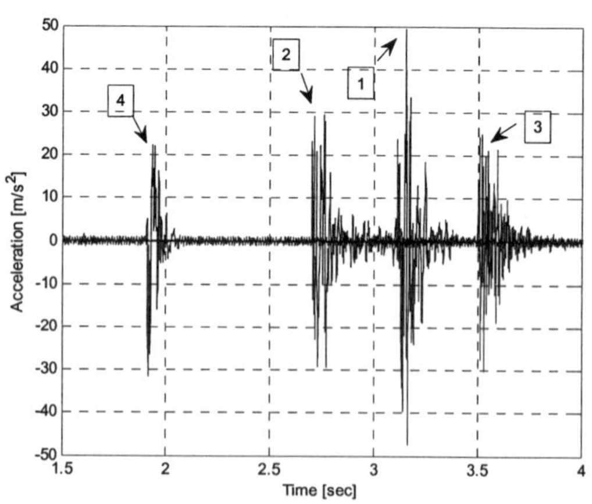

Figure 7. Signals of experimental acceleration. 1 = Passive; 2 = On/Off; 3 = Cont. Variable; 4 = Fuzzy.

4 Conclusion

This paper compared conventional control algorithms for automotive active suspension, such as on/off and continuously variable.

The main contribution of this work is the proposal of an efficient control with fuzzy logic strategy, where the fuzzy rules were generated through qualitative analyses. The results of the usual control laws were compared with the fuzzy strategy through numeric simulations and experimental tests.

The proposed strategy with fuzzy logic was that showed the best dynamic characteristics in general. The fuzzy strategy rules were obtained from the concepts of continuously variable strategy and the best performance of the fuzzy strategy is explained due to fuzzy systems are universal approximations of functions.

5 Acknowledgements

CAMP acknowledges the foundation FAPEMIG (TEC 195/01) for their support (regarding first part of this work). AJFB acknowledges BASF AG for the agreement BASF – UNIFEI and their support to attend ERMR'06.

References

1. Williams, R.A. "Automotive active suspensions. Part 1: Basic principles"; *Proc. Inst. Mech. Eng.*, 211: pp. 415-426; 1997.
2. Bica, I. "Damper with magneto-rheological suspension"; *Journal of Magnetism & Magnetic Materials*, 241: pp. 196-200; 2002.
3. Yue et al. "Alternative control laws for automotive active suspensions"; *J. Dyn. Syst., Measur, & Control,* 111: pp. 286-291; 1989.
4. Barbieri, F. A. A., Neto, A. C. "Modeling and Control of vehicular dynamics" (in Portuguese). 11° C ongresso Int. Tecnolog.; 2002.
5. Miller, L. R., Nobles, C. M. "Methods to eliminate jerk and noise in suspensions semi-active." Lord Corp., SAE Paper 902284; 1997.
6. Ricardo, R. M. "Control semi-active of automotive suspensions." Master′ s Degree Dissertation. (in Portuguese), State University of Campinas; 1998.
7. Yoshimura et al. "Active suspension of passenger cars using linear and fuzzy-logic controls"; *Control Engineering Practice* 7: pp. 41-47; 1999.
8. Cal, B., Konik, D. "Intelligent vehicle active suspension control using fuzzy logic"; 10th IFAC, Triennal World Congress, pp. 51-56; 1996.
9. Lee, C. C. "Fuzzy logic control systems: fuzzy logic controller – Part I and II"; *IEEE Trans. Syst. Man Cybern.;* 20: pp 404-435; 1990.
10. Wang, L. X. "Adaptive fuzzy systems and control"; Prentice Hall; 1994.
11. Machado, J. B. "Comparisons between Fuzzy controllers and conventional controllers applied in the active suspension systems"; Scientific Initiation Work, Federal University of Itajubá, Brazil, 2004.

EXPERIMENTAL STUDIES ON MAGNETORHEOLOGICAL MODEL SUSPENSIONS

A.-M. TRENDLER AND H. BÖSE*

Fraunhofer-Institut für Silicatforschung ISC, Neunerplatz 2, D-97082 Würzburg, Germany
E-mail: boese@isc.fraunhofer.de

Magnetorheological model suspensions with different contents of iron particles have been investigated. The results show an increase of the shear stress by increasing the magnetic field and the solid content, which was mathematically modelled. The influence of the solid content is approximately exponential without magnetic field and linear in strong magnetic fields. The effect of the temperature and of the base oil viscosity on the shear stress is negligible in the field but decisive without field. The sedimentation behavior also strongly depends on the iron particle content, where MR fluids with a higher concentration settle with a lower intensity due to the higher sediment height. The response time of the magnetorheological model suspensions could not be clearly determined, but it is less than 10 milliseconds.

1 Introduction

In analogy to electrorheological (ER) fluids responsive to electric fields, magnetorheological (MR) suspensions consist of polarizable particles (usually carbonyl iron powder) dispersed in a carrier liquid (for example mineral oil, silicone oil, ester oil etc). Due to dipole interactions, when a magnetic field is applied, the particles form chains along the field lines. The result is a dramatic change in the rheological properties. This MR effect generates a high potential for various applications like adaptable vibration dampers, shock absorbers, clutches, brakes, haptic devices etc [1].

The application-relevant properties of MR suspensions like base viscosity, shear stress in the field and sedimentation stability depend on a variety of characteristics of their composition and of the particles used. Relevant parameters which affect the properties are the solid content, the composition and morphology of the particles, and their size distribution [2,3]. The influence of the solid content on the rheological properties without applied magnetic field is also a very important issue for suspensions of various materials in the ceramic [4], chemical and food industry.

In contrast to the ER-fluid research field, not many material-oriented scientific studies concerning the MR suspensions are known. However, a variety of studies have been done in terms of MR applications, whereby the MR suspensions used were commercially available materials [5]. A variation of the solid content of commercial MR suspensions in different base oils and also an empirical mathematical equation (yield stress depending on solid volume content and magnetic field strength) with regard to the magnetorheological device design is described in [6].

In this paper, the results of the investigations carried out with a series of model MR fluids are presented and analyzed. This study was performed in order to quantitatively determine the influence of the concentration of magnetic particles and the viscosity of the carrier liquid on the rheological behavior of the MR suspension with and without magnetic field, respectively, as well as the sedimentation behavior.

2 Methods

For this study, MR suspensions were prepared dispersing different volume contents of iron powder in hydrocarbon oils with various viscosities. The iron powder used was a carbonyl iron powder provided by BASF, with a mean particle size of approx. 5 μm. The iron concentration in the suspensions was varied between 10 and 35 % by volume. The viscosities of the carrier oil were in the range between 4.5 and 75 mPa s. The MR model suspensions did not contain any other additives.

The rheological properties were studied using a rheometer named MCR 300 produced by Paar Physica. For the measurements without field, a system of concentric cylinders and, for the studies in the magnetic field, a parallel-plate system, type M 003, where the field lines are perpendicular to the 1 mm gap between the plates was used. A detailed description of the magnetic cell is given in [7]. For the determination of the magnetic flux density in the gap between the parallel plates of the rheometer two methods were used. The first method is the experimental calibration of the flux density with a hall probe at different currents of the coil in the magnetic cell as well as solid contents of the MR suspensions. The second procedure included a computer simulation of the magnetic circuit and B-H curves of the MR suspensions obtained by a vibration magnetometer. The results show that the flux densities which were measured are in accordance with the simulated ones (less than 5% deviation); thereby the maximal current of 6 A corresponds to a flux density of 725 mT and a field strength of 600 kA/m in the case of the 35 vol. % iron MR suspension.

The sediment height (i. e. the interface between clear and turbid phases) of the suspension over a time period of 30 days was visually measured in graded cylinders at room temperature according to the sedimentation time.

Response time evaluations of different MR suspensions were performed in the magnetorheological cell of the rheometer. For this purpose an external current-driven power supply and for the recording of the response torque an oscilloscope were used.

3 Results and Discussion

3.1 *Effect of the applied magnetic field*

The dependence of the shear stress on the magnetic flux density and on the volume content of iron at a constant shear rate of 100 s^{-1} and at a temperature of 25 °C is depicted in Figure 1.

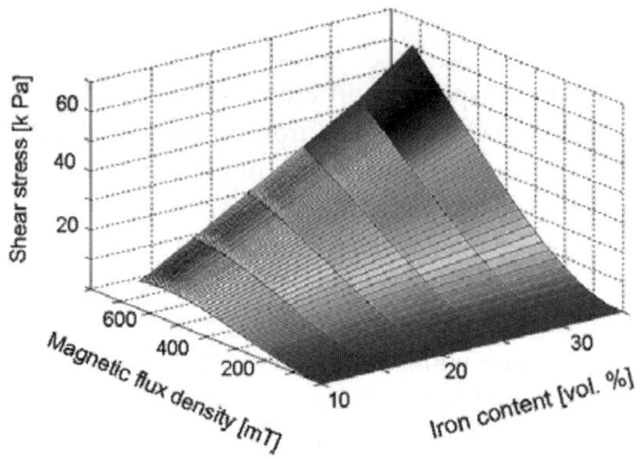

Figure 1. Dependence of the shear stress on the concentration of iron particles in the MR suspension and on the magnetic flux density at 100 s^{-1} and 25 °C

The shear stress of the MR suspension continuously increases with rising iron particle content and magnetic flux density. This dependence was mathematically modelled. The diagram (Figure 1) can be used as an orientation for the design of MR suspensions to be used in magnetorheological devices.

3.2 *Rheological behavior without magnetic field*

The flow curves of MR suspensions with various particle concentrations without magnetic field applied are given in Figure 2a.

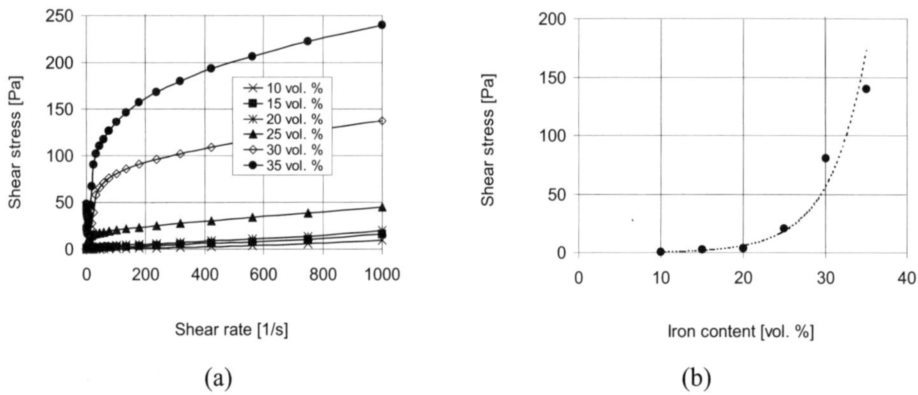

Figure 2. **(a)** Flow curves of MR suspensions with various iron particle concentrations without magnetic field at 25 °C; **(b)** Dependence of the shear stress on the iron volume content at a fixed shear rate of 100 s^{-1}

Considering the shear stress at a constant shear rate, as for example 100 s^{-1} (Figure 2b), the dependence on the iron particle content without magnetic field follows approximately an exponential function, in contrast to the corresponding relation with magnetic field where the dependence of the shear stress on the iron particle content is nearly linear.

3.3 *Effect of the temperature*

Figure 3 shows the influence of the temperature (range 25 – 125 °C) on the shear stress without field as well as in the magnetic field for four depicted magnetic flux densities. The volume content of iron particles is 35 %.

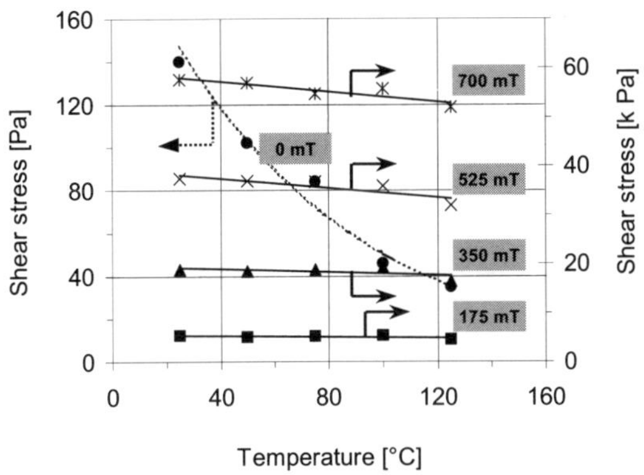

Figure 3. Dependence of the shear stress on the temperature at 100 s^{-1}, 25 °C and various magnetic flux densities (iron content 35 % by volume)

It could be observed that the temperature has a pronounced influence on the base viscosity of the MR suspension without applied magnetic field. This effect of the temperature corresponds to the exponential Arrhenius law. By applying the magnetic field the temperature has only a minor impact on the shear stress in the field because, under these conditions, the magnetic forces between the particles dominate the properties of the MR suspensions. Thereby the magnetorheological effect (i. e. the ratio of the shear stress with magnetic field to the shear stress without field) increases with a rising temperature due to the drastic decrease in the MR-suspension viscosity without field.

3.4 *Effect of the viscosity of the carrier liquid*

The influence of the base oil viscosity (range approx. 4.5 – 75 mPa s) on the rheological and magnetorheological properties is shown in Figure 4. The solid content is hold constant at 35 % by volume carbonyl iron powder.

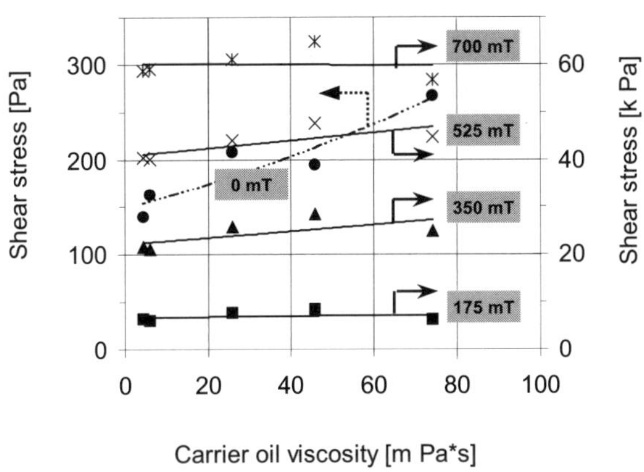

Figure 4. Influence of the viscosity of the carrier oil on the shear stress of the MR suspension at various magnetic flux densities, 100 s^{-1} and 25 °C (iron content 35 % by volume)

In analogy to the influence of temperature (Figure 3), the dependence of the base viscosity of the MR suspension without magnetic field on the carrier oil viscosity is relatively strong, whereas in the field the influence of the carrier oil is less pronounced (Figure 4). This behavior will be reflected in a decrease of the magnetorheological effect by using more viscous carrier oils.

3.5 Sedimentation stability

In order to investigate the sedimentation behavior, the lowering of the interface between the clear and turbid phases of MR fluids with various particle concentrations over 30 days was observed.

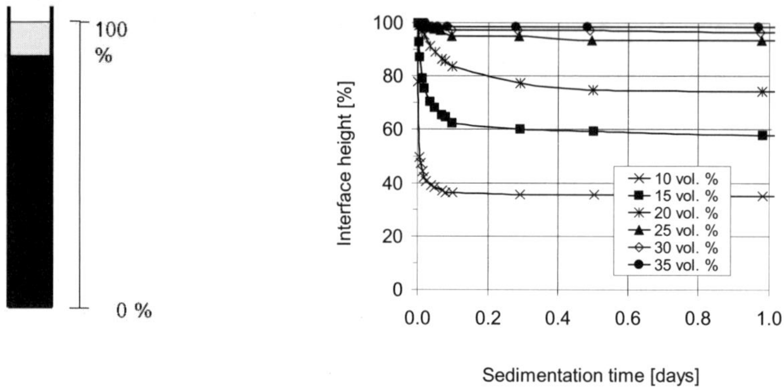

Figure 5. Sedimentation behavior of MR suspensions with different iron particle concentrations at 25 °C (change of the interface between clear and turbid phases)

The final state was nearly reached after one day (Figure 5). The sedimentation behavior strongly depends on the iron particle content, where MR suspensions with a higher concentration seem to settle with a lower intensity due to the higher sediment height.

3.6 Response time

Furthermore, preliminary measurements of the response time of different MR suspensions in the measuring cell of the rheometer were conducted. As an example for such experiments the increase of the torque with the MR suspension containing 35 % by volume solid is shown in Figure 6 for the application of different electric currents or magnetic flux densities, respectively.

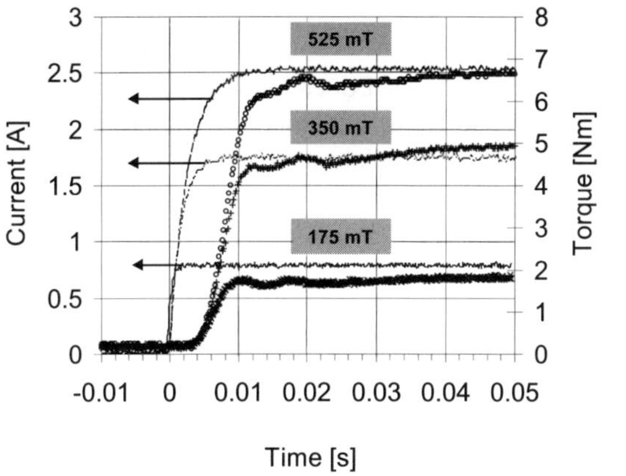

Figure 6. Response of the torque to a sudden increase of the current in the magnetorheological cell of the rheometer

It can be observed that the response time of the torque, i. e. the time of increase from the minimal to the maximal value, is approximately 10 ms. The initial delay time (about 2 ms) is independent of the shear rate, iron volume content, temperature and magnetic flux density. This constant time delay is attributed to the measurement procedure. However, the slope of the curves showing torque versus time increases with rising magnetic flux density. This behavior was also found for electrorheological fluids in [8]. Nevertheless, further studies have to be done in order to explain the magnetorheological response time of MR suspensions in the shear mode.

4 Conclusions

The dependence of relevant properties of MR suspensions on various parameters of composition (volume iron content, carrier oil viscosity) and experimental conditions (shear rate, temperature, magnetic flux density) were systematically investigated. The study has shown that the volume content of iron particles has a pronounced influence on the rheological properties of the MR suspensions.

The correlations between properties of the MR suspension, parameters of the composition and experimental conditions can be mathematically modelled and used for the application of the MR suspension in magnetorheological devices. The shear stress in strong magnetic fields increases nearly linearly with a rising iron particle content. However, the influence without field is even more pronounced (nearly exponential correlation). The temperature dependence of the shear stress is minor in the magnetic field and more relevant without field. Similarly, the dependence of the shear stress on the carrier liquid viscosity is negligible in the field and significant without field.

It can be concluded that systematic investigations of the dependence of properties of MR suspensions on the parameters of their composition, as performed in this study, are useful for the optimization of the application-relevant properties of these interesting smart materials.

5 Acknowledgements

Financial support from the German Federal Ministry of Education and Research (BMBF) is gratefully acknowledged. The authors thank Hans Schömig for preparing the MR suspensions of this study and Ramona Wolf for the simulations of magnetic flux density.

References

1. Bossis, G. (Ed.), *Proc. of the 8th Intern. Conf. of Electrorheological Fluids and Magnetorheological Suspensions, Nice 2001*, World Scientific Publishing, Singapore, (2002).
2. Trendler, A.-M., Böse, H., Influence of Particle Size on the Rheological Properties of Magnetorheological Suspensions. *Int. Journal of Modern Physics B*, **19**, 7-9 (2005) pp. 1416-22.
3. Böse, H., Trendler, A.-M., Rheological Properties of Magnetorheological Suspensions Containing Particles with Different Sizes. *Proc. 6th European Conference of Rheoly,* (2002) pp. 73–74.
4. Reed, J. S., Principles of Ceramics Processing. John Wiley & Sons Inc, New York (1995).
5. Carlson, J. D., Magnetorheological Fluid Actuators. Chapter in Adaptronics and Smart Structures (Ed. Janocha, H.), Springer-Verlag Berlin, (1999) pp. 180-95.
6. Carlson, J., D., MR Fluid and Devices in the Real World. *Int. Journal of Modern Physics B*, **19**, 7-9 (2005) pp. 1463-70.
7. Laeuger, J., Wollny, K., Stettin, H., Huck, S.; MR Fluid and Devices in the Real World. *Int. Journal of Modern Physics B*, **19**, 7-9 (2005) pp. 1463-70.
8. Zaun M., Design Concept for the Development of Cylinder Drives Based on Electrorheological Fluids, *Proc. of the 9th Intern. Conf. of Electrorheological Fluids and Magnetorheological Suspensions, Beijing 2004*, World Scientific Publishing, Singapore, (2005) pp. 915-21.

FUZZY MOVING SLIDING MODE CONTROL OF VEHICLE SUSPENSION FEATURING ER SHOCK ABSORBER

J. W. CHO, Y. S. PAENG, AND S. B. CHOI

Smart Structures and Systems Laboratory, Department of Mechanical Engineering, Inha University, Incheon 402-751, Korea
E-mail: seungbok@inha.ac.kr

K. S. KIM

Division of Mechanical and Automotive Engineering, Kongju National University, Chonan 330-717, Korea
E-mail: keysun@kongju.ac.kr

This paper presents a robust and superior control performance of a quarter-vehicle electrorheological (ER) suspension system. In order to achieve this goal, a moving sliding mode control algorithm is adopted, and its moving strategy is tuned by fuzzy logic. As a first step, ER damper is designed and manufactured for a commercial suspension system, and its field-dependent damping force is experimentally evaluated. The ER damper is then incorporated to a commercial quarter-vehicle suspension system, which is composed of sprung mass, spring and tire. After formulating the governing equation of motion for the quarter-vehicle ER suspension system, a stable sliding surface and moving algorithm based on fuzzy logic are formulated. The sliding mode controller is then synthesized and experimentally realized with a quarter-vehicle for the demonstration of a practical feasibility. Control performances such as vertical displacement and acceleration are evaluated in time and frequency domains under various road conditions such as bump and random road.

1 Introduction

Recently, vibration suppression of a vehicle system using semi-active suspension has been significantly increased. Though the passive suspension system featuring conventional oil damper provides design simplicity and cost-effectiveness, performance limitations are inevitable due to the uncontrollable damping force. On the other hand, the active suspension system can provide high control performance in wide frequency range. However, the active suspension requires high power consumption, many sensors, servo-valves, and sophisticated control logic. Consequently, one way to resolve these problems is to adopt the semi-active suspension system. The semi-active suspension system offers a desirable performance generally enhanced in the active mode without requiring large power consumption and expensive hardware. For the last decade, very attractive and effective semi-active suspension system featuring an electrorheological(ER) fluid has been proposed by many investigators [1-4].

However, in most of previous works on vehicle suspension system associated with ER fluid the dynamic bandwidth of the ER damper has not been considered for modeling and controller formulation. This is mainly due to that the response time of currently available ER fluids is fast enough to take into account for the body (1~2*Hz*) and wheel (10~ 15*Hz*) modes of a passenger vehicle. However, it is generally known that the dynamic bandwidth of the ER damper depends upon not only response time of the ER fluid itself but also design parameters of the hardware such as electrode gap and length. Therefore, the identification of the dynamic characteristics of the ER damper needs to be undertaken prior to formulating control algorithm. In addition, the sprung mass variation of the quarter vehicle should be incorporated with controller design in order to consider loading conditions as system uncertainty. Consequently, the robustness of the control performance should be guaranteed for the ER suspension system subjected to the dynamic characteristics of the ER damper and the sprung mass perturbation.

The robustness of the sliding mode controller (SMC) can be improved by shortening the time required to attain the sliding mode, or may be guaranteed during whole intervals of control action by eliminating the reaching phase. One easy way to minimize the reaching phase is to employ a large control input. However this will cause higher chattering which is undesirable in physical system, and also extreme sensitivity to unmodelled dynamics. Choi *et al.*[5] suggested a moving sliding mode controller (MSMC) to improve the performance and the robustness of the SMC. There are many superiorities of the SMC with the MSMC such as fast tracking performance, short reaching phase, and so on. However, the movement of the MSMC proposed by Choi *et al.*[5] depends on exact calculations of the surface equations to make the surface function have a value in a certain range. Hence, the generalization of the MSMC to the higher order systems is restricted by the complexity of the equations. To resolve this problem, a simplified MSMC based on the fuzzy theory (fuzzy moving sliding mode controller, FMSMC) is proposed [6].

In this study, we first design and manufacture a cylindrical ER damper, which is applicable for medium-sized vehicle. In addition, by regarding the sprung mass variation due to the loading conditions as system uncertainty, a moving sliding mode controller based on fuzzy logic which takes into account for parameter perturbations of the proposed quarter-vehicle ER suspension system is then designed. In order to demonstrate the robustness and performance of the proposed FMSMC, bump and random road responses are presented with respect to sprung mass perturbation in time and frequency domains. Furthermore, the control responses are compared with that of the conventional SMC.

2 Electrorheological Suspension System

2.1 Dynamic Modeling of Electrorheological Damper

A cylindrical type of ER damper shown in figure 1 is designed and manufactured in this study. In the absence of the electric field, the ER damper produces damping force only

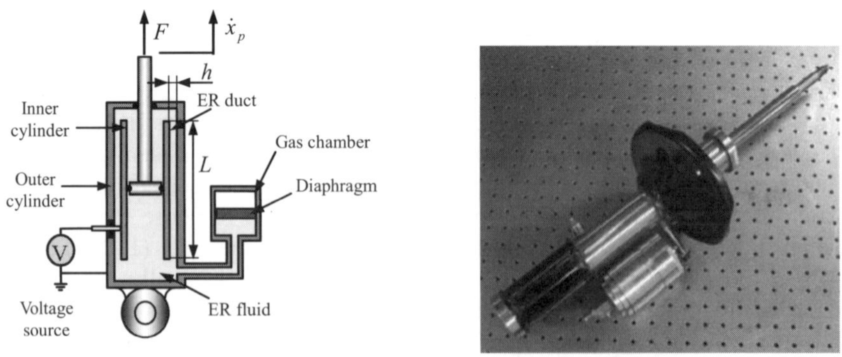

(a) schematic configuration (b) photograph

Figure 1. The proposed ER damper for passenger vehicle.

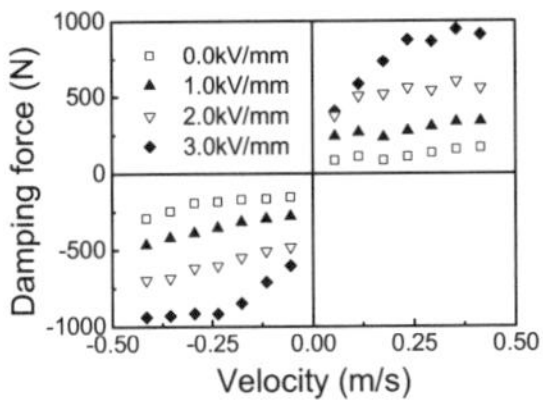

Figure 2. The field-dependent damping force

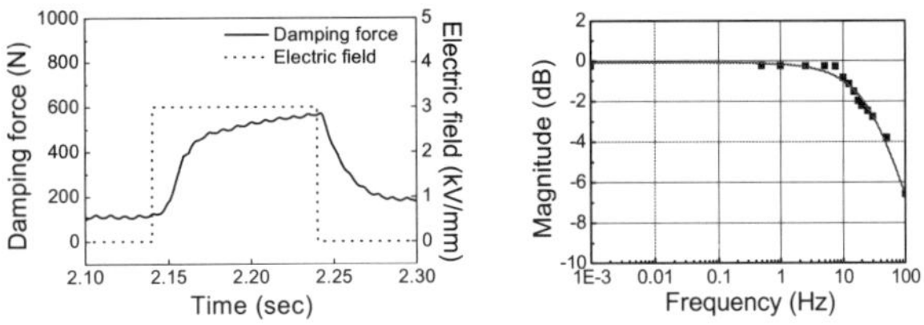

Figure 3. Damping force response in time and frequency domain

caused by the fluid viscous resistance. However, if a certain level of the electric field is supplied to the ER fluid domain, the ER damper produces additional damping force owing to the yield stress of the ER fluid. This damping force of the ER damper can be continuously tuned by controlling the intensity of the electric field. The damping force of the proposed ER damper can be obtained as follows [4].

$$F = k_e x_p + c_e \dot{x}_p + F_{ER} \tag{1}$$

where, k_e is the effective stiffness due to the gas pressure, c_e is the effective damping due to the fluid viscosity, x_p is the excitation displacement, and F_{ER} is the field-dependent damping force which is tunable as a function of applied electric field. The controllable damping force F_{ER} can be expressed by

$$F_{ER} = (A_p - A_r) 2 \frac{L}{h} \alpha E^{\beta} \operatorname{sgn}(\dot{x}_p) \tag{2}$$

where, A_p and A_r represent piston head and piston rod areas, respectively. $\operatorname{sgn}(\cdot)$ is a sign function, L is the electrode length, h is the electrode gap, and E is the electric field. The α and β are intrinsic values of the ER fluid to be experimentally determined. In this study, the field-dependent yield stress of the ER fluid is experimentally obtained by $876.3E^{1.38}$ Pa (Here the unit of E is kV/mm).

Figure 2 presents the damping force with respect to the piston velocity at various electric fields. This plot is obtained by calculating the maximum damping force at each velocity. The detailed experimental apparatus and procedure to obtain this result have been well described in [7]. It is clearly observed that the damping force is increased as the electric field increases. As a specific case, the damping force of 293*N* at piston velocity of 0.414*m/s* is increased up to 938*N* by applying the input voltage of 3.0*kV/mm*.

Figure 3 shows the measured step and dynamic response characteristic of the ER damper. It is obtained by applying constant input voltage (3.0kV/mm). It is identified that the response time of the ER damper is about 19ms and the dynamic response bandwidth of the ER damper is about 28Hz at -3dB. Thus, both the vehicle body mode (1~2Hz) and the wheel mode (10~13Hz) of the passenger vehicle can be effectively controlled by employing the proposed ER damper. From the results shown in figure 3, the static expression of the damping force and input field given by equation (2) should be modified to take into account for the dynamic characteristic as follow.

$$\tau \frac{d}{dt} F_{ER} + F_{ER} = (A_p - A_r) 2 \frac{L}{h} \alpha E^{\beta} \operatorname{sgn}(\dot{x}_p) \tag{3}$$

where, τ is the time constant of the ER damper. The dynamic model of the ER damper is to be incorporated with the suspension model.

2.2 Modeling of Quarter Vehicle ER Suspension System

In this study, a quarter-vehicle ER suspension system is established to evaluate the performance and robustness of control system. Figure 4 shows the quarter car model of the semi-active ER suspension system that has two degree-of-freedom. m_s and m_u represent the sprung mass and unsprung mass, respectively. The spring for the suspension is assumed to be linear and the tire is also modeled as linear spring component. Now, by considering the dynamic relationship, the state space model is expressed for the quarter-vehicle ER suspension system as follows:

$$\begin{aligned} \dot{\boldsymbol{x}} &= \boldsymbol{Ax} + \boldsymbol{B}u + \boldsymbol{L}z_r \\ y &= \boldsymbol{Cx} \end{aligned} \tag{4}$$

where,

$$\mathrm{x} = \begin{bmatrix} z_s & \dot{z}_s & z_u & \dot{z}_u & F_{ER} \end{bmatrix}^T$$

$$A = \begin{bmatrix} 0 & 1 & 0 & 0 & 0 \\ -\dfrac{k_s}{m_s + \Delta m_s} & -\dfrac{c_s}{m_s + \Delta m_s} & \dfrac{k_s}{m_s + \Delta m_s} & \dfrac{c_s}{m_s + \Delta m_s} & -\dfrac{1}{m_s + \Delta m_s} \\ 0 & 0 & 0 & 1 & 0 \\ \dfrac{k_s}{m_u} & \dfrac{c_s}{m_u} & -\dfrac{k_t + k_s}{m_u} & -\dfrac{c_s}{m_u} & \dfrac{1}{m_u} \\ 0 & 0 & 0 & 0 & -\dfrac{1}{\tau} \end{bmatrix}$$

$$B = \begin{bmatrix} 0 & 0 & 0 & 0 & \dfrac{1}{\tau} \end{bmatrix}^T, \quad C = \begin{bmatrix} 1 & 0 & 0 & 0 & 0 \end{bmatrix}, \quad L = \begin{bmatrix} 0 & 0 & 0 & \dfrac{k_t}{m_u} & 0 \end{bmatrix}^T,$$

$$u = (A_p - A_r) 2 \frac{L}{h} \alpha E^{\beta} \operatorname{sgn}(\dot{x})$$

where, Δm_s is the sprung mass perturbation. k_s is the total stiffness coefficient of the suspension, and it includes the effective stiffness k_e in equation (1). c_s is the damping coefficient of the suspension and it is assumed to be equal to c_e. In addition, k_t is the stiffness coefficient of the tire. z_s, z_u and z_r are the vertical displacement of sprung mass, unsprung mass, and excitation, respectively.

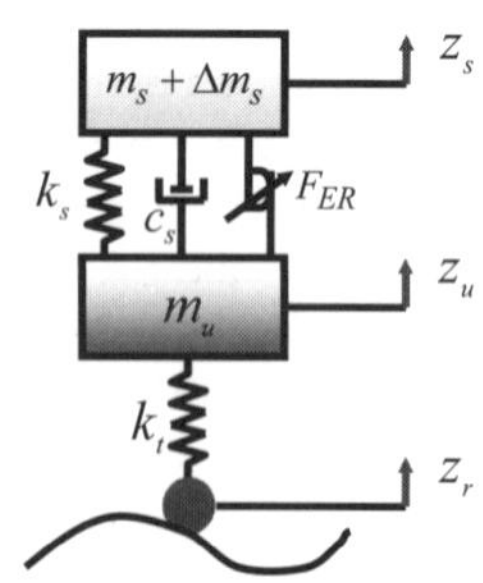

Figure 4. Mechanical model of the quarter-vehicle ER suspension system.

3 Controller Design

In this study, a FMSMC is adopted to suppress the vibration level. As a first step, we defined the following sliding surface s:

$$s = c_1 x_1 + c_2 x_2 + c_3 x_3 + c_4 x_4 \tag{5}$$

where, c_i (i =1,2,3,4,5) are coefficients of sliding surface and have fixed values satisfied stable sliding surface. It is required to reduce reaching phase to improve the performance and robustness of the SMC. Reaching time can be successfully reduced in case that sliding surface is rotated or shifted by considering reaching phase. In order to accomplish this goal, a FMSMC, which can change the coefficients and intercepts of sliding surface by fuzzy turning which takes into account for location of reaching phase, is proposed.

In this study, the coefficient of sprung mass displacement x_1 is considered with priority while other coefficients of sliding surface are fixed. Rotating algorithm, which can change the sliding surface c_1 as a function of x_1, is applied to reduce the reaching phase. However, it is hard to expect that the robustness is enhanced in case that initial condition of reaching phase is located in quadrant 1 or quadrant 3. Hence, shifting algorithm, which can change the intercept α as a function of x_1, is adopted and determined according to the relative place of sliding surface, and this will sufficiently reduce the reaching phase. Therefore, equation (5) can be rewritten as follows [8]:

$$s = c_1(x_1)x_1 + c_2 x_2 + c_3 x_3 + c_4 x_4 + \alpha(x_1) \tag{6}$$

where, c_1 and α are determined by turning of fuzzy logic which takes into account for displacement and direction of sprung mass displacement of x_1. In this case, the sliding surface should maintain stability, although c_1 and α is changed according to time. In order to reduce reaching time in the rotating algorithm, c_1 should be small in case that absolute value of x_1 is large. Also, c_1 should be high in case of the opposite case. In the letter case, the sprung displacement x_1 can be converged toward 0 more quickly. In the shifting control algorithm, absolute value of α should be high in case that absolute value of x_1 is large and vice versa. Furthermore, sliding surface is moved upward for positive x_1 and sliding surface is moved downward for the opposite case.

Consequently, we can formulate the following control input of the proposed FMSMC, which is combined rotating and shifting algorithm [7]:

$$\begin{aligned} u(t) = -\frac{\tau}{c_5}\Bigg\{ & \left(c_1 - \frac{\Delta c_1 \chi_{1r} \sinh(\chi_{1r})}{\cosh^2(\chi_{1r})} - \frac{(\alpha_{\max}/\sigma_s)}{\cosh^2(\chi_{1s})}\right)x_2 + \frac{c_2}{m_s}(-k_s x_1 - c_s x_2 + k_s x_3 + c_s x_4 - x_5) \\ & + c_3 x + \frac{c_4}{m_u}(k_s x_1 + c_s x_2 - (k_t + k_s)x_3 - c_s x_4 + x_5) + \frac{c_5}{\tau}x_5 \Bigg\} - k\,\mathrm{sgn}(s) \end{aligned} \tag{7}$$

where, k stands for discontinuous gain which is positive number. The control input determined from the FMSMC is to be applied to the ER damper depending upon the motion of suspension travel.

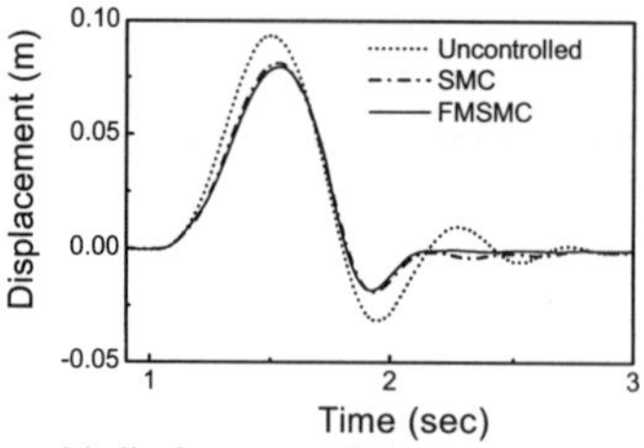

(a) displacement of the sprung mass

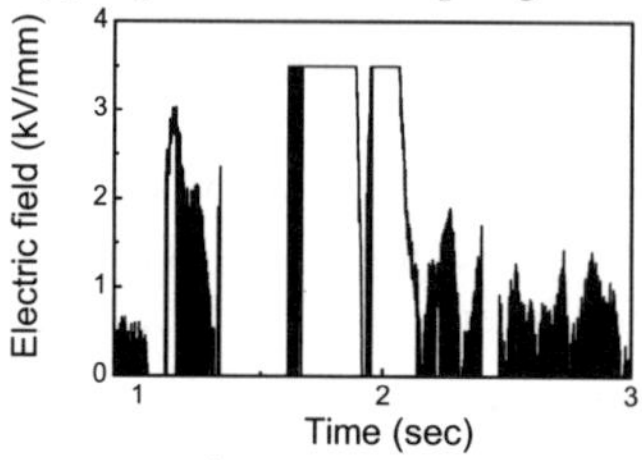

(b) electric field

Figure 5. Bump response of the ER suspension system with sprung mass:310kg.

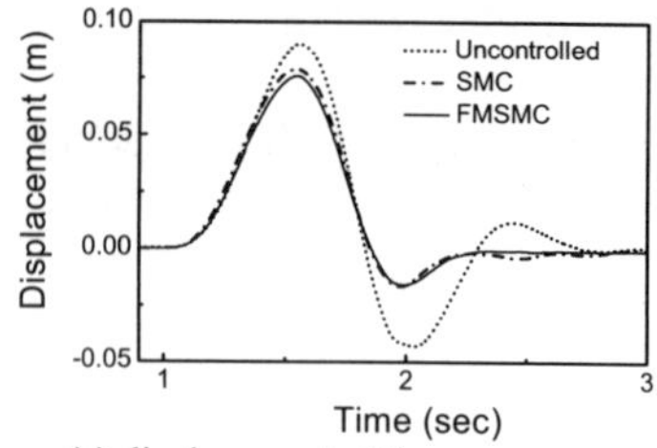

(a) displacement of the sprung mass

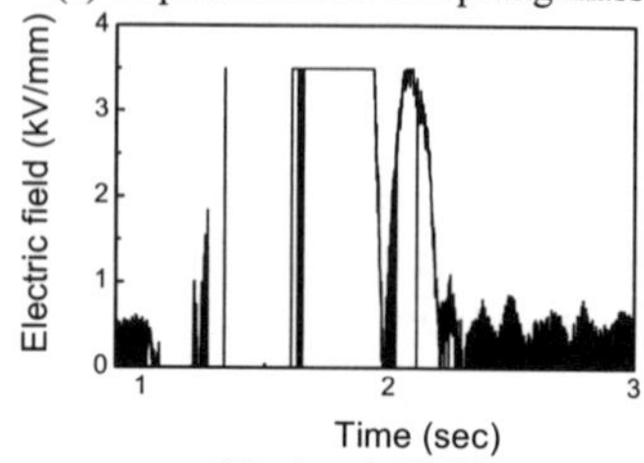

(b) electric field

Figure 6. Bump response of the ER suspension system with sprung mass:380kg.

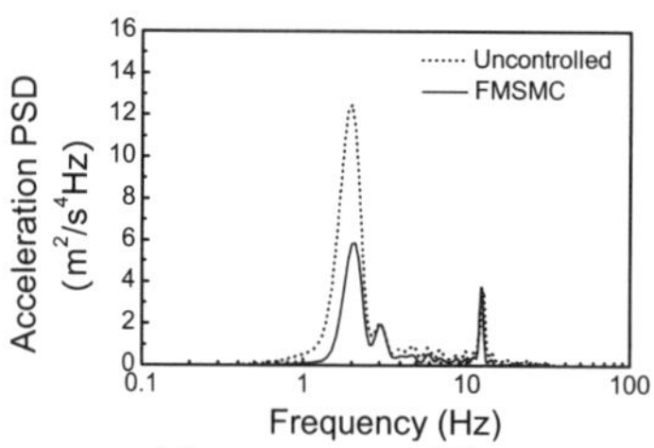

(a) sprung mass:310kg

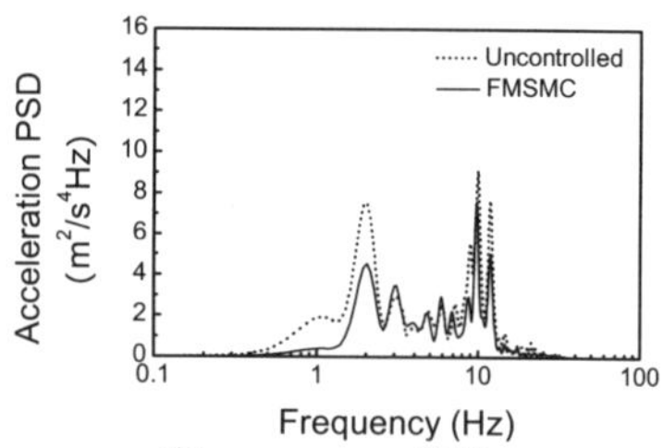

(b) sprung mass:380kg

Figure 7. Random road responses of the ER suspension system.

4 Results and Discussion

Figure 5 and 6 present time responses of the ER suspension system for the bump excitation. In this study, the system parameters of the quarter-vehicle ER suspension system are chosen on the basis of the conventional suspension system for a passenger vehicle. Also, it is assumed that the sprung mass without any additional mass (m_s =310*kg*) is the nominal value for the proposed control system. On the other hand, in order to investigate the robustness of the proposed FMSMC, the sprung mass is added the mass of a driver by 70*kg* (m_s =380*kg*) for perturbed quarter-vehicle system in consideration of the mass of a driver. As shown in figure 5, it can be obviously found that unwanted vibrations induced from the bump excitation are well suppressed by adopting the proposed FMSMC to the ER suspension system better than one of the SMC. This is mainly due to that reaching phase can be decreased

by applying the FMSMC. Figure 6 shows time responses of the proposed system with parameter perturbations. From the results it is seen that the vertical displacement is well reduced by employing the control electric field determined from proposed controller even in the presence of parameter perturbations. Figure 7 presents controlled frequency responses using the FMSMC under random road excitation. The frequency responses are obtained from power spectral density (PSD) for the vertical acceleration of the sprung mass. As expected, the PSD for the vertical acceleration is considerably reduced in the neighborhood of body resonance (1-2*Hz*) by applying control input.

5 Conclusions

Vibration control performance of the ER suspension system for a passenger vehicle was evaluated via fuzzy moving sliding mode controller (FMSMC). A cylindrical ER damper was designed and manufactured. After evaluating the field-dependent damping characteristics of the ER damper, a quarter-vehicle suspension system was then constructed and its governing equations of motion were derived. In order to obtain a favorable control performance of the ER suspension system subjected to parameter perturbations, a FMSMC was designed and realized to quarter-vehicle ER suspension system. The FMSMC can improve the performance and robustness of the SMC by reducing reaching phase. It is demonstrated vibration control performance under various road conditions in time and frequency domains. The results presented in this work are self-explanatory justifying that the proposed FMSMC is very effective for vibration isolation of a passenger vehicle with ER suspension system.

Acknowledgement

This work was supported by Inha University Research Grant. This financial support is gratefully acknowledged.

References

1. Petek, N. K., An Electronically controlled shock absorber using electro-rheological fluid. *SAE Technical Paper Series* 920275 (1992).
2. Sims, N. D., Stanway, R. and Beck, S. B. M., Proportional feedback control of an electro-rheological vibration damper. *J. of Intelligent Material Systems and Structures*, 8 (1997) pp. 426~ 433.
3. Choi, S. B., Choi, Y. T., Chang, E. G., Han, S. J. and Kim, C. S., Control characteristics of a continuously variable ER damper. *Mechatronics*, **8** (1998) pp. 143~ 161.
4. Choi, S. B., Choi, Y. T. and Park, D. W., A sliding mode control of a full-car electrorheological suspension system via hardware-in-the-loop simulation. *ASME J. of Dynamics, Measurement and Control*, **122** (2000) pp. 114~ 121.
5. Choi, S. B., Cheong, C. C. and Park, D. W., Moving switching surfaces for robust control of second-order variable structure systems. *International Journal of Control*, **58** (1993) pp. 229~ 247.
6. Ha, Q. P., Rye, D. C. and Durrent-Whyte, H. F., Fuzzy moving sliding mode control with application to robotic manipulators. *Automatica*, **35** (1999) pp. 607~ 616
7. Cho, J. W., Fuzzy moving sliding mode control of vehicle suspension featuring ER shock absorber. *M. S. Dissertation*, Inha University, 2006.
8. Park, D. W. and Choi, S. B., Moving sliding surface for high-order variable structure systems. *International Journal of Control*, **72** (1999) pp. 960~ 970.

COLE-COLE ANALYSIS ON DIELECTRIC SPECTRA OF ELECTRORHEOLOGICAL SUSPENSIONS

H. J. CHOI AND C. H. HONG

Department of Polymer Engineering and Science, Inha University, Incheon 402-751, Korea
E-mail: hjchoi@inha.ac.kr

M. S. JHON

Department of Chemical Engineering, Carnegie Mellon University, Pittsburgh 15213, USA

Along with rheological properties, dielectric spectra of the ER fluids were found to provide additional information on both analyzing their electrical polarization properties and interpreting their flow behavior. Dielectric properties of three different material systems such as spherical-monodisperse polymer microshperes (PAPMMA) consisting of a poly(methyl methacrylate) (PMMA) core and a polyaniline (PANI) shell, single-doped and mesoporous-doped TiO_2 system and copper phthalocyanine-doped mesoporous TiO_2 suspensions based ER fluid were analyzed via dielectric spectra of permittivity and loss factor as a function of the frequency, and Cole-Cole plots. Polarizability which is defined as a difference of dielectric constant at both zero frequency and infinite frequency was found to be a proper parameter to understand a better ER fluid. Using the Cole-Cole plot we were able not only to fit the dielectric spectra, but also to deduce the achievable polarizability by explaining the higher ER performance of the ER fluids.

1 Introduction

Electrorheological (ER) fluids are commonly suspensions of particles having higher dielectric constant and/or conductivity than that of suspending medium fluids with a low dielectric constant and a low viscosity. Rheological properties of ER fluids are drastically changed by an applied electric field from liquid-like to solid-like [1-5]. This effect is reversible, implying the fibril structures are not maintained when the electric field is turned off. ER activity has been related to interfacial polarization via the Wagner model, which is connected with the characteristics of dielectric spectra, i.e., the dependence of dielectric constant and loss factor or loss tangent on frequency. According to the polarization model, the suspending particles are polarized due to dielectric mismatch between the particles and the suspending fluid under an applied electric field. The interaction among the polarized particles causes them to arrange themselves in the form of strings along the electric field direction [2].

Khusid and Acrivos [6] related thermodynamic properties to the dielectric spectra based on microstructures of ER fluids. Although some discrepancies with the experimental results still exist and the rheological study of ER fluids was not explicitly treated in their report, their theory may explain the currently observed ER phenomena, especially the ones that cannot be described by previous polarization and conduction models. Furthermore, Hao et al. [7] showed that the slow polarization, especially the interfacial polarization is responsible for the conductive, dielectric, and surface properties of dry-base polymer or inorganic material based ER fluids. A large interfacial polarization would facilitate the particles to attain a large amount of charges on the surface, then leading to the turn of particle along the direction of an electric field to form a fibrillation structure. Therefore, the strength of fibrillation chains is determined by the particle polarization force of particle dielectric constant [7]. Considering previous experimental or theoretical works, the performance of an ER fluid crucially depends on the electrical properties of the suspended particles [8]. Dielectric spectra thereby become one of the important characteristics to analyze the ER performance.

On the other hand, in the theoretical treatment on ER fluids, suspended ER particles are normally assumed to be monodisperse and spherical. However, the experimentally used particles in ER fluids are commonly polydisperse and irregular. These morphological factors of the dispersed phase are one of the most critical parameters in the development of the theoretical modeling. Certain previous experiments used these idealized particles, including metal particles having nonconductive metal particles having nonconductive metal oxide shell [9], PMMA with double layered shell [10], and semiconductive particles having silicate shell [11]. Nonetheless, their essential ER characteriziations along with dielectric spectra were not fully analyzed.

Previously studied ER performances of monodisperse and spherical particles, which are only conductive on the surface, dispersed in silicone oil were reanalyzed based on Cole-Cole plot along with other two different inorganic systems. Three different sizes of PMMA microspheres were used as core material, and the PANI shell was prepared by in situ polymerization of aniline in the presence of core articles. In addition to our own system of PAPMMA, we analyzed dielectric spectra of two other systems of doped TiO_2 based ER fluids [12] and CuPC-doped mesoporous TiO_2 ER fluids [13].

2 ER Fluid Systems

1) Conducting Polyaniline Shell and PMMA Microsphere as Core System

As we previously reported [14], three different sizes of PMMA microparticles (diameter=2, 4.5 and 9μm) were prepared via a dispersion polymerizaion method. The monodisperse particles obtained were washed sufficiently by methanol and then dispersed in water prior to the PANI polymerization. ER fluids were prepared using the dried PAPMMA particles dispersed in silicone oil. Conductivities and dielectric constants of each PAPMMA were measured using pressed disks of PAPMMA samples. The thickness of the PANI coating in PAPMMAs was obtained to be 52, 131, and 261nm for PAPMMA2, PAPMMA4.5, and PAPMMA9, respectively. The DC conductivity and dielectric constant measured at 1 kHz were 8.02×10^{-9}, 2.54×10^{-9}, 5.14×10^{-9}S/cm, and 15.8, 20.7, 35.9 for PAPMMA2, PAPMMA4.5, and PAPMMA9, respectively. ER measurements were carried out using a rotational rheometer (Physica MC 120) with a Couette geometry. Dielectric relaxations of all prepared ER fluids were observed using the HP 4284A precision LCR meter with HP 16452A liquid test fixture at room temperature.

2)Single-doped and mesoporous-doped TiO_2 System

Yin and Zhao [12] prepared single-doped and mesoporous-doped TiO_2 based ER system, starting from Pluronic F-127 ($(PEO)_{106}(PPO)_{70}(PEO)_{106}$) dissolved in ethanol by stirring. TiCl4 and dilute water containing CeCl3 (Ce/Ti=7 mol%) were added with further stirring to obtain sol. The sol was dried in an open dish, and further condensed, and finally calcined at 450°C to remove the template.

3) Copper phthalocyanine-doped mesoporous TiO_2 suspensions

CuPC-doped mesoporous TiO_2 was synthesized by Di et al. [13]. Cetyltrimethylammonium bromide (CTAB) was employed as cationic template and tetrabutyl titabate ($Ti(OBu)_4$) was employed as the Ti source. To gain the mesoporous structure, the mole ratio of Ti to CTAB was about 3. $Ti(OBu)_4$–$TEAH_3$–CTAB–NaOH solution was stirred at room temperature for aging 12h. Thereafter, the solid material was filtered off, washed with deionized water.

3 Dielectric Analysis

Both orientational polarization (Debye polarization) and the interfacial polarization (Wagner polarization) were reported to be generated in ER suspension under an electric field, depending on the experimental condition [7]. Hao *et al.* [7] stated that they retard with an electric-field frequency, and contribute to the dielectric loss of the whole system. Therefore, the high dielectric loss to some extent means a large proportion of the orientational polarization and the interfacial polarization.

Meanwhile, dielectric relaxation processes have been generally described by a following Havriliak-Negami equation [15]:

$$\varepsilon^* = \varepsilon' + i\varepsilon'' = \varepsilon_\infty + \frac{\Delta\varepsilon}{\left\{1 + (i\omega\lambda)^{1-\alpha}\right\}^{\beta}} \tag{1}$$

where ε_0 is ε' as $\omega \to 0$ ($\varepsilon_0 = \lim_{\omega\to 0}\varepsilon^*(\omega)$), ε_∞ is ε' as $\omega \to \infty$ ($\varepsilon_\infty = \lim_{\omega\to\infty}\varepsilon^*(\omega)$) and $\Delta\varepsilon = \varepsilon_0 - \varepsilon_\infty$ is the polarizability. β and (1-α) describe the 'skewdness' and the 'depression' of the function or the broadness of the relaxation time distribution, respectively. Special cases can be deduced from this equation are the Cole-Cole function (β =1), the Davidson-Cole function (α =0), and the Debye function (α =0 and β =1) which has a single relaxation time [15]. The Cole-Cole function is expressed as follows:

$$\varepsilon^* = \varepsilon_\infty + \frac{\Delta\varepsilon}{1 + (i\omega\lambda)^{1-\alpha}} \tag{2}$$

Note that the Cole-Cole plot has been successively adopted to study the dielectric process not only for various systems of magnetic fluid [16], tobacco mosaic virus [17], and resin beads [18] but also for ER suspensions [19]. Our approach of using Cole-Cole interpretation along with the polarizability in this study has been also expanded to other various previously reported ER fluids such as PANI, microencapsulated PANI, mesoporous material and inorganic particles [20].

On the other hand, the dielectric relaxation time (λ) resulting from the interfacial polarization for an isolated process or its mean value for a distributed process can be calculated from the critical frequency, fmax, at which the value of ε'' is a maximum or at which the value of ε' shows the steepest descent with frequency [15]. The connection between the dielectric measurement made in the frequency domain and that made in the time domain is given by the Fourier transform relation shown in Eq. (3):

$$\lambda = \frac{1}{2\pi f_{max}} \tag{3}$$

The relaxation times calculated from Eq. (3) are given in Table 1. These are typical results for the interfacial polarization of suspensions including ER fluids consisting of polarizable-dispersed phases in an insulating media.

4 Results and Discussion

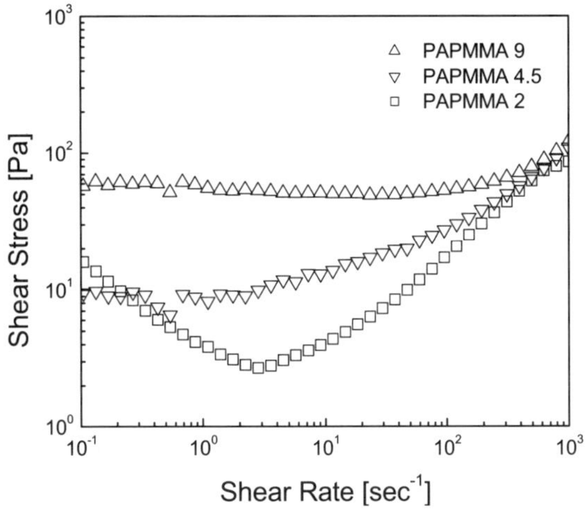

Figure 1. Flow curves for three PAPMMA based ER fluids (10 vol%) at an electric field strengths of 3 kV/mm.

The synthesized PAPMMA suspensions showed typical ER behaviors exhibiting a yield stress under an applied electric field. The flow curves for three different ER fluids with a particle concentration of 10 vol% under an applied electric field strength of 3 kV/mm. The enhancement of shear stress becomes larger with the size of PAPMMA particles within a broad range of shear rate.

The particle size effect on ER properties was also analyzed via dielectric spectra shown in Figure 2 with permittivity (ε′) and loss factor (ε") as a function of the frequency and Cole-Cole plots for PAPMMA based ER fluids. These are typical results for the interfacial polarizable, dispersed phases in an insulating media. Lines in Figure 2 are fits of the Cole-Cole formula (Eq. 2).

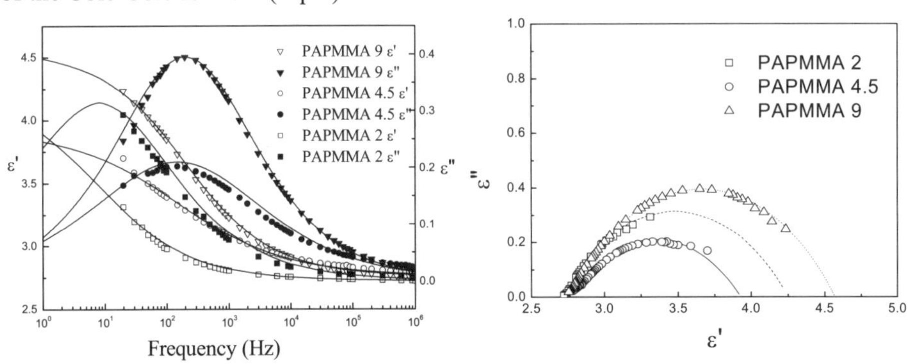

Figure 2. Dielectric spectra of ER fluid (left) and Cole-Cole fitting curves of ER fluids (right).

The parameters for each ER fluid are summarized in Table 1, in which $\Delta\varepsilon = \varepsilon_0 - \varepsilon_\infty$ is an achievable polarizability in ER fluids. Relaxation time (λ) for interfacial polarization of ER fluids is known to be related to yield stress and stress enhancement under the applied electric field. As the λ within adequate range gets smaller, the higher stress enhancement is achieved. In addition, the Δε dependence on ER properties also provides a positive effect on ER performance.

Table 1. Various parameters from Cole-Cole fitting.

Sample	ε_0	ε_∞	$\Delta\varepsilon = \varepsilon_0 - \varepsilon_\infty$	λ (ms)	α
PAPMMA 2	4.25	2.72	1.53	20	0.50
PAPMMA 4.5	4.05	2.76	1.29	1.4	0.61
PAPMMA 9	4.57	2.75	1.82	0.79	0.48
Mesoporous-doped TiO_2	22.0	5.17	16.83	0.16	0.32
Single-doped TiO_2	13.3	4.41	8.89	1.6	0.42
CuPC-doped TiO2 (0.10)	105.0	70	35	2.76	0.38
CuPC-doped TiO2 (0.15)	133.2	75.5	57.7	0.051	0.46

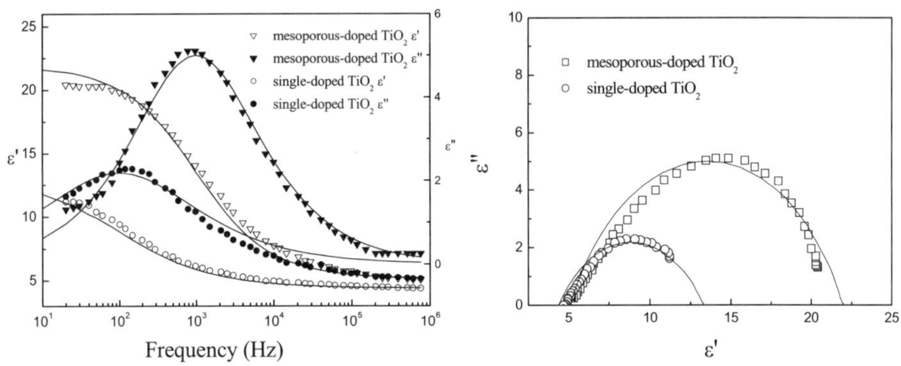

Figure 3. Dielectric spectra of mesoporous-doped TiO_2 ER fluid and single-doped TiO_2 ER fluid (left) and Cole-Cole fitting curves of mesoporous-doped and single-doped TiO_2 (right).

In a recent report, Yin and Zhao [12] showed that both single-doped and mesoporous-doped TiO_2 based ER fluids are superior to those of their counterparts without doping and explained their ER mechanism based on dielectric spectra. It is found that the adopting Cole-Cole plot not only fits the dielectric spectra of both mesoporous-doped TiO_2 templated ER fluid and single-doped TiO_2 ER fluid but also provides a better explanation on their polarizability [21] (see Figure 3) .

Table 1 indicate that the λ of mesoporous-doped TiO_2 is obtained to be smaller than that of single-doped TiO_2, corresponding to their yield stress characteristics. In addition, the $\Delta\varepsilon$ shows the higher value in mesoporous-doped TiO_2 based ER fluids which shows the higher yield stress. From the higher $\Delta\varepsilon$, the stiff chain fibrillar structure shows the better ER behavior under electric field strengths.

In addition, another novel water-free ER system consisting of conductive molecules of CuPC-doped mesoporous TiO_2 suspended in insulating oil reported, was also found to be fitted well with Eq. 2, explaining its ER mechanism based on the interfacial particle polarization along with the dielectric spectra [22].

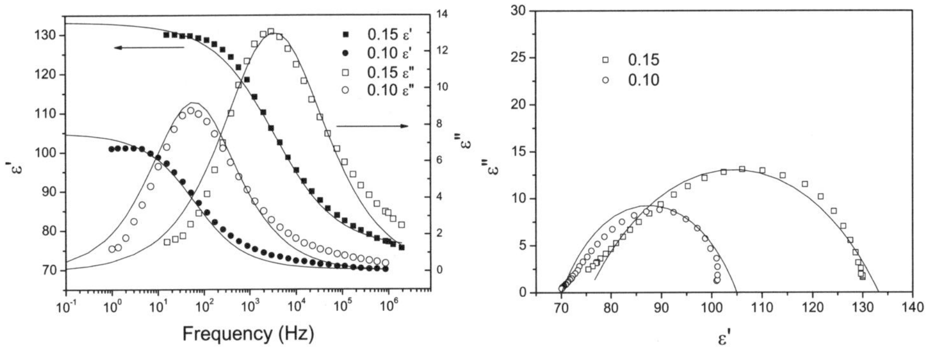

Figure 4. Dielectric spectra of CuPC-doped TiO_2 (0.10 and 0.15) ER fluid and single-doped TiO_2 ER fluid (left) and Cole-Cole fitting curves of CuPC -doped TiO_2 (right).

5 Conclusion

Monodisperse and spherical PMMA particles coated by conducting PANI were investigated. ER properties of PAPMMA particles increased with the PMMA particle size. Dielectric spectra confirmed the different flow behavior for the PAPMMA2 system. In addition, it is found that the dielectric polarization that controls the ER effect via the interfacial polarization can be fitted well using the Cole-Cole plot. Furthermore, single-doped and mesoporous-doped TiO_2 system and copper phthalocyanine-doped mesoporous TiO_2 suspensions based ER fluid were analyzed via dielectric spectra of permittivity and loss factor as a function of the frequency, and Cole-Cole plots. It was found that using the Cole-Cole plot we were able not only to fit the dielectric spectra, but also to deduce the achievable polarizability by explaining the higher ER performance of the ER fluids.

6 Acknowledgements

This study was supported by KOSEF via the Applied Rheology Center, Korea.

References

1. Zhao X. P. and Yin J. B., Advances in Electrorheological Fluids based on Inorganic Dielectric Materials, *J. Ind. Eng. Chem.* **12** (2006) pp. 184-198.
2. Hao T., Electrorheological fluids, *Adv. Mater.* **13** (2001) pp. 1847-1857.
3. Cho M. S., Kim J. W., Choi H. J., Webber R. M., and Jhon M. S., Electrorheological characteristics of polyaniline and its copolymer suspensions with ionic and nonionic substituents, *Colloid Polym. Sci.* **278** (2000) pp. 61-64.
4. Quadrat O. and Stejskal J., Polyaniline in Electrorheology, *J. Ind. Eng. Chem.* 12 (2006) pp. 353-361.
5. Choi H. J., Cho M. S., Kim J. W., Kim C. A., and Jhon M. S., A Yield Stress Scaling Function for Electrorheological Fluids, *Appl. Phys. Lett.* **78** (2001) pp. 3806-3808.

6. Khusid B. and Acrivos A., Effects of Conductivity in Electric-Field-Induced Aggregation in Electrorheological Fluids, *Phys. Rev. E* **52** (1995) pp. 1669-1693.
7. Hao T., Kawai A., and Ikazaki F., Mechanism of the Electrorheological Effect: Evidence from the Conductive, Dielectric, and Surface Characteristics of Water-Free Electrorheological Fluids, *Langmuir* **14** (1998) pp. 1256-1262.
8. Boissy C., Atten P., and Foulc J. N., On the Role of Conductivities and Frequency in the Electrorheological Effect, *J. Intell. Mater. Syst. Struct.* 7 (1996) pp. 599-603.
9. Davis L. C., The Metal-Particle Insulating Oil System - an Ideal Electrorheological Fluid, *J. Appl. Phys.* **73** (1993) pp. 680-683.
10. Saito T., Anzai H., Kuroda S., and Osawa Z., Electrorheological Particles Composed of Polymer Core with Controlled Diameter and Electro-conductive/ Nonconductive Double Layers Shell, *Int. J. Mod. Phys. B* **13** (1999) pp. 1689-1696.
11. Sakurai R., See H., Saito T., Asai S., and Sumita M., Suspension of Layered Particles: an optimum electrorheological fluid for dc applications, *Rheol. Acta* **38** (1999) pp. 478-483.
12. Yin J. B. and Zhao X. P., Giant Electrorheological Activity of High Surface Area Mesoporous Cerium-doped TiO_2 Templated by Block Copolymer, *Chem. Phys. Lett.* **398** (2004) pp. 393-399.
13. Di K., Zhu Y., Yang X. and Li C., Electrorheological behavior of copper phthalocyanine-doped mesoporous TiO_2 suspensions, *J. Colloid Interface Sci.* **294** (2006) pp. 499-503.
14. Cho M. S., Cho Y. H., Choi H. J., and Jhon M. S., Synthesis and Electrorheological Characteristics of Polyaniline-Coated Poly(methyl methacrylate) Microsphere: Size Effect, *Langmuir* **19** (2003) pp. 5875-5881.
15. Asami K., Characterization of Heterogenous Systems by Dielectric Spectroscopy, *Prog. Polym. Sci.* **27** (2002) pp. 1617-1659.
16. Fannin P. C., Characterisation of Magnetic Fluids, *J. Alloys Compounds* **369** (2004) pp. 43-51.
17. Ermolina I., Morgan H., Green N. G., Milner J. J., and Feldman Yu., Dielectric Spectroscopy of Tabacco Mosaic Virus, *Bioch. Biophys. Acta* **1622** (2003) pp. 57-63.
18. Chen Z. and Zhao K. S., Dielectric Analysis of Macroporous Anion-exchange Resin Beads Suspensions, *J. Colloid Interface Sci.* **276** (2004) pp. 85-91.
19. Lengalova A., PavlinekV, Saha P., Stejskal J., Kitano T., and Quadrat O, The Effect of Dielectric Properties on the Electrorheology of Suspensions of Silica Particles Coated with Polyaniline, *Physica A* **321** (2003) pp. 411-424.
20. Fang F. F., Sung J. H., and Choi H. J., Shear Stress and Dielectric Characteristics of Polyaniline/TiO_2 Composite-Based Electrorheological Fluid, *J. Macromol. Sci.:B Phys.*, **45** (2006) pp. 923-932.
21. Park B. J., Sung J. H., Lee I. S. and Choi H. J., Comment on "Giant electrorheological activity of high surface area mesoporous cerium-doped TiO_2 templated by block copolymer", *Chem. Phys. Lett.* **414** (2005) pp. 525-526
22. Hong C. H., Park B. J. and Choi H. J., Comment on "electrorheological behavior of copper phthalocyanine-doped mesoporous TiO_2 suspensions", *J. Colloid Interface Sci.* **300** (2006) pp. 818-819.

MAGNETORHEOLOGY OF CARBONYL-IRON SUSPENSION WITH IRON/ MCM-41 ADDITIVE

F. F. FANG, B. J. PARK, AND H. J. CHOI

Department of Polymer Science and Engineering, Inha University, Incheon 402-751, Korea
E-mail: hjchoi@inha.ac.kr

W. S. AHN

Department of Chemical Engineering, Inha University, Incheon 402-751, Korea

Iron containg hexagonal mesoporous silica particle (Fe-MCM-41) was prepared and adopted into carbonyl-iron (CI) based magnetorheological (MR) suspension as an additive to improve the sedimentation problem of the CI based MR fluid. Structural properties and morphology of the synthesized Fe-MCM-41 particles were observed using SEM. Their MR properties such as oscillation characteristics and flow response (shear stress and shear viscosity) were examined via a rotational rheometer in parallel plate geometry equipped with a magnetic field supplier under external magnetic field strengths ranging from 0 to 257 kA/m. The addition of Fe-containing mesoporous particles into CI suspension was found to improve not only MR behaviors but also sedimentation problem of the CI based MR fluid.

1 Introduction

Magnetorheological (MR) fluids are considered as smart colloidal suspensions which consist of ferro- or ferri-magnetic particles with a size on the order of several microns dispersed in a nonmagnetic fluid medium. When exposed to an external magnetic field, the fluid transforms from a freely-flowing liquid-like state to a solid-like state within milliseconds, because the magnetic dispersed particles attract each other to form fibril structures aligning along the magnetic field direction, consequently inhibiting the fluid flow [1, 2]. The unique focus has been placed on the rapidness, reversibility and tunability of the phase transition process. Meanwhile, the rheological properties (shear viscosity, yield stress, storage modulus, etc.) of the fluid flow are changed dramatically. The yield stress which is investigated to be a function of magnetic field strength represents the maximum in the shear stress versus strain rate relationship, since the gel-like structure will collapse when the shear stress reaches to this maximum. Many researches on the MR fluid focused on the yield stress as a function of an applied magnetic field strength for the optimum formulation in each specific application. As it is well known, the MR fluids are similar to electrorheological (ER) fluids regarding on the mechanism and their changes of structure [3-5]. However, the MR fluid is proved to be much stronger, stable and facile to be utilized. Thereby increasing research efforts in employing MR fluid to control mechanical elements have been undertaken since its first discovery [6]. As for the special merits of MR fluid in large mechanical loads, the potential applicability of the MR fluid has stimulated considerable research activities in developing controllable shock absorbers, automotive damping devices, and other vibration control systems [7-11]. However, many crucial negative factors of MR fluid, such as serious sedimentation problem, abrasion and lower saturated magnetization ought to be resolved to satisfy its commercial applications.

Meanwhile, much attention has been paid on finding high performance MR materials [12, 13], and several merits of magnetic carbonyl iron (CI) particles make them be competent as a standard MR material [14, 15]. However, their high density leads to undesirable sedimentation problem during the MR experiment. Furthermore, abrasion on both the device and the magnetic particles also becomes a main issue. To improve all these drawbacks, not only some additives, such as thixotropic agents or surfactant but also the encapsulation technology using polymeric materials is often introduced to resolve the severe sedimentation problem via being dispersed between the micron-sized CI spheres.

Recently, MCM-41 particles with hexagonal mesoporous structure have been well documented for their various catalytic reactions and other emerging technologies including solid state nanoreactors for the preparation of metal and metal oxide [16] and ER application [17-19]. These silica phases possess hexagonally ordered system of mesochannels with a diameter ranging from 2 to 50 nm and a length achieving several micrometers. In this study, Fe-MCM-41 mesoporous particles containing Fe were synthesized and applied into CI based MR suspension as an additive to improve the sedimentation problem. The oscillatory performances and flow responses (shear stress and shear viscosity) of the system with and without Fe-MCM-41 additive were investigated. The sedimentation problem was also examined by adding the mesoporous Fe-MCM-41 particles into CI suspension.

2 Experimental

Cationic surfactant of CTMABr (cetyltrimethyl ammonium bromide, Aldrich) as the structure directing agent, and iron(Ⅲ) nitrate, fumed silica, and sodium hydroxide as the substrates were used. The colloidal silica and fumed silica were adopted as the silica source and iron(Ⅲ) nitrate ($Fe(NO_3)_3 9H_2O$, 98%, Aldrich) were used as the iron source. The pH of the medium was maintained using NaOH. The aqueous solution of CTMABr was prepared by dissolving it in deionized water and stirring the mixture for 15 min. The molar ratio of synthesis substrates was taken as SiO_2: CTMABr: NaOH: $Fe(NO_3)_3$=1: 0.2: 0.5: 0.04. After $Fe(NO_3)_3$ added into the solution, it was heated to 423 K. After 7 days of hydrothermal heating in oven, the batch was cooled to room temperature and pH of this mixture was adjusted to 9 using acetic acid. The sample was dried for 24 h at room temperature. To remove the surfactant, the product was calcined for 5 h at 823 K with O_2 flow [20].

In order to prepare MR fluid, we adopted CI (average particle size: 4.5-5.2 μm, density: 7.86 g/cm^3) as dispersion phase. For the additive component and nonmagnetic suspending medium, Fe-MCM-41 and mineral oil (MO, Aldrich, Milwaukee, WI, USA) were utilized. Two types of MR fluids with and without Fe-MCM-41 particles at a fixed CI concentration (80 wt% or 30 v/v%) were prepared, and the amount of additive was adjusted to 3 wt% with respect to the continuum phase (MO and Fe-MCM-41).

MR characterizations were performed at 20°C using a rotational rheometer (Physica MCR 300, Stuttgart, Germany) equipped with a magnetorheological device (MRD 180). A parallel-plate measuring system was made of non-magnetic metal to prevent the occurrence of radial magnetic force components on the shaft of the measuring system. The magnetic field direction was set to be perpendicular to the flow direction.

3 Results and discussion

The morphologies of pure Fe-MCM-41 particles are depicted in Figure 1. The size of primary Fe-MCM-41 particle was less than 2 micron even though that of the agglomerate of the Fe-MCM-41 was around 8 micron, showing a shape of twisted rectangular bar [21]. The right picture appeared in Fig. 1 describes the surface characteristics of the Fe-MCM-41 particles in a magnified view. Some layered-like structure as well as embedded spherical iron oxide particles which have been oxidized from Fe at the surface was observed. The CI particles possess regular spherical shape, while these Fe-MCM-41 particles demonstrate irregular shape. Compared to those spherical CI particles with an average size of 4.5-5.2 μm, these magnetic hexagonal mesoporous silica phases occupy the interspaces disorderedly among the CI particles due to their relatively small size. In this way, the added Fe-MCM-41 particles restrain the direct contact of the CI particles, and consequently prevent the serious cake problem, by which they sustain the stability of the MR fluid [22]. As it is well-known, the MR fluids exposed to an applied magnetic field will undergo a chain rupture and reformation process. Therefore, under this condition, adding the additives help to enhance the strength of the column and to reform columns from the broken chain structure under an external shear force.

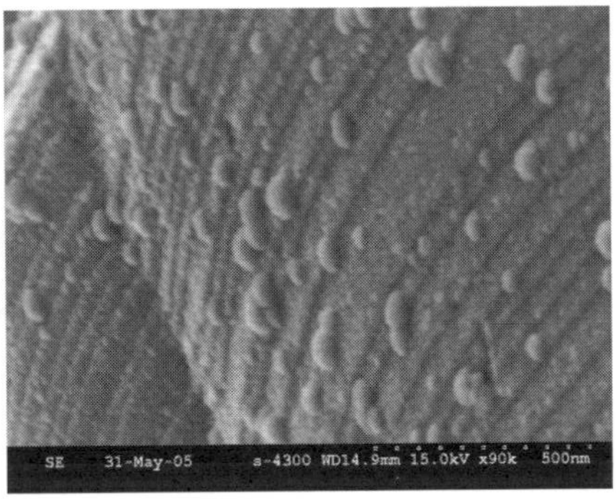

Figure 1. SEM images of pristine Fe-MCM-41 particles (left image) and their enlarged surfaces (right image).

The flow curves of these two suspensions with and without the additives were examined via a controlled shear rate mode. Figure 2 shows the resulting flow responses measured as a function of shear rate for four different magnetic field strengths ranging from 0 to 257 kA/m. Both of the two systems represent a plateau behavior of shear stress over the whole range of applied shear rate. The obvious increase of shear stress with the magnetic field strength is the typical feature of MR fluid with the formation of robust chain structure which is similar with the phenomenon observed for ER fluids [23, 24]. The dynamic yield stress was estimated up to 30 kPa at 257 kA/m by extrapolating the flow curve to a zero shear rate limit [3]. It is found that the CI/Fe-MCM-41 suspension indicates better MR effect than that of the pristine CI particles based MR fluid. We attribute this observation to the additives which contain some magnetic iron oxide particles as mentioned in the SEM images. Thereby, when exposed to magnetic field, these mesoporous Fe-MCM-41 additive particles pad the gaps among the CI particles, causing the large particles to form more chainlike aggregates leading to an enhanced MR performance. When the external field is absent, we can see that the system with additives also indicates a higher shear stress compared with the pristine CI suspension. It may be caused by these Fe-MCM-41 particles with relative small size which helps to make the MR suspension more polydisperse. By this way, the suspension with additives will produce higher shear stress [25]. In addition, it was also found that the suspension of CI/Fe-MCM-41 showed a typical shear thinning behavior which suggests that the particle structures developed under external magnetic fields remains over the whole shear range during measurement [26].

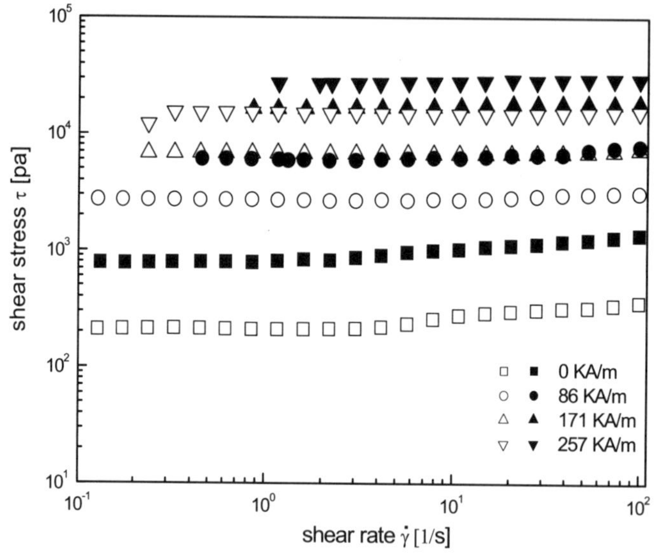

Figure 2. Shear stress of pure CI particle (open symbol) and CI/Fe-MCM41 particles (closed symbol) based MR fluid.

The conventional rotational tests as mentioned above can reveal some crucial informations on the influence of a magnetic field, however, they could not directly detect the strength of the structure at rest as a function of the magnetic field strength. Therefore, rotational tests are regarded not enough to describe the viscoelastic behavior as well as the process of chain formation for MR fluids precisely [15].

In order to decide a critical deformation over which the chainlike structure will be destroyed leading to a nonlinear oscillatory test, an amplitude sweep test at $\omega=10\ s^{-1}$ was performed. From the amplitude sweep test, it was observed that the linear viscoelastic range gets widened due to the enhanced chainlike structure when we increased the strength of magnetic field. The critical strain at about 10^{-4} % was chosen to carry out the magneto sweep. If the amplitude is higher than this value, the structure of the sample will be irreversibly changed or even completely ruined, while smaller than this value, the structure of the sample will be sustained.

The changes of storage modulus and loss modulus are shown in Figure 3. It is apparent that at a fixed magnetic strength, both the storage modulus and loss modulus vary weakly as a function of frequency, suggesting that the sample possesses a very strong solid-like structure rather than a liquid-like structure [27]. However, it is worthy to note that the storage modulus is always higher than the loss modulus over the whole range of frequency. This observation also helps to give an evidence of the predominant elastic property of the MR fluids.

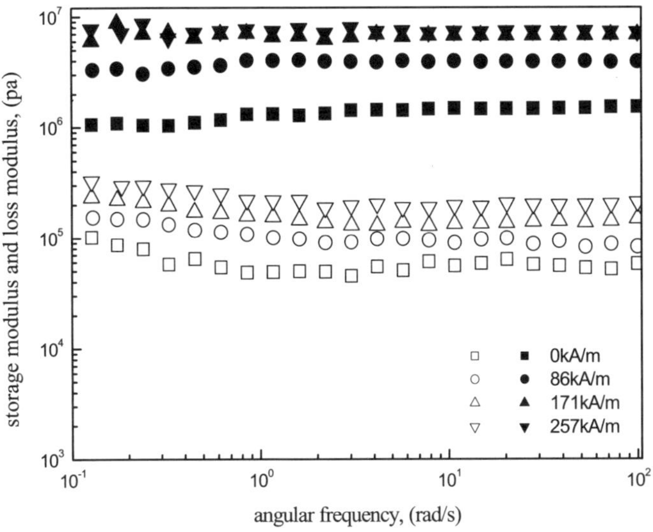

Figure 3. Magneto sweep for CI/Fe-MCM-41 suspension storage modulus (closed symbol), loss modulus (open symbol) ($\gamma = 10^{-4}$)

Figure 4 indicates the sedimentation measurement of the two systems of MR fluids with and without an additive as a function of time. Photos were taken before sedimentation (left-hand side picture) and after sedimentation (right-hand side picture) for CI particles (left vial) and CI/Fe-MCM-41 particles (right vial) dispersed in mineral oil. We can find

Figure 4. Photos taken before sedimentation (left) and after sedimentation (right) for CI particles (left vial) and CI/Fe-MCM-41 particles (right vial) dispersed in mineral oil.

that the sedimentation problem gets improved because the additives take up the gaps between the CI particles, by which way the CI particles lose much opportunity to contact [28, 29]. The right hand side picture is a final profile of the sedimentation test from both CI particles and CI/Fe-MCM-41 particles based MR fluids.

4 Conclusions

MR characteristics of two MR fluids with and without adding hexagonal Fe-MCM-41 particles were examined. The additives were found not only to improve sedimentation problem but also to enhance MR property. Flow curves indicated that the magnetic additives help to make the MR suspension polydisperse leading to a higher shear stress under an external field. The oscillatory tests also showed dominant elastic characteristics of the sample due to the strong solid-like structure under an applied magnetic field.

5 Acknowledgement

This study was supported by research grants from the KOSEF (# R01-203-000-10382-0).

References

1. Brand H. R. and Pleiner H., Origin of the slow wave in a magnetorheological slurry, *Phys. Rev. Lett.* **86** (2000) pp. 1385.
2. Klingenberg D. J., Magnetorheology: applications and challenges, *AIChE J.* 47 (2001) pp. 246-249.
3. Zhao X. P. and Yin J. B., Advances in electrorheological fluids based on inorganic dielectric materials, *J. Ind. Eng. Chem.,* **12** (2006) pp. 184-198.
4. Quadrat O. and Stejskal J., Polyaniline in electrorheology, *J. Ind. Eng. Chem.,* **12** (2006) pp. 352-361.
5. See H., Advances in electro-rheological fluids: materials, modeling and allpications, *J. Ind. Eng. Chem.* **10** (2004) pp. 1132-1145.
6. Rabinow J., The magnetic fluid clutch, *AIEE Trans.* **67** (1948) pp. 1308-1315.
7. Kuzhir P., Bossis G. and Bashtovoi V., Optimization of magnetorheological fluid valves, *Int. J. Mod. Phys. B* **19** (2005) pp.1229-1235.
8. Carlson J. D., MR fluids and devices in the real world, *Int. J. Mod. Phys. B* **19** (2005) pp. 1463-1470.
9. Zhu C. S., The response time of a rotor system with a disk-type magnetorheological fuid damper, *Int. J. Mod. Phys. B* **19** (2005) pp.1506-1512.
10. Hu Y., Wang Y. L., Gong X. Q., Gong X. L, Zhang X. Z., Jiang W. Q, Zhang P. Q. and Chen Z. Y., Magnetorheological elastomeres based on polyurethane/Si-rubber hybrid, *Int. J. Mod. Phys. B* **19** (2005) pp.1114-1120.
11. Gong X. L, Zhang X. Z. and Zhang P. Q., Study of mechanical behavior and microstructure of magnetorheological elastomers, *Int. J. Mod. Phys. B* **19** (2005) pp.1304-1310.

12. Jang I. B., Kim H. B., Lee J. Y., You J. L., Choi H. J. and Jhon M. S., Role of organic coating of carbonyl iron suspended particles in magnetorheological fluids, *J. Appl. Phys.* **97** (2005) pp. 10Q912.
13. Bombard A. J. F., Alcantara M. R., Knobel M. and Volpe P. L. O., Experimental study of MR suspensions of carbonyl iron powders with different particles sizes, *Int. J. Mod. Phys. B* **19** (2005) pp.1332-1338.
14. Lim S. T., Cho M. S., Jang I. B., Choi H. J. and Jhon M. S., Magnetorheology of carbonyl iron suspensions with submicron-sized filler, *IEEE Trans. Magn.* **40** (2004) pp. 3033-3035.
15. Lim S. T., Choi H. J. and Jhon M. S., Magnetorheological characterization of carbonyl iron-organic suspensions, *IEEE Trans. Magn.* **41** (2004) pp. 3745-3747.
16. Choi J. S., Yoon S. S., Jang S. H. and Ahn W. S., Phenol hydroxylation using Fe-MCM-41 catalysts, *Catal. Today* **111** (2006) pp. 280-287.
17. Cho M. S., Choi H. J. and Ahn W. S., Enhanced electrorheology of conducting polyaniline confined in MCM-41 channels, *Langmuir* **20** (2004) pp. 202-207.
18. Cho M. S., Choi H. J., Kim K. Y. and Ahn W. S., Synthesis and characterization of polyaniline/mesoporous SBA-15 nanocomposite, *Macromol. Rapid Commun.* **23** (2002) pp. 713-716.
19. Maity A. and Biswas M., Recent progress in conducting polymer, mixed polymer-inorganic hybrid nanocomposites, *J. Ind. Eng. Chem.* **12** (2006) pp. 311-351.
20. Pak C., Joo S. H. and Haller G. L., Nitrogen adsorption of cobalt-incorporated MCM-41 materials after compression at high pressure, *J. Ind. Eng. Chem.* **11** (2005) pp. 756-761.
21. Choi H. J., Cho M. S., Kang K. K. and Ahn W. S., Electrorheological properties of a suspension of a mesoporous molecular sieve (MCM-41), *Micropor. Mesopor. Mater.* **19** (2000) pp. 19-24.
22. Cho M. S., Lim S. T., Jang I. B., Choi H. J. and Jhon M. S., Encapsulation of spherical iron-particle with PMMA and its magnetorheological particles, *IEEE Trans. Magn.* **40** (2004) pp. 3036-3038.
23. Cho Y. H., Cho M. S., Choi H. J., and Jhon M. S., Electrorheological characterization of polyaniline-coated poly(methyl methacrylate) suspensions, *Colloid Polym. Sci.* **280** (2002) pp. 1062-1066.
24. Wen W. and Sheng P., Two- and three-dimensional ordered structures formed by electro-magnetorheological colloids, *Physica B* **338** (2003) pp. 343-346.
25. Lim S. T., Cho M. S., Jang I. B. and Choi H. J., Magnetorheological characterization of carbonyl iron based suspension stabilized by fumed silica, *J. Magn. Magn. Mater.* **282** (2004) pp. 170-173.
26. Wollny K., Läuger J. and Huck S., Magneto sweep-a new method for characterizing the viscoelastic properties for MR fluids, *Appl. Rheol.* **12** (2002) pp. 25-31.
27. Kittipoomwong D. and Klingenberg D. J., Dynamic yield stress enhancement in bidisperse magnetorheological fluids, *J. Rheol.* **49** (2005) pp. 1521-1538.
28. Gandhi F. and Bullough W. A., On the phenomenological modeling of electrorheological and magnetorheological fluid preyield behavior, *J. Int. Mater. Sys. Struct.* **16** (2005) pp. 237-247.
29. Facey W. B., Rosenfeld N. C., Choi Y. T., Wereley N. M., Choi S. B. and Chen P., Design and testing of a compact magnetorheological damper for high impulsive loads, *Int. J. Mod. Phys. B,* **19** (2005) pp.1549-1555.

SPEED CONTROL OF DC MOTOR USING ELECTRORHEOLOGICAL BRAKE SYSTEM

S. B. CHOI, J. Y. YOOK, M. K. CHOI AND Q. H. NGUYEN

Smart Structures and Systems Laboratory, Department of Mechanical Engineering, Inha University, Incheon 402-751, Korea

E-mail: seungbok@inha.ac.kr

This paper presents robust control performance of a direct current (DC) motor with brake system adopting the giant electrorheological (GER) fluid, whose distinguished feature is an extremely high value of yield stress. As a first step, Bingham characteristics of the GER fluid is experimentally investigated using the Couette type electroviscometer. A cylindrical type of ER brake is then devised based on the Bingham model, and its braking torque is experimentally evaluated. The ER break is then incorporated with a DC motor. After formulating the governing equation of motion for the DC motor with ER brake system, a sliding mode control algorithm, which is very robust to external disturbances and parameter uncertainties, is synthesized and experimentally realized in order to achieve desired rotational speed trajectories. The tracking responses of the control strategy are then evaluated for various sinusoidal trajectories. In addition, their tracking errors are evaluated and compared with those obtained from traditional PID controller.

1 Introduction

In modern industry, direct current (DC) motors are widely used with well-known advantages of simple design and reliable performance in speed and torque. Besides expensive servomotors, simple speed control mechanism for DC motors is one of the important issues. In conventional approaches, rotational speed of DC motors can be controlled by varying the voltage applied to the motor solenoid. However, it can't be reliably controlled at low speeds. Furthermore, mechanical response of the controlled DC motors is not dynamically fast enough to output torque responses in order to tackle the cogging torques generated. This makes difficulties to control speed of DC motors at high frequency. In recent years, there have been some attentions on applying brakes featuring electrorheological (ER) fluid to control speed and/or torque of DC motor [1-3]. However, in these researches, there exist some problems such as large size of electrodes to generate enough braking torque, system sensitivity to external disturbances and uncertainties.

ER fluid is a class of colloidal suspension whose rheological properties, such as yield stress, can be changed by field intensity. In general, ER fluid have fast response characteristics and hence wide control bandwidth. But their yield stress level is around $1KPa$ at maximum electric field up to $5KV/mm$ (normally limited by current density). Recently, a new type of ER fluid, called giant electrorheological (GER) fluid, has been reported by Weijia Wen et al [4, 5]. The GER fluid, consisting of urea-coated $20nm$ diameter nano particles (barium titanyl oxalate) suspended in insulating oil, e.g., silicone oil, can reach a yield stress above $200KPa$ at an electric field of $5kV/mm$. This actually provides many benefits to the ER brake such as relative lower voltage, faster response time and smaller electrode size.

Consequently, the main contribution of this paper is to show speed control performance of DC motors using GER brake. As a first step, a cylindrical type of the brake is designed and manufactured on the basic of Bingham model of the GER fluid. In order to achieve a desired rotational speed, sliding mode controller (SMC) is synthesized and experimentally realized. Then, speed control performance is evaluated for various trajectories in time domain.

2 Electrorheological Brake System

A cylindrical ER brake is derived on the basis of Bingham characteristics of the employed GER fluid. In general, ER fluid behaves as the Bingham fluid [6] whose constitutive equation is given by.

$$\tau = \tau_y(E) + \eta\dot{\gamma} \text{ , } \tau_y(E) = \alpha E^{\beta} \tag{1}$$

where τ is the shear stress [Pa], η is the dynamic viscosity [$Pa{\cdot}s$], $\dot{\gamma}$ is the shear rate [s^{-1}] and $\tau_y(E)$ is the dynamic yield stress of the ER fluid. It has been proved that $\tau_y(E)$ is a function of the electric field E [kV/mm] and exponentially increases with respect to the electric field. The parameters of α and β are characteristic values of the ER fluid to be determined by experiment. In this study, the GER fluid is dissolved by silicon oil to achieve appropriate viscosity. Figure 1 shows the measured Bingham characteristics and figure 2 shows microscopic photograph of the employed GER fluid. From the results of curve fitting, α , β and η are calculated which are 617.1, 1.76 and 0.78, respectively.

Figure 3 shows the configuration of GER brake. The GER brake consists of a fixed cylinder (outer cylinder) and a rotating cylinder (inner cylinder). GER fluid is fully filled between the two cylinders. The two cylinders also serve as a pair of electrodes. The rotating cylinder is fitted with a shaft that driven by a DC motor. When a voltage is applied between the cylinders, an electric field is produced in the GER fluid, increasing yield stress of the fluid. Thus, the rotating cylinder is forced to slow down, or brake, to provide a desired rotation speed.

The braking torque T_b of the ER brake consists of two elements. Those are controllable field-dependent torque T_y and viscous friction torque T_η. By considering Bingham model, the braking torque can be expressed by.

$$T_b = T_y(E) + T_\eta \text{ ,} \tag{2}$$

where

$$T_y = 2\pi r_i^2 L_1 \alpha E^{\beta} \text{ ,} \qquad T_\eta = 4\pi\eta\dot{\theta}\left(L_1 \frac{r_i^4}{r_0^2 - r_i^2} + 2L_2 \frac{r_1^4}{r_0^2 - r_1^2}\right)$$

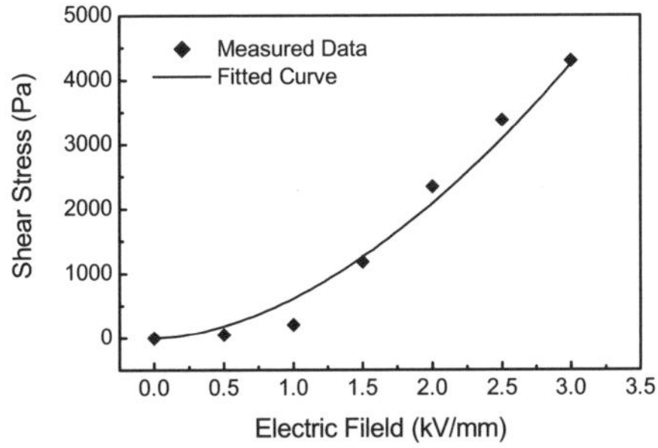

Figure 1. Bingham characteristics of GER fluid

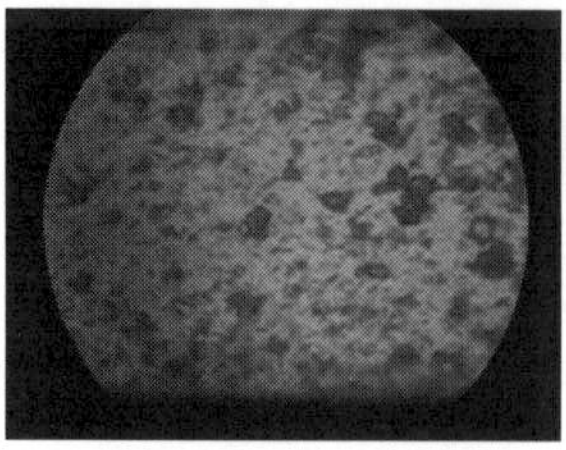

(a) without electric field

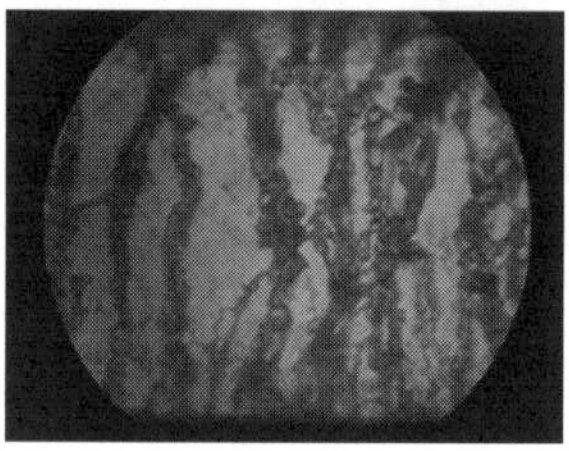

(b) with electric field

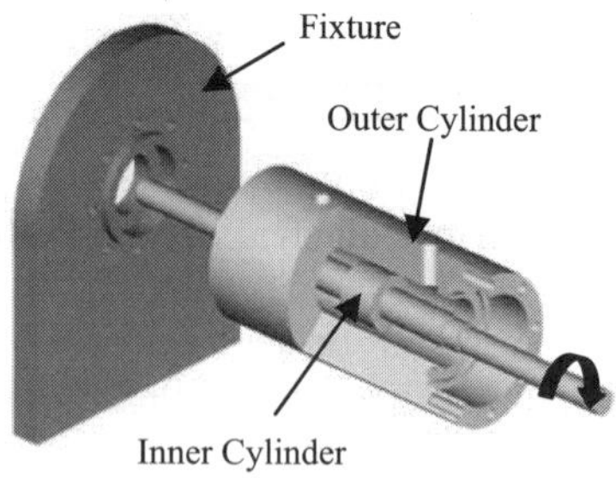

(a) configuration

(b) photograph

Figure 2. Microscopic photograph (X400) of the GER fluid.

Figure 3. Configuration of the proposed ER brake.

In above equation, r_0 and r_i are radius of outer and inner electrode respectively. L_1 and L_2 are the length of electrode and shaft mayor length. Table 1 shows mechanical dimensions of the proposed ER brake system which is determine by Eq. (2). By considering a rotating shaft by DC motor, the angular motion of θ is to be controlled by the proposed ER brake. Then, governing equation for the motion of θ is derived as follows:

$$J\ddot{\theta} + C\dot{\theta} = T_m - T_b \tag{3}$$

where J is rotational mass moment of inertia of the shaft, C is friction coefficient, θ is angular position of the shaft, T_m is motor torque and T_b is braking torque.
Combining Eq. (2) and Eq. (3) yields the following governing equation of motion:

$$J\ddot{\theta} + (C + 4\pi\eta(L\frac{r_i^4}{r_0^2 - r_i^2} + 2L_1\frac{r_1^4}{r_0^2 - r_1^2})\dot{\theta} = T_m - 2\pi r_i L\alpha E^{\beta} \tag{4}$$

The simulation and experimental response of the system, in the case of no electric field, are shown in figure 4 (a) and (b), respectively. It's seen that the rotational speed in the experiment is smaller than that in simulation due to friction. Its braking torque is also experimentally evaluated as shown in figure 5.

Table 1. Mechanical dimension of the ER brake.

Name	Value	Name	Value
Inner electrode diameter	$d_i = 30mm$	Electrode length	$L_1 = 20mm$
Shaft mayor diameter	$d_1 = 18mm$	Shaft mayor length	$L_2 = 40mm$
Outer electrode diameter	$d_o = 18mm$	Shaft length	$L_s = 40mm$

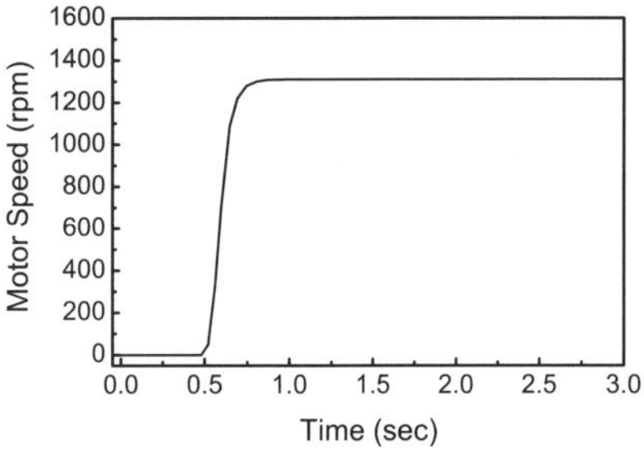

(a) simulation

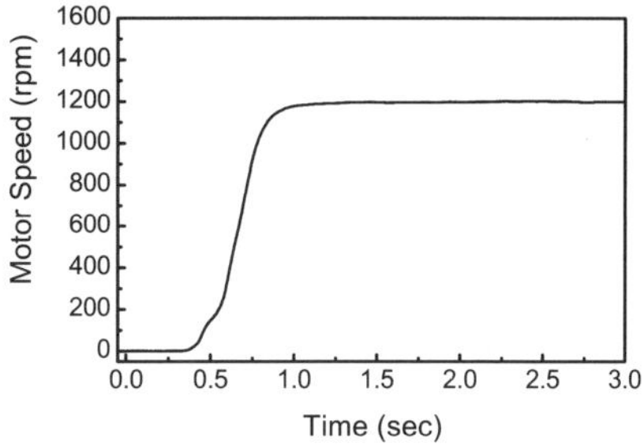

(b) experiment

Figure 4. Time response of the DC motor with ER brake System.

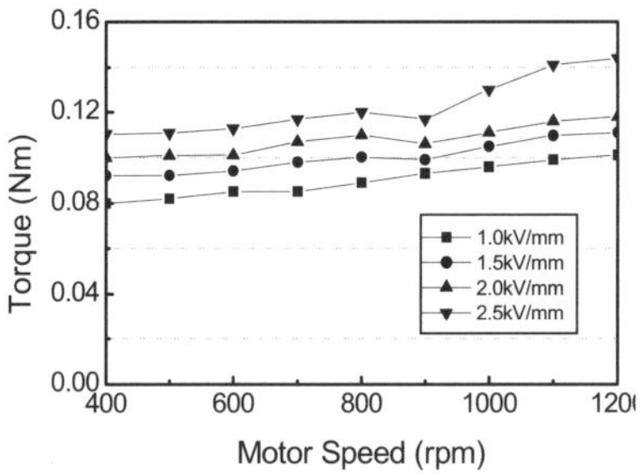

Figure 5. The measured braking torque.

3 Controller Design

The impending control issue of this work is to achieve the desired speed trajectories. However, there exist parameters uncertainties in the control model such as torque variation of the motor. This require to use robust control schemes such as sliding mode controller (SMC) which is very effective for nonlinear and parameter uncertain systems. In general, the control robustness to uncertainties is not guaranteed by applying PID controller.

The control design system (4) can be rewritten by considering uncertain parameter and mechanical dimension shown in Table 1 as follows:

$$\ddot{\theta}+15.92\dot{\theta}=45430(T_{m0}+\delta T)-792.23E^{1.757},\left|\delta T\right|<\alpha T_{m0} \tag{5}$$

where T_{m0} is the nominal motor torque. δT is the corresponding possible variation. The sliding surface on which guarantees stable sliding mode motion on is defined as follows.

$$s=\dot{e}+me=0\,, \qquad m>0 \tag{6}$$

where

$$e=\theta-\theta_d\,, \qquad \dot{e}=\dot{\theta}-\dot{\theta}_d$$

Then, a sliding mode exists on the sliding surface whenever the following sliding mode condition is satisfied.

$$s\cdot\dot{s}<0 \tag{7}$$

By applying sliding mode condition to Eq. (5), SMC control input is as follows:

$$E(t) = \left[\frac{1}{792.23m} (\ddot{\theta} - \ddot{\theta}_d + m(45430T_m - 15.92\eta\dot{\theta} - \ddot{\theta}_d) + k\,\mathrm{sgn}(s) \right]^{\frac{1}{1.757}} \tag{8}$$

It's straightforward to show that the uncertain system (5) with the proposed sliding mode controller (8), satisfies the sliding mode condition (7).

In practice, it is not desirable to use the discontinuous control law in equation (19) due to the chattering. Therefore, the discontinuous control law is normally approximated by a continuous one inside boundary layer width. This is accomplished by replacing $\mathrm{sgn}(s)$ in Eq. (8) by a saturation function [7,8].

$$sat(s) = \begin{cases} s/\varepsilon, & |s| \le \varepsilon \\ \mathrm{sgn}(s), & |s| \ge \varepsilon \end{cases} \tag{9}$$

4 Results and Discussions

Figure 6 shows the experimental apparatus and instrumentation for speed tracking control. The rotation speed was measured by the DC tachometer generator (SANYO DENKI, 104-8011-1) and was fed back to LABVIEW Real Time System (controller) via A/D converter. Then, the control electric field determined from the controller was applied to the ER Brake System through D/A converter and the high voltage amplifier (TREK10/10A). Figure 7 presents the measured tracking control response via PID controller for the desired sinusoidal trajectories at frequency of 1 Hz and 3 Hz, respectively. It is observed that the desired speed trajectory is favorably achieved, but their tracking error up to 8%. Figure 8 shows tracking control response via SMC where the desired trajectories are same as the case of PID control. It's clearly seen that the proposed SMC well follows the desired speed with small tracking error of 5% which is much smaller than 8% of PID control as shown in Figure 9. But input magnitude of sliding mode control is little higher than that of PID control because of the effect of parameter variation.

5 Conclusion

In this study, an ER brake system was proposed for speed control of DC motor. The ER Brake has been designed and manufactured by employing giant electrorheological (GER)

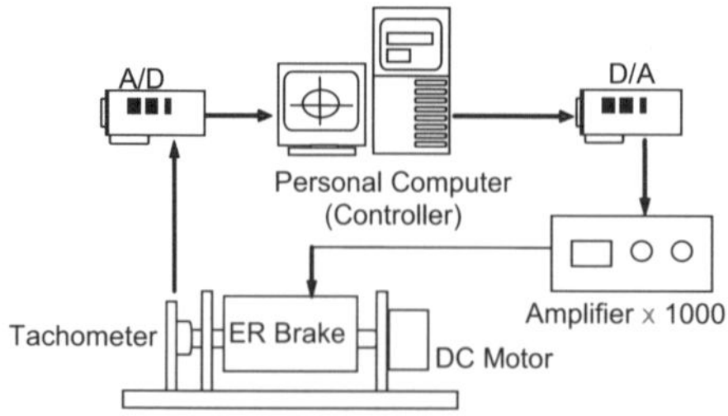

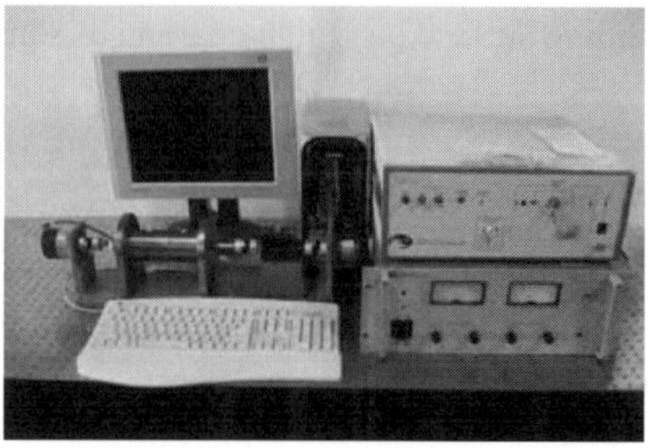

(a) schematic diagram (b) photograph of experimental set up

Figure 6. Experimental apparatus for speed tracking control.

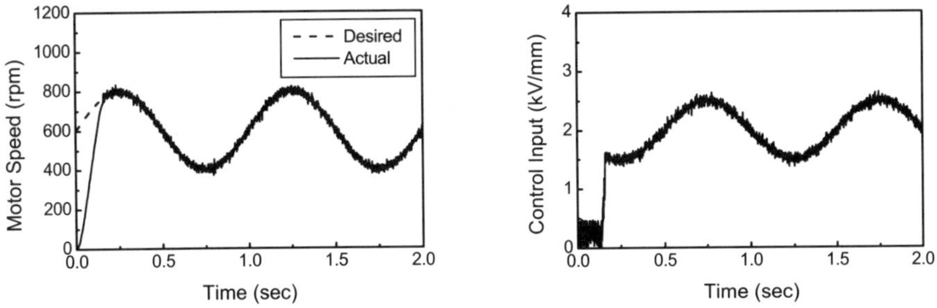

(a) 1*Hz* desired trajectory

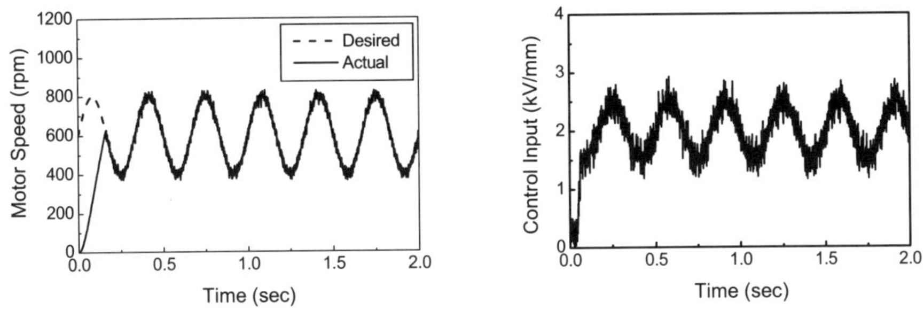

(b) 3*Hz* desired trajectory

Figure 7. Speed control response via PID controller.

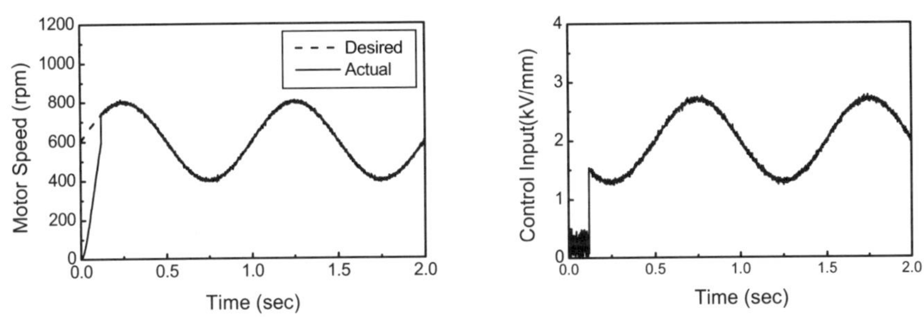

(a) 1 *Hz* desired trajectory

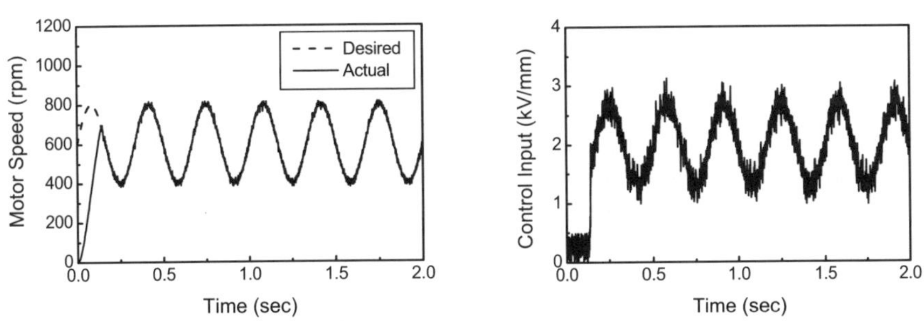

(b) 3 *Hz* desired trajectory

Figure 8. Speed control response via SMC

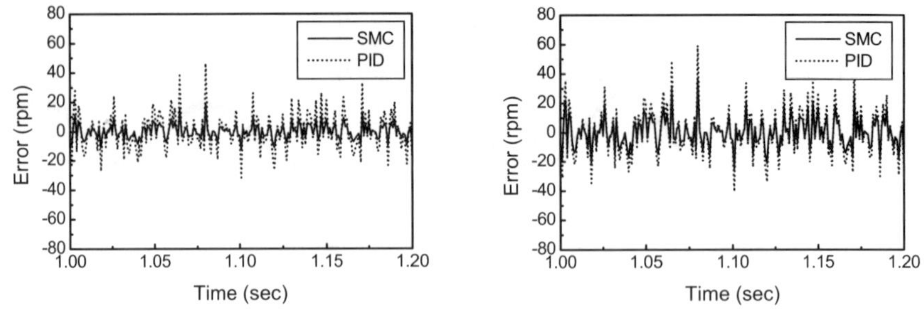

(a) 1*Hz* desired trajectory (b) 3*Hz* desired trajectory

Figure 9. Comparison of tracking errors

fluid which is characterized by high yield stress. In order to achieve desired speed trajectories, a sliding mode controller (SMC) is formulated by considering torque variation of the DC motor uncertainty, and experimentally realized. Subsequently, it has been demonstrated that the desired speed is well achieved by adopting the ER brake and SMC

6 Acknowledgement

The ER fluid used in this work was supplied by China Patent Investment Limited in Hong Kong. The authors would like to thank the president Mr. Benjamin Fok and the professor Ping Sheng at Hong Kong University of Science and Technology.

References

1. Kikuchi, T. and Furusho, J., Development of isokinetic machine using ER brake. *Proceedings of the 2003 IEEE International Conference on Robotics & Automation.* (Taipei, Taiwan, 2003) pp.214-219.
2. Nikitczuk, J., Das, A. Vyas, H., Weinberg, B. and Mavroidis, C, Adaptive torque control of electrorheological fluid brakes used in active knee rehabilitation devices. *Proceedings of the 2006 IEEE International Conference on Robotics and Automation*, (Orlando, FL, 2006) pp.211-218.
3. Furusho, J., Sakaguchi, M., Takesue, N. And Koyanagi, K., Development of ER Brake and Its Application to Passive Force Display. *Journal of Intelligent Material Systems and Structures* (2002) pp.425-450.
4. Wen, W., Huang, X. and Sheng, P., Particle size scaling of the giant electrorheological effect. *Applied Physics Letters* **85** (2004) pp.299-301.
5. Wen W., Huang X., Yang S., Lu K. and Sheng P., The giant electrorheological effect in suspensions of nanoparticles. *Nature Materials* **2** (2003) pp.727-730.
6. Shames, L. H. and Cozzarelli H. A., Elastic and inelastic stress analysis. *Prentice Hall* (England Cliffs, New Jersey, 1992).
7. Choi, Y. T., Chol, J. U., Choi, S. B. and Wereley, N. M., Constitutive models of electrorheological and magnetorheological fluids using viscometers. *Smart Mater. Struct.* **14** (2005) pp. 1025-1036.
8. Choi, S. B., Choi, Y. T. and Park, D. W., A sliding mode control of a full-car electrorheological suspension system via hardware-in-the-loop simulation. *ASME Journal of Dynamic Systems, Measurement and Control* **122** (2000) pp.114-121.

MAGNETORHEOLOGICAL BYPASS DAMPER EXPLOITING FLOW THROUGH A POROUS CHANNEL

EUGENE COOK, WEI HU, AND NORMAN M. WERELEY
Smart Structures Laboratory, Alfred Gessow Rotorcraft Center
Department of Aerospace Engineering, University of Maryland, College Park MD 20742
E-mail: ecook@umd.edu, weihu@umd.edu, wereley@umd.edu

A magnetorheological (MR) damper with a bypass valve containing porous media is developed. Movement of the piston forces MR fluid to flow from one side of the piston to the other, through an externally mounted bypass packed with magnetic spheres. The passageways between the spheres provide narrow, tortuous channels that act as a controllable valve when subjected to a magnetic field. A stationary magnetic coil is wrapped around the so called porous valve, isolated from the MR fluid and external to the device. While flowing through the tortuous channels in the porous media created by the packed spheres, the MR fluid experiences varied local flow directions, and thus an external field can be applied in any direction and still achieve variable damping. Additionally, the use of long channels tightly packed to create narrow passageways allows a very high damping force to be generated by a relatively compact damper. Furthermore, use of an externally mounted coil allows for maintenance accessibility, and the magnetic flux return path can be either empty (air) or any high permeability material. The damper is tested for steady state sinusoidal displacements, and equivalent viscous damping and complex stiffness are computed. It is shown that the porous bypass valve MR damper can provide high controllable damping force and a wide force range.

1 Introduction

Dampers that use a magnetorheological fluid as the working medium can provide controllable damping force by the means of a magnetic field to control motion, shock or vibration. Descriptions of typical MR dampers can be found in Refs. [1-5]. Most MR dampers currently published or patented use a piston made of ferrous metal with an integral magnetic coil that produces magnetic flux in and around the piston. In these MR damper configurations, an MR fluid flow channel is either moved with the piston [1-3, 5] or fixed relative to the damper body [4]. The channels are typically straight or of simple geometry, and are perpendicular to the magnetic field. The requirement for a field perpendicular to the channel often leads to placement of the coil within the piston head, where access is difficult for maintenance because the coil is usually immersed in the MR fluid, which reduces the life of the coil and contributes to poor heat dissipation. Additionally, the volume of the magnetic coil reduces the active length of the flow channel, and the configuration of the flow channel and flux return guide must follow constraints imposed by magnetic field saturation and precise flow gap width requirements.

A promising alternative is to exploit the tortuous channels that naturally exist in porous media. With the fluid flowing in varied local directions through the network of narrow, tortuous passages in the medium, a uniform field applied in any direction will cause a substantial variable damping effect, even for an externally mounted, accessible coil. Shulman [6] proposed spiral channels or channels packed with particles placed inside a solenoid. Kuzhir et al. [7] developed a hydraulic device for the investigation of MR fluid flow through porous media in a magnetic field parallel to the overall flow. Their measurements demonstrated that a packed bed of magnetic particles gave a much higher field-dependent force range than spiral channels. The objective of this study is to explore the practical design and implementation of a magnetorheological valve containing a porous media for a controllable or semi-active damper. MR dampers could utilize a porous valve placed in a bypass flow path, or incorporated into the damper cylinder. Taking advantage of a porous valve allows the active length of the flow channel to be increased. Also, the porous valve can be activated by a magnetic coil external to the device, allowing for easy maintenance and effective dissipation of the thermal energy generated through resistance heating of the solenoid.

The current research presents a prototype MR damper with a porous bypass valve, developed to evaluate the effects of a porous valve configuration on damper performance. The damper was tested for steady state sinusoidal displacements, and equivalent viscous damping and complex stiffness are computed. Advantages of the MR damper with porous valves over the conventional MR dampers are demonstrated by the experimental data at different amplitudes and frequencies.

2 Experimental Study

2.1 Damper description

Fig.1 is a schematic of the porous bypass damper prototype studied in this paper. Fig. 1 shows the overall damper, consisting of a hydraulic cylinder connected to a porous bypass valve. The piston head divides the MR fluid into two chambers, which communicate with each other through the bypass when the piston is driven. A double rod configuration was used to avoid volume changes due to a single rod entering and exiting the damper. The porous valve consists of a valve body, a magnetic solenoid, and hydraulic tubing. Spherical beads are randomly packed inside the hollow nonmagnetic stainless steel tube comprising the valve body. The magnetic coil is wrapped around the stainless steel tube and is stationary relative to the damper cylinder. The hydraulic tubing connects the porous valve to the piston chambers. In operation, when the piston moves relative to the cylinder, the piston head pushes the MR fluid from one chamber through the porous valve into the other. In traversing the porous valve, the MR fluid passes through the bed of packed spheres, in which the yield stress and viscosity of the MR fluid are controlled by an applied magnetic field.

The properties of the tortuous channels in the porous media through which the fluid must flow determine the damper efficiency and effectiveness. Fig. 2 shows a schematic of a tortuous channel in the valve, subjected to a magnetic field parallel to the porous valve axis. The inclined angle between the channel axis and magnetic field direction varies along the channel, and stochastically the average angle of the channel relative to the

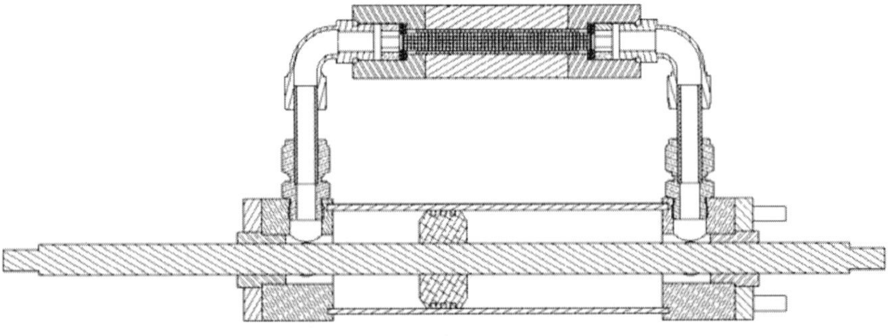

Figure 1 – Schematic of the MR damper with porous valve bypass

magnetic field is a function of the tortuosity (defined as the ratio of the fluid path length to the porous valve length) of the packed beads. By definition, as the tortuosity increases, the active length of the tortuous channel increases, leading to a larger controllable damping range.

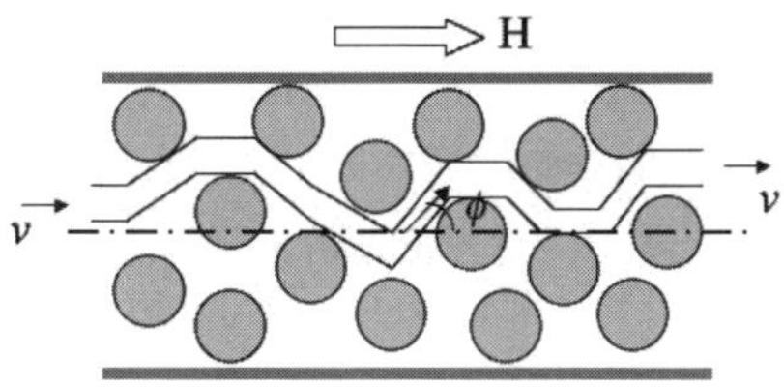

Figure 2 – Tortuous channel in the porous media

Both the MR fluid and the porous media are placed in the center of the solenoid and function as a magnetic flux guide, so the mean values of the magnetic field strength applied to the MR fluid depend on the material properties and the geometry of the porous media, allowing room for design flexibility. Comparatively, in conventional MR dampers, the fluid channel must convey MR fluid perpendicularly to magnetic field, which places geometric constraints on the damper and magnetic coil design, increasing damper cost.

The prototype constructed for this study was built from a commercially available hydraulic cylinder with a 50.8 mm (2 inch) bore and a 152.4 mm (6 inch) stroke. The bypass was built around a stainless steel tube with an inner diameter of 10 mm. Non-magnetic stainless steel was used to encourage the filed to be stronger in the fluid rather than in the bypass wall. The tube was packed with magnetic steel spheres of 3.5 mm diameter, secured in place by woven wire mesh. A solenoid with 50 ohms of resistance is wrapped around the outside of the bypass, coaxial with the stainless steel tube. The ends of the bypass were welded to machined steel end parts, which connect to the cylinder via standard tube fittings. A ball valve was also attached for filling the damper with MR fluid.

2.2 Damper testing

To evaluate the performance of the damper, measurements of damper response under steady-state sinusoidal displacements were taken using a 25 kN load frame (MTS 810) with a servo-hydraulic actuator, and on a scotch yoke shock dynamometer. The damper testing set up is shown in Fig. 3

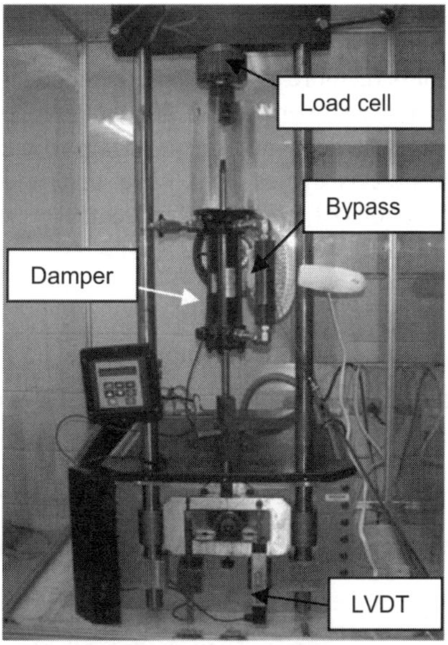

Figure 3 - Damper testing apparatus

An LVDT and a load cell were used to record displacement and force, respectively. All tests were conducted at room temperature, in displacement control, in which the excitation amplitude was fixed at 12.7 mm (1 inch peak to peak) and the frequency was varied from 0.05 Hz to 1.4 Hz. To reduce the noise of the sinusoidal displacement signal, a Fourier series was used to reconstruct the input displacement. The reconstructed simple harmonic displacement signal was then differentiated in the frequency domain to obtain the velocity. For each test signal, current ranging from 0.0 amps to 1.5 amps was supplied by a DC power supply.

3 Damper Characterization

3.1 Hysteresis behavior

Fig. 4 shows force response due to sinusoidal displacement and velocity excitation of 25.4 mm (1 inch) peak to peak amplitude at 0.2 Hz, where the applied current is varied from 0 to 1.5 A. Fig. 4(a) shows the force vs. piston displacement behavior, and Fig. 4(b) shows the force vs. piston velocity behavior of the damper. Clearly, the hysteresis cycle of the porous MR damper is similar to that of a conventional MR damper. The total energy dissipated by the damper over one cycle is represented by the area within the hysteresis

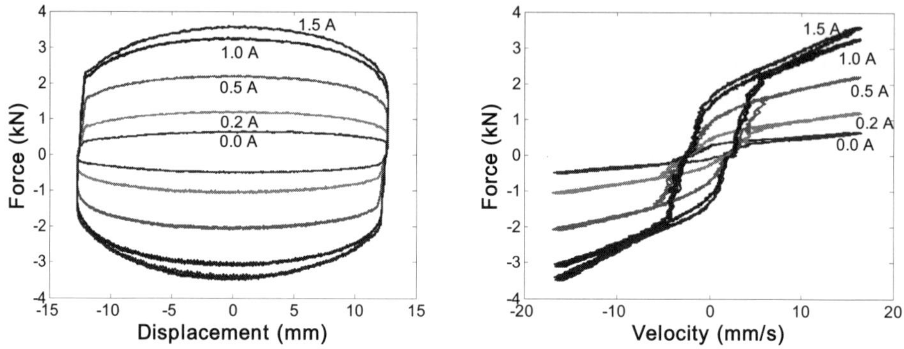

(a) Force vs. displacement (b) Force vs. velocity

Figure 4 - Typical hysteresis behavior of the MR damper with porous valve at 25.4 mm stroke and 0.2 Hz

cycles on the force vs. displacement plot in Fig. 4(a). As expected for MR dampers, applied current causes the damping force to increase over the cycle, which in turn increases the enclosed area. As higher current and, hence a stronger magnetic field is applied, more energy is dissipated by the damper. However, as the applied current is increased over 1.5 A, the damper approaches saturation, meaning that further current increase does not result in significant damping force gains. From force vs. velocity hysteresis cycles, it is clear that both yield force and post-yield viscous damping are functions of the applied current.

3.2 Complex modulus

A typical quantitative approach used for characterizing linearized viscoelastic behavior is the calculation of the complex stiffness, K^*, which is described by the in-phase or spring stiffness, K', and the quadrature or loss stiffness, K'', such that

$$K^* = K' + jK'' = K'(1 + j\eta) \tag{1}$$

where η is the loss factor. For a single frequency sinusoidal displacement

$$x(t) = X_c \cos\Omega t + X_s \sin\Omega t \tag{2}$$

the linearized viscoelastic force is given by

$$f = K^* x = F_c \cos\Omega t + F_s \sin\Omega t \tag{3}$$

Substituting Eq. (1) and (2) into (3) and equating the sine and cosine terms yields the in-phase and quadrature stiffness as

$$K' = \frac{F_c X_c + F_s X_s}{X_c^2 + X_s^2} \qquad K'' = \frac{F_c X_s - F_s X_c}{X_c^2 + X_s^2} \tag{4}$$

This characterization method allows the MR damper to be represented by an equivalent linear viscoelastic device, where the higher harmonic (nonlinear) terms in the force response are neglected. Figure 5 shows a comparison of the complex stiffness characterization to the experimental data at 0.1 Hz and 0.75 Hz. These comparisons demonstrate that the characterization deviates measured damper behavior at a lower frequency (Fig. 5a, 5b). However, at higher frequencies (Fig. 5c, 5d), the complex stiffness captures the characteristics of the damper better, since at a higher loading rate the stiffness due to the compressibility of the fluid contributes to a more significant equivalent stiffness, and the post-yield viscous damping dominates over the yield force.

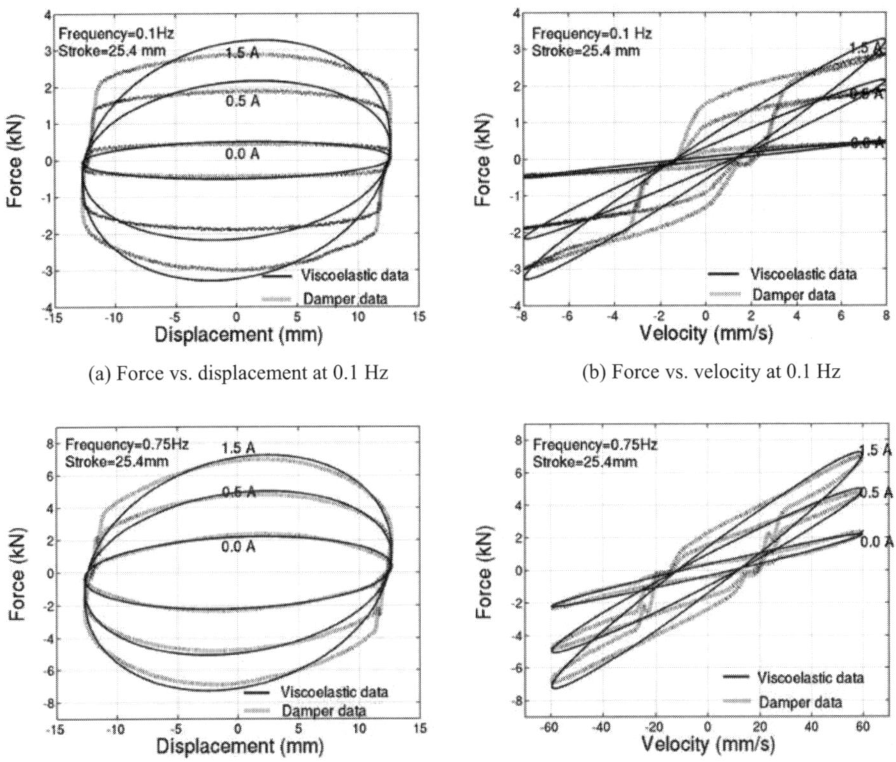

(a) Force vs. displacement at 0.1 Hz

(b) Force vs. velocity at 0.1 Hz

(c) Force vs. displacement at 0.75 Hz

(d) Force vs. velocity at 0.75 Hz

Figure 5 – Linearized complex modulus model compared to experiment at 25.4 mm stroke

The in-phase stiffness and quadrature stiffness of the damper are shown in Fig. 6. The quadrature stiffness is higher than the in-phase stiffness since the damper mainly provides damping force. The complex modulus is also a function of the applied current and excitation frequency. In-phase and quadrature stiffness both increase as the applied current increases. Similarly, increasing frequency leads to a higher stiffness since fluid compressibility and viscous damping become significant as the loading velocity is higher.

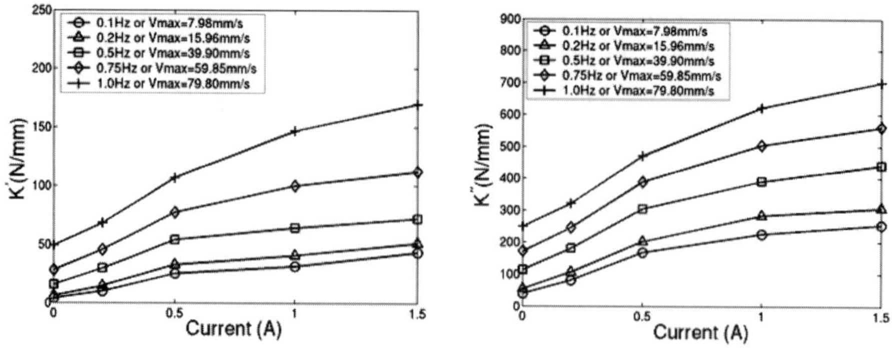

(a) In-phase stiffness (b) Quadrature stiffness

Figure 6. Complex stiffness as a function of current and frequency

3.3 *Equivalent damping*

Equivalent viscous damping is used to quantify the damping augmentation effect demonstrated by the MR damper. Equating the energy dissipated by the damper over a cycle to that of a viscous damper, results in an equivalent viscous damping coefficient:

$$C_{eq} = \frac{E}{\pi \Omega X_0^2} \qquad (5)$$

where Ω and X_0 are the frequency and amplitude of the displacement input. The energy dissipated over one cycle, E, is computed numerically by:

$$E = \oint F(x)dx \qquad (6)$$

where, F is the damper force over one cycle, and x is the displacement.

Fig. 7 shows the relationship between the controllable damping and the applied current and velocity. As the applied current increases (from 0 A to 1.5 A), so does the equivalent viscous damping. However, the variation of the equivalent viscous damping is more significant at lower frequencies than at higher frequencies. To evaluate the performance of the MR damper, the dynamic range, C_{eq}/C_0, is introduced as the ratio of field-on to field-off damping. Fig. 8 shows the equivalent viscous damping coefficient as a function of peak velocity at field off and field on (1.5 A) conditions. For each case the dynamic range is

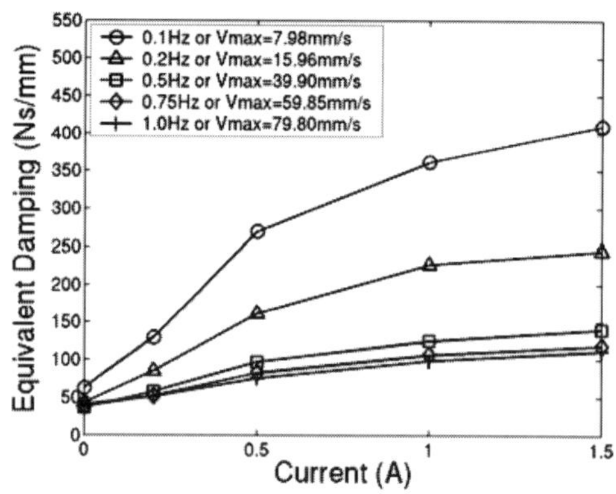

Figure 7 - Controllable equivalent damping with respect to applied frequency (damper velocity)

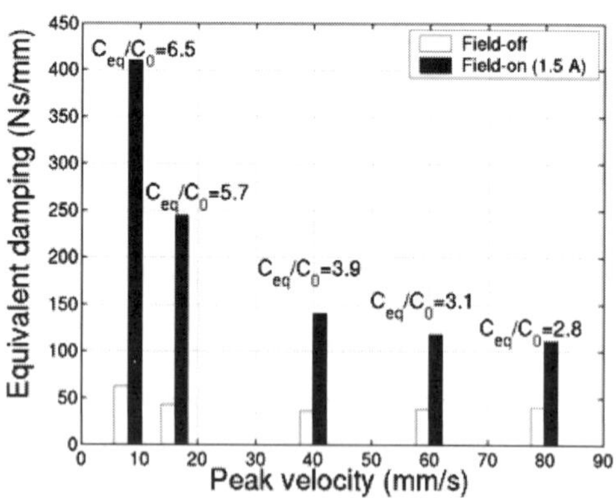

Figure 8 - Controllable dynamic range as a function of peak velocity

listed. The maximum dynamic range is 6.5 at an excitation frequency of 0.1 Hz (corresponding to a maximum velocity 7.98 mm/s). As the frequency increases, the dynamic range of the damper decreases. The minimum dynamic range is 2.8 at a frequency of 1 Hz (corresponding to a maximum velocity 79.80 mm/s). It can likewise be shown dynamic range decreases as displacement amplitude increases if the frequency is kept constant. In other words, the dynamic range is a function of the applied velocity. It is demonstrated that the damper behaves as a combination of a viscous and a friction damper. However, the MR damper with a porous valve differs from a conventional flow mode MR damper in that not only the friction or yield force but also the post-yield behavior is a function of the applied field. Thus, the porous valve configuration can provide a substantial damping control range even over a larger velocity range, which is desirable for a damper in high speed or shock resistance applications.

4 Conclusions

A magnetorheological damper with porous valve was developed which forces fluid to flow through a bypass packed with steel spheres rather than a typical gap. While flowing through the tortuous channels in the porous media created by the packed spheres, the fluid experiences varied local flow directions, and thus an external field can be applied in any direction and still attain variable damping. This approach allows the coil to be located externally, easily accessible for maintenance. Additionally, the use of long channels tightly packed to create narrow passageways allows a very high damping force to be generated by a relatively compact damper. The damper is tested for steady state sinusoidal displacements of 25.4 mm peak to peak amplitude and frequencies ranging from 0.05 Hz to 1.4 Hz while currents of 0 to 1.5 amps are applied to the coil. Equivalent viscous damping and complex stiffness are computed. As with conventional MR dampers, the equivalent damping is higher at low frequencies than at higher frequencies, and increases dramatically with applied current. Equivalent damping ranges from 50 N-s/mm to 400 N-s/mm. The maximum dynamic range is 6.5, at the lowest test frequency of 0.1 Hz, and reduces to 2.0 at 1.0 Hz.

References

1. Calson J.D. and Chrzan M.J., Magnetorheological fluid dampers, U.S. Pat. No. 5277281 (1994).
2. Lisenker I., Hopkins P.N., Baudendistel F. and Oliver M.L., Magnetorheological fluid damper, U.S. Pat. No. 6279700 (2001).
3. Hopkins P.N., Fehring J.D., Lisenker I., Longhouse R.E., Kruckemeyer W.C., Oliver M.L., Robinson, F.M. and Alexandridis, A.A., Magnetorheological fluid damper, U.S. Pat. No. 6311810 (2001).
4. Chen P.C. and Wereley N.M., Magnetorheological damper and energy dissipation method, U.S. Pat. No. 6694856 (2004).
5. Anderfaas E.N. and Banks D., Magnetorheological damper system, U.S. Pat. No. 6953108 (2005).
6. Shulman Z., Magnetorheological systems and their application, *Magnetic Fluids and Applications Handbook* (1996) pp. 188-229.
7. Kuzhir P., Bossis G., Bashtovoi V. and Volkova O., Flow of magnetorheological fluid through porous media, Euro. J. Mech. B/Fluids 22 (2003) pp. 331-343

ADAPTIVE TUNED VIBRATION ABSORBER BASED ON MAGNETORHEOLOGICAL ELASTOMER

H.X.DENG AND X.L.GONG*

CAS Key Laboratory of Mechanical Behavior and Design of Materials, Department of Mechanics and Mechanical Engineering, University of Science and Technology of China, Hefei 230027, China

**Email: gongxl@ustc.edu.cn*

This paper presents an adaptive tuned vibration absorber (ATVA) which based on magnetorheological elastomer (MRE). Traditional dynamic absorber has limited its application and vibration absorption capacity for its narrow working frequency bandwidth. MRE is a kind of smart material whose modulus can be controlled by applied magnetic field. Based on MREs, an adaptive tuned vibration absorber which works on shear mode is proposed in this paper. After the vibration mode shapes of the ATVA are analyzed, the mechanical structure of the ATVA is brought forward. Then the magnetic circuit of the ATVA is identified by ANSYS software. By using of a modified dipole model, the shift-frequency properties of the ATVA versus magnetic field and strains are theoretically analyzed and simulated. Further more, by employing a beam with two ends supported, its shift-frequency property and vibration absorption capacity are experimentally justified. The experimental results demonstrate that the designed ATVA has better performance than traditional passive absorber in terms of frequency-shift property and vibration absorption capacity.

1 Introduction

Most of mechanical, civil and construction systems encounter undesirable vibrations. To avoid damage, fatigue failure and to increase the system lifetime, many researches have been done to attenuate the unwanted vibration. Dynamic vibration absorber (DVA) which was invented by H.Frahm [1] in 1900s is a classic solution to suppress vibrations of machines and structures. Traditional DVA is typically composed of an oscillator, a spring element and a damping element. However, this kind of passive absorber theoretically brings the object base to rest at a single excitation frequency, the resonance frequency of the DVA. For many practical systems, the working condition may change over time and cause the absorber to become inefficient and potentially aggravate the base vibration. Adaptive tuned vibration absorber (ATVA) is developed to solve this problem, which is a device similar to DVA but holds adaptive element to trace the working condition. Generally, ATVA can vary its natural frequency to track uncertain or time-varying excitation frequencies by employing the adaptive spring element which is able to alter its stiffness. According to the method of changing the stiffness, the ATVAs can be sorted into two major groups: one is to use variable geometries [2-3] and the other is to use smart materials [4-6].

Magnetorheological elastomers (MREs) are smart materials where polarizable particles are suspended in a non-magnetic solid or gel-like matrix. Typically, magnetic fields are applied to the polymer composite during crosslinking so that chainlike structures can be formed and fixed in the matrix after curing. The unique characteristic of MRE is that its shear modulus can be controlled by the external magnetic field rapidly, continuously, and reversibly [7]. Such properties make MREs to have promising in many applications such as ATVAs, stiffness tunable mounts and suspensions, and variable impedance surfaces. Watson [8] applied a patent using MREs for a suspension bushing in 1997; Later, Ginder et al. [9, 10] constructed and tested tunable automotive mounts and bushings based on MR elastomers. Ginder and his coworkers [11] did pioneer work on the development of an adaptive tunable vibration absorber (ATVA) using MREs.

The objective of this paper is to investigate the application of MREs to adaptive tuned vibration absorber. The vibration mode shapes have been analyzed to ensure that the ATVA works by shear mode. Another problem to be confirmed is the magnetic circuit; a closed C-shape structure has been identified by ANSYS software. Further more, the shift-frequency properties of the ATVA versus magnetic field and strains are analyzed and simulated by using the modified magnetic dipole model. It is also experimentally verified together with its attenuation ability.

2 Material Preparation

The MRE materials consist of natural rubber as a matrix, additives and carbonyl iron particles with size of 3 to 5 microns. At the first step of the fabricating process, all ingredients are thoroughly mixed by mixing roll and then the mixture is compressed into a mold and placed in a self-developed magnet-heat coupled device. During the pre-cured stage, the particles are magnetized and form chains aligned along the magnetic field direction. 30 minutes later, the magnetic field is turned off and the temperature is raised to 153℃ to finish the sulfuration. The MRE properties under various magnetic fields are evaluated by a modified Dynamic Mechanical Analyzer (DMA) from the Triton Co. It can be seen from Fig.1 that the shear modulus shows an increasing trend with magnetic field intensity. However, the increasing slope decreases with the increment of magnetic fields, which is due to the magnetic saturation. In addition, the magneto-induced modulus also shows an increasing trend with the strain decreasing while the zero-field modulus is nearly the same with the strain change.

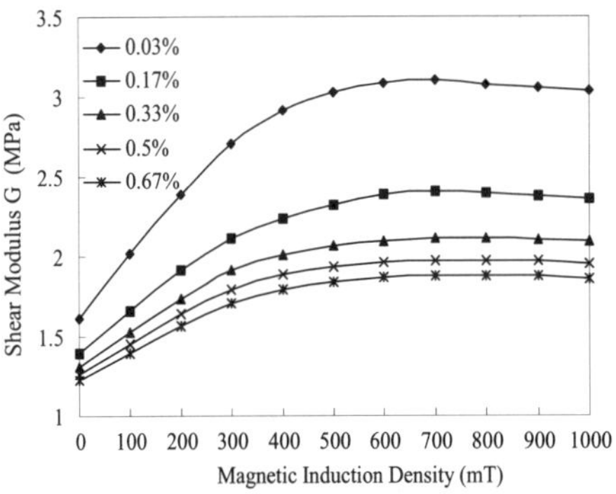

Fig.1 Shear modulus with magnetic field under different strains load

3 Design of ATVA

3.1 *Simulation of Mode Shapes*

This paper focused on the ATVA which works by shear mode. So the first step of the design is to analyze the vibration mode shape of the assumed ATVA and make sure the work mode of ATVA is shear. To this end, ANSYS software is employed and a popular sandwich structure is simulated. In this structure, dynamic mass is at the center and through MREs connecting the static mass. The results of vibration mode shapes analyses are shown in Fig.2. The left picture shows that the first mode shape of ATVA is swing while the right picture shows the second mode is shear in plumb direction. It means that the sandwich structure with this shape would easily swing instead of shear. To avoid this happening, some structures to make the ATVA work by shear mode should be leaded in, such as guiderod, sliding chute and linear bearing.

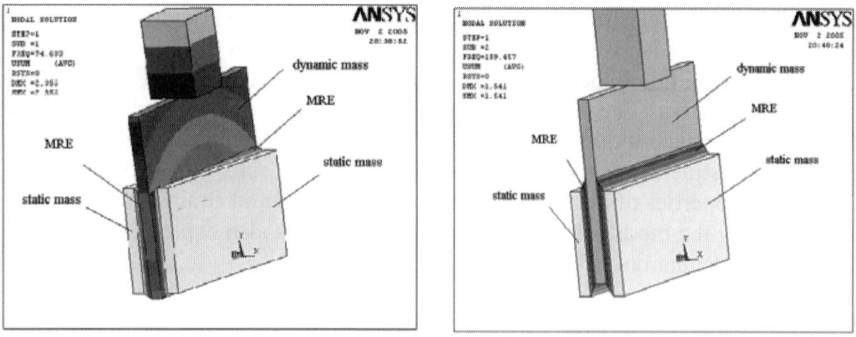

Fig.2 Mode shapes of sandwich structure

3.2 ATVA Structure

The efficient component for ATVA to absorb vibration is its oscillator or dynamic mass. As shown in Fig.3, the developed ATVA consists of three main parts: dynamic mass, static mass and smart spring element with MREs. The electromagnets and magnetic conductors form a closed C-shape magnetic circuit, which are assembled at mounting shell. These components together construct the dynamic mass. This development makes the ATVA more compact and more efficient because no additional oscillator and nearly most components can be considered as dynamic mass. The static mass consists of the shear plate and the base. Through the shear plate, the smart spring elements (MREs) connect the dynamic mass and the static mass. Further more, as analyzed above, a guiderod and a linear bearing are employed to ensure the ATVA works by the shear mode.

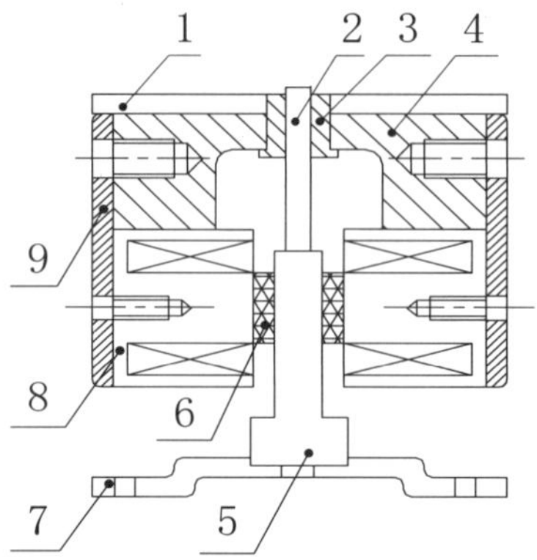

Fig.3 The scheme of designed ATVA
1.Cover; 2.Guiderod; 3.Linear bearing; 4.Magnetic conductor; 5.Shear plate; 6.MREs; 7.Base; 8.Electromagnet; 9.Mounting Shell

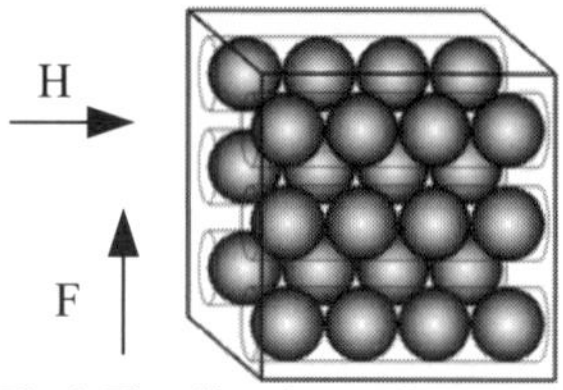

Fig.4 The directions of magnetic field and shear force

3.3 *Magnetic Circuit Analyses*

Magnetic conductor which forms a magnetic circuit with MREs is the vital component of the ATVA based on MREs. As shown in Fig.4, for shear mode, the direction of magnetic field is parallel with the direction of particle chains while the shear force is perpendicular to the particle chains. The purpose of analyzing the magnetic circuit is to optimize the turn number of coils, the circuit and the distribution of magnetic lines of flux, et al. The avail tool of the magnetic circuit analyses is ANSYS software. Fig. 5 is a result of the designed ATVA. For a coil with 2200 rounds, when applied current is 0.5A, the magnetic field through MRE can reach 0.9 Tesla, which is enough to utilize as shown in Fig. 1.

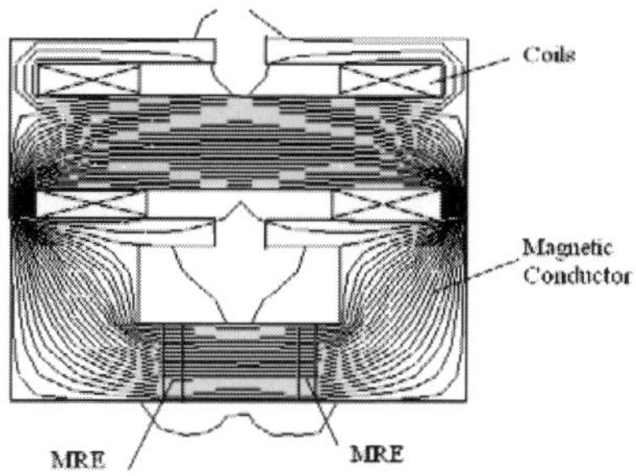

Fig.5 The magnetic circuit of a C-shape structure

4 Theoretic Analysis of Shift-Frequency Property

The MRE's shear modulus G consists of two terms, zero-field modulus G_0 and magneto-induced modulus ΔG_d :

$$G = G_0 + \Delta G_d \tag{1}$$

The natural frequency of the ATVA can be expressed as:

$$f = f_0 + \Delta f \tag{2}$$

where f_0 is the initial natural frequency, Δf is the magneto-induced frequency:

$$f_0 = \frac{1}{2\pi}\sqrt{\frac{G_0 A}{mh}} \tag{3}$$

$$\Delta f = \frac{1}{2\pi}\sqrt{\frac{G_0 \times A}{m \times h}} \bullet (\sqrt{1+\frac{\Delta G_d}{G_0}} - 1) \tag{4}$$

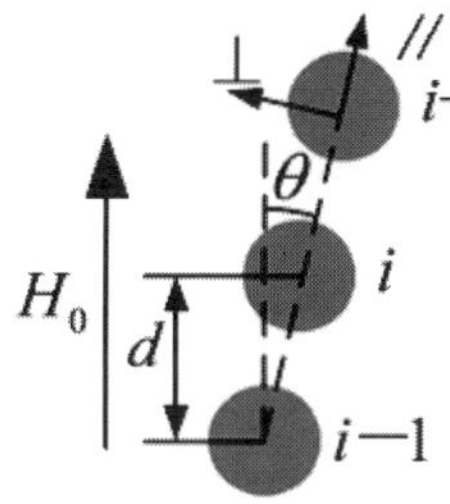

Fig.6 The model of particle chain

From Eqn. (3), the initial resonant frequency of the ATVA can be designed to match the primary system. Equation 4 reveals that the frequency-shift capacity is not only related to the MR effect but also related to the initial shear modulus. Large initial modulus with the same MR effect will cause wide frequency-shift bandwidth. To theoretically analyze the frequency-shift property with magnetic field, a modified magnetic dipoles model considering the interaction between all particles at the same chain is employed. The model of a single chain is shown in Fig.6, where d is the distance between two neighbor particles before deformation, and θ is the shear angle. For small deformation, $\theta \to 0$, the magneto-induced modulus can be expressed as

$$\Delta G = 3\phi\mu_m\mu_0\beta^2 H_0^2\left(\frac{R}{d}\right)^3\zeta\left(\left(\frac{10}{A^2}+\frac{2}{B^2}\right)+\frac{48\beta\zeta}{A^3}\left(\frac{R}{d}\right)^3\right) \tag{5}$$

where 3 β is the effective susceptibility and it approaches 3 for ferromagnetic particles, μ_0, μ_m correspond to the permeability in vacuum and the relative one of the matrix respectively, ϕ is the volume fraction of particles and R is the radius of the particle, $A = 1 - 4\beta\cos^3\theta(R/d)^3\zeta$, $B = 1 + 2\beta\cos^3\theta(R/d)^3\zeta$ and $\zeta = \sum_{k=1}^{\infty} 1/k^3 \approx 1.202$. Define the shear strain to be ε, the magneto-induced modulus can be represented as

$$\Delta G = 3\phi\mu_m\mu_0\beta^2 H_0^2\left(\frac{R}{d}\right)^3\zeta\left(\frac{12a^7\left(1+2ba^3+14b^2a^6-8b^3a^9\right)}{\left(1-4ba^3\right)^3\left(1+2ba^3\right)^2}\right) \tag{6}$$

where $a = (1+\varepsilon^2)^{-\frac{1}{2}}, b = \beta\zeta\left(\frac{R}{d}\right)^3$. Obtained from Eqn. (5), Fig.7 shows that the natural frequency of ATVA has increasing trend with magnetic field. In the initial stage, natural frequency accelerates and then the rate of slope drives to stabilization before saturation. From Eqn. (6), the influence of shear strain to the shift-frequency property can be obtained and shown in Fig. 8 where the magnetic field is fixed as 800 KA/m. The frequency decreases while the strain increase and gets stabilization when shear strain is 100%.

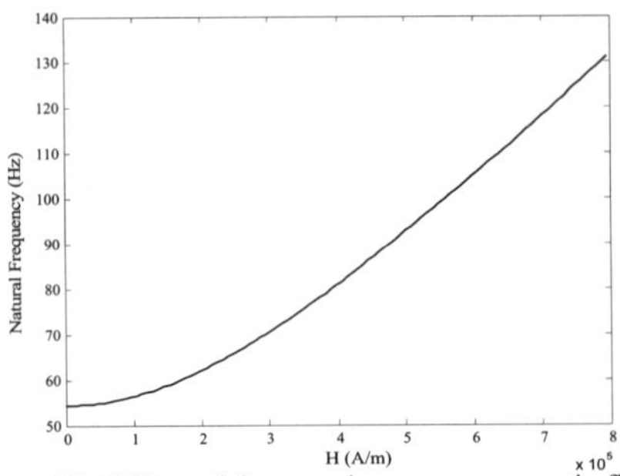

Fig.7 Natural frequencies versus magnetic field

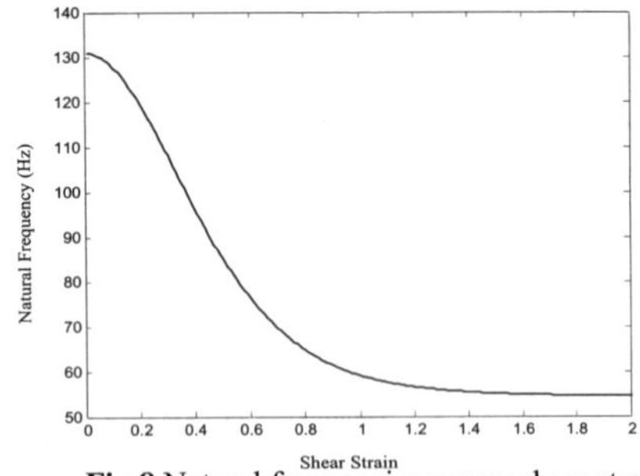

Fig.8 Natural frequencies versus shear strain

It can be seen that strain influences the shift-frequency property so hugely that the amplitude of the ATVA must be evaluated. The model of ATVA coupled with primary system can be simplified as Fig. 9, where m_1 is the mass of primary system, k_1 is the stiffness of primary system, ω_1 is the natural frequency of primary system, and x_2 is the vibration amplitude of primary system, m_2 is the mass of ATVA, k_2 is the stiffness of ATVA, ω_2 is the natural frequency of ATVA, and x_2 is the vibration amplitude of ATVA. When external excitation is $p_1 sin\omega t$, the complex amplitude $\overline{x}_2$ can be expressed as:

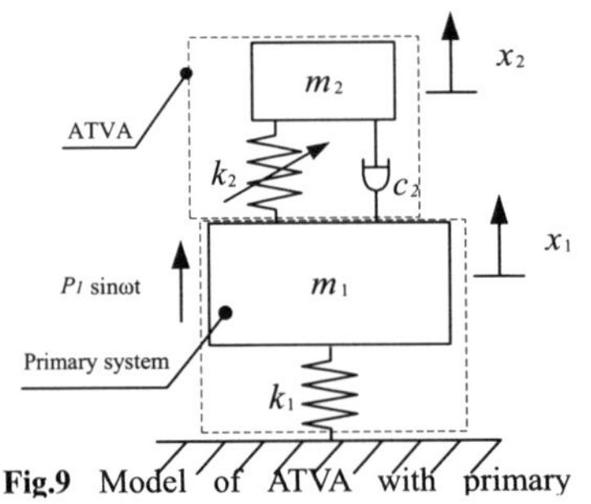

Fig.9 Model of ATVA with primary

$$\overline{x}_2 = \frac{p_1(k_2 + ic\omega)}{-k_2^2 + ic\omega(m_1\omega_1^2 - \frac{m_1}{m_2}k_2 - k_2)} \tag{7}$$

The maximum shear displacement B_2 equal:

$$B_2 = \delta_{st} + |x_2| = \frac{g}{\omega^2} + |x_2| \tag{8}$$

where δ_{st} is static deflection of ATVA. Fig. 10 is a classic example of calculated shear displacement with and without static deflection. The shear displacement without static deflection is relatively small compared to the displacement with static deflection. It means that the static deflection has vast contribution to the shear displacement especially in the low frequencies. So limiting the static deflection is a method to enhance the tuned frequency band. It also can be seen that the shear displacement decreases sharply with the excitation frequency increasing. When the excitation frequency is higher than 40 Hz in this example, the shear displacement trends to a stable level and has little influence on tuned frequency of ATVA. Thus, ATVA working with MRE should work at relatively high frequency band, because the stability of the shift-frequency property would have problem at low frequency band.

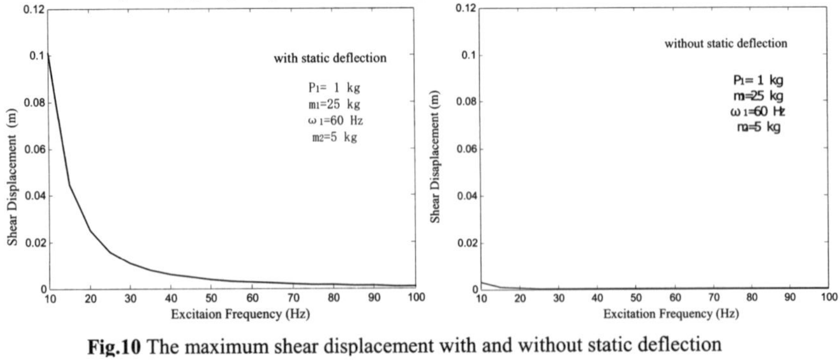

Fig.10 The maximum shear displacement with and without static deflection

5 Experimental Evaluation of ATVA

5.1 Frequency-shift Property

To investigate the frequency-shift capability of the MREs a beam with both ends supported is employed. The developed ATVA is placed at the centre of the beam. White noise excitation applied to the beam is generated by an exciter. Two accelerometers are placed on the oscillator and the base beam to measure their responses, respectively. The measured signals are sent to the Dynamic Signal Analyzer. The transfer function can be obtained by using FFT analysis and the testing results are shown in Fig. 11. It shows that the curves of transfer function move rightward with the increment of the magnetic field. The resonance frequency can be obtained by reading the peak-to-peak value. It is shown in Fig. 12. Its trend agrees well with the theoretical results. The resonance frequency increases from 41Hz at 0 A to 63.75 Hz at 0.6 A and its relative frequency change is as high as 155%.

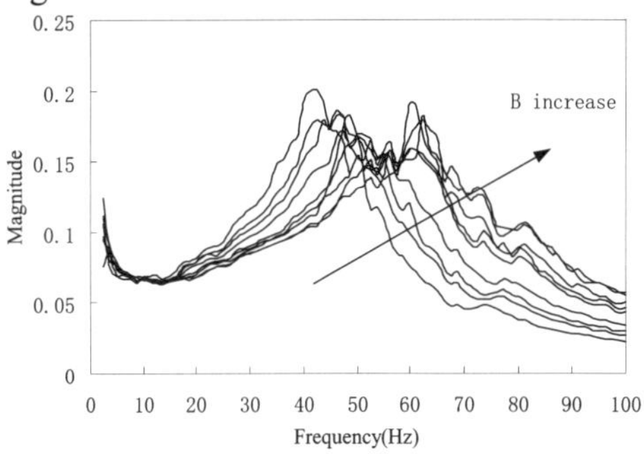

Fig.11 The transfer function at various magnetic fields

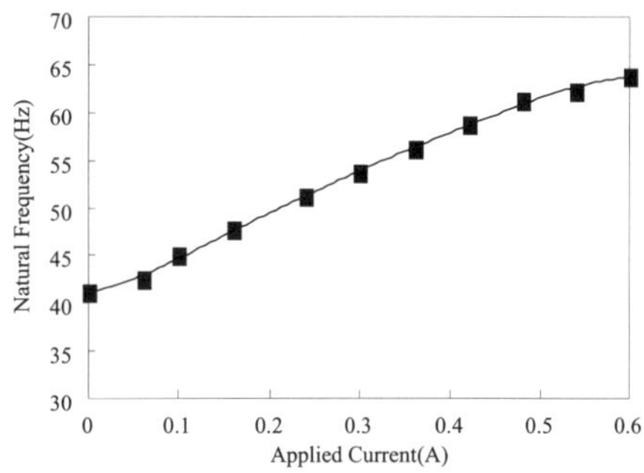

Fig.12 The relationship between applied current and resonance frequency

5.2 Vibration Attenuation

Compared with the frequency-shift experiment, the difference of system to evaluate vibration absorption is that an impedance head is induced. The impedance head connects with the beam and the exciter. The exciter provides sinusoidal excitation to the beam with a linear frequency scan. The base point impedance with various magnetic fields can be measured by the dynamic signal analyzer. To evaluate the vibration absorption capacity, a ratio γ of base point admittance H_A with TVA and that H_o without TVA is employed to reflect the effect of vibration absorption:

$$\gamma=20(|H_A / H_O|) \tag{9}$$

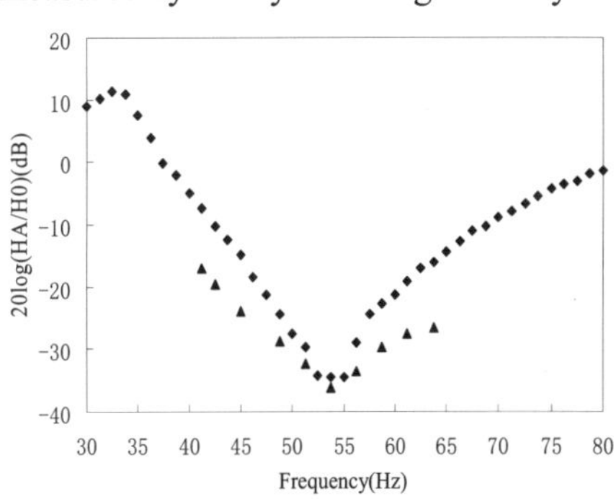

▲adaptive TVA; ◆ passive TVA

Fig.13 Comparison of vibration attenuation between ATVA and TVA

The comparison of vibration absorption capacity between the ATVA and the passive DVA is shown in Fig. 13. The upper data are the results of passive DVA which is realized by fixing the current of ATVA and the lower data are the results of ATVA whose frequency is tuned to trace the excitation frequency. For the passive DVA, the best vibration attenuation efficiency occurs at the natural frequency of the primary system. The effect become worse sharply while the excitation frequency is apart from beam natural frequency and a new resonance hump occurs at 34Hz. For ATVA, in tunable frequency band, it has better absorbing effects than passive DVA.

6 Conclusion

In this paper, a shear mode ATVA based on MRE was designed, simulated and evaluated. After the vibration mode shapes of the ATVA are analyzed and the magnetic circuit of the ATVA is identified by ANSYS software, a compact and efficient ATVA is proposed. Further more, the shift-frequency properties of the ATVA versus magnetic field and strains are analyzed and simulated by using the vibration theory and the modified magnetic dipole model. It is also experimentally verified together with its attenuation ability. The experimental results demonstrate that the designed ATVA has better performance than traditional passive absorber in terms of frequency-shift property and vibration absorption capacity.

7 Acknowledgements

This study is granted by BRJH Project of Chinese Academy of Sciences and Specialized Research Fund for the Doctoral Program of Higher Education (No. 20050358010).

References

1. Frahm H., Device for Damping Vibrations of Bodies, *U.S.Patent* (1909) No. 989958.
2. Walsh P.L., Lamancusa J.S., A variable stiffness vibration absorber for minimization of transient vibrations, *J. Sound & Vibration* **158** (1992) pp. 195-211.
3. Li J.F. and Gong X.L. et al., The research of dynamic properties of adaptive tuned vibration absorber. *Journal of Experimental Mechanics* **20** (2005) pp. 507-514.
4. Williams K.A., Chiu G.T.-., Bernhard R.J., Passive-adaptive vibration absorbers using shape memory alloys, *Proc. SPIE* **3668** (1999) pp.630-641.
5. Davis C.L., Lesieutre G.A., An actively tuned solid-state vibration absorber using capacitive shunting of piezoelectric stiffness, *J. Sound & Vibration* **3** (2000) pp.601–617.
6. Flatau A.B., Dapino M.J., Calkins F.T., High bandwidth tenability in a smart vibration absorber, *Proc. SPIE* **3327** (1998) pp. 463–473.
7. Gong X.L., Zhang X.Z., Zhang P.Q., Fabrication and characterization of isotropic magnetorheological elastomers, *Polymer Testing* **24**(2005) pp.669-676.
8. Watson J. R., Method and apparatus for varying the stiffness of a suspension bushing, *US Patent* (1997) 5,609,353.
9. Stewart W. M., Ginder J. M., Elie L.D., and Nichols M. E., Method and apparatus for reducing brake shudder, U.S. Patent (1998) No.5,816,587.
10. Ginder J.M., Nichols M.E., et al., Controllable-stiffness components based on magnetorheological elastomers, *Proc. SPIE* **3985** (2000) pp.418-425.
11. Ginder J.M., Schlotter W.F., and Nichols M.E., Magnetorheological elastomers in tunable vibration absorbers, *Proc. SPIE* **4331** (2001) pp.103-110.

ER EFFECT OF LOW MOLECULAR LIQUID CRYSTAL ON ONE-SIDED PATTERNED ELECTRODES

TAKEHITO KIKUCHI
Osaka University, 2-1 Yamadaoka, Suita, Osaka, Japan
E-mail: kikuchi@mech.eng.osaka-u.ac.jp

AKIO INOUE
ERtec Co., Ltd., 2-1-31 Sakuragaoka, Minoh, Osaka, Japan

JUNJI FURUSHO AND RYOHEI KAWAMUKI
Osaka University, 2-1 Yamadaoka, Suita, Osaka, Japan

Recently many researches on ER fluids (ERF) are developed for applications of clutches, brakes, dampers and other kind of mechatronics technologies. In many applications, these devices are constructed with bilateral electrodes to apply electric field in ERF. However, the electric field of several kV/mm may be necessary to generate an ER effect sufficiently for practical purposes. The gap between a pair of electrodes should be, therefore, maintained narrowly and exactly for fears of short-circuit. At the same time, this electrode system also requires an interconnection on driving parts. To dissolve these disadvantages on the conventional configuration of the electrodes, we have been proposing an ``one-sided patterned electrode" (OSPE) system. In this report, we confirmed the flow characteristics of low molecular liquid crystal (LMLC) on OSPE. Next, we also confirmed the different characteristics depending on the pattern type. According to results of electro-static analysis, we conclude that such a difference may results from directors of LC molecules derived by electric field.

1 Introduction

Recently many researches on ER fluids (ERF) are developed for applications of clutches, brakes, dampers and other kind of mechatronics technologies [1]-[4]. And ER technology have also adapted to the process technology like polishing [5]. In many case, particular-type ERFs have been used because of its responsibility and dynamic range of force. On the other hand, homogeneous-type ERFs have been applied in same devices. Liquid crystalline materials are also the homogeneous ERFs. In addition, they are classified into liquid crystalline polymer (LCP) and low molecular liquid crystal (LMLC). Particle-type ERFs and LCP-type ERFs have large ER effect. Therefore, many researches and applications have applied them. On the other hand, ER effect of LMLC is generally so small that there is few application of LMLC to mechatronics fields. However, when it comes to applications for lubricant, LMLC might have advantage because LMLC has stable properties and low viscosity at base condition (without electric field).

In this paper, ER effect of LMLC is verified experimentally and consider about the difference of the characteristics of LMLC on some difference electrode type with the result of electro-static analysis. We describe about two type of electrode as follows:

1) Bilateral electrodes shown in figure 1 (a), which have a pair of electrodes opposing each other and ERF is filled between the electrodes,

2) One-sided patterned electrode (OSPE) shown in figure 1(b), in which anodes and cathodes are patterned on one side.

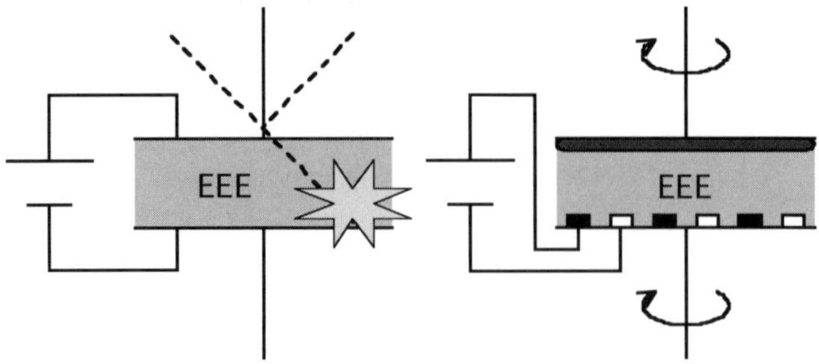

(a) Bilateral configuration (b) One-sided configuration

Fig.1 Configurations of electrodes

In most studies, the ERF has been placed between a pair of opposite electrodes or ``bilateral electrode system". However, the electric field of several kV/mm may be necessary to generate ER effect sufficiently for practical purposes. Therefore, the gap between the electrodes should be maintained exactly for a fear of short-circuit. At the same time, this electrode system requires interconnection on not only fixed parts but also driving parts. It may take any costs and lack maintainability. To dissolve these disadvantages, we proposed ``one-sided patterned electrode [6]" (OSPE). In another report, ER effects of LCP are investigated experimentally and analytically [6], [7]. Moreover, it depends on pattern types of OSPE. However, ER effects of LMLC on OSPE have never been reported.

2 Methods

2.1 Sample material

In this paper, P-type nematic liquid crystal (JC-5012LA, Chisso CO. LTD., Japan) is used as sample material. This material is generally used in LCD. The characteristics of this material are shown in table 1. T_{NI} is the transition temperature from nematic phase to isotropic liquid phase, called "clearing point". $\Delta\varepsilon$, $\varepsilon_{//}$, $\varepsilon_{\perp}$ are a dielectric anisotropy, a dielectric constant parallel to the director of liquid crystalline molecules, a dielectric constant vertical to the director, respectively , and following relation is formed as $\Delta\varepsilon = \varepsilon_{//} - \varepsilon_{\perp}$. It is easily estimated that the ER effect of liquid crystal is related to $\Delta\varepsilon$. But there is few report about the relationship between them.

Table 1. Characteristics of liquid crystal

Transition temperature (℃)	T_{NI}	77.7
Viscosity (at 20 ℃) (mPa· s)	η	24.2
Dielectric anisotropy (at 20 ℃, 1kHz) Parallel dielectric constant Vertical dielectric constant	$\Delta\varepsilon$ $\varepsilon_{//}$ $\varepsilon_{\perp}$	5.7 9.4 3.7
Specific resistivity (Ω· cm)	ρ	$1.8*10^{13}$

2.2 Experimental setup and evaluation method

An appearance of experimental setup is shown in figure 2. ERF is filled between the upper disk and lower disk. The upper disk is rotated with a servomotor (geared motor). Torque generated by the ERF is measured with a torque sensor built in a base plate. Signals from the torque sensor are into shear stresses by calculations depending on the diameter of the disks and the gap between them.

In this paper, bilateral electrodes (Fig.1 (a)) and OSPE (Fig.1 (b)) are used. Two types of OSPE shown in figure 3 are used. Figure 3 (a) illustrates a concentric circular pattern and figure 3 (b) illustrates a radial pattern, respectively. Table 2 shows size parameters of them.

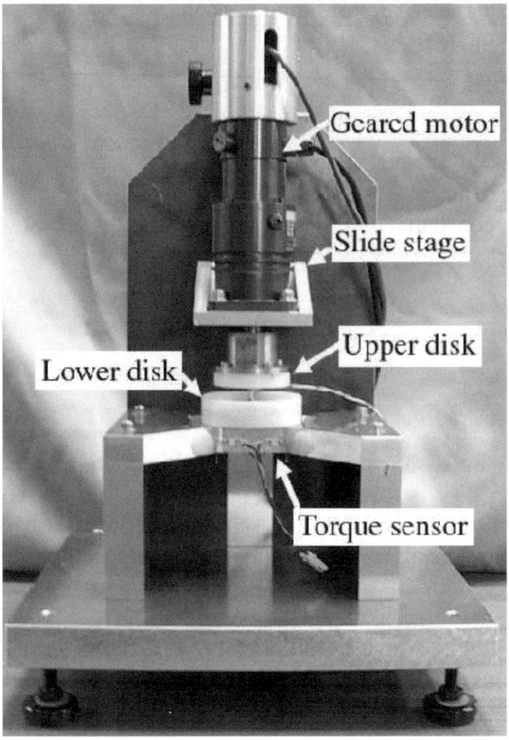

Figure 2. Experimental setup

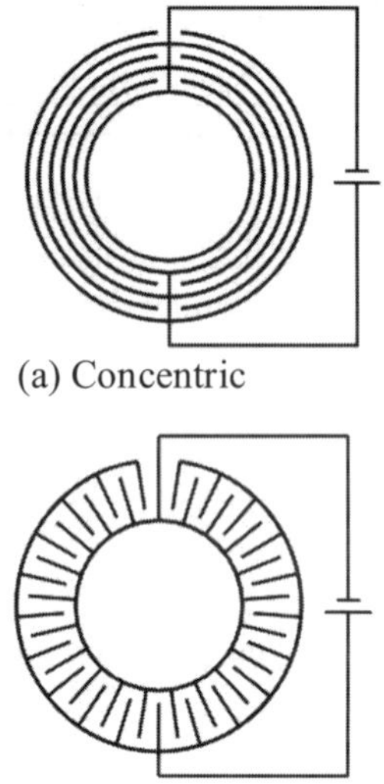

(a) Concentric

(b) Radial pattern

Figure 3. Pattern type of OSPE

Table 2. Size parameters of patterned electrodes

Outer diameter (mm)	60
Electrode gap (μ m)	200
Electrode width (μ m)	Circular : 200 Radial : 120 ~ 470
Electrode thickness (μ m)	40

In the case of experiment with bilateral electrodes, disks masked whole surface with copper are set on the upper and lower disks. In the case of experiment with OSPE, each type of OSPE is set on the lower disk and plastics disk is set on the upper disk.

Rotational speed is measured with rotary encoder built in the geared motor and shear rate is defined as follows;

$$\dot{\gamma} = \frac{R \cdot \omega}{h}$$

where $\dot{\gamma}$ is shear rate (1/s) , R is radius of upper disk (mm), ω is rotational speed of the disk (rad/s) and h is the gap between the upper disk and the lower disk (mm). Next, shear stress of LMLC is defined as follows;

$$\tau = \frac{3T}{2\pi R^3}$$

where τ is shear stress (Pa) and T is torque measured with the sensor (Nm). Considering the rotational parallel-plate type viscometer like this experimental setup, actual distribution of shear stress is not constant. However, we assume that the shear stress is constant overall area of the disk as above equations, because we do not have exact model about shear rate-shear stress relationship of LMLC.

3 Results

Viscosity changes of LMLC were measured by using the experimental setup explained above. Environmental temperature is about 25℃ through all experiments. We evaluate the characteristics of LMLC with shear rate- shear stress relationship.

At first experimental result with the bilateral electrodes is shown in figure 4. The disk gap is 0.1mm, applied electric field are 0.0, 1.0, 2.0, 3.0kV/mm, respectively. The relationship is straight line with constant electric field, and its slope changes depending on the strength of electric field applied to LMLC.

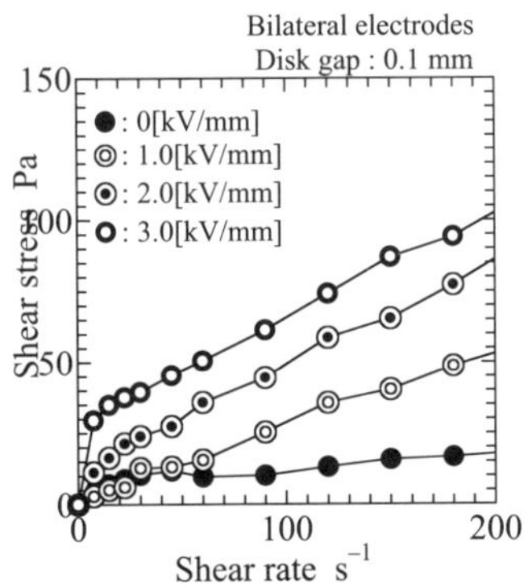

Figure 4. Flow curve of LMLC on bilateral electrodes

Secondly, the results with OSPE are shown in figures 5. Figures 5 (1-a)-(1-c) are the flow curve on circular pattern and (2-a)-(2-c) are the ones on radial pattern. The disk gaps are 0.05, 0.1, and 0.2 mm, respectively. On the radial pattern, the distribution of electric field is not constant on the whole area. Therefore we inscribed the value of electric field dividing the applied voltage by the central value of electrode gap (0.2mm).

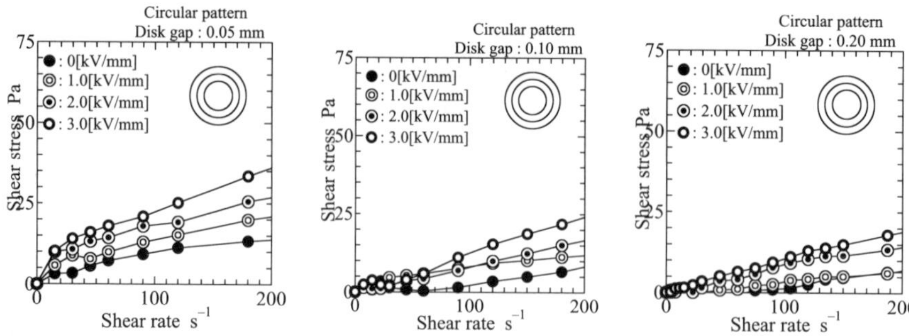

(1-a) On circular pattern (disk gap : 0.05mm) (1-b) On circular pattern (disk gap : 0.1mm) (1-c) On circular pattern (disk gap : 0.2mm)

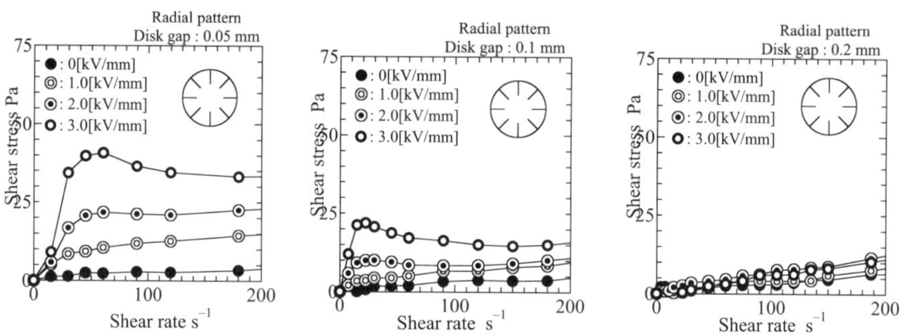

(2-a) On radial pattern (disk gap : 0.05mm) (2-b) On radial pattern (disk gap : 0.1mm) (2-c) On radial pattern (disk gap : 0.2mm)

Figure 5. Flow curve of LMLC on OSPE

It is confirmed that the larger the disk gap becomes, the smaller the ER effect becomes. In the case that the disk gap is 0.2mm, the disk gap and electrode gap are equal to each other. In this case, it is considered that the upper disk is out of the effective range of the viscosity change of the ERF and visible changes of shear stress did not generate. In the case that the disk gap is 0.05mm, these characteristics have grate difference between the cases of the concentric electrode and radial electrode. In the case of the concentric electrode shown in figure 8 (1-a), flow curve seems like Newtonian flow. On the other hand, flow curve of radial electrode shown in figure 8 (2-a) seems like Plastic flow.

In the case of using the radial pattern electrode with 0.05mm disk gap, large ER effect was shown. Although this effect is limited at low shear rate, the change ratio exceeds ten times. In conventional researches, such a large change of LMLC is not reported. Therefore, this result is highly important.

4 Discussion

To estimate the distribution of the director of LMLC, electro-static analyses are carried out. At first, figure 6 shows the result of the electro-static analysis on the concentric circular electrode. In the case of the circular electrode, 2-D analysis is sufficient because a shape of electrode is rotational symmetry. Each parameter is set as follows; electrode width is 200μm, electrode gap is 200μm, electrode height is 40μm and applied voltage is 400V. At second, figure 7 shows the result of the analysis on the radial patterned electrode. This type of electrode is not rotational symmetry; therefore, 3-D analysis was used. For analysis target, a pair of bosses of anode and cathode is adopted as a minimal unit. Size parameter is referenced form table 2. Applied voltage was 400V.

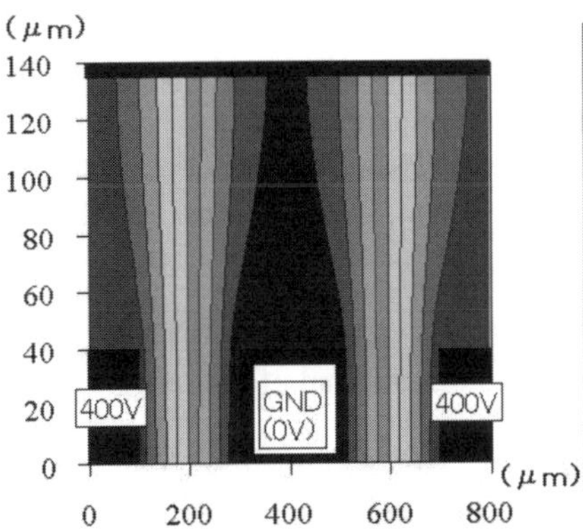

Figure 6. Electric potential on circular pattern

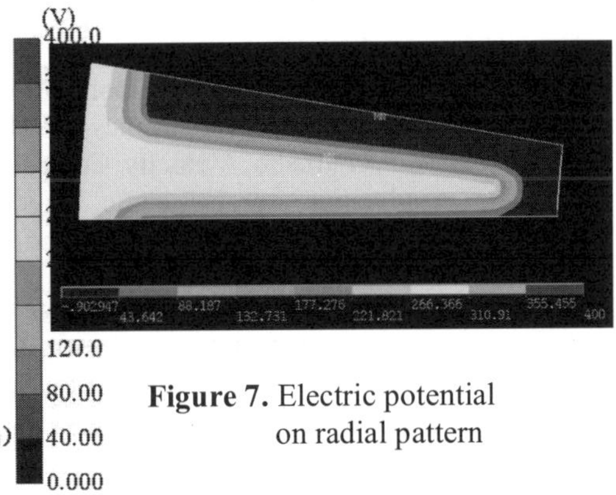

Figure 7. Electric potential on radial pattern

From these result of electro-static analyses, the distribution of electric field vector or the differential vector of the electric potential consists of the horizontal vector from anode to cathode. If the directors of LMLC follow toward this vector, the director makes the distribution shown in figure 8. It is considered that the different characteristics depending on the type of electrode result from such a difference of director of liquid crystal molecules on the electrodes.

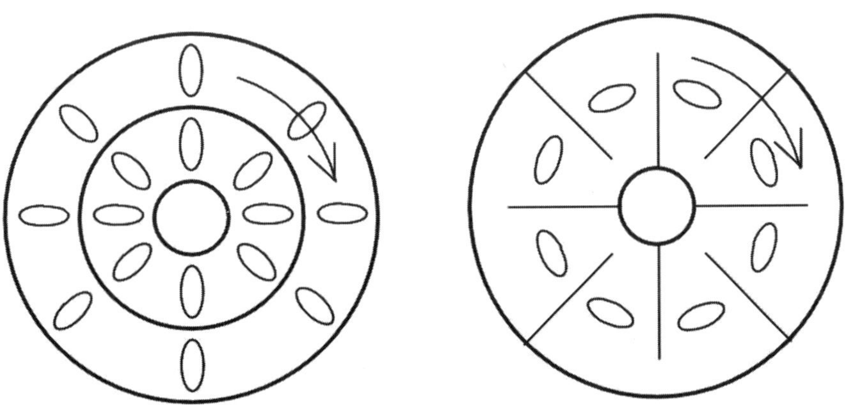

(a) On circular pattern (b) On radial pattern

Figure 8. Distributions of directors of liquid crystalline molecules

5 Acknowledgements

This work was financially supported by the Grant-in-Aid for Scientific Research of Japan (Younger scholar: No. 17760113). The authors gratefully acknowledge.

References

1. M. Sakaguchi, G. Zhang and J. Furusho, Modeling and motion control of ER actuator, Journal of Robotics and Mechatronics, Vol.13, No.1, pp. 80-87 (2001).
2. T.Kikuchi and J.Furusho, Velocity Control of Brake Using Particle-Type ER Fluid and Its Application to Isokinetic Exercise System, Proceedings of the 6th International Conference on Motion and Vibration Control, Vol.2, pp. 837-842 (2002).
3. K. Koyanagi and J. Furusho, DIRECT-DRIVE MOTOR SYSTEM WITH PARTICLE-TYPE ER FLUID DAMPER, International Journal of Modern Physics B, Vol. 19, Nos. 7-9, pp.1661-1667 (2005).
4. J. Zhang, J. Furusho, M. Uda and T. Yamaguchi, DEVELOPMENT OF A NEW LINEAR GUIDE SYSTEM WITH ELECTRICALLY CONTROLLABLE DAMPER USING LIQUID CRYSTALLINE POLYMER, International Journal of Modern Physics B, Vol. 19, Nos. 7-9, pp.1648-1654 (2005).
5. Takehito KIKUCHI, Junsuke FUJIWARA, Junji FURUSHO and Akio INOUE, Polishing Using ER Slurry on One-sided Patterned Electrodes, International Journal of Modern Physics B, Vol. 19, Nos. 7-9, pp.1682-1688 (2005).
6. N.TAKESUE , J.FURUSHO , and A.INOUE , J.Appl.Phys., Vol.91, No.3, pp.1618-1623 (2002).
7. N.TAKESUE , J.FURUSHO , and A.INOUE , J.Appl.Phys., Vol.94, No.8, pp.5367-5373 (2002).

DESIGN OF THE HIGH-PERFORMANCE MR BRAKE AND ITS CHARACTERISTICS

TAKEHITO KIKUCHI, JUNJI FURUSHO, YUHEI YAMAGUCHI AND SHINYA KIMURA

Osaka University, 2-1 Yamadaoka, Suita, Osaka, Japan
E-mail: kikuchi@mech.eng.osaka-u.ac.jp

MR Fluid (MRF) is suitable material for some mechatronics device, for example clutches or brakes because of its rapid response and large viscosity change. However, total response speed will grow worse due to the affect of eddy current at magnetic circuit. Realization of high torque devices requires large area of magnetic circuit. However, the larger the area becomes, the larger the effect of eddy current grows with reduction of electric resistance. To dissolve this problem, we propose to use laminated silicon steel sheets (LSSS) for yokes of electro-magnets. By using LSSS, we developed high-torque (> 30Nm) and high-speed (< several m illiseconds) MRBrake (MRB). In this paper, the design guideline of the MRB is described. In this guideline, we use FEM analysis to design shapes and location of yokes. In addition, experimental results of MRB are presented. As a result, high performance MRB, whose maximum output torque is 36Nm and output response speed is several milliseconds, was realized.

1 Introduction

We have developed some devices using ER fluid (Electro-rheological fluid, ERF) or MR fluid (Magneto-rheological fluid, MRF), for example clutches, brakes and dumpers [1]-[5]. These mechatronics devices have grate responsibility. In particular clutch-type- devices using ERF or MRF have some advantages of low-inertia and high-safety [5]. However, in the past researches to design the MR devices, we had a problem that the output response did not have prospective rapidity. According to some reports [6], the response time of MRF is several milliseconds. However, the response time of the devices using MRF had been several ten milliseconds or more. In the later researches, it became clear that the reason of this problem was a work of eddy current. The electric current to coils generates the eddy current at magnetic circuit. The eddy current grows up if the electric resistance of the magnetic circuit is small. MRF needs about 1T [tesla] of magnetic flux density to realize sufficient changing of viscosity. Therefore, it is so important to make the magnetic resistance of the magnetic circuit small. However, in generally, materials with low electric resistance have also low magnetic resistance like iron. It is important to make the electric resistance of magnetic circuit small, at the same time, to make the magnetic resistance of it large.

In this paper, we develop new brake using MRF. We simply call it MRB in this paper. This MRB has grate performance of output response by using laminated silicon steel sheets (LSSS) as the magnetic circuits. LSSS reduces the affect of eddy current and improve the responsibility of MRB to several milliseconds.

2 Methods

2.1 MR brake

The "MR clutch" is a mechanism using MRF for torque transmission. Figure 1 represents the basic structure of MR clutch. Application of electric current to the MRF controls the transmitted torque. MR clutch has higher responsibility, higher torque/inertia ratio and lower inertia than other conventional clutches, for example, a powder clutch and electromagnetic clutch. "MR actuator" consists of a MR clutch, a servomotor for input, and a reduction system. Figure 2 shows cross sectional view of a MR brake developed in a past report [7]. On the other hand, when the input part of MR clutch is fixed, we call it "MR brake".

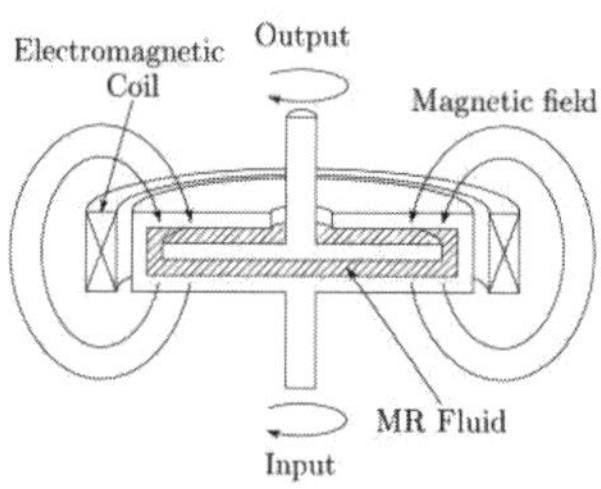

Figure 1. Basic structure of MR clutches

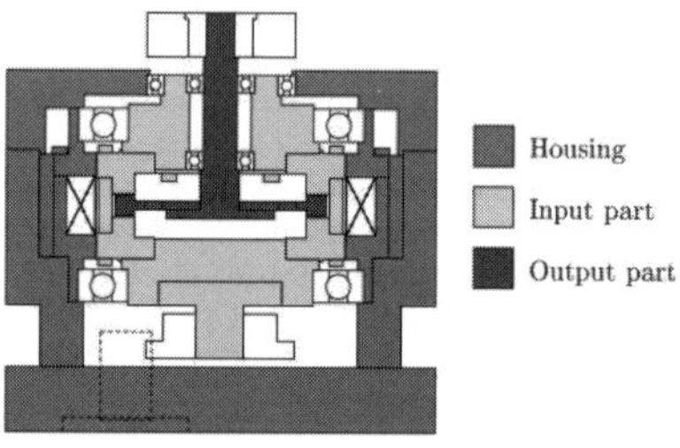

Figure 2. Cross section of MR actuator

Generally, the electromagnetic brakes and the powder brakes have time delay of several tens milliseconds or more. Therefore, it is difficult to use them as servomechanisms. On the other hand, ER brakes and MR brakes have so fast time constant (several milliseconds) that they are suitable for servomechanisms. However, we should consider the delay time of magnetic flux by eddy current of the MR brake.

2.2 Improvement of magnetic circuit

Changing of the magnetic flux crossing into the magnetic circuits made of conductors generates the eddy current. If the magnetic circuits have a wide width like shown in the left drawing of figure 3, electric resistance becomes small and eddy current grows up. On the other hand, if we use laminated silicon steel sheets (LSSS) shown in the right drawing of figure 3 as a yoke (or magnetic circuit), we can reduce the eddy current generated in the yoke. Therefore, we adopted C-Type-yoke with LSSS to the magnetic circuit of MR brake.

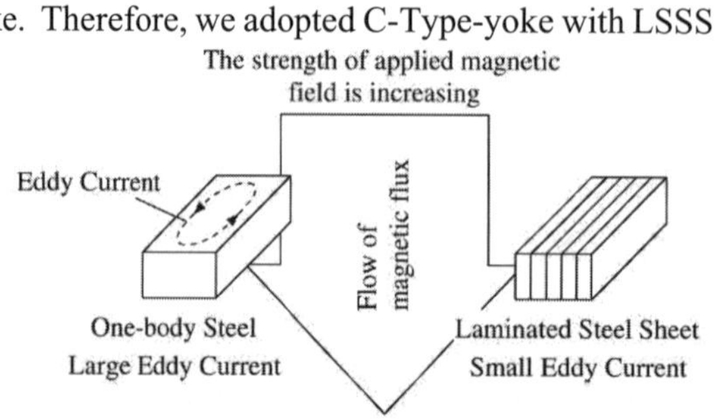

Figure 3. Laminated silicon steel sheet (LSSS)

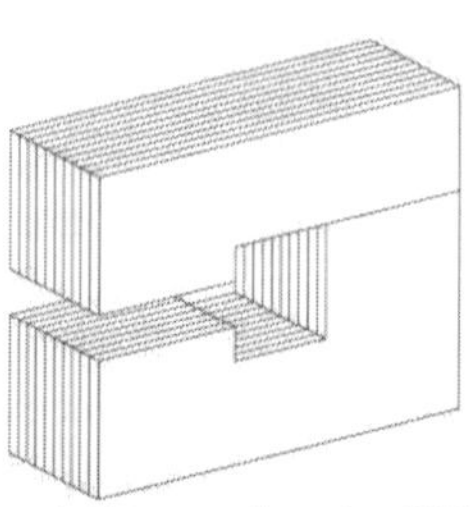

Figure 4. C-type yoke using LSSS

2.3 Magnetostatic analysis

Magnetostatic analysis was carried out for confirmation of the magnetic flux density of the designed brake. Figure 5 shows the basic components of MRB. This brake has a rotational disk and 4-yokes around the disk. MRF is filled into a case in which the disk is inserted. The yokes of the electro-magnets are made of LSSS.

Figure 6 shows a FEM (finite elemental method) model for the magnetostatic analysis. This model consists of a quarter shape of the new MRB. 3-D analysis was carried out using ANSYS/University Low Option, Ver.6.1. Material of a shaft and a disk are steel, a case is plastics and yokes are silicon steel, respectively. B-H curve of each material is set for a non-linear analysis.

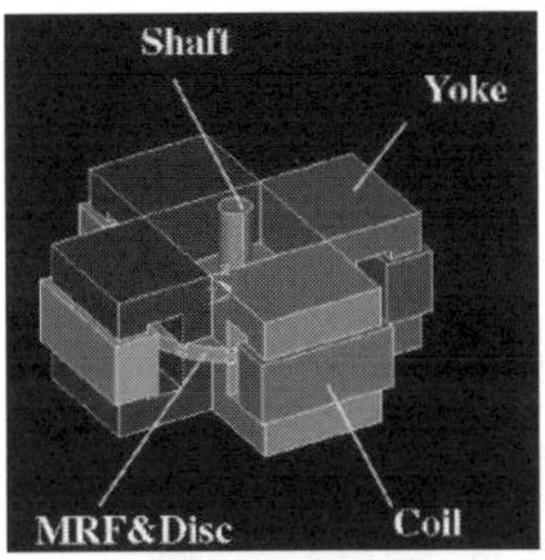

Figure 5. Basic components of MRB

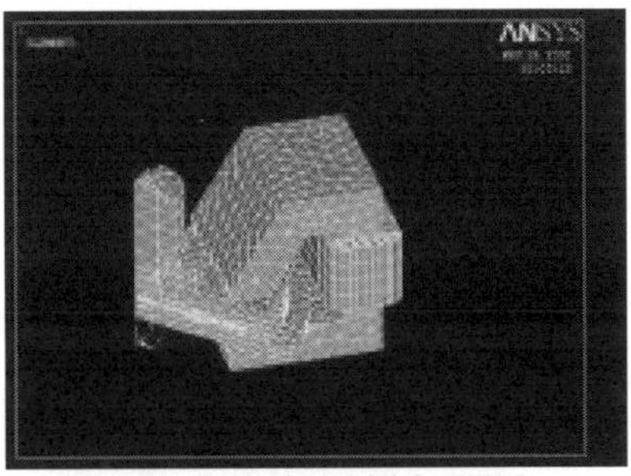

Figure 6. Analysis model for FEM

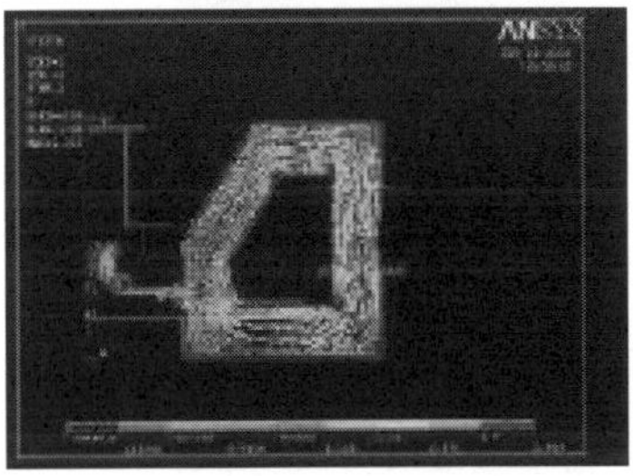

Figure 7. Result of analysis (Magnetic flux in the C-Type yoke)

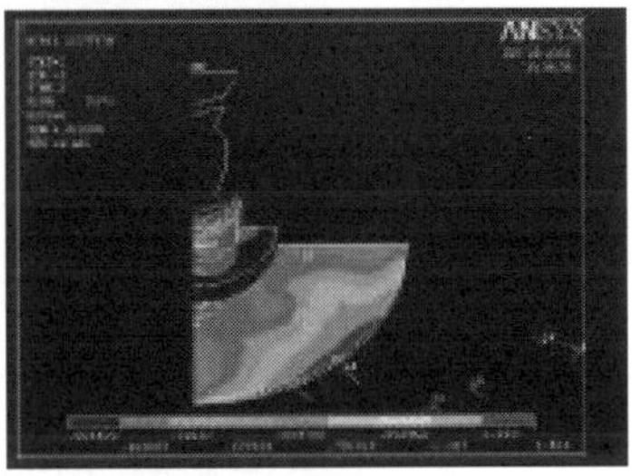

Figure 8. Result of analysis (Magnetic flux density on the disk)

Figure 7 and 8 show the results of the magnetostatic analysis. The goal of this analysis is to determine a value of current to generate about 1 [T] of magnetic field around the disk or MRF. According to the results, it was confirmed that we need to apply about 2 [A] of current to 4 coils whose number of winding are 400 each.

3 Results

3.1 Developed MR brake with C-type yokes

MRF-132LD (Lord Co.) is used as working fluid. Figure 9 shows a cross sectional view, and figure 10 shows a picture of the MRB. Table 1 shows specifications of it.

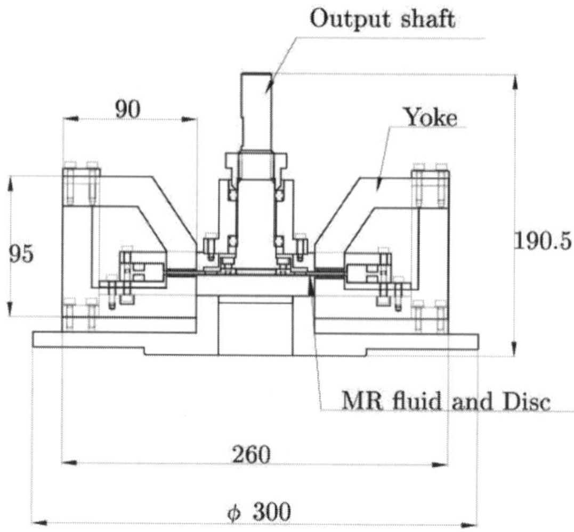

Figure 9. Sectional view of New MRB

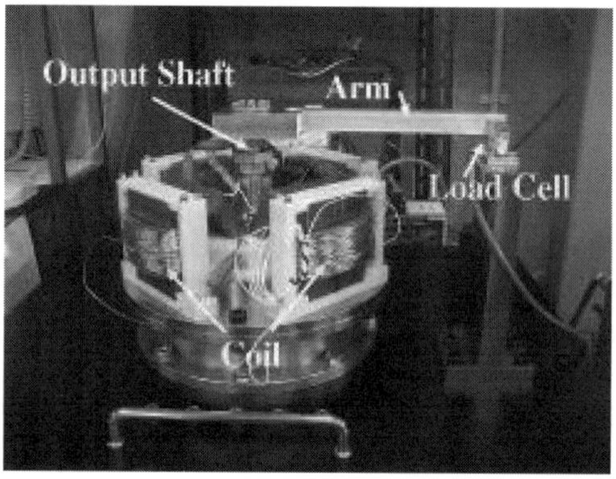

Figure 10. New MR Brake and torque sensor system

Table 1. Specifications of New MRB

Diameter of a casing (mm)	300
Height (mm)	190.5
Diameter of a disc (mm)	120
Disk gap (mm)	0.4
Number of electric-magnets	4
Mass (kg)	18.0
Maximum current (A)	2

3.2 Experimental confirmation for the output characteristics of new MRB

Some basic experiments on the MRB were carried out. Figure 11 and 12 show a experimental setup. A rotary table rotates MRB and an arm connected to the output-shaft pushes a force sensor. Then we can calculate the transmitted torque by the MRB. A servo-amplifier and a personal computer control the current applied to the coils.

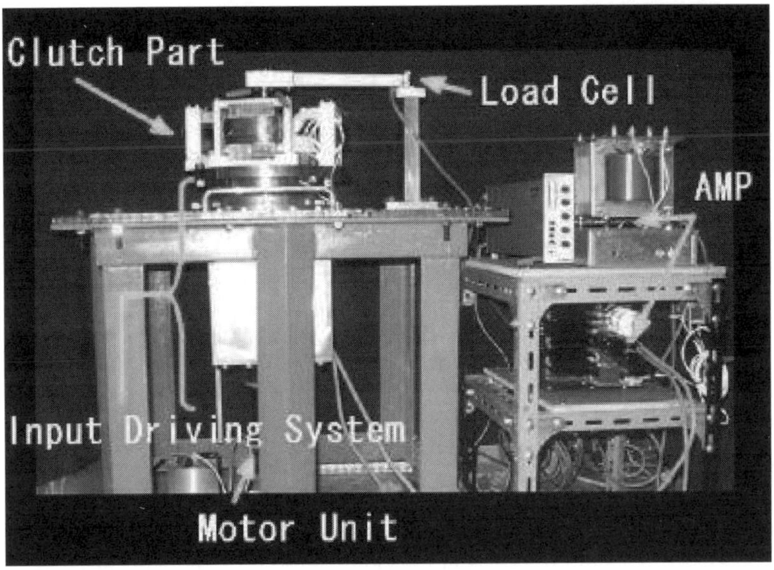

Figure 11. Experimental setup for new MRB

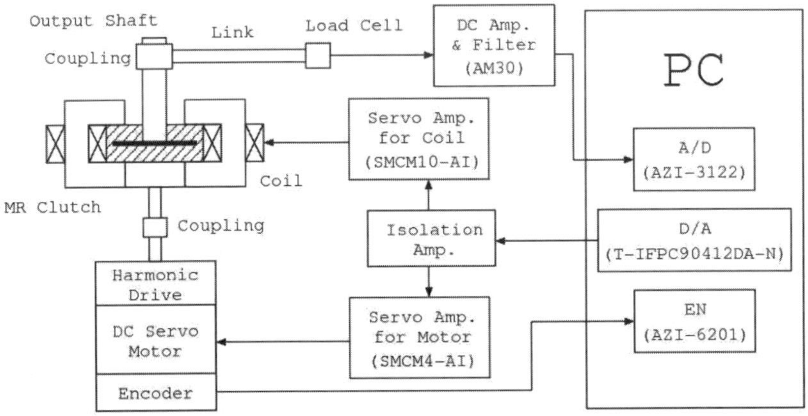

Figure 12. Control system of experimental setup for new MRB

Figure 13, 14 show dynamic and static characteristics of MRB respectively. In these experiments, the input rotating speed is 0.5 [rad/s]. Time constant of the MRB is several milliseconds and the maximum torque of it is about 36 [Nm]. Base torque or the torque generated by base viscosity of MR fluid is 0.1 [Nm]. Therefore, controllable range of output torque is about 360 times by the ratio for the base torque.

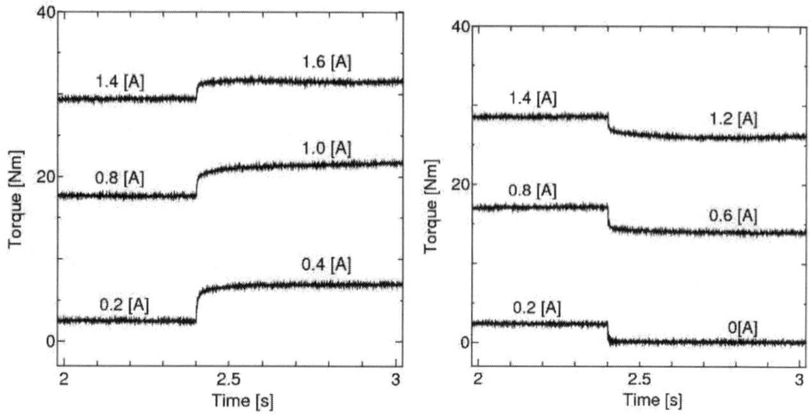

(a) Rising edge (b) Falling edge

Figure 13. Step response of new MRB (Current input)

4 Conclusion

In this paper, we discuss about a design method of a brake using MR fluid. The output response of MR fluid is so fast but the eddy current of magnetic circuit prevent it. Therefore, we propose laminated silicon steel sheets (LSSS) for yokes of electro-magnets. We confirmed that LSSS yokes could improve the responsibility of MRB experimentally.

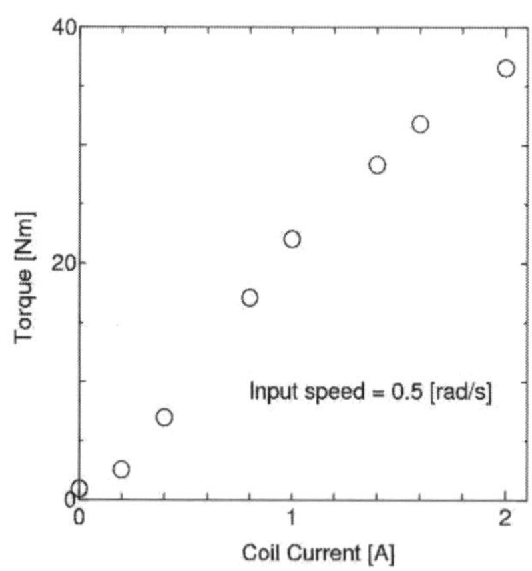

Figure 14. Static response of new MRB

We developed the new MRB which has high torque performance (Min. 0.1 [Nm], Max. 36[Nm]) and high responsibility (several milliseconds).

References

1. M. Sakaguchi, G. Zhang and J. Furusho, Modeling and motion control of ER actuator, Journal of Robotics and Mechatronics, Vol.13, No.1, pp. 80-87 (2001).
2. T.Kikuchi and J.Furusho, Velocity Control of Brake Using Particle-Type ER Fluid and Its Application to Isokinetic Exercise System, Proceedings of the 6th International Conference on Motion and Vibration Control, Vol.2, pp. 837-842 (2002).
3. K. Koyanagi and J. Furusho, DIRECT-DRIVE MOTOR SYSTEM WITH PARTICLE-TYPE ER FLUID DAMPER, International Journal of Modern Physics B, Vol. 19, Nos. 7-9, pp.1661-1667 (2005).
4. J. Zhang, J. Furusho, M. Uda and T. Yamaguchi, DEVELOPMENT OF A NEW LINEAR GUIDE SYSTEM WITH ELECTRICALLY CONTROLLABLE DAMPER USING LIQUID CRYSTALLINE POLYMER, International Journal of Modern Physics B, Vol. 19, Nos. 7-9, pp.1648-1654 (2005).
5. Y. Yamaguchi, J. Furusho, S. Kimura and K. Koyanagi, DEVELOPMENT OF HIGH-PERFORMANCE MR ACTUATOR AND ITS APPLICATION TO 2-D FORCE DI SPLAY, International Journal of Modern Physics B, Vol. 19, Nos. 7-9, pp.1485-1491 (2005).
6. Carlson, J. D. and Jolly, M. R., MR fluid, foam and elastomer devices, Mechatronics, Vol.10, 555-569 (2000).
7. J. Lin, H. Asaoka, M. Sakaguchi, G. Zhang and J. Furusho, Applying Study Connected with Development of New Actuator Using MR Fluid, Proceedings of the 2000 Japan-USA Flexible Automation Conference, pp.413-416 (2000).

THE INFLUENCE OF MAGNETIC FIELD ON THE SPEED OF FILLING THE CYLINDRICAL CAPILLARIES WITH MAGNETORHEOLOGICAL FLUID

E. V. KOROBKO, N. A. ZHURAVSKI

A. V. Luikov Heat and Mass Transfer Institute, National Academy of Sciences of Belarus, 15 P. Brovka Street, Minsk, 220072, Belarus

E-mail: srgor@hmti.ac.by

The results of experimental research of the process of capillary rise of magnetorheological suspensions in the vertical cylindrical capillary in the magnetic field were presented. It was shown that under small magnetic fields the speed of capillary filling increased. Under big values of magnetic induction this process is slowed down. In the capillaries of a small diameter the influence of magnetic field leads to the decrease of the height of capillary rise which is due to the overlapping of wall menisci and the change of the coefficient of surface tension.

1 Introduction

Effective controlling the form of the free surface and the kinetics of fluid absorption in the capillary by means of external magnetic impacts by regulating the surface properties is an actual problem. For example, the decrease of surface tension allows to change the height of the capillary rise of the electroconductive fluid by means of the impact of external electric signal [1]. The change of the rim wetting angle is known to allow to control the process of drop spread of electroconductive and electrorheological fluids [2, 3] and to stabilize liquid microstructures [3].

The capillary absorption of magnetic fluids was earlier determined to be influenced by magnetic field: under the influence of inhomogeneous magnetic field the column of magnetic fluid is directed towards the areas of higher magnetic fields and the speed of the process grows up [4]. In the homogeneous field the height of the rise is increased with the induction growth. The speed and the time of a rise depend on the orientation of magnetic field and on the influence of gravity [5].

The problems of the capillary flow of magnetorheological suspensions (MRS) that have the ability for strong structuring under magnetic fields (in comparison to magnetic fluids) were not considered before.

2 Methods

The aim of this paper is to do an experimental research on the influence of magnetic field on the capillary rise of MRS in the vertical capillary. A number of suspension compounds based on the carbonyl iron (dispersion medium — oil MOBIL) and MRS based on the chrome dioxide (dispersion medium — oil PMS–100) are used as the object of investigation.

On the first stage of experiments the rheological properties of MRS under and without magnetic field are determined by means of rotational viscometry. After that the capillary rise of these media is investigated in the cylindrical capillary. The experimental method is the following: in the initial time moment the capillary is put down by the lower face in MRS that is in the plain vessel. At that, the rise of the MRS column starts in the capillary because of capillary forces. As the surface area of the vessel is much bigger than the area of the capillary cross-section, the suspension level in the vessel did not change dramatically at a MRS rise. The capillary and the vessel are put in magnetic field, created by the electromagnet, which induction vector is directed parallel to the capillary. At certain periods of time the capillary is photographed, which allows to fix the motion kinetics and the form of MRS meniscus. Glass capillaries with the inner diameters of 0.27, 0.45, 0.6 and 0.8 mm were used in the experiments.

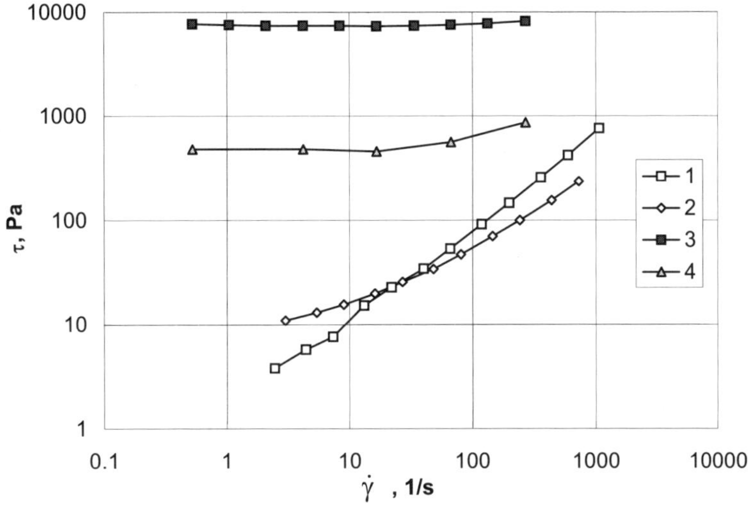

Figure 1. MRS flow curves: 1, 3 – MRS based on the carbonyl iron and oil Mobil, 2, 4 – MRS-14P; 1, 2 – B=0, 3, 4 – B=250 mT (B – magnetic field induction).

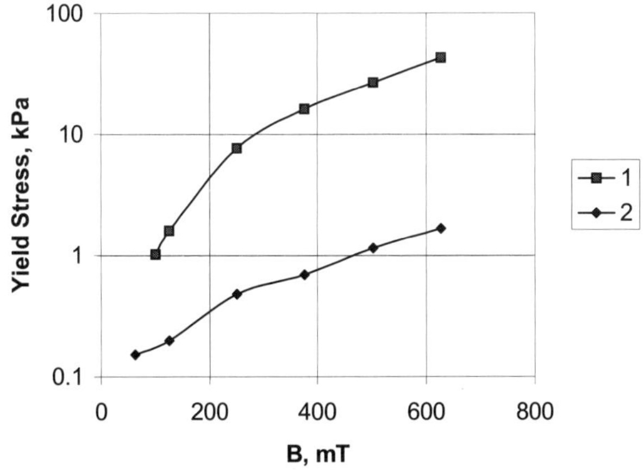

Figure 2. Dependence of MRS fluidity limit on the magnetic field induction: 1 – MRS based on the carbonyl iron and oil Mobil, 2 – MRS-14P.

3 Rheological properties

Flow curves of MRS based on the carbonyl iron with the mass concentration of the dispersed medium 85.8% and on the chrome dioxide with mass concentration 10% (MRS-14P) determined on the rotational viscometer Rheotest 2.1 are shown in Fig. 1. With the increase of magnetic field the flow curves rise along the axis. A substantial fluidity limit appears for this media in magnetic field. (Fig. 2). The increase of the MRS-14P fluidity limit is smaller in the magnetic field as chrome dioxide particles have smaller magnetic susceptibility than the carbonyl iron particles. More over the volume concentration of the dispersed phase of the second suspension is by order smaller.

4 Capillary raising

The experimental result analysis has shown that the investigated MRS wet partially the capillary walls and the wetting angle θ is acute. The raising curves (the diagrams of the column height change in the capillary with time) of MRS based on carbonyl iron in the capillary of a diameter of 0.27 mm under and without magnetic field at induction values 5 and 14 mT are shown in Fig. 3.

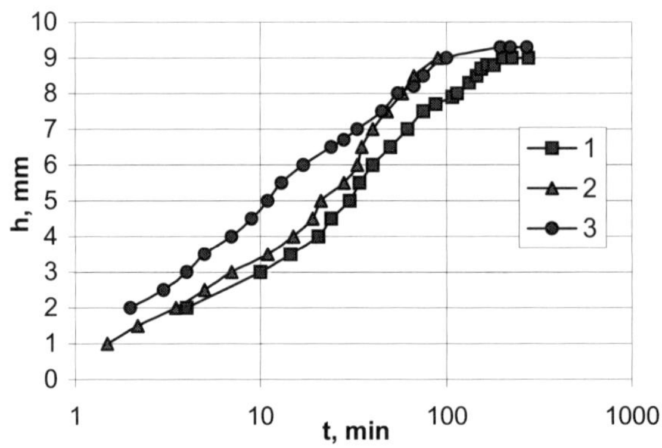

Figure 3. Capillary rise curves of MRS based on the carbonyl iron in the capillary of the diameter 0.27 mm: 1 – without magnetic field; 2 – B=5 mT; 3 – 14.

The process of MRS capillary rise up to the maximum height is found to develop slowly. The characteristic rise time [5] was not bigger than 20 s for magnetic fluids, for MRS the process lasted for 2 – 3 h.

The influence of the induction value of magnetic field on the maximum rise height is not found. In the investigated range of values of magnetic field induction the height was 9 mm in all cases. As the fluid raising height in the capillary is

$$h = \frac{2\sigma\cos\theta}{\rho g r}, \tag{1}$$

where r — capillary radius, σ – surface tension coefficient, then the calculated value σcosθ is 0.022 N/m. That corresponds to the capillary pressure $\Delta p = \rho g h = 3.4 \cdot 10^2$ Pa.

In the capillary of big inner diameter of 0.8 mm the maximum raising height h is 3 mm. That corresponds to the value of $\sigma\cos\theta$, equal to 0.023 N/m, which is in well agreement with experiments, carried out on the capillary with a diameter 0.27 mm. The capillary pressure is $1.1{\cdot}10^2$ Pa respectively.

It was determined, that the acceleration of the meniscus motion of the MRS takes place in the capillary under magnetic field. The dependence of rising time on the magnetic field induction is shown in Fig. 4. As the characteristic raising time the time when the column reaches 90 % of maximum height is taken. At magnetic field induction about 10 mT the MRS rising time decreases in two times.

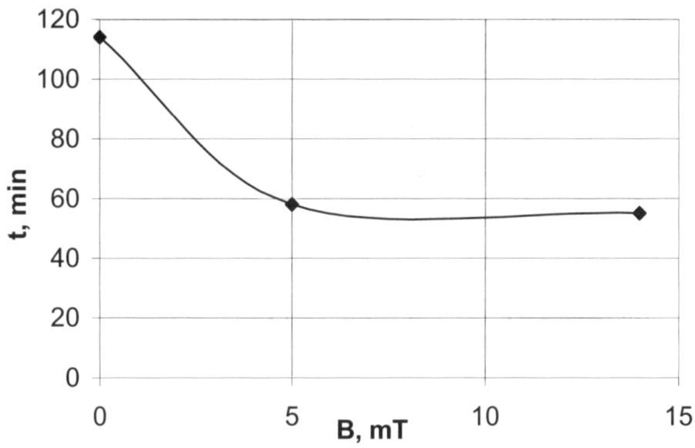

Figure 4. Dependence of the capillary rising time of MRS based on the carbonyl iron on the magnetic field induction.

The range of magnetic field induction values at which the capillary rise is possible is limited. Any substantial MRS rise was not found in the fields with induction 20 mT and higher as the fluidity limit exceeds the capillary pressure in this case.

Suspensions based on the chrome dioxide have smaller effective viscosity and density (about 1.0 g/cm^3 compared to 3.8 g/cm^3), than MRS based on the carbonyl iron. Therefore there are capable to rise on higher as it follows from (1). Besides, the process of rising takes place with higher speed, characteristic time of rise is 20 – 40 min. Capillary rise curves for a suspension based on the chrome dioxide (MRS-14P) in the capillary with the inner diameter of 0.6 mm are shown in Fig. 5. The value of capillary pressure in this case is 98 Pa. For this suspension the value $\sigma\cos\theta$ is 0.014 N/m. As MRS based on the chrome dioxide have other rheological propertied, the rise takes place at bigger values of magnetic field induction than in the first case, where the rage of B is limited by 20 mT.

According to rheological experiment data, the time of rise increased in power dependence with the increase of effective viscosity of MRS for all compositions under investigation. The dependence of capillary rising time on the effective viscosity of MRS (without magnetic field) is shown in Fig. 6.

For MRS based on the chrome dioxide the magnetic field also does not practically influence the final rising height in the range of under 35 mT in the capillaries with diameters of 0.45 and 0.6 mm. In magnetic fields with the induction of over 35 mT some decrease of the capillary rise height is seen in such capillaries (for example, in the capillary with a diameter of 0.6 mm – from 10 down to 8 mm in the field with induction of 60 mT, Fig. 5, curve 6). In the capillaries with a diameter of 0.27 mm the rising height of MRS is lowered substantially at the increase of magnetic filed induction, starting with 15 mT (from 20 down to 7.5 mm),which can be linked to the overlapping of wall menisci with the change of σ.

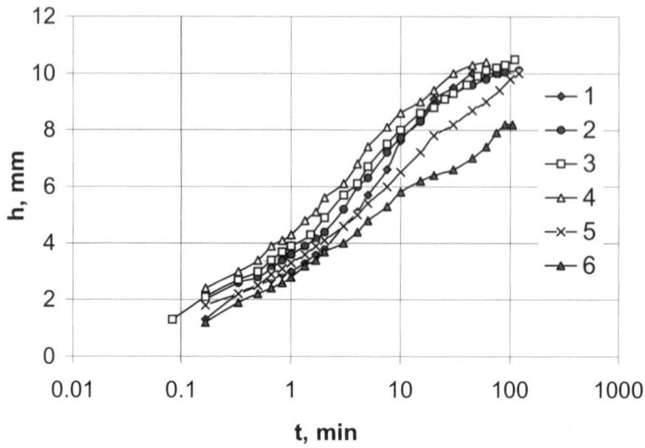

Figure 5. Capillary rise curves of MRS-14P in the capillary of the diameter 0.6 mm: 1 – without magnetic field; 2 – B=12 mT; 3 – 17.5; 4 – 25; 5 – 35; 6 – 60.

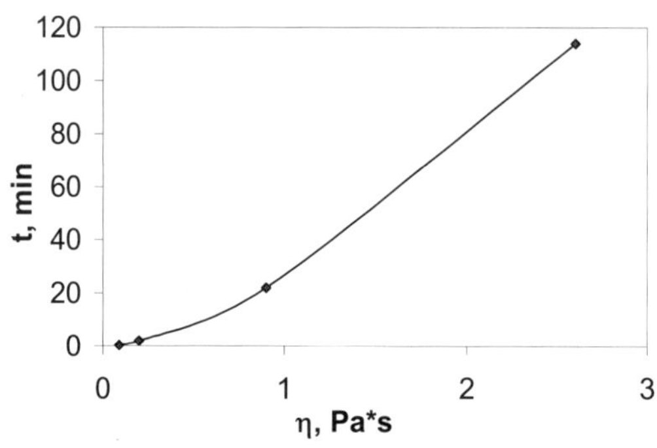

Figure 6. Dependence of the capillary rising time on the effective viscosity of MRS (B=0).

The rising height is inversely proportional to the capillary diameter. This is illustrated in Fig. 7, where the curves of capillary rise of MRS-14P are given under no magnetic field in the capillaries of different diameters.

The field changes the speed of capillary filling. The characteristic rising time of MRS-14P, depending on the magnetic field induction is shown in Fig. 8. In magnetic fields with induction up to 25 mT an acceleration of meniscus motion takes place, the rising time lowers from 1.5 up to 4 times in the capillaries of different diameters, then at big values of magnetic field the change of rheological properties of MRS under magnetic field starts influencing the capillary process. In the magnetic field with values of induction over 30 mT the slowing down of the rising speed is seen, in the fields with the induction over 60 мT a stratification of the suspension in the capillaries takes place.

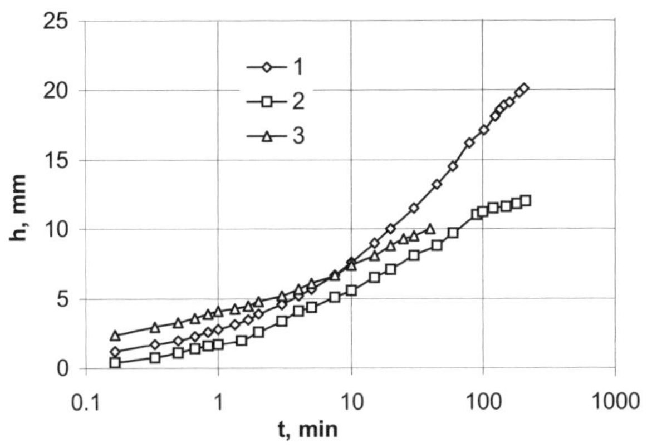

Figure 7. Capillary rise curves for MRS-14P without magnetic fielding the capillaries of different diameters: 1 – 0.27 mm; 2 – 0.45; 3 – 0.6.

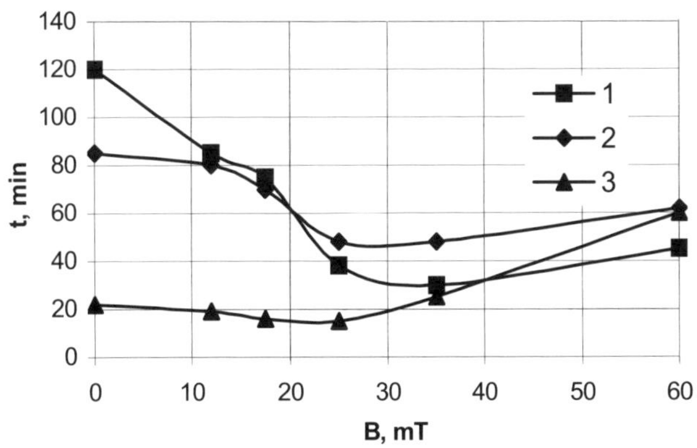

Figure 8. Dependence of the capillary rising time of MRS-14P on the induction of magnetic field: 1 – capillary diameter 0.27 mm; 2 – 0.45; 3 – 0.6.

5 Conclusion

Experimental study of the process of capillary rise of magnetorheological suspensions based on the carbonyl iron and chrome dioxide allowed to determine in the magnetic field, that the impact of the magnetic fields results in the increase of the capillary filling speed. The characteristic rising time is lowered in two times for the MRS based on the carbonyl iron under magnetic fields with the induction of about 10 – 15 mT. For a suspension based on the chrome dioxide the increase of rising speed from 1.5 up to 4 times takes place in the magnetic field at the value of induction about 25 mT. It was determined that 1—2 h are needed to reach the final rising height for a suspension based on the carbonyl iron and 20—40 min for a suspension based on the chrome dioxide. The capillary pressure of the meniscus is about 10^2 Pa.

It was found, that the rising time increases in the power dependence with the increase of effective viscosity of MRS. It was also determined that at big values of magnetic field induction the change of rheological properties of MRS leads to the decrease of capillary rising speed under the influence of the field.

References

1. Prins M. W. J., Wetters W. J. and Weekamp J. W., Fluid Control in Multichannel Structures by Electrocapillary Pressure, *Science* **291** (2001) pp. 277–280.
2. Korobko E. V., Gorodkin R. G., Goncharova N. A. and Baikow V. I., Some Features of the Composition Effect on the Rheology and Characteristics of ERF Spreading *Intern. J. of Applied Mechanics and Engineering* **8** (2003) pp. 67–72.
3. Mugele F. and Herminghaus S., Electrostatic Stabilization of Fluid Microstructures, *Applied Physics Letters* **81** (2002) pp. 2303–2305.
4. Bashtovoi V. G, Kuzhir P. P., Reks A. G., Statistics of magnetic fluids in the cylindrical capillaries. In *New "smart" materials, electro- and m,agnetorheological fluids and their application for energy effective technologies*: *Proceedings of Int. Conf. and workshop* (Minsk, "Technoprint", 2001) pp. 59 – 71.
5. Bashtovoi V., Kuzhir P., Reks A., Bossis G. and Volkova O., The Uniform Magnetic Field Influence on the Magnetic Fluid Meniscus Motion in the Cylindrical Capillary, *Intern. J. of Modern Physics B* **16** (2002) pp. 2590–2596.

CONTROL RESPONSES OF COMMERCIAL MAGNETORHEOLOGICAL DAMPER FOR PASSENGER VEHICLES

H. G. LEE

Department of Automotive Engineering, Daeduk College, Daejeon 305-715, Korea
E-mail: leehg@ddc.ac.kr

K. G. SUNG AND S. B. CHOI

Smart Structures and Systems Laboratory, Department of Mechanical Engineering, Inha University, Incheon 402-751, Korea
E-mail: seungbok@inha.ac.kr

K. W. MIN AND S. H. LEE

Department of Architectural Engineering, Dankook University, Seoul 140-714, Korea

This paper presents control characteristics of a commercially available magnetorheological (MR) damper which can be directly applied to passenger vehicles. After evaluating the field-dependent damping force, the damping force controllability is experimentally investigated. Subsequently, a quarter-car model which is composed of sprung mass, MR damper, spring and tire is constructed and the governing equation of motion is derived in order to investigate vibration control of a vehicle. A discrete-time fuzzy sliding mode control algorithm is then designed by considering system uncertainties such as the variation of the sprung mass. The controller consists of a discontinuous part and an equivalent part. In addition, the fuzzy algorithm is adopted to increase system robustness to the uncertainties and achieve faster reaching to the sliding surface. Control performances of the proposed sliding mode controller are experimentally evaluated under various road conditions and compared with continuous-time sliding mode controller.

1 Introduction

One of the most attractive candidates to formulate successful semi-active suspension is to use electrorheological (ER) or magnetorheological (MR) fluid damper. Nakano constructed a quarter car suspension system model using ER damper and investigated performance characteristics through two control strategies: constant control voltage and square root of control voltage proportional to absolute velocity of unsprung mass [1]. Petek *et al.* constructed a semi-active full suspension system consisting of four ER dampers and experimentally demonstrated that unwanted vehicle motions can be favorably suppressed using the skyhook control algorithm [2]. More recently, Choi *et al.* proposed a sliding mode control methodology for vibration control of vehicle suspension system installed with ER dampers [3]. They demonstrated superior performance characteristics such as significant reduction of vertical acceleration via hardware-in the-loop simulation(HILS). On the other hand, many investigators have proposed MR damper for a vehicle suspension system. Carlson *et al.* proposed a commercially available MR damper, which is applicable to on-and-off-highway vehicle suspension system [4]. They experimentally demonstrated that sufficient levels of damping force and also superior control capability of the damping force by applying control magnetic field could be achieved. Spencer Jr. *et al.* proposed dynamic model for prediction of damping force of a MR damper [5]. They compared the measured damping forces with the predicted ones in time domain. Kamath *et al.* proposed a semi-active MR lag mode damper [6].

They proposed dynamic model and verified its validity by comparing the predicted damping force with the measured one.

Vehicle suspension system can be subjected to parameter variations and disturbances. Hence robust control algorithm is required such as sliding mode controller (SMC) and H_∞ controller. In general, the sliding mode controller has been designed based on the continuous-time system. However, its implementation of an electronic control unit (ECU) for vehicle requires certain sampling processes. It is impossible that the sampling process is exactly realized in the system with the continuous-time sliding mode controller (CSMC) because sampling time of ECU is normally 0.01*s*. This leads to the study of the discrete-time sliding mode control (DSMC) system. The main discrepancy between the DSMC and the CSMC is the determination of the existence condition of the sliding mode. Sarpturk *et al.*[7] have presented the existence condition to determine a stable SMC for the discrete-time system. This condition requires a controller gain to have upper and lower bounds. But if the system has uncertainties, there exists a region in the neighborhood of the sliding surface where the feedback gain can't be defined.

In this work, the main contribution of this work is to present a new DFSMC for the MR suspension system. The damping force responses of a commercially available MR damper, which can be directly applied to passenger vehicles, are investigated. And then, a discrete-time sliding mode controller is then formulated by employing a fuzzy algorithm to appropriately determined principal control parameters. The Proposed controller consists of a discontinuous part and an equivalent part. In order to demonstrate the robustness and performance of the proposed DFSMC, bump and random road responses are presented in time and frequency domains. Furthermore, the control responses are compared with that of CSMC.

2 Modeling of Quarter Vehicle MR Suspension System

In this study, for the MR damper, the commercial product (MagneRide) from GM Cadillac is used and shown in figure 1. In order to verify the damping force controllability of commercial MR damper, the damping force with respect to the piston velocity at various magnetic fields is experimentally investigated as shown in figure 2. This plot is obtained by calculating the maximum damping force at each velocity. The piston velocity is changed by increasing the excitation frequency from 0.4*Hz* to 3.0*Hz*, while the excitation amplitude is maintained to be constant by ± 20*mm*. It is clearly observed that the damping force is increased as the magnetic field increases. As a specific case, the damping force of 316*N* at piston velocity of 0.38*m/s* is increased up to 2617*N* by

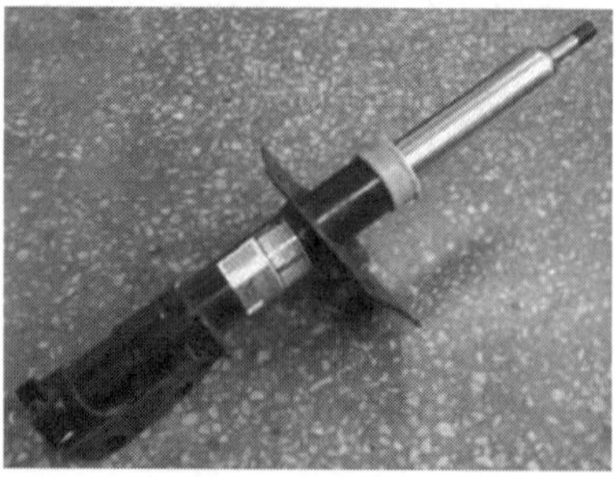

Figure 1. The commercial MR damper for passenger vehicle.

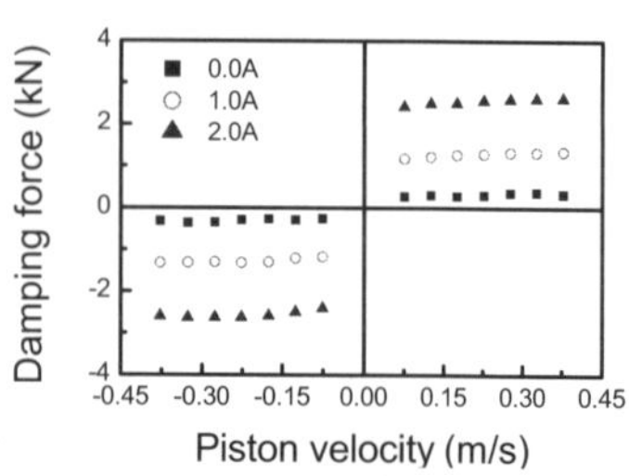

Figure 2. The field dependent damping force characteristics of MR damper

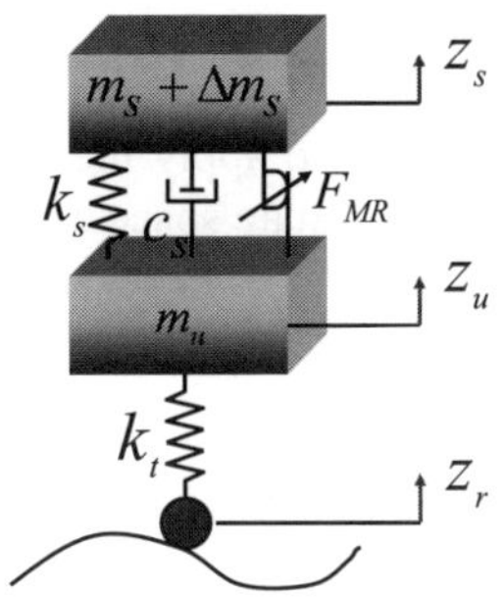

Figure 3. Mechanical model of the quarter-vehicle MR suspension system.

applying thc input current of $2.0A$.

To evaluate the performance and robustness of control system using commercial MR damper, a quarter-vehicle MR suspension system is established. Figure 3 shows the quarter car model of the semi-active MR suspension system that has two degree-of-freedom. m_s and m_u represent the sprung mass, sprung mass perturbation and unsprung mass, respectively. The spring for the suspension is assumed to be linear and the tire is also modeled as linear spring component. Now, we can construct the governing equation for the quarter-vehicle MR suspension system as follows.

$$\begin{aligned} m_s \ddot{z}_s &= -c_s(\dot{z}_s - \dot{z}_u) - k_s(z_s - z_u) - F_{ER} \\ m_u \ddot{z}_{us} &= c_s(\dot{z}_s - \dot{z}_{us}) + k_s(z_s - z_{us}) - k_t(z_{us} - z_r) + F_{ER} \end{aligned} \tag{1}$$

where, k_s is the total stiffness coefficient of the suspension. c_s is the damping coefficient of the suspension. In addition, k_t is the stiffness coefficient of the tire. z_s, z_u, and z_r are the vertical displacement of sprung mass, unsprung mass, and excitation, respectively.

Now, substituting equation (1) into state-space equation yields following uncertain state dynamic model:

$$\dot{\boldsymbol{x}}(t) = (\boldsymbol{A}_0 + \Delta\boldsymbol{A}(t))\boldsymbol{x}(t) + \boldsymbol{B}u(t) + \boldsymbol{f}(t) = \boldsymbol{A}(t)\boldsymbol{x}(t) + \boldsymbol{B}u(t) + \boldsymbol{f}(t) \tag{2}$$

By using the zero-order-hold method [8], the continuous-time system (2) can be written by following discrete-time system:

$$\boldsymbol{x}(k+1) = \Phi(k)\boldsymbol{x}(k) + \Gamma(k)u(k) + \boldsymbol{d}(k) \tag{3}$$

where,

$$\Phi(k) \cong \exp\left(\boldsymbol{A}\left(\frac{t_k + t_{k+1}}{2}\right)T \right), \quad \Gamma(k) \cong \int_0^T \exp\left(\boldsymbol{A}\left(\frac{t_k + t_{k+1}}{2}\right)\tau \right) d\tau \boldsymbol{B}\left(\frac{t_k + t_{k+1}}{2}\right),$$

$$\boldsymbol{d}(k) \cong \int_0^T \exp\left(\boldsymbol{A}\left(\frac{t_k + t_{k+1}}{2}\right)\tau \right) \boldsymbol{f}((k+1)T - \tau) d\tau$$

where, T and k are the sampling time and sampling number, respectively. Equation (3) can be rewritten using the nominal part and the uncertain part as follows:

$$\boldsymbol{x}(k+1) = (\Phi_0 + \Delta\Phi)\boldsymbol{x}(k) + (\Gamma_0 + \Delta\Gamma)u(k) + \boldsymbol{d}(k) \tag{4}$$

3 Controller Design

The sliding mode should be occurred to apply the SMC theory in equation (4). At first, we may design a sliding surface in the state space as follows:

$$s(k) = \boldsymbol{C}\boldsymbol{x}(k) \tag{5}$$

where, $\boldsymbol{C} = [c_1 \cdots c_n]$ is surface gradient vector. The designed sliding surface should be satisfied the Lyapunov Function, equation (5), to be stable.

$$\begin{aligned} F(k) &= |s(k)| > 0 \;\; for\ all\ k \\ \Delta F(k) &= F(k+1) - F(k) = |s(k+1)| - |s(k)| < 0 \;\; for\ all\ k \end{aligned} \tag{6}$$

From equation (6), we can get sliding mode existence conditions.

$$(s(k+1)) - s(k))\text{sgn}(s(k)) < 0 \tag{7}$$

$$(s(k+1)) + s(k))\text{sgn}(s(k)) > 0 \tag{8}$$

Equation (7) is sliding condition and equation (8) is convergence condition of the DSMC.

The DSMC is consists of the equivalent controller and discontinuous controller. However, if the control system has the uncertainty, sliding region could exist around the sliding surface. In this study, the β-equivalent controller in the sliding region is proposed to achieve robust control performance for the system.

The equivalent and β-equivalent controller can be designed using equation (5) and the nominal system,

$$\boldsymbol{x}(k+1) = \Phi\boldsymbol{x}(k) + \Gamma u(k) \tag{9}$$

When,

$$\Delta s = s(k+1) - s(k) = 0 \tag{10}$$

$$\Delta s = s(k+1) - \beta s(k) = 0 \tag{11}$$

The equivalent and β-equivalent controller are obtained as follows,

$$u_{eq}(k) = -(\boldsymbol{C}\Gamma_0)^{-1}(\boldsymbol{C}\Phi_0 - \boldsymbol{C})\boldsymbol{x}(k) \tag{12}$$

$$u_{eq,\beta}(k) = F_{eq,\beta}x(k) = -(\boldsymbol{C}\Gamma_0)^{-1}(\boldsymbol{C}\Phi_0 - \beta\boldsymbol{C})\boldsymbol{x}(k),\ 0 < \beta < 1 \tag{13}$$

In the sliding mode, the motion of the system is then subjected to following equations:

$$\boldsymbol{x}(k+1) = (\Phi_0 + \Gamma_0\boldsymbol{F}_{eq})\boldsymbol{x}(k) = \Phi_{eq}\boldsymbol{x}(k), \quad s(k) = \boldsymbol{C}\boldsymbol{x}(k) = 0 \tag{14}$$

Here, one eigenvalue of Φ_{eq} is 1 and $\boldsymbol{C}$ should be chosen that the others $(n-1)$ become predetermined desired eigenvalues(less than 1) [9]. Consequently, the nominal system with the equivalent controller equation (5) is marginally stable in the sliding mode.

The discontinuous controller is needed to guarantee that the control system is stable. The discontinuous controller is defined as follows,

$$u_d = -h(k)\,\mathrm{sgn}(\boldsymbol{C\Gamma}_0 s(k))\sum_{i=1}^{n}|x_i(k)| \tag{15}$$

where,

$$\begin{cases} h_s(k) \le h(k) \le h_c(k) & :\textit{outside the sliding region} \\ h(k) = 0 & :\textit{inside the sliding region} \end{cases} \tag{16}$$

From the sliding existence condition, the upper bound $h_c(k)$ and lower bound $h_s(k)$ are defined as follows,

$$h_s(k) = \frac{\sup(H_2(k))}{H_1(k)} \tag{17}$$

$$h_c(k) = \frac{2|s(k)| - \sup(H_2(k))}{H_1(k)} \tag{18}$$

The prominent advantage of the fuzzy controller is that it can effectively control complex ill-defined systems having nonlinearities, parameter variations and disturbances. The upper bound $h_c(k)$ and the lower bound $h_s(k)$ of the discontinuous feedback gain are determined by fuzzy algorithm. These bounds, $h_c(k)$ and $h_s(k)$, are equal to the boundary of the sliding region. Thus, we can obtain the information of this boundary and define the following function:

$$s_{bound}(k) \equiv \frac{h_s(k)\cdot\sup\big(H_1(k)\big)+\sup\big(H_2(k)\big)}{2} \tag{19}$$

The fuzzy controller consists of three components; a fuzzification interface, a decision making logic and a defuzzification interface. In the fuzzification interface, σ_β, $\Delta\sigma_\beta$ or σ_h, $\Delta\sigma_h$ are estimated as follows,

$$\sigma_\beta = |s(k)|/s_{bound}(k),\ \ \Delta\sigma_\beta = \sigma_\beta(k) - \sigma_\beta(k-1) \tag{20}$$

$$\sigma_h = s_{bound}(k)/|s(k)|,\ \ \Delta\sigma_h = \sigma_h(k) - \sigma_h(k-1) \tag{21}$$

and these are modified into linguistic values $\tilde{\sigma}$ and $\Delta\tilde{\sigma}$.

In the defuzzification interface, fuzzy output β or $h(k)$ decided from the decision making logic is changed into a numerical value for the purpose of using real control input. Hence, the β-equivalent controller and the discontinuous controller can be formulated as follows:

$$u_{eq}(k) = -(\boldsymbol{C\Gamma}_0)^{-1}(\boldsymbol{C\Phi}_0 - \beta(\tilde{\sigma},\Delta\tilde{\sigma})\boldsymbol{C})\boldsymbol{x}(k) \tag{22}$$

$$u_d = -h(\tilde{\sigma},\Delta\tilde{\sigma})\,\mathrm{sgn}(\boldsymbol{C\Gamma}_0 s(k))\sum_{i-1}^{n}|x_i(k)| \tag{23}$$

It is evident that only the β-equivalent controller (22) is activated inside the sliding region, while the controllers (6) and (23) are activated outside the sliding region.

4 Results and Discussion

From the practical point of view, the MR suspension system should be installed on the full vehicle and it's effectiveness for the ride quality needs to be tested under various road conditions. However, it is very expensive to completely build the entire full-vehicle system with hardware from the start. Therefore, we adopt a quarter-vehicle system, which is well corresponding with field test. An experimental apparatus for the quarter-vehicle test is established as shown in figure 4 to evaluate the effectiveness of vibration isolation of the proposed suspension system. The commercial MR damper, sprung mass, spring and tire are installed on the hydraulic system and the sprung mass displacement and suspension travel are measured by two LVDTs.

Figure 5 presents time responses of the MR suspension system for the bump excitation. In this study, the system parameters of the quarter-vehicle MR suspension system are chosen on the basis of the conventional suspension system for a passenger vehicle. As shown in figure 5, it can be obviously found that unwanted vibrations induced from the bump excitation are well suppressed by adopting the proposed DFSMC to the MR suspension system. In addition, the vibration control performance of the DFSMC is enhanced better than one of the CSMC. Figure 6 presents controlled frequency responses using the DFSMC under random road excitation. The frequency responses are obtained from power spectral density (PSD) for the vertical acceleration of the sprung mass. As expected, the PSD for the vertical acceleration is considerably reduced in the neighborhood of body resonance (1-2*Hz*) by applying the current input.

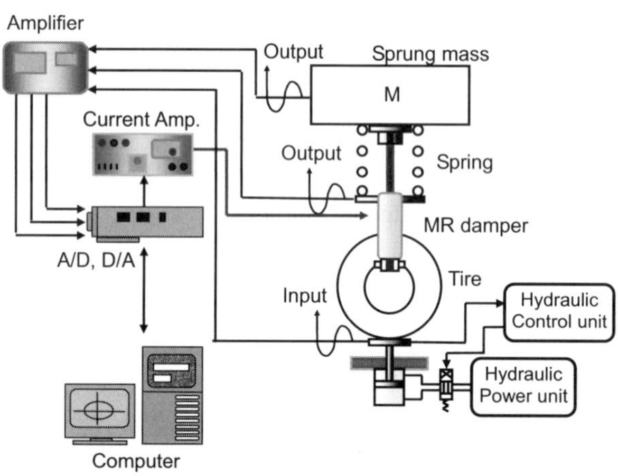

Figure 4. Experimental apparatus for performance evaluation of MR suspension system.

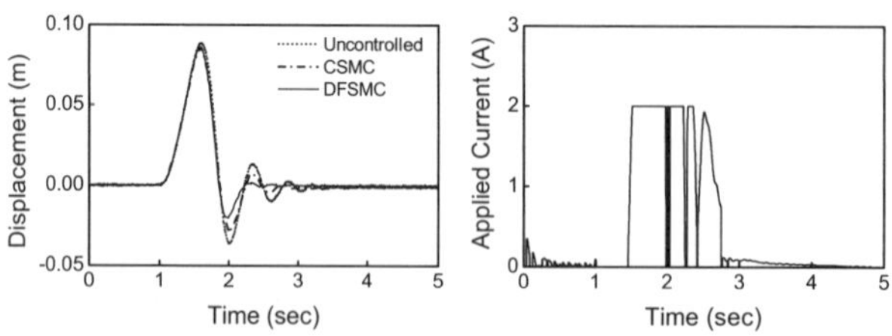

Figure 5. Comparison of CSMC and DFSMC of bump response.

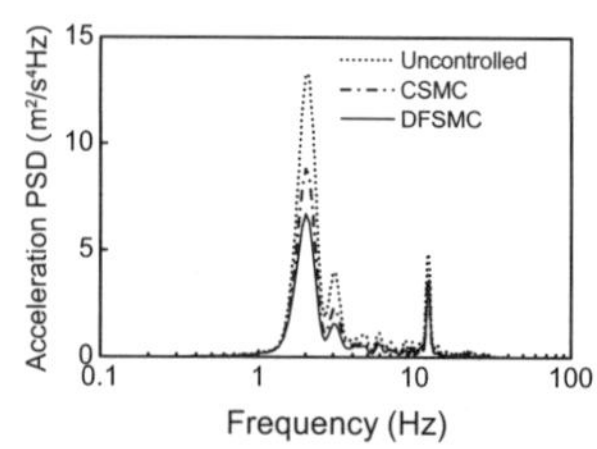

Figure 6. Comparison of CSMC and DFSMC of random road response.

5 Conclusions

Vibration control performance of the MR suspension system using commercially available MR damper was evaluated via discrete-time fuzzy sliding mode controller (DFSMC). In order to verify the damping force controllability of commercial MR damper, the damping force characteristics are experimentally investigated, and then a quarter-vehicle suspension system was constructed and its governing equations of motion were derived by using the zero-order-hold method. In order to obtain a favorable control performance of the MR suspension system with discrete-time controller, a DFSMC tuned by fuzzy algorithm was designed and realized to the quarter-vehicle MR suspension system. The proposed controller consists of a discontinuous part and an equivalent part. Bump and random road responses were presented in time and frequency domains. Furthermore, the control responses are compared with that of CSMC. The results presented in this work are self-explanatory justifying that the proposed DFSMC is very effective for vibration isolation of a passenger vehicle with the commercial MR damper.

Acknowledgement

This work was partially supported by GM-Korea. This support is gratefully acknowledged.

References

1. Nakano, M., A novel semi-active control of automotive suspension using an electrorheological shock absorber. *Proc. of the 5th Int. Conference on ER Fluid*, MR Suspensions and Associated Technical Paper Series 920275 (1995).
2. Petek, N. K., Romstadt, D. J., Lizell, M. B. and Weyenberg, T. R., Demonstration of an automotive semi-active suspension using electrorheological fluid. *SAE Technical Paper Series* 950586 (1995).
3. Choi, S. B., Choi, Y. T. and Park, D. W., A sliding mode control of a full-car electrorheological suspension system via hardware-in-the-loop simulation. *ASME J. of Dynamics, Measurement and Control*, **122** (2000) pp. 114~ 121.
4. Carlson, J. D., Cantanzarite, D. M. and St. Clair, K. A., Commercial magneto-rheological fluid devices. *Proc. of the 5th International Conference on ER Fluids*, MR Suspensions and Associated Technology, (1995) pp. 20-28.
5. Spencer Jr., B. F., Dyke, S. J., Sain, M. K. and Carlson, J. D., Phenomenological model for a magnetorheological damper. *Journal of Engineering Mechanics*, ASCE, **123** (1997) pp. 230-238.
6. Kamath, G. M., Wereley, N. M. and Jolly, M. R., Characterization of semi-active magnetorheological fluid lag mode dampers. *SPIE Conference on Smart Structures and Integrated Systems*, , San Diego CA (1998) pp. 3329-37.
7. Sarpturk, S. Z., Istefanopulos, Y. and Kaynak, O., On the stability of discrete-time sliding mode control systems. *IEEE Transactions on Automatic Control*, **32** (1983) pp. 930-932.
8. Franklin, G. F., Powell, J. D. and Workman, M. L., Digital control of dynamic systems. *Addison-Wesley Publishing*, New York, (1990).

9. Chern, T. L. and Wu, Y., C., Design of integral variable structure controller and application to electrohydraulic velocity servomechanism. *IEEE Proceedings-D*, **138** (1991) pp. 439-444.

RESPONSE CONTROL OF A REAL-SCALED FIVE-STORY STRUCTURE USING MAGNETO-RHEOLOGICAL DAMPER

S.H. LEE, E.C. PARK, K.J. YOUN, K.W. MIN, AND L.CHUNG

Department of Architectural Engineering, Dankook University,San-8, Han-nam dong, Young-San Gu, Seoul, KOREA, E-mail: lshyun00@dankook.ac.kr

S.B. CHOI, AND K.G. SUNG

Department of Mechanic Engineering, Inha University

H.G. LEE

Department of Automotive Engineering, Daeduk College

This paper presents vibration control of a real-scaled five-story steel structure subjected to horizontal excitation using a semi-active magneto-rheological (MR) damper. A large-sized double-rod MR damper, which is applicable for vibration control of the real-scaled five-story structure, is designed and manufactured on the basis of the field-dependent Bingham model of the MR fluid. The damping force of the MR damper is experimentally evaluated with respect to the excitation frequency at various magnetic fields. After formulating the governing equation of motion for the five-story structure associated with the MR damper, displacement and acceleration responses of the structure are discussed under pseudo earthquake excitation in which a hybrid mass damper is designed to reproduce seismically excited structural responses. The controllers which require only structural response of the damper installed floor for calculating the input current of the MR damper are designed to effectively suppress unwanted structural vibrations.

1 Introduction

Researches on vibration mitigation of large scale civil structures under dynamic loads by using damping devices have been extensively conducted. Generally, since earthquake load induces a civil structure to have significantly large vibratory energy, the damping devices controlling seismic responses of the civil structures should be designed to have large capacity corresponding to the induced structural energy. Active control devices have been used in last two decades to control seismically or wind excited structure due to their outstanding performance. However, they require additional large electric power supply in order to realize optimal control force, which is difficult to be secured when real earthquake occurs. As an alternative for active devices, semi-active damping devices such as Magnetho-rheological (MR) fluid dampers, are considered because they can generate large force with small power supply. MR dampers have been practically applied to real-scale structures such as a cable stayed bridge structure. Lord Co. developed 20-ton MR damper and Yang proposed various hysteretic models for describing the damping force characteristics of the 20-ton MR damper [1]. Since it is practically impossible to experimentally evaluate the performance of the MR damper for controlling dynamic response of a real scale civil structure, numerical analyses or small scale laboratory experiments have been generally conducted. Jansen and Dyke compared various semi-active control algorithms through numerical analysis of a five-story building structure installed with the MR damper [2].

In this study, the control performance of the practically manufactured MR dampers installed at a real scaled five-story steel structure is numerically evaluated as a preliminary study for performing experimental test. Numerical analysis is conducted under the expecting conditions in the experiment that Chevron braces are used to install MR dampers between the base and the first floor, and hybrid mass damper (HMD) at the fifth floor excites the building structure. Pseudo-earthquake excitation test, in which the input signal of the HMD is given for the HMD to reproduce identical floor responses to seismically induced ones, is supposed to be conducted for replacing conventional shaking table test. The response mitigation of the five story building with the MR damper is investigated according to the variation of the input voltage to the MR damper and the performance of the semi-active MR damper is evaluated.

2 Modeling of Damper and Building

2.1 MR Damper Design and Modeling

Figure 1(a) presents a schematic configuration of the proposed mixed mode two-way(double-rod) MR damper and its photograph is shown in figure 1(b). The piston head is moving to right and left sides. The MR damper is divided into the upper and lower chambers by the piston, and it is fully filled with the MR fluid. In order to simplify the analysis of the MR damper, it is assumed that the MR fluid is incompressible and friction force between oil seals and fluid inertia are negligible. By the motion of the piston, the MR fluid flows through the electrode at both ends from one chamber to the other. Thus, the damping force of the MR damper can be expressed as follows [3]:

$$\begin{aligned} F_d &= F_\eta + F_{MR}\,\mathrm{sgn}(\dot{x}_p) \\ &= (A_p - A_r)^2 \frac{12\eta L}{2\pi r h^3}\dot{x}_p + 4(A_p - A_r)\frac{2L}{h}\alpha H^\beta\,\mathrm{sgn}(\dot{x}_p) \end{aligned} \tag{1}$$

In the above, $\dot{x}_p$ is the piston velocity, r is the outer radius of the inner cylinder, A_p is the piston area, A_r is the rod area and the magnetic field is related by $H = NI/2h$. Here, N is the number of the coil turns and I is the input current. In this work, for the MR fluid, the commercial product (MRF-132-DG) from Load Corporation is used.

The size and level of required damping force adopted in this work are chosen on the basis of simulation results for a vibration control of five-story structure. The damping force is evaluated with respect to the magnetic field in order to determine design parameters. The designed MR damper has the following specifications; gap between magnetic poles(h): 1mm, electrode length(L): 30mm, area of the piston head(A_p): 4712mm^2, number of coil turn(N): 200, and diameter of the coil(copper wire): 0.8mm. It is noted that the material for the inner and outer piston is mild steel featuring a strong magnetic property, while the material for the piston rod and cylinder is aluminum featuring relatively low permeability.

Figure 2 presents the dynamic force with respect to the piston velocity at various magnetic fields. The magnetic field is applied to the fluid domain through the coil from 0.0A to 3.0A. This plot is obtained by calculating the maximum damping force at each velocity. The piston velocity is changed by increasing the excitation frequency from 0.5 to 2.0Hz, while the excitation amplitude is maintained to be constant by $\pm$30mm. This directly indicates that a certain desired damping force can be achieved by just controlling the input current regardless of the piston velocity. It is identified that the damping force of 476N at piston velocity of 0.377m/s is increased up to 1524N by applying the input current of 3.0A. Specifically, the damping force of 818N at 1.0A and 1122N at 2.0A are obtained respectively. The inherent damping characteristic of the MR damper needs to be considered for practical implementation of the building structure.

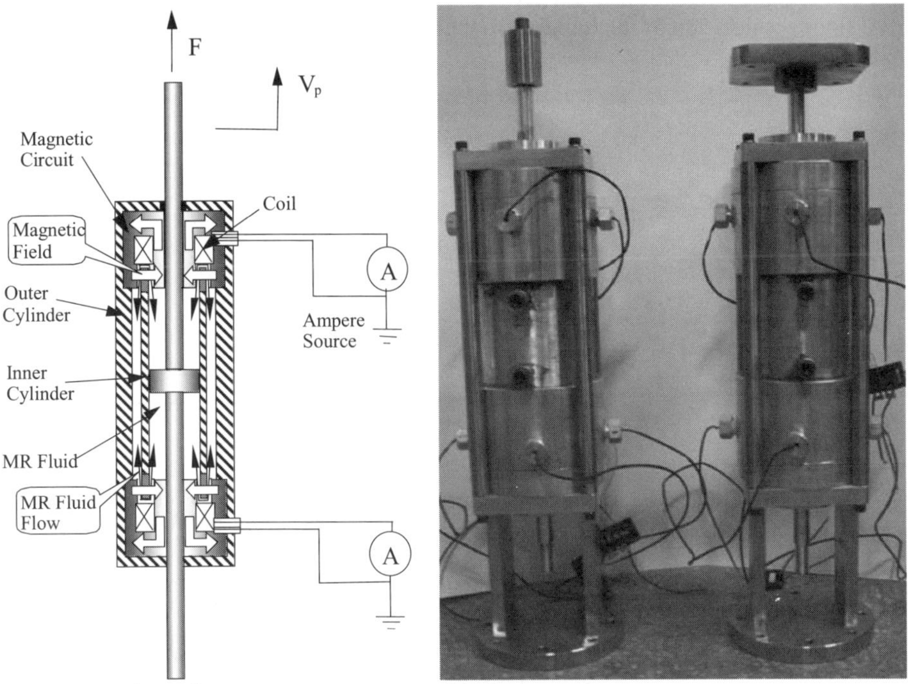

(a) configuration (b) photograph

Figure 1. The proposed mixed mode MR dampers

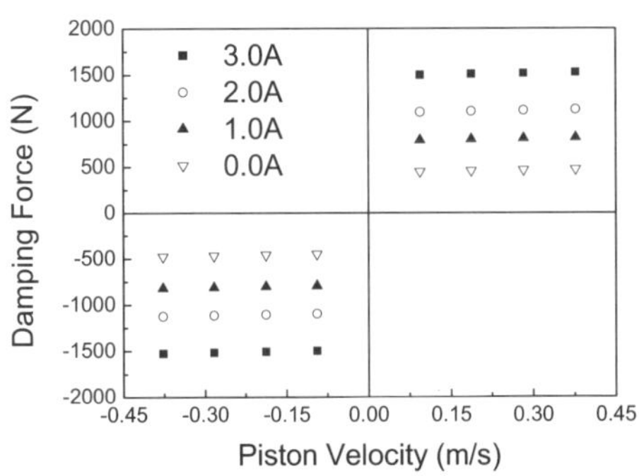

Figure 2. Field-dependent damping force characteristics of the MR Damper.

2.2 *Building Model*

An experimental model, which is shown in Figure 3, is a full-scale five-story steel structure which has story height of 6m, plan of 6m × 6m, and story mass of 20ton. The MR dampers introduced in section 2.1will be installed between the first floor and the base using the Chevron braces. Each floor is composed of four identical wide-flange type steel columns. A HMD on the fifth floor excites the model structure before MR dampers are installed. The fundamental frequency and damping ratio were found experimentally to be 0.51Hz and 1.98%, respectively [4]. Since the structure was assumed to be a shear building which had story weight concentrated on the floor and story stiffness was uniformly distributed, the story stiffness could be estimated by fitting the fundamental frequency, which was 2440kN/m. Figures. 3 and 4 show the five story steel building and the transfer function of the roof floor acceleration of the building to the HMD input.

Figure 3. Five-story steel structure

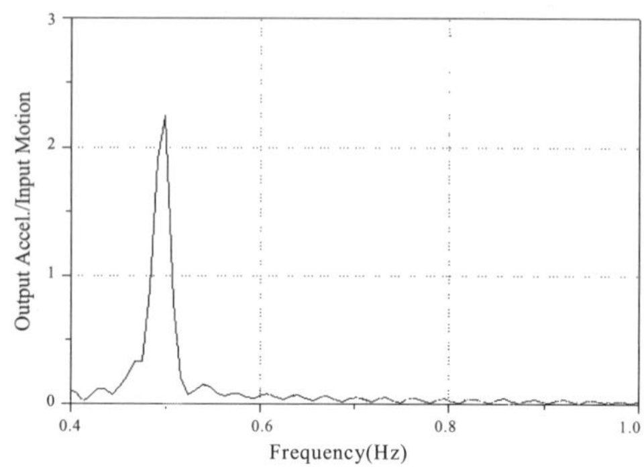

Figure 4. Transfer function from HMD loads to the acceleration of the roof floor

2.3 *Building-HMD Model*

Excitation force is generated by the HMD to vibrate the structure in the elastic range. To control the HMD, the linear oscillating actuator (LOA) using electromagnetic force as an exciter was adopted. Figure 5 shows the HMD and Figure 6 presents the control system for the HMD including a moving mass, inverter, and control computer. The relationship between computer input voltage e_c and inverter output voltage e is given as follows

$$\dot{\eta} = A_{in}\eta + B_{in}e_c\,,\ e = C_{in}\eta \tag{2}$$

in which, $\eta = \begin{bmatrix} \eta_1 \\ \eta_2 \end{bmatrix}$, $A_{in} = \begin{bmatrix} -10 & -22 \\ 1 & 0 \end{bmatrix}$, $B_{in} = \begin{bmatrix} 1 \\ 0 \end{bmatrix}$, $C_{in} = \begin{bmatrix} 177 & 780 \end{bmatrix}$

The HMD using the LOA system consists of a moving coil and a permanent magnet generating electromotive thrust that moves the HMD mass. The equation for the output voltage of the inverter inputted to the coil and the displacement of the moving mass is

$$L\dot{i} + Ri + k_E\dot{x}_r = e \tag{3}$$

where i, and x_r are the coil current, and the relative displacement of the moving mass to the HMD installed floor displacement, respectively. Also L, R, and k_E are the inductance, resistance, and induced electromotive force constant, respectively.

The thrust is a current multiplied by a constant k_T, which is applied to the HMD mass as follows:

$$m_h\ddot{x}_r + +c_h\dot{x}_r + k_hx_r = k_Ti \tag{4}$$

where m_h, k_h, and c_h are the mass, stiffness, and viscosity coefficient of the HMD, respectively. The parameters mentioned above were already identified through system experiments. The equation of motion of an n degree of freedom structure and the HMD system is given as

$$M_a\ddot{X}_a + C_a\dot{X}_a + K_aX_a = B_ak_Ti \tag{5}$$

in which

$$X_a = \begin{bmatrix} X_s \\ x_h \end{bmatrix}, M_a = \begin{bmatrix} M_s & 0 \\ 0 & m_h \end{bmatrix}, C_a = \begin{bmatrix} C_s + c_h B_h B_h^T & -c_h B_h \\ -c_h B_h^T & c_h \end{bmatrix},$$

$$K_a = \begin{bmatrix} K_s + k_h B_h B_h^T & -k_h B_h \\ -k_h B_h^T & c_h \end{bmatrix}, B_a = \begin{bmatrix} -B_h \\ 1 \end{bmatrix}$$

in which, X_s is the structural displacement vector, and M_s, C_s, and K_s are, respectively, mass, damping, and stiffness matrices of the structure. B_h denotes the HMD location vector and is $B_h = [0 \quad 0 \quad 0 \quad 1 \quad 0]^T$. The relationship between the HMD displacement relative to base x_h and x_r is $x_r = -B_h^T X_s + x_h = B_a X_a$

Figure 5. Photograph of the HMD

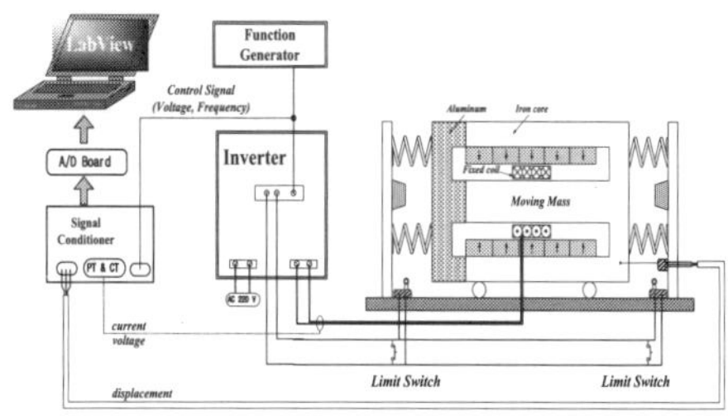

Figure 6. Control system for the HMD

Finally, the state-space form equation of the structure-HMD system is obtained as

$$\dot{z} = Az + Be_c = \begin{bmatrix} A_1 & B_1 C_{in} \\ 0 & A_{in} \end{bmatrix} z + \begin{bmatrix} 0 \\ B_{in} \end{bmatrix} e_c, \quad y = Cz \tag{6}$$

in which z, A_1 and B_1 are

$$z = \begin{bmatrix} X_a \\ \dot{X}_a \\ i \\ \eta \end{bmatrix}, \quad A_1 = \begin{bmatrix} 0_{(n+1)\times(n+1)} & I_{(n+1)\times(n+1)} & 0_{(n+1)\times 1} \\ -M_a^{-1} K_a & -M_a^{-1} C_a & k_T M_a^{-1} B_a \\ 0_{1\times(n+1)} & -L^{-1} k_E B_a & -L^{-1} R \end{bmatrix}, B_1 = \begin{bmatrix} 0_{(n+1)\times 1} \\ 0_{(n+1)\times 1} \\ L^{-1} \end{bmatrix} \tag{7}$$

3 Pseudo-Earthquake Excitation

In order to experimentally evaluate the seismic performance of a building structure with accuracy, shaking table test should be conducted. However, it is practically impossible and very expensive to excite a real scale building using the shaking table. Yu presented a linear shaker system to simulate structural seismic response by using inverse transfer function of the structural response induced by the linear shaker [5]. In this paper, the input signal of the HMD was obtained for the HMD to keep the structure in the target response trajectory calculated using numerical analysis of the structure excited by earthquake load. An inverse transfer function of a structural response to the input voltage of the HMD is obtained using a state space form governing equation of the structure-HMD model, and the discrete Fourier transform is performed. The output transfer function to e_c is given by

$$T_y(s) = \frac{Y(s)}{e_c(s)} = C(sI - A)^{-1}B \tag{8}$$

where the scalar s is a Laplace transformation variable, and $Y(s)$ and $e_c(s)$ are, respectively, Laplace transform of y and e_c. The e_c reproducing the earthquake induced structural response y_T , which is calculated by performing seismic analysis of the building model, can be obtained using the inverse of T_y.

$$e_c(s) = T_y^{-1}(s)Y_T(s) \tag{9}$$

Figure 7 shows the magnitude of the transfer function T_y, time history of input voltage, and HMD generated force, which reproduce the identical roof floor acceleration induced by El Centro (NS component, 1940) scaled to have peak value of 0.05g. It is observed in figure 7(a) that the transfer function has frequency intervals in which the magnitude is as small as zero and thus the magnitude of the inverse transfer function increases infinitely. Because the input force is calculated by the product of the inverse transfer function and the output signal, significant input force is obtained in order to realize the small magnitude of output components corresponding to the intervals. In this study, the input signal corresponding to the frequency intervals in which the magnitude of T_y is smaller than 5% of the peak value is removed using band stop filter. Figure 8 compares the HMD and earthquake induced roof, fifth, and first second floor accelerations. It is observed that the responses caused by different type of excitation match well with each other.

Figure 9 shows the floor distribution of the peak and RMS values of displacement and acceleration responses. Semi-active control is simply designed for the input current to be proportional to the first floor velocity. It is observed that passive-on control generating maximum damping force with input ampere of 3.0A provides the best control performance in reducing displacement response although the reduction magnitude is slightly larger than one by semi-active MR. Semi-active MR reduces floor acceleration most significantly and gives the almost equivalent displacement mitigation performance to the passive-on control.

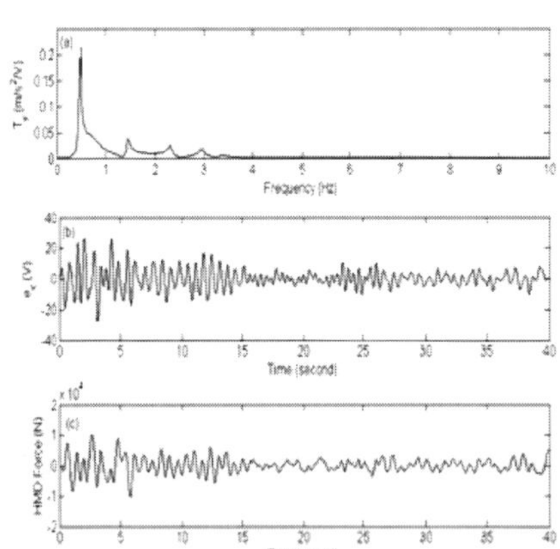

Figure 7. Pseudo-earthquake induction system

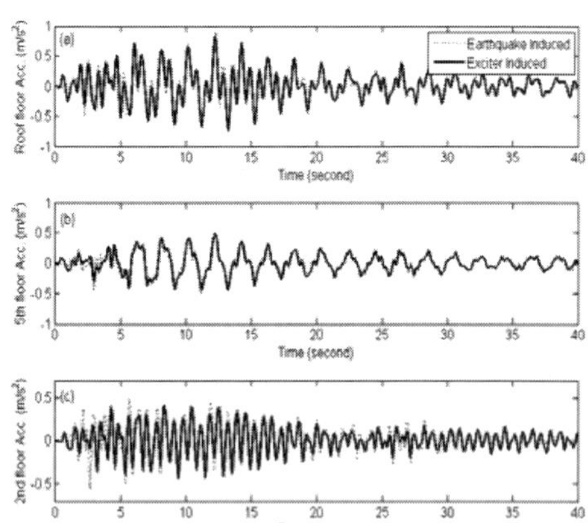

Figure 8. Exciter and earthquake induced acceleration

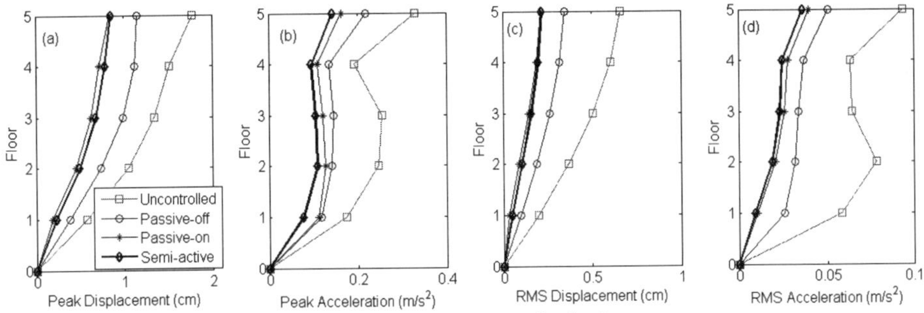

Figure 9. Floor response distribution

4 Conclusion

In this study, the control performance of the practically manufactured MR dampers is numerically evaluated for a real scaled five-story steel structure as a preliminary study for performing experimental test. The numerical results from pseudo-earthquake excitation using a HMD which reproduces the base-excited responses, indicate that the manufactured MR damper is effective in the vibration mitigation of the five-story steel structure. Also, it is observed that the semi-active MR damper reduces the floor acceleration most significantly and gives the almost equivalent displacement mitigation performance to the passive-on control. Efforts are currently underway to perform the experimental verification.

5 Acknowledgements

The work presented in this paper was partially supported by the National Research Laboratory Program (Project No. M1-0203-00-0068) from the Ministry of Science and Technology in Korea and the Second Stage of BK21 through the Ministry of Education & Human Resources Development through in Korea.

Reference

1. Yang G., Large-scale magnetorheological fluid damper for vibration mitigation: modeling, testing and control, (Ph.D thesis, Univ. of Notre Dame, Indiana, 2001)
2. Jansen, L.and M, Dyke S. J., Semiactive control strategies for MR dampers: comparative study. *Journal of Engineering Mechanics*, ASCE, **126** (2000), pp.795-803
3. Choi, S. B., Choi, Y. T., Chang, E. G., Han, S. J. and Kim, C. S., Control characteristics of a continuously variable ER damper. *Mechatronics*, **8** (1998) pp. 143-161.
4. Min, K. W., Kim, J., and Lee, S. H., Vibration tests of 5-storey steel frame with viscoelastic dampers, *Engineering Structures*, **26** (2004) pp.831-839
5. Yu E., Whang D.H., Conte J.P., Stewart J.P., and Wallace J.W., Forced vibration testing of buildings using the linear shaker seismic simulation (LSSS) testing method, *Earthquake Engng Struct. Dyn.*, **34** (2005) pp.737–761

PERFORMANCE EVALUATION OF MR DAMPER IN BUILDING STRUCTURES CONSIDERING DYNAMIC SOIL-STRUCTURE INTERACTION

S.-K. LEE AND J.-H. PARK

Remodeling Research Center, Dankook University, Seoul 140-714, Korea

S.-H. LEE, K. W. MIN, L. CHUNG AND B.-W. MOON

Department of Architectural Engineering, Dankook University, Seoul 140-714, Korea
E-mail: kwmin@dankook.ac.kr

K.-S. KANG

Department of Architectural Engineering, Tongmyong University, Pusan, Korea

S.-B. CHOI

Department of Mechanic Engineering, Inha University, Incheon, Korea

Since the force generated by a MR damper has large nonlinearity, the performance of the MR damper is dependent on the response characteristics such as frequency and amplitude. Soil-structure interaction (SSI) is generally known to have large effect on the seismic response of building structure. In this study, the performance of MR damper in mitigating seismic response of a building structure is evaluated considering SSI effect. Firstly, the performance variance of a MR damper due to the change of natural period is investigated by constructing its normalized response spectrum through the numerical analysis for many earthquake wave records and the natural period of structure. The variable friction force of a MR damper is normalized by the structural base shear force, and its amplitude and the decrement of response are quantitatively evaluated. Then, the response characteristics of SSI system due to the lengthening of structural natural period and various soil conditions are numerically evaluated based on the response spectrum analysis. Finally, the numerical results with and without considering the SSI effects are comparatively evaluated for the building structure with a MR damper. The comparison results show that the SSI effect should be considered in order that the undesirable effect of MR dampers on the structural control should not be neglected.

1 Introduction

The magnetorheological (MR) damper has been increasingly applied to the control of civil structures subjected to earthquake or wind load (Dyke et al. 1996). MR fluid has both frictional and viscous damping terms of which friction force can be modulated by varying the magnitude of the magnetic field applied to the fluid. The seismic response characteristics of building structures are strongly governed by the fundamental structural period, and identical frictional force may well provide different control performance according to the structural period. Accordingly, the effect of the MR damper should be evaluated considering structural period.

General seismic analysis of building structures is performed by assuming that the soil around the base shows rigid motion and the ground motion is directly transferred to the structure. However, soil has dynamic stiffness and damping in frequency domain and the soil-structure interaction (SSI) lengthens the structural period. Accordingly, the SSI should be considered for a structure of which seismic responses are strongly influenced by the period variation.

In this study, the performance variance of a MR damper due to the change of natural period is investigated by constructing its normalized response spectrum through the numerical analysis for many earthquake wave records and the natural period of structure. The earthquake-response control performance of a MR damper is also evaluated by considering the soil-structure interaction (SSI) effect, by which the period characteristic of the mode dominating structural behaviors is affected. The variable friction force of a MR damper is normalized by the structural base shear force, and its amplitude and the decrement of both period and response are quantitatively evaluated. Finally, the numerical results with and without considering the SSI effects are comparatively evaluated for the building structure with a MR damper.

2 Building Structure and MR Dampers Modeling

Many dynamic models were developed for reproducing the nonlinear hysteretic behavior of MR dampers; Bingham, biviscous, hysteretic biviscous models and Bouc-Wen model in the form of differential equation, by which arbitrary types of hysteretic curves can by produced [1]. Lee et. al. showed that the differences according to the respective models can be neglected in evaluating the control performance of building structures [2]. Accordingly, Bingham model that has mathematically simple form, which comprises coulomb friction force and linear viscosity as represented by Eq. (1), is used in this study. The differential equation of the mass-normalized single degree-of-freedom (SDOF) system with a MR damper is given by

$$m_s\ddot{x}_s + c_s\dot{x}_s + k_s x_s = -m_s\ddot{x}_g - f_f\,\mathrm{sgn}(\dot{x}_s) - c\dot{x}_s \tag{1}$$

where, x_s denotes the relative structural displacement to the ground motion, x_g. m_s, k_s, and c are, respectively, structural mass, stiffness, and damping, and f_f and c are friction and viscous forces of the MR damper. The viscosity of a MR damper is not considered in this study, since it is not variable and simply plays a role in increasing the linear viscosity of SDOF structure. The friction force of a MR damper, f_f, is normalized with the following equation.

$$\rho = f_f / f_s \tag{2}$$

where, f_s is the maximum base shear force of the original structure before the installation of a MR damper

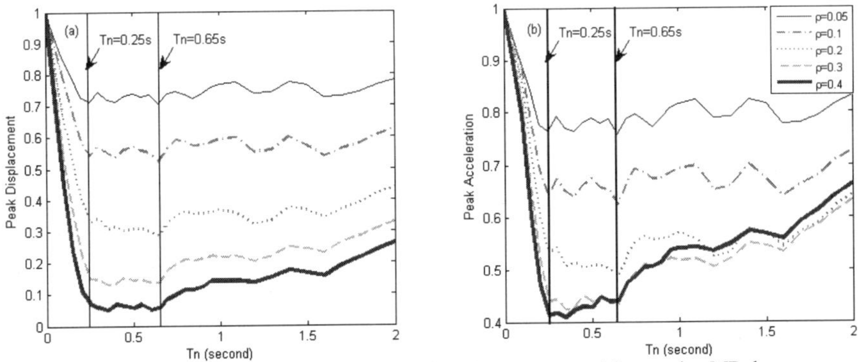

Figure 1. Normalized response spectrum of SDOF structures with a passive MR damper

To investigate performance of passive MR dampers with respect to the normalized friction force, ρ, and derive a design equation for optimal ρ, nonlinear time history analysis is conducted for ρ = 0.05, 0.1, 0.2, 0.3 and 0.4. 20 Ground acceleration data recorded at rock sites are applied [3]. For each ground acceleration input, peak responses of relative displacement and absolute acceleration are obtained. Then, those responses are normalized with uncontrolled responses. Finally, normalized response spectra are constructed using average normalized responses for 20 ground acceleration data.

Figure 1 shows variation of normalized peak responses with respect to ρ for the relative displacement and the absolute acceleration. From figure 1, different trends of response reduction by increasing ρ are observed for different range of the natural period. For $T_n \leq 0.25$ sec, reduction of responses is achieved little at $T_n = 0.0$ sec and increases with increasing T_n. For $0.25 \leq T_n \leq 0.65$ sec, the natural period does not affect the reduction ratio of the responses. Last, for $0.65 \leq T_n \leq 2.0$ sec, response reduction effect for a single ρ decreases linearly as the natural period increases, and this tendency becomes prominent for large ρ's. In particular, loss of control effectiveness becomes more serious for the absolute acceleration rather than the relative displacement. As a result, ρ higher than a certain level affects the absolute acceleration adversely.

A fixed value of friction force yields different normalized friction force for diverse base shear force of the structure. This can be confirmed in figure 2. Average base shear of SDOF structures are plotted in figure 2 (a) and corresponding variations of ρ for two specific of friction force, i.e. 1.0 and 2.0 N, are plotted in figure 2 (b). It can be observed that base shear reduction due to elongation of the natural period increases ρ for a fixed friction force. Therefore, performance of a passive MR damper with a fixed friction force is anticipated to change if it is installed to structures with different natural periods. Especially, it is estimated that increase of the absolute acceleration in long-period structures induced by a large normalized friction force, ρ, requires the elongation of the natural frequency due to soil-structure interaction to be taken into account in the design of MR dampers

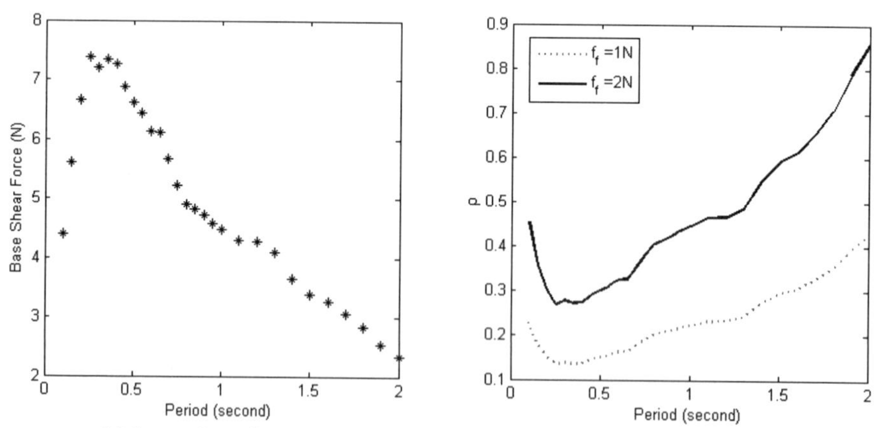

(a) Average base shear

(b) Variation of normalized friction force

Figure 2. Average base shear and normalized friction force for a fixed friction force

3 Numerical Evaluation of SSI System

The numerical computation is performed based on the substructure method for the SSI system with the squared foundation laid on half-space soil, as shown in figure 3.

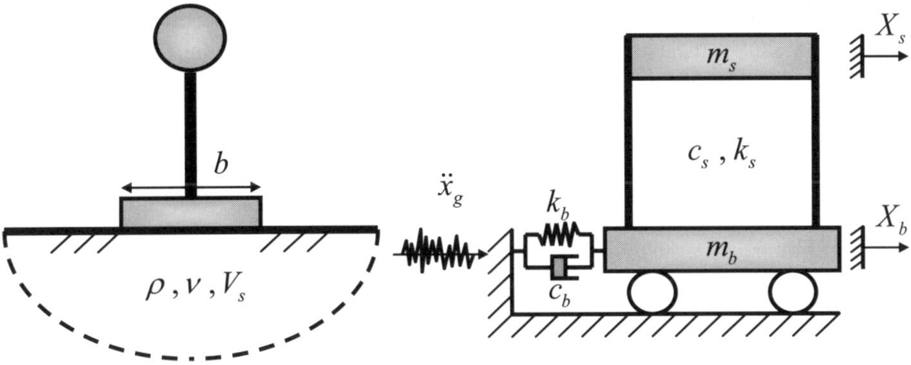

Figure 3. Numerical modeling of SSI system based on the substructure method

Eq. (3) represents the equation of motion the SSI system, which is formulated based on the substructure method.

$$\begin{bmatrix} m_s & 0 \\ 0 & m_b \end{bmatrix} \begin{Bmatrix} \ddot{x}_s \\ \ddot{x}_b \end{Bmatrix} + \begin{bmatrix} c_s & -c_s \\ -c_s & c_s \end{bmatrix} \begin{Bmatrix} \dot{x}_s \\ \dot{x}_b \end{Bmatrix} + \begin{bmatrix} k_s & -k_s \\ -k_s & k_s \end{bmatrix} \begin{Bmatrix} x_s \\ x_b \end{Bmatrix} = - \begin{Bmatrix} m_s \ddot{x}_g \\ m_b \ddot{x}_g \end{Bmatrix} - \begin{Bmatrix} 0 \\ r_b \end{Bmatrix} \tag{3}$$

where, $x_s = X_s - x_g$ and $x_b = X_b - x_g$ are the structural and foundation displacements relative to the input motion, x_g, respectively, and m_b is the mass of foundation. r_b represents the interaction force expressed by

$$r_b = c_g \dot{x}_b + k_g x_b \tag{4}$$

where, c_g and k_g are the damping and stiffness coefficients of soil, which are given by the following Parmelee's soil stiffness based on the exact solution for the circular rigid foundation with underlying half-space soil [4].

$$c_b = \frac{6.21}{2.54 - \nu} \cdot \rho \cdot V_s \cdot r^2 \quad , \quad k_b = \frac{6.77}{1.79 - \nu} \cdot \rho \cdot V_s^2 \cdot r \tag{5}$$

where, ρ and ν are the density and Poisson's ratio of soil medium, respectively. V_s denotes the velocity of shear wave transmitted through soil and r is the equivalent radius of circular foundation with the same area as the foundation under consideration.

Table 1. Numerical parameters of superstructure and soil

Superstructure parameters		Soil profile	
m_s (kg)	510204	ν	0.45
m_b (kg)	510204	ρ (kg/m^3)	1500
r (m)	10	V_s (m/s)	100, 200, 300

SSI system with the parameters such as Table 1 was excited by El Centro earthquake wave with the maximum amplitude of 0.35g in the response spectrum analysis. The structural stiffness was adjusted to correspond to the change of natural period of superstructure for the specified value of structural mass. Also, in order to investigate the effect of soil condition on the superstructure, the shear wave velocities were varied: the larger value of shear wave velocity varies, the relatively stiffer soil becomes.

Figure 4 shows the elongation of structural natural period in the SSI system. This tendency is obviously increased with the soft soil condition and in the region of short natural period. As the structural period is lengthened, the natural period of SSI system , T_s , approaches to that of fixed-base system, T_n .

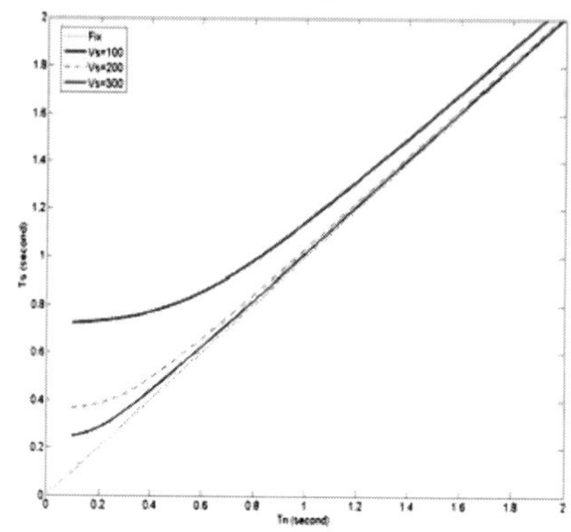

Figure 4. Period of SSI system vs. period of fixed-base system

Figure 5 shows the peak acceleration response spectrums for the SSI system with different soil conditions due to the change in shear wave velocity, V_s , in addition to those for fixed-base system without SSI effect. It is observed from figure 5 that as the soil condition becomes soft, the sharp peaks in the spectrum shape for the fixed-base system is smoothened. This phenomenon is because that the structural damping is additionally increased by the SSI effect, which is conspicuous in the soft soil condition.

Figure 6 expresses the normalized spectrums with different soil conditions, in which the response of SSI system was divided by that of fixed-base system. It is noted that in the range of short and moderate natural periods, the response of SSI system is grown with the increase of shear velocity and locally exceeds that of fixed-base system. However, in the case of long-period structure, $T_n \geq 1.4$, the response of SSI system excesses that of fixed-base system on the whole. This tendency is remarkable in the soft soil condition.

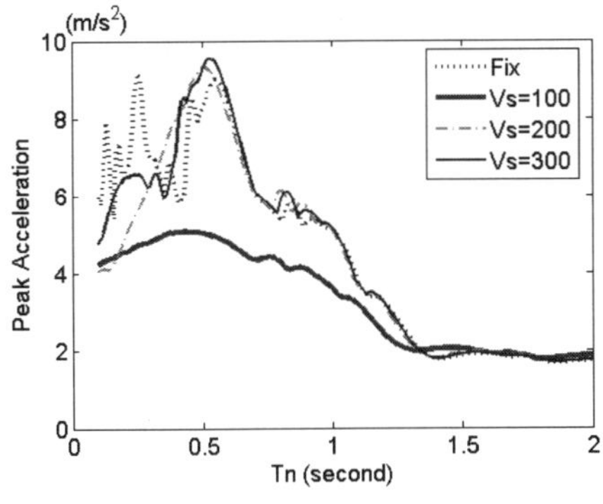

Figure 5. Acceleration response spectrum

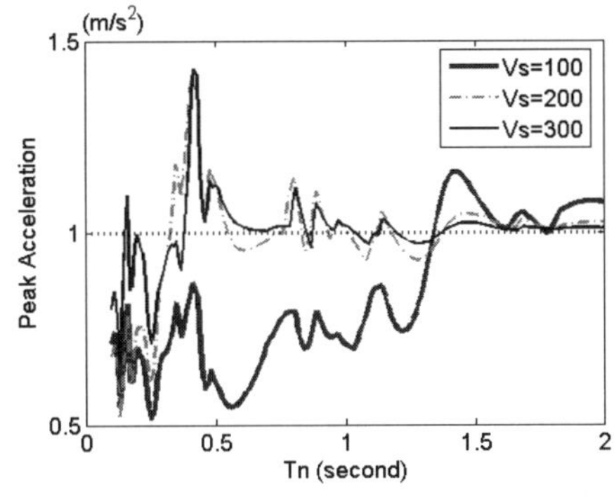

Figure 6. Normalized acceleration response spectrum

4 Building Structure with a MR Damper Considering SSI Effect

Eq. (6) represents the equation of motion of building structure with a MR damper considering the SSI effect.

$$\begin{bmatrix} m_s & 0\\ 0 & m_b\end{bmatrix}\begin{Bmatrix}\ddot{x}_s\\ \ddot{x}_b\end{Bmatrix}+\begin{bmatrix} c_s & -c_s\\ -c_s & c_s\end{bmatrix}\begin{Bmatrix}\dot{x}_s\\ \dot{x}_b\end{Bmatrix}+\begin{bmatrix}k_s & -k_s\\ -k_s & k_s\end{bmatrix}\begin{Bmatrix}x_s\\ x_b\end{Bmatrix}=-\begin{Bmatrix}m_s\ddot{x}_g\\ m_b\ddot{x}_g\end{Bmatrix}-\begin{Bmatrix}0\\ r_b\end{Bmatrix}-\begin{Bmatrix}f_f \operatorname{sgn}(\dot{x}_s-\dot{x}_b)+c(\dot{x}_s-\dot{x}_b)\\ 0\end{Bmatrix} \quad (6)$$

An MR damper was installed to the single degree-of-freedom structure with the mass of 10000ton, the natural period of 0.5 sec and also with the damping ratio of 5 %. The same numerical soil parameter as the softest soil condition in the previous section was designated for the numerical evaluation of SSI system with the MR damper. Figures 7 and 8 show the time histories of acceleration responses of fixed-base and SSI systems, respectively. The friction force generated by the MR damper was determined by the value ρ =0.05 and ρ =0.3 for the passive-off and passive-on control algorithms, respectively, and also for the semi-active algorithm, friction force was assumed to proportionally varies to the structural velocity of the structure.

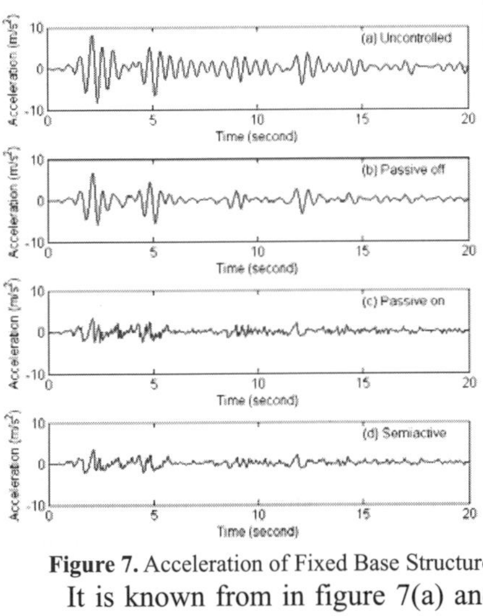

Figure 7. Acceleration of Fixed Base Structure

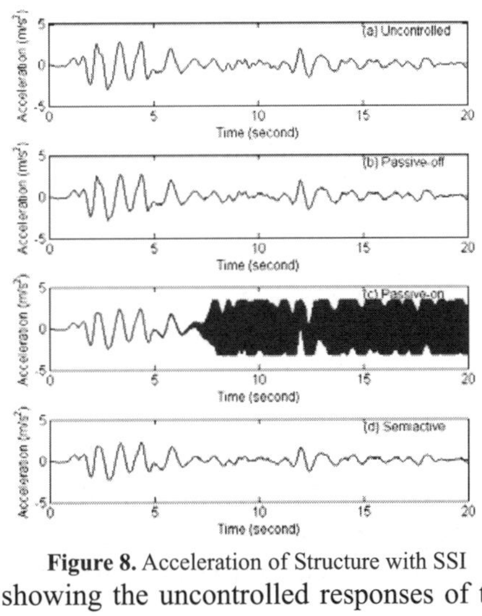

Figure 8. Acceleration of Structure with SSI

It is known from in figure 7(a) and 8(a) showing the uncontrolled responses of the structure that the structural natural period is lengthened by considering the SSI effect, and subsequently the acceleration response is decreased. Also, it can be indicated in figure 8(c) showing passive-on cases that the chattering-phenomenon, which is undesirable for general structures, occurs in considering the SSI effect, while in case of the fixed-base system (figure 7(c)) the passive-on controlled response shows the nearly same control performance as the semi-active controlled one (figure 7(d)). This implies that the control performance was rather deteriorated, since the amplitude of maximum control force, which was generated by the passive-on algorithm, significantly exceeded the proper value required in the control process. It is noted that the acceleration response was reduced by the semi-active control even when the SSI effect was considered. Consequently, the SSI effect should be considered in evaluating the control efficiency for the acceleration response of the structure with the MR damper, otherwise the detrimental effect such as the chattering phenomenon in the passive-on control would be neglected.

5 Conclusions

In this study, the performance of MR damper in reducing the seismic response of a building structure was numerically evaluated considering SSI effect. Firstly, the performance variance of a MR damper due to the change of natural period was investigated by constructing its normalized response spectrum through the numerical analysis for many earthquake wave records and the natural period of structure. The variable friction force of a MR damper was normalized by the structural base shear force, and its amplitude and the decrement of both period and response were quantitatively evaluated. Then, the response characteristics of SSI system due to the lengthening of structural natural period and various soil conditions were numerically evaluated based on the response spectrum analysis. Through these successive quantitative evaluations for the respective numerical examples, the increase of acceleration response due to the additional extension of natural period of long-period structure was commonly observed. Finally, the numerical results with and without considering the SSI effects were comparatively evaluated for the building structure with a MR damper. The comparison results showed that the SSI effect should be considered in order that the undesirable effect of MR dampers on the structural control such as chattering problem in acceleration response should not be neglected.

6 Acknowledgements

The work presented in this paper was partially supported by the National Research Laboratory Program (Project No. M1-0203-00-0068) from the Ministry of Science and Technology in Korea and the Second Stage of BK21 through the Ministry of Education & Human Resources Development through in Korea.

References

1. Dyke, S. J, Spencer Jr. B. F, Sain M. K, Carlson J. D. "Modeling and control of magnetorheological dampers for seismic response reduction", Smart Materials and Structures, Vol. 5, 1996, pp. 565 - 575.

2. Jansen, L. M, Dyke S. J. "Semiactive control strategies for MR dampers: comparative study". Journal of Engineering Mechanics, ASCE, Vol. 126, No. 8, 2000, pp.795-803
3. Yang G., "Large - scale magnetorheological Fluid Damper for Vibration Mitigation: Modeling, Testing and Control", Ph.D thesis, University of Notre Dame, Indiana, 2001
4. Applied Technology Council, Seismic evaluation and retrofit of concrete buildings, Report ATC-40, 1996
5. Parmelee, R.A., The Influence of Foundation Parameters on the Seismic Response of Interaction Systems, Proceedings of the 3rd Japan Earthquake Engineering Symposium, 1970

DISCRETE-TIME FUZZY SLIDING MODE CONTROL OF VEHICLE SUSPENSION SYSTEM FEATURING ELECTRORHEOLOGICAL DAMPER

K. H. LIM AND S. B. CHOI

Smart Structures and Systems Laboratory, Department of Mechanical Engineering, Inha University, Incheon 402-751, Korea
E-mail: seungbok@inha.ac.kr

Y. S. LEE AND M. S. HAN

Faculty of Mechanical and Automotive Engineering, Keimyung University, Daegu 704-701, Korea
E-mail: lysgd@kmu.ac.kr

This paper presents control characteristics of a quarter-vehicle Electrorheological (ER) suspension system by adopting a discrete-time fuzzy sliding mode control algorithm. A cylindrical ER damper is designed and manufactured by incorporating Bingham model of the ER fluid. After evaluating the field-dependent damping characteristics of the ER damper, a quarter-vehicle ER suspension system is constructed which is composed of sprung mass, ER damper, spring and tire. After deriving the governing equation of motion for the quarter-vehicle ER suspension system, a discrete-time control model with system uncertainties is formulated in a state space. A stable sliding surface is then designed followed by the formulations of a discrete-time sliding mode controller which consists of a discontinuous part and an equivalent part. In the controller formulation, the fuzzy control algorithm is also adopted to increase system robustness to the uncertainties and achieve faster reaching to the sliding surface. The controller is then experimentally realized and control performance is evaluated under various road conditions.

1 Introduction

The unwanted vehicle vibration needs to be attenuated under various road conditions. This is normally accomplished by employing suspension system. So far, three types of suspension have been proposed and successfully implemented; passive, active, and semi-active. The semi-active suspension system offers a desirable performance generally enhanced in the active mode without requiring large power sources and expensive hardware. Recently, very attractive and effective semi-active suspension system featuring an electrorheological (ER) fluid has been proposed by many investigators [1-3].

Vehicle suspension system can be subjected to parameter variations and disturbances. Hence robust control algorithm is required such as sliding mode controller (SMC). In general, the sliding mode controller has been designed based on the continuous-time system. However, its implementation by an electronic control unit (ECU) of vehicle requires certain sampling processes. It is impossible that the sampling process is exactly realized in the continuous-time sliding mode control (CSMC) system because sampling time of ECU is normally 0.01s. This leads to the study of the discrete-time sliding mode control (DSMC) system. The main discrepancy between the DSMC and the CSMC is the determination of the existence condition of the sliding mode. Sarpturk *et al.*[4] have presented the existence condition to determine a stable SMC for the discrete-time system. This condition requires a controller gain to have upper and lower bounds. But if the system has uncertainties, there exists a region in the neighborhood of the sliding surface where the feedback gain can' t be defined.

Consequently, the main contribution of this work is to present a new DFSMC for the ER suspension system. We first design and manufacture a cylindrical ER damper, which is applicable for medium-sized vehicle. Its damping force responses are evaluated. And then, a discrete-time sliding mode controller which takes into account for various sampling rate is then formulated by employing a fuzzy technique to appropriately determined principal control parameters. In order to demonstrate the robustness and performance of the proposed DFSMC, bump and random road responses are presented in time and frequency domains. Furthermore, the control responses are compared with that of CSMC.

2 Electrorheological Suspension System

2.1 Dynamic Modeling of Electrorheological Damper

A cylindrical type of ER damper shown in figure 1 is designed and manufactured in this study. The damping force of the proposed ER damper can be obtained as follows [3].

$$F = k_e x_p + c_e \dot{x}_p + F_{ER} \tag{1}$$

where, k_e is the effective stiffness due to the gas pressure, c_e is the effective damping due to the fluid viscosity, x_p is the excitation displacement, and F_{ER} is the field-dependent damping force which is tunable as a function of applied electric field. The controllable damping force F_{ER} can be expressed by

$$F_{ER} = (A_p - A_r)2\frac{L}{h}\alpha E^{\beta}\,\mathrm{sgn}(\dot{x}_p) \tag{2}$$

where, A_p and A_r represent piston head and piston rod areas, respectively. $\mathrm{sgn}(\cdot)$ is a sign function, L is the electrode length, h is the electrode gap, and E is the electric field. The α and β are intrinsic values of the ER fluid to be experimentally determined.

Figure 2 presents the damping force with respect to the piston velocity at various electric fields. This plot is obtained by calculating the maximum damping force at each velocity. The detailed experimental apparatus and procedure to obtain this result have been well described in [5].

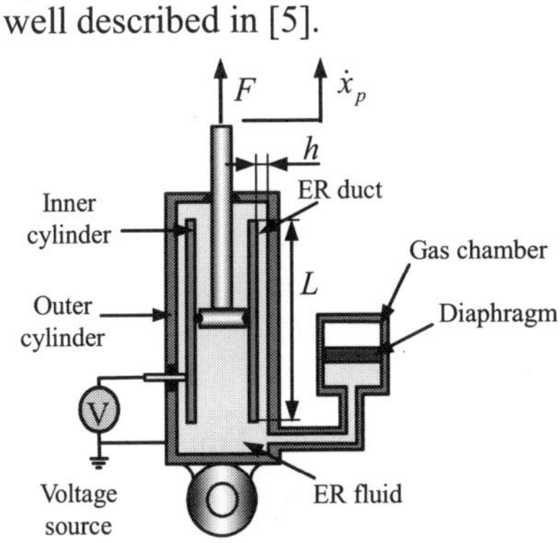

Figure 1. The proposed ER damper for passenger vehicle.

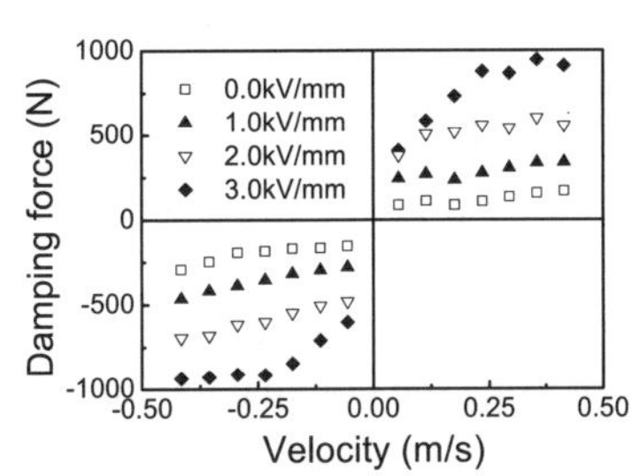

Figure 2. The field dependent damping force

2.2 Modeling of Quarter Vehicle ER Suspension System

In this study, a quarter-vehicle ER suspension system is established to evaluate the performance and robustness of control system. Figure 3 shows the quarter car model of the semi-active ER suspension system that has two degree-of-freedom. m_s and m_u represent the sprung mass, sprung mass perturbation and unsprung mass, respectively. The spring for the suspension is assumed to be linear and the tire is also modeled as linear spring component. Now, we can construct the governing equation for the quarter-vehicle ER suspension system as follows.

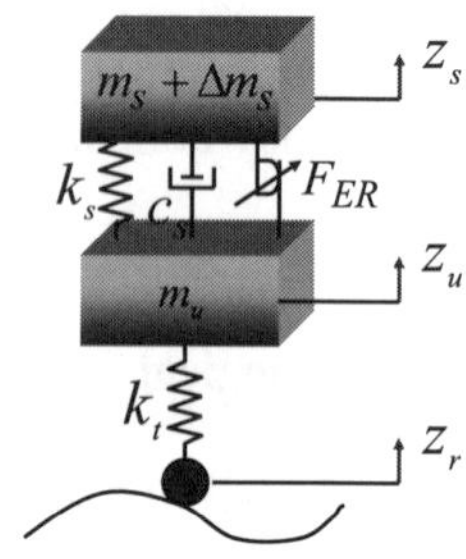

Figure 3. Mechanical model of the quarter-vehicle ER suspension system.

$$\begin{aligned} m_s \ddot{z}_s &= -c_s(\dot{z}_s - \dot{z}_u) - k_s(z_s - z_u) - F_{ER} \\ m_u \ddot{z}_{us} &= c_s(\dot{z}_s - \dot{z}_{us}) + k_s(z_s - z_{us}) - k_t(z_{us} - z_r) + F_{ER} \end{aligned} \tag{3}$$

where, k_s is the total stiffness coefficient of the suspension, and it includes the effective stiffness k_e in equation (1). c_s is the damping coefficient of the suspension and it is assumed to be equal to c_e. In addition, k_t is the stiffness coefficient of the tire. z_s, z_u, and z_r are the vertical displacement of sprung mass, unsprung mass, and excitation, respectively. Now, substituting equation (3) into state-space equation yields following uncertain state dynamic model:

$$\dot{\boldsymbol{x}}(t) = (\boldsymbol{A}_0 + \Delta\boldsymbol{A}(t))\boldsymbol{x}(t) + \boldsymbol{B}u(t) + \boldsymbol{f}(t) = \boldsymbol{A}(t)\boldsymbol{x}(t) + \boldsymbol{B}u(t) + \boldsymbol{f}(t) \tag{4}$$

Here, we assume that all elements of the system and input matrices are piecewise smooth functions. By using the zero-order-hold method [6], the continuous-time system (4) can be written by following discrete-time version:

$$\boldsymbol{x}(k+1) = \Phi(k)\boldsymbol{x}(k) + \Gamma(k)u(k) + \boldsymbol{d}(k) \tag{5}$$

where,

$$\Phi(k) = \exp\left(\int_{t_k}^{t_{k+1}} \boldsymbol{A}(\tau)d\tau\right) \cong \exp\left(\boldsymbol{A}\left(\frac{t_k + t_{k+1}}{2}\right)T\right),$$

$$\Gamma(k) = \int_{t_k}^{t_{k+1}} \exp\left(\int_{\tau}^{t_{k+1}} \boldsymbol{A}(\alpha)d\alpha\right)\boldsymbol{B}(\tau)d\tau \cong \int_0^T \exp\left(\boldsymbol{A}\left(\frac{t_k + t_{k+1}}{2}\right)\tau\right)d\tau\boldsymbol{B}\left(\frac{t_k + t_{k+1}}{2}\right),$$

$$\boldsymbol{d}(k) \cong \int_0^T \exp\left(\boldsymbol{A}\left(\frac{t_k + t_{k+1}}{2}\right)\tau\right)\boldsymbol{f}((k+1)T - \tau)d\tau$$

where, T is a sampling time and k is a sampling number. Equation (5) can be rewritten using the nominal part and the uncertain part as follows:

$$\boldsymbol{x}(k+1) = (\Phi_0 + \Delta\Phi)\boldsymbol{x}(k) + (\Gamma_0 + \Delta\Gamma)u(k) + \boldsymbol{d}(k) \tag{6}$$

3 Controller Design

The sliding mode should be occurred to apply the SMC theory in equation (6). At first, we may design a sliding surface in the state space as follows:

$$s(k) = \boldsymbol{C}\boldsymbol{x}(k) \tag{7}$$

where, $\boldsymbol{C} = [c_1 \cdots c_n]$ is surface gradient vector. The designed sliding surface should be satisfied the Lyapunov Function, equation (7), to be stable.

$$\begin{aligned} &F(k) = |s(k)| > 0 \ \ \textit{for all } k \\ &\Delta F(k) = F(k+1) - F(k) = |s(k+1)| - |s(k)| < 0 \ \textit{for all } k \end{aligned} \tag{8}$$

From equation (8), we can get sliding mode existence conditions.

$$(s(k+1)) - s(k))\operatorname{sgn}(s(k)) < 0 \tag{9}$$

$$(s(k+1)) + s(k))\operatorname{sgn}(s(k)) > 0 \tag{10}$$

Equation (9) is sliding condition and equation (10) is convergence condition of the DSMC.

The DSMC is consists of the equivalent controller and discontinuous controller. However, if the control system has the uncertainty, sliding region could exist around the sliding surface. In the sliding region, the discontinuous controller, then, creates chattering at control input which makes the control system be unstable and the equivalent controller may not deliver the state trajectory to the sliding surface that the robustness of the control system gets worse. So, this study proposes the β-equivalent controller in the sliding region. At the outside of the sliding region, equivalent controller and discontinuous controller deliver the state trajectory to the sliding region. At the inside of the sliding region, the β-equivalent controller delivers the state trajectory to the origin point through the sliding surface.

The equivalent and β-equivalent controller can be designed using equation (7) and the nominal system,

$$\boldsymbol{x}(k+1) = \Phi \boldsymbol{x}(k) + \Gamma u(k) \tag{11}$$

When,

$$\Delta s = s(k+1) - s(k) = 0 \tag{12}$$

$$\Delta s = s(k+1) - \beta s(k) = 0 \tag{13}$$

The equivalent and β-equivalent controller are obtained as follows,

$$u_{eq}(k) = -(\boldsymbol{C}\Gamma_0)^{-1}(\boldsymbol{C}\Phi_0 - \boldsymbol{C})\boldsymbol{x}(k) \tag{14}$$

$$u_{eq,\beta}(k) = F_{eq,\beta}x(k) = -(\boldsymbol{C}\Gamma_0)^{-1}(\boldsymbol{C}\Phi_0 - \beta\boldsymbol{C})\boldsymbol{x}(k), 0 < \beta < 1 \tag{15}$$

In the sliding mode, the motion of the system is then subjected to following equations:

$$\boldsymbol{x}(k+1) = (\Phi_0 + \Gamma_0 \boldsymbol{F}_{eq})\boldsymbol{x}(k) = \Phi_{eq}\boldsymbol{x}(k), \quad s(k) = \boldsymbol{C}\boldsymbol{x}(k) = 0 \tag{16}$$

Here, one eigenvalue of Φ_{eq} is 1 and $\boldsymbol{C}$ should be chosen that the others ($n-1$) become predetermined desired eigenvalues(less than 1) [7]. Consequently, the nominal system with the equivalent controller equation (7) is marginally stable in the sliding mode.

The discontinuous controller is needed to guarantee that the control system is stable. The discontinuous controller is defined as follows,

$$u_d = -h(k)\mathrm{sgn}(\boldsymbol{C\Gamma}_0 s(k))\sum_{i=1}^{n}|x_i(k)| \tag{17}$$

where,

$$\begin{cases} h_s(k) \le h(k) \le h_c(k) : \textit{outside the sliding region} \\ h(k) = 0 \qquad\qquad\quad : \textit{inside the sliding region} \end{cases} \tag{18}$$

From the sliding existence condition, the upper bound $h_c(k)$ and lower bound $h_s(k)$ are defined as follows,

$$h_s(k) = \frac{\sup(H_2(k))}{H_1(k)} \tag{19}$$

$$h_c(k) = \frac{2|s(k)| - \sup(H_2(k))}{H_1(k)} \tag{20}$$

The prominent advantage of the fuzzy controller is that it can effectively control complex ill-defined systems having nonlinearities, parameter variations and disturbances. When we design a fuzzy controller, how to select appropriate fuzzy control rules is the first problem to be resolved. For this, we consider the upper bound $h_c(k)$ and the lower bound $h_s(k)$ of the discontinuous feedback gain. These bounds, $h_c(k)$ and $h_s(k)$, are equal to the boundary of the sliding region. Thus, we can obtain the information of this boundary and define the following function:

$$s_{bound}(k) \equiv \frac{h_s(k)\cdot \sup(H_1(k)) + \sup(H_2(k))}{2} \tag{21}$$

The fuzzy controller consists of three components; a fuzzification interface, a decision making logic and a defuzzification interface. In the fuzzification interface, σ_β, $\Delta\sigma_\beta$ or σ_h, $\Delta\sigma_h$ are estimated as follows,

$$\sigma_\beta = |s(k)|/s_{bound}(k),\ \Delta\sigma_\beta = \sigma_\beta(k) - \sigma_\beta(k-1) \tag{22}$$

$$\sigma_h = s_{bound}(k)/|s(k)|,\ \Delta\sigma_h = \sigma_h(k) - \sigma_h(k-1) \tag{23}$$

and these are modified into linguistic values $\tilde{\sigma}$ and $\Delta\tilde{\sigma}$.

The decision making logic is the kernel of the fuzzy controller since it has the capability of both simulating decision making based on fuzzy concepts and inferring fuzzy control actions. In the defuzzification interface, fuzzy output β or $h(k)$ decided from the decision making logic is changed into a numerical value for the purpose of using real control input. Hence, the β-equivalent controller and the discontinuous controller can be formulated as follows:

$$u_{eq}(k) = -(\boldsymbol{C\Gamma}_0)^{-1}(\boldsymbol{C\Phi}_0 - \beta(\tilde{\sigma},\Delta\tilde{\sigma})\boldsymbol{C})\boldsymbol{x}(k) \tag{24}$$

$$u_d = -h(\tilde{\sigma},\Delta\tilde{\sigma})\operatorname{sgn}(\boldsymbol{C\Gamma}_0 s(k))\sum_{i-1}^{n}|x_i(k)| \tag{25}$$

It is evident that only the β-equivalent controller (24) is activated inside the sliding region, while the controllers (8) and (25) are activated outside the sliding region.

4 Results and Discussion

From the practical point of view, the ER suspension system should be installed on the full vehicle and it's effectiveness for the ride quality needs to be tested under various road conditions. However, it is very expensive to completely build the entire full-vehicle system with hardware from the start. Therefore, we adopt a quarter-vehicle system, which is well corresponding with field test. An experimental apparatus for the quarter-vehicle test is established as shown in figure 4 to evaluate the effectiveness of vibration isolation of the proposed suspension system. The ER damper, sprung mass, spring and tire are installed on the hydraulic system and the sprung mass displacement and suspension travel are measured by two LVDTs.

Figure 5 presents time responses of the ER suspension system for the bump excitation. In this study, the system parameters of the quarter-vehicle ER suspension system are chosen on the basis of the conventional suspension system for a passenger vehicle. As shown in figure 5, it can be obviously found that unwanted vibrations induced from the bump excitation are well suppressed by adopting the proposed DFSMC to the ER suspension system. In addition, the vibration control performance of the DFSMC is enhanced better than one of the CSMC.

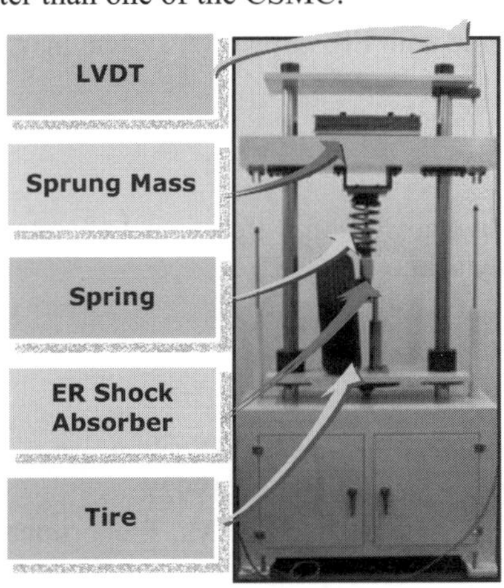

Figure 4. Experimental apparatus for performance evaluation of ER suspension system.

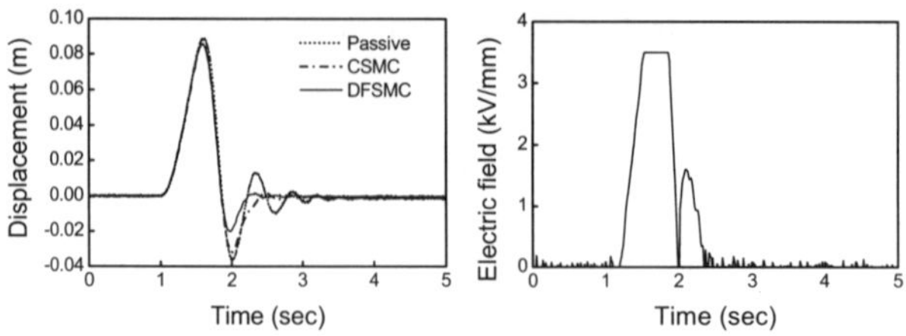

Figure 5. Comparison of CSMC and DFSMC of bump response.

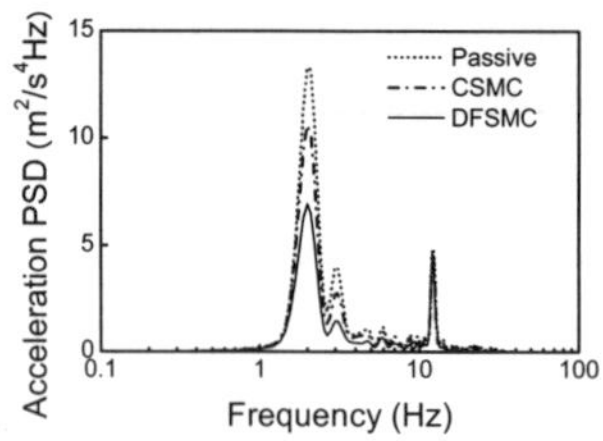

Figure 6. Comparison of CSMC and DFSMC of random road response.

Figure 6 presents controlled frequency responses using the DFSMC under random road excitation. The frequency responses are obtained from power spectral density (PSD) for the vertical acceleration of the sprung mass. As expected, the PSD for the vertical acceleration is considerably reduced in the neighborhood of body resonance (1-2*Hz*) by applying control input.

5 Conclusions

Vibration control performance of the ER suspension system for a passenger vehicle was evaluated via discrete-time fuzzy sliding mode controller (DFSMC). A cylindrical ER damper was designed and manufactured. After evaluating the field-dependent damping characteristics of the ER damper, a quarter-vehicle suspension system was then constructed and its governing equations of motion were derived by using the zero-order-hold method. In order to obtain a favorable control performance of the ER suspension system with discrete-time controller, a DFSMC was designed and realized to quarter-vehicle ER suspension system and tuned by fuzzy logic. Bump and random road responses were presented in time and frequency domains. Furthermore, the control responses are compared with that of CSMC. The results presented in this work are self-explanatory justifying that the proposed DFSMC is very effective for vibration isolation of a passenger vehicle with ER suspension system in real field.

Acknowledgement

This work was supported by the National Research Laboratory (NRL) program. This financial support is gratefully acknowledged.

References

1. Petek, N. K., An Electronically controlled shock absorber using electro-rheological fluid. *SAE Technical Paper Series* 920275 (1992).
2. Sims, N. D., Stanway, R. and Beck, S. B. M., Proportional feedback control of an electro-rheological vibration damper. *J. of Intelligent Material Systems and Structures*, 8 (1997) pp. 426~433.
3. Choi, S. B., Choi, Y. T. and Park, D. W., A sliding mode control of a full-car electrorheological suspension system via hardware-in-the-loop simulation. *ASME J. of Dynamics, Measurement and Control*, **122** (2000) pp. 114~121.

4. Sarpturk, S. Z., Istefanopulos, Y. and Kaynak, O., On the stability of discrete-time sliding mode control systems. *IEEE Transactions on Automatic Control*, **32** (1983) pp. 930-932.
5. Lim, K. H., Discrete-time fuzzy sliding mode control of vehicle suspension system featuring Electrorheological damper. *M. S. Dissertation*, Inha University, (2006).
6. Franklin, G. F., Powell, J. D. and Workman, M. L., Digital control of dynamic systems. *Addison-Wesley Publishing*, New York, (1990).
7. Chern, T. L. and Wu, Y., C., Design of integral variable structure controller and application to electrohydraulic velocity servomechanism. *IEEE Proceedings-D*, **138** (1991) pp. 439-444.

SEDIMENTATION RATE MEASUREMENT OF MAGNETORHEOLOGICAL FLUIDS WITH PARTIAL SUBSTITUTION OF MICROPARTICLES WITH NANOPARTICLES

G. NGATU[1], N. WERELEY[1], N. NERSSESIAN[2] AND B. WOODS[2]

[1]*Department of Aerospace Engineering, University of Maryland, College Park, MD 20742*
e-mail: wereley@umd.edu

[2]*Techno-Sciences, 11750 Beltsville Dr, 3rd Floor, Calverton, MD 20706*

Magnetorheological (MR) fluids consist of suspensions of micron scale magnetic particles (iron or cobalt) in a fluid medium such as silicone oil or mineral oil. MR fluids are used as dampers in vibration absorption applications as well as active valves in hydraulic applications. Nevertheless, one limitation preventing the broader use of MR fluids is the gradual sedimentation of the particles as a function of time. Typical MR fluids utilize micron scale particles at high weight fractions (as much as nominally 80wt% (app. 34vol%)). A reduction in sedimentation rate can be achieved by substituting nanometer scale particles in bi-disperse MR fluids (i.e., a mixture of micron-scale and nanometer scale particles). The nanometer scale particles reduce sedimentation rate, because nanometer scale particles fill in the voids between the micron scale particles, thereby reducing creeping flow. In this study we investigate and quantify the effect of substitution of micro-particles with nanometer scale particles on the sedimentation rate of the MR fluid. Different MR fluids varying from pure micron scale to nanometer scale particle loading of 10% and 15% are investigated. In order to measure sedimentation, a sedimentation rate measurement device was constructed. The device measures sedimentation magnetically utilizing the reduction in permeability with sedimentation. The flow curves of the MR fluids are determined using a Paar-Physica Rheometer and data is fitted with the Bingham-Plastic flow model to estimate yield stress.

1 Introduction

Magnetorheological fluids consist of suspended micron sized magnetic particles, such as iron or cobalt, in a fluid medium, such as silicone oil and hydraulic oil. Their main advantage is that their rheological properties, yield stress and viscosity, can be continuously varied and controlled by applying a varying magnetic field. In the presence of a magnetic field, magnetic dipoles of the suspended particles align themselves roughly parallel to the applied field and form chains which act as a flow barrier. A finite stress must be developed to yield these chained structures, thus giving rise to the so-called yield stress of MR fluids. MR fluids are gaining a wider application in active and semi-active systems, including dampers and clutches.

Typically, MR fluids contain micron scale particles suspended in non-magnetic carrier fluids [1-5]. These fluids have a high yield stress (20kPa to 100kPa) but tend to settle under low or no magnetic field and without constant mixing. Density differences between particles (~7 .5g/cm^3) and suspension fluid (~ 1.0 g/cm^3) causes particles to sediment and agglomerate at the bottom of the container, thus losing their rheological properties. To regain their properties, the particles must be effectively re-dispersed into the suspension.

To alleviate the sedimentation problem, partial substitution of micron scale particles with nanometer scale particles to form bidisperse MR fluids have been established [1-4,6]. The nanometer scale particles in MR fluids tend to re-disperse themselves due to thermal convection [3]. They act as plugs, filling in the voids between the micro scale particles [7], there by reducing creeping flow and enable the dispersion to maintain a uniform mixture. However, nanometer scale particles tend to reduce the yield stress of the MR fluid. To help quantify this trade off, two types of experiments were performed in this investigation: one measured the sedimentation rate and the other estimated the yield stress of the MR fluids.

To quantify the improvement in sedimentation velocity of MR fluids with nanometer scaled particles, an inductance based solenoid sensor was constructed to track mudline location of the fluid [8, 9]. The mudline is the boundary which separates the settling bottom part of the MR fluid from the upper clarified suspension (Fig. 1). As the MR fluid sediments, the mudline travels downwards, until all the iron particles in the suspension are fully deposited at the bottom of the container. The magnetic permeability of the MR fluid, thus the inductance of the sensor, is highly dependent on the volume fraction of iron particles in the dispersion. For MR fluid placed in the sensor, sedimentation of the fluid, which reduces the iron concentration of the fluid enclosed by the sensor, results in lowering the permeability of the fluid and inductance of the sensor in parallel. Thus, by measuring the rate of the monotonically decreasing magnetic inductance resulting from the settling MR fluid, we can track and estimate the rate of change of the mudline location as it traverses through the sensor. The Inductance measuring instrument comprises of an Inductance meter (with data acquisition system), a sensor assy (coil, cylinder for coil winding, and flux return), a vertically translating height gage and a base for fluid positioning. Ref. Fig.1. The main parameters measured are the change in sedimentation velocity as a function of change in fluid composition (wt % of micron vs. nanometer scale particles).

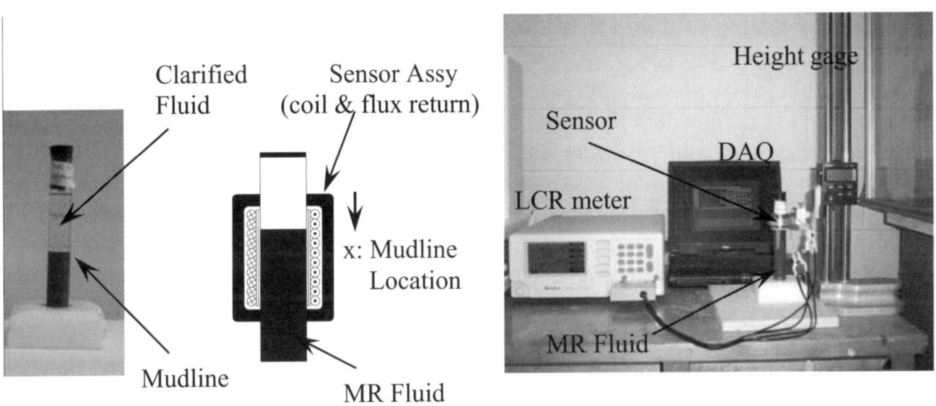

Fig. 1 Settling MR Fluid, Sensor Assembly and Sedimentation Test Set Up

The yield stresses of the fluids were determined using a Paar Physica MCR300 Parallel Disc Rheometer, Fig 2. This instrument is equipped with an MR cell capable of continuously applying a varying magnetic field to the MR fluid. A standard gap of 1 mm between the parallel plates was maintained where a sample of 3ml was poured in for each test run. The magnetic field lines are oriented normal to the parallel plates and MR fluid sample. The tests conducted were rotational tests to determine the flow curve (shear stress vs shear rate). The data obtained was then fitted with the Bingham-Plastic flow model given by:

$$\tau = \tau_y + \mu\dot{\gamma}$$

where τ_y is the yield stress, μ is the post-yield viscosity and $\dot{\gamma}$ is the shear rate. The model predicts that in the pre-yield region, where the stress is below the yield stress (τ_y), the fluid exhibits a rigid behavior. In the post-yield region, the fluid demonstrates an idealviscous flow. The Bingham-Plastic model was fitted to the data using a weighted least squares error minimization by selecting the parameters τ_y and $\dot{\gamma}$. High shear stress values at high shear rates were chosen to better fit the model and estimate the yield stress of the MR fluids, Fig 2.

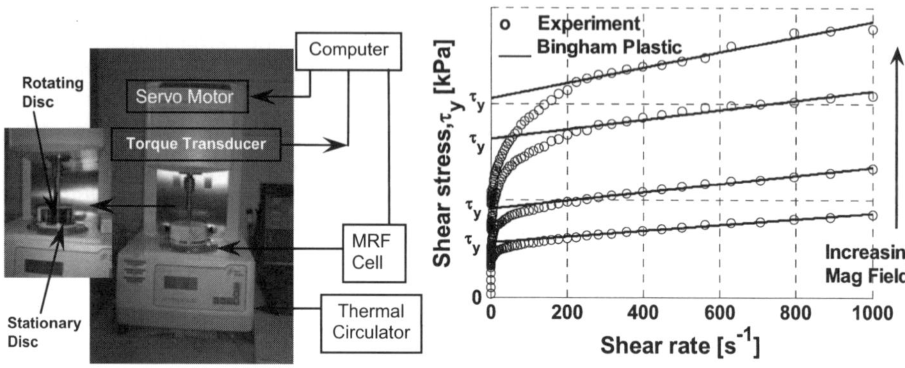

Fig. 2 Paar Physica MCR300 Rheometer and Bingham Plastic Model

2 Magnetorheological Fluid Synthesis and Sample Types

A patented microwave-based process was developed by Material Modification Inc to synthesis nano-scale powders [US Patent 6,409,851]. In this microwave plasma synthesis technique, microwave energy was employed to generate plasma through ionization, disassociation and recombination of gas molecules. The high temperature vaporized the precursors, which promoted chemical reactions at the molecular level in the presence of microwaves. These vapors were rapidly cooled in an inert atmosphere to form powders. Nanometer scale iron powders were synthesized from iron carbonyl precursors per the following chemical reaction:

$$\mathbf{Fe\ (CO)_5 \longrightarrow Fe\ (s) + 5CO\ (g)}$$

The average size of nanometer scale particles was 40 nm. Micron scale iron powder was commercially purchased with average size between 2 and 10 μm. Silicone oil was chosen as the carrier fluid to prepare stable MR fluids. Lecithin was used as a surfactant for producing nanofluidic dispersions. It was mixed in the oil using high speed emulsifier at nearly 11000 rpm. Both Nanometer scale and micron scale iron particles were then added to the silicone oil and mixing continued until a uniformly dispersed mixture is obtained. See Table 1.

Table 1: MR Fluid Samples

Total Fe Loading	%wt	50	50	50	60	60	60	70	70	70	80	80
Micron scale particle	%	100	90	85	100	90	85	100	90	85	100	90
	%wt	50	45	42.5	60	54	51	70	63	59.5	80	72
Nanometer scale particle	%	0	10	15	0	10	15	0	10	15	0	10
	%wt	0	5	7.5	0	6	9	0	7	10.5	0	8

3 Measurement of Sedimentation Rate

The test procedure comprises of two parts. The first one determined the relation between inductance and mudline location within the sensor. This was accomplished by mounting the fluid in the test set up (Fig.1) and measuring the inductance as the sensor was moved up along the bottle. First, the top edge of the sensor was aligned with the mudline and inductance was measured. This was maximum inductance where sensor fully enclosed MR fluid. Then the sensor was moved up by 0.05 in and measurement was taken again. This procedure was repeated at the same interval until the lower end of sensor assembly was above the mudline and sensor enclosed only air and/or clarified fluid, i.e., minimum inductance. Fig.3 shows mudline location in sensor versus inductance. We assumed the permeability of air and non-magnetic fluid to be same as vacuum.

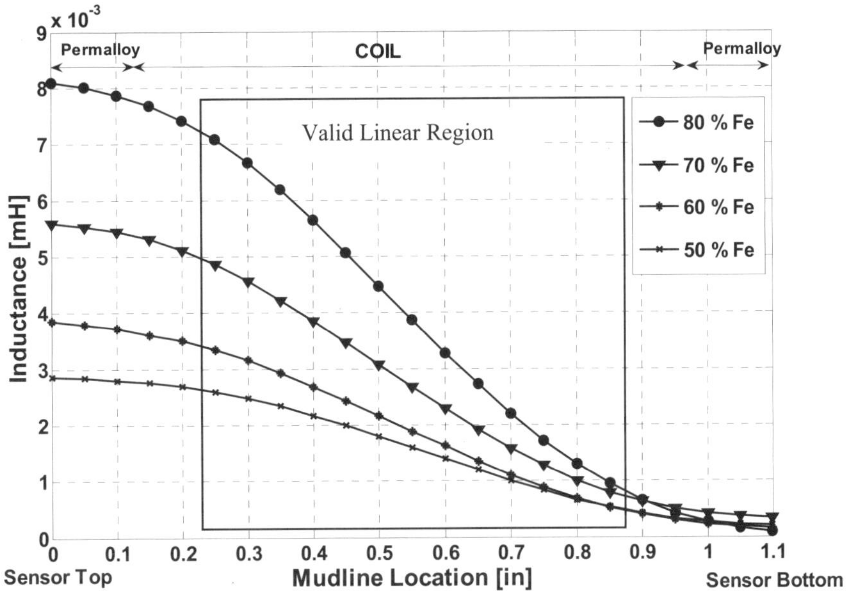

Fig. 3 Inductance Vs Mudline Position for Various Iron Loading

The plot shows inductance is directly proportional to location of mudline in sensor. This linearity holds well when the mudline was located between 0.20 in and 0.90 in within the sensor and disappears as the mudline approaches the edges of the sensor. For sedimentation measurement, we held the sensor at a stationary position and tracked the mudline as it traveled down the sensor due to fluid sedimentation. For instance, after leveling sensor top with the mudline and setting this as the zero position, the sensor was moved up 0.4 in using the height gage and the inductance, $L_{0.4}$, was measured. Then the sensor was further pushed up 0.2 in and inductance measurement was retaken, $L_{0.6}$. Then sensor was moved down and fixed just above the 0.4 in position. At this position, automated inductance measuring was started and inductance, L_i, was continuously measured as the mudline traveled down the sensor. When L_i reached a value below $L_{0.6}$, the test was terminated. The mudline location is estimated by:

Distance Traveled by Mudline: $x = b\dfrac{(L_{0.4} - L_i)}{L_{0.4} - L_{0.6}}$

where $L_{0.4}$ is initial inductance, $L_{0.6}$ final inductance and L_i is inductance at time t and b is the total distance traveled by the mudline.

4 Results

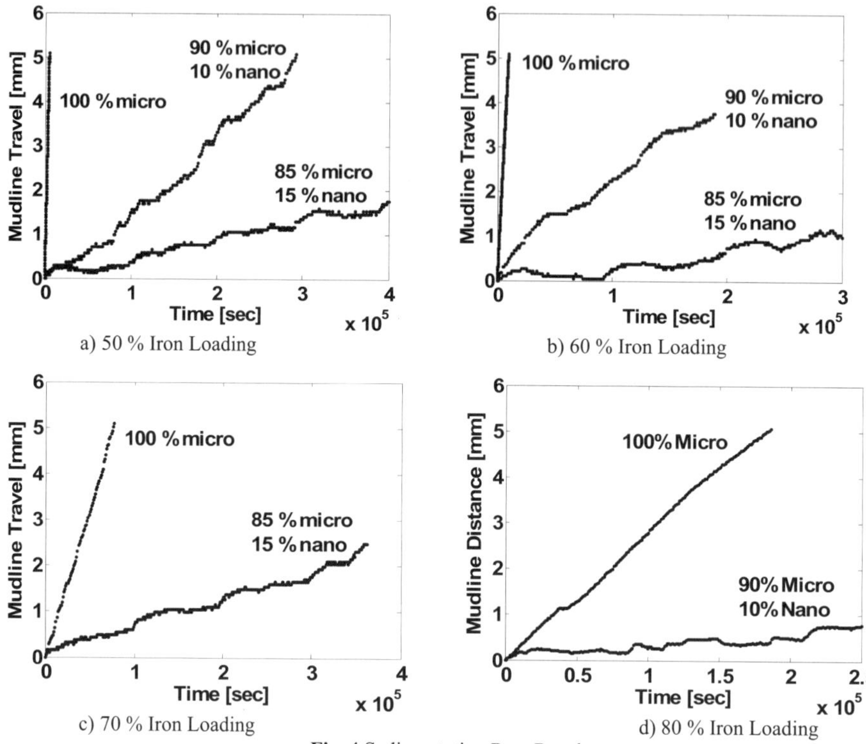

Fig. 4 Sedimentation Rate Results

Figure 4(a) through 4(d) show the mudline travel as obtained from inductance measurements for all MR fluid samples in this study. A linear least-squares fitting was used to determine the sedimentation velocity. Figure 5(a) shows the plot of the estimated sedimentation velocity for the MR fluid samples. These figures show an improvement in sedimentation as the micro particles are substituted with nanometer scale particles. For instance, for 50wt% iron loaded MR fluid, the estimated sedimentation velocity with 0% nanometer scale particles was 1.05 μm/s while with 10% nanometer scale particle substitution, sedimentation dropped to 0.017 μm/s. And for 70wt% iron loading, sedimentation dropped from 0.07 μm/s to 0.0058 μm/s with 10% nanometer scale particle substitution. We also noticed that the highest drop rate in sedimentation was achieved with 10% nanometer scale particle substitution.

Fig 5(b) shows the yield stress results. There was a significant decrement in yield stress as the concentration of nanometer scale particles substituting the micron scale particles increased in the MR fluid. For 50wt% iron loaded MR fluid, the yield stress dropped from 14.7kPa to 11.3kPa when substituting 10% of the micron scale particles with nanometer scale particles. In comparison, for 70wt% iron loading, the yield stress dropped from 34.4kPa to 26.1kPa when substituting 10% micron scale particles with nanometer scale particles.

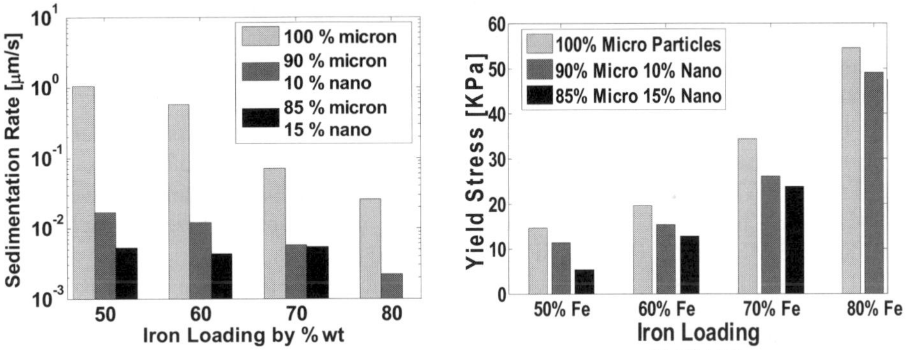

a) Sedimentation Rate b) Yield Stress

Fig. 5 Sedimentation Rate and Yield Stress

5 Conclusion

This investigation quantified the change in sedimentation rate and yield stress of MR fluids as nanometer scale particles are substituted for micron scale particles in the suspension. Substitution of nanometer scale particles reduces the settling rate enabling the MR fluid to maintain a uniform dispersion without marked sedimentation for a longer period. A settling velocity reduction of 99% was achieved by substituting micron scale particles with nanometer scale particles for the 50wt% iron loaded fluid. For 70wt% iron loading, a 91% sedimentation rate decrement was achieved. A two orders of magnitude reduction in sedimentation rate was achieved when substituting 10% of the micron scale particles with nanometer scale particles for all of the MR fluids. Further substitution of micron scale particles above 10% did not result in a proportional reduction in sedimentation.

Even though substitution of micron scale particles with nanometer scale particles in bidisperse MR fluid markedly reduced sedimentation rate, it was shown that nanometer scale particle substitution has a detrimental effect on the yield stress of the MR fluid. The penalty in yield stress increased as the concentration of nanometer scale particles of the MR fluid increased. The yield stress of the 50wt% iron loaded MR fluid dropped by 23% with 10% nanometer scale particle substitution and dropped by 64% with 15% nanometer scale particle substitution. At 70wt% iron loading, the drop in yield stress was about 23% with 10% nanometer scale particle substitution and 30% with 15% nanometer scale particle substitution. The results from the two experiments show that we can mitigate sedimentation with a relatively small yield stress penalty by substituting 10% of the micron scale particles with nanometer scale particles.

Acknowledgement

The authors would like to thank W. I. Kordonski for his thoughtfull suggestion on the experimental set up and R. Radhakrishnan at MMI (Fairfax,VA) for providing the MR fluid samples.

References

1. N. M. Wereley, A. Chaudhuri, J. –H. Yoo, S. John, S. Kotha, A. Suggs, R. Radhakrishnan, B. Love and T. Sudarshan, "Bidisperse Magnetorheological Fluids Using Fe Particles at Nanometer and Micron Scale," J. Intel. Mat. Syst. Str. **17** (2006) pp. 393.
2. A. Chaudhuri, G. Wang and N. Wereley, "Substitution of Micron by Nanometer Scale Powders in MR Fluids," Int. J. Mod. Phys. B **19** (2005) pp. 1374.
3. J. Trihan, J. Yoo, N. Wereley, S. Kotha, A. Suggs, R. Radhakrishnan, B. Love and T. Sudarshan, "Impact of Varying Concentrations of Nanometer Sized Particles in a Bidisperse MR Fluid," Proc. of SPIE **5052** (2003) pp. 175.
4. E. Lemaire, G. Bossis and A. Meunierl, "Influence of Particle Size on the Rheology of Magnetorheological Fluids," J Rheol. **39** (1995) pp. 1011.
5. G. Bossis, S. Lacis, A. Meunier and O. Volkova, "Magnetorheological Fluids," J. Magn. Magn. Mater. **252** (2002) pp.224-228.
6. C. Kormann, H. Laun and H. Richter, "MR Fluids with Nano-Sized Magnetic Particles," Int. J. Mod. Phys. B **10** (1996) pp. 3167.
7. E. F. Burguera, B. J. Love, R. Sahul, G. Ngatu, N.M. Wereley, "The Physical Basis for Stability in Bimodal Dispersions Including Micrometer-Sized Particles and Nanoparticles," Submitted.
8. S. R. Gorodkin, W. I. Kordonski, E. Medvedeva, Z. Novikova, A. Shorey, and S. Jacobs, "A Method and Device for Measurement of a Sedimentation Constant of Magnetorheological Fluids," Rev. Sci. Instrum. **71** (2000) pp. 2476.
9. L. Chen, D. Chen, "Permalloy Inductor Based Instrument that Measures the Sedimentation Constant of MR Fluids," Rev. Sci. Instrum. **74** (2003) pp. 3566.

PREPARATION AND MAGNETORHEOLOGICAL CHARACTERIZATION OF CI/PVB CORE/SHELL PARTICLE SUSEPENDED MR FLUIDS

J. L. YOU, B. J. PARK, H. J. CHOI, AND S. B. CHOI*

*Department of Polymer Science and Engineering and *Department of Mechanical Engineering, Inha University, Incheon 402-751, Korea* *Email: hjchoi@inha.ac.kr*

M. S. JHON

Department of Chemical Engineering, Carnegie Mellon University, Pittsburgh 15213, USA

Hybrid magnetic particles of carbonyl iron (CI) /poly(vinyl butyral) (PVB) with core/shell microstructure (CI-PVB) were prepared in order to enhance the dispersion stability of the magnetorheological (MR) fluids. Since the composite particles of CI-PVB have a lower density than that of the pristine CI particles, they are regarded to improve the sedimentation problem of magnetic particles in the MR fluid when the particles are dispersed in a mineral oil and to make easy redispersion after caking. The PVB coating layers were found to play an important role in the steric repulsion between the relatively large CI particles. Morphology and composition of the CI-PVB particles were observed via SEM and TGA, respectively. Flow properties of both CI and CI-PVB based MR fluids were examined via a rotational rheometer in parallel plate geometry equipped with a magnetic field supplier.

1 Introduction

Magnetorheological (MR) fluid is a stable colloidal particle suspension dispersed in a viscous or viscoelastic non-mangetic carrier fluid, exhibiting rapid changes of rheological properties under external magnetic fields, i.e. a gel-like state under a magnetic field with enhanced shear viscosity and viscoelastic characteristics. Such interesting phenomenon of magnetic particles building up chain-like structure with a magnetic field was found to be due to magnetic dipole-dipole interaction between particles [1-6], in which the MR fluids along with electrorheological (ER) fluids [7-10] are regarded as smart materials. MR fluid demonstrates a high yield stress up to 10-100 kPa under a strong magnetic field [11]. Controlling the intensity of the magnetic field applied in the MR fluid can adjust the apparent viscosity and other rheological properties of the fluid. Based on this tunable characteristic, these fluids can be widely used in designing damping devices such as shock absorbers, clutches and vibration dampers [12-15]. Carrier phase of the MR fluids can be silicone oil, mineral oil, synthetic hydrocarbon compound or other suitable organic liquid. Soft ferromagnetic materials such as magnetite and iron are generally adopted as the MR materials [16, 17], since these particles are easily magnetized under magnetic field. Among various MR materials, MR fluid based on carbonyl iron (CI) has attracted much attention with its strong magnetic properties. However, most magnetic fluids including MR suspension have not only sedimentation and aggregation problems due to their high density and the remnant magnetization, but also abrasivity problem of the particles. Therefore, the dispersion stability and the re-dispersibiliy of the MR fluids need to be improved for future engineering applications of the MR fluids. In order to improve the stability of the MR fluids, the surface of magnetic particles was treated with polymers or surfactants [18, 19]. To compensate these disadvantages of the pristine CI particles, a light and soft poly(vinyl butyral) (PVB) coating was applied to the surface of the CI particles. The PVB coating prevents the cake formation among the CI particles and protects the device from abrasion. Rheological properties of the MR suspensions under applied magnetic field strengths were then investigated. The flow behavior and viscoelastic properties were measured by the steady state and dynamic oscillation experiments, respectively.

2 Experimental

2.1 Materials

As for the suspended particles in the MR systems, we used carbonyl iron (CI, standard CM grade, BASF, Germany) with average particle size of about 3㎛. Poly(vinyl butyral) (PVB) (Mowital B30HH) was adopted for polymeric coating of the magnetic particle, and chloroform as a solvent due to its high dissolving capacity of PVB and fast evaporation rate at room temperature. Both poly(vinyl alcohol) (PVA) and triblock copolymer ($PEG_{15}PPG_{53}PEG_{15}$, Aldrich) were used as stabilizers, while sodium dodecyl sulfate (SDS, Aldrich) was used as an emulsifier.

2.2 Preparation of CI-PVB composite microspheres

Composite magnetic particles of CI-PVB were prepared in spherical shape with a PVB coating via a solvent evaporation technique [20]. PVB was initially dissolved in chloroform, and the CI particles were quickly added into the solution mixture of stabilizer (PVA), emulsifier (SDS) and triblock copolymer ($PEG_{15}PPG_{53}PEG_{15}$) dissolved in deionized water, separately. The dispersion medium was mixed using a mechanical stirrer. After 2 days, all of the chloroform used as a solvent was evaporated, leaving the CI particles completely coated with PVB. The coated CI particles were washed with deionized water to remove the physically adsorbed reactant at the surface of the composites.

To prepare the MR fluid, these composite particles were dispersed in a mineral oil (MO, Aldrich). Particle concentration of the CI-PVB composite particle dispersed in mineral oil was set to be 30 wt%.

2.3 Characterization

Diameter and surface characteristics of the prepared CI-PVB particles were observed with scanning electron microscope (SEM, JSM-6700F, JEOL, Tokyo, Japan). The thermogravimetric analysis (TGA) was carried out in the temperature range of 30-800℃ at a heating rate of 20℃/min under nitrogen atmosphere. The CI content in the CI-PVB microspheres was calculated from residues of the TGA diagram. Magnetic properties of the particles were examined using a vibration sample magnetometer (VSM) at room temperature. MR fluid was prepared by dispersing 30wt% of polymer coated carbonyl iron particles in mineral oil. Rheological properties of MR suspensions of the CI-PVB microspheres were examined under a magnetic field using a parallel plate typed rheometer (MCR300, Physica, Germany) at 25℃. The coil current and magnetic field strength using separate control unit and rheometer software (Rheopuls 2.65, Physica, Germany) were applied.

3 Results and discussion

Figure 1 shows the size and surface morphology of the CI-PVB particles observed from the SEM, in which the surface of the coated CI particles turns rugged as a result of the PVB coating. The average particle size was about 4㎛ which became slightly larger than that of pristine CI particles due to the encapsulation of PVB. After coating surface of the CI particles with PVB, particle size distribution becomes narrower than that of pure CI particles.

Figure 1. SEM image of CI-PVB particles

Magnetization curves of both polymer coated CI particles and pure CI particles are shown in Fig. 2. It can be found that the polymer coated CI possesses similar magnetic properties comparing with pure carbonyl iron particles. The saturation magnetizations of pure carbonyl iron and CI-PVB particles were found to be 203.36 emu/g and 191.66 emu/g, respectively. About 5 % decrease of the saturation magnetization of the CI-PVB particles compared with pure carbonyl iron particles, due to the polymeric coating was detected. That is to say, the magnetization can only be attained at the magnetic particle amount. Ideally, this means that a MR suspension based on magnetic powder will have a stronger MR effect than CI-PVB microspheres, if the magnetization was the only determining factor for the MR effect. The amount of PVB in the CI-PVB microspheres was indirectly analyzed using a thermal analysis. Figure 3 shows the TGA thermograms for both CI-PVB and pure PVB particles. The 23 wt% weight loss of organic material up to 400℃ and a constant weight up to 700℃ were observed, indicating that the coated magnetic particles contained 77wt% carbonyl iron particles.

MR fluids of CI-PVB microspheres dispersed in a mineral oil with a concentration of 20wt% were prepared and their steady shear properties under various magnetic fields are represented in Fig. 4. It shows increased shear stresses with increased magnetic fields. As the magnetic density is increased, the yield shear stress of the MR fluid also increases, exhibiting typical MR behavior. Both MR fluids exhibit non-Newtonian behavior in the absence of the magnetic field, because of remnant magnetization of the polymer coated carbonyl iron particles even though it is relatively small as shown in Fig. 2 and relatively high particle concentration [21].

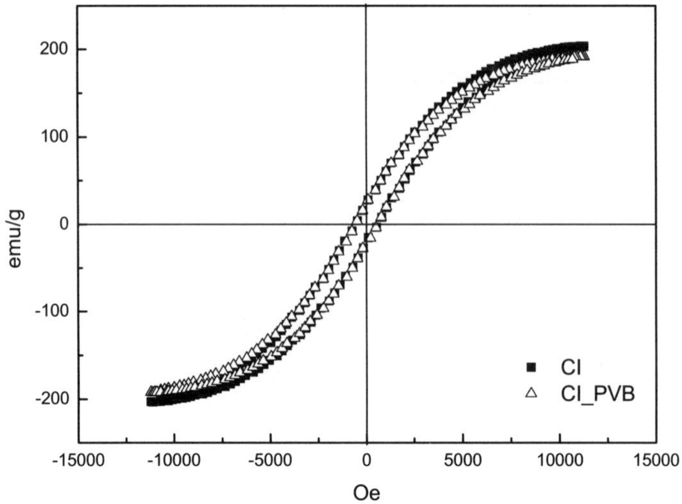

Figure 2. VSM image of pure CI and CI-PVB particles

The MR fluid of pure CI particles showed slightly larger yield stress compared with that of the CI-PVB microspheres due to the stronger magnetic property. Magnetorheological fluid changed flow property from a liquid-like to a solid-like state under magnetic field strength, and magnetic particles building up chain-like structure with an applied magnetic field was due to dipole-dipole interaction, causing increased shear viscosities [22-24].

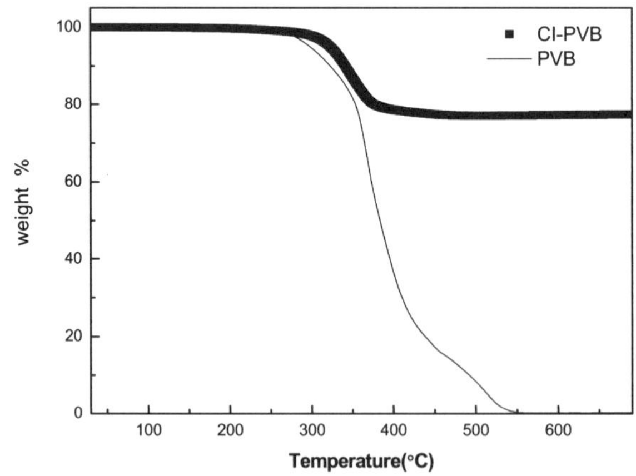

Figure 3. TGA thermogram of CI-PVB and PVB particles

Typical behavior of MR fluids under a magnetic field can be characterized by a Bingham plastic model as shown in Fig. 4. This means that a minimum yield stress is needed to start the fluid flow at a zero shear rate limit. This yield stress is one of the important factors that determine the MR effect. In general, higher yield stresses are required for a specific application of the MR fluid [6].

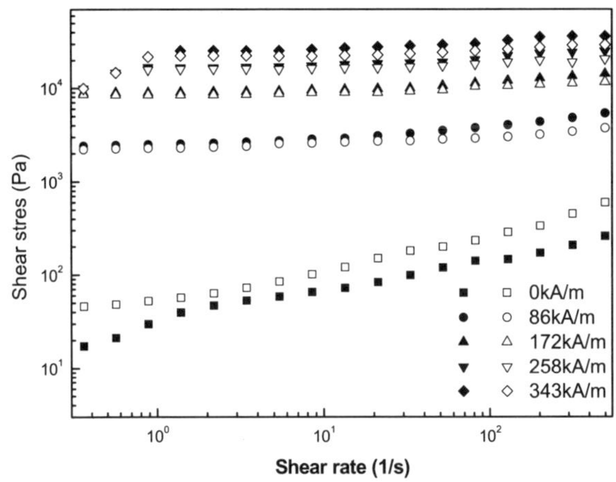

Figure 4. Shear stress (a) and viscosity(b) versus shear rate of CI-PVB particles (open: CI-PVB, close : CI)

Figure 5 shows storage modulus (a) and loss modulus (b) as a function of frequency for the CI-PVB under applied magnetic field strengths. The storage modulus (G') is the measurement of the elasticity and loss modulus G" response is the property related to the shear viscosity. The storage modulus G' of MR fluid showed slight increase when increasing the applied magnetic field. These results suggest that both G' and G" responses get enhanced with increased magnetic field strength. In addition, the G' value of storage modulus is obviously higher than that of loss modulus G", indicating that the sample presents well the elastic property rather than the viscous property [25-30].

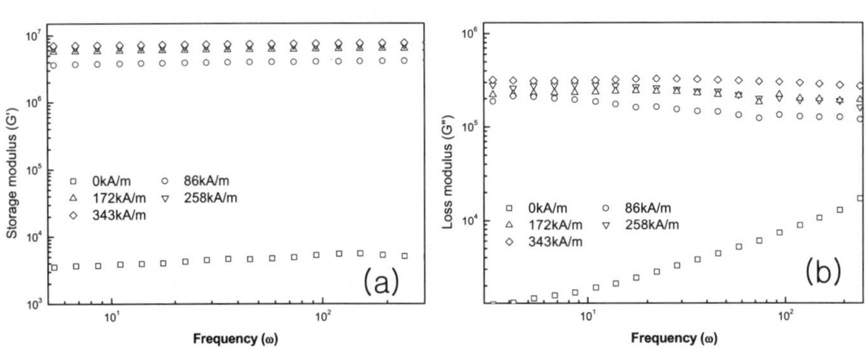

Figure 5 Storage modulus (a) and loss modulus (b) ad a function of frequency for CI-PVB based MR fluid

4 Conclusion

The polymer coated carbonyl iron particles were prepared by solvent evaporation techniques using PVB, while the magnetic properties of the CI-PVB particles slightly decreased compared with pure CI particles because of the encapsulation of polymer on the CI particle. MR properties of the polymer coated magnetic particles dispersed in mineral oil were studied, and shear stress was observed to be increased with an applied magnetic field. The more magnetic field strength was applied, the higher yield stress of the MR fluid was observed, exhibiting typical MR behavior. When the magnetic field strength increased, the storage modulus and loss modulus value were also increased indicating very strong elastic properties. It was also observed that the sedimentation was improved via polymeric coating.

5 Acknowledgements

This study was supported by research grants from the NRL program (2005).

References

1. Brand H. R. and Pleiner H., Origin of the Slow Wave in a Magnetorheological Slurry, *Phys. Rev. Lett.* **86** (2001) pp. 1385.
2. Lim S. T., Cho M. S., Jang I. B., Choi H. J. and Jhon M. S., Magnetorheology of carbonyl–iron suspensions with submicron-sized filler, *IEEE Trans. Magn.* **40** (2004) pp. 3033-3035.
3. Claracq J., Sarrazin J. and Montfort J. P., Viscoelastic properties of magnetorheological fluids, *Rheol Acta.* **43** (2004) pp. 38-49.
4. Wollny K., Läuger J. and Huck S., Mangeto sweep – a method for characterizing the viscoelastic properties of magneto-rheological fluids, *Appl. Rheol.* **12** (2002) pp. 25-31.
5. Lita M., Popa N. C., Velescu C. and Vékás L. N., Investigations of a Magnetorheological Fluid Damper, *IEEE Trans. Magn.* **40** (2004) pp. 469-472.
6. Carlson J. D., MR fluids and devices in the real world, *Int. J. Mod. Phys. B,* **19** (2005) pp. 1463-1470.
7. Zhao X. P. and Yin J. B., Advances in electrorheological fluids based on inorganic dielectric materials, *J. Ind. Eng. Chem.,* **12** (2006) pp. 184-198.
8. Quadrat O. and Stejskal J., Polyaniline in electrorheology, *J. Ind. Eng. Chem.,* **12** (2006) pp. 352-361.
9. Sung J. H., Park D. P., Park B. J. and Choi H. J., Universal yield stress characteristics of potato starch phosphate-based electrorheological fluid, *J. Ind. Eng. Chem.,* **12** (2006) pp. 301-305.
10. Han Y. M., Choi S. B. and Choi H. J., Preisach model of ER fluids considering temperature variations, *Int. J. Mod. Phys. B,* **19** (2005) pp.1325-1331.
11. Park J. H. and Park O. O., Electrorheology and magnetorheology, *Korea-Australia Rheol. J.*, **13** (2001) pp. 13-17.
12. Phulé P. P. and Ginder J. M., Synthesis and properties of novel magnetorheological fluids having improved stability and redispersibility, *Int. J. Mod. Phys. B*, **13** (1999) pp. 2019-2027.

13. Zhu C. S., The response time of a rotor system with a disk-type magnetorheological fluid damper, *Int. J. Mod. Phys. B,* **19** (2005) pp.1506-1512.
14. Sung K. G., Choi S. B., Lee H. G., Min K. W. and Lee S. H., Performance comparison of MR dampers with three different working modes: Shear, flow and mixed mode, *Int. J. Mod. Phys. B,* **19** (2005) pp.1556-1562.
15. Facey W. B., Rosenfeld N. C., Choi Y. T., Wereley N. M., Choi S. B. and Chen P., Design and testing of a compact magnetorheological damper for high impulsive loads, *Int. J. Mod. Phys. B,* **19** (2005) pp.1549-1555.
16. Kormann C., Laun H. M. and Richter H. J., MR fluids with nanosized magnetic particles, *Int. J. Mod. Phys. B* **10** (1996) pp. 23-24.
17. Bica I., The influence of the magnetic field on the electrical magnetoresistance of magnetorheological suspensions, *J. Magn. Magn. Mater.,* **299** (2006) pp. 412-418.
18. Lee S. U. and Harris M. T., Surface modification of magnetic nanoparticles capped by oleic acids: characterization and colloidal stability in polar solvents, *J. Colloid Interf. Sci.* **293** (2006) pp. 401-408.
19. Jun J. B., Kim J. W., Lee J. W. and Suh K. D., Spherical polarization body: synthesis of monodisperse micron-sized polyaniline composite particles, *Macromol. Rapid Commun.* **22** (2001) pp.937-940.
20. Maity A. and Biswas M., Recent progress in conducting polymer, mixed polymer-inorganic hybrid nanocomposites, *J. Ind. Eng. Chem.,* **12** (2006) pp. 311-351.
21. Choi H. J., Kim C. A., Kwon T. M. and Jhon M. S., Viscosity of magnetic particle suspensions, *J. Magn. Magn. Mater.* **209** (2000) pp. 228-230.
22. Lim S. T., Cho M. S., Jang I. B. and Choi H. J., Mangetorheological characterization of carbonyl iron based suspension stabilized by fumed silica, *J. Magn. Magn. Mater.,* **282** (2004) pp. 170-173.
23. Jang I. B., Kim H. B., Lee J. Y., You J. L., Choi H. J. and Jhon M. S., Role of organic coating on carbonyl iron susepended particles in magnetorheological fluids, *J. Appl. Phys.* **97** (2005) 10Q912.
24. Niraula B., King T. C. and Misran M., Rheology properties of dodecyl-β -D-maltoside stabilized mineral oil-in-water emulsions, *Colloid Surf. A-Physicochem. Eng. Asp.* **231** (2003) pp.159-172.
25. Ginder J. M., Davis L. C. and Elie L. D., Rheology of magnetorheological fluids: models and measurements, *Int. J. Mod. Phys. B,* **10** (1996) pp. 3293-3303.
26. Rankin P. J., Horvath A. T. and Klingenberg D. J., Magnetorheology in viscoplastic media, *Rheol. Acta*, **38** (1999) pp. 471-477.
27. Kuzhir P., Bossis G. and Bashtovoi V., Optimization of magnetorheological fluid valves, *Int. J. Mod. Phys. B,* **19** (2005) pp.1229-1235.
28. Lim S. T., Cho M. S., Choi H. J. and Jhon M. S., Magnetorheological characterization of organoclay added carbonyl-iron suspensions, *Int. J. Mod. Phys. B,* **19** (2005) pp.1142-1148.
29. Gordaninejad F., Kavlicoglu B. M. and Wang X., Friction factor of magneto-rheological fluid flow in grooved channels, *Int. J. Mod. Phys. B,* **19** (2005) pp.1297-1303.
30. Trendler A. M. and Bose H., Influence of particle size on the rheological properties of magnetorheological suspensions, *Int. J. Mod. Phys. B,* **19** (2005) pp.1416-1422.

PREPARATION AND ELECTRORHEOLOGICAL CHARACTERISTICS OF PANI/MWNT NANOCOMPOSITE

J. Y. LIM, S. T. KIM, B. J. PARK AND H. J. CHOI

Department of Polymer Science and Engineering, Inha University, Incheon 402-751, Korea
E-mail: hjchoi@inha.ac.kr

We prepared polyaniline (PANI) /multi-walled carbon nanotube (MWNT) nanocomposites of both PANI pierced MWNT nanocomposites synthesized using poly(vinyl alcohol) as a polymeric stabilizer and PANI/MWNT nanocables via an in-situ oxidative polymerization directed by a cationic surfactant of cetyltrimethylammonium bromide (CTAB). In addition to chemical analysis of the products, the formation of nanocomposite nanocable and their morphology in which the MWNTs were entirely encapsulated by amorphous PANI, were confirmed by scanning electron microscopy (SEM) and transmission electron microscopy (TEM). Electrorheological (ER) fluids were prepared by dispersing both of these synthesized MWNT/PANI nanocomposites in insulating silicone oil and their ER properties were investigated by controlling applied DC electric field strengths. The synthesized nanocomposite particles have enhanced interparticle interactions, and the shear stresses increase with the electric field strength for a broad range of shear rate. Among three different constitutive equations adopted, the Cho-Choi-Jhon model was found to fit the flow curves quite well.

1 Introduction

Polymer/carbon nanotube (CNT) nanocomposites have attracted significant attentions, because of their engineering potential applications with extraordinary mechanical and electrical properties [1, 2]. Many polymers such as polystyrene, poly(methyl methacrylate), and poly(ethylene oxide) [3] have been used as matrix materials in the CNT/polymer composites for target applications from various preparation methods of solution casting, melt mixing, grafting macromolecules to the CNTs, electrochemical and in situ polymerization [4, 5]. The CNTs can be also used as a conductive filler for several systems such as CNT/poly(phenylene-vinylene) [6] and CNT/polypyrrole [7] composites. These conjugated polymers with the incorporation of CNTs can lead to new composite materials possessing the properties of each component with a synergistic effect that would be useful in particular applications.

Meanwhile, PANI has been extensively studied as one of promising conducting polymers because of not only its stability and high conductivity [8], but also the fact that its doping level can be readily controlled through an acid/base doping/depoing process [9]. Furthermore, it has been extensively adopted as an ER material [10-13], when dispersed in an insulating fluid, showing a rapid and reversible change in shear viscosity under an applied electric field [14].

In this study, the MWNT and PANI nanocomposites were introduced as ER materials, in order to investigate the synergistic effects of CNT/conducting polymer nanocomposites. Recently, several studies on MWNT/PANI composites were reported, including electrochemical response of MWNT/PANI composite film [15] and CNT/PANI core-shell nanowires prepared by in situ inverse microemulsion [16].

We synthesized PANI/MWNT nanocomposites of both PANI pierced MWNT nanocomposites prepared using poly(vinyl alcohol) as a polymeric stabilizer [17] and PANI/MWNT nanocables by in situ oxidative polymerization directed by the cationic surfactant of CTAB [18]. After preparation of polymer nanocomposites, we examined both their dispersion stability and ER properties. ER fluids were prepared by dispersing both of these synthesized MWNT/PANI nanocomposites in insulating silicone oil and their ER properties were investigated by controlling applied DC electric field strengths.

2 Experimental

2.1 Materials

The MWNT with purity of ~95 wt% synthesized by a thermal chemical vapor deposition method (Iljin Nanotec Co., Korea) was used in this study. To improve purity of the MWNT prior to use, it was treated in 3M HNO_3 and refluxed for 24h at 60 ℃, and refluxed again in 5M HCl solution for 6h at 120 ℃. The purity of the acid-treated MWNT was measured to be 99% using thermogravimetric analysis (TGA, Polymer Lab., TGA1000, UK). Aniline (DC Chemical Co., Ltd, Korea), ammonium persulfate (APS) (Daejung, Korea) and poly(vinyl alcohol) (PVA, DC Chemical Co., Ltd, Korea) were used without further purification.

2.2 *Synthesis of PANI/MWNT nanocomposites*

The PANI/MWNT nanocomposite was synthesized via an oxidative dispersion polymerization with PVA as a polymeric stabilizer. PVA possesses hydrophilic and hydrophobic group, in which the PVA helped hydrophobic MWNT be dispersed into water. PVA was dissolved in deionized water at 90℃. The treated MWNT (0.1 wt% for aniline), aniline monomer, and 1.0 M HCl were dispersed in PVA solution by ultrasonic generation (G2806, Kyungil Ultrasonic Co., Korea), with a nominal frequency of 28kHz and power of 600 W, for 5h at 25℃. The deionized water containing APS was added to the suspension. The PANI/MWNT composite was polymerized with stirring and ultrasonication for 24 h at 20℃. After polymerization, the PANI/MWNT composites were washed by deionized water.

2.3 *Synthesis of PANI/MWNT nanocables*

PANI/MWNT nanocable was synthesized via in-situ chemical oxidation polymerization with cationic surfactant of CTAB. The acid-treated MWNT and CTAB were dispersed in 1.0 M HCl solution using the same ultrasonic generator with a power of 600 W for 2h at 25℃. Well dispersed suspensions were cooled down to -5℃. Aniline monomer and 1.0 M HCl containing APS were added to the above suspension. The suspension was sonicated for 2min, then left standing in the refrigerator at -5℃ for 24h. The PANI/MWNT composite was filtered and rinsed with deionized water and methanol for several times. The remaining PANI/MWNT nanocable was dried under a vacuum at room temperature for 24h.

2.4 *Characterization*

After polymerization of both samples, the PANI/MWNT composites were characterized by a Fourier transform infrared spectrometer (FT-IR, Perkin Elmer System, spectrum

2000 explorer sperctrometer, USA) for the evaluation of chemical structures using a KBr pellet. The surface morphology of the composites was observed by a SEM (Hitachi S-4300, Japan), while the inside of the composite observed by a TEM (Phillips CM200, NL), using the samples prepared by solvent casting method. The dc electrical conductivities of the PANI/MWNT composite and PANI/MWNT nanocable were measured by four pin probe method (Loresta GP, Mitsubishi Chemical, Japan).

3 Results and discussion

SEM pictures given in Fig. 1 show surface morphologies of both PANI/MWNT nanocomposites prepared with a polymeric stabilizer (a) and a cationic surfactant (b). PANI seems to be pierced with MWNT like a pearl necklace structure (Fig. 1(a)). These MWNT/PANI particles were well dispersed in water and no sedimentation was observed for a few weeks, suggesting that the PVA was stably adsorbed onto the particle surface. Meanwhile, Fig. 1(b) shows that the PANI/MWNT composites have thicker diameter compared to the raw MWNT. This result represents that the PANI coated surface of the MWNT.

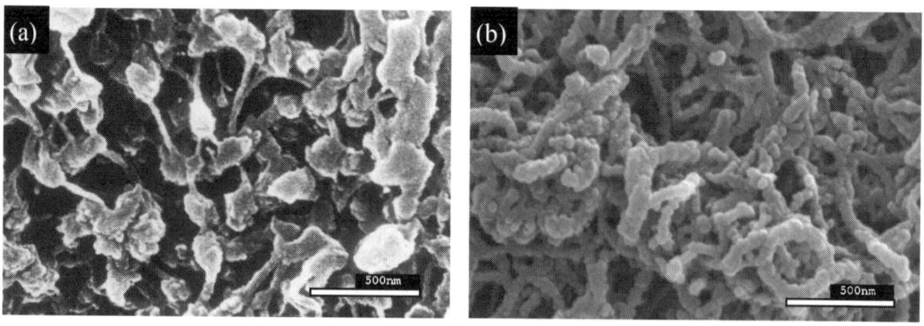

Figure 1. The SEM images of PANI/MWNT nanocomposite by in situ polymerization with (a) polymeric stabilizer (PVA) and (b) cationic surfactant (CTAB)

Figure 2 represents TEM images of the PANI/MWNT nanocomposite prepared from two different methods. The PANI droplets are observed to be formed in the middle of the MWNT chain with the pearl-necklace morphology (Fig. 2(a)). This morphology is considered to be due to the interfacial energy difference. If the interfacial energy between PANI and MWNT is smaller than the surface energy of MWNT, the PANI should be located on the MWNT surface. Furthermore, PANI showed spherical bead-shapes because of PVA presence. Figure 2(b) shows cable-like morphology of MWNT/PANI with the core/shell structure. The outer layer has an average thickness about 10~20 nm. The mechanism of these nanocable formations were described by Zhang et al. [18], explaining that the surfactant, CTAB plays an important role.

Figure 3 shows the FT-IR spectrum of MWNT, PANI, and PANI/MWNT nanocomposite. Even though the MWNT does not possess characteristic peaks, the spectrum of PANI

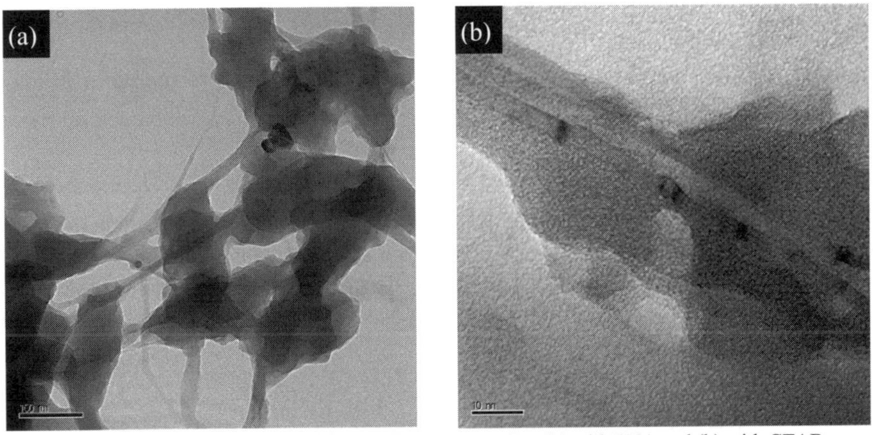

Figure 2. TEM images of PANI/MWNT nanocomposites (a) with PVA and (b) with CTAB

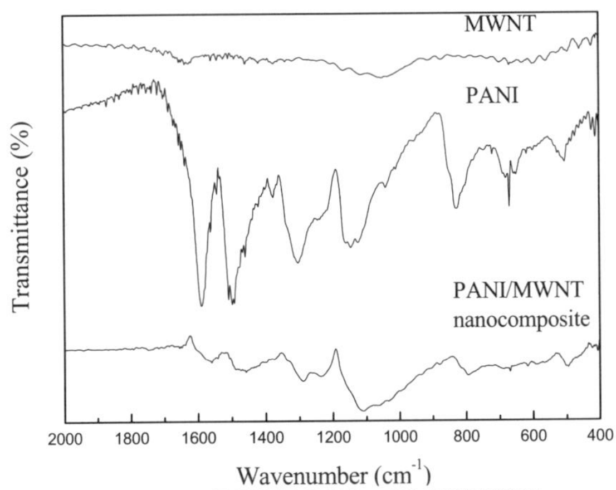

Figure 3. FT-IR spectrum of MWNT, PANI and PANI/MWNT nanocomposite

shows all characteristic peaks such as 1589 cm^{-1} indicating C=C stretching of the quinoid rings, 1458 cm^{-1} of C=C stretching of benzenoid rings, 1303 cm^{-1} of C-N stretching , and 1145 cm^{-1} of N=Q=N (where Q represents the quinoid ring) stretching. The spectrum of PANI/ MWNT nanocomposite shows all characteristic peaks of PANI, while all of the peaks in the PANI/MWNT nanocomposite shift to right about 30cm^{-1} than the PANI peaks.

Figure 4 represents the dependence of shear stress as a function of shear rate for different electric field strengths for the PANI/MWNT nanocomposite dispersed in silicone oil. While the electrostatic force of inter-particle dominates at low shear rate region, the hydrodynamic force of each particle does at high shear rate region. The plateau region describes the overcome of electrostatic force over an external shear force. In order to analyze this plateau region, three different models are adopted. In general, the plateau region has been described using a following Bingham fluid-like behavior with a non-vanishing value of yield stress in the presence of an applied electric field.

$$\tau = \tau_0 + \eta_0 \dot{\gamma} \qquad \tau \geq \tau_0$$

$$\dot{\gamma} = 0 \qquad \tau < \tau_0 \tag{1}$$

In addition, another commonly used form of the De Kee-Turcotte model is following [19]:

$$\tau = \tau_0 + \eta_1 \dot{\gamma} e^{-t_1 \dot{\gamma}} \tag{2}$$

In Fig. 4, dotted line and solid line (normal) show the fitting by Bingham and De Kee-Turcotte model, respectively. However, none of these can fit the decrease of shear stress especially at low shear region. In order to describe complicated flow curve of such ER system, recently Cho-Choi-Jhon (CCJ) proposed the following constitutive equation [20, 21]:

$$\tau = \tau_0 \left(\frac{1}{1 + (t_1 \dot{\gamma})^{\alpha}} \right) + \eta_\infty \left(1 + \frac{1}{(t_2 \dot{\gamma})^{\beta}} \right) \dot{\gamma} \tag{3}$$

The CCJ model can cover the stress decrease phenomena at a low shear rate region and provide an accurate value for the real yield stress. τ_0 is the dynamic yield stress defined as the extrapolated stress from low shear rate region, α is related to the decline in the shear stress at a low shear rate region, and the exponent β ranges from $0 < \beta \leq 1$, because $d\tau / d\dot{\gamma} \geq 0$ above the critical shear rate at which the shear stress becomes a minimum. t_1 and t_2 are time constants, in which t_1 is considered to the inverse of the shear rate at which the shear stress shows a minimum at a low shear rate region and t_2 is related to the inverse of the shear rate at which a pseudo-Newtonian behavior starts. η_∞ is the viscosity at an infinite shear rate.

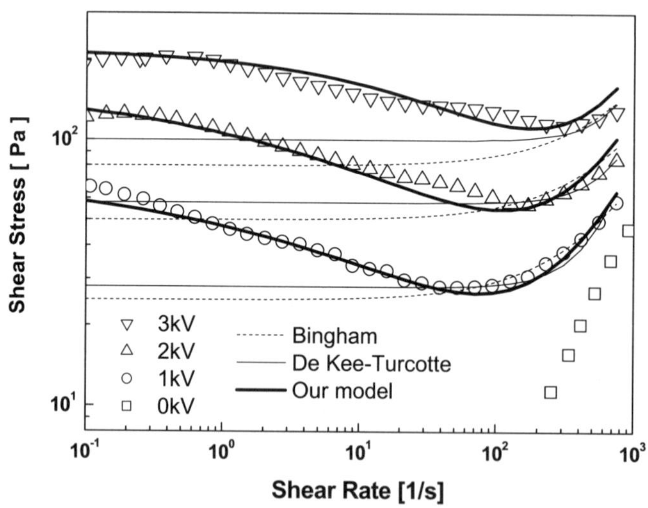

Figure 4. Flow curve of PANI/MWNT composite based ER fluid

Figure 5 demonstrates the ER phenomenon of MWNT/PANI composite with nanocable structure. The orientation of fabricated particles in the silicone oil toward the electrode under the external electrical field was observed via an optical microscopy. The randomly dispersed composite particles in the medium (a) transformed to fibrillated chains structure based on interfacial polarization (b) after application of electric voltages [22, 23].

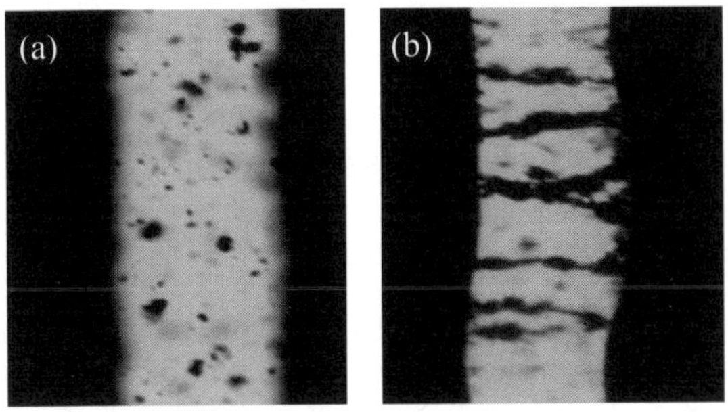

Figure 5. Optical microscopic image of the PANI/MWNT composite dispersed in silicone oil before (a) and after the application of an electric field (right)

4 Conclusions

We fabricated both MWNT/PANI nanocomposites with pearl-necklace structure and core-shell nanocable structure through in situ oxidation polymerization by using PVA as a polymeric stabilizer and by using the cationic surfactant CTAB, respectively. Chemical structures and ER characteristics of the MWNT/PANI composite particles were also investigated. The synthesized MWNT/PMMA particles exhibited typical ER responses, i.e., shear stress increased with the increase of an applied DC electric field strength.

5 Acknowledgments

This work was supported by research grants from KOSEF (R01-2006-000-10062-0), Korea.

References

1. Grossiord N., Loos J., Regev O. and Koning C. E., Toolbox for dispersing carbon nanotubes into polymers to get conductive nanocomposites. *Chem. Mater.* **18** (2006) pp. 1089-1099.
2. Park S. J., Cho M. S., Lim S. T., Choi H. J. and Jhon M. S., Synthesis and dispersion characteristics of multi-walled carbon nanotube composites with poly(methyl methacrylate) prepared by in-situ bulk polymerization. *Macromol. Rapid Commun.* **24** (2003) pp. 1070–1073.
3. Park S. J., Cho M. S., Lim S. T., Choi H. J. and Jhon. M. S., Electrorheology of multiwalled carbon nanotube/poly(methyl methacrylate) nanocomposites. *Macromol. Rapid Commun.* **26** (2005) pp. 1563–1566.
4. Jin H. J., Choi H. J., Yoon S. H., Myung S. J. and Shim. S. E., Carbon nanotube-adsorbed polystyrene and poly(methyl methacrylate) microspheres. *Chem. Mater.* **17** (2005) pp. 4034-4037.

5. Tang W. Z., Santare M. H.and Advani S. G., Melt processing and mechanical property characterization of multi-walled carbon nanotube/high density polyethylene (MWNT/HDPE) composite films. *Carbon* **41** (2003) pp. 2779-2785.
6. Coleman J. N., Curran S., Dalton A. B., Davey A. P., Mc Carthy B., Blau W. and Barklie R. C., Physical dopping of a conjugated polymer with carbon nanotubes. *Synth. Met.* **102** (1999) pp.1174-1175.
7. Chen G. Z., Shaffer M. S. P., Coleby D., Dixon G., Zhou W., Fray D. J. and Windle A. H., Carbon nanotube and polypyrrole composites: coating and doping. *Adv. Mater.* **12** (2000) pp.522-526.
8. MacDiarmid A. G. "Synthetic metals": a novel role for organic polymers. *Curr. Appl. Phys.* **1** (2001) pp.269-279.
9. Maity A. and Biswas M., Recent progress in conducting polymer, mixed polymer-inorganic hybrid nanocomposites. *J. Ind. Eng. Chem.* **12** (2006) pp. 311-351.
10. Park S. J., Park S. Y., Cho M. S., Choi H. J., and Jhon M. S., Synthesis and electrorheology of multi-walled carbon nanotube/polyaniline nanoparticles. *Synth. Met.* **152** (2005) pp. 337-340.
11. Choi H. J., Cho M. S., Lee Y. H., Cho Y. H and Jhon M. S., Electrorheology of conducting polyaniline composite particles. *e-Polymers* (2004) Art. 030.
12. Quadrat O. and Stejskal J., Polyaniline in electrorheology, *J. Ind. Eng. Chem.* 12 (2006) pp. 352-361.
13. Lu J. and Zhao X., Electrorheological properties of a polyaniline-montmorillonite clay nanocomposite suspension. *J. Mater. Chem.* **12** (2002) pp. 2603-2605.
14. Sim I. S., Kim. J. W., Choi H. J, Kim C. A. And Jhon M. S., Preparation and electrorheological characteristics of poly(p-phenylene)-based suspension. *Chem. Mater.* **13** (2001) pp. 1243-1247.
15. Guo D. and Li. H. Well-dispersed multi-walled carbon nanotube/polyaniline composite films. *J. Solid State Electrochem.* **9** (2005) pp.445-449.
16. Yu Y., Che B., Si Z., Li L., Chen W. and Xue G., Carbon nanotube/polyaniline core-shell nanowires prepared by in situ inverse microemulsion. *Synth. Met.* **150** (2005) pp.271-277.
17. Choi H. J., Park S. J., Kim S. T. and Jhon M. S., Electrorheological application of polyaniline/multi-walled carbon nanotube composites. *Diamond Relat. Mater.* 14 (2005) pp. 766-769.
18. Zhang X., Zhang J., Wang R. And Liu Z., Cationic surfactant directed polyaniline/ CNT nanocables: synthesis, characterization, and enhanced electrical properties. *Carbon* **42** (2004) pp.1455-1461.
19. Zhu H., Kim Y. D. and De Kee D., Non-Newtonian fluids with a yield stress. *J. Non-Newtonian Fluid Mech.,* **129** (2005) pp. 177-181.
20. Cho M. S., Choi H. J. and Jhon M. S., Shear stress analysis of a semiconducting polymer based electrorheological fluid system, *Polymer* **46** (2005) pp. 11484-11488.
21. Hong C. H., Choi H. J. and Jhon M. S., Comment on "preparation and enhanced electrorheological activity of TiO_2 doped with chromium ion". *Chem. Mater.* **18** (2006) pp. 2771-2772.
22. See H., Advances in electro-rheological fluids: materials, modeling and applications. *J. Ind. Eng. Chem.* **10** (2004) pp. 1132-1145.
23. Lengalova A., PavlinekV, Saha P., Stejskal J., Kitano T., and Quadrat O, The effect of dielectric properties on the electrorheology of suspensions of silica particles coated with polyaniline. *Physica A* **321** (2003) pp. 411-424.

VIBRATION CONTROL OF ENGINE MOUNT USING ER FLUID AND PIEZOELECTRIC ACTUATOR

H. J. SONG AND S. B. CHOI

Smart Structures and Systems Laboratory, Department of Mechanical Engineering, Inha University, Incheon 402-751, Korea
E-mail: seungbok@inha.ac.kr

Y. S. JEON

Department of Tool and Mold Design, Yuhan College, Gyeonggi-Do 442-749, Korea
E-mail: drjys@hanmail.net

N. M. WERELEY

Smart Structures Laboratory, Department of Aerospace Engineering, University of Maryland, College Park, MD 20742, U. S. A.
E-mail: wereley@umd.edu

This paper presents vibration control responses of a hybrid engine mount utilizing ER fluid and piezoelectric actuator to isolate the vibration of wide frequency range with various amplitudes. The function of ER fluid is to isolate vibration from engine unbalance force for low frequency range with large amplitude, while the piezoelectric actuator for high frequency range with small amplitude. The ER fluid part operating under the flow and shear modes is devised and the piezoelectric part is integrated with the magnification mechanism. After establishing the dynamic model of the proposed engine mount, both field-dependent dynamic stiffness of the engine mount are evaluated for various conditions. Then, in order to investigate the effectiveness of the proposed hybrid mount, one degree of freedom (DOF) system is established. The governing equation of motion of 1DOF system is formulated and then robust controller to attenuate vibration of wide range frequency component is designed. The vibration control performances such as displacement transmissibility are evaluated and presented in frequency domains.

1. Introduction

Noise and vibration problems of a passenger vehicle are caused by the engine excitation and considered more importantly. One of the sources of noise and vibration is engine excitation forces generated by the gas pressure and inertia force. These noise and vibrations bring about the passenger uncomfortable and structural problems. Therefore, various types of engine mounts have been proposed and developed to attenuate unwanted noise and vibration from the engine. Passive type engine mount exhibits performance limitation, showing good performance only narrow band for specially restricted region [1, 2]. One of attractive approaches to attenuate unwanted vibration is to utilize smart materials (electro-rheological (ER) fluid, magneto-rheological (MR) fluid and piezoelectric material) [3]. These materials are efficiently controllable by the intensity of electric field or magnetic field, respectively. Semi-active engine mount systems are featuring an ER fluid or MR fluid [4-6]. These are used to effectively isolation of vibration to the vehicle engine mount system. However, ER fluid and MR fluid have a property of semi-active that has insufficient performance to over the resonance range. The other type is active engine mount which is solenoid type utilizing electro-magnetic unit and force magnification type utilizing piezoelectric material of stack type [7, 8]. In the case of using piezoelectric material, it needs a magnification mechanism because displacement of actuator is very small and it can be generated high control force in a wide frequency range.

The hybrid engine mount proposed in this work has multi flow channel and mixed mode type featuring ER fluid and piezoelectric material. ER fluid and piezoelectric material are used to attenuation large displacement in low frequency range and small displacement in high frequency range, respectively. The hybrid engine mount then incorporated to the 1 degree-of-freedom system. The governing equation of motion is derived. A robust controller is designed to attenuate vibration. Control responses are presented in the frequency and time domain.

2. Dynamic Modeling of Hybrid Engine Mount

The hybrid engine mount consist of main rubber part that is support the engine mass and attenuate the vibration, ER fluid part which has multi cylindrical electrodes for large displacement and piezoelectric part which is located in lower part and generate control force. Then hybrid engine mount operated in wide frequency range. Figure 1 presents configuration of the hybrid engine mount. The flow path (electrodes) located between upper chamber and lower chamber. Fluid flow generated from the excitation input.

3.2 Dynamic Modeling of the ER Fluid Part

Figure 2 shows ER fluid part. Main rubber makes a role of supporting engine mass from stiffness and attenuating vibration from low damping. Flow channel is located between upper chamber and lower chamber. Fluid flow is generated by the relative motion of each chamber, then fluid moves on flow channel. In the case of without electric field, the damping force is produced by only fluid resistance. If electric field is applied, yield stress is produced by rheological behavior (Bingham property) between negative and positive electrodes. It is assumed that ER fluid is incompressible and the pressure in the chamber is uniform, rubber has linear stiffness constant and damping coefficient.

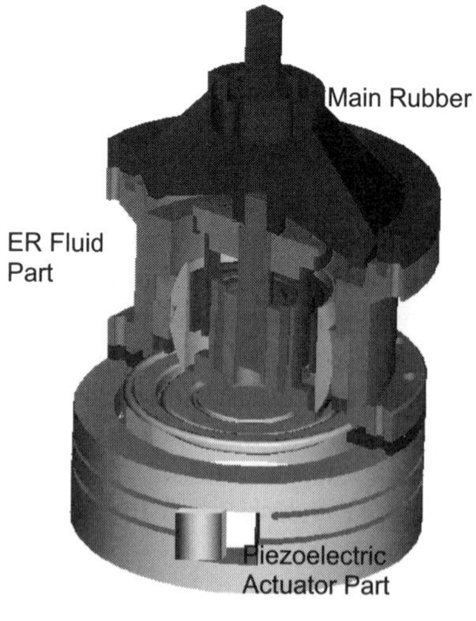

Figure 1. Configuration of hybrid engine mount.

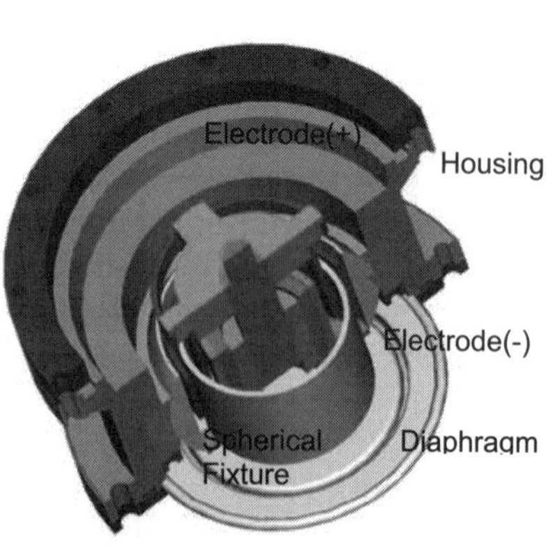

Figure 2. Schematic diagram of the ER fluid part.

Figure 3 presents the hydraulic model of ER part. The dynamic equation of ER part of the hybrid engine mount is derived from the hydraulic model shown in Fig.3 (a) and given by

$$m_m \ddot{x}_m = -k_R(x_m - x_a) - c_R(\dot{x}_m - \dot{x}_a) + A_p P_1 \tag{1}$$

where, m_m is the mass of the 1 degree-of-freedom system, k_R is the stiffness of the rubber in the vertical direction, c_R is the damping constant of the rubber in the vertical direction, A_p is the effective piston area, P_1 is the pressure of upper chamber, x_m and x_a are the displacement of mass and base, respectively.

When the motion of flow generated for the base excitation, continuity equation and the pressure drop of upper and lower chamber occurred by fluid flow resistance and ER fluid effect are obtained as follows.

$$C_1 \dot{P}_1 = A_p(\dot{x}_m - \dot{x}_b) + Q \tag{2}$$

$$C_2 \dot{P}_2 = -Q \tag{3}$$

$$P_2 - P_1 = \frac{12\eta l}{bh^3} Q + 2\frac{l}{h}\tau_y \operatorname{sgn}(Q) \tag{4}$$

where τ_y is the yield stress of the ER fluid, P_2 is the pressure in the lower chamber, η is the ER fluid viscosity, l is the electrode length, C_1 and C_2 is the compliance of the upper and lower chamber, Q is the flow from pressure difference between upper and lower chamber and defined as fluid flow through the flow channel, x_i is the displacement of the fluid in the flow channel, sgn(Q) is the signum function for ER fluid flow, respectively. Figure 2 (b) presents the mechanical model of the ER fluid part. We can be rewritten equation (1) from the Eq. (1)-(4) as follows

$$m_m \ddot{x}_m + k_x(x_m - x_a) + c_x(\dot{x}_m - \dot{x}_a) = -F_{ERt} \tag{5}$$

where, each parameters are defined as, $k_1 \equiv A_p^2/(C_1+C_2)$, $k_x = k_R + k_1$,

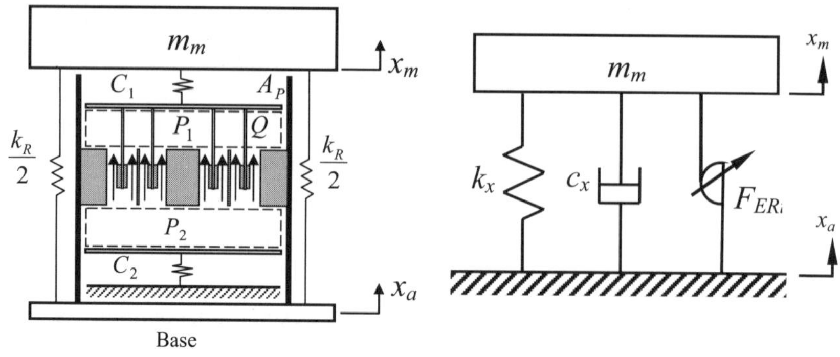

(a) hydraulic model (b) mechanical model

Figure 2. Dynamic model of ER fluid part.

$c_1 \equiv \left(A_p C_2/(C_1+C_2)\right)^2$, $c_1 \equiv R_f A_p A_s C_2/(C_1+C_2)$, $c_x = c_R + c_1 - c_2$,

$$F_{ERt} \cong \left(A_e + 2\frac{A_p C_2}{C_1+C_2}\frac{l}{h} \right)\tau_y$$

here, A_e is effective area, R_f is the flow resistance.

3.2 Dynamic Modeling of the Piezoelectric Actuator Part

The schematic configuration of piezoelectric actuator part is shown in Fig.3 (a). From this configuration, the mechanical model is shown as Fig.3 (b). Governing equation of motion derived as follows

$$m_{im}\ddot{x}_a + c_p\dot{x}_a + k_p x_a = F_a \tag{6}$$

here, m_{im} is the intermediate mass between ER fluid part and piezoelectric actuator, $F_a(=\alpha V)$ is the actuating force, k_p and c_p are the stiffness constant and damping coefficient of the piezoelectric actuator, respectively.

3.2 Dynamic Modeling of the Hybrid Engine Mount

Mechanical model of the hybrid engine mount can be represents to Fig.4 that is including ER fluid and piezoelectric actuator part. Equation of motion of hybrid engine mount expressed by

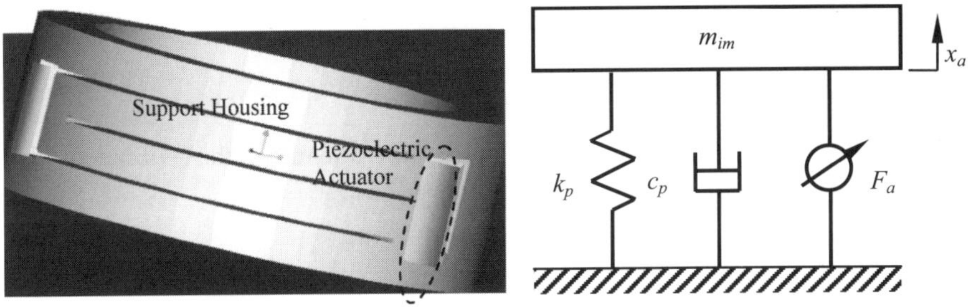

(a) Schematic configuration (b) mechanical model

Figure 3. Dynamic model of piezoelectric actuator part.

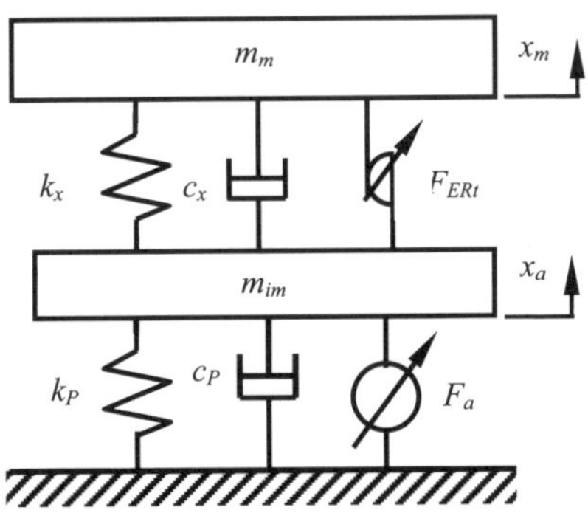

Figure 4. Mechanical model of the hybrid engine mount.

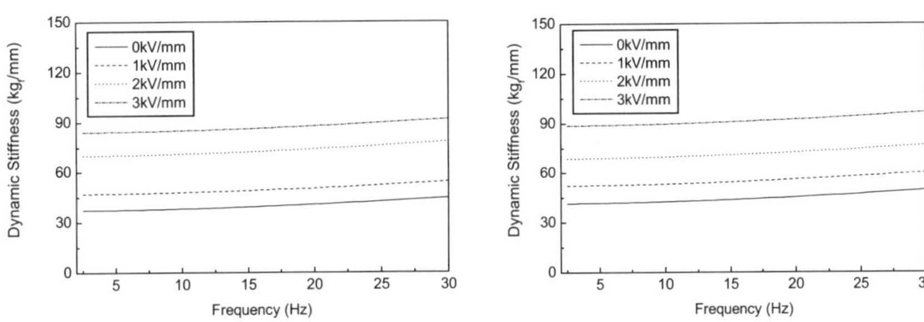

(a) ±0.5mm (b) ±1.0mm

Figure 5. Dynamic stiffness of the ER fluid part.

$$m_m\ddot{x}_m + k_x(x_m - x_a) + c_x(\dot{x}_m - \dot{x}_a) = -F_{ERt} \tag{7}$$

$$m_{im}x_a + c_x(\dot{x}_a - \dot{x}_m) + k_x(x_a - x_m) + c_p\dot{x}_a + k_px_a = -F_a(t)$$

Figure 5 presents the dynamic stiffness for the ER fluid part. A magnitude of sinusoidal excitation is ±0.5mm, ±1.0mm. Frequency range and applied electric field is from 1 to 30Hz and maximum 3kV/mm, respectively. In the case of, no electric field, dynamic stiffness has 38kg$_f$/mm at 10Hz, also, the value is increase as frequency increases. Dynamic stiffness is increased as the intensity of the applied electric field increases. Also, the result has same phenomena to ±1.0mm.

3. Control Characteristics

3.2 Controller Design

Sliding mode controller (SMC) is adopted to attenuate the vibration for 1 degree-of-freedom system. We can define sliding surface

$$s = Cx + \dot{x},\ \ C > 0 \tag{8}$$

where, C is the sliding surface gradient. The condition for existence of the sliding motion of sliding surface, $s\dot{s} < 0$, is satisfied. Prior to formulating the SMC, parameter variations of stiffness and damping coefficient are imposed as follows

$$k = k_0 + \delta k,\ \ |\delta k| < \alpha k_0 \tag{9}$$

$$c = c_0 + \delta c,\ \ |\delta c| < \alpha c_0 \tag{10}$$

The sliding mode controller is designed for the uncertain system as follows

$$u = mC\dot{x} - k_0 x - c_0 \dot{x} + k_s \,\mathrm{sgn}(s) \tag{11}$$

where, k_s means discontinuous gain which has positive number.

The controller given by Equation (11) is designed in an active actuating manner. However, in the case of ER fluid part, it is a semi-active actuator. Thus, the control input force should be applied to the ER fluid part according to the following actuating condition [9].

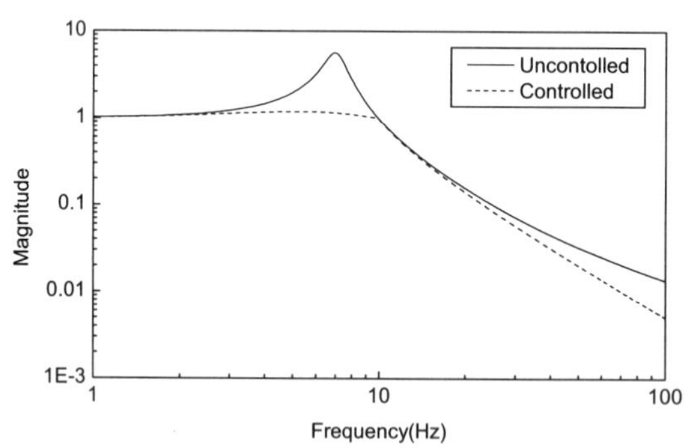

Figure 6. Control response of the hybrid engine mount.

$$u = \begin{cases} u_{ER} & for\ \ u(\dot{x}_m - \dot{x}_a) > 0 \\ 0 & for\ \ u(\dot{x}_m - \dot{x}_a) \le 0 \end{cases} \tag{12}$$

In this study, the maximum control electric field applied to the ER fluid and piezoelectric actuator is limited to 3kV/mm and 200V, respectively.

On the other hand, control input for each actuator should be applied depending upon the excitation frequency. In this work, the frequency criterion is fixed at 10Hz. Thus, the control input can be expressed as follows :

$$u = \begin{cases} u_{ER} & for \quad f_{ex} < 10Hz \\ u_P & for \quad f_{ex} \geq 10Hz \end{cases} \tag{13}$$

here, f_{ex} is the excitation frequency, u_{ER} and u_P are control input of the ER fluid part and piezoelectric actuator part, respectively.

3.2 Control Result

Control characteristics for vibration attenuation of the 1 degree-of-freedom hybrid engine mount system achieved using sliding mode controller. Control strategy separated two parts, one is low frequency range to 10Hz (resonance range) which is operate only ER fluid part, the other is high frequency range over 10Hz (over resonance range) which is operate only piezoelectric actuator. Figure 6 presents the frequency domain responses of the hybrid engine mount. The frequency response is obtained from transmissibility for the mass. It is clearly seen that frequency response is well reduced by adopting the proposed sliding model controller except neighborhood of 10Hz.

4. Conclusion

A hybrid engine mount was proposed and its dynamic characteristics and vibration control performances were verifying that dynamic stiffness can be controlled by the intensity of electric field. In order to obtain a favorable control performance of the hybrid engine mount, a robust sliding mode controller was designed. Then control characteristics for vibration attenuation of 1 degree-of-freedom were evaluated. The magnitude of transmissibility was well reduced by adopting the robust sliding mode controller.

5. Acknowledgement

This work was supported by National Research Laboratory (NRL) program directed by Korea Ministry of Science and Technology. This financial support is gratefully acknowledged.

References

1. Singh, R., Kim, G. and Ravinda, R. V., Linear analysis of automotive hydro-mechanical mount with emphasis on decoupler characteristics. *J. of Sound and Vibration* **158**(2) (1992) pp. 219-243.
2. Rivin, E. I., Passive engine mount - some directions for further development. *SAE Paper* **850481** (1985).
3. Mel Schwatrz, *Encyclopedia of smart materials* (Wiley-Interscience, 2001).
4. Choi, S. B. and Song, H. J., Vibration control of a passenger vehicle utilizing a semi-active ER engine mount. *Vehicle System Dynamics* **37**(3) (2002) pp. 193-216.

5. Williams, E. W., Rigby, S. G., Sproston, J. L. and Stanway, R., Electrorheological fluids applied to an automotive engine mount. *J. Non-Newtonian Fluid Mech.* **47**(1993) pp. 221-238.
6. Choi, S. B., Song, H. J., Lee, H. H., Lim, S. C., Kim, J. H. and Choi, H. J., Vibration control of a passenger vehicle featuring magnetorheological engine mounts. *Int. J. of Vehicle Design* **33**(1) (2003) pp. 2-16.
7. Matsuoka, H., Mikasa, T. and Nemoto, H., NV Countermeasure technology for a cylinder-on-demand engine - development of active control engine mount. *SAE Paper* **2004-01-0413** (2004).
8. Shibayama, T., Ito K., Gami, T. and Oku, T., Active engine mount for a large amplitude of idling vibration. *SAE Paper* **951298** (1995) pp. 531-538.
9. Leitmann, G., Semi-active control for vibration attenuation. *J. of Intelligent Material Systems and Structures* **5** (1994), pp. 841-846.

OPTIMUM DESIGN OF ONE-SIDED PATTERNED ELECTRODES FOR GEL-STRUCTURED ER FLUIDS BY MEANS OF A NUMERICAL ANALYSIS

HIROAKI WATANABE[1], YASUHIRO KAKINUMA[1],
TOJIRO AOYAMA[1] AND HIDENOBU ANZAI[2]

[1] Dept. of System Design Engineering, Keio University,
3-14-1 Hiyoshi, Kohoku-ku, Yokohama 223-8522, Japan
[2] FUJIKURA KASEI CO., LTD.
kakinuma@sd.keio.ac.jp

Electro-Rheological Fluids (ERFs) are functional fluids that change their apparent viscosity by applying electric field. ERF is composed of ER particles and silicon oil. ERF shows reversibility and high responsiveness of ER effects, but has a defect that passage of time causes sedimentation of ER particles. Electro-Rheological Gel (ERG) has been developed to overcome the defect of ERF. ERG is composed of ER particles and silicon gel. A flat plate placed on the ERG sheet clings to the gel by applying an electric field and ERG produces the shear force to the plate. The one-sided pattern of electrodes, which has been developed for ERF devices, is a very useful idea to free sliding or rotating parts from wiring arrangements. In this study, the optimum design of one-sided pattern electrodes for ERG is carried out by means of a numerical analysis. The general-purpose tool for the analysis of electric and magnetic field 'ANSYS Emag' is used. The influences of the electrodes pattern on the performances of ERG are analyzed and the yield forces are estimated from electric field intensity at the interface between the electrodes and ERG. Some experiments are also carried out to examine the validity of the numerical analysis.

1 Introduction

Electro-Rheological Fluids (ERFs) are functional fluids [1] that change their viscoelastic properties by applying electric field. ERF is composed of ER particles and silicon oil. ERF shows reversibility and high responsiveness of ER effects, but has a defect that the sedimentation of ER particles occurs in long time use. The Electro-Rheological Gel (ERG) [2, 3] has been developed to overcome the defect of ERF. ERG is composed of ER particles and silicon gel. ERG also produces shear force by applying electric field. The mechanism of shear force formation is shown in Fig. 1. ER particles polarize electrically by applying an electric field. The polarized ER particles attract each other and ER particles existing at ERG surface slightly move into silicon gel. A flat plate placed on the ERG sheet, therefore, clings to the gel. By the adhesion of the gel, ERG produces the shear force to the plate. Generated shear force is in proportion to the applied electric field because the adhesive area of ERG to a flat plate increases linearly with electric field.

In recent study, the ERG is driven by the "bilateral electrode system" shown in Fig. 2(a). But "bilateral electrode system" has some disadvantages in practical use. This electrode system requires the wiring to both sides of sliding or rotating parts. In order to attain the higher applicability of ERG to machine parts, "one-sided pattern electrodes system" [4] has been proposed. In the case of "one-sided pattern electrodes system", the anode and cathode are arranged alternately on one side as illustrated in Fig. 2(b).

In this study, the optimum design of one-sided pattern electrodes is carried out by means of a numerical analysis. The influences of the electrodes pattern on the performances of ERG are analyzed and the yield forces are estimated from the distribution of electric field intensity at the interface between the electrodes and ERG.

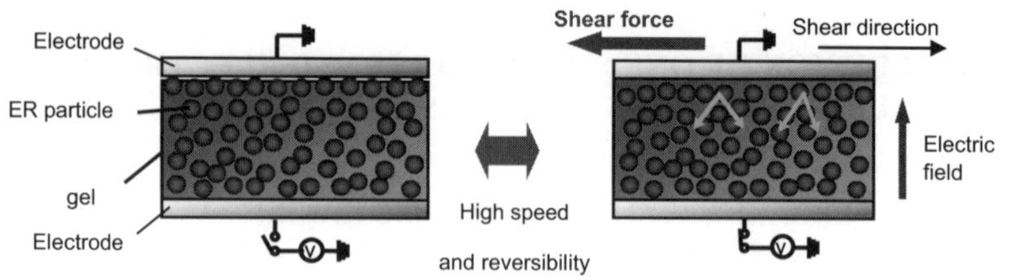

(a) No electric field (b) With electric field

Figure 1. Mechanism of shear force formation

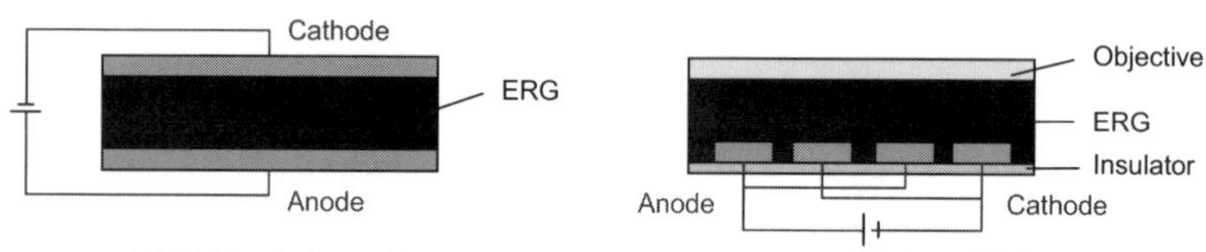

(a) Bilateral electrode (b) One-sided pattern electrode

Figure 2. Configuration of the electrodes

And, some experiments are also carried out to evaluate the validity of the numerical analysis.

2 Numerical analysis

Shear stress produced by ERG has a strongly dependence on the electric field intensity. The electric field applied by "one-sided pattern electrode system" shows more complicated distribution than the case of bilateral electrode. In this study, the general-purpose tool for the analysis of electric and magnetic field 'ANSYS 9.0' is used.

2.1 Model

The model for the electric field analysis is illustrated in Fig. 3. The model is two-dimensional and the conductive plate is placed on the ERG sheet with 0.3 mm thickness which is formed on the one-sided pattern electrodes. The behavior of electric field at the ERG surface is analyzed for various electrode width (Ew) and electrode gap (Eg). The dimension of an electrode set with 1.0mm of Ew and Eg is defined as standard model. Nine different cases that Ew or Eg is varied from standard model are prepared. Applied voltages to the anodes are 900V and the cathodes are earthed.

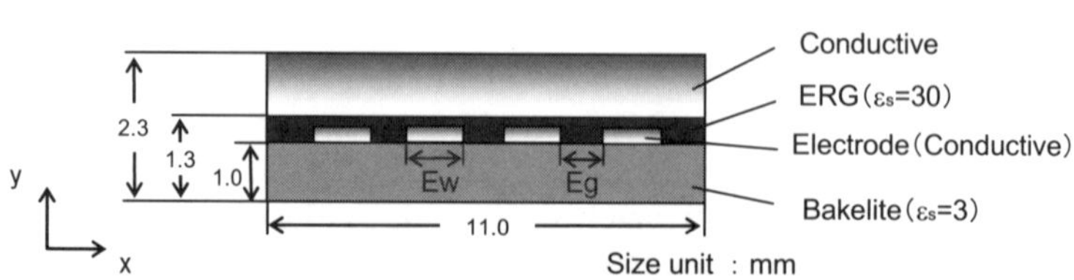

Figure 3. Two-dimensional analysis model

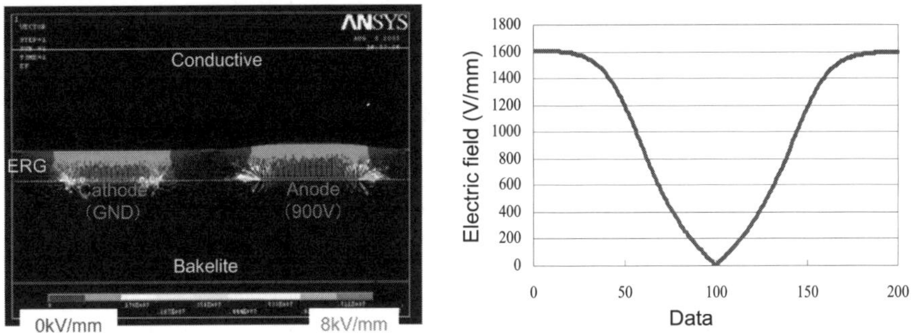

(a) A vector view of electric field (b) Electric field intensity at the ERG surface

Figure 4. Distribution of Electric field (E_W=1.0mm,E_G=1.0mm)

2.2 *Electric field vector*

The result of the standard model (Ew =1.0[mm], Eg =1.0[mm]) is shown in Fig. 4(a). The mechanism of shear force formation suggests that the electric field intensity at ERG surface is considerably related to the shear force. The distribution of the electric field intensity at ERG surface is shown in Fig. 4(b). The electric field intensity above the electrodes is strong and at the middle point between two electrodes is almost 0V/mm.

2.3 *Electric field at the ERG surface*

Average value of the electric field intensity at the ERG surface between the centers of two electrodes are evaluated for various dimensions of Ew and Eg. The relationships of the average value of the electric field intensity to Ew and to Eg are shown in Fig. 5(a) and Fig. 5(b) respectively.

As Ew becomes wider, the average electric field intensity increases. On the contract, as Eg becomes wider, the electric field intensity decreases. These results indicate that it is necessary to design the one-sided electrodes with the wide electrode and narrow electrode gap in order to obtain the higher electric field at the interface.

2.4 *Estimation of yield stress*

The yield stress is estimated from the relationship between electric field intensity and

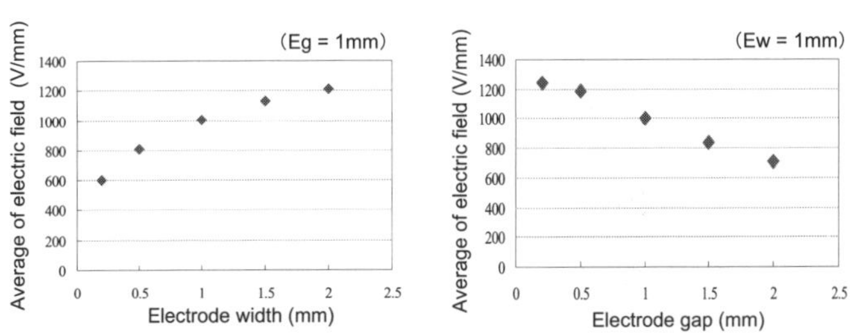

(a) Influence of electrode width (b) Influence of electrode gap

Figure 5. Influence of electrode dimension on electric field intensity at the interface

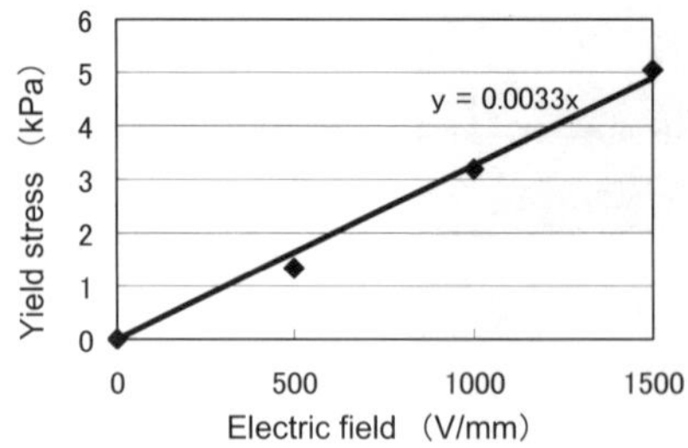

Figure 6. Characteristic of ERG in bilateral electrodes

yield stress measured in the experiment for "bilateral electrode system". The experimental result for "bilateral electrode system" is shown in Fig. 6. The yield stress changes linearly according to the applied electric field intensity, and the relation between these two physical quantities is formulated as follows,

$$\sigma y = 0.0033 \times Ey \qquad (1)$$

where σy is yield stress [kPa] and Ey is electric field intensity [V/mm]. Estimated yield stress is given by substituting the average value of electric field intensity obtained from the numerical analysis into equation (1).

3 Experiment

3.1 Experimental condition

The standard one-sided pattern electrode (Ew =1.0[mm], Eg =1.0[mm]) is shown in Fig. 7. In addition to standard type, five kinds of one-sided pattern electrodes with different dimensions of Ew or Eg are fabricated. An aluminum plate is placed on ERG as a sliding plate. The applied voltage to the anodes is 900V, and the cathode is earthed. While shear movement is applied to the aluminum plate, the displacement is measured by an eddy-current sensor and the shear force is measured by a dynamic strain gauge force sensor. For instance, behavior of shear stress in standard type of the one-sided electrodes is presented in Fig. 8. As the displacement of the plate increases up to yield point, the shear stress also increases. The shear stress, however, is almost constant beyond yield point. The point of transition from elastic deformation mode to sliding mode is defined as yield point.

Figure 7. One-sided pattern electrodes

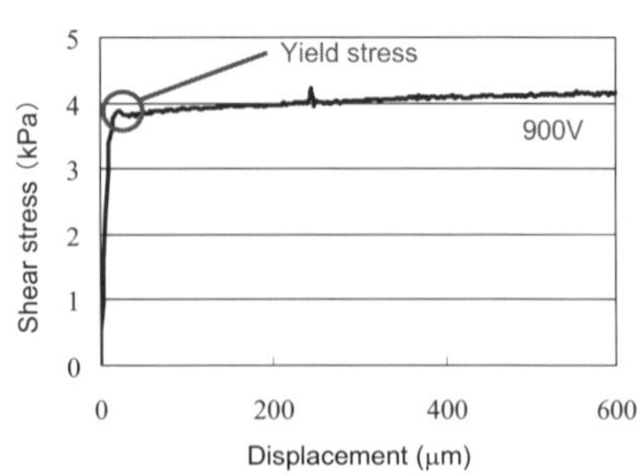

Figure 8. Behavior of shear stress in standard model (Ew=1.0, Eg=1.0)

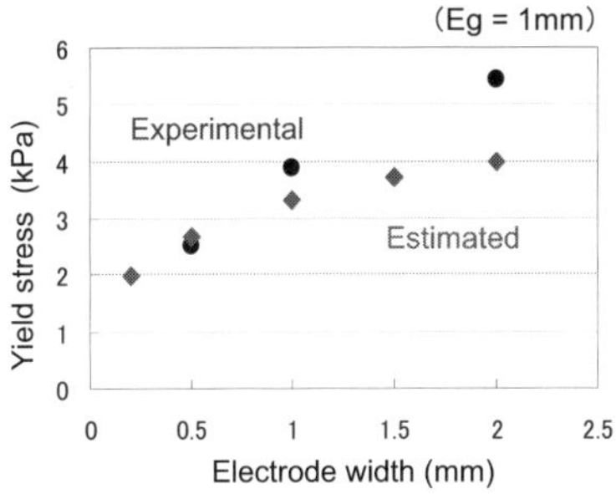

(a) Influence of electrode width

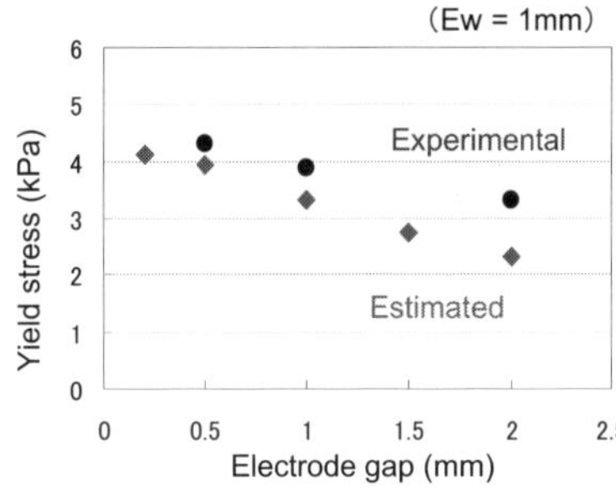

(b) Influence of electrode gap

Figure 9. Influence of electrode dimension on the generated yield stress

3.2 Results of experiment

Influence of electrode pattern on the generated yield stress is investigated. The relationships of yield stress to Ew and to Eg are shown in Fig. 9(a) and Fig. 9(b) respectively. The estimated values from analysis given by Eq. (1) are also plotted. As Ew becomes wider, the yield stress increases. Also, as Eg becomes wider, the yield stress decreases. The behavior of yield stress is consistent with that of electric field obtained by numerical analysis. However, a little difference is observed between the experimental values and estimated values of yield stresses.

4 Application to torque transmission device

Torque transmission device [5] is considered as one of the typical applications of ERG. The one-sided pattern electrode system is applied to the ERG torque transmission device as shown in Fig. 10. The two different patterns of electrodes (Concentric and Radial) are considered as illustrated in Fig. 11. The numerical analysis with two-dimensional model is impossible for the radial pattern. Therefore, the three-dimensional analysis is employed.

4.1 Model

The model for 3D numerical analysis is illustrated in Fig. 12. Two models of concentric circle type and radial type are prepared. In both model, the dimensions Ew and Eg is

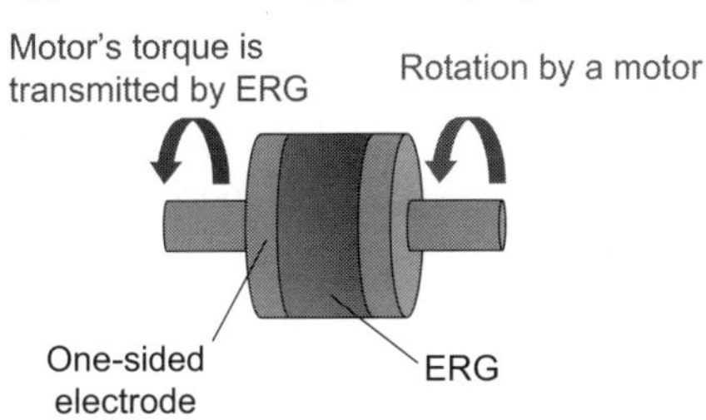

(a) Basic structure

(b) Appearance

Figure 10. Developed torque transmission device

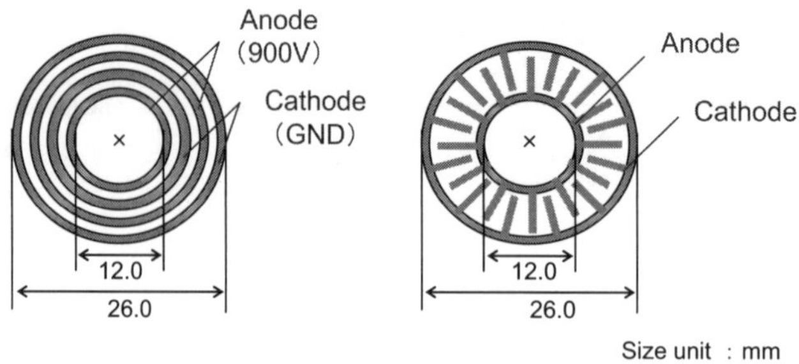

(a) Concentric circle type (b) Radial type

Figure 11. Electrode configurations for torque transmission device

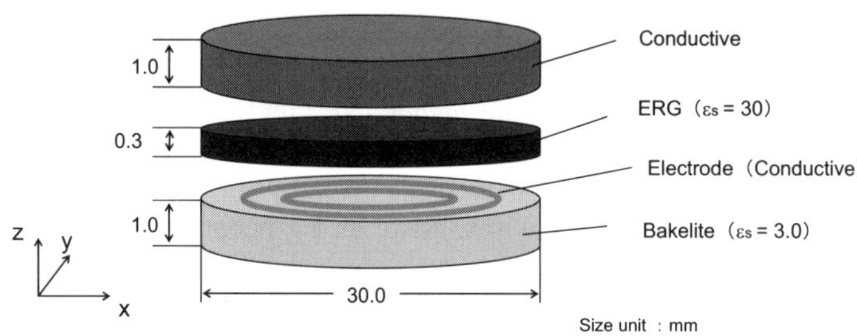

Figure 12. Three-dimensional model

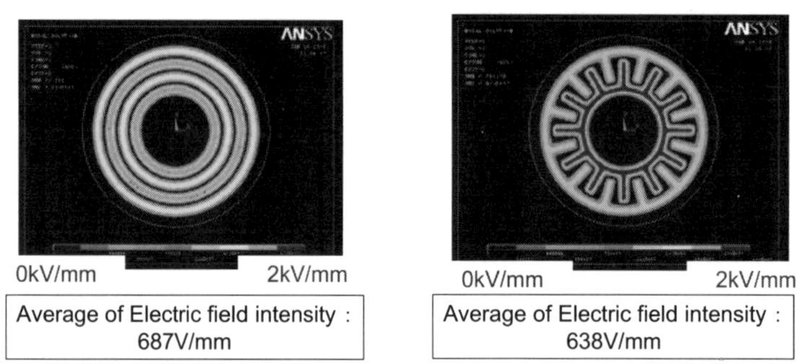

(a) Concentric circle type (b) Radial type

Figure 13. Contour view of electric field intensity at ERG surface

1.0[mm]. The total areas of two electrodes are equal. The closest electrode to the center is defined as the anode in the concentric circle model. Anodes and cathodes are arranged alternatively. The inner electrode is used as the anode while the outer electrode is provides as the cathode in the radial model. Applied voltage to the anodes is 900V, and the cathode is earthed.

4.2 Electric field at ERG surface

The distributions of electric field intensity at ERG surface are shown in Fig. 13. High electric field intensity is obtained according to the electrode pattern. The average of electric field intensity of concentric circle model is higher than the case of radial model.

From this result, it is predicted that the concentric circle electrode provides better performance of torque transmission device.

5 Conclusion

ER effect in ERG can be obtained in case of applying the one-sided electrodes. The influence of the electrode pattern on the shear stress generated by ER effect is investigated by numerical analysis of electric field and the experiment of shear test. The yield stress is related to the electric field at the interface between ERG and upper plate. When electrode width is wider and electrode gap is narrower, the ER effect is obtained efficiently. Based on the results of the numerical and experimental analysis, an empirical equation which can be used for rough estimation of ERG performance from numerical analysis is proposed.

Moreover, the model of one-sided electrodes is enhanced to 3 dimensions and the pattern to obtain the higher ER effect is estimated by numerical analysis. It is predicted that concentric circle model can produce higher shear stress than radial type.

6 Acknowledgements

This work was partially supported by a Grant-in-Aid for Young Scientists (B), The Ministry of Education, Culture, Sports, Science and Technology.

Reference

1. R.Stanway: Smart fluid -current and future development-, Materials Science and Technology, vol.20, p.931-939 (2004)
2. Y. Kakinuma, T. Aoyama, H. Anzai, K. Sakurai, K. Isobe and K. Tanaka, Basic Properties of Gel-structured Electro-rheological Fluids, International Journal of Modern Physics B, **19,** p. 1339-1345 (2005)
3. Y. Kakinuma, T. Aoyama, H. Anzai, K. Sakurai, K. Isobe and K. Tanaka, Development of Gel-Structured Electro-rheological Fluids and their Applications to the Control of Positioning, Proceedings of 8th International Research/Expert Conference, p. 459-462 (2004)
4. T.Kikuchi, J.Furusho, A.Inoue: Total force calculation method for ER effect of liquid crystalline polymer on one-sided pattern electrode system, ERMR2004, p.242-248 (2004)
5. Avi Fisch, Jason Nikitczuk, Brian Weinberg, Juan Melli-Huber, Constantinos Mavroidis, Charies Wampler: Development of an electro-rheological fluid actuator and haptic systems for vehicular instrument control, Proceedings of IMECE'03, IMECE2003-43514, p.759-768 (2003)

FRONTAL CRASH MITIGATION USING MR IMPACT DAMPER FOR CONTROLLABLE BUMPER

D. WOO AND S. B. CHOI

Smart Structures and Systems Laboratory, Department of Mechanical Engineering, Inha University, Incheon 402-751, Korea
E-mail: seungbok@inha.ac.kr

Y. T. CHOI AND N. M. WERELEY

Smart Structures Laboratory, Department of Aerospace Engineering, University of Maryland, College Park, MD 20742, U. S. A.
E-mail: wereley@umd.edu

This paper presents performance characteristics of a MR impact damper for controllable bumper in vehicle systems. Recently, several mechanisms are proposed in order to minimize the injury of vehicle occupants during frontal collision. One of promising candidates is to adopt MR fluid which undergoes significant instantaneous reversible changes in material characteristics when subjected to magnetic field. Using this salient property, a new type of MR impact damper is devised in this work. The proposed MR impact damper is integrated with bellows to induce the flow motion and the motion is operated under flow mode. The field-dependent damping force is evaluated by computer simulation with various conditions. In order to investigate the effectiveness of the proposed impact damper, a lumped parameter mathematical model of frontal vehicle crash system including MR impact damper is developed and realized in order to evaluate acceleration peak reductions and transmitted force in the frontal collision.

1. Introduction

Recently, it is essentially required that driver and sufferer is secured against vehicle collision. Therefore, there are demands on bumpers to absorb collision force and systems to prevent accidents. For this, the research is actively advanced about not only the system that occurs a braking force in advance related to breaking devices but bumper that protects vehicle and driver from collision. There are a variety of research such as security of the driver who uses air bag and safety belt, and reduction of collision throughout braking efficiency improvement using ABS [1]. Mizuno tried to observe both the collision characteristic from various directions in type of cars and collision accident analyses of pedestrians [2]. Witteman et al. proposed friction damper to guarantee security of vehicle through reducing impact force and demonstrated this by means of vehicle model analysis using FEM [3]. Wagstrom observed that the stiffness change of the vehicle could improve the stability in various conditions through vehicle model which was combined one degree of freedom system using computer simulation [4]. Jawad et al. considered smart bumper composed before collision controller which change stiffness beforehand utilizing hydraulic system [5]. Through this, efficiency of impact force decrement about different collision conditions was considered.

The research about a shock reduction is actively accomplished using smart fluid which has reversible properties with applied electric or magnetic field. Wereley et al. suggested MR damper to reduce shock transmitted to a helicopter. Then, he designed the controller through the mathematics modeling and verified controller performance through application to actual system [6]. Ahmadian et al. suggested shock reduction mechanism using the MR damper when fires a bomb [7]. Choi et al. verified controller performance using the MR damper through application to vehicle suspension [8, 9]. Song et al. proposed the shock damper to reduce impact and substantiated impact control performance by means of decrement of acceleration as composing one degree of freedom system [10]. From this researches, the sufficient possibilities were presented about reduction of impact using the smart fluid has reversible properties and the research of the various forms has been advanced.

In this study, MR impact damper is proposed to reduce transmitted force to vehicle chassis when the vehicle is occurred frontal collision. The governing equation of motion of MR impact damper is derived using hydraulic model. The dynamic equation includes a damping force term which is the field dependent damping force achieved from Bingham model of MR fluid. The performance of collision mitigation is then evaluated by computer simulation using vehicle model that is include occupant model.

2. Dynamic Modeling of MR Impact Damper

Figure 1 shows the configuration of MR impact damper. MR impact damper installed at the bumper structure. It generate the damping force when occur frontal collision. MR impact damper consists of upper chamber, magnetic circuit to produce magnetic field, lower chamber located on under the magnetic circuit and diaphragm. MR fluid filled in upper chamber, lower chamber and flow channel. Diaphragm makes a role of containing and accumulating the fluid flow. Magnetic circuit installed at inner cylinder. Flow channel is located between upper chamber and lower chamber. Fluid flow is generated for the motion of each chamber, then fluids move on flow channel. Bellows located at the position of upper chamber that have a good property of compression to large deformation and a good property of volume change to fluid flow. It has oil resistance, moisture resistant, pressure resistance and airtight characteristics for filled MR fluid. The damping force is produced by only fluid resistance of MR fluid without magnetic field. If magnetic field is applied to magnetic pole, the MR impact damper produces damping force on the magnetic pole. It can be continuously generated by adjusting the intensity of the magnetic field.

In advance, conventional finite element analysis program ANSYS is used to verifying magnetic flux in the magnetic pole. We can verify the uniform magnetic flex in the flow channel. Then shape of the flow channel and composition of the magnetic pole are determined. It is assumed for analysis that MR fluid is incompressible, MR impact damper has motion of one direction.

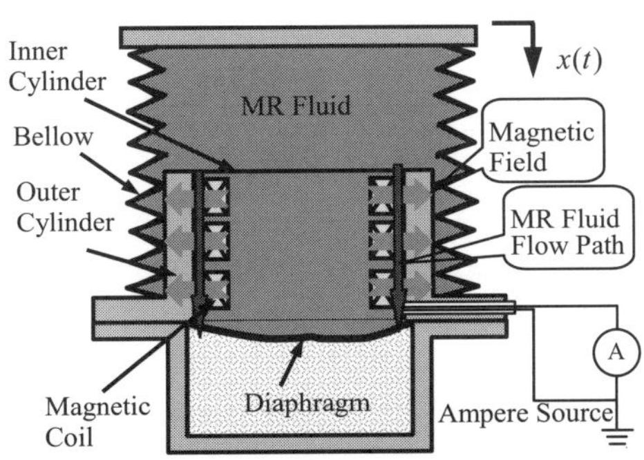

Figure 1. Configuration of MR impact damper.

It is assumed that diaphragm has linear property of compliance. Figure 3 shows hydraulic model of MR impact damper. The transmitted force of the proposed MR impact damper to the chassis body is obtained as follows.

$$F_{cb} = -\left(k_b x(t) + c_b x(t) - A_p P_1(t)\right) \tag{1}$$

where, k_b and c_b are the stiffness and damping constant of the bellows, P_1 is the pressure of the upper chamber, A_p is the effective piston area of the upper chamber, $A_p P_1(t)$ is produced damping force from pressure drop.

The pressure drop is consists of fluid resistance ΔP_{fp} and yield stress ΔP_{MR} of MR fluid. The fluid resistance is generated by fluid flow without magnetic field. The yield stress in produced by intensity of magnetic field in the position of magnetic pole. The pressure drop is derived as follows :

$$P_2(t) - P_1(t) = \Delta P_{fp} + \Delta P_{MR} \tag{2}$$

where ΔP_{fp} is represent as $R_{fp} Q_{fp}(t)$, R_{fp} is the fluid resistance, $Q_{fp}(t)$ is fluid flow at the flow path, $Q_{fp}(t)$ is defined as fluid flow $A_{fp}\dot{x}_{fp}(t)$, A_{fp} is the area of flow path, $\dot{x}_{fp}(t)$ is the velocity of fluid at the flow path, ΔP_{MR} is represent as $2L/h \cdot \tau_y(H)$, L is the length of magnetic pole (50mm), h and H are the gap size of magnetic pole (1mm), magnetic field, respectively.

When the motion of flow generated by input excitation $x(t)$, continuity equation of upper and lower chamber occurred by fluid flow resistance and MR fluid effect are obtained as follows.

$$\begin{aligned} C_1 \dot{P}_1(t) &= Q_{fp}(t) - A_p \dot{x}(t) \\ C_2 \dot{P}_2(t) &= -Q_{fp}(t) \end{aligned} \tag{3}$$

The transmitted force to body represents as follows from equation (1)-(3).

$$\begin{aligned} F_{cb} &= -k_b x(t) - c_b \dot{x}(t) - k_i (x(t) - x_{fp}(t)) \\ k_i (x(t) - x_{fp}(t)) - k_d x_{fp}(t) &= c_{fp} \dot{x}_{fp}(t) + \mathrm{sgn}(\dot{x}_{fp}(t)) F_{MR} \end{aligned} \tag{4}$$

where, each parameters are defined as $k_i \equiv A_p^2 / C_1$, $k_d \equiv A_p^2 / C_2$, $c_{fp} \equiv A_p^2 / R_{fp}$,

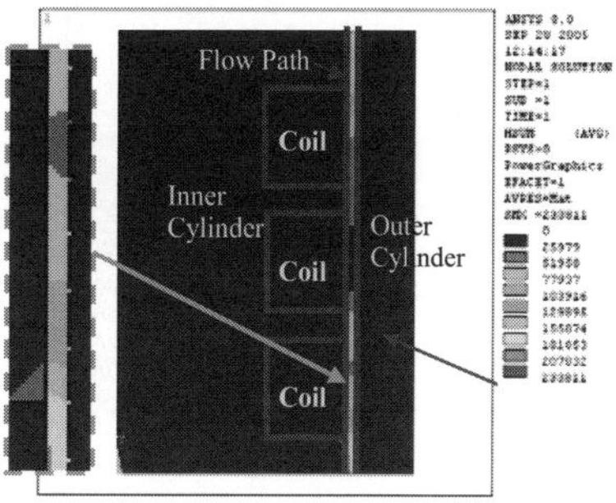

Figure 2. FEM result of the magnetic flux

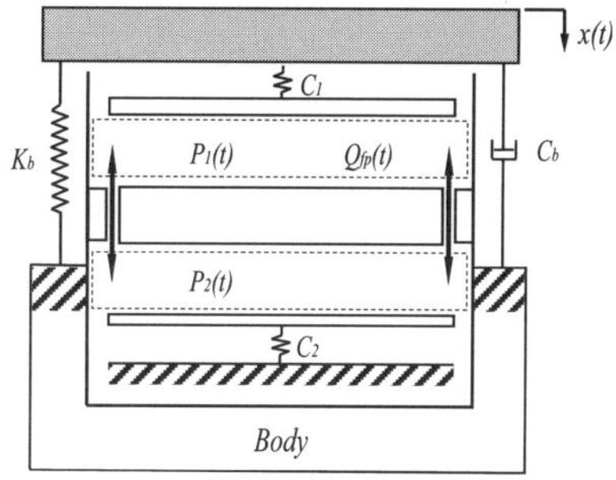

Figure 3. Hydraulic model of the MR impact damper

F_{MR} is field-dependent damping force due to yield stress of the MR fluid, C_1 and C_2 is the compliance of the upper and lower chamber.

The damping force F_{MR} can be expressed by

$$F_{MR} = A_p 2\frac{L}{h}\alpha H^{\beta} \tag{5}$$

Figure 4 presents the damping force with various excitation displacement and magnetic field. The displacement of input is Fig.4 (a) ±20mm, (b) ±50mm, (c) ±100mm, respectively. Fig.4 (d) is the damping force at ±50mm. It is clearly observed that the damping force is increased as the intensity of applied magnetic field increases.

3. Collision Mitigation Properties of the Vehicle System

3.1 Dynamic Modeling of the Vehicle System

Vehicle system is simplified to chassis, engine room and bumper. The motion of vehicle is considered only one direction of drive. Figure 5 shows the vehicle model with MR impact damper that has 3 degree-of-freedom [11, 12]. The occupant model is presented left side of k_3. It has 2 degree-of-freedom considering the upper and lower part of the body [13]. The vehicle model included occupant model as follow

$$M_1\ddot{x}_1 = -2k_1(x_1 - x_2) + 2F_{cb} \tag{6}$$

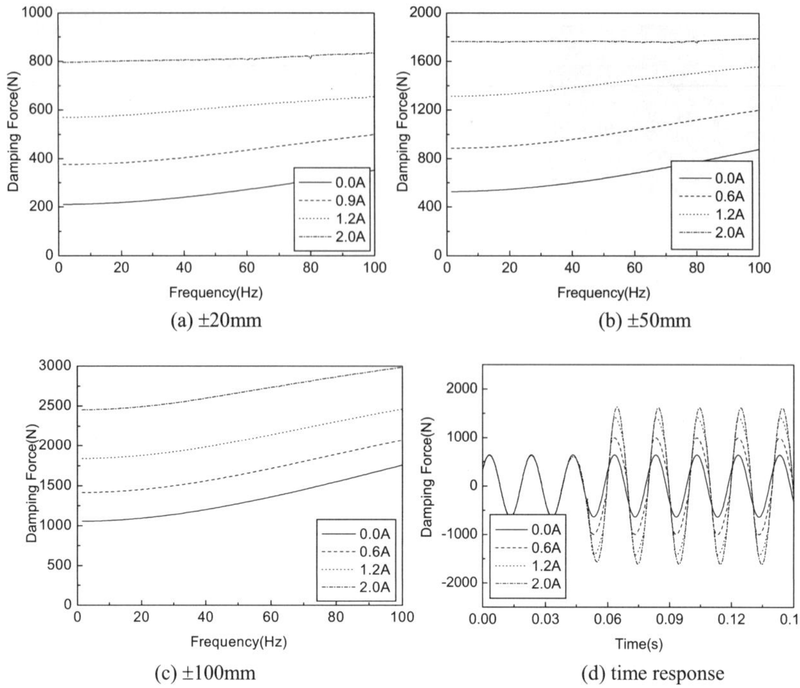

Figure 4. Dynamic characteristics of the MR impact damper.

$$M_2\ddot{x}_2 = 2k_1(x_1 - x_2) - 2k_2(x_2 - x_3) \tag{7}$$

$$M_3\ddot{x}_3 = 2k_2(x_2 - x_3) - 2k_3(x_3 - x_4 - R\sin\theta) \tag{8}$$

$$M_4\ddot{x}_4 = k_3(x_3 - x_4 - R\sin\theta) - c_2(\dot{x}_4 - \dot{x}_3) - M_5L\ddot{\theta}\cos\theta \tag{9}$$

$$M_5L^2\ddot{\theta} = k_3(x_3 - x_4 - R\sin\theta)R\cos\theta + M_5gL\sin\theta \tag{10}$$

where, M_1 is the mass of the bumper, M_2 is the mass of engine room, M_3 is the mass of the frame, M_4 and M_5 are the lower part of the body, head mass, respectively.

3.2 Performance Evaluation of the Vehicle System

The performance of vehicle system with MR impact damper is evaluated by computer simulation. Initial conditions are as follows

$$\theta(0) = 0\ , x_1(0) = 0, x_2(0) = 0, x_3(0) = x_4(0) = 0, \dot{x}_1(0) = \dot{x}_2(0) = \dot{x}_3(0) = \dot{x}_4(0) = v_0$$

The proposed MR impact damper produced damping force by intensity of magnetic field. Responses of vehicle system investigated when MR impact damper produced constant damping force. Damping force separated three steps that is crash initiation phase, airbag deployment phase and occupant contact phase. These conditions are represented as follows

$$F_{cb} = \begin{cases} 45\ kN & if\ \ 0 < t \leq 30\ \ (m\sec) \\ 15\ kN & if\ \ 30 < t \leq 70\ \ (m\sec) \\ 30\ kN & if\ \ t > 70\ \ (m\sec) \end{cases} \tag{11}$$

The vehicle crash severity index (VCSI) and frame deformation are used as a performance criterion. VCSI presents vehicle collision hazard index that is calculated by deceleration data. VCSI defined as follows

$$VCSI = \frac{1}{T}\int_0^T (-\ddot{x})^2 dt \tag{12}$$

Figure 6 presents the VCSI responses of the standard vehicle without MR impact damper and the vehicle with MR impact damper for the various vehicle speeds. In the case of vehicle with MR impact damper (48km/h), the value of deceleration decreased from

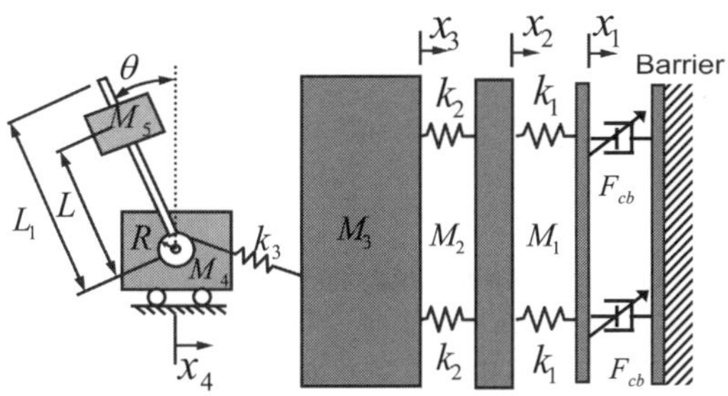

Figure 5. Mathematical model of the vehicle system.

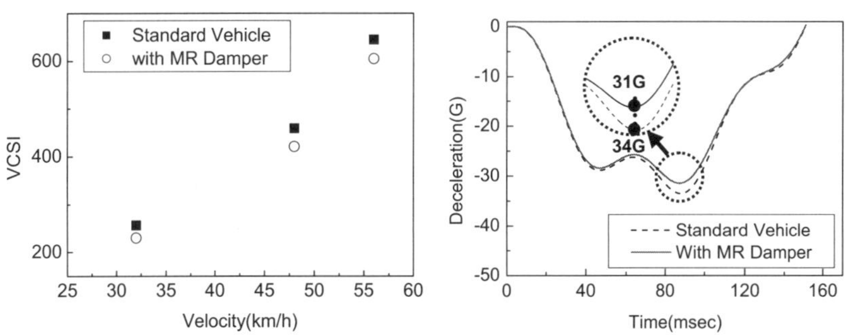

(a) VCSI (b) deceleration of the occupant (48km/h)

Figure 6. Comparison of the vehicle crash severity index.

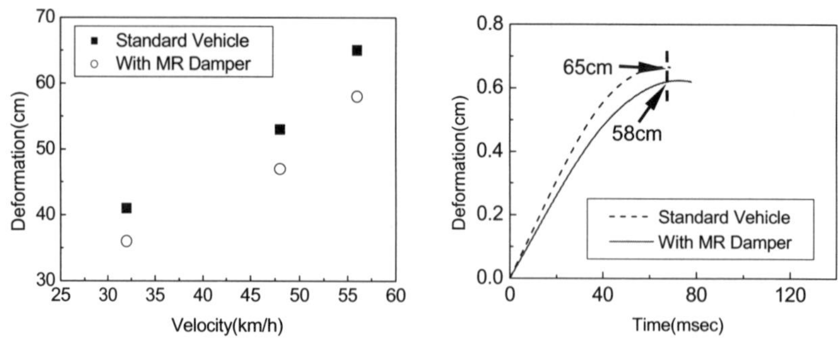

(a) frame deformation (b) 52km/h

Figure 7. Response comparison with and without MR damper.

34G to 31G. Also, VCSI value decreased from 459 to 421. Figure 7 presents simulated results. The largest value of deformation takes place at 52km/h and its value is decreased from 65cm to 58cm. From these results, it is clearly observed that VCSI and frame deformation is decreased by imposing constant damping force for each time steps.

4. Conclusion

A MR impact damper was proposed and its dynamic characteristics and collision mitigation performances were evaluated by considering variations of speeds and occupants model. After verifying that the damping force can be controlled by the intensity of magnetic field, the vehicle model including occupant model was constructed. It has been simulated that VCSI and frame deformation can be effectively reduced by employing the damping force determined control strategy. The performance of MR impact damper will be experimentally evaluated for the various test conditions.

5. Acknowledgement

This work was supported by National Research Laboratory (NRL) program directed by Korea Ministry of Science and Technology. This financial support is gratefully acknowledged.

References

1. Witteman, W., Improved vehicle crashworthiness design by control of the energy absorption for different collision situations. *PhD. Dissertation, Eindhoven University of Technology, Automotive Engineering & Product Design Technology, Eindhoven, Netherlands* (1999).
2. Mizuno. K. and Kajzer. J., 1999, Compatibility problems in frontal, side, single car collisions and car-to-pedestrian accidents in Japan. *Accident Analysis Prev.* **31**(4) (1999) pp. 381-391.

3. Witteman, W., Adaptive frontal structure design to achieve optimal deceleration pulses. *Proc. of the 19th Int. Technical Conf. on the Enhanced Safety of Vehicles* (2005) pp. 8.
4. Wågström, L., Thomson, R., and Pipkorn, B., Structural adaptivity for acceleration level reduction in passenger car frontal collisions. *Int. J. of Crash* **9**(2) (2004) pp. 121-127.
5. Jawad, S. A. W., Intelligent hydraulic bumper for frontal collision mitigation. *Crashworthiness and Occupant Protection in Transportation Systems ASME* **218** (1996) pp. 181-189.
6. Lee, D. Y., Choi, Y. T. and Wereley, N. M., Performance analysis of ER/MR impact damper systems using Herschel-Bulkley model. *J. of Intelligent Material Systems and Structures* **13**(7-8) (2002) pp. 525-531.
7. Ahmadian, M., Appleton R. and Norris, J. A., An analytical study of fire out of battery using magnetorheological dampers. *Shock and Vibration* **9** (2002) pp.129-142.
8. Lee, H. S. and Choi, S. B., Control and response characteristics of a magneto-rheological fluid damper for passenger vehicles. *J. of Intelligent Material Systems and Structures* **11**(1) (2000) pp. 80-87.
9. Lee, H. S., Choi, S. B. and Lee, S. K., Vibration control of a passenger vehicle featuring MR suspension units. *Transactions of the Korean Society for Noise and Vibration Eng.* **11**(1) (2001) pp. 41-48.
10. Song, H. J., Choi, S. B., Kim, J. H. and Kim, K. S., Performance evaluation of ER shock damper subjected to impulse excitation. *J. of Intelligent Material Systems and Structures* **13** (2004) pp. 625-628.
11. Elmarakbi, A. M. and Zu, J. W., Dynamic modeling and analysis of vehicle smart structures for frontal collision improvement. *Int. J. of Automotive Technology* **5**(4) pp.244-255.
12. Kamal, M., Vibration analysis and simulation of vehicle-to-barrier impact. *SAE Transactions* **700414** (1979).
13. Weaver, J. R., A simple occupant dynamics model. *J. of Biomechanics* **1** (1968) pp. 185-191.

MANUFACTURE OF BULK MAGNETORHEOLOGICAL ELASTOMERS USING VACUUM ASSISTED RESIN TRANSFER MOLDING

B. K.S. WOODS, N. WERELEY, AND R. HOFFMASTER

University of Maryland, 3181 Glenn L. Martin Hall, College Park MD 20742, USA E-mail: bewoods@umd.edu and wereley@umd.edu

N. NERSESSIAN

Techno-Sciences, 11750 Beltsville Dr, 3rd Floor Beltsville, MD, 20705-3194, USA E-mail: nersesse@technosci.com

Magnetorheological elastomers (MREs) consist of ferromagnetic particles embedded in a compliant matrix (i.e. elastomer). Due to the magnetic interaction of the ferromagnetic particles, MREs exhibit field dependent physical properties. Very significant changes in the modulus and loss factor of the elastomer can be realized. This makes MREs a promising candidate for active vibration control mechanisms. One factor currently limiting the implementation of this technology is the lack of an efficient manufacturing method that is practical for mass production. Most of the specimens created for previous MRE research were made using simple casting or mechanical mixing methods that are not ideal. In this research a new methodology for producing MREs using Vacuum Assisted Resin Transfer Molding (VARTM) was investigated. The method was used with a range of iron particles sizes and silicon elastomer systems and found to be effective within certain limits of applicability. The specimens produced were tested in compression under a range of magnetic fields to validate the presence of the MR effect. Changes in compressive modulus ranging from 35% to 150% (depending on volume fraction) under fields of around 0.3T were observed.

1. Introduction

Magneto Rheological Elastomers (MREs) are a promising group of enabling active materials for a wide range of vibration isolation and control problems. MREs are a relatively new field of research and therefore they have yet to be widely implemented into commercial products. One constraint on the commercial viability of MREs is the need for a low cost, near net shape and scalable method of manufacture. Previous MRE specimens have been created using two primary methods, simple casting [1,2,3] and mechanical mixing [4,5]. Neither method is truly suitable for mass production. VARTM is a technology that has come into wide use in the composites industry due to its versatility and scalability. These traits make it an excellent candidate for MRE production. This paper examines the adaptation of VARTM technologies to MRE manufacture.

The success of a VARTM process depends on careful selection of the resin system to be used. Dow Corning two-part thermoset Silicon elastomers were chosen for this testing because their long working times, low viscosities, and low curing temperatures are ideal for the VARTM process, and because they are available in a fairly wide range of stiffnesses. Previous research has shown that the absolute MR effect on the stiffness of an MRE is independent of the matrix stiffness [6], which implies that a softer matrix should show a larger relative MR effect [3]. In order to maximize the MR effect in our samples, silicon elastomers with durometers below 60 Shore A (medium hard – equivalent to tire tread rubber)[7] were used.

2. Methods

2.1 Manufacture of axially aligned MRE specimens

The magnetorheological elastomer specimens were manufactured using a Vacuum-Assisted Resin Transfer Molding (VARTM) process.

A three part mold (see figure 1 below) containing an inner cavity with dimensions of 0.4" x 0.4" x 2" (approximately 1 cm x 1 cm x 5.1 cm) was machined from 6061-T6 aluminum. Rubber o-rings were run in channels between the three parts of the mold so that they compressed tight when the mold was bolted together, which insured a tight seal when the vacuum was pulled.

Lur-Lok instant fittings were threaded into tapped holes on two opposite corners of the mold to allow for the attachment of a syringe on one end and a vacuum hose on the other. The syringe was filled with the mixed and desiccated elastomer resin, after which it was screwed into the side of the mold. When the vacuum was pulled the resin traveled into the mold at a controlled rate, filling the cavity and displacing any air that might otherwise have caused voids in the final specimen.

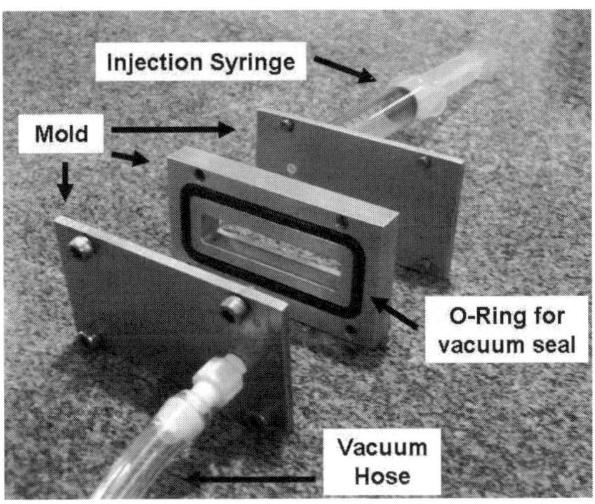

Figure 1. VARTM Mold.

This setup allowed for specimens with varying volume fractions of iron particles to be easily produced. This was done by placing a precisely measured mass of iron particles into the mold before it was sealed. Since the cavity volume and material densities were known, the mass of iron required for a given volume fraction is given by:

$$m_{fe} = V_{tot} * \varphi * \rho_{fe}$$

Here $\boldsymbol{\varphi}$ is the desired volume fraction of iron, $\boldsymbol{\rho_{fe}}$ is the density of iron, and $\mathbf{V_{tot}}$ is the total cavity volume (0.32 in^3 in this case).

The mold was then placed into an axial magnetic field created by two large (2"x 2"x 1") Neodymium permanent magnets. This setup is shown in figures 2 and 3 below. The field lines run along the major axis of the specimens so as to align the iron particles into long chains to strengthen the magnetorheological effect [8]. Using a hall probe, a field of approximately 0.3T was seen during curing, with some variance depending on the permeability of the sample, which is controlled by the volume fraction of iron.

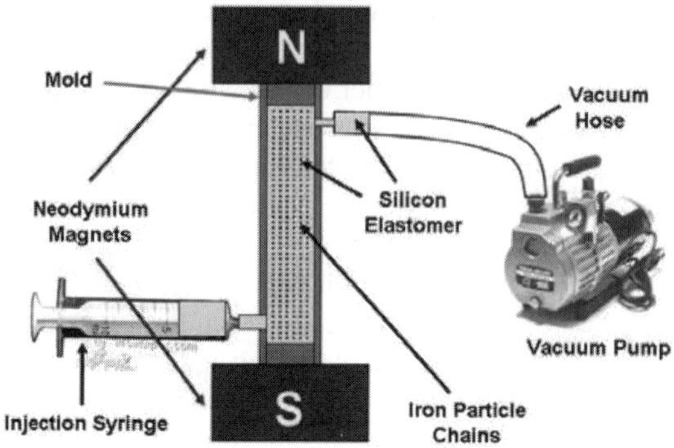

Figure 2. VARTM mold in magnetic field.

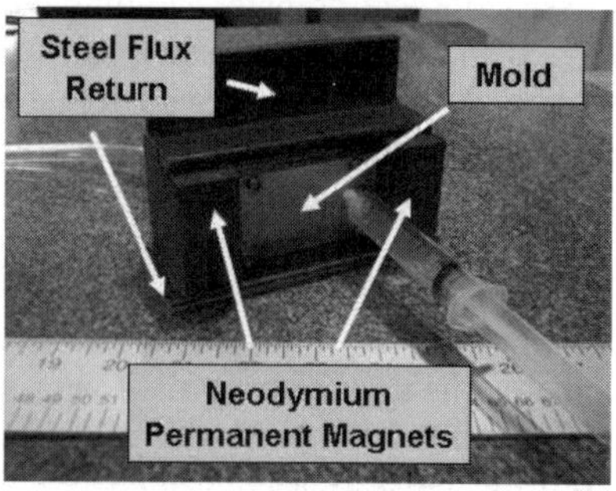

Figure 3. Block Diagram of VARTM system

The chosen elastomer was injected into the mold and the whole assembly was then placed in an oven where the elastomer cured at 70°C for three hours. The field was maintained during the curing process to insure that the particles do not become misaligned. The 70°C cure temperature was well below the Curie temperature of the magnets (310°C) which ensured they where not damaged.

Once cured, the mold was disassembled and the specimen was carefully removed. Applying a mold release agent (Freekote Exitt) to the interior faces of the mold before it is filled was found to simplify this step. Several of the specimens created using this VARTM process can be seen below in figure 4. From left to right these samples are; 1. Slygard 160, -100 mesh, $\varphi = 0.33$, 2. Slygard 170, -20 mesh, $\varphi = 0.20$, 3. Slygard 182, -20 mesh, $\varphi = 0.35$, 4. Slygard 182, $\varphi = 0$.

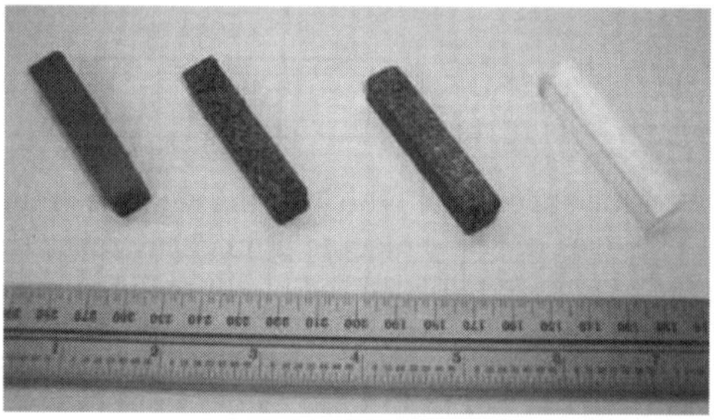

Figure 4. Sample specimens.

Using this process we were able to make a wide range of MRE specimens. Four different silicon systems were successfully used with durometers ranging from 48 Shore OO to 60 Shore A. The iron particles employed ranged in size from -20 mesh (particles fit through a 0.033"mesh) to -325 mesh (0.0017")[9]. Additionally, we were able to reliably make specimens with volume fractions between 0.2 and 0.35, although it must be noted that for the smaller particle sizes, the success rate for $\varphi = 0.35$ specimens was low.

2.2 Manufacturing Challenges

In order for this VARTM process to be practical, it must be able to produce with a high rate of success specimens having sound mechanical properties and an even distribution of well formed iron particle chains. Several significant challenges arose during the development of the VARTM process.

The most common problem was the presence of voids in the cured samples. While we were investigating different techniques for applying the vacuum to the mold and for injecting the elastomer into the cavity, it was noticed that the samples would often have large empty spaces inside of them that created inconsistent material properties, making them unsuitable for use. We realized that the size of the voids was dependent on how long the vacuum was maintained after elastomer injection was completed. We theorized that the vacuum was expanding any gas bubbles present in the elastomer either from the mixing process or from the curing reaction. We were able to create specimens with little to no visible voids simply by minimizing the amount of time the vacuum was pulled after the mold cavity was determined to be filled with elastomer.

Another challenge encountered with this method was incomplete particle wetting. This occurred when the elastomer was not able to fill all of the air space between the iron particles, resulting in portions of the sample disintegrating after removal from the mold. Initial attempts to mitigate this problem focused on pulling the vacuum for a longer period of time to ensure that the elastomer in these regions had sufficient time to fully infiltrate the mold. However, this was not particularly effective at ensuring full specimen wetting, and what's more it increased the chances of voids forming. A more practical solution was to simply limit the volume fraction of iron in the sample. Increasing the volume fraction of the sample increased the total surface area of the iron particles inside the mold. This resulted in higher injection pressure being needed to overcome the viscous forces generated by movement of the elastomer over the surface of the particles. Adding more than 35% iron by volume to the sample was found to increase the viscous forces to the point where regions of incomplete wetting were likely to occur. Limiting particle volume fraction $\varphi < 0.35$ was not seen as a problem because previous research by Lokander et al. [10] indicated that increasing the volume fraction past 30% lead to deterioration in the mechanical properties of the composite. Particle size also influences this problem. It was found in our experimentation that specimens made with smaller particles; particularly the -200 to -325 mesh (0.0029" to 0.0017") sizes, were more likely to not wet out completely. This was also likely due to an increase in viscous forces, since a given mass of smaller particles will have more surface area than the same mass of larger particles.

It is mentioned in the previous section that the lower bound of the range of volume fractions we can obtain with this process is 0.2. This limit is primarily set by the occurrence of uneven particle distribution at lower concentrations of iron. Several specimens with $\varphi = 0.15$ were attempted, but they all experienced clumping of the iron particles at one end of the mold. The iron concentration was too low to maintain particle chains over the entire length of the mold cavity. This could be eliminated by increasing the volume fraction, or by increasing the strength and uniformity of the curing field.

Based on our experience, $0.2 < \varphi < 0.35$ appeared to be the feasible volume fraction range for this VARTM process.

2.3 Testing Apparatus

Compressive testing of these specimens was carried out using an Instron DynaMight 8841 hydraulic axial test machine. An apparatus was manufactured to hold the specimens and apply the desired field levels. It consisted of three major parts. The sample was supported by a pair of grips that bolted on to the load cell and actuator arm of the Instron machine. These grips were machined from high permeability 1018 steel to ensure a strong, well directed field through the specimen. In one of these grips, at the MRE interface, a hall sensor was imbedded to allow for measurements of flux density through the MRE. Surrounding these grips and the MRE was an aluminum bobbin supporting an electromagnetic coil constructed from 5280 turns of 20 gauge copper wire. This coil was aligned to produce field lines along the major axis of the specimen. This provided for the strongest possible MR effect [11]. The third element of the test setup was an outer flux return path consisting of a 3.5" outer diameter 1018 steel pipe ¼" thick with two 0.6" thick 1018 steel discs at the ends. This assembly fits snugly around the bobbin and has a 1.4" hole bored out at each end to allow the sample grips to freely slide during testing. A cross-section view of this assembly appears in Figure 5 below. A picture of the Instron test machine, with the test rig mounted is shown in Figure 6.

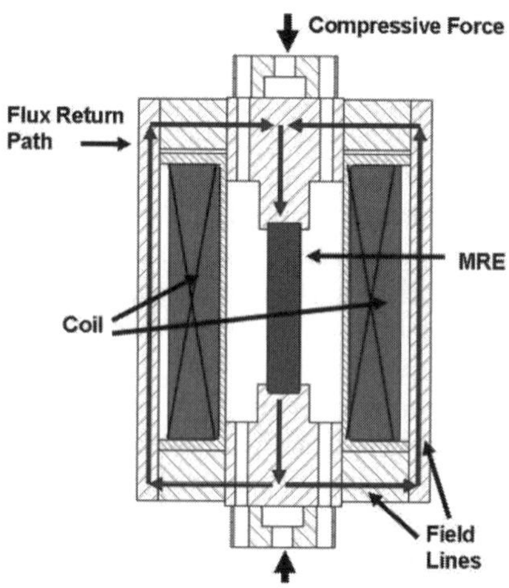

Figure 5. Test rig Cross Section

Figure 6. Instron DynaMight with test rig installed

Due to the long specimen length (2") and low elastomer permeability ($\mu_{el} \approx \mu_0$) this setup was only able to generate maximum flux densities on the order of 0.26T to 0.3T, with the exact maximum applied field being dependent on the volume fraction of the specimen being tested.

3. **Results**

3.1 Validation of field dependent stiffness

Each specimen was tested at 6 different values of field strength over a range of 0T to about 0.3T. At each field value a load ramp was performed that compressed the samples up to 20N (4.17lb) in 4 seconds, held the load for 2 seconds, then returned to the no load condition over another 4 seconds. The forces were kept small to ensure that no buckling occurred. The force ramps were performed slowly to minimize strain rate dependence of the stiffness parameter [12]. Finally, to remove the effects of temperature from the results, the coil was actively cooled using compressed air to maintain a temperature less than 85°F (~30°C)

A set of typical results from this testing is shown below in figure 7. This particular data set is for Dow Corning Sylgard 182 silicon elastomer (50 Shore A durometer) filled with various volume fractions of -20 mesh iron particles (Alfa Aesar). These results clearly show that the MR effect is present in these specimens. Note that the compressive stiffness parameter increases with increasing field strength for all volume fractions. Note also that the magnitude of the change in modulus is dependent on the volume fraction of the specimen.

With this data set we found modulus changes ranging from 35% ($\varphi = 0.2$) to 150% ($\varphi = 0.25$). These results are of similar magnitude to those presented in other research, where values of 20-60% change in modulus for a saturated MRE are typical [5,13].

Note that these results show an increase in the inherent (0 T) stiffness of the specimens with increasing volume fraction. The net stiffness of a MRE composite is governed both by the stiffness of its parts and by the way they interact. Specifically, the manner in which the elastomer wets and bonds to the iron particles has rather complex effects [3]. This results in the total MRE stiffness being more complicated than a simple weighted sum of its parts, causing non-linear increases in stiffness with increasing volume fraction.

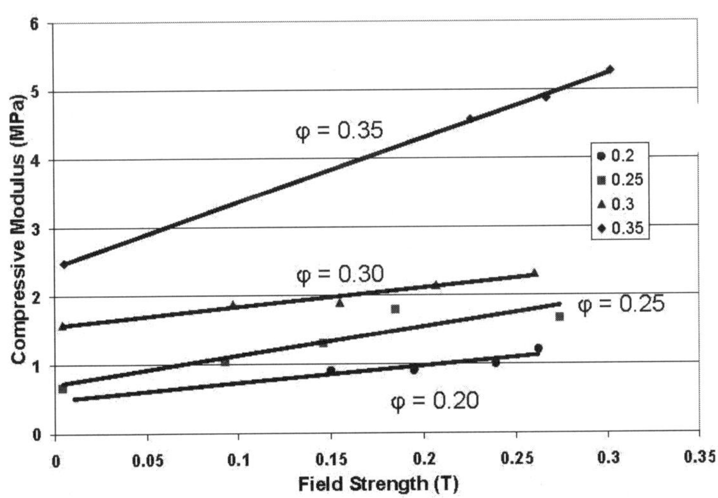

Figure 7. Measured compressive modulus as a function of magnetic flux density (Slygard 182, -20 mesh)

4. Conclusion

In summary, we were able to successfully fabricate axially aligned magnetorheological elastomer specimens using a VARTM composite manufacturing process. This process allowed for the efficient production of specimens with a wide range of mechanical and field-dependent properties. By altering the host elastomer, the iron particle size, and the iron particle volume fraction, MREs with tailored properties were produced. By following the procedures and manufacturing guidelines described in this study, this VARTM method can reliably create bulk MRE specimens that are suitable for vibration absorbing applications. Additionally, we were able to validate the MR effect in these materials, with changes in compressive modulus from 35% to 150% being observed.

$\varphi = 0.20$
$\varphi = 0.25$
$\varphi = 0.30$
$\varphi = 0.35$

References

1. Jolly, M., Carlson, J., and Munoz, B., 1996, A model of the behaviour of magnetorheological materials. Smart Materials and Structures. Vol. 5, pp. 607-614.
2. Zhou, G.Y., Li, J.R., 2003, Dynamic behavior of a magnetorheological elastomer under uniaxial deformation: I. Experiment. Smart Materials and Structures. Vol. 12, pp. 859-872.
3. Kallio M., The elastic and damping properties of magnetorheological elastomers, VTT Publications 565 pp. 38-39.
4. Ginder, J.M., *et al.*, 1999, Magnetorheological Elastomers: Properties and Applications. SPIE Smart Materials Technologies. Vol. 3675, pp. 131-138.
5. Shen, Y., Golnaraghi, M.F. and Heppler, G.R. 2004. Experimental research and modeling of magnetorheological elastomers. J. Int. Material Systems and Structures. Vol. 15, pp. 27-35.
6. Lokander, M. and Stenberg, B. 2003, Improving the magnetorheological effect in isotropic magnetorheological rubber materials. Polymer testing 22, pp. 677-680.
7. Durometer Hardness Scale, Online technical data, McMaster-Carr, accessed from http://www.mcmaster.com/
8. Farshad, M. and Benine, A. 2004. Magnetoactive elastomer composites, Polymer Testing 23, pp. 347-353.
9. Powder Particle Size, Online technical data, ESPI Metals, accessed from http://www.espimetals.com/tech/mesh.htm on 06/03/06.
10. Lokander, M. and Stenberg, B. 2003. Performance of isotropic magnetorheological rubber materials. Polymer Testing 22, pp. 245-251.
11. Ginder, J.M., Clark, S.M., Schlotter, W.F. and Nichols, M.E. 2002, Magnetostrictive phenomena in magnetorheological elastomers. Int. J. Modern Phys. B, Vol. 16, pp. 17-18.
12. Bergstrom, J.S. and Boyce, M.C. 1998. Constitutive modeling of the large strain time-dependent behavior of elastomers. J. Mech. Phys. Solids., Vol. 46. pp. 931-954.
13. Zhou, G.Y., 2003 Shear properties of a magnetorheological elastomer. Smart Materials and Structures. Vol. 12, pp. 139-146.

SQUEEZE FLOW OF ELECTRORHEOLOGICAL FLUIDS UNDER CONSTANT VOLUME

ERNEST CARL MCINTYRE

Department of Macromolecular Science and Engineering
University of Michigan Ann Arbor, MI, USA
E-mail: ecarlm@umich.edu

FRANK E. FILISKO

Department of Macromolecular Science and Engineering
University of Michigan Ann Arbor, MI, USA
E-mail: fef@umich.edu

Compression Studies of electrorheological (ER) fluids have been impeded by not knowing the concentration of particles held between the plates. This paper presents a technique that calls for using a constant volume setup that solves the problem. This paper also includes results on the effects of changing concentration of particles and changing the viscosity of the oils used in compression of an ER fluid. The conclusions of this paper were that the constant volume squeeze flows of ER fluids were significantly different than what has been predicted with constant volume squeeze flow theories.

1 Introduction

Compression studies of electrorheological fluids have been impeded by the "sealing effect". Using a constant area parallel plate apparatus this sealing effect is responsible for squeezing out an unknown quantity of particles from between the plates while the rest are held between the plates because of the electric field. Previous studies by Ahn, Chu and Lee have dealt with this problem by assuming that all particles remain in between the plates. This is an approximation that gets worse as the viscosity of the fluid is increased.

The present study avoids these restrictions by using a constant volume apparatus. In a constant volume apparatus instead of knowing that the area of the plate in contact with the fluid is constant throughout the test, instead the area of the plates in contact with the fluid is changing throughout the test, but the volume of the fluid contained in between the plates remains constant throughout the tests. The experiment and equation developed for this type of squeeze flow was done by Diennes and Klemm.

By using a constant volume apparatus for conducting this experiment the concentration of particles that stay in between the plates is known throughout the experiment. Because there is already a model for homogeneous Newtonian fluids in constant volume squeeze flow, the results for compression of a electrorheological fluid could be compared with this model and an analysis of both the effects of the ER fluid composition on squeeze flow and the effects of geometry and squeeze speed could be made. This study is concerned with the effects of the concentration of particles in the dispersion.

2 Materials and Methods

2.1 Apparatus

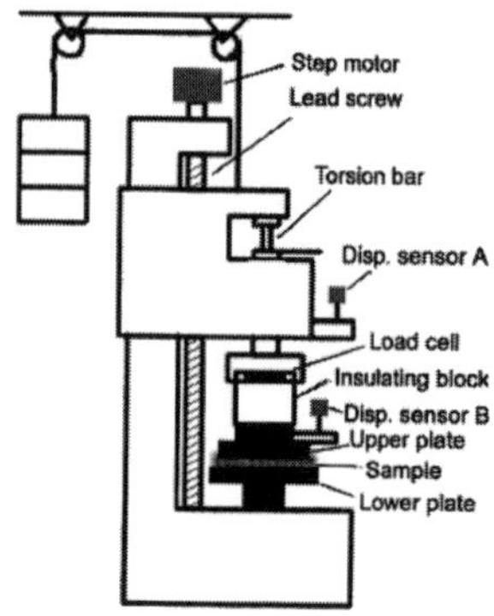

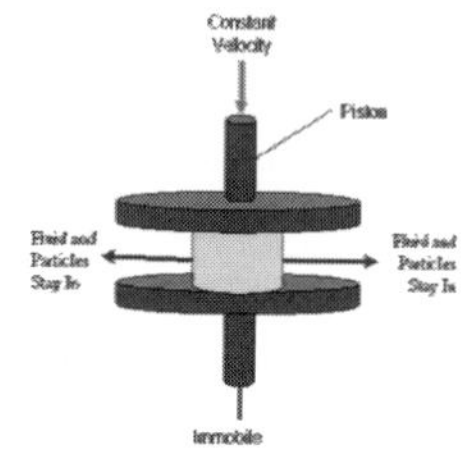

A Weissenberg rheometer with an added step motor-leadscrew loading system and an LVDT displacement sensor was used to make the measurements. For each test a syringe was used to measure a constant volume of 0.5 mL of ER fluid which was placed between the plates. Before compression, a DC voltage was imposed on the sample and was kept constant in the whole period of compression. The squeeze speed was 16.7 microns/second for all tests. A PC and a laptop computer were used to record the output of the load cell and the displacement sensor.

2.2 Materials Preparation

The electrorheological fluids were prepared using dried aluminosilicate powder (SG 1.1 @ 25 °C) in silicon oil. This study examined both a high viscosity-1,000 cSt (SG 1.27 @ 25°C) and a low viscosity-10 cSt (SG 0.963 @ 25°C) silicon oil. By weight 10, 20, 30, and 40% solutions were prepared by weighing the aluminosilicates out and adding the silicon oil dropwise into vials. The solutions were immediately mixed and capped.

3 Results

Figure 1 shows how the compression force develops during compression of the ER fluids containing the low viscosity silicon oil without an electric field. The compression force shows that the fluid squeezes down until there is a monolayer of particles where the force increases at the very end. There isn't a large difference in the way the squeeze force increases as the concentration is increased for the ER fluid containing the low viscosity silicon oil.

Figure 2 shows that the compression force in squeeze flow when an electric field (E= 1,000 V) is applied. The electric field causes an increase in the compression force for all the concentrations. The electrorheological effect is greater for larger concentrations. For example at a gap separation of 0.1 mm the force generated by 10% and 40% ER fluid is around 0-3 N without an electric field. When the electric field is applied at 0.1 mm the force is 200 N for 10% and >>450 N for the 40% ER fluid.

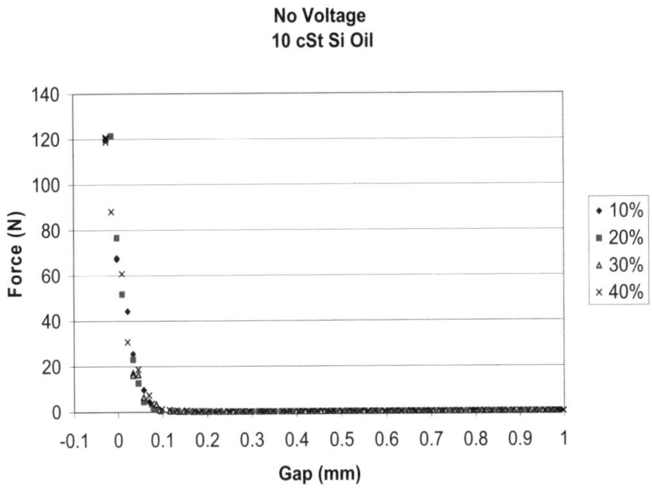

Figure 1. Force vs. Gap Chart using constant volume to show the effect of concentration of particles in low viscosity fluid with no electric field

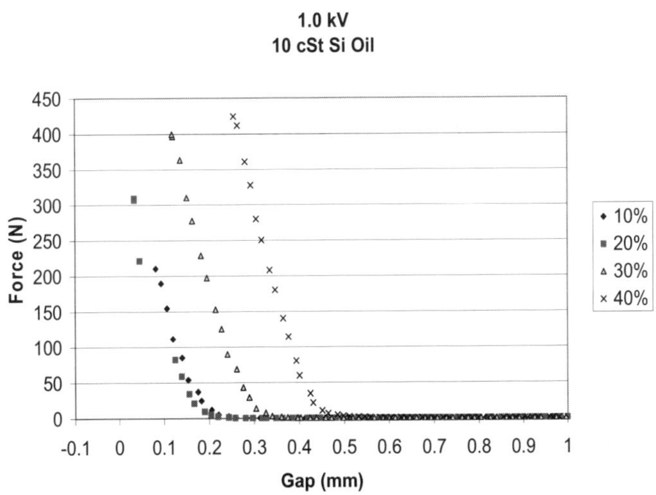

Figure 2. Force vs. Gap Chart using constant volume to show the effect of concentration of particles in low viscosity fluid under 1.0 kV electric field.

Figure 3 like Figure 1 shows how the compression force develops during compression of the ER fluids without an electric field, except this ER fluid contains the high viscosity silicon oil. Even though these fluids are much thicker and the forces generated by thicker fluids are larger even without an electric field, the curves still fall almost on top of each other with the exception of the most concentrated suspension 40%.

Figure 4 shows the compression force for the ER fluid with the high viscosity silicon oil under an electric field (E=1,000 V). Again the electric field causes an increase in the compression force for all the concentrations. The electrorheological effect is greater for larger concentrations. If a gap separation of 0.3 mm is taken for figures 3 and 4, for 10% and 40% ER fluid the force generated at that gap is around 5-10 N. Under an electric field the force generated at 0.3 mm is 15 N at 10% and 200 N at 40%.

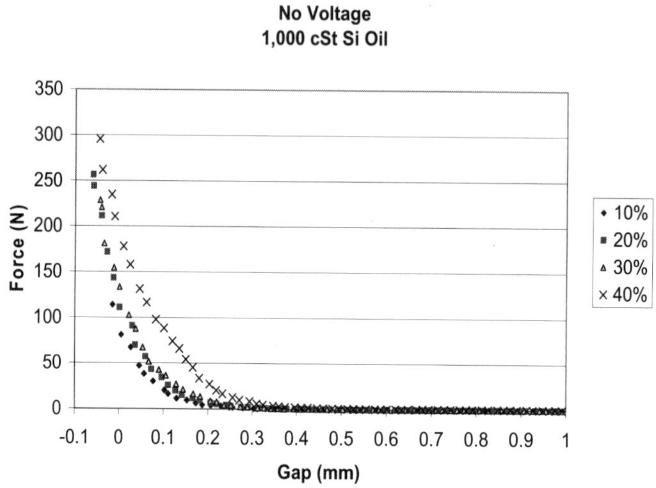

Figure 3. Force vs. Gap Chart using constant volume to show the effect of concentrationof particles in high viscosity fluid with no electric field.

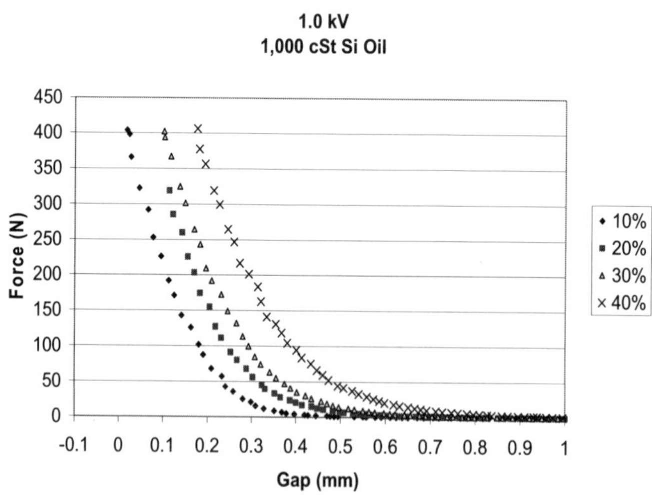

Figure 4. Force vs. Gap Chart using constant volume to show the effect of concentration of particles in high viscosity fluid with 1.0 kV.

Figure 5 gives the log-log plot data for the ER fluid containing the low viscosity silicon oil under an electric field. This figure illustrates how the results from the constant volume squeeze flow of an ER fluid matches squeeze flow theories. According to squeeze flow theories the log-log plot should produce a line with a slope of -5. The diagonal line on the graph gives a slope of -5. The data shows a dependence greater than -5. Figure 6 gives a similar plot for the ER fluid containing the high viscosity silicon oil under an electric field. Again the slopes of the data curves are greater than -5.

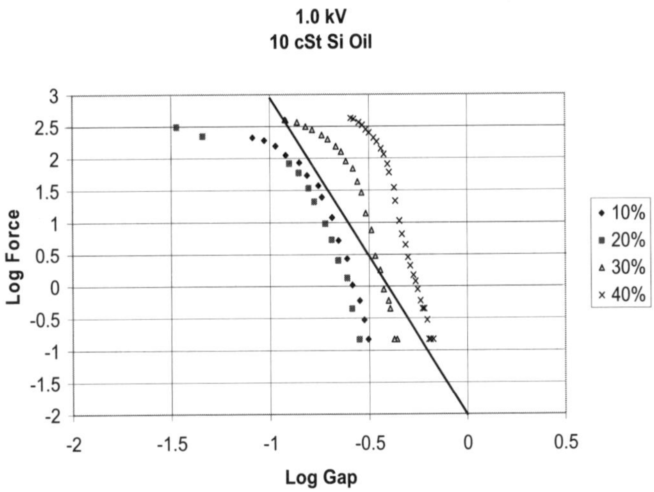

Figure 5. A log-log chart of Force vs. Gap for the low viscosity ER fluid at different concentrations of particles. The line represents the slope that would be predicted by Stefan's law for a constant volume Newtonian fluid.

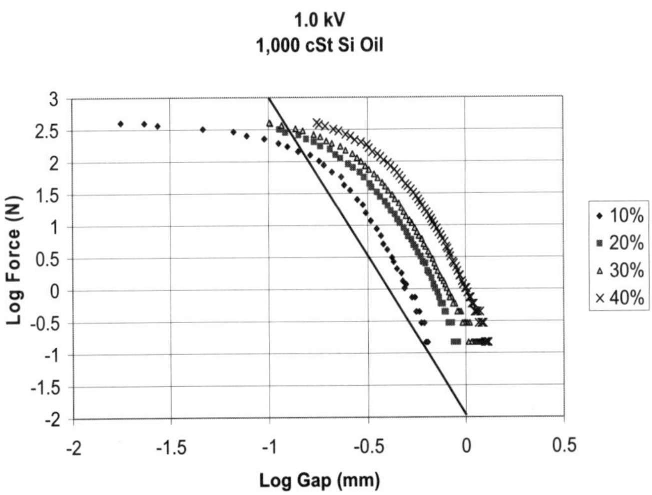

Figure 6. A log-log chart of Force vs. Gap for the high viscosity ER fluid at different concentrations of particles. The line represents the slope that would be predicted by Stefan's law for a constant volume Newtonian fluid.

4 Discussion

The figures show that the effect of concentration on the force is greater under an electric field than without one. This was true for both the high and the low viscosity fluids. Because this experiment was done using a constant volume setup where the concentration of particles between the plates was known throughout the test this suggests that the concentration of particles are a strong factor in the compressive forces generated by an ER fluid. While this result was implied in similar studies done using a constant area setup this is the first experiment where the concentration of particles is known and the experimental setup is unaffected by the "sealing effect"

Comparing figures 2 and 4 the effect of the continuous phase of the ER fluid can also be better understood.. In previous studies the effect of the concentration was unable to be isolated. In fact the concentration was not even able to be accurately determined once the particles started squeezing between the plates. Therefore any attempt to assess the effect of the continuous phase by such studies would have been contaminated by the inability to pin down the effects of the changing concentration of the ER fluid.

One observation that was made for the compression experiments on the viscosity of the continuous phase. For the low viscosity oils it was observed that the fluids containing low viscosity oils, only pure oil is exuded radially during compression. Whereas for fluids containing high viscosity oils the powder and oil were exuded radially during the compression.

The effect of the viscosity is best illustrated in plots 5 and 6 where the dependence of the force on the gap is shown in a log-log plot. Using the extension of Stefan's law for constant volume for a log-log plot of force vs. gap it should give a linear plot with a slope of -5. Looking at both figures 5 and 6 the data for the ER fluids containing both the high and low viscosity oils gave slopes that were larger then -5. The ER fluid containing the high viscosity oil had a slope closer to -5 than the ER fluid containing the low viscosity. As stated in the observation made in the previous paragraph this can be the result of particles staying in the center of the fluid for the low viscosity sample.

References

1. Ahn, K.H., S. Chu and S.J. Lee "An Experimental Study on the squeezing flow of electrorheological suspensions" *J. Rheol.* **44** (2000) pg. 105-120
2. Diennes G.J. and H.F. Klemm "Theory and Application of the Parellel Plate Plastomer" *Appl. Phys.* **17** (1946) pg. 458-471
3. Monkman, G.J. "Electrorheological Effect Under Compressive Stress" *J. Phys. D: Appl. Phys.* **28** (1995) pg. 588-593
4. Viera, S.L., M. Nakano, R. Oke, T. Nagata "Mechanical Properties of an ER fluid in Tensile, Compressive, and Oscillatory Squeeze Tests" **15** (2001) pg. 714-722
5. Filisko, F. and Meng Y. "Unidirectional Compression of Electrorheological Fluids in Electric Fields" J. Appl. Phys. **98** (2005)
6. Collomb, J., F. Chaari and M. Chaouche "Squeeze Flow of Concentrated Suspensions of Spheres in Newtonian and Shear Thinning Fluids" J. Rheol. **48** (2004) pg. 405-416

ANALYSIS OF CONDENSED SPECIES FROM ELECTRIC ARC DEPOSITION

A.M. KNAPP AND PROF. F.E. FILISKO

Materials Science and Engineering Department, University of Michigan, Ann Arbor, MI 48109 E-mail: *fef@umich.edu*

Carbonaceous condensed species, which include single-walled carbon nanotubes, produced by electric arc deposition (EAD) are subjected to thermogravimetric analysis at varying heating rates. It will be shown that the heating rate is critical when predicting the types and amounts of single-walled carbon nanotubes present in the sample. Examination of EAD products (from various applied currents) with TGA plots concludes that the 70-90 A current regime is the optimum operating parameter and the web-lid faction is the richest region of SWNT concentration.

1 Introduction

Characterization of the amount of single-walled carbon nanotubes (SWNT) formed during EAD as well as the area in the chamber with the highest yield is a primary concern when evaluating the efficiency of the EAD operation. EAD experiments were conducted over a range of currents applied to the electrodes to determine the optimum conditions for SWNT production (Fig. 1a) [1].

Numerous literature papers use thermogravimetric analysis (TGA) or temperature programmed oxidation (TPO) to quantify the amount of SWNT produced. It will be shown however that TGA alone cannot accurately determine the amount of SWNT produced.

2 Experimental

TGA was conducted on condensed carbonaceous species produced with EAD (Fig. 1b). Condensed species collected from the collar and web-lid regions of the EAD chamber were analyzed with a simultaneous TGA-DTA instrument SDT2560. Samples of 5-10 mg were placed in an alumina crucible and subjected to a dynamic air atmosphere (60 sccm) at various ramp rates from room temperature to 1000°C.

FTIR was conducted using a Perkin-Elmer 1740 Infrared Fourier Transform Spectrometer, while commercial software, Spectrum, allowed for quantitative evaluation of the material.

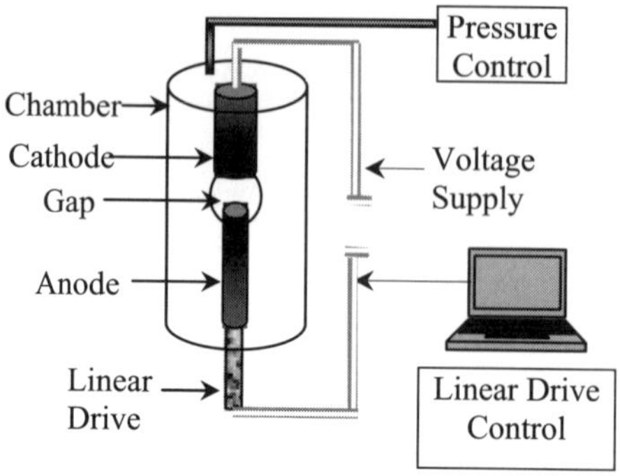

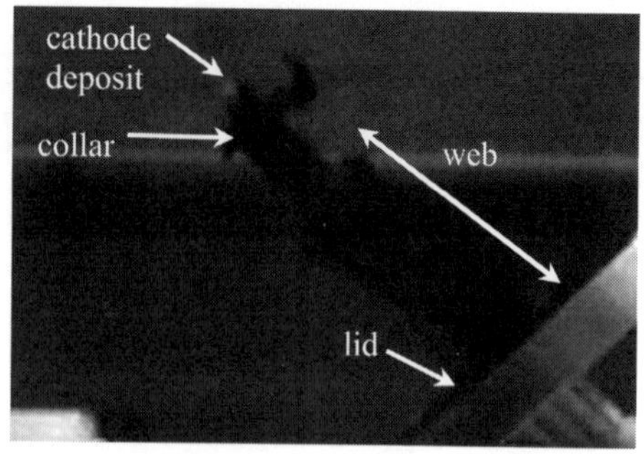

Fig. 1: a) Schematic of the EAD operation. b) Image of carbonaceous species collected on the upper chamber lid.

3 TGA Characterization of Condensable Species

During oxidation of organic materials, carbon solids are converted into CO and CO_2 gases with a corresponding decrease in mass. [2,3] TGA ascertains the temperature range of mass loss and the amount of residual material remaining, if any, after the experiment is complete. The derivative of the TGA data, the ratio of dm/dT evaluates the mass loss, dm, and changes in temperature, dT, over the experimental temperature regime to determine the regions of maximum rate of oxidation. Using this derivative in conjunction with Lorentzian curve fitting has been used by several researchers to quantify the amount of SWNT originally in the sample [4]. In addition, TGA of the EAD samples in dynamic oxygen rich environments may be used to determine the concentration of metal originally in the sample. Likewise, refined SWNT may also be evaluated with TGA to determine the efficiency of purification methods.

Besides SWNT, the EAD method also produces the unwanted products of amorphous carbon, fullerenes, and nanographitic particles. Pang et al have investigated the maximum rate of oxidation for C_{60}, graphite, and a mixture of nanotubes and nanoparticles which they reported to be 420°C, 645°C, and 695°C respectively [3]. Various papers in literature associate the weight loss event at 350-450°C with the oxidation of amorphous carbon and consider C_{60} to be amorphous in nature [2,4]. In characterizing the EAD produced material, the temperature regions as reported by Pang should be considered loosely as a guide since it has been reported that metal catalyst particles present have a tendency to promote the oxidation of SWNT below 500°C; [2,5] purified SWNT are resistive to oxidation until 600°C [2,6,7].

3.1 58-62 A Actual Current

TGA plots of the 58-62 A actual current web-lid faction at heating rates of 10, 7, and 5°C/min show a slight increase in weight of ~1% between 150-225°C attributed to the adsorption of oxygen by the sample (Fig. 2). In general, a single step weight loss is observed between 300-500°C for each heating rate. While Lorentzian fits of the data are not performed to specifically determine the different materials being oxidized, changes in the slope of the normalized ΔT curves may be used to estimate the number of species present. Accordingly, from these curves, it is predicted that two carbon polymorphs are present, amorphous carbon and SWNT, which agrees well with the literature.

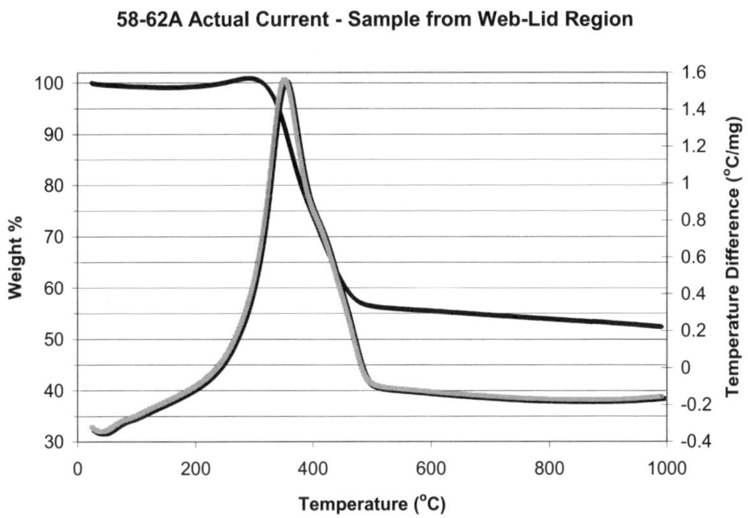

Fig. 2: TGA plot of sample from web-lid region with a heating rate of 10°C/min. This plot is consistent and representative of the three heating rates examined.

Further examination of the TGA plots determines that approximately 60 wt.% of the material remains after oxidation. Based upon the "recipe" used to pack the anode for the EAD experiments, the residual oxides of NiO and Y_2O_3 should compose 30 wt.% of the initial TGA sample weight. It is therefore unreasonable to conclude that the 60 wt.% of material remaining is composed entirely of NiO and Y_2O_3. It is highly possible that some polymorph of carbon is present that is resistant to oxidation at the temperatures examined and is supported by the gray color of the TGA residue suggesting the presence of carbon. If the sample did not contain any carbon species, the residue would be greenish in color evidence of NiO. FTIR was subsequently performed on the residue, which displayed peaks characteristic of C-C bonds between 1380-1600 cm^{-1}.

Plots of samples from the collar at the same heating rates again show a slight increase in weight of ~1% between 150-225°C as well as a single step weight loss observed between 300-500°C (Fig. 3). From examination of the ΔT curves, it is predicted that three species are present, amorphous carbon, C_{60}, and SWNT. Each ΔT curve for the samples from the collar have a smaller intensity than those for corresponding heating rate in the web-lid faction, suggesting that less heat is being generated during oxidation i.e. smaller exothermic reactions.

For all of the plots at this experimental current, at temperatures greater than 600°C, it appears that 3-5% weight loss may be possible. However, this weight loss is insignificant and attributed to baseline drift from the equipment.

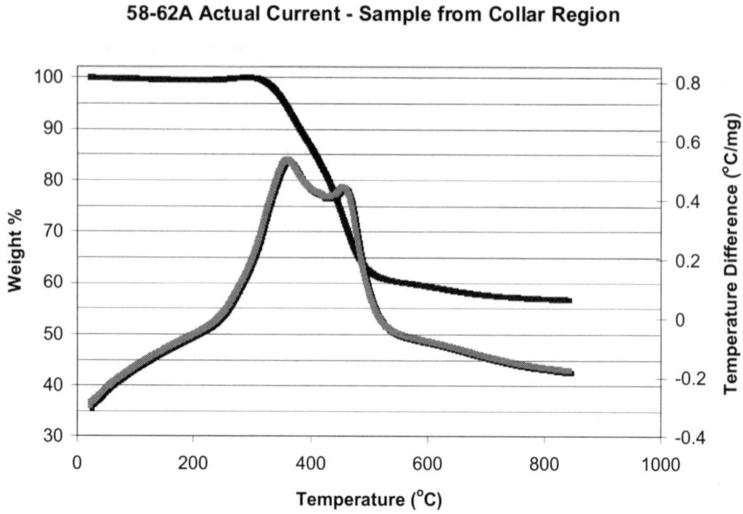

Fig. 3: TGA plot of sample from collar region with a heating rate of 7°C/min. This plot is also representative of the 5°C/min heating rate.

3.2 60-66 A Actual Current

Examinations of these plots with the corresponding heating rate from the 58-62 A current are quite similar with exception of the collar samples at the 5 and 7°C/min heating rates. At these conditions, the evolution of species is quite distinct in the 58-62 A and less so in the 60-66 A plots. It is theorized that more SWNT are present causing the overlap in the ΔT curves and concealing the third species. The mass remaining after TGA is abnormally large; the presence of carbon is once more suspect as each of the residues are gray to black in color. Possibly the amount of nanographite and/or graphite produced is not being eliminated during oxidation.

3.3 60-70 A Actual Current

In the first TGA experiment on a sample from the web-lid faction and at a heating rate of 10°C/min, an exothermic reaction with 3.3°C/min amplitude is glaringly visible at 420°C with two smaller exotherms of 1.2 and 0.15°C/min present at 360 and 460°C respectively (Fig. 4). Combustion of the material at 420°C expelled considerable energy resulting in an overall loss of heat by the sample which then reduced its temperature by ~ 5°C. Thoughts as to why the massive exotherm occurred included the compression of material as it was packed into the alumina crucible for TGA or the presence of a substantial quantity of SWNT, which efficiently transported the heat throughout the sample and caused spontaneous combustion.

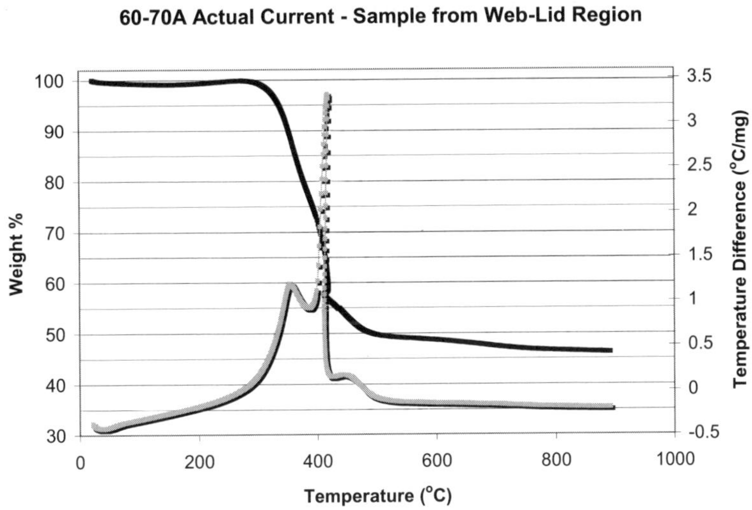

Fig. 4: TGA plot of sample from web-lid region with a heating rate of 10°C/min.

Harutyunuan et al reported similar TGA data on commercially obtained SWNT made by the EAD process [2]. Though a heating rate of 5°C/min was used, three exotherms are present at 350, 395, and 420°C, with the reaction at 395°C having an amplitude ~ 4.0°C/min. Harutyunuan further states the oxidized species are amorphous carbon, SWNT, and multishell carbons for each successive exotherm. Dillon et al attribute the final exotherm in similar data to surviving SWNT as opposed to multishell carbons while Landi et al assign it as nanostructure carbon or "carbonaceous impurities of comparable morphology to be stable above 500°C" [7,8].

Additional TGA experiments on the web-lid faction were conducted at heating rates of 3 and 1°C/min (Fig. 5). Curiously, the exotherm indicating combustion of material is absent in each subsequent plot. The derivatives of the weight loss curves for the latter plots depict only two carbon species present in each sample. However, as all three samples tested were acquired from the same experiment and collected from the same region of the chamber, it is expected that latter plots actually contain three species.

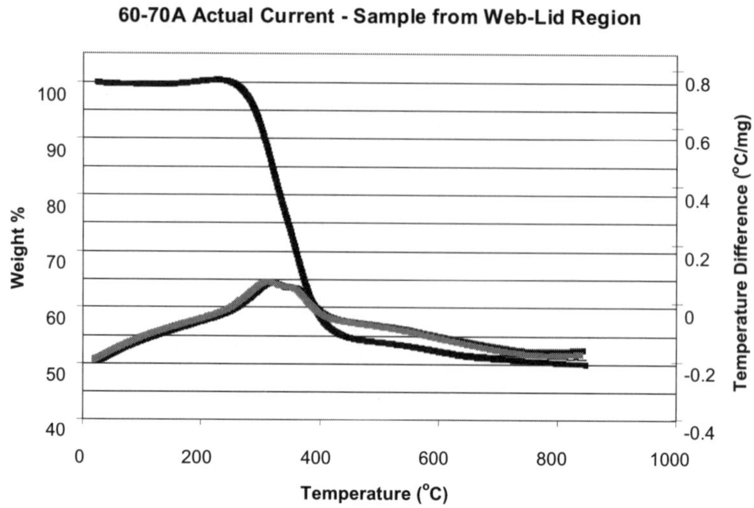

Fig. 5: TGA plot of sample from web-lid region with a heating rate of 1°C/min. This plot is also representative of the heating rate of 3°C/min.

TGA performed at 10°C/min on a sample from the collar does not behave as the corresponding heating rate of the web-lid faction (Fig. 6). While the ΔT curve does indicate the presence of three carbonaceous species, the exotherm at 420°C is not separate nor spontaneous and has a maximum amplitude of only 0.7°C/mg as compared to the 3.3°C/mg amplitude observed in the combustion of material from the web-lid. Also dissimilar is that the nanostructured carbons, a termed adopted from Landi, from the collar region are not oxidized until 600-800°C in the derivative curve [8]. Speculation of the combustion that occurred in the web-lid faction allows us theorize that incineration reaction changed the morphology of the nanostructured carbons to a simpler geometry which oxidized below 500°C as opposed to the more stable nanostructured carbons present in the collar.

Heating rates of 3 and 1°C are quite similar to their web-lid counterparts. Again it is proposed that three carbonaceous species were present despite the ΔT curve depicting two exotherms overlapping one another. While it is known that carbonaceous species remain in the residue due to its grayish color, the amount has decreased from the 60-66 A samples; approximately 50-55 wt.% remains after oxidation.

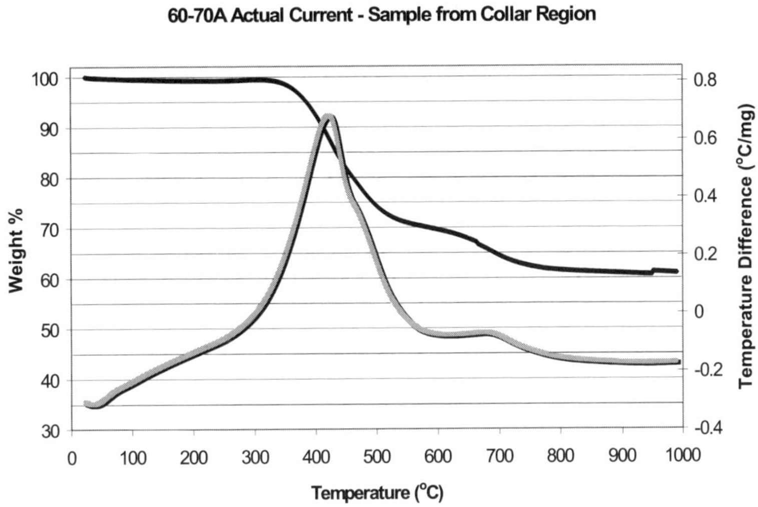

Fig. 6: TGA plot of sample from collar region with a heating rate of 10°C/min.

3.4 70-90 A Actual Current

Spontaneous combustion occurs yet again at the 10°C/min heating rate with an amplitude of 3.3°C/mg at 400°C (Fig. 7). Intriguingly, only two distinct exotherms are present corresponding to the oxidation of SWNT (3.6°C/mg) and nanostructured carbons (0.11°C/mg). While it is known that amorphous carbon is also produced during the EAD process, the exotherm denoting this species is obscured by the width of the single, sharp exotherm. It could be argued that a sufficient number of smaller diameter SWNT are present in the sample which oxidize at the same temperature regime as the amorphous carbon. This theory is supported by the work of Landi et al in which they concluded that thermal oxidation of raw soot involves the degradation of SWNT at each temperature that mass loss occurs and its oxidation is catalyzed by the presence of the metal impurities [8].

The 5°C/min heat rate likewise displays the distinct combustion exotherm at 400°C as well as a smaller exotherm at 350°C attributed to amorphous carbon (Fig. 8). Though the heating rate is smaller, the amplitude of the combustion exotherm is stronger having an amplitude of nearly 4°C/mg. It is curious that the corresponding shoulder for the nanostructured carbon and in the ΔT is masked by the SWNT exotherm. Also of note is the metal oxide content is 30-40 wt.% of the sample as predicted by the calculations in section 3.1 as well as the greenish color of the residue.

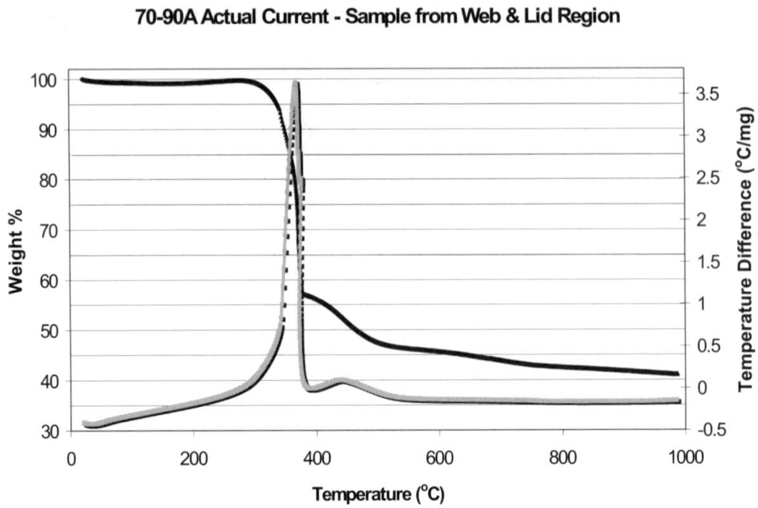

Fig. 7: TGA plot of sample from web-lid region with a heating rate of 10°C/min.

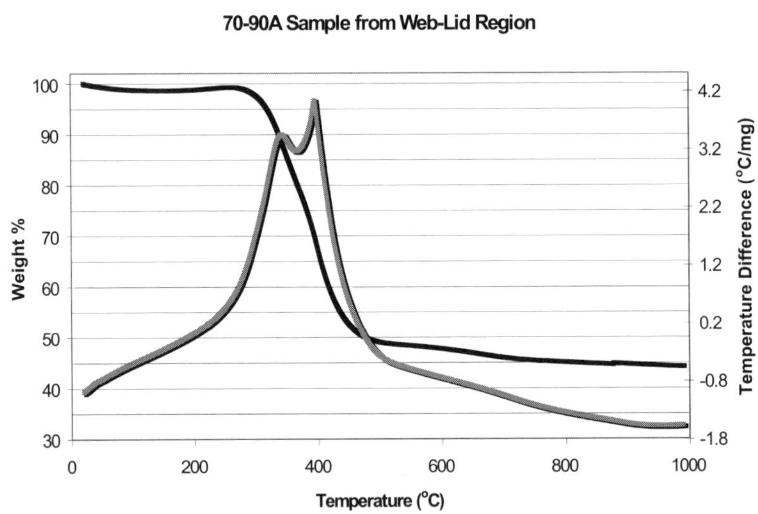

Fig. 8: TGA plot of sample from web-lid region with a heating rate of 5°C/min.

3.5 78-96 & 100-116 A Actual Current

Heating rates of 10 and 5°C/min for both the web-lid and collar factions display the expected three exotherms; weight loss above 500 °C due to nanostructured carbon is again apparent. Consistent with the high percentage of material remaining after oxidation, the color of the TGA residue is gray denoting the presence of unoxidize carbon.

3.6 Aldrich

Despite the fact that a two stage weight loss is observed, the amount of nanostructured carbon is minimal and completely removed by 700°C. The commercially obtained raw soot is quite consistent with the 70-90 A actual current TGA results in terms of residual metal content, unmistakable combustion exotherm, and its corresponding maximum amplitude. It is tentatively predicted that the 70-90 A current regime is the "sweet spot" for production with the EAD apparatus at the University of Michigan.

4 TGA conclusions

Slower heating rates allow one to observe exotherms of three condensed species, yet only the faster heating rates exhibit the combustion exotherm related to SWNT oxidation. In the 58-62 and 60-66 A actual currents, the samples from web-lid faction do not demonstrate the presence of nanostructured carbon but samples from the collar are obvious in their presence. The 60-70 A currents present evidence of nanostructured carbons in web-lid, but more so in the collar faction. The 70-90 A regime is the cautiously predicted as the optimum operating condition for the EAD apparatus as few nanostructured carbons are produced according to TGA analysis. Increasing the current to the 78-96 and 100-116 current regime does not appear to improve the yield of SWNT as more nanostructured carbon subsist within the samples and are not completely removed even after experiencing temperatures of 1000 °C.

References:

1. Waldorff E.I., Waas A.M., Friedmann P.P., and Keidar M., Characterization of Carbon Nanotubes Produced by Arc Discharge: Effect of the Background Pressure, *J. Appl. Phys.* **95** [5] (2004) pp. 2749-2754.
2. Harutyunuan A.R., Pradhan B.K., Chang J., Chen G., and Eklund P.C., Purification of Single-Wall Carbon Nanotubes by Selective Microwave Heating of Catalyst Particles, *J. Phys. Chem. B* **106** (2002) pp. 8671-8675.
3. Pang L.S.K., Saxby J.D., and Chatfield S.P., Thermogravimetric Analysis of Carbon Nanotubes and Nanoparticles, *J. Phys. Chem.* **97** [27] (1993) pp. 6941-6942.
4. Shi Z., Lian Y., Liao F., Zhou X., Gu Z., Zhang Y., and Iijima S., Purification of Single-Wall Carbon Nanotubes, *Solid State Comm.* **112** (1999) pp. 35-37.
5. Zhou W., Ooi Y.H., Russo R., Papanek P., Luzzi D.E., Fischer J.E., Bronikowski M.J., Willis P.A., and Smalley R.E., Structural Characterization and Diameter-Dependent Oxidative Stability of Single Wall Carbon Nanotubes Synthesized by the Catalytic Decomposition of CO, *Chem. Phys. Lett.* **350** (2001) pp. 6-14.
6. Chiang I.W., Brinson B.E., Smalley R.E., Margrave J.L., and Hauge R.H., Purification and Characterization of Single-Wall Carbon Nanotubes, *J. Phys. Chem. B* **105** (2001) pp.1157-61.
7. Dillon A.C., Gennett T., Jones K.M., Alleman J.L., Parilla P.A., and Heben M.J., A Simple and Complete Purification of Single-Walled Carbon Nanotube Materials, *Adv. Mater.* **11** [16] (1999) pp.1354-58.

8. Landi B.J., Cress C.D., Evans C.M., and Raffaelle R.P., Thermal Oxidation Profiling of Single-Walled Carbon Nanotubes, *Chem. Mater.* **17** (2005) pp. 6819-6834.

PERFORMANCE TESTING AND HYSTERESIS MODELING OF A POSITION-FEEDBACK MR ACTUATOR

B. LIU, W. H. LI, P. B. KOSASIH AND X. Z. ZHANG

School of Mechanical, Materials and Mechatronic Engineering, University of Wollongong, Wollongong, Northfields Avenue, NSW 2522, Australia

E-mail: weihuali@uow.edu.au

This paper presents the development and performance testing and modeling of a position-feedback MR actuator. The MR actuator, in disc-shape, composes of rotary shaft and plate, an electromagnetic coil, MR fluids, and casings. The working principle of the actuator was discussed and the transmitted torque equation was employed by using the Bingham plastic model. The optimal dimensions of the actuator were obtained by finite element analysis using COSMOSEMS package. Following the manufacturing and fabrication of the actuator prototype, the steady-state performance of the MR actuator was measured using a force gauge. The experimental results show that the actuator exhibit hysteresis behavior. A sub-hysteresis model was then proposed and the model parameters were identified using the experimental results. The comparison between experimental results and model-prediction value demonstrated that the model could accurately predict the steady-state performances of the MR actuator.

1 Introduction

MR devices have also found commercial application as exercise equipments [1]. The Lord Corporation developed rotary brakes (model: MRB-2107-3), which adopts a rotor that spins through a chamber filled with MR fluid. The magnetic field generated by the current flowing in the coil converts the fluid inside the chamber into a semisolid, thus increasing the shear stress and torque required to turn the rotor. The brake promises high controllability, fast response time (<10 m s), andwide torque range (0.3-5.6 Nm) with very low power requirement. This project aims to investigate the application of MR brakes in virtual realities. To fulfill various virtual reality applications where large forces are needed, the developed MR actuator must provide a larger range of resistant torque which is at least twice of that of MRB-2107-3. To this end, an optimization algorithm [2] was employed to design the MR actuator. The optimization analysis was carried out using finite element analysis. Based on the obtained optimal geometrical dimensions, the MR actuator was fabricated and its static performance was tested subsequently. The developed MR actuator exhibited hysteresis behaviros, where the resistant torque was not one-one matching with the coil current. A parametric subhysteresis model was proposed to describe the actuator performance.

2 MR Actuator Design

2.1 Working principle

The schematic of the MR actuator is shown in Figure 1. The main components of the MR actuator, as shown in Fig. 1, encompass a compact shear disk MR actuator and a low resistance linear potentiometer, which is attached concentrically and can be rotated synchronously with the shaft of the actuator. Obviously, the actuator operates in a direct-shear mode, shearing the MR fluid contained in the space between the two surfaces, the disk surface and the plate, that move with respect to one another.

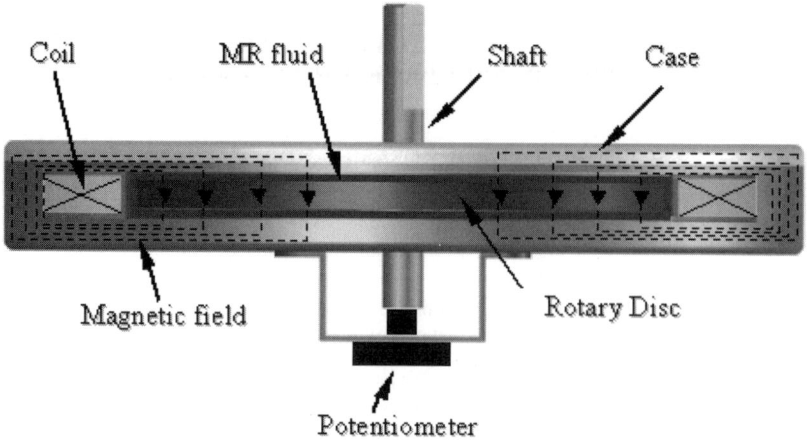

Figure 1. Schematic of the MR actuator.

Based on the Bingham plastic model, the resistance torque, T, is given by [3]

$$T = \frac{4\pi}{3}\tau_y\left(R_o^3 - R_i^3\right) + \frac{\pi\eta\omega}{h}\left(R_o^4 - R_i^4\right) \tag{1}$$

where τ_y is the dynamic yield stress, R_o and R_i are inner and outer radius, η is the fluid viscosity, ω is angular velocity of the rotary disc, h is the gap. As can be seen from equation (1), the resistance torque the resistance torque consists of two terms: one is due to MR effect, and the other is due to viscous flow. The MR actuator is designed for virtual reality application, where the viscous contribution is very small compared with the MR effect. So the mechanical design is only based on the first term.

2.2. Optimization Design

The MR fluid used in the actuator is MRF-132AD, supplied by the Lord Corporation. The maximum yield stress in the linear working area is τ_y = 30 kPa and the corresponding magnetic field is H = 100 kAmp/m. The shaft diameter is chosen as 8 mm or R_i = 4mm. The resistant torque is chosen as T=15 Nm. Substitute these parameters into equation (1) and let $\omega = 0$, the outer radius is determined as R_o = 50mm.

At the design stage, except for the inner and outer radius, many parameters including fluid gap, disc thickness, casing thickness, as well as the coil turns, should be considered. To achieve an efficient MR actuator, the flux density in the fluid gap should be maintained constant. The fluid gap range is recommended from 0.25 to 2 mm for ease of manufacture and assembly [4,5]. In this study, the gap is set to 1.0 mm. The other dimensions will be determined by using finite analysis with the help of COSMOSEMS package. This software is powerful in simulating 3D magnetic field for low frequency electromagnetic or electromechanical applications.

The flux density distributions in the gap along the plate radius direction at various coil currents are shown in Figure 2. Obviously, the distribution of the magnetic flux density is uniform except at the end, which is close to the coil. The discrepancy near the plate edge is probably due to the edge effect of the magnetic field and higher magnetic induction closer to the coil. Furthermore, at the coil current of 3 A, the flux densities at the MR fluid gap and the MILD steel plate are 0.6 Tesla and 1.4 Tesla, respectively. These two flux density values both correspond to operating points of MR fluids and MILD steels. Therefore, both the MR fluids and the MILD reach their saturation points at the same coil currents 3 Amps. The dimensions of the MR actuator are optimized and listed in Table 1.

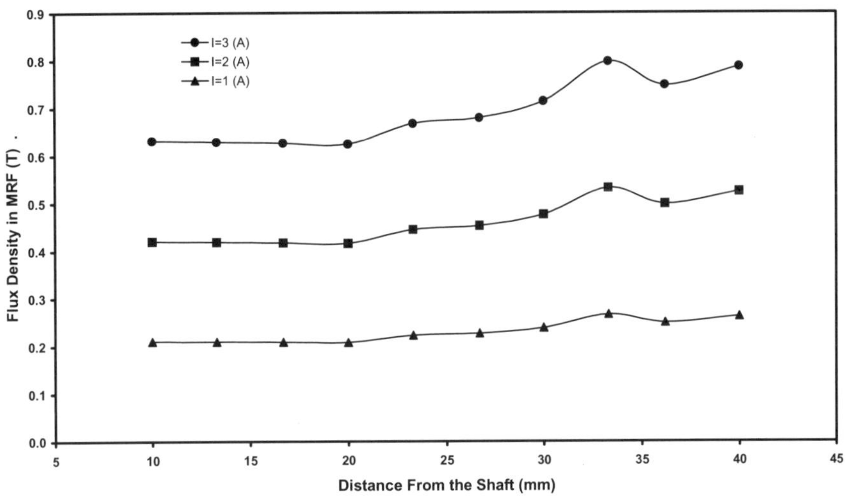

Figure 2. Magnetic flux density distribution.

Table 1. Dimensions of the MR actuator.

Items	**Dimensions**
Outer radius of the disc	50 mm
MR actuator radius	78 mm
Thickness of the case	10 mm
Thickness of the disc	5 mm
MRF gap	1 mm
Thickness of the plate	7 mm
Coil turns	95

3 Testing and Modeling of the MR Actuator

3.1 Steady-State Performance Testing

The resistant torque of the MR actuator under both clockwise and anticlockwise rotary conditions was characterized by using an IMADA digital torque gauge (model: HTG2). Figure 3 shows testing results as well as testing conditions. For each rotary direction, a series of torque measurements were conducted by varying the coil currents from 0 Amp to a certain value, such as 1.0 A or 2.0 A, and then back to 0 Amp. The current increasing or decreasing step is 0.1 A. the results of the clockwise and counter-clockwise tests are drawn side by side. The negative signs on the left part represent the opposite direction of the rotation instead of different electric current. As can be seen from this figure, the plots were symmetrical, which means that steady-sate performances of the actuator were not influenced by the direction of the shaft rotation. Therefore, only counter-clockwise experimental results were used for modeling approaches.

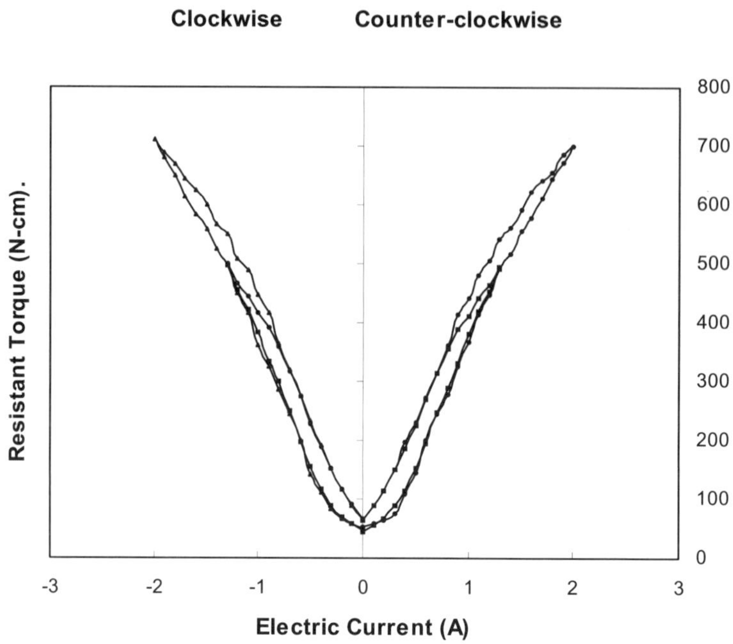

Figure 3. Experimental results under both clockwise and counter-clockwise test.

3.2 Subhysteresis Model

As shown in Figure 3, the resistance torques are not one-one matching with currents. There exists a hysteresis behavior. Generally, when a ferromagnetic material is magnetized in one direction, it will not relax back to zero magnetization when the imposed magnetizing field is removed. This phenomenon is called hysteresis. To further study the hysteresis behavior, a lot more tests were carried out under varied electric current conditions, including 0-0.5-0A, 0-1-0A, 0-3-0A, and 0-2-1-2A. As shown in Figure 4, all the testing hysteresis loops are limited in a largest loop boundary, which corresponds to the case of 0-3-0A. For convenience, the largest boundary hysteresis is defined as the main hysteresis loop. The main hysteresis loop is approximately in parallelogram shape, where the upper and bottom sides are called current shift. Further examinations of experimental results for other cases have found that all other hysteresis loops are enveloped in the main hysteresis loop. Therefore, the MR actuator behaves as subhysteretic behaviors.

To model such behaviors, a subhysteretic model is proposed based on the following three assumptions: (a) all the subhysteresis are limited in the main hysteresis domain; (b) all the electric current shifts of the subhysteresises are constant and equal to the electric shift of the main hysteresis; and (c) the top slopes of the subhysteresises are constant in the linear area. The schematic of the subhysteretic model is shown in Figure 5. In this figure, the red arrows are the paths of the resistant torque when electric current increases. The blue arrows indicate the paths of the resistant torque when electric current decreases. These paths and the superimposed red and blue paths form the subhysteresis model combining with the pure red and blue paths. In this model, T_0 is the resistant torque exhibited in the absence of electric current; I_s is the constant difference of electric current; S_1 and S_2 are the two constant slopes of the model; I_t and I_b are the two vertexes and their values vary when the torque moves along the red or blue arrow. However, the difference between I_t and I_b is constant, i.e., I_b - I_t = constant.

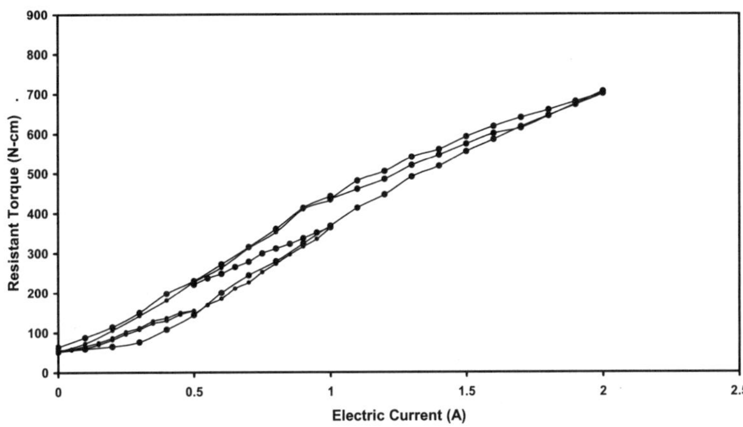

Figure 4. Subhysteresis testing of the MR actuator.

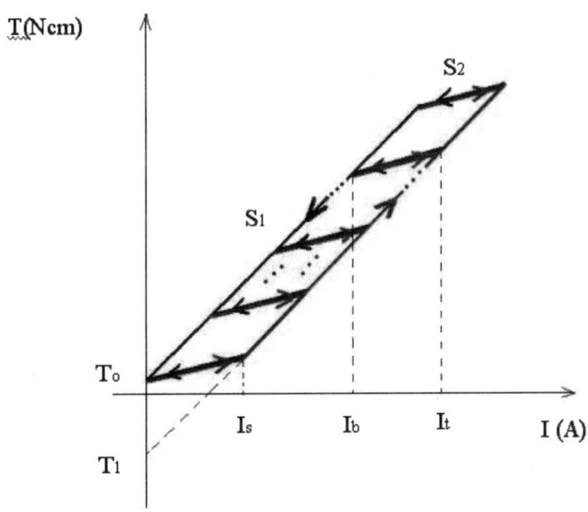

Figure 5. Subhysteresis model.

The resistant torque can be calculated according to the input electric current based on this subhysteresis model. The proposed subhysteresis model contains three modules: (a) the initialization module; (b) the torque track location module; and (c) the torque determination module. The program flowcharts for these three modules are shown in Figure 6.

Module 1: Initialization	$I_0 \le I_s$	$I_b=0;\ I_t=I_s;$ $T = S_2\ I_0+T_0$
	$I_0>I_s$	$I_b=I_0-I_s;\ I_t=I_0;$ $T = S_1\ I_0+T_1$

Module 2: Torque track location	$I_t-I \ge I_s$	Dec=1; Inc=0
	$0<I_b-I<I_s$ or $0<I-I_t<I_s$	Dec=0; Inc=0
	$I-I_b \ge I_s$	Dec=0; Inc=1

Module 3: Resistive torque calculation	$\dot{I}>0$	Inc=1	$I_t=I\ ;I_b=I-I_s;T = S_1\ I+T_1$
		Inc=0	$T = S_1\ I_b+T_0+S_2(I-I_b)$
	$\dot{I}<0$	Dec=1	$I_b=I\ ;I_t=I+I_s;T = S_1\ I+T_0$
		Dec=0	$T = S_1\ I_t+T_1+S_2(I-I_t)$
	$\dot{I}=0$		$T = T$

Figure 6. Program flowchart of the three modules.

The first module, the initialization module, is to find the initial resistance torque of the actuator for any arbitrary given initial coil current I_0. The relationships between the initial resistance torque versus the initial current are summarized in the initialization module, where the value of I_s is used as a critical value to change the slope of resistive torque.

The second module, the torque track location, is used to determine the torque track position for the given current I. As shown in Figure 5, there are three torque regions: torque increasing region (the purely blue arrows region), torque decreasing region (the purely red arrows region), and the subhestersis region (mixture of blue and red arrow regions). To distinguish or model three regions, two Boolean variables, Dec and Inc, are introduced to represent the following three cases:

- Case 1: Dec = 1 AND Inc = 0: the torque track in the purely blue line
- Case 2:Dec = 0 AND Inc = 1: the torque track in the purely red line
- Case 3: Dec = 0 AND Inc = 0: the torque track in red/blue mixture line

The last module is to determine resistive torques based on the current gradient and module 2 analyses. For both Case 1 and Case 2, resistive torques could easily be determined. For case 3, resistive torques depends on the current trend. Therefore, module 3 discussed three different cases in terms of the current gradient signal, i.e., $\dot{I}$ = dI/dt >0, =0, <0.

3.3 Comparison between Model-Prediction Results and Experimental Results

As discussed in Section 3.2 and shown in Figure 6, the model consists of four parameters, namely I_s, S_1, S_2, and T_0. By using experimental results, these parameters were identified as I_s=0.6, S_1=387.5, S_2=233.3 and T_0=45. These parameters were then used to re-construct model predictions. The comparison between experimental results and model-prediction values is shown in Figure 7. As can be seen from this figure, the proposed subhysteresis model can accurately predict steady-state performances of the MR actuator behaviors.

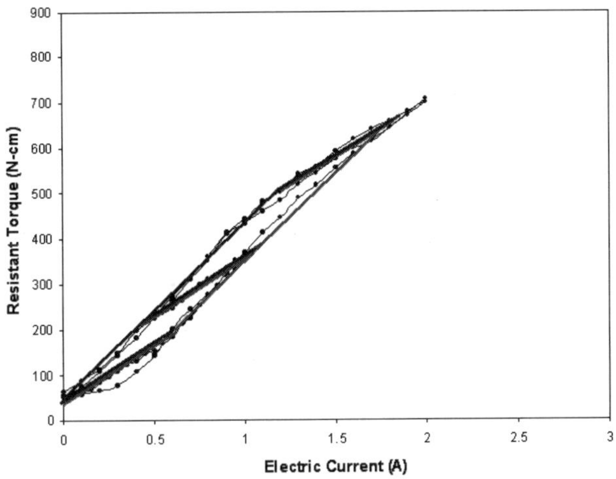

Figure 7. Program flowchart of the three modules

4 Conclusion

An effective and compact position-feedback MR actuator based on the disc-shaped MR model was developed in this project. In the design stage, finite element analysis was adopted to optimize the behavior of the actuator in terms of magnetic field distribution and materials selection. The steady state performances of the MR actuator was measured by using a torque meter. The experimental results indicated that MR actuator exhibited hysterestic behaviors. A subhysteresis model was proposed to accurately predict the steady-sate performances of the MR actuator.

References

1. Carlson J. D. and Jolly M. R., MR fluid, foam and elastomer devices. *Mechatronics* **6** (2000) pp.555–569.
2. Li W. H., Du H. and Guo N. Q., Finite Element analysis and simulation evaluation of a MR valve. *Int. J. Adv. Manuf. Tech.* **21** (2003) pp.438-445.
3. Li W. H. and Du H., Design and experimental evaluation of a magnetorheological brake. *Int. J. Adv. Manuf. Tech.* **13** (2002) pp.679-685
4. Lindler J. and Wereley N. M., Quasi-steady Bingham plastic analysis of an electrorheological flow mode bypass damper with piston bleed. *Smart Mater. Struct.* **21** (2003) pp.508-515.
5. Yoo J. H. and Wereley N. M., Design of a high-efficiency MR valve. *J. Intell. Mater. Syst. Struct.* **13** (2002) pp. 679–685.

INFLUENCE OF PARTICLE SHAPE ON THE PROPERTIES OF MAGNETORHEOLOGICAL FLUIDS

R.C. BELL, E.D. MILLER, J.O. KARLI, A.N. VAVRECK, D.T. ZIMMERMAN
The Pennsylvania State University, Altoona College, Altoona, PA 16601, USA
E-mail: rcb155@psu.edu

We present preliminary studies of the rheological properties and dispersion stability of MR fluids as a function of particle shape by comparing fluids made with uniform nickel spheres to those employing nickel microrods suspended in silicon oil. The rods were fabricated using template-based electrodeposition having diameters in the range 300 ± 30 nm and lengths in the range of 5 – 25 μm. The properties of these rods were compared to commercial nickel carbonyl spheres (1–10 μm dia.). Qualitatively, the off-state (field off) viscosity of fluids containing only rods was found to be substantially greater than those that contain only spherical particles. Rheological measurements of the on-state viscosities were conducted using a custom rheometer equipped with an electromagnet capable of magnetic fields up to 0.6 T. Placed in this field, we observed yield stresses of 1.88 ± 0.23 and 1.86 ± 0.26 kPa, respectively for 7.6% volume fraction of pure nickel spheres and rods, respectively. Due to limitations of the testing apparatus, only the on-state properties of these fluids were investigated. For this reason, the yield stress values reported do not account for the difference in off-state viscosity. While the fluids containing spherical nickel particles tended to settle rather quickly (< 20 minutes), those containing only rods remained suspended even after 48 hours.

1 Introduction

Conventional magnetorheological (MR) fluids are composed of roughly spherical, micron-sized ferro- or ferri-magnetic particles that are suspended in aqueous or organic carrier fluids [1,2]. By application and removal of an externally applied magnetic field, this smart material has the unique ability to rapidly (~few ms) change its effective viscosity in a nearly reversible manner. These materials operate on the principle that when subjected to a magnetic field of sufficient strength (on-state), the particles acquire a magnetic polarization and attract one another, forming chain-like structures that in turn join to form columnar structures along the field. In the off-state, conventional MR suspensions are viscous fluids (~0.1 – 1.0 Pa s^{-1}). However, with the application of an external magnetic field, they convert to a semi-solid with a substantial change in the viscosity (fractional change ~10^5 – 10^6) and concomitant large yield stress (~20 –100 kPa) [1]. There are a large variety of applications ideally suited to exploit this material's uniquely large change in physical properties in which the field-dependent yield stress (and apparent viscosity) is controllable over a continuous range. For example, MR fluids are currently used in commercial devices such as dampers, clutches, and optical polishing instruments [3-5].

Conventional MR fluids have had some difficulty finding widespread commercial use, owing to relatively high manufacturing costs, as well as a limitation on the number of particulates suspended in the fluid in order to maintain its low off-state viscosity (it is desirable to have a large number of particles to enhance the yield stress). In addition, the settling [6] and wear of the particulates [7] tend to reduce the efficacy over time and eventually to device failure. The shear strength depends on several factors including particle size, composition, the magnetic moment, magnetization saturation of the suspended particles, and the strength of the applied magnetic field.

A disadvantage of conventional MR fluids is that the particles suspended in the fluids tend to settle in the absence of constant mixing [6]. In time, they settle into a hard "cake" making re-dispersion of the particles difficult. One of the contributing factors is that the remanence of the particles causes them to stick together. It has been shown that the size of the particles can greatly influence the shear strength of MR fluids; the shear strength of the fluids decreases as particle size is decreased [8,9]. However, the larger the particle, the quicker the particles tend to settle; above 10 μm, these particles tend to settle rather quickly. The sedimentation can be avoided by using particles smaller than approximately 0.1 μm for which Brownian motion keeps them dispersed in the carrier fluid. However, this same phenomenon also hinders the formation of the chain-like structures in the energized fluids, thus reducing the yield stress of the material. Several methods to prevent settling and enhance dispersion have been explored, such as thixotropic additives [10], electrostatic additives, the addition of non-spherical particles, [11,12] or the addition of smaller sized particles [13,14] to the suspensions. MR fluids composed of a combination of different sized particles have exhibited slower settling and facilitate re-dispersion of the particles [14]. However, the addition of these particles has been shown to either increase or decrease the shear strength depending on the composition of particle sizes [13-16].

Control of the magnetic anisotropy of these materials may be useful in such applications as chemical sensors and thermal switches [17,18]. The way in which the particles pack together is believed to play an essential role in the magnetic anisotropy of these materials [12,19]. Here, the relevant contributing factors include the crystalline magnetic anisotropy and the shape anisotropy of the particles in the field-oriented composites [11]. It has been shown that particle shape plays an important role in the properties of electrorheological (ER) fluids. An increase in dynamic modulus and yield stress [20,21] has been shown using ellipsoidal particles. For these fluids, the dynamic modulus was found to increase almost linearly with the length-to-diameter ratio of the particles [17]. These studies also showed that mixtures of elongated and spherical particles displayed an enhanced yield stress compared to those fluids that simply contained spherical particles [18].

In this preliminary investigation, we are examining the rheological properties and settling characteristics of fluids containing micron-sized rods with diameters of 300 ± 30 nm with an average length of 5 – 25 μm suspended in silicon oil. We studied the settling and re-dispersibility behavior of these rod-shaped particles suspended in MR fluid oils. Using a custom rheometer, the rheological properties of these rods were compared to commercial nickel carbonyl spheres (range of 1-10 μm dia.) in the same fluids.

2 Methods

The commercially available nickel spheres (Novamet) used in this study are synthesized by the thermal decomposition of nickel carbonyl. The micron-sized nickel rods (nearly monodisperse) were grown using template-based, direct-current electrodeposition. We used commercially available anodized alumina membranes (Whatman Anodiscs) as our templates. The bottom sides of the templates were coated with a Ga/In eutectic, which provides an electrical contact between the bottom of the channels and the electrolyte solution. The membranes were mounted in cells in which electrodes constructed of 99.5 % nickel foil (Alfa Aesar) were suspended approximately 2 cm above the templates. Each cell was filled with an electrolyte solution containing 1.1 M $NiSO_4$, 0.19 M $NiCl_2$, and 0.6 M H_3BO_3, with the pH of the solution adjusted to 3.5 using concentrated H_2SO_4. The diameter of the rods was fixed by the diameter of the channels within the membrane and the length of the rods was controlled by the current and deposition time. In these studies, 50 mA of current at a potential of approximately 2 VDC was used. The rods were recovered by dissolving the template in a solution of 1.0 *M* sodium hydroxide at room temperature while in an ultrasonic bath.

Characterization of the nickel spheres and microrods was performed on a Hitachi S570 scanning electron microscope (SEM) with a LaB_6 filament at an acceleration voltage of 20 kV. The microscope is equipped with an Orion PCI digital imaging system. The SEM provides critical information about diameter and length of the rods, as well as the size distribution of the particles. The microrods were analyzed on a Scintag X2 automated diffractometer (XRD) equipped with a Peltier solid-state detector to examine the crystal structure and approximate grain size of the crystallites formed within the rods. Using Scherrer's formula, the approximate crystallite size for un-annealed nickel wires was determined to be 23.8 nm The purity of the rods (greater than 99.4 % nickel) was confirmed using an induction-coupled plasma mass spectrometer (ICP).

MR fluids tested in this study were composed of 7.6 vol. % nickel wires suspended in silicon oil (polydimethylsiloxane, Alfa Aesar). Fluids containing 7.6, 17.5, and 40.0 % volume fraction nickel spheres were also tested for comparison. A custom built rheometer was constructed to test the yield stress of the energized fluids. This device consists of a C-shaped electromagnet capable of generating fields up to 0.6 T as measured by a Hall probe between the magnet poles. The sample was then placed tightly between the poles. To measure effective viscosity, a thin aluminum plate is submersed 13 mm into the fluid with 1 mm of the MR fluid on either side of the plate. With the magnetic field applied, we use a transducer (range of 0.02 – 45 N) to measure the force required to pull the plate out of the energized fluid. However, due to the poor resolution of this force transducer, the maximum field of 0.6 T was required to measure the yield stress within a reasonable error. Thus, the low sensitivity of this instrument did not allow for the determination of the off-state viscosity.

3 Results

Figure 1 shows an SEM image of 5.8 μm long nickel rods produced from a single template. The rods for these samples were made from approximately 100 separate templates, which were then mixed together.

The magnetic anisotropy of spheres is significantly smaller than that of rods. In work by Qi *et al.* [22], rods composed of nickel coated glass fibers were found to line up parallel to a static, planar magnetic field. To confirm whether pure nickel rods behaved similarly, the particles were suspended in a drop of water, placed in a uniform magnetic field while the water was allowed to evaporate, and then viewed under an SEM. This allowed us to see the frozen-in, energized state of the particles. (The structure of the fluid both in the presence and absence of a magnetic field has been presented in the literature by suspending the particles in heated paraffin or epoxy, then cooling to lock in the position of the particles [23,24].) We processed water dispersions of nickel spheres in the same manner and observed patterns similar to spheres suspended in wax. As seen in the SEM image shown in Fig. 2., we find that the rods and spheres indeed line up nearly parallel with the magnetic field.

Figure 1. Nickel microrods with 300 nm diameters and average lengths of 5.8 μm with no magnetic filed present.

MR fluids were prepared using 7.6 vol. % rods in silicon oil. At this percentage, the fluids have a consistency of hand cream. Addition of more rods quickly results in a "dry" fluid. Due to the limitations of our custom rheometer, the off-state viscosity of these fluids could not be determined quantitatively. Fluids containing the nickel spheres at 7.6 % vol. % changed the viscosity of the silicon oil very little, but the spheres quickly (< 20 min) settled out of the fluids. Addition of 1 vol. % of lecithin had little effect on preventing the settling of spheres from the low volume percentage fluid.

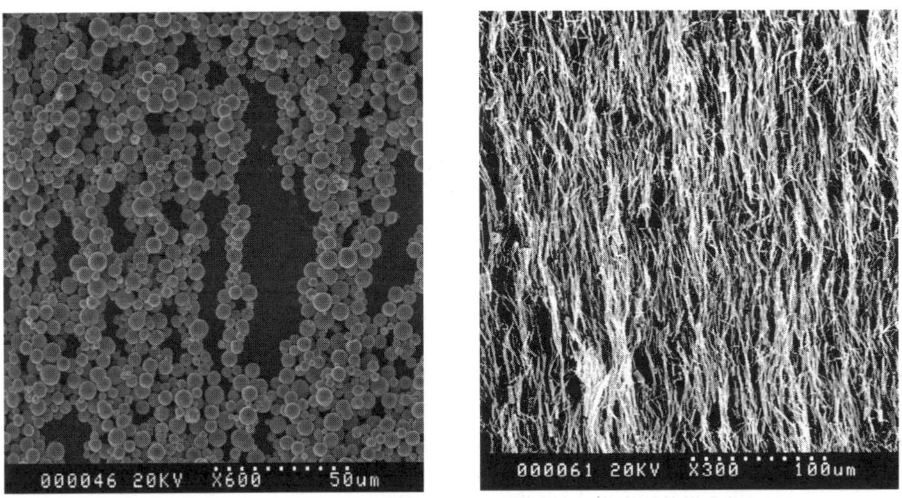

Figure 2. Nickel spheres (left) and microrods (right) in the presence of a planar magnetic field.

As shown in Fig. 3, the rheology tests indicate that both the spheres and rods produced the same approximate yield stress of 1.88 ± 0.23 and 1.86 ± 0.26 kPa, respectively, by application of an external field of 0.6 T. These graphs show the displacement of the aluminum plate versus the force required to pull it out of the fluid. As expected, the curves display a nearly linear reduction in force as the plate is removed from the fluid. The approximate maximum yield stress is determined by dividing the maximum force by the area of the submersed part of the plate. The approximate yield stress was also determined for fluids consisting of 17.5 and 40 vol. % spheres, which resulted in approximate yield stresses of 2.63 kPa and 11.3 kPa, respectively.

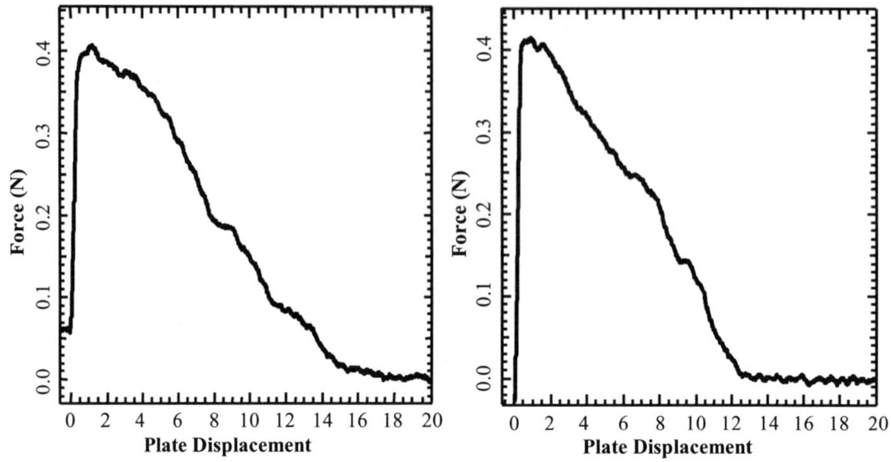

Figure 3. Force versus plate displacement for MR fluids containing 7.6 vol. % nickel microrods (left) and 7.6 vol. % nickel spheres (right) suspended in silicon oil with a field of 0.6 T.

4 Discussion

This preliminary study was intended to explore the feasibility of creating a new type of MR fluid that contains either pure microrods or a combination of spheres and rods to increase the apparent yield stress of the materials (as observed in ER fluids [20,21]). These studies also indicate that the addition of rods to fluids containing spheres will also reduce the sedimentation and enhance the re-dispersion properties of the fluid.

The coercivity, remanence, and magnetic saturation field all depend on the various anisotropies associated with changing diameters and lengths of the microrods [25]. The diameter of the rods is crucial for the desired magnetic properties. The coercivity and remanence of the rods must be as small as possible for the fluid to have a low off-state viscosity. Studies on the magnetic properties of nanowires and microrods have shown a distinct pattern for the decrease in both coercivity and remanence as the diameter of the rods are increased. For example, nickel rods with 30 nm diameters have a coercivity of 378 Oe perpendicular to their long axis [26]. This decreases to 273 Oe and 72 Oe for 50 nm and 200 nm diameters, respectively. If following this same trend, wires with 300 nm diameters should have a coercivity of ~32 Oe. Nielsch *et al.* [27], have shown that 25 nm wires have a remanence of 100 % of the hysteresis that drops to 30 % upon doubling the diameter of the wires. In our case, the use of commercial alumina templates limits us to growing rods with a maximum diameter of 300 nm.

The apparent yield stress of fluids containing either spheres or rods (both at a 7.6 vol. %) are comparable. However, these results were determined with field strengths that resulted in the magnetic saturation of the particles. At these field strengths, it is likely that any effects due to the shape anisotropy would be masked. Once the magnetization of the particle is saturated, increasing the strength of the applied field does not further increase the shear strength of the fluid [28].

The most striking difference between these fluids was found in the off-state viscosity and sedimentation properties. At 7.6 vol. %, the nickel spheres have an off-state viscosity qualitatively similar to that of the pure silicon oil. These particles settle in a relatively short period of time (< 20 minutes). This was only slightly improved with the addition of lecithin. The fluids containing only microrods at this same volume percentage displayed an off-state viscosity similar to hand cream. This resulted in negligible settling over a 48 hour period.

Future studies using a conventional torque rheometer will be carried out on fluids in which both volume percentages and sphere-to-rod ratios are varied to better understand the rheological properties of MR fluids containing rods. Studies performed at various field strengths up to saturation should exemplify the differences in rheological properties that shape anisotropies should have on these materials. The settling characteristics will also be investigated using more quantitative methods.

5 Acknowledgements

We would like to thank Inco Special Products (Wyckoff, New Jersey, USA) for supplying the Novamet spherical nickel powders used in this study. We acknowledge the efforts of undergraduate student Patrick J. Dougherty for his work collecting the rheological measurements. This work was supported by a seed grant from The Pennsylvania State University, Altoona College.

References

1. He J. M. and Huang J., Magnetorheological fluids and their properties, *Int. J. Mod. Phys. B* **19** (2005) pp. 593–596.
2. Rabinow J., The magnetic fluid clutch, *AIEE Trans.* **67** (1948) pp. 1308–1315.
3. Zipser L., Richter L. and Lange U., Magnetorheologic fluids for actuators, *Sens. Actuators A* **92** (2001) pp. 318–325.
4. Ramallo J. C., Johnson E. A. and Spencer Jr. B. F., 'Smart' Base Isolation Systems, *J. Eng. Mech.* **128** (2002) pp. 1088–1099.
5. Harland N. R., Mace B. R. and Jones R. W., Adaptive-passive control of vibration transmission in beams using electro/magnetorheological fluid filled inserts, *IEEE Trans. on Control Syst. Technol.* **9** (2001) pp. 209–220.
6. Chen L. S. and Chen D. Y., Permalloy inductor based instrument that measures the sedimentation constant of magnetorheological fluids, *Rev. Scien. Instru.* **74** (2003) pp. 3566–3568.
7. Carlson J. D., Critical factors for MR fluids in vehicle systems, *Int. J. Vehicle Des.* **33** (2003) pp. 207–217.
8. de Gans B. J., Duin N. J., van den Ende D. and Mellema J., The influence of particle size on the magnetorheological properties of an inverse ferrofluid, *J. Chem. Phys.* **113** (2000) pp. 2032–2042.
9. Genç S and Phulé P. P., Rheological properties of magnetorheological fluids, *Smart Mater. Struct.* **11** (2002) pp. 140–146.
10. Weiss K. D., Nixon D. A., Carlson J. D. and Margida A. J., Thixotropic magnetorheological materials, US Patent No. 5,645,752 (1997).
11. Chin B. D., Park J. H., Kwon M. H. and Park O. O., Rheological properties and dispersion stability of Magnetorheological (MR) suspensions, *Rheol. Acta*, **40** (2001) pp. 211–219.

12. Phulé P. P. and Ginder J. M., Synthesis and properties of novel magnetorheological fluids having improved stability and redispersibility, *Int. J. Mod. Phys. B* **13** (1999) pp. 2019–2027.
13. Chaudhuri A., Wang G., Wereley N. M., Tasovksi V. and Radhakrishnan R., Substitution of micron by nanometer scale powders in magnetorheological fluids, *Int. J. Mod. Phys. B* **19** (2005) pp. 1374–1380.
14. Wereley N. M., Chaudhuri A., Yoo J.–H., John S., Kotha S., Suggs A., Radhakrishnan R., Love B. J. and Sudarshan T. S., Bidisperse magnetorheological fluids using Fe particles at nanometer and micro scale, *J. Intell. Mater. Syst. Struct.* **17** (2006) pp. 393–401.
15. Wu C. W. and Conrad H., Influence of mixed particle size on electrorheological response, *J. Appl. Phys.* **83** (1998) pp. 3880–3884.
16. Lemaire E., Meunier A., Bossis G., Liu J., Felt D. W., Bashtovoi P. and Matoussevitch N., Influence of the particle size on the rheology of magnetorheological fluids, *J. Rheol.* **39** (1995) pp. 1011–1020.
17. Kanu R. C. and Shaw M. T., Enhanced electrorheological fluids using anisotropic particles, *J. Rheol.* **42** (1998) pp. 657–670.
18. Asano K., Suto H. and Yatsuzuka K., Influence of the particle configuration on electrorheological effect, *J. Electrostatics* **40-41** (1997) pp. 573–578.
19. Rankin P. J., Horvath A. T. and Klingenberg D. J., Magnetorheology in viscoplastic media, *Rheol. Acta* **38** (1999) pp. 471–477.
20. Kormann C., Laun H. M. and Richter H. J., MR fluids with nano-sized magnetic particles, *Int. J. Mod. Phys. B* **10** (1996) pp. 3167–3172.
21. Margida A. J., Weiss K. D. and Carlson J. D., Magnetorheological materials based on iron alloy particles, *Int. J. Mod. Phys. B* **10** (1996) pp. 3335–3341.
22. Qi Y., Zhang L. and Wen W., Anisotropy properties of magnetic colloidal materials, *J. Phys. D: Appl. Phys.* **36** (2003) pp. L10–L14.
23. Henley S. and Filisko F. E., Lamellar structures induced by coupled electromagnetic and shear fields: a defining character of ER/MR activity, *Int. J. Mod. Phys. B* **16** (2002) pp. 2286–2292.
24. Martin J. E., Venturini E., Odinek J. and Anderson R. A., Anisotropic magnetism in field-structured composites, *Phys. Rev. E* **61** (2000) pp. 2818–2830.
25. Ross C. A., Hwang M., Shima M., Smith H. I., Farhoud M., Savas T. A., Schwarzacher W., Parrochon J., Escoffier W., Bertram H. N., Humphrey F. B. and Redjdal M., Magnetic properties of arrays of electrodeposited nanorods, *J. Magn. Magn. Mater.* **249** (2002) pp. 200–207.
26. Lin S. W., Chang S. C., Liu R. S., Hu S. F. and Jan N. T., Fabrication and magnetic properties of nickel nanowires, *J. Magn. Magn. Mater.* **282** (2004) pp. 28–31.
27. Nielsch K., Wehrspohn R. B., Fischer S. F., Kronmuller H., Barthel J., Kirschner J. and Gösele U., Magnetic properties of 100 nm-period nickel nanorod arrays obtained from ordered porous-alumina templates, *Mat. Res. Soc. Symp. Proc.* **636** (2001) pp. D1.9.1–6.
28. Jones T. B. and Saha B., Nonlinear interactions of particles in chains, *J. Appl. Phys.* **68** (1990) pp. 404–410.

EXPERIMENTAL VALIDATION OF A SIGNUM FUNCTION BASED DAMPER CONTROLLER FOR MR FLUID DAMPERS

D. H. WANG AND G. YUAN

Key Laboratory of Optoelectronic Technology and Systems of the Ministry of Education of China, Chongqing University, Chongqing, China 400044
E-mail: dhwang@cqu.edu.cn

W. H. LIAO

Smart Materials and Structures Laboratory, Department of Automation and Computer-Aided Engineering, The Chinese University of Hong Kong, Hong Kong, China
E-mail: whliao@cuhk.edu.hk

Recently, Wang and Liao (2005) proposed a new damper controller for MR fluid dampers based on signum function and the control algorithm and simulation results are analyzed. In this paper, the experimental setup to validate the signum function based damper controller is established and the testing results are presented. The experimental results show that the signum function based damper controller for MR dampers can let the controlled damping force track the desired damping force very well and the effectiveness of the damper controller based on the signum function is verified.

1 Introduction

MR fluid dampers, which utilize the advantages of MR fluids, are semi-active control devices that are capable of generating the magnitude of force sufficient for large-scale applications, while requiring only a battery for power [1,2]. Already, a 20 ton MR fluid damper prototype has been developed and tested in the laboratory [3] and the applications of MR fluid dampers can be found from civil engineering (such as buildings and bridges [2,3,4]) to automobiles [5], railway vehicles [6,7,8], and aerospace applications [9,10].

The schematic of a semi-active control system based on MR fluid dampers is illustrated in Figure 1 [11]. The MR damper based semi-active control system consists of a system controller and a damper controller. The system controller generates the desired damping force according to the dynamic responses of the plant while the damper controller adjusts the command voltage to the power driver to track the desired damping force. The damping force of the smart fluid damper should be monitored and/or predicted and then fed to the damper controller to generate the command voltage or current according to the desired damping force generated by the system controller.

Referring to Figure 1, damper controllers for MR fluid dampers affect the effectiveness of the overall system. Several kinds of damper controllers have been proposed as follows: Heaviside function method [2], continuous-state controller [12], direct inverse function method [13,14], neural network based damper controller [15], and signum function based damper controller [11].

In this paper, the experiments on the damper controller based on the signum function are carried out to validate the expression and effectiveness of the developed signum function based damper controller.

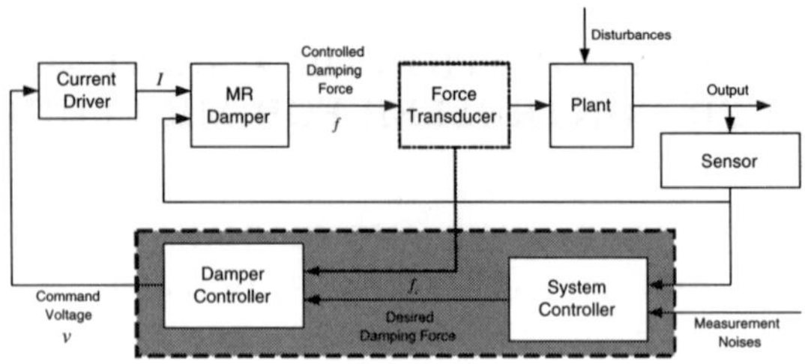

Figure 1. Semi-active control system for a plant integrated with an MR fluid damper.

2 Signum function based damper controller

According to Wang and Liao [11], the following issues are of great importance when designing the damper controllers for MR fluid dampers: (1) tracking ability of damping force, (2) energy requirement for MR fluid dampers, and (3) high frequency characteristics. In order to deal with the above problems, a damper controller for MR fluid dampers based on signum function is proposed as follows [11]

$$v_{Sign} = \frac{V_{max}}{2N}\left\{\sum_{0\le j\le N-1}\left\{\mathrm{sgn}\left\{\left[f_c-(1+kj)f\right]f\right\}+1\right\}\right\}\times$$
$$\left\{1-\left[\frac{\mathrm{sgn}\left(v_{Sign1}-V_{max}\right)+1}{2}\right]\bigcap\left[\frac{1-\mathrm{sgn}(f\dot{f})}{2}\right]\right\} \tag{1}$$

where sgn(·) is the signum function; $\bigcap$ is the logical AND; f_c and f are the desired and controlled (measured) damping forces, respectively; V_{max} is the maximum voltage to the current driver associated with the saturation of the magnetic field in the MR fluid damper; N is an integer with $0 \le j \le N-1$; k is a small constant; $\dot{f}$ is the time derivative of f; and v_{Sign1} is given by

$$v_{Sign1} = \frac{V_{max}}{2N}\left\{\sum_{0\le j\le N-1}\left\{\mathrm{sgn}\{\left[f_c-(1+kj)f\right]f\}+1\right\}\right\} \tag{2}$$

Let

$$v_{Sign2} = 1-\left[\frac{\mathrm{sgn}(v_{sign1}-V_{max})+1}{2}\right]\bigcap\left[\frac{1-\mathrm{sgn}(f\dot{f})}{2}\right] \tag{3}$$

then according to Eqs. (2) and (3), Eq. (1) can be rewritten as

$$v_{Sign} = v_{Sign1}\times v_{Sign2} \tag{4}$$

It should be noted that v_{sign1} alone can be used as a damper controller, but v_{sign2} itself can not do that (v_{sign2} is used to adjust v_{sign1} here).

According to Wang and Liao, the feasible ranges for k and N are given by [11]

$$-10^{-1} \le k \le -10^{-4};\ N \ge 5 \tag{5}$$

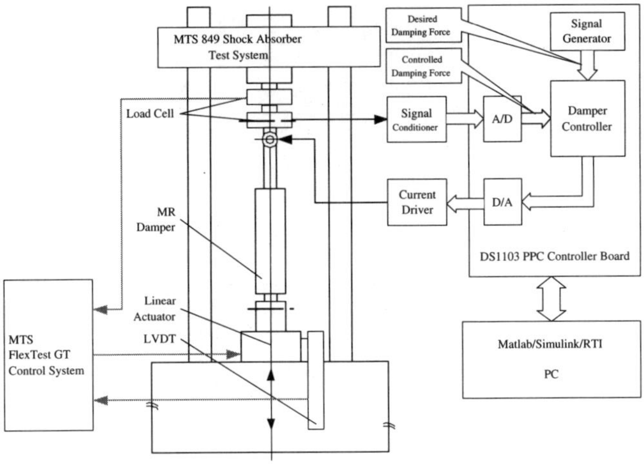

Figure 2. Schematic of the experimental setup.

3 Experimental validation

In this section, the validating experiments on the damper controller based on the signum function are presented.

3.1 Experimental setup

In order to validate the signum function based damper controller for MR fluid dampers, the experiment is set up as shown in Figure 2. The testing setup consists of two parts: (1) the damper controller based on the rapid control prototype (RCP) technique and (2) the measuring system for the force vs displacement/velocity of the MR fluid damper with/without control by the damper controller. The photo of the experimental setup is shown in Figure 3. In the experiment, the MR damper (Type: RD-1005) manufactured by Lord Corporation is tested.

The rapid control prototype of the damper controller for the MR fluid damper consists of the DS1103PPC controller board from dSPACE GmbH, the controllable current driver for the MR fluid damper [16], the 849 Shock Absorber Test System from the MTS, the load cell B (Type: MTS 661.19F-01, range: 0~5.00kN) and its signal conditioning system (DC strain amplifier 6M72, San.ei). It is worthy to note that the load cell B, which is installed in-series with the MR damper and the load cell A (used by the 849 Shock Absorber Test System, type: MTS 661.19F-01, range: 0~5.00kN) as shown in Figure 3, and its signal conditioning system are extra provided to monitor and feedback the damping force to the damper controller. The signum function based damper controller for the MR fluid damper is realized with Matlab/Simulink, then compiled and downloaded into the DS1103PPC controller board, which composes the rapid prototype of the signum function based damper controller. The Simulink block diagram of the signum function based damper controller is shown in Figure 4. Observing Figure 4, the sinusoidal signal output from

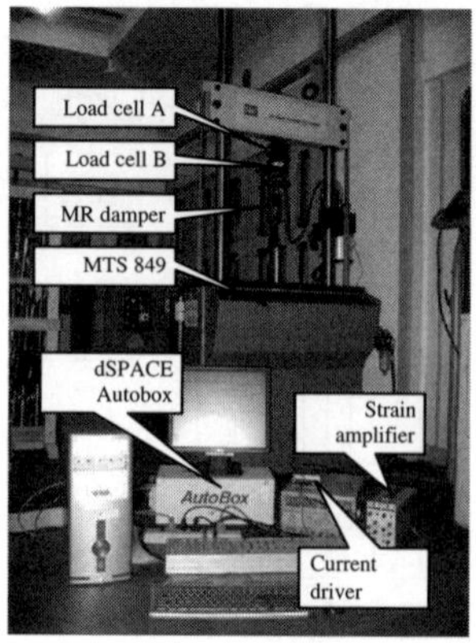

Figure 3. Photo of the experimental setup.

the unit "Sine Wave", which is realized by the DS1103 controller board, produces the desired damping force f_c. The controlled damping force f is digitized and input from the A/D port of the DS1103 controller board to the host computer. Based on the desired and controlled damping forces, the command voltage can be determined through the signum function based damper controller with the MATLAB/SIMULINK/RTI and input to the current driver for the MR fluid damper through the D/A port of DS1103. The damper controller can control the damping force of the MR fluid damper through adjusting the command voltage to the current driver.

The measuring system for the force vs displacement/velocity of the MR fluid damper with/without control by the damper controller relies on the 849 Shock Absorber Test System, which also provides the displacement excitation for the tested MR fluid damper. The damping force of the tested MR damper is monitored by the load cell A and its signal conditioning system provided by the MTS. The corresponding displacement and velocity of the piston of the tested MR fluid damper are monitored by the LVDT provided by the MTS. The tested MR fluid damper is installed on the 849 Shock Absorber Test System, as shown in Figure 3, and the rod of the damper moves relatively to the cylinder under the actuation of the hydraulic actuator of the 849 Shock Absorber Test System.

3.2 Open loop testing of the MR damper

Before testing the signum function based damper controller, the open loop test of the MR damper is carried out. When the MR damper is excited by a sinusoidal displacement with an amplitude of ±15.00mm and a frequency of 1.00Hz and driven by the current 0.00A, 0.50A, 1.00A, 1.50A, and 2.00A, the tested damping forces versus the displacement and velocity are shown in Figures 5(a) and 5(b), respectively. For each case with a driving current, three tests are carried out to verify the repeatability. The tested results are plotted in Figure 5, which indicates the repeatability of the measured damping forces. Observing Figure 5, the maximum

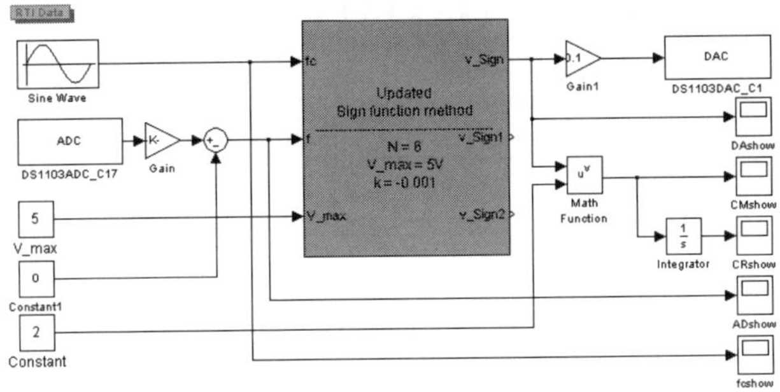

Figure 4. Simulink block diagram of the signum function based damper controller for MR dampers.

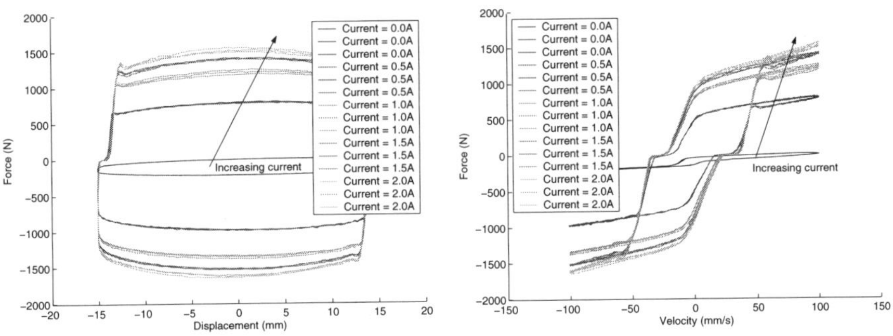

(a) (b)

Figure 5. Testing results of the MR damper: (a) force vs displacement and (b) force vs velocity.

damping force of the MR damper is less than 200.00N when the driving current is 0.00A and becomes larger than 1500.00N when the driving current is 2.00A. The damping force is increased when the driving current increases. Besides, the damping force is also increased with increasing the excitation velocity given a fixed driving current. The force-velocity diagram also exhibits highly nonlinear hysteresis.

3.3 Closed loop testing of the MR damper based on signum function controller

The time history of the controlled damping force under the signum function based damper controller accompanying with the desired damping force is shown in Figure 6(a) and the corresponding command voltage is shown in Figure 6(b). The MR damper is not controlled from 0.00 to 1.25s through turning off the current driver. When the damper is not controlled from 0.00 to 1.25s, the damper controller is still at work and generates the command voltage, which can be seen from Figure 6(b). However, the generated command voltage can not control the current driver because it is turned off from 0.00 to 1.25s. After 1.25s, the current driver is turned on

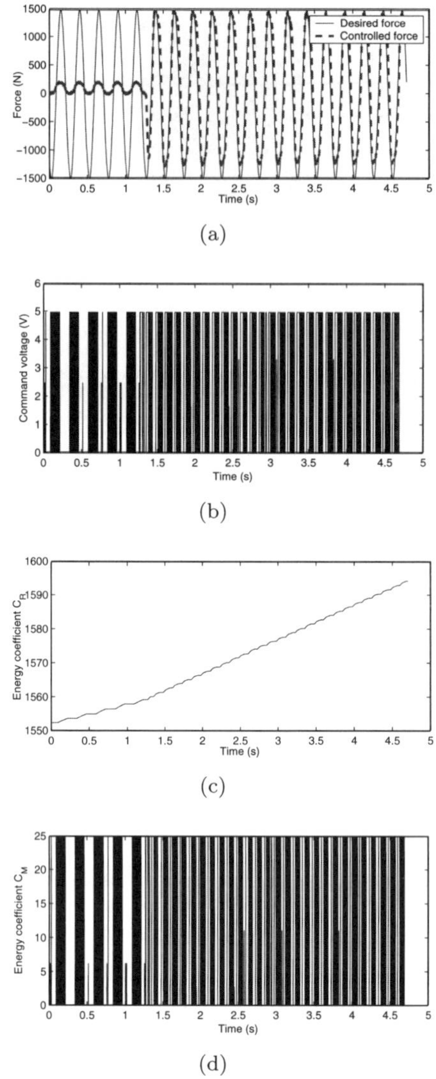

Figure 6. Time history before and after control: (a) controlled and desired damping forces, (b) command voltage, (c) energy coefficient related to the resistance of the electromagnetic circuit, and (d) energy coefficient related to the inductance of the electromagnetic circuit.

and the MR damper is controlled. The controlled damping force tracks the desired damping force very well when the MR damper is controlled.

The energy coefficients [11] ($C_R = \int_{t_0}^{t_1} v^2 dt$ and $C_M = v^2$) related to the resistance and inductance of the electromagnetic circuit are shown in Figures 6(c) and 6(d), respectively. It should be noted that the trends of C_R and C_M are consistent with the simulation results [11].

4 Conclusions

In this paper, the experimental setup to validate the signum function based damper controller is established and the testing results are presented. The experimental results show that the signum function based damper controller can let the controlled damping force track the desired damping force very well and the effectiveness of the damper controller based on the signum function is verified.

Acknowledgments

The authors wish to acknowledge the financial supports by a Foundation for the Author of National Excellent Doctoral Dissertation of PR China (Grant No. 200132) and a Program for New Century Excellent Talents in University (Grant No. NCET-05-0765). The authors also wish to acknowledge the help from the Predetection Center of Zongshen Motorcycle Group R&D.

References

1. J. B. F. Spencer *et al*, Journal of Engineering Mechanics **123**, 230 (1997).
2. S. J. Dyke *et al*, Smart Materials and Structures **5**, 565 (1996).
3. J. B. F. Spencer *et al*, in *Proceedings of the Second World Conference on Structural Control (2WCSC)*, (Proc. WCSC 1, 1998).
4. G. W. Housner *et al*, Journal of Engineering Mechanics **123**, 897 (1997).
5. A. H. F. Lam and W. H. Liao, International Journal of Vehicle Design **33**, 50 (2003).
6. W. H. Liao and D. H. Wang, Journal of Intelligent Material Systems and Structures **14**, 161 (2003).
7. Y. K. Lau and W. H. Liao, Journal of Rail and Rapid Transit, Proceedings of IMechE Part F **219**, 261 (2005).
8. D. H. Wang and W. H. Liao, in *Proceedings of Smart Structures and Materials 2003: Smart Structures and Integrated Systems*, ed. A. M. Baz (Proc. SPIE 5056, 2003).
9. F. Gandhi *et al*, Smart Materials and Structures **10**, 96 (2001).
10. N. M. Wereley *et al*, Journal of Intelligent Material Systems and Structures **10**, 624 (1999).
11. D. H. Wang and W. H. Liao, Journal of Intelligent Material Systems and Structures **16**, 983 (2005).
12. N. D. Sims *et al*, Smart Materials and Structures **8**, 601 (1999).
13. H. S. Lee and S. B. Choi, Journal of Intelligent Material Systems and Structures **11**, 80 (2000).
14. S. B. Choi *et al*, Vehicle System Dynamics **38**, 341 (2002).
15. D. H. Wang and W. H. Liao, Smart Materials and Structures **14**, 111 (2005).
16. G. Yuan *et al*, Journal of Functional Materials **37**, 736 (2006).

THE EFFECT OF POLYPYRROLE LOADING ON THE ELECTRORHEOLOGICAL PROPERTIES OF POLYPYRROLE/SBA-15 SUSPENSIONS

QILIN CHENG, VLADIMIR PAVLINEK, TOMAS BELZA AND ANEZKA LENGALOVA
Polymer Centre, Faculty of Technology, Tomas Bata University in Zlin, 762 72 Zlin, Czech Republic
E-mail: chengql@ecust.edu.cn

YING HE AND CHUNZHONG LI
Key Laboratory for Ultrafine Materials of Ministry of Education, School of Materials Science and Engineering, East China University of Science and Technology, 200237 Shanghai, China
E-mail: czli@ecust.edu.cn

Encapsulated polypyrrole (PPy) with different amount in the channels of mesoporous silica (SBA-15) was synthesized. The XRD and N_2 adsorption/desorption isotherms analysis show that PPy with different loadings in the channels does not affect the ordered hexagonal structure of resultant PPy/SBA-15 nanocomposites. Further, the electrorheological (ER) properties of the nanocomposites with different PPy loadings were investigated by steady and oscillatory shear experiments. The results reveal that PPy/SBA-15 fluids exhibit typical ER and viscoelastic behavior under the applied electric field and ER response depends mainly on PPy loadings and there is an optimum PPy loading for strong ER effect.

1 Introduction

Mesoporous materials have attracted wide attention in the past decade due to their potential applications in catalysis, separation, drug delivery, and so on. These materials exhibit large pores (2-50 nm) with a narrow pore-size distribution and a high surface area. Based on above mentioned characteristics, special attention has been paid to mesoporous materials in recent years as promising application in preparation of ER fluids. For example, Zhao et al. [1,2] reported giant ER activity of high surface area mesoporous cerium-doped TiO_2 templated by block copolymer and developed a new kind of mesoporous rare-earth –doped TiO_2 with strong ER effect. Choi et al.[3,4] found that semiconducting polyaniline confined in mesoporous MCM-41 or SBA-15 can significantly enhance electrorheology of the composite system. Cheng et al. [5,6] synthesized conducting PPy in mesoporous silica MCM-41 and discussed its ER properties. ER behavior of copper phthalocyanine-doped (or urea-doped) mesoporous TiO_2 suspensions was also investigated by Zhu et al. [7,8]. These experimental studies suggested that large interfacial polarization was mainly responsible for the higher ER effect.

It is well-known that ER fluids consist of dielectric particles and insulating liquids. Their rheological properties can be changed rapidly and reversibly upon an external electric field. Therefore, ER fluids have promising applications in engineering fields. However, low yield stress is still the main obstacle hindering their practice application. In an effort to overcome it, synthesis and design of high-performance ER materials has been an active topic so far. The discovery and development of mesoporous materials offer the possibility of using them as host for making some advanced ER materials [1,9]. In our previous study of polypyrrole (PPy) confined in mesoporous SBA-15 [10], we have found that incorporation of PPy in the channels of SBA-15 can enhance the ER effect of the material remarkably. It is of interest whether the ER behavior would be affected by the

PPy loading in the nanocomposite. The investigation of this issue is the aim of this study, as a part of our continuous research work in PPy-based ER materials.

2 Experimental

2.1 *Synthesis of nanocomposites*

SBA-15 materials were synthesized according to the literature, using tri-block copolymer poly (ethylene glycol)-block-poly (propylene glycol)-block-poly (ethylene glycol) (Pluronic P123, Aldrich) as a template in acid conditions [11]. To obtain PPy/SBA-15 nanocomposites, the calcined SBA-15 was dried under vacuum at 250 °C for 6 h to remove air and water from the channels. Then the host was exposed to equilibrium vapors from liquid pyrrole (98%, Aldrich, distilled under reduced pressure before use) in a flask for 24 h at 20 °C, 40 °C and 50 °C, respectively. The SBA-15-containing pyrrole was immersed in a 0.25 M aqueous solution of $FeCl_3 \cdot 6H_2O$ with stirring at room temperature for 24 h. Finally, the PPy/SBA-15 nanocomposites with different PPy loadings (designated as PS1, PS2, PS3) were washed several times with water and acetone, and then dried at room temperature under reduced pressure.

2.2 *Characterization of materials*

Small-angle X-ray diffraction patterns were determined with Rigaku D/MAX 2550V diffractometer using Cu Kα radiation (λ = 1.5406 Å). N_2 adsorption/desorption isotherms at about -195.6 °C were measured using Micromeritics ASAP 2010 system. Thermogravimetric analysis was carried out with a TGA/SDTA 851^e at a heating rate of 10°C/min under a purging atmosphere of air.

2.3 *Electrorheological measurements*

Dried PS1, PS2 and PS3 particles were dispersed in silicone oil (Fluid 200, Dow Corning, UK; viscosity η_c = 108 mPa s, density d_c = 0.965 g cm^{-3}) in the concentration of 10% (w/w). Measurements of rheological properties were carried out under controlled shear-rate mode using a coaxial cylinder viscometer (Bohlin GEMINI, Malvern Instruments, UK). The suspensions were placed in the Couette cell with a rotating inner cylinder of 14 mm diameter and the outer cylinder separated by a 0.7 mm gap. They were connected to a DC power supply producing a field strength E = 0.5 – 3 kV mm^{-1}. Dynamic viscoelastic tests were performed by dynamic strain sweeps and frequency sweeps. The strain sweep was carried out with applied strains of 0.0001 to 1 at a frequency of 10 Hz under an electric field to determine the linear viscoelastic region. The rheological parameters were then obtained from the frequency sweep tests (0.1 to 50 Hz) at fixed strain amplitude in the linear viscoelastic region. All experiments were performed at 25 °C.

3 Results and Discussion

3.1 Characterizatin of PPy/SBA-15 nanocomposites

PPy chains encapsulated in the channels of SBA-15 were confirmed by the different experimental methods in our previous study [10]. The amount of PPy loading in the PPy/SBA-15 nanocomposites (i.e.PS1, PS2 and PS3) was estimated by thermogravimetric analysis to be about 27, 40 and 47 wt.%, respectively. Figure 1 shows the small-angle XRD patterns of PS, PS2 and PS3. The three well-resolved diffraction peaks indexed as 100, 110 and 210 indicate that the hexagonal pore structure is remained after incorporation of PPy in the channels. It is worth noting that the intensity of the peak (100) from PS1, PS2 and PS3 decreases significantly with increasing PPy loadings because of PPy chains in the channels.

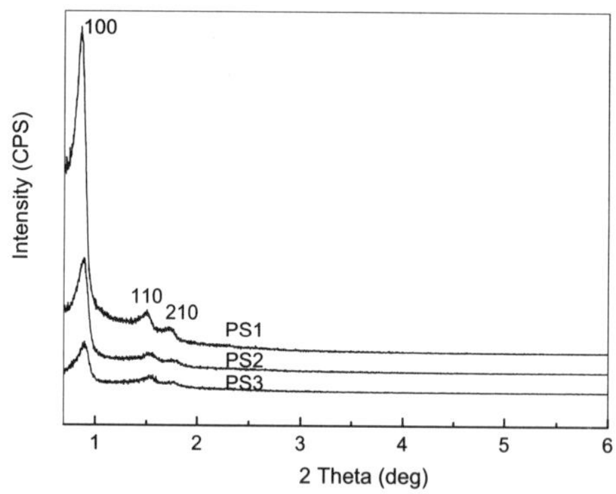

Figure 1. Small-angle XRD patterns of PS1, PS2 and PS3

N_2 adsorption/desorption isotherms for SBA-15, PS1, PS2 and PS3 are depicted in Fig. 2. All the isotherms belong to type IV with H_1 hysteresis loops, according to the IUPAC classification, suggesting the typical characteristic of mesoporous materials. The isotherms for PS1, PS2 and PS3 are similar in the shape to that of pure SBA-15. Clearly, with the amount of PPy loading increased, the N_2 adsorption decreases quickly and the inflection point of the step shifts to lower P/P_0 , which indicates that the inclusion of PPy into SBA-15 leads to a significant change in channel size [12].

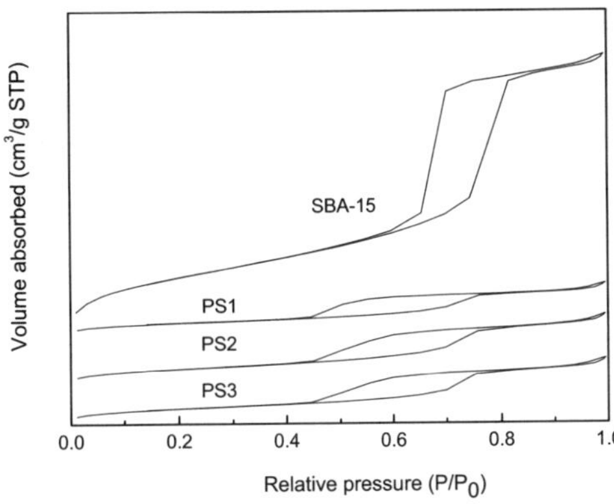

Figure 2. N_2 adsorption-desorption isotherms of SBA-15, PS1, PS2 and PS3

3.2 ER properties of PPy/SBA-15 suspension

As a representative of PPy/SBA-15 nanocomposites, Fig. 3 shows typical flow curves for PS1 suspension under different electric field strengths. In the absence of electric field, the suspension exhibits a slight departure from Newtonian fluid behavior. When an electric field is applied, the shear stress increases abruptly with electric field and the fluid shows a significant yield stress, suggesting that the fluid is solidified strongly by the external electric field. The shear stress exhibits either a plateau region or a slight decrease with shear rate up to a critical shear rate, and afterwards, it starts to increase with shear rate similarly to non-Newtonian fluids [13]. This is mainly ascribed to the competition between hydrodynamic and electrostatic forces.

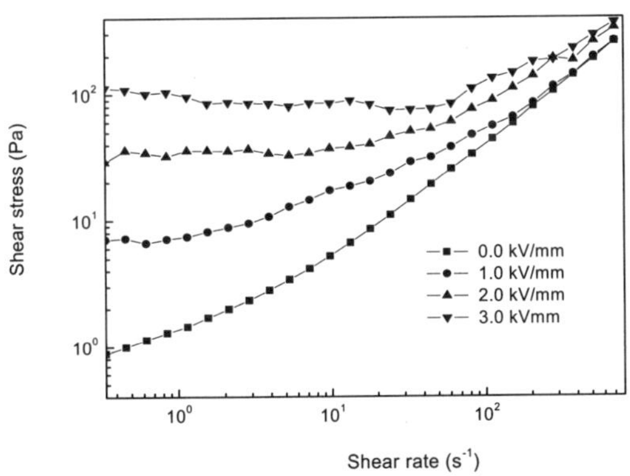

Figure 3. Shear stress as a function of shear rate for 10 wt.% PS1 fluid

To follow the effect of PPy loadings in the channels of SBA-15, the shear stress versus shear rate dependence of PS1, PS2 and PS3 fluids were plotted in one graph at electric field strength of 2.0 kV/mm (Fig. 4). It can be found that the yield stresses of PS1, PS2 and PS3 are approximately 35.8, 142.6 and 106.9 Pa, respectively. Moreover, the PS2 fluid shows highest shear stress over the broad shear rate region, which supports the fact that PS2 suspension displays the strongest ER effect among PPy/SBA-15 based ER fluids.

Based on the above results, we consider that ER effect of PPy/SBA-15 fluids does not increase linearly with the amount of PPy loadings in the channels of SBA-15. There is a critical PPy loading in which the composites show the strongest ER activity. Choi et al. [3] proposed that the combined effect from the polarization within polyaniline/MCM-41 particles and the electrical anisotropy of polyaniline/MCM-41 influences the ER properties of polyaniline/MCM-41 fluid. Hence, encapsulated PPy affects not only the structural properties but also polarization and electrical properties of PPy/SBA-15. On the other hand, according to widely accepted studies of Block, Tao and other researchers [14-16], the large polarization ability (related to high dielectric constant) and the suitable polarization response (related to suitable dielectric loss or conductivity) mainly dominate high ER effect [17]. The different amount of PPy in the channels may change the polarization and conductivity of PPy/SBA-15 particles, which in turn makes PS1, PS2 and PS3 fluids show different ER activity.

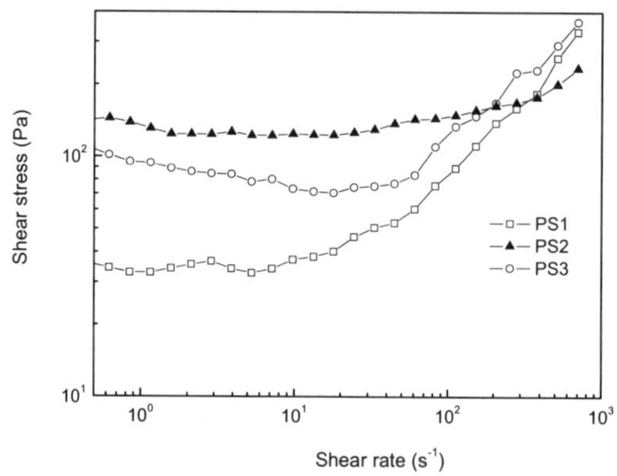

Figure 4. The relation between shear stress and shear rate of PS1, PS2 and PS3 based fluids ($E = 2.0$ kV/mm)

Dynamic oscillatory measurements, which can help us further understand the interactions among PPy/SBA-15 particles forming the columnar structures under an electric field, were also performed to study the viscoelastic properties of the solidified PPy/SBA-15 ER fluids under an applied electric field. Figure 5 represents storage modulus (G') and loss modulus (G'') of PS1, PS2 and PS3 versus strain amplitude at an electric field strength of 2.0kV/mm. In the linear viscoelastic region, for PS1, PS2 and PS3, $G' > G''$, and beyond the linear region, the reverse phenomena can be observed. This can be explained that PPy/SBA-15 fluids are strongly solidified with an applied electric field and the particles form well-developed chains spanning the gap between the electrodes. Meanwhile, it also indicates that these fluids become stiffer showing a larger modulus, but become more fragile and is easily fractured at relatively small strains [18]. In addition, it can be found from Fig. 5 that G' for PS2 suspension is the largest and G' for PS1is the lowest in the linear viscoelastic region, suggesting that PS2 fluid becomes the most elastic than the PS1 and PS3 fluids, which is mainly due to the stiffest particle chain structure of PS2 under an electric field.

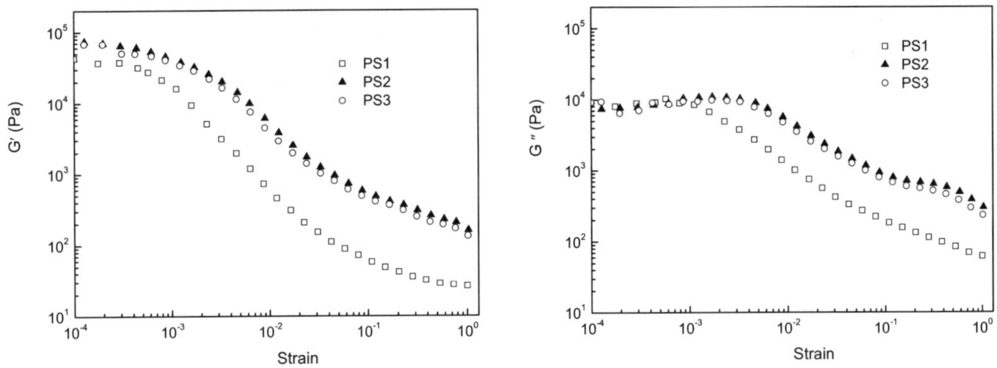

Figure 5. G′ and G″ as a function of strain for PS1, PS2 and PS3 fluids at an electric field strength of 2 kV/mm

Figure 6 shows the *G′* as a function of frequency at a strain of 5×10^{-4} for PS1, PS2 and PS3 fluids at an electric field strength of 2.0 kV/mm. For PS2 fluid, *G′* exhibits a plateau over a wide range of frequency, but *G′* for PS1 and PS3 fluids shows a relatively slight fluctuation. The results imply that the particular microstructures formed in PS2 behaves the most elastic and stiffest showing solid-like behavior, while relatively weak particle chain structure formed in PS1 and PS3 fluids can not undergo larger deformation.

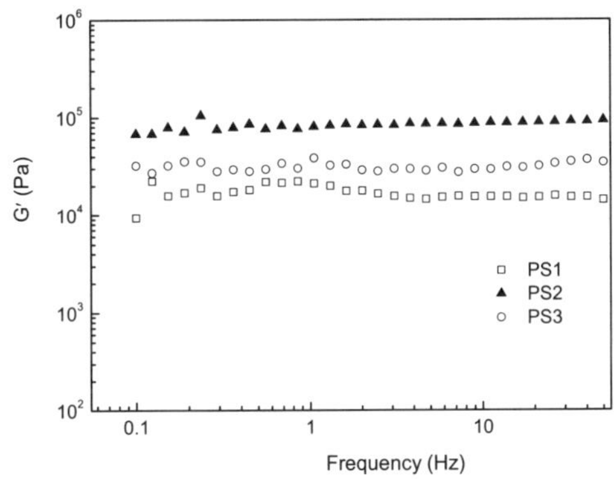

Figure 6. G′ as a function of frequency for PS1, PS2 and PS3 fluids at an electric field strength of 2 kV/mm

4 Conclusion

A new nanocomposite with different amount of polypyrrole confined in the channels of mesoporous SBA-15 has been synthesized by in situ oxidative polymerization. The nanocomposites still possess well-defined mesoporous structure with various PPy loadings. The nanocomposite particles can be used as a dispersed phase in ER fluids, and they show typical ER behavior. ER activity of PPy/SBA-15 fluids is significantly influenced by the PPy loadings. The viscoelastic properties of ER fluids further suggest that PPy/SBA-15 containing 40 wt.% PPy fluid exhibits the most elastic within the linear viscoelastic region due to the stiffest fibrillar structure under an applied electric field.

5 Acknowledgement

This work was supported by Ministry of Education, Youth and Sports of the Czech Republic (MSM 70 88352101) and Grant Agency of the Czech Republic (202/06/0419). The authors also wish to thank to the National Science Foundation of China (20236020, 20176009); the major basic research project of Shanghai (04DZ14002) and the special project for nanotechnology of Shanghai.

References

1. Yin J. B. and Zhao X. P., Giant electrorheological activity of high surface area mesoporous cerium-doped TiO_2 templated by block copolymer, *Chem. Phys. Lett.* **398** (2004) pp. 393-399.
2. Yin J. B. and Zhao X. P., Preparation and electrorheological activity of mesoporous rare-earth-doped TiO_2, *Chem. Mater.* **14** (2002) pp. 4633-4640.
3. Cho M. S., Choi H. J. and Ahn W. S., Enhanced electrorheology of conducting polyaniline confined in MCM-41 channels, *Langmuir* **20** (2004) pp. 202-207.
4. Cho M. S., Choi H. J., Kim K. Y. and Ahn W. S., Synthesis and characterization of polyaniline/mesoporous SBA-15 nanocomposite, *Macromol. Rapid Commun.* **23** (2002) pp. 713-716.
5. Cheng Q., Pavlinek V., Li C., Lengalova A., He Y. and Saha P., Synthesis and characterization of new mesoporous material with conducting polypyrrole confined in mesoporous silica, *Mater. Chem. Phys.* **98** (2006) pp. 504-508.
6. Cheng Q., Pavlinek V., Lengalova A., Li C., Belza T. and Saha P., Electrorheological properties of new mesoporous material with conducting polypyrrole in mesoporous silica, *Microporous Mesoporous Mater.* In press.
7. Di K., Zhu Y. H., Yang X.L. and Li C. Z., Electrorheological behavior of copper phthalocyanine-doped mesoporous TiO_2 suspensions, *J. Colloid Interf. Sci.* **294** (2006) pp. 499-503.
8. Di K, Zhu Y. H., Yang X.L. and Li C. Z., Electrorheological behavior of urea-doped mesoporous TiO_2 suspensions, *Colloid Surface A* **280** (2006) pp. 71-75.
9. Choi H. J., Cho M. S. and Ahn W. S., Synthesis of polyaniline/MCM-41 nanocomposite and its electrorheological property, *Synthetic Met.* **135-136** (2003) pp. 711-712.
10. Cheng Q., Pavlinek V., Lengalova A., Li C., He Y. and Saha P., Conducting polypyrrole confined in ordered mesoporous silica SBA-15 channels: Preparation and its electrorheology, *Microporous Mesoporous Mater.* In press.
11. Zhao D., Huo Q., Feng J., Chemelka B. F. and Stucky G. D., Nonionic Triblock and Star Diblock Copolymer and Oligomeric Surfactant Syntheses of Highly Ordered, Hydrothermally Stable, Mesoporous Silica Structures, *J. Am. Chem. Soc.* **120** (1998) pp. 6024-6036.
12. Li N., Li X., Geng W., Zhang T., Zuo Y. and Qiu S., Synthesis and humidity sensitivity of conducting polyaniline in SBA-15, *J. Appl. Polym. Sci.* **93** (2004) pp. 1597-1601.
13. Woo D. J., Suh M. H., Shin E. S., Lee C. W. and Lee S. H., Electrorheological behavior of suspensions of a substituted polyaniline with long alkyl pendants, *J. Colloid Interf. Sci.* **288** (2005) pp. 71-74.
14. Hao T., Kawai A. and Ikazaki F., Dielectric criteria for the electrorheological effect, *Langmuir* **15** (1999) pp. 918-921.

15. Block H. and Kelly J.P., Electro-rheology, *J. Phys. D: Appl. Phys.* **21** (1988) pp. 1661-1677.
16. Davis L.C, Polarization forces and conductivity effects in electrorheological fluids, *J. Appl. Phys.* **72** (1992) pp.1334-1340.
17. Zhao X. P and Yin J. B, Preparation and electrorheological characterizatics of rare-earth-doped TiO_2 suspensions, *Chem. Mater.* **14** (2002) pp. 2258-2263.
18. Hiamtup P., Sirivat A. and Jamieson A. M., Electrorheological properties of polyaniline suspensions: Field-induced liquid to solid transition and residual gel structure, *J. Colloid Interf. Sci.* **295** (2006) pp. 270-278.

EFFECTS OF EXCITATION CHARACTERISTICS ON THE EQUIVALENT LINEAR SYSTEM OF A BUILDING STRUCTURE WITH MR DAMPERS

J.-H.PARK AND S.-K.LEE

Remodeling Research Center, Dankook University, Seoul 140-714, Korea

S.-H.LEE, K.-W.MIN AND B.-W.MOON

Department of Architectural Engineering, Dankook University, Seoul 140-714, Korea
E-mail: kwmin@dankook.ac.kr

S.-B. CHOI

Department of Mechanic Engineering, Inha University, Incheon, Korea

Seismic control performance of MR dampers, which have severe nonlinearity, changes with respect to the dynamic characteristics of earthquake excitations such as magnitude and frequency contents. In this study, effects of excitation characteristics on the equivalent linear system represented by an equivalent damping ratio for single-degree-of-freedom (SDOF) systems with a MR damper are investigated through extensive numerical analysis for various natural frequencies of the structures and design parameters of the MR damper. In addition, to implement an equivalent linearization procedure considering non-stationarity and frequency contents of the earthquake excitation, seismic response reduction factors for artificial earthquake ground motions are proposed using regression analysis of linear structural responses obtained from numerical analysis. Equivalent linearization results show that the relative magnitude of the excitation compared with the friction force of the MR damper and the frequency contents of the excitation affect the equivalent damping ratio considerably, and appropriate combination of the friction and the viscous damping produces additional damping effect.

1 Introduction

Magnetorheological (MR) fluid is a kind of fluid that can have solid state by magnetic field. It has been applied to vibration control and shock-absorbing devices in various fields, since it is discovered in the late 1940's. In particular, MR dampers have been utilized as a promising semi-active control device for civil and building structures, because its control force can be changed using small amount of electric power and the stability of the controlled structure can be guaranteed [1]. Many algorithms, such as clipped optimal control, Lyapunov control, decentralized bang-bang control etc., have been applied to structural control using MR dampers [2].

However, severe non-linearity of MR dampers makes analysis and design of them difficult and time-consuming. So, modeling of a structure with MR dampers as an equivalent linear system can facilitate application of the MR dampers, especially in the early stage of design. That is, first, required damping ratio is determined. Then, design parameters of the MR dampers producing the required equivalent damping ratio are found. However, the relation between design parameters of the MR dampers and the corresponding equivalent damping ratio is affected by the dynamic characteristics of the structure and excitation. Therefore, this study addresses effects on the equivalent damping ratio of the MR dampers caused by the natural frequency of the structure, and the magnitude and frequency contents of the earthquake excitation through extensive numerical analysis. In addition, to implement an equivalent linearization procedure considering non-stationarity and frequency contents of the earthquake excitation, seismic response reduction factors for artificial earthquake ground motions are proposed by regression analysis of linear structural responses obtained from numerical analysis.

2 Modeling of MR Dampers and Structures

For numerical modeling of MR dampers, this study adopted the Bingham model, which represents mechanical properties of the MR damper with only two parameters, friction force and damping coefficient, and simplifies parametric study [3]. The structures, in which MR dampers are installed, are assumed to be linear elastic SDOF systems. Idealized model of a SDOF structure with a MR damper is illustrated in Fig. 1, where m_s, c_s and k_s denote structural mass, damping coefficient and stiffness, respectively, and c_d and f_d denote viscous damping coefficient and friction force of the MR damper, respectively. Properties of the structures and MR dampers applied to numerical analysis are presented in Table 1. The friction force of the MR damper is normalized with inertia force induced by the peak ground acceleration and denoted by R_f in Table 1. Also, the viscous damping coefficient of the MR damper is normalized with the critical damping coefficient of the structure and denoted by β_d.

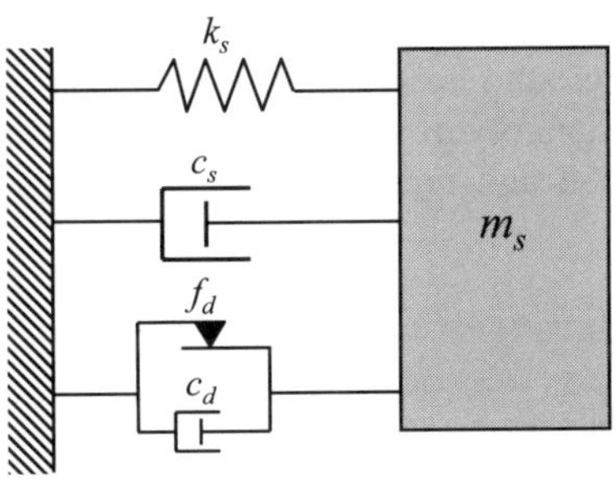

Figure 1. SDOF system with a MR damper

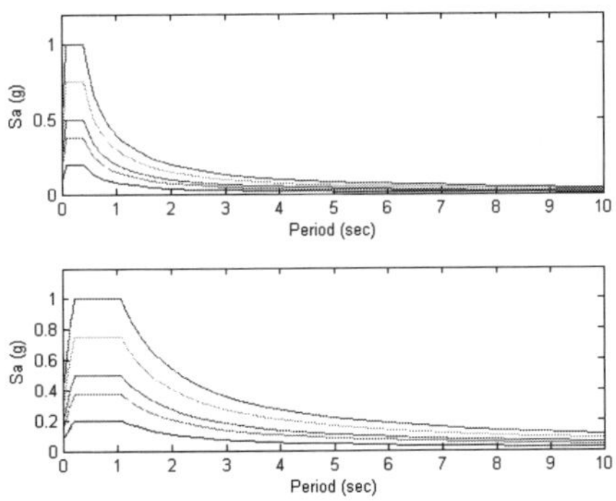

Figure 2. Elastic response sepectra of pseudo-acceleration

Table 1. Properties of structures and MR dampers for numerical analysis

Structure	Mass, m_s (kg)	1.0
	Natural period, T_n (sec)	0.2, 0.4, 0.7, 1.0, 1.5, 2.0, 3.0
	Damping ratio, β_s (%)	5
MR damper	Normalized friction force, R_f	From 0.01 to 0.1(increment = 0.01) From 0.1 to 0.2 (increment = 0.02)
	Viscous damping, β_d (%)	0, 2, 5

3 Earthquake Excitations

Artificial earthquakes are used as dynamic loads for parameter study. Linear elastic response spectra adopted in ATC-40 are used to generate ground acceleration data and are defined by

$$S_a = \begin{cases} C_A(1+1.5T/T_A) & (T < T_A) \\ 2.5C_A & (T_A < T < T_S) \\ C_V/T & (T > T_S) \end{cases} \tag{1}$$

where

$$T_S = C_V/(2.5C_A); \quad T_A = 0.2T_S \tag{2}$$

and C_A and C_V are seismic coefficients related to the spectral response for short and long natural periods, respectively [4]. Various seismic coefficients are chosen for numerical analysis to investigate effects of magnitude and frequency contents of earthquakes on the equivalent linear damping ratio of the structure with a MR damper. Ten response spectra used in numerical analysis are represented in Fig. 2. Five C_A's, which are defined in ATC-40 for seismic intensity ≤ 0.40 and soil profile type A, are adopted to represent difference in excitation magnitude. Peak ground acceleration of the generated ground acceleration is set equal to C_A, which means the peak pseudo-acceleration of a rigid structure as can be seen in Eq. (1). Two seismic coefficient ratios, C_V/C_A, 1.0 and 2.67 are employed to represent difference in frequency contents of excitation as shown in Fig. 2. In ATC-40, C_V/C_A equal to 1.0 is applied to stiff soil profile type and near seismic source, and C_V/C_A equal to 2.67 is applied to soft soil profile type and distant seismic source [4]. For each combination of seismic coefficients, 10 ground acceleration time histories are generated using SIMQKE [5]. To simulate non-stationarity of the ground acceleration, an envelope function proposed by Jennings is applied and represented in Fig. 3 with a sample ground acceleration time history [6].

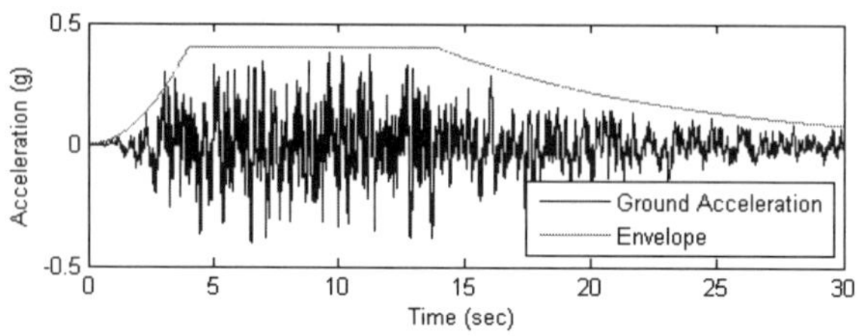

Figure 3. Artificial ground acceleration and envelope function

4 Equivalent Damping Ratio

If a MR damper behaves in the exact manner of Coulomb friction and viscous damping in parallel, as in the Bingham model, the control force depends on only velocity response at the connections to the structure. This results in the increase of the damping ratio without change in the natural period. Therefore, the equivalent linear system for a linear SDOF structure with a MR damper can be defined using only an equivalent damping ratio.

Newmark and Hall proposed amplification factors for design spectra using ground motion recorded on firm ground [7]. Response reduction factors of SDOF systems for arbitrary damping ratios on the basis of the response of 5%-damped structure were derived using those amplification factors, and utilized for capacity spectrum method proposed in ATC-40 [4]. Using equations of those response reduction factors, the equivalent damping ratio of a nonlinear structure can be calculated inversely. However, artificial ground motions generated in this study may have different properties from those used in Newmark and Hall's study, and covers soft ground condition. Therefore, new response reduction factors are proposed for those artificial ground motions through least square regression and given as the following equation.

$$R = a_0 + a_1 \log \beta + a_2 (\log \beta)^2 \tag{2}$$

where β is the damping ratio, and a_0, a_1 and a_2 are constants, which are represented in Table 2. As can be seen from Table 2, proposed response reduction factors have an additional second order term, and different coefficients are proposed with respect to frequency contents represented by C_V/C_A. Two damping ratios are calculated using both response reduction ratios that belong to acceleration- and velocity- sensitive frequency ranges, and the smaller one is chosen.

Table 2. Constants in response reduction factors

	Period Range	a_0	a_1	a_2
Newmark & Hall	Acceleration-sensitive	1.51	-0.321	0
	Velocity-sensitive	1.40	-0.249	0
This Study (C_V/C_A =1.00)	Acceleration-sensitive	1.84	-0.620	0.0603
	Velocity-sensitive	1.65	-0.446	0.0273
This Study (C_V/C_A =2.67)	Acceleration-sensitive	1.76	-0.550	0.0477
	Velocity-sensitive	1.58	-0.386	0.0172

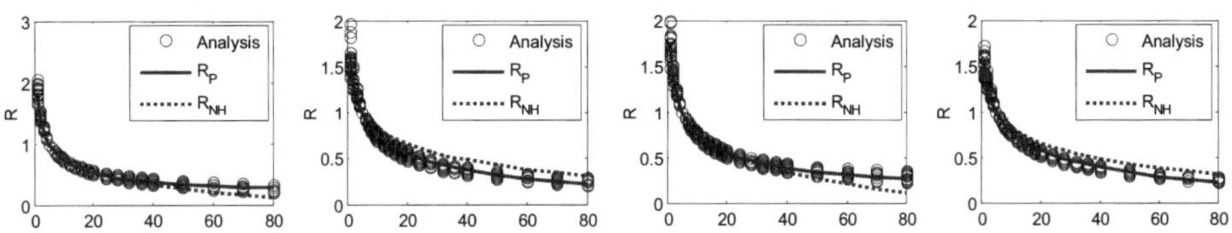

(a) For acceleration-sensitive period range (C_V/C_A=1.00)
(b) For velocity-sensitive period range (C_V/C_A=1.00)
(c) For acceleration-sensitive period range (C_V/C_A=2.67)
(d) For velocity-sensitive period range (C_V/C_A=2.67)

Figure 4. Response reduction factors (β_s =0.00, R_{NH}: Newmark & Hall, R_P: proposed)

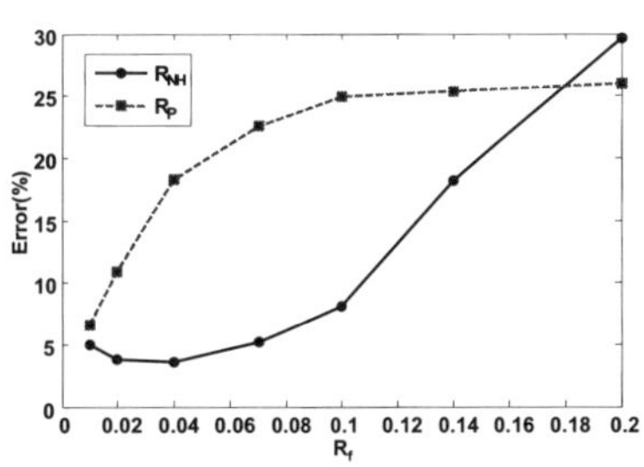

Figure 5. Maximum peak displacement errors with respect to the normalized friction force

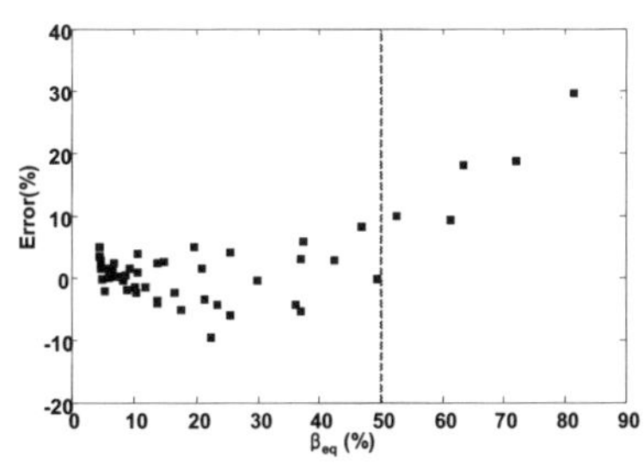

Figure 6. Peak displacement errors with respect to the equivalent damping ratio

Response reduction factors are compared in Figure 4, in which R_{NH} and R_P are response reduction factors proposed by Newmark and Hall and present study, respectively. Proposed response reduction factors agree with analysis result well, while Newmark and Hall's response reduction factors have large errors for higher damping ratios. In particular, since the slopes of the curves are small for the range of higher damping ratios, small error in response reduction ratio can result in large error in the damping ratio.

To evaluate accuracy in response prediction using equivalent linear systems, response of the equivalent linear systems and corresponding original systems with a MR damper are compared. Peak displacement errors with respect to the normalized friction force, R_f, are represented in Fig. 5. Errors increase as R_f increases due to strong nonlinearity. However, errors for the proposed response reduction ratios are restricted under 10 %, for $R_f \leq 0.1$. Peak displacement errors with respect to the equivalent damping ratio are represented in Fig. 6. For equivalent damping ratio smaller than 50 %, errors are limited to bellow 10 %. As a result, above limitations on the normalized friction force and equivalent damping ratio can be used as a criterion to determine whether nonlinear analysis is needed or not to evaluate the performance of a MR damper.

5 Effects of Excitation Characteristics on Equivalent Damping Ratios

To investigate effects of excitation magnitude on the equivalent damping ratio, equivalent damping ratios for short and long period structures subjected to the seismic coefficient ratio of 1.0 and 2.67, respectively, are represented in Fig. 7. Equivalent damping ratios are not dependent on the absolute excitation magnitude represented by the seismic coefficient C_A, but on the normalized friction force, the ratio between the excitation magnitude and friction force. This tendency is common for both seismic coefficient ratios describing frequency contents of the excitation, and can be found for structures with other natural periods and seismic coefficients than in Fig. 7.

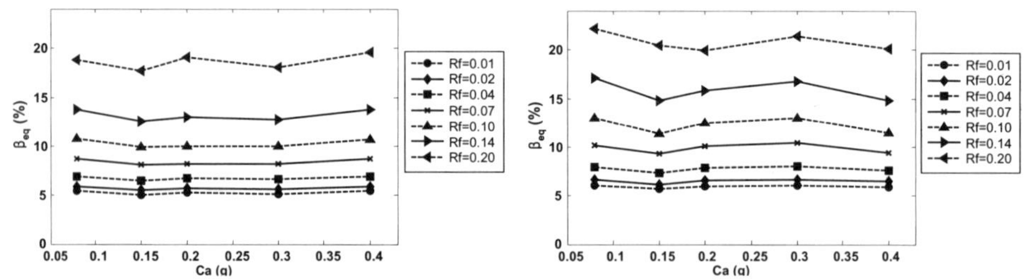

(a) $T_n = 0.2$ sec, $C_V/C_A = 1.0$ (b) $T_n = 2.0$ sec, $C_V/C_A = 2.67$

Figure 7. Equivalent damping ratio with respect to the excitation magnitude ($\beta_d = 0$)

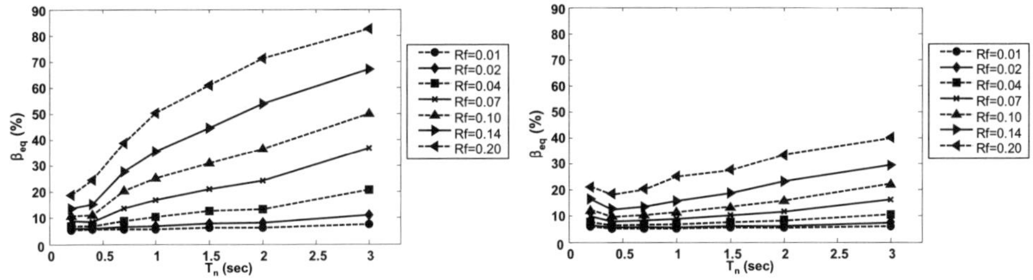

(a) $C_V/C_A = 1.0$ (b) $C_V/C_A = 2.67$

Figure 8. Equivalent damping ratio with respect to the natural period ($C_A = 0.08$, $\beta_d = 0$)

Next, variation of the equivalent damping ratio with respect to the natural period is examined. Fig. 8 represents equivalent damping ratios for C_A = 0.08 and β_d = 0 %. In Fig.8 it is observed that, for C_V/C_A = 1.0, long period structures achieve much larger equivalent damping ratios than short period ones with the same normalized friction force. On the other hand, for C_V/C_A = 2.67, the equivalent damping ratio does not increase by the period elongation as much as for C_V/C_A = 1.0. Moreover, enhancement of the equivalent damping ratio by increasing the normalized friction force is less significant for C_V/C_A = 2.67 than for C_V/C_A = 1.0. For other C_A's and β_d's than in Fig. 8, the same tendency is obtained.

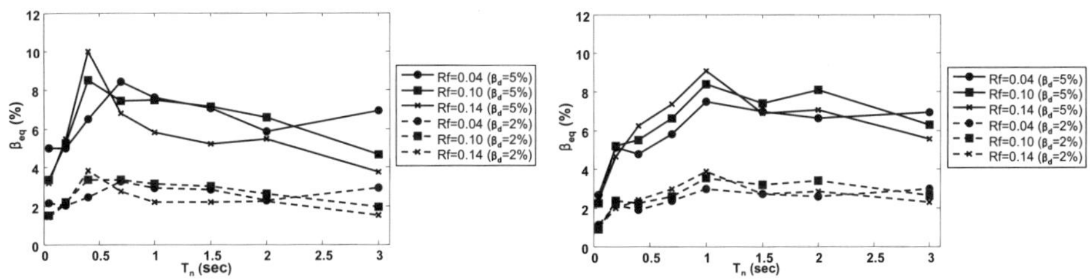

(a) $C_V/C_A = 1.0$ (b) $C_V/C_A = 2.67$

Figure 9. Increment of equivalent damping ratio by the viscous damping of the MR damper (C_V=0.08)

In Table 1, β_d means the increment of the structural damping ratio by the viscous damping part of the MR damper, because it is a viscous damping coefficient normalized by the the critical damping ratio of the structure. To investigate if the equivalent damping ratio increment by the viscous damping of the MR damper is equal to β_d, equivalent damping ratio increments by β_d = 2 % and 5 % are represented in Fig. 9. It is observed that the equivalent damping ratio increment is similar to β_d for short period of 0.2 sec. However, for several numbers of natural periods longer than 0.2 sec, the difference between the equivalent damping ratio increment and β_d is considerably large. The maximum difference for each R_f occurs at T_n = 0.4 or 0.7 for C_V/C_A = 1.0 and at T_n = 1.0 for C_V/C_A = 2.67, and amounts to 80~100 % of β_d for R_f = 0.14.

For those natural periods, R_f has positive effects in increasing the equivalent damping ratio. On the other handm R_f affects the equivalent damping ratio increment adversely for the long natural period of 3.0 sec. Therefore, appropriate combination of the friction force and viscous damping is required to enhance the structural damping effectively.

6 Concluding Remarks

In this study, seismic response reduction factors based on regression analysis for artificial earthquakes are proposed and employed in the equivalent linearization procedure for SDOF systems with a MR damper. The equivalent damping ratios are examined for extensive natural periods and excitation characteristics such as magnitude and frequency contents of the ground motion. Main results obtained from numerical analysis are summarized as follows.

1. The equivalent damping ratio does not depend on the absolute magnitude of the excitation, but depends on the ratio between the excitation magnitude and the friction force of the MR damper.
2. Change in the frequency contents of the ground acceleration with the same peak values results in the difference of the equivalent damping ratios, which is more remarkable for the structures with a long natural period.
3. For the same friction force of the MR damper, long-period structures have lower equivalent damping ratios for excitations with more frequency contents near its resonance frequency, because of limited energy dissipation capacity of the MR damper and increase of strain energy in the structure.
4. Linear viscous damping of the MR damper provides additional equivalent damping ratio to the one calculated from its damping coefficient mathematically.

7 Acknowledgements

The work presented in this paper was supported by the Ministry of Education & Human Resources Development through the Second Stage of BK21 and Research Funds of the National Research Laboratory Program (Project No.M1-0203-00-0068) from the Ministry of Science and Technology in Korea.

References

1. Jung, H.-J., Spencer, Jr, B.F., Ni, Y.Q., Lee, I.-W., State-of-the-art of semiactive control systems using MR fluid dampers in civil engineering applications, *Structural Engineering and Mechanics*, **17** (2004) pp. 493-526.
2. Jansen, L.M., Dyke, S.J., Semiactive control strategies for MR dampers: Comparative study, *J. of Eng. Mech.*, **126** (2000) pp. 795-803.
3. Spencer Jr., B.F., Dyke, S.J., Sain, M.K., Carlson, J.D., Phenomenological model for magnetorheological dampers, *J. of Eng. Mech.*, 123 (1997) pp. 230-238.
4. ATC, Seismic evaluation and retrofit of concrete buildings, ATC-40 Report, Applied Technology Council, Redwood City, California, (1996).
5. Vanmarcke, E.H., Gasparini, D.A., SIMQKE User's Manual and Document, Division of Advanced Environmental Research and Technology, MIT, (1976).
6. Jennings, P.C., Housner, G.W., Tsai, N.C., Simulated earthquake motions, California Institute of Technology, Pasadena, (1968).
7. Newmark, N.M., Hall, W.J., Earthquake Spectra and Design, Earthquake Engineering Research Institute, Berkeley, (1982).

ELECTROSTRUCTURED SENSITIVITY OF INORGANIC AND NATURAL FILLERS FOR ERF

E.V. KOROBKO, N. A. BEDZIK
A. V. Luikov Heat and Mass Transfer Institute, National Academy of Sciences of Belarus, 15 P. Brovka str., Minsk, 220072, Belarus
E-mail: eva@itmo.by

L. S. ESCHENKO, G. M. ZHUK
The Belarussian State Technological University, Minsk, Belarus

Electrorheological activity of suspensions, which have inorganic natural (glauconite, kaolin) and synthetic (frame aluminosilicates and hydrated aluminum oxides) fillers as dispersed medium, has been investigated. The dependences of electrorheological activity of the fillers on their composition, structure and the presence of different types of water have been found. It was assumed that the presence of solid phase of the coordinated water in the structure of synthetic compounds leads to the its protolytic dissociation (under the influence of metal ions) with the formation of H^+ protons, responsible for the appearance of electrorheological effect of the electrorheological fluids investigated.

1 Introduction

Analysis of effectiveness of known ERF compositions [1-3] and their application for different technologies [4] has shown, that all of them have these or that disadvantages. The main disadvantage is a not high ER –sensitivity. Besides, many ERF compositions, containing adsorbed water (on the basis of oxides and natural fillers, as usual) change their characteristics in the long run, have the residual effect and could be used only in a specific range of temperatures and humidity, which lowers their exploitation possibilities.

The aim of this paper is to develop the ways of optimizing electrorheological activity of the dispersed phase, as it is responsible for a display of the ER-effect in most cases. The comparison analysis of ERF on the basis of natural, synthetic silicates and hydrated aluminum oxides, which ER-activity has been improved by means of variation of different kinds of treatment, is made.

2 Experimental part

Inorganic natural and synthetic materials, containing a certain quantity of bonded activator (water) have been investigated as the ERF fillers. These are natural silicate materials with the layer structure: kaolin and glauconite, dried and thermal treated at temperatures needed. Crystalline aluminosilicates with a frame structure, in particular, zeolites NaA, NaY, modified by phosphoric acid and the products of their dehydration and hydrated aluminum oxides were regarded as synthetic materials.

Investigation of processes, happening at heating silicate and aluminum hydroxides, is carried out by means of thermographimetric analysis on the derivatograph OD-102 by firm «MOM» (Hungary) at heating speed 10 deg/min, mass of the hinge-plate 0.5–1.0 g at sensitivity DTA 1/10. The measuring method is based on the determining the difference between temperatures in the investigated sample and in the standard, measured by the differential thermo pair. Heating curves represent the dependence of the temperature difference on heating temperature. The phase compositions of the investigated samples were determined by the X-ray analysis on the diffractometer «DRON-3» with copper radiation. The detection speed was 1–2 deg/min in the range of angles $2\Theta = 10–70°$. Mass loss of the samples was carried out in the range of temperatures 20—850°C in the polythermal regime in the following way. Crucibles with a hinge-plate of a sample were put into the muffle oven and were heated to a given temperature, whereupon they were taken out, cooled and weighed. The evaluation of Al_2O_3, SO_4^{2-} content was carried out by standard procedures [5]. Experimental evaluation of electrorheological sensitivity of ERF was made on a rotational viscometer with a working site of a co-axial cylindrical type with a radial slot between cylinders equal to 0.7 mm. It was modified for the application of electric field from the voltage source in the range of voltages U = 0-2.5 kV. The milliampermeter was used for measuring the current in the slot between cylinders. Maximum measuring error is not higher than 10%. In the experiments a shear stress τ, shear rate $\dot{\gamma}$, current strength I were measured and a comparison evaluation of rheological indices of sensitivity by the values of effective viscosity $\eta = \frac{\tau}{\dot{\gamma}}$ was done.

The chemical composition of the kaolin and enriched glauconite under investigation, that are used as dispersed phase, is presented in the Table 1. The average size of a particle of enriched glauconite and kaolin was 4 and 7 μm respectively. In the result of measurements was found that the mass loss curves of glauconite and kaolin are characterized by presence of endothermic effects in the range of 70-550 °C due to detaching of water that is present in the form of adsorbed as well as in coordinated molecules and in the bridge hydroxyl groups. According to the mass loss curve the main quantity of water from kaolin and glauconite is removed in the range of temperatures 440-650 °C that could be an evidence of presence in them chemical fast bound water, being in the form of bridge OH-groups. Based on data from mass loss curve analysis, the conditions of getting the products of dehydration of glauconite and kaolin that differ in composition of combined activator — chemical bound water, were found. The composition of the samples obtained is presented in the Table 2

Frame aluminosilicates with zeolite structure with the type of NaA and NaY were investigated as synthetic silicate materials. Chemical composition of these compounds is described by brutto-formulas $Al_2O_3 \cdot 2Na_2O \cdot 2SiO_2 \cdot 80H_2O$ and $Al_2O_3 \cdot 4Na_2O \cdot 8SiO_2 \cdot 160H_2O$, respectively. Synthetic zeolites are characterized by mono dispersiveness with the particle diameter in the range of 0–5 μm.

Table 1. Chemical composition of kaolin and glauconite.

Sample	Content, mass.%												
	SiO_2	Al_2O_3	TiO_2	Fe_2O_3	CaO	MgO	Na_2O	K_2O	FeO	MnO	SO_3	P_2O_5	p.o.d.
Glauconite	52.77	10.90	1.08	6.51	1.24	2.99	0.12	5.27	1.75	0.04	0.24	0.03	10.06
Kaolin	47.80	36.80	1.20	0.45	0.63	0.30	0.10	0.44	--	--	--	--	12.23

Table 2. Content of H_2O in glauconite and kaolin samples.

Sample	Temperature, °C	Full mass loss at burning, %	Content of H_2O in samples, %	
			Hydrated	Chemically bonded
Glauconite	22*	11.2	3.2	8.0
	105	7.7	0.5	7.2
	450	4.7	--	4.7
Kaolin	22	13.0	0.8	12.2
	105	12.2	0.3	11.9
	400	11.9	--	11.9

*) Dried at room temperature.

3 Results

Thermograms of these compounds show that the main quantity of water from zeolites vanishes in the range of temperatures 70–750 °C. That allows to assume that coordinated molecules are the majority in the frame aluminosilicates.

It is interesting to mark that there are some differences between a thermogram of the zeolite NaA, modified by phosphoric acid, from a thermogram of the zeolite NaA that can be caused by introduction of phosphate groups in structure of the zeolite and by a change of binding energy of molecules of water in it. The dependencies of shear stress on the intensity of electric field for ERF with synthetic and natural silicate materials are submitted in Figure 1. In case of use of zeolites in ERF essential distinctions in their electrorheological activity are observed, depending on their composition, structure and presence of various forms of water in it. According to Figure 1 an increase of shear stress with the increase of the intensity of electric field by constant shift speed 41.7 s^{-1} was observed for the ERF on the basis of NaY and heat treated at temperature 350^{0}C. On the other hand, the ERF on the basis of the zeolite NaA, heat treated at the same temperature, almost does not show ER - activity. This distinction is possible to explaine by the difference in structures of crystal lattices and in arrangement of the cations, molecules of water and OH – groups in them. So the comparison of the thermograms of NaA and NaY shows their difference in the area of endothermic effect, which is linked to deleted molecules of water and the hydroxyl groups. As it follows from the thermogram analysis data, the quantity of the chemically bound water in the zeolite NaY is higher, than in the structure of the NaA. As it is known, in the zeolites the hydroxyl groups with the mobile hydrogen are present. Besides, thanks to electrostatic field, cations promote the dissociation of the coordinating bound water with the formation of the $MeOH^+$ and H^+. The formation of the mobile protons and their number in the structure of the zeolites can be one of the reasons of the polarization and the electrorheological effect observed.

According to [6] ion conductivity of the zeolites can be induced by cation compositions. It is noted, the electrorheological activity of ERF based on zeolites, containing different cations differs significantly. This fact is explained by authors [7 – 8]

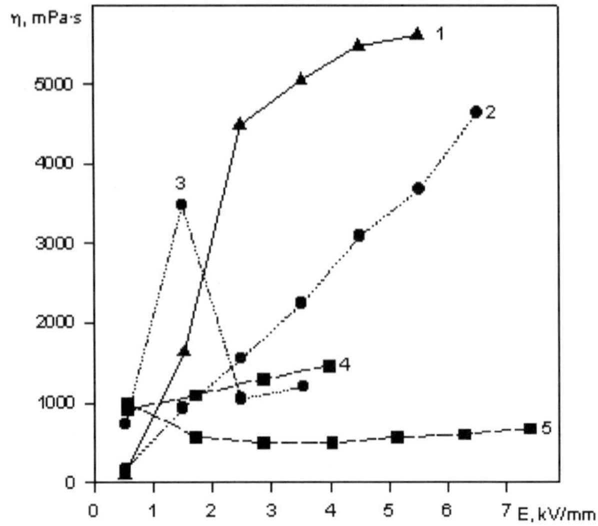

Figure 1. The dependence of effective viscosity on the value of the applied electric field for ERF based on the natural and synthetic silicates: 1 – glauconite (T t.h. = 22°C); 2 – zeolite NaY (T t.h. = 350°C); 3 – zeolites, modified by phosphoric acid; 4 – kaolin (T t.h. = 105°C); 5 - kaolin (T t.h. = 22°C). C=20%. $\dot{\gamma}$=41,7 s^{-1}.

by the different placement of the cations in the structure of zeolite and their mobility. Besides, electrorheological activity can be also induced by the fact that the nature of a cation influences its electrostatic field and therefore influences the bound water dissociation process and the number of mobile hydrogen atoms that cause polarization.

The research of ERF based on the modified zeolites with molar ratio P_2O_5/Al_2O_3 – 0.2 – 0.3 as disperse phase, has shown (fig. 1), that ERF – effect increase at the intensity of an electric field equal 0.5 kV. As intensity increases, the shear stress sharply fell, that can be connected to the formation of phosphate HPO_4^{2-}, H_2PO^- — groups that connect mobile H^+ proton.

Electrorheological activity of ERF on the basis of glauconite and the products of its dehydration is apparently determined by the presence of the mobile H^+-protons in their structure. As it follows from the experimental data suspensions based on glauconite, dried under ambient atmosphere at 22 °C till the constant mass and that contained all types of water have demonstrated the biggest electrorheological effect. At the same time detaching of adsorbed as well as coordinated water from glauconite lowers its electrosensitivity and therefore the effectiveness of its application as disperse phase in ERF, in particular, the one based on transform oil. In the case of thermal treatment of glauconite at 230 °C it contains up to 5.5% of chemically bound water that is present in the form of bridge hydroxyl groups. Apparently, the presence of coordinated water in the structure of glauconite that detaches according to mass loss curve analysis in the range of 70-160 °C, results in its protolytic dissociation under action of metal ions forming H^+-protons, that are responsible for electrorheological effect of ERF.

As the results of research have shown that ERF based on kaolin presents insignificant ER-effect. It is necessary to mention, that suspensions, prepared with the use of kaolin and products of its thermal treatment, are characterized by high viscosity that could influence electrosensitivity, because it differs from the movement of particles of a disperse phase in super viscous dispersed medium. On the other hand, kaolin does not practically contain crystal hydrated water; water, that is a part of a structure of kaolin, and is detached at rather high temperatures and is chemically fast bound.

Table 3. Conditions of getting aluminum hydroxides and the results of their investigation.

№ Experiment	pH sedimentation	Conditions: aging, Temperature, °C	Conditions: aging, Time, h	Conditions: Thermal treatment of the samples, °C	Chemical composition, mole $Al_2O_3 : H_2O$	Type of the structure
1	9.0-9.2	20	24	100	1.0 : 2.86	Weakly crystallized Pseudo boehmite
				170	1.0 : 2.09	
				350	1.0 : 1.04	
				500	1.0 : 0.23	
2		100	8	100	1.0 : 2.21	Pseudo boehmite
				170	1.0 : 1.66	
				350	1.0 : 1.01	
				500	1.0 : 0.35	
3		150	7	100	1.0 : 2.08	boehmite
				170	1.0 : 1.44	
				350	1.0 : 1.22	
				500	1.0 : 0.24	
4		180	7	100	1.0 : 2.07	boehmite
				170	1.0 : 1.51	
				350	1.0 : 1.13	
				500	1.0 : 0.23	

The investigation results of natural (kaolin and glauconite) and synthetic (zeolites) silicates have shown the correlation between electrorheological activity and a type of structure of the studied compounds. This way a inequality of energetic states of water molecules and OH-groups, forming silicate crystalline lattice was found. The data from DTA and TG-curves prove this fact. The formation of the mobile proton, appearing under impact of metal ions, leads to the electrorheological effect. It means that different cations will influence differently the process of electrolytic dissociation of water molecules and OH-groups in the silicate composition (for example in zeolites). Our aim was to investigate the correlation between the ER-effect and different types of water in the filler compositions for ERF. Hence, the next stage of the work was investigation of electrorheological sensitivity of ERF based on the modified aluminum oxides (only the quantity of water and water type in the compound with the same cation Al^{3+} was varied). The investigation of rheological properties of ERF based on the dispersed phase of 3 types has been carried out. The first type was an ERF based on the hydrated aluminum oxides, containing all three types of water adsorbed, interlayer and in the form of OH-groups (air-dried samples № 1 and № 2) (see Table 3) (see Fig. 2, curve I). To the second group, ERF obtained with the use of hydrated aluminum oxides, containing interlayer water and water in the form of OH-groups were included (samples № 1 и № 2, heated up to 170 °C). The third group of ERF is characterized by the presence of the filler in which the water is in the form of fast chemically bonded OH-groups, vanishing in the

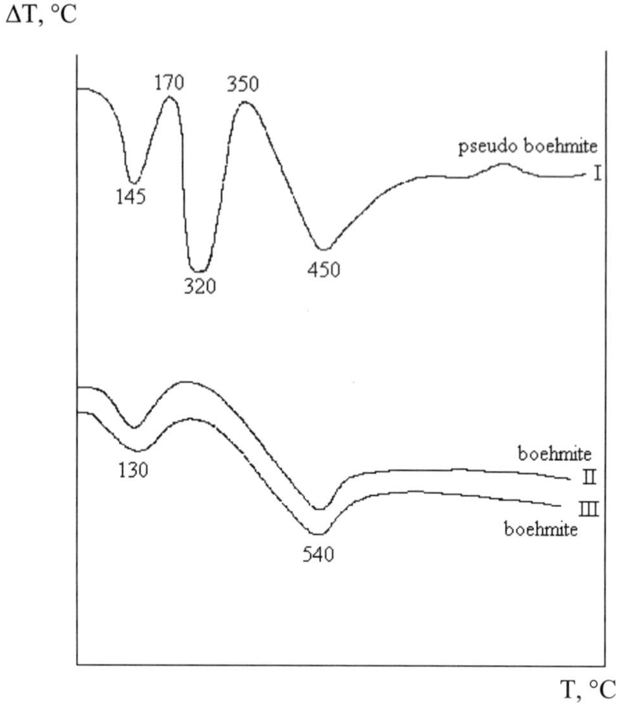

Figure 2. Curves of aging product heating of the segregated aluminum oxides.

range of temperatures 350–520 °C (samples № 1 – 4, heated up to 350 °C) (Fig.2, curve III). The dependence of effective viscosity on the intensity value of the electric field is shown in Fig. 3 for the ERF, in which abovementioned samples of hydrated aluminum oxides were used as a dispersed phase. As it follows from the experimental data the type of changing effective viscosity in the electric field is different for ERF with different filler types.

In Fig 3 is shown that the biggest change of effective viscosity (η ~1700–4000 mPa·s) is seen for ERF from the second group, containing in their composition up to 30% of introduced interlayer water. (curves 5-8).

In the case of use of ERF from the third group (curves 9–12), the growth of the effective viscosity is much lower (η ~ 40–280 mPa s). Such insignificant electrorheological effect is probably due to the fact that water molecules that are the part of aluminum hydroxides in the form of OH-group are weak activator. For suspensions of the first group the growth of effective viscosity is 200–1500 mPa · s (curves 1–4) depending on the aging conditions. For example, for aluminum hydroxides, aged at 100 and 20 °C, the values of effective viscosity are 700–1550 mPa · s respectively (curves 4, 2). At that it is important to note that for the latter high current was detected (I = 2.64mA), compared to other samples (the current was from 0.001 up to 0.011 mA). The observed difference in particle activity in the electric field could be linked to the fact that the samples aged at 20 °C contain significant quantity of free water molecules and have a weakly ordered structure that influences the ER-effect.

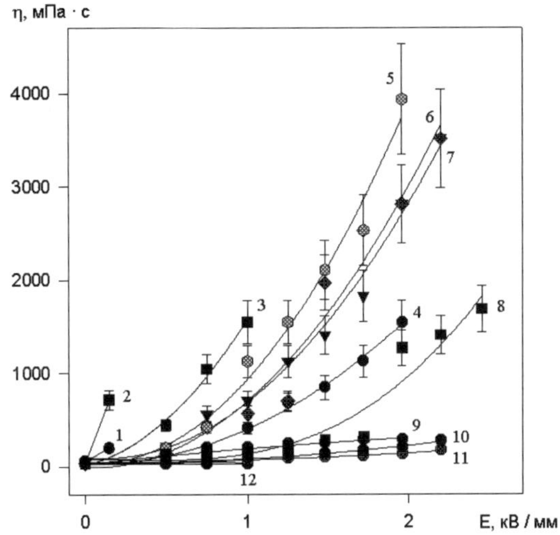

Figure 3. The dependence of effective viscosity on the value of applied electric field for aluminum hydroxide:
1 - №3 (T t.h.=105°C), 2 - №6 (T t.h.=105°C), 3 - №1 (T t.h.=105°C), 4 - №2 (T t.h.=105° C), 5 - №3 (T t.h.=170° C), 6 - №4 (T t.h.=170° C), 7 - №4 (T t.h.=170° C, 8 - №2 (T t.h.=170°C), 9 - №3 (T t.h.=350°C), 10 - №4 (T t.h.=350°C), 11 - №2 (T t.h.=350°C).

4 Conclusion

The analysis of results, obtained at investigating ERF based on the synthetic aluminum hydroxide allows to conclude that the growth of effective viscosity is firstly due to the presence of interlayer water, forming hydrogen bridges. One can assume that this type of bonding water molecules that are in the filler provides the mobility of protons H^+ in the electric field and therefore proton conductivity, that causes the ER-effect. At the same time water molecules, forming fast chemical bonding Al – O – H with aluminum, are different by the relative low activity due to low content of mobile protons. Free water molecules i.e. non chemically bound, do not practically undergo photolytic and therefore are not sources of protons H^+. Considering abovementioned one can assume that the mechanism of ER-effect in water containing ERF, consisted of aluminum hydroxides and mineral oil, is based on the proton conductivity but not on the ion polarization, initiated by water molecules.

References

1. Bloodworth R., Electrorheological Fluids Based on Polyurethane Dispersions. In *Proceedings of the 4 th Int. Conf. on Electro-Rheological Fluids, Magneto-Rheological Suspensions and Associated Technology*. (10-14 July 1995. Sheffield, UK. Ed. W. A. Bullough) pp. 118-131.
2. Coulter J. P., Weiss D., Carison J. D., Electrorheological Materials and their usage in Intelligent Material Systems and Structures. Part II: Applications. In *Recent Advances in Sensory and Adoptive Materials and Their Applications* (C.A. Rogers, Ed., Technomic Public. Lancaster PA).
3. Weiss D., Coulter J. P., Carison J. D., Electrorheological Materials and their Usage in Intelligent Material Systems and Structures. Part I: Mechanisms, Formulations and Properties. In *Recent Advances in Sensory and Adoptive Materials and Their Applications* (C.A. Rogers, Ed., Technomic Public. Lancaster, PA).
4. Korobko E.V. Electrostructured (electrorheological) fluids: the peculiarities of hydromechanics and the possibilities of application (in rus.).- Minsk: HMTI AS Belarus.– 1996.– 189 p.
5. Analysis methods of phosphate raw materials, phosphate and complex fertilizing, forage phosphates. – M.: Chemistry, 1975. – 215 p.
6. Negita K., Ohsawa Y., Dielectric Properties of Electroreological Fluids Containang Zeolite Particals, *J. Phys. II France,* **№ 5** (1995), pp. 2249.
7. Bose H., Investigations on Zeolite – Based ER Fluid Supported by Experimental Design. In *Proceedings of the 6th Int. Conf. on Electroreological Fluids, Magneto – Reological Fluids and Their Applications* (22-25 July 1997, Yonesawa, Japan. Eds. M. Nakano. K. Oyama, World Scientific Publishing Co., Singapore) pp. 240-247.
8. Subramanian V., Seff K. A., Near Zero Coordinate Sodium Ion in Dehedrated Zeolite 4F, Na12-A, *J. Phys. Chem.* **№ 81** (1977) pp. 2249.